ENGINEERING MECHANICS

P C Dumir

Formerly, Professor, Department of Applied Mechanics
Indian Institute of Technology Delhi

S Sengupta

Formerly, General Manager, Engineers India Limited
and Visiting Professor
Indian Institute of Technology Delhi

Srinivas V Veeravalli

Professor, Department of Applied Mechanics
Indian Institute of Technology Delhi

Universities Press

ENGINEERING MECHANICS

UNIVERSITIES PRESS (INDIA) PRIVATE LIMITED

Registered Office
3-6-747/1/A & 3-6-754/1, Himayatnagar, Hyderabad 500 029, Telangana, India
info@universitiespress.com; www.universitiespress.com

Distributed by
Orient Blackswan Private Limited

Registered Office
3-6-752 Himayatnagar, Hyderabad 500 029, Telangana, India

Other Offices
Bengaluru / Bhopal / Chennai / Guwahati / Hyderabad / Jaipur
Kolkata / Lucknow / Mumbai / New Delhi / Noida / Patna / Visakhapatnam

First published 2020

ISBN 978 93 89211 57 3

Front cover image courtesy of Wikimedia
Portrait of Leonhard Euler (1707–1783) by artist Jakob Emanuel Handmann (1718–1781), given by Rudolf Bischoff-Mrian, 1849

Typeset in Times New Roman 11/13.2

501695

Printed at
B. B. Press, Noida

Published by
Universities Press (India) Private Limited
3-6-747/1/A & 3-6-754/1, Himayatnagar, Hyderabad 500 029, Telangana, India

Contents

Preface ix

1 KINEMATICS 1
- 1.1 Mechanics 1
- 1.2 Vectors 3
- 1.3 Problem Solving Process 8
- 1.4 Mathematical Model of Classical Mechanics 12
- 1.5 Position Vectors of Locations 15
- 1.6 Kinematics of a Moving Point 16
- 1.7 Kinematics of a Point in Cartesian Coordinates 19
- 1.8 Kinematics of a Point in Cylindrical Polar Coordinates 21
- 1.9 Kinematics of a Point in Path Coordinates 24
- 1.10 Relative Velocity and Acceleration 30
- 1.11 Rectilinear Motion 32
- 1.12 Illustrative Problems 35
- 1.13 Concept Review Questions 1A 52
- 1.14 Practice Problems 1A 56
- 1.15 Index Notation 59
- 1.16 Angular Velocity of Frame m Relative to Frame F 61
- 1.17 Angular Velocity of m if $\vec{e}_{3|F} = \vec{0}$ 64
- 1.18 Relations between $\vec{v}_{P|F}$, $\vec{a}_{P|F}$ and $\vec{v}_{P|m}$, $\vec{a}_{P|m}$ 66
- 1.19 Kinematics Relative to Translating Frame T 68
- 1.20 Rigid Body Kinematics 70
- 1.21 Composition of Angular Velocities and Angular Accelerations 73
- 1.22 Rolling Without Slip 75
- 1.23 Gear, Belt and Chain Drives 79
- 1.24 Examples Involving Moving Frames 80
- 1.25 Concept Review Questions 1B 103
- 1.26 Practice Problems 1B 109

2 AXIOMS AND FORCE SYSTEMS **115**
2.1 Force and Moment . . . 115
2.2 Mass and Centre of Mass . . . 121
2.3 Euler's Axioms . . . 131
2.4 Equivalent / Resultant Force Systems . . . 134
2.4.1 Centre of parallel forces . . . 141
2.5 Newton's Third Law of Motion . . . 153
2.6 Kinetic Energy, Power and Work . . . 154
2.7 Expressions of some Entities in Terms of C . . . 158
2.8 $\dot{\vec{H}}_A - \vec{M}_A$ Relation for Arbitrary Point A . . . 162
2.8.1 $\dot{\vec{H}}_C - \vec{M}_A$ Relation for arbitrary point A . . . 164
2.8.2 $\dot{\vec{H}}_A - \vec{M}_A$ Relation for any point A for a translating frame T . 165
2.9 Equations of Equilibrium . . . 166
2.10 Impulse and Angular Impulse . . . 168
2.11 Impulse–Momentum Relations . . . 169
2.12 Angular Impulse–Moment of Momentum Relations . . . 171
2.13 Work–Energy Relation for Centre of Mass C of any System . . . 172
2.14 Friction . . . 172
2.14.1 Axiom of dry friction or Coulomb friction . . . 173
2.14.2 Rolling friction . . . 178
2.15 Axiom of Electromagnetic Force . . . 180
2.16 Axiom of Gravitational Force . . . 181
2.16.1 Model of earth's gravitational field . . . 183
2.17 Free Body Diagram and Reactions of Supports . . . 184
2.18 Belt Friction . . . 199
2.19 Frictional Torque at a Thrust Bearing and Frictional Clutches . . . 203
2.20 Frictional Torque for a Square-Threaded Screw . . . 205
2.21 Conservative Forces . . . 206
2.21.1 Potential energies of some conservative forces . . . 209
2.22 Workless Force Systems . . . 219
2.23 Deformable Body, Rigid Body and Mass-point Models . . . 223
2.24 Concept Review Questions 2A . . . 227
2.25 Practice Problems 2A . . . 232
2.26 Work–Energy Relation for C for Conservative Forces . . . 238
2.27 Dynamics of a Translating System . . . 239
2.27.1 Virtual work approach for equivalent force systems . . . 242
2.28 Dynamics of a Rigid Body with Mass Concentrated at C . . . 243

2.29 Projectiles with Resistance 244
2.29.1 General motion under $\vec{R}_f = -mk\vec{v}$ 245
2.29.2 Body falling vertically under $\vec{R}_f = -mk\vec{v}$ 247
2.29.3 Body falling vertically under $R_f = \frac{1}{2}C_D\rho Av^2$ 248
2.29.4 Motion of a shell fired up under $R_f = \frac{1}{2}C_D\rho Av^2$ 249
2.30 Equation of Motion of C Relative to a Translating Frame 250
2.31 Examples on Motion of C and Translating Systems 250
2.32 Concept Review Questions 2B 284
2.33 Practice Problems 2B 289

3 DYNAMICS OF A RIGID BODY 300

3.1 $\vec{H}_A$ about a Point A of a Rigid Body 301
3.2 Inertia Tensor at A 303
3.2.1 Principal axes of inertia at A 306
3.2.2 Parallel axes theorem in terms of C 309
3.2.3 Inertia elements for some symmetric uniform bodies 311
3.3 Concept Review Questions 3A 321
3.4 Practice Problems 3A 322
3.5 Angular Impulse–Moment of Momentum Relation for a Rigid Body 324
3.6 Moment of Momentum Equations of Motion of a Rigid Body 335
3.6.1 Motion of a rigid body with $\vec{e}_3 =$ constant, i.e., $\vec{\omega} = \omega\,\vec{e}_3$ 336
3.6.2 Euler's equations for three-dimensional motion of a rigid body 338
3.6.3 Direct equations of moment of momentum if $I_{11}^A = I_{22}^A$ 340
3.7 Kinetic Energy Expression of a Rigid Body 345
3.8 Work–Energy Relation for a Rigid Body 346
3.8.1 Work–energy relation for interconnected system of rigid bodies 347
3.8.2 Conditions of equilibrium of a rigid body 348
3.8.3 Virtual work relation for inter-connected rigid bodies in motion 350
3.8.4 Dynamics of a rigid body relative to a translating frame 351
3.9 Balancing of Rotors 352
3.10 Centre of Percussion 356
3.11 Plane Motion with z-Axis as a Principal Axis of Inertia at C 360
3.12 Illustrative Examples 365
3.13 Concept Review Questions 3B 403
3.14 Practice Problems 3B 413
3.15 Impact of Rigid Bodies 428
3.16 Concept Review Questions 3C 461
3.17 Practice Problems 3C 463

3.18 Torque-Free Motion of a Rigid Body 467
3.18.1 Torque-free motion of rigid body with $I_{11}^O = I_{22}^O \neq I_{33}^O$ 468
3.19 Stability of Uniform Rotation of a Rigid Body 473
3.19.1 Flying saucers are more stable than flying cigars . . . 475
3.20 Steady Precession of Gyroscope 476
3.20.1 Spinning projectile 480
3.20.2 Navigational gyroscopes 480
3.20.3 Gyrocompass 482
3.21 General Motion of a Gyroscope 484
3.22 Spinning Top 485
3.23 Rolling Without Slip of a 'Disc' 491
3.24 Motion of a Billiard Ball 494
3.25 Concept Review Questions 3D 496
3.26 Practice Problems 3D 497

4 STATICS **500**
4.1 Remarks on the Equations of Equilibrium 500
4.1.1 Statically determinate and statically indeterminate systems . . 502
4.2 Examples of One Rigid Body 504
4.3 Analysis of Frames and Mechanisms 512
4.4 Concept Review Question 4A 520
4.5 Practice Problems 4A 523
4.6 Problems Involving Friction 528
4.6.1 Impending motion with tipping or slipping 533
4.6.2 Impending rolling of a cylinder with contact at two or more surfaces 534
4.6.3 Rigid body in plane / line contact under 3D loads 538
4.6.4 Friction grips and friction locking 541
4.6.5 Wedge 542
4.7 Screw 546
4.7.1 Square thread 546
4.7.2 V-thread 548
4.8 Flexible Cables on Rough Surfaces 552
4.9 Rough Pin Joint / Journal Bearing 555
4.10 Rolling Friction 560
4.11 Concept Review Questions 4B 563
4.12 Practice Problems 4B 565
4.13 Internal Force Resultants in Bars 574
4.14 Equations of Equilibrium of a Straight Bar 575
4.15 Trusses 577

4.15.1 Degrees of freedom of a truss 578
4.15.2 Determination of forces in members of a truss 580
4.15.3 Simple and complex trusses 582
4.15.4 Examples on trusses 585
4.16 Flexible Cables 594
4.16.1 Equations of equilibrium of coplanar cable 596
4.16.2 Distributed force $q_0(x)$ per unit projected length 597
4.16.3 Distributed load $\vec{q}(s)$ per unit length 605
4.16.4 Cable subjected to discrete transverse loads 611
4.17 Beams 614
4.17.1 Useful results for drawing BM and SF diagrams 617
4.17.2 Non-straight beams and three-dimensional loads 627
4.18 Concept Review Questions 4C 633
4.19 Practice Problems 4C 636

5 VARIATIONAL MECHANICS **643**
5.1 Hamilton's Principle 643
5.2 Lagrange's Equations for a System with Finite DOF 646
5.2.1 Illustrative examples 648
5.3 Lagrange's Equations for Impulsive Forces 654
5.4 Principle of Virtual Work 656
5.4.1 Illustrative examples 659
5.5 Stability of Equilibrium Configuration 666
5.5.1 Stability of a conservative system of rigid bodies 667
5.5.2 Conditions of stability of 1-dof conservative system 668
5.5.3 Conditions of stability for 2-dof systems 683
5.6 Concept Review Questions 5 688
5.7 Practice Problems 5 692

6 SPECIAL TOPICS **699**
6.1 Motion Under a Central Force 699
6.1.1 Attractive linear central force 701
6.2 Motion Under an Inverse Square Gravitational Force 702
6.2.1 Equation of the trajectory 703
6.2.2 Time period for a closed orbit 706
6.2.3 Parameters of a satellite's orbit in terms of initial r_0, v_0, α_0 . . 707
6.2.4 Geostationary satellite 709
6.2.5 Transfer orbit 710

6.2.6 Effect of atmospheric drag on the orbit of a satellite 710
6.3 Motion of a Charged 'Particle' in an Electromagnetic Field . . . 719
6.3.1 Motion in a uniform magnetic field 719
6.3.2 Motion in an uniform electromagnetic field 721
6.3.3 Effect of a weak magnetic field on motion in an electric field . . 723
6.4 Motion Relative to a Non-Inertial Frame 724
6.5 Effect of Earth's Rotation 724
6.6 Concept Review Questions 6 A, C 734
6.7 Practice Problems 6 A, C 736
6.8 Equations of Motion for a Control Volume 738
6.9 Rocket 743
6.9.1 Single-stage rocket 744
6.9.2 Two-stage rocket 746
6.10 Concept Review Questions 6 D 760
6.11 Practice Problems 6 D 762
6.12 Equation of Motion 765
6.13 Free Vibration of an Undamped System 766
6.14 Free Vibration of a Damped System 771
6.15 Response to Harmonic Excitation 776
6.15.1 Response of a rotating unbalanced rotor 780
6.15.2 Response to the harmonic motion of the support 781
6.15.3 Force transmitted to the foundation 784
6.16 Concept Review Questions 6 E 787
6.17 Practice Problems 6E 790
6.18 Concept of Stress Vector 793
6.19 Concept of a Fluid 794
6.20 Pascal's Law 795
6.21 Equations of Motion 796
6.21.1 Equations of equilibrium of fluids 796
6.22 Pressure Variation in a Fluid in Equilibrium 797
6.22.1 Incompressible fluids 797
6.22.2 Compressible ideal gas 800
6.23 Resultant Force on a Surface in a Fluid in Equilibrium 801
6.24 Buoyancy 807
6.24.1 Equilibrium conditions of floating bodies 808
6.24.2 Stability of partially submerged floating bodies 809
6.25 Concept Review Questions 6 F 813
6.26 Practice Problems 6 F 814
APPENDIXES 817

PREFACE

The subject of mechanics is as old as civilization. People used tools to shift, break and reshape stones, logs and other materials using the knowledge of mechanics inherited from generations of experience. The present status of knowledge of mechanics has been achieved after some false starts, confusion and incremental steps, through debate, experimentation and abstraction. The fuzzy state still exists in some textbooks where the theory is developed from a single particle through multiple-particles to a rigid body and wherein the transition from multi-particles to a continuous body is not rigorous. In mechanics of deformable bodies, the basic postulate of Cauchy consists of a force distribution across an internal surface and not forces between distant material points as postulated in the above development. Moreover, many concepts are not rigorously defined or elaborated, and introduction to variational mechanics is ignored. The present text attempts to reduce and, hopefully, eliminate these shortcomings.

In this book, the subject is developed in an integrated manner, in the spirit of modern continuum mechanics, since the classical model is physically valid only for systems of finite size. Euler's two axioms of momentum and moment of momentum form the basis of this theory. The logical sequence of kinematics, axioms, dynamics and statics has been followed. Statics has been treated as a special case of dynamics and presented later, so that the tendency to treat equations of dynamics like the equations of equilibrium, through pseudo-forces, is curbed. This logical approach is also economical. A chapter on variational mechanics, based on Hamilton's principle rather than D'Alembert's principle, is included to expose the students, early on, to an alternative formulation of mechanics which is the basis of engineering theories of beams, plates and shells and of modern computational methods.

The concepts are rigorously defined for the general case, prior to presenting their specific forms for particular cases. The purpose is to avoid ambiguities and inculcate more abstract thinking. This approach ensures that students avoid learning expressions for a special case first, which are not valid for the general case. For example, the angular velocity of a rigid body in plane motion reduces to the derivative of a finite angle, which is not so for general three-dimensional motion. The work done by a force acting at the same material point is related to the differential displacement of this material point, whereas for the general case it is the time integral of the rate of work done by the force, which itself is defined in terms of the velocity of its material point of application.

The book is replete with solved examples which illustrate the application of the underlying principles and governing equations to complex engineering problems and the manner in which they can be broken down for analysis. Special emphasis has been laid on drawing free-body diagrams before application of force or moment equations.

The students can gauge themselves on their understanding of the subject at three levels: (i) Through parts of the theory left as exercises, to concretise their understanding of the process of development of theory; (ii) through concept review questions that test their knowledge of basic concepts and the direct application of governing equations; and (iii) through practice problems which provide an opportunity to apply the knowledge gained to problems of greater complexity. Practice problems include elementary design problems (identified by a superscript ^{d} that accompanies their numbering) and challenging problems (indicated by a superscript *). Solving the latter will ensure that learning is deep-rooted.

Most of the questions and problems presented here are original or have been formulated afresh by us. However, several problems from standard existing texts on mechanics have also been included. In addition to those appearing in the printed book, 340 questions and problems, and 8 examples are included in a 110-page-long supplement available at www.universitiespress.com/EngineeringMechanics-DSV. The complete textbook, including the supplement, is also available in the e-book format.

The book is suitable as a text-book for a one semester undergraduate first course on mechanics mainly for engineering students but also for physics and mathematics students. Some topics should be given as reading assignments. Lagrange's equations and some applications in various chapters, especially part D of chapter 3 and some topics of chapter 6, may be excluded depending on the intended audience. Some parts of the book can be used as supplementary bridge material to an advanced course.

All learning, ultimately, is self-learning. One may forget the spoken word, retain parts of images seen, but grasps what one does or experiences. Students are therefore advised to read the book by working out the theory, exercises, most examples, questions and problems, including some challenging ones. Instructors will find the bank of problems, including those in the web supplement, large enough to choose from for ensuring distinct assignments for various groups of the same class and for assessment tests.

We thank the undergraduate students of IIT Delhi, since through teaching them, we have gained an understanding of mechanics and how best to teach it. We are indebted to several faculty members of the Applied Mechanics Department (IIT Delhi) for informative discussions and feedback on various versions of the text. We are very grateful to the reviewer for the conscientious reading of our manuscript and the many constructive suggestions. We also thank Mr. Pawan Tripathi for his help with the drawings in the book. We heartily thank Ms. Sreelatha Menon (Managing Editor, Universities Press) for all the cooperation, guidance and assistance in the execution of this publication. We would be grateful to our readers if typographical and other errors are brought to our notice. This book would not have been possible without the active encouragement of our families and their patient support through several ups and downs.

Based on our collective experience at IIT Delhi, we feel that the following sections can be easily discussed in a semester-long introductory course on Engineering Mechanics.

Sections 1.4 to 1.11; 1.15 to 1.23; 2.1 to 2.23; 2.26 to 2.28; 2.30; 3.1, 3.2, 3.5 to 3.11; 3.15; 3.20; 4.1; 4.3; 4.6; 4.8; 4.13 to 4.15; 4.17; 5.1 to 5.5; 6.1, 6.2, 6.4 and 6.5.

Some sections are elementary (e.g., 1.1 to 1.3), while others may be a little advanced for an introductory course (e.g., 2.29).

P C Dumir, S Sengupta and S V Veeravalli

CHAPTER 1

KINEMATICS

In this chapter, we discuss kinematics of a moving point with reference to two frames and kinematics of a rigid body.

1.1 Mechanics

Motion is inherent in the universe, from the subatomic to the cosmic scale. For example, motion of galaxies, planets, glaciers, ground water, winds, blood, vehicles, robots, electric motors, dust particles, gas molecules, electrons, etc. The diverse motions of various material systems, are partly governed by common physical principles. *Mechanics is the physical science dealing with the study of the mechanical actions (forces and moments) and their relation to the state of rest, motion, deformation or flow of material bodies of all kinds.* A material *body* is modelled as a continuous distribution of mass whose configuration at time t is given by the volume V enclosed in a surface S. Its points are called *material points.* Mechanics forms the basis of design, manufacture and operation of engineering components subject to mechanical actions.

Mechanics deals with problems of equilibrium configurations and their stability, transient and steady state dynamic response, wave propagation in bodies, etc. Its basic divisions are: kinematics, dynamics and statics. *Kinematics is the study of the geometry of motion of material systems without considering the forces and the mass involved in it.* It covers the study of relations between the position, displacement, velocity and acceleration of points, lines and bodies and the deformation of bodies. Kinematics is the basis of analysis and synthesis of mechanisms and is an essential first step for the study of dynamics. *Dynamics is the study of the interrelations between the applied forces, the inertial properties of the mechanical system and its resulting motion. Statics deals with material systems in equilibrium.* In moving systems wherein the inertia term is small compared to the individual applied forces, time-dependence is included, but the inertia term is neglected. Such an analysis is called *quasi-static.*

The mechanical behaviour of materials under applied forces differs considerably. The distance between material points of rigid bodies remains constant. Fluids continue to deform under arbitrarily small shear stresses. Elastic bodies instantaneously deform under load and instantaneously recover their original size on unloading. Thus, mechanics can also be classified into *mechanics of rigid bodies* and *mechanics of deformable bodies.* The latter is subdivided into *mechanics of solids* (with branches of elasticity, plasticity, etc.), *mechanics of fluids* (with branches of hydraulics, gas dynamics, etc.), and *rheology* (with branches of viscoelasticity, viscoplasticity, etc.).

The development of mechanics is linked to the development of the productive forces of society. The tools of the food gatherers, weapons of the hunters, ploughs of the agrarian people; building construction implements like – wheel, lever, pulley and winch for settled populations; vehicles for traders, boats of the sailors, etc., developed empirically over ages. This acquisition of data led to curiosity about *cause, effect and prediction.* The pyramids prove ancient man's knowledge of some basic principles of statics. This knowledge was summed up by Aristotle and Archimedes. The transition from handicrafts to mechanised methods of production led to the development of dynamics by Galileo, and Newton gave a definite shape to the principles of mechanics. Newton deduced the axiom of gravitational force from Kepler's three empirical laws of planetary motion. The discovery of the planet Neptune in 1846, as predicted by Newtonian mechanics was a triumph of theoretical mechanics. Euler modified and extended Newton's laws to all systems. Lagrange evolved an alternative variational formulation of Newtonian mechanics which was generalised by Hamilton's principle.

With more accurate instruments, it was found that classical or Newtonian mechanics wrongly predicts the motion of bodies travelling at speeds comparable to the speed of light (3×10^8 m/s), and also motion of bodies of atomic and subatomic size. For the former case, Einstein developed relativistic mechanics; and for the latter case, quantum mechanics was developed by Schrödinger, Heisenberg, etc.

A mathematical theory of a branch of physical science consists of the following.

i) A list of certain basic entities, called *primitives*, which are fundamental concepts ingrained in our experience about which we have an intuitive idea but whose precise definition cannot be given.

ii) A list of other entities, *defined* in terms of the primitives.

iii) Codification of physical observations into a *minimum* number of principles called *axioms*, which are expressed as mathematical relations between the defined entities and the primitives.

iv) Theorems derived from the axioms.

Quantum mechanics, relativistic mechanics and classical mechanics are three mathematical models for the motion of mechanical systems: *Quantum mechanics* is needed to explain the behaviour of bodies of atomic and subatomic size; *relativistic mechanics* is relevant to bodies of any size whose speeds approach the speed of light (3×10^8 m/s); *classical mechanics* is physically valid for the motion of bodies of size much greater than the atomic size, ranging from dust particles to galaxies, in a range of speed from zero to one-tenth the speed of light. The two formulations of classical mechanics are Newtonian and variational. The Newtonian formulation is based on forces and accelerations, and uses differential and integral calculus. The variational formulation is based on work and energy, and employs variational calculus. In this text, the emphasis is on Newtonian mechanics with a small introduction to variational mechanics.

Newton presented the foundations of classical mechanics in 1687 through three laws:

Law 1: Every body preserves in its state of rest or of uniform motion in a right line, unless it is compelled to change that state by forces impressed thereon.
Law 2: The alteration of motion is ever proportional to the motive force impressed, and is made in the direction of the right line in which that force is impressed.
Law 3: To every action, there is always opposed an equal reaction, or the mutual actions of two bodies upon each other are always equal and directed to contrary parts.

The laws, as stated, are imprecise and have been found to be physically invalid and inadequate. For example, a shell translating with uniform velocity under zero external force, may explode resulting in non-uniform motion of its parts. This invalidates the first law. The second law does not specify the point whose alteration of motion is proportional to the applied force. The frame of reference[1] for the first two laws is not specified. The third law can be derived from a more basic set of axioms of classical mechanics and so it is not a law in that model. The second law gives only the motion of the centre of mass of any system and the three laws together are insufficient to predict the motion of other points of the system, even for a rigid one. These laws do not even yield the complete system of equations of motion of a rigid body in translation. An exception is the case of an idealised system of mass-points, but classical mechanics is not physically valid for them.

Almost a century later, Euler modified and extended Newton's laws by adding the axiom of moment of momentum. His two axioms form the basis of classical mechanics.

1.2 Vectors

An entity that is completely characterised by its magnitude, which is independent of the frame of reference and the choice of the axes in it, is called a *scalar*. Length, area, volume, mass, density, temperature, time, work, energy and pressure are scalars.

[1] will be defined in Sec. 1.4

Entities such as velocity, force and momentum, are modelled by directed line segments.

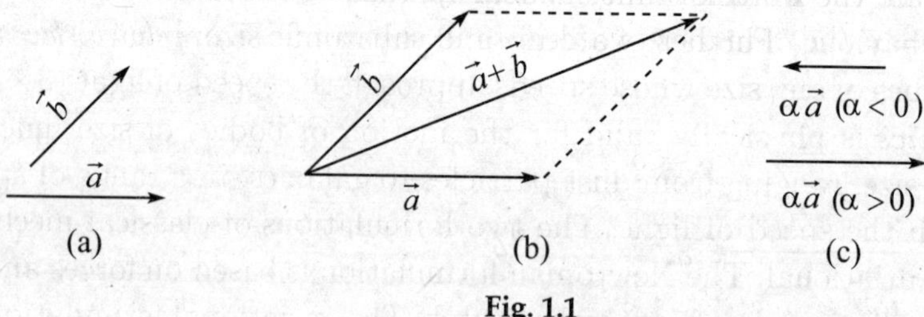

Fig. 1.1

We define the addition of two directed line segments (arrows) $\vec{a}$ and $\vec{b}$ (Fig. 1.1) by the parallelogram rule of addition, i.e., the sum $\vec{a}+\vec{b}$ is a directed line segment represented by the diagonal of a parallelogram, whose adjacent sides are $\vec{a}$ and $\vec{b}$. Additionally, we define the multiplication of $\vec{a}$ with a scalar α, denoted by $\alpha\,\vec{a}$, by a directed line segment of magnitude $|\alpha|$ times that of $\vec{a}$ with the direction same as or opposite to $\vec{a}$ depending on whether α is positive or negative. The set of all directed line segments, with these defined operations of addition and multiplication with a scalar, forms a **vector space of arrows**. Thus, a physical entity which has magnitude and direction, and which adds according to the parallelogram rule and gets multiplied with a scalar according to the aforementioned rule, is a vector. Position vector, displacement, velocity, acceleration, force, momentum, angular velocity, angular acceleration, moment and moment of momentum are vectors. The length of a directed line segment $\vec{a}$ is taken as the *length* or *magnitude* of the vector and is denoted as $|\vec{a}|$. A physical entity having magnitude and direction needs a unit to specify its magnitude and a coordinate system to specify its direction.

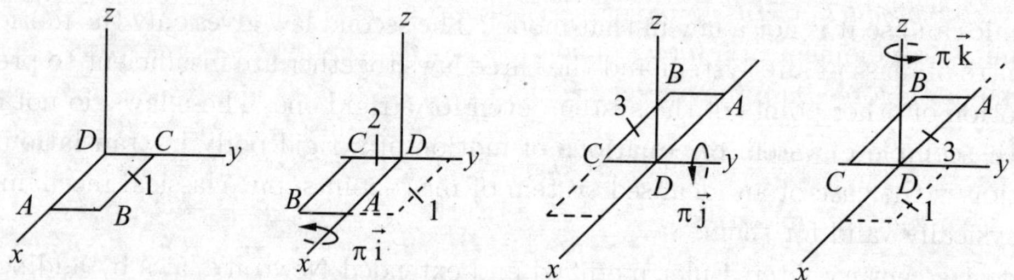

Fig. 1.2

Finite rotation has magnitude and direction, but it is not a vector since its addition is not according to the parallelogram law. Consider a lamina (Fig. 1.2) in position 1 in the xy-plane. A rotation of $\pi\vec{i}$ (π radians about the x-axis) displaces it to position 2. A further rotation of $\pi\vec{j}$ displaces it to position 3. A single rotation of $\pi\vec{k}$ displaces the lamina from position 1 to position 3. Thus, successive rotations of $\pi\vec{i}$ and $\pi\vec{j}$ add to a rotation of $\pi\vec{k}$ and not to a rotation of $\pi(\vec{i}+\vec{j})$ as per the parallelogram law.

A vector of unit magnitude is defined as a *unit vector*. It is often denoted by the letter $\vec{e}$ (in German, the word for unity is einheit). The unit vector $\vec{e}_a$ in the direction of vector $\vec{a}$ is $\vec{e}_a = \vec{a}\,/|\vec{a}|$. Further, we denote a vector of zero magnitude as $\vec{0}$ (zero vector).

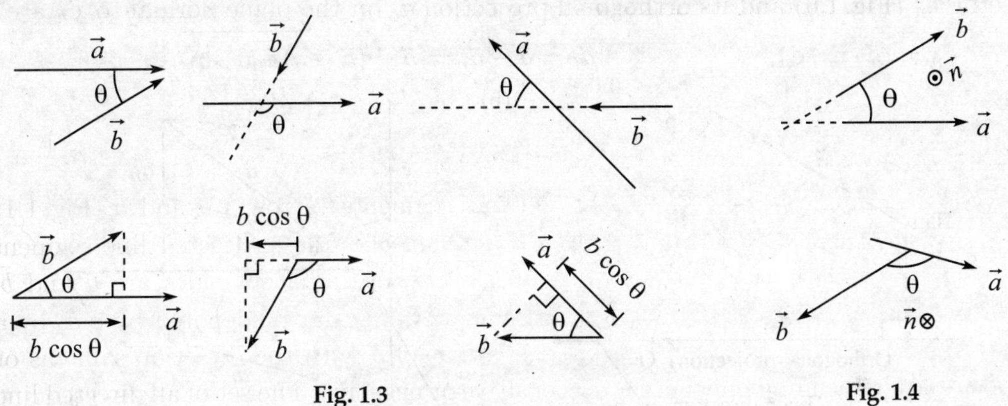

Fig. 1.3 **Fig. 1.4**

The dot product is useful for the definition of kinetic energy and work. The *dot product* or *scalar product* of two vectors $\vec{a}$ and $\vec{b}$, written as $\vec{a}\cdot\vec{b}$ is defined by

$$\vec{a}\cdot\vec{b} = \left\{ \begin{array}{ll} |\vec{a}||\vec{b}|\cos\theta & \text{if } a \neq \vec{0} \text{ and } \vec{b} \neq \vec{0} \\ 0 & \text{if } a = \vec{0} \text{ or } \vec{b} = \vec{0} \end{array} \right\}$$

where θ is the smaller interior angle between them (Fig. 1.3). Note that if $\vec{a}\cdot\vec{b} = 0$, then $\vec{a} = \vec{0}$ or $\vec{b} = \vec{0}$ or $\vec{a} \perp \vec{b}$. It follows that if $\vec{a}\cdot\vec{b} = \vec{c}\cdot\vec{b}$, i.e., $(\vec{a}-\vec{c})\cdot\vec{b} = 0$, then $\vec{a}-\vec{c} = \vec{0}$ or $\vec{b} = \vec{0}$ or $(\vec{a}-\vec{c}) \perp \vec{b}$. However, if $\vec{a}\cdot\vec{b} = \vec{c}\cdot\vec{b}\ \forall\, \vec{b}$, then $\vec{a} = \vec{c}$.

The *vector product* or *cross product* of vectors $\vec{a}$ and $\vec{b}$, denoted by $\vec{a}\times\vec{b}$, is useful for the definition of moment of a force, moment of momentum and the description of the electromagnetic force on a moving charge in a magnetic field. It is defined as follows.

1. If $\vec{a}$ and $\vec{b}$ are linearly independent vectors, then

   $$\vec{a}\times\vec{b} = |\vec{a}|\,|\vec{b}|\sin\theta\,\vec{n}$$

   where θ $(0 < \theta < \pi)$ is the smaller angle between $\vec{a}$ and $\vec{b}$ (Fig. 1.4). $\vec{n}$ is a unit vector normal to both $\vec{a}$ and $\vec{b}$ decided by the right-hand rule, i.e., if the fingers of the right hand are curled from $\vec{a}$ to $\vec{b}$, then $\vec{n}$ is in the direction of the thumb.
2. If $\vec{a}$ and $\vec{b}$ are linearly dependent vectors, i.e., $\theta = 0$ or π, then $\vec{a}\times\vec{b} = \vec{0}$.

The area of a parallelogram with adjacent sides $\vec{a}$ and $\vec{b}$ is $|\vec{a}\times\vec{b}|$. If $\vec{a}\times\vec{b} = \vec{0}$, then at least one of the following three conditions holds: $\vec{a} = \vec{0}$, $\vec{b} = \vec{0}$, $\vec{a} \parallel \vec{b}$. Hence, if $\vec{a}\times\vec{b} = \vec{c}\times\vec{b}$, then $(\vec{a}-\vec{c})\times\vec{b} = \vec{0}$ and at least one of the following three conditions holds: $\vec{a}-\vec{c} = \vec{0}$, $\vec{b} = \vec{0}$, $(\vec{a}-\vec{c}) \parallel \vec{b}$. However, if $\vec{a}\times\vec{b} = \vec{c}\times\vec{b}\ \forall\, \vec{b}$, then $\vec{a} = \vec{c}$.

The orthogonal projection or *rectangular component* or orthogonal component of a vector $\vec{a}$ in a given direction is a directed line segment $\overrightarrow{PQ}$ from the orthogonal projection P of the tail of $\vec{a}$ to the orthogonal projection Q of the tip of $\vec{a}$ on that direction (Fig. 1.5). The orthogonal component $\vec{a}_n$ of $\vec{a}$ in the direction of the unit vector $\vec{\mathrm{e}}_n$ (Fig. 1.6) and its orthogonal projection $\vec{a}_t$ on the plane normal to $\vec{\mathrm{e}}_n$ are

$$\vec{a}_n = (\vec{a} \cdot \vec{\mathrm{e}}_n)\,\vec{\mathrm{e}}_n, \qquad \vec{a}_t = \vec{a} - \vec{a}_n = \vec{a} - (\vec{a} \cdot \vec{\mathrm{e}}_n)\,\vec{\mathrm{e}}_n$$

Fig. 1.5

Fig. 1.6

A rotation about a directed line L is called *positive* if it bears to the direction of L the same relation as the rotation of a right-handed screw bears to its direction of motion. Three non-coplanar vectors at a point, say $\vec{\mathrm{b}}_1$, $\vec{\mathrm{b}}_2$, $\vec{\mathrm{b}}_3$, can be chosen as the basis vectors and any vector $\vec{v}$ can be uniquely expressed as a linear combination of these: $\vec{v} = v_1\vec{\mathrm{b}}_1 + v_2\vec{\mathrm{b}}_2 + v_3\vec{\mathrm{b}}_3$. v_is are called the components of $\vec{v}$ with respect to the bases $\vec{\mathrm{b}}_i$. Basis vectors, taken in some order, are called a *triad.* In an *orthogonal triad* the vectors $\vec{\mathrm{a}}_1$, $\vec{\mathrm{a}}_2$, $\vec{\mathrm{a}}_3$, are mutually orthogonal. It is called right-handed if $\vec{\mathrm{a}}_3$ is directed along $\vec{\mathrm{a}}_1 \times \vec{\mathrm{a}}_2$. Any other triad is called right-handed if it can be deformed continuously into a right-handed orthogonal triad without its vectors becoming coplanar at any stage. We choose a right-handed orthogonal triad of unit vectors, called an *orthonormal* triad, $\vec{\mathrm{i}}, \vec{\mathrm{j}}, \vec{\mathrm{k}}$ as a basis: $\vec{\mathrm{i}}\cdot\vec{\mathrm{j}} = \vec{\mathrm{j}}\cdot\vec{\mathrm{k}} = \vec{\mathrm{k}}\cdot\vec{\mathrm{i}} = 0$, $\vec{\mathrm{i}}\times\vec{\mathrm{j}} = \vec{\mathrm{k}}$, $\vec{\mathrm{j}}\times\vec{\mathrm{k}} = \vec{\mathrm{i}}$, $\vec{\mathrm{k}}\times\vec{\mathrm{i}} = \vec{\mathrm{j}}$. The rotation $\vec{\mathrm{i}}$ to $\vec{\mathrm{j}}$ is about $\vec{\mathrm{k}}$, the rotation $\vec{\mathrm{j}}$ to $\vec{\mathrm{k}}$ is about $\vec{\mathrm{i}}$ and the rotation $\vec{\mathrm{k}}$ to $\vec{\mathrm{i}}$ is about $\vec{\mathrm{j}}$.

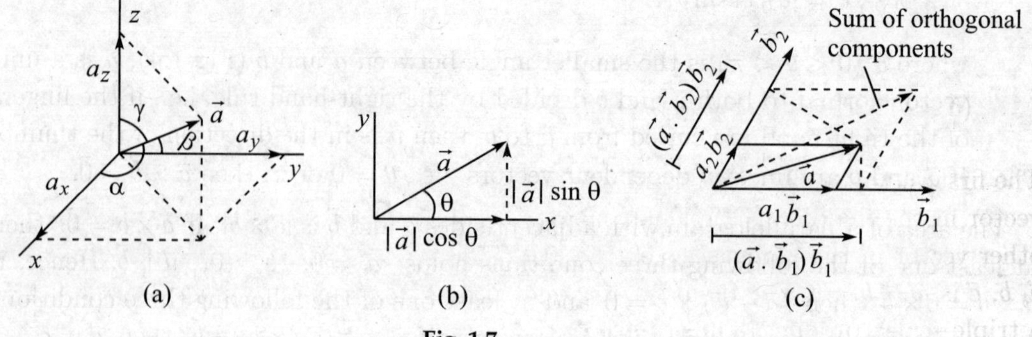

Fig. 1.7

A vector $\vec{a}$, making angles α, β, γ with $\vec{\mathrm{i}}$, $\vec{\mathrm{j}}$, $\vec{\mathrm{k}}$, is represented as $\vec{a} = a_x\,\vec{\mathrm{i}} + a_y\,\vec{\mathrm{j}} + a_z\,\vec{\mathrm{k}}$ (Fig. 1.7a), with $a_x = \vec{a}\cdot\vec{\mathrm{i}} = |\vec{a}|\,|\vec{\mathrm{i}}|\cos\alpha = |\vec{a}|\cos\alpha$, $a_y = |\vec{a}|\cos\beta$, $a_z = |\vec{a}|\cos\gamma$:

$$\vec{a} = |\vec{a}|(\cos\alpha\,\vec{\mathrm{i}} + \cos\beta\,\vec{\mathrm{j}} + \cos\gamma\,\vec{\mathrm{k}}) = |\vec{a}|(l\,\vec{\mathrm{i}} + m\,\vec{\mathrm{j}} + n\,\vec{\mathrm{k}})$$

where l, m, n are the direction cosines of $\vec{a}$ and the vector in the parenthesis is a unit vector in its direction. A vector $\vec{a}$ normal to $\vec{\mathrm{k}}$, at an angle θ to $\vec{\mathrm{i}}$ (Fig. 1.7b), is

$$\vec{a} = |\vec{a}|(\cos\theta\,\vec{\mathrm{i}} + \sin\theta\,\vec{\mathrm{j}}) = |\vec{a}|(l\,\vec{\mathrm{i}} + m\,\vec{\mathrm{j}})$$

Note that if the basis vectors $\vec{\mathrm{b}}_i$, $i = 1, 2, 3$, are of unit length but not orthogonal, then the components of vector $\vec{a}$ with respect to this basis are not given by $\vec{a}\cdot\vec{\mathrm{b}}_i$. However, $\vec{a}\cdot\vec{\mathrm{b}}_i$ are the orthogonal components of $\vec{a}$ along $\vec{\mathrm{b}}_i$ and their sum differs from the given vector. This is illustrated in Fig. 1.7c for the coplanar case.

Let $\vec{a} = a_x\,\vec{\mathrm{i}} + a_y\,\vec{\mathrm{j}} + a_z\,\vec{\mathrm{k}}$, $\vec{b} = b_x\,\vec{\mathrm{i}} + b_y\,\vec{\mathrm{j}} + b_z\,\vec{\mathrm{k}}$, and α be a real number. Then,

$$\begin{aligned}
\vec{a} + \vec{b} &= (a_x + b_x)\,\vec{\mathrm{i}} + (a_y + b_y)\,\vec{\mathrm{j}} + (a_z + b_z)\,\vec{\mathrm{k}} \\
\alpha\vec{a} &= (\alpha a_x)\,\vec{\mathrm{i}} + (\alpha a_y)\,\vec{\mathrm{j}} + (\alpha a_z)\,\vec{\mathrm{k}} \\
\vec{a}\cdot\vec{b} &= a_x b_x + a_y b_y + a_z b_z \\
|\vec{a}| &= (\vec{a}\cdot\vec{a})^{1/2} = (a_x^2 + a_y^2 + a_z^2)^{1/2} \\
\vec{a}\times\vec{b} &= \begin{vmatrix} \vec{\mathrm{i}} & \vec{\mathrm{j}} & \vec{\mathrm{k}} \\ a_x & a_y & a_z \\ b_x & b_y & b_z \end{vmatrix}
\end{aligned}$$

A vector equation in 3-dimensional space is, in general, equivalent to 3 scalar equations for the components. The scalar and vector triple products $\vec{a}\times\vec{b}\cdot\vec{c}$, $(\vec{a}\times\vec{b})\times\vec{c}$, $\vec{a}\times(\vec{b}\times\vec{c})$, can be expressed as

$$\vec{a}\times\vec{b}\cdot\vec{c} = \begin{vmatrix} a_x & a_y & a_z \\ b_x & b_y & b_z \\ c_x & c_y & c_z \end{vmatrix}$$

$$(\vec{a}\times\vec{b})\times\vec{c} = (\vec{a}\cdot\vec{c})\vec{b} - (\vec{b}\cdot\vec{c})\vec{a}, \qquad \vec{a}\times(\vec{b}\times\vec{c}) = (\vec{a}\cdot\vec{c})\vec{b} - (\vec{a}\cdot\vec{b})\vec{c}$$

The first vector on the right hand side of the triple vector product is a multiple of the vector in the middle on the left hand side and the second vector is a multiple of the other vector in the parenthesis on the left. The volume of a parallelepiped with sides $\vec{a}$, $\vec{b}$, $\vec{c}$ is $|\vec{a}\times\vec{b}\cdot\vec{c}|$. The moment of a force about a line can be expressed in terms of a triple scalar product. The triple cross product appears in the relation between the accelerations of two points of a rigid body.

§ **Exercise 1.1:** Let α_i be the components of a vector $\vec{a}$ with respect to any basis $\vec{b}_i$, $\vec{a} = \alpha_1\vec{b}_1 + \alpha_2\vec{b}_2 + \alpha_3\vec{b}_3$. By forming the triple scalar product $\vec{a} \cdot \vec{b}_2 \times \vec{b}_3$, show that the component $\alpha_1 = [\vec{a} \cdot (\vec{b}_2 \times \vec{b}_3)] / [\vec{b}_1 \cdot (\vec{b}_2 \times \vec{b}_3)]$.

The vector function $\vec{a}(u)$ is said to be *continuous* at u_0 if $\lim_{u \to u_0}[\vec{a}(u) - \vec{a}(u_0)] = \vec{0}$. The *derivative* of $\vec{a}(u)$ with respect to u is defined by

$$\frac{d\vec{a}}{du} = \lim_{\Delta u \to 0} \frac{\vec{a}(u + \Delta u) - \vec{a}(u)}{\Delta u}$$

The derivative of a vector is a vector. Higher derivatives are similarly defined.

§ **Exercise 1.2:** Show, by taking the derivative of $\vec{a} \cdot \vec{a}$, that the rate of change of a vector $\vec{a}(u)$ of constant magnitude is perpendicular to the vector itself, except in the trivial case of either $d\vec{a}/du$ or $\vec{a}$ being zero.

$\vec{A}(u)$ is defined as an *integral* or *anti-derivative* of $\vec{a}(u)$ if $d\vec{A}(u)/du = \vec{a}(u)$, and we write $\vec{A}(u) = \int \vec{a}(u)du$. This is an indefinite integral, as an arbitrary constant vector can be added to it. The *definite integral* of $\vec{a}(u)$ is defined as the limit of a sum. It is related to the indefinite integral by the fundamental theorem of integral calculus:

$$\frac{d\vec{A}(u)}{du} = \vec{a}(u) \;\; \forall\, u \in (u_1, u_2) \quad \Rightarrow \quad \int_{u_1}^{u_2} \vec{a}(u)du = \vec{A}(u_2) - \vec{A}(u_1)$$

The rules of differentiation and integration of vector functions are the same as those for scalar functions, except that the order of the cross product must be preserved. Linear differential equations involving differential coefficients of vectors with respect to a scalar variable are solved in exactly the same way as the differential equations of scalar variables, but the constants of integration are vectors.

It is significant that the relations among vectors are invariant with respect to translation and rotation of coordinates. Hence, it is ideal to present physical laws in vector form, since the experiments on which these laws are based and the laws of physics themselves remain unchanged when we rotate or translate the coordinate system.

1.3 Problem Solving Process

The first step to find the solution of a physical system in mechanics is to make its mathematical model by suitable abstractions/idealisations of the physical situation to have a simple model. For example, a rigid/elastic body, uniform density, uniform gravitational field, discrete forces, point contact, perfectly circular cylinder, etc. In this text, we shall mostly deal with physical systems with a given mathematical model.

The mathematical model of a physical system depends on the type of information required and the desired accuracy. For example, the earth is modelled as a mass-point for finding its orbital motion, as a rigid sphere for the problem of tides and lunar eclipses, as a rigid ellipsoid for the problem of precession of the equinoxes, and as an elastic sphere for the study of waves due to an earthquake.

The second step is to draw a sketch of the isolated mechanical system under consideration, showing the forces and moments exerted by the surroundings on it. Such a diagram is called a *free body diagram* (see Section 2.17 for details). For a vector entity with a known direction but an unknown sense, we do not impose an intuitive idea of its sense. The solution of the equations will yield the correct sign. An exception is the case of friction for impending or actual slip. The third step is to set up the governing equations of the system. Their solution is interpreted physically and compared with the observed facts, if any. In case of disagreement, the idealisations are modified.

An exact analytical algebraic solution is usually obtained for a simple physical problem using a simplified mathematical model. If the problem is complex, or if a more realistic complex model is used, then an approximate analytical solution or a numerical one is tried. The analytical solution permits a dimensional check and the numerical values are substituted only in the final expressions. The effect of a parameter in the equation on the solution is not directly revealed in a numerical solution, but it is directly observed in an analytical solution. Thus, one should attempt an exact, an approximate analytical or a numerical solution, in that order. Most problems in this text have exact solutions. Arrange the solution in an orderly fashion to reduce chance of errors and permit a quick recheck.

The accuracy of the model/governing equations/method of solution should conform to the accuracy of the available data and that of the desired results. The data in engineering problems are seldom known to have an accuracy greater than 0.2% and so an accuracy of solution greater than 0.2% is not necessary in most problems. The results of intermediate calculations should be recorded with at least 4 significant digits. More significant digits are needed when subtractions of almost equal numbers occur.

The import of this discussion is illustrated by the following anecdote. A visitor to a museum is told by a guide that the age of a particular dinosaur on display is 60 million and 3 years. On being questioned about this 'accurate' description, the guard explains that when he joined the museum 3 years back, he was told that the age of a dinosaur was 60 million years. Such 'accuracy' is absurd—it is both impossible and unnecessary to specify the age of the dinosaur to better accuracy than a few thousand years.

Example 1.1: Express fluid pressure p ($50\,\mathrm{kN/m^2}$) at a point A, on solid bodies immersed in a fluid (Fig. E1.1), as a vector in terms of Cartesian components.

Solution: If $\vec{n}$ is the outward normal to the body at point A, then $\vec{p} = -p\,\vec{n}$.

(a) $\vec{p} = -50(\cos\theta\,\vec{\mathrm{i}} + \sin\theta\,\vec{\mathrm{j}})\,\mathrm{kPa}$

(b) $\overrightarrow{PQ} = -2\vec{\mathrm{i}} + 3\vec{\mathrm{j}}, \qquad \overrightarrow{PR} = -2\vec{\mathrm{i}} + 6\vec{\mathrm{k}}$

$$\vec{n} = \frac{\overrightarrow{PQ} \times \overrightarrow{PR}}{|\overrightarrow{PQ} \times \overrightarrow{PR}|} = \frac{18\vec{i} + 12\vec{j} + 6\vec{k}}{6(3^2+2^2+1^2)^{1/2}} = 0.80178\vec{i} + 0.53452\vec{j} + 0.26726\vec{k}$$

$$\vec{p} = -50\,\vec{n} = (-40.09\vec{i} - 26.73\vec{j} - 13.36\vec{k})\,\text{kPa}$$

(c) $$\vec{n} = \frac{\vec{a}_1 \times \vec{a}_2}{|\vec{a}_1 \times \vec{a}_2|} = \vec{a}_1 \times \vec{a}_2$$

where $\vec{a}_2$, $\vec{a}_1$ are the orthogonal unit vectors in the peripheral direction and along the slant side of the cone at point A. The unit vector along OB is $\vec{a}_3$.

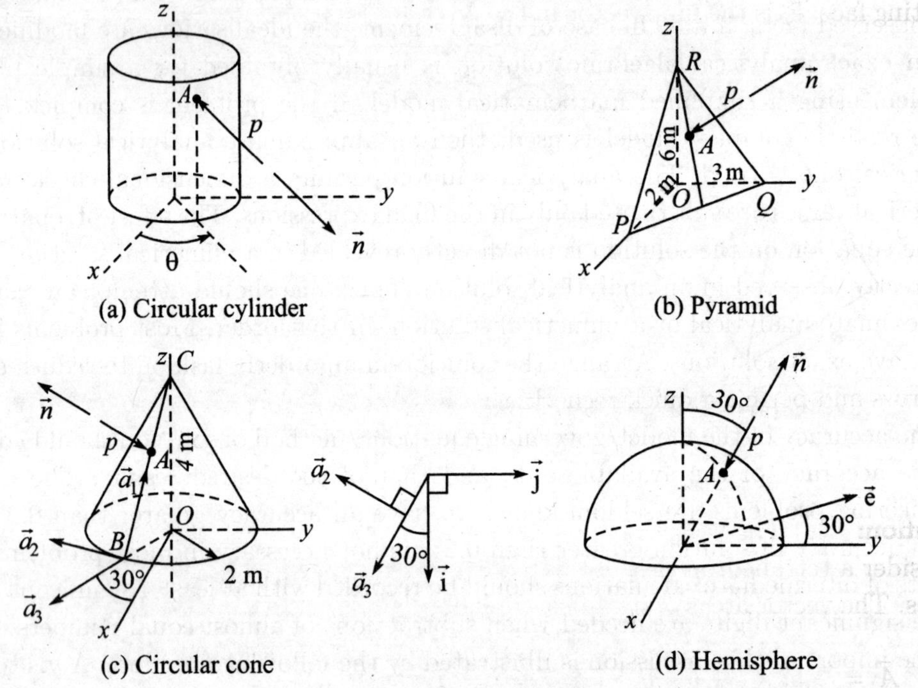

Fig. E1.1

$$\overrightarrow{BC} = -2\,\vec{a}_3 + 4\vec{k} = -2(\cos 30^\circ\,\vec{i} - \sin 30^\circ\,\vec{j}) + 4\vec{k} = -1.732\vec{i} + \vec{j} + 4\vec{k}$$

$$\vec{a}_1 = \frac{\overrightarrow{BC}}{|\overrightarrow{BC}|} = \frac{-1.732\vec{i} + \vec{j} + 4\vec{k}}{\sqrt{1.732^2 + 1^2 + 4^2}} = -0.3873\vec{i} + 0.2236\vec{j} + 0.8944\vec{k}$$

$$\vec{a}_2 = -(\sin 30^\circ\,\vec{i} + \cos 30^\circ\,\vec{j}) = -0.5\vec{i} - 0.866\vec{j}$$

$$\vec{n} = \vec{a}_1 \times \vec{a}_2 = (-0.3873\vec{i} + 0.2236\vec{j} + 0.8944\vec{k}) \times (-0.5\vec{i} - 0.866\vec{j})$$

$$= 0.7746\vec{i} - 0.4472\vec{j} + 0.4472\vec{k}$$

$$\vec{p} = -p\,\vec{n} = (-38.73\vec{i} + 22.36\vec{j} - 22.36\vec{k})\,\text{kPa}$$

(d) $\vec{p} = -p\vec{n} = -50(\cos 30° \vec{k} + \sin 30° \vec{e})$

$= -43.3\vec{k} - 25(\cos 30° \vec{j} - \sin 30° \vec{i}) = (12.5\vec{i} - 21.65\vec{j} - 43.3\vec{k})$ kPa

Example 1.2: Prove that (a) the total vector surface area of a tetrahedron is zero, (b) the total vector surface area of any solid is zero, (c) the projected area A^* of a surface with area $\vec{A}$ on a given plane with normal $\vec{n}$ is $\vec{A} \cdot \vec{n}$, and (d) if A is the area of a slant face of a tetrahedron (Fig. E1.2), then the area of the face normal to $\vec{e}_i$ is An_i, where n_i is the direction cosine with respect to $\vec{e}_i$ of the outward normal $\vec{n}$ to the slanting face. $\vec{e}_i$ is the unit vector in the direction of the coordinate axis x_i $(i = 1, 2, 3)$.

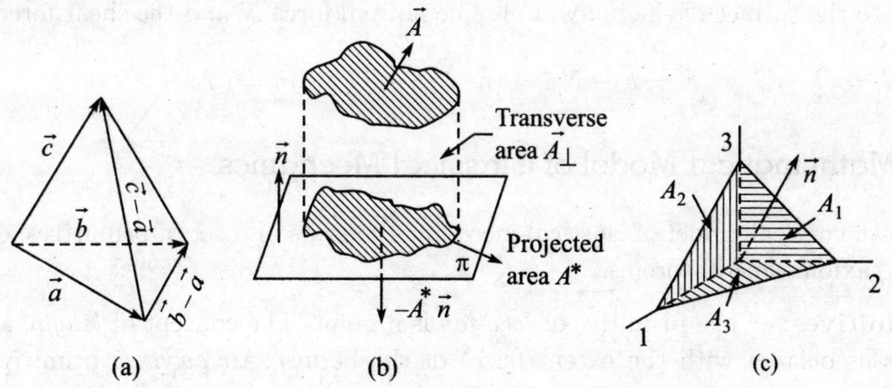

Fig. E1.2

Solution: (a) The vector area of a surface element is directed along outward normal. Consider a tetrahedron (Fig. E1.2a), with vectors $\vec{a}$, $\vec{b}$, $\vec{c}$ representing its 3 concurrent edges. The vector areas of its 4 faces are along their outward normal vectors with sum:

$$\vec{A} = \tfrac{1}{2}[\vec{a}\times\vec{c} + \vec{b}\times\vec{a} + \vec{c}\times\vec{b} + (\vec{b}-\vec{a})\times(\vec{c}-\vec{a})]$$

$$= \tfrac{1}{2}[\vec{a}\times\vec{c} + \vec{b}\times\vec{a} + \vec{c}\times\vec{b} + \vec{b}\times\vec{c} - \vec{b}\times\vec{a} - \vec{a}\times\vec{c} + \vec{a}\times\vec{a}] = \vec{0}$$

(b) Any solid can be split into a series of tetrahedrons. The sum of the vector areas of all the tetrahedrons is zero. Hence, the sum of the vector areas of the interfaces (they cancel out in pairs) and the vector area of the surface bounding the solid is zero. ⇒ The vector area of the surface bounding the solid is zero.

(c) Project every point of the given surface on the given plane π. This generates a cylinder with generators parallel to the normal $\vec{n}$ to π. Its end-faces have areas $\vec{A}$ and $-A^*\vec{n}$ (Fig. E1.2b), its transverse surface has area $\vec{A}_\perp \perp \vec{n}$. Its total vector area is zero

$$\vec{A} - A^*\vec{n} + \vec{A}_\perp = \vec{0} \qquad \Rightarrow \qquad (\vec{A} - A^*\vec{n} + \vec{A}_\perp)\cdot\vec{n} = 0$$

$$\Rightarrow \vec{A} \cdot \vec{n} - A^* \vec{n} \cdot \vec{n} + \vec{A}_\perp \cdot \vec{n} = 0 \quad \Rightarrow A^* = \vec{A} \cdot \vec{n}$$

(d) Let the area of a face in a coordinate plane perpendicular to $\vec{e}_i$ be A_i (Fig. E1.2c). The vector area of the tetrahedron is given by

$$A\vec{n} - \sum_{i=1}^{3} A_i \vec{e}_i = \sum_{i=1}^{3} (An_i - A_i) \vec{e}_i = \vec{0} \quad \Rightarrow \quad A_i = An_i$$

The force $\vec{F}$ acting at a point A on the surface of a body can be resolved into a component normal to the surface, called the normal force $\vec{N}$, and a component parallel to the surface, called the shear force $\vec{S}$. Let $\vec{n}$ be the unit vector along the outward normal to the surface of the body at A. The normal force N and the shear force $\vec{S}$ are

$$N = \vec{F} \cdot \vec{n}, \qquad \vec{S} = \vec{F} - N\vec{n} = (\vec{n} \cdot \vec{n})\vec{F} - (\vec{F} \cdot \vec{n})\vec{n} = (\vec{n} \times \vec{F}) \times \vec{n}$$

1.4 Mathematical Model of Classical Mechanics

The mathematical model of classical mechanics consists of a set of primitives, defined entities, axioms and theorems.

1. Primitives: The primitive of *location* is a point. The concept of *length*, a primitive, is associated with the extent (size) of the bodies. An *event*, a primitive, is a happening or an occurrence. *Time* is a primitive, which gives idea about the flow of events: past, present, future; and to specify the interval between them. The primitives of *force* and *moment* are mechanical actions which arise respectively from our ideas of muscular effort and twisting/bending action by hands. The property of a body due to which an external force is required to accelerate it is called *inertia* and its measure is a primitive called *mass*. The mass of a body also determines the gravitational force between it and other material bodies. The primitive of *energy* is related to its capacity to transform into various forms and thereby make things happen.

The set of all possible locations is a primitive, called *space*. *Space is modelled as homogeneous, isotropic, three-dimensional, without gaps, with the metric of Euclid and the axioms of Euclidean geometry as its properties.* It is 'absolute', i.e., its properties (e.g., distance between its locations) are independent of the material bodies occupying its locations or moving through it and also independent of its own motion.

2. Defined entities: *Newtonian unit of time is* **defined** *as the recurring interval of an ideal repeatable phenomenon / experiment*, called a *clock*. Thus, the time interval between two events is a real number $\in (-\infty, \infty)$. Time is modelled by real numbers.

A *reference frame* is **defined** as a *space*, moving in any manner, with an observer *having a clock*. In classical mechanics, time is *'absolute'*, i.e., it is independent of the reference frame and the presence or motion of the material bodies in it, and it is '*universal*', i.e., time interval measured between two events is the same in all reference frames, irrespective of their relative motion.

A material body is a collection of matter (mass), whose each point is called a *material point.* Matter is not continuously distributed over a volume, but is composed of very large number of protons, neutrons, electrons, etc., whose arrangement is not completely known. Statistical methods are required for the analysis of this discrete model. Pressure, density and temperature are the gross effects of the actions of many molecules and atoms. To *study the average of an entity in the small neighbourhood of a material point*, for mathematical convenience, the *continuum hypothesis* is postulated: There exists *a hypothetical distribution of mass and other non-inertial properties over the volume of a given body* with a volume density. Such an idealised continuous distribution of matter is called a *continuum*, in which exist– material points, material curves, material surfaces and material volume elements (internal/external/or both, big/small/infinitesimal). Integral calculus simplifies the analysis of this continuum. Despite smearing out of the discrete characteristics and disregard of the quantum nature of energy, the continuum model yields physically valid results for the effect of mechanical action on material bodies if their size is several orders of magnitude bigger than the atomic size.

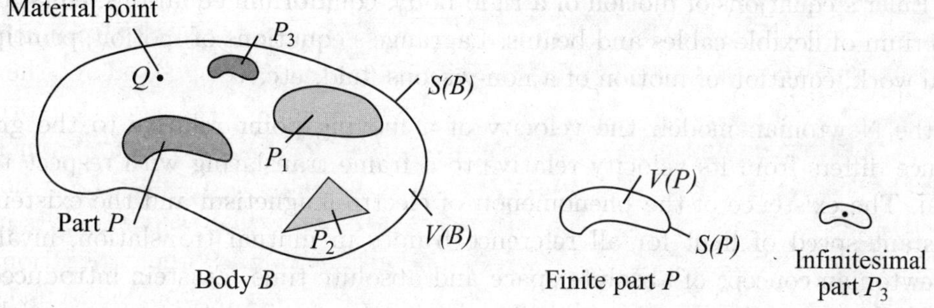

Fig. 1.8

In the mathematical model, *a body or a material body or a mechanical system or a system* B is a collection of matter (mass) of well-defined identity, whose configuration at a given time t is given by the volume $V(B)$ of space occupied by it, which is bounded by a piecewise smooth surface $S(B)$ (Fig. 1.8). A part P of the body is a subset of B, which at t occupies volume $V(P)$ bounded by a piecewise smooth surface $S(P)$. A part P (finite or infinitesimal) can itself be considered a body. A dust particle, a missile, a train, a lake and a galaxy are all bodies.

A *rigid body* is defined as a body in which the distance between any two material points remains the same for all time. A *rigid extension* of a rigid body means its massless extension such that the points in this extension move with the body as if they were rigidly attached to it.

Other defined entities are: position vector $\vec{r}$, displacement $\vec{d}$, velocity $\vec{v}$ and acceleration $\vec{a}$ of a point; $\vec{v}_{PA}$, $\vec{a}_{PA}$, angular velocity $\vec{\omega}$ and angular acceleration $\dot{\vec{\omega}}$ of a reference frame/rigid body; moment $\vec{M}_A$, M_{AB} of a force about a point A and a line AB; couple, resultant force-couple system, internal force resultants; momentum $\vec{p}$, moment of momentum $\vec{H}_A$ about A, kinetic energy T; impulse $\vec{I}$, angular impulse $\vec{I}_{\text{ang}_A}$ about A, power $\dot{W}$ and work done W due to a force $\vec{F}$, conservative force, potential energy, centre of mass C, inertia matrix, variation, first variation, virtual work, etc.

3. Axioms:
Axiom of mass: The mass $m(B)$ of a body $B = \cup P_i$ is a positive scalar whose value is independent of the external actions on the body and its motion, and $m(B) = \sum m(P_i)$.
Euler's axioms of momentum, moment of momentum: $\dot{\vec{p}} = \vec{F}$, $\dot{\vec{H}}_O = \vec{M}_O$ for fixed O. Hamilton's principle, Coulomb's axiom of dry friction, axiom of rolling resistance, Coulomb's law of electrostatic forces, Lorentz law of electromagnetic forces, axiom of gravitational force, Cauchy and Euler's axiom of stress vector, Hook's law for linear elastic solids, Stokes law of viscosity for fluids and other axioms for material behaviour.

4. Theorems derived from the axioms: Newton's 3rd law, impulse-momentum and angular impulse-moment of momentum principles, work-energy principle for C, $\vec{F} = m\vec{a}_C$, Euler's equations of motion of a rigid body, equilibrium equations, equations of equilibrium of flexible cables and beams, Lagrange's equations of motion, principle of virtual work, equation of motion of a non-viscous fluid, etc.

In the Newtonian model, the velocity of a moving point relative to the ground reference differs from its velocity relative to a frame translating with respect to the ground. The existence of the phenomenon of electro-magnetism and the existence of a constant speed of light for all reference frames in mutual translation, invalidate the Newtonian concept of absolute space and absolute time. Einstein introduced the concept of a space–time continuum in his special theory of relativity. Space and time being the forms of existence of matter, their properties are related to the properties of the moving material bodies in his mathematical model. The interrelation between space and time is significant for $\beta^2 = v^2/c^2 \to 1$, where v and c are the speed of the body and the speed of light. At the speeds encountered in most practical situations, the Newtonian concepts of absolute space and absolute time are physically meaningful, since for speeds less than 30000 km/s, space and time are very weakly dependent on the properties of the material bodies moving in it as $\beta^2 < 0.01$.

An infinite number of reference frames exist since any number of rigid spaces moving relative to one another can be considered. For convenience, any number of coordinate systems can be embedded in a given reference frame, but all of them with a device to measure time constitute the same reference frame. Reference frames can be attached to any rigid body. The reference frames will be denoted by alphanumeric symbols, for example, the frames $F, G, T, I, m, 1, 2, 3$, shown in Fig. 1.9. We shall often represent a reference frame as a rigid body to emphasise that the axes have been embedded in it.

Rectangular Cartesian axes x, y, z can be embedded in a given frame F, with arbitrary origin and orientation. Let $\vec{\mathrm{i}}, \vec{\mathrm{j}}, \vec{\mathrm{k}}$ be the unit vectors in the directions x, y, z. These vectors being constant (fixed) in F, have relative to F, $\frac{d\vec{\mathrm{i}}}{dt}|_F = \frac{d\vec{\mathrm{j}}}{dt}|_F = \frac{d\vec{\mathrm{k}}}{dt}|_F = \vec{0}$. The time derivative of a vector $\vec{A}(t) = a_x(t)\,\vec{\mathrm{i}} + a_y(t)\,\vec{\mathrm{j}} + a_z(t)\,\vec{\mathrm{k}}$ with respect to frame F (Fig. 1.9), called the *frame derivative* of $\vec{A}$ relative to F, is defined by

$$\dot{\vec{A}}_{|F} = \left.\frac{d\vec{A}}{dt}\right|_{\vec{\mathrm{i}},\vec{\mathrm{j}},\vec{\mathrm{k}}\text{ fixed}} = \lim_{\Delta t \to 0} \left.\frac{\vec{A}(t+\Delta t) - \vec{A}(t)}{\Delta t}\right|_{\vec{\mathrm{i}},\vec{\mathrm{j}},\vec{\mathrm{k}}\text{ fixed}}$$

$$= \dot{a}_x\,\vec{\mathrm{i}} + \dot{a}_y\,\vec{\mathrm{j}} + \dot{a}_z\,\vec{\mathrm{k}} \tag{1.1}$$

This derivative depends only on F and not on the choice of the origin and orientation of the axes x, y, z. However, the components of $\dot{\vec{A}}_{|F}$ can be given with respect to any basis vectors. The rules of frame derivatives of vectors are the same as those for scalars.

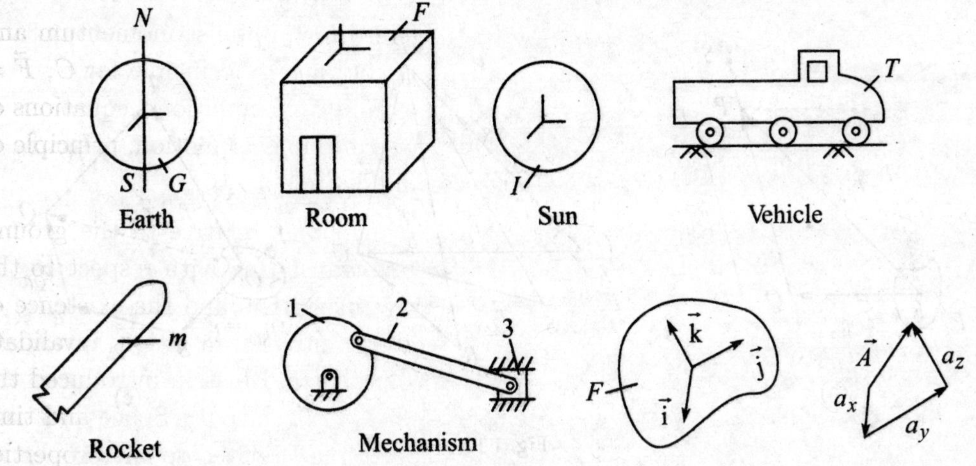

Fig. 1.9

For a time-dependent scalar function $f(t)$ like temperature, since there is no orientation involved and time t is the same for all frames: $\dot{f}_{|F} = \dot{f}$; independent of F.

1.5 Position Vectors of Locations

Consider locations P, Q, R, S of a reference frame F. The *position vector of P relative to Q*, denoted by $\vec{r}_{PQ}$, is the directed line segment to P from Q, $\vec{r}_{PQ} = \overrightarrow{QP}$ (Fig. 1.10a). The position vectors of all points relative to a particular (fixed) reference location O of the reference frame are termed their *position vectors.* The position vector of P is denoted as $\vec{r}_P$ (Fig. 1.10b). It follows from Fig. 1.10a,b that

$$\vec{r}_P = \vec{r}_{PO} = \overrightarrow{OP} \tag{1.2}$$

$$\vec{r}_{QP} = -\vec{r}_{PQ} \qquad \text{and} \qquad \vec{r}_{PR} = \vec{r}_{PQ} + \vec{r}_{QR} \tag{1.3}$$

$$\Rightarrow \vec{r}_{PQ} = \vec{r}_{PR} - \vec{r}_{QR} = \vec{r}_{PO} - \vec{r}_{QO} = \vec{r}_P - \vec{r}_Q = \vec{r}_{PS} - \vec{r}_{QS} \tag{1.4}$$

i.e., $\vec{r}_{PQ}$ is the difference of the position vectors of locations P and Q with respect to *any* location, in particular, the difference of the position vectors relative to the reference location O. The chain rule for relative position vectors (Fig. 1.10c) follows from eq. (1.3):

$$\vec{r}_P = \vec{r}_{PQ} + \vec{r}_Q, \qquad \vec{r}_{PO} = \vec{r}_{PQ} + \vec{r}_{QR} + \vec{r}_{RS} + \vec{r}_{SO} \tag{1.5}$$

The relative position vector, between any two material points of a body, is defined as the relative position vector of the locations of the space occupied by them, and the distance between the material points is defined as the length of this vector.

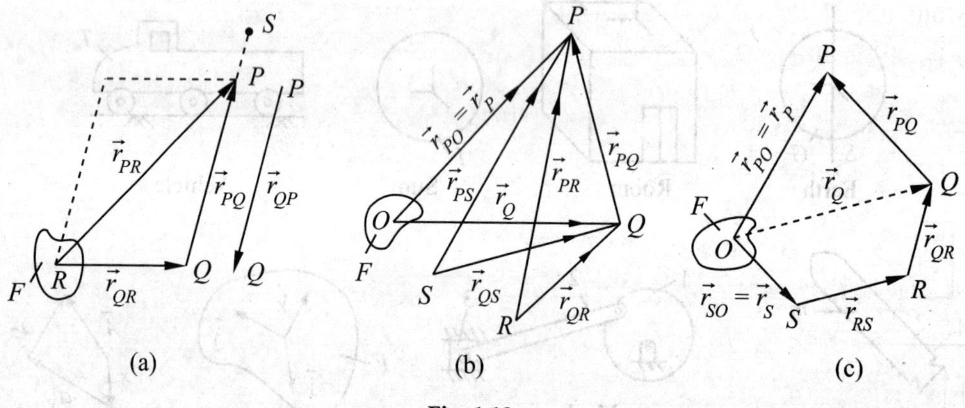

Fig. 1.10

A location in a three-dimensional space can be prescribed by giving an ordered set of three numbers, called its coordinates, with respect to a triplet set of coordinate lines embedded in the space. The coordinate lines may be rectilinear or curvilinear and mutually orthogonal or non-orthogonal.

1.6 Kinematics of a Moving Point

A point is said to be at *rest* relative to a reference frame F if it continues to occupy the same location of F, otherwise it is said to be in *motion* relative to F. The centre C of the front wheel 1, of a car moving without slip straight ahead, is at rest relative to the reference frames of wheel 1 and main body 3 of the car, but is in motion relative to the reference frames of rear wheel 2 and ground 4 (Fig. 1.11).

The *position vector of a moving point P* in frame F at time t is defined as the position vector of the location occupied by it at time t relative to a reference location O of F and is denoted by $\vec{r}_P(t)$ (Fig. 1.12a). The function $\vec{r}_P(t)$ is defined as a *continuous function of time.* Let a moving point P occupy locations P_1, P_2 of reference frame F at times t_1, t_2. Its *displacement vector* with respect to frame F from time t_1 to t_2, denoted as $\vec{d}_{|F}$, is defined as the directed line segment from P_1 to P_2 (Fig. 1.12a):

$$\vec{d}_{|F} = \overrightarrow{P_1P_2} = \vec{r}_{P_2P_1} = \vec{r}_{P_2} - \vec{r}_{P_1} \tag{1.6}$$

The displacement vector with respect to frame F is independent of the choice of the reference location (origin) O for the position vectors (Fig. 1.12a). However, it does depend on the reference frame. Consider a material point A of the rear wheel 2 of the car moving without slip ,which is the contact point on ground in configuration I. As the car moves from configuration I at t_1 to configuration II at t_2, it occupies locations A_1, A_2, A_3, A_4 of frames 1, 2, 3, 4 at t_1 and occupies locations A_1', A_2', A_3', A_4' of frames 1, 2, 3, 4 at t_2 (Fig. 1.11). A_i are coincident at t_1 but not at t_2, and A_i' are coincident at t_2 but not at t_1. The displacement vectors of A are $\overrightarrow{A_1A_1'} = 2b\,\vec{\imath}_1$, $\overrightarrow{A_2A_2'} = \vec{0}$, $\overrightarrow{A_3A_3'} = 2r\,\vec{\jmath}_3$, $\overrightarrow{A_4A_4'} = d\,\vec{\imath}_4 + 2r\,\vec{\jmath}_4$ relative to frames 1, 2, 3, 4.

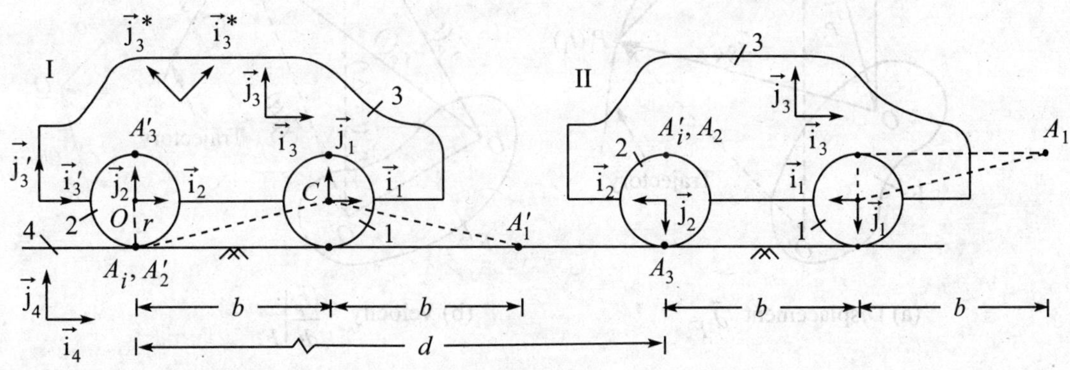

Fig. 1.11

The curve (locus of the set of locations of frame F occupied by P) traced by the moving point P in the frame F is called its *path* or *trajectory* in that frame. The parametric equation of the trajectory is given by $\vec{r} = \vec{r}_P(t)$. The trajectory of a moving point may be different for different frames, for example, the trajectory of A with respect to frames 2, 3, 4, in Fig. 1.11 is respectively a single point, a circle of radius r and a cycloid. The trajectory of O with respect to frames 1, 2, 3, 4 in Fig. 1.11 is a circle of radius b, a single point, a single point and a straight line. The motion is called *spatial* or *planar* according as the trajectory is a spatial or a planar curve. The motion is *rectilinear* if the trajectory is a straight line; else it is *curvilinear*, which may be spatial or planar. A point on a nut being tightened on a bolt has spatial curvilinear motion since it advances in the direction of the bolt axis and moves around this axis.

Let a moving point P occupy locations P and P' (Fig. 1.12b) of frame F at times t and $t+\Delta t$. The displacement $\Delta\vec{r}_P$ in time Δt is $\Delta\vec{r}_P = \overrightarrow{PP'}$. The velocity $\vec{v}_{P|F}(t)$ of point P with respect to frame F is defined as the rate of change of displacement with respect to frame F at time t, and the acceleration $\vec{a}_{P|F}(t)$ of point P with respect to frame F is defined as the rate of change of $\vec{v}_{P|F}(t)$ with respect to frame F at time t:

$$\begin{aligned} \vec{v}_{P|F}(t) &= \lim_{\Delta t\to 0}\frac{\overrightarrow{PP'}}{\Delta t} = \lim_{\Delta t\to 0}\frac{\Delta\vec{r}_P}{\Delta t} = \dot{\vec{r}}_{P|F} \\ \vec{a}_{P|F}(t) &= \dot{\vec{v}}_{P|F}(t) = \ddot{\vec{r}}_{P|F}(t) \end{aligned} \tag{1.7}$$

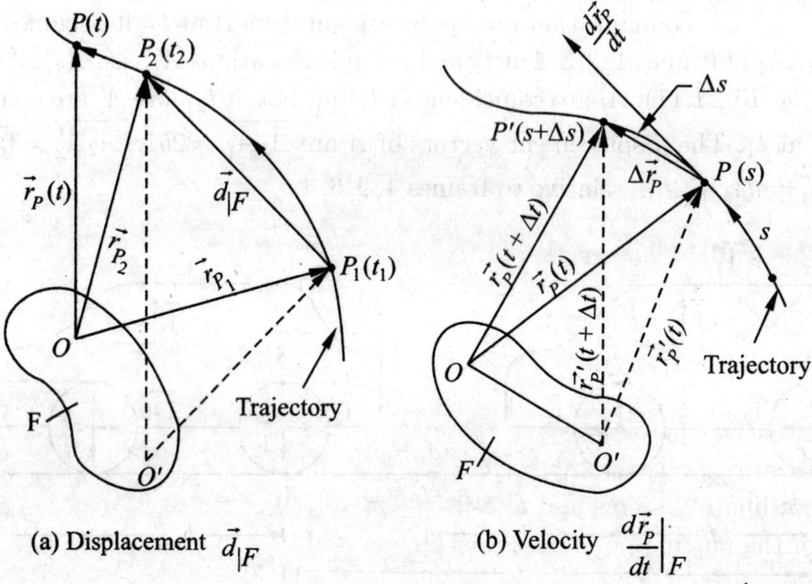

Fig. 1.12

The velocity vector is always tangent to the trajectory since in the limit as $\Delta t \to 0$, the direction of the chord $\Delta\vec{r}_P$ is along the tangent. The position vectors depend on the choice of the origin O, but the relative position vectors $\overrightarrow{P_1P_2}$ and $\overrightarrow{PP'}$ (Fig. 1.12b) are independent of O and the orientation of the axes x, y, z embedded in the frame F. It follows from eqs (1.6) and (1.7) that *displacement* $\vec{d}_{|F}$, *velocity* $\vec{v}_{P|F}$ *and acceleration* $\vec{a}_{P|F}$ of P *depend on* F *but are independent of the choice of the origin and the orientation of the axes* x, y, z embedded in the frame F. For example, the displacement, velocity and acceleration of point A with respect to frame 3 (Fig. 1.11) do not depend on whether the axes $\vec{\mathrm{i}}_3, \vec{\mathrm{j}}_3$; $\vec{\mathrm{i}}'_3, \vec{\mathrm{j}}'_3$, or $\vec{\mathrm{i}}^*_3, \vec{\mathrm{j}}^*_3$ are taken as the basis vectors.[2]

P is at rest in frame F if $\vec{r}_P(t) = \vec{r}_P(0)\ \forall\, t$, i.e., P is fixed in F iff $\vec{v}_{P|F} = \vec{0}\ \forall\, t$. P is at *instantaneous rest* in F at t_1 if $\vec{v}_{P|F}(t_1) = \vec{0}$ and $\vec{a}_{P|F}(t_1)$ may not be zero.

We shall drop the suffix P and even $(\ )_{|F}$ when these are clear from the context:

$$\vec{v} = \dot{\vec{r}}, \qquad \vec{a} = \dot{\vec{v}} = \ddot{\vec{r}} \tag{1.8}$$

their directions being $\vec{v}\,/|\vec{v}\,|$ and $\vec{a}\,/|\vec{a}\,|$. In general, $|\vec{a}\,| \neq \frac{d}{dt}|\vec{v}\,|$. Integrating eq. (1.8):

$$\begin{aligned} \vec{v}(t) &= \vec{v}(0) + \int_0^t \vec{a}(t)\, dt \\ \vec{r}(t) &= \vec{r}(0) + \int_0^t \vec{v}(t)\, dt \\ v^2(t) - v^2(0) &= \int_{v^2(0)}^{v^2(t)} d(v^2) = \int_0^t \frac{d}{dt}(\vec{v}\cdot\vec{v}\,)dt = 2\int_0^t \dot{\vec{v}}\cdot\vec{v}\,dt = 2\int_{\vec{r}(0)}^{\vec{r}(t)} \vec{a}\cdot d\vec{r} \end{aligned} \tag{1.9}$$

For *uniform acceleration*, $\vec{a}(t) = \vec{a}_0$, where $\vec{a}_0$ is a constant vector, eq. (1.9) $\Rightarrow$

$$\begin{aligned} \vec{v}(t) &= \vec{v}(0) + \vec{a}_0 t \\ \vec{r}(t) &= \vec{r}(0) + \vec{v}(0)t + \tfrac{1}{2}\vec{a}_0 t^2 \\ v^2(t) - v^2(0) &= 2\vec{a}_0\cdot[\vec{r}(t) - \vec{r}(0)] \end{aligned} \tag{1.10}$$

Choosing $\vec{r}(0) = \vec{0}$ in the expression for $\vec{r}(t)$ shows that the trajectory is a plane curve in the plane of $\vec{v}(0)$ and $\vec{a}_0$. For *uniform motion*, $\vec{a}_0 = \vec{0}$ and eq. (1.10) yields $\vec{v}(t) = \vec{v}(0)$, and $\vec{r}(t) = \vec{r}(0) + \vec{v}(0)t$ which is a straight path.

We next derive the expressions for $\vec{v}$ and $\vec{a}$ in *extrinsic coordinate systems* whose coordinates are *a priori* fixed such as the Cartesian (x, y, z) and cylindrical polar (r, ϕ, z) coordinate systems and also in an *intrinsic coordinate system* determined by the path of the moving point using path coordinate s along the path.

1.7 Kinematics of a Point in Cartesian Coordinates

Let $x(t)$, $y(t)$, $z(t)$ (Fig. 1.13) be the Cartesian coordinates of the moving point P,

$$\begin{aligned} \vec{r}(t) &= x(t)\,\vec{\mathrm{i}} + y(t)\,\vec{\mathrm{j}} + z(t)\,\vec{\mathrm{k}} \\ \vec{v}(t) &= \dot{x}\,\vec{\mathrm{i}} + \dot{y}\,\vec{\mathrm{j}} + \dot{z}\,\vec{\mathrm{k}} \\ \vec{a}(t) &= \ddot{x}\,\vec{\mathrm{i}} + \ddot{y}\,\vec{\mathrm{j}} + \ddot{z}\,\vec{\mathrm{k}} \end{aligned} \tag{1.11}$$

[2] For biological systems, acceleration is usually expressed as a multiple of the acceleration g due to gravity on the surface of the earth since their endurance limit is related to the g which they are used to.

$$\begin{aligned} v_x &= \dot{x}, & v_y &= \dot{y}, & v_z &= \dot{z} \\ a_x &= \ddot{x} = \dot{v}_x, & a_y &= \ddot{y} = \dot{v}_y, & a_z &= \ddot{z} = \dot{v}_z \\ v &= (\dot{x}^2 + \dot{y}^2 + \dot{z}^2)^{1/2}, & & & a &= (\ddot{x}^2 + \ddot{y}^2 + \ddot{z}^2)^{1/2} \end{aligned} \tag{1.12}$$

In general, $a_x = \ddot{x}$ may be a function of $x,\ y,\ z,\ \dot{x},\ \dot{y},\ \dot{z}, t$, so that 3 coupled differential equations have to be solved to get $x(t),\ y(t),\ z(t)$ if $\vec{a}$ is given. If $a_x = \ddot{x} = f_1(x, \dot{x}, t)$, then it can be integrated, independently of the other two, as for rectilinear motion.

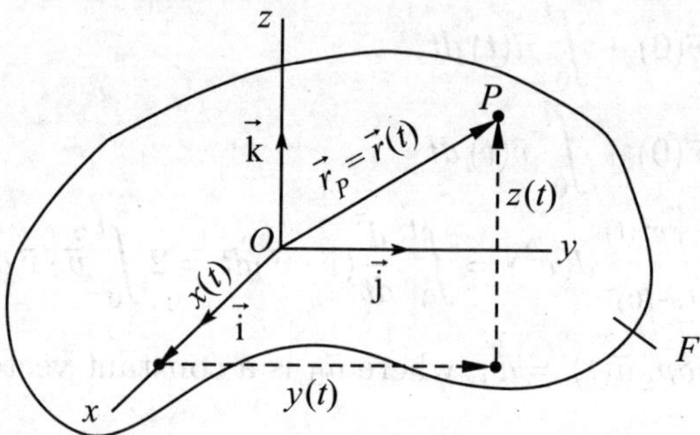

Fig. 1.13

If the point has a planar motion, then a good choice of the coordinate system is such that the xy-plane is the plane of motion, and for rectilinear motion, a good choice for the x-axis is along the path.

The trajectory can also be described implicitly as a function of the coordinates as

$$g(x, y, z) = 0, \qquad h(x, y, z) = 0 \tag{1}$$

which do not contain t explicitly. The trajectory is simply the curve of intersection of the two surfaces described by eq. (1). The relations between the velocity and acceleration components are obtained by differentiating eqs (1) twice with respect to t. For example, for the plane trajectory $f(x, y) = 0$:

$$\frac{\partial f}{\partial x}\dot{x} + \frac{\partial f}{\partial y}\dot{y} = 0$$

$$\left(\frac{\partial^2 f}{\partial x^2}\dot{x} + \frac{\partial^2 f}{\partial y \partial x}\dot{y}\right)\dot{x} + \frac{\partial f}{\partial x}\ddot{x} + \left(\frac{\partial^2 f}{\partial x \partial y}\dot{x} + \frac{\partial^2 f}{\partial y^2}\dot{y}\right)\dot{y} + \frac{\partial f}{\partial y}\ddot{y} = 0$$

yield $\dot{y},\ \ddot{y}$, given the values of $\dot{x},\ \ddot{x}$.

An important case consists of the velocity being a known function of the position (x, y, z) of the moving point and time t. This happens in fluid mechanics where velocities are measured at fixed locations. Thus, given $\vec{v}(x, y, z, t)$, the acceleration components of a material point with $\vec{r}(t) = x\,\vec{\mathrm{i}} + y\,\vec{\mathrm{j}} + z\,\vec{\mathrm{k}}$ are obtained as

$$\begin{aligned} a_x &= \frac{d}{dt}[v_x(x,y,z,t)] = \lim_{\Delta t \to 0} \frac{v_x[\vec{r}(t+\Delta t), t+\Delta t] - v_x[\vec{r}(t), t]}{\Delta t} \\ &= \frac{\partial v_x}{\partial x}\frac{dx}{dt} + \frac{\partial v_x}{\partial y}\frac{dy}{dt} + \frac{\partial v_x}{\partial z}\frac{dz}{dt} + \frac{\partial v_x}{\partial t} \quad \Rightarrow \end{aligned}$$

$$a_x = \frac{\partial v_x}{\partial t} + \frac{\partial v_x}{\partial x} v_x + \frac{\partial v_x}{\partial y} v_y + \frac{\partial v_x}{\partial z} v_z$$

Similarly, $$a_y = \frac{\partial v_y}{\partial t} + \frac{\partial v_y}{\partial x} v_x + \frac{\partial v_y}{\partial y} v_y + \frac{\partial v_y}{\partial z} v_z \tag{1.13}$$

and $$a_z = \frac{\partial v_z}{\partial t} + \frac{\partial v_z}{\partial x} v_x + \frac{\partial v_z}{\partial y} v_y + \frac{\partial v_z}{\partial z} v_z$$

Note that $\frac{d}{dt}$ is a *material time derivative* for a specific material point, whereas $\frac{\partial}{\partial t}$ is a time derivative at a fixed spatial location.

1.8 Kinematics of a Point in Cylindrical Polar Coordinates

Let $r(t)$, $\phi(t)$, $z(t)$ (Fig. 1.14) be the cylindrical polar coordinates of the moving point P at time t. At this position $\vec{e}_r$ is the *radial* outward unit vector in the direction of increasing r, tangent to the radial coordinate line $r\colon \phi =$ constant, $z =$ constant; $\vec{e}_\phi$ is the *transverse/circumferential* unit vector, in the direction of increasing ϕ, tangent to the circumferential coordinate line $\phi\colon r =$ constant, $z =$ constant; and $\vec{e}_z$ is the *axial* unit vector in the direction of increasing z, tangent to the axial coordinate line $z\colon r =$ constant, $\phi =$ constant. These form a right-handed triad with

$$\begin{aligned} \vec{e}_r \times \vec{e}_\phi &= \vec{e}_z, \qquad \vec{e}_\phi \times \vec{e}_z = \vec{e}_r, \qquad \vec{e}_z \times \vec{e}_r = \vec{e}_\phi \\ \vec{e}_r(t) &= \cos\phi\,\vec{\mathrm{i}} + \sin\phi\,\vec{\mathrm{j}} \\ \vec{e}_\phi(t) &= -\sin\phi\,\vec{\mathrm{i}} + \cos\phi\,\vec{\mathrm{j}} \\ \dot{\vec{e}}_{r|F} &= (-\sin\phi\,\vec{\mathrm{i}} + \cos\phi\,\vec{\mathrm{j}})\dot{\phi} = \dot{\phi}\,\vec{e}_\phi \\ \dot{\vec{e}}_{\phi|F} &= (-\cos\phi\,\vec{\mathrm{i}} - \sin\phi\,\vec{\mathrm{j}})\dot{\phi} = -\dot{\phi}\,\vec{e}_r \\ \vec{e}_z(t) &= \vec{\mathrm{k}}, \qquad \dot{\vec{e}}_{z|F} = \vec{0} \end{aligned} \tag{1.14}$$

In what follows, $\dot{(\,)} = \dot{(\,)}_{|F}$ is implied. Differentiating $\vec{r}$ with respect to F and using eq. (1.14) yields $\vec{v}, \vec{a}$:

$$\vec{r} = r(t)\,\vec{e}_r(t) + z(t)\,\vec{e}_z(t) \qquad \Rightarrow \qquad \vec{v} = \dot{\vec{r}} = \dot{r}\,\vec{e}_r + r\,\dot{\vec{e}}_r + \dot{z}\,\vec{e}_z + z\,\dot{\vec{e}}_z$$

$$\vec{v} = \dot{r}\,\vec{e}_r + r\dot{\phi}\,\vec{e}_\phi + \dot{z}\,\vec{e}_z \tag{1.15}$$

$$\vec{a} = \dot{\vec{v}} = \ddot{r}\,\vec{e}_r + \dot{r}\,\dot{\vec{e}}_r + (\dot{r}\dot{\phi} + r\ddot{\phi})\,\vec{e}_\phi + r\dot{\phi}\,\dot{\vec{e}}_\phi + \ddot{z}\,\vec{e}_z + \dot{z}\,\dot{\vec{e}}_z$$

$$\vec{a} = (\ddot{r} - r\dot{\phi}^2)\,\vec{e}_r + (2\dot{r}\dot{\phi} + r\ddot{\phi})\,\vec{e}_\phi + \ddot{z}\,\vec{e}_z \tag{1.16}$$

$$v_r = \dot{r}, \qquad v_\phi = r\dot{\phi}, \qquad v_z = \dot{z}, \quad v = (\dot{r}^2 + r^2\dot{\phi}^2 + \dot{z}^2)^{1/2} \tag{1.17}$$

$$a_r = \ddot{r} - r\dot{\phi}^2, \quad a_\phi = 2\dot{r}\dot{\phi} + r\ddot{\phi}, \quad a_z = \ddot{z}, \quad a = (a_r^2 + a_\phi^2 + a_z^2)^{1/2} \tag{1.18}$$

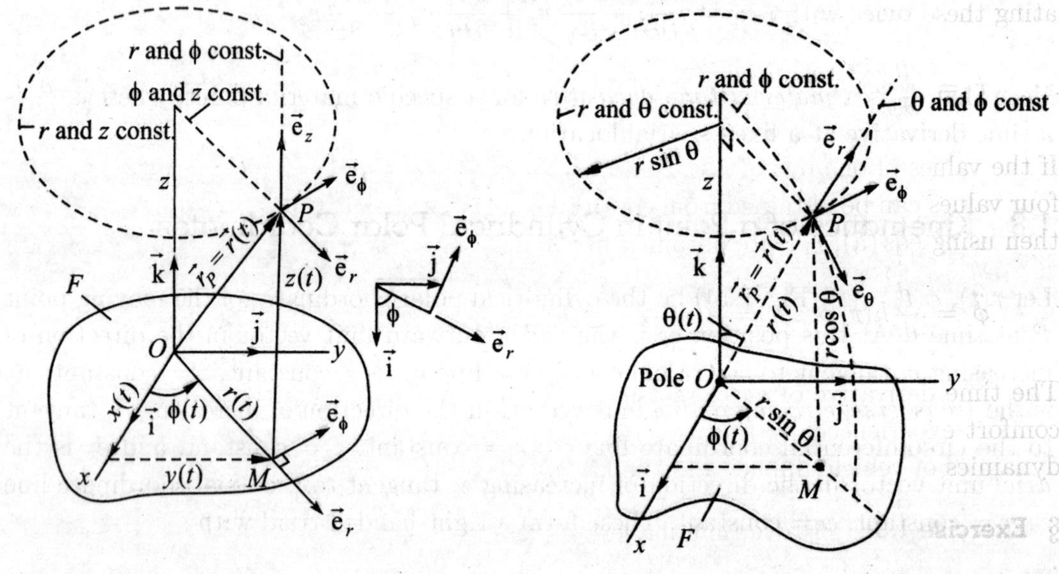

Fig. 1.14 Fig. 1.15

The dependence of $\vec{r}$ on ϕ is implicit through the dependence of $\vec{e}_r$ and $\vec{e}_\phi$ on ϕ. Note that $a_\phi = \frac{1}{r}\frac{d}{dt}(r^2\dot{\phi}) = \frac{1}{r}\frac{d}{dt}(rv_\phi)$ and $a_r \neq \dot{v}_r$, $a_\phi \neq \dot{v}_\phi$ since $\vec{e}_r$, $\vec{e}_\phi$ are not constant vectors in frame F. The quantity $\dot{\phi}$ is termed the *'angular speed'* of the *projected line OM* (Fig. 1.14) about the z-axis, its *angular velocity being* $\dot{\phi}\,\vec{k}$ *and angular acceleration* $\ddot{\phi}\,\vec{k}$. However, a moving point has no angular velocity as it has no orientation. The integrals of a_r, a_ϕ with respect to time do not yield v_r, v_ϕ, but integral of ra_ϕ yields:

$$\int_0^t ra_\phi\,dt = \int_0^t \frac{d}{dt}(rv_\phi)\,dt = r(t)v_\phi(t) - r(0)v_\phi(0) \tag{1.19}$$

For planar motion in plane $z =$ constant, the axial components v_z, a_z are zero and

$$\vec{v} = \dot{r}\,\vec{e}_r + r\dot{\phi}\,\vec{e}_\phi, \qquad \vec{a} = (\ddot{r} - r\dot{\phi}^2)\,\vec{e}_r + (2\dot{r}\dot{\phi} + r\ddot{\phi})\,\vec{e}_\phi \tag{1.20}$$

The cylindrical and Cartesian coordinates (Fig. 1.14) are related by

$$\begin{aligned} r &= (x^2 + y^2)^{1/2}, \qquad & \phi &= \tan^{-1}(y/x) \\ x &= r\cos\phi, & y &= r\sin\phi \end{aligned} \tag{1.21}$$

The cylindrical components are relevant in problems involving motion around an axis of symmetry, e.g., in turbines and centrifugal pumps, the change in the axial and transverse components of velocity leads respectively to axial thrust and axial moment.

If the path is prescribed not as parametric equations in time, but as $r = f(\phi)$, $z = g(\phi)$, then the relations between time derivatives of r, ϕ, z are obtained by differentiating these twice with respect to the time, using chain rule:

$$\dot{r} = \frac{df}{d\phi}\dot{\phi}, \quad \dot{z} = \frac{dg}{d\phi}\dot{\phi}; \quad \ddot{r} = \frac{d^2 f}{d\phi^2}\dot{\phi}^2 + \frac{df}{d\phi}\ddot{\phi}, \quad \ddot{z} = \frac{d^2 g}{d\phi^2}\dot{\phi}^2 + \frac{dg}{d\phi}\ddot{\phi} \tag{a}$$

If the values of one out of $\dot{r}, \dot{\phi}, \dot{z}$ and one out of $\ddot{r}, \ddot{\phi}, \ddot{z}$ are known, then the remaining four values can be obtained from eqs (a). For example, if $\dot{\phi}$ is known as $\dot{\phi} = h(r, \phi, z, t)$, then using eqs (a), $\dot{r}, \dot{z}$ are obtained first followed by $\ddot{r}, \ddot{z}$, where $\ddot{\phi}$ is obtained by:

$$\ddot{\phi} = \frac{d}{dt}h(r, \phi, z, t) = \frac{\partial h}{\partial r}\dot{r} + \frac{\partial h}{\partial \phi}\dot{\phi} + \frac{\partial h}{\partial z}\dot{z} + \frac{\partial h}{\partial t}$$

The time derivative of acceleration, $\dot{\vec{a}}$, is defined as *jerk*. It is a measure of the discomfort experienced by the passengers in moving vehicles. Jerk is of interest in the dynamics of vehicle impact and the kinematics of cams and followers.

§ **Exercise 1.3:** Prove that the jerk $\dot{\vec{a}}$ is given by

$$\dot{\vec{a}} = (\dddot{r} - 3\dot{r}\dot{\phi}^2 - 3r\dot{\phi}\ddot{\phi})\,\vec{e}_r + [3\ddot{r}\dot{\phi} + 3\dot{r}\ddot{\phi} + r\dddot{\phi} \; - r\dot{\phi}^3]\,\vec{e}_\phi + \dddot{z}\,\vec{e}_z$$

§ **Exercise 1.4:** The *spherical polar coordinates* of a moving point P at time t are $r(t), \theta(t), \phi(t)$ (Fig. 1.15) with $r \geq 0$, $0 \leq \theta \leq \pi$, $0 \leq \phi < 2\pi$ and $\vec{r}_P = r\,\vec{e}_r$, where at this position, $\vec{e}_r$ is the *radial* outward unit vector, $\vec{e}_\theta$ is the *meridional* unit vector in the direction of increasing θ, and $\vec{e}_\phi$ is the *peripheral* unit vector in the direction of increasing ϕ. These form a right-handed triad with

$$\vec{e}_r \times \vec{e}_\theta = \vec{e}_\phi, \qquad \vec{e}_\theta \times \vec{e}_\phi = \vec{e}_r, \qquad \vec{e}_\phi \times \vec{e}_r = \vec{e}_\theta$$

Prove that spherical and Cartesian coordinates and their base vectors are related by

$$r = (x^2 + y^2 + z^2)^{1/2}, \quad \theta = \tan^{-1}[(x^2 + y^2)^{1/2}/z], \quad \phi = \tan^{-1}\frac{y}{x}$$

$$x = r\sin\theta\cos\phi, \qquad y = r\sin\theta\sin\phi, \qquad z = r\cos\theta$$

$$\vec{e}_r(t) = \sin\theta(\cos\phi\,\vec{i} + \sin\phi\,\vec{j}\,) + \cos\theta\,\vec{k}, \qquad \vec{e}_\phi(t) = -\sin\phi\,\vec{i} + \cos\phi\,\vec{j}$$

$$\vec{e}_\theta(t) = \vec{e}_\phi \times \vec{e}_r = \cos\theta(\cos\phi\,\vec{i} + \sin\phi\,\vec{j}\,) - \sin\theta\,\vec{k} \quad \Rightarrow$$

$$\sin\theta\,\vec{e}_r + \cos\theta\,\vec{e}_\theta = \cos\phi\,\vec{i} + \sin\phi\,\vec{j}$$

$$\dot{\vec{e}}_{r|F} = \dot\theta\,\vec{e}_\theta + \dot\phi\sin\theta\,\vec{e}_\phi, \qquad \dot{\vec{e}}_{\theta|F} = -\dot\theta\,\vec{e}_r + \dot\phi\cos\theta\,\vec{e}_\phi$$

$$\dot{\vec{e}}_{\phi|F} = -\dot\phi(\sin\theta\,\vec{e}_r + \cos\theta\,\vec{e}_\theta)$$

and show that velocity and acceleration are given by

$$\vec{v} = \dot r\,\vec{e}_r + r\dot\theta\,\vec{e}_\theta + r\dot\phi\sin\theta\,\vec{e}_\phi$$

$$\vec{a} = (\ddot r - r\dot\theta^2 - r\dot\phi^2\sin^2\theta)\,\vec{e}_r + (r\ddot\theta + 2\dot r\dot\theta - r\dot\phi^2\sin\theta\cos\theta)\,\vec{e}_\theta$$

$$+[(r\ddot\phi + 2\dot r\dot\phi)\sin\theta + 2r\dot\theta\dot\phi\cos\theta]\,\vec{e}_\phi$$

The choice of the origin and orientation of the coordinate axes in a reference frame is not unique. It depends on the particular problem under study, so as to exploit the inherent geometrical symmetries to obtain an easier solution. The coordinate lines in Cartesian, cylindrical and spherical cases are orthogonal with the ϕ and θ coordinate lines being circular and the rest rectilinear. For specific geometry of bodies, more general coordinate systems with non-orthogonal curvilinear coordinate lines are used.

1.9 Kinematics of a Point in Path Coordinates

We first study some aspects of the geometry of curves in space. Let $s(t)$ (Fig. 1.16a) be the coordinate along the trajectory of the moving point P at time t. A right-handed triad of unit vectors $\vec{e}_t$, $\vec{e}_n$, $\vec{e}_b$ at the location s is defined later by eq. (1.22) as follows.

$$(ds)^2 = d\vec{r}\cdot d\vec{r} = \frac{d\vec{r}}{ds}ds\cdot\frac{d\vec{r}}{ds}ds \qquad \Rightarrow \qquad \left|\frac{d\vec{r}}{ds}\right| = 1$$

Thus, $\vec{e}_t = \frac{d\vec{r}}{ds} = \lim_{\Delta s\to 0}\frac{\Delta\vec{r}}{\Delta s}$ is defined as the unit tangent vector to the curve in the direction of increasing s since in the limit as $\Delta s \to 0$, the direction of the chord $\Delta\vec{r}$ becomes the tangent. Note that

$$\frac{d}{ds}(\vec{e}_t\cdot\vec{e}_t) = \frac{d}{ds}(1) = 0 \quad \Rightarrow \quad 2\frac{d\vec{e}_t}{ds}\cdot\vec{e}_t = 0 \quad \Rightarrow \quad \frac{d\vec{e}_t}{ds} \perp \vec{e}_t \ \text{ if } \ \frac{d\vec{e}_t}{ds} \neq \vec{0}$$

Therefore, the unit vector $\vec{e}_n$ *in the direction of* $d\vec{e}_t/ds$ (if $d\vec{e}_t/ds \neq \vec{0}$) is defined as the *principal normal vector* of the curve, there being an infinite set of unit vectors normal to the curve at P. *The magnitude of* $d\vec{e}_t/ds$ is defined as the *curvature* $1/\rho$, and ρ is called the *radius of curvature* of the curve at P. $\vec{e}_n$ is normal to $\vec{e}_t$ and *directed* towards the *centre of curvature* C ($\vec{r}_C = \vec{r}_P + \rho\,\vec{e}_n$) of the curve. Thus, $d\vec{e}_t/ds = (1/\rho)\,\vec{e}_n$. The vector $\vec{e}_b = \vec{e}_t \times \vec{e}_n$ is defined as the *binormal vector* as it is normal to both $\vec{e}_t$ and $\vec{e}_n$.

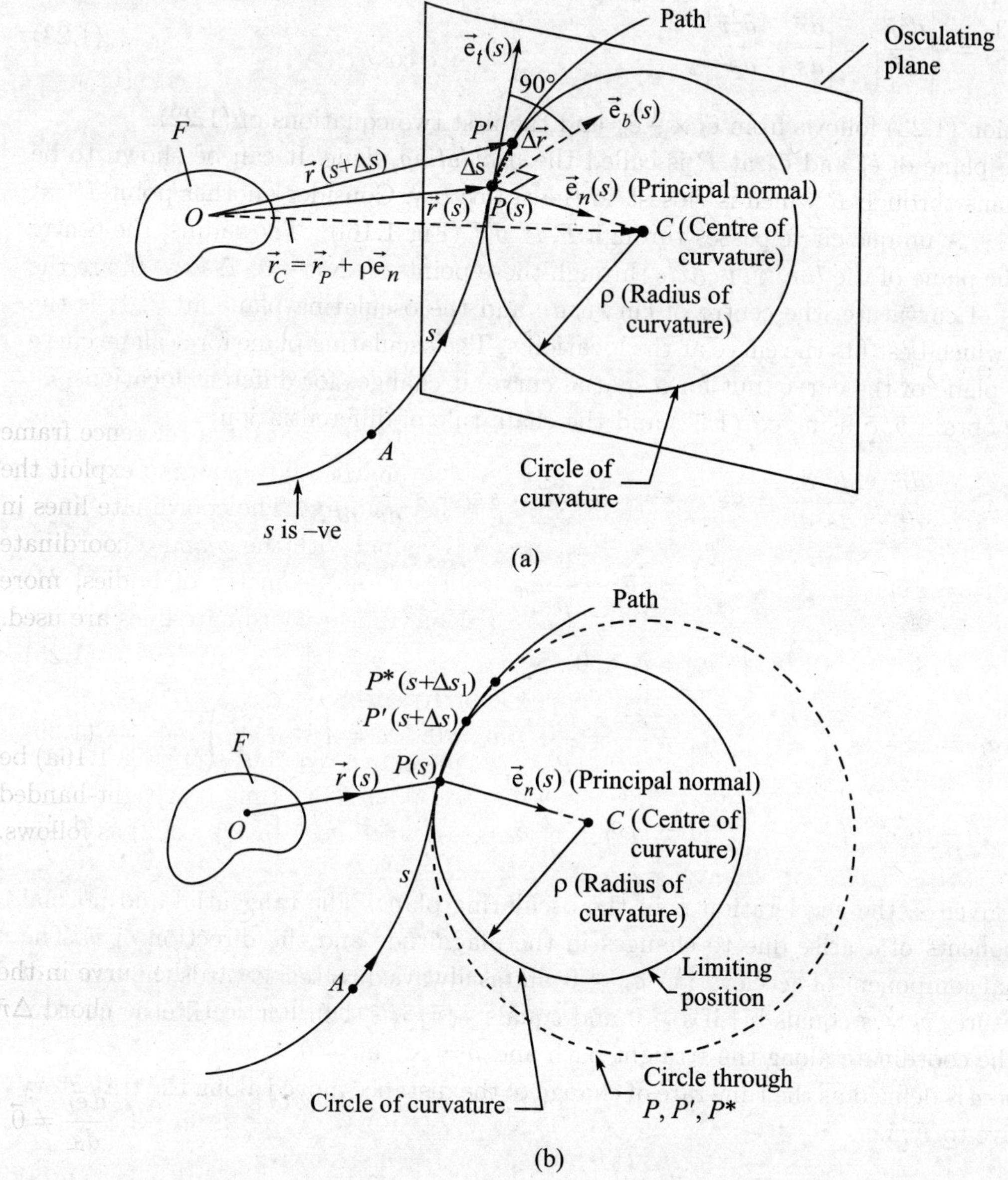

Fig. 1.16

We re-state the definitions of $\vec{e}_t$, $\vec{e}_n$, $\vec{e}_b$, ρ:

$$\vec{e}_t = \frac{d\vec{r}}{ds} = \lim_{\Delta s \to 0} \frac{\Delta \vec{r}}{\Delta s}, \qquad \frac{d\vec{e}_t}{ds} = \frac{d^2\vec{r}}{ds^2} = \frac{1}{\rho}\vec{e}_n, \qquad \vec{e}_b = \vec{e}_t \times \vec{e}_n \tag{1.22}$$

$$\vec{e}_t \times \vec{e}_n = \vec{e}_b, \qquad \vec{e}_n \times \vec{e}_b = \vec{e}_t, \qquad \vec{e}_b \times \vec{e}_t = \vec{e}_n$$

$$\frac{1}{\rho} = \left|\frac{d^2\vec{r}}{ds^2}\right| = \left|\frac{d\vec{r}}{ds} \times \frac{d^2\vec{r}}{ds^2}\right| \tag{1.23}$$

Equation (1.23) follows from $\vec{e}_t \times \frac{1}{\rho}\vec{e}_n$ and the first two equations of (1.22).

The plane of $\vec{e}_t$ and $\vec{e}_n$ at P is called the *osculating plane.* It can be shown to be the plane through P which is closest to the trajectory. Consider another point P^* at $s + \Delta s_1$. A unique circle passes through P, P', P^* (Fig. 1.16b). The radius, the centre and the plane of the *limiting circle* through these points as $\Delta s \to 0,\ \Delta s_1 \to 0$, are the radius of curvature, the centre of curvature and the osculating plane at P. It is the circle which best fits the curve at the location s. The osculating plane for a plane curve is the plane of the curve, but for a spatial curve, it changes for different locations s.

We obtain $\vec{v}$, $\vec{a}$ using eq. (1.22) and the chain rule of differentiation:

$$\vec{v} = \frac{d\vec{r}}{dt} = \frac{d\vec{r}}{ds}\frac{ds}{dt} = \dot{s}\,\vec{e}_t \quad \Rightarrow \quad \vec{a} = \dot{\vec{v}} = \ddot{s}\,\vec{e}_t + \dot{s}\,\frac{d\vec{e}_t}{ds}\frac{ds}{dt}$$

$$\vec{v} = \dot{s}\,\vec{e}_t, \qquad \vec{a} = \ddot{s}\,\vec{e}_t + \frac{\dot{s}^2}{\rho}\,\vec{e}_n \tag{1.24}$$

$$v_t = \dot{s}, \qquad v_n = v_b = 0, \qquad v = |\dot{s}| \tag{1.25}$$

$$a_t = \ddot{s} = \dot{v}_t, \qquad a_n = \frac{\dot{s}^2}{\rho} = \frac{v_t^2}{\rho}, \quad a_b = 0, \quad a = \left(\ddot{s}^2 + \frac{\dot{s}^4}{\rho^2}\right)^{1/2} \tag{1.26}$$

$$\vec{v} = v_t\,\vec{e}_t, \qquad \vec{a} = \dot{v}_t\,\vec{e}_t + \frac{v_t^2}{\rho}\,\vec{e}_n \tag{1.27}$$

At a given s, the acceleration is in the osculating plane. The tangential and normal components of $\vec{a}$ arise due to changes in the magnitude and the direction of $\vec{v}$. The normal component of acceleration $a_n \geq 0$ and is always directed towards the centre of curvature. $v_t = \dot{s}$ equals $|\vec{v}|$ if $\dot{s} \geq 0$ and equals $-|\vec{v}|$ if $\dot{s} < 0$. For rectilinear motion, s is the coordinate along the straight path and $\rho = \infty,\ a_n = 0$.

Speed is defined as the time rate of change of the distance moved along the trajectory relative to F, i.e.,

$$\text{speed} = |\dot{s}| = v = |\vec{v}| = |\dot{\vec{r}}| = |v_t| \tag{1.28}$$

The distance moved in frame F from time t_0 to t and the displacement are given by

$$\text{distance moved} \;=\; \int |ds| = \int_{t_0}^{t} |\dot{s}|\,dt = \int_{t_0}^{t} |\vec{v}(t)|\,dt \tag{1.29}$$

$$\text{displacement} \;=\; \Delta\vec{r} = \vec{r}(t) - \vec{r}(t_0) = \int_{t_0}^{t} \vec{v}(t)\,dt \tag{1.30}$$

The fact that a_n is proportional to $1/\rho$ is taken into account in the design of structures and mechanisms such as aeroplane wings, rail tracks and cams since sudden changes of curvature leading to sudden changes of acceleration are detrimental to equipment and passengers. Wing profile is designed without any sudden change in its curvature to avoid sudden change in the acceleration of the material points of air flowing past it. Special transition curves in rail tracks (and in high-speed cams) are used to make a smooth transition from a straight section of infinite radius of curvature to a circular section of finite radius of curvature.

If the trajectory is given parametrically as $\vec{r} = \vec{r}(\tau)$, then

$$|\vec{e}_t| = \left|\frac{d\vec{r}}{ds}\right| = \left|\frac{d\vec{r}}{d\tau}\frac{d\tau}{ds}\right| = \left|\frac{d\vec{r}}{d\tau}\right|\left|\frac{d\tau}{ds}\right| = 1 \quad\Rightarrow\quad \left|\frac{d\tau}{ds}\right| = 1\Big/\left|\frac{d\vec{r}}{d\tau}\right|$$

$$\frac{d\vec{r}}{ds} = \frac{d\vec{r}}{d\tau}\frac{d\tau}{ds} \quad\Rightarrow\quad \frac{d^2\vec{r}}{ds^2} = \frac{d^2\vec{r}}{d\tau^2}\left(\frac{d\tau}{ds}\right)^2 + \frac{d\vec{r}}{d\tau}\frac{d^2\tau}{ds^2}$$

$$\text{eq. (1.23)} \Rightarrow \quad \frac{1}{\rho} = \left|\frac{d\vec{r}}{ds} \times \frac{d^2\vec{r}}{ds^2}\right| = \left|\frac{d\vec{r}}{d\tau} \times \frac{d^2\vec{r}}{d\tau^2}\right|\left|\frac{d\tau}{ds}\right|^3 = \frac{\left|\dfrac{d\vec{r}}{d\tau} \times \dfrac{d^2\vec{r}}{d\tau^2}\right|}{\left|\dfrac{d\vec{r}}{d\tau}\right|^3}$$

$$\Rightarrow \quad \frac{1}{\rho} = \frac{|\vec{v} \times \vec{a}|}{|\vec{v}\,|^3}, \qquad \rho = \frac{|\vec{v}\,|^3}{|\vec{v} \times \vec{a}\,|} \tag{1.31}$$

using parameter $\tau = t$. ρ at s depends only on the shape and size of the trajectory, and not on how fast the point moves over it. Substituting expressions of $\vec{v}$ and $\vec{a}$ in path coordinates in eq. (1.31) reveals that it is an identity independent of $\dot{s}$ and $\ddot{s}$.

§ **Exercise 1.5:** Prove eq. (1.31) directly using $\vec{v}\times\vec{a}$ in path coordinates: $\vec{v}\times\vec{a} = \frac{\dot{s}^3}{\rho}\,\vec{e}_b$. Hence, show that for a given $\vec{v}$, $\vec{a}$: $\vec{e}_n = \vec{e}_b \times \vec{e}_t = \dfrac{\vec{v} \times \vec{a}}{|\vec{v} \times \vec{a}|} \times \dfrac{\vec{v}}{|\vec{v}\,|}$.

Depending on the functional form of a_t and v_t, the following equations are used to obtain v_t and s by separating the variables and integration. For $a_t(t)$ and $a_t(v_t)$:

$$a_t = \frac{dv_t}{dt} \quad\Rightarrow\quad v_t(t) - v_t(0) = \int_0^t a_t(t)\,dt, \qquad t = \int_{v_t(0)}^{v_t(t)} \frac{1}{a_t(v_t)}\,dv_t \tag{1.32}$$

For $a_t(s)$ and $a_t(v_t)$: $\quad a_t = \dfrac{dv_t}{dt} = \dfrac{dv_t}{ds}\dfrac{ds}{dt} = v_t\,\dfrac{dv_t}{ds} \quad\Rightarrow$

$$v_t^2(s) - v_t^2(0) = 2\int_0^s a_t(s)\,ds, \qquad s = \int_{v_t(0)}^{v_t(s)} \frac{v_t}{a_t(v_t)}\,dv_t \tag{1.33}$$

For $v_t(t)$ and $v_t(s)$:

$$v_t = \frac{ds}{dt} \quad \Rightarrow \quad s(t) - s(0) = \int_0^t v_t(t)\,dt, \qquad t = \int_{s(0)}^{s(t)} \frac{1}{v_t(s)}\,ds \tag{1.34}$$

The integrals are evaluated in closed form or by numerical integration. For *uniform tangential acceleration*, $a_t(t) = a_0$ (a constant), eqs $(1.32)_1$, $(1.34)_1$, $(1.33)_1$ yield

$$v_t(t) = v_t(0) + a_0 t, \quad s(t) = s(0) + v_t(0)t + \tfrac{1}{2}a_0 t^2, \quad v_t^2(s) - v_t^2(0) = 2a_0 s \tag{1.35}$$

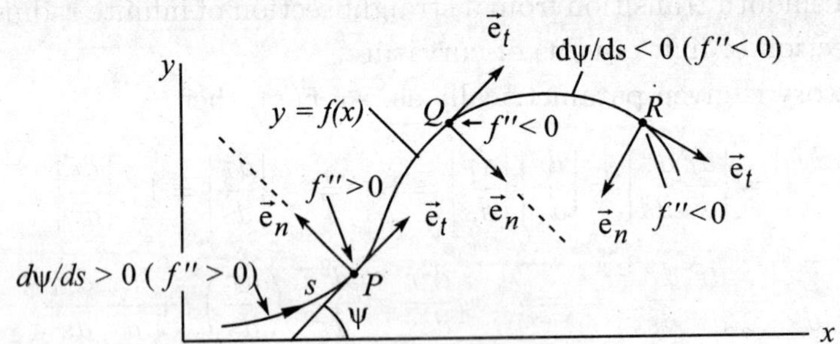

Fig. 1.17

For a *plane trajectory* (Fig. 1.17), $\vec{r} = x\vec{\mathrm{i}} + f(x)\vec{\mathrm{j}}$, ρ is obtained using $\tau = x$ in eq. $(1.31)_1$. To distinguish between time and space derivatives, we denote $\frac{d(\,)}{dx}$ by $(\,)'$:

$$\frac{d\vec{r}}{dx} = \vec{\mathrm{i}} + f'\,\vec{\mathrm{j}}, \qquad \frac{d^2\vec{r}}{dx^2} = f''\,\vec{\mathrm{j}}, \qquad \left|\frac{d\vec{r}}{dx}\right| = (1 + f'^2)^{1/2}$$

$$\left|\frac{d\vec{r}}{dx} \times \frac{d^2\vec{r}}{dx^2}\right| = |f''|,$$

$$\text{eq. } (1.31)_1 \;\Rightarrow\; \rho = \frac{[1 + f'^2]^{3/2}}{|f''|} \tag{1.36}$$

Further,

$$\tan\psi = \frac{dy}{dx} = f', \qquad \vec{\mathrm{e}}_t = \cos\psi\,\vec{\mathrm{i}} + \sin\psi\,\vec{\mathrm{j}}$$

$$\frac{d\vec{\mathrm{e}}_t}{ds} = (-\sin\psi\,\vec{\mathrm{i}} + \cos\psi\,\vec{\mathrm{j}})\frac{d\psi}{ds}, \qquad \frac{1}{\rho} = \left|\frac{d\vec{\mathrm{e}}_t}{ds}\right| = \left|\frac{d\psi}{ds}\right|$$

$$\vec{\mathrm{e}}_n = \left(\frac{d\vec{\mathrm{e}}_t}{ds}\right)\Big/\left|\frac{d\vec{\mathrm{e}}_t}{ds}\right| = (-\sin\psi\,\vec{\mathrm{i}} + \cos\psi\,\vec{\mathrm{j}})\,\frac{d\psi/ds}{|d\psi/ds|} = \pm(-\sin\psi\,\vec{\mathrm{i}} + \cos\psi\,\vec{\mathrm{j}})$$

Here the +ve sign is to be used in $\vec{e}_n$ for point P in Fig. 1.17 where $d\psi/ds > 0$ and −ve sign is to be used in $\vec{e}_n$ for points Q and R, where $d\psi/ds < 0$. It is observed that in each case, $\vec{e}_n$ is directed to the centre of curvature of the trajectory.

Interchanging x and y in the analysis leading to eq. (1.36) yields:

$$\rho = \frac{[1 + (dx/dy)^2]^{3/2}}{|d^2x/dy^2|} \tag{1.37}$$

§ **Exercise 1.6:** Show that for a plane trajectory $r = r(\phi)$, the radius of curvature $\rho = (r^2 + r'^2)^{3/2}/|r^2 + 2r'^2 - rr''|$, where $(\ )' = d(\ )/d\phi$.

§ **Exercise 1.7:** Prove the expressions of ρ for spatial trajectories

1. $y = y(x)$, $z = z(x)$, with $(\ )' = d(\ )/dx$:

$$\rho = (1 + y'^2 + z'^2)^{3/2}/[y''^2 + z''^2 + (y'z'' - y''z')^2]^{1/2}$$

2. $r = r(\phi)$, $z = z(\phi)$, with $(\ )' = d(\ )/d\phi$:

$$\rho = \frac{(r^2 + r'^2 + z'^2)^{3/2}}{[(rz'' - 2r'z')^2 + (r'z'' - r''z' + rz')^2 + (r^2 + 2r'^2 - rr'')^2]^{1/2}}$$

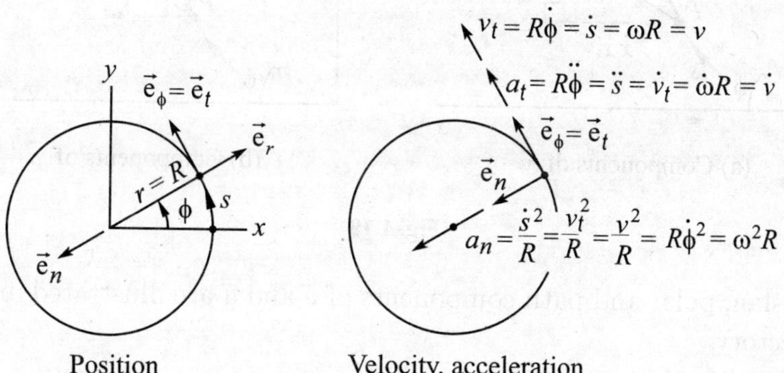

Fig. 1.18

For a *circular trajectory* of radius R (Fig. 1.18), using eqs (1.15, 16, 24, 25), we get:

$$r = R, \quad z = 0, \quad \dot{r} = \dot{z} = 0, \quad \ddot{r} = \ddot{z} = 0, \quad \rho = R, \quad \vec{e}_t = \vec{e}_\phi, \quad \vec{e}_n = -\vec{e}_r$$

$$\omega = \dot{\phi}, \quad \dot{\omega} = \ddot{\phi}, \quad s = R\phi, \quad v_t = \dot{s} = R\dot{\phi} = \omega R, \quad \dot{v}_t = \ddot{s} = R\ddot{\phi} = \dot{\omega}R$$

$$\vec{v} = \dot{s}\,\vec{e}_t = R\dot{\phi}\,\vec{e}_\phi = \omega R\,\vec{e}_\phi$$

$$\vec{a} = \ddot{s}\,\vec{e}_t + \frac{\dot{s}^2}{R}\,\vec{e}_n = -R\dot{\phi}^2\,\vec{e}_r + R\ddot{\phi}\,\vec{e}_\phi = -\omega^2 R\,\vec{e}_r + \dot{\omega}R\,\vec{e}_\phi \tag{1.38}$$

The radial and transverse components of $\vec{a}$, viz., $-r\dot{\phi}^2\,\vec{e}_r$ $(-\omega^2 R\vec{e}_r)$ and $r\ddot{\phi}\,\vec{e}_\phi$ $(\dot{\omega}R\,\vec{e}_\phi)$ are called the *centripetal* and *tangential* components.

§ **Exercise 1.8:** A pulley rotates at angular velocity ω and angular acceleration $\dot{\omega}$ relative to the ground. Its axis of rotation is at a distance b from a point P fixed to the ground. Find $\vec{v}_P$ and $\vec{a}_P$ of P relative to the reference frame of the pulley.

When the path is known a priori and information is known or desired for speed and rate of change of speed, then it is useful to express $\vec{v}$ and $\vec{a}$ in terms of the path coordinates. Path components are mostly used for problems of plane motion because it is generally easier to describe spatial motion in extrinsic coordinates.

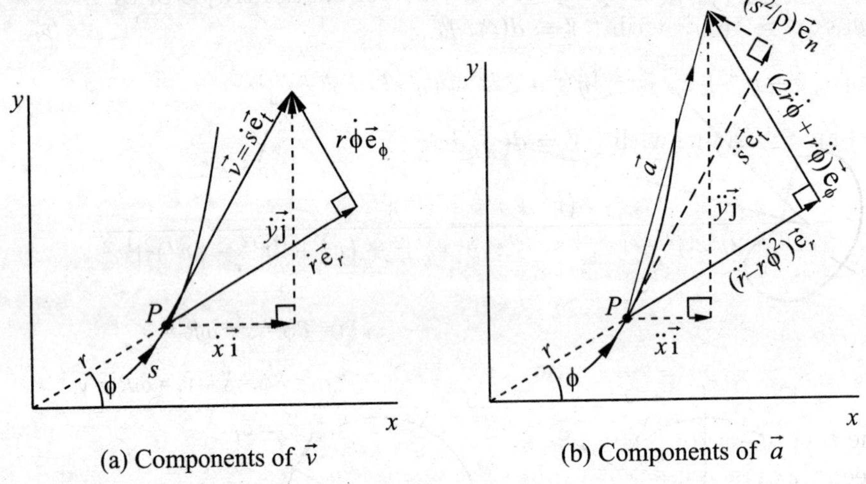

Fig. 1.19

The Cartesian, polar and path components of $\vec{v}$ and $\vec{a}$ are illustrated in Fig. 1.19 for a plane trajectory.

§ **Exercise 1.9:** Prove that for a plane trajectory, in path coordinates: $\vec{J} = \dot{\vec{a}} = (\dddot{s} - \dot{s}^3/\rho^2)\,\vec{e}_t + [3\dot{s}\ddot{s}/\rho - \dot{s}^3(d\rho/ds)/\rho^2]\,\vec{e}_n$.

1.10 Relative Velocity and Acceleration

We define *relative position vector* $\vec{r}_{PA|F}$, *relative velocity* $\vec{v}_{PA|F}$ and *relative acceleration* $\vec{a}_{PA|F}$ of a moving point P relative to a moving point A (Fig. 1.20) in frame F by

$$\begin{aligned} \vec{r}_{PA|F} &= \vec{r}_{P|F} - \vec{r}_{A|F} \\ \vec{v}_{PA|F} &= \dot{\vec{r}}_{PA|F} = \dot{\vec{r}}_{P|F} - \dot{\vec{r}}_{A|F} = \vec{v}_{P|F} - \vec{v}_{A|F} \end{aligned} \tag{1.39}$$

$$\vec{a}_{PA|F} = \dot{\vec{v}}_{PA|F} = \dot{\vec{v}}_{P|F} - \dot{\vec{v}}_{A|F} = \vec{a}_{P|F} - \vec{a}_{A|F} \tag{1.40}$$

The chain rule for composition of relative position, velocity and acceleration vectors with respect to frame F (dropping $(\,)_{|F}$ for simplicity), follows from Fig. 1.20:

$$\begin{aligned} &\vec{r}_P = \vec{r}_{PA} + \vec{r}_A, && \vec{r}_P = \vec{r}_{PA} + \vec{r}_{AB} + \vec{r}_B, && \vec{r}_{PB} = \vec{r}_{PA} + \vec{r}_{AB} \\ &\vec{v}_P = \vec{v}_{PA} + \vec{v}_A, && \vec{v}_P = \vec{v}_{PA} + \vec{v}_{AB} + \vec{v}_B, && \vec{v}_{PB} = \vec{v}_{PA} + \vec{v}_{AB} \\ &\vec{a}_P = \vec{a}_{PA} + \vec{a}_A, && \vec{a}_P = \vec{a}_{PA} + \vec{a}_{AB} + \vec{a}_B, && \vec{a}_{PB} = \vec{a}_{PA} + \vec{a}_{AB} \end{aligned} \tag{1.41}$$

and $\vec{a}_{PS} = \vec{a}_{PA} + \vec{a}_{AB} + \vec{a}_{BC} + \vec{a}_{CS}$.

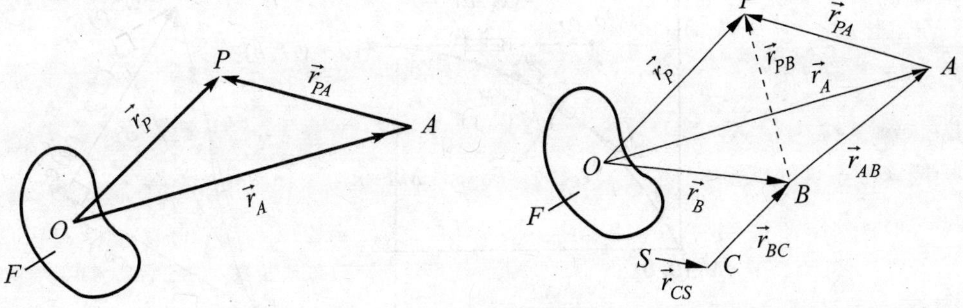

Fig. 1.20

The *rate of separation* of points A and P is the rate at which the distance d $(= AP)$ between them increases with time, the *rate of approach* being its negative. The distance $d = AP$ measured at a given time is the same for all reference frames. Hence, the *rate of separation* $\dot{d}$ *and its derivative* $\ddot{d}$ *are independent of the reference frame.* These are obtained using

$$\begin{aligned} & d^2 = \vec{r}_{PA} \cdot \vec{r}_{PA} && \Rightarrow \quad 2d\dot{d} = \vec{v}_{PA} \cdot \vec{r}_{PA} + \vec{r}_{PA} \cdot \vec{v}_{PA} \\ \Rightarrow \quad & d\dot{d} = \vec{v}_{PA} \cdot \vec{r}_{PA} && \Rightarrow \quad \dot{d} = \vec{v}_{PA} \cdot \vec{e}_{PA} \\ & d\ddot{d} + \dot{d}^2 = \vec{a}_{PA} \cdot \vec{r}_{PA} + \vec{v}_{PA} \cdot \vec{v}_{PA} && \Rightarrow \quad \ddot{d} = \vec{a}_{PA} \cdot \vec{e}_{PA} + (v_{PA}^2 - \dot{d}^2)/d \end{aligned}$$

$\dot{d}$ is the component of $\vec{v}_{PA}$ in the direction of $\vec{r}_{PA}$ $(\vec{e}_{PA} = \vec{r}_{PA}/d)$, but $\ddot{d}$ is *not* the component of $\vec{a}_{PA}$ in the direction of $\vec{r}_{PA}$.

Example 1.3: A missile M moves at velocity v_M towards its target aircraft A moving at velocity v_A (Fig. E1.3). The initial positions are M_0, A_0. Find the time of capture T. (Note: The missile continuously changes its orientation so that it always moves directly towards the target.).

Solution: Use the rate of decrease of separation r at $v_M - v_A \cos\phi$ and $v_{MA_x} = v_M \cos\phi - v_A$, to set up equations for $(a^2+b^2)^{1/2}$ and a:

$$\int_0^T (v_M - v_A \cos\phi)\, dt = -\int_{\sqrt{a^2+b^2}}^0 dr = (a^2+b^2)^{1/2}$$

$$\int_0^T (v_M \cos\phi - v_A)\, dt = \int_{-a}^0 d(x_M - x_A) = a$$

Eliminate $\int_0^T \cos\phi\, dt$ to get: $\quad T = [a v_A + (a^2+b^2)^{1/2} v_M]/(v_M^2 - v_A^2)$.

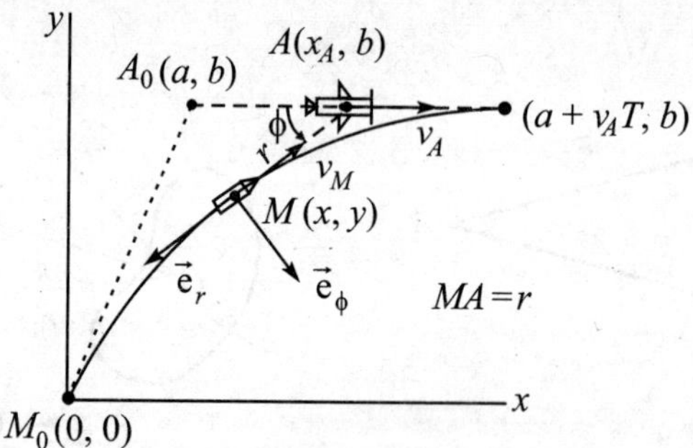

Fig. E1.3

1.11 Rectilinear Motion

For rectilinear motion, choose x-axis along the path with $s = x$. In this section and the next, we denote the velocity and acceleration along $\vec{i}$ by v and a:

$$\vec{r} = x\,\vec{i}, \qquad \vec{v} = \dot{x}\,\vec{i} = v\,\vec{i}, \qquad \vec{a} = \ddot{x}\,\vec{i} = \dot{v}\,\vec{i} = a\,\vec{i} \tag{1.42}$$

$$v = \dot{x} = \dot{s} = v_t, \qquad a = \ddot{x} = \dot{v} = v\frac{dv}{dx} = \ddot{s} = a_t \tag{1.43}$$

The speed and magnitude of acceleration are *not* v and a but $|v|$ and $|a|$. If a and v have the same sign, then the speed $|v|$ increases; and if they have the opposite signs, then the speed decreases and the magnitude of acceleration for this case is called *deceleration*. In general, a may be a function of x, v, t:

$$\ddot{x} = a = a(x, v, t) \tag{1.44}$$

The two constants of integration in the solution of this second order differential equation are generally evaluated from the two *initial conditions* of position and velocity at

$t = 0$: $x(0) = x_0$, $\dot{x}(0) = v_0$. The two given conditions could also be prescribed values of $x(0)$, $x(t_1)$.

If a is a known function of only one of the variables x, v, t; then eq. (1.44) can be integrated by separation of variables. The relations for x, v, t are the same as eqs (1.32) to (1.35) with x, v, a replacing s, v_t, a_t. The curves of x-t, v-t, a-t for rectilinear motion and of s-t, v_t-t, a_t-t for curvilinear motion are called motion curves. The following relations hold:

$$v = \frac{dx}{dt} = \text{slope of } x\text{-}t \text{ curve}, \qquad a = \frac{dv}{dt} = \text{slope of } v\text{-}t \text{ curve}$$

$$\Delta x = x(t_2) - x(t_1) = \int_{t_1}^{t_2} v\, dt = \text{algebraic area under } v\text{-}t \text{ curve}$$

$$\Delta v = v(t_2) - v(t_1) = \int_{t_1}^{t_2} a\, dt = \text{algebraic area under } a\text{-}t \text{ curve}$$

The area in the half plane of positive v in v-t plane, and of positive a in a-t plane, is considered as positive, and the rest of the area as negative. The distance d_{12} travelled from t_1 to t_2 is given by

$$d_{12} = \int_{t_1}^{t_2} |v|\, dt = \text{sum of the magnitudes of area under the } v\text{-}t \text{ curve}$$

The relations for motion curves for curvilinear motion are similar.

The relations between the translational motion of two or more bodies, connected by inextensible cables through pulleys or rigid rods, are established by writing equations for the constant length of each cable or rod in terms of the displacement parameters of the bodies. Two successive differentiations of these equations with respect to t yield relations between their velocities and accelerations. For the cases shown in Fig. 1.21:

a. $s_A = s_B$, $\quad v_A = v_B$, $\quad a_A = a_B$

b. $x_A^2 + y_B^2 = L^2$, $\quad x_A \dot{x}_A + y_B \dot{y}_B = 0$, $\quad \dot{x}_A^2 + x_A \ddot{x}_A + \dot{y}_B^2 + y_B \ddot{y}_B = 0$

c. $2x_1 + x_2 = \text{const.}$, $\quad 2v_1 + v_2 = 0$, $\quad 2a_1 + a_2 = 0$

d. $x_1 + 2(x_1 - y - d) + 2y + x_2 = \text{const.}$, $\quad 3v_1 + v_2 = 0$, $\quad 3a_1 + a_2 = 0$

e. $x_1 + 2x_2 = \text{const.}$, $\quad v_1 + 2v_2 = 0$, $\quad a_1 + 2a_2 = 0$

$x_3 + (x_3 - x_2) = \text{const.}$, $\quad 2v_3 - v_2 = 0$, $\quad 2a_3 - a_2 = 0$

It is assumed in case d. that θ is small throughout the motion.

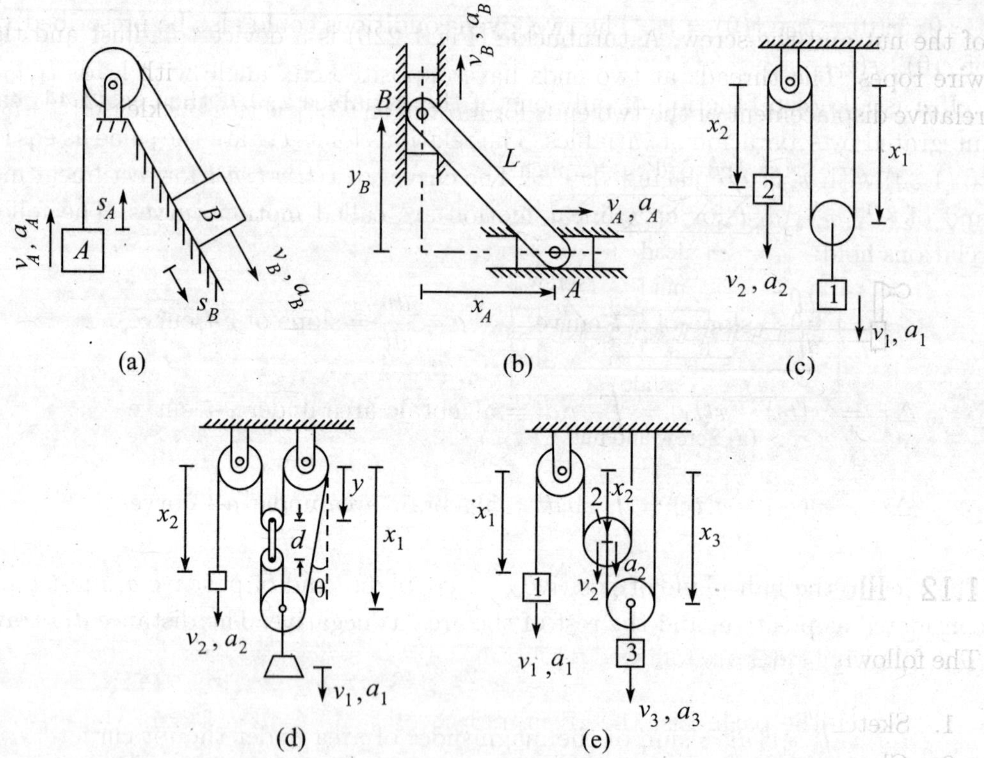

Fig. 1.21

Rotation is often transformed into rectilinear translation by a screw and a nut as in the lead screw of a lathe, parallel vise, screw-jack and hand drill. Screw 2 in Fig. 1.22a rotates in fixed bearings and carries a nut 1 which slides in guides but cannot turn. If p is the *pitch* (axial distance between two successive threads) and the screw has n threads per cm, then $p = (1/n)$ cm. The lead l of the screw is the axial distance between corresponding points of the helix of a thread which have a difference of 2π radians in their polar angle with respect to the screw axis. For a single start thread $l = p$, and for a screw with k starts of thread $l = kp$. l is the axial advance of a point on the screw moving in fixed bearings in one revolution. It is also the axial advance of a point on a nut moving on a fixed screw in one revolution. The axial distance s moved by a nut on a screw rotating in fixed bearings (Fig. 1.22a) for a rotation θ of the screw is given by

$$s = l\theta/2\pi \tag{1.45}$$

In general, mechanisms involving screws are not convenient if the direction of motion changes during motion, because of the inevitable backlash due to gap on the threads of the nut and the screw. A turnbuckle (Fig. 1.22b) is a device to adjust and tighten wire ropes. The threads at two ends have opposite helix angle with leads l_1, l_2. The relative displacement of the two ends for a rotation θ of the turnbuckle is $\theta(l_1+l_2)/2\pi$.

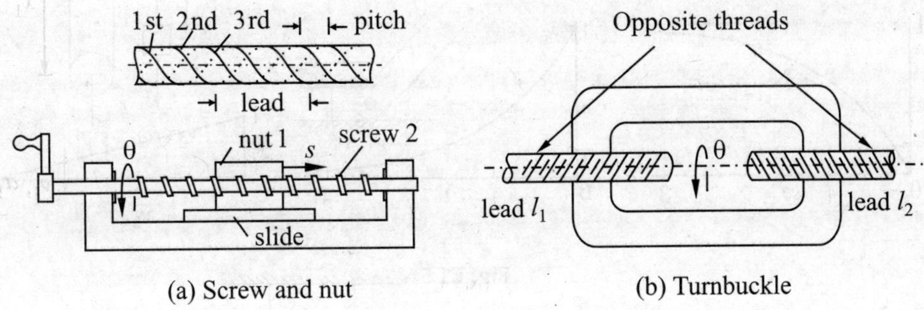

(a) Screw and nut

(b) Turnbuckle

Fig. 1.22

1.12 Illustrative Problems

The following steps are followed to obtain the solution of problems of point kinematics.

1. Sketch the path. List the given information and the variables to be solved.
2. Choose the set of coordinates (Cartesian/cylindrical/path) which best fit the information given and the results required. Show the unit vectors in the sketch. Write geometric relations between lengths and angles in terms of coordinates.
3. Write kinematic relations between unknowns and the given data.
4. Solve the equations for the required variables.

Example 1.4: The ionosphere is a region of electrically neutral gas composed of positively charged ions and negatively charged electrons. The region surrounds the earth at a height of approximately 200 km. Find the motion of an electron of charge $-e$ and mass m which is initially at rest and has acceleration of $(-eE/m)\sin\omega t$ under an electric field $E\sin\omega t$ due to the passage of a radio wave.

Solution: Choose the origin at the initial position: $x(0)=0,\ v(0)=0$.

$$v(t) = v(0) + \int_0^t a(t)\,dt = -\frac{eE}{m}\int_0^t \sin\omega t\,dt = \frac{eE}{m\omega}(\cos\omega t - 1)$$

$$x(t) = x(0) + \int_0^t v(t)\,dt = \frac{eE}{m\omega}\int_0^t (\cos\omega t - 1)\,dt = \frac{eE}{m\omega}\left[\frac{\sin\omega t}{\omega} - t\right]$$

The first term in $x(t)$ represents the jiggling motion around the motion with uniform velocity represented by the second term.

Example 1.5: A gun, recoiling under viscous damping, has an acceleration $a = -c\,v$, where c is a constant. The initial velocity is v_0. Find $v(t)$, $x(t)$, $v(x)$. Find the distance moved by the gun and the time elapsed before its speed reduces to (i) $0.5v_0$, (ii) $0.01v_0$.

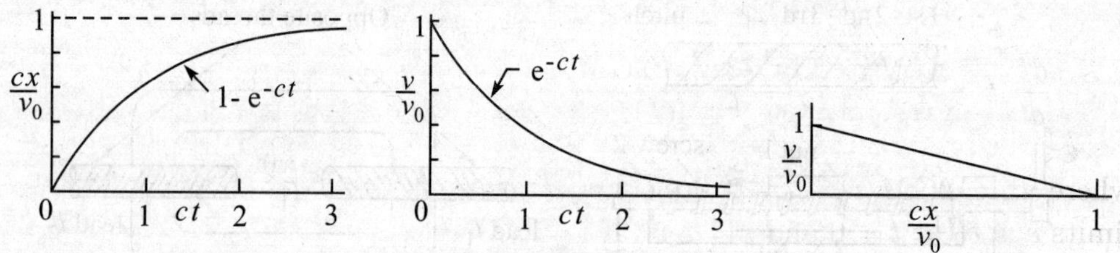

Fig. E1.5

Solution: The solution is obtained by separation of variables:

1. $v(t)$: $$\frac{dv}{dt} = -c\,v \quad \Rightarrow \quad \int_{v_0}^{v} \frac{1}{v}\,dv = -c\int_0^t dt \quad \Rightarrow \quad v = v_0 \mathrm{e}^{-ct} \qquad (1)$$

2. $x(t)$: $$\frac{dx}{dt} = v_0\mathrm{e}^{-ct} \quad \Rightarrow \quad \int_0^x dx = v_0\int_0^t \mathrm{e}^{-ct}\,dt, \qquad x = \frac{v_0}{c}(1-\mathrm{e}^{-ct}) \qquad (2)$$

Hence, v decays to zero exponentially and $x \to v_0/c$ as $t \to \infty$.

3. $v(x)$: $$\frac{dv}{dt} = v\frac{dv}{dx} = -c\,v \quad \Rightarrow \quad \int_{v_0}^{v} dv = -c\int_0^x dx, \quad v = v_0 - c\,x \qquad (3)$$

which could have also been obtained by eliminating t from eqs (1) and (2). Hence, $x \to v_0/c$ as $v \to 0$. For $v = 0.5v_0$, eqs (1) and (3) yield $t = \ln 2\,/c = 0.693/c$ and $x = v_0/2c$. Similarly, for $v = 0.01v_0$, eqs (1) and (3) yield $t = \ln 100\,/c = 4.605/c$ and $x = 0.99v_0/c$. The motion curves are shown in Fig. E1.5.

Example 1.6: A mass-point at distance r_0 from the centre O of the earth is projected radially out with velocity v_0. Its acceleration at distance r from O is GM/r^2 directed to O. GM is a +ve constant. Show that it returns if $v_0 < (2GM/r_0)^{1/2}$. In this case, find $r_{\max}$, the velocity with which it hits the ground at $r = R$ and the time of travel.

Solution: The velocity and acceleration are: $v = \dot{r},\ a = \ddot{r}$.

$$a = \frac{dv}{dt} = \frac{dv}{dr}\dot{r} = v\frac{dv}{dr} = -\frac{GM}{r^2}$$

$$\Rightarrow \int_{v_0}^{v} v\,dv = -GM\int_{r_0}^{r}\frac{1}{r^2}\,dr \quad \Rightarrow \quad v^2 = \left(v_0^2 - \frac{2GM}{r_0}\right) + \frac{2GM}{r} \qquad (1)$$

$\Rightarrow\ r_{\max} = \infty$ if corresponding $v_\infty = v_0^2 - 2GM/r_0 \geq 0$, i.e., if $v_0^2 \geq 2GM/r_0$.

$\Rightarrow$ The body returns if $v_0 < (2GM/r_0)^{1/2}$. In this cae, finite $r_{\max}$ occurs for $v = 0$: at $r_{\max} = 2GM/(2GM/r_0 - v_0^2)$ using eq. (1). The speed v_1 with which it hits the ground at $r = R$ is obtained from eq. (1) as $v_1 = [v_0^2 + 2GM(1/R - 1/r_0)]^{1/2}$. The time of travel t_1 is obtained by expressing the acceleration in terms of v using eq. (1):

$$\frac{dv}{dt} = -\frac{GM}{r^2} = -\frac{[v^2 - v_0^2 + 2GM/r_0]^2}{4GM} = -\frac{(v^2 + c^2)^2}{4GM}\,.$$

where $c^2 = 2GM/r_0 - v_0^2 > 0$. Separating the variables and integrating between the limits $v = v_0$ at $t = 0$ and $v = -v_1$ at $t = t_1$, by the substitution $v = c\tan\theta$, we get

$$\begin{aligned} t_1 &= \int_0^{t_1} dt = -4GM \int_{v_0}^{-v_1} \frac{1}{(v^2 + c^2)^2}\, dv \\ &= -4GM \left[\frac{v}{2c^2(v^2 + c^2)} + \frac{1}{2c^3} \tan^{-1}\left(\frac{v}{c}\right) \right] \Bigg|_{v_0}^{-v_1} \end{aligned}$$

Example 1.7: The knowledge of the acceleration of material points of earth during an earthquake is needed for the design of foundations of heavy machines. The acceleration recorded by an acceleration measuring device in the North to South direction is shown in Fig. E1.7a with scales, t: 1 cm = 0.1 s, a: 1 cm = $0.3g$. Find (a) the speed at 0.6 s, (b) the maximum speed in the first 0.6 s, (c) the displacement in 0.6 s.

Solution: (a) $v(0) = 0$. The change in velocity equals the area under the *a*-*t* curve.

$$v(0.6) = \text{area } 0 \text{ to } 0.6\,\text{s} = (1 - 2.5 + 0.5 - 0.2) \times (0.3g \times 0.1) = -0.3532 \text{ m/s}$$

(b) The cumulative area A_c under the *a*-*t* curve is plotted in Fig. E1.7b.

t(s)	0	0 to 0.2	0.2 to 0.4	0.4 to 0.5	0.5 to 0.6
Additional area	0	1	−2.5	0.5	−0.2
Total area A_c cm^2	0	1	−1.5	−1.0	−1.2

The maximum speed $|v|_{\max}$ occurs when $|A_c|_{\max} = 1.5\,\text{cm}^2$ at $t = 0.4s$. $\Rightarrow$ $|v|_{\max} = 1.5(0.3g)0.1 = 0.4415\,\text{m/s}$ and is directed towards the North since $A_c = -1.5 < 0$.
(c) In the *moment–area method* to obtain x directly from the *a*-*t* curve, the area under the *v*-*t* curve (Fig. E1.7c) is found from the horizontal strips of area $(t_1 - t)dv = (t_1 - t)a\,dt$, which is the first moment of the elementary area $a\,dt$ under the *a*-*t* curve about the line $t = t_1$:

$$x_1 - x_0 = \int_{t_0}^{t_1} v\,dt = v_0(t_1 - t_0) + \int_{t_0}^{t_1} (t_1 - t)a\,dt$$

$$\Rightarrow \qquad x_1 = x_0 + v_0(t_1 - t_0) + A(t_1 - t_C)$$

where A is the area under the a-t curve from t_0 to t_1 and C is its centroid.

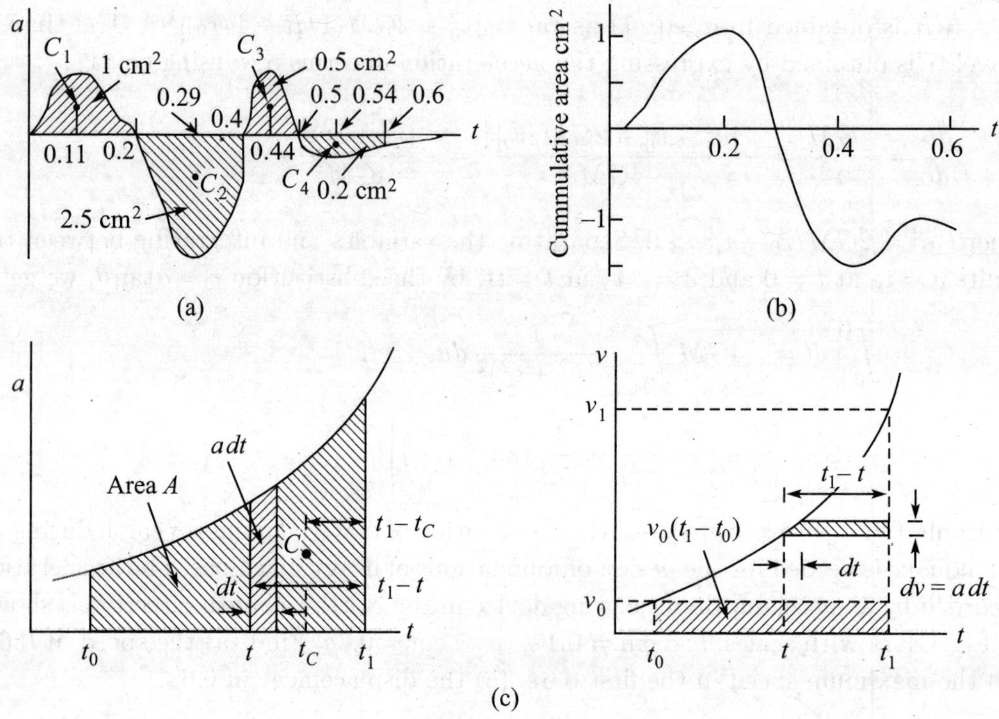

Fig. E1.7

Apply the moment–area method from $t_0 = 0$ to $t_1 = 0.6\,\text{s}$, with $x_0 = 0, v_0 = 0$:

$$\begin{aligned} x(0.6) &= x_0 + v_0(t_1 - t_0) + A(t_1 - t_C) = \sum A_i(t_1 - t_{C_i}) \\ &= [1(0.6 - 0.11) - 2.5(0.6 - 0.29) + 0.5(0.6 - 0.44) \\ &\quad -0.2(0.6 - 0.54)](0.3g \times 0.1) = -0.06386\,\text{m} \end{aligned}$$

i.e., the displacement in 0.6 s is 0.06386 m to the North.

Example 1.8: For an elevator, find the minimum time to travel 81 m from rest on the ground floor to the top floor and $v_{\max}$, for the following cases. (a) The acceleration and deceleration of the elevator is limited to $4.5\,\text{m/s}^2$ and the maximum speed is limited to $6.75\,\text{m/s}$. (b) The maximum magnitude of jerk is limited to $0.9\,\text{m/s}^3$. (c) The magnitudes of jerk and acceleration are limited to $0.9\,\text{m/s}^3$ and $2.7\,\text{m/s}^2$. (d) The magnitudes of jerk, acceleration and velocity are limited to $0.9\,\text{m/s}^3$, $4.5\,\text{m/s}^2$ and $6.75\,\text{m/s}$.

Solution: For the *a-t*, *v-t* diagrams in Fig. E1.8a, b, c, d: $v(0) = v_0 = 0,\ v(t_1) = 0$.

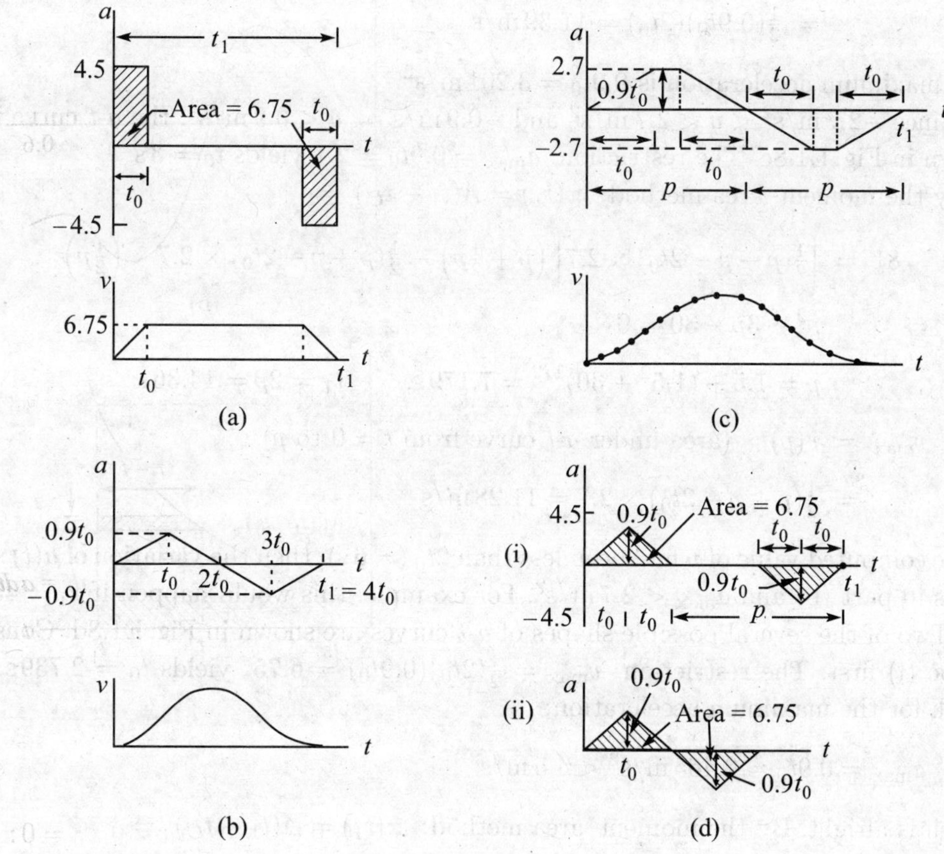

Fig. E1.8

(a) $\quad v(t_0) = 4.5t_0 = v_{\max} = 6.75 \quad \Rightarrow \quad t_0 = 1.5\,\text{s}.$

By the moment–area method,

$$x(t_1) = x_0 + v_0(t_1 - 0) + A(t_1 - t_C)$$

$$\Rightarrow \quad 81 = 6.75(t_1 - 0.5t_0) - 6.75 \times 0.5t_0 \quad \Rightarrow \quad t_1 = 13.5\,\text{s}$$

(b) The *a-t* curve has to be continuous as the jerk is finite. Its maximum slope is: $|\dot{a}| = 0.9\,\text{m/s}^3$. The area under the *a-t* curve is zero since the final velocity is zero. Hence, the *a-t* curve has the shape shown in Fig. E1.8b. By the moment–area method,

$$x(t_1) = A(t_1 - t_C) \quad \Rightarrow \quad 81 = \tfrac{1}{2}(2t_0)(0.9t_0)(3t_0) - \tfrac{1}{2}(2t_0)(0.9t_0)t_0$$

$$\Rightarrow \quad t_0 = 3.557\,\text{s}, \qquad t_1 = 4t_0 = 14.23\,\text{s},$$

$$v_{max} = v(2t_0) = (\text{area under } a\text{-}t \text{ curve from } t = 0 \text{ to } 2t_0)$$
$$= \tfrac{1}{2}(0.9t_0)(2t_0) = 11.39\,\text{m/s}$$

The maximum acceleration is $0.9t_0 = 3.201\,\text{m/s}^2$.
(c) Since $-2.7\,\text{m/s}^2 < a < 2.7\,\text{m/s}^2$ and $-0.9\,\text{m/s}^3 < \dot{a} < 0.9\,\text{m/s}^3$, the *a-t* curve is as shown in Fig. E1.8c. The restriction $a_{max} = 0.9t_0 = 2.7$ yields $t_0 = 3\,\text{s}$.

By the moment–area method: $x(t_1) = A(t_1 - t_C) \Rightarrow$

$$81 = \left[\tfrac{1}{2}(p + p - 2t_0) \times 2.7\right]\left(p + \tfrac{1}{2}p\right) - \tfrac{1}{2}(p + p - 2t_0) \times 2.7 \times \left(\tfrac{1}{2}p\right)$$
$$\Rightarrow \quad p^2 - 3p - 30 = 0$$
$$p = 1.5 + (1.5^2 + 30)^{1/2} = 7.179\,\text{s}, \qquad t_1 = 2p = 14.36\,\text{s}$$
$$v_{max} = v(p) = (\text{area under } a\text{-}t \text{ curve from } t = 0 \text{ to } p)$$
$$= \tfrac{1}{2}(p + p - 2t_0) \times 2.7 = 11.28\,\text{m/s}$$

If the computed value of p had been less than $2t_0$ ($= 6\,\text{s}$), then the variation of $a(t)$ shall be as in part (b) and $a_{max} < 2.7\,\text{m/s}^2$. For example, this would happen if $x_1 = 45\,\text{m}$.
(d) Two of the several possible shapes of *a-t* curves are shown in Fig. E1.8d. Consider shape (i) first. The restriction $v_{max} = \tfrac{1}{2}(2t_0)(0.9t_0) = 6.75$ yields $t_0 = 2.739\,\text{s}$. We check for the maximum acceleration:

$$a_{max} = 0.9t_0 = 2.465\,\text{m/s}^2 < 4.5\,\text{m/s}^2$$

which is alright. By the moment–area method: $x(t_1) = A(t_1 - t_C) \Rightarrow$

$$81 = 6.75(p + t_0) - 6.75t_0 \quad \Rightarrow \quad p = 12\,\text{s}, \qquad t_1 = p + 2t_0 = 17.48\,\text{s}$$

Hence, the *a-t* curve (i) of Fig. E1.8d is the solution.

Example 1.9: The rotary motion of a wheel (Fig. E1.9a) is converted into translation of the rod by: (i) Scotch yoke mechanism, (ii) cam with tangent follower, (iii) cam with roller follower. In the given position, the disc has angular speed ω and angular acceleration $\dot{\omega}$. Find the velocity $\dot{x}$ and acceleration $\ddot{x}$ of the follower rod in each case.

Solution: The relation of position x of the rod with ϕ is set up for the general configuration (Fig. E1.9b). $\dot{x}$ and $\ddot{x}$ are obtained by differentiating this relation.

(i) $\quad x = R\cos\phi \quad \Rightarrow \quad \dot{x} = -R\sin\phi\,\dot{\phi} = -\omega R\sin\phi$

$$\ddot{x} = -\omega R\cos\phi\,\dot{\phi} - \dot{\omega}R\sin\phi = -\omega^2 R\cos\phi - \dot{\omega}R\sin\phi$$

(ii) $\quad x = R + e\cos\phi \quad \Rightarrow \quad \dot{x} = -e\sin\phi\,\dot{\phi} = -\omega e\sin\phi$

$$\ddot{x} = -\omega e \cos\phi\,\dot{\phi} - \dot{\omega} e \sin\phi = -\omega^2 e \cos\phi - \dot{\omega} e \sin\phi$$

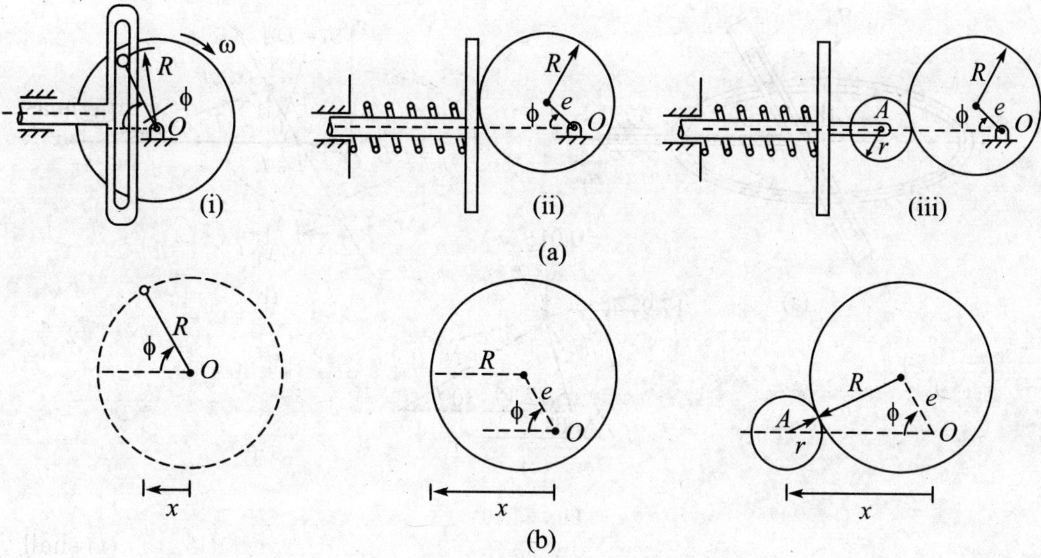

Fig. E1.9

(iii) By cosine law, $$2xe\cos\phi = x^2 + e^2 - (R+r)^2 \tag{1}$$

It can be solved for x. Differentiating eq. (1) with respect to time twice, yields $\dot{x}$, $\ddot{x}$:

$$2\dot{x}e\cos\phi - 2xe\sin\phi\,\dot{\phi} = 2x\dot{x} \quad\Rightarrow\quad \dot{x} = -\omega ex\sin\phi/(x - e\cos\phi) \tag{2}$$

$$\ddot{x}e\cos\phi - \dot{x}e\sin\phi\,\dot{\phi} - \dot{x}e\sin\phi\,\dot{\phi} - xe(\cos\phi\,\dot{\phi}^2 + \sin\phi\,\ddot{\phi}) = \dot{x}^2 + x\ddot{x} \tag{3}$$

Substituting for $\dot{x}$ from eq. (2) in eq. (3) and simplifying, we get

$$\ddot{x} = \frac{ex}{x - e\cos\phi}\left[\frac{\omega^2}{x - e\cos\phi}\left\{e(1+\sin^2\phi) - x\cos\phi - \frac{xe\sin^2\phi}{x - e\cos\phi}\right\} - \dot{\omega}\sin\phi\right]$$

Example 1.10: Pins A and B move in a fixed elliptical slot and in a straight slot which translates with acceleration $a_1 = -\omega^2 x_0 \cos\omega t$ (Fig. E1.10a). The position and velocity of the point D of the slot at $t = 0$ are x_0 and zero. Find the (a) velocity and acceleration of the centres of pins with respect to (1) ground reference G and (2) translating body T, (b) rate at which the centres of the pins separate and the rate of change of this relative speed of separation with respect to G and T, (c) reactions from the smooth slots on the pin A of mass 0.01 kg; when point D of the slot is for the first time at a distance of $0.6x_0$ to the right of the centre of the ellipse. Given: $\theta = 60°$,

$c = 20\,\text{cm}$, $d = 10\,\text{cm}$, $x_0 = 10\,\text{cm}$ and $\omega = 80\,\text{rad/s}$.

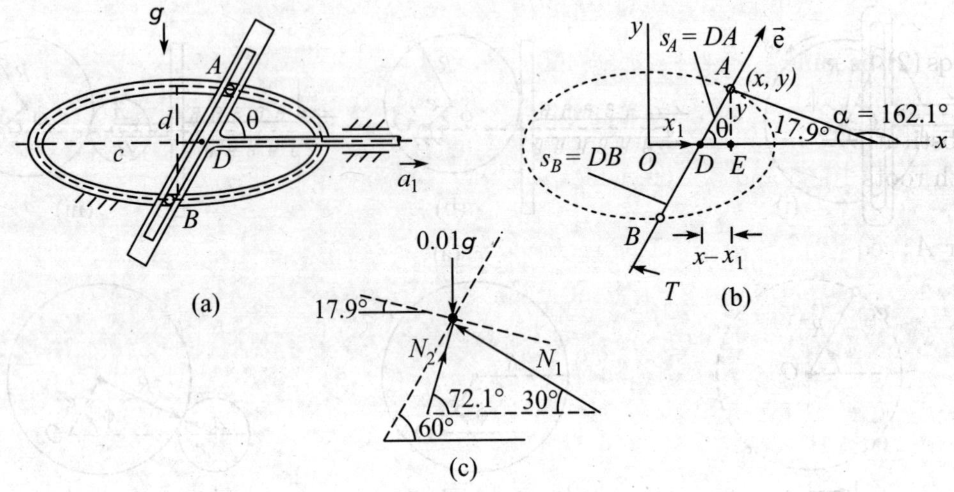

Fig. E1.10

Solution: Let (x, y) be the position of A (Fig. E1.10b) and x_1 be the position of D at time t. For D, we have $x_1(0) = x_0$, $v_1(0) = 0$, $a_1 = -\omega^2 x_0 \cos\omega t$. Hence,

$$v_1 = \dot{x}_1 = v_1(0) + \int_0^t a_1(t)\,dt = -\omega x_0 \sin\omega t \tag{1}$$

$$x_1 = x_1(0) + \int_0^t v_1\,dt = x_0 \cos\omega t$$

The coordinates x, y are related by the equations of the ellipse and the straight slot:

$$\frac{x^2}{c^2} + \frac{y^2}{d^2} = 1, \qquad y = (x - x_1)\tan\theta \qquad (AE = DE\tan\theta) \tag{2}$$

Equations (2) are solved for x, y. Differentiating eqs (2) yield relations between $\dot{x}, \dot{y}$:

$$\frac{x\dot{x}}{c^2} + \frac{y\dot{y}}{d^2} = 0, \qquad \dot{y} = (\dot{x} - \dot{x}_1)\tan\theta \tag{3}$$

Equations (3) are solved for $\dot{x}, \dot{y}$. Differentiating eqs (3) yields relations between $\ddot{x}, \ddot{y}$:

$$\frac{\dot{x}^2 + x\ddot{x}}{c^2} + \frac{\dot{y}^2 + y\ddot{y}}{d^2} = 0, \qquad \ddot{y} = (\ddot{x} - \ddot{x}_1)\tan\theta \tag{4}$$

Equations (4) are solved for $\ddot{x}, \ddot{y}$. For the given data, at the position of interest:

$$x_1 = x_0\cos\omega t = 0.6x_0 = 6\,\text{cm} \quad \Rightarrow \quad \cos\omega t = 0.6, \quad \sin\omega t = 0.8$$

$$\text{eqs (1)} \Rightarrow \quad \dot{x}_1 = -80(10)(0.8) = -640\,\text{cm/s}$$

$$\ddot{x}_1 = -(80)^2(10)(0.6) = -38400\,\text{cm/s}^2$$

$$\text{eqs (2) become} \qquad x^2 + 4y^2 = 400, \qquad y = (x-6)\sqrt{3} \tag{5}$$

Substituting y from the 2nd equation into the 1st equation yields, $13x^2 - 144x + 32 = 0$ with roots 10.85 and 0.2269, which are the x-coordinates of A and B, respectively.

For A: eqs (2) $\Rightarrow$ $y = (10.85 - 6)\sqrt{3} = 8.400$

$$\text{eqs (3)}: \qquad 10.85\dot{x}/400 + 8.4\dot{y}/100 = 0, \qquad \dot{y} = (\dot{x} + 640)\sqrt{3}$$

$$\Rightarrow \qquad \dot{x} = -539.4\,\text{cm/s}, \qquad \dot{y} = 174.2\,\text{cm/s}$$

$$\text{eqs (4)}: \qquad (539.4^2 + 10.85\ddot{x})/400 + (174.2^2 + 8.4\ddot{y})/100 = 0$$

$$\ddot{y} = (\ddot{x} + 38400)\sqrt{3}$$

$$\Rightarrow \qquad \ddot{x} = -38338\,\text{cm/s}^2, \qquad \ddot{y} = 107.99\,\text{cm/s}^2$$

(a)

$$\vec{v}_A = \dot{x}\,\vec{\imath} + \dot{y}\,\vec{\jmath} = -539.4\,\vec{\imath} + 174.2\,\vec{\jmath}\ \text{cm/s}$$

$$\vec{a}_A = \ddot{x}\,\vec{\imath} + \ddot{y}\,\vec{\jmath} = -38338\,\vec{\imath} + 107.99\,\vec{\jmath}\ \text{cm/s}^2$$

Similarly, by taking $x = 0.2269$, we can obtain velocity and acceleration of point B:

$$\vec{v}_B = -642.1\,\vec{\imath} - 3.64\,\vec{\jmath}\ \text{cm/s}, \qquad \vec{a}_B = -32554\,\vec{\imath} + 10125\,\vec{\jmath}\ \text{cm/s}^2$$

Relative to frame T, the position vector of A is $s_A\vec{e}$, where $\vec{e}$ is the unit vector along AB, with $s_A = y_A \operatorname{cosec}\theta$,

$$\dot{s}_A = \dot{y}_A \operatorname{cosec}\theta = 174.2 \operatorname{cosec} 60^\circ = 201.1\,\text{cm/s}$$

$$\dot{s}_B = \dot{y}_B \operatorname{cosec}\theta = -3.64 \operatorname{cosec} 60^\circ = -4.2\,\text{cm/s}$$

$$\ddot{s}_A = \ddot{y}_A \operatorname{cosec}\theta = 107.99 \operatorname{cosec} 60^\circ = 124.7\,\text{cm/s}^2$$

$$\ddot{s}_B = \ddot{y}_B \operatorname{cosec}\theta = 10125 \operatorname{cosec} 60^\circ = 11691\,\text{cm/s}^2$$

$$\vec{v}_{A|T} = \dot{s}_A\,\vec{e} = 201.1\,\vec{e}\ \text{cm/s}, \qquad \vec{a}_{A|T} = \ddot{s}_A\,\vec{e} = 124.7\,\vec{e}\ \text{cm/s}^2$$

$$\vec{v}_{B|T} = \dot{s}_B\,\vec{e} = -4.2\,\vec{e}\ \text{cm/s}, \qquad \vec{a}_{B|T} = \ddot{s}_B\,\vec{e} = 11691\,\vec{e}\ \text{cm/s}^2$$

(b) The position vector of A relative to B is $s\vec{e}$ with the separation:

$$s = AB = (y_A - y_B)\operatorname{cosec}\theta$$

$$\dot{s} = (\dot{y}_A - \dot{y}_B)\operatorname{cosec}\theta = 201.1 - (-4.2) = 205.3\,\text{cm/s}$$

$$\ddot{s} = (\ddot{y}_A - \ddot{y}_B)\operatorname{cosec}\theta = 124.7 - 11691 = -11566\,\text{cm/s}^2$$

The rates of separation with respect to translating frame T and with respect to ground reference are equal as these are the same for all reference frames.

(c) Let the tangent to the ellipse at (x_A, y_A) make an angle α with the x-axis.

$$\frac{x^2}{400} + \frac{y^2}{100} = 1 \quad \Rightarrow \quad \frac{2x}{400} + \frac{2yy'}{100} = 0, \quad y' = \frac{dy}{dx} = -\frac{x}{4y}$$

$$\Rightarrow \quad \tan\alpha = y'_A = -\frac{x_A}{4y_A} = -\frac{10.85}{4 \times 8.4} \quad \Rightarrow \quad \alpha = 162.1^\circ$$

The FBD (free body diagram) of the pin is shown in Fig. E1.10c. The equation of motion of pin A: $\vec{F} = m\vec{a}_A \Rightarrow$

$$N_1(-\cos 30^\circ\,\vec{i} + \sin 30^\circ\,\vec{j}) + N_2(\cos 72.1^\circ\,\vec{i} + \sin 72.1^\circ\,\vec{j}) - 0.01g\,\vec{j}$$
$$= 0.01(-383.38\,\vec{i} + 1.0799\,\vec{j})$$

$$\vec{i}: \qquad -0.8660N_1 + 0.3074N_2 = -3.8338$$

$$\vec{j}: \qquad 0.5N_1 + 0.9516N_2 = 0.1089$$

$\Rightarrow N_1 = 3.765\,\text{N}$, $N_2 = -1.864\,\text{N}$. The negative sign in N_2 implies that the actual direction of reaction N_2 is opposite to that shown in Fig. E1.10c.

Example 1.11: (a) The acceleration of a point moving in the xy-plane is proportional to its distance from a fixed point and is directed towards it, i.e., with origin at this point, $\vec{a} = -\omega^2\vec{r} = -\omega^2(x\,\vec{i} + y\,\vec{j})$, where ω is a constant. Find its trajectory. (b) Generalise for the case $\vec{a} = -\omega_1^2 x\,\vec{i} - \omega_2^2 y\,\vec{j}$, where ω_1, ω_2 are constants.

Solution: (a) Let $\vec{r}(0) = x(0)\vec{i} + y(0)\vec{j}$, $\quad \vec{v}(0) = \dot{x}(0)\vec{i} + \dot{y}(0)\vec{j}$.

$$\vec{a} = \ddot{x}\,\vec{i} + \ddot{y}\,\vec{j} = -\omega^2(x\,\vec{i} + y\,\vec{j}) \quad \Rightarrow \quad \ddot{x} + \omega^2 x = 0 \qquad (1)$$

$$\ddot{y} + \omega^2 y = 0 \qquad (2)$$

The solution of eq. (1) is $x = A\cos\omega t + B\sin\omega t$. The constants A, B are found from the initial conditions: $x(0) = A,\ \dot{x}(0) = \omega B \ \Rightarrow$

$$x = x(0)\cos\omega t + [\dot{x}(0)/\omega]\sin\omega t \quad \Rightarrow \quad x = A_1\cos(\omega t - \phi_1) \tag{3}$$

with $A_1 = [x^2(0) + \{\dot{x}(0)/\omega\}^2]^{1/2}$, $\quad \phi_1 = \tan^{-1}[\dot{x}(0)/\omega x(0)]$. Similarly,

$$y = A_2\cos(\omega t - \phi_2) \tag{4}$$

with $A_2 = [y^2(0) + \{\dot{y}(0)/\omega\}^2]^{1/2}$, $\phi_2 = \tan^{-1}[\dot{y}(0)/\omega y(0)]$. The trajectory is obtained by eliminating t from eqs (4) and (3) using $t - \phi_1/\omega = t^*$, $\phi = \phi_1 - \phi_2$:

$$y = A_2\cos(\omega t^* + \phi) = A_2\cos\omega t^*\cos\phi - A_2\sin\omega t^*\sin\phi, \quad x = A_1\cos\omega t^*$$

$$\Rightarrow \quad \cos\omega t^* = x/A_1, \qquad \sin\omega t^* = [(x/A_1)\cos\phi - y/A_2]\,/\sin\phi$$

and $\sin^2\omega t^* + \cos^2\omega t^* = 1$ yields

$$A_2^2 x^2 - 2A_1A_2\cos\phi\, xy + A_1^2 y^2 = A_1^2A_2^2\sin^2\phi \tag{5}$$

Equations (3) and (4) imply that $|x| \le A_1,\ |y| \le A_2$, i.e., the trajectory lies in a rectangle. Hence, the trajectory of eq. (5) is a conic section which is an ellipse (Fig. E1.11a). The motion is periodic with a period of $2\pi/\omega$. If $\phi = 0$ or π, then eq. (5) reduces to $(A_2x \mp A_1y)^2 = 0$, i.e., $A_2x = \pm A_1y$, which are straight lines.

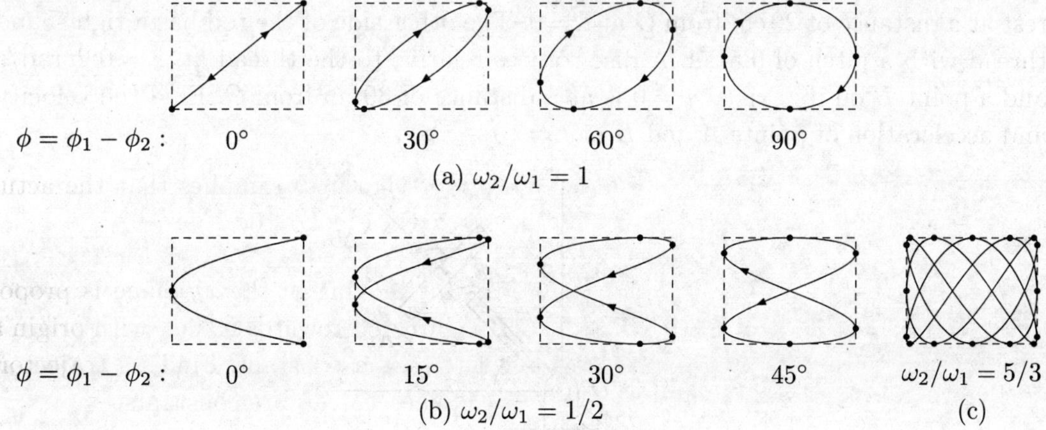

Fig. E1.11

(b) The general solution is similar to eqs (3) and (4) with ω replaced by ω_1 and ω_2:

$$x = A_1\cos(\omega_1 t - \phi_1), \qquad y = A_2\cos(\omega_2 t - \phi_2) \tag{6}$$

The paths obtained by eliminating t from eqs (6) are called *Lissajous figures* (first studied by J.A. Lissajous). These lie in a rectangle with sides $x = \pm A_1$, $y = \pm A_2$.

If ω_1/ω_2 is a rational number, say m/n, where m, n are integers with no common factors, then the periods $T_1 = 2\pi/\omega_1, T_2 = 2\pi/\omega_2$ of x, y displacements are related by

$$\frac{T_2}{T_1} = \frac{\omega_1}{\omega_2} = \frac{m}{n} \qquad \Rightarrow \qquad mT_1 = nT_2 = T \tag{7}$$

T is the periodic time which is the lowest common multiple of T_1 and T_2. In one period, the trajectory touches n $(= T/T_2)$ times the edge $y = A_2$ of the rectangle, and touches m times the edge $x = A_1$. Some trajectories for $\omega_1/\omega_2 = m/n = 2/1$ are plotted in Fig. E1.11b. The property of tangency of the trajectories is illustrated in Fig. E1.11c.

To find experimentally the frequency of a signal, a transducer is used to produce a proportional voltage, which is fed to one set of plates of a cathode ray oscilloscope. An input is applied to the other set and its frequency is adjusted until a simple Lissajous figure, such as an ellipse, appears. From the numbers n and m of points of tangency and the known frequency, the unknown frequency is found using eq. (7).

If ω_1/ω_2 is not a rational number, then the trajectory is an open curve. After a long time, the moving point would have passed through every point inside the rectangle, the moving point never passing twice through a given point with the same velocity.

Example 1.12: A rod with a threaded hub rotates on a right-hand screw of pitch 2 cm with $\phi = 0.4t^2$ rad (Fig. E1.12), where the time t is in seconds. One side of the rod has a telescopic arm which has an outward acceleration of $0.4t$ cm/s^2, with its tip A at rest at a distance of 20 cm from O at $t = 0$. The other side of the rod has a right-hand thread with a pitch of 0.5 cm. A disc rotates relative to the thread at $\omega = 0.2t$ rad/s and a point B on its axis at $t = 0$ is at a distance of 30 cm from O. Find the velocity and acceleration of points A and B at $t = 10$ s.

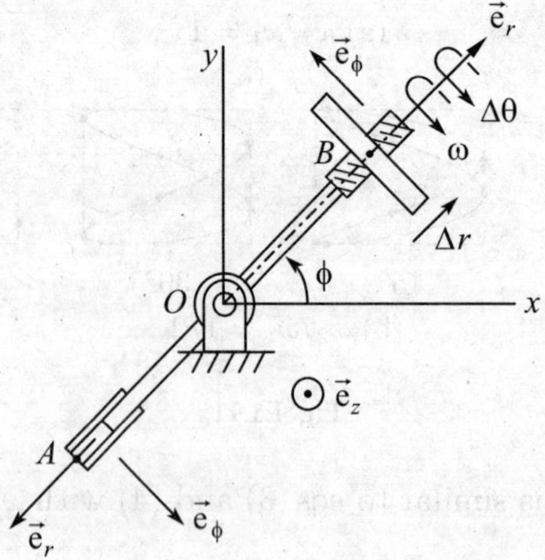

Fig. E1.12

Solution: We use cylindrical components as r, $\dot{r}$, $\ddot{r}$, ϕ, $\dot{\phi}$, $\dot{z}$, $\ddot{z}$ can be easily computed.

For A: $\quad \ddot{r} = 0.4t\,\text{cm/s}^2 \quad \Rightarrow \quad \dot{r}(t) = \dot{r}(0) + \int_0^t 0.4t\,dt = 0.2t^2\,\text{cm/s}$

$$\Rightarrow \quad r(t) = r(0) + \int_0^t 0.2t^2\,dt = 20 + \tfrac{1}{3}(0.2t^3)\,\text{cm}$$

$$\phi = 0.4t^2\,\text{rad}, \qquad \dot{\phi} = 0.8t\,\text{rad/s}, \qquad \ddot{\phi} = 0.8\,\text{rad/s}^2$$

For one rotation (2π), the hub advances axially by one pitch (2 cm). Hence, for rotation $\Delta\phi$, the axial advance Δz of the hub is given by

$$\Delta z = \frac{2}{2\pi}\Delta\phi = \frac{\Delta\phi}{\pi}\,\text{cm} \quad \Rightarrow \quad \dot{z} = \frac{\dot{\phi}}{\pi}\,\text{cm/s}, \qquad \ddot{z} = \frac{\ddot{\phi}}{\pi}\,\text{cm/s}^2$$

At $t = 10\,\text{s}$:

$$r = 20 + 0.2(10)^3/3 = 86.67\,\text{cm}, \qquad \dot{r} = 0.2(10)^2 = 20\,\text{cm/s}$$

$$\ddot{r} = 0.4(10) = 4\,\text{cm/s}^2, \qquad \dot{\phi} = 0.8(10) = 8\,\text{rad/s}, \qquad \ddot{\phi} = 0.8\,\text{rad/s}^2$$

$$\dot{z} = 8/\pi = 2.546\,\text{cm/s}, \qquad \ddot{z} = 0.8/\pi = 0.2546\,\text{cm/s}^2$$

$$\vec{v}_A = \dot{r}\,\vec{e}_r + r\dot{\phi}\,\vec{e}_\phi + \dot{z}\,\vec{e}_z = 20\,\vec{e}_r + 693.4\,\vec{e}_\phi + 2.546\,\vec{e}_z\,\text{cm/s}$$

$$\vec{a}_A = (\ddot{r} - r\dot{\phi}^2)\,\vec{e}_r + (2\dot{r}\dot{\phi} + r\ddot{\phi})\,\vec{e}_\phi + \ddot{z}\,\vec{e}_z$$

$$= -5543\,\vec{e}_r + 389.3\,\vec{e}_\phi + 0.2546\,\vec{e}_z\,\text{cm/s}^2$$

where $\vec{e}_r$, $\vec{e}_\phi$, $\vec{e}_z$ are as shown at point A in Fig. E1.12.

For a rotation $\Delta\theta$ of the disc, the increase Δr in the radial coordinate of B is:

$$\Delta r = \frac{0.5\Delta\theta}{2\pi}\,\text{cm} \quad \Rightarrow \quad \dot{r} = \frac{0.5\dot{\theta}}{2\pi} = \frac{0.5\,\omega}{2\pi}\,\text{cm/s}, \qquad \ddot{r} = \frac{0.5\dot{\omega}}{2\pi}\,\text{cm/s}^2$$

$$\omega = 0.2t\,\text{rad/s}, \qquad \dot{\omega} = 0.2\,\text{rad/s}^2$$

$$r(t) = r(0) + \int_0^t \frac{0.5\omega}{2\pi}\,dt = 30 + \int_0^t \frac{0.5(0.2t)}{2\pi}\,dt = 30 + \frac{0.025t^2}{\pi}\,\text{cm}$$

$$\Rightarrow \text{ at } t = 10\,\text{s}: \; r = 30.80\,\text{cm}, \qquad \omega = 2\,\text{rad/s},$$

$$\dot{r} = \frac{0.5(2)}{2\pi} = 0.1592\,\text{cm/s}, \qquad \ddot{r} = \frac{0.5(0.2)}{2\pi} = 0.01592\,\text{cm/s}^2$$

The values of $\dot{\phi}$, $\ddot{\phi}$, $\dot{z}$, $\ddot{z}$ for point B are the same as for point A. Hence,

$$\vec{v}_B = \dot{r}\,\vec{e}_r + r\dot{\phi}\,\vec{e}_\phi + \dot{z}\,\vec{e}_z = 0.1592\,\vec{e}_r + 246.4\,\vec{e}_\phi + 2.546\,\vec{e}_z\,\text{cm/s}$$

$$\vec{a}_B = (\ddot{r} - r\dot{\phi}^2)\,\vec{e}_r + (2\dot{r}\dot{\phi} + r\ddot{\phi})\,\vec{e}_\phi + \ddot{z}\,\vec{e}_z = -1971\,\vec{e}_r + 27.19\,\vec{e}_\phi + 0.2546\,\vec{e}_z \text{ cm/s}^2$$

where $\vec{e}_r$, $\vec{e}_\phi$, $\vec{e}_z$ are as shown at point B in Fig. E1.12.

Example 1.13: The velocity of a point of a fluid is $\vec{v} = c(\cos\phi\,\vec{e}_r + \sin\phi\,\vec{e}_\phi)/r^2$, where c is a constant. Find its acceleration and its tangential component, its trajectory and ρ.

Solution: $\vec{v} = \dot{r}\,\vec{e}_r + r\dot{\phi}\,\vec{e}_\phi + \dot{z}\,\vec{e}_z = \dfrac{c}{r^2}(\cos\phi\,\vec{e}_r + \sin\phi\,\vec{e}_\phi)$

$$\Rightarrow \quad \dot{r} = \frac{c}{r^2}\cos\phi, \qquad \dot{\phi} = \frac{c}{r^3}\sin\phi, \qquad \dot{z} = 0$$

$$\Rightarrow \quad \ddot{r} = -\frac{2c}{r^3}\dot{r}\cos\phi - \frac{c}{r^2}\sin\phi\,\dot{\phi} = -\frac{c^2}{r^5}(2\cos^2\phi + \sin^2\phi)$$

$$\ddot{\phi} = -\frac{3c}{r^4}\dot{r}\sin\phi + \frac{c}{r^3}\cos\phi\,\dot{\phi} = -\frac{2c^2}{r^6}\sin\phi\cos\phi, \qquad \ddot{z} = 0$$

$$\Rightarrow \quad \vec{a} = (\ddot{r} - r\dot{\phi}^2)\,\vec{e}_r + (2\dot{r}\dot{\phi} + r\ddot{\phi})\,\vec{e}_\phi + \ddot{z}\,\vec{e}_z$$

$$= -\frac{c^2}{r^5}(2\cos^2\phi + \sin^2\phi + \sin^2\phi)\,\vec{e}_r = -\frac{2c^2}{r^5}\,\vec{e}_r$$

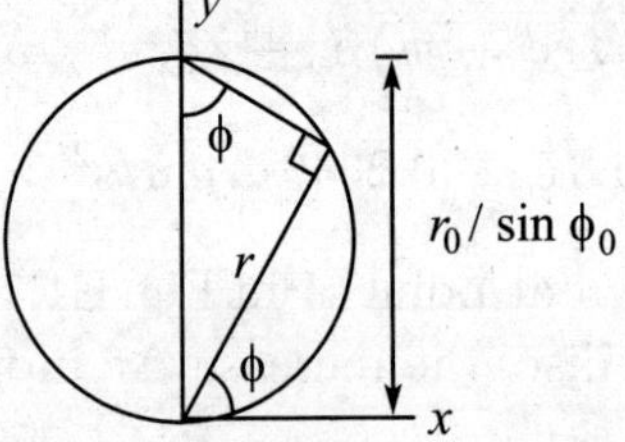

Fig. E1.13

The path $r(\phi)$ is obtained by integrating the expression for $dr/d\phi$:

$$\frac{dr}{d\phi} = \frac{\dot{r}}{\dot{\phi}} = r\frac{\cos\phi}{\sin\phi} \quad \Rightarrow \quad \int_{r_0}^{r}\frac{1}{r}\,dr = \int_{\phi_0}^{\phi}\frac{\cos\phi}{\sin\phi}\,d\phi$$

$$\Rightarrow \quad \ln\frac{r}{r_0} = \ln\frac{\sin\phi}{\sin\phi_0} \quad \Rightarrow \quad r = r_0\frac{\sin\phi}{\sin\phi_0}$$

which is the equation of a circle of radius $R = r_0/2\sin\phi_0$ centred on the y-axis (Fig. E1.13). Choosing $\vec{e}_t$ in the direction of $\vec{v}$, we obtain a_t and ρ from $\vec{v}$ and $\vec{a}$ as:

$$a_t = \vec{a}\cdot\vec{e}_t = \vec{a}\cdot\frac{\vec{v}}{|\vec{v}|} = -\frac{2c^2}{r^5}\left(\frac{c}{r^2}\cos\phi\right)\Big/\left(\frac{c}{r^2}\right) = -\frac{2c^2}{r^5}\cos\phi$$

$$\rho = \frac{|\vec{v}|^3}{|\vec{v}\times\vec{a}|} = \frac{c^3/r^6}{|(2c^2/r^5)(c/r^2)\sin\phi\,\vec{e}_z|} = \frac{r}{2|\sin\phi|}$$

ρ is independent of c, as it depends only on the shape and size of the trajectory.

Example 1.14: The tip of a cutting tool, shown in Fig. E1.14, describes an elliptic path: $x^2/a^2 + y^2/b^2 = 1$ with $a > b$. (a) Find the acceleration of the tool tip when it is at an end of the major axis where its speed is v and is increasing at the rate of $\dot{v}$. (b) If the tool tip has uniform speed v, what is its maximum acceleration?

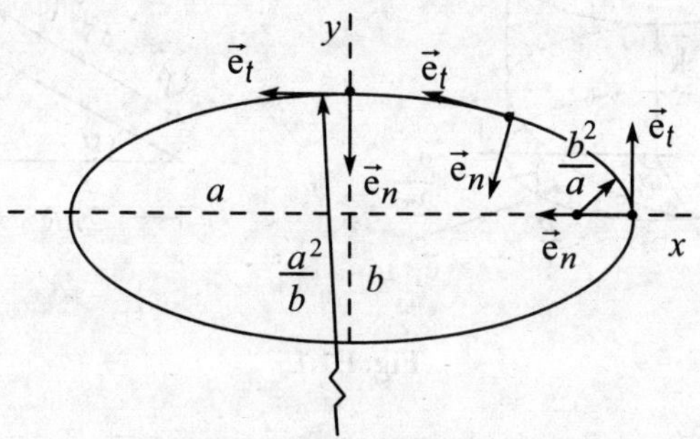

Fig. E1.14

Solution: Since the path and $v, \dot{v}$ are given, it is convenient to use path description of acceleration. The parametric equations of the ellipse are: $x = a\cos\tau,\ y = b\sin\tau$.

$$\frac{dy}{dx} = \frac{dy/d\tau}{dx/d\tau} = \frac{b\cos\tau}{-a\sin\tau} = -\frac{b}{a}\cot\tau$$

$$\Rightarrow \quad \frac{d^2y}{dx^2} = \frac{b}{a}\operatorname{cosec}^2\tau\,\frac{d\tau}{dx} = -\frac{b}{a^2}\operatorname{cosec}^3\tau$$

$$\rho = \frac{[1+(dy/dx)^2]^{3/2}}{|d^2y/dx^2|} = \frac{a^2[1+b^2\cot^2\tau/a^2]^{3/2}}{b|\operatorname{cosec}^3\tau|}$$

$$= [a^2\sin^2\tau + b^2\cos^2\tau]^{3/2}/ab = [(a^2-b^2)\sin^2\tau + b^2]^{3/2}/\,ab$$

(a) At an end of the major axis, $x = a\cos\tau = a \Rightarrow \cos\tau = 1,\ \sin\tau = 0,\ \rho = b^2/a,$

$$\vec{a} = \dot{v}\,\vec{e}_t + \frac{v^2}{\rho}\,\vec{e}_n = \dot{v}\,\vec{e}_t + \frac{av^2}{b^2}\,\vec{e}_n$$

(b) The minimum value of ρ is for $\sin\tau = 0$, i.e., at the end of the major axis with $\rho_{\min} = b^2/a$. As $\dot{v} = 0,\ \vec{a} = (v^2/\rho)\,\vec{e}_n \ \Rightarrow\ |\vec{a}|_{\max} = v^2/\rho_{\min} = v^2a/b^2.$

Example 1.15: A point P moves along a circular helix $\vec{r} = R\cos\phi\,\vec{\mathrm{i}} + R\sin\phi\,\vec{\mathrm{j}} + c\,\phi\,\vec{\mathrm{k}}$, with ϕ increasing with time (Fig. E1.15). The constant c is related to pitch p by $c = p/2\pi$. Its speed at ϕ is v and is decreasing at the rate of a. Express its velocity and acceleration in (a) Cartesian coordinates, (b) cylindrical coordinates and (c) path components using the relevant expressions directly.

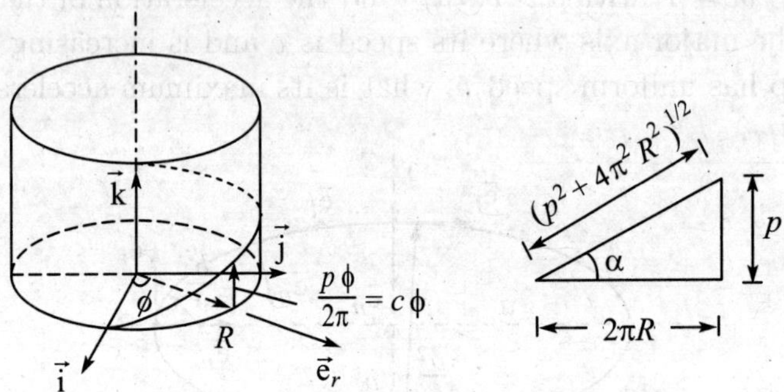

Fig. E1.15

Solution: (a) $\dot{\phi} > 0, \qquad \dot{v} = -a$ and $\qquad \vec{r} = R\cos\phi\,\vec{\mathrm{i}} + R\sin\phi\,\vec{\mathrm{j}} + c\,\phi\,\vec{\mathrm{k}},$

$$\vec{v} = \dot{\vec{r}} = (-R\sin\phi\,\vec{\mathrm{i}} + R\cos\phi\,\vec{\mathrm{j}} + c\,\vec{\mathrm{k}})\dot{\phi}$$

$$v = |\vec{v}| = (R^2 + c^2)^{1/2}\dot{\phi}, \qquad \dot{\phi} = v/(R^2 + c^2)^{1/2}$$

$$\Rightarrow \quad \vec{v} = (-R\sin\phi\,\vec{\mathrm{i}} + R\cos\phi\,\vec{\mathrm{j}} + c\,\vec{\mathrm{k}})v/(R^2 + c^2)^{1/2}$$

$$\vec{a} = \dot{\vec{v}} = -(-R\sin\phi\,\vec{\mathrm{i}} + R\cos\phi\,\vec{\mathrm{j}} + c\,\vec{\mathrm{k}})a/(R^2 + c^2)^{1/2}$$
$$-(R\cos\phi\,\vec{\mathrm{i}} + R\sin\phi\,\vec{\mathrm{j}})v^2/(R^2 + c^2)$$

We can obtain $\rho = |\vec{v}|^3/|\vec{v} \times \vec{a}|, \quad \vec{\mathrm{e}}_n = \vec{\mathrm{e}}_b \times \vec{\mathrm{e}}_t = \dfrac{\vec{v} \times \vec{a}}{|\vec{v} \times \vec{a}|} \times \dfrac{\vec{v}}{|\vec{v}|}$ from $\vec{v}, \vec{a}$.

(b) $\vec{r} = R\,\vec{\mathrm{e}}_r + c\,\phi\,\vec{\mathrm{e}}_z, \quad r = R, \quad \dot{r} = 0, \quad \ddot{r} = 0, \quad z = c\,\phi, \quad \dot{z} = c\,\dot{\phi}, \quad \ddot{z} = c\,\ddot{\phi},$

$$\vec{v} = \dot{r}\,\vec{\mathrm{e}}_r + r\dot{\phi}\,\vec{\mathrm{e}}_\phi + \dot{z}\,\vec{\mathrm{e}}_z = R\dot{\phi}\,\vec{\mathrm{e}}_\phi + c\,\dot{\phi}\,\vec{\mathrm{e}}_z$$

$$\Rightarrow \quad \dot{\phi} = v/(R^2 + c^2)^{1/2} \qquad \Rightarrow \qquad \ddot{\phi} = -a/(R^2 + c^2)^{1/2}$$

$$\vec{v} = (R\,\vec{\mathrm{e}}_\phi + c\,\vec{\mathrm{e}}_z)v/(R^2 + c^2)^{1/2}$$

$$\vec{a} = (\ddot{r} - r\dot{\phi}^2)\,\vec{\mathrm{e}}_r + (2\dot{r}\dot{\phi} + r\ddot{\phi})\,\vec{\mathrm{e}}_\phi + \ddot{z}\,\vec{\mathrm{e}}_z = -R\dot{\phi}^2\,\vec{\mathrm{e}}_r + R\ddot{\phi}\,\vec{\mathrm{e}}_\phi + c\,\ddot{\phi}\,\vec{\mathrm{e}}_z$$
$$= -[Rv^2/(R^2 + c^2)]\,\vec{\mathrm{e}}_r - [R\,\vec{\mathrm{e}}_\phi + c\,\vec{\mathrm{e}}_z]a/(R^2 + c^2)^{1/2}$$

(c) $d\phi/ds > 0, \qquad \vec{e}_t = \dfrac{d\vec{r}}{ds} = (-R\sin\phi\,\vec{i} + R\cos\phi\,\vec{j} + c\,\vec{k})\dfrac{d\phi}{ds} \Rightarrow$

$$|\vec{e}_t| = (R^2+c^2)^{1/2}\left|\frac{d\phi}{ds}\right| = 1 \quad \Rightarrow \quad \frac{d\phi}{ds} = \frac{1}{(R^2+c^2)^{1/2}}$$

$$\vec{e}_t = \frac{-R\sin\phi\,\vec{i} + R\cos\phi\,\vec{j} + c\,\vec{k}}{(R^2+c^2)^{1/2}}$$

$$\frac{d\,\vec{e}_t}{ds} = \frac{-R\cos\phi\,\vec{i} - R\sin\phi\,\vec{j}}{(R^2+c^2)^{1/2}}\frac{d\phi}{ds} = \frac{-R\cos\phi\,\vec{i} - R\sin\phi\,\vec{j}}{R^2+c^2}$$

$$\frac{1}{\rho} = \left|\frac{d\,\vec{e}_t}{ds}\right| = \frac{R}{R^2+c^2} \quad \Rightarrow \quad \rho = \frac{R^2+c^2}{R} \neq R$$

$$\vec{e}_n = \frac{\dfrac{d\,\vec{e}_t}{ds}}{\left|\dfrac{d\,\vec{e}_t}{ds}\right|} = -\cos\phi\,\vec{i} - \sin\phi\,\vec{j} = -\,\vec{e}_r$$

$$\vec{v} = v\,\vec{e}_t, \qquad \vec{a} = -a\,\vec{e}_t + \frac{v^2}{\rho}\,\vec{e}_n = -a\,\vec{e}_t + \frac{v^2R}{R^2+c^2}\,\vec{e}_n$$

The angle α between $\vec{e}_t$ and $\vec{e}_\phi$ $(= -\sin\phi\,\vec{i} + \cos\phi\,\vec{j})$ is called the helix angle:

$$\cos\alpha = \vec{e}_t \cdot \vec{e}_\phi = R/(R^2+c^2)^{1/2} \quad \Rightarrow \quad \tan\alpha = c/R = p/2\pi R$$

$\vec{e}_n$ at P is radial. The osculating plane of $\vec{e}_t$ and $\vec{e}_n$ is at an angle $\pi/2 - \alpha$ to the axis and rotates about it as a screw in a fixed nut. The axial displacement for one rotation about the axis is called the pitch p of the helix: $p = 2c\pi$. The given circular helix has constant values of pitch p, helix angle α and radius of curvature ρ.

Review 1 A

1. A directed entity, which adds according to the parallelogram law, is a vector.
2. Finite rotation has magnitude and direction, but it is *not* a vector.
3. $\vec{a}\cdot\vec{b} = \vec{b}\cdot\vec{a}$, $\vec{a}\times\vec{b} = -\vec{b}\times\vec{a}$.
4. In $\vec{A}\cdot\vec{B}\times\vec{C}$, dot and cross can be interchanged; and the sequence of vectors in cyclic order of $\vec{A}, \vec{B}, \vec{C}$ can also be interchanged.
5. $\vec{A}\times(\vec{D}\times\vec{C}) = (\vec{A}\cdot\vec{C})\,\vec{D} - (\vec{A}\cdot\vec{D})\vec{C}$ is in terms of vectors in the parenthesis with the positive term for the vector $\vec{D}$ in the middle of the triple cross-product.
6. For a vector $\vec{A}$ of constant magnitude, $\dot{\vec{A}} \perp \vec{A}$ or $\dot{\vec{A}} = \vec{0}$.
7. Displacement, velocity and acceleration depend on the reference frame, but are independent of the choice of the origin and the orientation of the axes in it.

8. $\vec{r} = x\vec{i} + y\vec{j} + z\vec{k}, \quad \vec{v} = \dot{x}\vec{i} + \dot{y}\vec{k} + \dot{z}\vec{k}, \quad \vec{a} = \ddot{x}\vec{i} + \ddot{y}\vec{j} + \ddot{z}\vec{k}, \; a_x = \dot{v}_x$
9. $\vec{e}_r$ is radially out from the z-axis, $\vec{e}_\phi = \vec{e}_z \times \vec{e}_r, \quad d\vec{e}_r/d\phi = \vec{e}_\phi, \; d\vec{e}_\phi/d\phi = -\vec{e}_r$.
$\vec{r} = r\vec{e}_r + z\vec{e}_z, \quad \vec{v} = \dot{r}\vec{e}_r + r\dot{\phi}\vec{e}_\phi + \dot{z}\vec{e}_z, \quad \vec{a} = (\ddot{r} - r\dot{\phi}^2)\vec{e}_r + (2\dot{r}\dot{\phi} + r\ddot{\phi})\vec{e}_\phi + \ddot{z}\vec{e}_z$
$a_r = \ddot{r} - r\dot{\phi}^2 \neq \dot{v}_r, \quad a_\phi = 2\dot{r}\dot{\phi} + r\ddot{\phi} \neq \dot{v}_\phi, \quad a_z = \ddot{z} = \dot{v}_z, \quad a_\phi = \frac{1}{r}\frac{d}{dt}(rv_\phi)$
10. $\vec{e}_t = d\vec{r}/ds, \; d\vec{e}_t/ds = \vec{e}_n/\rho, \; \vec{e}_n$ is directed inwards to the centre of curvature.
$\vec{v} = \dot{s}\,\vec{e}_t, \quad \vec{a} = \ddot{s}\,\vec{e}_t + \frac{\dot{s}^2}{\rho}\vec{e}_n, \quad a_t = \ddot{s} = \dot{v}_t, \quad a_n = \dot{s}^2/\rho \neq \dot{v}_n.$
11. $\dfrac{1}{\rho} = \left|\dfrac{d\vec{r}}{ds} \times \dfrac{d^2\vec{r}}{ds^2}\right| = \left|\dfrac{d\vec{r}}{d\tau} \times \dfrac{d^2\vec{r}}{d\tau^2}\right| \Big/ \left|\dfrac{d\vec{r}}{d\tau}\right|^3 = |\vec{v} \times \vec{a}| \,/\, |\vec{v}|^3.$ For a plane curve:

$$\rho = \frac{\left[1 + \left(\dfrac{dx}{dy}\right)^2\right]^{3/2}}{\left|\dfrac{d^2x}{dy^2}\right|} = \frac{\left[1 + \left(\dfrac{dy}{dx}\right)^2\right]^{3/2}}{\left|\dfrac{d^2y}{dx^2}\right|} = \frac{(r^2 + r'^2)^{3/2}}{|r^2 + 2r'^2 - rr''|}, \quad \text{with } (\;)' = \frac{d(\;)}{d\phi}.$$

1.13 Concept Review Questions 1A

1. Find the x and y components of each force given in Fig. Q1.1.
   [(a) $-F(\cos\alpha, \sin\alpha)$ (b) $mg(\sin\theta, -\cos\theta)$ (c) $-48, -20$ (d) $-40\sqrt{5}, 80\sqrt{}$

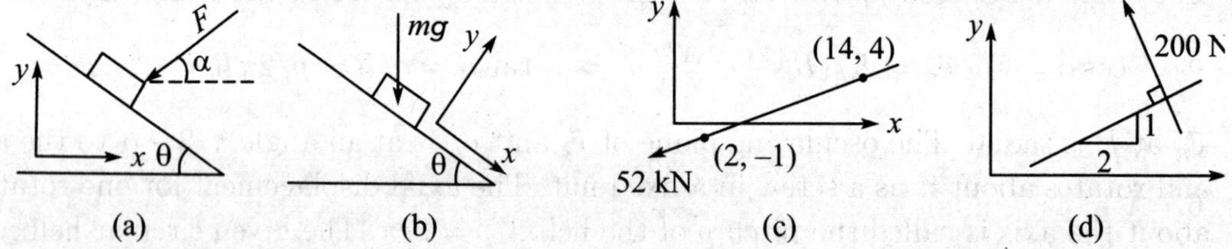

Fig. Q1.1

2. Express the forces $\vec{F_i}$ (Fig. Q1.2) in terms of unit vectors $\vec{i}, \vec{j}, \vec{k}$. $\cos\phi = 0.6$, $F_1 = 70\,\text{kN}$, $F_2 = 40\,\text{kN}$, $F_3 = F_4 = F_5 = 700\,\text{kN}$.

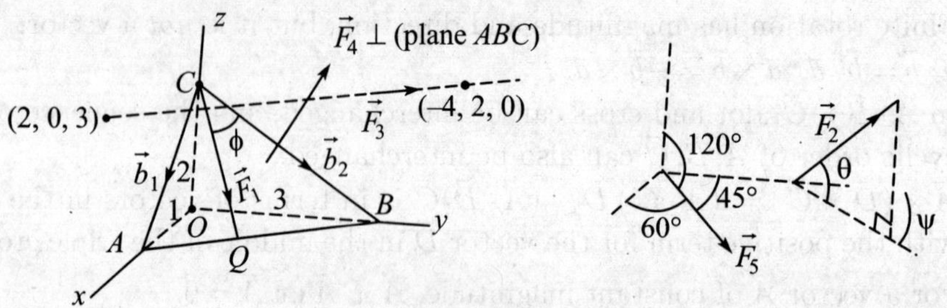

Fig. Q1.2

$[\vec{F_1} = 28.85\vec{i} + 8.320\vec{j} - 63.24\vec{k}, \qquad \vec{F_2} = 40\cos\theta(\sin\psi\,\vec{i} + \cos\psi\,\vec{j}) + 40\sin\theta\,\vec{k},$

$\vec{F}_3 = 200\vec{j} - 600\vec{i} - 300\vec{k}$, $\vec{F}_4 = 600\vec{i} + 200\vec{j} + 300\vec{k}$, $\vec{F}_5 = 350\vec{i} + 495\vec{j} - 350\vec{k}$ kN]

3. Two persons 1 and 2, standing on a rotating platform, observe the motion of a projectile P and an ant D moving relative to the ground along OE at the speed v which is increasing at the rate a (Fig. Q1.3). C is a point of the platform. In the *reference frame of the platform*, (a) are $\vec{v}_P$, $\vec{a}_P$ observed by them the same? (b) Find $\vec{v}_D$, $\vec{a}_D$ of the ant D observed by them. (c) Find $\vec{v}_{BA}$, and (d) $\vec{v}_C$, $\vec{a}_C$.
[(a) Yes (b) $v\,\vec{e}_r - \omega d\,\vec{e}_\phi$, $(a - \omega^2 d)\,\vec{e}_r - (2\omega v + \dot{\omega} d)\,\vec{e}_\phi$ (c) $\vec{0}$ (d) $\vec{0}$, $\vec{0}$]

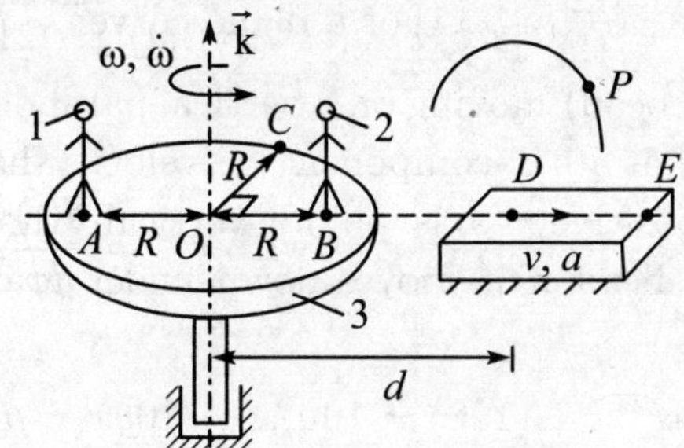

Fig. Q1.3

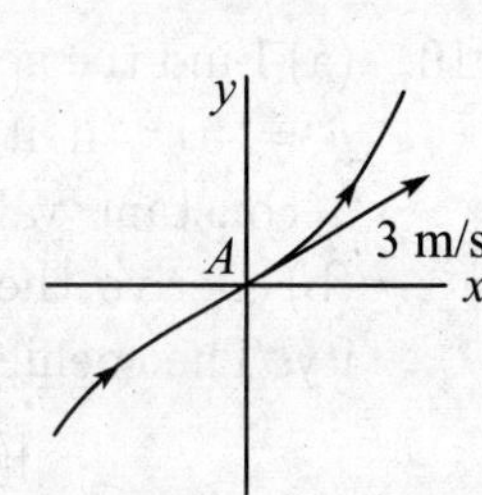

Fig. Q1.5

Fig. Q1.6

4. Which of the following are possible? (a) $\vec{v} = -2\,\vec{e}_t$ (b) $\vec{a} = -3\,\vec{e}_t$
(c) $\vec{a} = 3\,\vec{e}_t + 5\,\vec{e}_n$ (d) $\vec{a} = 3\,\vec{e}_t - 5\,\vec{e}_n$ (e) $\vec{a} = -3\,\vec{e}_t + 5\,\vec{e}_n$ (f) $\vec{a} = 5\,\vec{e}_n$. [a, b, c, e, f]

5. Cam 1 has velocity v and acceleration a (Fig. Q1.5). Find the velocity and acceleration of 2. A cam is a rotating body in contact with another body, called a follower, to which it imparts prescribed motion. $[-f'v\,\vec{j},\ (v^2 f'' - af')\,\vec{j}]$

6. A moving point at location A of its plane trajectory $y - x = (x + y)^3$ (x, y in metres) has speed 3 m/s which is decreasing at the rate of 2 m/s^2 (Fig. Q1.6). Find its acceleration in this position. $[-1.414(\vec{i} + \vec{j})\ \text{m/s}^2]$

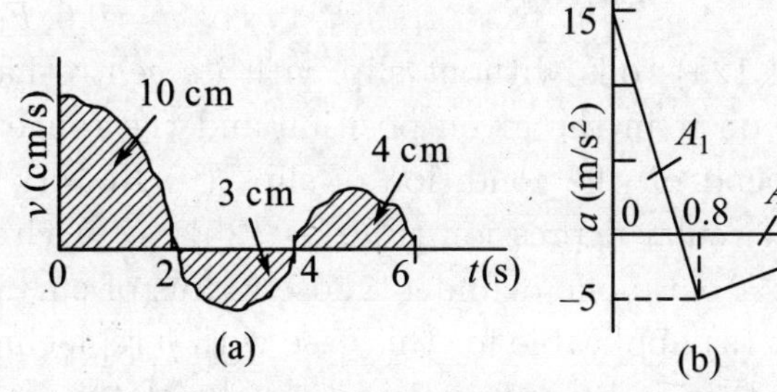

Fig. Q1.7

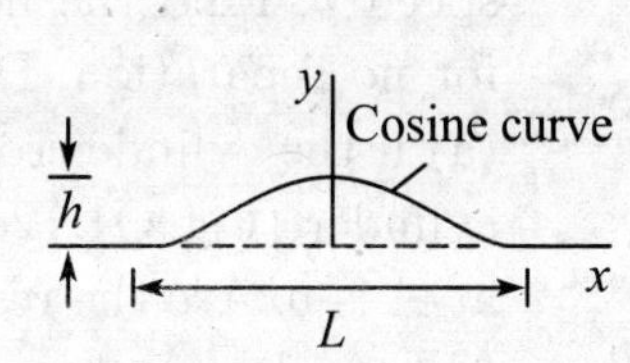

Fig. Q1.8

7. (a) The velocity of a point in rectilinear motion is shown in Fig. Q1.7a. Find its displacement in the first six seconds. (b) The acceleration of an object due

to a blast is approximated by the curve shown in Fig. Q1.7b. Its initial velocity is 0.5 m/s and it comes to rest at time t_1. Find t_1, the displacement and the maximum velocity. [(a) 11 cm (b) 2.6 s, 5.767 cm, 5 cm/s]

8^d. Design the height h of the cosine curve shaped road bumper (Fig. Q1.8) such that when a wheel of radius, r, moves over it at a constant speed v, the maximum acceleration at its centre does not exceed ng ($n \sim 1$). [$h < L^2/2\pi^2(r + v^2/ng)$]

9. The velocity of a material point P of a fluid at location x, y, z of the reference frame at time t is $\vec{v} = bt\vec{\imath} + (ct^2/x)\vec{\jmath} - 2t\vec{k}$. Find the acceleration of P at x, y, z at time t. [$b\vec{\imath} + [2ct/x - (ct^2/x^2)(bt)]\vec{\jmath} - 2\vec{k}$]

10. (a) Find the acceleration of a mass-point (a bead) moving on a vertical parabola, $y = 5x^2$, if it is moved in such a way that its x-component of velocity has a constant value of 4 m/s. (Note: The positive y-axis points vertically up.) (b) Derive the equation of motion of the bead if it moves down under gravity. The coefficient of friction is μ.

[(a) $160\vec{\jmath}$ m/s^2 (b) $(1 + 100x^2)\ddot{x} + (100x - 10\mu)\dot{x}^2 + 10gx = \mu g$]

11. A nozzle discharges a stream of water in air with speed v_0 at the exit at angle of elevation θ to the horizontal. Find the radius of curvature of the stream and the rate of change of speed at the exit and at its highest point.

[$v_0^2/g\cos\theta$, $-g\sin\theta$; ${v_0}^2\cos^2\theta/g$, 0]

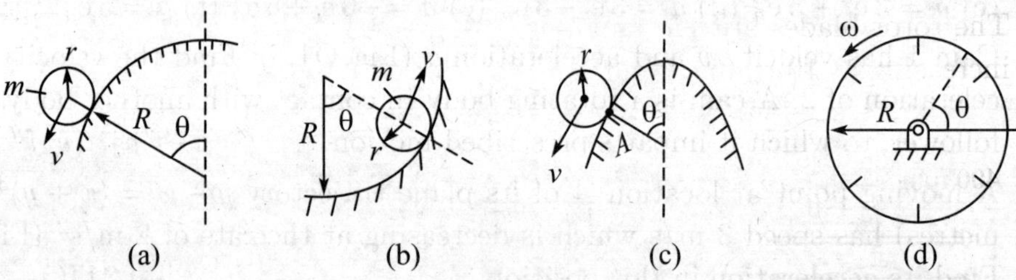

Fig. Q1.12

12. (a) A circular cylinder (Fig. Q1.12a) rolls without slip with its centre having speed v. Find the normal force on it in the given position and the range of v for no separation. Do these depend on the condition of slip? (b) Rework part (a) if the cylinder rolls inside a circular depression (Fig. Q1.12b). (c) A circular cylinder (Fig. Q1.12c) rolls over a parabolic cylinder whose radius of curvature at A is ρ. Are the results of part (a) applicable for this case with ρ replacing R? (d)^d A clothes drier (Fig. Q1.12d) is to be designed so that the clothes fall off the cylindrical surface at angle θ. Small radial vanes prevent slippage till loss of contact. Find ω. [(a) $m[g\cos\theta - v^2/(r + R)]$, $v < [(R + r)g\cos\theta]^{1/2}$, no (b) $m[g\cos\theta + v^2/(R - r)]$, $\forall\, v$, no (c) Yes (d) $(g\sin\theta/R)^{1/2}$]

13*. Bodies 1 and 2 are connected by an inextensible cable passing around small pulleys. Set up kinematic relations between v_1, a_1 and v_2, a_2 for the given position for the systems shown in Fig. Q1.13.

[(a) $v_1 \cos\theta_1 = v_2 \cos\theta_2, \quad a_1 \cos\theta_1 + v_1^2 \sin^2\theta_1/s_1 = a_2 \cos\theta_2 - v_2^2 \sin^2\theta_2/s_2$

(b) $v_1(2+\cos\theta) = v_2, \quad a_1(2+\cos\theta) + v_1^2 \sin^2\theta/r = a_2$

(c) $v_1 rs = 4v_2 Lh \sin\theta$, with $s^2 = L^2 + h^2 - 2Lh\cos\theta$,

$r^2(v_1^2 - 4sa_1) = 16Lh(v_2^2 \cos\theta - a_2 r \sin\theta)$]

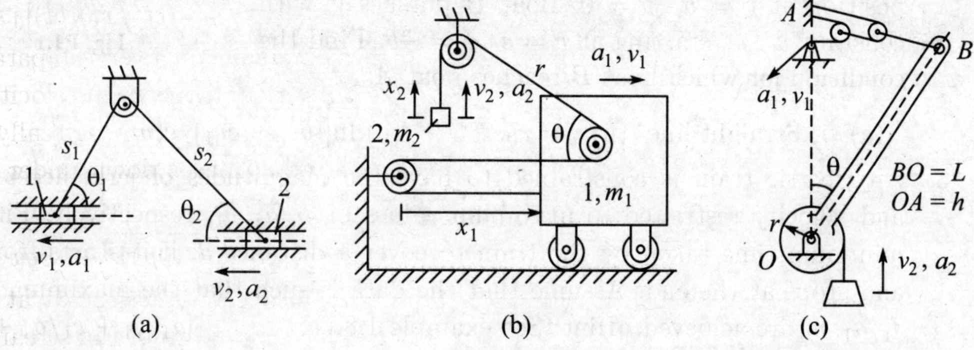

Fig. Q1.13

14. A helicopter is ascending at 8 m/s which is decreasing at 1.5 m/s^2 (Fig. Q1.14). The rotor blades are 4 m long and are rotating at a speed of 400 rpm which is increasing at 10 rpm/s. Find $\vec{a}_P$. [$-7019\vec{e}_r + 4.189\vec{e}_\phi - 1.5\vec{e}_z$ m/s^2]

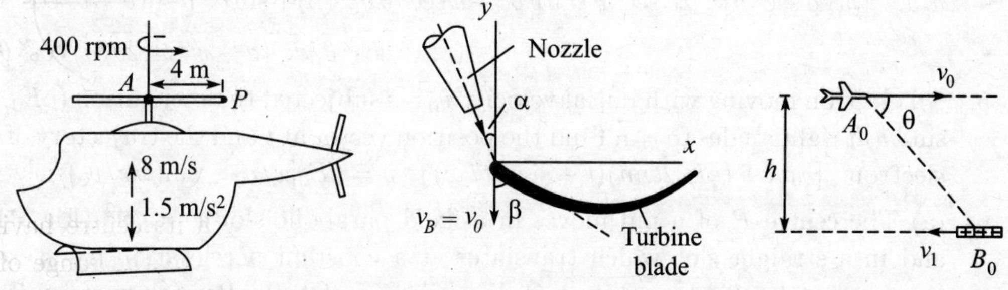

Fig. Q1.14 **Fig. Q1.15** **Fig. Q1.16**

15[d]. In a steam turbine, steam leaves the nozzle at speed v_A and point B of the turbine blade is moving at speed v_0 (Fig. Q1.15). Find v_0/v_A if steam flows tangentially relative to B in the direction tangent to the blade at B. [$\sin(\beta - \alpha)/\sin\beta$]

16. Find the angle θ for a bomber flying with velocity v_0 to score a direct hit on a train moving with opposite velocity v_1 in the same vertical plane (Fig. Q1.16). The initial positions are A_0 and B_0. [$\cot\theta = \sqrt{2}(v_0 + v_1)/\sqrt{gh}$]

1.14 Practice Problems 1A

1. Express force $\vec{F}$ in terms of $\vec{\mathrm{i}}, \vec{\mathrm{j}}, \vec{\mathrm{k}}$ (Fig. P1.1). $OABC$ is a rectangular plate with OC in the xy-plane and $F = 20\sqrt{5}$ kN. [$-28.66\vec{\mathrm{i}} - 29.64\vec{\mathrm{j}} + 17.32\vec{\mathrm{k}}$ kN]

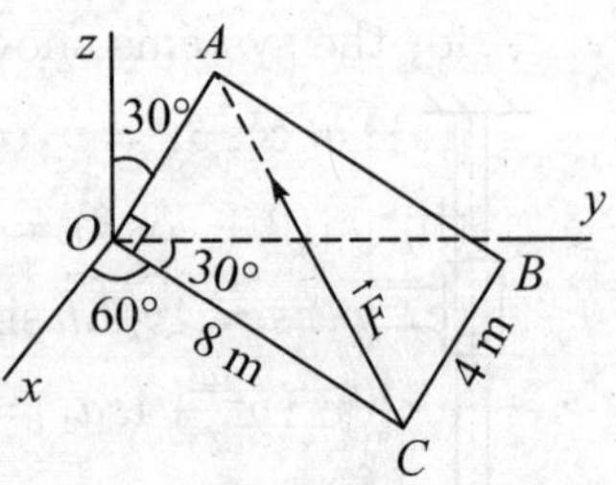

Fig. P1.1

2. (a) Find the trajectory of a point which moves in a plane (i) with constant v_x, v_y, and (ii) with constant v_r, v_ϕ. (b) A boat A moves radially at v_1 with initial position at $r = b$, $\phi = 0$. Boat B pursues it with constant v_r, v_ϕ starting at $r = c$, $\phi = -\theta$. Find the condition for which boat B reaches boat A.

   [(a) (i) Straight line (ii) $r = r_0 e^{\phi v_r / v_\phi}$ (b) $\ln[(bv_r - cv_1)/c(v_r - v_1)] = v_r\theta/v_\phi$]

3. An electric train is constrained to have the magnitudes of jerk, acceleration and velocity restricted to maximum values of j_1, a_1, v_1, respectively. Find the minimum time taken by the train to cover a distance d, if it starts from rest and stops at the end. Assume that the data is such that the maximum values j_1, a_1, v_1 are achieved. [Hint: See example 1.8.] [$a_1/j_1 + v_1/a_1 + d/v_1$]

4. A rocket is launched vertically up and is observed by a telescope at a distance d from the launch site. The angle of elevation $\theta(t)$ of the telescope is recorded. (a) Find the velocity v and acceleration a of the rocket. (b) Find the rate of separation of the distance between the rocket and the telescope in terms of v, a. Find the time derivative of this rate of separation. (c) Express $\dot{\theta}$ and $\ddot{\theta}$ in terms of v, a. [(a) $d\sec^2\theta\,\dot{\theta}$, $2d\sec^2\theta\tan\theta\,\dot{\theta}^2 + d\sec^2\theta\,\ddot{\theta}$ (b) $v\sin\theta$, $a\sin\theta + v^2\cos^3\theta/d$

   (c) $v\cos^2\theta/d$, $(a - v^2\sin 2\theta/d)\cos^2\theta/d$]

5. An electron moving with initial velocity $v_0\vec{\mathrm{i}}$ is subjected to acceleration (eE_0/m) $\sin\omega t$ at right angles to $v_0\vec{\mathrm{i}}$. Find the position vector at t and the trajectory of the electron. [$v_0 t\vec{\mathrm{i}} + (eE_0/\omega m)(t - \sin\omega t/\omega)\vec{\mathrm{j}}$, $y = eE_0[x/v_0 - \sin(\omega x/v_0)/\omega]/\omega m$]

6. (a) The centre P of a pin moves in a fixed parabolic slot $x = cy^2$ (Fig. P1.6) and in a straight slot which translates at a constant acceleration a_0, starting from rest at $t = 0$ when P is at O. Find $\vec{v}_P$, $\vec{a}_P$ at t. (b) Rework part (a) if the parabolic slot has an acceleration $bt\vec{\mathrm{i}}$ and is at rest at $t = 0$.

   [(a) $ca_0^2t^3\vec{\mathrm{i}} + a_0t\vec{\mathrm{j}}$, $3ca_0^2t^2\vec{\mathrm{i}} + a_0\vec{\mathrm{j}}$ (b) $(\frac{1}{2}bt^2 + ca_0^2t^3)\vec{\mathrm{i}} + a_0t\vec{\mathrm{j}}$, $(bt + 3ca_0^2t^2)\vec{\mathrm{i}} + a_0\vec{\mathrm{j}}$]

7. A slotted arm OA rotates about the horizontal axis at O (Fig. P1.7). A pin of mass 0.1 kg can slide in the slot and is kept in contact with the fixed disc due to the spring force. In the given position: (a) find $\vec{a}_P$, (b) find the normal components of the reactions from the plate and the slot on the pin. The free length of the spring is 1 m and $k = 200$ N/m. For the plate and the slot $\mu_k = 0.2$. Assume

that the pin slips on the plate and the slot in the directions of velocity of P relative to them. $OA = 0.8\,\text{m}$. [(a) $-4.456\,\vec{e}_r - 1.880\,\vec{e}_\theta\ \text{m/s}^2$ (b) 227.8 N, 152.7 N]

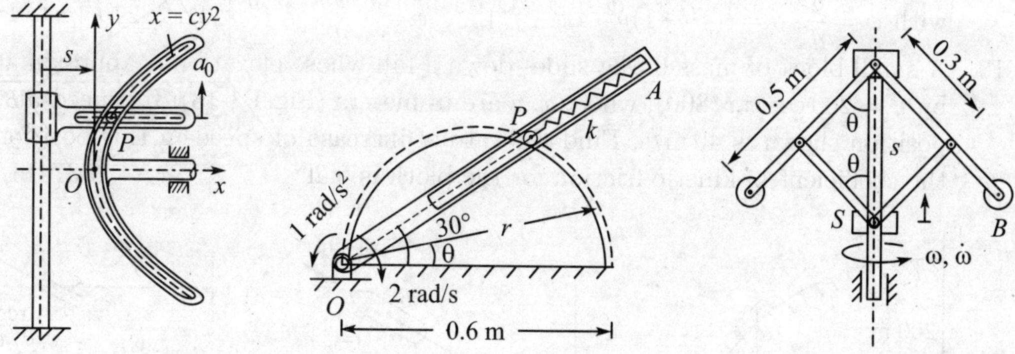

Fig. P1.6 **Fig. P1.7** **Fig. P1.9**

8. (a) A point moves along a plane curve $r = c\,e^\phi$ with $\dot{\phi} = b/r^2$, where c, b are constants. Find $\vec{v}$, $\vec{a}$, ρ at any position r. (b) Rework part (a) if $\dot{\phi} = at^2$. (c) Rework part (a) if v is constant. [$\rho = \sqrt{2}r\ \forall$ cases (a) $(b/r)(\vec{e}_r + \vec{e}_\phi)$, $(-2b^2/r^3)\,\vec{e}_r$ (b) $rat^2(\vec{e}_r + \vec{e}_\phi)$, $2atr[\vec{e}_r + (1 + at^3)\,\vec{e}_\phi]$ (c) $v(\vec{e}_r + \vec{e}_\phi)/\sqrt{2}$, $(v^2/2r)(-\vec{e}_r + \vec{e}_\phi)$]
9. The balls of the fly-ball speed governor (Fig. P1.9) move out as the sleeve S moves up on the vertical shaft. At the given instant, $\omega = 2\,\text{rad/s}$, $\dot{\omega} = 0.4\,\text{rad/s}^2$, $\theta = 40°$ and S is moving up with a speed of $2\,\text{m/s}$ which is decreasing at $0.1\,\text{m/s}^2$. Find the axial force and shear force from the ball B on the rod at the given instant. The mass of each ball is 0.2 kg. [4.357 N, 2.702 N]
10. A point moves on a plane curve $r = \phi + e^\phi$. At $\phi = 0$, the speed is $\sqrt{5}\,\text{m/s}$ and is increasing at $2\sqrt{5}\,\text{m/s}^2$. Angle ϕ increases with time. Find $\vec{v}$ and $\vec{a}$ in (a) cylindrical coordinates, (b) Cartesian coordinates and (c) path coordinates. Also find the radius of curvature of the path at this position.

    [$2\,\vec{e}_r + \vec{e}_\phi$, $2.4\,\vec{e}_r + 5.2\,\vec{e}_\phi$; $2\,\vec{i} + \vec{j}$, $2.4\,\vec{i} + 5.2\,\vec{j}$; $\sqrt{5}\,\vec{e}_t$, $2\sqrt{5}\,\vec{e}_t + 1.6\sqrt{5}\,\vec{e}_n$; $5\sqrt{5}/8$]
11. The automatic controller of a high-speed train slows it as per $v = v_0 \cos(\pi s/6)$, where s is in kilometers. The train is moving on a circular track of radius 36 km. Its initial speed $v_0 = 180\,\text{km/h}$ when the brakes are applied. Find (i) the acceleration at $s = 2\,\text{km}$, (ii) the position when the rate of change of speed with respect to time is maximum and (iii) the position when the magnitude of acceleration is maximum.

    [(i) $-7346\,\vec{e}_t + 225\,\vec{e}_n\ \text{km/h}^2$ (ii) $s = 1.5\,\text{km}$ (iii) $s = 1.497\,\text{km}$]
12. An army truck is going up an axi-symmetric hill: $r = f(z)$ on a helix of pitch p. Find its velocity and acceleration at a height z where its speed is v and decreasing

at the rate a. $[\{f'p\vec{e}_r/2\pi + f\,\vec{e}_\phi + p\vec{e}_z/2\pi\}\dot{\phi}$,

$$\{f'p\ddot{\phi}/2\pi + (f''p^2/4\pi^2 - f)\dot{\phi}^2\}\,\vec{e}_r + (2f'p\dot{\phi}^2/2\pi + f\ddot{\phi})\,\vec{e}_\phi + p\ddot{\phi}\,\vec{e}_z/2\pi]$$

with $\dot{\phi} = \dfrac{v}{d}$, $d = [(f'^2{+}1)p^2/4\pi^2{+}f^2]^{1/2}$, $\ddot{\phi} = -\dfrac{a}{d}{-}[f'f''p^2/2\pi^2{+}2ff']pv^2/4\pi d^4$

13. A small block of mass 0.5 kg slides down a hill whose elevation is approximated by $y = 200\cos(\pi x/800)$, where x, y are in meters (Fig. P1.13). Its speed at the position shown is 40 m/s. Find the rate of increase of speed in this position if the coefficient of kinetic friction for the block is 0.2. [4.374 m/s²]

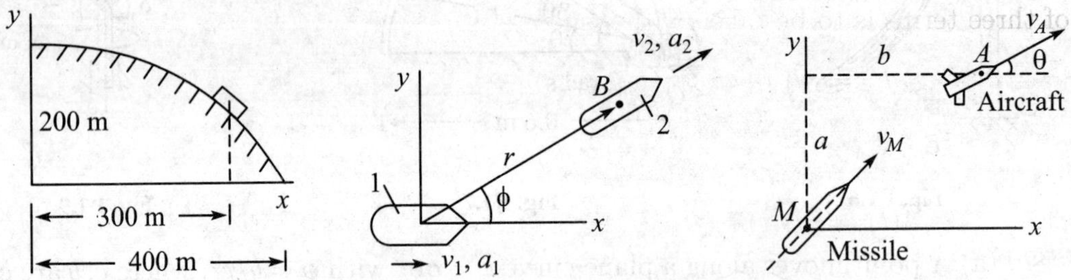

Fig. P1.13 Fig. P1.14 Fig. P1.15

14. Ships 1 and 2 translate relative to the ground F with constant acceleration. The point B is observed by a telescope from ship 1 and its coordinates are (r, ϕ) (Fig. P1.14). Find $\dot{r}, \dot{\phi}, \ddot{r}, \ddot{\phi}$ at the given position where $r = 10$ km, $\phi = \cos^{-1} 0.8$. Relative to F: $v_1 = 8$ km/h, $v_2 = 10$ km/h, $a_1 = 0$, $a_2 = 1.8$ km/h². Draw the trajectory of point B relative to frame 1 and show the profile of the ship on it for two positions of B.

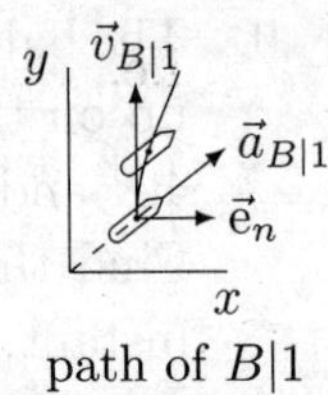

path of $B|1$

[3.6 km/h, .48 rad/h, 4.104 km/h², −.3456 rad/h², $\vec{e}_t$ at 53.13° to $\vec{e}_r$, $\rho = 25$ km]

15. A missile is launched at the position shown and is programmed to follow a pursuit curve so that its velocity is always directed towards the high altitude spy aircraft (Fig. P1.15) in rectilinear flight. Find the time taken to hit the aircraft assuming that the aircraft and the missile move with constant speeds v_A, v_M. $[\{v_M\sqrt{a^2+b^2} + v_A(b\cos\theta + a\sin\theta)\}/(v_M^2 - v_A^2)]$

16*. (a) Four men A, B, C, D stand at the corners of a square of side s. They start moving with speed v so that A pursues B, B pursues C, C pursues D and D pursues A. Find the position where they meet, the path followed, the time taken and the distance moved. [Hint: Use r, ϕ with centre of the square as origin.]
(b) Rework part (a) if 6 men stand at the corners of a regular hexagon of side s and each pursues the person in front at constant speed v.

[(a) at centre, $r = se^{-\phi}/\sqrt{2}$, s/v, s (b) at centre, $r = se^{-\phi/\sqrt{3}}$, $2s/v$, $2s$]

1.15 Index Notation

In the development of theory, it is convenient to use the *index notation*, i.e., to denote the coordinates x, y, z, the components a_x, a_y, a_z of a vector, and vectors $\vec{\mathrm{i}}$, $\vec{\mathrm{j}}$, $\vec{\mathrm{k}}$, by x_1, x_2, x_3 (x_i); a_1, a_2, a_3 (a_i) and $\vec{\mathrm{e}}_1$, $\vec{\mathrm{e}}_2$, $\vec{\mathrm{e}}_3$ $(\vec{\mathrm{e}}_i)$, respectively (Fig. 1.23). For brevity, we use the following *summation convention*: unless specified otherwise, a repeated algebraic index, such as j, in a *term* of an expression, implies summation over the range of that index. For example, the indices in the following examples range over values 1, 2, 3 and a repeated algebraic index, such as j, in a term implies that the sum of three terms is to be taken with j successively taking values 1, 2, 3. Thus,

$$d_{ii} = d_{11} + d_{22} + d_{33}, \qquad a_j b_j = a_1 b_1 + a_2 b_2 + a_3 b_3$$

$$\vec{\mathrm{e}}_i \cdot \vec{\mathrm{e}}_i = \vec{\mathrm{e}}_1 \cdot \vec{\mathrm{e}}_1 + \vec{\mathrm{e}}_2 \cdot \vec{\mathrm{e}}_2 + \vec{\mathrm{e}}_3 \cdot \vec{\mathrm{e}}_3 = 3$$

$$\vec{A}(t) = a_1(t)\,\vec{\mathrm{e}}_1 + a_2(t)\,\vec{\mathrm{e}}_2 + a_3(t)\,\vec{\mathrm{e}}_3 = \sum_{i=1}^{3} a_i(t)\,\vec{\mathrm{e}}_i = a_i\,\vec{\mathrm{e}}_i$$

$$\vec{r} = x_1\,\vec{\mathrm{e}}_1 + x_2\,\vec{\mathrm{e}}_2 + x_3\,\vec{\mathrm{e}}_3 = \sum_{i=1}^{3} x_i\,\vec{\mathrm{e}}_i = x_i\,\vec{\mathrm{e}}_i$$

$$b_i + d_{ij} c_j = a_i \qquad \Rightarrow \qquad b_i + d_{i1} c_1 + d_{i2} c_2 + d_{i3} c_3 = a_i \tag{a}$$

i.e.,

$$b_1 + d_{11} c_1 + d_{12} c_2 + d_{13} c_3 = a_1$$

$$b_2 + d_{21} c_1 + d_{22} c_2 + d_{23} c_3 = a_2$$

$$b_3 + d_{31} c_1 + d_{32} c_2 + d_{33} c_3 = a_3$$

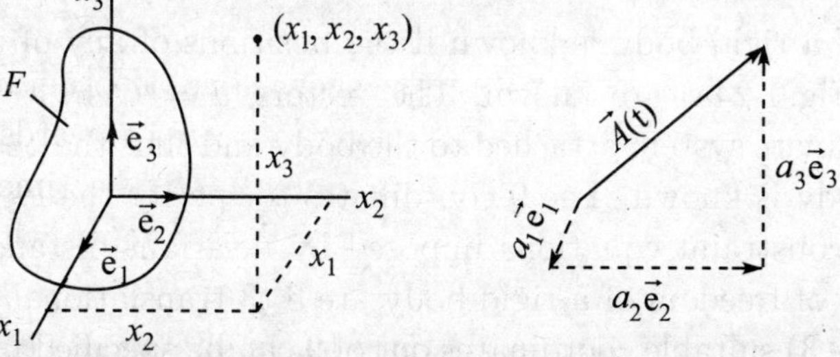

Fig. 1.23

A non-repeated index i is a *free index* which can be successively assigned values 1, 2, 3 to obtain three equations. The repeated index is a *dummy index* and can be replaced by any other algebraic index, e.g., eq. (a) can be written as: $b_i + d_{ik} c_k = a_i$. No index is permitted to occur in a given term three or more times. The dot products of $\vec{\mathrm{e}}_i$ can be expressed in terms of nine elements of Kronecker's delta δ_{ij}, defined as follows:

$$\delta_{ij} = \begin{cases} 1 & \text{if } i = j \\ 0 & \text{if } i \neq j \end{cases} \tag{1.46}$$

$$\vec{e}_i \cdot \vec{e}_j = \delta_{ij}$$

The *degrees of freedom* (abbreviated as dof) of a mechanical system are defined as the number of *independent* parameters (coordinates) needed to specify its configuration at the given time. The dof of a point moving in three-dimensional space, a point constrained to move on a given surface and a point constrained to move on a given curve are respectively 3 (x, y, z), 2 (surface coordinates ξ, η), and 1 (path coordinate s).

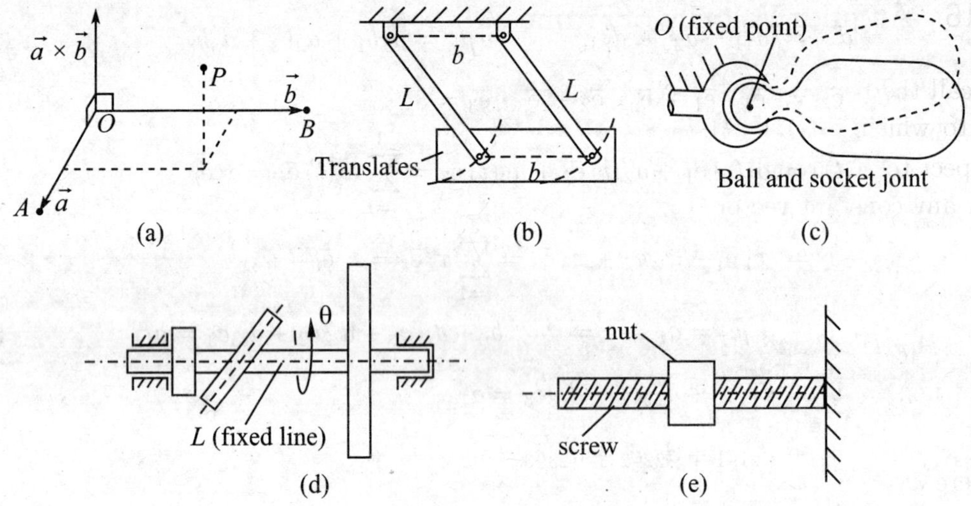

Fig. 1.24

The position of a rigid body is known if the positions of any of its 3 non-collinear points O, A, B (Fig. 1.24a) are known. The vectors $\vec{a} = \overrightarrow{OA}$, $\vec{b} = \overrightarrow{OB}$ and $\vec{a} \times \vec{b}$ determine a coordinate system attached to the body and thus the position of any other point P of the body is known. The 9 coordinates to specify the locations of O, A, B are related by 3 constraint equations imposed by invariant distances OA, AB, OB. Thus, the degrees of freedom of a rigid body are 6 (3 translational and 3 rotational), since only 6 $(= 9 - 3)$ suitable coordinates out of 9 can be specified independently. The basic modes of motion of a rigid body are: translation, rotation about a fixed point and general motion. Its motion is called *translation* (3 dof), if all its material points have the same displacement in F, or equivalently if its orientation remains fixed in F (Fig. 1.24b). Its motion is called *rotation about a point* (3 dof) fixed in F if one material point of the body occupies the same location in F for all time (Fig. 1.24c). A particular case of this is a *rotation about a fixed line* L (1 dof) in F in which all material points of the body on this line L do not undergo any displacement (Fig. 1.24d).

A rigid body is said to have *general motion* relative to F if (i) its orientation changes relative to F and (ii) there does not exist any point in it, or in its rigid extension, which remains fixed in F. The motion of a nut on a fixed screw (Fig. 1.24e) is an example of general motion of a rigid body. By **Euler's kinematical theorem**: *an arbitrary displacement of a rigid body with a fixed point is equivalent to a single rotation about a certain line through that point.* **Chasle's theorem** follows from this: *'The most general displacement of a rigid body in spatial motion is equivalent to a translation followed by a rotation about some base point and vice versa.'*

1.16 Angular Velocity of Frame m Relative to Frame F

Recall the basis vectors $\vec{e}_r(t)$, $\vec{e}_\phi(t)$, of the cylindrical coordinates. Consider a frame m to which $\vec{e}_r(t), \vec{e}_\phi(t), \vec{e}_z$ are attached. $\vec{e}_r(t), \vec{e}_\phi(t)$ change their orientation with respect to F through $\phi(t)$, but $\vec{e}_z$ does not. Vectors $\vec{i}, \vec{j}, \vec{k}$ are embedded in frame F. For any constant vector $\vec{A}$ in frame m, its components a_r, a_ϕ, a_z are constant with

$$\begin{aligned}
\vec{A}(t) &= a_r\,\vec{e}_r(t) + a_\phi\,\vec{e}_\phi(t) + a_z\,\vec{e}_z(t) \\
\dot{\vec{A}}_{|F}(t) &= a_r\,\dot{\vec{e}}_r + a_\phi\,\dot{\vec{e}}_\phi + a_z\,\dot{\vec{e}}_z = a_r\dot{\phi}\,\vec{e}_\phi(t) - a_\phi\dot{\phi}\,\vec{e}_r(t) \\
&= \dot{\phi}\,\vec{e}_z \times [a_r\,\vec{e}_r(t) + a_\phi\,\vec{e}_\phi(t) + a_z\,\vec{e}_z(t)] = \vec{\omega}(t) \times \vec{A}(t) \quad \forall\, \vec{A}
\end{aligned}$$

where $\vec{\omega}(t) = \dot{\phi}(t)\,\vec{e}_z$ is a measure of the rate of change of orientation of m with respect to F. We shall establish that this result is a particular case of the general result valid for any motion of m, in which $\vec{\omega}$ *need not be the derivative of a finite angle of rotation.*

We now take up the case of general motion of frame m relative to frame F. Consider right-handed unit triads: $\vec{E}_i$ fixed to F and $\vec{e}_i$ fixed to m (Fig. 1.25). The rate of change of orientation of m relative to F at time t is governed by $\dot{\vec{e}}_{i|F}(t)$, which are not independent as

$$\begin{aligned}
\vec{e}_i(t)\cdot\vec{e}_j(t) &= \delta_{ij} \quad \Rightarrow \quad \dot{\vec{e}}_{i|F}(t)\cdot\vec{e}_j(t) + \vec{e}_i(t)\cdot\dot{\vec{e}}_{j|F}(t) = 0 \Rightarrow \\
\dot{\vec{e}}_{i|F}(t)\cdot\vec{e}_j(t) &= -\,\vec{e}_i(t)\cdot\dot{\vec{e}}_{j|F}(t) \quad \text{(Six relations in all for distinct } i, j) \qquad (a)
\end{aligned}$$

$$[\vec{e}_i(t) \times \dot{\vec{e}}_{i|F}(t)] \times \vec{A}(t) = \{\vec{A}(t)\cdot\vec{e}_i(t)\}\,\dot{\vec{e}}_{i|F}(t) - \{\vec{A}(t)\cdot\dot{\vec{e}}_{i|F}(t)\}\,\vec{e}_i(t) \qquad (b)$$

Consider a constant vector $\vec{A} = a_i\,\vec{e}_i$ in m, i.e., all a_i are constant. With respect to F, $\vec{A}(t) = a_i\,\vec{e}_i(t)$ and using eq. (a), we express $\dot{\vec{A}}_{|F}$ in terms of $\vec{e}_i(t)$, $\dot{\vec{e}}_{i|F}(t)$ and $\vec{A}(t)$:

$$\dot{\vec{A}}_{|F}(t) = a_i\, \dot{\vec{e}}_{i|F}(t) \tag{c}$$

$$\begin{aligned}
&= [\vec{A}(t)\cdot\vec{e}_i(t)]\,\dot{\vec{e}}_{i|F}(t) \overset{(b)}{=} [\vec{e}_i(t)\times\dot{\vec{e}}_{i|F}(t)]\times\vec{A}(t) + \{\vec{A}(t)\cdot\dot{\vec{e}}_{i|F}(t)\}\,\vec{e}_i(t)\\
&\qquad\qquad\qquad\qquad = [\vec{e}_i(t)\times\dot{\vec{e}}_{i|F}(t)]\times\vec{A}(t) + \{a_j\,\vec{e}_j(t)\cdot\dot{\vec{e}}_{i|F}(t)\}\,\vec{e}_i(t)\\
&\qquad\qquad\qquad\qquad \overset{(a)}{=} [\vec{e}_i(t)\times\dot{\vec{e}}_{i|F}(t)]\times\vec{A}(t) - \{\vec{e}_i\cdot\dot{\vec{e}}_{j|F}(t)a_j\}\,\vec{e}_i(t)\\
&\qquad\qquad\qquad\qquad \overset{(c)}{=} [\vec{e}_i(t)\times\dot{\vec{e}}_{i|F}(t)]\times\vec{A}(t) - \{\vec{e}_i\cdot\dot{\vec{A}}_{|F}(t)\}\,\vec{e}_i(t)
\end{aligned}$$

$$\Rightarrow\quad \dot{\vec{A}}_{|F} = \tfrac{1}{2}[\vec{e}_i(t)\times\dot{\vec{e}}_{i|F}(t)]\times\vec{A}(t) \tag{d}$$

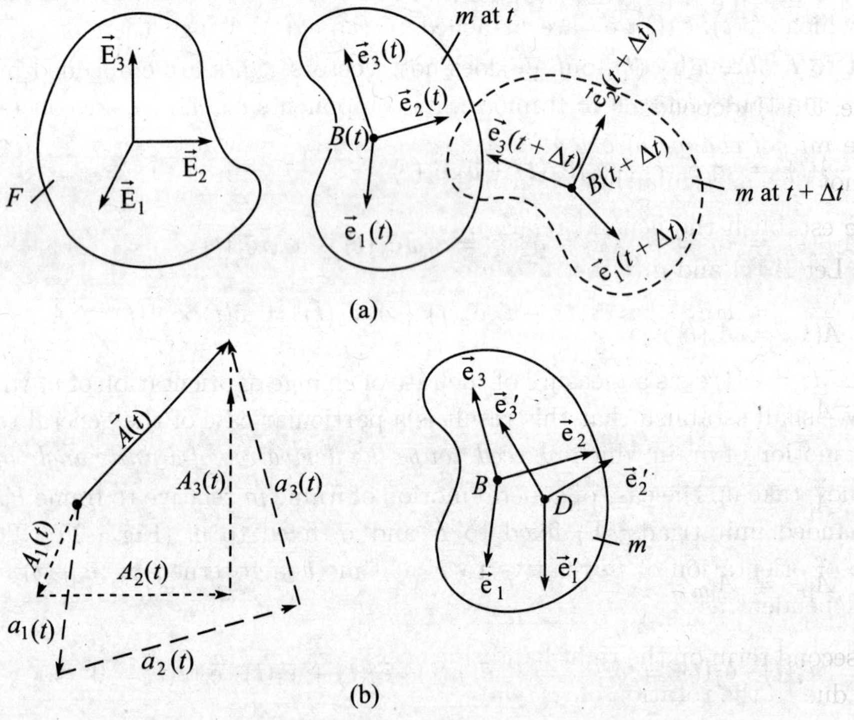

(a)

(b)

Fig. 1.25

Hence, we *define* the angular velocity $\vec{\omega}_{m|F}(t)$ of frame m relative to frame F at time t in terms of $\dot{\vec{e}}_{i|F}(t)$ by

$$\vec{\omega}(t) = \tfrac{1}{2}\,\vec{e}_i(t)\times\dot{\vec{e}}_{i|F}(t) \tag{1.47}$$

$$= \tfrac{1}{2}[\,\vec{e}_1(t)\times\dot{\vec{e}}_{1|F}(t) + \vec{e}_2(t)\times\dot{\vec{e}}_{2|F}(t) + \vec{e}_3(t)\times\dot{\vec{e}}_{3|F}(t)]$$

It is a measure of the rate of change of orientation of m. From eq. (d) we infer that for all constant vectors $\vec{A}$ in m, there exists a unique $\vec{\omega}(t)$, defined above, such that

$$\dot{\vec{A}}_{|F} = \vec{\omega} \times \vec{A} \tag{1.48}$$

In particular, for constant vectors $\vec{e}_i$ in m:

$$\dot{\vec{e}}_{i|F} = \vec{\omega} \times \vec{e}_i \tag{1.49}$$

We prove the *consistency* of the definition of $\vec{\omega}$. Consider another orthogonal right-handed unit triad $\vec{e}'_i$ fixed to $m \Rightarrow \vec{e}'_i$ are constant in m. Let $\vec{\omega} = \omega_i \vec{e}_i = \omega'_i \vec{e}'_i$ and

$$\begin{aligned} \vec{\omega}' &= \tfrac{1}{2}\vec{e}'_i \times \dot{\vec{e}}'_{i|F} = \tfrac{1}{2}[\vec{e}'_i \times (\vec{\omega} \times \vec{e}'_i)] \qquad \text{[Using eq. (1.48) for } \vec{A} = \vec{e}'_i] \\ &= \tfrac{1}{2}[(\vec{e}'_i \cdot \vec{e}'_i)\vec{\omega} - (\vec{e}'_i \cdot \vec{\omega})\vec{e}'_i] = \tfrac{1}{2}[3\vec{\omega} - \vec{\omega}] = \vec{\omega} \end{aligned}$$

since $\vec{e}'_i \cdot \vec{e}'_i = \vec{e}'_1 \cdot \vec{e}'_1 + \vec{e}'_2 \cdot \vec{e}'_2 + \vec{e}'_3 \cdot \vec{e}'_3 = 3$ and $(\vec{e}'_i \cdot \vec{\omega})\vec{e}'_i = \omega'_i \vec{e}'_i = \vec{\omega}$. Hence, $\vec{\omega}$ is independent of the location or the directions of $\vec{e}_i$ in m. It is a measure of the *rate of change of orientation of the frame m as a whole relative to the frame F* and not of a particular line element in it.

We establish the general relation between $\dot{\vec{A}}_{|m}$ and $\dot{\vec{A}}_{|F}$ for an **arbitrary vector** $\vec{A}(t)$. Let $A_i(t)$ and $a_i(t)$ be its components along $\vec{E}_i$ and $\vec{e}_i$ as shown in Fig. 1.25:

$$\begin{aligned} \vec{A}(t) &= A_i(t)\,\vec{E}_i = a_i(t)\,\vec{e}_i \\ \dot{\vec{A}}_{|F} &= \dot{\vec{A}}\Big|_{\vec{E}_i=\text{const.}} = \dot{A}_i\,\vec{E}_i, \qquad \dot{\vec{A}}_{|m} = \dot{\vec{A}}\Big|_{\vec{e}_i=\text{const.}} = \dot{a}_i\,\vec{e}_i \\ \dot{\vec{A}}_{|F} &= \overline{(a_i\,\vec{e}_i)}^{\,\cdot}_{|F} = \dot{a}_i\,\vec{e}_i + a_i\dot{\vec{e}}_{i|F} = \dot{\vec{A}}_{|m} + a_i\vec{\omega} \times \vec{e}_i = \dot{\vec{A}}_{|m} + \vec{\omega} \times a_i\,\vec{e}_i \;\Rightarrow \\ \dot{\vec{A}}_{|F} &= \dot{\vec{A}}_{|m} + \vec{\omega} \times \vec{A}, \qquad \text{i.e.,} \qquad (\dot{\vec{\ }})_{|F} = (\dot{\vec{\ }})_{|m} + \vec{\omega}_{m|F} \times (\vec{\ }) \end{aligned} \tag{1.50}$$

The second term on the right hand side of eq. (1.50) represents the change of $\vec{A}$ relative to F due to the rotation of m relative to F.

Using eq. (1.50), the time derivative of $\vec{\omega}_{m|F}$ with respect to frame F is given by

$$\dot{\vec{\omega}}_{|F} = \frac{d\,\vec{\omega}_{m|F}}{dt}\Big|_F = \overline{(\omega_{m|F})}^{\,\cdot}_{|m} + \vec{\omega} \times \vec{\omega} = \dot{\vec{\omega}}_{|m} \tag{1.51}$$

$\dot{\vec{\omega}} = \dot{\vec{\omega}}_{|F} = \dot{\vec{\omega}}_{|m}$ is defined as the *angular acceleration* of frame m relative to frame F. Do not confuse $\dot{\vec{\omega}}_{|m} = \overline{(\omega_{m|F})}^{\,\cdot}_{|m}$ with $\overline{(\vec{\omega}_{m|m})}^{\,\cdot}_{|m}$ which is zero. Three cases are described for obtaining $\dot{\vec{\omega}}$.

1. If $\vec{\omega}_{m|F}(t) = \omega_i(t)\,\vec{e}_i$, then $\dot{\vec{\omega}}_{m|F} = \overline{(\vec{\omega}_{m|F})}^{\,\cdot}_{|m} = \dot{\omega}_i\,\vec{e}_i$.
2. If $\vec{\omega}_{m|F}(t) = \Omega_i(t)\,\vec{E}_i$, then $\dot{\vec{\omega}}_{m|F} = \overline{(\vec{\omega}_{m|F})}^{\,\cdot}_{|F} = \dot{\Omega}_i\,\vec{E}_i$.
3. If $\vec{\omega}_{m|F}(t) = \Omega_i^*(t)\,\vec{E}_i + \omega_i^*(t)\,\vec{e}_i$, then

$$\dot{\vec{\omega}}_{m|F} = \overline{(\vec{\omega}_{m|F})}^{\,\cdot}_{|F} = \dot{\Omega}_i^*\,\vec{E}_i + \dot{\omega}_i^*\,\vec{e}_i + \vec{\omega}_{m|F} \times \omega_i^*\,\vec{e}_i$$

§ **Exercise 1.10:** Prove that for a constant vector $\vec{A}$ in frame m:

$$\ddot{\vec{A}}_{|F} = \frac{d}{dt}(\dot{\vec{A}}_{|F})\Big|_F = \dot{\vec{\omega}} \times \vec{A} + \vec{\omega} \times (\vec{\omega} \times \vec{A}).$$

1.17 Angular Velocity of m if $\dot{\vec{e}}_{3|F} = \vec{0}$

A rigid body m is said to have *plane motion* relative to F if all its material points have plane trajectories in parallel planes. Choose $\vec{e}_3$ attached to m in a direction normal to the planes of trajectories, then relative to frame F, $\vec{e}_3$ moves parallel to itself and $\dot{\vec{e}}_{3|F} = \vec{0}$. Most real-life systems have plane motion, e.g., a rigid body rotating about a fixed axis and a trolley rolling without slip on a plane. Plane motion is much simpler to analyse mathematically than spatial motion (the one which is not a plane motion). The dof of a rigid body in plane motion are 3 – two translational and one rotational.

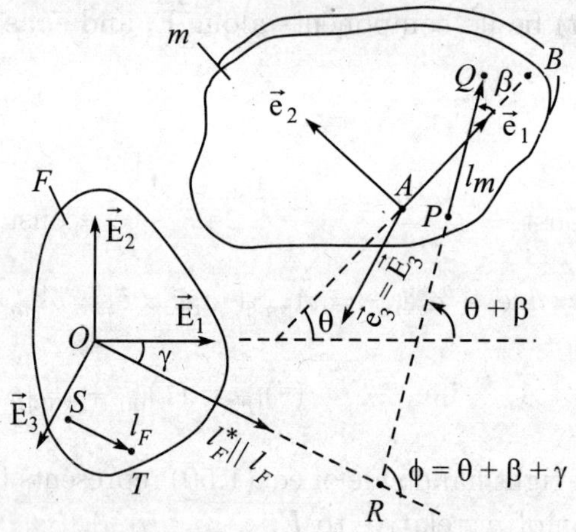

Fig. 1.26

For plane motion, $\dot{\vec{e}}_{3|F} = \vec{0}$, but the converse is not true. For example, the motion of a rigid nut on a fixed screw F is not a plane motion, but $\vec{e}_3$ attached to the nut in the direction of the screw axis does not change its orientation relative to F, i.e., $\dot{\vec{e}}_{3|F} = \vec{0}$.

For cases such as the screw motion and plane motion since $\dot{\vec{e}}_{3|F} = \vec{0}$, we choose $\vec{E}_3 = \vec{e}_3$ (Fig. 1.26). Then, $\vec{e}_1$, $\vec{e}_2$ have no components along $\vec{E}_3$. Let $\theta(t)$ be the angle between $\vec{e}_1$ and $\vec{E}_1$ at time t:

$$
\begin{aligned}
\vec{e}_1(t) &= \cos\theta(t)\,\vec{E}_1 + \sin\theta(t)\,\vec{E}_2, \qquad\qquad \vec{e}_3(t) = \vec{E}_3 \\
\vec{e}_2(t) &= -\sin\theta(t)\,\vec{E}_1 + \cos\theta(t)\,\vec{E}_2, \qquad\qquad \dot{\vec{e}}_{3|F} = \vec{0} \\
\dot{\vec{e}}_{1|F} &= (-\sin\theta\,\vec{E}_1 + \cos\theta\,\vec{E}_2)\dot{\theta} = \dot{\theta}\,\vec{e}_2 \\
\dot{\vec{e}}_{2|F} &= (-\cos\theta\,\vec{E}_1 - \sin\theta\,\vec{E}_2)\dot{\theta} = -\dot{\theta}\,\vec{e}_1
\end{aligned}
$$

and eq. (1.47) $\Rightarrow$

$$
\begin{aligned}
\vec{\omega} &= \tfrac{1}{2}(\vec{e}_i \times \dot{\vec{e}}_{i|F}) = \tfrac{1}{2}[\vec{e}_1 \times \dot{\vec{e}}_{1|F} + \vec{e}_2 \times \dot{\vec{e}}_{2|F} + \vec{e}_3 \times \dot{\vec{e}}_{3|F}] \\
&= \tfrac{1}{2}[\vec{e}_1 \times \dot{\theta}\,\vec{e}_2 + \vec{e}_2 \times (-\dot{\theta}\,\vec{e}_1) + \vec{e}_3 \times \vec{0}] \\
\Rightarrow \quad \vec{\omega} &= \dot{\theta}\,\vec{e}_3, \qquad \vec{\omega} = \omega\,\vec{e}_3, \qquad \text{with} \qquad \omega = \dot{\theta} \qquad (1.52)
\end{aligned}
$$

$\vec{\omega}$ is not only the rate of rotation of $\vec{e}_1$ relative to $\vec{E}_1$, but also the rate of rotation of **any line** $l_m \perp \vec{e}_3$ fixed in m, (e.g., $\overrightarrow{PQ}$) relative to **any line** $l_F \perp \vec{E}_3$ fixed in F (e.g., $\overrightarrow{ST}$) since $\dot{\phi} = \dot{\theta} + \dot{\beta} + \dot{\gamma} = \dot{\theta}$. In this special case, $\vec{\omega}$ is the rate of change of a *finite angle* θ. For the general spatial motion of a rigid body, an infinitesimal rotation is *not* an infinitesimal increment of *a finite angular measure* and ω is not equal to the rate of change of a finite angle.

The angular position θ, angular velocity $\vec{\omega}$, angular acceleration $\dot{\vec{\omega}}$ are related by

$$
\begin{aligned}
\vec{\omega} &= \omega\,\vec{e}_3 = \dot{\theta}\,\vec{e}_3 \qquad \Rightarrow \qquad \dot{\vec{\omega}} = \dot{\omega}\,\vec{e}_3 = \ddot{\theta}\,\vec{e}_3 = \alpha\,\vec{e}_3 \qquad \Rightarrow \\
\alpha &= \ddot{\theta} = \dot{\omega} = \frac{d\omega}{d\theta}\frac{d\theta}{dt} = \omega\frac{d\omega}{d\theta} \qquad (1.53)
\end{aligned}
$$

Equations (1.53) relating $\alpha, \omega, \theta, t$ are identical to the ones relating a_t, v_t, s, t. In general, α may be a function of θ, ω, t:

$$
\ddot{\theta} = \alpha = \alpha(\omega, \theta, t) \qquad (1.54)
$$

The two constants of integration in the solution of this second order differential equation are evaluated from the two *initial conditions* of θ and ω at $t = 0$: $\theta(0) = \theta_0$, $\dot{\theta}(0) = \omega_0$. If α is a known function of only one of the variables θ, ω, t, eq. (1.54) can be integrated by expressing α in a suitable form to separate the variables. These relations are the same as eqs (1.32) to (1.35) with θ, ω, α replacing s, v_t, a_t:

$$
\alpha(t),\ \alpha(\omega),\ \alpha = \frac{d\omega}{dt} \ \Rightarrow\ \omega(t) - \omega(0) = \int_0^t \alpha(t)\,dt, \qquad t = \int_{\omega(0)}^{\omega(t)} \frac{1}{\alpha(\omega)}\,d\omega \qquad (1.55)
$$

$$
\alpha(\theta),\ \alpha(\omega),\ \alpha = \omega\frac{d\omega}{d\theta} \ \Rightarrow\ \omega^2(\theta) - \omega^2(0) = 2\int_0^\theta \alpha(\theta)\,d\theta, \quad \theta = \int_{\omega(0)}^{\omega(\theta)} \frac{\omega}{\alpha(\omega)}\,d\omega \qquad (1.56)
$$

$$\omega(t),\ \omega(\theta),\ \omega = \frac{d\theta}{dt} \quad \Rightarrow \quad \theta(t) - \theta(0) = \int_0^t \omega(t)\,dt, \qquad t = \int_{\theta(0)}^{\theta(t)} \frac{1}{\omega(\theta)}\,d\theta \quad (1.57)$$

For $\alpha(t) = \text{const.} = \alpha_0$:

$$\omega(t) = \omega(0) + \alpha_0 t, \qquad \theta(t) = \theta(0) + \omega(0)t + \tfrac{1}{2}\alpha_0 t^2, \qquad \omega^2(\theta) - \omega^2(0) = 2\alpha_0\theta \quad (1.58)$$

For *uniform rotation*: $\alpha(t) = 0, \qquad \omega(t) = \omega(0), \qquad \theta(t) = \theta(0) + \omega(0)t.$

1.18 Relations between $\vec{v}_{P|F}$, $\vec{a}_{P|F}$ and $\vec{v}_{P|m}$, $\vec{a}_{P|m}$

These relations are needed for (i) application of equations of dynamics, where acceleration relative to an inertial frame I is required whereas it may be conveniently described relative to a non-inertial frame m, (ii) a simple visualisation of motion relative to a frame m moving relative to the ground frame F, e.g., the flow of gases through the nozzle of a rocket engine has a simpler description relative to the frame m attached to the rocket than relative to the ground frame F, (iii) study of kinematics of a point on a slider moving in a slot in a rigid body m which is in motion relative to the ground reference F.

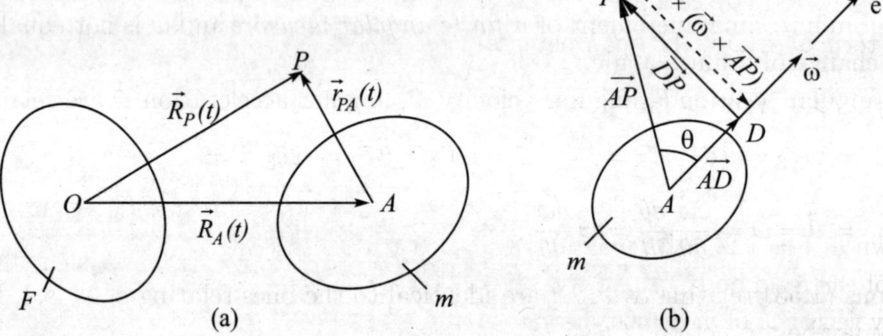

Fig. 1.27

For a point P in motion, $\vec{v}_{P|F}$, $\vec{a}_{P|F}$ are related to $\vec{v}_{P|m}$, $\vec{a}_{P|m}$ through $\vec{\omega}_{m|F}$, $\dot{\vec{\omega}}_{m|F}$ of m; $\vec{v}_{A|F}$, $\vec{a}_{A|F}$ of a point A of m and $\vec{r}_{PA}$. We denote the position vectors with respect to frames F and m by $\vec{R}$ and $\vec{r}$, respectively (Fig. 1.27a). Let O be fixed to F. A is fixed to m. Thus,

$$\vec{R}_P(t) = \vec{R}_A(t) + \vec{r}_{PA}(t) \qquad (a)$$

We take the time derivative of equation (a) with respect to frame F and use eq. (1.50) to yield the velocity relation:

$$\dot{\vec{R}}_{P|F} \ = \ \dot{\vec{R}}_{A|F} + \dot{\vec{r}}_{PA|F} = \dot{\vec{R}}_{A|F} + [\dot{\vec{r}}_{PA|m} + \vec{\omega} \times \vec{r}_{PA}] \quad \Rightarrow$$

$$\vec{v}_{P|F} = \vec{v}_{A|F} + \vec{\omega} \times \vec{r}_{PA} + \vec{v}_{P|m} \tag{1.59}$$

We take the time derivative of eq. (1.59) with respect to frame F and use eq. (1.50) to yield the acceleration relation:

$$\begin{aligned} \vec{a}_{P|F} &= \vec{a}_{A|F} + \dot{\vec{\omega}}_{|F} \times \vec{r}_{PA} + \vec{\omega} \times \dot{\vec{r}}_{PA|F} + \frac{d}{dt}(\vec{v}_{P|m})\Big|_F \\ &= \vec{a}_{A|F} + \dot{\vec{\omega}} \times \vec{r}_{PA} + \vec{\omega} \times \left[\dot{\vec{r}}_{PA|m} + \vec{\omega} \times \vec{r}_{PA}\right] + \left[\frac{d}{dt}(\vec{v}_{P|m})\Big|_m + \vec{\omega} \times \vec{v}_{P|m}\right] \end{aligned}$$

$$\vec{a}_{P|F} = \vec{a}_{A|F} + \dot{\vec{\omega}} \times \vec{r}_{PA} + \vec{\omega} \times (\vec{\omega} \times \vec{r}_{PA}) + 2\vec{\omega} \times \vec{v}_{P|m} + \vec{a}_{P|m} \tag{1.60}$$

Writing $\vec{r}_{PA} = \overrightarrow{AP}$, eqs (1.59) and (1.60) can be written as

$$\vec{v}_{P|F} = \vec{v}_{A|F} + \vec{\omega} \times \overrightarrow{AP} + \vec{v}_{P|m} \tag{1.61}$$

$$\vec{a}_{P|F} = \vec{a}_{A|F} + \dot{\vec{\omega}} \times \overrightarrow{AP} + \vec{\omega} \times (\vec{\omega} \times \overrightarrow{AP}) + 2\vec{\omega} \times \vec{v}_{P|m} + \vec{a}_{P|m} \tag{1.62}$$

In eqs (1.59) to (1.62), it is understood that $\vec{\omega} = \vec{\omega}_{m|F}$ and $\dot{\vec{\omega}} = \dot{\vec{\omega}}_{m|F}$.

The term $\vec{\omega} \times (\vec{\omega} \times \overrightarrow{AP})$ can be expressed as:

$$\begin{aligned} \vec{\omega} \times (\vec{\omega} \times \overrightarrow{AP}) &= (\vec{\omega} \cdot \overrightarrow{AP})\vec{\omega} - \omega^2 \overrightarrow{AP} = -\omega^2[\overrightarrow{AP} - (AP)\cos\theta\, \vec{e}] \\ &= -\omega^2(\overrightarrow{AP} - \overrightarrow{AD}) = -\omega^2 \overrightarrow{DP} \end{aligned}$$

as shown in Fig. 1.27b. Its magnitude equals the product of the square of the angular speed of the rigid body m and the perpendicular distance of point P from the angular velocity vector $\vec{\omega}$ of m drawn through A and is always directed radially in towards this vector. Hence, this term $\vec{\omega} \times (\vec{\omega} \times \overrightarrow{AP})$ is called *centripetal acceleration* (centripetal means centre-seeking). The term $2\vec{\omega} \times \vec{v}_{P|m}$ is called *Coriolis acceleration* in honour of G. Coriolis who first derived it.

Note that for special motion of frame m with $\vec{e}_3 = \vec{E}_3$, $\vec{\omega} = \omega\, \vec{e}_3$ and for points A and P in the same plane normal to $\vec{e}_3$, $\vec{\omega} \cdot \overrightarrow{AP} = 0$ and $\vec{\omega} \times (\vec{\omega} \times \overrightarrow{AP}) = -\omega^2 \overrightarrow{AP}$.

§ **Exercise 1.11:** (a) A is a point of frame m. Show that the jerk of P is given by

$$\begin{aligned} \dot{\vec{a}}_{P|F} &= \dot{\vec{a}}_{A|F} + \ddot{\vec{\omega}}_{|m} \times \vec{r}_{PA} + \vec{\omega} \times (\dot{\vec{\omega}} \times \vec{r}_{PA}) - \omega^2(\vec{\omega} \times \vec{r}_{PA}) \\ &\quad + 2\dot{\vec{\omega}} \times (\vec{\omega} \times \vec{r}_{PA}) + 3\dot{\vec{\omega}} \times \vec{v}_{P|m} + 3\vec{\omega} \times \vec{a}_{P|m} + 3\vec{\omega} \times (\vec{\omega} \times \vec{v}_{P|m}) + \dot{\vec{a}}_{P|m} \end{aligned}$$

(b) In the computerised calculation of eqs (1.59) and(1.60), it is easier to work with matrix products. Writing any vector $\vec{B}$ as 3×1 column matrix $[\vec{B}]$, prove that

$[\vec{\omega} \times \vec{B}] = [\Omega][\vec{B}]$, where $[\Omega]$ is a skew-symmetric matrix associated with $\vec{\omega}$:

$$[\Omega] = \begin{bmatrix} 0 & -\omega_z & \omega_y \\ \omega_z & 0 & -\omega_x \\ -\omega_y & \omega_x & 0 \end{bmatrix}$$

Thus, show that eqs (1.59) and (1.60) can be put in matrix form as:

$$\begin{aligned} [\vec{v}_{P|F}] &= [\vec{v}_{A|F}] + [\Omega][\vec{r}_{PA}] + [\vec{v}_{P|m}] \\ [\vec{a}_{P|F}] &= [\vec{a}_{A|F}] + [\dot{\Omega}][\vec{r}_{PA}] + [\Omega][\Omega][\vec{r}_{PA}] + 2[\Omega][\vec{v}_{P|m}] + [\vec{a}_{P|m}] \end{aligned}$$

1.19 Kinematics Relative to Translating Frame T

A frame T is said to *translate* with respect to F if displacements of all its points A, B, C, ... are the same in F (Fig. 1.28a). Hence, the velocities and accelerations of all points of T with respect to F are the same. $\Rightarrow$

$$\vec{v}_{BA|F} = \dot{\vec{R}}_{BA|F} = \vec{v}_{B|F} - \vec{v}_{A|F} = \vec{0} \qquad \Rightarrow \qquad \vec{R}_{BA} \text{ is a constant vector in } F$$

and $\vec{a}_{BA|F} = \vec{0}$. The relative displacement, velocity and acceleration between any two points of T with respect to frame F are zero. The motion of just one point of T is necessary and sufficient to describe the motion of T. All the directed line elements of T are constant in F. Therefore, $\dot{\vec{e}}_{i|F}(t) = \vec{0}$ and eqs (1.47) and (1.50) yield

$$\vec{\omega}(t) = \tfrac{1}{2}[\vec{e}_i(t) \times \dot{\vec{e}}_{i|F}(t)] = \vec{0} \qquad \Rightarrow \qquad \dot{\vec{\omega}} = \vec{0}, \qquad (\dot{\vec{\ }})_{|F} = (\dot{\vec{\ }})_{|T} \tag{1.63}$$

Conversely, if $\vec{\omega}(t) \equiv \vec{0}$, then for any two locations A, B of T: $\dot{\vec{R}}_{BA} = \vec{\omega} \times \vec{R}_{BA} = \vec{0} \Rightarrow \vec{R}_{BA}(t) = \vec{R}_{BA}(0) \Rightarrow \vec{R}_B(t) - \vec{R}_B(0) = \vec{R}_A(t) - \vec{R}_A(0) \Rightarrow$ displacements of all locations of T are the same $\Rightarrow$ T is in translation relative to F. Hence, a rigid body m is in translation relative to frame F if and only if $\vec{\omega}_{m|F}(t) \equiv \vec{0}$.

Using eq. (1.63), eqs (1.59) and (1.60) yield $\vec{v}_{P|F}$, $\vec{a}_{P|F}$ in terms of a point A of T:

$$\vec{v}_{P|F} = \vec{v}_{A|F} + \vec{v}_{P|T}, \qquad \vec{a}_{P|F} = \vec{a}_{A|F} + \vec{a}_{P|T} \tag{1.64}$$

$\vec{v}_{A|F}$, $\vec{a}_{A|F}$ being the same for every point of T, eq. (1.64) is re-written in terms of velocity $\vec{v}_{T|F}$ and acceleration $\vec{a}_{T|F}$ of frame T translating relative to frame F:

$$\vec{v}_{P|F} = \vec{v}_{P|T} + \vec{v}_{T|F}, \qquad \vec{a}_{P|F} = \vec{a}_{P|T} + \vec{a}_{T|F} \tag{1.65}$$

$\vec{v}_{A|F}$, $\vec{v}_{P|T}$, $\vec{a}_{A|F}$, $\vec{a}_{P|T}$ can be expressed in Cartesian/cylindrical/path components and added by expressing them in the same basis. It follows (Fig. 1.28b) from eq. (1.65) that

$$\vec{v}_{PQ|F} = \vec{v}_{P|F} - \vec{v}_{Q|F} = \vec{v}_{P|T} + \vec{v}_{T|F} - \vec{v}_{Q|T} - \vec{v}_{T|F} = \vec{v}_{P|T} - \vec{v}_{Q|T} = \vec{v}_{PQ|T}$$

and similarly, $\vec{a}_{PQ|F} = \vec{a}_{PQ|T}$, i.e., the relative velocity and relative acceleration of point P relative to Q in all reference frames in mutual translation are the same.

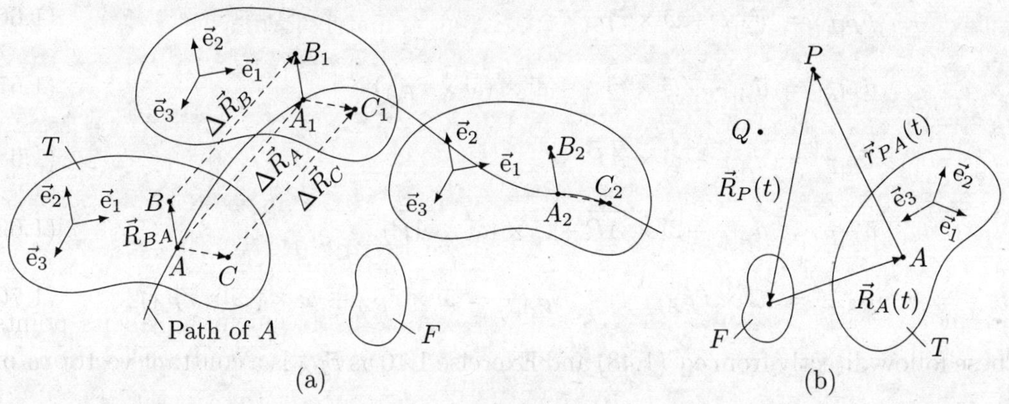

Fig. 1.28

Relations (1.65) can be compounded for several translating frames, e.g., if frame T_2 translates relative to frame T_1 and T_1 translates relative to frame F (Fig. 1.29a), then

$$\vec{v}_{P|F} = \vec{v}_{P|T_2} + \vec{v}_{T_2|T_1} + \vec{v}_{T_1|F}, \qquad \vec{a}_{P|F} = \vec{a}_{P|T_2} + \vec{a}_{T_2|T_1} + \vec{a}_{T_1|F}$$

(a) (b)

Fig. 1.29

The dof of a translating system are 3. If a material point of T is constrained to move on a given surface or a given curve, then the dof of T are respectively 2 or 1. If the path of a point of T is a plane curve or a straight line, then it is called respectively, a plane translation or a rectilinear translation. In a locomotive running along a straight level track (Fig. 1.29b), the main body 1, piston 2, piston rod 3 and cross-head 4 have rectilinear and the parallel coupler rods 5 and 6 have plane curvilinear translation.

1.20 Rigid Body Kinematics

For material points P and A of a rigid body m (Fig. 1.30), $\vec{v}_{P|m} = \vec{0}$, $\vec{a}_{P|m} = \vec{0}$ and eqs (1.59) to (1.62) yield

$$\vec{v}_{P|F} = \vec{v}_{A|F} + \vec{\omega} \times \vec{r}_{PA} \tag{1.66}$$

$$\vec{a}_{P|F} = \vec{a}_{A|F} + \dot{\vec{\omega}} \times \vec{r}_{PA} + \vec{\omega} \times (\vec{\omega} \times \vec{r}_{PA}) \tag{1.67}$$

$$\vec{v}_{P|F} = \vec{v}_{A|F} + \vec{\omega} \times \overrightarrow{AP} \tag{1.68}$$

$$\vec{a}_{P|F} = \vec{a}_{A|F} + \dot{\vec{\omega}} \times \overrightarrow{AP} + \vec{\omega} \times (\vec{\omega} \times \overrightarrow{AP}) \tag{1.69}$$

$$\Rightarrow \quad \vec{v}_{PA|F} = \vec{\omega} \times \vec{r}_{PA}, \qquad \vec{a}_{PA|F} = \dot{\vec{\omega}} \times \vec{r}_{PA} + \vec{\omega} \times (\vec{\omega} \times \vec{r}_{PA}) \tag{1.70}$$

These follow directly from eq. (1.48) and Exercise 1.10 as $\vec{r}_{PA}$ is a constant vector in m.

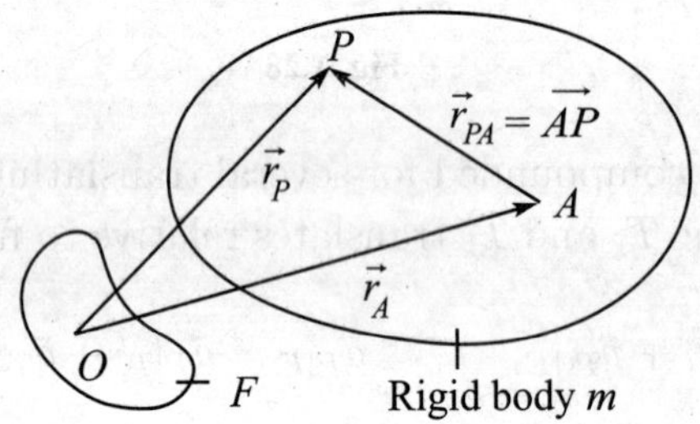

Fig. 1.30

§ **Exercise 1.12:** Using eq. (1.66), prove the following results for a rigid body. (i) The components of velocity of all points on a given line, in the direction of the line are equal. (ii) At a given instant, the component of velocity of all points in the direction of $\vec{\omega}$ are the same. (iii) At a given instant, the set of points with velocity equal to that of a material point A lie on a line through A in the direction of $\vec{\omega}$.

Note that for plane motion of a rigid body with $\vec{e}_z = \vec{e}_3 = \vec{E}_3$, $\vec{\omega} = \omega\,\vec{e}_z$: for its material points A and B in the same plane normal to $\vec{e}_z$, with $\overrightarrow{AB} = r\,\vec{e}_r$ (Fig. 1.31):

$$\begin{aligned}\vec{v}_{BA} &= \vec{\omega} \times \overrightarrow{AB} = \omega\,\vec{e}_z \times r\,\vec{e}_r = \omega r\,\vec{e}_\phi \\ \vec{a}_{BA} &= \dot{\vec{\omega}} \times \overrightarrow{AB} + \vec{\omega} \times (\vec{\omega} \times \overrightarrow{AB}) = \dot{\omega}\,\vec{e}_z \times r\,\vec{e}_r - \omega^2 r\,\vec{e}_r \\ &= -\omega^2 r\,\vec{e}_r + \dot{\omega} r\,\vec{e}_\phi\end{aligned} \tag{1.71}$$

i.e., the relative values correspond to a circular motion of B relative to A with AB rotating at the rate $\vec{\omega}$.

A rigid body is said to have an *instantaneous axis of rotation* at time t if all its points (its material points or points on its rigid extension) on such an axis have zero velocity at that instant. *It does not always exist.* For the given angular velocity $\vec{\omega}$ of the body and velocity $\vec{v}_A$ of its point A at time t, the instantaneous axis of rotation exists if we can find a point B such that

$$\vec{v}_B = \vec{v}_A + \vec{\omega} \times \vec{r}_{BA} = \vec{0}, \qquad \text{i.e.,} \quad \text{if} \quad \vec{v}_A = -\vec{\omega} \times \vec{r}_{BA} \tag{1}$$

The solution of eq. (1) exists if $\vec{v}_A \cdot \vec{\omega} = -(\vec{\omega} \times \vec{r}_{BA}) \cdot \vec{\omega} = 0$. Hence, the instantaneous axis of rotation exists *only if* $\vec{v}_A \cdot \vec{\omega} = 0$, i.e., only if $\vec{v}_A = \vec{0}$ or $\vec{v}_A \perp \vec{\omega}$.

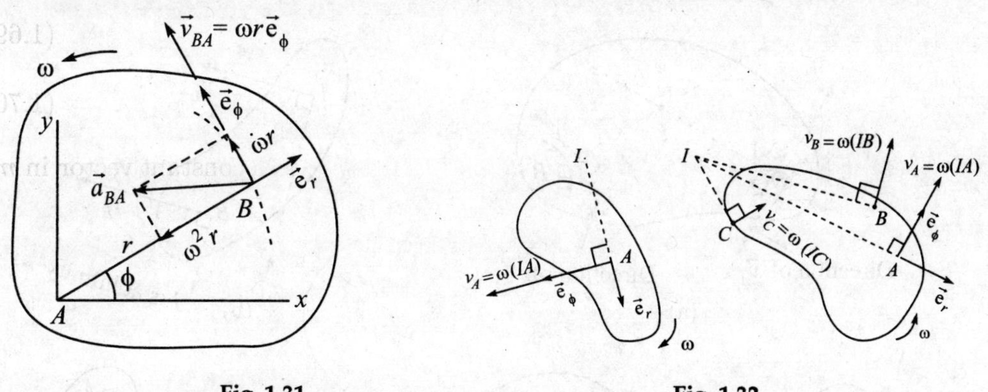

Fig. 1.31 **Fig. 1.32**

The *instantaneous axis of rotation always exists for plane motion* of a rigid body since $\vec{v}_A \perp \vec{\omega}$. The point I at which this axis intersects the plane of motion of A is called the *instantaneous centre* and it is obtained using $\vec{v}_I = \vec{v}_A + \vec{\omega} \times \vec{r}_{IA} = 0 \Rightarrow$

$$\vec{\omega} \times \vec{v}_A = -\vec{\omega} \times (\vec{\omega} \times \vec{r}_{IA}) = \omega^2 \vec{r}_{IA} \qquad \Rightarrow \qquad \vec{r}_{IA} = \frac{\vec{\omega} \times \vec{v}_A}{\omega^2}$$

For instantaneous centre I: $\vec{v}_I = \vec{0}$, but in general $\vec{a}_I \neq \vec{0}$ since, in general, a different point of the body is the instantaneous centre at different instants.

The velocity of points A, B, C, whose plane of motion is through I, can be expressed in terms of their radial distances IA, IB, IC from I (Fig. 1.32):

$$\vec{v}_A = \vec{v}_I + \vec{\omega} \times \overrightarrow{IA} = \omega \vec{e}_z \times (IA)\vec{e}_r = \omega(IA)\vec{e}_\phi \quad \Rightarrow$$

$$\frac{|\vec{v}_A|}{IA} = \frac{|\vec{v}_B|}{IB} = \frac{|\vec{v}_C|}{IC} = |\vec{\omega}| \tag{1.72}$$

$|\vec{v}_A|$ of a point A equals the product of its radial distance IA from I and $|\vec{\omega}|$. $\vec{v}_A$ is directed along normal $\vec{e}_\phi$ to $\overrightarrow{IA}$ obtained by rotating IA about I in the sense of $\vec{\omega}$. Equation (1.72) $\Rightarrow$ if $\vec{\omega} = \vec{0}$ then $IA = \infty$, i.e., I is at infinity. For plane translation, I remains at infinity since $|\vec{\omega}| \equiv 0$.

For a rigid body in plane motion, if the directions of velocities of its two points A and B, having the same plane of motion, are known at a given instant, then the instantaneous centre I can be located as follows.

1. If $\vec{v}_A \nparallel \vec{v}_B$, then I is at the intersection of normals to $\vec{v}_A$, $\vec{v}_B$ at A, B (Fig. 1.33a).
2. If $\vec{v}_A \parallel \vec{v}_B$ and $AB \not\perp \vec{v}_A$, then I is at infinity and $\vec{v}_A = \vec{v}_B$ (Fig. 1.33b).
3. If $\vec{v}_A \parallel \vec{v}_B$ and $AB \perp \vec{v}_A$, then I is on $AB, IA/IB = |\vec{v}_A| \,/\, |\vec{v}_B|$ and it follows from similar triangles (Fig. 1.33c) that I lies at the intersection of AB with the line joining the tips of $\vec{v}_A$ and $\vec{v}_B$ drawn at A and B. In particular, for $\vec{v}_A = \vec{v}_B$, I is at infinity (Fig. 1.33d).

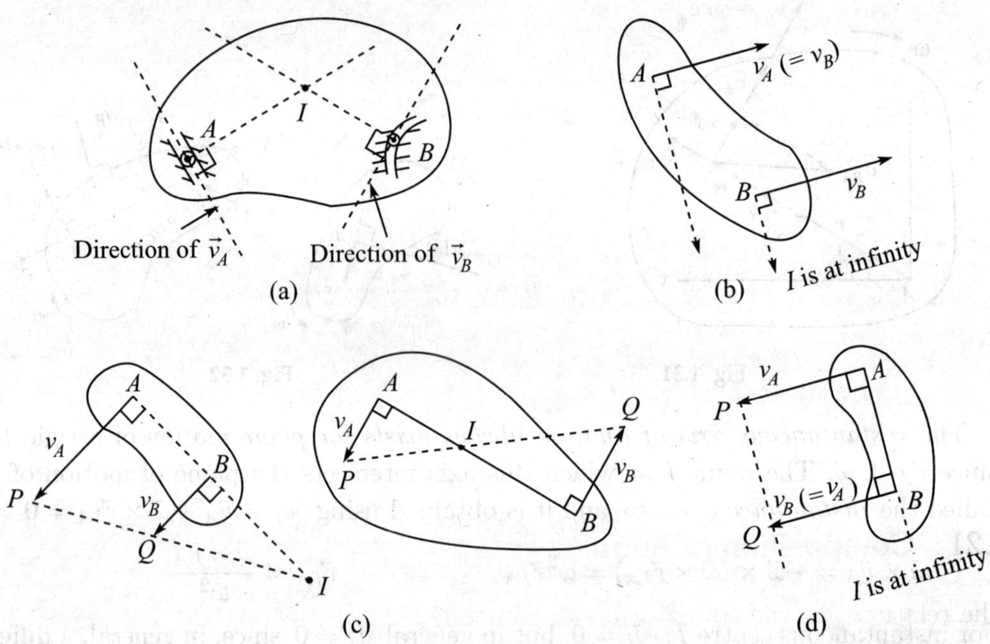

Fig. 1.33

The instantaneous centres I_k of bodies k are shown in Fig. 1.34. If I can be located easily, then it is sometimes useful for finding the velocities, but not the accelerations.

§ **Exercise 1.13:** Prove that for plane motion of a rigid body (excluding translation), at any given instant, a material point D called the acceleration centre, exists in the plane of motion of A for which $\vec{a}_D = \vec{a}_A + \dot{\vec{\omega}} \times \vec{r}_{DA} - \omega^2 \vec{r}_{DA} = \vec{0}$. Prove the existence by showing explicitly, $\vec{r}_{DA} = [\omega^2 \vec{a}_A + \dot{\vec{\omega}} \times \vec{a}_A] \,/\, (\omega^4 + \dot{\omega}^2)$ using $\dot{\vec{\omega}} \times \vec{a}_D$.

The acceleration centre D differs from I, except for rotation about a fixed axis. For a cylinder rolling without slip (Fig. 1.34) with uniform ω, the instantaneous centre I_1 is at the point of contact and the acceleration centre D is at the centre C.

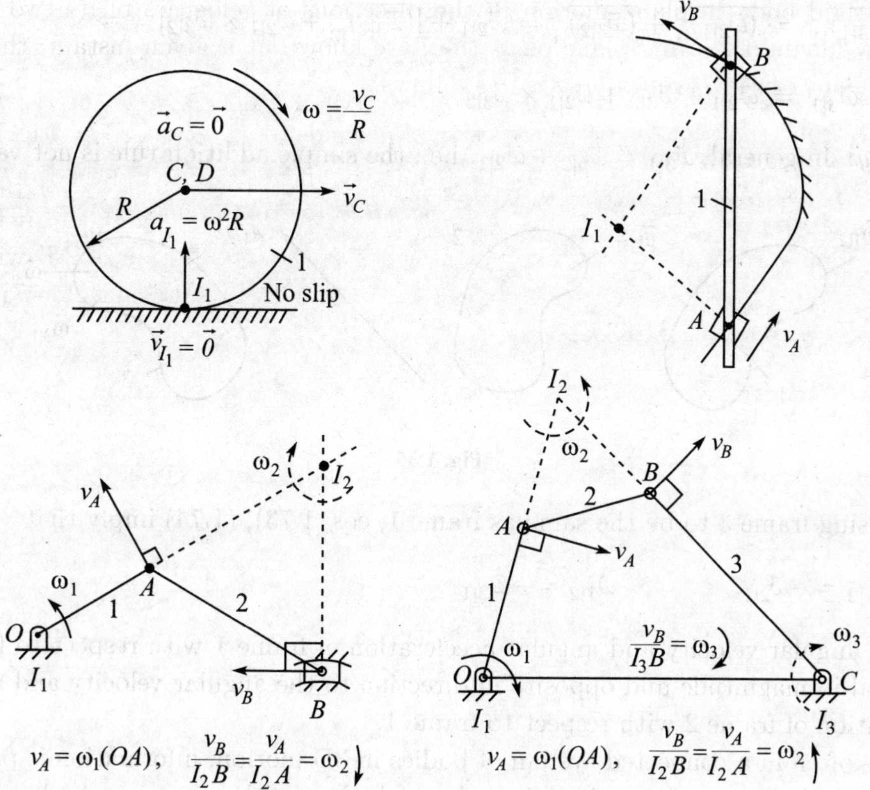

Fig. 1.34

1.21 Composition of Angular Velocities and Angular Accelerations

The relations for the composition of angular velocities and angular accelerations are useful for interconnected rigid bodies *executing spatial motion* in which, the rotational motion of a body is described successively relative to other bodies in the system. We prove the simple addition relation for *composition of angular velocities* using eq. (1.50). Let $\vec{\omega}_{3|1}$, $\vec{\omega}_{3|2}$, $\vec{\omega}_{2|1}$ be the angular velocities of frame 3 relative to frame 1, frame 3 relative to frame 2, frame 2 relative to frame 1, respectively (Fig. 1.35). Therefore,

$$\dot{\vec{A}}_{|1} = \dot{\vec{A}}_{|3} + \vec{\omega}_{3|1} \times \vec{A} = \dot{\vec{A}}_{|2} + \vec{\omega}_{2|1} \times \vec{A}$$

$$= (\dot{\vec{A}}_{|3} + \vec{\omega}_{3|2} \times \vec{A}) + \vec{\omega}_{2|1} \times \vec{A}$$

$$= \dot{\vec{A}}_{|3} + (\vec{\omega}_{3|2} + \vec{\omega}_{2|1}) \times \vec{A}, \qquad \forall \vec{A} \Rightarrow$$

$$\vec{\omega}_{3|1} = \vec{\omega}_{3|2} + \vec{\omega}_{2|1}, \qquad \text{i.e.,} \quad \vec{\omega}_{3|1} = \vec{\omega}_{2|1} + \vec{\omega}_{3|2} \tag{1.73}$$

$$\overline{(\vec{\omega}_{3|1})_{|1}}^{\cdot} = \overline{(\vec{\omega}_{2|1})_{|1}}^{\cdot} + \overline{(\vec{\omega}_{3|2})_{|1}}^{\cdot} = \dot{\vec{\omega}}_{2|1} + [\overline{(\vec{\omega}_{3|2})_{|2}}^{\cdot} + \vec{\omega}_{2|1} \times \vec{\omega}_{3|2}] \quad \Rightarrow$$

$$\dot{\vec{\omega}}_{3|1} = \dot{\vec{\omega}}_{2|1} + \dot{\vec{\omega}}_{3|2} + \vec{\omega}_{2|1} \times \vec{\omega}_{3|2} \tag{1.74}$$

Warning: In general, $\dot{\vec{\omega}}_{3|1} \neq \dot{\vec{\omega}}_{3|2} + \dot{\vec{\omega}}_{2|1}$, i.e., the simple addition rule is not valid.

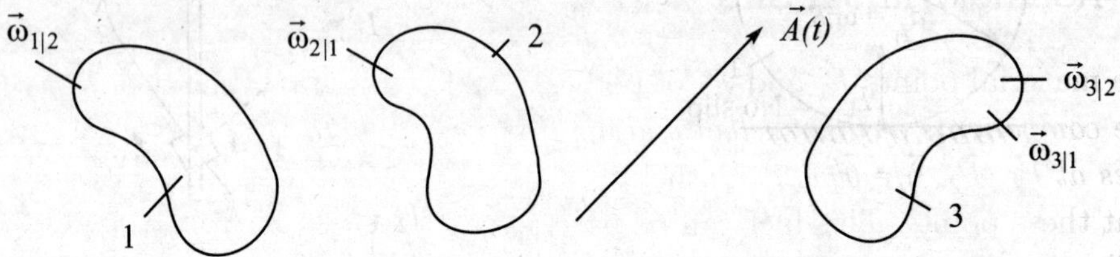

Fig. 1.35

Choosing frame 3 to be the same as frame 1, eqs (1.73), (1.74) imply that

$$\vec{\omega}_{1|2} = -\vec{\omega}_{2|1}, \qquad \dot{\vec{\omega}}_{1|2} = -\dot{\vec{\omega}}_{2|1} \tag{1.75}$$

i.e., the angular velocity and angular acceleration of frame 1 with respect to frame 2 are equal in magnitude and opposite in direction to the angular velocity and angular acceleration of frame 2 with respect to frame 1.

Quite often in a connected system of bodies in 3D motion, information is provided for the relative rates of rotation $\vec{\omega}_{i|j}$, $\dot{\vec{\omega}}_{i|j}$ of body i with respect to body j, about an axis fixed in body j. It is preferable to name the body with respect to which the information is sought as F. The body whose rotation is given with respect to F is called body 1, the body whose rotation is given with respect to body 1 is named body 2, and so on. Then,

$$\vec{\omega}_{2|F} = \vec{\omega}_{1|F} + \vec{\omega}_{2|1}, \qquad \dot{\vec{\omega}}_{2|F} = \dot{\vec{\omega}}_{1|F} + (\dot{\vec{\omega}}_{2|1} + \vec{\omega}_{1|F} \times \vec{\omega}_{2|1})$$

$$\vec{\omega}_{3|F} = \vec{\omega}_{2|F} + \vec{\omega}_{3|2}, \qquad \dot{\vec{\omega}}_{3|F} = \dot{\vec{\omega}}_{2|F} + (\dot{\vec{\omega}}_{3|2} + \vec{\omega}_{2|F} \times \vec{\omega}_{3|2})$$

$$\vec{\omega}_3 = \vec{\omega}_1 + \vec{\omega}_{2|1} + \vec{\omega}_{3|2}$$

$$\begin{aligned}\dot{\vec{\omega}}_3 &= \dot{\vec{\omega}}_1 + (\dot{\vec{\omega}}_{2|1} + \underline{\vec{\omega}_1} \times \vec{\omega}_{2|1}) + (\dot{\vec{\omega}}_{3|2} + \underline{\vec{\omega}_2} \times \vec{\omega}_{3|2}) \\ &= \dot{\vec{\omega}}_1 + (\dot{\vec{\omega}}_{2|1} + \underline{\vec{\omega}_1} \times \vec{\omega}_{2|1}) + [\dot{\vec{\omega}}_{3|2} + \underline{(\vec{\omega}_1 + \vec{\omega}_{2|1})} \times \vec{\omega}_{3|2}]\end{aligned}$$

where, for convenience, F has been dropped for the entities with respect to F. Study carefully the *underlined* first terms in the cross-products. The first terms in the cross-products in $\dot{\vec{\omega}}$ are the *angular velocities relative to the basic frame* F and **not relative to the previous body**. If the bodies are named as described and the angular velocities are added up starting with body 1, body 2 with respect to 1, body 3 with respect to 2, etc., then the first term, in the cross-product appearing in the derivative of the nth term, equals the sum of all the terms to the left of the nth term in the $\vec{\omega}$ expression. This procedure has been followed in all the examples.

1.22 Rolling Without Slip

Let the material points P_1 and P_2 of bodies 1 and 2 (Fig. 1.36a) be in contact at time t_1. The *components of their velocity along the normal $\vec{n}$, to the tangent plane π of the surfaces at P_1, P_2, are equal*, as otherwise, the bodies will separate or penetrate each other at these points. This fact can be used to find the velocity relation between cams and followers. However, if body 1 rolls/slides on body 2, in general, the components of $\vec{a}_{P_1}$ and $\vec{a}_{P_2}$ along $\vec{n}$ differ, since different points make contact at different instants.

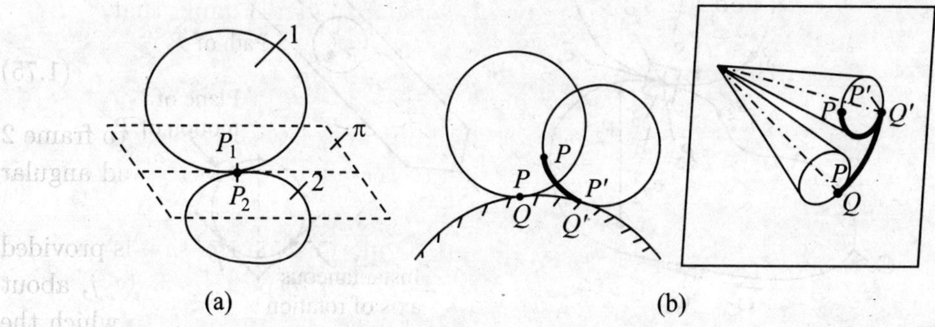

Fig. 1.36

For no slip between P_1 and P_2 in the tangent plane π, their velocity components in this plane must be equal $\Rightarrow \vec{v}_{P_1P_2|F}(t_1) = \vec{0}$, i.e., $\vec{v}_{P_1|F}(t_1) = \vec{v}_{P_2|F}(t_1)\ \forall$ frames F.

The three possible kinematic conditions at contact are defined as follows:

1. slip at t_1 if $\vec{v}_{P_1|F}(t_1) \neq \vec{v}_{P_2|F}(t_1)$,
2. no slip at t_1 if $\vec{v}_{P_1|F}(t_1) = \vec{v}_{P_2|F}(t_1)$ at t_1 and also subsequently $\vec{v}_{P_1^+|F}(t_1^+) = \vec{v}_{P_2^+|F}(t_1^+)$ for the corresponding different material points P_1^+, P_2^+ in contact at t_1^+. It does not imply that $\vec{a}_{P_1|F}(t_1) = \vec{a}_{P_2|F}(t_1)$.
3. impending slip at t_1 if $\vec{v}_{P_1|F}(t_1) = \vec{v}_{P_2|F}(t_1)$ at t_1 but this equality does not hold for the corresponding points P_1^+, P_2^+ in contact at t_1^+: $\vec{v}_{P_1^+|F}(t_1^+) \neq \vec{v}_{P_2^+|F}(t_1^+)$, where F is *any* reference frame. If velocities are equal or unequal for frame F, then by eq. (1.59) the same holds for other frames m with $\vec{\omega}_{m|F}$, as $\vec{r}_{P_1P_2} = \vec{0}$ at t_1.

§ **Exercise 1.14:** Prove that if $\overrightarrow{P_1P_2} = \vec{0}$, then $\vec{v}_{P_1P_2|F} = \vec{v}_{P_1P_2|m}$.

If body 2 is at rest, then $\vec{v}_{P_2|F}(t_1) = \vec{0}$ in the above definitions and for no slip, $\vec{v}_{P_1|F}(t_1) = \vec{0}$. Hence, if 1 is a rigid body, then P_1 is on its instantaneous axis of rotation and all additional points of contact of body 1 on 2 must lie on this axis parallel to $\vec{\omega}_1$.

An equivalent definition of rolling without slip is that the curves on the surfaces in rolling contact be tangential and the arcs traversed on them between successive points of contact be of equal length. For rolling without slip in Fig. 1.36b: $\widehat{QQ'} = \widehat{PP'}$.

If body 1 moves relative to body 2 in contact with it such that there is no slip at all points of contact, then the two bodies are said to be rolling without slip, e.g., wheel on a road, roller bearing in the bearing race, billiard ball on a table.

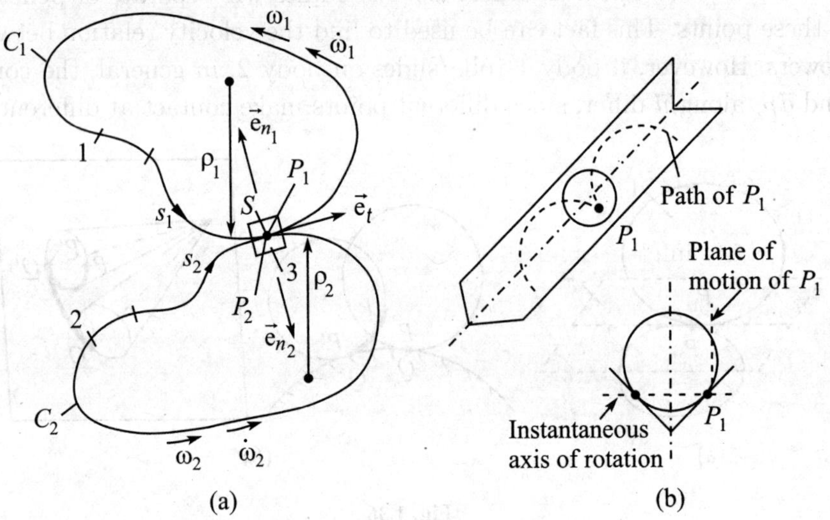

Fig. 1.37

We establish the equivalence of two definitions of rolling without slip, for the special case of *rolling without slip of a cylindrical rigid body 1 in relative plane motion on another cylindrical body 2* with the plane of motion normal to their generators (Fig. 1.37a). Relative to frame of body 2 (i.e., $\vec{\omega}_2 = \vec{0}$), let $\vec{\omega}_1 = \omega_1 \vec{k}$. The bounding curve C_1 of body 1 rolls over the bounding curve C_2 of 2. At the given instant a material point P_1 of body 1 makes contact with a material point P_2 of 2. Consider a common slider S which slides on C_1 and C_2 and which at the given instant is coincident with P_1 and P_2: $\overrightarrow{P_1S} = \overrightarrow{P_2S} = \vec{0}$. C_1 and C_2 are the trajectories of S relative to bodies 1 and 2 with path coordinates s_1, s_2. We find expressions of $\vec{v}_{P_1P_2}$, $\vec{a}_{P_1P_2}$:

$$\vec{v}_S = \vec{v}_{P_1} + \vec{\omega}_1 \times \overrightarrow{P_1S} + \vec{v}_{S|1} = \vec{v}_{P_2} + \vec{v}_{S|2} \Rightarrow$$

$$\vec{v}_{P_1P_2} = \vec{v}_{P_1} - \vec{v}_{P_2} = \vec{v}_{S|2} - \vec{v}_{S|1} = (\dot{s}_2 - \dot{s}_1)\,\vec{e}_t \tag{1}$$

$$\vec{a}_S = \vec{a}_{P_1} + \dot{\vec{\omega}}_1 \times \overrightarrow{P_1S} + \vec{\omega}_1 \times (\vec{\omega}_1 \times \overrightarrow{P_1S}) + 2\vec{\omega}_1 \times \vec{v}_{S|1} + \vec{a}_{S|1} = \vec{a}_{P_2} + \vec{a}_{S|2} \Rightarrow$$

$$\begin{aligned}\vec{a}_{P_1P_2} &= \vec{a}_{P_1} - \vec{a}_{P_2} = \vec{a}_{S|2} - \vec{a}_{S|1} - 2\omega_1 \dot{s}_1\, \vec{k} \times \vec{e}_t \\ &= (\ddot{s}_2 - \ddot{s}_1)\, \vec{e}_t - (\dot{s}_1^2/\rho_1 + \dot{s}_2^2/\rho_2)\, \vec{e}_{n_1} - 2\omega_1 \dot{s}_1\, \vec{k} \times \vec{e}_t \qquad (2)\end{aligned}$$

$$\Rightarrow \quad \text{its component in the tangent plane} \qquad \vec{a}_{P_1P_2} \cdot \vec{e}_t = \ddot{s}_2 - \ddot{s}_1 \qquad (3)$$

We consider three cases at the instant $t = t_1$.

1. There is slip if $\vec{v}_{P_1P_2}(t_1) \neq \vec{0}$ and eq. (1) $\Rightarrow$ $\dot{s}_2(t_1) \neq \dot{s}_1(t_1)$, i.e., $ds_2 \neq ds_1$ at t_1.
2. There is no slip if $\vec{v}_{P_1P_2}(t_1) = \vec{0}$ and also at $t = t_1^+$, eq. (1) $\Rightarrow$ $\dot{s}_2(t_1) = \dot{s}_1(t_1)$, and also at $t = t_1^+$, i.e., $ds_2 = ds_1$ at t_1 and $\ddot{s}_2(t_1) = \ddot{s}_1(t_1)$. Hence, *for the case of no slip*, eqs (3), (2) imply that $\vec{a}_{P_1P_2}$ *has no component in the tangential plane of contact* but can have a component normal to it.
3. There is impending slip if $\vec{v}_{P_1P_2}(t_1) = \vec{0}$ but is just violated at $t = t_1^+$, eq. (1) $\Rightarrow$ $\dot{s}_2(t_1) = \dot{s}_1(t_1)$, but is just violated at $t = t_1^+$, i.e., $|\dot{s}_2(t_1^+) - \dot{s}_1(t_1^+)| \neq 0$ $\Rightarrow$ $\ddot{s}_2(t_1) \neq \ddot{s}_1(t_1)$. Hence, *for impending slip*, eq. (3) implies that $\vec{a}_{P_1P_2}$ *has a non-zero component in the tangential plane of contact.*

Thus, the two definitions of rolling without slip are equivalent for this special case.

Note: The above inferences for the component of $\vec{a}_{P_1P_2}$ in the tangential plane do not hold for the case of only one point of each body making contact on the tangential plane. For example, if a sphere rolls without slip on a wedge-shaped channel (Fig. 1.37b), then the vertical plane of the cycloidal path of P_1 is not normal to the tangent plane. The vertical acceleration of P_1 has a component in the tangential plane of contact.

Example 1.16: A cylinder 1 rolls without slip on a fixed cylinder 2 (Fig. E1.16a). Find the velocity and acceleration of material points A, O, C, D on cylinder 1. Find the radii of curvature of the paths of C and D in this position.

Solution: Let point B of body 2 be in contact with point A of body 1. Choose $\vec{i}, \vec{j}$ in the plane of motion. Let v_O and $\dot{v}_O = a_O$ be the speed and rate of change of speed of point O. O moves in a circle of radius $\rho_O = R + r$. Hence,

$$\vec{v}_B = \vec{0}, \qquad \vec{a}_B = \vec{0}, \qquad \vec{\omega}_1 = \omega_1 \vec{k}, \qquad \dot{\vec{\omega}}_1 = \dot{\omega}_1 \vec{k}$$

$$\overrightarrow{OA} = -r\,\vec{i}, \qquad \overrightarrow{OD} = r\,\vec{i}, \qquad \overrightarrow{OC} = -r\,\vec{j}, \qquad \rho_O = R + r$$

We use velocity and acceleration relations of two points on the rigid body 1 taking the *geometric centre* as the reference point. Since the plane of motion of all points of interest is the same, $\vec{\omega} \times (\vec{\omega} \times \vec{r}) = -\omega^2 \vec{r}$. The kinematic condition of no slip, $\vec{v}_A = \vec{v}_B$:

$$\vec{v}_A = \vec{v}_O + \vec{\omega}_1 \times \overrightarrow{OA} = v_O\,\vec{j} + \omega_1 \vec{k} \times (-r\,\vec{i}) = (v_O - \omega_1 r)\,\vec{j} = \vec{v}_B = \vec{0} \Rightarrow$$

$$v_O = \omega_1 r \quad \forall\, t \qquad \Rightarrow \qquad a_O = \dot{v}_O = \dot{\omega}_1 r, \qquad \vec{v}_O = \omega_1 r\,\vec{j} \qquad (1)$$

$$\vec{a}_O = a_O\, \vec{e}_t + (v_0^2/\rho_O)\, \vec{e}_n = -[\omega_1^2 r^2 / (R + r)]\,\vec{i} + \dot{\omega}_1 r\,\vec{j} \qquad (2)$$

$$\vec{a}_A = \vec{a}_O + \dot{\vec{\omega}}_1 \times \overrightarrow{OA} - \omega_1^2 \overrightarrow{OA}$$

$$= \vec{a}_O + \dot{\omega}_1 \vec{k} \times (-r\vec{i}) - \omega_1^2(-r\vec{i}) = \frac{\omega_1^2}{[1/R + 1/r]}\vec{i} \quad (3)$$

Fig. E1.16

As expected, $\vec{a}_A$ is normal to the plane of contact but is independent of $\dot{\omega}_1$. Hence, it is more convenient to express *accelerations in terms of centre C rather than A.*

$$\vec{v}_D = \vec{v}_O + \vec{\omega}_1 \times \overrightarrow{OD} = \omega_1 r\vec{j} + \omega_1 \vec{k} \times r\vec{i} = 2\omega_1 r\vec{j}$$

$$\vec{v}_C = \vec{v}_O + \vec{\omega}_1 \times \overrightarrow{OC} = \omega_1 r\vec{j} + \omega_1 \vec{k} \times (-r\vec{j}) = \omega_1 r\vec{i} + \omega_1 r\vec{j}$$

$$\vec{a}_D = \vec{a}_O + \dot{\vec{\omega}}_1 \times \overrightarrow{OD} - \omega_1^2 \overrightarrow{OD} = \vec{a}_O + \dot{\omega}_1 \vec{k} \times r\vec{i} - \omega_1^2 r\vec{i}$$

$$= -[\omega_1^2 r(R+2r)/(R+r)]\vec{i} + 2\dot{\omega}_1 r\vec{j}$$

$$\vec{a}_C = \vec{a}_O + \dot{\vec{\omega}}_1 \times \overrightarrow{OC} - \omega_1^2 \overrightarrow{OC} = \vec{a}_O + \dot{\omega}_1 \vec{k} \times (-r\vec{j}) - \omega_1^2(-r\vec{j})$$

$$= [\dot{\omega}_1 r - \omega_1^2 r^2/(R+r)]\vec{i} + (\dot{\omega}_1 r + \omega_1^2 r)\vec{j}$$

$$\Rightarrow \rho_D = \frac{v_D^3}{|\vec{v}_D \times \vec{a}_D|} = \frac{8\omega_1^3 r^3}{|2\omega_1^3 r^2 (R+2r)\vec{k}/(R+r)|} = \frac{4r(R+r)}{R+2r} \quad (4)$$

$$\rho_C = \frac{v_C^3}{|\vec{v}_C \times \vec{a}_C|} = \frac{2\sqrt{2}\omega_1^3 r^3}{|\omega_1^3 r^2 (R+2r)\vec{k}/(R+r)|} = \frac{2\sqrt{2}r(R+r)}{R+2r} \quad (5)$$

For **impending slip**: $\vec{v}_A = \vec{v}_B$, *and the components of* $\vec{a}_A$ *and* $\vec{a}_B$ *in the tangential plane of contact are not equal at that instant*, i.e., $v_O = \omega_1 r$ and $a_O \neq \dot{\omega}_1 r$.

We now consider some particular cases of rolling without slip.

1. Cylinder 1 on a fixed flat surface (Fig. E1.16b), $R = \infty$, eqs (1)–(5) $\Rightarrow$

$$\vec{v}_O = \omega_1 r\,\vec{\jmath}, \qquad \vec{a}_O = \dot{\omega}_1 r\,\vec{\jmath}, \qquad \vec{a}_A = \omega_1^2 r\,\vec{\imath}, \qquad \rho_D = 4r, \qquad \rho_C = 2\sqrt{2}r$$

2. Cylinder 1 inside a fixed hollow cylinder 2 of radius R_1 (Fig. 1.16c), $R = -R_1$ as centres of 2 and 1 are on the same side of the tangent plane, eqs (1)–(3) $\Rightarrow$

$$\vec{v}_O = \omega_1 r\,\vec{\jmath}, \qquad \vec{a}_O = \frac{\omega_1^2 r^2}{R_1 - r}\,\vec{\imath} + \dot{\omega}_1 r\,\vec{\jmath}, \qquad \vec{a}_A = \frac{\omega_1^2}{1/r - 1/R_1}\,\vec{\imath}$$

3. Body 1 rolls without slip on a fixed cylinder 2 (Fig. E1.16d), $r = \infty$ and eq. (3) yields $\vec{a}_A = \omega_1^2 R\,\vec{\imath}$. [*Prove this result abinitio without using* eq. (3)!] For D,

$$\begin{aligned}\vec{v}_D &= \vec{v}_A + \vec{\omega}_1 \times \overrightarrow{AD} = \omega_1\,\vec{k} \times d\,\vec{\imath} = \omega_1 d\,\vec{\jmath}\\ \vec{a}_D &= \vec{a}_A + \dot{\vec{\omega}}_1 \times \overrightarrow{AD} - \omega_1^2\,\overrightarrow{AD} = \omega_1^2 (R - d)\,\vec{\imath} + \dot{\omega}_1 d\,\vec{\jmath}\end{aligned}$$

4. Cylinder 1 turns around a corner at A (Fig. E1.16e), $R = 0$ and eqs (1)–(5) $\Rightarrow$

$$\vec{v}_O = \omega_1 r\,\vec{\jmath}, \qquad \vec{a}_O = -\omega_1^2 r\,\vec{\imath} + \dot{\omega}_1 r\,\vec{\jmath}, \qquad \vec{a}_A = \vec{0}, \qquad \rho_D = 2r, \qquad \rho_C = \sqrt{2}r$$

5. Ring 1 on fixed cylinder 2 (Fig. E1.16f), $r = -r_1$ and eqs (1), (2) and (3) $\Rightarrow$

$$\vec{v}_O = -\omega_1 r_1\,\vec{\jmath}, \qquad \vec{a}_O = \frac{\omega_1^2 r_1^2}{r_1 - R}\,\vec{\imath} - \dot{\omega}_1 r_1\,\vec{\jmath}, \qquad \vec{a}_A = \frac{\omega_1^2}{1/R - 1/r_1}\,\vec{\imath}$$

If a rigid cylinder 1 rolls without slip on a moving rigid cylinder 2, then ω_1, $\dot{\omega}_1$, $\vec{v}$, $\vec{a}$ in this example would be relative to the reference frame of body 2.

1.23 Gear, Belt and Chain Drives

The no slip condition for belt and chain drives, and at the pitch circles of gears in mesh, takes a very *simple form if it is written with respect to a frame in which the material points* of the gears, pulleys and sprockets have *circular motion.* Consider the five pairs of gears rotating about fixed axes in Fig. 1.38a with the gear number i having angular velocity ω_i relative to the ground as shown. The no slip condition yields:

$$\omega_1 R_1 = -\omega_2 R_2, \qquad \omega_3 R_3 = \omega_4 R_4 \tag{1.76}$$

The radii can be replaced by the number of teeth on the gears. If the gears were *mounted on the same rotating body* m, then eq. (1.76) applies with ω_i being the angular velocity of the ith gear relative to body m. For example, in Fig. 1.38b, bent shaft 1 and bevel gear 2 are rotating at ω_1 and ω_2 with respect to the ground. Gear 3 is rotating at angular velocity Ω relative to body 1. The axes of the gears 2 and 3 are on body 1 and the contact points A and B have circular motion relative to body 1. Let $\vec{v}_{A|1}$, $\vec{v}_{B|1}$ be in the direction inward to the plane of the figure. The condition of no slip yields:

$$\vec{v}_{A|1} = \vec{v}_{B|1} \quad \Rightarrow \quad (\omega_2 - \omega_1)R_2 = \Omega R_3 \tag{1}$$

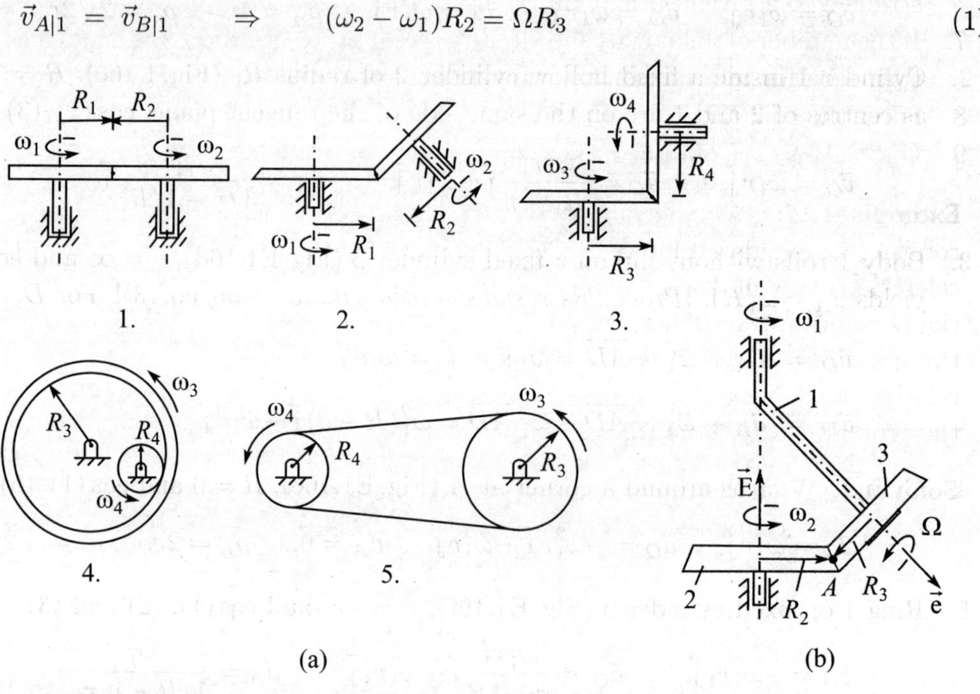

Fig. 1.38

1.24 Examples Involving Moving Frames

The following steps are used to apply eqs (1.61) and (1.62) to solutions of problems.

1. Draw a sketch showing the axes of rotation, the angular motions and mark constrained points and common points of connected rigid bodies. Choose $\vec{e}_i$ such that the position vectors and angular velocities can be expressed conveniently.
2. List the given data and the unknowns. Choose frames, attached to rigid bodies or otherwise, so that $\vec{v}$, $\vec{a}$ of some point(s) relative to them can be expressed easily.
3. Write the kinematic restrictions on all constrained points.
4. For 3D problems, compute successively $\vec{\omega}$, $\dot{\vec{\omega}}$ of rigid bodies using $\dot{\vec{A}}_{|F} = \dot{\vec{A}}_{|m} + \vec{\omega}_{m|F} \times \vec{A}$. Express these and the relative position vectors in terms of $\vec{e}_i$.
5. Succesively obtain $\vec{v}$, $\vec{a}$ of points on the axes of rotation of 2 relative to 1, 3 relative to 2, etc. If P is a point of body i or its motion relative to it is known,

then in terms of a point Q of body i, whose $\vec{v}_Q$, $\vec{a}_Q$ have already been found:
$\vec{v}_P = \vec{v}_Q + \vec{\omega}_i \times \overrightarrow{QP} + \vec{v}_{P|i}$, $\vec{a}_P = \vec{a}_Q + \dot{\vec{\omega}}_i \times \overrightarrow{QP} + \vec{\omega}_i \times (\vec{\omega}_i \times \overrightarrow{QP}) + 2\vec{\omega}_i \times \vec{v}_{P|i} + \vec{a}_{P|i}$
with $\vec{v}_{P|i} = \dot{s}_{P|i}\,\vec{e}_t$, $\vec{a}_{P|i} = \ddot{s}_{P|i}\,\vec{e}_t + (\dot{s}^2_{P|i}\,/\,\rho_P)\,\vec{e}_n$. The last term in $\vec{v}_P$ and the last *two* terms in $\vec{a}_P$ are included only if P has motion relative to body i.

6. Write in two ways $\vec{v}$, $\vec{a}$ of a fixed point/common point of two bodies/point constrained to a given path/point of contact of bodies that roll without slip.
7. If the number of unknowns and the number of scalar equations set up are unequal, then look for information/relations not used.
8. Solve the equations.
9. Once $\vec{\omega}_i$, $\dot{\vec{\omega}}_i$ are obtained, the required $\vec{v}$, $\vec{a}$ are obtained using eqs (1.61), (1.62).

Example 1.17: A vane pump (Fig. E1.17), is rotating at angular velocity ω and angular acceleration $\dot{\omega}$ about a fixed vertical axis. A material point P of water flows out along a vane with the speed of v and rate of increase of speed of $\dot{v}$ relative to the vane. Find $\vec{v}_P$ and $\vec{a}_P$ relative to the ground. Neglect the thickness of the vanes.

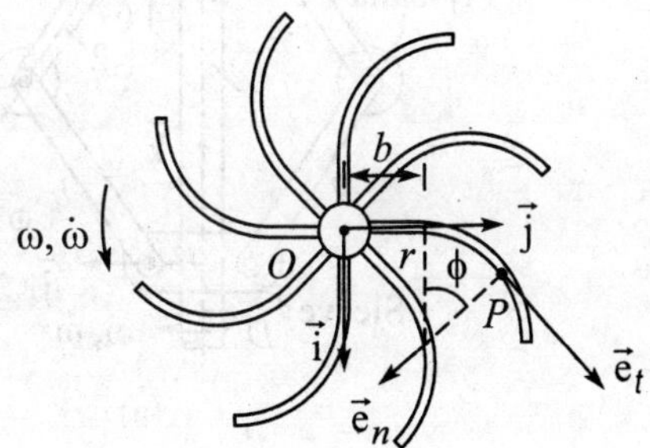

Fig. E1.17

Solution: With the choice of axes (Fig. E1.17),

$$\vec{\omega}_1 = \omega\,\vec{k}, \quad \dot{\vec{\omega}}_1 = \dot{\omega}\,\vec{k}, \quad \vec{v}_O = \vec{0}, \quad \vec{a}_O = \vec{0}$$

$$\overrightarrow{OP} = r(1-\cos\phi)\,\vec{i} + (b + r\sin\phi)\,\vec{j}$$

$$\vec{e}_t = \sin\phi\,\vec{i} + \cos\phi\,\vec{j}, \quad \vec{e}_n = \cos\phi\,\vec{i} - \sin\phi\,\vec{j}$$

The path of P relative to vane 1 is a circle of radius r. Hence,

$$\vec{v}_{P|1} = v\,\vec{e}_t, \quad \vec{a}_{P|1} = \dot{v}\,\vec{e}_t + (v^2/r)\,\vec{e}_n$$

$$\vec{v}_P = \vec{v}_O + \vec{\omega}_1 \times \overrightarrow{OP} + \vec{v}_{P|1}$$

$$\vec{a}_P = \vec{a}_O + \dot{\vec{\omega}}_1 \times \overrightarrow{OP} + \vec{\omega}_1 \times (\vec{\omega}_1 \times \overrightarrow{OP}) + 2\vec{\omega}_1 \times \vec{v}_{P|1} + \vec{a}_{P|1} \Rightarrow$$

$$\vec{v}_P = [-\omega(b + r\sin\phi) + v\sin\phi]\,\vec{i} + [\omega r(1-\cos\phi) + v\cos\phi]\,\vec{j}$$

$$\begin{aligned}\vec{a}_P = {} & [-\dot{\omega}(b + r\sin\phi) - \omega^2 r(1-\cos\phi) - 2\omega v\cos\phi + \dot{v}\sin\phi + (v^2/r)\cos\phi]\,\vec{i} \\ & + [\dot{\omega} r(1-\cos\phi) - \omega^2(b + r\sin\phi) + 2\omega v\sin\phi + \dot{v}\cos\phi - (v^2/r)\sin\phi]\,\vec{j}\end{aligned}$$

Example 1.18: (a) A *fly-ball governor* (Fig. E1.18a) is used to control the speed of a body rotating about a vertical axis at ω, $\dot{\omega}$. The displacement of the sleeve controls

ω by a connection to a feedback device. Find the angular acceleration of bar AB and $\vec{v}_A$, $\vec{v}_B$ at $t = \pi/4p$, if $\theta = \theta_0 + \theta_1 \sin pt$. (b) A helicopter turns in a horizontal circle (Fig. E1.18b), with O moving in a circle of radius R at speed v, which is increasing at the rate of $\dot{v}$ at $t = \pi/4p$. The helicopter blades oscillate with $\theta = \theta_0 + \theta_1 \sin pt$. Find $\vec{v}_{BO}$, $\vec{a}_{BO}$ at $t = \pi/4p$. (c) A spacecraft is revolving about its fixed axis OD, at a rate ω (Fig. E1.18c). Its solar panels are opening at a programmable rate given in the graph. Find $\vec{v}_B$, $\vec{a}_B$ and $\dot{\vec{\omega}}_2$ just before and just after θ equals θ_0.

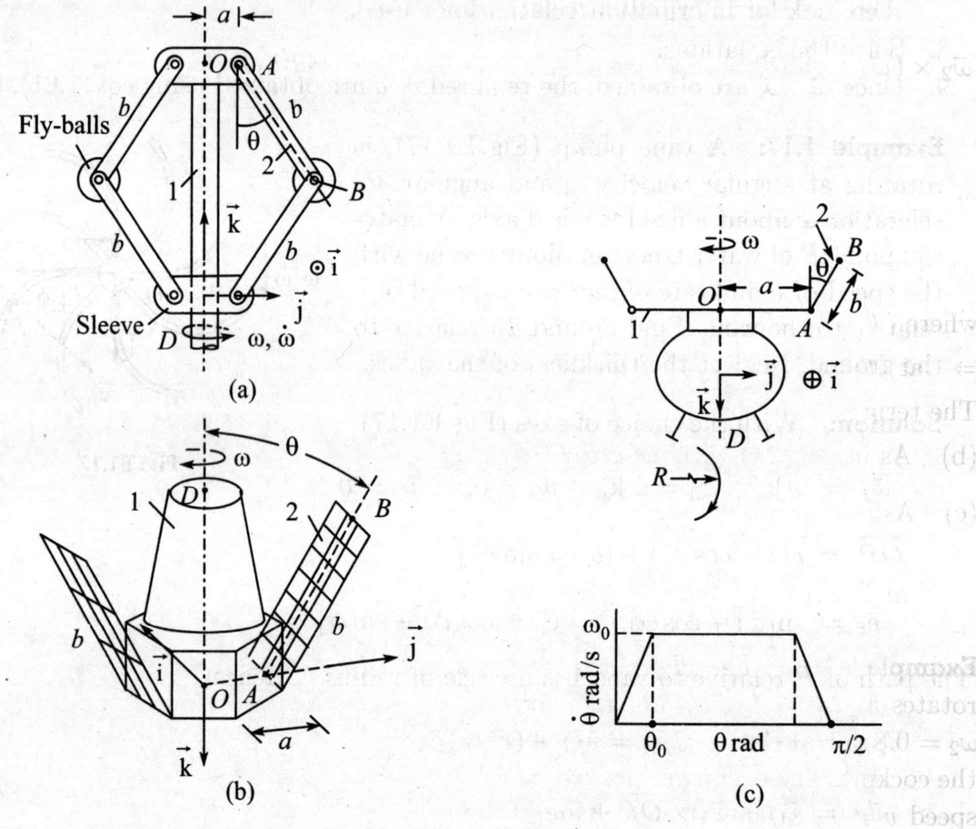

Fig. E1.18

Solution: (a) The three physical problems are kinematically the same. Let $\vec{\mathrm{i}}, \vec{\mathrm{j}}, \vec{\mathrm{k}}$ be fixed to body 1. Body 1 rotates with respect to the ground about the fixed axis $\vec{\mathrm{k}}$. Body 2 rotates with respect to body 1 about axis $\vec{\mathrm{i}}$ fixed to 1.

$$\vec{v}_O = \vec{0}, \qquad \vec{a}_O = \vec{0}, \qquad \overrightarrow{OA} = a\,\vec{\mathrm{j}}, \qquad \overrightarrow{AB} = b(\sin\theta\,\vec{\mathrm{j}} - \cos\theta\,\vec{\mathrm{k}})$$

$$\vec{\omega}_1 = \omega\,\vec{\mathrm{k}}, \qquad \dot{\vec{\omega}}_1 = \dot{\omega}\,\vec{\mathrm{k}}$$

$$\vec{\omega}_2 = \vec{\omega}_1 + \dot{\theta}\,\vec{\mathrm{i}} = \dot{\theta}\,\vec{\mathrm{i}} + \omega\,\vec{\mathrm{k}}$$

$$\dot{\vec{\omega}}_2 = \dot{\vec{\omega}}_1 + (\ddot{\theta}\vec{\mathrm{i}} + \vec{\omega}_1 \times \dot{\theta}\vec{\mathrm{i}}) = \dot{\omega}\vec{\mathrm{k}} + \ddot{\theta}\vec{\mathrm{i}} + \omega\vec{\mathrm{k}} \times \dot{\theta}\vec{\mathrm{i}} = \ddot{\theta}\vec{\mathrm{i}} + \omega\dot{\theta}\vec{\mathrm{j}} + \dot{\omega}\vec{\mathrm{k}}$$

Points O, A are on body 1 and points A, B are on body 2. We successively compute

$$\vec{v}_A = \vec{v}_O + \vec{\omega}_1 \times \overrightarrow{OA}, \qquad \vec{a}_A = \vec{a}_O + \dot{\vec{\omega}}_1 \times \overrightarrow{OA} + \vec{\omega}_1 \times (\vec{\omega}_1 \times \overrightarrow{OA})$$

$$\vec{v}_B = \vec{v}_A + \vec{\omega}_2 \times \overrightarrow{AB}, \qquad \vec{a}_B = \vec{a}_A + \dot{\vec{\omega}}_2 \times \overrightarrow{AB} + \vec{\omega}_2 \times (\vec{\omega}_2 \times \overrightarrow{AB})$$

$$\vec{\omega}_2 \times \overrightarrow{AB} = -\omega b \sin\theta\,\vec{\mathrm{i}} + \dot{\theta} b \cos\theta\,\vec{\mathrm{j}} + \dot{\theta} b \sin\theta\,\vec{\mathrm{k}} \Rightarrow$$

$$\vec{\omega}_2 \times (\vec{\omega}_2 \times \overrightarrow{AB}) = -\omega\dot{\theta} b \cos\theta\,\vec{\mathrm{i}} - (\dot{\theta}^2 + \omega^2) b \sin\theta\,\vec{\mathrm{j}} + \dot{\theta}^2 b \cos\theta\,\vec{\mathrm{k}}$$

$$\vec{v}_B = -(a + b\sin\theta)\omega\,\vec{\mathrm{i}} + b\dot{\theta}(\cos\theta\,\vec{\mathrm{j}} + \sin\theta\,\vec{\mathrm{k}})$$

$$\vec{a}_B = -[2\omega\dot{\theta} b\cos\theta + \dot{\omega}(a + b\sin\theta)]\,\vec{\mathrm{i}} + (b\dot{\theta}^2\cos\theta + b\ddot{\theta}\sin\theta)\,\vec{\mathrm{k}}$$
$$+[-\omega^2(a + b\sin\theta) - b\dot{\theta}^2\sin\theta + b\ddot{\theta}\cos\theta]\,\vec{\mathrm{j}}$$

where $\theta = \theta_0 + \theta_1 \sin pt$, $\quad \dot{\theta} = p\theta_1 \cos pt$, $\quad \ddot{\theta} = -p^2\theta_1 \sin pt$.
$\Rightarrow$ at $t = \pi/4p$: $\quad \theta = \theta_0 + \theta_1/\sqrt{2}$, $\quad \dot{\theta} = p\theta_1/\sqrt{2}$, $\quad \ddot{\theta} = -p^2\theta_1/\sqrt{2}$.
The term $-2\omega b\dot{\theta}\cos\theta\,\vec{\mathrm{i}}$ in $\vec{a}_B$ is not an obvious one.

(b) As in part (a) with $\omega = v/R$, $\dot{\omega} = \dot{v}/R$ and $\vec{v}_B$, $\vec{a}_B$ replaced by $\vec{v}_{BO}$, $\vec{a}_{BO}$.

(c) As in part (a) with $\theta = \theta_0$, $\dot{\theta} = \omega_0$, and $\ddot{\theta} = \dfrac{d\dot{\theta}}{d\theta}\dfrac{d\theta}{dt} = \omega_0\dfrac{d\dot{\theta}}{d\theta}$ $\Rightarrow$ using graph:

$$\text{at } \theta = \theta_0^- : \quad \frac{d\dot{\theta}}{d\theta} = \frac{\omega_0}{\theta_0}, \quad \ddot{\theta} = \frac{\omega_0^2}{\theta_0}; \qquad \text{at } \theta = \theta_0^+ : \quad \frac{d\dot{\theta}}{d\theta} = 0, \quad \ddot{\theta} = 0$$

Example 1.19: The main arm (OB = 10 m) of a device used in amusement parks rotates at ω_1 = 1 rad/s about a fixed axis (Fig. E1.19). A cockpit on it rotates at $\omega_2 = 0.8$ rad/s relative to it. Point A is on the ear of an occupant strapped to a seat in the cockpit. The occupant observes that a light spot S moves in a parabolic path with speed $v = 0.4$ m/s as shown which is decreasing at 0.1 m/s². Find $\dot{\vec{\omega}}_2$, $\vec{v}_A$, $\vec{a}_A$, $\vec{v}_S$, $\vec{a}_S$ at the given instant when $\dot{\omega}_1 = 0.5$ rad/s², $\dot{\omega}_2 = -0.4$ rad/s², $\cos\theta = 0.6$

Solution: Let arm = body 1, cockpit = body 2, motion of S relative to 2 in terms of s,

$$\vec{e}_t = \vec{\mathrm{i}}, \qquad \vec{e}_n = -\vec{\mathrm{j}}, \qquad \dot{s} = 0.4\,\mathrm{m/s}, \qquad \ddot{s} = -0.1\,\mathrm{m/s^2}$$

$$y_1 = 2x_1^2 \qquad \Rightarrow \qquad y_1' = \frac{dy_1}{dx_1} = 4x_1, \qquad y_1'' = \frac{d^2y_1}{dx_1^2} = 4$$

$$\Rightarrow \text{ at } (0,0): \qquad y_1' = 0, \qquad \rho = \frac{(1 + y_1'^2)^{3/2}}{|y_1''|} = 0.25\,\mathrm{m}$$

$$\vec{v}_{S|2} = \dot{s}\,\vec{e}_t = 0.4\vec{i}\text{ m/s}, \qquad \vec{a}_{S|2} = \ddot{s}\,\vec{e}_t + (\dot{s}^2/\rho)\,\vec{e}_n = -0.1\vec{i} - 0.64\vec{j}\text{ m/s}^2$$

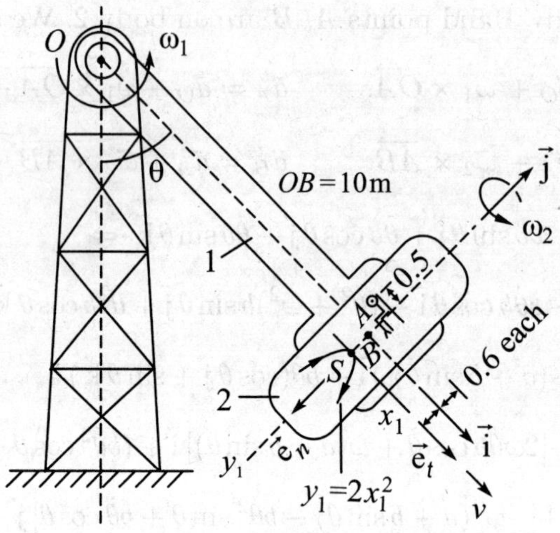

Fig. E1.19

We compose angular velocities and angular accelerations relative to the ground:

$$\vec{\omega}_1 = \vec{k}\text{ rad/s}, \qquad \dot{\vec{\omega}}_1 = 0.5\vec{k}\text{ rad/s}^2$$

$$\vec{\omega}_2 = \vec{\omega}_1 + 0.8\vec{j} = 0.8\vec{j} + \vec{k}\text{ rad/s}$$

$$\dot{\vec{\omega}}_2 = \dot{\vec{\omega}}_1 - 0.4\vec{j} + \vec{\omega}_1 \times 0.8\vec{j} = 0.5\vec{k} - 0.4\vec{j} + \vec{k} \times 0.8\vec{j}$$

$$= -0.8\vec{i} - 0.4\vec{j} + 0.5\vec{k}\text{ rad/s}^2$$

Points O, B are on body 1 with $\vec{v}_O = \vec{0}$, $\vec{a}_O = \vec{0}$, $\overrightarrow{OB} = 10\vec{i}$ m,

$$\vec{v}_B = \vec{v}_O + \vec{\omega}_1 \times \overrightarrow{OB} = \vec{k} \times 10\vec{i} = 10\vec{j}\text{ m/s}$$

$$\vec{a}_B = \vec{a}_O + \dot{\vec{\omega}}_1 \times \overrightarrow{OB} + \vec{\omega}_1 \times (\vec{\omega}_1 \times \overrightarrow{OB})$$

$$= 0.5\vec{k} \times 10\vec{i} + \vec{k} \times (\vec{k} \times 10\vec{i}) = -10\vec{i} + 5\vec{j}\text{ m/s}^2$$

Points B, A are on body 2 with $\overrightarrow{BA} = -0.5\vec{i} + 0.6\vec{j}$ m,

$$\vec{\omega}_2 \times \overrightarrow{BA} = (0.8\vec{j} + \vec{k}) \times (-0.5\vec{i} + 0.6\vec{j}) = -0.6\vec{i} - 0.5\vec{j} + 0.4\vec{k}\text{ m/s}$$

$$\dot{\vec{\omega}}_2 \times \overrightarrow{BA} = (-0.8\vec{i} - 0.4\vec{j} + 0.5\vec{k}) \times (-0.5\vec{i} + 0.6\vec{j})$$

$$= -0.3\vec{i} - 0.25\vec{j} - 0.68\vec{k}\text{ m/s}^2$$

$$\vec{\omega}_2 \times (\vec{\omega}_2 \times \overrightarrow{BA}) = (0.8\vec{j} + \vec{k}) \times (-0.6\vec{i} - 0.5\vec{j} + 0.4\vec{k})$$

$$= 0.82\vec{i} - 0.6\vec{j} + 0.48\vec{k} \text{ m/s}^2$$

$$\vec{v}_A = \vec{v}_B + \vec{\omega}_2 \times \overrightarrow{BA} = -0.6\vec{i} + 9.5\vec{j} + 0.4\vec{k} \text{ m/s}$$

$$\vec{a}_A = \vec{a}_B + \dot{\vec{\omega}}_2 \times \overrightarrow{BA} + \vec{\omega}_2 \times (\vec{\omega}_2 \times \overrightarrow{BA})$$

$$= -9.48\vec{i} + 4.15\vec{j} - 0.2\vec{k} \text{ m/s}^2$$

S moves relative to 2 $\Rightarrow$ we express $\vec{v}_S$, $\vec{a}_S$ in terms of point B of 2: $\overrightarrow{BS} = -0.6\vec{j}$ m,

$$\vec{\omega}_2 \times \overrightarrow{BS} = (0.8\vec{j} + \vec{k}) \times (-0.6\vec{j}) = 0.6\vec{i} \text{ m/s}$$

$$\dot{\vec{\omega}}_2 \times \overrightarrow{BS} = (-0.8\vec{i} - 0.4\vec{j} + 0.5\vec{k}) \times (-0.6\vec{j}) = 0.3\vec{i} + 0.48\vec{k} \text{ m/s}^2$$

$$\vec{\omega}_2 \times (\vec{\omega}_2 \times \overrightarrow{BS}) = (0.8\vec{j} + \vec{k}) \times (0.6\vec{i}) = 0.6\vec{j} - 0.48\vec{k} \text{ m/s}^2$$

$$2\vec{\omega}_2 \times \vec{v}_{S|2} = 2(0.8\vec{j} + \vec{k}) \times 0.4\vec{i} = 0.8\vec{j} - 0.64\vec{k} \text{ m/s}^2$$

$$\vec{v}_S = \vec{v}_B + \vec{\omega}_2 \times \overrightarrow{BS} + \vec{v}_{S|2} = \vec{i} + 10\vec{j} \text{ m/s}$$

$$\vec{a}_S = \vec{a}_B + \dot{\vec{\omega}}_2 \times \overrightarrow{BS} + \vec{\omega}_2 \times (\vec{\omega}_2 \times \overrightarrow{BS}) + 2\vec{\omega}_2 \times \vec{v}_{S|2} + \vec{a}_{S|2}$$

$$= -9.8\vec{i} + 5.76\vec{j} - 0.64\vec{k} \text{ m/s}^2$$

Example 1.20: A box rotates about a fixed axis and a yoked arm rotates relative to it (Fig. E1.20). A disc rotates relative to the arm. Points P and Q on sliders moving in straight and circular slots ($\cos\theta = 0.8$) have $v = 2$ cm/s decreasing at 1 cm/s^2 relative to the disc. Find velocity and acceleration of points P, Q and a point A of the disc.

Solution: Let $\vec{i}$, $\vec{j}$, $\vec{k}$ be fixed to the box. We successively name the box 1 rotating about a fixed axis $\vec{k}$, the yoked arm 2 with its axis of rotation $\vec{i}$ relative to box 1 being fixed to box 1, and the disc 3 with its axis of rotation $\vec{e}$ relative to the arm 2 being fixed to the arm 2. At the given instant, $\vec{e} = \vec{j}$. We successively compose the angular velocities and angular accelerations of bodies 1, 2, 3.

$$\vec{\omega}_1 = 5\vec{k} \text{ rad/s}, \qquad \dot{\vec{\omega}}_1 = -\vec{k} \text{ rad/s}^2, \qquad \vec{\omega}_2 = \vec{\omega}_1 - 4\vec{i} = -4\vec{i} + 5\vec{k} \text{ rad/s}$$

$$\dot{\vec{\omega}}_2 = \dot{\vec{\omega}}_1 - 2\vec{i} + \vec{\omega}_1 \times (-4\vec{i}) = -\vec{k} - 2\vec{i} + 5\vec{k} \times (-4\vec{i})$$

$$= -2\vec{i} - 20\vec{j} - \vec{k} \text{ rad/s}^2$$

$$\vec{\omega}_3 = \vec{\omega}_2 - 3\vec{j} = -4\vec{i} - 3\vec{j} + 5\vec{k} \text{ rad/s}$$

$$\begin{aligned}\dot{\vec{\omega}}_3 &= \dot{\vec{\omega}}_2 + \vec{j} + \vec{\omega}_2 \times (-3\vec{j}) \\ &= -2\vec{i} - 20\vec{j} - \vec{k} + \vec{j} + (-4\vec{i} + 5\vec{k}) \times (-3\vec{j}) \\ &= 13\vec{i} - 19\vec{j} + 11\vec{k} \text{ rad/s}^2\end{aligned}$$

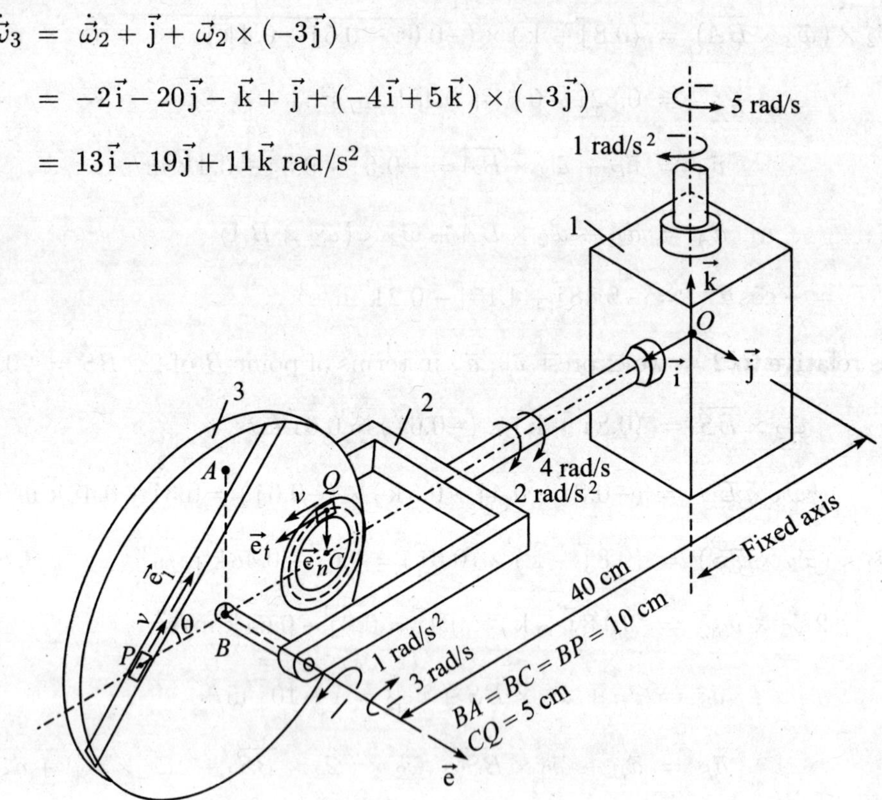

Fig. E1.20

Points O, B are on body 2 with $\vec{v}_O = \vec{0}$, $\vec{a}_O = \vec{0}$, $\overrightarrow{OB} = 40\vec{i}$ cm,

$$\begin{aligned}\vec{v}_B &= \vec{v}_O + \vec{\omega}_2 \times \overrightarrow{OB} = (-4\vec{i} + 5\vec{k}) \times 40\vec{i} = 200\vec{j} \text{ cm/s} \\ \vec{a}_B &= \vec{a}_O + \dot{\vec{\omega}}_2 \times \overrightarrow{OB} + \vec{\omega}_2 \times (\vec{\omega}_2 \times \overrightarrow{OB}) \\ &= (-2\vec{i} - 20\vec{j} - \vec{k}) \times 40\vec{i} + (-4\vec{i} + 5\vec{k}) \times 200\vec{j} \\ &= -1000\vec{i} - 40\vec{j} \text{ cm/s}^2\end{aligned}$$

Points B, A are on body 3 with $\overrightarrow{BA} = 10\vec{k}$ cm,

$$\begin{aligned}\vec{\omega}_3 \times \overrightarrow{BA} &= (-4\vec{i} - 3\vec{j} + 5\vec{k}) \times 10\vec{k} = -30\vec{i} + 40\vec{j} \text{ cm/s} \\ \dot{\vec{\omega}}_3 \times \overrightarrow{BA} &= (13\vec{i} - 19\vec{j} + 11\vec{k}) \times 10\vec{k} = -190\vec{i} - 130\vec{j} \text{ cm/s}^2 \\ \vec{\omega}_3 \times (\vec{\omega}_3 \times \overrightarrow{BA}) &= (-4\vec{i} - 3\vec{j} + 5\vec{k}) \times (-30\vec{i} + 40\vec{j}) \\ &= -200\vec{i} - 150\vec{j} - 250\vec{k} \text{ cm/s}^2\end{aligned}$$

$$\begin{aligned}
\vec{v}_A &= \vec{v}_B + \vec{\omega}_3 \times \overrightarrow{BA} = -30\,\vec{\imath} + 240\,\vec{\jmath}\ \text{cm/s} \\
\vec{a}_A &= \vec{a}_B + \dot{\vec{\omega}}_3 \times \overrightarrow{BA} + \vec{\omega}_3 \times (\vec{\omega}_3 \times \overrightarrow{BA}) \\
&= -1390\,\vec{\imath} - 320\,\vec{\jmath} - 250\,\vec{k}\ \text{cm/s}^2
\end{aligned}$$

P moves *relative to 3* $\Rightarrow$ we express $\vec{v}_P$, $\vec{a}_P$ in terms of point B of 3: $\overrightarrow{BP} = 10\,\vec{\imath}$ cm,

$$\begin{aligned}
\vec{e}_1 &= -\cos\theta\,\vec{\imath} + \sin\theta\,\vec{k} = -0.8\,\vec{\imath} + 0.6\,\vec{k} \\
\vec{v}_{P|3} = 2\,\vec{e}_1 &= -1.6\,\vec{\imath} + 1.2\,\vec{k}\ \text{cm/s} \\
\vec{a}_{P|3} = -\,\vec{e}_1 &= 0.8\,\vec{\imath} - 0.6\,\vec{k}\ \text{cm/s}^2 \\
\vec{\omega}_3 \times \overrightarrow{BP} &= 50\,\vec{\jmath} + 30\,\vec{k}\ \text{cm/s} \\
\dot{\vec{\omega}}_3 \times \overrightarrow{BP} &= 110\,\vec{\jmath} + 190\,\vec{k}\ \text{cm/s}^2 \\
\vec{\omega}_3 \times (\vec{\omega}_3 \times \overrightarrow{BP}) &= -340\,\vec{\imath} + 120\,\vec{\jmath} - 200\,\vec{k}\ \text{cm/s}^2 \\
2\,\vec{\omega}_3 \times \vec{v}_{P|3} &= 2(-4\,\vec{\imath} - 3\,\vec{\jmath} + 5\,\vec{k}) \times (-1.6\,\vec{\imath} + 1.2\,\vec{k}) \\
&= -7.2\,\vec{\imath} - 6.4\,\vec{\jmath} - 9.6\,\vec{k}\ \text{cm/s}^2 \\
\vec{v}_P &= \vec{v}_B + \vec{\omega}_3 \times \overrightarrow{BP} + \vec{v}_{P|3} = -1.6\,\vec{\imath} + 250\,\vec{\jmath} + 31.2\,\vec{k}\ \text{cm/s} \\
\vec{a}_P &= \vec{a}_B + \dot{\vec{\omega}}_3 \times \overrightarrow{BP} + \vec{\omega}_3 \times (\vec{\omega}_3 \times \overrightarrow{BP}) + 2\,\vec{\omega}_3 \times \vec{v}_{P|3} + \vec{a}_{P|3} \\
&= -1346.4\,\vec{\imath} + 183.6\,\vec{\jmath} - 20.2\,\vec{k}\ \text{cm/s}^2
\end{aligned}$$

Q moves *with respect to body 3* $\Rightarrow$ we express $\vec{v}_Q$, $\vec{a}_Q$ in terms of point B *of body 3*: $\overrightarrow{BQ} = -10\,\vec{\imath} + 5\,\vec{k}$ cm, $\quad \vec{e}_t = \vec{\imath}$, $\quad \vec{e}_n = -\,\vec{k}$,

$$\begin{aligned}
\dot{s}_{Q|3} &= 2\,\text{cm/s}, \qquad \ddot{s}_{Q|3} = -1\,\text{cm/s}^2, \qquad \vec{v}_{Q|3} = \dot{s}_{Q|3}\,\vec{e}_t = 2\,\vec{\imath}\ \text{cm/s} \\
\vec{a}_{Q|3} &= \ddot{s}_{Q|3}\,\vec{e}_t + (\dot{s}^2_{Q|3}\,/\,5)\,\vec{e}_n = -\,\vec{\imath} - 0.8\,\vec{k}\ \text{cm/s}^2 \\
\vec{\omega}_3 \times \overrightarrow{BQ} &= -15\,\vec{\imath} - 30\,\vec{\jmath} - 30\,\vec{k}\ \text{cm/s} \\
\dot{\vec{\omega}}_3 \times \overrightarrow{BQ} &= -95\,\vec{\imath} - 175\,\vec{\jmath} - 190\,\vec{k}\ \text{cm/s}^2 \\
\vec{\omega}_3 \times (\vec{\omega}_3 \times \overrightarrow{BQ}) &= 240\,\vec{\imath} - 195\,\vec{\jmath} + 75\,\vec{k}\ \text{cm/s}^2 \\
2\,\vec{\omega}_3 \times \vec{v}_{Q|3} &= 2(-4\,\vec{\imath} - 3\,\vec{\jmath} + 5\,\vec{k}) \times 2\,\vec{\imath} = 20\,\vec{\jmath} + 12\,\vec{k}\ \text{cm/s}^2
\end{aligned}$$

$$\vec{v}_Q = \vec{v}_B + \vec{\omega}_3 \times \overrightarrow{BQ} + \vec{v}_{Q|3} = -13\vec{\imath} + 170\vec{\jmath} - 30\vec{k}\ \text{cm/s}$$

$$\begin{aligned}\vec{a}_Q &= \vec{a}_B + \dot{\vec{\omega}}_3 \times \overrightarrow{BQ} + \vec{\omega}_3 \times (\vec{\omega}_3 \times \overrightarrow{BQ}) + 2\,\vec{\omega}_3 \times \vec{v}_{Q|3} + \vec{a}_{Q|3} \\ &= -856\vec{\imath} - 390\vec{\jmath} - 103.8\vec{k}\ \text{cm/s}^2\end{aligned}$$

Example 1.21: A *universal joint* is used to transmit power from one shaft to another whose axes intersect *at a small angle* α (Fig. E1.21). It consists of a rigid cross-link $ABCD$ with $AB \perp CD$. AB is hinged to drive shaft 1 and CD to shaft 2. Given ω_1, find the angular velocities of shaft 2 and cross-link 3. The axes of the shafts are in the yz-plane and the arm AB is in the xy-plane at angle ψ with the x-axis.

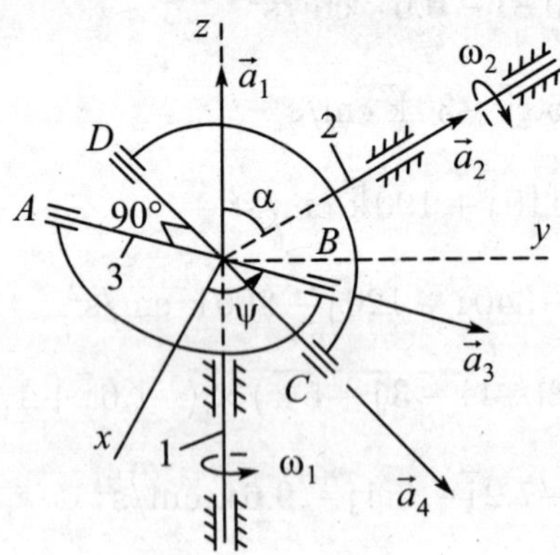

Fig. E1.21

Solution: For the unit vectors $\vec{a}_i$,

$$\vec{a}_1 \cdot \vec{a}_2 = \cos\alpha, \quad \vec{a}_1 \cdot \vec{a}_3 = 0, \quad \vec{a}_3 \cdot \vec{a}_4 = 0, \quad \vec{a}_4 \cdot \vec{a}_2 = 0, \quad \vec{a}_1 = \vec{k}$$

$$\vec{a}_2 = \sin\alpha\,\vec{\jmath} + \cos\alpha\,\vec{k}, \qquad \vec{a}_3 = \cos\psi\,\vec{\imath} + \sin\psi\,\vec{\jmath}$$

$$\Rightarrow \quad \vec{a}_2 \cdot \vec{a}_3 = \sin\alpha\sin\psi$$

Let $\vec{\omega}_{3|1} = \omega_{31}\vec{a}_3$, $\quad \vec{\omega}_{3|2} = \omega_{32}\,\vec{a}_4$. $\qquad \vec{\omega}_3 = \vec{\omega}_{3|1} + \vec{\omega}_1 = \vec{\omega}_{3|2} + \vec{\omega}_2 \quad \Rightarrow$

$$\omega_{31}\vec{a}_3 + \omega_1\vec{a}_1 = \omega_{32}\,\vec{a}_4 + \omega_2\,\vec{a}_2 \tag{1}$$

Eliminate $\omega_{32}\,\vec{a}_4$ from eq. (1) by forming dot products with $\vec{a}_3$ and $\vec{a}_2$ orthogonal to it

$$\omega_{31} = \omega_2(\vec{a}_2 \cdot \vec{a}_3), \qquad \omega_{31}(\vec{a}_3 \cdot \vec{a}_2) + \omega_1\cos\alpha = \omega_2$$

$$\Rightarrow \quad \omega_{31} = \omega_2\sin\alpha\sin\psi, \qquad \omega_{31}\sin\alpha\sin\psi + \omega_1\cos\alpha = \omega_2$$

$$\Rightarrow \quad \omega_2 = \frac{\omega_1\cos\alpha}{1 - \sin^2\alpha\sin^2\psi}, \qquad \omega_{31} = \frac{\omega_1\cos\alpha\sin\alpha\sin\psi}{1 - \sin^2\alpha\sin^2\psi}$$

$$\vec{\omega}_3 = \omega_{31}\vec{a}_3 + \omega_1\vec{a}_1 = \frac{\omega_1 \cos\alpha \sin\alpha \sin\psi}{1-\sin^2\alpha\sin^2\psi}(\cos\psi\,\vec{\mathrm{i}} + \sin\psi\,\vec{\mathrm{j}}) + \omega_1\,\vec{\mathrm{k}}$$

The variation of ω_2 from $\omega_1 \cos\alpha$ at $\psi = 0$ to $\omega_1/\cos\alpha$ at $\psi = \pi/2$ is small if α is small, i.e., if $\sin^2\alpha \simeq \alpha^2 \ll 1$. Integrating the $\omega_2 - \omega_1$ relation, we get the rotation θ_2 of shaft 2 as $\theta_2 = \cot^{-1}[\cot\psi/\cos\alpha]$.

Example 1.22: (a) The gimbal support of a *gyroscope* rotates about a fixed axis AB (Fig. E1.22a). The frame m supporting the shaft of the rotor rotates relative to the gimbal. The rotor rotates relative to frame m. Find at this instant ($\theta = \cos^{-1} 0.8$), the angular velocity and angular acceleration of the rotor and $\vec{v}_P$, $\vec{a}_P$, $\vec{v}_Q$, $\vec{a}_Q$ relative to the ground. The kinematically equivalent problems of a spinning top and part of a robotic device are shown in Figs E1.22b,c. (b) The telescopic arm is increasing its length at the given rates (Fig. E1.22d). For a point Q on it, find $\vec{v}_Q$, $\vec{a}_Q$.

Solution: (a) In order of successive rotations, we name the gimbal support, the frame m and the rotor as bodies 1, 2 and 3. Choose axes $\vec{\mathrm{i}}$, $\vec{\mathrm{j}}$, $\vec{\mathrm{k}}$ attached to body 1, with $\vec{\mathrm{k}}$ along its fixed axis of rotation, and $\vec{\mathrm{i}}$ along the axis of rotation of body 2 relative to body 1. We successively compose the angular velocities, angular accelerations, velocities and accelerations. The rotor spins about the axis $\vec{e} = \sin\theta\,\vec{\mathrm{j}} + \cos\theta\,\vec{\mathrm{k}} = 0.6\,\vec{\mathrm{j}} + 0.8\,\vec{\mathrm{k}}$.

$$\begin{aligned}
\vec{\omega}_1 &= 4\,\vec{\mathrm{k}}\ \mathrm{rad/s}, \qquad \dot{\vec{\omega}}_1 = -0.5\,\vec{\mathrm{k}}\ \mathrm{rad/s^2}\\
\vec{\omega}_2 &= \vec{\omega}_1 - 0.2\,\vec{\mathrm{i}} = -0.2\,\vec{\mathrm{i}} + 4\,\vec{\mathrm{k}}\ \mathrm{rad/s}\\
\dot{\vec{\omega}}_2 &= \dot{\vec{\omega}}_1 - 0.1\,\vec{\mathrm{i}} + \vec{\omega}_1 \times (-0.2\,\vec{\mathrm{i}})\\
&= -0.5\,\vec{\mathrm{k}} - 0.1\,\vec{\mathrm{i}} + 4\,\vec{\mathrm{k}} \times (-0.2\,\vec{\mathrm{i}}) = -0.1\,\vec{\mathrm{i}} - 0.8\,\vec{\mathrm{j}} - 0.5\,\vec{\mathrm{k}}\ \mathrm{rad/s^2}\\
\vec{\omega}_3 &= \vec{\omega}_2 + 20\,\vec{e} = (-0.2\,\vec{\mathrm{i}} + 4\,\vec{\mathrm{k}}) + 20(0.6\,\vec{\mathrm{j}} + 0.8\,\vec{\mathrm{k}}) = -0.2\,\vec{\mathrm{i}} + 12\,\vec{\mathrm{j}} + 20\,\vec{\mathrm{k}}\ \mathrm{rad/s}\\
\dot{\vec{\omega}}_3 &= \dot{\vec{\omega}}_2 - 2\,\vec{e} + \vec{\omega}_2 \times 20\,\vec{e}\\
&= (-0.1\,\vec{\mathrm{i}} - 0.8\,\vec{\mathrm{j}} - 0.5\,\vec{\mathrm{k}}) - 2(0.6\,\vec{\mathrm{j}} + 0.8\,\vec{\mathrm{k}}) + (-0.2\,\vec{\mathrm{i}} + 4\,\vec{\mathrm{k}}) \times 20(0.6\,\vec{\mathrm{j}} + 0.8\,\vec{\mathrm{k}})\\
&= -48.1\,\vec{\mathrm{i}} + 1.2\,\vec{\mathrm{j}} - 4.5\,\vec{\mathrm{k}}\ \mathrm{rad/s^2}
\end{aligned}$$

Points P and O are on body 2 with $\vec{v}_O = \vec{0}$, $\vec{a}_O = \vec{0}$,

$$\begin{aligned}
\overrightarrow{OP} &= 0.5\,\vec{e} = 0.5(0.6\,\vec{\mathrm{j}} + 0.8\,\vec{\mathrm{k}}) = 0.3\,\vec{\mathrm{j}} + 0.4\,\vec{\mathrm{k}}\ \mathrm{m}\\
\vec{\omega}_2 \times \overrightarrow{OP} &= (-0.2\,\vec{\mathrm{i}} + 4\,\vec{\mathrm{k}}) \times (0.3\,\vec{\mathrm{j}} + 0.4\,\vec{\mathrm{k}}) = -1.2\,\vec{\mathrm{i}} + 0.08\,\vec{\mathrm{j}} - 0.06\,\vec{\mathrm{k}}\ \mathrm{m/s}\\
\vec{\omega}_2 \times (\vec{\omega}_2 \times \overrightarrow{OP}) &= (-0.2\,\vec{\mathrm{i}} + 4\,\vec{\mathrm{k}}) \times (-1.2\,\vec{\mathrm{i}} + 0.08\,\vec{\mathrm{j}} - 0.06\,\vec{\mathrm{k}})\\
&= -0.32\,\vec{\mathrm{i}} - 4.812\,\vec{\mathrm{j}} - 0.016\,\vec{\mathrm{k}}\ \mathrm{m/s^2}
\end{aligned}$$

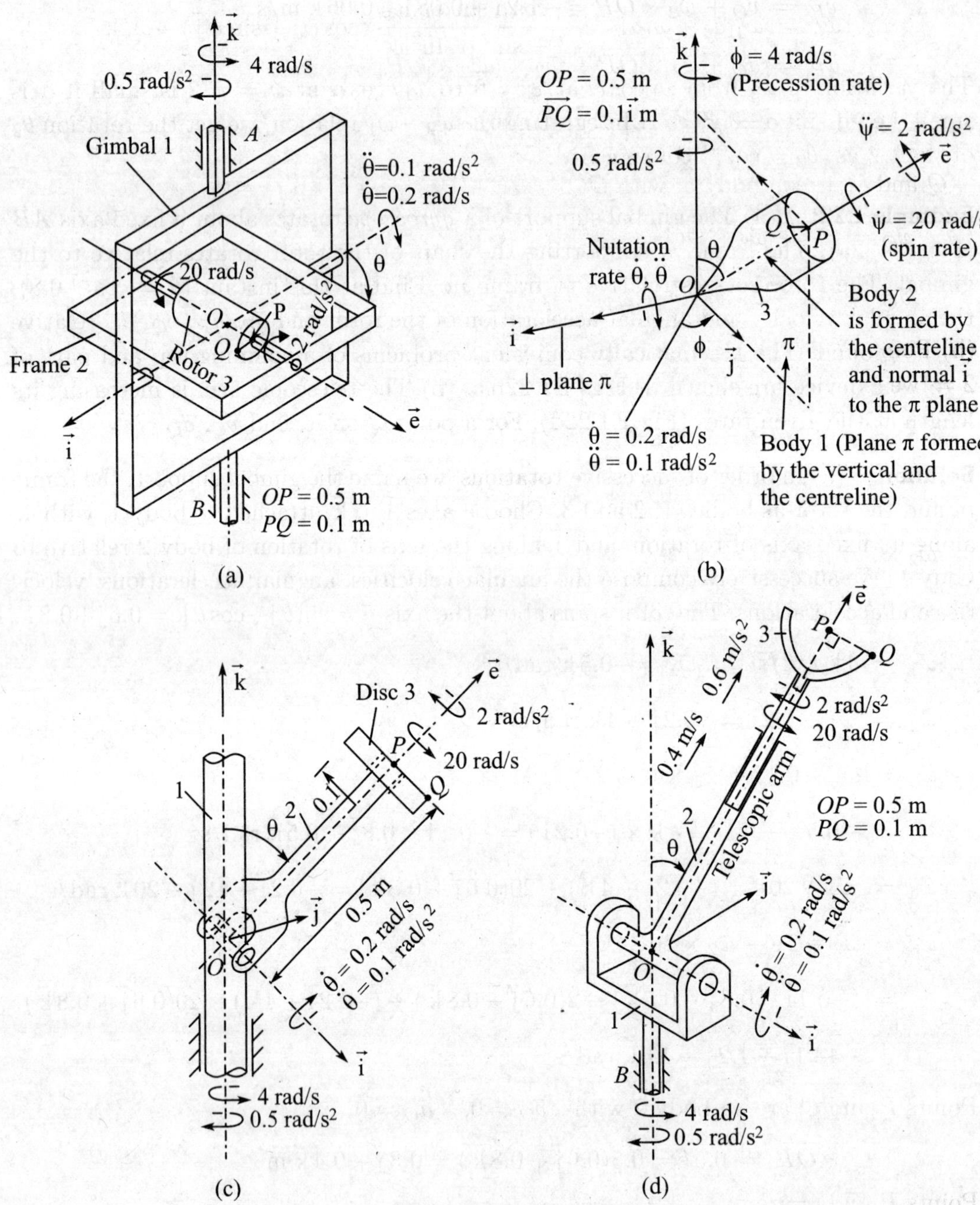

Fig. E1.22

$$\dot{\vec{\omega}}_2 \times \overrightarrow{OP} = (-0.1\vec{\mathrm{i}} - 0.8\vec{\mathrm{j}} - 0.5\vec{\mathrm{k}}) \times (0.3\vec{\mathrm{j}} + 0.4\vec{\mathrm{k}})$$

$$= -0.17\vec{\mathrm{i}} + 0.04\vec{\mathrm{j}} - 0.03\vec{\mathrm{k}}\ \mathrm{m/s^2}$$

$$\vec{v}_P = \vec{v}_O + \vec{\omega}_2 \times \overrightarrow{OP} = -1.2\vec{i} + 0.08\vec{j} - 0.06\vec{k} \text{ m/s}$$

$$\vec{a}_P = \vec{a}_O + \dot{\vec{\omega}}_2 \times \overrightarrow{OP} + \vec{\omega}_2 \times (\vec{\omega}_2 \times \overrightarrow{OP})$$

$$= -0.49\vec{i} - 4.772\vec{j} - 0.046\vec{k} \text{ m/s}^2$$

Q and O are on body 3, with $\overrightarrow{OQ} = \overrightarrow{OP} + \overrightarrow{PQ} = 0.5\vec{e} + 0.1\vec{i} = 0.1\vec{i} + 0.3\vec{j} + 0.4\vec{k}$,

$$\vec{v}_Q = \vec{v}_O + \vec{\omega}_3 \times \overrightarrow{OQ} = -1.2\vec{i} + 2.08\vec{j} - 1.26\vec{k} \text{ m/s}$$

$$\vec{a}_Q = \vec{a}_O + \dot{\vec{\omega}}_3 \times \overrightarrow{OQ} + \vec{\omega}_3 \times (\vec{\omega}_3 \times \overrightarrow{OQ}) = -54.89\vec{i} - 5.462\vec{j} - 0.566\vec{k} \text{ m/s}^2$$

(b) Point P, on the rigid extension of arm 3, has given motion with respect to body 2 $\Rightarrow$ we express $\vec{v}_P$, $\vec{a}_P$ in terms of point O of body 2 with $\vec{v}_O = \vec{0}$, $\vec{a}_O = \vec{0}$,

$$\overrightarrow{OP} = 0.5\vec{e} = 0.5(0.6\vec{j} + 0.8\vec{k}) = 0.3\vec{j} + 0.4\vec{k} \text{ m}$$

$$\vec{\omega}_2 \times \overrightarrow{OP} = (-0.2\vec{i} + 4\vec{k}) \times (0.3\vec{j} + 0.4\vec{k}) = -1.2\vec{i} + 0.08\vec{j} - 0.06\vec{k} \text{ m/s}$$

$$\vec{\omega}_2 \times (\vec{\omega}_2 \times \overrightarrow{OP}) = (-0.2\vec{i} + 4\vec{k}) \times (-1.2\vec{i} + 0.08\vec{j} - 0.06\vec{k})$$

$$= -0.32\vec{i} - 4.812\vec{j} - 0.016\vec{k} \text{ m/s}^2$$

$$\dot{\vec{\omega}}_2 \times \overrightarrow{OP} = (-0.1\vec{i} - 0.8\vec{j} - 0.5\vec{k}) \times (0.3\vec{j} + 0.4\vec{k})$$

$$= -0.17\vec{i} + 0.04\vec{j} - 0.03\vec{k} \text{ m/s}^2$$

$$\vec{v}_{P|2} = 0.4\vec{e} = 0.4(0.6\vec{j} + 0.8\vec{k}) = 0.24\vec{j} + 0.32\vec{k} \text{ m/s}$$

$$\vec{a}_{P|2} = 0.6\vec{e} = 0.6(0.6\vec{j} + 0.8\vec{k}) = 0.36\vec{j} + 0.48\vec{k} \text{ m/s}^2$$

$$2\vec{\omega}_2 \times \vec{v}_{P|2} = 2(-0.2\vec{i} + 4\vec{k}) \times (0.24\vec{j} + 0.32\vec{k})$$

$$= -1.92\vec{i} + 0.128\vec{j} - 0.096\vec{k} \text{ m/s}^2$$

$$\vec{v}_P = \vec{v}_O + \vec{\omega}_2 \times \overrightarrow{OP} + \vec{v}_{P|2} = -1.2\vec{i} + 0.32\vec{j} + 0.26\vec{k} \text{ m/s}$$

$$\vec{a}_P = \vec{a}_O + \dot{\vec{\omega}}_2 \times \overrightarrow{OP} + \vec{\omega}_2 \times (\vec{\omega}_2 \times \overrightarrow{OP}) + 2\vec{\omega}_2 \times \vec{v}_{P|2} + \vec{a}_{P|2}$$

$$= -2.41\vec{i} - 4.284\vec{j} + 0.338\vec{k} \text{ m/s}^2$$

Points P and Q are on body 3 with $\overrightarrow{PQ} = 0.1\vec{i}$ m,

$$\vec{\omega}_3 \times \overrightarrow{PQ} = (-0.2\vec{i} + 12\vec{j} + 20\vec{k}) \times 0.1\vec{i} = 2\vec{j} - 1.2\vec{k} \text{ m/s}$$

$$\vec{\omega}_3 \times (\vec{\omega}_3 \times \overrightarrow{PQ}) = (-.2\vec{i} + 12\vec{j} + 20\vec{k}) \times (2\vec{j} - 1.2\vec{k}) = -54.4\vec{i} - .24\vec{j} - .4\vec{k} \text{ m/s}^2$$

$$\dot{\vec{\omega}}_3 \times \overrightarrow{PQ} = (-48.1\vec{\imath} + 1.2\vec{\jmath} - 4.5\vec{k}) \times 0.1\vec{\imath} = -0.45\vec{\jmath} - 0.12\vec{k}\ \mathrm{m/s^2}$$

$$\vec{v}_Q = \vec{v}_P + \vec{\omega}_3 \times \overrightarrow{PQ} = -1.2\vec{\imath} + 2.32\vec{\jmath} - 0.94\vec{k}\ \mathrm{m/s}$$

$$\vec{a}_Q = \vec{a}_P + \dot{\vec{\omega}}_3 \times \overrightarrow{PQ} + \vec{\omega}_3 \times (\vec{\omega}_3 \times \overrightarrow{PQ}) = -56.81\vec{\imath} - 4.974\vec{\jmath} - 0.182\vec{k}\ \mathrm{m/s^2}$$

Example 1.23: For the mechanism shown in Fig. E1.23a, find ω_2, $\dot{\omega}_2$ of wheel 2 and the instantaneous centre of body 3 at the given instant. Where is the instantaneous centre of wheel 2 with respect to the reference frame of body 1?

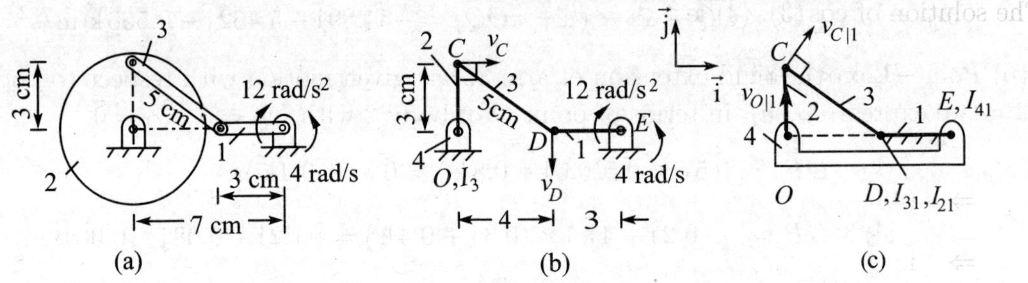

Fig. E1.23a,b,c

Solution: This mechanism (Fig. E1.23b) is called a *four bar chain* as it consists of 4 members—members 1 and 2 have fixed axis of rotation, the connecting member 3 has general plane motion and the 4th member is the fixed body supporting 1 and 2. We write $\vec{v}_C$, $\vec{a}_C$ of common point C (Fig. E1.23b) of bodies 2 and 3 in two ways in terms of other convenient coplanar points of these bodies. $\Rightarrow$ $\vec{\omega} \times (\vec{\omega} \times \vec{r}) = -\omega^2 \vec{r}$.

$$\vec{v}_O = \vec{v}_E = \vec{0}, \quad \vec{a}_O = \vec{a}_E = \vec{0}, \quad \vec{\omega}_1 = 4\vec{k}, \quad \dot{\vec{\omega}}_1 = -12\vec{k}$$

$$\vec{\omega}_2 = \omega_2\vec{k}, \quad \dot{\vec{\omega}}_2 = \dot{\omega}_2\vec{k}, \quad \vec{\omega}_3 = \omega_3\vec{k}, \quad \dot{\vec{\omega}}_3 = \dot{\omega}_3\vec{k}$$

$$\vec{v}_C = \vec{v}_D + \vec{\omega}_3 \times \overrightarrow{DC}$$

$$= (\vec{v}_E + \vec{\omega}_1 \times \overrightarrow{ED}) + \vec{\omega}_3 \times \overrightarrow{DC} = \vec{v}_O + \vec{\omega}_2 \times \overrightarrow{OC} \qquad (1)$$

$$\vec{a}_C = \vec{a}_D + \dot{\vec{\omega}}_3 \times \overrightarrow{DC} - \omega_3^2\,\overrightarrow{DC}$$

$$= (\vec{a}_E + \dot{\vec{\omega}}_1 \times \overrightarrow{ED} - \omega_1^2\,\overrightarrow{ED}) + \dot{\vec{\omega}}_3 \times \overrightarrow{DC} - \omega_3^2\,\overrightarrow{DC}$$

$$= \vec{a}_O + \dot{\vec{\omega}}_2 \times \overrightarrow{OC} - \omega_2^2\,\overrightarrow{OC} \qquad (2)$$

$$OD = (5^2 - 3^2)^{1/2} = 4\,\mathrm{cm}, \qquad \overrightarrow{DC} = -4\vec{\imath} + 3\vec{\jmath}\ \mathrm{cm}$$

$$\overrightarrow{ED} = -3\vec{\imath}\ \mathrm{cm}, \qquad \overrightarrow{OC} = 3\vec{\jmath}\ \mathrm{cm}$$

The unknowns ω_2, ω_3 are solved from the two scalar equations corresponding to eq. (1)

$\dot{\omega}_2$, $\dot{\omega}_3$ are solved from the two scalar equations corresponding to eq. (2). Equation (1):

$$4\vec{k} \times (-3\vec{i}) + \omega_3 \vec{k} \times (-4\vec{i} + 3\vec{j}) = \omega_2 \vec{k} \times 3\vec{j}$$

$$\Rightarrow \quad -3\omega_3 \vec{i} - (12 + 4\omega_3)\vec{j} = -3\omega_2 \vec{i}$$

$$\Rightarrow \vec{i}: \quad -3\omega_3 = -3\omega_2 \tag{3}$$

$$\vec{j}: \quad -12 - 4\omega_3 = 0 \tag{4}$$

The solution of eqs (3), (4) is $\omega_2 = \omega_3 = -3\,\text{rad/s}$. Equation (2):

$$-12\vec{k} \times (-3\vec{i}) - 4^2(-3\vec{i}) + \dot{\omega}_3 \vec{k} \times (-4\vec{i} + 3\vec{j}) - (-3)^2(-4\vec{i} + 3\vec{j})$$
$$= \dot{\omega}_2 \vec{k} \times 3\vec{j} - (-3)^2(3\vec{j})$$

$$\Rightarrow (48 - 3\dot{\omega}_3 + 36)\vec{i} + (36 - 4\dot{\omega}_3 - 27)\vec{j} = -3\dot{\omega}_2 \vec{i} - 27\vec{j}$$

$$\Rightarrow \vec{i}: \quad 48 - 3\dot{\omega}_3 + 36 = -3\dot{\omega}_2 \tag{5}$$

$$\vec{j}: \quad 36 - 4\dot{\omega}_3 - 27 = -27 \tag{6}$$

The solution of eqs (5), (6) is $\dot{\omega}_2 = -19\,\text{rad/s}^2$, $\dot{\omega}_3 = 9\,\text{rad/s}^2$.

The instantaneous centre I_3 of body 3 is at the intersection O of the normals to $\vec{v}_D$ and $\vec{v}_C$ (Fig. E1.23b). ω_2 and ω_3 can also be obtained using I_3:

$$\frac{v_D}{I_3D} = \frac{v_C}{I_3C} = |\omega_3| \quad \Rightarrow \quad \frac{|\omega_1|(DE)}{I_3D} = \frac{|\omega_2|(OC)}{I_3C} = |\omega_3|$$

$$\Rightarrow \quad \frac{4(3)}{4} = \frac{|\omega_2|(3)}{3} = |\omega_3| \quad \Rightarrow \quad |\omega_2| = 3, \quad |\omega_3| = 3\,\text{rad/s}$$

The direction of $\vec{v}_D$ implies that 3 rotates about I_3, in the direction from $\vec{j}$ to $\vec{i}$, i.e., about $-\vec{k}$. This yields the direction of $\vec{v}_C$ as shown in Fig. E1.23b. Hence, 2 rotates from $\vec{j}$ to $\vec{i}$, i.e., about $-\vec{k}$. Therefore, ω_2 and ω_3 are negative.

With respect to the reference frame of body 1: the instantaneous centre I_{31} of body 3 is at D and the instantaneous centre I_{41} of body 4 is at E. Hence, directions of $\vec{v}_{C|1}$ and $\vec{v}_{O|1}$ are as shown in Fig. E1.23c. The instantaneous centre I_{21} of body 2 with respect to body 1 is at the intersection D of the normals to $\vec{v}_{C|1}$ and $\vec{v}_{O|1}$.

The above procedure of writing $\vec{v}$ and $\vec{a}$ of a point in two different ways applies for the mechanisms shown in Fig. E1.23d. For mechanisms in Fig. E1.23d(i):

$$\vec{v}_B = \vec{v}_A + \omega_2 \vec{k} \times \overrightarrow{AB} = (\vec{v}_O + \omega_1 \vec{k} \times \overrightarrow{OA}) + \omega_2 \vec{k} \times \overrightarrow{AB} = v_B \vec{e}_t \tag{7}$$

$$\vec{a}_B = \vec{a}_A + \dot{\omega}_2 \vec{k} \times \overrightarrow{AB} - \omega_2^2 \overrightarrow{AB}$$

$$= (\vec{a}_O + \dot{\omega}_1 \vec{k} \times \overrightarrow{OA} - \omega_1^2 \overrightarrow{OA}) + \dot{\omega}_2 \vec{k} \times \overrightarrow{AB} - \omega_2^2 \overrightarrow{AB} = \dot{v}_B \vec{e}_t + (v_B^2/\rho_B) \vec{e}_n \quad (8)$$

where $\vec{v}_O = \vec{0}$, $\vec{a}_O = \vec{0}$ and $\rho_B = \infty$ if path of B is a straight line. Equation (7) is solved for ω_2, v_B and eq. (8) for $\dot{\omega}_2$, $\dot{v}_B$. $|\omega_3| = |v_B|/R$, $\dot{\omega}_3 = |\dot{v}_B|/R|$ and their signs are found from the directions of $v_B \vec{e}_t$, $\dot{v}_B \vec{e}_t$, knowing that I_3 is at the point of contact.

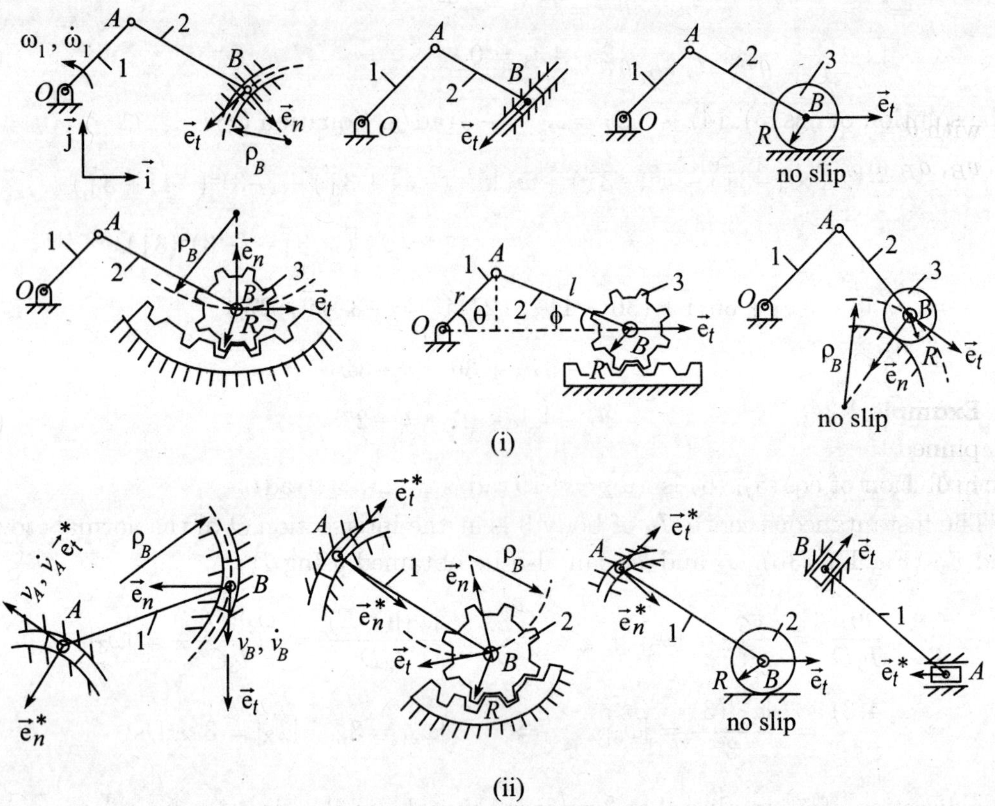

Fig. E1.23d

For the mechanisms in Fig. E1.23d(ii), let the radius of curvature of path of A be ρ_A and of path of B be ρ_B:

$$\vec{v}_B = v_A \vec{e}_t^* + \omega_1 \vec{k} \times \overrightarrow{AB} = v_B \vec{e}_t \quad (9)$$

$$\vec{a}_B = [\dot{v}_A \vec{e}_t^* + (v_A^2/\rho_A) \vec{e}_n^*] + \dot{\omega}_1 \vec{k} \times \overrightarrow{AB} - \omega_1^2 \overrightarrow{AB} = \dot{v}_B \vec{e}_t + (v_B^2/\rho_B) \vec{e}_n \quad (10)$$

where $\rho_B = \infty$ if path of B is a straight line and $\rho_A = \infty$ if path of A is a straight line. Equation (9) is solved for ω_1, v_B and eq. (10) for $\dot{\omega}_1$, $\dot{v}_B$. Then, $|\omega_2| = |v_B|/R$, $|\dot{\omega}_2| = |\dot{v}_B|/R$ and their signs are obtained from the directions of $v_B \vec{e}_t$, $\dot{v}_B \vec{e}_t$, knowing that

I_2 is at the point of contact.

Note : Sometimes, it is convenient to work out kinematics using geometrical relations, e.g., for the mechanism of $(i)_5$ (Fig. E1.23d), we can obtain ω_2, v_B, $\dot{\omega}_2$, a_B as follows:

$$r \sin\theta = l \sin\phi$$

$$\Rightarrow \qquad r\cos\theta\,\dot{\theta} = l\cos\phi\,\dot{\phi} \tag{11}$$

$$-r\sin\theta\,\dot{\theta}^2 + r\cos\theta\,\ddot{\theta} = -l\sin\phi\,\dot{\phi}^2 + l\cos\phi\,\ddot{\phi} \tag{12}$$

with $\dot{\theta} = \omega_1$, $\ddot{\theta} = \dot{\omega}_1$, $\dot{\phi} = -\omega_2$, $\ddot{\phi} = -\dot{\omega}_2$. ω_2, $\dot{\omega}_2$ are obtained from eqs (11) and (12). v_B, a_B are then obtained by differentiating x_B :

$$x_B = r\cos\theta + l\cos\phi$$

$$v_B = \dot{x}_B = -r\sin\theta\,\omega_1 + l\sin\phi\,\omega_2$$

$$a_B = \ddot{x}_B = -r\sin\theta\,\dot{\omega}_1 - r\cos\theta\,\omega_1^2 + l\sin\phi\,\dot{\omega}_2 - l\cos\phi\,\omega_2^2$$

Example 1.24: Disc 1 rolls without slip (Fig. E1.24). It is pinned to rod 2 which is pinned to slider 3. Pin P is welded to plate 4 hinged at Q. Pin P can slide in the slot in 1. Find the acceleration of point E and the angular acceleration of plate 4.

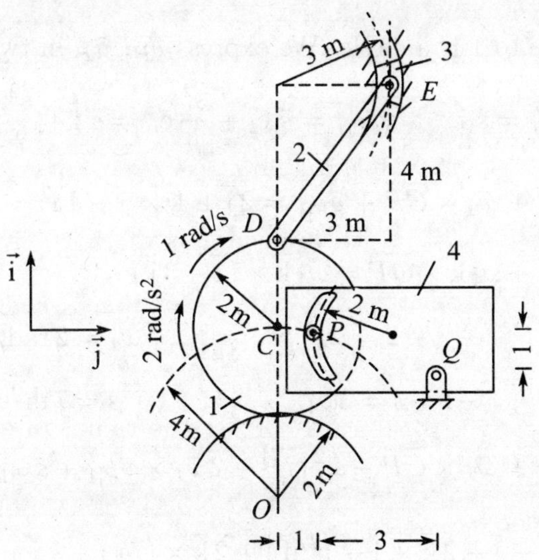

Fig. E1.24

Solution: $\vec{\omega}_1 = 1\,\vec{k}$ rad/s, $\dot{\vec{\omega}}_1 = 2\,\vec{k}$ rad/s^2. From no slip condition: $\dot{s}_C = 2\omega_1 \;\forall\, t$
$\Rightarrow \quad \ddot{s}_C = 2\dot{\omega}_1$: $\quad \dot{s}_C = 2(1) = 2\,$m/s, $\quad \ddot{s}_C = 2(2) = 4\,$m/s^2, $\quad \vec{v}_C = \dot{s}_C\,\vec{e}_t^{\,C} = 2\vec{j}$,

$$\vec{v}_D = \vec{v}_C + \vec{\omega}_1 \times \overrightarrow{CD} = 2\vec{j} + \vec{k} \times 2\vec{i} = 4\vec{j}\,\mathrm{m/s}$$

$$\vec{v}_E = \vec{v}_D + \omega_2 \vec{k} \times \overrightarrow{DE} = 4\vec{j} + \omega_2 \vec{k} \times (3\vec{j} + 4\vec{i}) = v_E \vec{e}_t^{\,E} = v_E \vec{i}$$

$$\Rightarrow \quad \vec{j}: \quad 4 + 4\omega_2 = 0 \quad \Rightarrow \quad \omega_2 = -1\,\mathrm{rad/s}$$

$$\vec{i}: \quad -3\omega_2 = v_E \quad \Rightarrow \quad v_E = 3\,\mathrm{m/s}$$

$$\vec{a}_C = \ddot{s}_C \vec{e}_t^{\,C} + \frac{\dot{s}_C^2}{\rho_C} \vec{e}_n^{\,C} = 4\vec{j} + \frac{2^2}{4}(-\vec{i}) = 4\vec{j} - \vec{i}\,\mathrm{m/s^2}$$

$$\vec{a}_D = \vec{a}_C + \dot{\vec{\omega}}_1 \times \overrightarrow{CD} - \omega_1^2 \overrightarrow{CD} = 4\vec{j} - \vec{i} + 2\vec{k} \times 2\vec{i} - 1^2 \times 2\vec{i} = 8\vec{j} - 3\vec{i}\,\mathrm{m/s^2}$$

$$\vec{a}_E = \vec{a}_D + \dot{\omega}_2 \vec{k} \times \overrightarrow{DE} - \omega_2^2 \overrightarrow{DE}$$

$$= 8\vec{j} - 3\vec{i} + \dot{\omega}_2 \vec{k} \times (3\vec{j} + 4\vec{i}) - 1^2(3\vec{j} + 4\vec{i}) = \dot{v}_E \vec{e}_t^{\,E} + \frac{v_E^2}{\rho_E} \vec{e}_n^{\,E} = \dot{v}_E \vec{i} + \frac{3^2}{3}(-\vec{j})$$

$$\Rightarrow \quad \vec{j}: \quad 8 + 4\dot{\omega}_2 - 3 = -3 \quad \Rightarrow \quad \dot{\omega}_2 = -2\,\mathrm{rad/s^2}$$

$$\vec{i}: \quad -3 - 3\dot{\omega}_2 - 4 = \dot{v}_E \quad \Rightarrow \quad \dot{v}_E = -1\,\mathrm{m/s^2}$$

$$\vec{a}_E = \dot{v}_E \vec{i} + \frac{3^2}{3}(-\vec{j}) = -\vec{i} - 3\vec{j}\,\mathrm{m/s^2}$$

The path of P relative to 1 is a circle. We express $\vec{v}_P$, $\vec{a}_P$ in two ways:

$$\vec{v}_{P|1} = \dot{s}\,\vec{e}_t = \dot{s}\,\vec{i}, \qquad \vec{a}_{P|1} = \ddot{s}\,\vec{e}_t + \frac{\dot{s}^2}{\rho}\vec{e}_n = \ddot{s}\,\vec{i} + \frac{\dot{s}^2}{2}\vec{j}$$

$$\vec{v}_P = \vec{v}_C + \vec{\omega}_1 \times \overrightarrow{CP} + \vec{v}_{P|1} = 2\vec{j} + \vec{k} \times \vec{j} + \dot{s}\,\vec{i}$$

$$= \vec{v}_Q + \omega_4 \vec{k} \times \overrightarrow{QP} = \omega_4 \vec{k} \times (\vec{i} - 3\vec{j})$$

$$\Rightarrow \quad \vec{j}: \quad 2 = \omega_4 \quad \Rightarrow \quad \omega_4 = 2\,\mathrm{rad/s}$$

$$\vec{i}: \quad -1 + \dot{s} = 3\omega_4 \quad \Rightarrow \quad \dot{s} = 7\,\mathrm{m/s}$$

$$\vec{a}_P = \vec{a}_C + \dot{\vec{\omega}}_1 \times \overrightarrow{CP} - \omega_1^2 \overrightarrow{CP} + 2\vec{\omega}_1 \times \vec{v}_{P|1} + \vec{a}_{P|1}$$

$$= 4\vec{j} - \vec{i} + 2\vec{k} \times \vec{j} - 1^2(\vec{j}) + 2\vec{k} \times (7\vec{i}) + \ddot{s}\,\vec{i} + \frac{7^2}{2}\vec{j}$$

$$= \vec{a}_Q + \dot{\omega}_4 \vec{k} \times \overrightarrow{QP} - \omega_4^2 \overrightarrow{QP} = \dot{\omega}_4 \vec{k} \times (\vec{i} - 3\vec{j}) - 2^2(\vec{i} - 3\vec{j})$$

$$\Rightarrow \quad \vec{j}: \quad 4 - 1 + 14 + 24.5 = \dot{\omega}_4 + 12 \quad \Rightarrow \quad \dot{\omega}_4 = 29.5\,\mathrm{rad/s^2}$$

$$\vec{i}: \quad -1 - 2 + \ddot{s} = 3\dot{\omega}_4 - 4 \quad \Rightarrow \quad \ddot{s} = 87.5\,\mathrm{m/s^2}$$

Example 1.25: *Quick return mechanism* (Fig. E1.25a) is used in metal cutting to speed up the idling return stroke. The pin P is fixed to wheel 1. When $\theta = \cos^{-1}(0.8)$, find (a) the angular velocity and angular acceleration of link AB, (b) $\vec{v}_D$, $\vec{a}_D$. (c) Find the ratio OP/OA, if the ratio of the times for forward and return strokes is β.

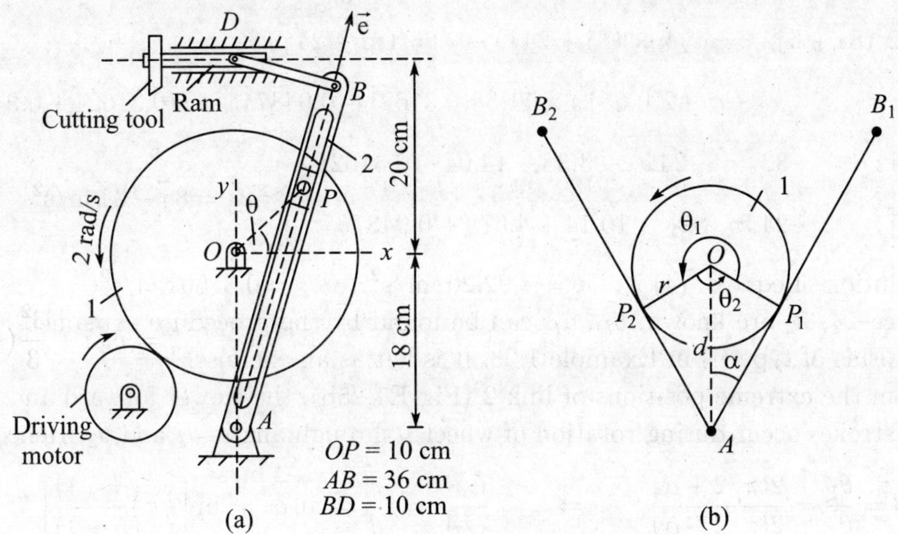

Fig. E1.25

Solution: $\cos\theta = 0.8$, $\sin\theta = 0.6$. For centre P of the pin

$$\overrightarrow{OP} = 10(\cos\theta\,\vec{i} + \sin\theta\,\vec{j}) = 8\vec{i} + 6\vec{j} \text{ cm}$$

$$\overrightarrow{AP} = \overrightarrow{AO} + \overrightarrow{OP} = 18\vec{j} + \overrightarrow{OP} = 8\vec{i} + 24\vec{j} \text{ cm}$$

$$\vec{e} = \frac{\overrightarrow{AP}}{|\overrightarrow{AP}|} = \frac{8\vec{i} + 24\vec{j}}{(8^2 + 24^2)^{1/2}} = 0.3162\vec{i} + 0.9487\vec{j}$$

$$\vec{\omega}_1 = 2\vec{k}, \qquad \dot{\vec{\omega}}_1 = \vec{0}, \qquad \vec{\omega}_2 = \omega_2\vec{k}, \qquad \dot{\vec{\omega}}_2 = \dot{\omega}_2\vec{k}$$

$$\vec{v}_O = \vec{v}_A = \vec{0}, \qquad \vec{a}_O = \vec{a}_A = \vec{0}$$

The path of P relative to body 2 is the straight line $AB \Rightarrow \vec{v}_{P|2} = v\vec{e}$, $\vec{a}_{P|2} = \dot{v}\vec{e}$. ω_2, v are obtained by writing $\vec{v}_P$ in two ways. $\dot{\omega}_2$, $\dot{v}$ are obtained by writing $\vec{a}_P$ in two ways. We consider points in the same plane of motion $\Rightarrow \vec{\omega} \times (\vec{\omega} \times \vec{r}) = -\omega^2\vec{r}$.

$$\vec{v}_P = \vec{v}_O + \vec{\omega}_1 \times \overrightarrow{OP} = \vec{v}_A + \vec{\omega}_2 \times \overrightarrow{AP} + \vec{v}_{P|2}$$

$$\Rightarrow \quad 2\vec{k} \times (8\vec{i} + 6\vec{j}) = \omega_2\vec{k} \times (8\vec{i} + 24\vec{j}) + v(0.3162\vec{i} + 0.9487\vec{j})$$

$$\Rightarrow \quad \vec{i}: \qquad -12 = -24\omega_2 + 0.3162v \tag{1}$$

$$\vec{j}: \quad 16 = 8\omega_2 + 0.9487v \tag{2}$$

The solution of eqs (1), (2) is: $v = 11.38\,\text{cm/s}, \quad \omega_2 = 0.650\,\text{rad/s}.$

$$\vec{a}_P = \vec{a}_O + \dot{\vec{\omega}}_1 \times \overrightarrow{OP} - \omega_1^2\overrightarrow{OP} = \vec{a}_A + \dot{\vec{\omega}}_2 \times \overrightarrow{AP} - \omega_2^2\overrightarrow{AP} + 2\vec{\omega}_2 \times \vec{v}_{P|2} + \vec{a}_{P|2}$$

$$\Rightarrow -2^2(8\vec{i}+6\vec{j}) = \dot{\omega}_2\vec{k} \times (8\vec{i}+24\vec{j}) - 0.65^2(8\vec{i}+24\vec{j})$$

$$+2(0.65\vec{k}) \times 11.38(0.3162\vec{i}+0.9487\vec{j}) + \dot{v}(0.3162\vec{i}+0.9487\vec{j})$$

$$\Rightarrow \vec{i}: \quad -32 = -24\dot{\omega}_2 - 3.38 - 14.04 + 0.3162\dot{v} \tag{3}$$

$$\vec{j}: \quad -24 = 8\dot{\omega}_2 - 10.14 + 4.678 + 0.9487\dot{v} \tag{4}$$

The solution of eqs (3), (4) is: $\dot{v} = -22.20\,\text{cm/s}^2, \quad \dot{\omega}_2 = 0.3150\,\text{rad/s}^2.$
(b) Once ω_2, $\dot{\omega}_2$ are known, $\vec{v}_D$, $\vec{a}_D$ can be found by the procedure explained for the mechanisms of type (i) in Example 1.23. It is left as an exercise.
(c) From the extreme positions of link 2 (Fig. E1.25b), the slower forward and faster return strokes occur during rotation of wheel 1 through angles θ_1 and θ_2. Hence,

$$\beta = \frac{\theta_1}{\theta_2} = \frac{2(\pi/2+\alpha)}{2(\pi/2-\alpha)} \quad \Rightarrow \quad \frac{OP}{OA} = \frac{r}{d} = \sin\alpha = \sin\left[\tfrac{1}{2}\pi\frac{(\beta-1)}{(\beta+1)}\right]$$

Example 1.26: A planetary gear train consists of a sun gear S which rotates about a fixed axis at ω_1, $\dot{\omega}_1$ (Fig. E1.26a). Three planetary gears P mesh with S and the ring gear R. R rotates at ω_2, $\dot{\omega}_2$. The bearings of the planets are mounted on a frame F called the spider. (a) Find the angular velocity and angular acceleration of the planets and the spider. (b) Find the acceleration of points A, Q on a planet and point B on the ring gear. (c) Find the relation between ω_1 and ω_2 so that the planet 4 translates. What is the trajectory of Q in this case? Planetary gears are used in transmission systems where power transmission has to be altered, by stopping one or more gears.

Solution: (a) The axes of rotations of sun 1, ring gear 2 and the planets 4 are mounted on spider 3. Let the angular velocities of S and P be ω_3 and ω_4. The angular velocities of bodies 1, 2, 4 relative to body 3 are $\omega_1-\omega_3$, $\omega_2-\omega_3$, $\omega_4-\omega_3$. Since points A, B, D, E have circular motion relative to 3, the no slip conditions at A (internal and external gears in mesh) and D (external gears in mesh) are $\vec{v}_{B|3} = \vec{v}_{A|3}$, $\vec{v}_{D|3} = \vec{v}_{E|3}$:

$$(\omega_2-\omega_3)(R_1+2R_2) = (\omega_4-\omega_3)R_2$$

$$(\omega_1-\omega_3)R_1 = -(\omega_4-\omega_3)R_2$$

$$\Rightarrow \quad \omega_3 = [\omega_1 R_1 + \omega_2(R_1+2R_2)]\,/\,2(R_1+R_2) \tag{1}$$

$$\omega_4 = [\omega_2(R_1+2R_2) - \omega_1 R_1]\,/\,2R_2 \tag{2}$$

$$\Rightarrow \quad \dot{\omega}_3 = [\dot{\omega}_1 R_1 + \dot{\omega}_2(R_1 + 2R_2)] \,/\, 2(R_1 + R_2)$$

$$\dot{\omega}_4 = [\dot{\omega}_2(R_1 + 2R_2) - \dot{\omega}_1 R_1] \,/\, 2R_2$$

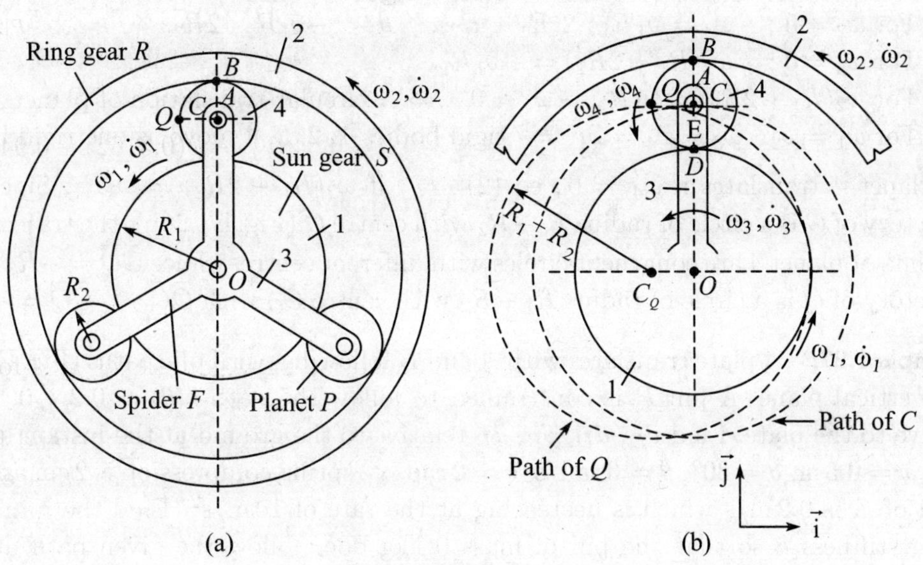

Fig. E1.26

(b) Since all points of interest have the same plane of motion, $\vec{\omega} \times (\vec{\omega} \times \vec{r}) = -\omega^2 \vec{r}$. As O and C are points of body 3 with $\overrightarrow{OC} = (R_1 + R_2)\,\vec{\mathrm{j}}$, $\vec{a}_C$ is obtained from

$$\vec{a}_C = \vec{a}_O + \dot{\omega}_3\,\vec{\mathrm{k}} \times \overrightarrow{OC} - \omega_3^2\,\overrightarrow{OC} \tag{3}$$

The accelerations of points Q, A on body 4 are obtained in terms of point C of 4:

$$\vec{a}_Q = \vec{a}_C + \dot{\omega}_4\,\vec{\mathrm{k}} \times \overrightarrow{CQ} - \omega_4^2\,\overrightarrow{CQ}, \qquad \vec{a}_A = \vec{a}_C + \dot{\omega}_4\,\vec{\mathrm{k}} \times \overrightarrow{CA} - \omega_4^2\,\overrightarrow{CA} \tag{4}$$

with $\overrightarrow{CQ} = -R_2\,\vec{\mathrm{i}}$, $\overrightarrow{CA} = R_2\,\vec{\mathrm{j}}$. Equations (1) to (4) $\Rightarrow$

$$\vec{a}_Q = [\omega_4^2 R_2 - \dot{\omega}_3(R_1 + R_2)]\,\vec{\mathrm{i}} - [\dot{\omega}_4 R_2 + \omega_3^2(R_1 + R_2)]\,\vec{\mathrm{j}}$$

$$\vec{a}_A = -\dot{\omega}_2(R_1 + 2R_2)\,\vec{\mathrm{i}} - \left[\frac{\{\omega_1 R_1 + \omega_2(R_1 + 2R_2)\}^2}{4(R_1 + R_2)} + \frac{\{\omega_2(R_1 + 2R_2) - \omega_1 R_1\}^2}{4R_2}\right]\vec{\mathrm{j}}$$

The points O and B are on body 2 with $\overrightarrow{OB} = (R_1 + 2R_2)\,\vec{\mathrm{j}}$. Hence,

$$\vec{a}_B = \vec{a}_O + \dot{\omega}_2\,\vec{\mathrm{k}} \times \overrightarrow{OB} - \omega_2^2\,\overrightarrow{OB} = -\dot{\omega}_2(R_1 + 2R_2)\,\vec{\mathrm{i}} - \omega_2^2(R_1 + 2R_2)\,\vec{\mathrm{j}}$$

Note that $\vec{a}_A \neq \vec{a}_B$, but the components of $\vec{a}_A$, $\vec{a}_B$ along the tangent direction $\vec{\mathrm{i}}$ are the same, as they should be, to satisfy the condition of rolling without slip.

Particular cases: The following results follow from relations (1) and (2).

1. For sun at rest, $\omega_1 = 0$, $\quad \omega_3 = \dfrac{\omega_2(R_1 + 2R_2)}{2(R_1 + R_2)}$, $\quad \omega_4 = \dfrac{\omega_2(R_1 + 2R_2)}{2R_2}$.
2. For $\omega_2 = 0$: $\quad \omega_3 = \omega_1 R_1 \,/\, 2(R_1 + R_2)$, $\quad \omega_4 = -\omega_1 R_1 \,/\, 2R_2$.
3. For $\omega_3 = 0$: $\quad \omega_2(R_1 + 2R_2) = -\omega_1 R_1$.
4. For $\omega_2(R_1 + 2R_2) = \omega_1 R_1$: $\quad \omega_4 = 0 \quad \Rightarrow \quad$ circular translation of planets.
5. For $\omega_1 = \omega_2$: $\omega_3 = \omega_4 = \omega_1 \;\Rightarrow\;$ rigid bodies 1, 2, 3, 4 move as one rigid body.

(c) Planet 4 translates if $\omega_4 = 0$: eq. (2) $\Rightarrow$ if $\omega_2(R_1 + 2R_2) = \omega_1 R_1$. Since the trajectory of C is a circle of radius $R_1 + R_2$ with centre O (Fig. E1.26b), the trajectories of points of planet 4 are congruent circles with different centres. Since $\overrightarrow{CQ} = -R_2\,\vec{\imath}$, the trajectory of Q is a circle of radius $R_1 + R_2$ with centre C_Q with $\overrightarrow{OC_Q} = \overrightarrow{CQ} = -R_2\,\vec{\imath}$.

Example 1.27: A plate translates with A and B following parabolic paths (Fig. E1.27) in a vertical plane. A pin P is constrained to follow the path $r(\theta) = 0.2 + 0.1\cos\theta$ relative to the plate. Find $\vec{v}_D$, $\vec{a}_D$, $\vec{v}_P$, $\vec{a}_P$ relative to the ground at the instant shown when $x = 0.5\,\mathrm{m}$, $\theta = 30°$, $\dot{\theta} = 3\,\mathrm{rad/s}$, $\ddot{\theta} = 2\,\mathrm{rad/s^2}$, spring compression is 2 cm and the speed of A is $0.2\,\mathrm{m/s}$ which is decreasing at the rate of $10\,\mathrm{m/s^2}$. Find the minimum spring stiffness k so that the pin of mass 0.1 kg does follow the given path at this instant. The slot and the curved support of the pin are smooth.

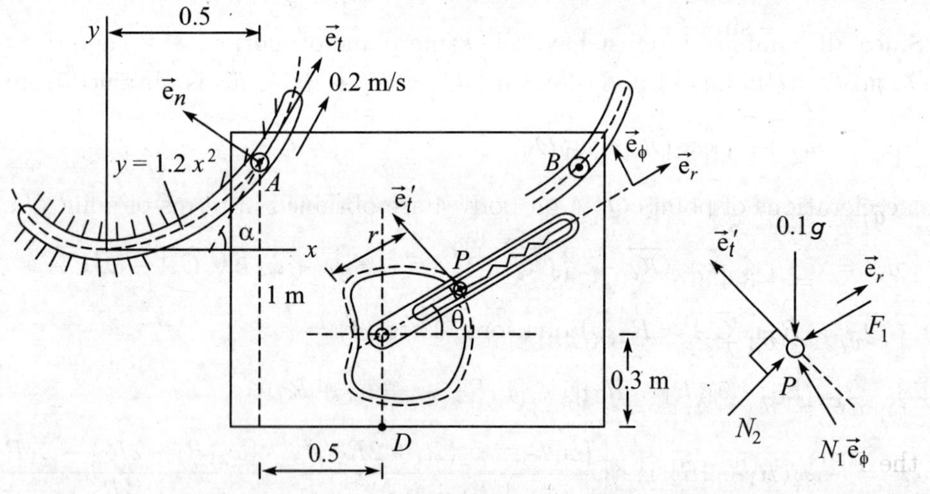

Fig. E1.27

Solution: We use path components for $\vec{a}_A$ and cylindrical components for $\vec{a}_{P|T}$ relative to the translating frame T of the plate. Since $\vec{a}_P$ relative to the ground is the sum of these accelerations, we express each of them in a common basis, say $\vec{\imath}$, $\vec{\jmath}$. For A:

$$\dot{s} = 0.2\,\mathrm{m/s}, \qquad \ddot{s} = -10\,\mathrm{m/s^2}, \qquad y = 1.2x^2, \qquad y' = 2.4x, \qquad y'' = 2.4\,\mathrm{m^{-1}}$$

At $x = 0.5\,\text{m}$:

$$\begin{aligned}
y' &= \tan\alpha = 1.2, \qquad \alpha = 50.19^\circ \\
\rho &= (1+y'^2)^{3/2}\,/\,|y''| = 1.588\,\text{m} \\
\vec{\text{e}}_t &= \cos\alpha\,\vec{\text{i}} + \sin\alpha\,\vec{\text{j}} = 0.6402\,\vec{\text{i}} + 0.7682\,\vec{\text{j}} \\
\vec{\text{e}}_n &= -\sin\alpha\,\vec{\text{i}} + \cos\alpha\,\vec{\text{j}} = -0.7682\,\vec{\text{i}} + 0.6402\,\vec{\text{j}} \\
\vec{v}_A &= \dot{s}\,\vec{\text{e}}_t = 0.2\,\vec{\text{e}}_t\,\text{m/s} = 0.1280\,\vec{\text{i}} + 0.1536\,\vec{\text{j}}\,\text{m/s} = \vec{v}_D \\
\vec{a}_A &= \ddot{s}\,\vec{\text{e}}_t + (\dot{s}^2/\rho)\,\vec{\text{e}}_n = -10\,\vec{\text{e}}_t + 0.02519\,\vec{\text{e}}_n = -6.421\,\vec{\text{i}} - 7.666\,\vec{\text{j}}\,\text{m/s}^2 = \vec{a}_D
\end{aligned}$$

as A, D are points of T. For $\vec{a}_{P|T}$: $\dot{z} = \ddot{z} = 0, \quad \phi = \theta, \quad \dot{\phi} = \dot{\theta} = 3, \quad \ddot{\phi} = \ddot{\theta} = 2,$

$$\begin{aligned}
r &= 0.2 + 0.1\cos\theta\,\text{m}, \qquad \dot{r} = -0.1\sin\theta\,\dot{\theta}\,\text{m/s} \\
\ddot{r} &= -0.1\cos\theta\,\dot{\theta}^2 - 0.1\sin\theta\,\ddot{\theta}\,\text{m/s}^2
\end{aligned}$$

At $\theta = 30^\circ$:

$$\begin{aligned}
r &= 0.2866\,\text{m}, \qquad \dot{r} = -0.15\,\text{m/s}, \qquad \ddot{r} = -0.8794\,\text{m/s}^2 \\
\vec{\text{e}}_r &= \cos\phi\,\vec{\text{i}} + \sin\phi\,\vec{\text{j}} = 0.866\,\vec{\text{i}} + 0.5\,\vec{\text{j}} \\
\vec{\text{e}}_\phi &= -\sin\phi\,\vec{\text{i}} + \cos\phi\,\vec{\text{j}} = -0.5\,\vec{\text{i}} + 0.866\,\vec{\text{j}} \\
\vec{v}_{P|T} &= \dot{r}\,\vec{\text{e}}_r + r\dot{\phi}\,\vec{\text{e}}_\phi + \dot{z}\,\vec{\text{e}}_z = -0.15\,\vec{\text{e}}_r + 0.8598\,\vec{\text{e}}_\phi \\
&= -0.5598\,\vec{\text{i}} + 0.6696\,\vec{\text{j}}\,\text{m/s} \\
\vec{a}_{P|T} &= (\ddot{r} - r\dot{\phi}^2)\,\vec{\text{e}}_r + (2\dot{r}\dot{\phi} + r\ddot{\phi})\,\vec{\text{e}}_\phi + \ddot{z}\,\vec{\text{e}}_z \\
&= -3.459\,\vec{\text{e}}_r - 0.3268\,\vec{\text{e}}_\phi\,\text{m/s}^2 = -2.832\,\vec{\text{i}} - 2.013\,\vec{\text{j}}\,\text{m/s}^2 \\
\vec{v}_P &= \vec{v}_A + \vec{v}_{P|T} = -0.4318\,\vec{\text{i}} + 0.8232\,\vec{\text{j}}\,\text{m/s} \\
\vec{a}_P &= \vec{a}_A + \vec{a}_{P|T} = -9.253\,\vec{\text{i}} - 9.679\,\vec{\text{j}}\,\text{m/s}^2
\end{aligned}$$

In the FBD of the pin (Fig. E1.27), F_1, N_1, N_2 (= 0 given) are compressive spring force, normal reaction from the slot and normal reaction from the curved support.

$$\vec{F} = -F_1\,\vec{\text{e}}_r + N_1\,\vec{\text{e}}_\phi - 0.1g\,\vec{\text{j}} = m\vec{a}_P = 0.1(-9.253\,\vec{\text{i}} - 9.679\,\vec{\text{j}}) \tag{1}$$

F_1, N_1 can be found from the $\vec{\text{i}}$, $\vec{\text{j}}$ components, or F_1 is found directly by eq. (1) $\cdot\,\vec{\text{e}}_r$:

$$-F_1 - 0.1(9.81)(0.5) = 0.1[(-9.253)(0.866) + (-9.679)(0.5)]$$

$\Rightarrow F_1 = 0.7948\,\text{N}$. For compression $\delta = 0.02\,\text{m}$, $k = F_1/\delta = 0.7948/0.02 = 39.74\,\text{N/m}$.

Example 1.28: Set up equations to find the velocity and acceleration of the reciprocating/oscillating followers of cams (Fig. E1.28).

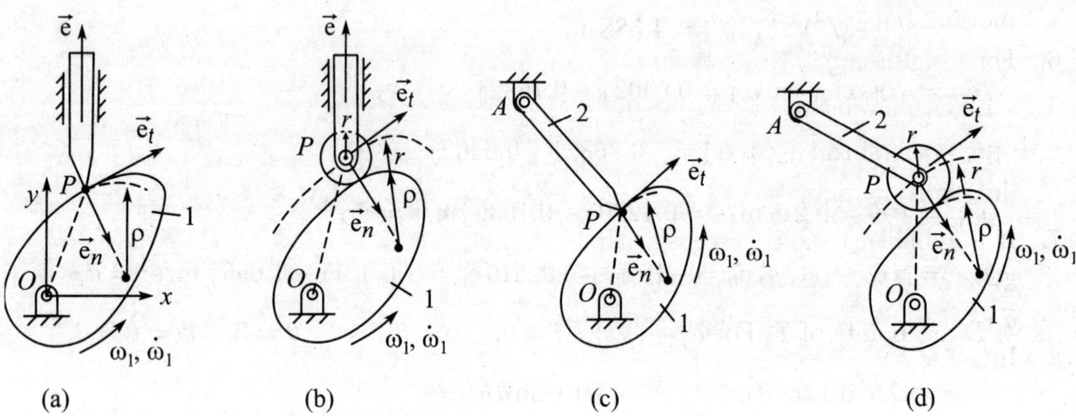

Fig. E1.28

Solution: (a), (b) The path of point P with respect to cam 1 is the outer profile of the cam in case (a) and a curve at distance r from this profile in case (b). Hence,

$$\vec{v}_{P|1} = \dot{s}\,\vec{e}_t, \qquad \vec{a}_{P|1} = \ddot{s}\,\vec{e}_t + \frac{\dot{s}^2}{\rho_1}\,\vec{e}_n$$

where $\rho_1 = \rho$ for case (a) and $\rho_1 = \rho + r$ for case (b). We write $\vec{v}_P$, $\vec{a}_P$ in two ways:

$$\vec{v}_P = \omega_1 \vec{k} \times \overrightarrow{OP} + \dot{s}\,\vec{e}_t = v_P\,\vec{e}$$

$$\vec{a}_P = \dot{\omega}_1 \vec{k} \times \overrightarrow{OP} - \omega_1^2\,\overrightarrow{OP} + 2\omega_1 \vec{k} \times \dot{s}\,\vec{e}_t + \ddot{s}\,\vec{e}_t + \frac{\dot{s}^2}{\rho_1}\,\vec{e}_n = a_P\,\vec{e}$$

$\dot{s}$, v_P and $\ddot{s}$, a_P are obtained by comparing $\vec{i}, \vec{j}$ components.
(c), (d) The only change is to replace the unknowns v_P, a_P by ω_2, $\dot{\omega}_2$, and $v_P\,\vec{e}$, $a_P\,\vec{e}$ by $\omega_2 \vec{k} \times \overrightarrow{AP}$ and $\dot{\omega}_2 \vec{k} \times \overrightarrow{AP} - \omega_2^2\,\overrightarrow{AP}$.

Review 1 B

1. $\vec{\omega}_{m|F} = \frac{1}{2}(\vec{e}_i \times \dot{\vec{e}}_{i|F})$, $\quad \dot{\vec{A}}_{|F} = \dot{\vec{A}}_{|m} + \vec{\omega} \times \vec{A}$, $\quad$ i.e., $\quad \dot{(\vec{\ })}_{|F} = \dot{(\vec{\ })}_{|m} + \vec{\omega}_{m|F} \times (\vec{\ })$
2. $\vec{\omega}_{2|F} = \vec{\omega}_{1|F} + \vec{\omega}_{2|1}$, $\quad \dot{\vec{\omega}}_{2|F} = \dot{\vec{\omega}}_{1|F} + \dot{\vec{\omega}}_{2|1} + \vec{\omega}_{1|F} \times \vec{\omega}_{2|1}$.
3. $\frac{d}{dt}(\vec{\omega}_{m|F})\Big|_F = \frac{d}{dt}(\vec{\omega}_{m|F})\Big|_m$, $\quad \vec{\omega}_{1|2} = -\,\vec{\omega}_{2|1}$, $\quad \dot{\vec{\omega}}_{1|2} = -\dot{\vec{\omega}}_{2|1}$.
4. If P is a point of body i or its motion relative to it is known, then $\vec{v}_P$, $\vec{a}_P$ are found in terms of a point Q of body i, whose $\vec{v}_Q$, $\vec{a}_Q$ have already been found:

$$\vec{v}_P = \vec{v}_Q + \vec{\omega}_i \times \overrightarrow{QP} + \vec{v}_{P|i}$$

$$\vec{a}_P = \vec{a}_Q + \dot{\vec{\omega}}_i \times \overrightarrow{QP} + \vec{\omega}_i \times (\vec{\omega}_i \times \overrightarrow{QP}) + 2\,\vec{\omega}_i \times \vec{v}_{P|i} + \vec{a}_{P|i}$$

with $\vec{v}_{P|i} = \dot{s}_{P|i}\,\vec{e}_t$, $\vec{a}_{P|i} = \ddot{s}_{P|i}\,\vec{e}_t + (\dot{s}^2_{P|i}\,/\,\rho_P)\,\vec{e}_n$. The last term in $\vec{v}_P$ and the last *two* terms in $\vec{a}_P$ being included only if P has motion relative to body i.

5. An instantaneous axis of rotation does not always exist for a rigid body in spatial motion, but it always exists for plane motion.
6. For a rigid body T in *translation* with respect to frame F,
   i. Displacement, $\vec{v}_{|F}$, $\vec{a}_{|F}$ of all points of T relative to F are equal. $\vec{\omega}_{T|F}(t) \equiv \vec{0}$.
   ii. For any vector $\vec{A}(t)$, $\dot{\vec{A}}_{|T} = \dot{\vec{A}}_{|F}$.
   iii. For any points P, Q: $\vec{v}_{PQ|T} = \vec{v}_{PQ|F}$, $\vec{a}_{PQ|T} = \vec{a}_{PQ|F}$.
7. If a cylindrical body 1 rolls without slip on another such body 2, with their generators in contact and point A of body 1 in contact with body 2, then $\vec{v}_{A|2} = \vec{0}$. $\vec{a}_{A|2}$ *may not be* zero. $\vec{a}_{A|2} \perp$ common tangent plane and is independent of $\dot{\vec{\omega}}_{1|2}$.
8. In general, the acceleration of the instantaneous centre is *not* zero.

1.25 Concept Review Questions 1B

17. (a) A bird flies at uniform velocity relative to the ground reference G. Find the shape of its trajectory observed from a jeep moving at uniform acceleration relative to G. (b) Points A_i, B_i are the material points of rigid body i, $i = 1, 2, 3, 4$. $\vec{\omega}_{i|j}$ denotes the angular velocity of body i relative to body j and $\vec{v}_{A_i|j}$ denotes the velocity of A_i relative to the reference frame of body j. $\vec{A}(t)$ is an arbitrary vector. Fill up the blanks by adding the appropriate angular velocity/position vectors/velocity vectors/$\vec{0}$.

$$\dot{\vec{A}}_{|4} = \dot{\vec{A}}_{|2} + (\ldots) \times \vec{A}, \qquad \vec{v}_{B_2|6} = \ldots + (\ldots) \times (\ldots) + \vec{v}_{B_2|4},$$

$$\vec{\omega}_{2|4} = \vec{\omega}_{2|6} + \vec{\omega}_{6|4} + (\ldots) \times (\ldots), \quad \vec{v}_{A_4|2} = \vec{v}_{A_3|2} + (\ldots) \times \overrightarrow{A_3A_4} + (\ldots),$$

$$\vec{v}_{A_4|1} = \vec{v}_{B_4|1} + (\ldots) \times \overrightarrow{B_4A_4} + (\ldots)$$

(c) The base B of an antenna A rotates relative to G, and the antenna rotates relative to the base B about the axis of the pin at O in direction $\vec{k}$ (Fig. Q1.17a). Find the angular acceleration of the antenna A relative to G. (d) A cylinder rolls without slip with C moving at constant speed v (Fig. Q1.17b). At the given position, find the radii of curvature of the trajectories of A and B. Will these change if C has an acceleration?

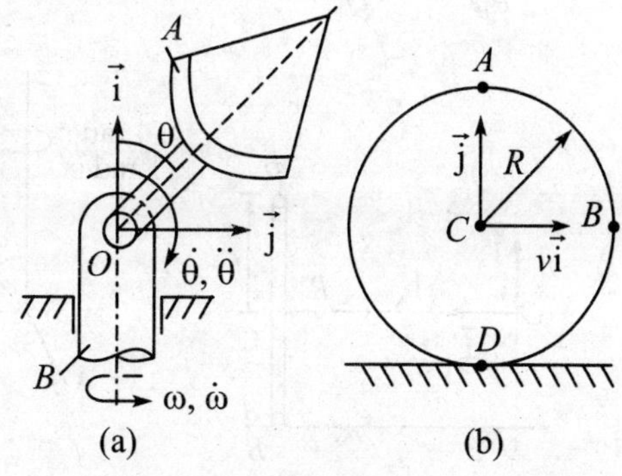

Fig. Q1.17

[(a) Parabola/straight line (b) $\vec{\omega}_{2|4}$, $\vec{v}_{B4|6} + \vec{\omega}_{4|6} \times \overrightarrow{B_4B_2}$ or $\vec{v}_{A4|6} + \vec{\omega}_{4|6} \times \overrightarrow{A_4B_4}$, $\vec{\omega}_{6|4} \times \vec{\omega}_{2|6}$, $\vec{\omega}_{3|2}$ & $\vec{v}_{A4|3}$, $\vec{\omega}_{4|1}$ & $\vec{0}$ (c) $\dot{\omega}\,\vec{\mathrm{i}} - \omega\dot{\theta}\,\vec{\mathrm{j}} + \ddot{\theta}\,\vec{\mathrm{k}}$ (d) $4R$, $2\sqrt{2}R$, no]

18. Write in long hand, $H_{A_i} = I^A_{ij}\omega_j$, $I^A_{ns} = I^A_{ij} n_i s_j$, $a_i b_i$, $\vec{\mathrm{e}}_i \cdot \vec{\mathrm{e}}_i$, $\frac{1}{2} I^C_{ij}\omega_i\omega_j$.

$[I^A_{i1}\omega_1 + I^A_{i2}\omega_2 + I^A_{i3}\omega_3,\ \ I^A_{11}n_1s_1 + I^A_{22}n_2s_2 + I^A_{33}n_3s_3 + I^A_{12}n_1s_2 + I^A_{21}n_2s_1 + I^A_{23}n_2s_3$
$+ I^A_{32}n_3s_2 + I^A_{31}n_3s_1 + I^A_{13}n_1s_3,\ \ a_1b_1 + a_2b_2 + a_3b_3,\ \ \vec{\mathrm{e}}_1\cdot\vec{\mathrm{e}}_1 + \vec{\mathrm{e}}_2\cdot\vec{\mathrm{e}}_2 + \vec{\mathrm{e}}_3\cdot\vec{\mathrm{e}}_3,$
$\frac{1}{2}\{I^C_{11}\omega_1^2 + I^C_{22}\omega_2^2 + I^C_{33}\omega_3^2 + (I^C_{12} + I^C_{21})\omega_1\omega_2 + (I^C_{23} + I^C_{32})\omega_2\omega_3 + (I^C_{31} + I^C_{13})\omega_3\omega_1\}]$

19. $\vec{\mathrm{E}}_i$ and $\vec{\mathrm{e}}_i$ are the bases in frames F and m respectively. The angular velocity and angular acceleration of m relative to F are $\vec{\omega}$ and $\dot{\vec{\omega}}$. A vector $\vec{A} = a_i\,\vec{\mathrm{e}}_i = A_i\,\vec{\mathrm{E}}_i$. (a) Express $\dot{\vec{A}}_{|m}$ in terms of $\vec{\mathrm{e}}_i$. (b) Express $\dot{\vec{A}}_{|F}$ in terms of $\vec{\mathrm{E}}_i$. (c) Express $\dot{\vec{A}}_{|F}$ in terms of $\vec{\mathrm{e}}_i$. (d) If $\vec{\omega} = \omega_i\,\vec{\mathrm{e}}_i$, find $\dot{\vec{\omega}}$. (e) If $\vec{\omega} = \Omega_i\,\vec{\mathrm{E}}_i + \omega_2\,\vec{\mathrm{e}}_2$, find $\dot{\vec{\omega}}$.

[(a) $\dot{a}_i\,\vec{\mathrm{e}}_i$ (b) $\dot{A}_i\,\vec{\mathrm{E}}_i$ (c) $\dot{a}_i\,\vec{\mathrm{e}}_i + \vec{\omega}\times a_i\,\mathrm{e_i}$ (d) $\dot{\omega}_i\,\mathrm{e_i}$ (e) $\dot{\Omega}_i\,\vec{\mathrm{E}}_i + \dot{\omega}_2\,\vec{\mathrm{e}}_2 + \vec{\omega}\times\omega_2\,\vec{\mathrm{e}}_2$]

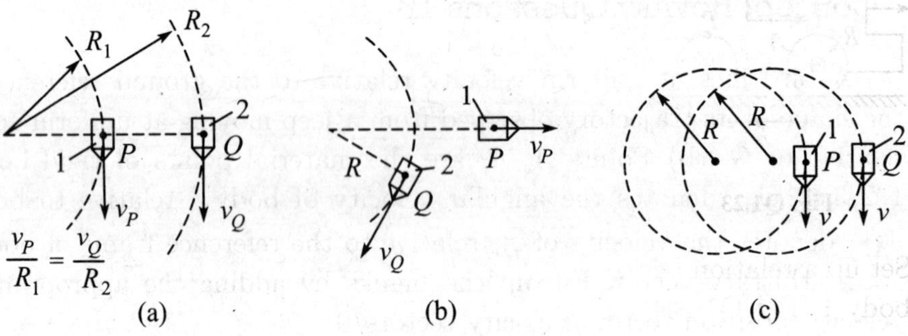

Fig. Q1.20

20. Two ships are moving as shown in Fig. Q1.20. The trajectories of points P and Q fixed to them are shown. In each case, state whether at the given instant, $\vec{v}_{P|2} = -\vec{v}_{Q|1}$. [(a) True (b) False (c) True]

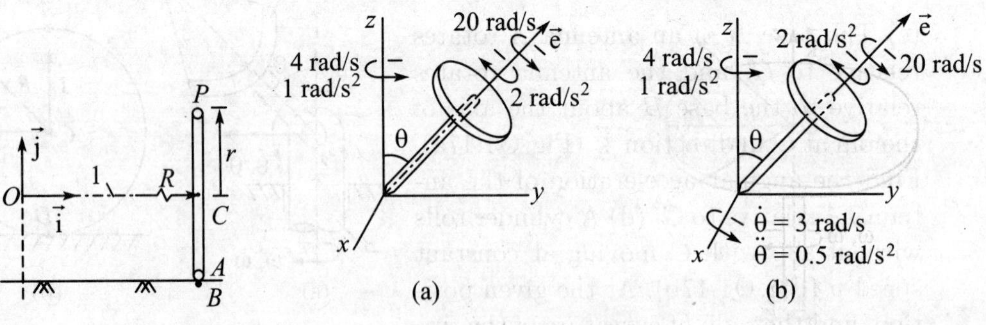

Fig. Q1.21 **Fig. Q1.22**

21. A wheel rolls without slip such that its centre describes a horizontal circle of radius R (Fig. Q1.21) at constant speed v. Find the acceleration of its points P and A. $[-v^2(3\vec{\mathrm{i}}/R+\vec{\mathrm{j}}/r),\ v^2(\vec{\mathrm{i}}/R+\vec{\mathrm{j}}/r)]$

22. (a) A spinning top spins about its axis (Fig. Q1.22a) and its axis precesses about the vertical axis at a fixed angle $\theta = 60°$. Find the angular acceleration of the top at the given instant when the axis is in the yz-plane. (b) Rework part (a) if the top also has a nutation (Fig. Q1.22b).
$[(a)\ -69.28\vec{\mathrm{i}} - 1.732\vec{\mathrm{j}}\ \mathrm{rad/s^2}\ \ (b)\ -69.78\vec{\mathrm{i}} + 16.27\vec{\mathrm{j}} - 51.96\vec{\mathrm{k}}\ \mathrm{rad/s^2}]$

23. A truck moves to the left with velocity v and acceleration a (Fig. Q1.23). Assume that its axles have no motion relative to the main body of the truck and the wheels roll without slip. Find $\vec{v}_{P|1}$, $\vec{a}_{P|1}$. $[-bv\vec{\mathrm{j}}/R,\ bv^2\vec{\mathrm{i}}/R^2 - ba\vec{\mathrm{j}}/R]$

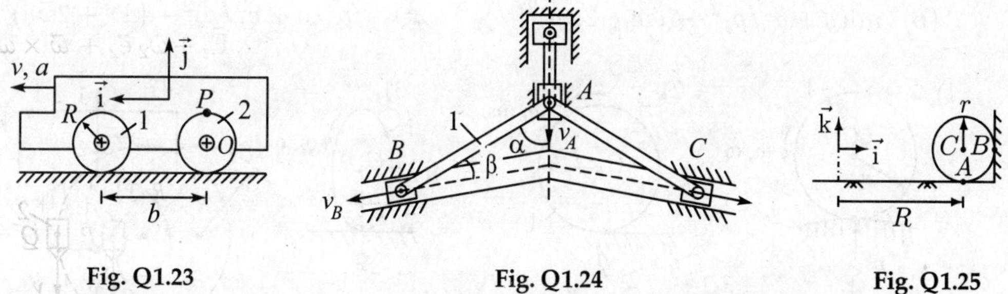

Fig. Q1.23 Fig. Q1.24 Fig. Q1.25

24. Set up a relation between v_A, v_B, α, β and find the angular velocity of the rigid body 1 (Fig. Q1.24). $[v_A \cos\alpha = v_B \cos\beta,\ (v_A \sin\alpha + v_B \sin\beta)/(AB)]$

25*. In a cylindrical cavity, a sphere rolls without slip at A and B (Fig. Q1.25). In the position shown, $\vec{v}_C = v\vec{\mathrm{j}}$ and v is increasing at the rate a. Find the angular velocity and angular acceleration of the sphere. Find $\vec{a}_A$. Why has $\vec{a}_A$ a component in the tangent plane? $[-v(\vec{\mathrm{i}}+\vec{\mathrm{k}})/r,\ -a(\vec{\mathrm{i}}+\vec{\mathrm{k}})/r - v^2\vec{\mathrm{j}}/rR,$ $v^2(\vec{\mathrm{k}}-\vec{\mathrm{i}})/r$, as instantaneous axis of rotation AB is not in this plane$]$

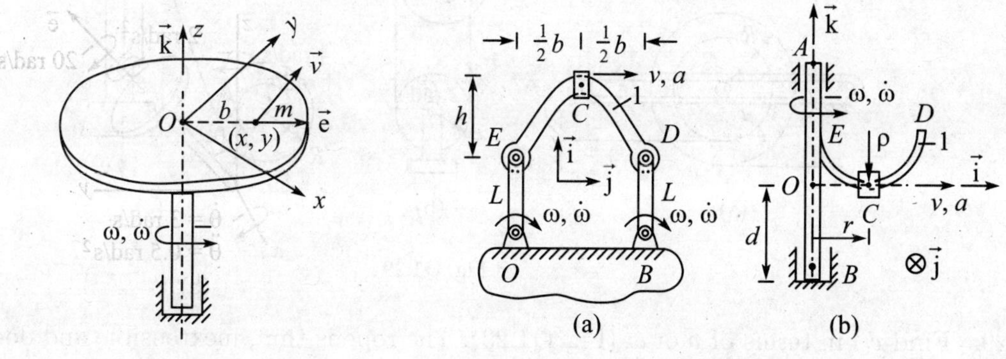

Fig. Q1.26 Fig. Q1.27

26*. A disc rotates about the vertical axis (Fig. Q1.26). A small slider moves with velocity $\vec{v}$ relative to the disc. Find the acceleration of the slider relative to the disc. The coefficient of friction between the slider and the disc is μ. Hence, derive the governing equations for the position (x, y) of the slider relative to the axes x, y attached to the disc. $[\omega^2 b\vec{e} - \dot{\omega}\vec{k} \times b\vec{e} - 2\omega\vec{k} \times \vec{v} - \mu g\vec{v}/|\vec{v}|$,

$\ddot{x} - 2\omega\dot{y} - \dot{\omega}y - \omega^2 x = -\mu g\dot{x}/(\dot{x}^2+\dot{y}^2)^{1/2}$, $\ddot{y} + 2\omega\dot{x} + \dot{\omega}x - \omega^2 y = -\mu g\dot{y}/(\dot{x}^2+\dot{y}^2)^{1/2}]$

27. A bead C of mass m moves over a parabolic rod ED (Fig. Q1.27). The radius of curvature of the parabola at the position of C is ρ and its coefficient of friction is μ. C moves with respect to the rod at speed v which is increasing at an unknown rate a. Find the normal reaction N from the rod on the bead and a.

[(a) $m|g - \omega^2 L - v^2/\rho|$, $-\mu|g - \omega^2 L - v^2/\rho| - \dot{\omega}L$

(b) $m[(g + v^2/\rho)^2 + (\dot{\omega}r + 2\omega v)^2]^{1/2}$, $\omega^2 r - \mu\{(g + v^2/\rho)^2 + (\dot{\omega}r + 2\omega v)^2\}^{1/2}]$

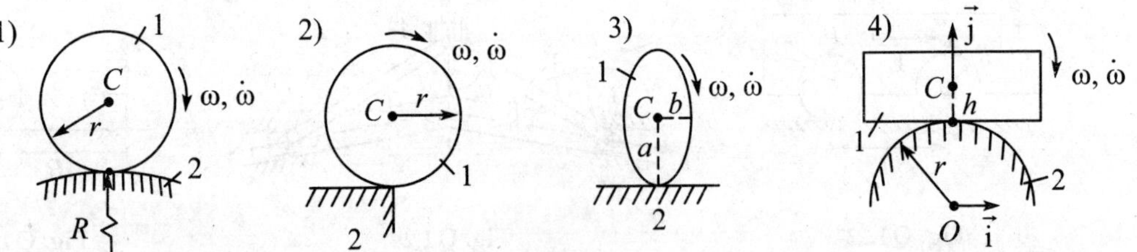

Fig. Q1.28

28. Body 1 rolls without slip on fixed body 2 with angular velocity ω and angular acceleration $\dot{\omega}$ as shown in Fig. Q1.28. Find $\vec{a}_C$. [1. $\dot{\omega}r\,\vec{e}_t + [\omega^2 r^2/(R + r)]\,\vec{e}_n$

2. $\dot{\omega}r\,\vec{e}_t + \omega^2 r\,\vec{e}_n$ 3. $\dot{\omega}a\,\vec{i} + \omega^2(b^2 - a^2)\,\vec{j}/a$ 4. $\dot{\omega}h\,\vec{i} + \omega^2(r - h)\,\vec{j}]$

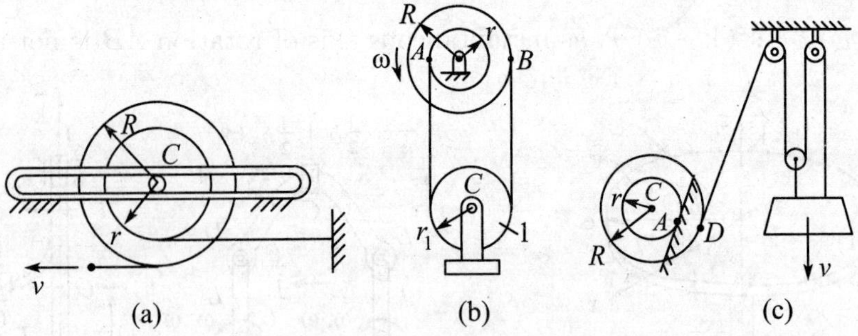

Fig. Q1.29

29. Find $\vec{v}_C$ in terms of v or ω (Fig. Q1.29). The rope is thin, inextensible and does not slip. There is no slip at A for case (c).

[(a) $vr/(R-r)$ to right (b) $\omega(R-r)/2$ up (c) $3vr/(R-r)$ down]

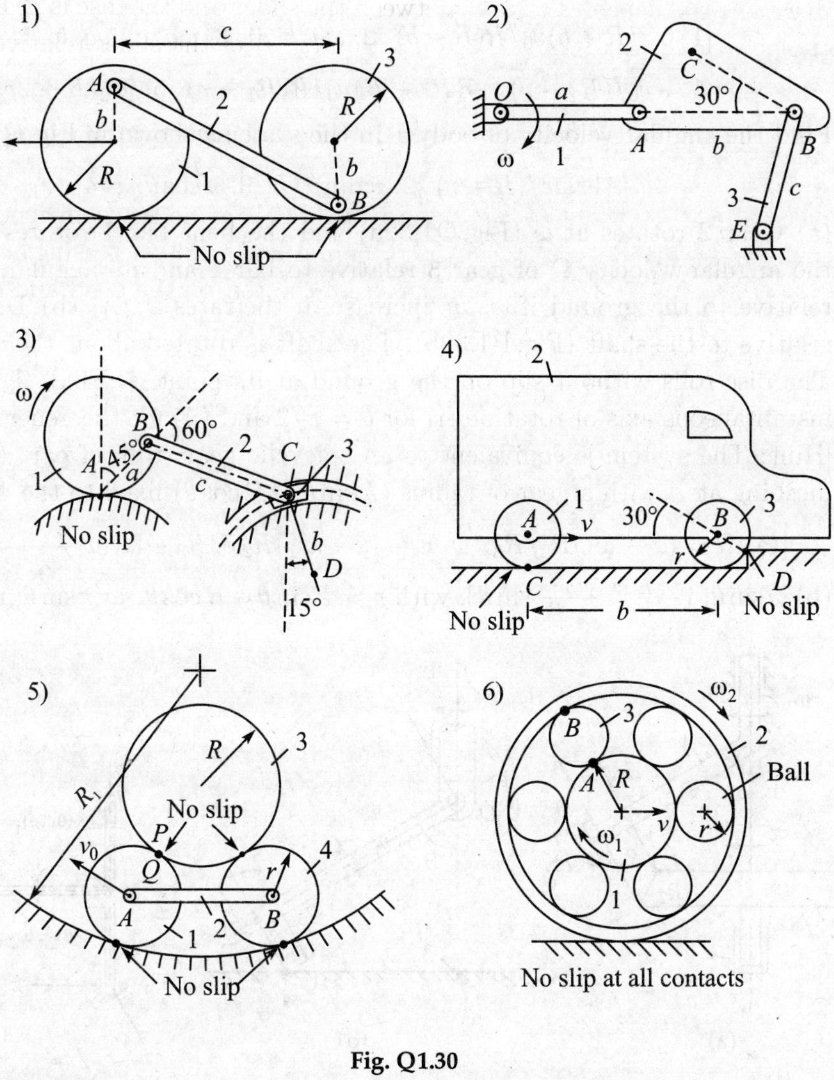

Fig. Q1.30

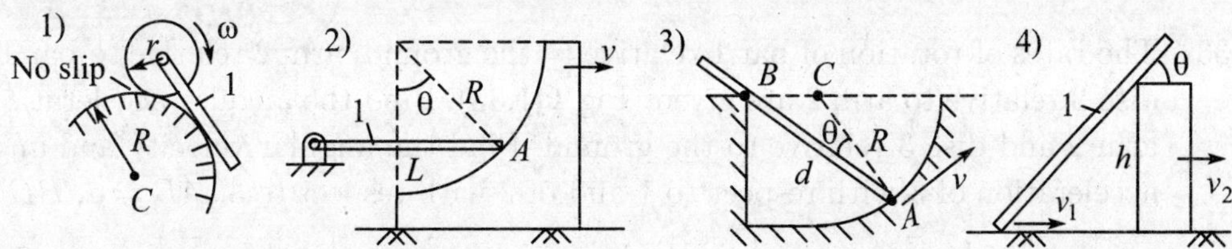

Fig. Q1.31

30. For the systems of rigid bodies shown in Fig. Q1.30, find the angular velocities of body 2 (if unknown) and of body 3.

[1. 0, $(R+b)v_0/R(R-b)$ 2. $\omega a/b$, 0 3. $\omega a/c$, $\omega a/b$ 4. $\sqrt{3}v/b$, $2v/r$ 5. $-v_0/(R_1-r)$, $-(R_1+R)v_0/R(R_1-r)$ 6. $[\omega_2(R+2r)-\omega_1 R]/2r$]

31. Find the angular velocity of body 1 in the systems shown in Fig. Q1.31.

[1. $\omega r/(R+r)$ 2. $v\tan\theta/L$ 3. $v\cos\theta/d$ 4. $(v_1-v_2)\sin^2\theta/h$]

32. (a) Gear 2 rotates at ω (Fig. Q1.32a) and the bent bar 1 rotates at ω_1. Find the angular velocity Ω of gear 3 relative to bar 1 and its angular acceleration relative to the ground if ω, ω_1 increase at the rates $\dot{\omega}$, $\dot{\omega}_1$. (b) Disc 2 rotates relative to the shaft (Fig. P1.32b). The shaft is rotated about the vertical axis. The disc rolls without slip on the ground at its point A. Find $\dot{\vec{\omega}}_2$, $\vec{a}_A$ and its instantaneous axis of rotation (i) for $\theta=\pi/2$ and (ii) for the general value of θ. [Hint: The system is equivalent to an epicyclic gear train of part (a) with disc meshing at A with a gear of radius $(L\sin\theta - R\cos\theta)$ fixed to the ground.]

[(a) $\Omega=(\omega-\omega_1)R_2/R_3$, $\dot{\omega}_1\vec{E}+(\dot{\omega}-\dot{\omega}_1)R_2\vec{e}/R_3+\omega_1(\omega-\omega_1)R_2\vec{E}\times\vec{e}/R_3$ (b) $\dot{\omega}\sin\theta(\vec{j}-\frac{L}{R}\vec{i})+\frac{\omega^2 r}{R}\sin\theta\,\vec{k}$ with $r=L\sin\theta-R\cos\theta$, $\omega^2 r\sin\theta(\vec{i}+\frac{L}{R}\vec{j})$, OA]

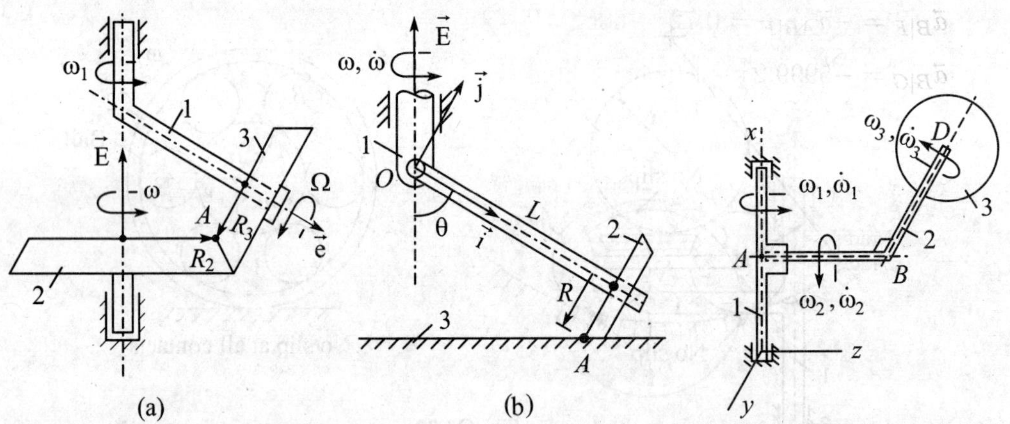

Fig. Q1.32 Fig. Q1.33

33. The rates of rotation of bar 1 relative to the ground, arm 2 relative to bar 1 and disc 3 relative to arm 2 are given (Fig. Q1.33). Find the angular accelerations of arm 2 and disc 3 relative to the ground. Find the angular velocity and angular acceleration of 3 with respect to 1 and of 2 with respect to 3. $AB=d$, $BD=a$.

[$\dot{\omega}_1\vec{i}-\omega_1\omega_2\vec{j}+\dot{\omega}_2\vec{k}$, $(\dot{\omega}_1-\omega_2\omega_3)\vec{i}+(\dot{\omega}_3-\omega_1\omega_2)\vec{j}+(\dot{\omega}_2+\omega_1\omega_3)\vec{k}$; $\omega_2\vec{k}+\omega_3\vec{j}$, $-\omega_2\omega_3\vec{i}+\dot{\omega}_3\vec{j}+\dot{\omega}_2\vec{k}$; $-\omega_3\vec{j}$, $-\dot{\omega}_3\vec{j}$]

1.26 Practice Problems 1B

17. A rigid tee 1 rotates about a fixed axis (Fig. P1.17). A rigid disc 2 mounted on the arm OC rotates relative to it about an axis at C. A block slides relative to the disc at 1 m/s which is increasing at 2 m/s^2. Find, at the given instant, (a) $\vec{\omega}_2$, $\dot{\vec{\omega}}_2$, (b) $\vec{a}_C$ and (c) $\vec{a}_P$. (d) Find the result of part (c) if the slot is a circle with centre at D. [(a) $\vec{k}+2\vec{i}$ rad/s, $3\vec{i}+2\vec{j}+2\vec{k}$ rad/s^2 (b) $-4\vec{i}-2\vec{j}$ m/s^2 (c) $-2\vec{i}-3\vec{j}$ m/s^2 (d) $-2\vec{i}-3\vec{j}-0.5\vec{k}$ m/s^2]

18. The landing gear of an aircraft is being retracted by rotating the gear support arm AD about a hinge at A in the fuselage, at rates of rotation relative to it as shown in Fig. P1.18 (aircraft as viewed from the front). The landing wheel is spinning at the given rates relative to the arm AD. The aircraft is gaining height in a vertical loop of radius $R_1 = 1000$ m and is at the bottom of the loop. The speed of the aircraft is v and is increasing at $\dot{v}$. Find $\vec{v}_B$, $\vec{a}_B$, $\vec{v}_{AB}$, $\vec{a}_{AB}$ (1) relative to frame F of the aircraft and (2) relative to the ground reference. $\theta = 30°$, $v = 60$ m/s, $\dot{v} = 15$ m/s^2, $\omega_\alpha = 0.8$ rad/s, $\dot{\omega}_\alpha = 0.2$ rad/s^2, $\omega_\beta = 100$ rad/s, $\dot{\omega}_\beta = -10$ rad/s^2, $r = 0.6$ m, $R = 3$ m.

[$\vec{v}_{B|F} = -\vec{v}_{AB|F} = 2.88\vec{j} - 60\vec{k}$ m/s, $\vec{v}_{B|G} = 2.88\vec{j} + 0.187\vec{k}$ m/s,

$\vec{a}_{B|F} = -\vec{a}_{AB|F} = 0.72\vec{j} - 6002.3\vec{i} + 6\vec{k}$, $\vec{a}_{AB|G} = 5996.1\vec{i} + 2.874\vec{j} - 5.874\vec{k}$ m/s^2

$\vec{a}_{B|G} = -5999.2\vec{i} - 1.0744\vec{j} + 20.874\vec{k}$ m/s^2, $\vec{v}_{AB|G} = -2.88\vec{j} + 59.813\vec{k}$ m/s,]

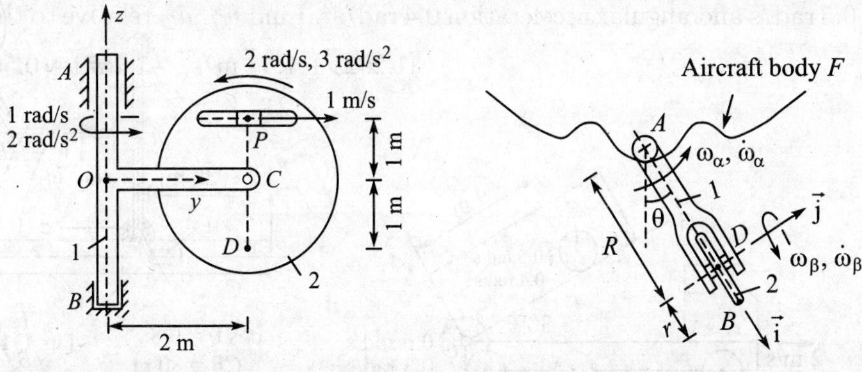

Fig. P1.17 Fig. P1.18

19*. A robotic arm, to pour molten metal into molds (Fig. P1.19), consists of link DA fixed to the base plate, D, and a link AB connected to the bucket. Given $DA = 7$ m, $AB = 2.5$ m, $\alpha = 60°$, $\beta = 45°$, $\omega_b = 0.4$ rad/s, $\dot{\omega}_b = 0.2$ rad/s^2, $\omega_a = 0.8$ rad/s, $\dot{\omega}_a = 0.2$ rad/s^2. The bucket is rotated relative to link AB so that it remains vertically upright and the molten metal does not spill. Find the angular velocity and angular acceleration of the bucket relative to link AB and relative to the ground. Also find $\vec{v}_P$, $\vec{a}_P$. [$-0.8\vec{k}$ rad/s, $-0.2\vec{k}$ rad/s^2; $0.4\vec{j}$ rad/s, $0.2\vec{j}$ rad/s^2, $1.414(\vec{i}+\vec{j})-3.376\vec{k}$ m/s, $-2.127\vec{i}+1.485\vec{j}-2.817\vec{k}$ m/s^2]

20*. In an amusement park ride, the cockpit containing the passengers rotates relative to the telescopic arm ACB (Fig. P1.20). The length of the arm ACB increases as shown and it swings relative to the arm AO. The arm AO rotates relative to the fixed vertical axis. Point P is at the centre of the eye of a passenger strapped to a seat fixed in the cockpit. Find $\vec{a}_P$. [$0.04\vec{i} - 2.602\vec{j} + 2.978\vec{k}$ m/s^2]

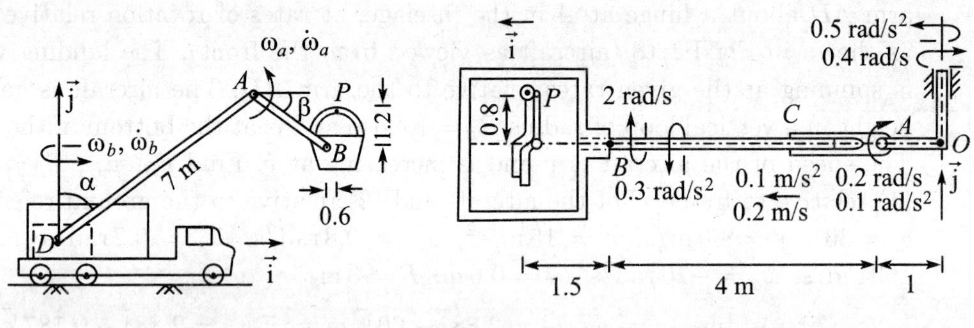

Fig. P1.19 Fig. P1.20

21. The loading compartment of a truck (Fig. P1.21) rotates as shown. A cylinder of radius 0.4 m rolls without slip relative to the compartment at angular velocity 0.5 rad/s and angular acceleration 0.4 rad/s^2. Find $\vec{v}_C$, $\vec{a}_C$ relative to the ground. [$1.292\vec{i} + 1.4\vec{j}$ m/s, $-1.287\vec{i} + 0.566\vec{j}$ m/s^2]

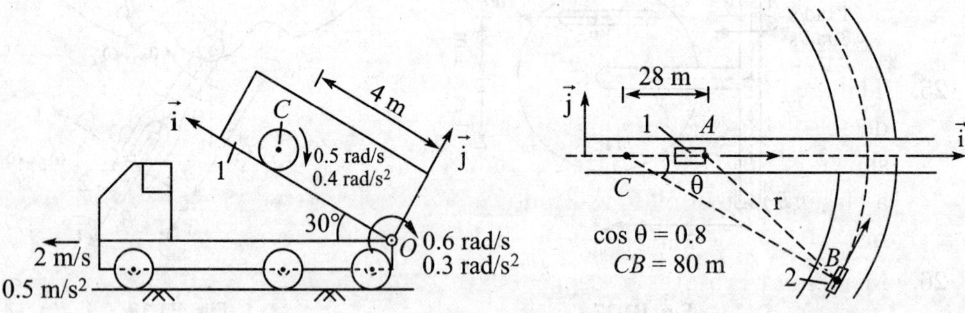

Fig. P1.21 Fig. P1.22

22. Car 1 approaches an over-pass at a speed of 10 m/s which is decreasing at 2 m/s^2 (Fig. P1.22). Point B on car 2 moves at a constant speed of 20 m/s in a circle of radius 80 m with $\cos\theta = 0.8$. The ground reference is G. Find (a) $\vec{v}_{BA|G}$, $\vec{a}_{BA|G}$, (b) $\vec{v}_{B|1}$, $\vec{a}_{B|1}$, (c) $\vec{v}_{A|2}$, $\vec{a}_{A|2}$. (d) Find derivatives $\dot{r}$, $\ddot{r}$ of the separation $AB = r$

at the given position as measured in reference frames $G, 1, 2$.

[(a) $2\vec{i}+16\vec{j}$, $-2\vec{i}+3\vec{j}$ (b) $2\vec{i}+16\vec{j}$, $-2\vec{i}+3\vec{j}$ (c) $10\vec{i}-7\vec{j}\,$m/s, $-3.75\vec{i}-5\vec{j}\,$m/s^2
(d) $\dot{r}=-11.6\,$m/s, $\ddot{r}=-1.509\,$m/s^2 relative to frames G, 1, 2]

23. The radar equipment on board ship 1 follows ship 2 by measuring the distance R and angle ϕ with respect to the longitudinal axis of ship 1 (Fig. P1.23). Assume that both ships are moving in a plane. Ship 1 is taking a right turn having radius R_1 and its speed is constant at v_1. At the instant shown, the radar indicates the values of R, $\dot{R}$, $\ddot{R}$, ϕ, $\dot{\phi}$, $\ddot{\phi}$. Find $\vec{v}_P$ and $\vec{a}_P$ with respect to the ground.

[$\dot{R}\,\vec{e}_R + v_1\,\vec{e}_t + R(\dot{\phi}-v_1/R_1)\,\vec{e}_\phi$,
$(v_1^2/R_1)\,\vec{e}_n + (\ddot{R}-R\dot{\phi}^2+2v_1R\dot{\phi}/R_1-v_1^2R/R_1^2)\,\vec{e}_R + (2\dot{R}\dot{\phi}+R\ddot{\phi}-2v_1\dot{R}/R_1)\,\vec{e}_\phi$]

24. An arm OA rotates at angular velocity ω_1 and angular acceleration $\dot{\omega}_1$ at the instant shown in Fig. P1.24. Find $\vec{v}_P$ and $\vec{a}_P$ at the given instant.

[$\omega_1 r\,\vec{i}+\omega_1 r\,\vec{j}$, $(\dot{\omega}_1 r-2\omega_1^2 r)\,\vec{i}+(\dot{\omega}_1 r-0.5\omega_1^2 r)\,\vec{j}$]

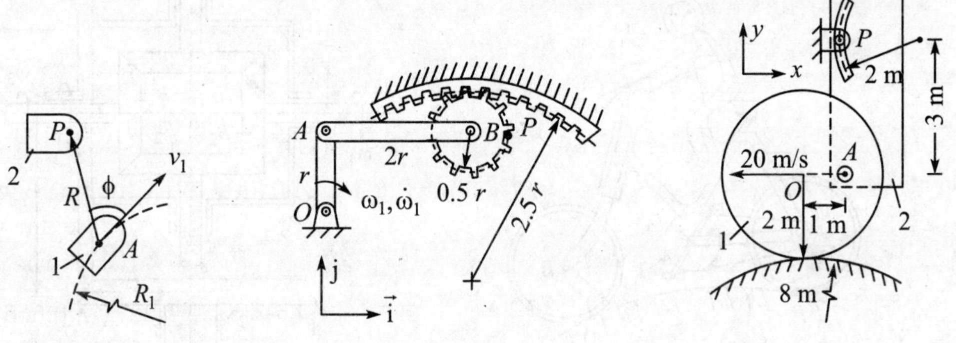

Fig. P1.23 **Fig. P1.24** **Fig. P1.25**

25. The centre O of disc 1, rolling without slip, has a speed of 20 m/s which is decreasing at the rate of 4 m/s^2 at the instant when A is vertically below P as shown in Fig. P1.25. Plate 2 is pinned to the disc at A. A fixed pin P can slide in a circular slot in plate 2. Find the angular acceleration of the plate at the given instant. [$-59.78\,\vec{k}$ rad/s^2]

26. Member OA rotates at a constant angular velocity of 2 rad/s when in the position shown in Fig. P1.26. Find the angular velocity and angular acceleration of body 3, velocity of point B and the velocity of body AB relative to body 3.

[$0.64\,\vec{k}$ rad/s, $1.984\,\vec{k}$ rad/s^2, $8.96\vec{i}-6.72\vec{j}$ cm/s]

27. The drive wheel of a tape-recorder has a constant angular velocity ω_1. The thickness of the tape is t. Find the angular acceleration of the driven wheel at the instant shown in Fig. P1.27. [$\omega_1^2(1+r_1^2/r_2^2)t/2\pi r_2$]

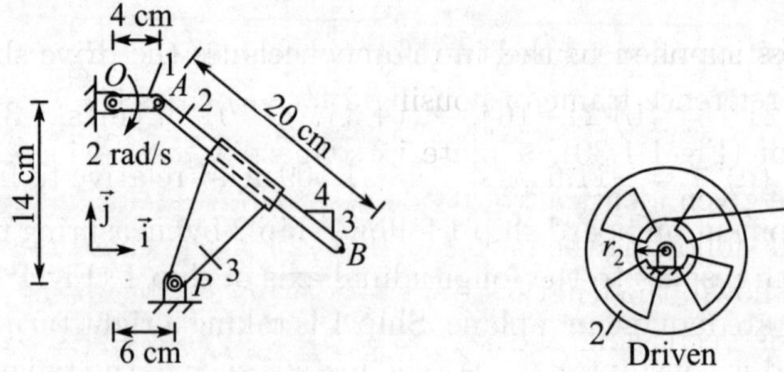

Fig. P1.26 Fig. P1.27

28. A cycle (Fig. P1.28) is moving without slip over a cylindrical mound with no rotation of the handle with respect to the main frame. Find the angular velocity of the wheels relative to the ground if the angular velocity of the axle of the paddles relative to the ground is ω. $[\omega r_1(R+r)/(rr_1+Rr_2)]$

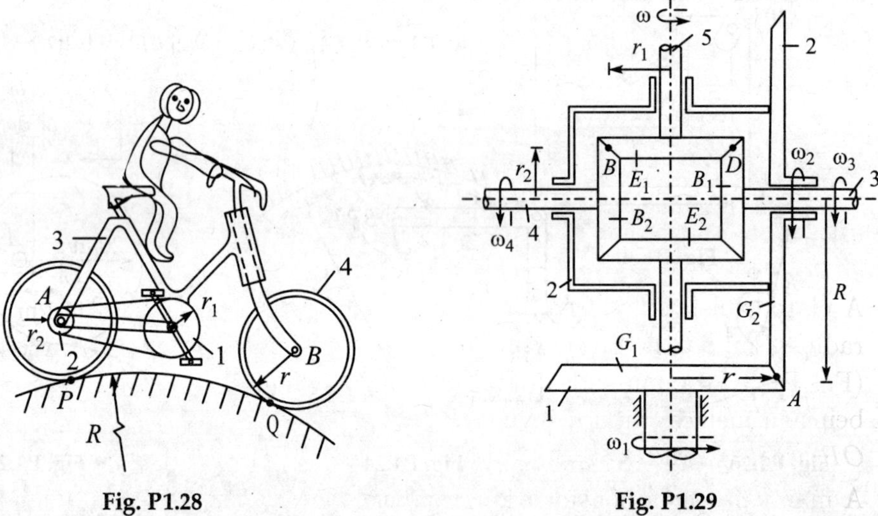

Fig. P1.28 Fig. P1.29

29. When an automobile turns around a sharp curve, the centres of the inner and outer tyres have different velocities as they move over curves of different radii. Thus, both tyres cannot roll without slip if they are mounted on a rigid shaft. To avoid this, the axle is made in two parts and driven through a *differential gear* (Fig. P1.29). Drive shaft 1 rotates at angular velocity ω_1 relative to the vehicle and drives the housing 2, through bevel gears G_1 and G_2, with angular velocity ω_2 relative to the vehicle. Two or four identical epicyclic gears E_1, E_2 are free to rotate about axes fixed to housing 2. These gears mesh with bevel gears B_1, B_2 fixed to the rear wheel shafts 3 and 4. Let ω_3, ω_4 be the angular velocities of the shafts 3, 4 relative to the vehicle and ω be the angular velocity of E_1 relative to housing 2. Using no slip conditions at points D and B show that $\omega_2 = (\omega_3+\omega_4)/2 = \omega_1 r/R$, $\omega = (\omega_4-\omega_3)r_2/2r_1$. Neglecting losses, show

that the torques supplied to the two rear wheels by the drive shaft are equal. [Hint: Use the reference frame of housing 2 for velocities.]

30. In a mechanism (Fig. P1.30), a plate is connected to two identical telescopic arms, which are given identical rotations. A pin moves relative to a smooth slot, with its centre P following the path $x = 0.9 - 2y^2 + y^3$, where x, y are in metres. At the instant shown, the lengths of both the arms are increasing at 0.8 m/s and this rate itself is decreasing at the rate of 0.2 m/s^2. The location of P at this instant is at $y = 0.3$ m and its speed is $v = 2$ m/s increasing at an unknown rate $\dot{v}$ relative to the plate. Find $\vec{a}_D$ and $\dot{v}$. $[0.235\vec{\imath} + 0.713\vec{\jmath},\ -7.546\,\text{m/s}^2]$

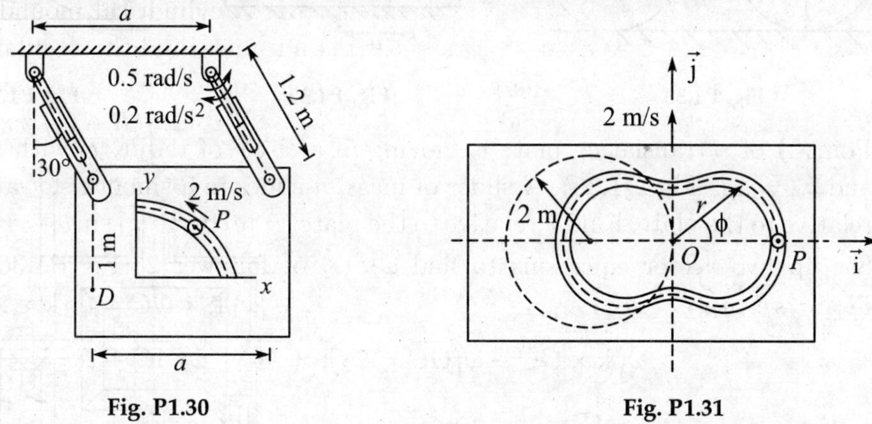

Fig. P1.30

Fig. P1.31

31. A plate translates relative to the ground such that point O describes a circle of radius 2 m. At the given instant, $\vec{v}_O = 2\vec{\jmath}$ m/s and v_O is decreasing at 0.5 m/s^2 (Fig. P1.31). A pin slides in a curved slot $r = 2 + \cos 2\phi$ in the plate with r being in metres. Find $\vec{a}_P$ when P is on the x-axis and the rates of rotation of OP are $\dot{\phi} = 2$ rad/s, $\ddot{\phi} = -3$ rad/s^2. $[-30\vec{\imath} - 9.5\vec{\jmath}\ \text{m/s}^2]$

32. A man can row a boat at 6 km/h in still water. The water, in a river of width 0.5 km, flows at 3 km/h. (a) In which direction should the boat be headed, if he wants to reach a point directly opposite his starting point? Find the time taken to cross. (b) Find the time taken to row across to a point 3 km downstream and then back to the starting point. (c) In what direction should he head the boat if he wants to cross the river in the shortest time? Where would he land in this case? (d) The man can run at 10 km/h. He wants to get to a point across the stream and 3 km downstream of his starting point in the least time. Find the downstream distance to be covered by rowing. [(a) 120° to flow, 0.09622 h (b) 1.346 h (c) 90° to flow, 0.25 km downstream (d) 1.318 km downstream]

33. Plank 3 rests on top of cylinders 1 and 2 (Fig. P1.33). There is no slip at any contact. Point O has velocity v and acceleration a. Find the angular velocity and angular acceleration of 2 and 3. Find $\vec{v}_C$, $\vec{a}_C$.

$[-v\vec{k}/c,\ \vec{0};\ -a\vec{k}/c,\ \vec{0};\ (1+\cos\theta)v\vec{i}-v\sin\theta\vec{j},\ (1+\cos\theta)a\vec{i}-a\sin\theta\vec{j}]$

34*. A thrust ball bearing of a vertical shaft (Fig. P1.34) is to be designed so that there is no slip at A, B, C. Prove that the required relation between a, r and β is $a/r = (1+\sin\beta)/(\cos\beta - \sin\beta)$.

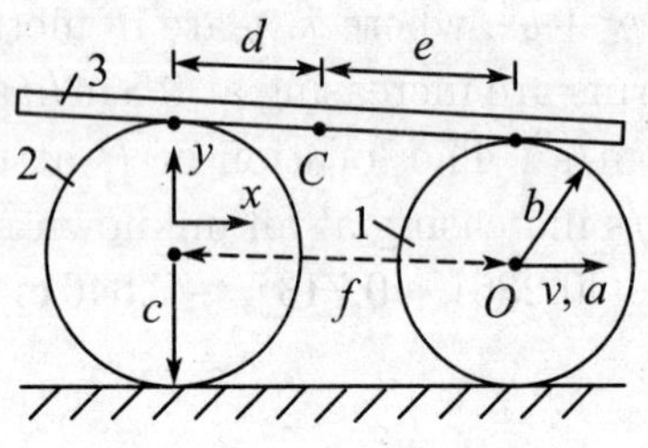

Fig. P1.33

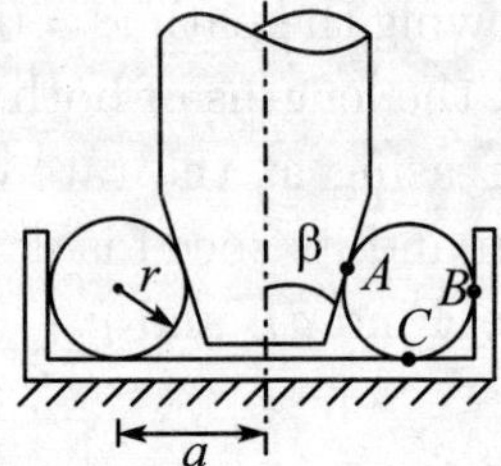

Fig. P1.34

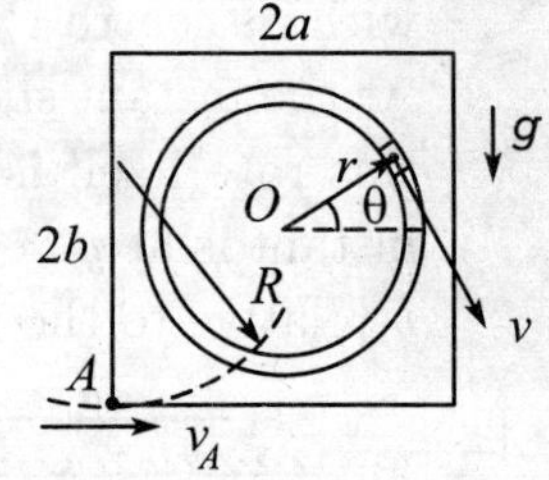

Fig. P1.35

35. Point A of a translating plate is moving in a circle of radius R with speed v_A and $\dot{v}_A = a_A$ (Fig. Q1.35). A slider of mass m slides in a smooth slot at speed v relative to the plate. Find $\dot{v}$ relative to the plate. $[g\cos\theta - a_A\sin\theta + v_A^2\cos\theta\,/R]$

36. Set up two vector equations to find ω_2, $\dot{\omega}_2$ of follower 2 (Fig. P1.36). [Hint: $\vec{v}_{P|2} = \dot{s}\,\vec{e}$, $\vec{a}_{P|2} = \ddot{s}\,\vec{e}$] $[\omega_1\vec{k}\times\overrightarrow{OC} = \omega_2\vec{k}\times\overrightarrow{AC} + \dot{s}\,\vec{e}$,

$\dot{\omega}_1\vec{k}\times\overrightarrow{OC} - \omega_1^2\overrightarrow{OC} = \dot{\omega}_2\vec{k}\times\overrightarrow{AC} - \omega_2^2\overrightarrow{AC} + 2\omega_2\vec{k}\times\dot{s}\,\vec{e} + \ddot{s}\,\vec{e}]$

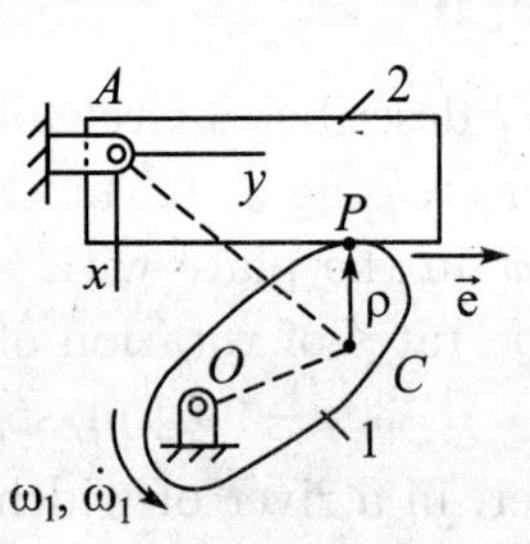

Fig. P1.36

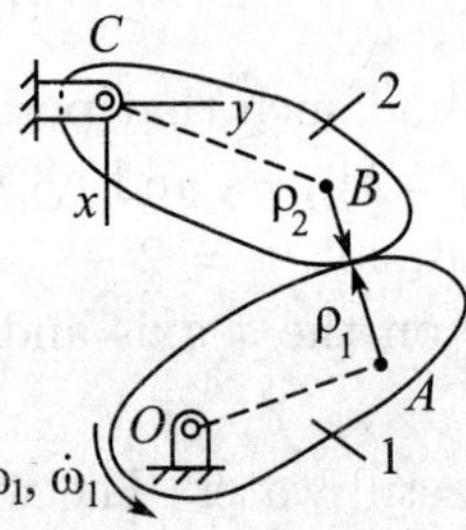

Fig. P1.37

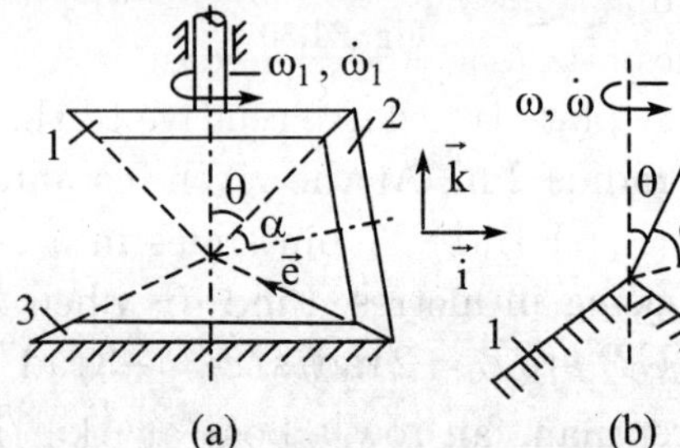

Fig. P1.38

37*. Set up two vector equations to find ω_2, $\dot{\omega}_2$ of follower 2 (Fig. P1.37). [Hint: This system is equivalent to a 4-bar mechanism consisting of links OA, AB, BC]

$[\omega_2\vec{k}\times\overrightarrow{CB} = \omega_1\vec{k}\times\overrightarrow{OA} + \Omega\vec{k}\times\overrightarrow{AB}$,

$\dot{\omega}_2\vec{k}\times\overrightarrow{CB} - \omega_2^2\overrightarrow{CB} = \dot{\omega}_1\vec{k}\times\overrightarrow{OA} - \omega_1^2\overrightarrow{OA} + \dot{\Omega}\vec{k}\times\overrightarrow{AB} - \Omega^2\overrightarrow{AB}]$

38*. (a) For the bevel gear system in Fig. P1.38a, find $\vec{\omega}_2$, $\dot{\vec{\omega}}_2$. (b) Cone 2 rolls without slip on cone 1 (Fig. P1.38b), with its centreline rotating at ω, $\dot{\omega}$. Find $\vec{\omega}_2$, $\dot{\vec{\omega}}_2$.

$[(a)\ \omega_1\sin\theta\,\vec{e}/\sin 2\alpha,\ \ \frac{\sin\theta}{\sin 2\alpha}[\dot{\omega}_1\vec{e} - \{\omega_1^2\sin(2\alpha+\theta)\sin\theta\sin\alpha/\sin 2\alpha\sin(\theta+\alpha)\}\vec{j}\,]$

$(b)\ \omega\sin(\theta+\alpha)\,\vec{e}/\sin\alpha,\quad \{\dot{\omega}\,\vec{e} - \omega^2\sin(\theta+2\alpha)\vec{j}\,\}\sin(\theta+\alpha)/\sin\alpha]$

CHAPTER

2

AXIOMS AND FORCE SYSTEMS

IN this chapter, force and mass are introduced. Moment of a force, work, kinetic energy, momentum and moment of momentum are defined. Euler's axioms are presented. Equivalent force systems and resultant of a force system are defined. The axioms of Coulomb friction, rolling friction, electromagnetic and gravitational forces are presented. The principles of impulse–momentum and angular impulse–moment of momentum are derived. The work–energy relation for centre of mass is established. Free body diagrams and reactions are discussed. Conservative forces and their potential energies are presented. The dynamics of the centre of mass of any system is covered.

2.1 Force and Moment

The concept of force (a primitive) arises from muscular effort. Force is a *quantitative measure of the mechanical interaction of material bodies in contact or at a distance.* It is represented by an arrow, whose length gives its magnitude, direction gives its line of action, arrow-head gives its sense and the tail usually represents its point of application P (Fig. 2.1a). It is denoted by $\vec{F}$ with magnitude $F = |\vec{F}|$. Its SI unit is newton (N). The forces acting on a body (Fig. 2.1b) may be modelled as *discrete* forces, $\vec{F}_i$ N acting at point i with position vector $\vec{r}_i$, or *distributed* forces distributed over a volume, a surface or a line. These may be functions of time t. The force distributed over a volume is called *body force* and is described by a *body force density* $\vec{f}_b\,\mathrm{N/m^3}$, for example, gravitational force. The force distributed over a surface is called *surface force* and is described by a *surface force density* $\vec{f}_s\,\mathrm{N/m^2}$, for example, pressure. The force distributed over a line is called *line force* and is described by a *line force density* $\vec{f}_l\,\mathrm{N/m}$, for example, load per unit length on a beam. A discrete force is an idealisation of a physical force distributed over a 'small' contact area or over a 'small' volume.

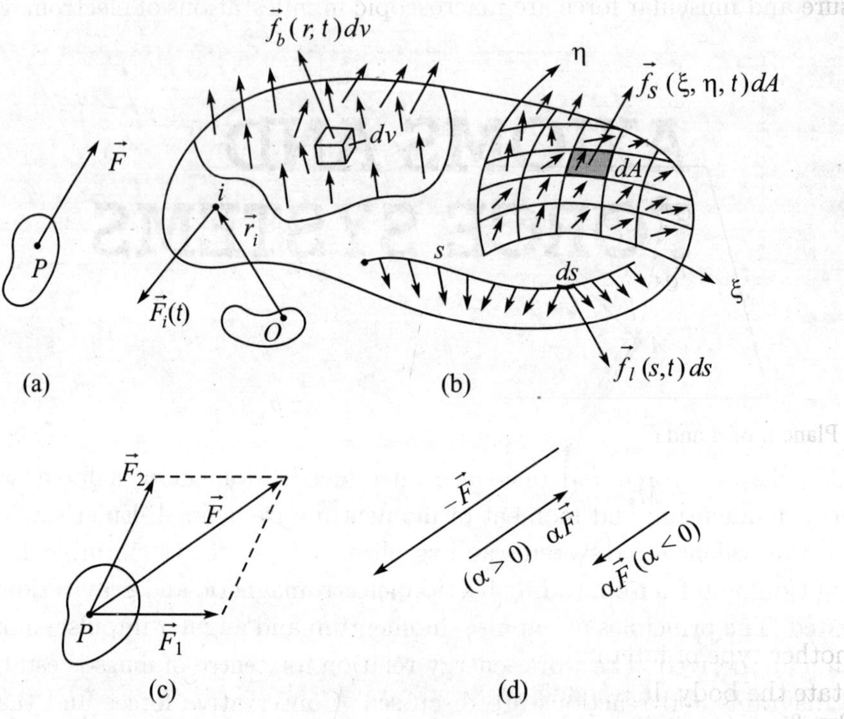

Fig. 2.1

Forces are frame-indifferent axiomatically, i.e., the interaction between material bodies is independent of the reference frame of observation. All *forces are governed by the parallelogram law (axiom) of forces*: 'the effect of forces $\vec{F}_1$ and $\vec{F}_2$ acting at the same point P of a body is the same as due to a single force $\vec{F}$ acting at the same point, which is represented by the diagonal arrow of a parallelogram formed by the two force arrows $\vec{F}_1, \vec{F}_2$, emanating at point P as adjacent sides (Fig. 2.1c). We write this addition as $\vec{F} = \vec{F}_1 + \vec{F}_2$. The product of a force $\vec{F}$ with a scalar α, denoted by $\alpha\vec{F}$, is a force of magnitude $|\alpha||\vec{F}|$ with the same point of application and line of action as $\vec{F}$ and sense the same as or opposite to that of $\vec{F}$ according as α is positive or negative (Fig. 2.1d). We define $-\vec{F} = (-1)\vec{F}$.

Four fundamental forces: gravitational, electromagnetic, weak nuclear and strong nuclear forces arise due to interaction between bodies. Electromagnetic forces of attraction and repulsion cancel each other to a large extent in large systems and gravity is predominant. Electromagnetic forces are much stronger than gravitational forces on the atomic scale and play a significant role in most chemical and biological processes. Nuclear forces are stronger compared to others for subatomic particles at distances of the order of 10^{-15} m and are negligible when the separation of the interacting particles is greater than 10^{-10} m. Frictional force, spring force, rope tension, hydrostatic pressure and muscular force are macroscopic manifestations of electromagnetic forces.

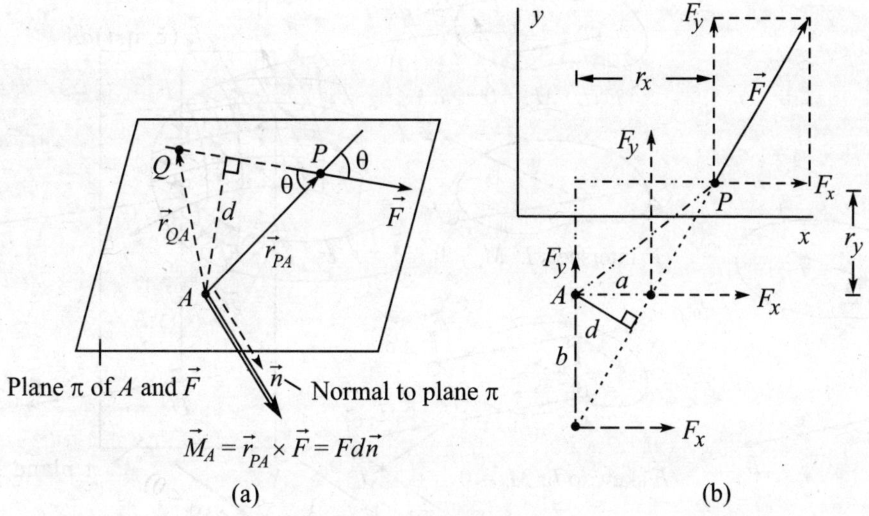

Fig. 2.2

Another type of interaction between bodies is called *moment* (a primitive). It tends to rotate the body. It is modelled as an arrow vector acting at a point with dimension $[ML^2T^{-2}]$ and with the operations of addition and scalar multiplication defined just as for forces. This concept finds application in polar media and dislocation theory. Closely related to moment is the entity of *moment* $\vec{M}_A$ of a force $\vec{F}$ acting at point P about a point A (Fig. 2.2a) with units of newton metre (N.m). It is defined by

$$\begin{aligned} \vec{M}_A &= \vec{r}_{PA} \times \vec{F} = \overrightarrow{AP} \times \vec{F} = (AP)\, F \sin\theta\, \vec{n} = F d\, \vec{n} \qquad (2.1) \\ &= \begin{vmatrix} \vec{\mathrm{i}} & \vec{\mathrm{j}} & \vec{\mathrm{k}} \\ r_x & r_y & r_z \\ F_x & F_y & F_z \end{vmatrix} \end{aligned}$$

where $\vec{F} = F_x \vec{\mathrm{i}} + F_y \vec{\mathrm{j}} + F_z \vec{\mathrm{k}}$, $\vec{r}_{PA} = r_x \vec{\mathrm{i}} + r_y \vec{\mathrm{j}} + r_z \vec{\mathrm{k}}$ and $\vec{n}$ is the unit vector normal to the plane of $\vec{F}$ and A, given by the cross product. Note that $\vec{M}_A = \overrightarrow{AQ} \times \vec{F}$, where $\overrightarrow{AQ}$ is the relative position vector of *any* convenient point Q on the line of action of $\vec{F}$, as

$$\overrightarrow{AQ} \times \vec{F} = (\overrightarrow{AP} + \overrightarrow{PQ}) \times \vec{F} = \overrightarrow{AP} \times \vec{F} = \vec{M}_A \qquad (\overrightarrow{PQ} \parallel \vec{F})$$

$\vec{M}_A$ is zero if $\vec{F}$ passes through A. For a force $\vec{F}$ coplanar with point A in the xy-plane (Fig. 2.2b), $\vec{n} = \pm \vec{\mathrm{k}}$ and eq. (2.1) yields

$$\vec{M}_A = (r_x F_y - r_y F_x)\, \vec{\mathrm{k}} = \pm |F_y a|\, \vec{\mathrm{k}} = \pm |F_x b|\, \vec{\mathrm{k}} = \pm F d\, \vec{\mathrm{k}} \qquad (2.2)$$

with +ve/−ve according as $\vec{F}$ tends to rotate the body about A in direction $\vec{\mathrm{k}}/-\vec{\mathrm{k}}$.

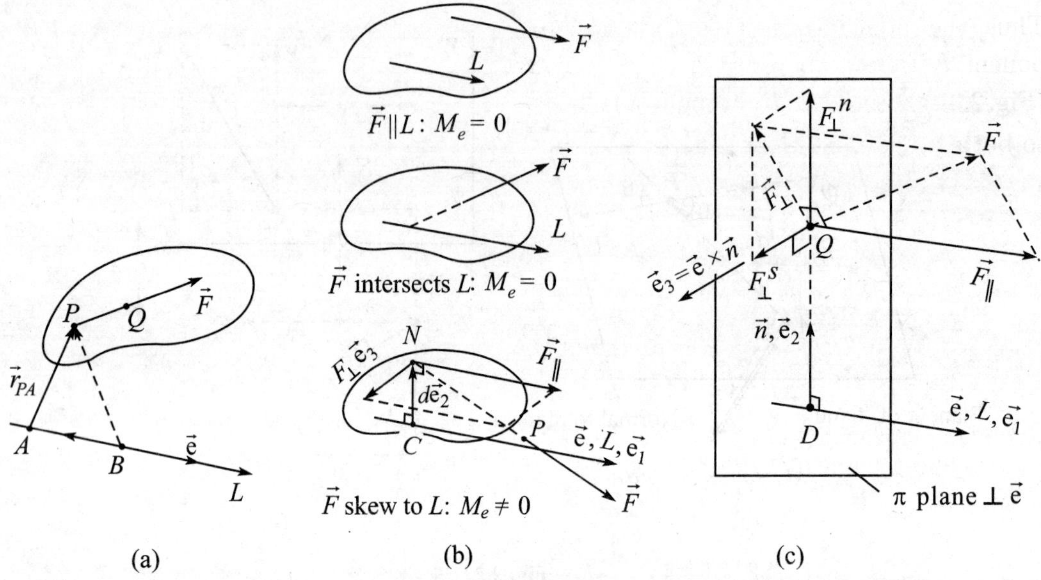

Fig. 2.3

The *moment* $\vec{M}_e$ *of the force* $\vec{F}$ *about a directed line* L (Fig. 2.3a) with unit vector $\vec{\mathrm{e}}$ along it is defined as the component along $\vec{\mathrm{e}}$ of its moment $\vec{M}_A$ about *any* point A on L:

$$M_e = \vec{M}_A \cdot \vec{\mathrm{e}} = \overrightarrow{AP} \times \vec{F} \cdot \vec{\mathrm{e}} = \begin{vmatrix} r_x & r_y & r_z \\ F_x & F_y & F_z \\ e_x & e_y & e_z \end{vmatrix}, \qquad \vec{M}_e = M_e\,\vec{\mathrm{e}} \tag{2.3}$$

where $\vec{\mathrm{e}} = e_x\,\vec{\mathrm{i}} + e_y\,\vec{\mathrm{j}} + e_z\,\vec{\mathrm{k}}$. $\vec{M}_e$ is well-defined since for any other point B on L,

$$\begin{aligned} \vec{M}_B \cdot \vec{\mathrm{e}} &= \overrightarrow{BP} \times \vec{F} \cdot \vec{\mathrm{e}} = (\overrightarrow{BA} + \overrightarrow{AP}) \times \vec{F} \cdot \vec{\mathrm{e}} \\ &= \overrightarrow{BA} \times \vec{F} \cdot \vec{\mathrm{e}} + M_e = M_e \qquad (\overrightarrow{BA} \parallel \vec{\mathrm{e}}) \end{aligned}$$

Equation (2.3) $\Rightarrow \vec{M}_e = \vec{0}$ iff $\vec{F}$ intersects L or is parallel to L (Fig. 2.3b). $\vec{M}_e \neq \vec{0}$ only if $\vec{F}$ and L are skew lines. Note that $\vec{M}_A$ is the vector sum of the moments of the force about three orthogonal axes through A.

Consider the plane π through *any* point Q on the line of action of $\vec{F}$ which is normal to $\vec{\mathrm{e}}$ (Fig. 2.3c). It intersects line L at D. The unit vector $\vec{n} = \vec{\mathrm{e}}_2$ along DQ is normal to $\vec{\mathrm{e}}$ and the unit vector $\vec{\mathrm{e}}_3 = \vec{\mathrm{e}} \times \vec{n}$ at Q is skew to vector $\vec{\mathrm{e}}$ through D. Resolve $\vec{F}$ into components: $\vec{F}_\parallel$, $\vec{F}_\perp$ which are parallel and normal to $\vec{\mathrm{e}}$. Resolve $\vec{F}_\perp$ into components $\vec{F}_\perp^n$, $\vec{F}_\perp^s$ along $\vec{n}$ and $\vec{\mathrm{e}}_3$, where $\vec{F}_\perp^n$ intersects line L and $\vec{F}_\perp^s$ is skew to line L.

$$\vec{M}_e = [(\vec{r}_{QD} \times \vec{F}) \cdot \vec{e}\,]\,\vec{e} = [\{DQ\,\vec{e}_2 \times (F_\parallel\,\vec{e}_1 + F_\perp^n\,\vec{e}_2 + F_\perp^s\,\vec{e}_3)\} \cdot \vec{e}_1\,]\,\vec{e} = (DQ)F_\perp^s\,\vec{e}$$

Thus, the moment of a force about a line is the moment of its transverse skew component $F_\perp^s$ about the line. Let $d = CN$ be the minimum distance between L and $\vec{F}$ (Fig. 2.3b). As $CN \perp \vec{F}$, in the resolution of $\vec{F} = \vec{F}_\parallel + \vec{F}_\perp$, $\vec{F}_\perp = F_\perp\,\vec{e}_3$ is perpendicular to both L and CN. Hence, $M_e = \vec{M}_C \cdot \vec{e} = d\,\vec{e}_2 \times (F_\parallel\,\vec{e}_1 + F_\perp\,\vec{e}_3) \cdot \vec{e}_1 = F_\perp d$.

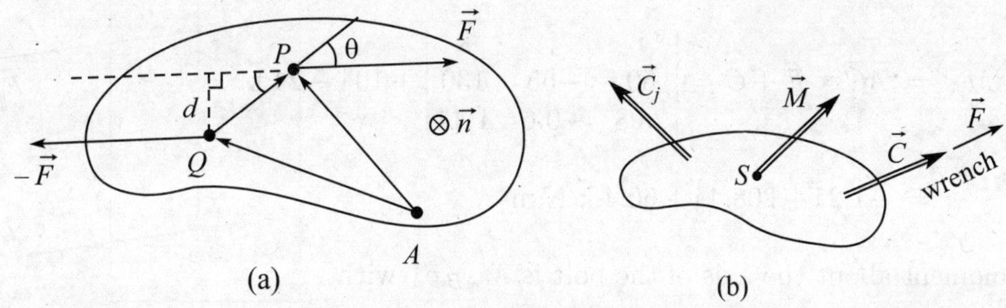

Fig. 2.4

A *couple* is a set of forces $\vec{F}$ and $-\vec{F}$ (Fig. 2.4a) with different lines of action. Its moment about every point is the same and is called the *moment* $\vec{C}$ *of the couple*:

$$\begin{aligned}\vec{M}_A &= \overrightarrow{AP} \times \vec{F} + \overrightarrow{AQ} \times (-\vec{F}) = (\overrightarrow{AP} - \overrightarrow{AQ}) \times \vec{F} = \overrightarrow{QP} \times \vec{F} \\ &= Fd\vec{n} = \vec{C}\end{aligned}$$

which is independent of A. The force sum of a couple is zero. Operationally, a couple will not merely signify a pair of forces $\vec{F}$ and $-\vec{F}$, but also a moment (a primitive independent of force). The moment $\vec{C}_1$ of a couple or a moment $\vec{M}$ is denoted by an arrow with double lines whose length gives its magnitude, direction gives its line of action, arrow-head gives its sense and the tail its point of application S (Fig. 2.4b).

A wrench is a system of force $\vec{F}$ and a moment (or moment of a couple) $\vec{C}$ whose direction is the same as or opposite to $\vec{F}$ (Fig. 2.4b). The force systems acting on a drill bit during drilling and on a screw being tightened or removed are wrenches.

Example 2.1: Find $\vec{M}_A$ and M_{AB} for the force system shown in Fig. E2.1.

Solution: Let $\vec{e}$, $\vec{e}^*$, $\vec{e}_1$ be the unit vectors along RC, AB, $\vec{F}$:

$$\vec{e} = (-5\,\vec{j} + 12\,\vec{k})/(5^2 + 12^2)^{1/2} = (-5\,\vec{j} + 12\,\vec{k})/13$$

$$\vec{e}^* = (-4\,\vec{i} + 3\,\vec{k})/(4^2 + 3^2)^{1/2} = -0.8\,\vec{i} + 0.6\,\vec{k}$$

$$\vec{F} = 2600\,\vec{e}_1 = 2600(3\,\vec{e} + 4\,\vec{i}\,)/(3^2 + 4^2)^{1/2} = 1560\,\vec{e} + 2080\,\vec{i}$$

$$= 1560(-5\vec{j} + 12\vec{k})/13 + 2080\vec{i} = (2.08\vec{i} - 0.6\vec{j} + 1.44\vec{k})10^3 \text{ N}$$

$$\overrightarrow{AC} = \overrightarrow{AP} + \overrightarrow{PQ} + \overrightarrow{QR} + \overrightarrow{RC} = -30\vec{j} + 50\vec{i} + 70\vec{k} + 65\vec{e} \text{ mm}$$

$$= 50\vec{i} - 55\vec{j} + 130\vec{k} \text{ mm} = (50\vec{i} - 55\vec{j} + 130\vec{k})10^{-3} \text{ m}$$

$$\vec{C} = -26\vec{e} = -26(-5\vec{j} + 12\vec{k})/13 = 10\vec{j} - 24\vec{k} \text{ N.m}$$

$$\vec{M}_A = \overrightarrow{AC} \times \vec{F} + \vec{C} = \begin{vmatrix} \vec{i} & \vec{j} & \vec{k} \\ 50 & -55 & 130 \\ 2.08 & -0.6 & 1.44 \end{vmatrix} + 10\vec{j} - 24\vec{k}$$

$$= -1.2\vec{i} + 208.4\vec{j} + 60.4\vec{k} \text{ N.m}$$

The moment about the axis of the bolt is $M_{AB}\,\vec{e}^*$ with

$$M_{AB} = M_{e^*} = \vec{M}_A \cdot \vec{e}^* = (-1.2)(-0.8) + 60.4 \times 0.6 = 37.2 \text{ N.m}$$

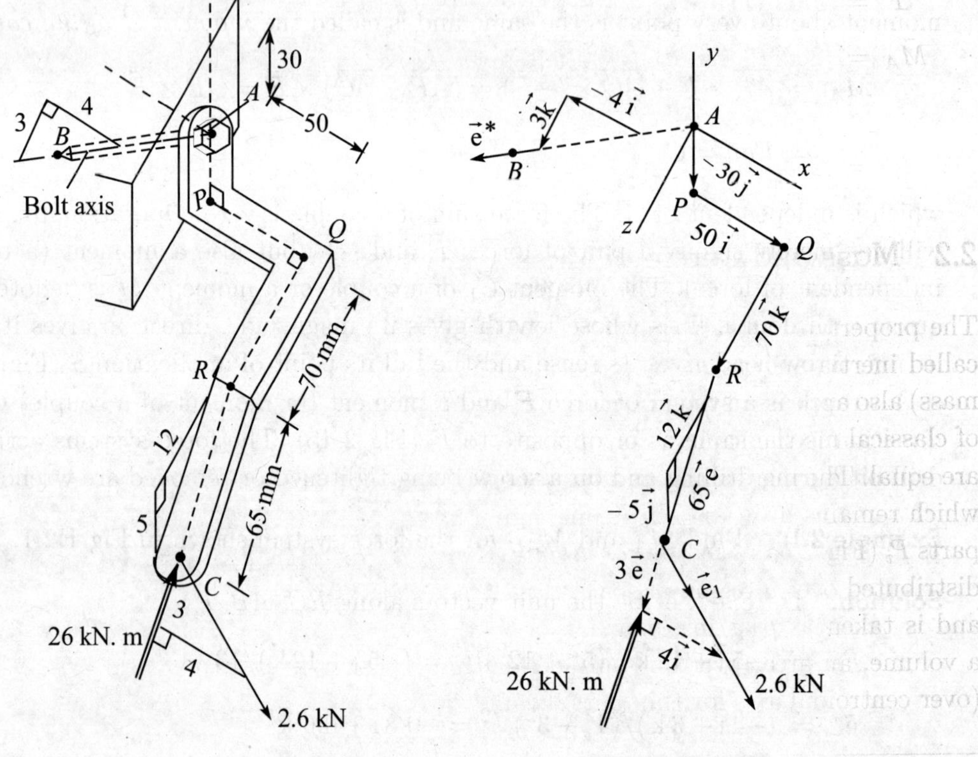

Fig. E2.1

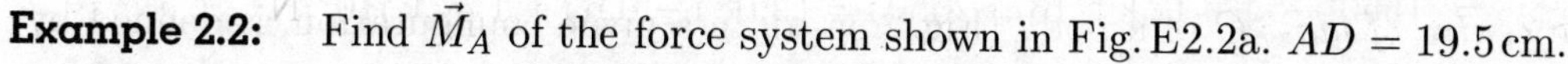

Example 2.2: Find $\vec{M}_A$ of the force system shown in Fig. E2.2a. $AD = 19.5\,\text{cm}$.

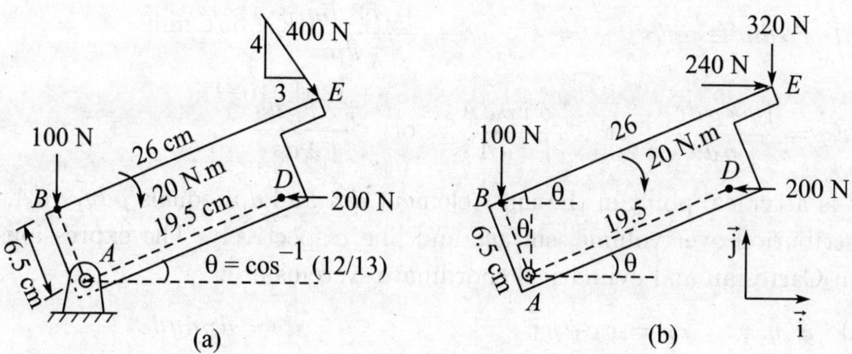

Fig. E2.2

Solution: For the coplanar force system, we resolve the forces and position vectors in the $\vec{i}$ and $\vec{j}$ components (Fig. E2.2b) and take their moment about A using the product of force and distance, with the sense x to y being $\vec{k}$. As $\cos\theta = 12/13 \Rightarrow \sin\theta = 5/13$.

$$\begin{aligned}
\vec{F} &= 400(-4\vec{j} + 3\vec{i}\,)/(4^2 + 3^2)^{1/2} = 240\,\vec{i} - 320\,\vec{j}\text{ N} \\
\vec{M}_A &= [200(19.5\sin\theta) + 100(6.5\sin\theta) - 240(6.5\cos\theta + 26\sin\theta) \\
&\qquad -320(26\cos\theta - 6.5\sin\theta)]10^{-2}\,\vec{k} - 20\,\vec{k} \\
&= [200(7.5) + 100(2.5) - 240(16) - 320(21.5)]10^{-2}\,\vec{k} - 20\,\vec{k} = -109.7\,\vec{k}\text{ N.m}
\end{aligned}$$

2.2 Mass and Centre of Mass

The property of a body due to which an external force is required to accelerate it is called inertia and its measure is called its mass (inertial mass). Mass (gravitational mass) also appears in the gravitational law of attraction. Experience within the domain of classical mechanics has shown that inertial mass and gravitational mass of a body are equal. The mass $m(B)$ of a body B is taken axiomatically as a positive real scalar, which remains invariant with time and equals the sum of the masses $m(P_i)$ of all its parts P_i (Fig. 2.5a)[1]. By the *continuum hypothesis*, mass is modelled to be continuously distributed over a region of space (Fig. 2.5a,b) and the *mass density* at a point exists and is taken as $\rho\,\text{kg/m}^3$, $\sigma\,\text{kg/m}^2$ or $\lambda\,\text{kg/m}$ for idealised distribution of mass over a volume, an area (over mid-surface for thin plates, membranes and shells) or a line (over centroidal axis for thin bars, beams, flexible cables and strings).

[1] We use the same symbol 'm' for mass and a moving frame. The meaning is usually clear from the context. In some situations (especially Chapter 3), we have used Ⓜ for the frame for greater clarity.

The *centre of mass* C of a body (Fig. 2.5a,b), in a given configuration, is defined by

$$\int_m \vec{r}_P\, dm \equiv m\vec{r}_C \qquad \Rightarrow \qquad \vec{r}_C \equiv \frac{\int_m \vec{r}_P\, dm}{\int_m dm} \tag{2.4}$$

$$\vec{r}_C = \frac{\int \rho\, \vec{r}_P\, dv}{\int \rho\, dv}, \quad \text{or} \quad \frac{\int \sigma\, \vec{r}_P\, dA}{\int \sigma\, dA}, \quad \text{or} \quad \frac{\int \lambda\, \vec{r}_P\, ds}{\int \lambda\, ds}$$

where P is a typical point in the mass element dm and dm equals ρdv, σdA, λds for mass distribution over volume, surface and line respectively. The expressions for $\vec{r}_P$ and dv in Cartesian and cylindrical coordinate systems are:

(a) $x, y, z: \quad \vec{r}_P = x\,\vec{\mathrm{i}} + y\,\vec{\mathrm{j}} + z\,\vec{\mathrm{k}}, \qquad dv = dx\,dy\,dz$

(b) $r, \phi, z: \quad \vec{r}_P = r\cos\phi\,\vec{\mathrm{i}} + r\sin\phi\,\vec{\mathrm{j}} + z\,\vec{\mathrm{k}}, \qquad dv = dr(r\,d\phi)dz$

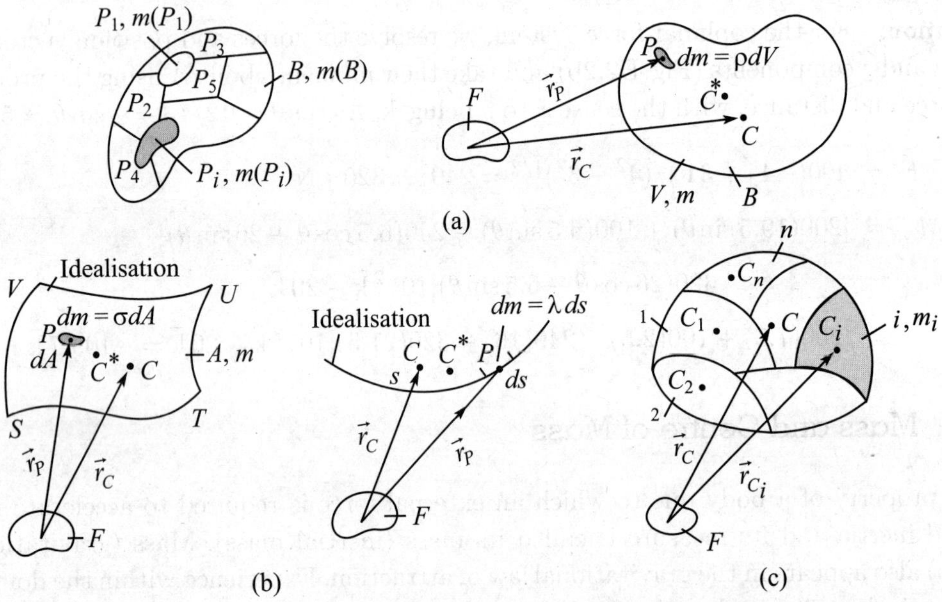

Fig. 2.5

The *centroids* C^* of a volume, a surface and a curve (Fig. 2.5a,b) are defined by

$$\vec{r}_{C^*} = \frac{\int \vec{r}_P\, dv}{\int dv}, \qquad \vec{r}_{C^*} = \frac{\int \vec{r}_P\, dA}{\int dA}, \qquad \vec{r}_{C^*} = \frac{\int \vec{r}_P\, ds}{\int ds} \tag{2.5}$$

C coincides with C^* for homogeneous bodies. C does not coincide with a specific material point for a deformable body.

It is easy to prove that:

1. For a body with a *centre of mass symmetry* O $[\rho(-\vec{r}) = \rho(\vec{r})]$, $C \equiv O$. (Fig. 2.6-1)

2. For a body with a *plane of mass symmetry*, say xy-plane $[\rho(x, y, -z) = \rho(x, y, z)]$, C lies on this plane. (Fig. 2.6-2)
3. For two planes of mass symmetry, C lies on their line of intersection. (Fig. 2.6-3)
4. A body is called a *body of revolution* if its volume is obtained by rotating an area about an axis (called *axis of revolution*), say z-axis, and its density $\rho = \rho(r, z)$. Then, C lies on the axis of revolution of the body. (Fig. 2.6-4)
5. For a body with 3 planes of mass symmetry not through one line, C lies at their point of intersection. (Fig. 2.6-5)

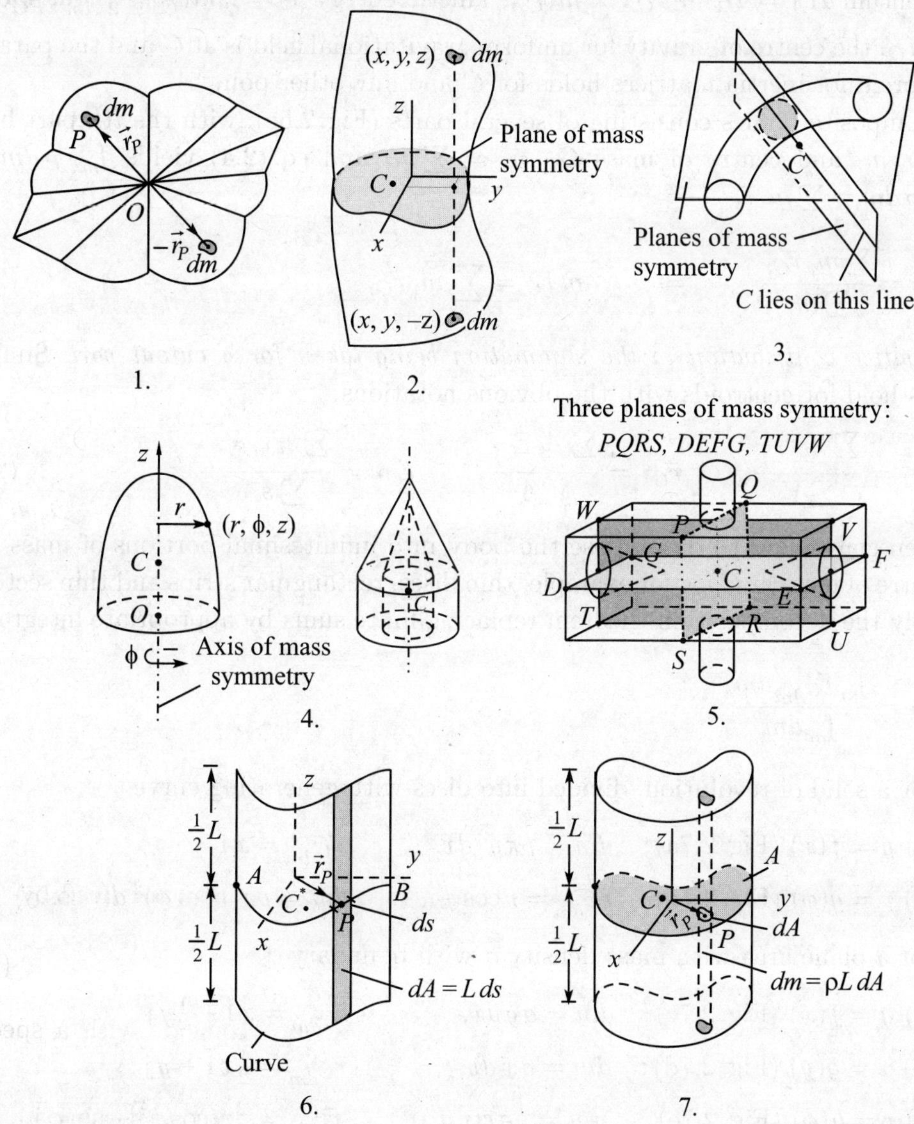

Fig. 2.6

6. For a cylindrical surface, C^* is at the centroid of its curve of intersection with its mid-plane of symmetry. (Fig. 2.6-6)
7. For a homogeneous cylindrical body, C is at the centroid of its cross-section in the mid-plane of symmetry. (Fig. 2.6-7)

The centroid is useful for finding the resultant force due to hydrostatic pressure on a plane area and stresses in thin bars. We shall find later that C is a point of vital importance in mechanics, for example, momentum $\vec{p} = m\vec{v}_C$, $\vec{F} = m\vec{a}_C$, moment of momentum $\vec{H}_A = \vec{H}_C + \vec{r}_{CA} \times m\vec{v}_{CA}$, kinetic energy $T = \frac{1}{2}mv_C^2 + \frac{1}{2}\int_m v_{PC}^2 dm$, $\vec{M}_C = \dot{\vec{H}}_C$, the centre of gravity for uniform gravitational field is at C and the parallel axis theorem for inertia matrices holds for C and any other point A.

For composite bodies consisting of several parts (Fig. 2.5c), with the ith part having mass m_i and centre of mass C_i, $m = \sum m_i$ and eq. (2.4) yields $\int_m \vec{r}_P\, dm = \sum \int_{m_i} \vec{r}_P\, dm = \sum m_i \vec{r}_{C_i} \quad \Rightarrow$

$$\vec{r}_C = \frac{\sum m_i \vec{r}_{C_i}}{\sum m_i} \qquad \Rightarrow \qquad m\vec{r}_C = \sum m_i \vec{r}_{C_i} \tag{2.6}$$

with negative contribution in the summation being taken for a cutout part. Similar relations hold for centroids with the obvious notations:

$$\vec{r}_{C^*} = \frac{\sum V_i \vec{r}_{C_i^*}}{\sum V_i}, \qquad \vec{r}_{C^*} = \frac{\sum A_i \vec{r}_{C_i^*}}{\sum A_i}, \qquad \vec{r}_{C^*} = \frac{\sum s_i \vec{r}_{C_i^*}}{\sum s_i} \tag{2.7}$$

It is often convenient to decompose the body into infinitesimal portions of mass dm with centre of mass at C_{dm}, for example, thin discs, rectangular strips and thin sectors, and apply the decomposition theorem replacing finite sums by appropriate integrals:

$$\vec{r}_C = \frac{\int_m \vec{r}_{C_{dm}}\, dm}{\int_m dm}$$

1. For a solid of revolution, divided into discs with generating curve

   (a) $y = f(x)$ (Fig. 2.7a): $\quad dm = \rho\pi y^2\, dx, \qquad \vec{r}_{C_{dm}} = x\vec{\imath}$

   (b) $r = g(\phi)$ (Fig. 2.7b): $\quad \vec{r}_{C_{dm}} = r\cos\phi\,\vec{\imath}, \qquad dm = \rho\pi(r\sin\phi)^2\, d(r\cos\phi)$

2. For a plane area with mass density σ with boundary

   (a) $y = f(x)$ (Fig. 2.7c): $\quad dm = \sigma y\, dx, \qquad \vec{r}_{C_{dm}} = x\vec{\imath} + \frac{1}{2}y\vec{\jmath}$

   (b) $x = g(y)$ (Fig. 2.7d): $\quad dm = \sigma x\, dy, \qquad \vec{r}_{C_{dm}} = \frac{1}{2}x\vec{\imath} + y\vec{\jmath}$

   (c) $r = h(\phi)$ (Fig. 2.7e): $\quad dm = \frac{1}{2}\sigma r(r\, d\phi), \qquad \vec{r}_{C_{dm}} = \frac{2}{3}r(\cos\phi\,\vec{\imath} + \sin\phi\,\vec{\jmath})$

3. For a plane curve with mass density λ, which is described as

$$\text{(a) } y = f(x) \text{ (Fig. 2.7f)}: \quad \vec{r}_P = x\,\vec{\mathrm{i}} + y\,\vec{\mathrm{j}}, \qquad dm = \lambda\left[1 + \left(\frac{dy}{dx}\right)^2\right]^{1/2} dx$$

$$\text{(b) } x = g(y) \text{ (Fig. 2.7f)}: \quad \vec{r}_P = x\,\vec{\mathrm{i}} + y\,\vec{\mathrm{j}}, \qquad dm = \lambda\left[\left(\frac{dx}{dy}\right)^2 + 1\right]^{1/2} dy$$

$$\text{(c) } r = h(\phi) \text{ (Fig. 2.7g)}: \quad \vec{r}_P = r\cos\phi\,\vec{\mathrm{i}} + r\sin\phi\,\vec{\mathrm{j}}, \quad dm = \lambda\left[\left(\frac{dr}{d\phi}\right)^2 + r^2\right]^{1/2} d\phi$$

$$\text{(d) } \phi = l(r) \text{ (Fig. 2.7g)}: \quad \vec{r}_P = r\cos\phi\,\vec{\mathrm{i}} + r\sin\phi\,\vec{\mathrm{j}}, \quad dm = \lambda\left[1 + r^2\left(\frac{d\phi}{dr}\right)^2\right]^{1/2} dr$$

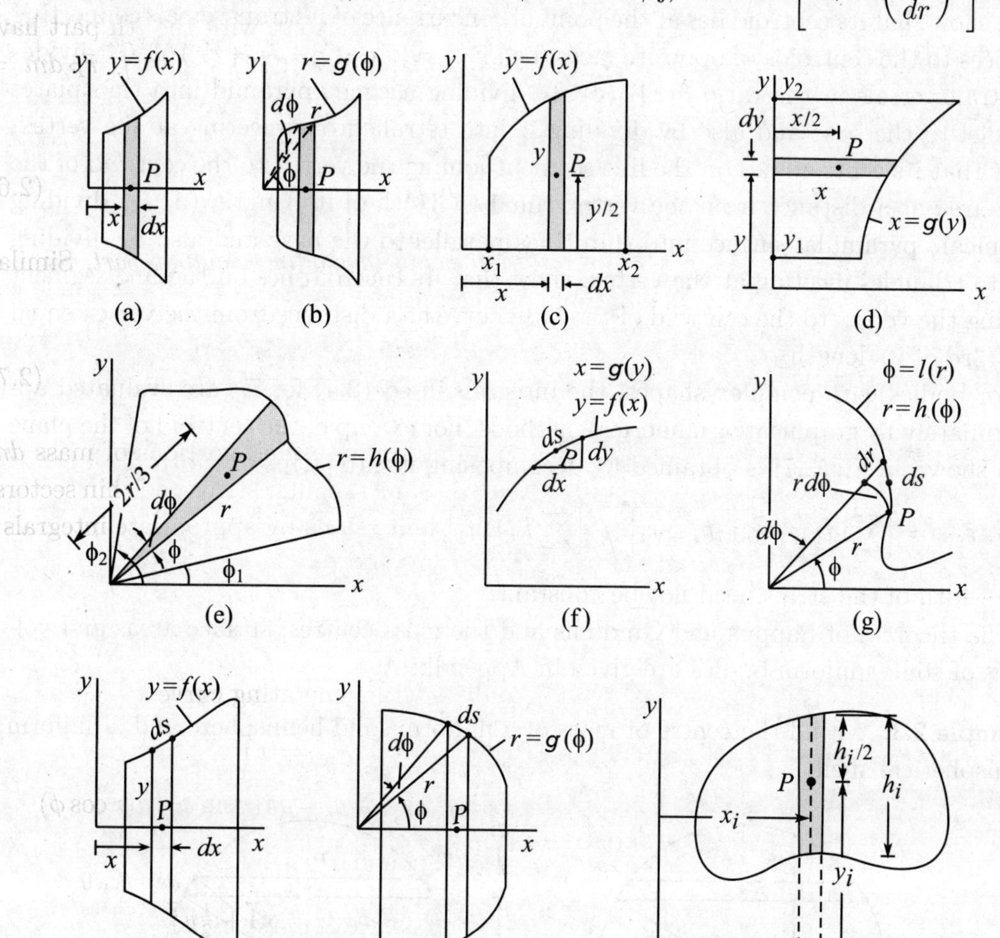

Fig. 2.7

4. A surface of revolution (σ) divided into rings, with generating curve

(a) $y = f(x)$ (Fig. 2.7h): $dm = 2\pi y\, ds\, \sigma = 2\pi\sigma y \left[1 + \left(\dfrac{dy}{dx}\right)^2\right]^{1/2} dx,\ \vec{r}_{C_{dm}} = x\vec{\imath}$

(b) $r = g(\phi)$ (Fig. 2.7i): $dm = 2\pi\sigma r \sin\phi \left[\left(\dfrac{dr}{d\phi}\right)^2 + r^2\right]^{1/2} d\phi,\ \vec{r}_{C_{dm}} = r\cos\phi\,\vec{\imath}$

§ **Exercise 2.1:** (a) By dividing a triangle ABC into strips parallel to its sides, show that its centroid lies at the point of intersection of its medians, i.e., $\vec{r}_{C^*} = (\vec{r}_A + \vec{r}_B + \vec{r}_C)/3$. (b) By dividing a tetrahedron $ABDE$ into thin plates parallel to a face, show that its centroid lies at the point of concurrence of its transversals connecting vertices to the centroids of opposite faces, i.e., $\vec{r}_{C^*} = (\vec{r}_A + \vec{r}_B + \vec{r}_D + \vec{r}_E)/4$. C^* divides the transversals in the ratio 3 : 1. (c) By dividing a cone / pyramid into thin plates parallel to the base and also by dividing it into tetrahedrons meeting at the vertex, show that its centroid lies on the line segment joining the vertex to the centroid of the base area at a distance from the vertex equal to 3/4th of its length. (d) By dividing a conical / pyramidal surface into thin rings parallel to the base and also by dividing it into triangles meeting at the vertex, show that its centroid lies on the line segment joining the vertex to the centroid of the base curve at a distance from the vertex equal to 2/3rd of its length.

For bodies with complex shapes, the integrals in eq. (2.4) for $\vec{r}_C$ are evaluated approximately by graphical or numerical methods. For example, the centroid of the plane area shown in Fig. 2.7j is obtained by decomposing it into strips of width Δx_i:

$$\vec{r}_{C^*} = \left[\sum (x_i\vec{\imath} + y_i\vec{\jmath}) h_i\, \Delta x_i\right] \Big/ \left[\sum h_i\, \Delta x_i\right]$$

The width of the strips need not be constant.

The theorem of Pappus and Guldinus and the mass centres, surface areas and volumes of some uniform bodies are given in Appendix A.

Example 2.3: Find the centre of mass of a uniform solid hemisphere and a uniform hemispherical shell.

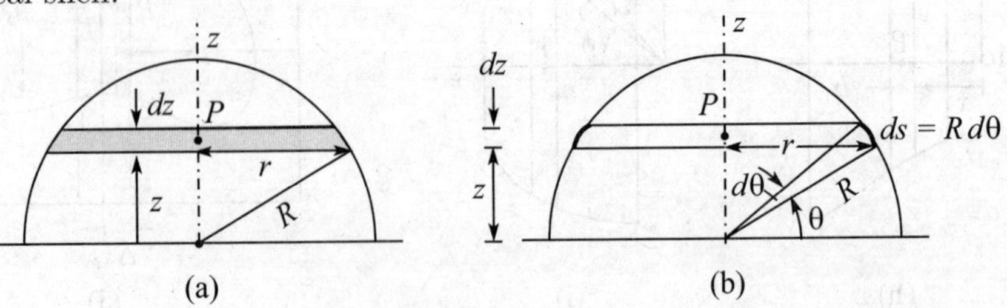

Fig. E2.3

Solution: Decompose the solid hemisphere into thin circular plates (Fig. E2.3a) with $dV = \pi r^2 dz = \pi(R^2 - z^2)dz, \quad \vec{r}_P = z\vec{k}$,

$$\vec{r}_C = \frac{\int_V \vec{r}_P\, dV}{V} = \frac{\int_0^R \pi(R^2 - z^2)z\, dz\, \vec{k}}{\int_0^R \pi(R^2 - z^2)\, dz} = \frac{3}{8}R\vec{k}$$

Decompose the shell into thin rings (Fig. E2.3b) with $dA = 2\pi r ds = 2\pi(R\cos\theta)(R\, d\theta)$,

$$\vec{r}_C = \frac{\int_A \vec{r}_P\, dA}{A} = \frac{\int_0^{\pi/2}(R\sin\theta\,\vec{k})2\pi R^2\cos\theta\, d\theta\,\vec{k}}{\int_0^{\pi/2} 2\pi R^2\cos\theta\, d\theta} = \tfrac{1}{2}R\vec{k}$$

Example 2.4: Find the centre of mass of (a) a thin right circular conical shell of base radius R, height h and density varying linearly with distance from the base, with σ_1 at the base to σ_0 at the vertex, and (b) a solid right circular cone of base radius R, height h and density varying linearly with distance from the axis of symmetry, with ρ_0 at the axis to ρ_1 at radius R. (c) Rework parts (a) and (b) for one-half of the body obtained by splitting with a plane through the axis of symmetry.

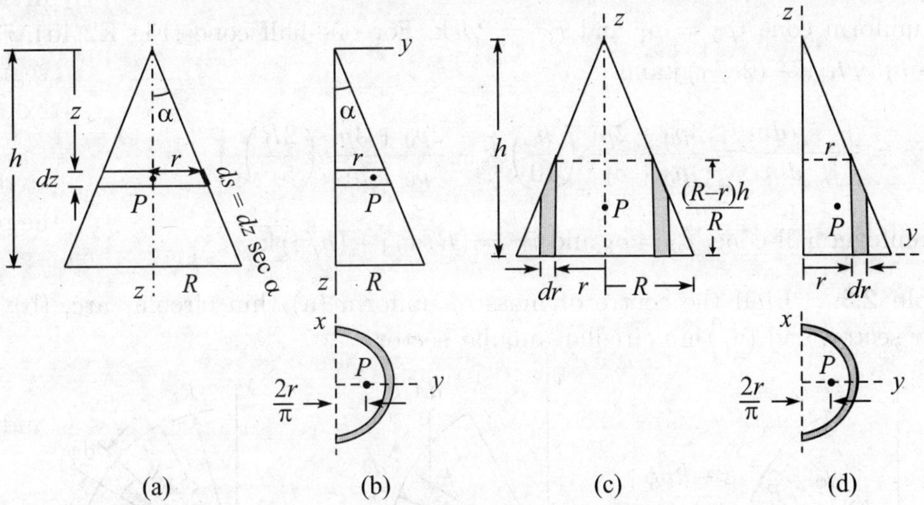

Fig. E2.4

Solution: (a) For the ring element shown in Fig. E2.4a,

$$ds = dz\sec\alpha, \qquad r = Rz/h, \qquad \vec{r}_P = z\vec{k}, \qquad \sigma = \sigma_0 + (\sigma_1 - \sigma_0)z/h$$

$$dm = 2\pi r\, ds\, \sigma = 2\pi R[\sigma_0 z + (\sigma_1 - \sigma_0)z^2/h]\sec\alpha\, dz\,/h$$

$$\vec{r}_C = \frac{\int_m \vec{r}_P\, dm}{\int_m dm} = \frac{\int_0^h[\sigma_0 z^2 + (\sigma_1 - \sigma_0)z^3/h]\, dz}{\int_0^h[\sigma_0 z + (\sigma_1 - \sigma_0)z^2/h]\, dz}\,\vec{k} = \frac{\sigma_0 + 3\sigma_1}{2(\sigma_0 + 2\sigma_1)}\,h\vec{k}$$

For a uniform shell, $\sigma_0 = \sigma_1$ and $\vec{r}_C = (2h/3)\,\vec{\mathrm{k}}$. For one-half shell (Fig. E2.4b), $\vec{r}_P = z\,\vec{\mathrm{k}} + (2r/\pi)\,\vec{\mathrm{j}} = z(2\tan\alpha\,\vec{\mathrm{j}}/\pi + \vec{\mathrm{k}})$ and hence

$$\vec{r}_C = \frac{\sigma_0 + 3\sigma_1}{2(\sigma_0 + 2\sigma_1)}\,h(2\tan\alpha\,\vec{\mathrm{j}}/\pi + \vec{\mathrm{k}})$$

$$\Rightarrow \quad \text{for } \sigma_0 = \sigma_1: \qquad \vec{r}_C = \frac{2}{3}(2R\,\vec{\mathrm{j}}/\pi + h\,\vec{\mathrm{k}})$$

(b) For an element of radius r, thickness dr and height $(R-r)h/R$ (Fig. E2.4c),

$$\vec{r}_P = [(R-r)h/2R]\,\vec{\mathrm{k}}, \qquad dm = [\rho_0 + (\rho_1 - \rho_0)r/R](2\pi r)\,dr\,(R-r)h/R$$

$$\vec{r}_C = \frac{\int_m \vec{r}_P\,dm}{\int_m dm} = \frac{h\int_0^R (R-r)[\rho_0 Rr + (\rho_1 - 2\rho_0)r^2 - (\rho_1 - \rho_0)r^3/R]\,dr}{2R\int_0^R[\rho_0 Rr + (\rho_1 - 2\rho_0)r^2 - (\rho_1 - \rho_0)r^3/R]\,dr}\,\vec{\mathrm{k}}$$

$$= \frac{3\rho_0 + 2\rho_1}{\rho_0 + \rho_1}\left(\frac{h}{10}\right)\vec{\mathrm{k}}$$

For a uniform cone, $\rho_0 = \rho_1$ and $\vec{r}_C = \frac{1}{4}h\,\vec{\mathrm{k}}$. For one-half cone (Fig. E2.4d), $\vec{r}_P = \frac{1}{2}[(R-r)h/R]\,\vec{\mathrm{k}} + (2r/\pi)\,\vec{\mathrm{j}}$ and

$$\vec{r}_C = \frac{\int_m \vec{r}_P\,dm}{\int_m dm} = \frac{3\rho_0 + 2\rho_1}{\rho_0 + \rho_1}\left(\frac{h}{10}\right)\vec{\mathrm{k}} + \frac{2\rho_0 + 3\rho_1}{\rho_0 + \rho_1}\left(\frac{2R}{5\pi}\right)\vec{\mathrm{j}}$$

For a uniform half-cone, $\rho_0 = \rho_1$ and $\vec{r}_C = (R/\pi)\,\vec{\mathrm{j}} + (h/4)\,\vec{\mathrm{k}}$.

Example 2.5: Find the centre of mass of uniform (a) thin circular arc, (b) thin circular sector, and (c) thin circular annular sector.

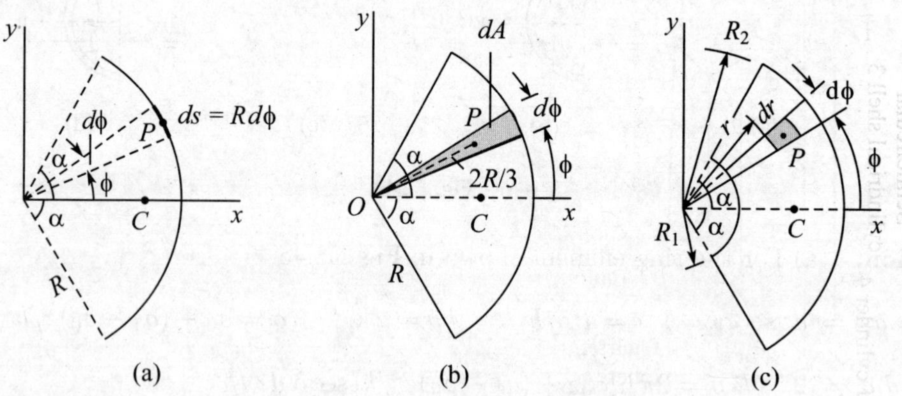

Fig. E2.5

Solution: We use polar coordinates since the boundaries are: r = constant and

$\phi =$ constant. The elements for integration are shown in Fig. E2.5.

$$\text{(a)}\quad \vec{r}_C = \frac{\int \vec{r}_P\, ds}{\int ds} = \frac{\int_{-\alpha}^{\alpha} R(\cos\phi\,\vec{\imath} + \sin\phi\,\vec{\jmath})R\,d\phi}{\int_{-\alpha}^{\alpha} R\,d\phi} = \frac{R\sin\alpha}{\alpha}\,\vec{\imath}$$

$$\text{(b)}\quad \vec{r}_C = \frac{\int \vec{r}_P\, dA}{\int dA} = \frac{\int_{-\alpha}^{\alpha} \frac{2}{3}R(\cos\phi\,\vec{\imath} + \sin\phi\,\vec{\jmath})(\frac{1}{2}R^2\,d\phi)}{\int_{-\alpha}^{\alpha} \frac{1}{2}R^2\,d\phi} = \frac{2R\sin\alpha}{3\alpha}\,\vec{\imath}$$

$$\text{(c)}\quad \vec{r}_C = \frac{\int \vec{r}_P\, dA}{\int dA} = \frac{\int_{-\alpha}^{\alpha}\int_{R_1}^{R_2} r(\cos\phi\,\vec{\imath} + \sin\phi\,\vec{\jmath})(r\,dr\,d\phi)}{\int_{-\alpha}^{\alpha}\int_{R_1}^{R_2} r\,dr\,d\phi} = \frac{2(R_2^3 - R_1^3)\sin\alpha}{3(R_2^2 - R_1^2)\alpha}\,\vec{\imath}$$

For a semi-circular thin plate, $R_1 = 0, \alpha = \frac{1}{2}\pi \Rightarrow \vec{r}_C = (4R_2/3\pi)\,\vec{\imath}$. For a semi-circular thin annular plate of mean radius R and thickness t, $\alpha = \frac{1}{2}\pi$, $R_1 = R - \frac{1}{2}t$, $R_2 = R + \frac{1}{2}t$, $\vec{r}_C = (2R/\pi)[1 + t^2/12R^2]\,\vec{\imath}$. The centroid of a semi-circular arc ($t = 0$) of radius R is at $(2R/\pi)\,\vec{\imath}$. The error e in approximating a semi-circular annular plate to a semi-circular arc is $e \simeq (100t^2/12R^2)\% \Rightarrow e < 0.1\%$ for $t/R < 0.1$ and $e < 1\%$ even for $t/R = 0.3$.

Example 2.6: Find the centre of mass of the body given in Fig. E2.6. The thin semi-circular ring, thin semi-circular cylindrical shell, and the solid top and base have mass densities λ, σ and ρ, respectively.

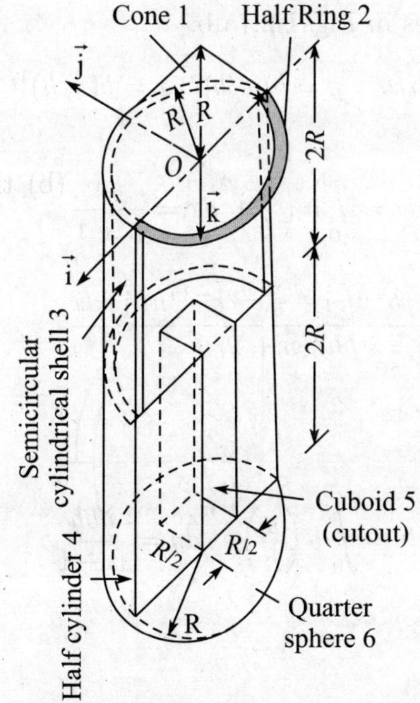

Fig. E2.6

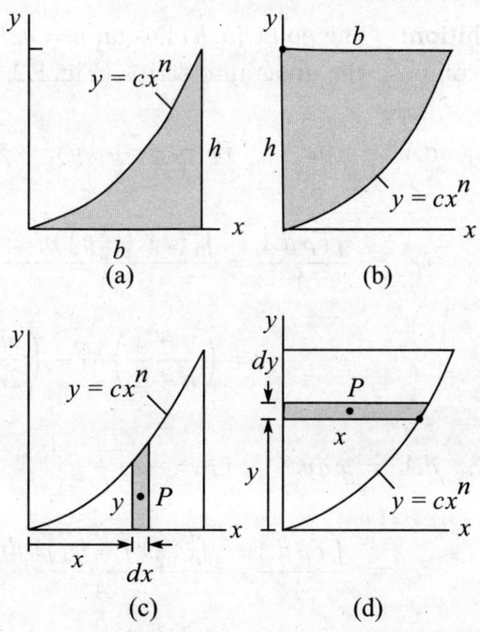

Fig. E2.7

Solution: We decompose the body into a cone 1, a thin semi-circular ring 2, a thin semi-circular cylindrical shell 3, a semi-circular solid cylinder 4, a deletion of a cuboid 5 of size $\frac{1}{2}R\times\frac{1}{2}R\times 2R$ and a quarter sphere 6. The contribution of the deleted cuboid 5 is taken as negative in the summations. The mass and centre of mass of each part are:

$$\begin{aligned}
m_1 &= (\pi R^2/3)R\rho = \pi R^3\rho/3, & \vec{r}_{C_1} &= -(R/4)\,\vec{k}\\
m_2 &= \pi R\lambda, & \vec{r}_{C_2} &= -(2R/\pi)\,\vec{j}\\
m_3 &= (\pi R)(2R)\sigma = 2\pi R^2\sigma, & \vec{r}_{C_3} &= (2R/\pi)\,\vec{j} + R\,\vec{k}\\
m_4 &= (\tfrac{1}{2}\pi R^2)(2R)\rho = \pi R^3\rho, & \vec{r}_{C_4} &= (4R/3\pi)\,\vec{j} + 3R\,\vec{k}\\
m_5 &= (\tfrac{1}{2}R)(\tfrac{1}{2}R)(2R)\rho = R^3\rho/2, & \vec{r}_{C_5} &= -(R/4)\,\vec{i} + (R/4)\,\vec{j} + 3R\,\vec{k}\\
m_6 &= [\tfrac{1}{4}(4\pi R^3/3)]\rho = \pi R^3\rho/3, & \vec{r}_{C_6} &= (3R/8)\,\vec{j} + (4R + 3R/8)\,\vec{k}
\end{aligned}$$

$$\begin{aligned}
\vec{r}_C &= \frac{\sum m_i \vec{r}_{C_i}}{\sum m_i} = [(R^4\rho/8)\,\vec{i} - \{2R^2\lambda - 4R^3\sigma - (29+3\pi)R^4\rho/24\}\,\vec{j}\\
&\quad + \{2\pi R^3\sigma + (35\pi - 12)R^4\rho/8\}\,\vec{k}]\;/\;[(5\pi/3 - 0.5)R^3\rho + 2\pi R^2\sigma + \pi R\lambda]
\end{aligned}$$

Example 2.7: Find the area and centroid of figures in Figs E2.7a,b.

Solution: The point (b, h) lies on $y = cx^n \Rightarrow c = h/b^n,\ y = h(x/b)^n,\ x = b(y/h)^{1/n}$. Decompose the areas into strips (Fig. E2.7c,d).

$$\text{(a)}\quad dA = y\,dx, \qquad \vec{r}_P = x\,\vec{i} + \tfrac{1}{2}y\,\vec{j}, \qquad A = \int_0^b y\,dx = \int_0^b h\left(\frac{x}{b}\right)^n dx = \frac{bh}{n+1}$$

$$\begin{aligned}
\vec{r}_C &= \frac{\int \vec{r}_P\,dA}{A} = \frac{\int_0^b (x\,\vec{i} + \frac{1}{2}y\,\vec{j})y\,dx}{A} = \frac{\int_0^b (x^{n+1}b^{-n}h\,\vec{i} + \frac{1}{2}x^{2n}b^{-2n}h^2\,\vec{j})\,dx}{bh/(n+1)}\\
&= \left(\frac{n+1}{n+2}\right) b\,\vec{i} + \left(\frac{n+1}{2n+1}\right)\frac{h}{2}\,\vec{j}
\end{aligned}$$

$$\text{(b)}\quad dA = x\,dy, \qquad \vec{r}_P = \tfrac{1}{2}x\,\vec{i} + y\,\vec{j}, \qquad A = \int_0^h x\,dy = \int_0^h b\left(\frac{y}{h}\right)^{1/n} dy = \frac{nbh}{n+1}$$

$$\begin{aligned}
\vec{r}_C &= \frac{\int \vec{r}_P\,dA}{A} = \frac{\int_0^h (\frac{1}{2}x\,\vec{i} + y\,\vec{j})x\,dy}{A}\\
&= \frac{\int_0^h (\frac{1}{2}b^2h^{-2/n}y^{2/n}\,\vec{i} + bh^{-1/n}y^{1/n+1}\,\vec{j})\,dy}{nbh/(n+1)} = \left(\frac{n+1}{n+2}\right)\frac{b}{2}\,\vec{i} + \left(\frac{n+1}{2n+1}\right)h\,\vec{j}
\end{aligned}$$

2.3 Euler's Axioms

The *momentum* $\vec{p}_{|F}$ and the *moment of momentum* $\vec{H}_{A|F}$ of a body B (a system) about a point A (which could be a material point of the body, a fixed location of frame F or a point of another body in motion with respect to F) with respect to frame F (Fig. 2.8a) are defined by

$$\vec{p}_{|F} = \int_m \vec{v}_{P|F}\, dm, \qquad \vec{H}_{A|F} = \int_m \vec{r}_{PA} \times \vec{v}_{PA|F}\, dm \tag{2.8}$$

The concept of $\vec{H}_A$ became more significant when its natural unit was discovered in quantum mechanics to be equal to Planck's constant h divided by 2π. It is a tiny unit, $h/2\pi = 1.054 \times 10^{-34}\,\text{kg.m}^2/\text{s}$, but it implies huge rotation speeds of the order of 10^{11} Hertz in systems of atomic size.

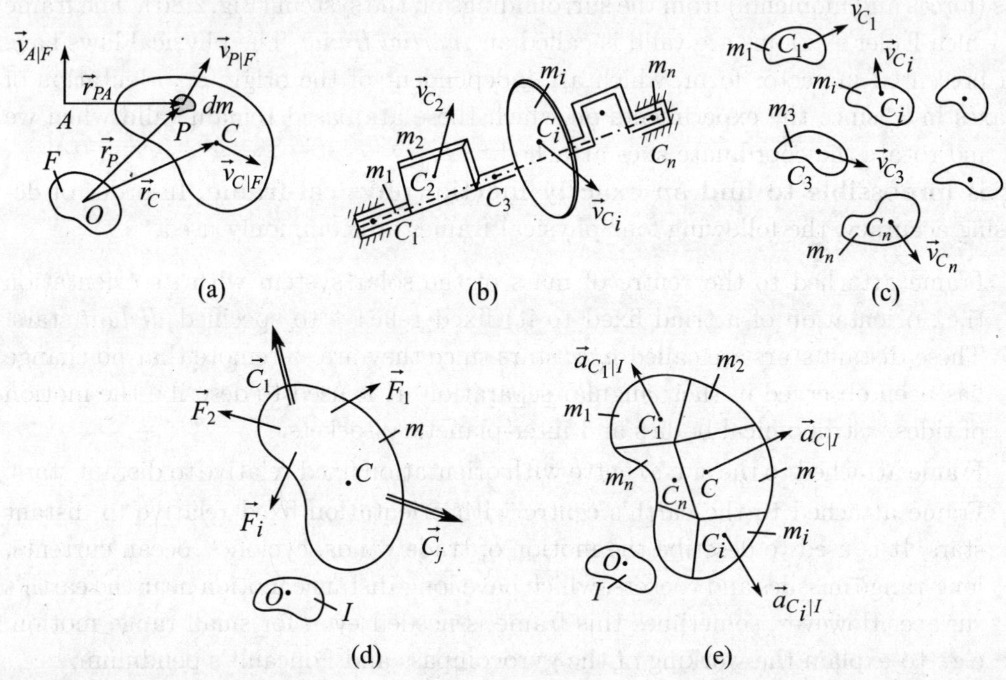

Fig. 2.8

The momentum of any system (Fig. 2.8a,b,c) can be expressed in terms of the velocity of C, since $\quad \vec{p}_{|F} = [\int \vec{r}_P\, dm]_{|F} = \overline{(m\vec{r}_C)}_{|F} \quad \Rightarrow$

$$\vec{p}_{|F} = m\vec{v}_{C|F} = \sum m_i \vec{v}_{C_i|F} \tag{2.9}$$

i.e., momentum of a body equals the momentum evaluated as if the total mass m of the body were concentrated at C and moving at its velocity $\vec{v}_{C|F}$.

§ **Exercise 2.2:** Frame T translates arbitrarily with respect to frame F. Using $\vec{v}_{P|F} = \vec{v}_{P|T} + \vec{v}_{T|F}$, prove that $\vec{p}_{|F} = \vec{p}_{|T} + m\vec{v}_{T|F}$ and $\vec{H}_{A|F} = \vec{H}_{A|T}$.

Euler's axioms relate $\dot{\vec{p}}$ and $\dot{\vec{H}}_O$ with the external forces and moments.

Euler's Axioms: There exists a frame I such that for *any system*

$$\dot{\vec{p}}_{|I} = \vec{F} \tag{2.10}$$

$$\dot{\vec{H}}_{O|I} = \vec{M}_O \tag{2.11}$$

where O **is a point fixed in** I, $\vec{F}$ is the sum of all the *external* forces from the surroundings on the system and $\vec{M}_O$ is the sum of the moments about O of all *external* loads (forces and moments) from the surroundings on the system (Fig. 2.8d). The frame I in which Euler's axioms are valid is called an *inertial frame.* The physical laws have been presented in vector form, which are independent of the origin or orientation of the axes in I, since the experiments on which these are based remain valid when we shift and rotate the coordinate axes attached to I.

It is impossible to find an exactly inertial physical frame. In order of decreasing accuracy, the following four physical frames are commonly used.

1. Frame attached to the centre of mass of the solar system with its orientation (i.e., orientation of a triad fixed to it) fixed relative to specified *distant* stars. These distant stars are called *fixed* stars since they are so remote that no change has been observed in their angular separation. It is used to describe the motion of tides, astronomical bodies and inter-planetary rockets.
2. Frame attached to the sun's centre with orientation fixed relative to distant stars.
3. Frame attached to the earth's centre with orientation fixed relative to distant stars. It is used to describe the motion of trade winds, cyclones, ocean currents, long-range missiles and rockets, which have long-distance motion near the earth's surface. However, sometimes this frame is needed even for small range motion, e.g., to explain the working of the gyrocompass and Foucault's pendulum.
4. Frame attached to the earth and rotating with it at constant angular velocity relative to frame in item 3. It yields excellent results for almost all dynamic problems in engineering, where the range of motion is small compared to the earth's radius, such as turbines, machine dynamics, automobiles and channel flows.

The six scalar equations corresponding to eqs (2.10) and (2.11), when applied to the whole system, are *not sufficient* for determining the motion of a *general system* which is deformable and hence possesses an infinite number of dof. However, these six scalar equations are *necessary and sufficient* for the complete determination of the motion of a *rigid body* since it has precisely six degrees of freedom.[2]

For the purpose of application of Euler's axioms to a system, it is necessary that

1. the system should be *well-identified* and its sketch be drawn *in isolation* from its surroundings, and
2. the *external forces and moments* exerted by the surroundings on the system should be drawn on it.

Such a diagram is called a **free body diagram** (FBD). The forces exerted by one part of the system on another part of the system are called *internal forces.* These should not be shown in an FBD since they do not appear in Euler's Axioms.

The **equation of motion of the centre of mass** C of *any system* is obtained using $\vec{p}_{|I}$ from eq. (2.9) in eq. (2.10) (Fig. 2.8e):

$$\vec{F} = m\vec{a}_{C|I} = \sum m_i \vec{a}_{C_i|I} \tag{2.12}$$

$$\begin{aligned} &F_x = m\ddot{x}_C, && F_y = m\ddot{y}_C, && F_z = m\ddot{z}_C \\ &F_r = m(\ddot{r}_C - r_C\dot{\phi}_C^2), && F_\phi = m(2\dot{r}_C\dot{\phi}_C + r_C\ddot{\phi}_C), && F_z = m\ddot{z}_C \\ &F_t = m\ddot{s}_C, && F_n = ma_{C_n} = m\dot{s}_C^2 \,/\, \rho_C, && F_b = 0 \end{aligned} \tag{2.13}$$

where the position of C with respect to frame I has Cartesian coordinates (x_C, y_C, z_C), cylindrical polar coordinates (r_C, ϕ_C, z_C), path coordinate s_C, and the radius of curvature of its path is ρ_C. The system can be a rigid body, a deformable body, a system of rigid bodies, a finite part or an infinitesimal part of a body. Equation (2.12) is the real content of Newton's second law.

If $\vec{F}(t) \equiv \vec{0}$, then $\vec{F} = m\vec{a}_{C|I} \Rightarrow$

$$\begin{aligned} &\vec{a}_{C|I} = \vec{0}, && \sum m_i \vec{a}_{C_i|I} = \vec{0} \quad \forall\, t \\ &\vec{v}_{C|I}(t) = \vec{v}_{C|I}(0), && \sum m_i \vec{v}_{C_i|I}(t) = \sum m_i \vec{v}_{C_i|I}(0) \end{aligned} \tag{2.14}$$

i.e., the centre of mass C of any system moves with uniform velocity relative to frame I if the total external force on it is zero. The other material points of the system may not have uniform velocity and the body can in fact deform, split or rotate. This is the precise content of Newton's first law. If in addition $\vec{p}\,(0) = \vec{0}$, then eq. (2.14) $\Rightarrow$

$$\begin{aligned} &\vec{v}_C(t) \equiv \vec{0}, && \sum m_i \vec{v}_{C_i}(t) \equiv \vec{0} \\ &\vec{r}_C(t) = \vec{r}_C(0), && \sum m_i \vec{r}_{C_i}(t) = \sum m_i \vec{r}_{C_i}(0) \end{aligned} \tag{2.15}$$

[2]Note, however, that if one wants to predict the motion of systems of rigid bodies, especially when friction is present, one could encounter *deterministic chaos*, i.e., although the system may be in principle deterministic, in practice prediction may not be possible beyond a short initial period because of the extreme sensitivity of the system to infinitesimal changes in initial and boundary conditions.

i.e., the centre of mass C of any system remains at rest relative to frame I, though parts of the system may move. If only a particular Cartesian component of $\vec{F}(t)$ is zero for all t, then eqs (2.14), (2.15) hold only for that component. However, these comments do not hold for $\vec{e}_r$ or $\vec{e}_\phi$ components if $F_r(t) \equiv 0$ or $F_\phi(t) \equiv 0$.

The kinematic variables $\vec{a}_{C|I}$ and $\vec{v}_{PA|I}$ in Euler's axioms have the same value in all inertial frames in uniform translation with respect to one another. Hence, the force $\vec{F}(\vec{r}, \dot{\vec{r}}, t)$ appearing in these axioms must be of the form $\vec{F}(\vec{r}_2 - \vec{r}_1, \dot{\vec{r}}_2 - \dot{\vec{r}}_1, t)$. All forces do in fact depend on the relative positions and/or velocities of interacting bodies.

§ **Exercise 2.3:** (a) Derive Euler's 2nd axiom from the 1st axiom for a mass-point subjected to a force $\vec{F}$ and no discrete moment. (b) In the theory of relativity, $m = m_0/[1-(v/c)^2)]^{1/2}$, where m_0 is the rest mass. Prove that the momentum principle for a mass-point $\vec{F} = d(m\vec{v})/dt$ becomes $\vec{F} = m\vec{a} + [(v/c)^2/\{1-(v/c)^2\}]m\dot{v}\,\vec{e}_t$. $\vec{F}$ differs from $m\vec{a}$ by less than 1% for $v/c < 0.1$.

2.4 Equivalent / Resultant Force Systems

To apply Euler's axioms to a system, we need to compute $\vec{F}$ and $\vec{M}_O$ of the external loads acting on it. We shall often refer to a load system consisting of forces and moments simply as a force system.

Two force systems are defined to be *equivalent* if they have the same total force $\vec{F}$ and the same total moment $\vec{M}_A$ about *one* point A. A *simpler force system* equivalent to a given force system is called its *resultant.*

The following inferences follow for equivalent force systems.

1. Equivalent force systems have equal total moment $\vec{M}_B$ about *any* point B.

   Proof: Consider a force system consisting of n discrete forces with the ith force $\vec{F}_i$ acting at point i, and m couples with moments $\vec{C}_j$, $j = 1, \ldots, m$ (Fig. 2.9a).

$$\begin{aligned}\vec{M}_B &= \sum_i \overrightarrow{Bi} \times \vec{F}_i + \sum_j \vec{C}_j = \sum_i (\overrightarrow{BA} + \overrightarrow{Ai}) \times \vec{F}_i + \sum_j \vec{C}_j \\ &= \overrightarrow{BA} \times \sum_i \vec{F}_i + \Big(\sum_i \overrightarrow{Ai} \times \vec{F}_i + \sum_j \vec{C}_j\Big) = \overrightarrow{BA} \times \vec{F} + \vec{M}_A\end{aligned}$$

   ⇒ The moment $\vec{M}_B$ for equivalent force systems is the same since $\vec{F}$ and $\vec{M}_A$ for them are equal. The proof can be extended for distributed forces and moments. Hence, *for finding the moment sum and force sum, a given force system can be replaced by its resultant.* For example, for applying Euler's axioms, a complex external force system can be replaced by its resultant.

2. Two equivalent force systems impart the same motion to a *single rigid body* since it is governed only by $\vec{F}$ and $\vec{M}_O$, which are the same for the two systems. But the *internal forces* induced are *different.* For the equivalent force systems in Fig. 2.9b,

the rigid rod has the same motion but the internal forces differ.

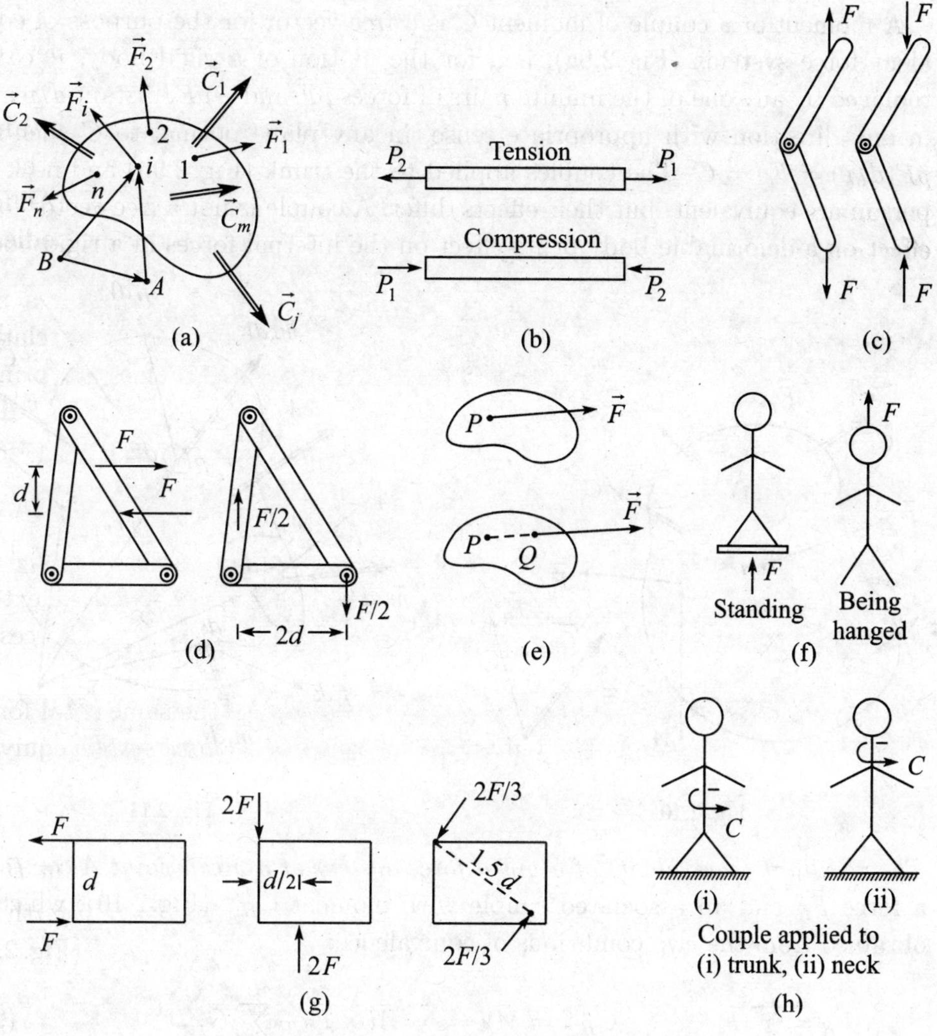

Fig. 2.9

3. Two equivalent force systems, in general, impart different motion to a deformable system and to a system of rigid bodies, as illustrated for equivalent force systems in Fig. 2.9c. However, if the system of rigid bodies together form a rigid system, then equivalent force systems acting on it produce the same motion (Fig. 2.9d).

Force is a transmissible vector for the purpose of equivalent force systems (Fig. 2.9e). Thus, the magnitude, line of action and sense of a force are significant for its effect on the motion of a rigid body and its point of application is immaterial. The force on the feet of a person (a deformable body) is equivalent to the force on the neck (Fig. 2.9f) but their effect is completely different! Force is not a transmissible vector for its effect on a deformable body.

A moment or a couple of moment $\vec{C}$ is a free vector for the purpose of equivalent force systems (Fig. 2.9g), i.e., for the motion of a rigid body, it can be replaced by any one of the infinite pairs of forces $p\vec{F}$ and $-p\vec{F}$, distant d/p apart in any direction with appropriate sense, in any plane normal to $\vec{C}$ such that $pF(d/p) = Fd = C$. The couples applied to the trunk (Fig. 2.9h) and neck of a person are equivalent, but their effects differ. A couple is not a free vector for its effect on a deformable body or its effect on the internal forces in a rigid body.

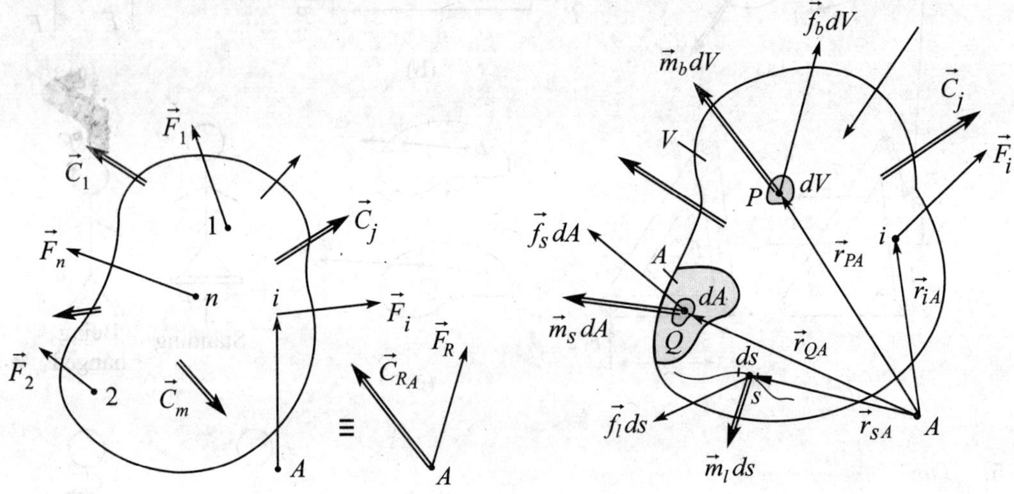

Fig. 2.10

Fig. 2.11

4. *The resultant (equivalent) of a given force system at a given point A* consists of a force $\vec{F}_R$ and an associated couple with moment $\vec{C}_{R_A}$ (Fig. 2.10), which are obtained from the two conditions of equivalence:

$$\vec{F}_R = \sum_i \vec{F}_i, \qquad \vec{C}_{R_A} = \vec{M}_A = \sum_i \overrightarrow{Ai} \times \vec{F}_i + \sum_j \vec{C}_j \tag{2.16}$$

The subscript A has not been used in $\vec{F}_R$ since it is independent of A. $\vec{C}_{R_A}$ depends on A. The integrals $\int_s \vec{f}_l\, ds$, $\int_A \vec{f}_s\, dA$, $\int_V \vec{f}_b\, dV$ and $\int_s (\vec{r}_{SA} \times \vec{f}_l + \vec{m}_l) ds$, $\int_A (\vec{r}_{QA} \times \vec{f}_s + \vec{m}_s) dA$, $\int_V (\vec{r}_{PA} \times \vec{f}_b + \vec{m}_b) dV$ are added to the right hand side of eqs $(2.16)_1$ and $(2.16)_2$ respectively, if the given force system includes distributed forces and distributed moments (Fig. 2.11) $\vec{m}_l, \vec{m}_s, \vec{m}_b$ distributed over a line, a surface and a volume as N.m/ m, N.m/m^2, N.m/m^3, respectively.

The direction cosines l, m, n of $\vec{F}_R$ can be obtained from

$$l = \cos\alpha = \frac{F_{Rx}}{F_R}, \qquad m = \cos\beta = \frac{F_{Ry}}{F_R}, \qquad n = \cos\gamma = \frac{F_{Rz}}{F_R}$$

The direction cosines of $\vec{C}_{R_A}$ are obtained similarly. For the case of $\vec{F}_R = F_{Rx}\vec{\imath} + F_{Ry}\vec{\jmath}$, angle α that $\vec{F}_R$ makes with x-axis is given by

$$\cos\alpha = \frac{F_{Rx}}{F_R}, \qquad \sin\alpha = \frac{F_{Ry}}{F_R}, \qquad \tan\alpha = \frac{F_{Ry}}{F_{Rx}}$$

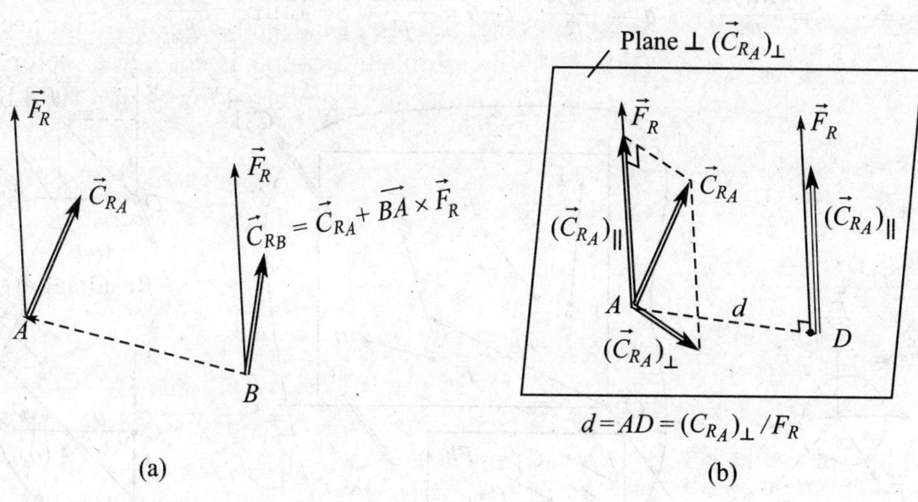

Fig. 2.12

5. *The simplest resultant (simplest equivalent) of a given force system is a wrench.*
   Proof: Let the resultant at point A be $\vec{F}_R$, $\vec{C}_{R_A}$. The resultant at point B is $\vec{F}_R$, $\vec{C}_{R_B}$ (Fig. 2.12a) with

$$\vec{M}_B = \vec{C}_{R_B} = \vec{C}_{R_A} + \overrightarrow{BA} \times \vec{F}_R = (\vec{C}_{R_A})_\parallel + (\vec{C}_{R_A})_\perp + \overrightarrow{BA} \times \vec{F}_R \qquad (a)$$

   where $(\vec{C}_{R_A})_\parallel$, $(\vec{C}_{R_A})_\perp$ are the components of $\vec{C}_{R_A}$ parallel and normal to $\vec{F}_R$. It is possible to choose an appropriate point B such that the last two terms on the right hand side of eq. (a) cancel each other, since $\overrightarrow{BA} \times \vec{F}_R \perp \vec{F}_R$. For such a point, say D, the resultant is a wrench consisting of $\vec{F}_R$, $(\vec{C}_{R_A})_\parallel$ (Fig. 2.12b).

It follows that if $(\vec{C}_{R_A})_\parallel = \vec{0}$, i.e., *if* $\vec{C}_{R_A} \cdot \vec{F}_R = 0$, *then the simplest resultant is a single force* $\vec{F}_R$ *if* $\vec{F}_R \neq \vec{0}$, *or a single couple* $\vec{C}_{R_A}$ *if* $\vec{F}_R = \vec{0}$, *or a null system if* $\vec{F}_R = \vec{0}$, $\vec{C}_{R_A} = \vec{0}$. Such a simplest resultant exists for the coplanar and parallel force systems, defined next, since the condition $\vec{C}_{R_O} \cdot \vec{F}_R = 0$ holds.

The simplest resultant of a concurrent force system (Fig. 2.13a) is a single force through the point of concurrence. *A coplanar force system* (Fig. 2.13b) consists of forces in one plane (say xy-plane) and couples with moments along the normal to this plane (i.e., parallel to z). For the origin O, $\vec{C}_{R_O} \cdot \vec{F}_R = M_{Oz}\,\vec{k} \cdot (F_x\vec{\imath} + F_y\vec{\jmath}) = 0$. For the case of $\vec{F}_R \neq \vec{0}$, the line of action of the simplest resultant $\vec{F}_R = F_x \vec{i} + F_y \vec{j}$ is at distance d from O and cuts the x and y axes at $A(a, 0, 0)$ and $B(0, -b, 0)$ such that

$$aF_y = bF_x = F_R d = M_{Oz} = \vec{C}_{RO} \cdot \vec{k}$$

$$\Rightarrow \quad d = \frac{M_{Oz}}{F_R}, \quad a = \frac{M_{Oz}}{F_y}, \quad b = \frac{M_{Oz}}{F_x}, \quad \tan\alpha = \frac{F_y}{F_x}$$

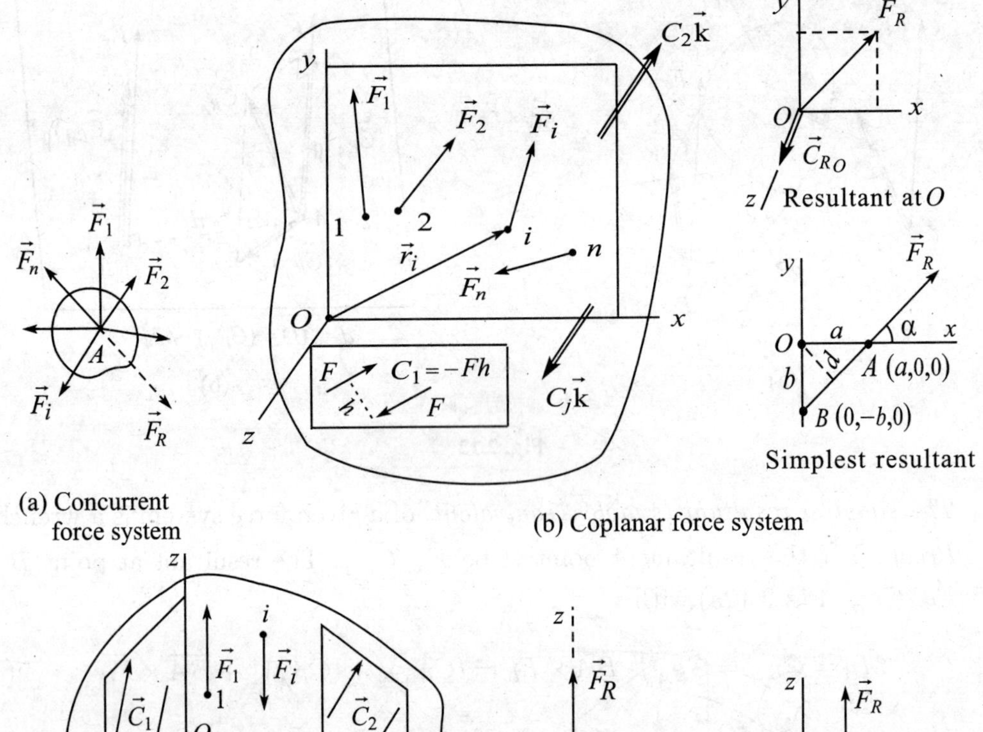

Fig. 2.13

A parallel force system (Fig. 2.13c) consists of forces parallel to a line (say z-axis) and couples with moments normal to this line (i.e., having components along $\vec{i}$ and $\vec{j}$). $\vec{C}_{RO} \cdot \vec{F}_R = (M_{Ox} \vec{i} + M_{Oy} \vec{j}) \cdot F_R \vec{k} = 0$. For $\vec{F}_R \neq \vec{0}$, the simplest resultant $\vec{F}_R = F_R \vec{k}$ intersects the xy-plane at $(a, b, 0) \ni$

$$(a\vec{i} + b\vec{j}) \times F_R \vec{k} = \vec{C}_{RO} = M_{Ox} \vec{i} + M_{Oy} \vec{j}$$

$$\Rightarrow \quad a = -\frac{M_{Oy}}{F_R}, \qquad b = \frac{M_{Ox}}{F_R}$$

The resultant of a force system may be computed for finding $\vec{F}$ and $\vec{M}_O$. Although the simplest resultant is a wrench, we usually use the resultant force–couple system at a convenient point A since the latter is easier to determine.

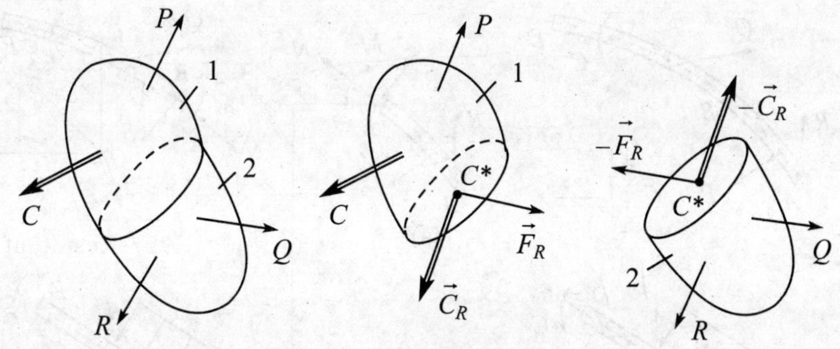

Fig. 2.14

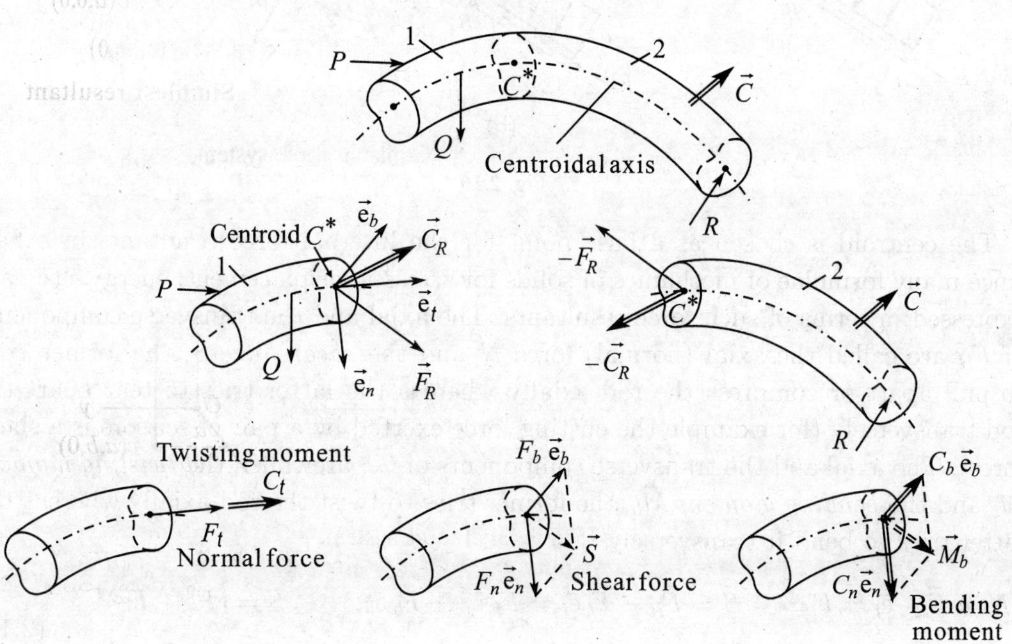

Fig. 2.15

Internal force resultants: For an internal section of a body, the distributed force across the section can be replaced by its resultant at the centroid C^* of the cross-section, consisting of a resultant force $\vec{F}_R$ and a resultant couple with moment $\vec{C}_R$ at C^* (Fig. 2.14). For the case of a *thin bar* (i.e., a body whose transverse dimensions are small compared to the axial length), these can be resolved (Fig. 2.15) in terms of the right-handed unit triad (tangent $\vec{e}_t$, normal $\vec{e}_n$, binormal $\vec{e}_b$) of its centroidal axis joining the centroids of its cross-sections:

$$\vec{F}_R = F_t\,\vec{e}_t + F_n\,\vec{e}_n + F_b\,\vec{e}_b, \qquad \vec{C}_R = C_t\,\vec{e}_t + C_n\,\vec{e}_n + C_b\,\vec{e}_b \tag{2.17}$$

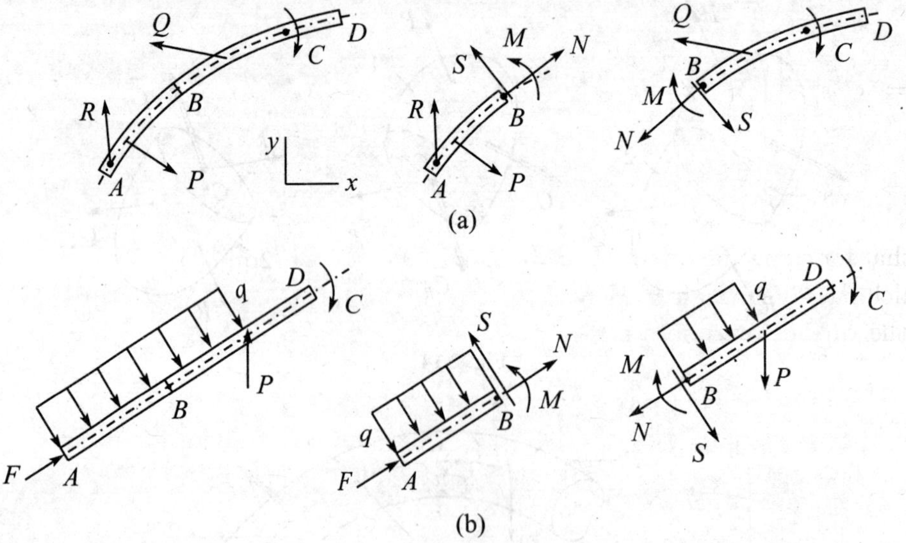

(a)

(b)

Fig. 2.16

The centroid is chosen as a base point for the internal force resultants in a bar, since many formulae of mechanics of solids for stresses, displacement, energy, etc., are expressed in terms of such force resultants. The axial and the transverse components of $\vec{F}_R$ are called the axial (normal) force N and the *shear force* $\vec{S}$, the former tries to pull apart or compress the rod axially whereas the latter tries to tear apart the rod transversely (for example the cutting force exerted by a pair of scissors is a shear force). The axial and the transverse components of $\vec{C}_R$ are called the *twisting moment* M_t and the *bending moment* $\vec{M}_b$, the former tries to twist the rod axially whereas the latter tries to bend it transversely.

$$\begin{aligned} N &= \vec{F}_R \cdot \vec{e}_t = F_t, & \vec{S} &= \vec{F}_R - F_t\,\vec{e}_t = F_n\,\vec{e}_n + F_b\,\vec{e}_b, & S &= (F_n^2 + F_b^2)^{1/2} \\ M_t &= \vec{C}_R \cdot \vec{e}_t = C_t, & \vec{M}_b &= \vec{C}_R - M_t\,\vec{e}_t = C_n\,\vec{e}_n + C_b\,\vec{e}_b, & M_b &= (C_n^2 + C_b^2)^{1/2} \end{aligned} \tag{2.18}$$

For a bar with a plane centroidal axis, say in the xy-plane, subjected to a coplanar force system in the plane of the centroidal axis, the components F_{Rz}, C_{Rx}, C_{Ry} of $\vec{F}_R$, $\vec{C}_R$ are zero, and the internal force resultants are often shown as a normal force N, a shear force S and a bending moment M (Fig. 2.16).

2.4.1 Centre of parallel forces

Consider a parallel force system (Fig. 2.17) consisting only of forces $F_i\vec{e}$ acting at $\vec{r}_i$, $i = 1, \ldots, n$ such that $\vec{F}_R \neq \vec{0}$. Its resultant $\vec{F}_R$ always passes through a point R called the *centre of parallel forces*, irrespective of the direction of $\vec{e}$, since

$$\vec{F}_R = \left(\sum_i F_i\right)\vec{e}, \qquad \vec{M}_O = \sum_i \vec{r}_i \times F_i\vec{e} = \vec{r}_R \times \vec{F}_R = \vec{r}_R \times \left(\sum_i F_i\right)\vec{e} \quad (1)$$

$$\Rightarrow \quad \left[\left(\sum_i F_i\right)\vec{r}_R - \left(\sum_i \vec{r}_i F_i\right)\right] \times \vec{e} = \vec{0}, \quad \forall\, \vec{e}$$

$$\Rightarrow \quad \vec{r}_R = \frac{\sum_i \vec{r}_i F_i}{\sum_i F_i}, \quad \forall\, \vec{e} \tag{2.19}$$

Note that for a specific direction $\vec{e} = \vec{e}_1$, eq. (1) yields the equation of a line, parallel to $\vec{e}_1$, which passes through R. Hence, eq. (2.19) yields the common point of intersection, R, of the various lines obtained by changing the direction of $\vec{e}$.

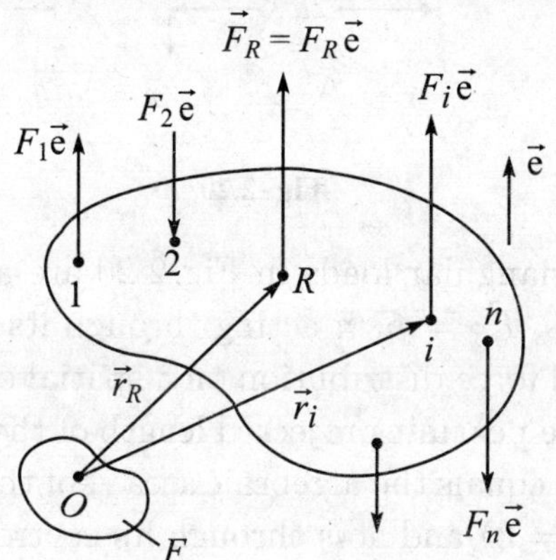

Fig. 2.17

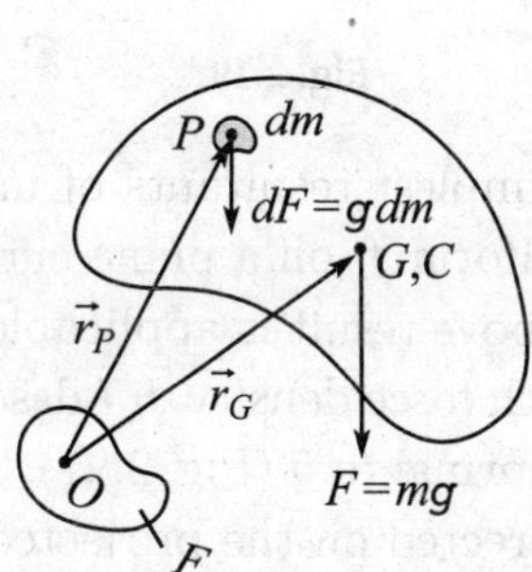

Fig. 2.18

For a distributed force, the summation in eq. (2.19) is replaced by integration. The centre of the distributed parallel *uniform gravitational force* $dF = g\,dm$ on mass element dm of a body of mass m, called *centre of gravity* G, coincides with C (Fig. 2.18):

$$\vec{r}_G = \vec{r}_R = \frac{\sum_i \vec{r}_i F_i}{\sum_i F_i} = \frac{\int_F \vec{r}_P\,dF}{\int_F dF} = \frac{\int_m \vec{r}_P\,g\,dm}{\int_m g\,dm} = \frac{\int_m \vec{r}_P\,dm}{\int_m dm} = \vec{r}_C$$

Consider a parallel distributed force $f_l\,\vec{n}$ (N/m) normal to a plane curve C_1. At every point of the curve, erect an ordinate parallel to $\vec{n}$ equal to f_l. The surface so generated is called a *loading surface* (Fig. 2.19). Its *simplest resultant* equals the algebraic *area* A *of the loading surface* (if $A \neq 0$) and *acts through its centroid* C^*:

$$dF = f_l\,ds = dA, \qquad \vec{F}_R = \int_F dF\,\vec{n} = \int_A dA\,\vec{n} = A\,\vec{n} \tag{1}$$

$$\vec{r}_{C^*} = \frac{\int_A \vec{r}_P\,dA}{A} = \frac{\int (\vec{r}_Q + 0.5 f_l\,\vec{n})\,dF}{F}$$

$$= \frac{\int \vec{r}_Q\,dF}{F} + (\ldots)\,\vec{n} = \vec{r}_R + (\ldots)\,\vec{n} \tag{2}$$

Fig. 2.19

Fig. 2.20

The simplest resultants of uniform and triangular loads in Fig. 2.20 are as shown. For a uniform f_l on a plane curve of length s, $\vec{F}_R = f_l s\,\vec{n}$ acting through its centroid.

The above result is applicable to a parallel force distribution on a spatial or a plane curve with force density $f_l\vec{n}$ described as force per unit projected length of the curve in a plane normal to $\vec{n}$ (Fig. 2.21). The resultant equals the algebraic area A of the loading surface erected on the projected curve (if $A \neq 0$) and acts through its centroid C^*.

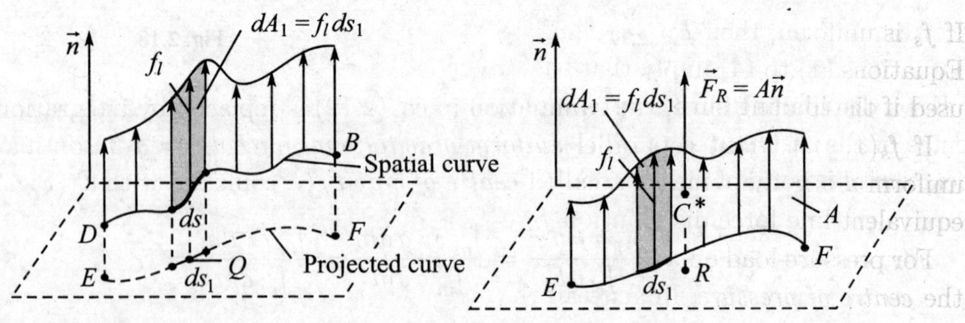

Fig. 2.21

Consider a parallel distributed force $f_s\,\vec{n}$ (N/m^2) normal to a plane area A. At every point of the area, erect an ordinate parallel to $\vec{n}$ equal to f_s. The solid so generated is called a *pressure space* (Fig. 2.22a). Its *simplest resultant* equals the algebraic *volume V of the pressure space* (if $V \neq 0$) and *acts through its centroid C^**.

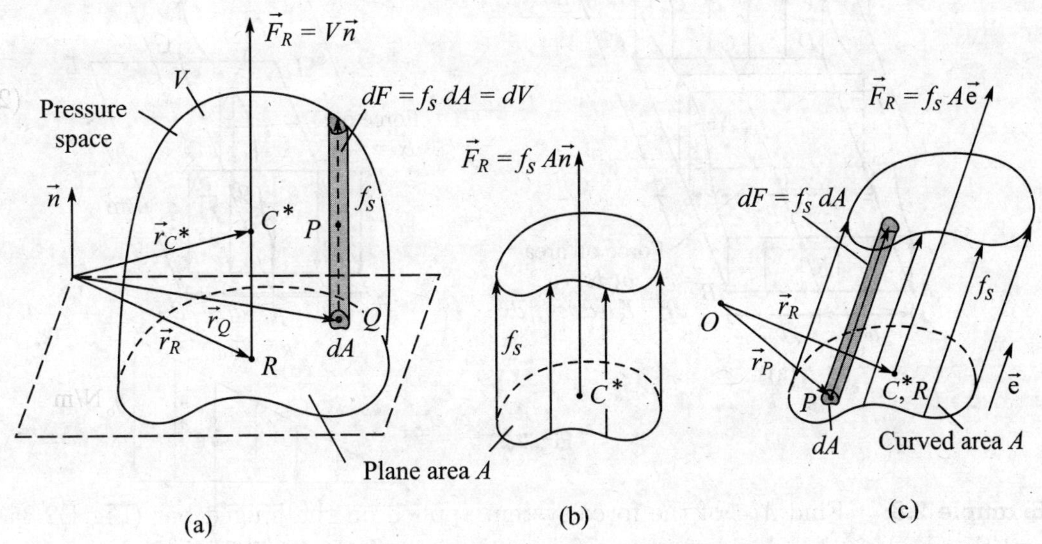

Fig. 2.22

$$dF = f_s\,dA = dV, \qquad \vec{F}_R = \int_F dF\,\vec{n} = \int_V dV\,\vec{n} = V\,\vec{n} \tag{3}$$

$$\vec{r}_{C^*} = \frac{\int_V \vec{r}_P\,dV}{V} = \frac{\int(\vec{r}_Q + 0.5 f_s\vec{n})\,dF}{F}$$
$$= \frac{\int \vec{r}_Q\,dF}{F} + (\ldots)\,\vec{n} = \vec{r}_R + (\ldots)\,\vec{n} \tag{4}$$

If f_s is uniform, then $\vec{F}_R = f_s A\,\vec{n}$ acts at the centroid of the plane area (Fig. 2.22b).[3] Equations (1) to (4) imply that the tabulated values of area, volume, centroids can be used if the loading surface/pressure space is composed of basic shapes.

If $f_s(x,y) = f_s(x)$ acts on a rectangular area in the xy-plane (Fig. 2.23), which is uniform along the width b, then the surface force distribution can be replaced by an equivalent line force distribution $f_l = f_s b$ on the centre line of the rectangle.

For pressure load on a plane area, the point R where the resultant force acts is called the *centre of pressure*. In general, the simplest resultant of a pressure distribution on a curved surface may be a wrench and the centre of pressure may not exist.

---

[3] Resultant of uniform distributed parallel force $f_s\,\vec{e}$ (N/m^2) (Fig. 2.22c) on area A (plane/curved) is $\vec{F}_R = \int_A f_s\,dA\,\vec{e} = f_s A\,\vec{e}$ through its centroid C^* as $\vec{r}_R = \int_A \vec{r}_P f_s\,dA / \int_A f_s\,dA = \int_A \vec{r}_P dA/A = \vec{r}_{C^*}$. In particular, this result applies to uniform tangential parallel force distribution.

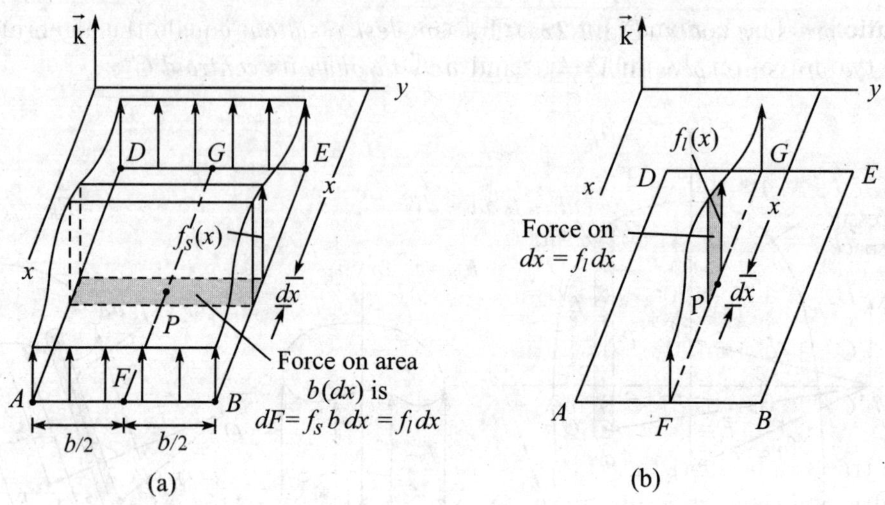

Fig. 2.23

Example 2.8: Find $\vec{M}_A$ of the force system applied on the hinged bar (Fig. E2.8a) and the resultant (equivalent) of this force system at A. $P = 5.773\,\text{kN}$.

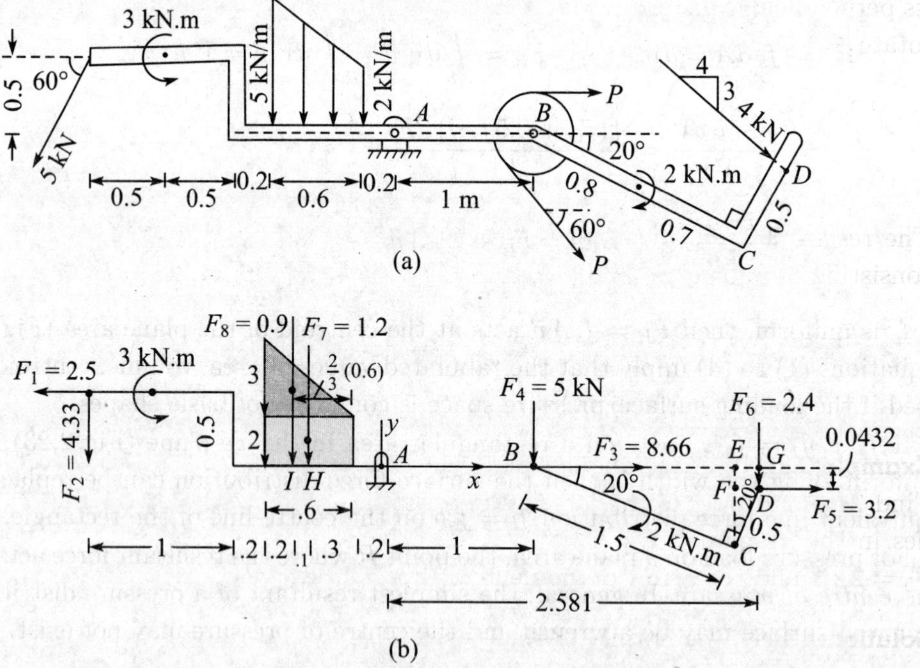

Fig. E2.8

Solution: The coplanar forces are resolved into suitable components (Fig. E2.8b). The two forces on the pulley are equivalent to F_3 and F_4 at B since their $\vec{M}_B = \vec{0}$.

$$F_1 = 5\cos 60^\circ = 2.5\,\text{kN}, \qquad F_4 = P\sin 60^\circ = 5\,\text{kN}$$

$$F_2 = 5\sin 60^\circ = 4.330\,\text{kN}, \qquad F_5 = 4[4/(3^2+4^2)^{1/2}] = 3.2\,\text{kN}$$

$$F_3 = P + P\cos 60^\circ = 8.660\,\text{kN}, \qquad F_6 = 4[3/(3^2+4^2)^{1/2}] = 2.4\,\text{kN}$$

$$BE = 1.5\cos 20^\circ = 1.410\,\text{m}, \qquad FD = 0.5\sin 20^\circ = 0.1710\,\text{m}$$

$$CE = 1.5\sin 20^\circ = 0.5130\,\text{m}, \qquad DG = EF = CE - CF = 0.0432\,\text{m}$$

$$CF = 0.5\cos 20^\circ = 0.4698\,\text{m}, \qquad AG = 1 + 1.410 + 0.1710 = 2.581\,\text{m}$$

The trapezoidal distributed force is decomposed into a rectangle and a triangle. The resultant of the rectangular distribution is $F_7 = \text{area} = 0.6(2) = 1.2\,\text{kN}$, and acts at H through its centroid with $AH = 0.2 + 0.6/2 = 0.5\,\text{m}$. The resultant of the triangular distribution is $F_8 = \text{area} = \frac{1}{2}(0.6)0.3 = 0.9\,\text{kN}$, and acts at I through its centroid with $AI = 0.2 + \frac{2}{3}(0.6) = 0.6\,\text{m}$.

For coplanar force system, instead of using cross product, it is easier to compute the moment of each force component about A as the product of its magnitude with its perpendicular distance from A and assign +ve or −ve sign according as it tends to rotate the bar from $x\,(\vec{\imath})$ to $y\,(\vec{\jmath})$, or $y\,(\vec{\imath})$ to $x\,(\vec{\jmath})$. The same applies to couples. Thus,

$$\begin{aligned}\vec{M}_A &= [2.5\times 0.5 + 4.330\times 2 + 8.660\times 0 - 5\times 1 + 3.2\times 0.0432\\ &\quad -2.4\times 2.581 + 1.2\times 0.5 + 0.9\times 0.6 + 3 - 2]\,\vec{k} = 0.9938\,\vec{k}\ \text{kN.m}\end{aligned}$$

The resultant (equivalent) of the given force system at A is a force–couple system consisting of a force $\vec{F}_R$ and a couple $\vec{C}_R$ with $\vec{C}_R = \vec{M}_A = 0.9938\,\vec{k}$ kN.m and

$$\begin{aligned}\vec{F}_R &= \sum_i \vec{F}_i = -2.5\vec{\imath} - 4.330\vec{\jmath} + 8.660\vec{\imath} - 5\vec{\jmath} + 3.2\vec{\imath} - 2.4\vec{\jmath} - 1.2\vec{\jmath} - 0.9\,\vec{\jmath}\\ &= 9.36\vec{\imath} - 13.83\vec{\jmath}\ \text{kN}\end{aligned}$$

Example 2.9: A rectangular plate (Fig. E2.9) of weight 2 kN is hinged along AB. Find $\vec{M}_A$ and M_{AB} of the force system shown and its resultant (equivalent) at A. G lies in the xz-plane. $\alpha = 120^\circ$, $\beta = 60^\circ$, $\gamma = 45^\circ$, $OG = 4\,\text{m}$, $DF = 3\,\text{m}$, weight $F_5 = 2\,\text{kN}$ and the arrow of moment 2 kN.m is in the plane $ABED$.

Solution: Let $\vec{n}$ be normal to the plane. The coordinates and position vectors are:

$$A(4,0,0), \quad B(0,0,3), \quad D(4,2,0), \quad E(0,2,3)$$

$$G(4\sin 30^\circ, 0, 4\cos 30^\circ) = (2, 0, 3.464), \quad H(0,1,3)$$

$$\vec{r}_A = 4\vec{i}, \quad \vec{r}_B = 3\vec{k}, \quad \vec{r}_D = 4\vec{i} + 2\vec{j}, \quad \overrightarrow{AB} = -4\vec{i} + 3\vec{k}$$

$$\vec{r}_E = 2\vec{j} + 3\vec{k}, \quad \vec{r}_G = 2\vec{i} + 3.464\vec{k}, \quad \vec{r}_H = \vec{j} + 3\vec{k}\text{ m}$$

$$\vec{r}_C = (\vec{r}_A + \vec{r}_B + \vec{r}_D + \vec{r}_E)/4 = 2\vec{i} + \vec{j} + 1.5\vec{k}\text{ m}$$

$$\vec{e} = \frac{\overrightarrow{AB}}{|\overrightarrow{AB}|} = \frac{-4\vec{i} + 3\vec{k}}{(4^2 + 3^2)^{1/2}} = -0.8\vec{i} + 0.6\vec{k}$$

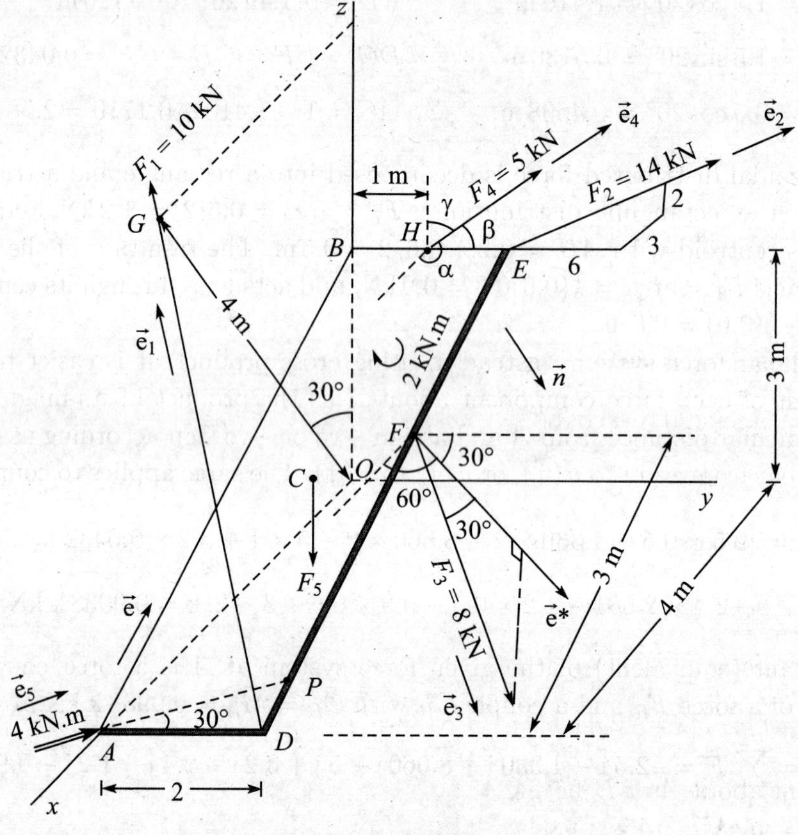

Fig. E2.9

$$\overrightarrow{AB} \times \overrightarrow{AD} = (-4\vec{i} + 3\vec{k}) \times 2\vec{j} = -6\vec{i} - 8\vec{k}$$

$$\overrightarrow{DG} = \vec{r}_G - \vec{r}_D = -2\vec{i} - 2\vec{j} + 3.464\vec{k}\text{ m}$$

$$\vec{r}_F = \vec{r}_D + 3\vec{e} = 4\vec{i} + 2\vec{j} + 3(-0.8\vec{i} + 0.6\vec{k}) = 0.6\vec{i} + 2\vec{j} + 1.8\vec{k}\text{ m}$$

$$\vec{n} = \frac{\overrightarrow{AB} \times \overrightarrow{AD}}{|\overrightarrow{AB} \times \overrightarrow{AD}|} = \frac{-6\vec{i} - 8\vec{k}}{(6^2 + 8^2)^{1/2}} = -0.6\vec{i} - 0.8\vec{k}$$

The two couples and the five forces are given by

$$\begin{aligned}
\vec{C}_1 &= 2\vec{n} = 2(-0.6\,\vec{\imath} - 0.8\,\vec{k}) = -1.2\,\vec{\imath} - 1.6\,\vec{k}\ \text{kN.m}\\
\vec{C}_2 &= 4\,\vec{e}_5 = 4(\cos 30^\circ\,\vec{\jmath} + \sin 30^\circ\,\vec{e}) = 4[0.866\,\vec{\jmath} + 0.5(-0.8\,\vec{\imath} + 0.6\,\vec{k})]\\
&= -1.6\,\vec{\imath} + 3.464\,\vec{\jmath} + 1.2\,\vec{k}\ \text{kN.m}\\
\vec{F}_1 &= 10\,\vec{e}_1 = 10\frac{\overrightarrow{DG}}{|\overrightarrow{DG}|} = 10\,\frac{-2\,\vec{\imath} - 2\,\vec{\jmath} + 3.464\,\vec{k}}{(2^2 + 2^2 + 3.464^2)^{1/2}}\\
&= -4.472\,\vec{\imath} - 4.472\,\vec{\jmath} + 7.746\,\vec{k}\ \text{kN}\\
\vec{F}_2 &= 14\,\vec{e}_2 = 14\,\frac{6\,\vec{\jmath} - 3\,\vec{\imath} + 2\,\vec{k}}{(6^2 + 3^2 + 2^2)^{1/2}} = -6\,\vec{\imath} + 12\,\vec{\jmath} + 4\,\vec{k}\ \text{kN}\\
\vec{F}_3 &= 8\,\vec{e}_3 = 8[\cos 30^\circ\,\vec{e}^* - \sin 30^\circ\,\vec{k}]\\
&= 8[0.866(\cos 60^\circ\,\vec{\imath} + \sin 60^\circ\,\vec{\jmath}) - 0.5\,\vec{k}] = 3.464\,\vec{\imath} + 6\,\vec{\jmath} - 4\,\vec{k}\ \text{kN}\\
\vec{F}_4 &= 5\,\vec{e}_4 = 5(\cos\alpha\,\vec{\imath} + \cos\beta\,\vec{\jmath} + \cos\gamma\,\vec{k})\\
&= 5(\cos 120^\circ\,\vec{\imath} + \cos 60^\circ\,\vec{\jmath} + \cos 45^\circ\,\vec{k}) = -2.5\,\vec{\imath} + 2.5\,\vec{\jmath} + 3.536\,\vec{k}\ \text{kN}\\
\vec{F}_5 &= -2\,\vec{k}\ \text{kN}
\end{aligned}$$

The position vectors of the points of application of the forces with respect to A are:

$$\overrightarrow{AD} = \vec{r}_D - \vec{r}_A = 2\,\vec{\jmath}\ \text{m}, \qquad \overrightarrow{AE} = \vec{r}_E - \vec{r}_A = -4\,\vec{\imath} + 2\,\vec{\jmath} + 3\,\vec{k}\ \text{m}$$

$$\overrightarrow{AH} = \vec{r}_H - \vec{r}_A = -4\,\vec{\imath} + \vec{\jmath} + 3\,\vec{k}\ \text{m}, \qquad \overrightarrow{AC} = \vec{r}_C - \vec{r}_A = -2\,\vec{\imath} + \vec{\jmath} + 1.5\,\vec{k}\ \text{m}$$

$$\overrightarrow{AF} = \vec{r}_F - \vec{r}_A = -2.4\,\vec{\imath} + 2\,\vec{\jmath} + 1.8\,\vec{k}\ \text{m}$$

The moment about A of $\vec{F}_i$ acting at I, is computed by $\overrightarrow{AI} \times F_i$:

$$\begin{aligned}
\vec{M}_A &= \overrightarrow{AD}\times\vec{F}_1 + \overrightarrow{AE}\times\vec{F}_2 + \overrightarrow{AF}\times\vec{F}_3 + \overrightarrow{AH}\times\vec{F}_4 + \overrightarrow{AC}\times\vec{F}_5 + \vec{C}_1 + \vec{C}_2\\
&= (15.49\,\vec{\imath} + 8.944\,\vec{k}) + (-28\,\vec{\imath} - 2\,\vec{\jmath} - 36\,\vec{k}) - 18.8\,\vec{\imath} - 3.365\,\vec{\jmath} - 21.33\,\vec{k}\\
&\quad + (-3.964\,\vec{\imath} + 6.644\,\vec{\jmath} - 7.5\,\vec{k}) + (-2\,\vec{\imath} - 4\,\vec{\jmath}) + (-1.2\,\vec{\imath} - 1.6\,\vec{k})\\
&\quad + (-1.6\,\vec{\imath} + 3.464\,\vec{\jmath} + 1.2\,\vec{k}) = -40.07\,\vec{\imath} + 0.743\,\vec{\jmath} - 56.29\,\vec{k}\ \text{kN.m}
\end{aligned}$$

The moment M_e of the force system about line AB is given by

$$M_e = \vec{M}_A \cdot \vec{e} = (-40.07)(-0.8) + (-56.29)(0.6) = -1.718\ \text{kN.m.}$$

The resultant (equivalent) at A is a force–couple system, consisting of a force $\vec{F}_R$ and a couple $\vec{C}_R$ given by $\vec{C}_R = \vec{M}_A = -40.07\,\vec{i} + 0.743\,\vec{j} - 56.29\,\vec{k}$ kN.m and

$$\begin{aligned}\vec{F}_R = \sum_i \vec{F}_i &= -4.472\,\vec{i} - 4.472\,\vec{j} + 7.746\,\vec{k} - 6\,\vec{i} + 12\,\vec{j} + 4\,\vec{k} \\ &\quad +3.464\,\vec{i} + 6\,\vec{j} - 4\,\vec{k} - 2.5\,\vec{i} + 2.5\,\vec{j} + 3.536\,\vec{k} - 2\,\vec{k} \\ &= -9.508\,\vec{i} + 16.03\,\vec{j} + 9.282\,\vec{k}\text{ kN}\end{aligned}$$

Example 2.10: Find the simplest resultant of the line and surface loads shown in Fig. E2.10a–f. The axisymmetric pressure p in case (e) and the circumferential force τ N/m² in case (f), on circular area of radius R, have been shown as a function of r.

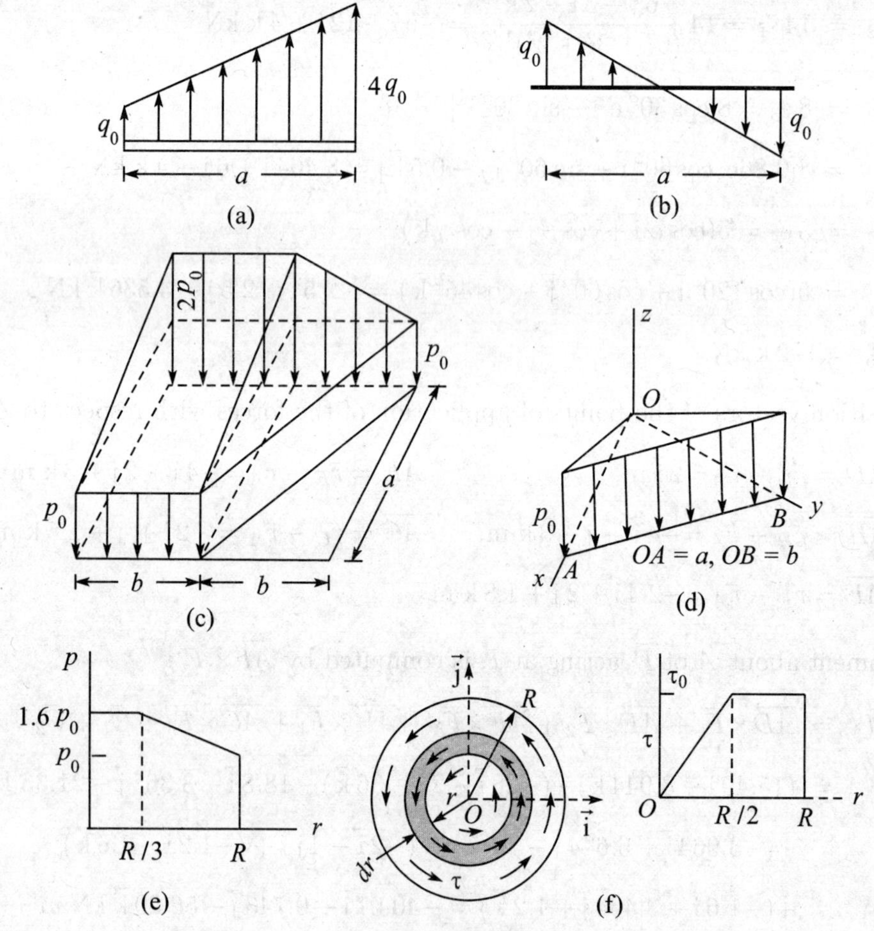

Fig. E2.10a-f

Solution: (a) Decompose the loading surface into a rectangular and a triangular area

(Fig. E2.10g). Their resultants are $F_1\,\vec{k}$ and $F_2\,\vec{k}$:

$$F_1 = (\text{area of rectangle}) = aq_0, \quad \text{through } C_1 \text{ at } \vec{r}_1 = \tfrac{1}{2}a\,\vec{i}$$

$$F_2 = (\text{area of triangle}) = \tfrac{1}{2}a(3q_0) = 1.5aq_0, \quad \text{through } C_2 \text{ at } \vec{r}_2 = \frac{2}{3}a\,\vec{i}$$

$$F_R = F_1 + F_2 = aq_0 + 1.5aq_0 = 2.5aq_0 \neq 0$$

The resultant is a force $\vec{F}_R = F_R\,\vec{k} = 2.5aq_0\,\vec{k}$ acting at point R with

$$\vec{r}_R = \frac{\vec{r}_1 F_1 + \vec{r}_2 F_2}{F_R} = [\tfrac{1}{2}a\,\vec{i}(aq_0) + \tfrac{2}{3}a\,\vec{i}(1.5aq_0)]/2.5aq_0 = 0.6a\,\vec{i}$$

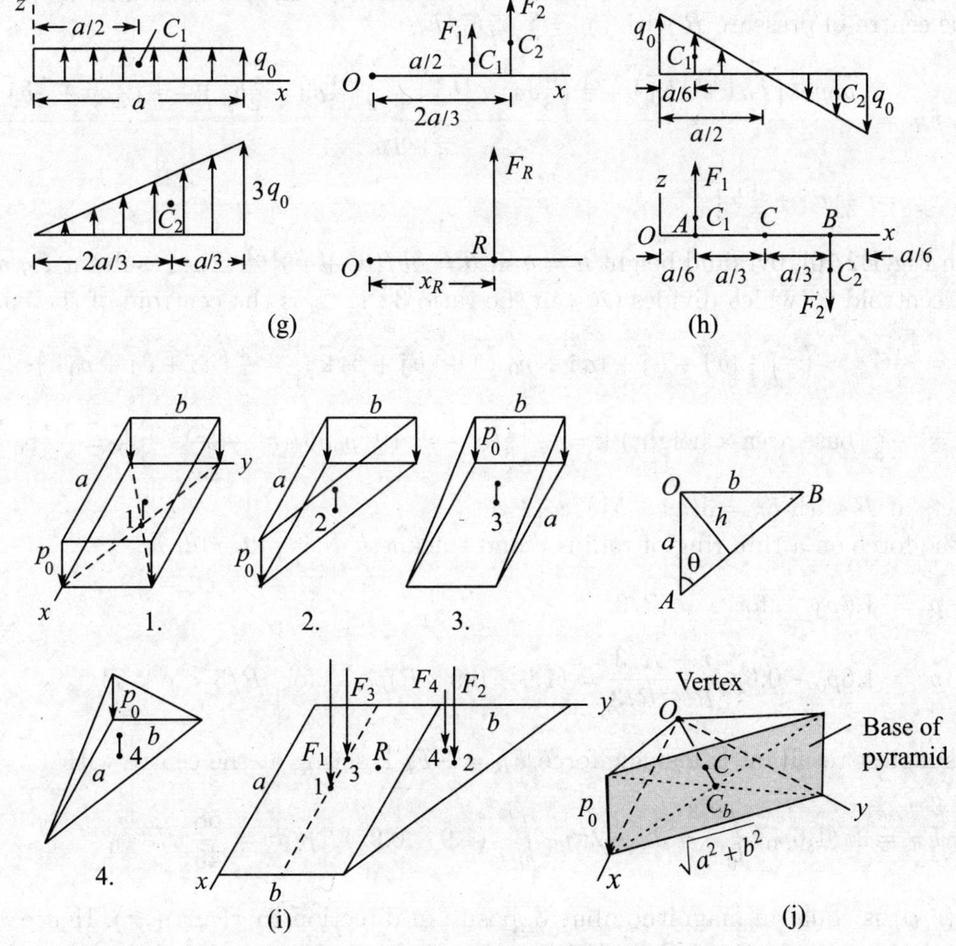

Fig. E2.10g-j

(b) The resultants of the two triangular loadings are $F_1 = F_2 = \frac{1}{2}(\frac{1}{2}a)q_0 = aq_0/4$, acting through their centroids at A and B. $AC = CB = \frac{2}{3}(\frac{1}{2}a) = \frac{1}{3}a$ (Fig. E2.10h). The

simplest resultant is a single couple of moment $\vec{C}_R = F_1(AB)\,\vec{\jmath} = \frac{1}{4}aq_0(\frac{2}{3}a)\,\vec{\jmath} = \frac{1}{6}q_0a^2\,\vec{\jmath}$.
(c) Resultant F_i = volume of pressure space (Fig. E2.10i) and acts through its centroid:

1. a cuboid $a \times b \times p_0$: $\quad F_1 = V_1 = abp_0 \quad$ at $\vec{r}_1 = \frac{1}{2}a\,\vec{\imath} + \frac{1}{2}b\,\vec{\jmath}$,

2. a triangular prism: $\quad F_2 = V_2 = \frac{1}{2}abp_0 \quad$ at $\vec{r}_2 = \frac{1}{3}a\,\vec{\imath} + (b + \frac{1}{3}b)\,\vec{\jmath} = \frac{1}{3}a\,\vec{\imath} + \frac{4}{3}b\,\vec{\jmath}$,

3. a triangular prism: $\quad F_3 = V_3 = \frac{1}{2}abp_0 \quad$ at $\vec{r}_3 = \frac{1}{3}a\,\vec{\imath} + \frac{1}{2}b\,\vec{\jmath}$,

4. a tetrahedron: $\quad F_4 = V_4 = \frac{1}{3}(\frac{1}{2}ab)p_0 = \frac{1}{6}abp_0$ through its centroid at 1/4th of its height from each base at $\vec{r}_4 = \frac{1}{4}a\,\vec{\imath} + (b + \frac{1}{4}b)\,\vec{\jmath} = \frac{1}{4}a\,\vec{\imath} + \frac{5}{4}b\,\vec{\jmath}$.

$F_R = \sum F_i = 13abp_0/6 \neq 0 \;\Rightarrow$ the resultant is $\vec{F}_R = -F_R\,\vec{k} = -(13abp_0/6)\,\vec{k}$ acting at the centre of pressure R with $\vec{r}_R = \sum \vec{r}_i F_i / F_R$:

$$\begin{aligned}\vec{r}_R &= \frac{abp_0\left[\left(\frac{1}{2}a\vec{\imath} + \frac{1}{2}b\vec{\jmath}\right) + \frac{1}{2}\left(\frac{1}{3}a\vec{\imath} + \frac{4}{3}b\vec{\jmath}\right) + \frac{1}{2}\left(\frac{1}{3}a\vec{\imath} + \frac{1}{2}b\vec{\jmath}\right) + \frac{1}{6}\left(\frac{1}{4}a\vec{\imath} + \frac{5}{4}b\vec{\jmath}\right)\right]}{\frac{13}{6}abp_0} \\ &= \tfrac{21}{52}\,a\,\vec{\imath} + \tfrac{3}{4}\,b\,\vec{\jmath}\end{aligned}$$

(d) In Fig. E2.10j, pyramid height $h = a\sin\theta = ab/(a^2+b^2)^{1/2}$. The resultant $\vec{F}_R$ acts thro' centroid C, which divides OC_b in the ratio 3 : 1. C_b is the centroid of the base.

$$\vec{r}_C = \tfrac{3}{4}\vec{r}_{C_b} = \left(\tfrac{3}{4}\right)\tfrac{1}{4}\left[a\vec{\imath} + b\vec{\jmath} + (a\vec{\imath} + p_0\vec{k}) + (b\vec{\jmath} + p_0\vec{k})\right] = \tfrac{3}{8}\left(a\vec{\imath} + b\vec{\jmath} + p_0\vec{k}\right)$$

$$\vec{F}_R = -\tfrac{1}{3}\,(\text{base area} \times \text{height})\,\vec{k} = -\left[\tfrac{1}{3}(a^2+b^2)^{1/2}p_0ab/(a^2+b^2)^{1/2}\right]\vec{k} = -\tfrac{1}{3}p_0ab\,\vec{k}$$

$\vec{F}_R$ acts at R with $\vec{r}_R = 3(a\,\vec{\imath} + b\,\vec{\jmath})/8$.
(e) The force on a thin ring of radius r and thickness dr is $p(2\pi r)dr$:

$$p = 1.6p_0 \qquad \text{for } r < R/3$$

$$p = 1.6p_0 - 0.6p_0\frac{r - R/3}{R - R/3} = (1.9 - 0.9r/R)p_0 \qquad \text{for } R/3 < r < R$$

The simplest resultant is a single force $\vec{F}_R = -F_R\,\vec{k}$ acting at the centre with

$$F_R = 2\pi(1.6p_0)\int_0^{R/3} r\,dr + 2\pi p_0\int_{R/3}^{R}(1.9 - 0.9r/R)r\,dr = \frac{58}{45}\pi R^2 p_0$$

(f) $\tau(r,\phi)$ is equal in magnitude but opposite in direction to $\tau(r,\phi+\pi)$. Hence, the resultant is a couple $\vec{C}_R = C_R\,\vec{k}$. Using a ring element of radius r and thickness dr:

$$C_R = M_{Oz} = \int_0^{\frac{1}{2}R} r\left(\frac{2\tau_0 r}{R}\right)(2\pi r)dr + \int_{\frac{1}{2}R}^{R} r\,\tau_0(2\pi r)dr = \frac{31}{48}\pi\tau_0 R^3$$

Example 2.11: Find the total force due to constant pressure p on an area A on the inside of a pressure vessel (Fig. E2.11a). Also, find the resultant force due to constant pressure p on the inside of spherical and cylindrical vessels with circular and rectangular cut-outs as shown in Fig. E2.11b,c.

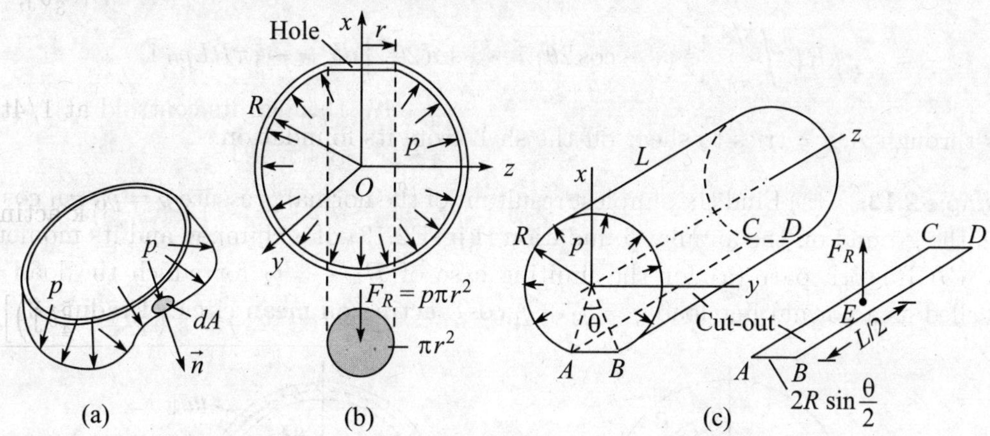

Fig. E2.11

Solution: The force $d\vec{F}$ on an area $d\vec{A} = dA\,\vec{n}$ with outward unit normal $\vec{n}$ is $d\vec{F} = p\,dA\,\vec{n} = p\,d\vec{A}$. The total force on area $\vec{A}$ is $\vec{F}_R = \int_{\vec{A}} p\,d\vec{A} = p\vec{A}$ $\Rightarrow$ the total force on a closed pressure vessel under constant pressure is zero since its $\vec{A} = \vec{0}$ (see Ex. 1.2).

The projected areas of the spherical vessel on yz, zx, xy planes are $-\pi r^2, 0, 0$, i.e., $\vec{A} = -\pi r^2\,\vec{\mathrm{i}}$. The simplest resultant is a force $\vec{F}_R = -p\pi r^2\,\vec{\mathrm{i}}$ through the centre O.

The projected areas of the cylindrical vessel on yz, zx, xy planes are $2R\sin(\theta/2)L, 0, 0$, i.e., $\vec{A} = 2RL\sin(\theta/2)\,\vec{\mathrm{i}}$. The simplest resultant is a force $\vec{F}_R = 2pRL\sin(\theta/2)\,\vec{\mathrm{i}}$ through the centroid E of the area $ABDC$.

Example 2.12: Find the simplest resultant of the distributed force on the cylindrical roof of a hangar in Fig. E2.12 due to wind pressure $p = p_0 \sin\theta$.

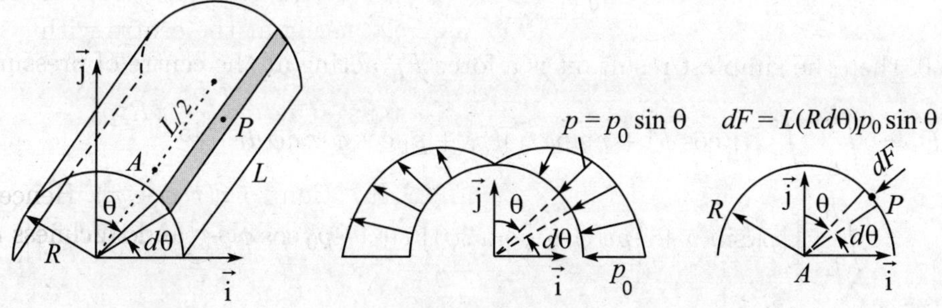

Fig. E2.12

Solution: The force on the area element $(R\,d\theta)L$ is a radial force $dF = p(R\,d\theta)L$ at P. These forces are concurrent at A at the middle of the axis. The resultant force $\vec{F}_R$:

$$\vec{F}_R = \int d\vec{F} = \int_{-\pi/2}^{\pi/2} RLp_0 \sin\theta(-\sin\theta\,\vec{\imath} - \cos\theta\,\vec{\jmath})\,d\theta$$

$$= -p_0 RL \int_{-\pi/2}^{\pi/2} \left[\tfrac{1}{2}(1-\cos 2\theta)\,\vec{\imath} + \tfrac{1}{2}\sin 2\theta\,\vec{\jmath}\right] d\theta = -\tfrac{1}{2}\pi RLp_0\,\vec{\imath}$$

acts through A. $\vec{F}_R$ tries to shear off the shell from its foundation.

Example 2.13: (a) Find the simplest resultant of the normal pressure $p = p_0 + p_1\cos\phi$ from the ground on the annular foundation (Fig. E2.13) of a chimney and its moment $\vec{M}_O$. (b) Rework part (a) for the limiting case of $R_1 \to R_2$ for which the load is modelled as a normal line load $q = q_0 + q_1\cos\phi$ acting on mean circle of radius R.

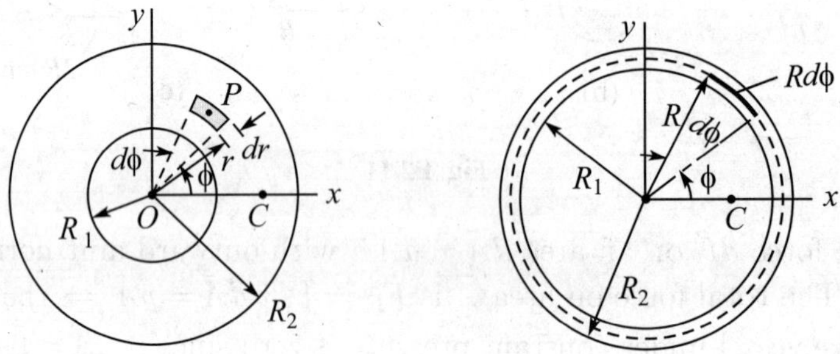

Fig. E2.13

Solution: The force on the foundation is a parallel force system with force $dF = p\,dA$ on an element of area $dA = r\,d\phi\,dr$. The total force $\vec{F}_R = F_R\,\vec{k}$ with

$$F_R = \int dF = \int_{R_1}^{R_2}\int_0^{2\pi} (p_0 + p_1\cos\phi)r\,d\phi\,dr$$

$$= \int_{R_1}^{R_2} (p_0\phi + p_1\sin\phi)\Big|_0^{2\pi} r\,dr = 2\pi p_0 \int_{R_1}^{R_2} r\,dr = \pi p_0(R_2^2 - R_1^2)$$

If $p_0 \neq 0$, then the simplest resultant is a force $\vec{F}_R$ acting at the centre of pressure C:

$$\int \vec{r}_P\,dF = \int_{R_1}^{R_2}\int_0^{2\pi} (r\cos\phi\,\vec{\imath} + r\sin\phi\,\vec{\jmath})(p_0 + p_1\cos\phi)r\,d\phi\,dr$$

$$= \int_{R_1}^{R_2} \left[\{p_0\sin\phi + \tfrac{1}{2}p_1(\phi + \tfrac{1}{2}\sin 2\phi)\}\,\vec{\imath} + (-p_0\cos\phi - \tfrac{1}{4}p_1\cos 2\phi)\,\vec{\jmath}\right]\Big|_0^{2\pi} r^2\,dr$$

$$= p_1\pi \int_{R_1}^{R_2} r^2\,dr\,\vec{\imath}$$

$$\vec{r}_C = \frac{\int \vec{r}_P\, dF}{F_R} = \frac{p_1(R_2^3 - R_1^3)}{3p_0(R_2^2 - R_1^2)}\,\vec{\mathrm{i}}$$

$$\vec{M}_O = \vec{r}_C \times \vec{F}_R = -\tfrac{1}{3}p_1\pi(R_2^3 - R_1^3)\,\vec{\mathrm{j}}$$

If $p_0 = 0$, then $\vec{F}_R = \vec{0}$ and the simplest resultant is a couple of moment $\vec{M}_O$.

For the parallel line force distribution $q = q_0 + q_1 \cos\phi$ on a circle of radius R, $dF = qR\,d\phi$ on an element of length $R\,d\phi$. The resultant $\vec{F}_R = F_R\,\vec{\mathrm{k}}$ acts at C with

$$F_R = \int_0^{2\pi} (q_0 + q_1 \cos\phi) R\,d\phi = 2\pi q_0 R$$

$$\vec{r}_C = \frac{\int \vec{r}_P\, dF}{F_R} = \frac{\int_0^{2\pi} R(\cos\phi\,\vec{\mathrm{i}} + \sin\phi\,\vec{\mathrm{j}})(q_0 + q_1\cos\phi)R\,d\phi}{2\pi q_0 R} = \frac{q_1 R}{2q_0}\,\vec{\mathrm{i}}$$

$$\vec{M}_O = \vec{r}_C \times \vec{F}_R = -\pi q_1 R^2\,\vec{\mathrm{j}}$$

2.5 Newton's Third Law of Motion

Loads (forces and moments) of interaction between two bodies (or their parts) have equal magnitudes and opposite directions. Let body B_1 interact with body B_2 at a contact point or over a contact surface or at a distance (Fig. 2.24a,b,c). Let the action of body B_1 on body B_2 have force sum $\vec{F}_{21}$ and moment sum $\vec{M}_{O_{21}}$ about O (a fixed point of an inertial frame), and the action of B_2 on B_1 have force sum $\vec{F}_{12}$ and moment sum $\vec{M}_{O_{12}}$ about O (Fig. 2.24d,e). Then,

$$\vec{F}_{12} = -\vec{F}_{21}, \qquad \vec{M}_{O_{12}} = -\vec{M}_{O_{21}} \tag{2.20}$$

$$\Rightarrow \quad \vec{M}_{A_{12}} = -\vec{M}_{A_{21}}, \qquad \forall\, A$$

Proof: Let external load due to surroundings of B_1 and B_2 be a force sum $\vec{F}_1^*$ and moment sum $\vec{M}_{O_1}^*$ about O on B_1, and a force sum $\vec{F}_2^*$ and moment sum $\vec{M}_{O_2}^*$ about O on B_2. Apply Euler's axioms to B_1 alone, to B_2 alone and to $B_1 \cup B_2$ (Fig. 2.24f):

$$\dot{\vec{p}}_{1|I} = \vec{F}_{12} + \vec{F}_1^*, \qquad \dot{\vec{H}}_{O_1} = \vec{M}_{O_{12}} + \vec{M}_{O_1}^* \tag{a}$$

$$\dot{\vec{p}}_{2|I} = \vec{F}_{21} + \vec{F}_2^*, \qquad \dot{\vec{H}}_{O_2} = \vec{M}_{O_{21}} + \vec{M}_{O_2}^* \tag{b}$$

$$\dot{\vec{p}}_{(B_1\cup B_2)|I} = \dot{\vec{p}}_{1|I} + \dot{\vec{p}}_{2|I} = \vec{F}_1^* + \vec{F}_2^* \tag{c}$$

$$\dot{\vec{H}}_{O\,(B_1\cup B_2)|I} = \dot{\vec{H}}_{O_1} + \dot{\vec{H}}_{O_2} = \vec{M}_{O_1}^* + \vec{M}_{O_2}^* \tag{d}$$

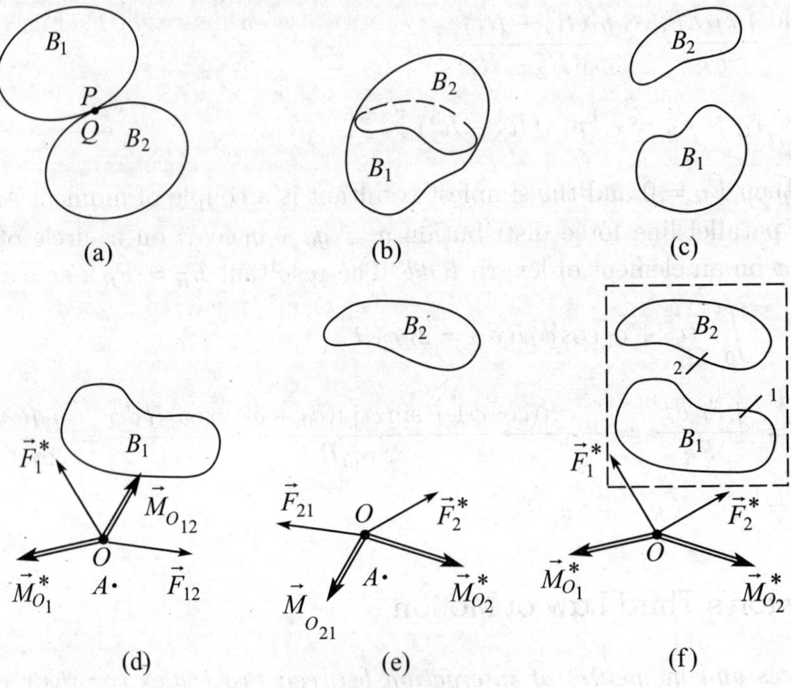

Fig. 2.24

The result is obtained by substituting from (a) and (b) in (c) and (d). Newton's 3rd law is not an axiom as it has been derived from Euler's axioms. B_1 and B_2 can be finite or infinitesimal parts of the same body. Hence, the loads of interaction have equal magnitudes and opposite directions both point-wise as well as globally in terms of the total force sum and the total moment sum of the loads of 'action' and 'reaction'. The terms 'action' and 'reaction' do not imply any cause and effect; rather, a mutual simultaneous interaction is implied. Thus, a single isolated force in nature cannot exist. In Fig. 2.14 the internal force resultants on part 1 are $\vec{F}_R$, $\vec{C}_R$ and on part 2 are $-\vec{F}_R$, $-\vec{C}_R$. The theorem of action and reaction is also valid for impulsive forces and moments.

2.6 Kinetic Energy, Power and Work

Energy is a primitive which can be transformed into various forms. Mass and energy are mutually inconvertible in classical mechanics.[4] The *kinetic energy* $T_{|F}$ of a body relative to frame F is defined by

$$T_{|F} \equiv \tfrac{1}{2}\int_m \vec{v}^{\,2}_{P|F}\, dm \tag{2.21}$$

$$= \tfrac{1}{2}\int_m (\dot{x}^2 + \dot{y}^2 + \dot{z}^2) dm = \tfrac{1}{2}\int_m (\dot{r}^2 + r^2\dot{\phi}^2 + \dot{z}^2) dm \tag{2.22}$$

[4]We simply define T here and avoid the usual circular definition of T as the *capacity to do work*.

(x, y, z) and (r, ϕ, z) are Cartesian and cylindrical coordinates of P (Fig. 2.25) in F.

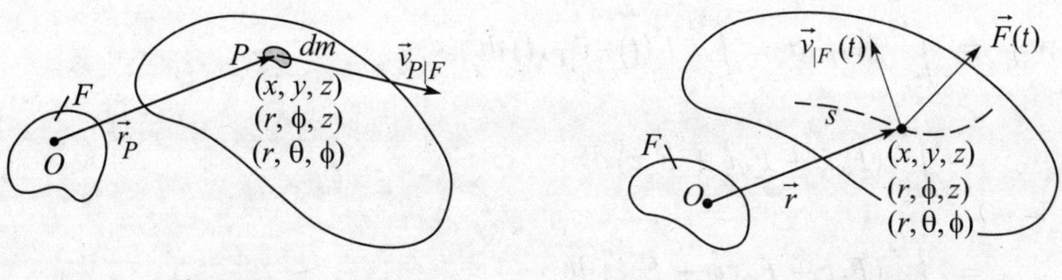

Fig. 2.25 **Fig. 2.26**

A body with kinetic energy has the capacity to do work as in pile-driving. The *power* $\dot{W}_{|F}(t)$ with respect to frame F of a force $\vec{F}(t)$ acting on a material point with velocity $\vec{v}_{|F}(t)$ and coordinates (x, y, z), (r, ϕ, z), (r, θ, ϕ), s (Fig. 2.26), is defined by

$$\dot{W}_{|F}(t) \equiv \vec{F}(t) \cdot \vec{v}_{|F}(t) \tag{2.23}$$

$$= F_x v_x + F_y v_y + F_z v_z = F_x \dot{x} + F_y \dot{y} + F_z \dot{z}$$

$$= F_r v_r + F_\phi v_\phi + F_z v_z = F_r \dot{r} + F_\phi r\dot{\phi} + F_z \dot{z} \tag{2.24}$$

$$= F_r v_r + F_\theta v_\theta + F_\phi v_\phi = F_r \dot{r} + F_\theta r\dot{\theta} + F_\phi r\dot{\phi}\sin\theta = F_t v_t = F_t \dot{s}$$

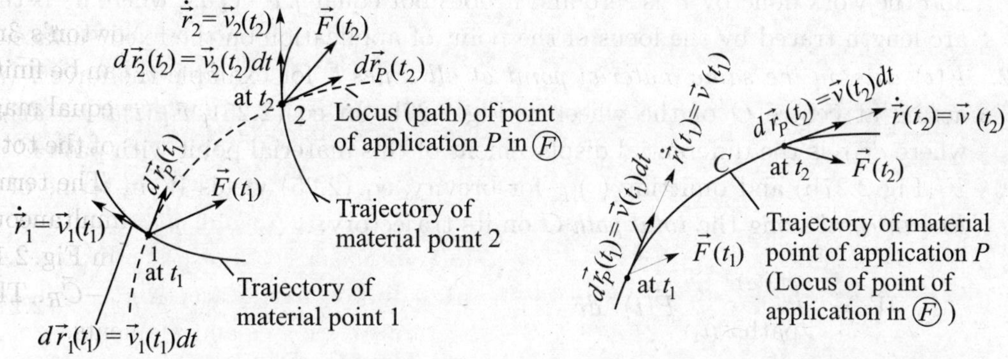

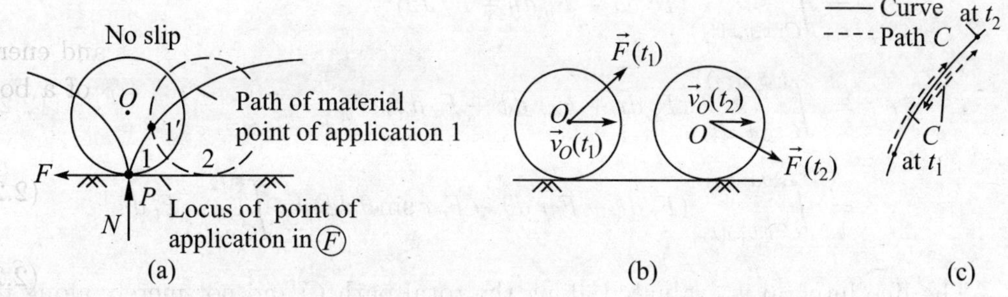

Fig. 2.27

The *work* done $W_{|F}$ with respect to frame F by a force $\vec{F}$ from t_1 to t_2 is defined by

$$W_{|F} \equiv \int_{t_1}^{t_2} \dot{W}_{|F}\,dt = \int_{t_1}^{t_2} \vec{F}(t)\cdot\vec{v}_{|F}(t)\,dt \tag{2.25}$$

$$= \int_{t_1}^{t_2} (F_x\dot{x} + F_y\dot{y} + F_z\dot{z})\,dt$$

$$= \int_{t_1}^{t_2} (F_r\dot{r} + F_\phi r\dot{\phi} + F_z\dot{z})\,dt \tag{2.26}$$

$$= \int_{t_1}^{t_2} (F_r\dot{r} + F_\theta r\dot{\theta} + F_\phi r\dot{\phi}\sin\theta)\,dt = \int_{t_1}^{t_2} F_t\dot{s}\,dt$$

We consider two cases:

1. $\vec{F}(t)$ *acts on different material points at different instants* (Fig. 2.27a). The locus of its point of application in F differs from the trajectories of the material points on which $\vec{F}$ acts at different instants. In eq. (2.25), $\vec{v}_{|F}(t)dt$ is *not equal to* $d\vec{r}$ of the same material point for all t or even $d\vec{r}_P$. For example, the contact point of a body rolling without slip on a fixed surface has zero velocity. Hence, the power of the friction force $\vec{F}$ acting at this material point is zero for all t (Fig. 2.27a) and the work done by $\vec{F}$ is zero and it does not equal $\int \vec{F}\cdot\vec{e}_t\,ds$, where ds is the arc length traced by the locus of the point of application on the fixed surface.
2. $\vec{F}(t)$ *acts on the same material point at all times* t, for example the force $\vec{F}(t)$ acting at centre O of the wheel in Fig. 2.27b. In eq. (2.25), $\vec{v}_{|F}(t)dt = d\vec{r}_{|F}$, where $d\vec{r}_{|F}$ is the differential displacement of this material point with path C in F (Fig. 2.27b) and omitting $(\ )_{|F}$ for brevity, eq. (2.25) yields W in terms of a *line integral* along the *total path* C on its trajectory:

$$W = \int_{\text{path}:\,\vec{r}(t_1)}^{\vec{r}(t_2)} \vec{F}(t)\cdot d\vec{r} \tag{2.27}$$

$$= \int_{C:\,x,y,z(t_1)}^{x,y,z(t_2)} (F_x\,dx + F_y\,dy + F_z\,dz)$$

$$= \int_{C:\,r,\phi,z(t_1)}^{r,\phi,z(t_2)} (F_r\,dr + F_\phi r\,d\phi + F_z\,dz) \tag{2.28}$$

$$= \int_{C:\,r,\theta,\phi(t_1)}^{r,\theta,\phi(t_2)} (F_r\,dr + F_\theta r\,d\theta + F_\phi r\sin\theta\,d\phi) = \int_{C:\,s(t_1)}^{s(t_2)} F_t\,ds$$

The line integral is evaluated along the total path C and not merely along the geometric curve along which it lies as a part of the curve may be traversed more than once (Fig. 2.27c). The two are identical only if the material point moves in one direction on its trajectory. If the force is time-dependent, then the work expression (2.27) is not strictly a line integral.

If the force acts on a specific material point at all times and depends only on its position vector $\vec{r}$, i.e., it is of the form $\vec{F}(\vec{r})$, then eq. (2.27) is often used to compute work, provided the path C is known or the force is conservative, i.e., the work is independent of the particular path C followed between positions $\vec{r}(t_1)$ and $\vec{r}(t_2)$. The line integrals in eq. (2.28) can be evaluated by converting them to ordinary definite integrals using the equation of the path C. For example, if the path is described as $\vec{r}(\tau)$ in terms of a parameter τ, then $\vec{F}(\vec{r}) = \vec{F}(\tau)$, $dx = (dx/d\tau)d\tau$, etc., and eq. (2.27) yields

$$W = \int_{C:\vec{r}(\tau_1)}^{\vec{r}(\tau_2)} \vec{F} \cdot d\vec{r} = \int_{\tau_1}^{\tau_2} \left[F_x(\tau)\frac{dx}{d\tau} + F_y(\tau)\frac{dy}{d\tau} + F_z(\tau)\frac{dz}{d\tau} \right] d\tau$$

In particular, if $\tau = x$, with $F_x = f(x, y, z) = f[x, y(x), z(x)] = F_x(x)$, etc., then

$$W = \int_{x_1}^{x_2} \left[F_x(x) + F_y(x)\frac{dy}{dx} + F_z(x)\frac{dz}{dx} \right] dx$$

Example 2.14: Find the total work done on the disc (Fig. E2.14a,b,c) by force P and frictional force F acting at the point of contact, in the time interval in which C moves a distance d and the conveyor moves a distance b relative to the ground. In case (b), F acts on the disc in the direction of motion of the conveyor. In case (c), C moves in a circle of radius R with speed $v_C(t)$ and the point of contact at time t has velocity $3\vec{v}_C(t)$. Find also the work done by the cable tension on the disc in case (a).

Solution: (a) The instantaneous centre of rotation of the disc (Fig. E2.14d) is at I, since the velocity of the part of the cable not in contact with the disc is zero. Hence, the tension T in it does no work. $v_A = \omega(IA) = (v_C/r)(R + r)$ downward. The work done by P acting downward at C and F acting upward at A on the disc is

$$\begin{aligned} W &= \int (\vec{P} \cdot \vec{v}_C + \vec{F} \cdot \vec{v}_A)\, dt = P \int v_C\, dt - F \int v_A\, dt \\ &= [P - F(R + r)/r] \int v_C\, dt = [P - F(R + r)/r]d \end{aligned}$$

(b) In Fig. E2.14e, $v_A = v,\ v_B = v_C + \omega R,\ v_A = v_C - \omega R \Rightarrow v_B = 2v_C - v_A = 2v_C - v$.

Work done by P and F on the disc is

$$\begin{aligned} W &= \int (\vec{P} \cdot \vec{v}_B + \vec{F} \cdot \vec{v}_A)\, dt = P \int v_B\, dt + F \int v_A\, dt \\ &= P \int (2v_C - v)\, dt + F \int v\, dt = P(2d - b) + Fb \end{aligned}$$

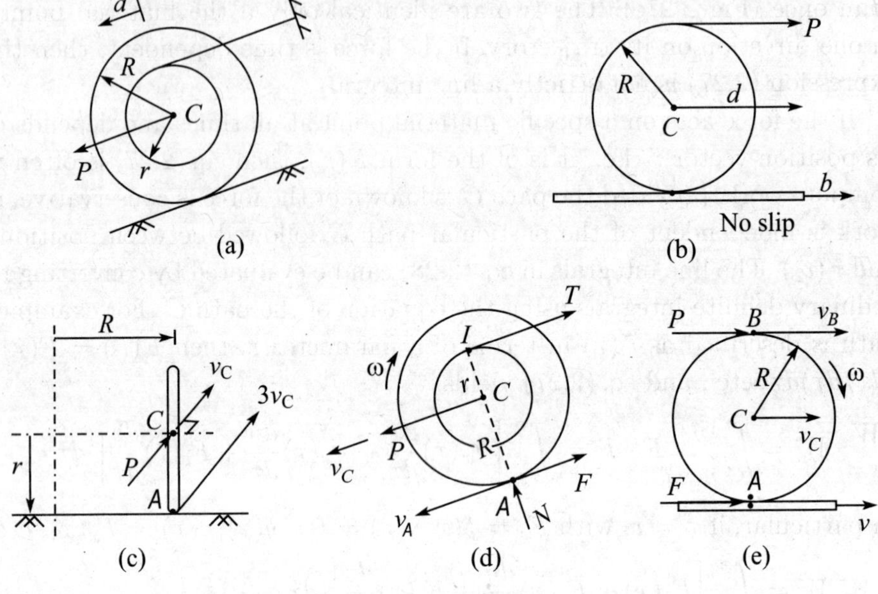

Fig. E2.14

(c) Work done by P and F (in the direction opposite to $\vec{v}_A$) on the disc is

$$W = \int (\vec{P} \cdot \vec{v}_C + \vec{F} \cdot \vec{v}_A)\, dt = \int [P v_C - F(3v_C)]\, dt = (P - 3F)d$$

2.7 Expressions of some Entities in Terms of C

$\vec{p}_{|F}$, $\vec{H}_{A|F}$, $T_{|F}$, $\dot{W}_{|F}$ can be expressed in terms of the centre of mass C (Fig. 2.28) as:

$$\vec{p}_{|F} = m\vec{v}_{C|F} \tag{2.29}$$

$$\vec{H}_{A|F} = \vec{H}_{C|F} + \vec{r}_{CA} \times m\vec{v}_{CA|F} \tag{2.30}$$

$$T_{|F} = \tfrac{1}{2} m v_{C|F}^2 + \tfrac{1}{2} \int_m v_{PC|F}^2\, dm \tag{2.31}$$

$$\dot{W}_{|F} = \vec{F} \cdot \vec{v}_{C|F} + \sum_i \vec{F}_i \cdot \vec{v}_{iC|F} \tag{2.32}$$

Thus, each of $\vec{H}_A, T$, and $\dot{W}$ is a sum of two contributions—the first one is as if the body were a hypothetical mass-point of mass m and moving with C and the second one is due to the motion relative to C.

For brevity, we omit $(\ )_{|F}$ in the proofs. Using $\int_m \vec{r}_{PC}\, dm = \vec{0}$, $\int_m \vec{v}_{PC}\, dm = \vec{0}$:

$$\int_m \vec{r}_{PC}\, dm = \int_m (\vec{r}_P - \vec{r}_C)\, dm = \vec{r}_C m - \vec{r}_C m = \vec{0} \quad \Rightarrow \quad \int_m \vec{v}_{PC}\, dm = \vec{0}$$

$$\begin{aligned}
\vec{H}_A &= \int_m \vec{r}_{PA} \times \vec{v}_{PA}\, dm = \int_m (\vec{r}_{PC} + \vec{r}_{CA}) \times (\vec{v}_{PC} + \vec{v}_{CA})\, dm \\
&= \int_m \vec{r}_{PC} \times \vec{v}_{PC}\, dm + \vec{r}_{CA} \times \int_m \vec{v}_{PC}\, dm \\
&\quad + \int_m \vec{r}_{PC}\, dm \times \vec{v}_{CA} + \vec{r}_{CA} \times \vec{v}_{CA} \int_m dm = \vec{H}_C + \vec{r}_{CA} \times m\vec{v}_{CA} \\
T &= \tfrac{1}{2}\int_m \vec{v}_P^{\,2}\, dm = \tfrac{1}{2}\int_m \vec{v}_P \cdot \vec{v}_P\, dm = \tfrac{1}{2}\int_m (\vec{v}_{PC} + \vec{v}_C)\cdot(\vec{v}_{PC} + \vec{v}_C)\, dm \\
&= \tfrac{1}{2}\int_m v_{PC}^2\, dm + \tfrac{1}{2}v_C^2 \int_m dm + \int_m \vec{v}_{PC}\, dm \cdot \vec{v}_C = \tfrac{1}{2}mv_C^2 + \tfrac{1}{2}\int_m v_{PC}^2\, dm \\
\dot{W} &= \sum_i \vec{F}_i \cdot \vec{v}_i = \sum_i \vec{F}_i \cdot (\vec{v}_{iC} + \vec{v}_C) = \sum_i \vec{F}_i \cdot \vec{v}_C + \sum_i \vec{F}_i . \vec{v}_{iC} \\
&= \vec{F} \cdot \vec{v}_C + \sum_i \vec{F}_i \cdot \vec{v}_{iC}
\end{aligned}$$

Fig. 2.28

The *rate of work* $\dot{W}$ *done by the external forces on a rigid body* (Fig. 2.29a) is:

$$\dot{W} = \sum_i \vec{F}_i \cdot \vec{v}_i = \sum_i \vec{F}_i \cdot (\vec{v}_C + \vec{\omega} \times \vec{r}_{iC}) = \sum_i \vec{F}_i \cdot \vec{v}_C + \vec{\omega} \cdot \sum_i \vec{r}_{iC} \times \vec{F}_i$$

$$\Rightarrow \qquad \dot{W} = \vec{F} \cdot \vec{v}_C + \vec{M}_C \cdot \vec{\omega} \tag{2.33}$$

Hence, two equivalent force systems perform the same work on a rigid body, as their $\vec{F}$ and $\vec{M}_C$ are the same. Thus, the work done by the distributed uniform gravitational force on a rigid body is the same as that due to its resultant mg acting at C. The rate of work done by a moment $\vec{C}$ (or a couple–moment) acting at point Q of a rigid body (Fig. 2.29b) is obtained from eq. (2.33) or is defined as

$$\dot{W} = \vec{M}_C \cdot \vec{\omega} = \vec{C} \cdot \vec{\omega} \tag{2.34}$$

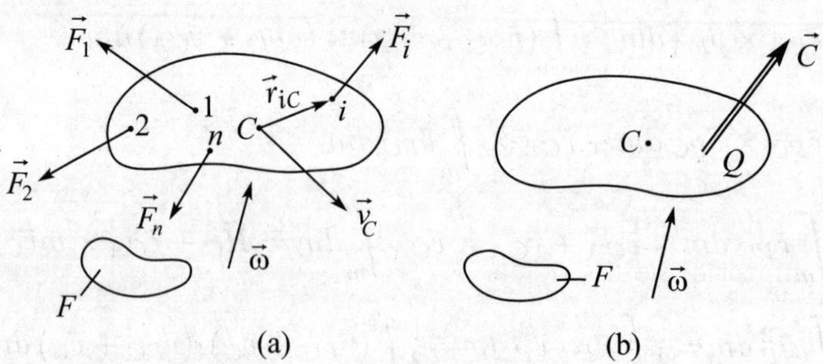

Fig. 2.29

§ **Exercise 2.4:** Prove that if A is a point of a rigid body, then $\dot{W} = \vec{F} \cdot \vec{v}_A + \vec{M}_A \cdot \vec{\omega}$.

Example 2.15: (a) A thin uniform disc of mass m moves with slip on a conveyor moving at velocity v_0 (Fig. E2.15a). (b) A thin non-uniform disc of mass m, centre of mass C and axial principal moment of inertia I, is rotating about a fixed axis at A (Fig. E2.15b). Find $\vec{p}, \vec{H}_C, \vec{H}_A, \vec{H}_B, \vec{H}_D$ for each case. Note: As will be shown in Chapter 3, $\vec{H}_A = I_{zz}^A \omega \vec{\mathrm{k}}$ for plane motion of a rigid body with $\vec{\omega} = \omega \vec{\mathrm{k}}$ and I_{zz}^A is a principal moment of inertia at material point A of the body.

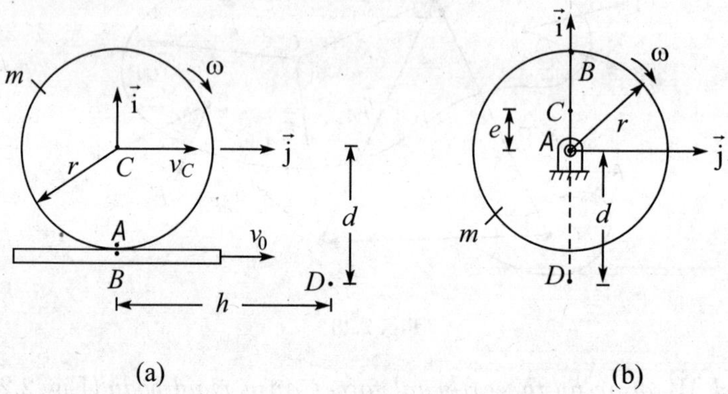

Fig. E2.15

Solution: (a) We use the parallel axes theorem for moment of inertia (see Chapter 3 for details). Note that A and C are material points of the disc, but B and D are not.

$$\vec{p} = m\vec{v}_C = mv_C\,\vec{\mathrm{j}}, \qquad \vec{H}_C = I_{zz}^C \omega\,\vec{\mathrm{k}} = \tfrac{1}{2}mr^2\omega\,\vec{\mathrm{k}}$$

$$\vec{H}_A = \vec{H}_C + \vec{r}_{CA} \times m\vec{v}_{CA} = \tfrac{1}{2}mr^2\omega\,\vec{\mathrm{k}} + r\,\vec{\mathrm{i}} \times m\omega r\,\vec{\mathrm{j}} = \tfrac{3}{2}mr^2\omega\,\vec{\mathrm{k}} = I_{zz}^A\omega\,\vec{\mathrm{k}}$$

$$\begin{aligned}\vec{H}_D &= \vec{H}_C + \vec{r}_{CD} \times m\vec{v}_{CD} = \tfrac{1}{2}mr^2\omega\,\vec{\mathrm{k}} + (d\,\vec{\mathrm{i}} - h\,\vec{\mathrm{j}}) \times mv_C\,\vec{\mathrm{j}} \\ &= (\tfrac{1}{2}mr^2\omega + mv_C d)\,\vec{\mathrm{k}} \neq I_{zz}^D\omega\,\vec{\mathrm{k}}\end{aligned}$$

$$\begin{aligned}\vec{H}_B &= \vec{H}_C + \vec{r}_{CB} \times m\vec{v}_{CB} = \tfrac{1}{2}mr^2\omega\,\vec{\mathrm{k}} + r\,\vec{\mathrm{i}} \times m(v_C - v_0)\,\vec{\mathrm{j}} \\ &= [\tfrac{1}{2}mr^2\omega + m(v_C - v_0)r]\,\vec{\mathrm{k}} \neq \vec{H}_A \end{aligned} \tag{1}$$

$\Rightarrow$ For no slip, $v_C = v_0 + \omega r$, $\vec{H}_B = \frac{3}{2}mr^2\omega\,\vec{\mathrm{k}} = \vec{H}_A$. Since $\vec{v}_B = \vec{v}_A$, B can be considered to be a point of the disc at that instant.
(b) A, B, C are material points of the disc, but D is not and $\vec{v}_C = e\omega\,\vec{\mathrm{j}}$.

$$\vec{p} = m\vec{v}_C = me\omega\,\vec{\mathrm{j}}, \qquad \vec{H}_A = I^A_{zz}\omega\,\vec{\mathrm{k}} = I\omega\,\vec{\mathrm{k}}$$

$$\begin{aligned}\vec{H}_C &= I^C_{zz}\omega\,\vec{\mathrm{k}} = (I^A_{zz} - me^2)\omega\,\vec{\mathrm{k}} = (I - me^2)\omega\,\vec{\mathrm{k}} \\ \vec{H}_B &= I^B_{zz}\omega\,\vec{\mathrm{k}} = [I^C_{zz} + m(r-e)^2]\omega\,\vec{\mathrm{k}} = [I - me^2 + m(r-e)^2]\omega\,\vec{\mathrm{k}} \\ \vec{H}_D &= \vec{H}_C + \vec{r}_{CD} \times m\vec{v}_{CD} = (I - me^2)\omega\,\vec{\mathrm{k}} + (d+e)\,\vec{\mathrm{i}} \times me\omega\,\vec{\mathrm{j}} \\ &= (I + med)\omega\,\vec{\mathrm{k}} \neq I^D_{zz}\omega\,\vec{\mathrm{k}}\end{aligned}$$

Example 2.16: (a) Find the kinetic energy of the chain of mass m of a tank relative to the tank frame 1 and the ground frame F (Fig. E2.16a). (b) Ring 1 rotates about a fixed axis (Fig. E2.16b) at ω relative to F. A thin rod 2 is pinned to the ring at C and rotates at Ω relative to 1. Two small sliders, each of mass m, slide relative to 2 at speed v. Find the kinetic energy of the pair of sliders with respect to frames 2, 1, F.

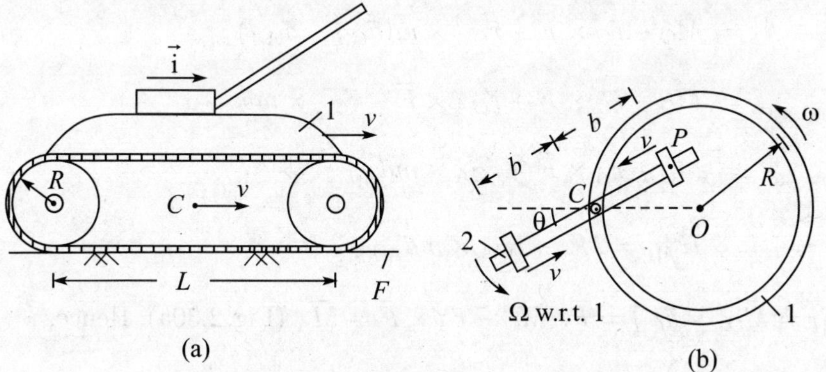

Fig. E2.16

Solution: (a) The centre of mass C of the chain is at rest relative to 1 and moves at v relative to F: $\vec{v}_{C|1} = \vec{0}$, $\vec{v}_{C|F} = v\,\vec{\mathrm{i}}$. For any point P of the chain, $|\vec{v}_{PC|1}| = |\vec{v}_{PC|F}| = v$.

$$T_{|F} = \tfrac{1}{2}mv^2_{C|F} + \tfrac{1}{2}\int_m v^2_{PC|F}\,dm = \tfrac{1}{2}mv^2 + \tfrac{1}{2}mv^2 = mv^2$$

$$T_{|1} = \tfrac{1}{2}mv^2_{C|1} + \tfrac{1}{2}\int_m v^2_{PC|1}\,dm = \tfrac{1}{2}mv^2$$

(b) The sliders have rectilinear motion in frame 2 with speed v: $T_{|2} = 2(\frac{1}{2}mv^2) = mv^2$. Using polar coordinates for sliders in frame 1 with origin at C, $\dot{\phi} = \Omega$, $r = b$, $\dot{r} = -v$,

$$T_{|1} = \tfrac{1}{2}m(\dot{r}^2 + r^2\dot{\phi}^2) \times 2 = m(v^2 + b^2\Omega^2)$$

The centre of mass C of the pair of sliders has $v_{C|F} = \omega R$. $\vec{\omega}_{2|F} = (\Omega + \omega)\,\vec{\mathrm{k}}$ and $\vec{r}_{PC} = b\,\vec{\mathrm{e}}_r$, $\vec{v}_{P|2} = -v\,\vec{\mathrm{e}}_r$. Hence, $\vec{v}_{PC|F} = \vec{\omega}_{2|F} \times \vec{r}_{PC} + \vec{v}_{P|2} = (\Omega + \omega)b\,\vec{\mathrm{e}}_\phi - v\,\vec{\mathrm{e}}_r$.

$$T_{|F} = \tfrac{1}{2}(2m)v_{C|F}^2 + \tfrac{1}{2}mv_{PC|F}^2 \times 2 = m[\omega^2R^2 + v^2 + (\Omega + \omega)^2b^2]$$

2.8 $\dot{\vec{H}}_A - \vec{M}_A$ Relation for Arbitrary Point A

Euler's second axiom takes the form $\dot{\vec{H}}_{A|I} = \vec{M}_A - \vec{r}_{CA} \times m\vec{a}_{A|I}$ for an arbitrary point A (Fig. 2.30) Using eq. (2.30): $\vec{H}_{A|I} = \vec{H}_{C|I} + \vec{r}_{CA} \times m\vec{v}_{CA|I}$,

$$\vec{H}_{O|I} = \vec{H}_{C|I} + \vec{r}_{CO} \times m\vec{v}_{CO|I} \quad \Rightarrow \quad \vec{H}_{C|I} = \vec{H}_{O|I} - \vec{r}_C \times m\vec{v}_{C|I}$$

$$\begin{aligned}
\Rightarrow \vec{H}_{A|I} &= \vec{H}_{O|I} - \vec{r}_C \times m\vec{v}_{C|I} + \vec{r}_{CA} \times m\vec{v}_{CA|I} \\
\dot{\vec{H}}_{A|I} &= \dot{\vec{H}}_{O|I} - \vec{v}_{C|I} \times m\vec{v}_{C|I} - \vec{r}_C \times m\vec{a}_{C|I} \\
&\quad + \vec{v}_{CA|I} \times m\vec{v}_{CA|I} + \vec{r}_{CA} \times m\vec{a}_{CA|I} \\
&= \vec{M}_O - \vec{r}_C \times \vec{F} + \vec{r}_{CA} \times m(\vec{a}_{C|I} - \vec{a}_{A|I}) \\
&= \vec{M}_O - \vec{r}_C \times \vec{F} + \vec{r}_{CA} \times \vec{F} - \vec{r}_{CA} \times m\vec{a}_{A|I} \\
&= \vec{M}_O - \vec{r}_A \times \vec{F} - \vec{r}_{CA} \times m\vec{a}_{A|I} \quad \Rightarrow
\end{aligned}$$

$$\dot{\vec{H}}_{A|I} = \vec{M}_A - \vec{r}_{CA} \times m\vec{a}_{A|I} \tag{2.35}$$

since $\dot{\vec{H}}_{O|I} = \vec{M}_O$, $m\vec{a}_{C|I} = \vec{F}$, $\vec{M}_O - \vec{r}_A \times \vec{F} = \vec{M}_A$ (Fig. 2.30a). Hence,

$$\dot{\vec{H}}_{A|I} = \vec{M}_A \tag{2.36}$$

provided point A satisfies at least one of the following three conditions:

1. $\vec{a}_{A|I} = \vec{0}$, i.e., A has zero acceleration in I,
2. $\vec{a}_{A|I} \parallel \vec{r}_{CA}$, i.e., acceleration of A is along AC,
3. $\vec{r}_{CA} = \vec{0}$, i.e., A coincides with C. It yields the most useful form

$$\dot{\vec{H}}_{C|I} = \vec{M}_C \tag{2.37}$$

If $\vec{M}_A(t) \equiv \vec{0}$, then eq. (2.36) yields $\vec{H}_A(t) = \vec{H}_A(0)$.

Equation 2.36)/(2.37) is derived using $\vec{F} = m\vec{a}_{C|I}$, $\dot{\vec{H}}_{O|I} = \vec{M}_O$, and so is not independent of these. Hence, once $\vec{F} = m\vec{a}_{C|I}$ and a moment equation have been written for a system, writing another moment equation does not yield an independent equation.

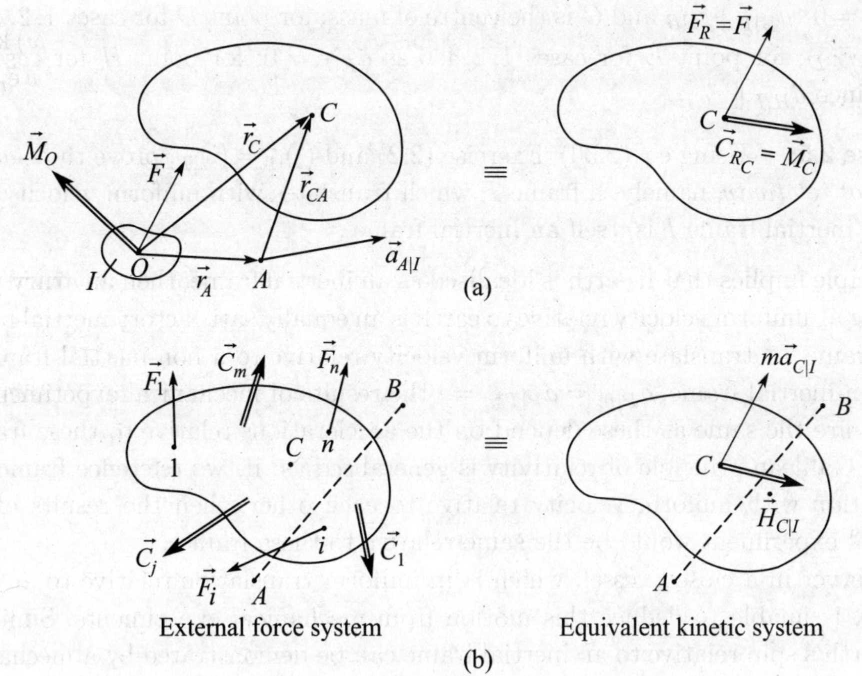

Fig. 2.30a,b

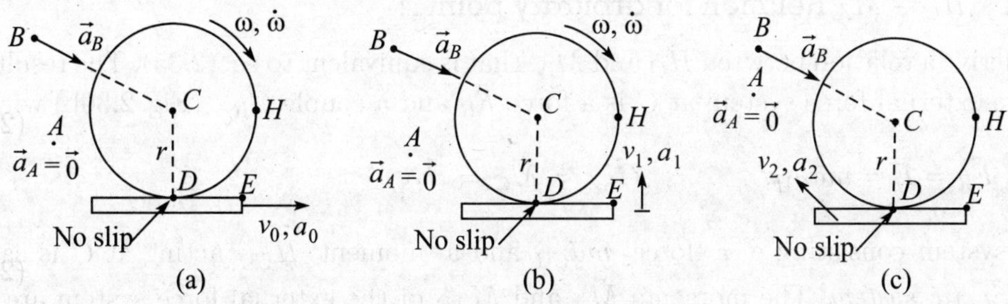

Fig. 2.31

Consider 7 types of motion of a uniform disc rolling without slip on

(a) a conveyor (Fig. 2.31a) with

1. $v_0 = 0,\ a_0 = 0$; 2. $v_0 \neq 0,\ a_0 = 0$; 3. $v_0 \neq 0,\ a_0 \neq 0$;

(b) an elevator (Fig. 2.31b) with

4. $v_1 \neq 0,\ a_1 = 0$; 5. $v_1 \neq 0,\ a_1 \neq 0$;

(c) the step of an escalator (Fig. 2.31c) with

6. $v_2 \neq 0,\ a_2 = 0$, and 7. $v_2 \neq 0,\ a_2 \neq 0$.

Euler's second axiom takes the form of eq. (2.36) for points A, B, C for all 7 cases since $\vec{a}_{A|I} = \vec{0}$, $\vec{a}_{B|I} \parallel \vec{r}_{CB}$ and C is the centre of mass; for point D for cases $1, 2, 4, 5, 6$ as $\vec{a}_{D|I} \parallel \vec{r}_{CD}$; for point E for cases $1, 2, 4, 6$ as $\vec{a}_{E|I} = \vec{0}$; for point H for case 5 if $a_1 = \dot{\omega} r$ since $\vec{a}_{H|I} \parallel \vec{r}_{CH}$.

§ **Exercise 2.5:** Using eq. (2.36), Exercise (2.2) and $(\dot{\ })_{|I} = (\dot{\ })_{|T}$, prove the *Galilean principle of relativity*, namely 'a frame T which translates with uniform velocity relative to an inertial frame I is itself an inertial frame'.

This principle implies that if earth is idealised as an inertial frame, then any rigid body translating at uniform velocity relative to earth is an equally satisfactory inertial frame.

Let a frame T^* translate with uniform velocity relative to a non-inertial frame m. T^* is a non-inertial frame. $\vec{a}_{P|m} = \vec{a}_{P|T^*} \Rightarrow$ the results of mechanical experiments in m and T^* are the same as these depend on the accelerations relative to these frames. Thus, the Galilean principle of relativity is generalised as: if two reference frames are in translation with uniform velocity relative to each other, then the results of any mechanical experiment would be the same relative to these frames.

An observer in a closed vessel, which is in uniform translation relative to an inertial frame, is unable to deduce this motion from mechanical experiments. Similarly, though earth's spin relative to an inertial frame can be demonstrated by a mechanical experiment on it (see Chapter 3 for gyrocompass), its translation is undetermined.

2.8.1 $\dot{\vec{H}}_C - \vec{M}_A$ Relation for arbitrary point A

We derive a relation between $\dot{\vec{H}}_C$ and $\vec{M}_A$, that is equivalent to eq. (2.35). The resultant of the external force system at C is a force $\vec{F}_R$ and a couple $\vec{C}_{R_C}$ (Fig. 2.30b) with:

$$\vec{F}_R = \vec{F} = m\vec{a}_{C|I}, \qquad \vec{C}_{R_C} = \vec{M}_C = \dot{\vec{H}}_{C|I}$$

The system consisting of a 'force' $m\vec{a}_{C|I}$ and a 'moment' $\dot{\vec{H}}_{C|I}$ acting at C is called the *kinetic system*. The moments $\vec{M}_A$ and M_{AB} of the external force system are the same as the moments of its equivalent kinetic system at C:

$$\vec{M}_A = \vec{M}_C + \vec{r}_{CA} \times \vec{F}_R = \dot{\vec{H}}_{C|I} + \vec{r}_{CA} \times m\vec{a}_{C|I} \tag{2.38}$$

$$\Rightarrow \quad M_{AB} = [\dot{\vec{H}}_{C|I} + \vec{r}_{CA} \times m\vec{a}_{C|I}] \cdot \vec{e} \tag{2.39}$$

These are more useful than eq. (2.36), as A can be chosen where many unknown forces act and AB as that line through which many unknown forces act, to eliminate them.

2.8.2 $\dot{\vec{H}}_A - \vec{M}_A$ Relation for any point A for a translating frame T

Consider a frame T translating relative to inertial frame I at $\vec{v}_T(t), \vec{a}_T(t)$ and body B of mass m in motion relative to T (Fig. 2.30c). $\vec{v}_{P|I} = \vec{v}_{P|T} + \vec{v}_T$, $\vec{a}_{P|I} = \vec{a}_{P|T} + \vec{a}_T$, $(\dot{\ })_{|I} = (\dot{\ })_{|T} \Rightarrow \vec{v}_{PA|I} = \vec{v}_{PA|T}$. For body B:

$$\vec{F} = m\vec{a}_{C|I} = m\vec{a}_{C|T} + m\vec{a}_T$$

$$\Rightarrow \quad m\vec{a}_{C|T} = \vec{F}' \tag{1}$$

where $\vec{F}' = \vec{F} + (-m\vec{a}_T)$ is the sum of forces of an *augmented force system* consisting of external loads on body augmented by an inertia 'force' $-m\vec{a}_T$ acting at C (Fig. 2.30d).

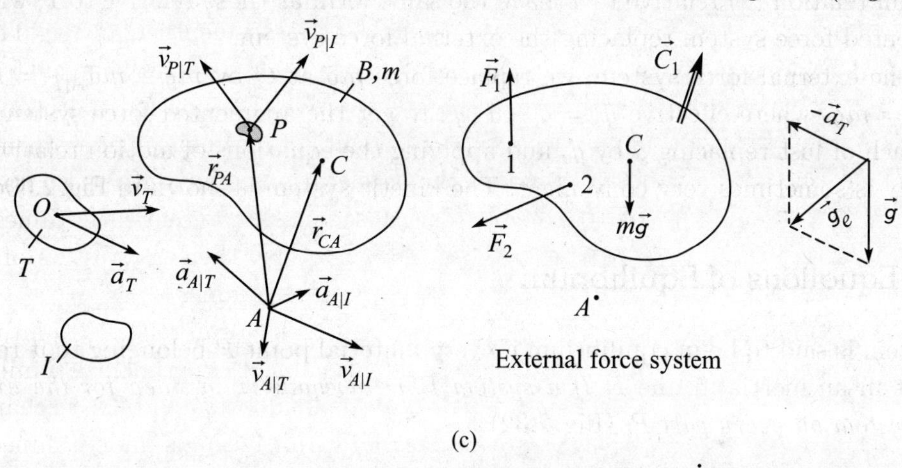

(c)

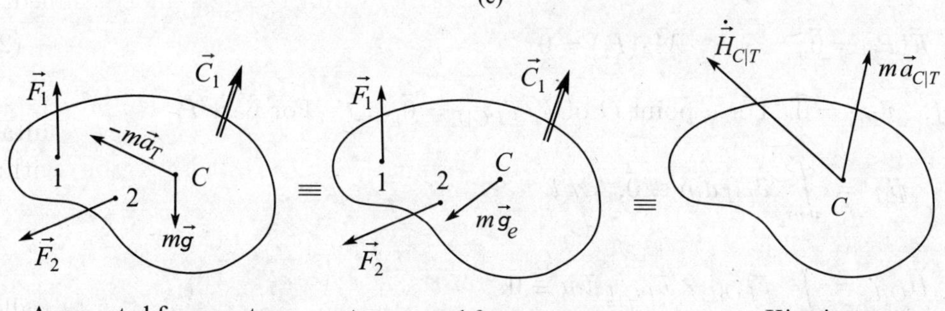

(d)

Fig. 2.30c,d

Let $\vec{M}'_A = \vec{M}_A + \vec{r}_{CA} \times (-m\vec{a}_T)$ be the moment of the augmented force system about A. Note that $\vec{M}'_C = \vec{M}_C$. For body B:

$$\vec{H}_{A|I} = \int_m \vec{r}_{PA} \times \vec{v}_{PA|I}\, dm = \int_m \vec{r}_{PA} \times \vec{v}_{PA|T}\, dm = \vec{H}_{A|T}$$

$$\begin{aligned}
\vec{H}_{A|T} &= \vec{H}_{A|I} = \vec{H}_{C|I} + \vec{r}_{CA} \times m\vec{v}_{CA|I} = \vec{H}_{C|I} + \vec{r}_{CA} \times m\vec{v}_{CA|T} \\
\dot{\vec{H}}_{A|T} &= \dot{\vec{H}}_{C|I} + \vec{r}_{CA} \times m\vec{a}_{CA|T} = \vec{M}_C + \vec{r}_{CA} \times m(\vec{a}_{C|T} - \vec{a}_{A|T}) \\
&= \vec{M}'_C + \vec{r}_{CA} \times (\vec{F}' - m\vec{a}_{A|T}) = \vec{M}'_A - \vec{r}_{CA} \times m\vec{a}_{A|T}
\end{aligned}$$

$$\Rightarrow \quad \dot{\vec{H}}_{A|T} = \vec{M}'_A \tag{2}$$

if point A is coincident with C / $\vec{a}_{A|T} = \vec{0}$ / $\vec{a}_{A|T}$ is along AC.

We conclude that the force–momentum relation (1) and moment–moment of momentum relation (2) relative to T have the same form as those relative to I, with the augmented force system replacing the external force system.

In the external force system, we replace force $m\vec{g}$ at C by $m\vec{g} - m\vec{a}_{A|T} = m(\vec{g} - \vec{a}_{A|T}) = m\vec{g}_e$ where effective $\vec{g}_e = \vec{g} - \vec{a}_{A|T}$, to get the augmented force system. This approach of just replacing $\vec{g}$ by $\vec{g}_e$ and applying the equations of motion relative to T directly is sometimes very convenient. The kinetic system is shown in Fig. 2.30d.

2.9 Equations of Equilibrium

A system is said to be in equilibrium if every material point P belonging to it remains at rest in an inertial frame I. *If a system B is in equilibrium, then for the external load system on every part P_i* (Fig. 2.32):

$$\vec{F}(P_i) = \vec{0}, \qquad \vec{M}_A(P_i) = \vec{0} \tag{2.40}$$

Proof: $\vec{v}_{P|I} = \vec{0}$. For a point O of I, $\vec{v}_{PO|I} = \vec{0}$, $\forall\, t$. For part P_i

$$\begin{aligned}
\vec{p}_{|I} &= \int_{m_i} \vec{v}_{P|I}\, dm \equiv \vec{0}, \quad \forall\, t \\
\vec{H}_{O|I} &= \int_{m_i} \vec{r}_{PO|I} \times \vec{v}_{PO|I}\, dm \equiv \vec{0}, \quad \forall\, t
\end{aligned}$$

Euler's axioms yield $\quad \vec{F}(P_i) = \dot{\vec{p}}_{|I} = \vec{0}, \qquad \vec{M}_O(P_i) = \dot{\vec{H}}_{O|I} = \vec{0} \Rightarrow$

$$\vec{M}_A(P_i) = \vec{M}_O(P_i) + \overrightarrow{AO} \times \vec{F}(P_i) = \vec{0}$$

A could be any point of the system or outside and may be in arbitrary motion. In particular, eq. (2.40) holds for the complete system B in equilibrium, i.e.,

$$\vec{F}(B) = \vec{0}, \qquad \vec{M}_A(B) = \vec{0} \tag{2.41}$$

Equations (2.41) are necessary conditions of equilibrium, but not sufficient conditions even for a single rigid body. For example, a uniform rigid circular cylinder rolls without slip with uniform angular velocity on a horizontal plane, and free vibrations occur in an unconstrained deformable body, even though $\vec{F}(B) = \vec{0},\ \vec{M}_A(B) = \vec{0}$.

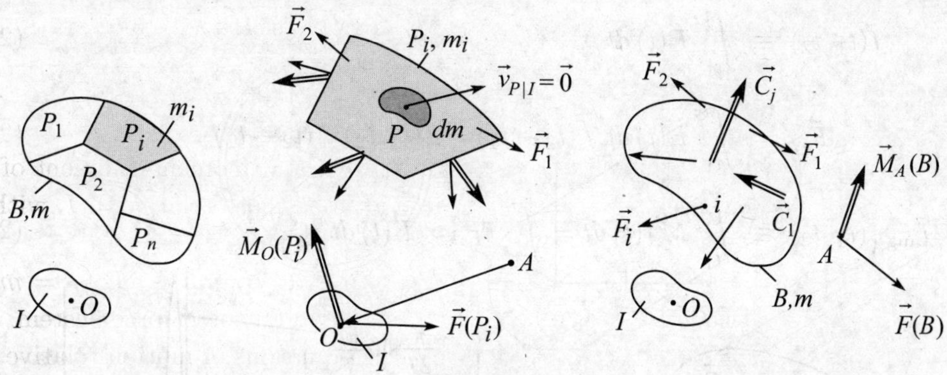

Fig. 2.32

Equations (2.40) are also the *equations of motion of an idealised inertia-less ('light') system*, since $m = 0 \Rightarrow$ for all parts $P_i:\ \vec{p}_{|I} = \vec{0},\ \vec{H}_{O|I} = \vec{0},\ \forall\, t$.

In statics, a system is said to be *light* if the gravitational force on it is very small (modelled as zero) compared to other forces acting on it. In dynamics, if the product of mass of a body and its acceleration is small compared to individual forces, over the interval of interest, then the mass is taken as zero. Such an analysis is called *quasi-static* since the equations of equilibrium hold. However, we should be careful while neglecting inertia of a system for impact problems, as large accelerations occur.

Two-force member: A two-force member is a member subjected to only two external forces and no couple. If a two-force member (Fig. 2.33a) is *in equilibrium* or a two-force *inertia-less* member is *in motion*, then the two forces have *equal magnitudes, opposite directions and act along the line joining their points of application.*

Proof: $\quad \vec{M}_A = \overrightarrow{AB} \times \vec{F}_2 = \vec{0} \ \Rightarrow\ \vec{F}_2$ acts through A along AB. Similarly, $\vec{F}_1$ acts along AB. $\quad \vec{F} = \vec{F}_1 + \vec{F}_2 = \vec{0} \ \Rightarrow\ \vec{F}_2 = -\vec{F}_1$ (Fig. 2.33b).

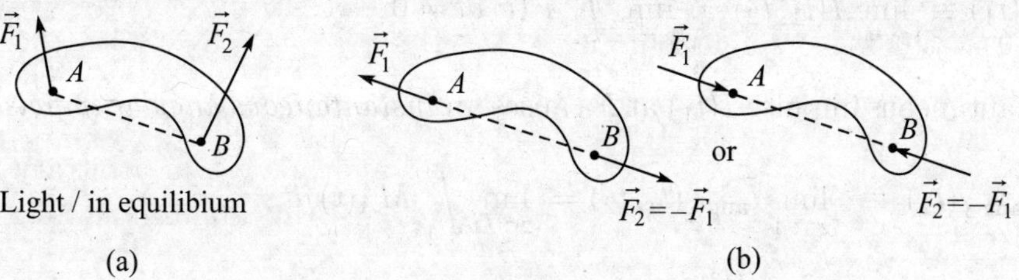

Fig. 2.33

2.10 Impulse and Angular Impulse

The *impulse* $\vec{I}(t_1, t_2)$ of a force $\vec{F}(t)$ acting at P (Fig. 2.34), the time average of $\vec{F}(t)$, and the *angular impulse* $\vec{I}_{\text{ang}_A}(t_1, t_2)$ of $\vec{F}$ about A for interval t_1 to t_2 are defined by

$$\vec{I}(t_1, t_2) \equiv \int_{t_1}^{t_2} \vec{F}(t)\, dt \tag{2.42}$$

$$\vec{F}_{\text{av}} = \int_{t_1}^{t_2} \vec{F}(t)\, dt \,/\, (t_2 - t_1) = \vec{I}(t_1, t_2) \,/\, (t_2 - t_1) \tag{2.43}$$

$$\vec{I}_{\text{ang}_A}(t_1, t_2) \equiv \int_{t_1}^{t_2} \vec{M}_A(t)\, dt = \int_{t_1}^{t_2} \vec{r}_{PA} \times \vec{F}(t)\, dt \tag{2.44}$$

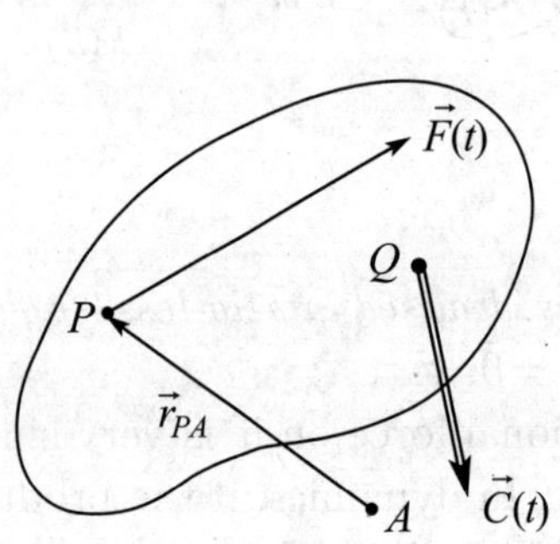

Fig. 2.34

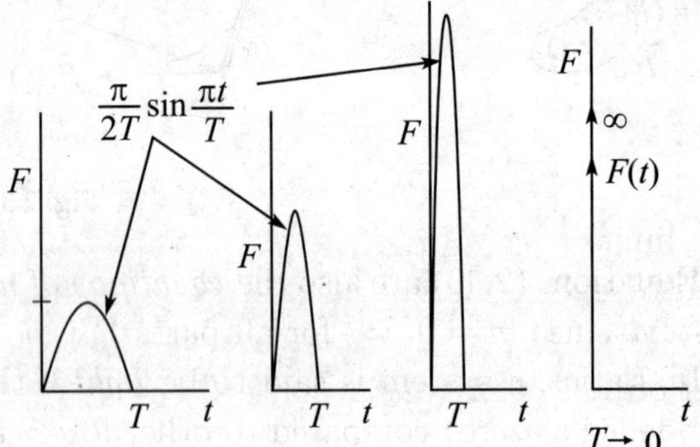

Fig. 2.35

The impulse $\vec{I}(t_1, t_2)$ of a moment (or couple–moment) $\vec{C}(t)$ (Fig. 2.34) acting at Q for interval t_1 to t_2 is zero and its angular impulse $\vec{I}_{\text{ang}_A}(t_1, t_2)$ about A for this interval is

$$\vec{I}_{\text{ang}_A}(t_1, t_2) = \int_{t_1}^{t_2} \vec{M}_A(t)\, dt = \int_{t_1}^{t_2} \vec{C}(t)\, dt \tag{2.45}$$

An *instantaneous impulse* $\vec{I}(t_1)$ (also called an *impulsive 'force'* with dimensions of impulse) is said to act at point P at time t_1 if the force $\vec{F}(t)$ acting at P is such that

$$\vec{I}(t_1) \equiv \lim_{t_2 \to t_1} \vec{I}(t_1, t_2) = \lim_{t_2 \to t_1} \int_{t_1}^{t_2} \vec{F}(t)\, dt \neq \vec{0} \tag{2.46}$$

An instantaneous impulse $\vec{I}(t_1)$ at P causes an *instantaneous angular impulse* $\vec{I}_{\text{ang}_A}(t_1)$

$$\begin{aligned}
\vec{I}_{\text{ang}_A}(t_1) &\equiv \lim_{t_2 \to t_1} \vec{I}_{\text{ang}_A}(t_1, t_2) = \lim_{t_2 \to t_1} \int_{t_1}^{t_2} \vec{M}_A(t)\, dt \\
&= \lim_{t_2 \to t_1} \int_{t_1}^{t_2} \vec{r}_{PA}(t) \times \vec{F}(t)\, dt = \vec{r}_{PA}(t_1) \times \vec{I}(t_1)
\end{aligned}$$

An *instantaneous angular impulse* (or *impulsive 'moment'* with dimensions of angular impulse) $\vec{I}_{\text{ang}_Q}(t_1)$ is said to act at Q at t_1 if the primitive of moment $\vec{C}(t)$ is such that

$$\vec{I}_{\text{ang}_Q}(t_1) = \lim_{t_2 \to t_1} \vec{I}_{\text{ang}_Q}(t_1, t_2) = \lim_{t_2 \to t_1} \int_{t_1}^{t_2} \vec{C}(t)\, dt \neq \vec{0} \tag{2.47}$$

Its instantaneous angular impulse about any point A is the same: $\vec{I}_{\text{ang}_A}(t_1) = \vec{I}_{\text{ang}_Q}(t_1)$.

Since $\vec{I}(t_1, t_2) = \vec{F}_{\text{av}}(t_2 - t_1)$, a finite force does not cause an instantaneous impulse. A step force, i,e., the sudden application or removal of a force, does not cause an instantaneous impulse. A force causes an instantaneous impulse only if it becomes infinite for an infinitesimal duration. Consider the force of Fig. 2.35:

$$\vec{F}(t) = \begin{cases} (\pi/2T)\sin(\pi t/T)\,\vec{e} & \text{for } 0 < t < T \\ \vec{0} & \text{for } t \notin (0, T) \end{cases} \tag{1}$$

$$\Rightarrow \qquad \vec{I}(0, T) = \int_0^T \vec{F}(t)\, dt = \frac{\pi}{2T}\int_0^T \sin\frac{\pi t}{T}\, dt\, \vec{e} = 1\,\vec{e} \;\; \forall\, T$$

In the limit as $T \to 0$, $|\vec{F}_{\text{av}}| \to \infty$, but $\vec{I}(0, t)$ has still the same value $1\,\vec{e}$. The force defined by eq. (1) for $T \to 0$ causes an instantaneous impulse $\vec{I}(0) = 1\,\vec{e}$ at $t = 0$.

Very large forces acting for a very short duration, such as those present during impact of bodies, are mathematically modelled as impulsive forces.

2.11 Impulse–Momentum Relations

All entities in this section are with respect to an inertial frame. The solution of $\vec{F} = m\vec{a}_C$ requires two integrations with respect to t. If $\vec{F}(t)$ is known and $\vec{v}_C(t)$ is required, then the first integral of $\vec{F} = m\vec{a}_C$, i.e., the impulse–momentum relation, is useful.

$$\begin{aligned} \vec{I}(t_1, t_2) &= \int_{t_1}^{t_2} \vec{F}\, dt = \int_{t_1}^{t_2} \dot{\vec{p}}\, dt = \int_{\vec{p}(t_1)}^{\vec{p}(t_2)} d\vec{p} \\ \vec{I}(t_1, t_2) &= \vec{F}_{\text{av}}(t_2 - t_1) = \vec{p}(t_2) - \vec{p}(t_1) \\ &= \Delta\vec{p} = m\Delta\vec{v}_C = \sum m_i \Delta\vec{v}_{C_i} \end{aligned} \tag{2.48}$$

i.e., the *impulse of the* **external** *force* $\vec{F}$ *equals the change in momentum.* For a pile-driver hammer, the impact interval $(t_2 - t_1)$ is kept small to have large $\vec{F}_{\text{av}}$. In contrast, devices meant to prevent injury in accidents, such as a helmet and a seat belt, are designed to increase the time $(t_2 - t_1)$ of absorption of the momentum to decrease $\vec{F}_{\text{av}}$.

Equation (2.48) holds for every Cartesian component, as $a_x = \dot{v}_x$, $a_y = \dot{v}_y$, $a_z = \dot{v}_z$,

e.g., $\quad I_x(t_1, t_2) = \int_{t_1}^{t_2} F_x\, dt = F_{x\text{av}}(t_2 - t_1) = \Delta p_x = m\Delta v_{Cx} = \sum m_i \Delta v_{C_{ix}}$

Such relations are not valid for $\vec{e}_r$, $\vec{e}_\phi$, $\vec{e}_n$ components, since $a_r \neq \dot{v}_r$, $a_\phi \neq \dot{v}_\phi$, $a_n \neq \dot{v}_n$.
The first integral of the equation of motion, $F_\phi = ma_{C_\phi} = m[\frac{1}{r_C}\frac{d}{dt}(r_C^2\dot{\phi}_C)] \Rightarrow$

$$m\Delta(r_C^2\dot{\phi}_C) = m\Delta(r_C v_{C_\phi}) = \int_{t_1}^{t_2} r_C F_\phi dt \tag{2.49}$$

In particular, $F_\phi \equiv 0$[5] $\Rightarrow$

$$r_C^2\dot{\phi}_C = r_C v_{C_\phi} = \text{constant} \tag{2.50}$$

For an *instantaneous impulse* at t_1, eq. (2.48) $\Rightarrow$

$$\begin{aligned} \vec{I}(t_1) &= \Delta\vec{p} = \vec{p}(t_1^+) - \vec{p}(t_1^-) = m[\vec{v}_C(t_1^+) - \vec{v}_C(t_1^-)] \\ &= \sum m_i[\vec{v}_{C_i}(t_1^+) - \vec{v}_{C_i}(t_1^-)] \end{aligned} \tag{2.51}$$

$\vec{p}$, $\vec{v}_C$, and $\vec{v}_{C_i}$ change instantaneously *without change in position* since $\vec{v}_{C\text{av}}$ is finite:

$$\vec{r}_C(t_1^+) - \vec{r}_C(t_1^-) = \lim_{t_2 \to t_1} \int_{t_1}^{t_2} \vec{v}_C(t)\, dt = \lim_{t_2 \to t_1} [\vec{v}_{C\text{av}}(t_2 - t_1)] = \vec{0}$$

If $\vec{I}(t_1, t_2) = \vec{0}$, then eq. (2.48) implies *conservation of momentum*:

$$\begin{aligned} &\vec{p}(t_2) = \vec{p}(t_1), \qquad \vec{v}_C(t_2) = \vec{v}_C(t_1) \\ &\sum m_i \vec{v}_{C_i}(t_2) = \sum m_i \vec{v}_{C_i}(t_1) \end{aligned} \tag{2.52}$$

Internal forces alone cannot change the velocity of C, though the velocities $\vec{v}_{C_i}$ of different parts can be changed. If a particular component of impulse in some fixed direction is zero, then the corresponding component of momentum is conserved, for example if only $I_y(t_1, t_2) = 0$, then the component $p_y = mv_{Cy} = \sum m_i v_{C_{iy}}$ is conserved.

Example 2.17: Discuss the impact of a rigid block of mass m attached to a spring of stiffness k, moving with velocity v_0 on a smooth horizontal plane, with a smooth rigid vertical wall (Fig. E2.17).

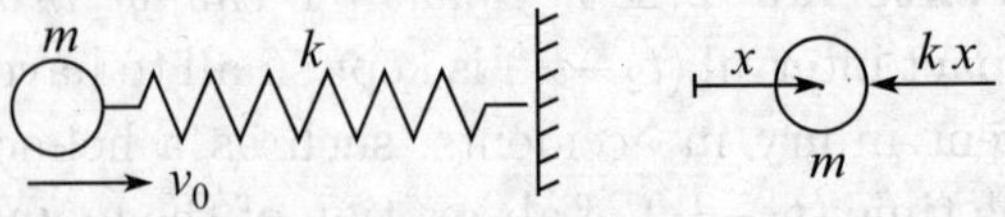

m k v0 x kx m

Fig. E2.17

[5]Note that for a mass-point m, $H_{O_z} = [(r\,\vec{e}_r + z\,\vec{e}_z) \times m(\dot{r}\,\vec{e}_r + r\dot{\phi}\,\vec{e}_\phi + \dot{z}\,\vec{e}_z)] \cdot \vec{e}_z = mr^2\dot{\phi} = mrv_\phi$.

Solution: Let x be the compression of the spring (Fig. E2.17). The initial conditions are: $x(0) = 0,\ \dot{x}(0) = v_0$. The equation of motion for $x \geq 0$ is

$$m\ddot{x} = -kx \quad \Rightarrow \quad x(t) = \frac{v_0}{\omega}\sin\omega t, \quad \text{with } \omega^2 = \frac{k}{m}, \quad x_{\max} = \frac{v_0}{\omega} = v_0\sqrt{m/k}$$

The impact ends at t_1 when the spring force becomes zero, i.e., when

$$x(t_1) = \frac{v_0}{\omega}\sin\omega t_1 = 0 \quad \text{at } t_1 = \pi/\omega = \pi\sqrt{m/k}, \quad \text{and} \quad \dot{x}(t_1) = -v_0$$

$$\vec{I}(0, t_1) = \Delta\vec{p} = m(-v_0\,\vec{\mathrm{i}} - v_0\,\vec{\mathrm{i}}) = -2mv_0\,\vec{\mathrm{i}}$$

This impulse on the block is independent of k. In the limit as $k \to \infty$ (stiff spring), $t_1 \to 0,\ x_{\max} \to 0 \ \Rightarrow$ there is an instantaneous change of $-2v_0\,\vec{\mathrm{i}}$ in velocity without change in position and the force from the wall, kx, tends to infinity in such a way that it exerts an instantaneous impulse $\vec{I}(0) = -2mv_0\,\vec{\mathrm{i}}$.

2.12 Angular Impulse–Moment of Momentum Relations

All entities in this section are with respect to an inertial frame. If A is a point such that $\vec{a}_A(t) \equiv \vec{0}$, or $A \equiv C$, or $\vec{a}_A(t)$ is along AC, $\forall\, t \in (t_1, t_2)$, then $\vec{M}_A = \dot{\vec{H}}_A$ yields the angular impulse–moment of momentum relation:

$$\vec{I}_{\text{ang}_A}(t_1, t_2) = \int_{t_1}^{t_2} \vec{M}_A\,dt = \int_{\vec{H}_A(t_1)}^{\vec{H}_A(t_2)} d\vec{H}_A$$

$$\vec{I}_{\text{ang}_A}(t_1, t_2) = \vec{H}_A(t_2) - \vec{H}_A(t_1) = \Delta\vec{H}_A \tag{2.53}$$

$\Rightarrow$ **1.** *The external angular impulse about A equals the change in the moment of momentum $\vec{H}_A$.*

2. If $\vec{I}_{\text{ang}_A}(t_1, t_2) = \vec{0}$, then eq. (2.53) implies that $\vec{H}_A$ is conserved.

Since the results 1 and 2 are valid for a point A fixed in I, if a component of $\vec{I}_{\text{ang}_A}(t_1, t_2)$ along a fixed direction in I is zero, then the component of $\vec{H}_A$ in that direction is conserved. Similarly, if a component of $\vec{I}_{\text{ang}_C}$ along a fixed direction of I is zero, then the component of $\vec{H}_C$ in that direction is conserved.

If an instantaneous angular impulse acts at t_1, then without change in position there is an instantaneous change in moment of momentum about the centre of mass C, about a point O fixed in I and *about any point A in arbitrary motion with respect to I whose velocity does not change instantaneously at t_1 (A may be a material point of the system or a point outside it)*, with

$$\vec{I}_{\text{ang}_C}(t_1) = \Delta\vec{H}_C, \qquad \vec{I}_{\text{ang}_O}(t_1) = \Delta\vec{H}_O, \qquad \vec{I}_{\text{ang}_A}(t_1) = \Delta\vec{H}_A \tag{2.54}$$

The last equation is valid (Fig. 2.36) since the positions of C, A and $\vec{v}_A$ remain the same, (i.e., $\Delta\vec{v}_A = \vec{0}$) and

$$\begin{aligned}\Delta\vec{H}_A &= \Delta[\vec{H}_C + \vec{r}_{CA} \times m\vec{v}_{CA}]\\ &= \Delta\vec{H}_C + \vec{r}_{CA} \times m\Delta\vec{v}_C\\ &= \vec{I}_{\text{ang}_C}(t_1) + \vec{r}_{CA} \times \vec{I}(t_1) = \vec{I}_{\text{ang}_A}(t_1)\end{aligned}$$

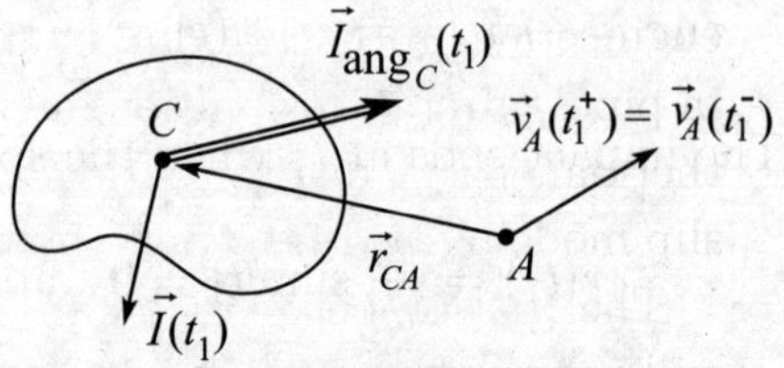

Fig. 2.36

But $\vec{I}_{\text{ang}_B}(t_1)$ is not equal to $\Delta\vec{H}_B$ for any other point B of the system or otherwise.

2.13 Work–Energy Relation for Centre of Mass C of any System

We define $\dot{W}^*$ and T_C by

$$\dot{W}^* \equiv \vec{F} \cdot \vec{v}_C = \text{power of all the } \textit{external} \text{ forces } \textit{as if} \text{ acting at } C \tag{2.55}$$

$$T_C \equiv \tfrac{1}{2}mv_{C|I}^2 = \text{kinetic energy } \textit{as if} \text{ all mass is concentrated at } C \tag{2.56}$$

In general, $\dot{W}^* \neq \dot{W}$, $T_C \neq T$.

$$\vec{F} = m\vec{a}_{C|I} \;\Rightarrow\; \vec{F} \cdot \vec{v}_{C|I} = m\vec{a}_{C|I} \cdot \vec{v}_{C|I} = \frac{d}{dt}\left[\tfrac{1}{2}m\vec{v}_{C|I} \cdot \vec{v}_{C|I}\right]_{|I} = \frac{d}{dt}\left[\tfrac{1}{2}mv_{C|I}^2\right]_{|I}$$

$$\Rightarrow \dot{T}_C = \dot{W}^* \qquad \Rightarrow \qquad T_{C_2} - T_{C_1} = W_{1-2}^* \tag{2.57}$$

T_{C_1}, T_{C_2} are the values of T_C in configurations 1 and 2 of the system and W_{1-2}^* is the work done from configuration 1 to 2 by the external forces as if they were acting at C. The work–energy relation is *not applicable when instantaneous impulsive loads act*, since at that instant $|\vec{a}_C| = \infty$ and the proof of eq. (2.57) requires $\vec{a}_C$ to remain finite.

2.14 Friction

The mechanical interaction at the surfaces of contact of two bodies is modelled by a distributed surface force whose components normal and tangent to the contact surface are called normal reaction and force of friction. Friction obstructs relative sliding between the bodies. Its direction on body 1 is opposite to the possible or existing relative velocity of point P_1 of body 1 relative to the point P_2 of body 2 with which it is in instantaneous contact (Fig. 2.37a). Since the bodies press against each other, the normal reaction on a body always points into its interior.

For the case of slip, friction causes energy dissipation, partially as thermal energy. This energy 'loss' can be high (20% for automobiles), but it is the resulting wear causing rough operation and premature failure of components, which is a principal engineering worry. Friction is undesirable in bearings, power screws, gears, fluid flow in pipes, aircraft propulsion, etc. Sometimes, the conversion of mechanical energy into thermal energy due to friction is beneficial, e.g., in brakes. Friction is utilised in its no slip mode in land transport, belt and clutch drives, holding and fastening devices, etc.

The friction between dry rubbing solid bodies is called *dry friction.* Fluid friction will not be covered in this text. Rolling friction occurs between bodies in rolling contact. The major cause of dry friction is bonding between the elements of the bodies over the microscopic roughness of the surfaces in contact. Interlocking microscopic protuberances and troughs of the contacting surfaces oppose relative motion giving rise to frictional resistance force. When sliding is present between such surfaces, some of these protuberances are sheared off or melt due to high local temperature, causing wear. Experiments broadly reveal that the frictional force at the instant of impending slip is proportional to the normal reaction and is independent of the surface area of contact. The friction force during continued slip, called kinetic friction, is less than that at impending slip because some of the interlocking protuberances have sheared off and there is no time to form new bonds between the elements of the bodies in contact. Over a limited range of speed, friction force is independent of the relative velocity between two points in contact. Friction force depends on local temperatures generated, adhesion at contact points, relative hardness of mating surfaces and presence of surface films of oxide, oil, dirt or other substances.

2.14.1 Axiom of dry friction or Coulomb friction

Consider two bodies with a surface contact (Fig. 2.37a). At a point P_1 of body 1, which at time t is in contact with point P_2 of body 2, the normal (*pressure*) and tangential (*frictional*) components, $\vec{p}$ and $\vec{\tau}$ respectively, of the surface force density (Fig. 2.37b) with magnitudes $p = |\vec{p}|$, and $\tau = |\vec{\tau}|$, are related as follows.

1. For no slip between P_1 and P_2 :

$$\tau < \mu_s p \tag{2.58}$$

$\vec{\tau}$ is found from the equation of motion.

2. For impending slip between P_1 and P_2 :

$$\tau = \mu_s p \tag{2.59}$$

and $\vec{\tau}$ is directed opposite to the direction of impending slip of P_1 relative to P_2, i.e., $\vec{\tau}$ is opposite to the component of $\vec{a}_{P_1P_2}(t^+)$ in the tangent plane.

3. For slip between P_1 and P_2 :

$$\tau = \mu_k p \tag{2.60}$$

and $\vec{\tau}$ is directed opposite to $\vec{v}_{P_1P_2}(t)$. $\mu_s\ (>0)$, $\mu_k\ (>0)$ are called the static and kinetic (dynamic) coefficients of friction with $\mu_k \leq \mu_s$. These are independent of p, $\vec{v}_{P_1P_2}$ and the area of contact. If a tentative assumption of no slip is made, then the direction and magnitude of τ can be shown arbitrarily in an FBD. However, it must be verified that the magnitude of τ, determined from the equation of motion, lies within the bound of $\mu_s p$. If the check holds, the assumption of no slip is correct, else there is slip. Often the direction of friction forces for the latter case are the same as those obtained with the assumption of no slip.

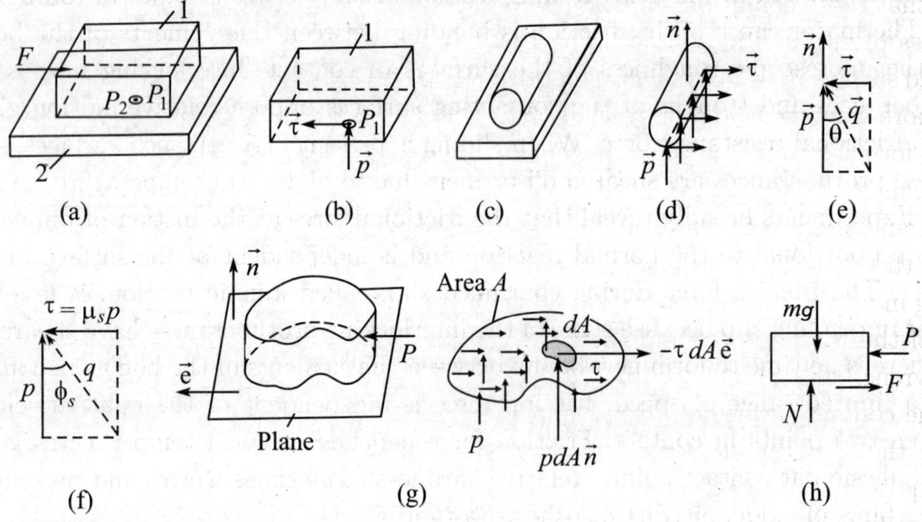

Fig. 2.37

For two bodies with idealised line contact (Fig. 2.37c), eqs (2.58) to (2.60) hold with p and τ being the normal and tangential (*frictional*) components of the line force density respectively (Fig. 2.37d).

In reality, μ is not exactly independent of p or the area of contact. The axiom is satisfactory for most metals in air. For diamond, rubber and plastics, μ decreases with increase in p. One should avoid using μ from tables. A variation of 25–100% can occur depending on the cleanliness, surface finish, pressure, lubrication and velocity. It is preferable to find the actual μ for the bodies to be used. The typical values are $\mu_s \simeq$: 1 for copper, cast iron, glass rubber tyres on road; 0.7 for mild steel, hemp rope on wood; 0.5 for leather on metal, wood on wood; 0.4 for steel on babbitt, brake lining on cast iron, rubber on metal / wood; 0.3 for hemp rope on metal, 0.2 for wire rope on steel, 0.1 for rubber on ice; 0.04 for teflon on steel/teflon, metal on ice. These data are for commercially available materials having some impurities. For pure metals with fine surface finish, the surfaces could bond and μ_s could become very large.

The resultant $\vec{q}$ of $\vec{p}$ and $\vec{\tau}$ makes angle $\theta = \tan^{-1}(\tau/p)$ with the normal (Fig. 2.37e).

The limiting value of θ for no slip (Fig. 2.37f) is called the *angle of static friction* ϕ_s:

$$\tan\phi_s = \mu_s, \qquad \tau/p \leq \mu_s \quad \Rightarrow \quad \tan\theta \leq \tan\phi_s \quad \Rightarrow \quad \theta \leq \phi_s \tag{2.61}$$

Most often, we work with $\vec{p}$ and $\vec{\tau}$ and not with their resultant.

If the frictional force $\tau = 0$ for no slip, impending slip and actual slip, i.e., $\mu_s = \mu_k = 0$, then the contact is defined to be a *perfectly smooth contact.* If the frictional force τ can be arbitrarily high, i.e., $\mu_s = \infty$ and slip between points in contact is not possible, then the contact is defined to be a *perfectly rough contact.* However, it should be noted that these definitions in our mathematical model are not based on the physical waviness of surfaces in contact. The coefficient of friction for a highly ground and polished surface (i.e., a physically smooth surface) is more than for an unground surface with more waviness (i.e., a physically rough surface). This is due to greater adhesion for the former case as the surface elements are much closer over most of the surface.

Like any other force system, the simplest resultants of the distributed contact force system and the distributed frictional force system are, in general, wrenches. For example, the simplest resultants of the contact force and the frictional force on a screw moving in a fixed nut (see Section 2.20) are wrenches (Fig. 2.61b). The simplest resultants of the contact force and the frictional force on an axisymmetric thrust bearing are a wrench and a single couple, respectively (Fig. 2.60).

If the contact surface of area A is plane (Fig. 2.37g,h) and the direction and sense of $\vec{\tau}$ is the same at all points (for example, when 1 translates with respect to 2), then eqs (2.58) to (2.60) imply that the resultant normal force $N = \int_A p\,dA$, and the resultant frictional force $F = \int_A \tau\,dA$, are related by

$$\text{for no slip: } F < \mu_s N, \qquad \text{impending slip: } F = \mu_s N, \qquad \text{for slip: } F = \mu_k N \tag{2.62}$$

Most often, $F \neq 0$ *for no slip,* and its direction is a priori arbitrary and is obtained from equations of motion. *In general,* N *does not act at the centroid of the contact area.*

Equations (2.62) are valid (i) for idealised straight line contact if the direction and sense of $\vec{\tau}$ is the same at all points, (ii) for idealised point contact (Fig. 2.38a,b) with the normal reaction N being in the direction of the normal to one (if the radius of curvature of one of the surfaces is zero) or both surfaces of contact. $N \geq 0$ for a one-sided constraint, but the sign of N is arbitrary for a two-sided constraint. For the one-sided constraint, check whether the computed value of N is non-negative. If this check fails, then rework the problem with $N = 0$.

We consider two cases of point contact.

1. A body makes contact at a point on a curved thin body (Fig. 2.38c) idealised as a curve C (for example, a small bead on a wire or a slider in a slot). Then,

$$\vec{F} = -F\,\vec{e}_t, \qquad \vec{N} = N_1\,\vec{e}_n + N_2\,\vec{e}_b,$$

$$N = (N_1^2 + N_2^2)^{1/2} \quad \text{and for no slip:} \quad |F| < \mu_s(N_1^2 + N_2^2)^{1/2}$$

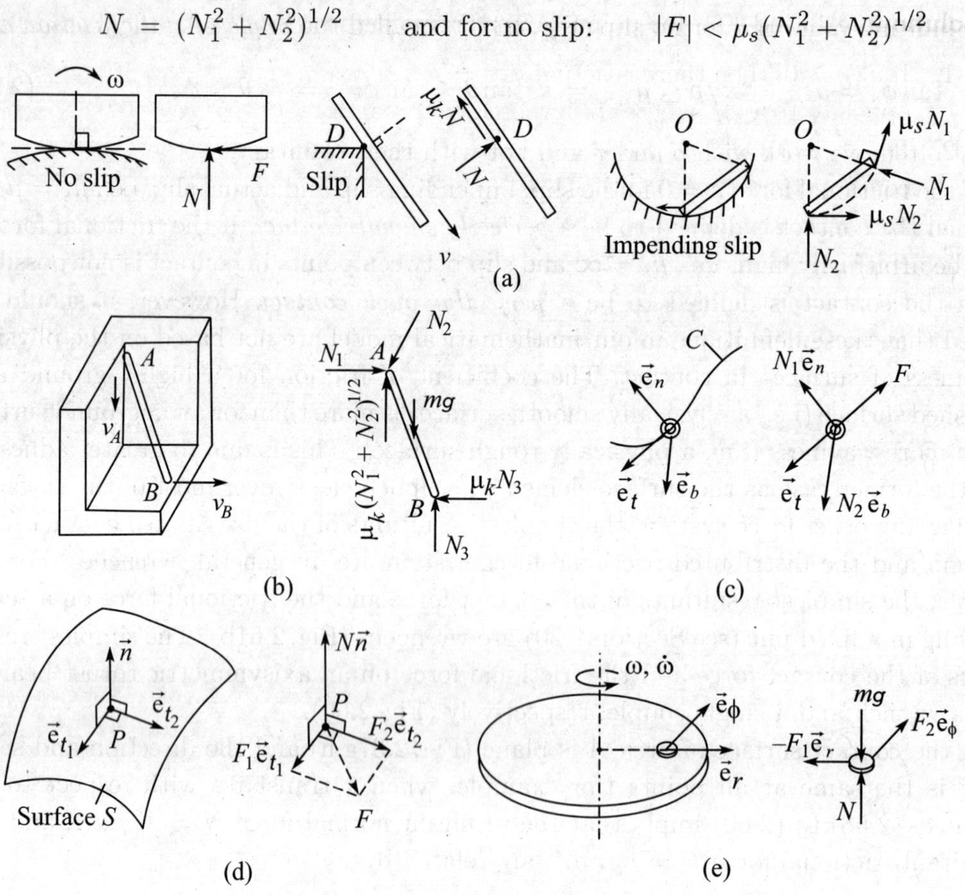

Fig. 2.38

2. A body makes point contact on a surface S (Fig. 2.38d) of another body (for example, a small slider on a cylinder),

$$\vec{F} = F_1\,\vec{e}_{t_1} + F_2\,\vec{e}_{t_2}, \qquad F = (F_1^2 + F_2^2)^{1/2}, \qquad \vec{N} = N\vec{n}$$

$$\text{and for no slip:} \qquad (F_1^2 + F_2^2)^{1/2} < \mu_s|N|$$

where $\vec{e}_{t_1}$, $\vec{e}_{t_2}$ are orthogonal unit vectors in the plane tangent to the surface S and $\vec{n}$ is its unit normal vector. For example, for impending slip of a small coin on a disc (Fig. 2.38e) rotating about a fixed vertical axis with angular acceleration, there are frictional force components in the radially inward and the circumferential directions, with $|\vec{F}| = \sqrt{F_1^2 + F_2^2} = \mu_s N$.

Example 2.18: Draw FBD of body 1 of mass m for each case shown in Fig. E2.18a. In case 3, there is no slip at A and B and body 3 translates at v. Given μ_s and μ_k.

Solution: Find $\vec{v}_{AB}$ for points A and B in contact. $\vec{F}$ at A for the 3 cases is:

1. If $\vec{v}_{AB} \neq \vec{0}$, then there is actual slip $\Rightarrow \vec{F}$ has magnitude $F = \mu_k N$ and is directed opposite to $\vec{v}_{AB}$, where N is the normal reaction.
2. If $\vec{v}_{AB} = \vec{0}$ and $\vec{a}_{AB} \cdot \vec{e}_t = 0$, where $\vec{e}_t$ is any vector in the tangent plane of contact, then there is no slip $\Rightarrow \vec{F}$ has arbitrary magnitude (bounded by $\mu_s N$) and arbitrary direction, which are obtained by solving the equations of motion.
3. If $\vec{v}_{AB} = \vec{0}$ and $\vec{a}_{AB} \cdot \vec{i} \neq 0$, where $\vec{i}$ is in the tangent plane of contact, then there is impending slip $\Rightarrow F = \mu_s N$ and is directed opposite to $(\vec{a}_{AB} \cdot \vec{i})\,\vec{i}$. $F = \mu_s N$ exists momentarily and becomes $\mu_k N$ immediately afterwards.

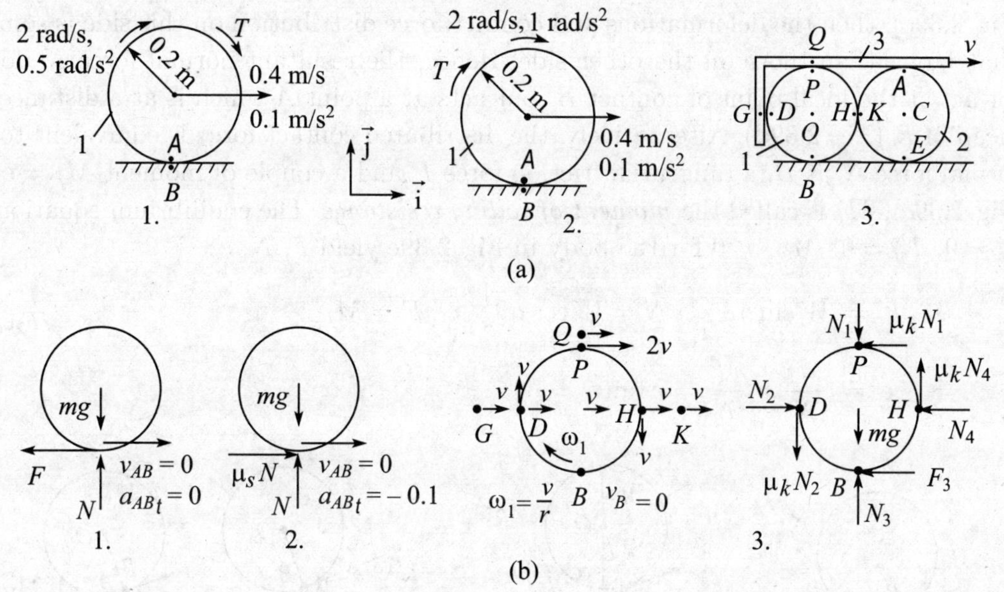

Fig. E2.18

Case 1: $v_{AB} = 0.4 - 2(0.2) = 0,\ a_{AB_t} = 0.1 - 0.5(0.2) = 0 \quad \Rightarrow \quad$ no slip. $|F| < \mu_s N$.

Case 1: $v_{AB} = 0.4 - 2(0.2) = 0,\ a_{AB_t} = 0.1 - 1(0.2) = -0.1 \Rightarrow$ impending slip to left.

Case 3: $\vec{v}_A = v\,\vec{i}, \quad \vec{v}_C = v\,\vec{i} \quad \Rightarrow \quad$ body 2 translates $\Rightarrow \vec{v}_K = v\,\vec{i}$.

$\vec{v}_B = \vec{0}, \quad \vec{v}_O = v\,\vec{i} \quad \Rightarrow \quad$ body 1 rotates at $\omega_1 = v/r$. $\vec{v}_Q = \vec{v}_G = v\,\vec{i}$.

$\vec{v}_P = \omega_1(2r)\,\vec{i} = 2v\,\vec{i}, \quad \vec{v}_{PQ} = v\,\vec{i} \quad \Rightarrow \quad P$ slips in direction $\vec{i}$.

$\vec{v}_D = \vec{v}_O + \omega_1 r\,\vec{j} = v\,\vec{i} + v\,\vec{j}, \quad \vec{v}_{DG} = v\,\vec{j} \quad \Rightarrow \quad D$ slips in direction $\vec{j}$.

$\vec{v}_H = \vec{v}_O + \omega_1 r(-\vec{j}) = v\,\vec{i} - v\,\vec{j}, \quad \vec{v}_{HK} = -v\,\vec{j} \quad \Rightarrow \quad H$ slips in direction $-\vec{j}$.

No slip at $B \Rightarrow |F_3| < \mu_s N_3$.

2.14.2 Rolling friction

Consider a wheel resting in equilibrium on an inclined plane of small inclination $\alpha \neq 0$ (Fig. 2.39a). Its FBD, on the assumption of line contact, is shown in Fig. 2.39b. There do not exist any values of F and N which can satisfy $F_x = 0,\ F_y = 0,\ M_{Oz} = 0 \Rightarrow$

$$F = W \sin\alpha, \qquad N = W\cos\alpha, \qquad rF = 0.$$

$\Rightarrow$ No matter how small α ($\neq 0$) is, and how large μ is, the wheel should roll. This contradicts our experience. Hence, our modelling of the contact surface as a line contact is not adequate. The bodies are not exactly rigid but deform enough near the contact region to have a small surface of contact. If the body tends to roll in the x-direction (Fig. 2.39c), then the deformations and contact force distribution on this side is somewhat larger than those on the other side. Hence, the resultant normal force N does not act at the ideal point of contact B, but acts at a point A which is at a distance e ahead of B (Fig. 2.39d). Alternatively, the distributed contact force is equivalent to a normal force N at B, a tangential friction force F and a couple of moment $M_f = eN$ (Fig. 2.39e). M_f is called the *moment of rolling resistance.* The equilibrium equations: $F_x = 0,\ F_y = 0,\ M_{Oz} = 0$ for the body in Fig. 2.39e yield $F,\ N,\ e$:

$$F = W\sin\alpha, \qquad N = W\cos\alpha, \qquad rF = M_f$$

$$\Rightarrow \quad e = \frac{M_f}{N} = \frac{rF}{N} = r\tan\alpha$$

Fig. 2.39

Experiments show that the wheel remains at rest only for small values of angle α, i.e., the eccentricity e of N has an upper limit e_1, i.e., $|e| \leq e_1$ and $M_f \leq e_1 N$. It is called the *inequality of rolling resistance* and e_1 is called the *length of rolling resistance* (also called the 'coefficient' of rolling friction). It has the dimensions of length and it may depend on the curvature of the contacting curves, speed of travel, elastic and plastic properties of the contacting materials and roughness of surfaces. Tests indicate little variation of e_1 with the wheel radius. So, e_1 (also denoted by a) is taken as independent of the radii of curvatures of the curves of the bodies rolling on each other at the point of contact. In case of an impending and actual rolling motion, the inequality of rolling resistance is replaced by an *equality of rolling resistance* with usually the same length of rolling resistance e_1 as in the inequality. We can thus state the following *axiom of rolling resistance.*

If a body tends to roll over another body, then the resultant normal force N acts through a point at a distance e ahead of the ideal line or point of contact in the direction in which impending or actual subsequent contact occurs, such that

$$\text{for no rolling:} \qquad |e| < e_1 \tag{2.63}$$

$$\text{and for impending or actual rolling:} \qquad |e| = e_1 \tag{2.64}$$

The range of values of e_1 for hardened steel on hardened steel (ball bearing) is 0.0005–0.0013 cm, for steel wheels on steel rails is 0.006–0.010 cm, for pneumatic tyre on smooth pavement is 0.05–0.07 cm, for wood on wood is 0.05–0.08 cm and for steel on wood is 0.15–0.25 cm. For the wheel of Fig. 2.39a, if both the conditions

1. $F/N = \tan\alpha < \mu_s$ (for no slip)
2. $e = M_f/N = r\tan\alpha < e_1$ (for no rolling)

are fulfilled, then it remains at rest. If α is gradually increased, then the wheel starts to slip at $\alpha = \tan^{-1}\mu_s$ if $\mu_s < e_1/r$, and starts to roll at $\alpha = \tan^{-1}(e_1/r)$ if $e_1/r < \mu_s$.

Example 2.19: A cylindrical roller of mass m is at rest. Find M to initiate rolling over the step (Fig. E2.19a). The length of rolling resistance is e. Find the minimum μ_s to ensure no slip.

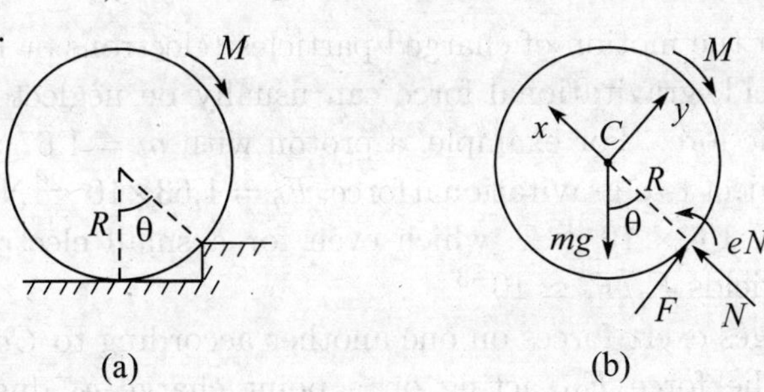

Fig. E2.19

Solution: N acts at a distance e ahead of the ideal contact point on the side which is about to make contact subsequently. Hence, in the FBD for impending rolling, the direction of moment eN is as shown (Fig. E2.19b). M, N, F are obtained from:

$$F_x = N - mg\cos\theta = 0 \quad \Rightarrow \quad N = mg\cos\theta$$

$$F_y = F - mg\sin\theta = 0 \quad \Rightarrow \quad F = mg\sin\theta$$

$$M_{Cz} = M - eN - FR = 0 \quad \Rightarrow \quad M = mg(e\cos\theta + R\sin\theta)$$

$$\mu_{s_{\min}} = F/N = \tan\theta$$

2.15 Axiom of Electromagnetic Force

Force $\vec{F}$ on a mass-point with charge q moving with velocity $\vec{v}$ in a static electromagnetic field with electric field vector $\vec{E}$ and a magnetic field vector $\vec{B}$ (Fig. 2.40a) is given by the Lorentz law as

$$\vec{F} = q\,\vec{E} + q\vec{v}\times\vec{B}. \tag{2.65}$$

$\vec{F}$ is in newtons if q is in coulombs (C), $\vec{E}$ is in N.C^{-1}, $\vec{B}$ is in N.C^{-1}.m^{-1}.s and $\vec{v}$ is in m.s^{-1} (1 C = 1 ampere second= 1 A.s).

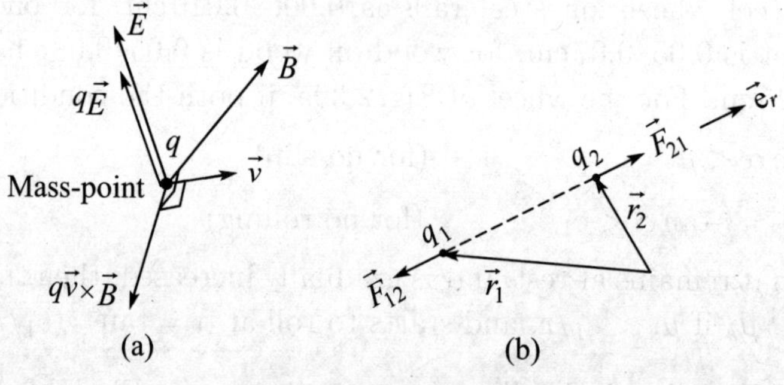

Fig. 2.40

For the study of the motion of charged particles (electrons or ionised atoms) in an electromagnetic field, gravitational force can usually be neglected in comparison to the electromagnetic force. For example, a proton with $m = 1.67\times10^{-27}$ kg, and $q = 1.6\times10^{-19}$ C is subject to a gravitational force, $F_g = 1.63\times10^{-26}$ N and an electrostatic force, $F_e = qE = 1.6\times10^{-19}E$, which even for a small electric field of strength, $E = 10^{-4}$N.C^{-1}, yields $F_g/F_e \simeq 10^{-3}$.

Stationary charges exert forces on one another according to *Coulomb's law of electrostatic forces.* The force F_{21} acting on a point charge q_2 due to a point charge q_1, shown in Fig. 2.40b, in an intervening medium of permittivity ϵ (for vacuum,

$\epsilon = 8.854 \times 10^{-12}\,\mathrm{C}^2.\mathrm{N}^{-1}.\mathrm{m}^{-2}$) is given by

$$\vec{F}_{21} = -\vec{F}_{12} = \frac{q_1 q_2}{4\pi\epsilon r^2}\,\vec{\mathrm{e}}_r = \frac{q_1 q_2}{4\pi\epsilon}\,\frac{(\vec{r}_2 - \vec{r}_1)}{|\vec{r}_2 - \vec{r}_1|^3} \tag{2.66}$$

2.16 Axiom of Gravitational Force

Newton's law of gravitation for mass-points (particles) is extended to finite bodies in the following axiom of gravitation: 'The total gravitational force $\vec{F}_{21}$ on body 2 of mass m_2 due to body 1 of mass m_1 (Fig. 2.41) is given by

$$\vec{F}_{21} = -\int_{m_2}\left[\int_{m_1} G\,\frac{\vec{r}}{r^3}\,dm_1\right]dm_2 \tag{2.67}$$

where $G = (6.672 \pm 0.004) \times 10^{-11}\,\mathrm{m}^3.\mathrm{kg}^{-1}.\,\mathrm{s}^{-2}$ is the universal gravitational constant.' The total force $\vec{F}_{21}$ is called the *weight* of body 2 due to body 1. The *simplest resultant* of the spatially distributed force $d\vec{F}_{21}$ is, in general, a *wrench.* If the simplest resultant is a single force, then the *centre of gravity* G of body 1 due to body 2 is defined as the unique point where the whole mass m_1 can be concentrated so that the force on it due to body 2 equals that on the actual distributed mass m_1. For a uniform sphere, G coincides with C and remains unchanged as the body moves. In general, the position of G relative to body 1 depends on its orientation and its distance from body 2 and it is distinct from C. In such cases, it is more useful to show an equivalent force–couple system at C.

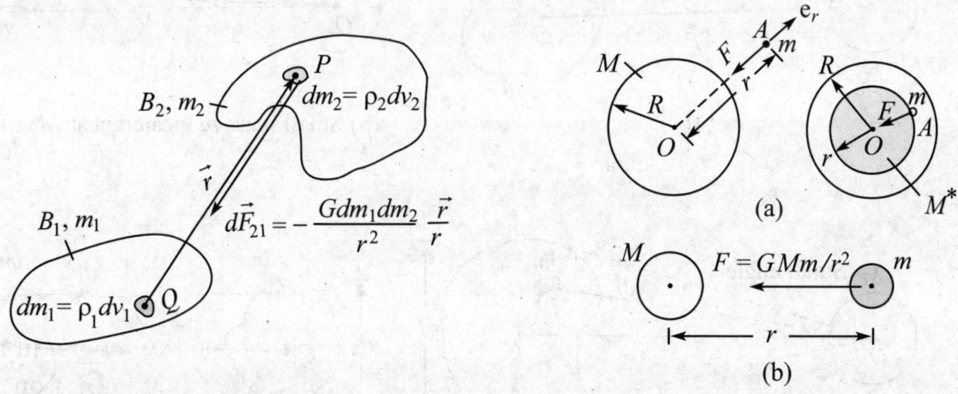

Fig. 2.41

Fig. 2.42

The following results can be proved from eq. (2.67).

1. For a *radially symmetric* [$\Rightarrow\ \rho(r)$] sphere of mass M, radius R and *a mass-point* of mass m (Fig. 2.42a) at distance r from its centre:

$$\text{for } m \text{ outside } M\text{, i.e., } r \geq R: \qquad F = \frac{GMm}{r^2} \tag{2.68}$$

$$\text{for } m \text{ inside } M, \text{ i.e., } r < R: \qquad F = \frac{GM^*m}{r^2} \tag{2.69}$$

where M^* is the mass inside a sphere of radius r. For a uniform sphere:

$$\frac{M^*}{M} = \frac{r^3}{R^3} \quad \Rightarrow \quad \text{for a uniform sphere:} \quad F = \frac{GMmr}{R^3} \quad \text{for } r < R \tag{2.70}$$

2. The resultant force between *two separate radially symmetric spheres* of mass M and m (Fig. 2.42b), with centres at a distance r apart, is $F = GMm/r^2$.
3. The *simplest resultant* of the concurrent gravitational force between *a radially symmetric sphere of mass M and an arbitrary body of mass m* (Fig. 2.43a) is a single force $\vec{F} = -GM\int_m(\vec{r}_P/r_P^3)\,dm$ through O. The *centre of gravity* G of m due to M exists and is given by

$$-GMm\frac{\vec{r}_G}{r_G^3} = -GM\int_m \frac{\vec{r}_P}{r_P^3}\,dm \tag{2.71}$$

In general, the location of G differs from C. If the *size of the body is small* compared to its distance from the sphere (Fig. 2.43b,c), then $\vec{r}_P = \vec{r}_{PC} + \vec{r}_C \simeq \vec{r}_C$ and eq. (2.71) $\Rightarrow \vec{r}_G \simeq \vec{r}_C$, i.e., model G to be at C and $\vec{F} = -GMm\vec{r}_C/r_C^3$.

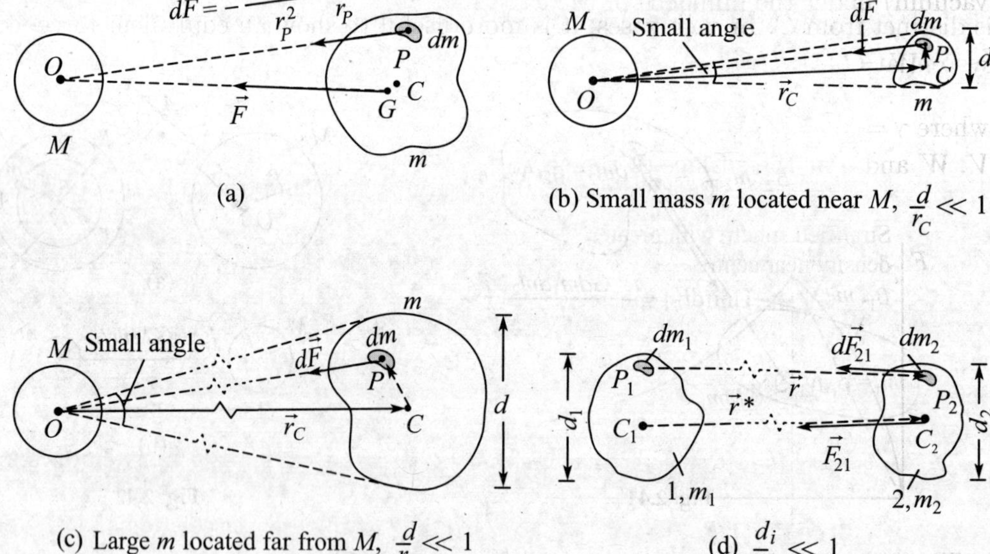

Fig. 2.43

Note: The distinction between C and G is rarely of importance in engineering applications on earth, but it is of interest in celestial mechanics and dynamics of space vehicles. In Skylab, 3 control moment gyros were installed to cancel $\vec{H}_C$ caused by moment $\vec{M}_C$ of a gravitational force of only a few N.m. Since the earth is not an exact sphere, the resultant of the gravitational force of the sun and the moon does not always pass through the same point. It exerts a variable moment about the centre of mass of the earth causing a gradual change in the direction of the earth's axis of rotation, which is called precession of the equinoxes.

4. For *two bodies of arbitrary shapes whose sizes* d_1, d_2 *are small* compared to the distance r^* separating them (Fig. 2.43d), since $\vec{r} = \overrightarrow{P_1P_2} = \overrightarrow{P_1C_1} + \overrightarrow{C_1C_2} + \overrightarrow{C_2P_2} \simeq \overrightarrow{C_1C_2} = \vec{r}^*$, eq. (2.67) yields $\vec{F}_{21} \simeq -Gm_1m_2\vec{r}^*/r^{*3}$.

2.16.1 Model of earth's gravitational field

For most practical problems, the variation of the earth's gravitational force due to its non-spherical shape, radially un-symmetric mass density and its motion can be neglected. Earth is modelled as a radially symmetric sphere of radius $R = 6395\,\text{km}$ with mass M equal to that of the physical earth. Its gravitational force F on a mass-point m at a distance r from its centre is given by eqs (2.68) and (2.69), and whose variation with r is shown in Fig. 2.44. The variation in the interior is not linear in r (as in a uniform sphere) since the density increases as its centre is approached. Consider a body of mass m whose dimensions are much smaller than its distance r from earth's centre. Its weight W and acceleration g of its centre of mass C in 'free fall' (i.e., in vacuum) under the influence of the earth's gravitational field (Fig. 2.45) are given by

$$W = GMm/r^2 = mg = \rho g V = \gamma V \quad \Rightarrow \quad g = GM/r^2 \tag{2.72}$$

where $\gamma = \rho g = \rho GM/r^2$ is the weight density of the body having density ρ and volume V. W and g at the surface of the earth ($r = R$) are: $W_0 = GMm/R^2$, $g_0 = GM/R^2$.

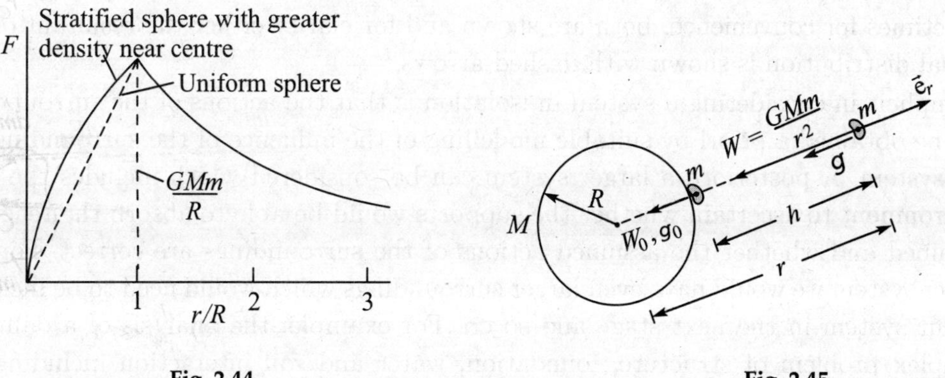

Fig. 2.44

Fig. 2.45

Equation (2.72) $\Rightarrow$ $W \simeq W_0$, $g \simeq g_0$ if $h/R \ll 1$, with only 1% error for $h < 30\,\text{km}$. For 1° variation in the direction of g_0 on the earth's surface, the corresponding arc-length $s = R(\pi/180) = 111.7\,\text{km} \Rightarrow$ the direction of g_0 varies by less than $0.01°$ for $s = 1\,\text{km}$. Thus, for a body of dimensions very much less than R, moving through small distances ($s \ll R$, $h \ll R$) near the surface of the earth, the distributed gravitational force is modelled as a parallel uniform field with force $d\vec{F} = dm\, g\, \vec{e}$ on mass element dm, where $\vec{e}$ is the local vertically downward direction. Their resultant is a constant force, $\vec{F} = mg\,\vec{e}$ acting at the centre of mass C.

Two uniform spheres of mass 5 kg each, with centres 10 cm apart, attract with a force of $16.68{\times}10^{-8}\,\text{N}$, which is extremely small compared to their weight of 49.05 N at the earth's surface. Hence, the mutual gravitational forces between objects on the earth's surface are negligible compared to their weights unless the masses involved are great.

2.17 Free Body Diagram and Reactions of Supports

To **apply Euler's axioms to a system** consisting of n connected/unconnected bodies or mass-points, or a body or a finite/an infinitesimal part of a body,

1. draw a sketch of the *well-identified* system *in isolation* from its surroundings, and
2. show the *external forces and moments* exerted by the surroundings on the system.

Such a diagram is called a **Free body diagram** (FBD). The forces exerted by one part of the system on another part P are called *internal forces* and are not shown in an FBD. Such forces appear in the FBD of part P as these are external forces for P.

Inertial forces like centrifugal force, Coriolis force, $-m\vec{a}_A$, $-m\vec{a}_C$, etc., should not appear in an FBD. Since the purpose of drawing an FBD is to obtain $\vec{F}$ and $\vec{M}_A$, we often replace some external forces by their simpler equivalent systems, for example, uniform gravitational force is often shown as mg at C. In general, both the actual force distribution and its equivalent should not be shown in the same FBD. However, sometimes for convenience, both are shown and for clarity either the resultant or the actual distribution is shown with dashed arrows.

Implicit in considering a system in isolation is that the actions of the surroundings can be obtained a priori by suitable modelling of the influence of the surroundings on the system. A posteriori, a larger system can be considered which includes the local environment to ascertain whether the supports would be able to absorb the reactions obtained and whether the assumed actions of the surroundings are correct. For this larger system we would have even larger surroundings which would need to be included in our system in the next stage and so on. For example, the analysis of a dam is a complex problem of structure, foundation, water and soil interaction including the effect of waves due to earthquakes whose intensity and frequency could be affected by the huge water reservoir and the dam. We need to consider the tectonic plates involved to address the problem of earthquakes. For the analysis of the dam, the actions of its surroundings are suitably modelled. The accuracy of a model can only be 'checked' a posteriori by actually measuring the response of the surrounding systems.

The loads exerted by the supports on the system are called *reactions*. The nature of the load system that body B_2 can exert on body B_1 through some connector or support depends on the type of constraint on their relative motion provided by it. It exerts a force component for each of the relative displacement components constrained by it, and a moment component for each of the relative rotation components constrained by it. To find out which relative displacement or rotational components are constrained, we mentally try to relatively translate or rotate the bodies in three convenient orthogonal directions. While drawing an FBD, *no attempt should be made to decide the direction of a constraint force/moment component, except for the friction force for impending slip/actual slip.* We consider several types of supports.

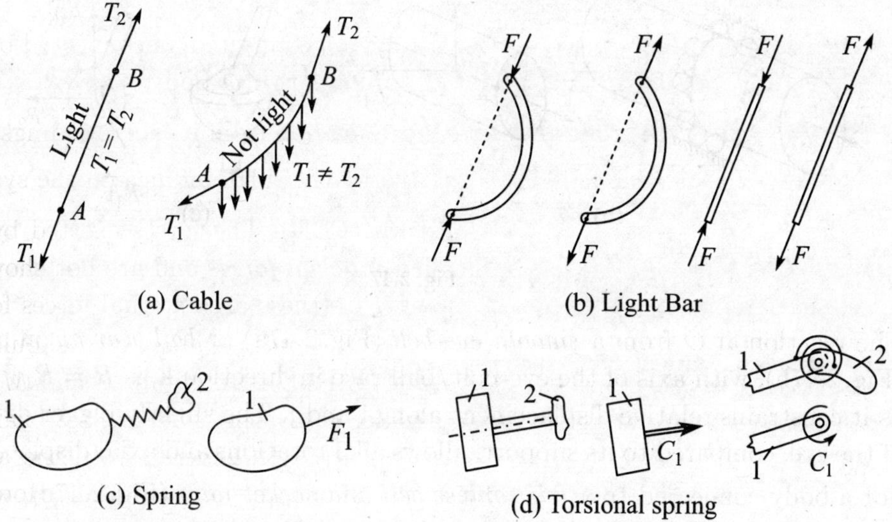

Fig. 2.46

1. A *flexible cable* (belt, chain, rope, string) is idealised as a perfectly flexible one-dimensional body with the internal force resultant being a tensile axial force if it is taut and a zero force if it is slack. At a point of fastening to a body, it provides a one-sided displacement constraint as it can only apply a pull in its axial direction and not a push (Fig. 2.46a). A *'light' cable* subjected to only two forces (pulls) at its ends is a two-force member. Hence, the end forces must be along the line joining the ends, the cable remaining straight. The cable is often modelled as *inextensible.* The forces in inextensible cables can change instantaneously.
2. A *light bar* subjected to only two forces at its points of connection is a two-force member and provides a two-sided displacement constraint at the points of connection along the line joining them, since it can apply a force F (as a pull or

as a push) in this direction (Fig. 2.46b).

3. For a body connected to an *ideal (inertia-less) spring* under extension e, a force $F_1(e)$ is exerted on the body along the axis of the spring (Fig. 2.46c).
4. For a body connected to an *ideal (inertia-less) torsional spring* under twist θ, a moment $C_1(\theta)$ is exerted on the body along the axis of the spring (Fig. 2.46d).

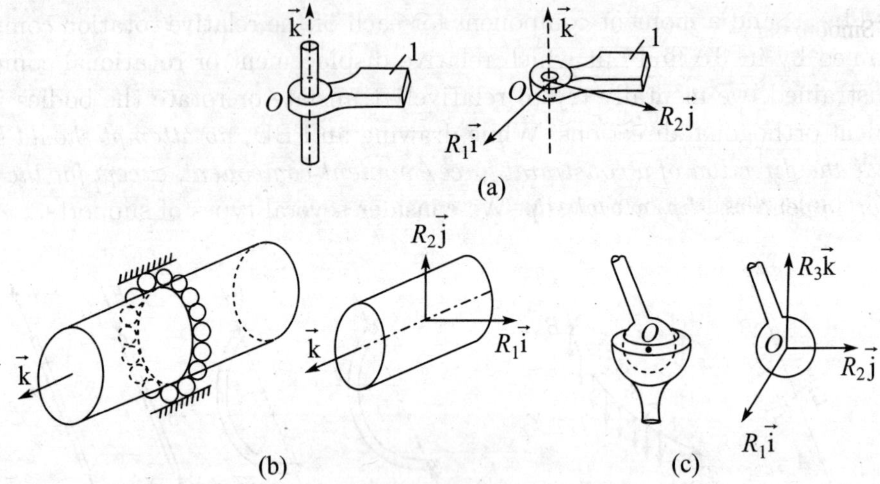

Fig. 2.47

5. The reaction at O from a *smooth eye-bolt* (Fig. 2.47a) or *ball bearing* on a body (Fig. 2.47b), with axis of the eye-bolt/ball race in direction $\vec{k}$, is $\vec{R} = R_1\,\vec{i} + R_2\,\vec{j}$, as it constrains relative displacement along $\vec{i}$ and $\vec{j}$. The slightly bigger diameter of the eye, compared to its support, allows all 3 rotations and axial displacement.
6. For a body connected to a *frictionless ball and socket joint* (Fig. 2.47c), relative rotation about all axes at O is allowed, but relative displacement at O is fully prevented. Hence, the reaction at O is $\vec{R} = R_1\,\vec{i} + R_2\,\vec{j} + R_3\,\vec{k}$.
7. In a *frictionless clevis joint* (Fig. 2.48a), the relative displacement along $\vec{k}$ and relative rotation about $\vec{j}$ and $\vec{k}$ are allowed. The reaction is $\vec{R} = R_1\,\vec{i} + R_2\,\vec{j}$ and $\vec{C} = C_1\,\vec{i}$. A similar reaction is exerted by the *rail on a flanged wheel* (Fig. 2.48b).
8. The reaction at O (Fig. 2.48c) from a *smooth rod on a collar*, with the axis of the rod in direction $\vec{k}$, is $\vec{R} = R_1\vec{i} + R_2\vec{j}$ and $\vec{C} = C_1\vec{i} + C_2\vec{j}$, since relative displacement along the rod axis $\vec{k}$ and relative rotation about it are allowed. A similar reaction is exerted by a *smooth roller bearing on a shaft* (Fig. 2.48d).
9. In a *frictionless pin and bracket* or a *frictionless hinge joint* or a *frictionless thrust bearing* with axis $\vec{k}$, relative displacement is constrained but relative rotation about axis $\vec{k}$ is allowed (Fig. 2.49a). The reaction at O is $\vec{R} = R_1\,\vec{i} + R_2\,\vec{j} + R_3\,\vec{k}$ and $\vec{C} = C_1\,\vec{i} + C_2\,\vec{j}$.

If the relative axial displacement of the pin is not constrained, then $R_3 = 0$, $\vec{R} = R_1\vec{i} + R_2\vec{j}$, $\vec{C} = C_1\vec{i} + C_2\vec{j}$ (Fig. 2.49b). This joint is same as a collar joint.

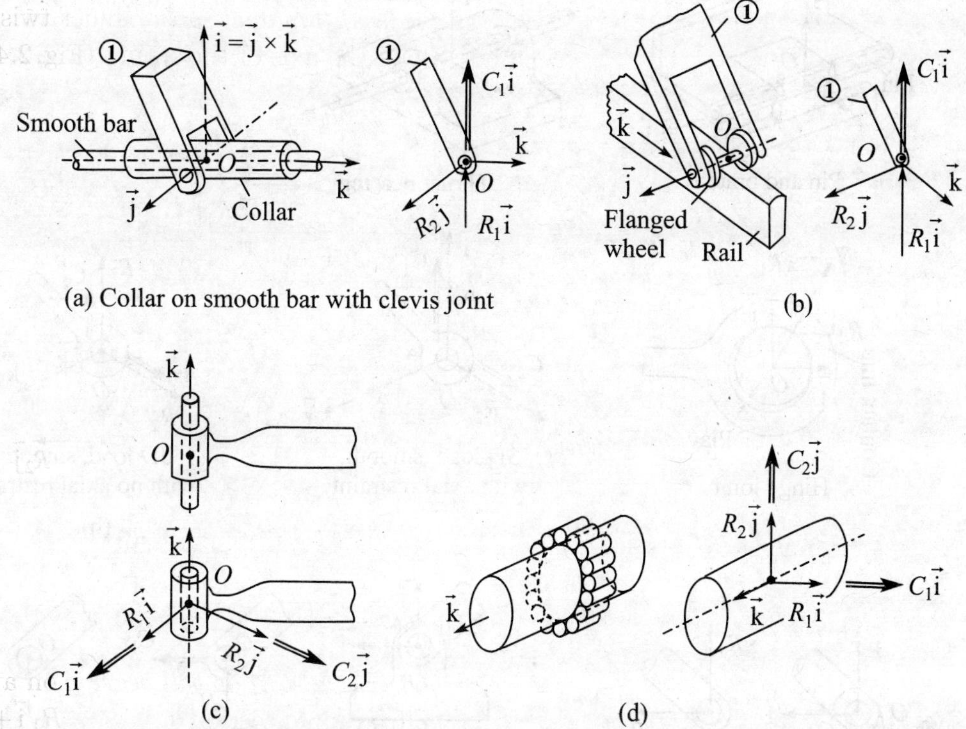

Fig. 2.48

If there are only coplanar loads in the mid-plane (xy-plane) of the joint, then $R_3\vec{k}$ and $C_1\vec{i} + C_2\vec{j}$ may not be of interest in the governing equations. Hence, for 2D loads, the relevant reaction is $\vec{R} = R_1\vec{i} + R_2\vec{j}$ (Fig. 2.49c).

For frictionless pin joints with a short pin, C_1, C_2, being of the order of force reaction times length of the pin, are small. Hence, only force reactions are included in the analysis unless it is found that C_1, C_2 are required to satisfy the equations of motion. For example, *the couple reactions are neglected for a shaft with two hinge joints, but are included for a shaft with only one hinge support.*

10. Consider a *smooth pin joint which itself is supported on smooth light spherical balls ('rollers')* on a plane (Fig. 2.49d) with normal $\vec{j}$. The pin support is free to move in directions $\vec{i}$ and $\vec{k}$, while the pin itself is free to rotate about the z axis relative to its support. Each ball is subjected to only normal reaction from the plane, as it is a two-force light member. The simplest resultant of these normal reactions is $N\vec{j}$ which has a moment $C_1\vec{i} + C_3\vec{k}$ about O. Reaction at O

is $\vec{F} = N\vec{\mathrm{j}}$ and $\vec{C} = C_1\vec{\mathrm{i}}$ since $C_3 = 0$ as pin allows free rotation about z axis.

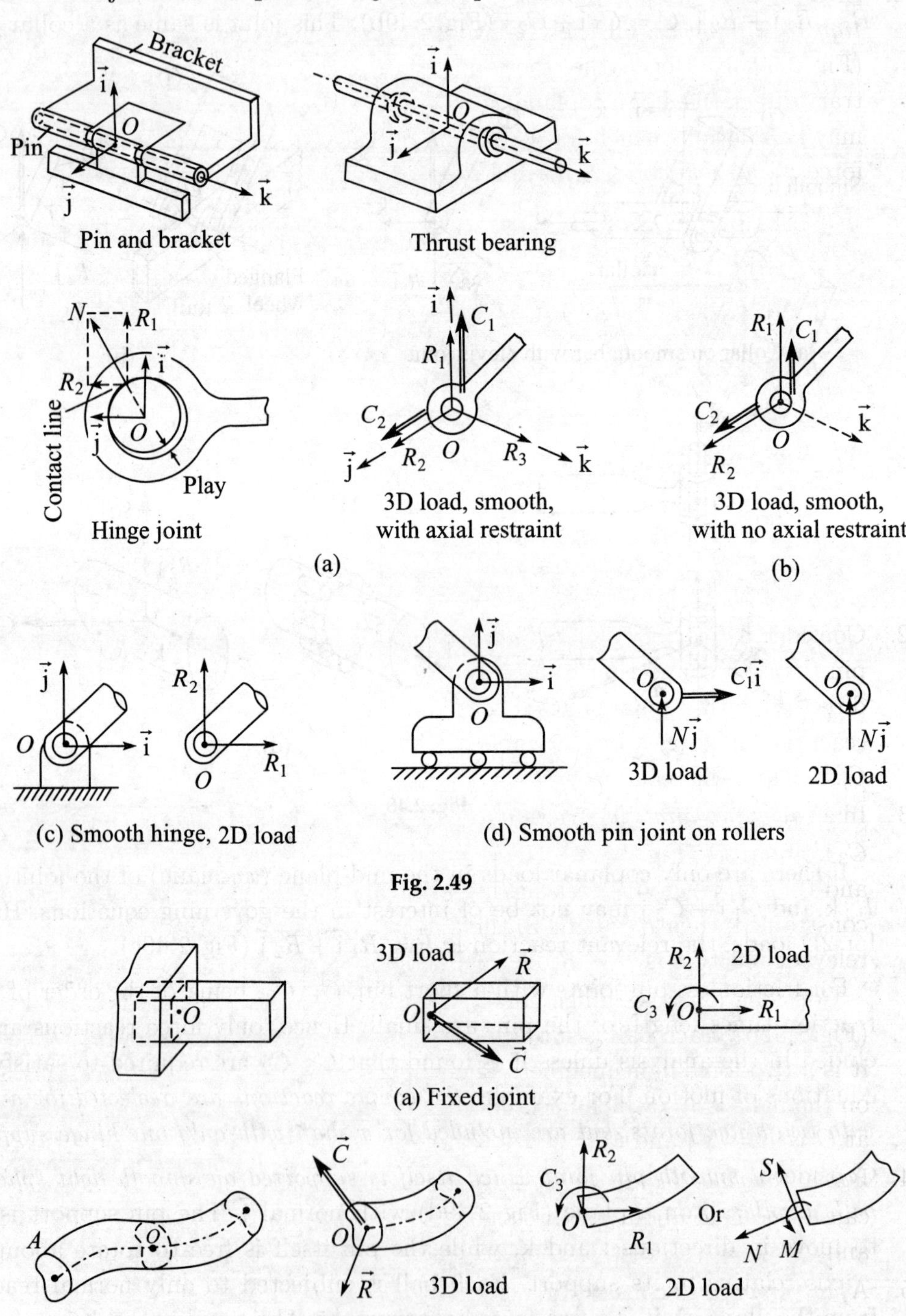

(c) Smooth hinge, 2D load

(d) Smooth pin joint on rollers

Fig. 2.49

(a) Fixed joint

(b) Internal force resultant in a bar

Fig. 2.50

11. In a *welded, 'fixed', 'built-in'* or *'encastre'* joint or a transverse *internal section of a thin bar*, all the relative displacements and rotations are constrained (Fig. 2.50a,b). Hence, the reaction is a force $\vec{R}$ and a moment $\vec{C}$ at O having arbitrary directions. For a coplanar force system acting on the body, only R_1, R_2, C_3, may be relevant, which for a straight/curved bar are usually shown as a normal force N, a shear force S normal to the axis and a bending moment M.

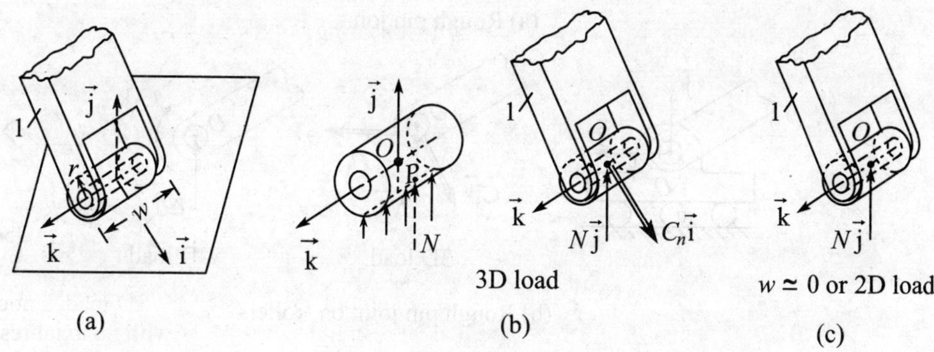

Fig. 2.51

12. Consider a *light circular cylindrical roller on a smooth surface* with a smooth pin (axis $\vec{k}$) connection to a body (Fig. 2.51a). It makes a line contact in the $\vec{k}$ direction with a plane normal to $\vec{j}$. Normal reaction on the pin is $N\vec{j}$ at $P \Rightarrow$ reaction at O is a force $N\vec{j}$ and a couple $C_n\vec{i}$ (Fig. 2.51b). If w is small, then $C_n \simeq 0$ and the reaction is $N\vec{j}$ (Fig. 2.51c). For 2D loads also, the reaction is $N\vec{j}$.
13. In a *rough pin joint*, the friction force $F\vec{e}_\phi$ at the line of contact causes a moment $C_3\vec{k}$ about O (Fig. 2.52a). Hence, the reaction at O is $\vec{R} = R_1\vec{i} + R_2\vec{j} + R_3\vec{k}$ and $\vec{C} = C_1\vec{i} + C_2\vec{j} + C_3\vec{k}$ having arbitrary directions. If the pin is not axially constrained, then R_3 is due to the axial friction force. For a 2D load system, the relevant reaction is $\vec{R} = R_1\vec{i} + R_2\vec{j}$ and $\vec{C} = C_3\vec{k}$.

    Consider a rough hinge joint supported on light spherical balls ('rollers') (Fig. 2.52b). Each ball is subjected to only normal reaction from the plane, since it is a two-force light member. The simplest resultant of these normal reactions on the balls is $N\vec{j}$ which has a moment $C_1\vec{i} + C_3\vec{k}$ about O. Hence, the reaction at O is $\vec{R} = N\vec{j}$ and $\vec{C} = C_1\vec{i} + C_3\vec{k}$. For 2D loads, it reduces to $N\vec{j}$ and $C_3\vec{k}$.
14. For a *rough ball and socket joint* (Fig. 2.52c), the friction force $F_\theta\vec{e}_\theta + F_\phi\vec{e}_\phi$ causes moment about all three axes. The reaction at O is $\vec{R} = R_1\vec{i} + R_2\vec{j} + R_3\vec{k}$ and $\vec{C} = C_1\vec{i} + C_2\vec{j} + C_3\vec{k}$. For 2D loads, it reduces to $\vec{R} = R_1\vec{i} + R_2\vec{j}$, $\vec{C} = C_3\vec{k}$.
15. A *universal joint* consists of a cross 3 whose two arms are pinned to shafts 1 and 2 (Fig. 2.53). Cross 3 can rotate relative to 1 about the $\vec{j}$ axis alone and 2 can rotate relative to 3 about the $\vec{i}$ axis alone. The point O of the cross is on the

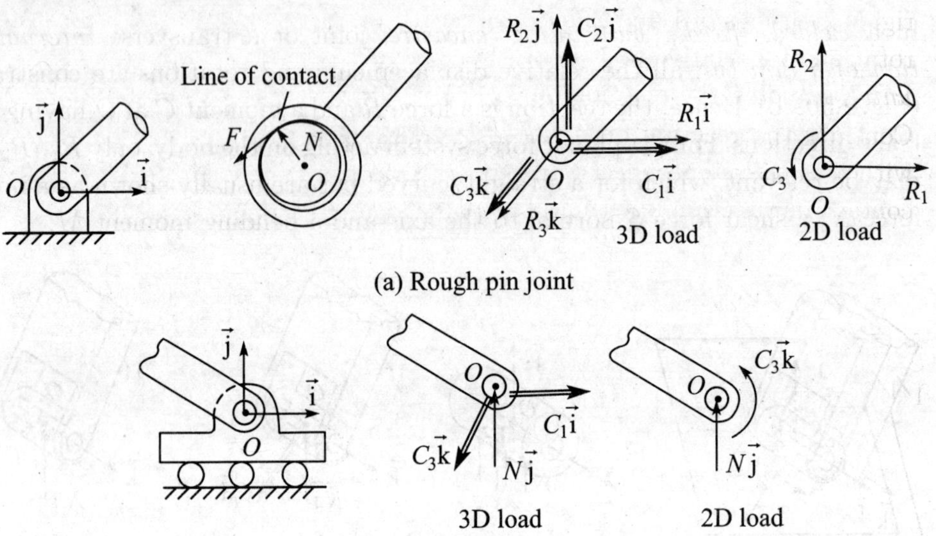

(a) Rough pin joint

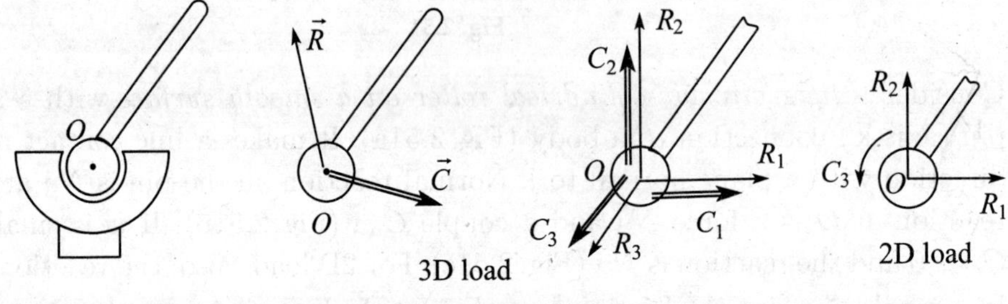

(b) Rough pin joint on 'rollers'

(c) Rough ball and socket joint

Fig. 2.52

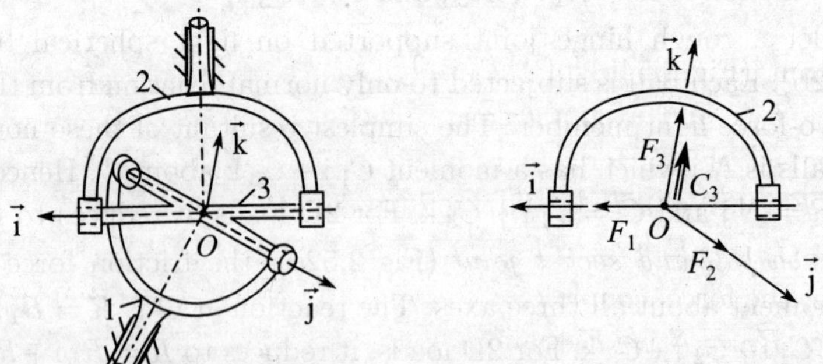

Fig. 2.53

rigid extensions of 1 and 2. It constrains translation of 2 relative to 1, but allows rotation of 2 relative to 1 along the axes $\vec{\mathrm{i}}$ and $\vec{\mathrm{j}}$ of the cross. For a *frictionless universal joint*, the reaction on 2 at O is $\vec{R} = F_1\vec{\mathrm{i}} + F_2\vec{\mathrm{j}} + F_3\vec{\mathrm{k}}$ and $\vec{C} = C_3\vec{\mathrm{k}}$.

16. Consider the reaction on a *light circular cylindrical roller on a rough surface* with a *smooth pin* connection to a body with axis $\vec{\mathrm{k}}$ (Fig. 2.54a). It makes a line contact in direction $\vec{\mathrm{k}}$ with a plane normal to $\vec{\mathrm{j}}$.

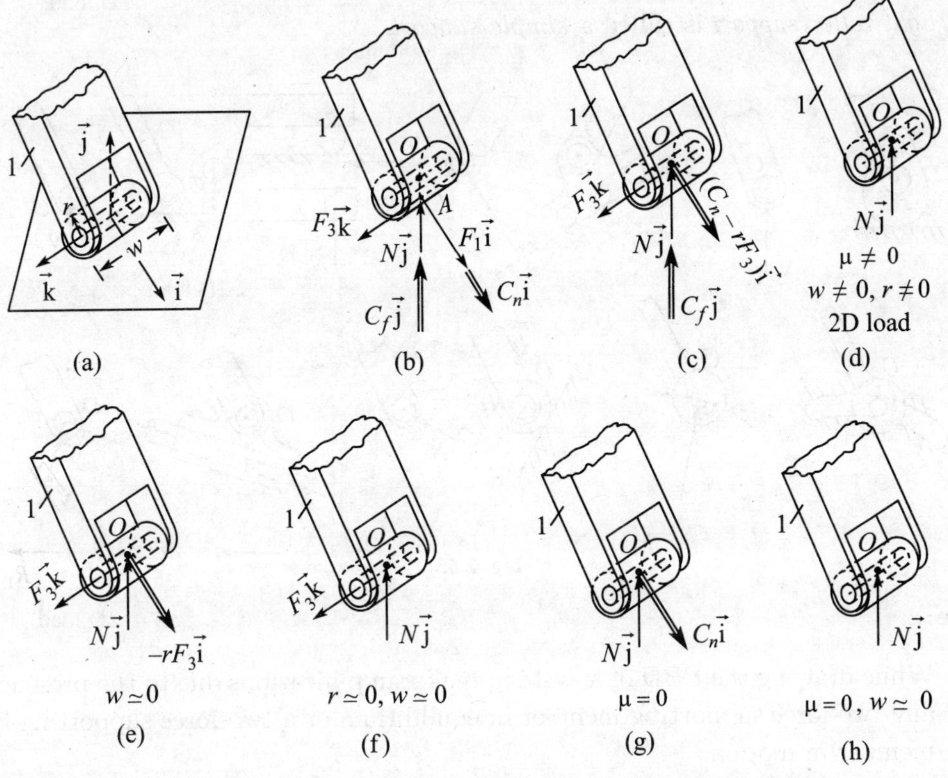

Fig. 2.54

The resultant (Fig. 2.54b) of the coplanar friction forces (in $\vec{\mathrm{i}}$-$\vec{\mathrm{k}}$ plane) at A is a force $F_1\vec{\mathrm{i}} + F_3\vec{\mathrm{k}}$ and a couple $C_f\vec{\mathrm{j}}$. The resultant of the coplanar normal force (in $\vec{\mathrm{j}}$-$\vec{\mathrm{k}}$ plane) at A is a force $N\vec{\mathrm{j}}$ and a couple $C_n\vec{\mathrm{i}}$. The force–couple reaction at O is $\vec{R}_O = F_1\vec{\mathrm{i}} + N\vec{\mathrm{j}} + F_3\vec{\mathrm{k}}$ and $\vec{C}_O = C_n\vec{\mathrm{i}} + C_f\vec{\mathrm{j}} - r\vec{\mathrm{j}} \times (F_1\vec{\mathrm{i}} + N\vec{\mathrm{j}} + F_3\vec{\mathrm{k}}) = (C_n - rF_3)\vec{\mathrm{i}} + C_f\vec{\mathrm{j}} + rF_1\vec{\mathrm{k}} \Rightarrow rF_1 = 0$, since the roller can rotate freely about the pin axis $\vec{\mathrm{k}}$. Hence, the force–couple reaction of the roller support on the body (Fig. 2.54c) at O is a force $\vec{R}_O = N\vec{\mathrm{j}} + F_3\vec{\mathrm{k}}$ and a couple $\vec{C}_O = (C_n - rF_3)\vec{\mathrm{i}} + C_f\vec{\mathrm{j}}$.

For 2D load system in xy plane, the relevant reaction at O is a force $\vec{R}_O = N\vec{\mathrm{j}}$ alone (Fig. 2.54d), irrespective of the size (r, w) of the roller and its coefficient of friction μ.

If the width w of the roller is small, then $C_f \simeq 0$, $C_n \simeq 0$ and the reaction at O

is $\vec{R}_O = N\,\vec{\mathrm{j}} + F_3\,\vec{\mathrm{k}}$ and $\vec{C}_O = -rF_3\,\vec{\mathrm{i}}$ (Fig. 2.54e). If the radius r of the roller is also small, it reduces to only a force reaction $\vec{R}_O = N\,\vec{\mathrm{j}} + F_3\,\vec{\mathrm{k}}$ at O (Fig. 2.54f).

If the roller is smooth, then the reaction at O is $N\,\vec{\mathrm{j}}$ and $C_n\,\vec{\mathrm{i}}$ (Fig. 2.54g). If the width of the smooth roller is small, then $C_n \simeq 0$ and the reaction is $N\,\vec{\mathrm{j}}$ (Fig. 2.54h).

For coplanar loads, for all the joints shown in Fig. 2.55, the relevant reaction is just a normal force N with no couple, irrespective of the size of the smooth or rough roller.

A *light roller support* is called a *simple support*.

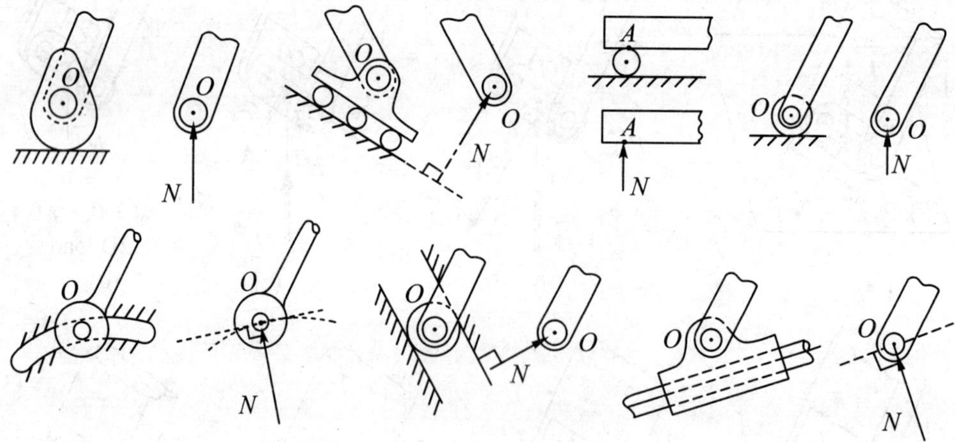

Fig. 2.55

Note:

1. While drawing the FBD of a system, make simplifications due to the presence of any two-force supporting member in equilibrium or a two-force supporting light member in motion.
2. Once support reactions on body B_1 due to body B_2 have been shown in the FBD of B_1, then in the FBD of B_2 show the same support reactions but with opposite sense. If these reactions are $\vec{R}$, $\vec{C}$ on B_1 then show $-\vec{R}$, $-\vec{C}$ on B_2.
3. Most often, the FBD of the pulley is not useful. It is more useful to draw the FBD of the system of the pulley and a part of the belt which overlaps it, since the internal contact force between the pulley and the belt does not appear in it.
4. If more than two members meet at a hinge joint, then either the FBD of the pin is drawn separately, or preferably, the FBDs of the members are drawn with the pin forming part of one of the members along with the load, if any, acting on it.
5. Consider FBDs of wheels of a vehicle, in which their inertia is included. One equation of motion is $M_{Cz} = I_{zz}^{C}\dot{\omega}$. Hence, the FBDs for the un-powered and un-braked wheels, shown in Fig. 2.56a, include a friction force to ensure angular

acceleration. For the case of no slip, the direction of friction force F follows from the equation $M_{Cz} = I_{zz}^{C}\dot{\omega}$, though it could be drawn in either direction.

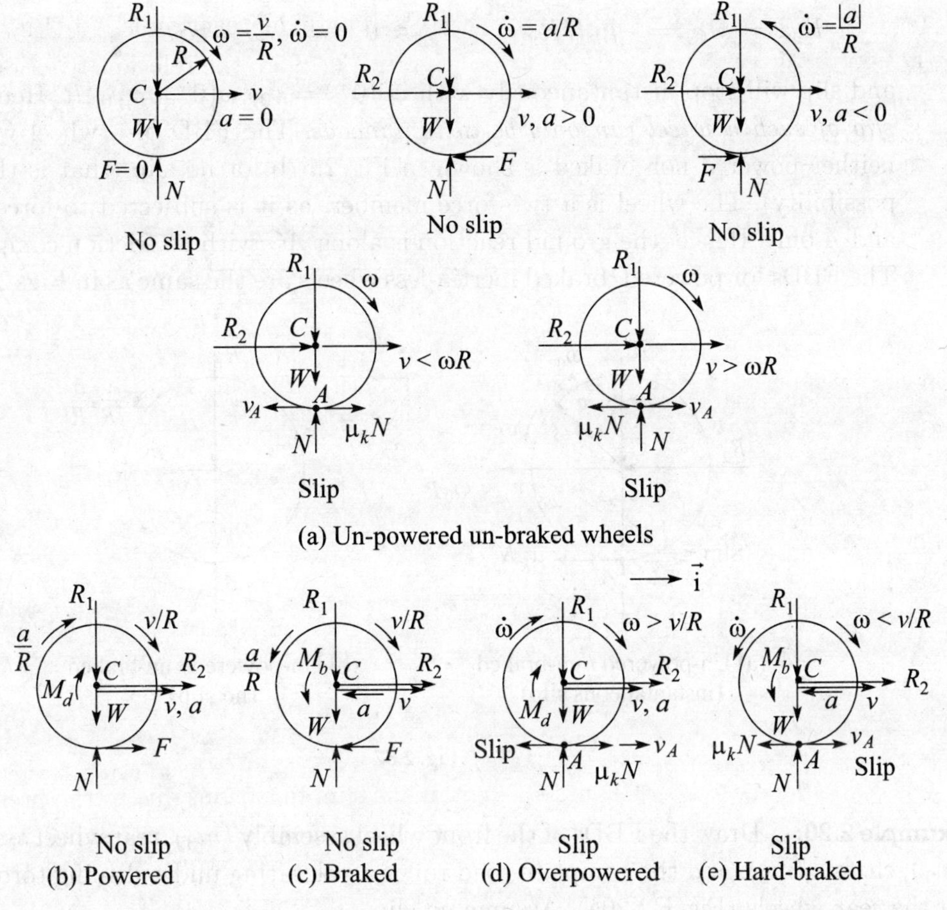

Fig. 2.56

The FBDs for the case of powered and braked wheels rolling without slip are shown in Figs 2.56b,c. F could be drawn in either direction since there is no slip. The corresponding FBDs for the case of over-powered and hard-braked wheels for which slip occurs are shown in Figs 2.56d,e. For the over-powered case, greater driving moment $\Rightarrow \dot{\omega} > a/R \Rightarrow \omega > v/R$, $v_A = v - \omega R < 0$, i.e., A slips in $-\vec{i}$ direction and $F = \mu_k N$ is in $\vec{i}$ direction. For the hard-braked case, since a and $\dot{\omega}$ are negative, and greater braking torque implies that $|\dot{\omega}| > |a/R| \Rightarrow \omega < v/R$, $v_A = v - \omega R > 0$, i.e., A slips in $\vec{i}$ direction and $F = \mu_k N$ is in $-\vec{i}$ direction.

6. Consider FBDs of wheels of a vehicle in which their inertia is neglected. First

consider the case of an un-powered, un-braked wheel (Fig. 2.57a) with $v_C(0) = v_0$, $\omega(0) = \omega_0$, $v_0 < \omega_0 R$. $\vec{v}_A(0) = -(\omega_0 R - v_0)\,\vec{\imath} \Rightarrow$ initial slip of A is in the $-\vec{\imath}$ direction. The FBD is shown in Fig. 2.57a.

$$I_{zz}^C \dot{\omega} = M_{Cz} = -\mu_k N R, \qquad I_{zz}^C \approx 0 \quad \Rightarrow \quad \dot{\omega} = \infty$$

and slip will stop instantaneously with $v_C(0^+) = v_0$, $\omega(0^+) = v_0/R$. Hence, the *slip of such a wheel can only be instantaneous.* The FBD of a wheel which is neither powered nor braked is shown in Fig. 2.57b for no slip (that is the only possibility). The wheel is a two-force member, as it is subjected to forces at C and A only. Hence, the ground reaction is along AC with no friction component. The FBDs for powered/braked inertia-less wheels are the same as in Figs 2.56b-e.

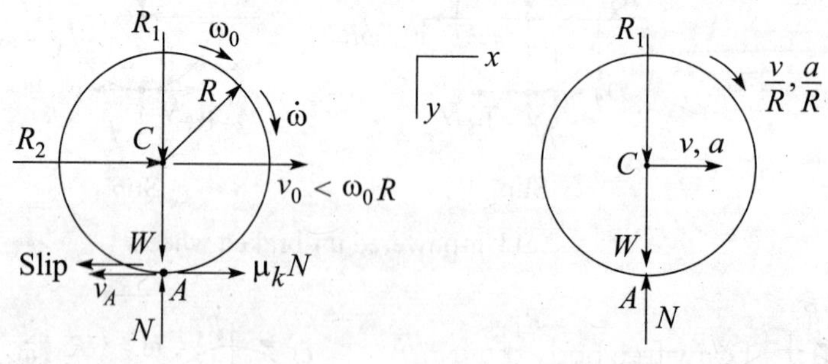

(a) Un-powered / un-braked (instantaneous slip)

(b) Un-powered / un-braked (no slip)

Fig. 2.57

Example 2.20: Draw the FBDs of the front wheel assembly (m_3), rear wheel assembly (m_1), chassis (m_2), and the complete road roller accelerating under driving torque M_0 on the rear wheels (Fig. E2.20a). Assume no slip.

Solution: The friction force F_2 on the front wheels (Fig. E2.20b) is in the backward direction to provide a moment about C_3 in the direction of its angular acceleration a/r. The reaction from the axle on it has a component R_3 in the forward direction to provide forward acceleration a (since F_2 is backward) and a component R_4 downward due to the weight of the chassis. In the FBD of the complete vehicle (Fig. E2.20c), the friction force F_1 on the rear wheels is in the forward direction to provide forward acceleration to the vehicle (since F_2 is backward). The driving torque, being internal to the system, does not appear in this FBD. In the FBD of the chassis (Fig. E2.20d), the reaction R_1 from the axle is in the forward direction to provide it a forward acceleration (since R_3 is in the backward direction) and the reaction R_2 is upwards to support its weight. The FBD of the rear wheel assembly (Fig. E2.20e) is now completed with the driving moment M_0 on the rear axle shown in the direction of its angular acceleration, and an opposite moment is shown as acting on the chassis (Fig. E2.20d).

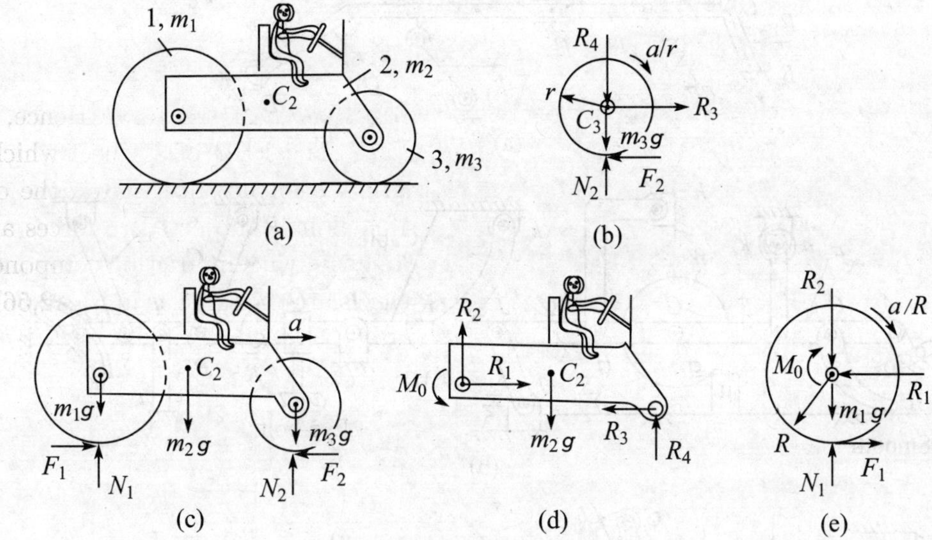

Fig. E2.20

Example 2.21: Draw free body diagrams of the following bodies for the systems given in Fig. E2.21a–d: (a) $ABOD$, $ABOE$, (b) AB, AE, (c) 1, 2, 3, $4+5$, 4, 5, AB, $1+2+3+4+5$, (d) 1, 2, 3. All contact surfaces are smooth. The belt and the cables are light. The members, for which no mass is shown, are light. Make simplifications, if any, due to the presence of supports which are two-force members. The loading on the systems in Fig. E2.21b–d is coplanar.

Solution: (a) The FBDs are shown in Fig. E2.21e. All relative rotations are allowed at an eye-bolt, but the relative displacements normal to the axis of the bolt are prevented. Hence, the eye-bolt exerts two components of force in the two directions normal to the axis of its eye. The supporting members PQ and RS, being two-force members, exert forces along PQ and RS only. However, the member TU exerts 3 force components, since it is not a two-force member. At the support B in the first FBD and at the internal section at E in the second FBD, three force components and three couple components are shown, because there is complete constraint on the relative rotation and relative displacement and the applied force system is three-dimensional. The reaction at the ball and socket joint D has three force components as complete constraint on translation occurs, but there is no rotational constraint. The hinge at H with axis y provides a force reaction $R_{14}\vec{\mathrm{i}} + R_{15}\vec{\mathrm{k}}$ and a moment reaction $C_4\vec{\mathrm{i}} + C_5\vec{\mathrm{k}}$, since it permits axial displacement and prevents rotation about $\vec{\mathrm{k}}$ and $\vec{\mathrm{i}}$ only.

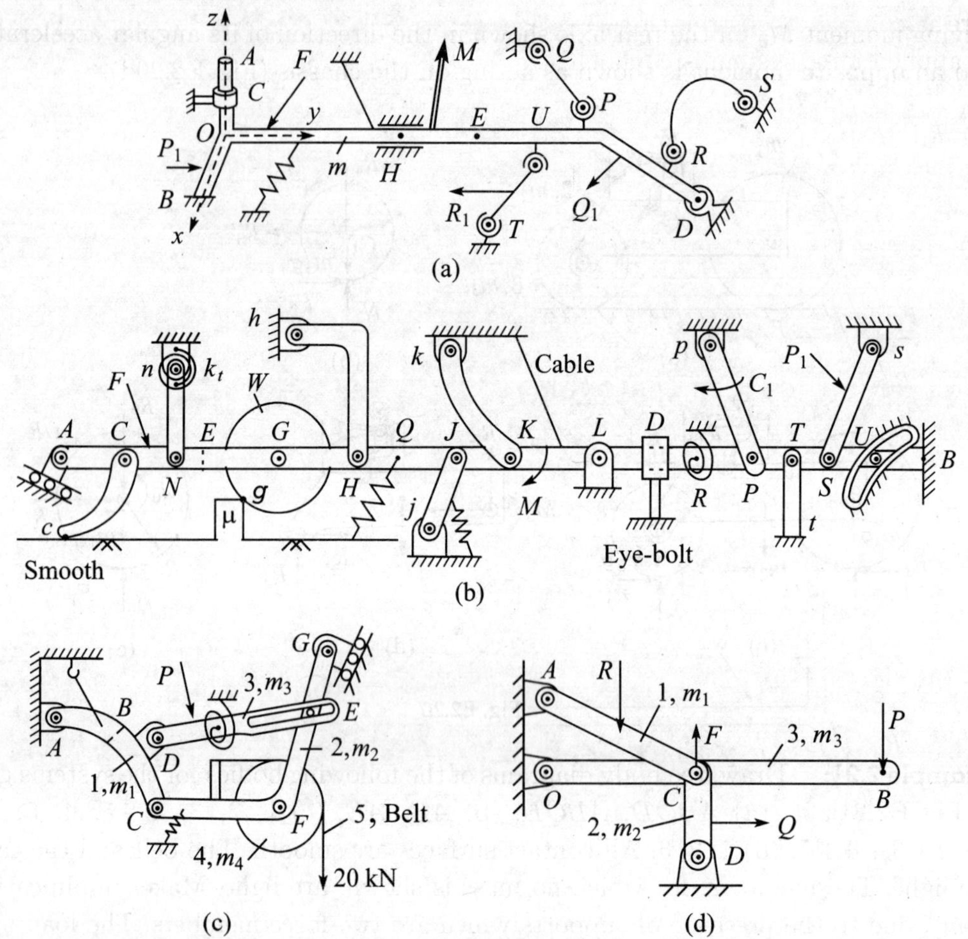

Fig. E2.21a-d

(b) The FBDs are depicted in Fig. E2.21f. The supporting members Kk and Hh, being two-force members, exert forces along Kk and Hh only. Member Cc is a two-force member, and the support reaction $N\vec{e}_n$ must be zero as it is not in the direction Cc. Thus, Cc does not apply any reaction at C on member AB. The member Gg is subjected to a force at G (pin reactions F_1, F_2 and weight W) and a force at g (normal and friction forces). Hence, it is a two-force member and reaction R_3 at g is along Gg only $\Rightarrow$ $\vec{W} + \vec{R}_3 - F_1\vec{\imath} - F_2\vec{\jmath} = \vec{0}$ $\Rightarrow$ $F_1\vec{\imath} + F_2\vec{\jmath} = \vec{W} + \vec{R}_3$. The members Ss, Tt, Pp, Jj, Nn are not two-force members since an additional force/couple acts on them in between or a couple also acts at one end. Each of them provides a force reaction on AB with two non-zero components in general, since the loading is coplanar. At the support B in the first FBD and at the internal section at E in the second FBD, two in-plane force components and one out-of-plane couple component are shown, because there is complete constraint on the relative rotation and relative displacement, and the applied force system is coplanar.

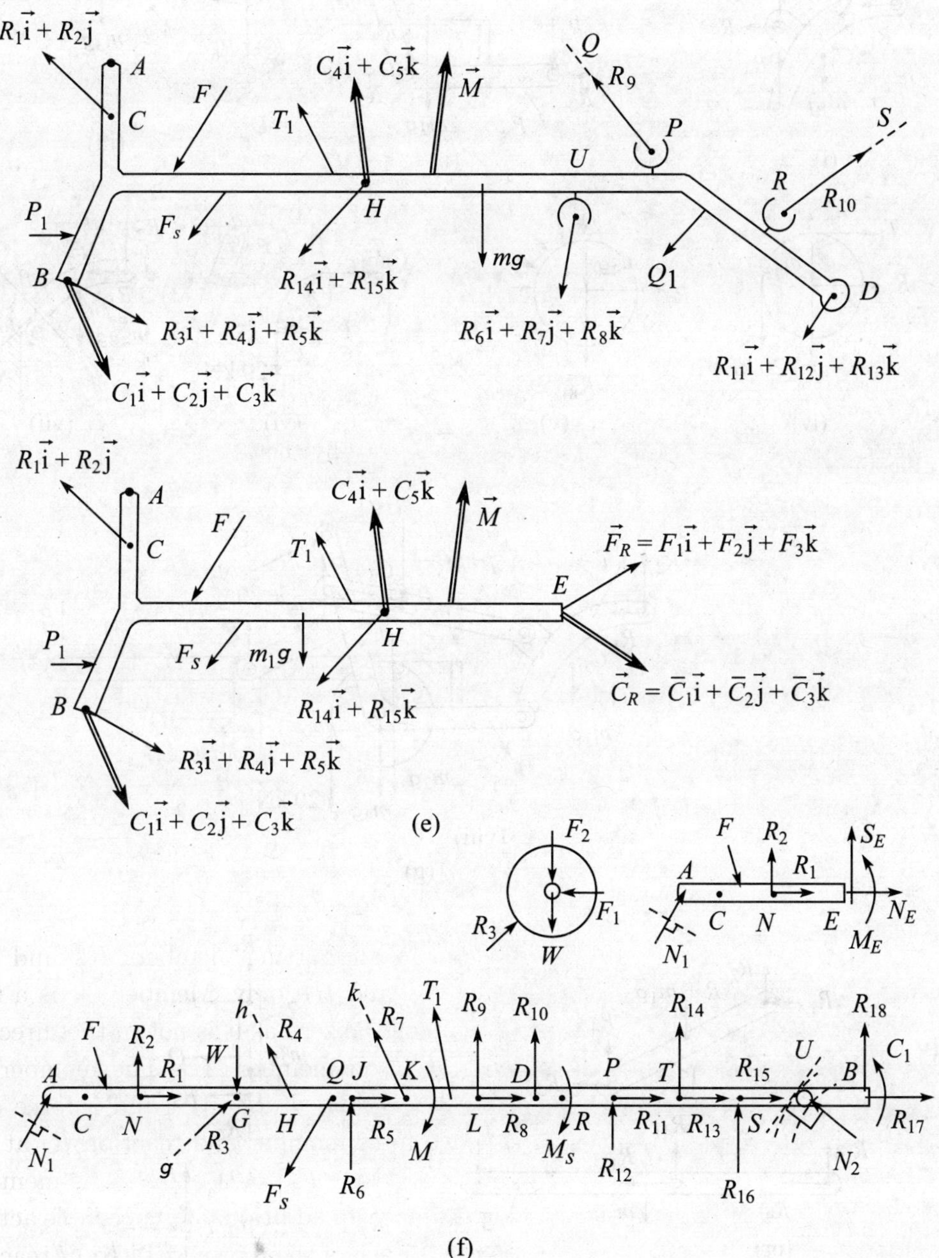

Fig. E2.21e,f

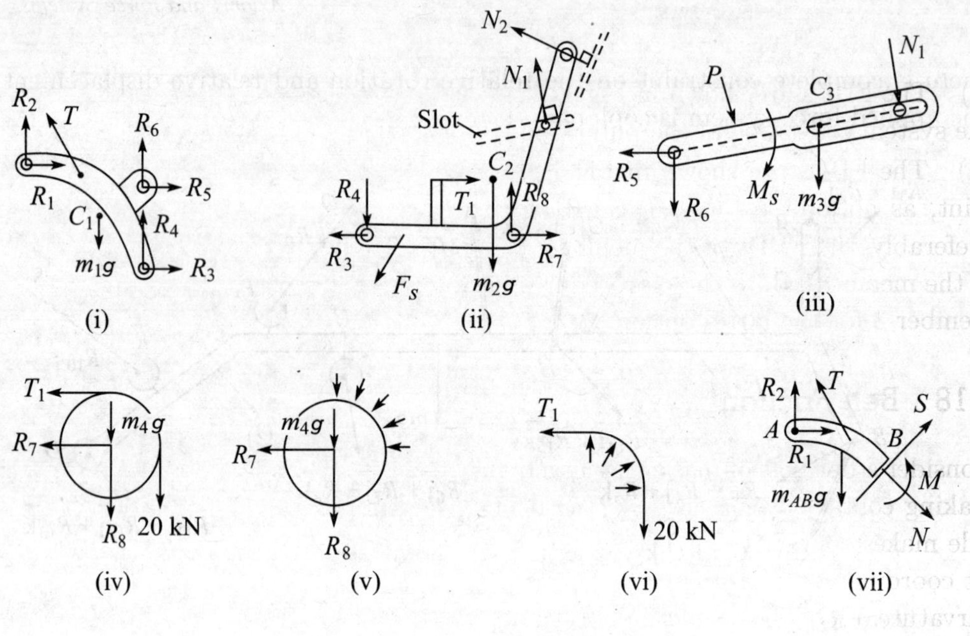

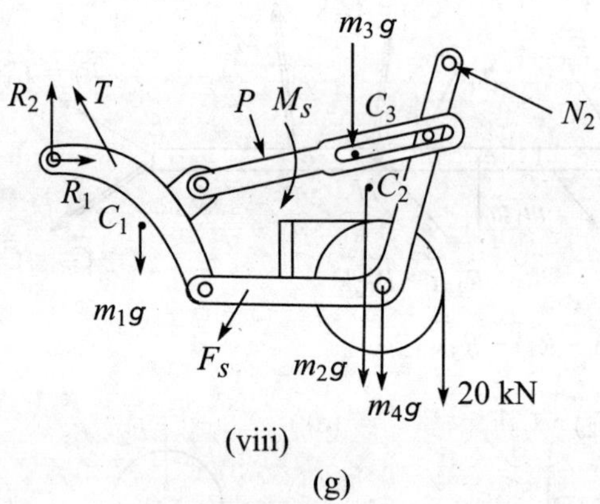

(g)

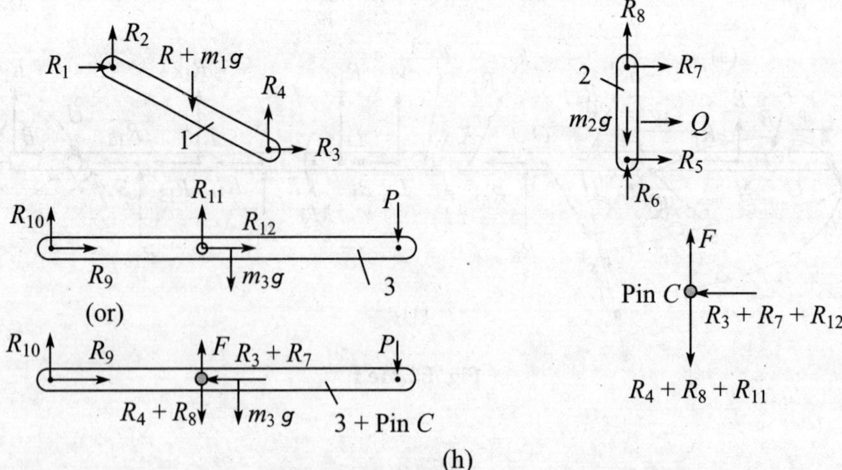

(h)

Fig. E2.21g,h

(c) The FBDs are shown in Fig. E2.21g. The FBD of the pulley has been drawn for the system consisting of the pulley and the part of the belt which overlaps it.

(d) The FBDs are shown in Fig. E2.21h. If more than 2 members meet at a hinge joint, as at joint C, then the FBD of the pin can either be drawn separately, or preferably, the FBDs of the members can be drawn, with the pin forming part of one of the members. Both these procedures have been illustrated with pin C included with member 3 for the latter option.

2.18 Belt Friction

Consider a flat belt of mass λ kg/m moving at speed v on a *fixed surface* S (Fig. 2.58a) making contact over a curve C (need not be a plane curve) from A where the slack side makes contact to B. Let T_1 and T_2 be the tensions on the tight and slack sides. At coordinate s measured from A, let $T(s)$ be the tension and $\rho(s)$ be the radius of curvature of C. The coefficients of friction are μ_k and μ_s. At location s, let the normal and frictional forces be p N/m and $\mu_k p$ N/m. In engineering applications, the tension in the belt is much larger than the weight of the belt. Hence, neglecting gravity, the FBD of an element of length Δs is shown in Fig. 2.58b with $\vec{e}_N$ being the outward normal to surface S at s. The equation of motion of this element is:

$$(T\,\vec{e}_t)|_{s+\Delta s} - (T\,\vec{e}_t)|_s + (p\,\vec{e}_N)_{\text{av}}\Delta s - (\mu_k p\,\vec{e}_t)_{\text{av}}\Delta s = \lambda\,\Delta s\Big(\frac{v^2}{\rho}\vec{e}_n\Big)_{\text{av}}$$

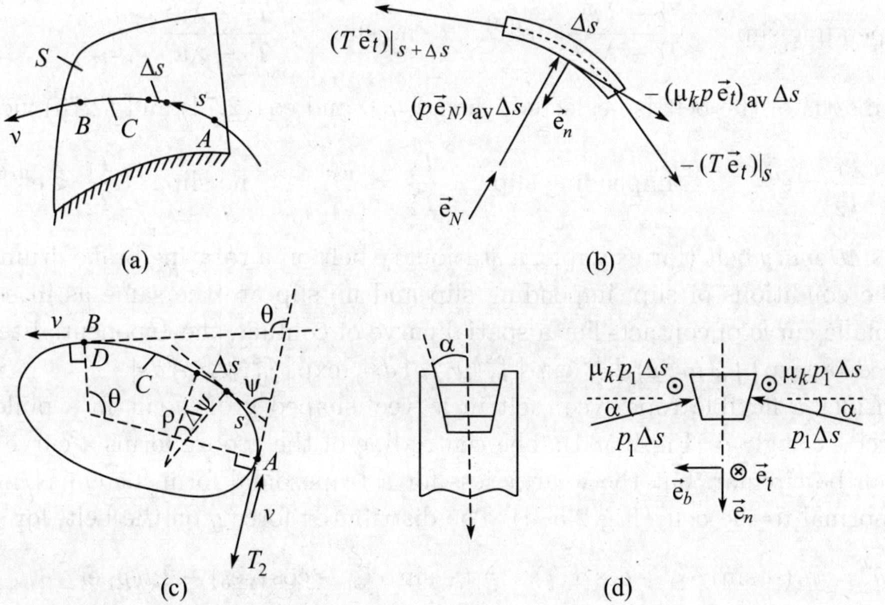

(a) (b) (c) (d)

Fig. 2.58

Divide by Δs, take limit as $\Delta s \to 0$ and use $d\vec{e}_t/ds = \vec{e}_n/\rho$, $\vec{e}_t \perp \vec{e}_N$:

$$\frac{d}{ds}(T\vec{e}_t) + p\,\vec{e}_N - \mu_k p\,\vec{e}_t = \frac{\lambda v^2}{\rho}\vec{e}_n \Rightarrow$$

$$\frac{dT}{ds}\vec{e}_t + \frac{T}{\rho}\vec{e}_n + p\,\vec{e}_N - \mu_k p\,\vec{e}_t = \frac{\lambda v^2}{\rho}\vec{e}_n$$

$$\vec{e}_n : \quad p = \frac{T - \lambda v^2}{\rho} \qquad \text{and} \qquad \vec{e}_N = -\vec{e}_n$$

$$\vec{e}_t : \quad \frac{dT}{ds} = \mu_k p = \mu_k \frac{T - \lambda v^2}{\rho} \qquad \Rightarrow \qquad \int_{T_2}^{T} \frac{1}{T - \lambda v^2}\, dT = \int_0^s \frac{\mu_k}{\rho}\, ds$$

$$\frac{T - \lambda v^2}{T_2 - \lambda v^2} = \exp\left[\int_0^s \frac{\mu_k}{\rho}\, ds\right], \quad p = \frac{T_2 - \lambda v^2}{\rho}\exp\left[\int_0^s \frac{\mu_k}{\rho}\, ds\right] \qquad (2.73)$$

For a plane curve C shown in Fig. 2.58c, the contact is over an arc AB, the belt changes its direction by angle θ, $ds/\rho = d\psi$ and eq. (2.73) reduces to

$$\frac{T - \lambda v^2}{T_2 - \lambda v^2} = e^{\mu_k \psi}, \quad p = \frac{T_2 - \lambda v^2}{\rho} e^{\mu_k \psi}, \quad \frac{T_1 - \lambda v^2}{T_2 - \lambda v^2} = e^{\mu_k \theta} \qquad (2.74)$$

where ψ and θ are in radians. $T_{\max}\,(= T_1)$ occurs at B, where the tension is in the direction of $\vec{v}_{BD}$. If the belt is on a *rotating pulley*, then eq. (2.74) holds for slip and

$$\text{for impending slip :} \quad \frac{T_1 - \lambda v^2}{T_2 - \lambda v^2} = e^{\mu_s \theta}, \qquad \text{no slip :} \quad \frac{T_1 - \lambda v^2}{T_2 - \lambda v^2} < e^{\mu_s \theta} \qquad (2.75)$$

If the inertia of the belt is neglected, then $\lambda = 0$ and eqs (2.74) and (2.75) yield for

$$\text{slip :} \quad \frac{T_1}{T_2} = e^{\mu_k \theta}, \qquad \text{impending slip :} \quad \frac{T_1}{T_2} = e^{\mu_s \theta}, \qquad \text{no slip :} \quad \frac{T_1}{T_2} < e^{\mu_s \theta} \quad (2.76)$$

For a *stationary* belt (for example, a stationary belt on a rotating brake drum), $v = 0$ and the conditions of slip, impending slip and no slip are the same as in eq. (2.76) for a plane curve of contact. For a spatial curve of contact, the exponential terms are replaced by $\exp[\int_0^s (\mu_k/\rho)\, ds]$, $\exp[\int_0^s (\mu_s/\rho)\, ds]$, $\exp[\int_0^s (\mu_s/\rho)\, ds]$.

Consider a flexible rope/'vee' belt in a 'vee' shaped groove cut in a pulley, with semi-vertex angle α (Fig. 2.58d). The centre line of the groove forms a curve C. The belt can be circular, but the wear is less for a trapezoidal form. Let p_1 N/m be the force normal to the belt (Fig. 2.58d). The distributed force $\vec{q}$ on the belt, for slip is

$$\begin{aligned} \vec{q} &= p_1(-\sin\alpha\,\vec{e}_n + \cos\alpha\,\vec{e}_b) + p_1(-\sin\alpha\,\vec{e}_n - \cos\alpha\,\vec{e}_b) - 2\mu_k p_1\,\vec{e}_t \\ &= -2\mu_k p_1\,\vec{e}_t - 2p_1 \sin\alpha\,\vec{e}_n \end{aligned}$$

The forces in $-\vec{e}_n$ and $-\vec{e}_t$ directions are $p = 2p_1 \sin\alpha$ and $2\mu_k p_1 = (\mu_k/\sin\alpha)p$ instead of $\mu_k p$ for a flat belt. Hence, eqs (2.73)–(2.76) are valid with an effective coefficient of friction $\mu_{\text{eff}} = \mu/\sin\alpha$ $(> \mu)$ replacing μ. For a 'vee' belt, eq. (2.76) becomes for

$$\text{slip: } \frac{T_1}{T_2} = e^{\mu_k\theta/\sin\alpha}, \quad \text{impending slip: } \frac{T_1}{T_2} = e^{\mu_s\theta/\sin\alpha}, \quad \text{no slip: } \frac{T_1}{T_2} < e^{\mu_s\theta/\sin\alpha}$$

A 'vee' belt is better for power transmission than a flat belt since $\mu_{\text{eff}} = \mu_k/\sin\alpha > \mu_k$. For example, for $\alpha = 18°$, let $(T_1/T_2)_f, (T_1/T_2)_v$ be the limiting tension ratios for a flat and a 'vee' belt. Then $\mu_{\text{eff}} = \mu_s/\sin 18° = 3.24\mu_s \Rightarrow (T_1/T_2)_v = e^{\mu_s\theta/\sin\alpha} = [e^{\mu_s\theta}]^{1/\sin\alpha} = (T_1/T_2)_f^{3.24} \gg (T_1/T_2)_f$.

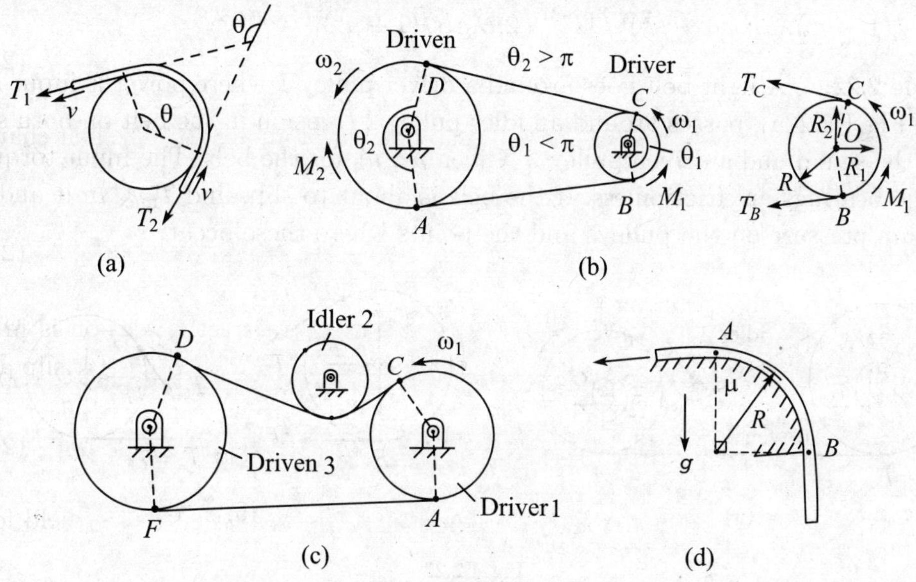

Fig. 2.59

For a circular pulley, angle θ is the angle of lap of the belt on the pulley (Fig. 2.59a). For power transmission, if μs are the same for both pulleys (Fig. 2.59b), then the *smaller* angle of lap θ_1 is used in the above expressions. Idler pulleys are used to increase this value (Fig. 2.59c). On a driving pulley, the input torque M_1 is in the same direction as ω_1, whereas on the driven pulley, the load torque M_2 is in a direction opposite to ω_2 (Fig. 2.59b). For uniform rotation, from the FBD of the driver, $M_{Oz} = M_1 + (T_C - T_B)R = 0 \Rightarrow T_B > T_C$. In Fig. 2.59b, the belt tension is maximum in the segment AB and minimum in the segment CD, and the maximum pressure on the pulleys is at B, where T is maximum and ρ is smaller. Let T_0 be the allowable tension in the belt. Substituting T_2 from eq. $(2.75)_1$, the power P transmitted is

$$P = T_1 v - T_2 v = (T_1 - \lambda v^2)v(1 - e^{-\mu_s\theta})$$

For maximum P, $\quad \dfrac{dP}{dv} = 0 \quad \Rightarrow \quad T_1 = 3\lambda v^2 = T_0.$

Hence, the maximum power is transmitted when $v^2 = T_0/3\lambda$.

§ **Exercise 2.6:** The speed v of the belt on a circular fixed pulley of radius R is increasing at the rate $\dot{v}$. Include acceleration a_t in the equation of motion of Δs to prove that eq. $(2.74)_3$ is replaced by $[T_1 - \lambda(v^2 - R\dot{v}/\mu_k)]/[T_2 - \lambda(v^2 - R\dot{v}/\mu_k)] = e^{\mu_k \theta}$.

§ **Exercise 2.7:** A heavy chain of mass λ kg/m is at rest (Fig. 2.59d) around a corner of radius R and coefficient of friction μ. The weight has to be included in the FBD of an infinitesimal element of the chain. Show that for impending upward motion $[T_A - (1-\mu^2)\lambda g R/(1+\mu^2)] \,/\, [T_B + 2\mu\lambda g R/(1+\mu^2)] = e^{\mu\pi/2}$.

Example 2.22: A light belt goes around a driver pulley 1 where power is input by a motor (Fig. E2.22a), passes around an idler pulley 2 (tension in the belt on both sides of it is the same) and a driven pulley 3. Given μ_s, μ_k for the belt. The input torque is M_1. The bearings are frictionless. If the belt is about to slip, find $T_{\max}$ in it and the maximum pressure on the pulleys and the points where these occur.

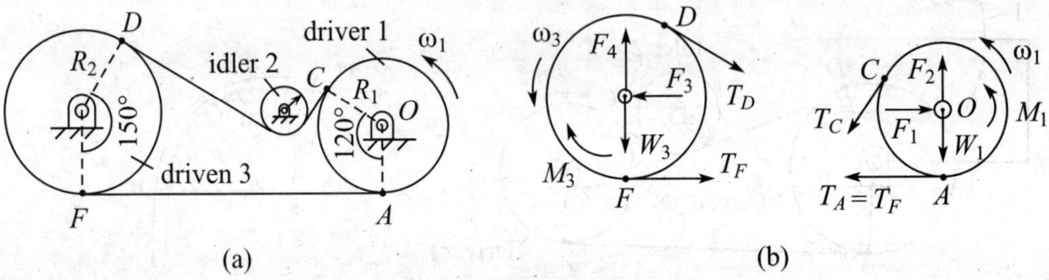

Fig. E2.22

Solution: In the FBDs (Fig. E2.22b), the driving torque is shown in the direction of ω_1 and the load torque is shown in the direction opposite to ω_3. For pulley 1:

$$M_{O_z} = M_1 + T_C R_1 - T_A R_1 = 0 \quad \Rightarrow \quad T_A = T_C + M_1/R_1 > T_C \tag{1}$$

Hence, part AF carries $T_{\max} = T_A$ and $p_{\max} = T_{\max}/R_1$ is at A where the radius of the pulley is smaller than at F. The angle of lap of 210° ($= 7\pi/6$ rad) on pulley 3 is less than that of 240° on pulley 1. Thus, the belt has impending slip on pulley 3:

$$\frac{T_A}{T_C} = \frac{T_F}{T_D} = \frac{T_{\max}}{T_{\min}} = e^{\mu_s \theta} = e^{7\pi\mu_s/6} \tag{2}$$

$$(1),\ (2) \Rightarrow \quad T_{\max} = T_A = \frac{M_1}{R_1\left(1 - e^{-7\pi\mu_s/6}\right)}, \qquad p_{\max} = \frac{T_A}{R_1} = \frac{M_1}{R_1^2\,(1 - e^{-7\pi\mu_s/6})}$$

2.19 Frictional Torque at a Thrust Bearing and Frictional Clutches

Consider an axisymmetric body 1 (Fig. 2.60a) in contact with another over an axisymmetric surface (conical, flat, spherical, etc.). The axial torque M needed for impending axial rotation of body 1 under an axial thrust P is to be determined. The FBD of body 1 is shown in Fig. 2.60b. Consider an elementary ring of the contact surface of width ds and radius r with its normal making an angle $\pi/2 - \alpha$ with the axis. Let the normal force be p N/m^2. The frictional force is $\mu_s p$ N/m^2 in the circumferential direction at a radial distance r from the axis. Equations of equilibrium for axial force and axial moment yield

$$P = \int_s (p \sin\alpha)(2\pi r)\, ds, \qquad M = \int_s (r\mu_s p)(2\pi r)\, ds$$

$$\Rightarrow \quad \frac{M}{P} = \frac{\int_s \mu_s r^2 p\, ds}{\int_s rp \sin\alpha\, ds} \tag{1}$$

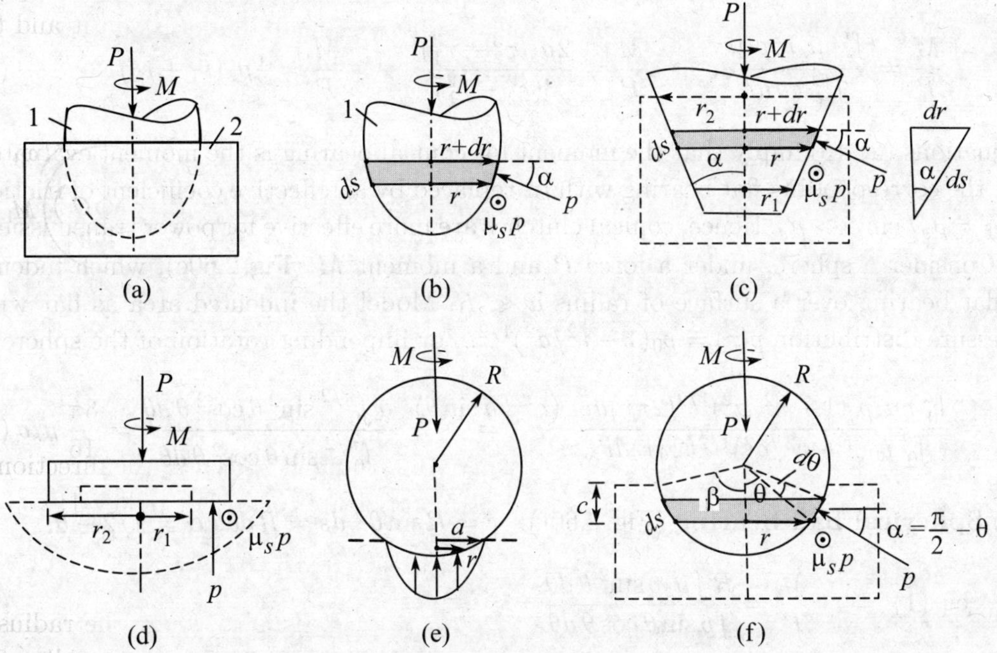

Fig. 2.60

(a) **Conical bearing** (Fig. 2.60c): $\alpha =$ constant, $ds = dr/\sin\alpha$, eq. (1)$\Rightarrow$

$$\frac{M}{P} = \frac{\int_{r_1}^{r_2} \mu_s r^2 p\, dr/\sin\alpha}{\int_{r_1}^{r_2} rp\, dr} \tag{2}$$

1. For uniform p: $$M/P = (2\mu_s/3\sin\alpha)(r_2^3 - r_1^3)/(r_2^2 - r_1^2) \quad (3)$$

2. For uniform wear: $p = C/r$, $$M/P = (\mu_s/\sin\alpha)(r_1 + r_2)/2 \quad (4)$$

This approximation of uniform wear is good if the bearing has been worn out, since the surfaces retain their relative shape and hence further wear is constant over the surface. The amount of wear at radius r, in n revolutions, is proportional to (i) the distance $2\pi rn$ through which slip takes place, and (ii) the intensity of the frictional force $\mu_s p$. Hence, for uniform wear, $2\pi rn\mu_s p$ =constant, i.e., rp =constant= C (say). Let the moment for the cases 1 and 2 be denoted by M_1 and M_2. Then,

$$M_1 - M_2 = \frac{\mu_s P(r_1 - r_2)^2}{6(r_1 + r_2)}\sin\alpha > 0 \quad \Rightarrow \quad M_1 > M_2$$

i.e., the resisting moment due to friction decreases when the bearing is worn out. The decrease is 25% for $r_1 = 0$.

(b) **Flat bearing** (Fig. 2.60d): $\alpha = \pi/2$, eqs (2)–(4) for various cases yield

$$\frac{M}{P} = \frac{\int_{r_1}^{r_2}\mu_s r^2 p\,dr}{\int_{r_1}^{r_2} rp\,dr}, \qquad \frac{M}{P} = \frac{2\mu_s(r_2^3 - r_1^3)}{3(r_2^2 - r_1^2)}, \qquad \frac{M}{P} = \tfrac{1}{2}\mu_s(r_1 + r_2) \quad (5)$$

Equations (2)–(5) imply that the moment for conical bearing is the moment evaluated for the corresponding flat bearing with μ_s replaced by an effective coefficient of friction $\mu_{\text{eff}} = \mu_s/\sin\alpha > \mu_s$. Hence, conical clutches are more effective for power transmission.

Consider a sphere, under a force P and a moment M (Fig. 2.60e), which indents a flat bearing over a surface of radius $a \ll R$. Model the indented area as flat with pressure distribution $p(r) = p_0(1 - r^2/a^2)^{1/2}$. For impending rotation of the sphere:

$$\frac{M}{P} = \frac{\int_0^a r\mu_s p_0(1 - r^2/a^2)^{1/2}(2\pi r)dr}{\int_0^a p_0(1 - r^2/a^2)^{1/2}(2\pi r)dr} \overset{(r = a\sin\theta)}{=} \frac{a\int_0^{\pi/2}\sin^2\theta\cos^2\theta\,d\theta}{\int_0^{\pi/2}\sin\theta\cos^2\theta\,d\theta} = \frac{3\pi}{16}\mu_s a \quad (6)$$

(c) **Spherical ball bearing** (Fig. 2.60f): $r = R\sin\theta$, $ds = R\,d\theta$, $\alpha = \pi/2 - \theta$,

$$\text{eq. (1)} \Rightarrow \quad \frac{M}{P} = \frac{R\int\mu_s p\sin^2\theta\,d\theta}{\int p\sin\theta\cos\theta\,d\theta} \quad (7)$$

For a ball bearing in a race, p is assumed to be either p = constant or $p = C/r$. For a sphere pressed into an elastic plate, we assume that p is proportional to normal indentation $c = R(\cos\theta - \cos\beta)$. For these three cases, M/P is obtained from eq. (7) as follows:

1. for p = constant: $$\frac{M}{P} = \frac{R\mu_s\int_0^\beta\sin^2\theta\,d\theta}{\int_0^\beta\sin\theta\cos\theta\,d\theta} = \frac{\mu_s R(2\beta - \sin 2\beta)}{1 - \cos 2\beta}$$

2. for $p=\dfrac{C}{r}=\dfrac{C}{R\sin\theta}$: $\qquad \dfrac{M}{P}=\dfrac{R\mu_s\int_0^\beta \sin\theta\, d\theta}{\int_0^\beta \cos\theta\, d\theta}=\mu_s R\tan\frac{1}{2}\beta$

3. for $p\propto c$, $p=AR(\cos\theta-\cos\beta)$: $\qquad \dfrac{M}{P}=\dfrac{R\mu_s\int_0^\beta \sin^2\theta(\cos\theta-\cos\beta)\, d\theta}{\int_0^\beta \sin\theta(\cos^2\theta-\cos\theta\cos\beta)\, d\theta}$

$$=\frac{\mu_s R(3\sin\beta-3\beta\cos\beta-\sin^3\beta)}{\cos^3\beta-3\cos\beta+2}$$

For actual slip, μ_k should replace μ_s.

2.20 Frictional Torque for a Square-Threaded Screw

Square-threaded screws are commonly used in screw-jacks and power-transmission (as in lead-screw of a lathe machine) and in testing machines and presses. We compute the axial torque M needed to impend the motion of a screw in a fixed nut against an axial thrust P. Let the length of the thread of the screw in contact with the nut be s_1 and its mean radius and lead be r and l respectively (Fig. 2.61a). The helix angle α is given by $\tan\alpha = l/2\pi r$ since the helix axially advances by l for a circumferential distance of $2\pi r$ (Fig. 2.61a). The FBD of the screw is shown in Fig. 2.61b. The nut exerts a distributed reaction force from one side of the thread of the screw. Since the width of the thread is small, this force is modelled as a line load at the mean radius r. This reaction extends from $s=0$ where the thread first makes contact with the nut to $s=s_1$ where the thread emerges from the nut.

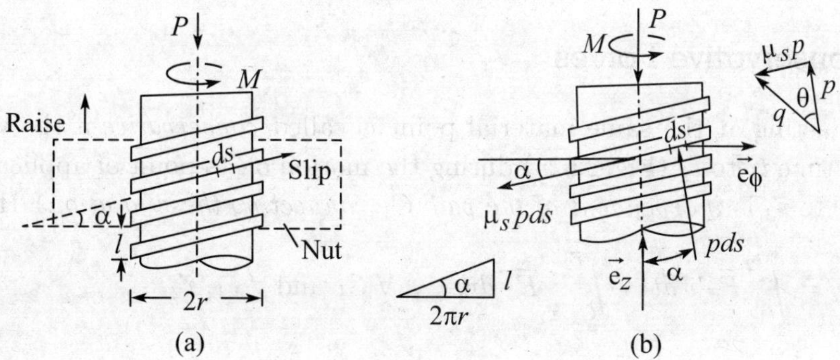

Fig. 2.61

Let the normal force (per unit length) on the thread be p. This makes an angle α with the axis as shown. The frictional force (per unit length) is $\mu_s p$ along the helix in the direction opposite to the direction of impending slip. Hence, the force $d\vec{F}$ on an element of length ds at a distance r from the axis is

$$d\vec{F} = p\,ds(-\sin\alpha\,\vec{e}_\phi + \cos\alpha\,\vec{e}_z) + \mu_s p\,ds(-\cos\alpha\,\vec{e}_\phi - \sin\alpha\,\vec{e}_z)$$

Equations of equilibrium for axial force and axial moment yield

$$P = \int_0^{s_1} (p\cos\alpha - \mu_s p \sin\alpha)ds, \qquad M = \int_0^{s_1} r(p\sin\alpha + \mu_s p\cos\alpha)ds$$

$$\Rightarrow \quad \frac{M}{P} = \frac{r(\sin\alpha + \mu_s\cos\alpha)}{\cos\alpha - \mu_s\sin\alpha} = \frac{r(\tan\alpha + \mu_s)}{1 - \mu_s\tan\alpha} = r\tan(\alpha+\theta) \tag{1}$$

$$q_{\text{av}} = \left[\int_0^{s_1} \frac{p}{\cos\theta}ds\right] / s_1 = \frac{P\sec\theta}{s_1(\cos\alpha - \mu_s\sin\alpha)} = P\sec(\alpha+\theta)/s_1$$

where q_{av} is the average force per unit length on the thread and $\tan\theta = \mu_s$. The threads will be sheared off if they are not strong enough to sustain q_{av}. q_{av} can be reduced by increasing s_1, i.e., by increasing the number of threads in contact with the nut. The opposite torque $-M_1\,\vec{e}_z$ needed for impending motion in the direction of the axial thrust P can be obtained from eq. (1) by replacing M with $-M_1$ and μ_s with $-\mu_s$ since the directions of the torque and the friction force in this case are opposite to those of the previous case:

$$\frac{M_1}{P} = \frac{r(\mu_s - \tan\alpha)}{1 + \mu_s\tan\alpha} = r\tan(\theta - \alpha) \tag{2}$$

The screw is *self-locking* if it does not advance under P in the absence of M_1. Hence, it is self-locking if $M_1 > 0$, i.e., if $\tan\alpha < \mu_s$.

2.21 Conservative Forces

A force $\vec{F}$ acting at the same material point is called *conservative* if the work done by it from time t_1 to t_2 (Fig. 2.62a) during the motion of its point of application from location $\vec{r}_1$ to $\vec{r}_2$ is *independent of the path* C_1 *connecting these locations.* Hence,

$$W_{1-2} = \int_{t_1}^{t_2} \vec{F}\cdot\vec{v}\,dt = \int_{C_1:\,\vec{r}_1}^{\vec{r}_2} \vec{F}\cdot d\vec{r}, \qquad \forall\, C_1 \text{ and } \forall\, \vec{r}_1, \vec{r}_2 \tag{2.77}$$

in the region R in which $\vec{F}$ is defined. R may be the whole of space or a part of it. Hence, a conservative force *acts on the same material point of the body and does not depend explicitly on its velocity and time,* i.e., $\vec{F} = \vec{F}(\vec{r})$. The force due to friction is non-conservative as its direction is velocity dependent in the case of dry friction and its magnitude and direction are velocity dependent if a lubricating film is present.

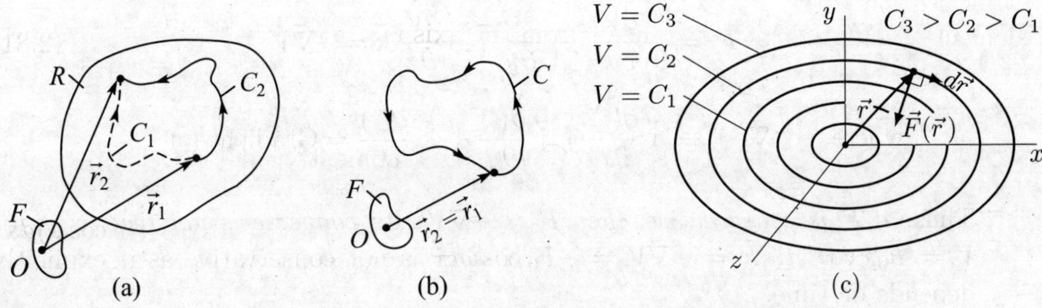

Fig. 2.62

The following inferences can be drawn for conservative forces.

1. Equation (2.77) implies that the integrand is a perfect differential of a position dependent single-valued function $V(\vec{r})$, called the *potential energy*, such that

$$dW = \vec{F} \cdot d\vec{r} = -dV \tag{2.78}$$

$$\Rightarrow \quad W_{1-2} = -\int_{V(\vec{r}_1)}^{V(\vec{r}_2)} dV = -[V(\vec{r}_2) - V(\vec{r}_1)] \tag{2.79}$$

i.e., *the work done from $\vec{r}_1$ to $\vec{r}_2$ equals negative of the change in V from $\vec{r}_1$ to $\vec{r}_2$.* The customary negative sign in the definition of V ensures that in work–energy relations [e.g., eq. (2.57)], when the work done by a conservative force is transferred to the left hand side, then we get the sum of kinetic and potential energies, rather than their difference.

2. Hence, the *work done by a conservative force in any closed path C ($\vec{r}_2 = \vec{r}_1$) is zero* (Fig. 2.62b).

3. Let $\vec{r}_O$ be the *datum* for V, i.e., $V(\vec{r}_O) = 0$. Equation (2.78) $\Rightarrow$

$$V(\vec{r}) = -\int_{\vec{r}_O}^{\vec{r}} \vec{F} \cdot d\vec{r} \tag{2.80}$$

i.e., potential energy at a given position $\vec{r}$ equals the negative of the work done from the datum to this position and also equals the *potential* (actual) work done from the given position $\vec{r}$ to the datum. For variable forces, the datum is usually chosen at the position where the force is zero.

4. Writing $\vec{F} \cdot d\vec{r} = -dV$ in terms of Cartesian components of $\vec{F}$ and $d\vec{r}$, we have:

$$F_x\,dx + F_y\,dy + F_z\,dz = -\frac{\partial V}{\partial x}dx - \frac{\partial V}{\partial y}dy - \frac{\partial V}{\partial z}dz \quad \forall\ dx, dy, dz.$$

$\Rightarrow$ The coefficients of dx, dy, dz on the two sides must be the same, i.e.,

$$F_x = -\frac{\partial V}{\partial x}, \qquad F_y = -\frac{\partial V}{\partial y}, \qquad F_z = -\frac{\partial V}{\partial z}$$

$$\Rightarrow \qquad \vec{F} = -\frac{\partial V}{\partial x}\vec{\mathrm{i}} - \frac{\partial V}{\partial y}\vec{\mathrm{j}} - \frac{\partial V}{\partial z}\vec{\mathrm{k}} = -\vec{\nabla}V \tag{2.81}$$

$$\text{where} \qquad \vec{\nabla}(\;) \equiv \vec{\mathrm{i}}\frac{\partial(\;)}{\partial x} + \vec{\mathrm{j}}\frac{\partial(\;)}{\partial y} + \vec{\mathrm{k}}\frac{\partial(\;)}{\partial z} = \text{Gradient}(\;)$$

Thus, *if $\vec{F}$ is conservative, then* $\vec{F} = -\vec{\nabla}V$. Its *converse is not true*, e.g., for $V = F_0 x\cos\omega t$, $\vec{F} = -\vec{\nabla}V = -F_0\cos\omega t\,\vec{\mathrm{i}}$ is not conservative as it explicitly depends on time.

For any $d\vec{r}$ along a contour surface $V(x, y, z) = \text{constant}$ (Fig. 2.62c):

$$dV = \frac{\partial V}{\partial x}dx + \frac{\partial V}{\partial y}dy + \frac{\partial V}{\partial z}dz = \vec{\nabla}V\cdot d\vec{r} = 0 \quad \Rightarrow \quad \vec{\nabla}V \perp d\vec{r}\;\forall\; d\vec{r} \;\Rightarrow$$

$\vec{F} = -\vec{\nabla}V$ *is along normal to contour surface of* V in the direction of decreasing V

Using the expressions for dW in cylindrical, spherical and path coordinates [eq. (2.28)], we have:

$$F_r = -\frac{\partial V}{\partial r}, \qquad F_\phi = -\frac{1}{r}\frac{\partial V}{\partial \phi}, \qquad F_z = -\frac{\partial V}{\partial z}$$

$$F_r = -\frac{\partial V}{\partial r}, \qquad F_\theta = -\frac{1}{r}\frac{\partial V}{\partial \theta}, \qquad F_\phi = -\frac{1}{r\sin\theta}\frac{\partial V}{\partial \phi} \tag{2.82}$$

$$F_t = -\frac{\partial V}{\partial s}$$

5. The force component F_e in direction $\vec{\mathrm{e}}$ is obtained as

$$F_e = \vec{F}\cdot\vec{\mathrm{e}} = -\vec{\nabla}V\cdot\vec{\mathrm{e}} = -(\text{directional derivative of } V \text{ along } \vec{\mathrm{e}}).$$

6. *If $\vec{F}$ is conservative, then* $\vec{\nabla}\times\vec{F} = \vec{0}$:

$$\vec{\nabla}\times\vec{F} = \begin{vmatrix} \vec{\mathrm{i}} & \vec{\mathrm{j}} & \vec{\mathrm{k}} \\ \dfrac{\partial}{\partial x} & \dfrac{\partial}{\partial y} & \dfrac{\partial}{\partial z} \\ F_x & F_y & F_z \end{vmatrix} = \left[\frac{\partial F_z}{\partial y} - \frac{\partial F_y}{\partial z}\right]\vec{\mathrm{i}} + \ldots = \left[-\frac{\partial^2 V}{\partial y\partial z} + \frac{\partial^2 V}{\partial z\partial y}\right]\vec{\mathrm{i}} + \ldots = \vec{0}$$

In cylindrical coordinates:

$$\vec{\nabla}\times\vec{F} = \begin{vmatrix} \dfrac{1}{r}\vec{\mathrm{e}}_r & \vec{\mathrm{e}}_\phi & \dfrac{1}{r}\vec{\mathrm{e}}_z \\ \dfrac{\partial}{\partial r} & \dfrac{\partial}{\partial \phi} & \dfrac{\partial}{\partial z} \\ F_r & rF_\phi & F_z \end{vmatrix} \tag{2.83}$$

7. The following theorem provides a simple condition to check whether a given force is conservative.

If $\vec{\nabla} \times \vec{F} = \vec{0}$ in the region R of the definition of $\vec{F}$, then $\vec{F}$ is conservative in every **simply-connected** *sub-region R_1 of R, provided $\vec{F}$ does not explicitly depend on velocity and time.*

Proof: Using Stoke's theorem, for any closed path C in R_1 (Fig. 2.63), $W = \oint_C \vec{F} \cdot d\vec{r} = \int_A \vec{\nabla} \times \vec{F} \cdot d\vec{A} = 0$, where A is any surface area in R_1 whose bounding curve is C.

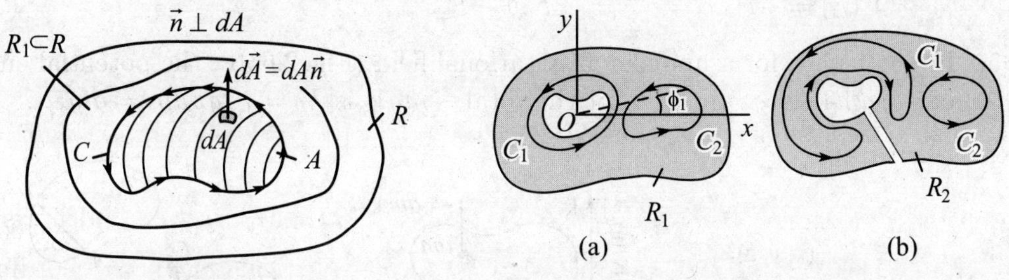

Fig. 2.63

Fig. 2.64

Consider a planar force field $\vec{F} = (c/r)\,\vec{e}_\phi$ which is not defined for $r = 0$. $\vec{\nabla} \times \vec{F} = (0/r)(\vec{e}_r + \vec{e}_z) = \vec{0}$ for $r \neq 0$ in the doubly-connected region R_1 in the xy-plane (Fig. 2.64a). $\oint_{C_1} \vec{F} \cdot d\vec{r} = \int_0^{2\pi} c\,d\phi = 2\pi c \neq 0$ for curves like C_1 which enclose the hole, and hence the force is non-conservative in R_1, with a potential $V = -c\phi$ that is multi-valued in R_1: $V(r, \phi, z) = -c\phi, -c(\phi + 2\pi), \ldots$. For any closed curve C_2 not enclosing the hole: $W = \int_{\phi_1}^{\phi_1} c\,d\phi = 0$. If the region is modified to a simply-connected region R_2 by making a slit in R_1 (Fig. 2.64b), then $\vec{F}$ is conservative in R_2.

2.21.1 Potential energies of some conservative forces

$V(\vec{r}) = -\int_{\vec{r}_O}^{\vec{r}} \vec{F} \cdot d\vec{r}$ is computed by integrating along a convenient path.

1. Integrating along the piecewise linear curves $C_1(OABD)$ and $C_2(OBD)$, shown in Fig. 2.65a, yields in general

$$V(\vec{r}) - V(\vec{0}) = -\int_{C_1:\,\vec{0}}^{\vec{r}} \vec{F} \cdot d\vec{r}$$

$$= -\int_0^x F_x(x, 0, 0)\,dx - \int_0^y F_y(x, y, 0)\,dy - \int_0^z F_z(x, y, z)\,dz \quad (2.84)$$

$$V(\vec{r}) - V(\vec{0}) = -\int_{C_2:\,\vec{0}}^{\vec{r}} \vec{F} \cdot d\vec{r}$$

$$= -\int_0^r F_r(r, \phi, 0)\,dr - \int_0^z F_z(r, \phi, z)\,dz \quad (2.85)$$

2. For $\vec{F} = F_x(x)\,\vec{\mathrm{i}} + F_y(y)\,\vec{\mathrm{j}} + F_z(z)\,\vec{\mathrm{k}}$, with its components dependent only on the corresponding coordinate, eq. (2.84) yields

$$V(\vec{r}) - V(\vec{0}) = -\int_0^x F_x(x)\,dx - \int_0^y F_y(y)\,dy - \int_0^z F_z(z)\,dz \tag{2.86}$$

3. For $\vec{F} = F_x(x)\,\vec{\mathrm{i}}$, eq. (2.86) $\Rightarrow$ $V(\vec{r}) - V(\vec{0}) = -\int_0^x F_x(x)\,dx$.
4. For a constant force $\vec{F}$, with datum at $\vec{0}$, eq. (2.80) $\Rightarrow$

$$V(\vec{r}) = -\vec{F}\cdot\vec{r} = -F_x x - F_y y - F_z z \tag{2.87}$$

5. In particular, for a uniform gravitational field (Fig. 2.65b), the potential energy of $-g\,dm\,\vec{\mathrm{k}}$ is $gz\,dm$ and for the total $-mg\,\vec{\mathrm{k}}$ is $V = \int_m gz\,dm = mgz_C$.

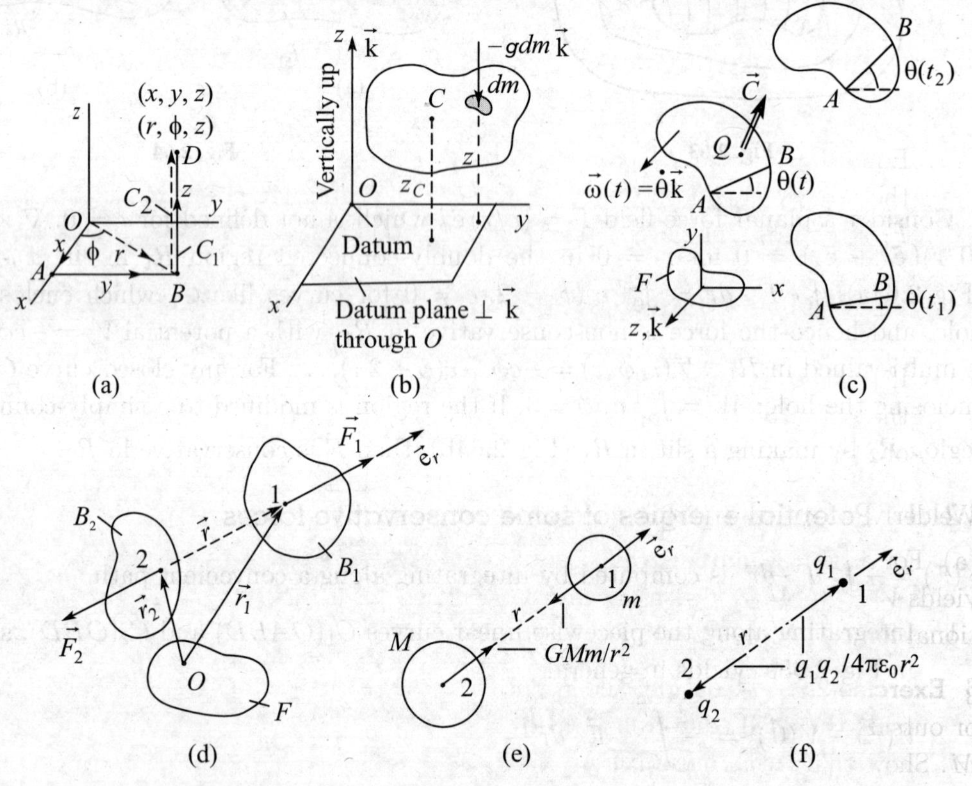

Fig. 2.65

6. The work done by a constant moment (or couple–moment) $\vec{C}$ acting at point Q of a rigid body (Fig. 2.65c) moving with $\vec{\omega}(t) = \omega(t)\,\vec{\mathrm{k}} = \dot{\theta}(t)\,\vec{\mathrm{k}}$ from orientation $\theta(t_1)$ at time t_1 to orientation $\theta(t_2)$ at time t_2 is given by

$$W = \int_{t_1}^{t_2} \vec{C}\cdot\vec{\omega}(t)\,dt = \vec{C}\cdot[\theta(t_2) - \theta(t_1)]\,\vec{\mathrm{k}} = C_z[\theta(t_2) - \theta(t_1)] \tag{1}$$

which is independent of the path followed by the body from its initial configuration to its final configuration. Hence, a constant moment acting on a rigid body, moving with angular velocity having only z-component, is conservative with potential energy $V(\theta) = -C_z\theta$ with datum at $\theta = 0$. However, a constant moment acting on a rigid body having arbitrary $\vec{\omega}$ is *non-conservative.*

§ **Exercise 2.8:** Prove the above result by considering two motions between the same initial and final configurations: (1) $\vec{\omega} = \omega\,\vec{\mathrm{i}}$ for time 0 to π/ω, followed by $\vec{\omega} = \omega\,\vec{\mathrm{j}}$ for time π/ω to $2\pi/\omega$, and (2) $\vec{\omega} = \frac{1}{2}\omega\,\vec{\mathrm{k}}$ for time 0 to $2\pi/\omega$.

7. For *a pair of mutual central forces* $\vec{F}_1$ *and* $\vec{F}_2$ (Fig. 2.65d), *with* $\vec{F}_1 = -\vec{F}_2 = F_1\,\vec{\mathrm{e}}_r = F(r,\theta,\phi)\,\vec{\mathrm{e}}_r$:

$$\begin{aligned} dW &= \vec{F}_1 \cdot d\vec{r}_1 + \vec{F}_2 \cdot d\vec{r}_2 = \vec{F}_1 \cdot d(\vec{r}_1 - \vec{r}_2) = F_1\,\vec{\mathrm{e}}_r \cdot d\vec{r} \\ &= F_1\,\vec{\mathrm{e}}_r \cdot (dr\,\vec{\mathrm{e}}_r + r\,d\theta\,\vec{\mathrm{e}}_\theta + r\sin\theta\,d\phi\,\vec{\mathrm{e}}_\phi) = F(r,\theta,\phi)\,dr \end{aligned} \qquad (1)$$

Equation (1) $\Rightarrow$ this pair of forces is conservative if $F(r,\theta,\phi) = F(r)$, i.e., if the forces depend only on the separation distance between points 1 and 2. For $\vec{F}_1 = F(r)\,\vec{\mathrm{e}}_r$, the potential energy of the pair of forces with datum at $r = r_0$ is:

$$V(r) = -\int_{r_0}^{r} F(r)\,dr \qquad (2.88)$$

If point 2 remains fixed, then the work done by $\vec{F}_2$ is zero and eq. (2.88) gives the potential energy of the single central force $\vec{F}_1 = F(r)\,\vec{\mathrm{e}}_r$ alone (with centre of force at point 2), which by itself is conservative.

We derive V of gravitational, electrostatic, spring forces and torsional spring moment.
(a) For a pair of mutual gravitation forces $F(r) = -GMm/r^2$ (Fig. 2.65e), eq. (2.88) yields $V = -GMm/r$ with datum at $r_0 = \infty$. This is also potential energy of gravitational force $\vec{F} = (-GMm/r^2)\,\vec{\mathrm{e}}_r$ on m due to a large mass M which remains fixed.

§ **Exercise 2.9:** Consider the gravitational force on a mass-point of mass m (inside or outside the shell) due to a fixed uniform thin spherical shell of radius a and mass M. Show that, with datum at $r_0 = \infty$, its potential energy $V = -GMm/r$ if $r > a$ and $V = -GMm/a$ if $r < a$.

(b) For a pair of mutual electrostatic forces $F(r) = q_1q_2/4\pi\epsilon_0 r^2$ on two point charges q_1 and q_2 (Fig. 2.65f) with datum at $r_0 = \infty$, eq. (2.88) yields $V = q_1q_2/4\pi\epsilon_0 r$. This is also the potential energy of a central electrostatic force $\vec{F} = (q_1q_2/4\pi\epsilon_0 r^2)\,\vec{\mathrm{e}}_r$ on q_1 due to a point charge q_2 which remains fixed.
(c) Consider a pair of forces due to an *ideal elastic* spring (Fig. 2.66a,b) such as a helical spring. Ideal means that its inertia is neglected. The spring is an inertia-less

two-force member and the two forces at its ends must be equal in magnitude, opposite in direction and act along the line joining the end-points. Elastic means that the spring recovers its un-deformed length on removal of the end-loads and the load-extension relation is the same during loading and unloading. Let L_0 be its un-stretched length.

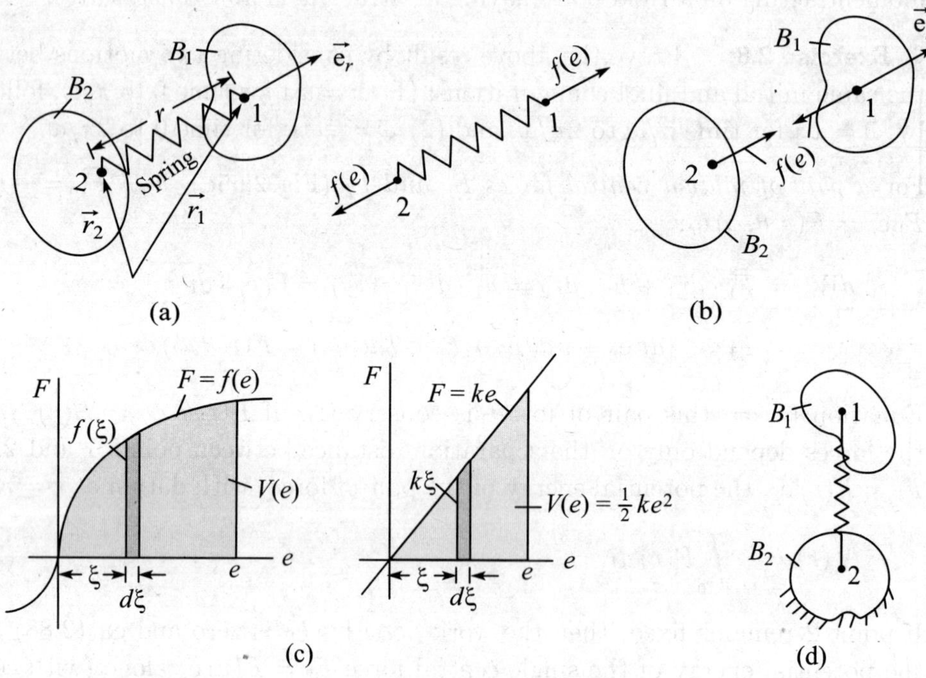

Fig. 2.66

Let the pull on the spring be $f(e)$ for an extension $e = r - L_0$. Hence, $F(r) = -f(e)$, $de = dr$ and eq. (2.88) yields potential energy of pair of spring forces

$$V(e) = \int_0^e f(e)\,de \tag{2.89}$$

with datum at un-stretched (un-deformed) configuration $r_0 = L_0$. The integral in eq. (2.89) represents the area under the force-extension curve (Fig. 2.66c). This is also the potential energy of a central spring force $\vec{F} = -f(e)\,\vec{e}_r$ acting on body B_1 with the other end of the spring fixed (Fig. 2.66d). For example, for

$$f(e) = k_1e + k_2e^3, \qquad V(e) = \tfrac{1}{2}k_1e^2 + \tfrac{1}{4}k_2e^4 \qquad (a)$$

For a *linear spring* of *stiffness* k: $\quad f(e) = ke, \qquad V = \frac{1}{2}ke^2 = \frac{1}{2}k\delta^2 \qquad$ (b)

with $\delta\ (= e)$ being the extension. The expressions (2.89), (a), (b) are valid for compression where e is negative since both $f(e)$ and de change sign.

(d) Consider a pair of moments due to an *ideal elastic torsional spring*, such as a torque bar and a spiral spring (Fig. 2.67a). Ideal means that its inertia is neglected. Hence, the end-moments must be $\pm M_t \vec{e}$ along the axis $\vec{e}$ of the spring. Elastic means that the spring recovers its un-deformed shape and size on removal of the end-moments and the moment–twist relation is the same during loading and unloading.

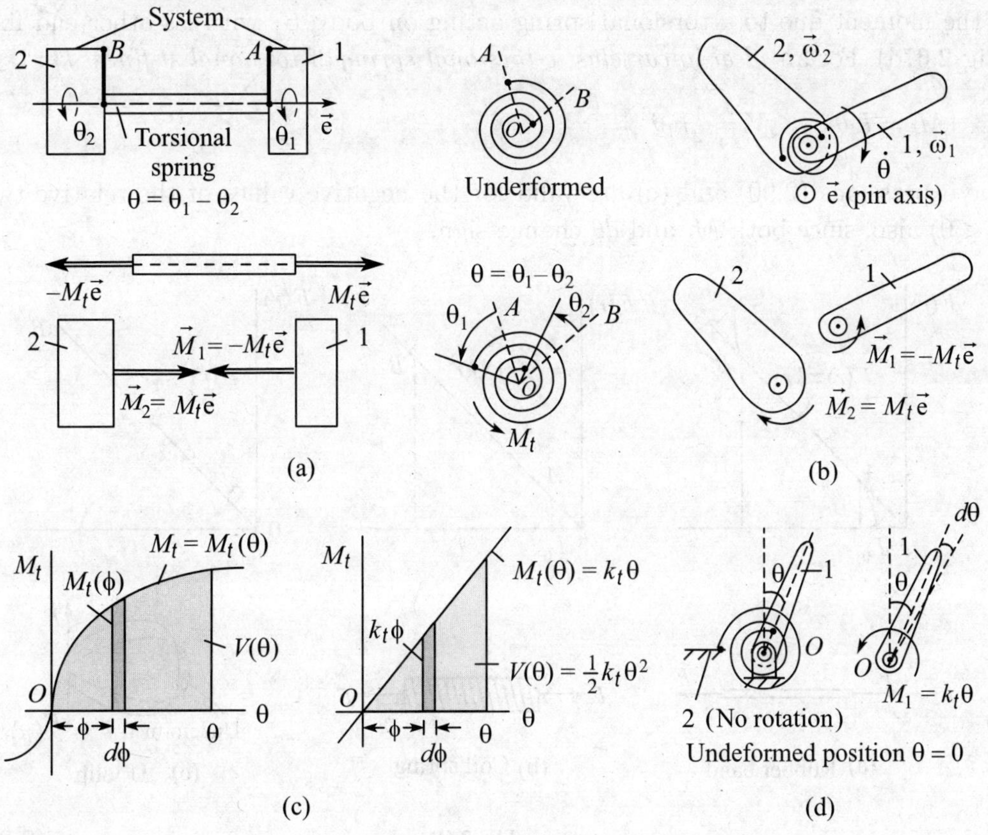

Fig. 2.67

Let the axial torsional (twisting) moment on the spring at the ends be $M_t(\theta)$ for axial twist (relative axial rotation of the two ends of the spring) θ, where for the un-deformed spring, $\theta = 0$ and $M_t(0) = 0$. Using $\vec{\omega}_1 = \vec{\omega}_2 + \dot{\theta}\,\vec{e}$ and $\vec{M}_1 = -M_t(\theta)\,\vec{e} = -\vec{M}_2$, the rate of work done by the pair of moments $\vec{M}_1$ and $\vec{M}_2$, acting on the rigid bodies 1 and 2 due to a torsional spring (Fig. 2.67b), is given by

$$\begin{aligned}\dot{W} &= \vec{M}_1 \cdot \vec{\omega}_1 + \vec{M}_2 \cdot \vec{\omega}_2 = \vec{M}_1 \cdot (\vec{\omega}_1 - \vec{\omega}_2) \\ &= -M_t(\theta)\,\vec{e} \cdot (\dot{\theta}\,\vec{e}) = -M_t(\theta)\,\dot{\theta} \quad \Rightarrow \quad dW = -M_t(\theta)\,d\theta\end{aligned}$$

Hence, the pair of moments together are conservative with potential energy

$$V(\theta) = \int_0^\theta M_t(\theta)\, d\theta \tag{2.90}$$

with the untwisted configuration $\theta = 0$ as the datum. The integral in eq. (2.90) is the area under the moment–twist curve (Fig. 2.67c). This is also the potential energy of the moment due to a torsional spring acting on body B_1 with its other end fixed (Fig. 2.67d). For an *ideal linear elastic torsional spring* of *torsional stiffness* k_t:

$$M_t = k_t\theta, \qquad V = \tfrac{1}{2}k_t\theta^2 \tag{c}$$

The expressions (2.90) and (c) are valid for the negative values of the relative twist ($\theta < 0$) also, since both M_t and $d\theta$ change sign.

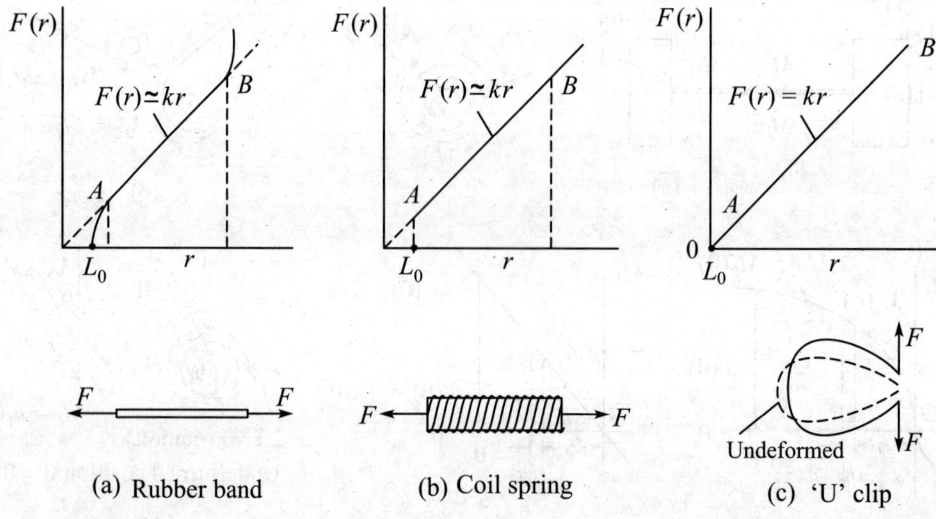

Fig. 2.68

(e) **'Zero-length' linear spring:** In some types of springs and their specific form of attachment, the relation of force F to extension e can be approximated as:

$$F(r) = f(e) = f(r - L_0) \simeq kr \quad \Rightarrow \quad V(r) = \tfrac{1}{2}kr^2$$

i.e., effectively it is a *'zero-length' linear spring.* For example, in range A to B of force-extension relation (Fig. 2.68): a rubber band, a pre-stressed spring with coils touching each other, a pinched 'U' clip , and springs with $r \gg L_0$ behave as zero-length springs.

Example 2.23: (a) Check whether the forces: (1) $\vec{F} = (-2x + y^2 + z^2 \sin x)\,\vec{\imath} + 2(x + z^2)y\,\vec{\jmath} + 2z(-\cos x + y^2)\,\vec{k}$, (2) $\vec{F} = (2x - y + yz^2)\,\vec{\imath} + (y^2x - x + xz^2)\,\vec{\jmath} + 2xyz\,\vec{k}$, are conservative force fields. (b) Find the work done by these forces in a closed path shown in Fig. E2.23. (c) Find the work done by these forces along the curve $C : \vec{r}(\tau) = 2\tau\,\vec{\imath} - \tau^3\,\vec{\jmath} + \tau^2\,\vec{k}$ from $\tau = 1$ to $\tau = 2$.

Solution: (1) $\vec{F}$ is independent of velocity and time. $F_x = -2x + y^2 + z^2 \sin x$, $F_y = 2(x + z^2)y$, $F_z = 2z(-\cos x + y^2)$

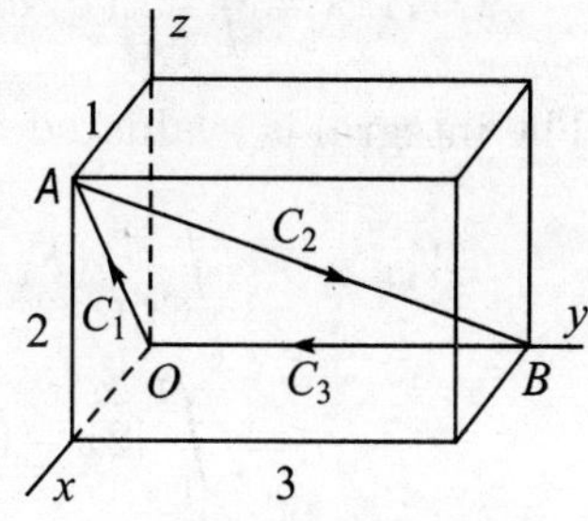

Fig. E2.23

(a) $\vec{\nabla} \times \vec{F} =$

$$\begin{vmatrix} \vec{\imath} & \vec{\jmath} & \vec{k} \\ \dfrac{\partial}{\partial x} & \dfrac{\partial}{\partial y} & \dfrac{\partial}{\partial z} \\ (-2x + y^2 + z^2 \sin x) & 2(x + z^2)y & 2z(-\cos x + y^2) \end{vmatrix}$$

$$= (4zy - 4zy)\,\vec{\imath} - (2z \sin x - 2z \sin x)\,\vec{\jmath} + (2y - 2y)\,\vec{k} = \vec{0}$$

Hence, this force is conservative for all simply-connected regions.

(b) The work done over the closed path is zero since $\vec{F}$ is conservative.

(c) $W = -\Delta V$. We choose the datum at $\vec{r}_0 = \vec{0}$ and use eq. (2.84) to find V of $\vec{F}$.

$$\begin{aligned} V(\vec{r}) &= -\int_0^x F_x(x, 0, 0)\, dx - \int_0^y F_y(x, y, 0)\, dy - \int_0^z F_z(x, y, z)\, dz \\ &= -\int_0^x (-2x)\, dx - \int_0^y 2xy\, dy - \int_0^z 2z(-\cos x + y^2)\, dz \\ &= x^2 - xy^2 - z^2(-\cos x + y^2) \end{aligned}$$

$C: \quad \vec{r}_1 = \vec{r}(1) = 2\,\vec{\imath} - \vec{\jmath} + \vec{k}, \quad \vec{r}_2 = \vec{r}(2) = 4\,\vec{\imath} - 8\,\vec{\jmath} + 4\,\vec{k},$

$$\begin{aligned} V(\vec{r}_1) &= V(2, -1, 1) = 2^2 - 2(-1)^2 - 1^2[-\cos 2 + (-1)^2] = 1 + \cos 2 \\ V(\vec{r}_2) &= V(4, -8, 4) = 4^2 - 4(-8)^2 - 4^2[-\cos 4 + (-8)^2] = -1264 + 16 \cos 4 \\ W(\vec{r}_1 \to \vec{r}_2) &= -[V(\vec{r}_2) - V(\vec{r}_1)] = 1265 + \cos 2 - 16 \cos 4 = 1275 \text{ units} \end{aligned}$$

(2) $\vec{F}$ is not explicitly dependent on $\vec{v}$ and t.

$$F_x = 2x - y + yz^2, \quad F_y = y^2 x - x + xz^2, \quad F_z = 2xyz$$

$$\text{(a)} \quad \vec{\nabla} \times \vec{F} = \begin{vmatrix} \vec{\imath} & \vec{\jmath} & \vec{k} \\ \dfrac{\partial}{\partial x} & \dfrac{\partial}{\partial y} & \dfrac{\partial}{\partial z} \\ (2x - y + yz^2) & (y^2 x - x + xz^2) & 2xyz \end{vmatrix} = y^2\,\vec{k} \neq \vec{0}$$

Hence, this force is non-conservative.

(b) C_1: $O(0,0,0) \to A(1,0,2)$, $y=0$, $(x-0)/(1-0)=(z-0)/(2-0)$

$$\Rightarrow \quad z=2x, \quad dz=2\,dx, \quad dy=0$$

$$W_{O-A}=\int_{C_1:O}^{A}(F_x\,dx+F_y\,dy+F_z\,dz)$$

The integral is evaluated after replacing y, z in terms of x for points on C_1:

$$\begin{aligned} W_{O-A} &= \int_{C_1:O}^{A}(F_x\,dx+F_y\,dy+F_z\,dz) \\ &= \int_0^1 [2x-0+(0)(2x)^2]\,dx+\int_0^1 [2x(0)(2x)]2\,dx=1 \end{aligned}$$

C_2: $A(1,0,2) \to B(0,3,0)$, $\quad \dfrac{x-1}{0-1}=\dfrac{y-0}{3-0}=\dfrac{z-2}{0-2} \Rightarrow$

$$y=-3x+3, \quad z=2x, \quad dy=-3\,dx, \quad dz=2\,dx.$$

$$\begin{aligned} \int_{C_2:A}^{B} F_x\,dx &= \int_{C_2:A}^{B}(2x-y+yz^2)\,dx \\ &= \int_1^0 [2x-(-3x+3)+(-3x+3)(2x)^2]\,dx=-0.5 \end{aligned}$$

$$\begin{aligned} \int_{C_2:A}^{B} F_y\,dy &= \int_{C_2:A}^{B}(y^2x-x+xz^2)\,dy \\ &= \int_1^0 [(-3x+3)^2x-x+x(2x)^2](-3)\,dx=3.75 \end{aligned}$$

$$\begin{aligned} \int_{C_2:A}^{B} F_z\,dz &= \int_{C_2:A}^{B}(2xyz)\,dz \\ &= \int_1^0 [2x(-3x+3)(2x)]2\,dx=\int_1^0 (24x^2-24x^3)\,dx=-2 \end{aligned}$$

$$W_{A-B} = \int_{C_2:A}^{B}(F_x\,dx+F_y\,dy+F_z\,dz)=-0.5+3.75-2=1.25$$

C_3: $B(0,3,0) \to O(0,0,0)$, $\quad x=0, \quad z=0 \quad \Rightarrow \quad dx=0, \quad dz=0,$

$$\begin{aligned} W_{B-O} &= \int_{C_3:B}^{O}(F_x\,dx+F_y\,dy+F_z\,dz)=\int_3^0 F_y\,dy \\ &= \int_3^0 \left[y^2(0)-0+0(0)^2\right]dy=0 \end{aligned}$$

$$W_{O-A-B-O} = W_{O-A} + W_{A-B} + W_{B-O} = 1 + 1.25 + 0 = 2.25 \text{ units}$$

The work done for this closed path is non-zero. The work done by a non-conservative force in a closed path is, in general, non-zero.

(c) C: $\vec{r}(\tau) = 2\tau\,\vec{\mathrm{i}} - \tau^3\,\vec{\mathrm{j}} + \tau^2\,\vec{\mathrm{k}} \quad \Rightarrow \quad x = 2\tau, \;\; y = -\tau^3, \;\; z = \tau^2,$

$$\Rightarrow \qquad dx = 2\,d\tau, \quad dy = -3\tau^2\,d\tau, \quad dz = 2\tau\,d\tau,$$

$$\begin{aligned} W &= \int_{C:\vec{r}(1)}^{\vec{r}(2)} (F_x\,dx + F_y\,dy + F_z\,dz) = \int_1^2 (2F_x - 3\tau^2 F_y + 2\tau F_z)\,d\tau \\ &= \int_1^2 \Big[2\left\{2(2\tau) - (-\tau^3) - \tau^3(\tau^2)^2\right\} - 3\tau^2\left\{(-\tau^3)^2(2\tau) - 2\tau + 2\tau(\tau^2)^2\right\} \\ &\qquad + 2\tau\left\{2(2\tau)(-\tau^3)(\tau^2)\right\}\Big]\,d\tau \\ &= \int_1^2 \left[8\tau + 8\tau^3 - 16\tau^7 - 6\tau^9\right] d\tau = -1081.8 \text{ units} \end{aligned}$$

Example 2.24: (a) Is the force $\vec{F} = (C/r^3)\,\vec{e}_r + b\,\vec{e}_\phi$ conservative? If so find its potential energy V. (b) Find the work done by $\vec{F}$ over the closed path $ABCD$ (Fig. E2.24a) and the quarter circular path from F to G. C and b are constants. (c) Is the force $\vec{F} = (C/r^3)\,\vec{e}_r + (b/r)\,\vec{e}_\phi$ conservative? If so find its potential energy V.

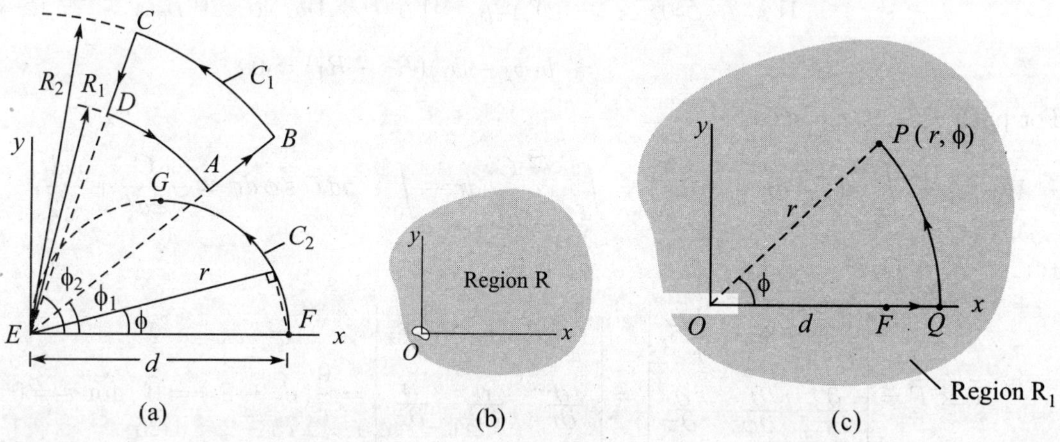

Fig. E2.24

Solution: (a)

$$\vec{\nabla} \times \vec{F} = \begin{vmatrix} \frac{1}{r}\,\vec{e}_r & \vec{e}_\phi & \frac{1}{r}\,\vec{e}_z \\ \frac{\partial}{\partial r} & \frac{\partial}{\partial \phi} & \frac{\partial}{\partial z} \\ F_r & rF_\phi & F_z \end{vmatrix} = \begin{vmatrix} \frac{1}{r}\,\vec{e}_r & \vec{e}_\phi & \frac{1}{r}\,\vec{e}_z \\ \frac{\partial}{\partial r} & \frac{\partial}{\partial \phi} & \frac{\partial}{\partial z} \\ C/r^3 & rb & 0 \end{vmatrix} = \frac{b}{r}\,\vec{e}_z \neq \vec{0}$$

$\Rightarrow$ $\vec{F}$ is non-conservative and V does not exist. $F_r = C/r^3$, $F_\phi = b$, $F_z = 0$.

Alternative method: Assume V exists and try to obtain V by indefinite integration.

$$F_r = -\frac{\partial V}{\partial r} \quad \Rightarrow \quad \frac{\partial V}{\partial r} = -\frac{C}{r^3} \quad \Rightarrow \quad V = \frac{C}{2r^2} + g(\phi, z)$$

$$F_\phi = -\frac{1}{r}\frac{\partial V}{\partial \phi} \quad \Rightarrow \quad \frac{\partial g}{\partial \phi} = -br \quad \Rightarrow \quad g = -br\phi + f(z)$$

This is an inconsistency as $g(\phi, z)$ is showing dependence on r $\Rightarrow$ V does not exist.

(b) $W = \int \vec{F} \cdot d\vec{r} = \int (F_r\, dr + F_\phi r\, d\phi + F_z\, dz) = \int [(C/r^3) dr + rb\, d\phi]$ (1)

$AB:\quad \phi = \phi_1, \quad d\phi = 0, \qquad W_{A-B} \overset{(1)}{=} \int_{R_1}^{R_2} \frac{C}{r^3}\, dr = -\frac{1}{2}C(R_2^{-2} - R_1^{-2})$

$BC:\quad r = R_2, \quad dr = 0, \qquad W_{B-C} \overset{(1)}{=} \int_{\phi_1}^{\phi_2} R_2 b\, d\phi = R_2 b(\phi_2 - \phi_1)$

Similarly, $\qquad W_{C-D} \overset{(1)}{=} \int_{R_2}^{R_1} \frac{C}{r^3}\, dr = -\frac{1}{2}C(R_1^{-2} - R_2^{-2})$

$$W_{D-A} \overset{(1)}{=} \int_{\phi_2}^{\phi_1} R_1 b\, d\phi = R_1 b(\phi_1 - \phi_2)$$

$$\begin{aligned} W_{A-B-C-D-A} &= W_{A-B} + W_{B-C} + W_{C-D} + W_{D-A} \\ &= b(\phi_2 - \phi_1)(R_2 - R_1) \neq 0 \end{aligned}$$

For path $FG:\quad r = d\cos\phi$,

$$W_{F-G} \overset{(1)}{=} \int_{\vec{r}_F}^{\vec{r}_G} \left(\frac{C}{r^3}\, dr + rb\, d\phi\right) = \int_d^{d/\sqrt{2}} \frac{C}{r^3}\, dr + \int_0^{\pi/4} bd\cos\phi\, d\phi = -\frac{C}{2d^2} + \frac{bd}{\sqrt{2}}$$

(c) $\quad F_r = C/r^3, \qquad F_\phi = b/r, \qquad F_z = 0,$

$$\vec{\nabla} \times \vec{F} = \begin{vmatrix} \frac{1}{r}\vec{e}_r & \vec{e}_\phi & \frac{1}{r}\vec{e}_z \\ \frac{\partial}{\partial r} & \frac{\partial}{\partial \phi} & \frac{\partial}{\partial z} \\ F_r & rF_\phi & F_z \end{vmatrix} = \begin{vmatrix} \frac{1}{r}\vec{e}_r & \vec{e}_\phi & \frac{1}{r}\vec{e}_z \\ \frac{\partial}{\partial r} & \frac{\partial}{\partial \phi} & \frac{\partial}{\partial z} \\ C/r^3 & b & 0 \end{vmatrix} = \frac{0}{r}(\vec{e}_r + \vec{e}_z) = \vec{0} \quad \text{for } r \neq 0$$

$\vec{F}$ is non-conservative in the doubly-connected region R, but is conservative in the simply-connected region R_1 (Fig. E2.24b,c). O cannot be datum as $O \notin R_1$. A simple path from the datum at F to $P(r, \phi)$ is radial from F to Q and circular from Q to P:

$$V_P = -\int_d^r (F_r|_{\phi=0})\, dr - \int_0^\phi rF_\phi\, d\phi = \frac{C}{2r^2} - \frac{C}{2d^2} - b\phi$$

Example 2.25: (a) Find $\dot{W}$ for the force $F = F_0 e^{-\alpha t}$ at time t when angular velocity of the rigid wheel is ω and velocity of O is v_O (Fig. E2.25). Angle ϕ and the distances h and b are constant. (b) Find the work done by $F = F_0$ ($\alpha = 0$), during the time O covers distance d. Assume no slip. Consider cases: $\phi = 0$, $\phi \neq 0$.

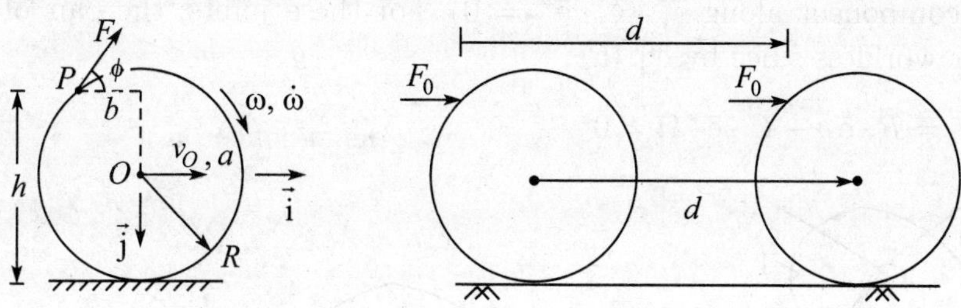

Fig. E2.25

Solution: (a) The force $\vec{F} = F_0 e^{-\alpha t}(\cos\phi\,\vec{\mathrm{i}} - \sin\phi\,\vec{\mathrm{j}})$ acts on different material points P at different instants. Hence, work is obtained by integrating $\dot{W}$ with respect to time.

$$\begin{aligned}\vec{v}_P &= \vec{v}_O + \vec{\omega} \times \overrightarrow{OP} = v_O\,\vec{\mathrm{i}} + \omega\,\vec{\mathrm{k}} \times [-b\,\vec{\mathrm{i}} - (h-R)\,\vec{\mathrm{j}}\,] \\ &= [v_O + \omega(h-R)]\,\vec{\mathrm{i}} - \omega b\,\vec{\mathrm{j}} \\ \dot{W} &= \vec{F}\cdot\vec{v}_P = F_0 e^{-\alpha t}[\{v_O + \omega(h-R)\}\cos\phi + \omega b \sin\phi] \qquad (1)\end{aligned}$$

(b) For the case of no slip, $\omega = v_O/R$ and for $\phi = 0, \alpha = 0$:

$$\dot{W} \overset{(1)}{=} F_0[v_O + \omega(h-R)] = \frac{hF_0 v_O}{R} \quad \Rightarrow \quad W = \frac{hF_0}{R}\int_0^t v_O\,dt = \frac{hF_0 d}{R}$$

The work done by constant force F_0: $W \neq F_0 d$ where d is the distance between the final and initial points of application of force. Similarly, for $\omega = v_O/R$, $\alpha = 0$, $\phi \neq 0$:

$$\dot{W} \overset{(1)}{=} (h\cos\phi + b\sin\phi)\,\frac{F_0 v_O}{R} \quad \Rightarrow \quad W = (h\cos\phi + b\sin\phi)\,\frac{F_0 d}{R}$$

2.22 Workless Force Systems

A force system which does no work on a body in any motion is called a workless system. Their identification is useful in the application of work–energy relations. The rate of work done by a pair of force reactions $\vec{R}, -\vec{R}$ and moment reactions $\vec{C}, -\vec{C}$ on rigid bodies 1 and 2 (Fig. 2.69a) at a joint/contact at A and B, or by $\vec{R}$ and $\vec{C}$ alone if body 2 is fixed, is given by

$$\dot{W} = \vec{R}\cdot(\vec{v}_A - \vec{v}_B) + \vec{C}\cdot(\vec{\omega}_1 - \vec{\omega}_2) \qquad (a)$$

We first prove three specific force systems to be workless, which are applied later.
(i) Consider frictionless joints (Fig. 2.69b) with A coinciding with B, and with constraint on relative displacement and relative rotation such that $\vec{v}_{AB} = \vec{v}_A - \vec{v}_B = v\,\vec{e}$, $\vec{\omega}_{1|2} = \vec{\omega}_1 - \vec{\omega}_2 = \Omega\,\vec{e}^*$. Hence, $\vec{R}$ has no component along $\vec{e}$ ($\vec{R}\cdot\vec{e} = 0$) and $\vec{C}$ has no component along $\vec{e}^*$ ($\vec{C}\cdot\vec{e}^* = 0$). For these joints, the pair of reactions are together workless since by eq. (a):

$$\dot{W} = \vec{R}\cdot\vec{e}\,v + \vec{C}\cdot\vec{e}^*\,\Omega = 0 \qquad (b)$$

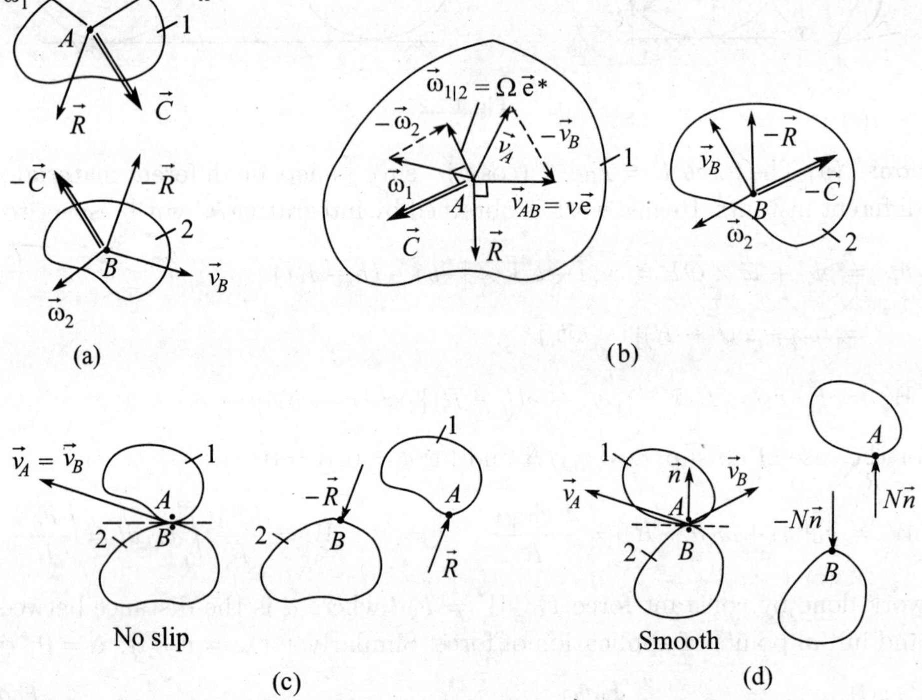

Fig. 2.69

(ii) For point contact (Fig. 2.69c) with no slip, $\vec{v}_A = \vec{v}_B$ and $\vec{C} = \vec{0}$, eq. (a) $\Rightarrow$ $\dot{W} = 0$.
(iii) For smooth point contact (Fig. 2.69d) with or without slip, $\vec{C} = \vec{0}$, $\vec{R} = N\vec{n}$, where $\vec{n}$ is normal to the tangent plane at the point of contact, and eq. (a) yields

$$\dot{W} = \vec{R}\cdot(\vec{v}_A - \vec{v}_B) = N\vec{n}\cdot(\vec{v}_A - \vec{v}_B) = N(v_{An} - v_{Bn}) = 0 \qquad (c)$$

as $N = 0$ for impending separation, or $v_{An} = v_{Bn}$ for no separation.

Using the expression (a) for $\dot{W}$, some examples of pairs of reactions/reactions which are not workless occur at: sliding contact with friction, joint with driving force, joint with driving torque, joint with frictional torque; and connections involving extensible cable, spring, damper, rotational spring or rotational damper.

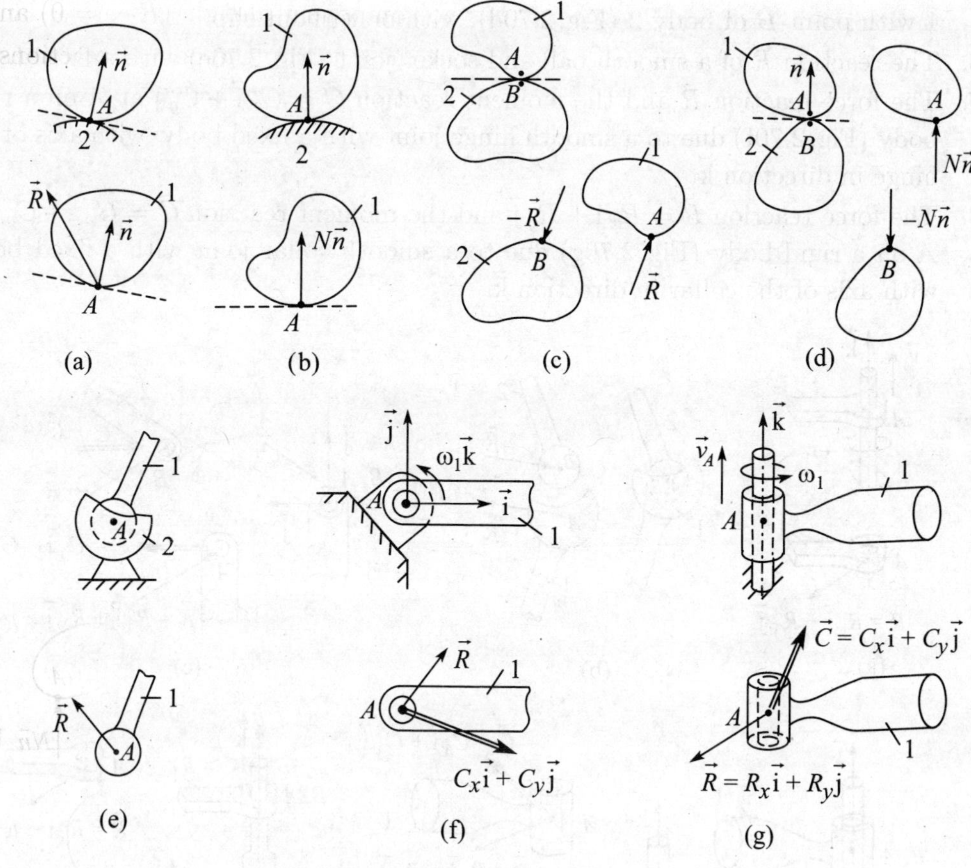

Fig. 2.70

Based on eqs (b) and (c), we list below in items 3 to 13 some force systems which are workless while acting on a body and some pairs of force systems acting on two bodies which together are workless, though individually each may perform work.

1. *Gyroscopic force* is a velocity dependent force $\vec{F}(\vec{r}, \vec{v})$, which always acts at right angles to the velocity $\vec{v}$ of its point of application. It is workless since $\dot{W} = \vec{F} \cdot \vec{v} = 0$. For example, the force $\vec{F} = q\vec{v} \times \vec{B}$ on a point charge q moving with velocity $\vec{v}$ in a magneto-static field $\vec{B}$ is workless.
2. The reaction $\vec{R}$ at contact A with a fixed body with no slip (Fig. 2.70a), is workless since $\dot{W} = \vec{R} \cdot \vec{v}_A = 0$ as $\vec{v}_A = \vec{0}$.
3. The reaction $\vec{R} = N\vec{n}$ at A with a smooth fixed body (Fig. 2.70b), with or without slip.

4. The pair of reactions $\vec{R}$ and $-\vec{R}$ at the contact of point A of body 1 with point B of body 2 (Fig. 2.70c) with no slip.
5. The pair of reactions $\vec{R} = N\vec{n}$ and $-\vec{R}$ at a smooth contact of point A of body 1 with point B of body 2 (Fig. 2.70d), with or without slip.
6. The reaction $\vec{R}$ of a smooth ball and socket joint (Fig. 2.70e) with a fixed body.
7. The force reaction $\vec{R}$ and the moment reaction $\vec{C} = C_x\vec{\mathrm{i}} + C_y\vec{\mathrm{j}}$ at A on a rigid body (Fig. 2.70f) due to a smooth hinge joint with a fixed body, with axis of the hinge in direction $\vec{\mathrm{k}}$.
8. The force reaction $\vec{R} = R_x\vec{\mathrm{i}} + R_y\vec{\mathrm{j}}$ and the moment reaction $\vec{C} = C_x\vec{\mathrm{i}} + C_y\vec{\mathrm{j}}$ at A on a rigid body (Fig. 2.70g) due to a smooth collar joint with a fixed body, with axis of the collar in direction $\vec{\mathrm{k}}$.

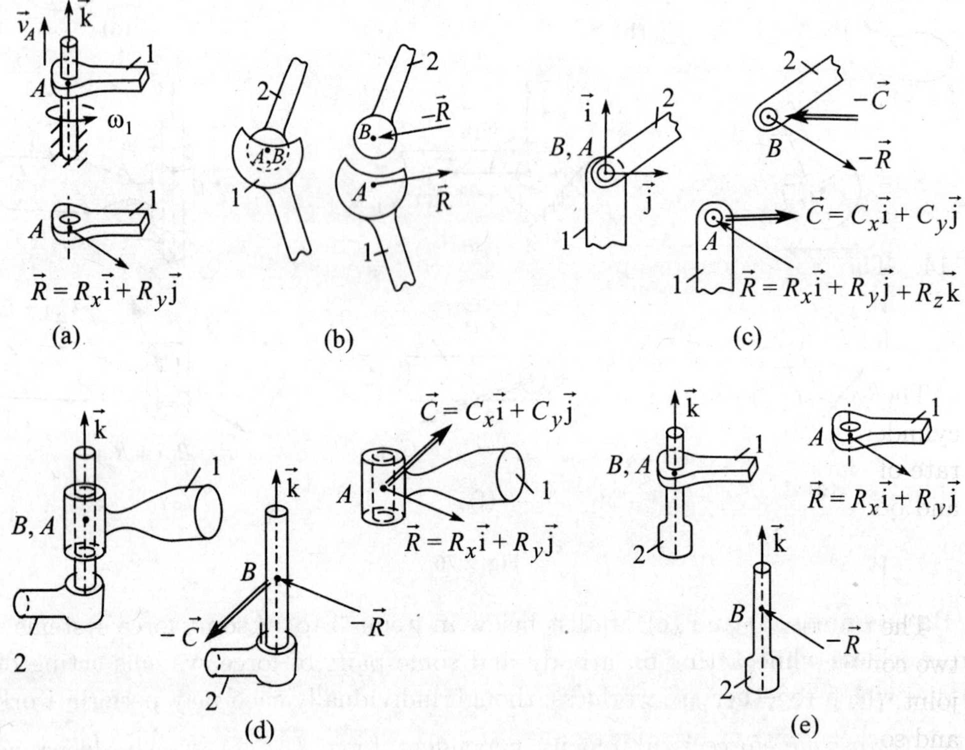

Fig. 2.71

9. The force reaction $\vec{R} = R_x\vec{\mathrm{i}} + R_y\vec{\mathrm{j}}$ at A on a rigid body (Fig. 2.71a), due to a smooth eye-bolt joint with a fixed body, with the bolt axis in direction $\vec{\mathrm{k}}$.
10. The pair of reactions $\vec{R}$ and $-\vec{R}$ on two bodies (Fig. 2.71b) due to a smooth ball and socket joint connecting them.
11. The pair of force reactions $\vec{R}$, $-\vec{R}$ and the pair of moment reactions $\vec{C} = C_x\vec{\mathrm{i}}+$

$C_y\vec{j}$, $-\vec{C}$ on two rigid bodies (Fig. 2.71c) due to a smooth hinge joint connecting them, with axis of hinge in direction $\vec{k}$.

12. The pair of force reactions $\vec{R} = R_x\vec{i}+R_y\vec{j}$, $-\vec{R}$ and the pair of moment reactions $\vec{C} = C_x\vec{i}+C_y\vec{j}$, $-\vec{C}$ on two rigid bodies (Fig. 2.71d) due to a smooth collar joint, with axis of the collar in direction $\vec{k}$.
13. The pair of reactions $\vec{R} = R_x\vec{i} + R_y\vec{j}$, $-\vec{R}$ on two bodies (Fig. 2.71e) due to a smooth eye-bolt joint connecting them, with axis of the eye-bolt in direction $\vec{k}$.

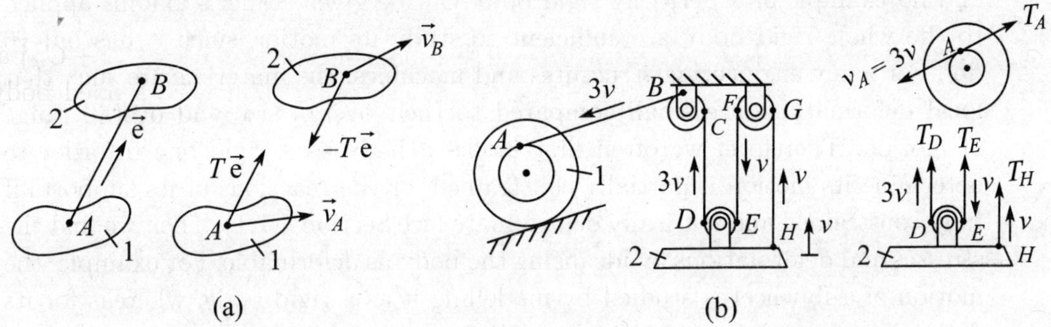

Fig. 2.72

14. The pair of tensions in a light inextensible cable (Fig. 2.72a) are workless since $\dot{W} = T(\vec{e}\cdot\vec{v}_A - \vec{e}\cdot\vec{v}_B) = 0$, as either $T = 0$ for impending slackness or $\vec{e}\cdot\vec{v}_A = \vec{e}\cdot\vec{v}_B$ for taut inextensible cable.

The system shown in Fig. 2.72b consists of 3 light pulleys, a light cable, a stepped cylinder 1 and body 2. $T_A = T_B = T_C = T_D = T_E = T_F = T_G = T_H$ and the total rate of work done by T_A, T_D, T_E, T_H, on the system of body 1 with cable around it and body 2 with the pulley on it and the cable around it, is given by

$$\dot{W} = T_A v_A + T_D v_D + T_E v_E + T_H v_H = T_A(-3v + 3v - v + v) = 0$$

The reactions on a rigid body connected to a fixed body, and the pair of reactions on two connected rigid bodies, are workless if the connection is by (i) a smooth universal joint, (ii) a clevis-type joint, (iii) a light rod connected by two hinge joints, two ball and socket joints or by a hinge joint and a ball and socket joint to the two bodies.

2.23 Deformable Body, Rigid Body and Mass-point Models

The 3 models for a physical system are: deformable body, rigid body and mass-point.

1. **Deformable body model:** The dof of a deformable body are infinite since the position of every material point is to be specified. Hence, Euler's axioms are applied to an infinitesimal volume which in the limit yield partial differential

equations of motion with the number of unknowns in them being more than the number of equations. The constitutive equations of the material of the body, relate the local internal force distributions (stresses) point-wise and the deformation measures (strains, strain-rates). The strain–displacement or strain-rate–velocity relations complete the set of governing equations.

2. **Rigid body model:** A rigid body has invariant distances between its material points for all time, for all motions and loadings. This is just an idealization and no example of a perfectly rigid body can be given. Euler's axioms applied to the whole rigid body are sufficient to study its motion since it has only 6 dof. For many engineering structures and machines, the materials are such that local deformations are small compared to their overall size and overall range of motion. Therefore, we often first model a body as a rigid one in order to determine its motion if partially constrained, or the reactions of its supports if fully constrained and statically determinate (see Section 4.1.1). Then we find the stresses and deformations by idealising the body as deformable. For example, the motion of a flywheel is studied by modelling it as a rigid body, whereas for its design for strength it is modelled as a deformable body. However, for statically indeterminate problems (see Section 4.1.1), it is not possible to separate the two analyses *even when the deformations are small* and the body has to be modelled as deformable.
3. **Mass-point model ('particle' or point-body):** $\vec{F} = m\vec{a}_{C|I}$ implies that the motion of the centre of mass C of a system is the same as that of a point-body of mass m under the action of a force $\vec{F}$. Thus, a *mass-point/point-mass/particle/point-body* is a model for a body in which the mass and the forces are lumped at its centre of mass C. It has a position and mass m but no size. Irrespective of the size of the system, the particle model is good if we only seek the motion of C, and the total external force on the system is not significantly affected by the motion of the system relative to C. For example, the motion of the centre of mass of a small lead pellet falling in still air can be determined by modelling it as a mass-point, but the motion of the centre of mass of a leaf falling in still air cannot be determined by modelling it as a mass-point since the total external force on the leaf depends on the aerodynamic forces on it, which are sensitive to the rotation of the leaf about its centre of mass. The leaf has to be modelled as a deformable body since it is also quite flexible. A missile falling in still air can be modelled as a rigid body since its deformations are small, but the aerodynamic forces on it also depend on its angular velocity. Hence, it cannot be modelled as a mass-point.

Consider a uniform rigid cylinder of mass m rolling without slip down an inclined plane at an angle α with the horizontal. It can be shown that irrespective of its size, i.e., its radius R, its centre has acceleration $2g\sin\alpha/3$ and the friction force is $mg\sin\alpha/3$. Slip occurs if the coefficient of friction $\mu < \tan\alpha/3$. These results differ from those of a cuboid which slips down the plane with acceleration $g(\sin\alpha - \mu\cos\alpha)$, and the friction force is $\mu mg\cos\alpha$ if $\mu < \tan\alpha$. Thus, a mass-point model of the cylinder to determine the motion of its centre C is wrong, even when R is very small. This is because the friction-force part of the total external force $\vec{F}$ depends on the rotational motion of the body.

The mass-point model may be convenient for a physical problem in which a body is very small in comparison with the distances involved. If we observe a system (a galaxy, a star or a spinning sphere) from a very great distance, it appears as a moving point, and the detailed local motions of rotation and vibration are not observable. Our only interest is about the *observable* motion of a body's centre of mass. This motion is independent of the way the external forces are distributed and the way its parts interact and move relative to each other, *provided the total external force $\vec{F}$ is the same.* Such a system can be modelled as a mass-point.

From tiny protein molecules, particles of pollutants, satellites, planets, stars, up to large galaxies, encompassing a wide range—10^{40} in length size and 10^{80} in mass—many physical systems are modelled as mass-points for the study of the motion of their centre of mass.

Review 2 A

1. The moment of $\vec{F}$ about a line parallel to it, and about a line intersecting it is 0.
2. The moment of a couple about all points is the same.
3. $\vec{p} = m\vec{v}_C = \sum m_i\vec{v}_{C_i}$, $\quad \vec{H}_A = \vec{H}_C + \vec{r}_{CA} \times m\vec{v}_{CA}$, $\quad T = \frac{1}{2}mv_C^2 + \frac{1}{2}\int_m v_{PC}^2\, dm$.
4. $\vec{F} = \dot{\vec{p}} = m\vec{a}_C = \sum m_i\vec{a}_{C_i}$, $\quad \dot{\vec{H}}_O = \vec{M}_O$ for O fixed in I.
5. Two equivalent force systems acting separately on the same material system with the same initial conditions cause the same $\vec{a}_{C|I}$. For *a rigid body,* they cause the same motion and $\dot{W}$, but may cause *different internal force distributions.*
6. The resultant/equivalent system at A is a force $\vec{F}_R = \vec{F}$ and a couple $\vec{C}_{R_A} = \vec{M}_A$.
7. The simplest resultant is a wrench. It simplifies to a force/couple/null system if $\vec{F}_R \cdot \vec{C}_{R_A} = 0$, which holds for concurrent, coplanar, and parallel force systems.
8. The centre R of parallel forces is located at $\vec{r}_R = \sum \vec{r}_i F_i / \sum F_i$.
9. For a distributed parallel force $\perp$ to a plane curve, $F_R =$ algebraic area of loading surface (if $\neq 0$) and acts through its C^*. For a distributed force $\perp$ to a plane area, $F_R =$ algebraic volume of pressure space (if $\neq 0$) and acts through its C^*.
10. $W = \int \vec{F}(t) \cdot \vec{v}(t)\, dt$. For $\vec{F}$ acting at same material point, $W = \int_{C:\,\vec{r}(t_1)}^{\vec{r}(t_2)} \vec{F}(t) \cdot d\vec{r}$.
11. A is a point of a rigid body, $\dot{W} = \vec{F} \cdot \vec{v}_A + \vec{M}_A \cdot \vec{\omega}$. For a couple $\vec{C}$: $\dot{W} = \vec{C} \cdot \vec{\omega}$.

12. $\dot{\vec{H}}_{A|I} = \vec{M}_A$ if $\vec{a}_{A|I} = \vec{0} \,/\, A \equiv C \,/\, \vec{a}_{A|I} \parallel \vec{r}_{CA}$.
13. For plane motion of a rigid body: $\vec{M}_A = \dot{\vec{H}}_A$ is *not always* valid for A on instantaneous axis of rotation ($\vec{e}$) in a plane through C normal to $\vec{e}$.
14. For any system and any point A: $\vec{M}_A = \dot{\vec{H}}_C + \vec{r}_{CA} \times m\vec{a}_C$.
15. For T translating at $\vec{a}_T$ relative to I, the equations of motion relative to T are identical to those relative to I if we replace $m\vec{g}$ by $m\vec{g}_e$ where $\vec{g}_e = \vec{g} - \vec{a}_T$.
16. For part P of a system in equilibrium/light system in motion: $\vec{F}(P) = \vec{0}$, $\vec{M}_A(P) = \vec{0}$
17. For a member in equilibrium/light member in motion: (i) If subjected only to $\vec{F}_1$ at A and $\vec{F}_2$ at B, then $\vec{F}_1 = -\vec{F}_2$ and both act along AB. (ii) If subjected to 3 forces, then these are coplanar, and concurrent/parallel.
18. $\vec{I} = \Delta\vec{p} = m\Delta\vec{v}_C = \sum m_i \Delta\vec{v}_{C_i}$.
19. $\vec{I}_{\mathrm{ang}_A} = \Delta\vec{H}_A$ if $\vec{a}_{A|I}(t) \equiv \vec{0} \,/\, A \equiv C \,/\, \vec{a}_{A|I}(t) \parallel \vec{r}_{CA}(t)$.
20. For impulsive loads at t_1: $\vec{I}_{\mathrm{ang}_C}(t_1) = \Delta\vec{H}_C$. $\vec{I}_{\mathrm{ang}_A}(t_1) = \Delta\vec{H}_A$ if $\Delta\vec{v}_A = \vec{0}$.
21. $\dot{W}^* = \vec{F} \cdot \vec{v}_C$, $T_C = \frac{1}{2} m v_C^2$, $\dot{T}_C = \dot{W}^*$, $T_{C_2} - T_{C_1} = W^*_{1-2}$.
22. For no slip, $\vec{F}$ is arbitrary in magnitude and direction but bounded by $|\vec{F}| < \mu_s N$.
23. Displacement constraint $\Rightarrow$ force reaction. Rotation constraint $\Rightarrow$ moment reaction. Smooth hinge with axis z, reactions are $R_1\vec{\imath} + R_2\vec{\jmath} + R_3\vec{k}$ and $C_1\vec{\imath} + C_2\vec{\jmath}$.
24. Impending slip of belt: $\dfrac{T_1 - \lambda v^2}{T_2 - \lambda v^2} = e^{\mu_s \theta / \sin\alpha}$, $p = \dfrac{T - \lambda v^2}{\rho}$. T_1 along slip.
25. Thrust bearing: $M/P = [\int_s \mu_s r^2 p\, ds] \,/\, [\int_s r p \sin\alpha\, ds]$, p at r at $\pi/2 - \alpha$ to axis.
26. Screw-jack: $M/P = r\tan(\alpha + \theta)$, where $\theta = \tan^{-1}\mu_s$. Self-locking if $\tan\alpha < \mu_s$.
27. Conservative $\vec{F} \Rightarrow \vec{F}(\vec{r})$: work path-independent, work in closed path is 0, $\vec{\nabla} \times \vec{F} = \vec{0}$, $dW = -dV$, $F_y = -\partial V/\partial y$, $F_r = -\partial V/\partial r$, $F_\phi = -\partial V/r\partial\phi$.
28. $V(\vec{r}) - V(\vec{0}) = -\int_0^x F_x(x,0,0)dx - \int_0^y F_y(x,y,0)dy - \int_0^z F_z(x,y,z)dz$
$= -\int_0^r F_r(r,\phi,0)dr - \int_0^z F_z(r,\phi,z)dz$.
29. For constant $\vec{F}$: $V = -\vec{F} \cdot \vec{r} = -F_x x - F_y y - F_z z$.
30. For constant couple $\vec{C}$ on a rigid body with $\vec{\omega}(t) = \dot{\theta}(t)\,\vec{k}$: $V = -C_z\theta$.
31. For $\vec{F} = F_r(r)\,\vec{e}_r$: $V(r) = -\int_{r_0}^r F_r(r)\,dr$.
32. Spring, extension e, $F(e)$, $V(e) = \int_0^e F(e)\,de$. Linear $k \Rightarrow V = \frac{1}{2}ke^2$.
33. Torsional spring, twist θ, $M(\theta)$, $V(\theta) = \int_0^\theta M(\theta)\,d\theta$. Linear $k_t \Rightarrow V = \frac{1}{2}k_t\theta^2$.
34. For pair of $\vec{R}, -\vec{R}$ and $\vec{C}, -\vec{C}$ on bodies 1, 2 at smooth joint/contact or at rough contact with no slip: $\dot{W} = 0$. Not so for internal driving moment/driving force.

2.24 Concept Review Questions 2A

1. Reference frame 1 is in translation with respect to frame 2. Which of the measurements of $\vec{p}$, $\dot{\vec{p}}$, $\vec{H}_A$, $\dot{\vec{H}}_A$, T, $\dot{T}$, W, $\dot{W}$ for a system made by two observers in these frames will be the same? [Only $\vec{H}_A$, $\dot{\vec{H}}_A$]

2. (a) The resultant of a force system at A is a force $\vec{F}_{R_A}$ and a couple $\vec{C}_{R_A}$. Find its resultant at B. (b) $\vec{F}_{R_A} = 4\vec{\imath} - 3\vec{\jmath} + 5\vec{k}$ kN, $\vec{C}_{R_A} = -2\vec{\imath} - 4\vec{\jmath} + \vec{k}$ kN.m. Find the couple of minimum magnitude to balance this force system in addition to an appropriate force. [(a) $\vec{F}_{R_A}$, $\overrightarrow{BA} \times \vec{F}_{R_A} + \vec{C}_{R_A}$ (b) $-0.72\vec{\imath} + 0.54\vec{\jmath} - 0.9\vec{k}$ kN.m]

3. Find the simplest resultant of the force systems shown in Fig. Q2.3a,b.

   [(a) $-10\vec{k}$ kN at $12.8\vec{\imath} - 11.6\vec{\jmath}$ m (b) $-80\vec{\imath} + 210\vec{\jmath}$ kN through $(0, 11.5)$ m]

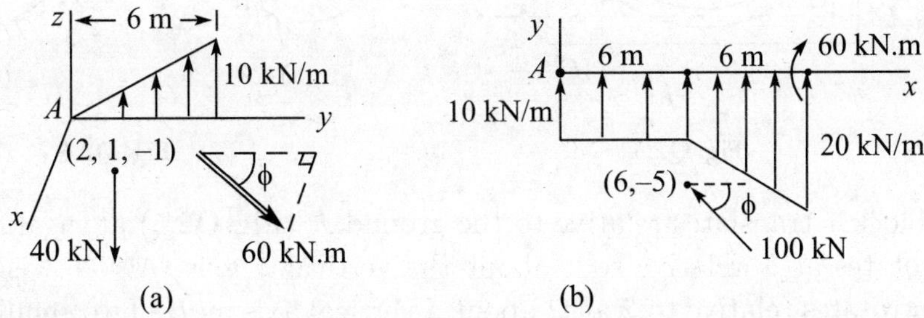

Fig. Q2.3

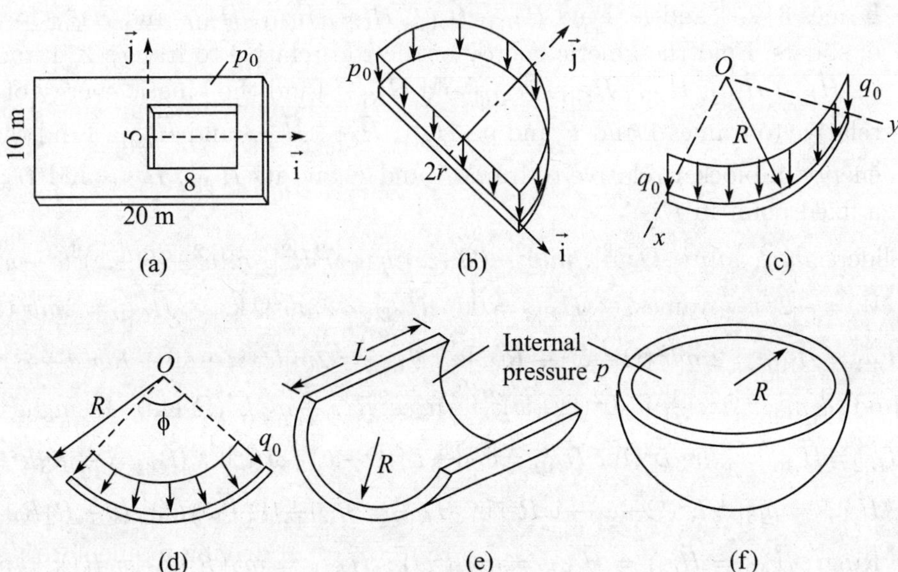

Fig. Q2.4

4. Find the simplest resultant of the uniform normal force distributions on plates (Fig. Q2.4a,b), rods (Fig. Q2.4c,d) and in shells (Fig. Q2.4e,f).

   [(a) $160p_0$ at $-\vec{\imath}$ m (b) $-\frac{1}{2}p_0\pi r^2\,\vec{k}$ at $(4r/3\pi)\vec{\jmath}$ (c) $-(q_0\pi R/2)\,\vec{k}$ at $(2R/\pi)(\vec{\imath}+\vec{\jmath})$ (d) $2q_0R\sin\frac{\phi}{2}\,\vec{e}_r$ thro' O bisecting ϕ (e) $2RLp$ at centre (f) πR^2p at centre]

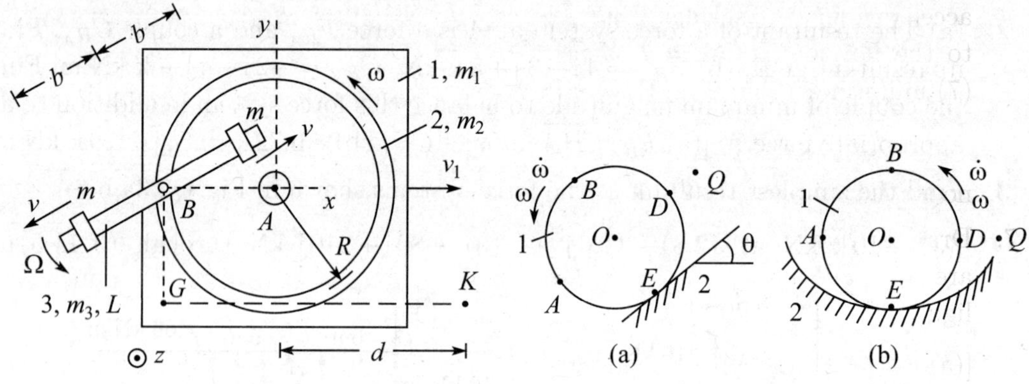

Fig. Q2.5

Fig. Q2.6

5. A block 1 translates relative to the ground F (Fig. Q2.5) at v_1 along x. A ring 2 rotates at ω relative to 1 about the vertical z axis. A rod 3 is pinned to 2 and rotates relative to 2 at Ω about a vertical axis at B. Two small sliders slide at v relative to 3 and are subjected to friction force f each. Find the kinetic energy of the pair of sliders, $\dot{W}$ by the friction forces on the pair relative to frames 3, 2, 1 and F. Find $\vec{H}_{B|3}$, $\vec{H}_{A|2}$, $\vec{H}_{G|1}$, $\vec{H}_{A|1}$, $\vec{H}_{A|F}$ and $\vec{H}_{K|F}$ for the pair of sliders. Find the kinetic energy of the rod relative to frames 2, 1 and F, and its $\vec{H}_{B|2}$, $\vec{H}_{A|2}$, $\vec{H}_{A|1}$, $\vec{H}_{G|1}$, $\vec{H}_{A|F}$ and $\vec{H}_{K|F}$. Find the kinetic energy of the ring relative to frames 1 and F and its $\vec{H}_{A|1}$, $\vec{H}_{G|1}$, $\vec{H}_{A|F}$ and $\vec{H}_{K|F}$. Find the kinetic energy of block 1 relative to frame 1 and F and its $\vec{H}_{A|1}$, $\vec{H}_{G|F}$ and $\vec{H}_{K|F}$. K is a fixed point in F.

[sliders: mv^2, $m[v^2+\Omega^2b^2]$, $m[v^2+(\Omega+\omega)^2b^2+\omega^2R^2]$, $m[v^2+(\Omega+\omega)^2b^2+\omega^2R^2+v_1^2]$
$\dot{W}=-2fv\ \forall$ frames, $\vec{H}_{B|3}=\vec{0}$, $\vec{H}_{A|2}=2mb^2\Omega\,\vec{k}$, $\vec{H}_{G|1}=2mb^2(\Omega+\omega)\,\vec{k}$,
$\vec{H}_{A|1}=\vec{H}_{A|F}=2m[b^2(\Omega+\omega)+R^2\omega]\vec{k}$, $\vec{H}_{K|F}=2m[b^2(\Omega+\omega)+R\omega(d+R)-v_1R]\vec{k}$
Rod: $\frac{1}{24}m_3L^2\Omega^2$, $\frac{1}{24}m_3L^2(\Omega+\omega)^2+\frac{1}{2}m_3\omega^2R^2$, $\frac{1}{24}m_3L^2(\Omega+\omega)^2+\frac{1}{2}m_3(\omega^2R^2+v_1^2)$
$\vec{H}_{B|2}=\vec{H}_{A|2}=\frac{1}{12}m_3L^2\Omega\vec{k}$, $\vec{H}_{A|1}=m_3[\frac{1}{12}L^2(\Omega+\omega)+\omega R^2]\vec{k}$, $\vec{H}_{G|1}=\frac{1}{12}m_3L^2(\Omega+\omega)\vec{k}$,
$\vec{H}_{A|F}=m_3[\frac{1}{12}L^2(\Omega+\omega)+\omega R^2]\,\vec{k}$, $\vec{H}_{K|F}=m_3[\frac{1}{12}(\Omega+\omega)L^2+(d+R)R\omega-v_1R]\,\vec{k}$
Ring: $\vec{H}_{A|1}=\vec{H}_{G|1}=\vec{H}_{A|F}=m_2R^2\omega\,\vec{k}$, $\vec{H}_{K|F}=m_2(R^2\omega-v_1R)\,\vec{k}$, $\frac{1}{2}m_2R^2\omega^2$,
$\frac{1}{2}m_2(R^2\omega^2+v_1^2)$, Block 1: 0, $\frac{1}{2}m_1v_1^2$, $\vec{H}_{A|1}=\vec{H}_{G|F}=\vec{0}$, $\vec{H}_{K|F}=-m_1v_1R\,\vec{k}$]

6. For which of the points A, B, D, E, O, Q is $\vec{M}_P = \dot{\vec{H}}_P$ valid for each of the uniform cylinders in Fig. Q2.6 rolling without slip with $\omega \neq 0$, $\dot{\omega} \neq 0$? Work out by considering i surface 2 is (1) fixed; (ii) surface 2 is fixed to a conveyor translating to the right with (2) uniform velocity, (3) uniform acceleration; (iii) surface 2 is fixed to an elevator translating up with (4) uniform velocity, (5) uniform acceleration; (iv) surface 2 is fixed to an escalator translating up at angle θ to the horizontal (angle θ same as in Fig. Q2.6a) with: (6) uniform velocity, (7) uniform acceleration. How will the answers change if $\omega = 0$?

[No for A, B, D; yes for O, Q; yes for E in cases (1), (2), (4), (6); no for E in cases (3), (7). For E in case (5): no for (a) & yes for (b). No change]

7. Draw FBD of member $ABCDEFJ$, making simplifications for supports which are two-force members (Fig. Q2.7). All contacts are smooth and all members are light. (a) The loading is coplanar. (b) The loading is three dimensional.

[(a) C_0, Q, triangle load, load left of J; R_1, R_2, C_1 at A; N, S, M at J; cable T, normal reactions at slot and roller, 1 reaction component from a straight and a curved 2-force member, 2 reaction components at remaining supports and D.

(b) At hinge of 2-force members of part (a) also couple $c_1\vec{\imath} + c_2\vec{\jmath}$ and force $F_3\vec{k}$. Other hinges: 3 reaction force components and couples $C_{i1}\vec{\imath} + C_{i2}\vec{\jmath}$, 3 reaction force components at E, 3 force reaction components and 3 couple components at A and J, cable tension, reactions at slot and roller support consist of a normal force N and a couple $\perp N$ in xy-plane and applied loads]

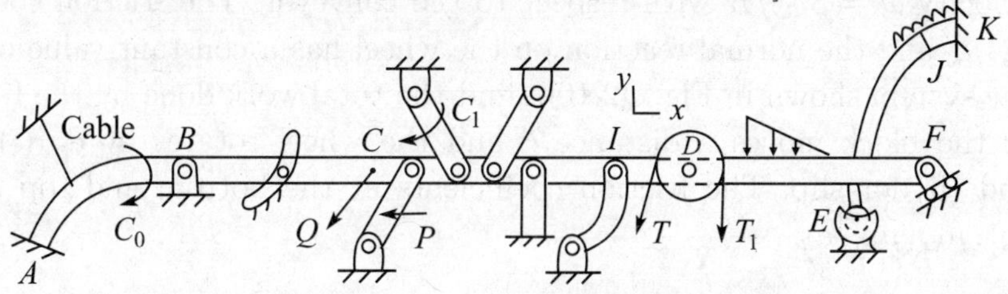

Fig. Q2.7

8^d. (a) Find the condition to be satisfied by μ_1, μ_2 so that the light wedges gripping a specimen ensure no slip (Fig. Q2.8). [Hint: As the specimen deforms under gripping force, the wedges will slip a little on their supports.] (b) A door stopper is a light link hinged to the door. It operates at an angle θ to the vertical. The coefficient of friction with the floor may vary from 0.1 to 0.8. Design the maximum value of θ for the stopper. [(a) $\mu_1 + \tan\alpha < \mu_2(1 - \mu_1 \tan\alpha)$ (b) 5.71°]

9^d. (a) Find h if the sluice gate (Fig. Q2.9a)is designed to open when the water depth equals d. The hydrostatic gauge pressure at a depth y is $\rho g y$. (b) Compare the efficacy of the two designs for supporting two loads P (Fig. Q2.9b). The load is transferred to the tie rods via nuts and washers. The criteria for design are the loads carried by nut connections and the tie rods. [(a) $\frac{1}{3}d$ (b) 1 is better]

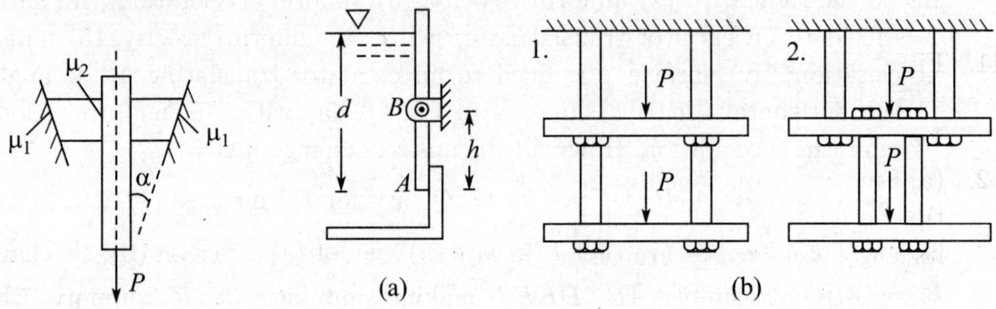

Fig. Q2.8

Fig. Q2.9

10. (a) A rigid wheel rolls without slip on a conveyor T (Fig. Q2.10a) translating relative to the ground G. Constant forces acting are: F_1 at C, Q at the end of a string wrapped around the axle, P at an angle θ with the vertical at a point at distance d above C and the friction force F in the direction of motion of the conveyor. A moment M acts as shown. Find the work done by P, Q, F_1, F, M relative to frames T and G in the period in which C advances a distance b relative to the conveyor and the conveyor moves a distance s in the same direction. (b) Rework part (a) if the wheel rolls with slip with velocity v_C and angular velocity $\omega = 3v_C/R$ with respect to the conveyor. The friction coefficients are μ_s, μ_k and the normal reaction on the wheel has a constant value of N. (c) For the system shown in Fig. Q2.10b, find the total work done by the friction forces as the block moves a distance d and the wheel rotates by θ: (i) for no slip and (ii) for slip. The friction coefficients at the bottom and top contacts are $\mu_{s_1}, \mu_{k_1}, \mu_{s_2}, \mu_{k_2}$.

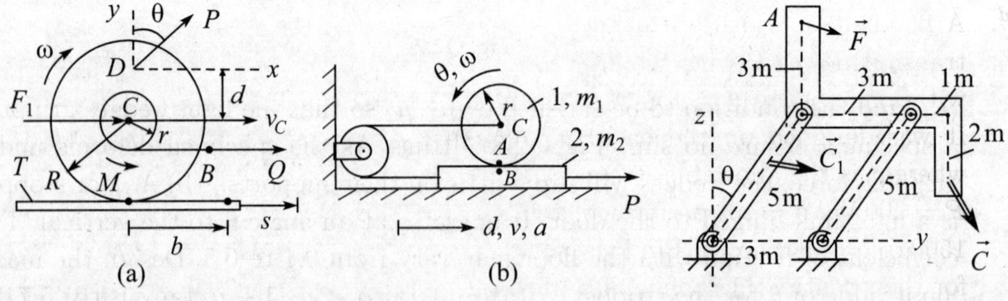

Fig. Q2.10

Fig. Q2.11

[(a) $W_{|T} = [Pb\sin\theta(d+R)/R,\ \ Q(R-r)b/R,\ \ F_1 b,\ \ 0,\ \ -Mb/R]$

(b) $W_{|T} = [Pb\sin\theta(1+3d/R),\ \ Q(1-3r/R)b,\ \ F_1 b,\ \ -2\mu_k Nb,\ \ -3Mb/R]$

For both (a) & (b): $W_{|G} = W_{|T} + [P\sin\theta s,\ \ Qs,\ \ F_1 s,\ \ Fs,\ \ 0]$

(c) (i) $-\mu_{k_1}(m_1+m_2)gd$ (ii) $-\mu_{k_1}(m_1+m_2)gd - \mu_{k_2} m_1 g|2d - r\theta|$]

11. Find the work done by $\vec{F} = 20\vec{\imath} + 5\vec{\jmath} - 4\vec{k}$ kN (Fig. Q2.11) and couples $\vec{C} = -3\vec{\imath}+2\vec{\jmath}-3\vec{k}$, $\vec{C}_1 = -4\vec{\imath}-2\vec{\jmath}-4\vec{k}$ kN.m when the system moves from position 1 at $\theta = \cos^{-1}0.8$ to position 2 at $\theta = \cos^{-1}0.6$. [10.14 kN.m]

12. (a) Find the work done by a force P acting in the circumferential direction at the tip of a telescopic link which is rotated at uniform ω during the interval in which its length L increases from L_1 to L_2 at: (i) uniform rate $\dot{L} = v$, and (ii) uniform $\ddot{L} = a$ with initial $\dot{L} = 0$. (b) Rework part (a) if the link has uniform $\dot{\omega}$ with initial $\omega = 0$. [(a) (i) $P\omega(L_2^2 - L_1^2)/2v$ (ii) $P\omega\sqrt{2(L_2-L_1)/a}(2L_1+L_2)/3$
(b) (i) $P\dot{\omega}(L_2-L_1)^2(L_1+2L_2)/6v^2$ (ii) $P\dot{\omega}(L_2^2-L_1^2)/2a$]

13. Find the work done by: (a) $\vec{F} = \phi(1/r + 1/r^3)\,\vec{e}_r + [(1+\cos\phi)/r^2]\,\vec{e}_\phi$ along the straight path from $A(r_1,\phi_1)$ to $B(r_2,\phi_1)$ followed by the circular path from $B(r_2,\phi_1)$ to $C(r_2,\phi_2)$, (b) $\vec{F} = (c_1\cos\phi/r)\,\vec{e}_r + (c_2/r)\,\vec{e}_\phi$ along the path $r = 1/(a+b\cos\phi)$ from $(1/(a+b), 0)$ to $(1/a, \pi/2)$, (c) $\vec{F} = x\vec{\imath} + yx^2\vec{\jmath} + 2z\vec{k}$ along the circular path of radius 1 m centred at $(2,0,0)$ from $A\ (1,0,0)$ to $B\ (2,1,0)$.
[(a) $\phi_1[\ln(r_2/r_1) - \frac{1}{2}(r_2^{-2} - r_1^{-2})] + [\phi_2 - \phi_1 + \sin\phi_2 - \sin\phi_1]/r_2$
(b) $\frac{1}{2}\pi c_2 + c_1[a\ln\{a/(a+b)\} + b]/b$ (c) 29/12]

14. Find the potential energy of the following forces, if conservative. (a) $4\vec{\imath}-3\vec{\jmath}+5\vec{k}$, (b) $f(x)\vec{\imath}+g(y)\vec{\jmath}+h(z)\vec{k}$, (c) $f[(x^2+y^2)^{1/2}](x\vec{\imath}+y\vec{\jmath})$, (d) $(x\vec{\imath}+y\vec{\jmath}+z\vec{k})g(r)/r$ with $r^2 = x^2+y^2+z^2$, (e) $c\vec{r}/r^5$, (f) $3\dot{x}\vec{\imath} - \dot{z}\vec{\jmath}$, (g) $(\alpha/r)[\cos 2\phi(\vec{e}_r + \cos\phi\,\vec{e}_\phi) + \sin\phi\,\vec{e}_\phi]$, (h) $\vec{A}\sin\omega t$.

[(a) $-4x+3y-5z$ (b) $-\int_0^x f(x)\,dx - \int_0^y g(y)\,dy - \int_0^z h(z)\,dz$ (c) $-\int_0^r rf(r)\,dr$
(d) $-\int_0^r g(r)\,dr$ (e) $c/3r^3$, (f) no (g) no (h) no]

15[d]. A flat belt of width b, thickness t and density $\rho = 1000\,\text{kg/m}^3$ is designed to transmit power P with the driver pulley of radius $r = 0.1$ m, rotating at n rpm, with angle of lap $\theta = 150°$ and $\mu = 0.25$. For belt: $T_{\max} = T_1$ and $T_{\min} = T_2$. It was designed on the basis of $(T_1 - T_2)r\omega = P$, $T_1/T_2 = e^{\mu\theta}$, $T_1 = \sigma bt$, where σ is the allowable stress. The belt designed for $n = 150$ rpm, $\sigma = 2$ MPa, $P = 200$ W, $b = 10t$ performed well. However, a belt designed similarly for $n = 1500$ rpm broke down. Find the dimensions of the two belts, analyse the reason for their contrasting behaviour, and hence re-design the belt for $n = 1500$ rpm.

[150 rpm: 3.641 mm × 36.41 mm. 1500 rpm: 1.151 mm × 11.51 mm, fails due to neglect of λv^2 in tension ratio in design, re-design: 1.23 mm × 12.3 mm]

2.25 Practice Problems 2A

1. Find $\vec{M}_O$, M_{OB} and the equivalent (resultant) force system at O (Fig. P2.1) for (a) tension 700 N in cable ED and couple $\vec{C} = 200\,\vec{i} + 300\,\vec{j} - 400\,\vec{k}$ N.m acting on a hinged plate, (b) force applied to a crank, and (c) weight mg acting at C, where AC has rotated by angle θ about OB from its equilibrium position in the vertical yz plane.

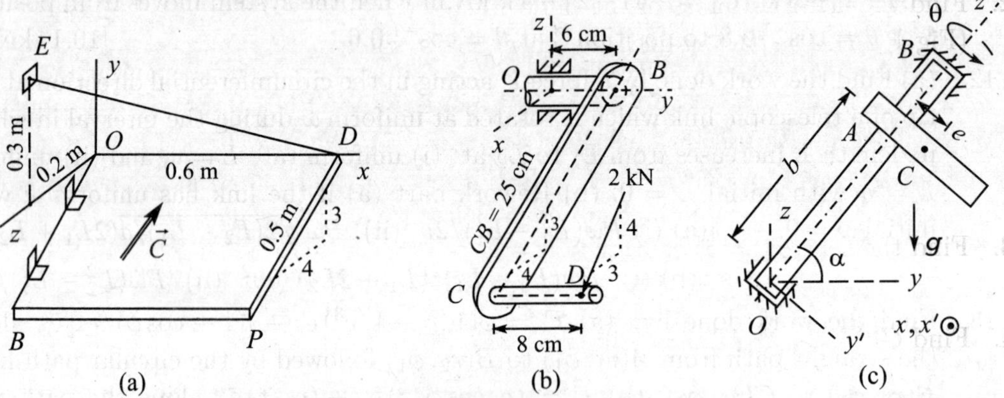

Fig. P2.1

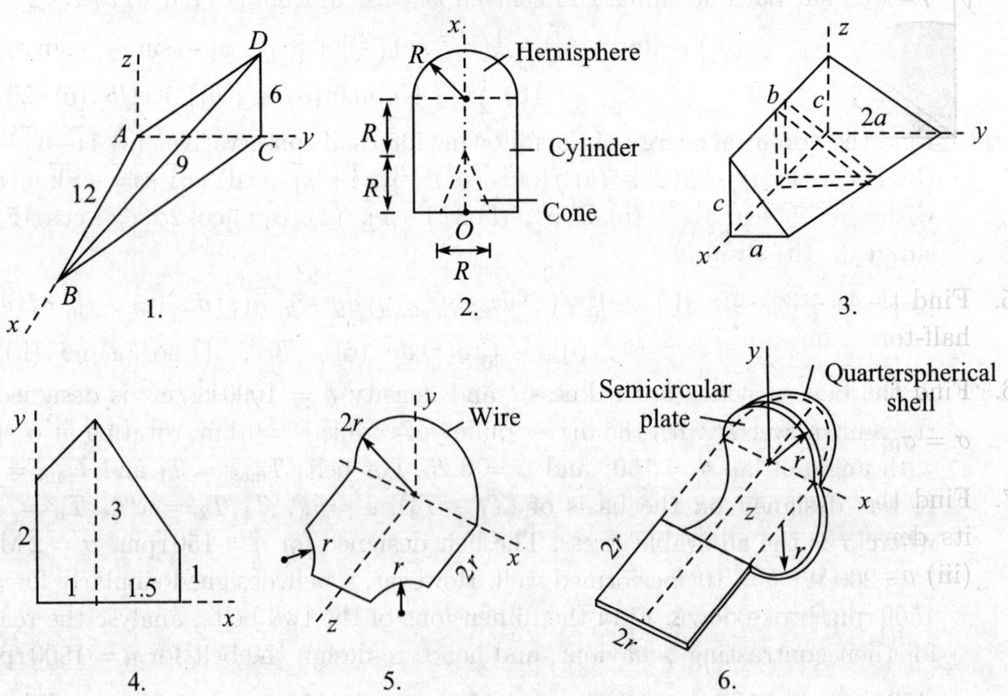

Fig. P2.2

[(a) $\vec{C}_{R_O} = \vec{M}_O = 200\vec{i}+180\vec{j}-220\vec{k},\ M_{OB} = -284\,\text{N.m},\ \vec{F}_R = -600\vec{i}+300\vec{j}+200\vec{k}$ N

(b) $\vec{C}_{R_O} = \vec{M}_O = -0.224\vec{i}+0.14\vec{j}-0.168\vec{k},\ M_{OB} = 0.14\,\text{kN.m},\ \vec{F}_R = 1.2\vec{i}-1.6\vec{k}$ kN

(c) $\vec{C}_{R_O} = \vec{M}_O = -(e\cos\theta\sin\alpha + b\cos\alpha)mg\,\vec{i}' - e\sin\theta\sin\alpha\, mg\,\vec{j}' - emg\sin\theta\cos\alpha\,\vec{k}'$,

$M_{OB} = -emg\sin\theta\cos\alpha,\ \vec{F}_R = -mg\vec{k}$]

2. Find the centre of mass of the uniform solids 1, 2, 3, plate 4, wire 5 and shell 6 (Fig. P2.2). For part 1, find the centre of mass for a thin box too.

[1. $3\vec{i}+4.5\vec{j}+1.5\vec{k}$, $3.434\vec{i}+4.131\vec{j}+1.434\vec{k}$ 2. $1.379R\vec{i}$
3. $\frac{1}{27}(12b\vec{i}+14a\vec{j}+9c\vec{k})$ 4. $1.129\vec{i}+1.167\vec{j}$ 5. $0.5186\,r\vec{j}+1.130\,r\vec{k}$
6. $-0.1175r\vec{j}+1.114r\vec{k}$]

3. Find the centroid of a quarter turn of a helical curve of radius R and pitch p.

[$(2R/\pi)(\vec{i}+\vec{j}) + p\vec{k}/8$]

4. Find C^* of the areas shown in Fig. P2.4.

[(a) $0.4194\, r_0\vec{i} + 0.3500\, r_0\vec{j}$ (b) $2.133\vec{i}+6.095\vec{j}$]

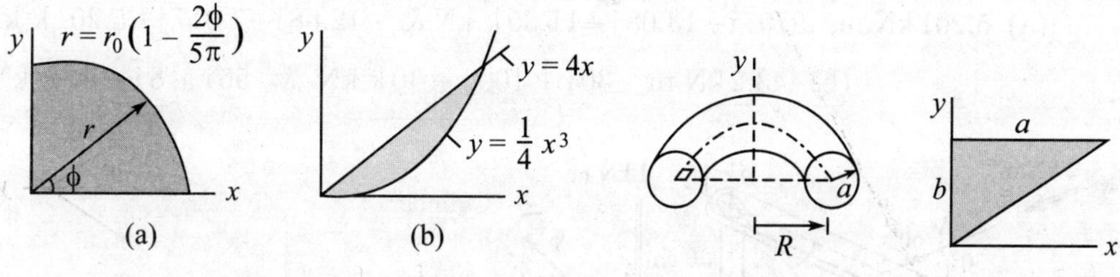

Fig. P2.4 Fig. P2.5 Fig. P2.6

5. Find the centroid of the volume enclosed and of the surface area of the solid half-torus shown in Fig. P2.5. [$(4R^2+a^2)\vec{j}/2\pi R,\ (2R^2+a^2)\vec{j}/(\pi R+a)$]

6. Find the centre of mass of a thin plate (Fig. P2.6) whose density σ varies as: $\sigma = \sigma_0[1+(x/a)(1+y^2/b^2)]$. [$\frac{55}{138}a\vec{i}+\frac{65}{92}b\vec{j}$]

7. Find the centre of mass of a hemisphere of radius R, shown in Fig. P2.7, if its density ρ varies as: (i) $\rho = \rho_0(1+cr/R)$, (ii) $\rho = \rho_0(1+cz/R)$ and (iii) $\rho = \rho_0(1+c\xi/R)$.

[(i) $\frac{3R(1+0.5333c)}{8(1+0.5890c)}\vec{k}$ (ii) $\frac{3R(1+0.5333c)}{8(1+0.375c)}\vec{k}$ (iii) $\frac{3R(1+0.8c)}{8(1+0.75c)}\vec{k}$]

8. Find the resultant force on (a) the bracket and (b) the antenna (Fig. P2.8) due to the applied forces shown. [(a) $22\vec{i} - 1.68\vec{j}$ kN (b) $5.410\vec{i} + 4\vec{j} - 25.27\vec{k}$ kN]

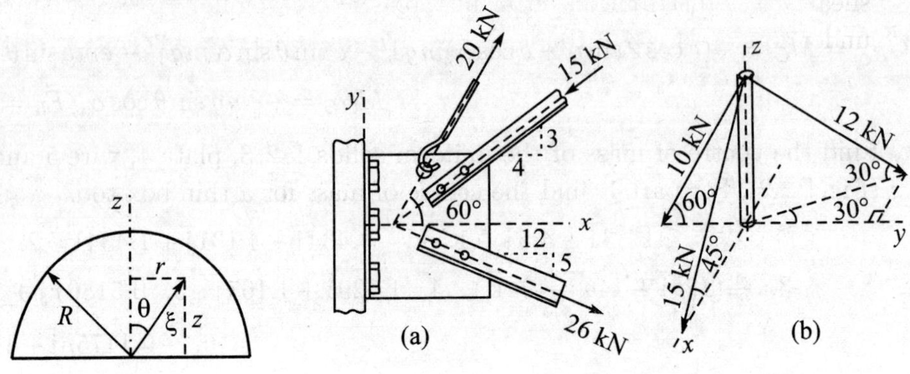

Fig. P2.7 **Fig. P2.8**

9. A column of a machine and a concrete anchor block are subjected to the forces shown in Fig. P2.9. In Fig. P2.9a, $\alpha = 60°$, $\beta = 45°$, $\gamma = 120°$, $BC = CD = 40$ cm, $EF = 20$ cm. Find M_{AB} and resultant at A for each force system.

[(a) 5.201 kN.m, $.9797\vec{i} + 13.08\vec{j} - 11.35\vec{k}$ kN & $-12.68\vec{i} + 7.017\vec{j} + 5.201\vec{k}$ kN.m

(b) 43.2 kN.m, $30\vec{i} + 100\vec{j} + 40\vec{k}$ kN & $56\vec{i} + 8\vec{j} - 12\vec{k}$ kN.m]

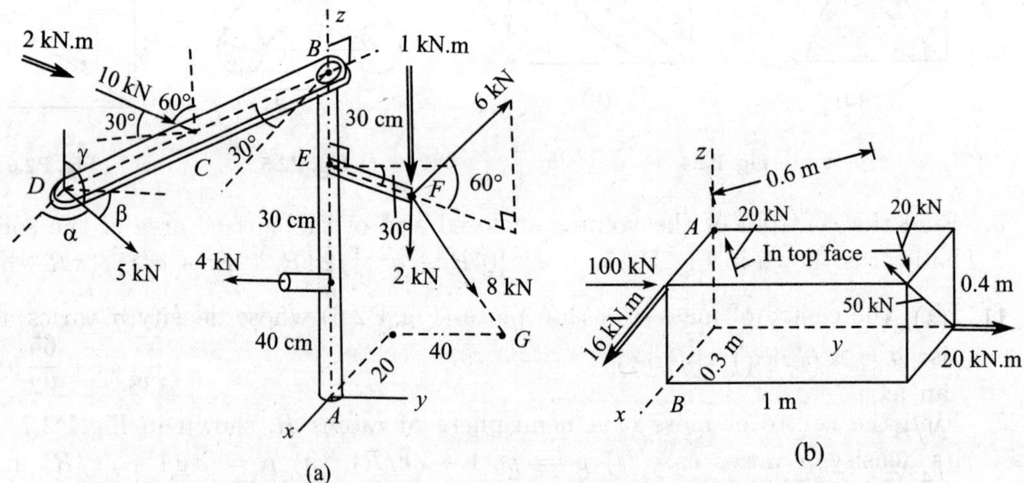

Fig. P2.9

10. (a) A hollow cylinder of radii R_1, R_2 is twisted by end loads distributed tangentially in the circumferential direction $\vec{e}_\phi$ as $\tau = c\,r\,\text{N/m}^2$ (Fig. P2.10a). Prove that its simplest resultant is a couple $\vec{C}_R = \frac{1}{2}\pi c(R_2^4 - R_1^4)\,\vec{k}$. (b) The residual

shear stress distribution on a circular section of a shaft after unloading is shown in Fig. P2.10b. Find its simplest resultant. (c) Find the simplest resultant of the shear stress distribution on a fully plastic rectangular cross-section of a shaft under torsion (Fig. P2.10c). [(b) null system (c) $C_R = (3b - a)a^2\tau_0/6$]

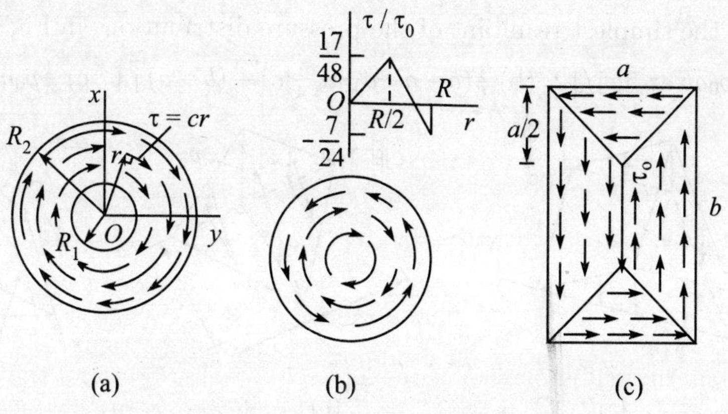

Fig. P2.10

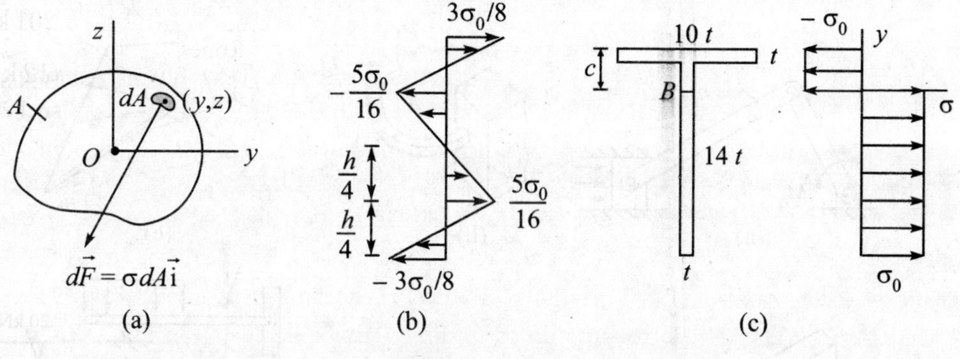

Fig. P2.11

11. (a) A normal stress distribution σ on the cross-section of a beam (Fig. P2.11a) is: $\sigma = a + by + cz$. Find its resultant at the centroid O. If this resultant is an axial force $N\vec{i}$ and a bending moment $M_y\vec{j} + M_z\vec{k}$, then show that $\sigma = N/A + [(I_{yz}M_y - I_{yy}M_z)y + (I_{zz}M_y - I_{yz}M_z)z] / (I_{yy}I_{zz} - I_{yz}^2)$, where $I_{yy} = \int_A z^2\, dA$, $I_{zz} = \int_A y^2\, dA$, $I_{yz} = -\int_A yz\, dA$ are the elements of the inertia matrix of the area of cross-section[6]. If the axes y, z are such that $I_{yz} = 0$, then $\sigma = N/A - M_z y / I_{zz} + M_y z / I_{yy}$ (according to engineering theory of bending of elastic beams.) (b) The residual normal stress, on a rectangular section of width b of a beam, across its depth h is shown in Fig. P2.11b. Find its simplest resultant.

[6]The inertia matrix of area is discussed in detail in Appendix C.

(c) The distribution of normal stress σ on a fully plastic T-section of a beam in pure bending is shown in Fig. P2.11c. Find c so that its simplest resultant is a couple and find its magnitude.

[(a) $\vec{F}_R = aA\,\vec{\imath}$ & $\vec{C}_{RO} = (I_{yy}c - I_{yz}b)\,\vec{\jmath} + (I_{yz}c - I_{zz}b)\,\vec{k}$ (b) null (c) $3t$, $99\sigma_0 t^2$]

12. Find the simplest resultant of the pressure distributions in Fig. P2.12.

[(a) $p_0 ab$ at $5a\,\vec{\jmath}/12$ (b) $\frac{1}{6}(b+a)cp_0$ at $\frac{1}{4}[c\,\vec{\imath} + (b-a)\,\vec{\jmath}]$ c) $\frac{1}{4}p_0 ab$ at $\frac{2}{3}(a\,\vec{\imath} + b\,\vec{\jmath})$]

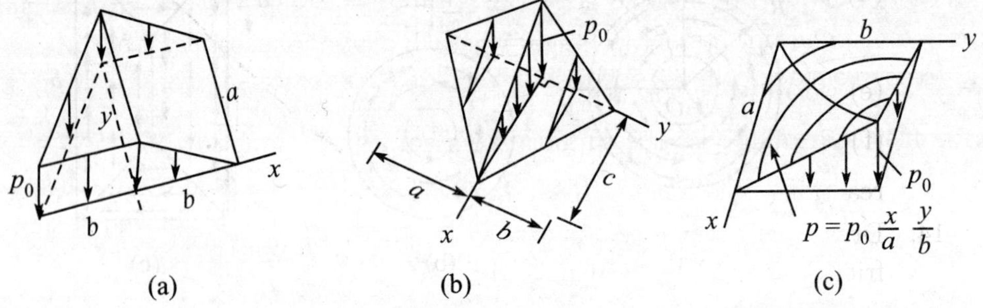

Fig. P2.12

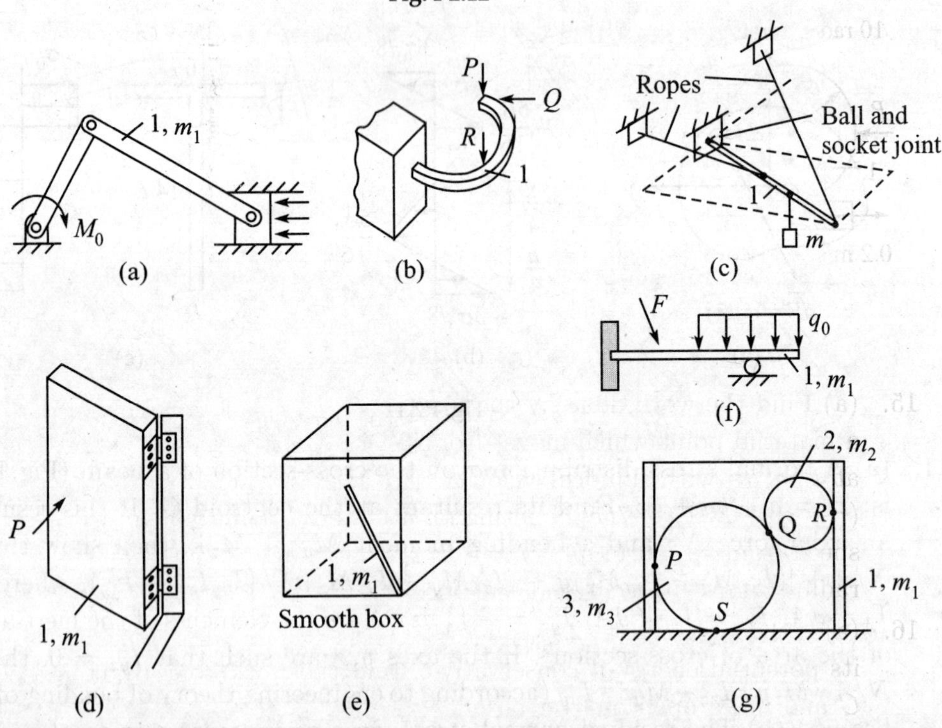

Fig. P2.13

13. Draw FBD of body 1 of mass m_1 in the systems in Fig. P2.13. The cables are

light. Assume all hinge and ball and socket joints and surfaces to be smooth. Make simplifications due to the presence of two-force members, if any. Rework case (a) if member 1 is light.

[(a) R_1, R_2; R_3, R_4 at ends, m_1g at C_1. For light case: only collinear F_1 at ends
(b) $\vec{R}$, $\vec{C}$ at support, weight m_1g at C_1 and given loads P, Q, R
(c) Cable tensions T_1, T_2; weights m_1g, mg; reactions R_1, R_2, R_3
(d) P, weight m_1g at C; $R_1, R_2, R_3, C_1\vec{\text{i}} + C_2\vec{\text{j}}$ at bottom & R_4, R_5, R_6, $C_3\vec{\text{i}} + C_4\vec{\text{j}}$ at top hinges where $\vec{\text{k}}$ along hinge axis. $C_1 = C_2 = C_3 = C_4 \simeq 0$
(e) weight m_1g; R_1, R_2, R_3 at bottom, $R_4\vec{\text{i}} + R_5\vec{\text{j}}$ at top ($\vec{\text{k}}$ vertical)
(f) R_1, R_2, C_1 at left support; N at roller, F, q_0, m_1g (g) weight m_1g, normal reactions N_2, N_3 on walls at P, R; N_1 on bottom at distance d from left corner]

14. Draw FBD of body 1 of mass m for each case in Fig. P2.14. The coefficients of friction are μ_s, μ_k. [Given loads, mg, normal reactions N's and (1) $|F| < \mu_s N$ (2) $\mu_s N$ to right (3) F at incline, $\mu_s N_1$ down (4) F at bottom, $\mu_k N_1$ down]

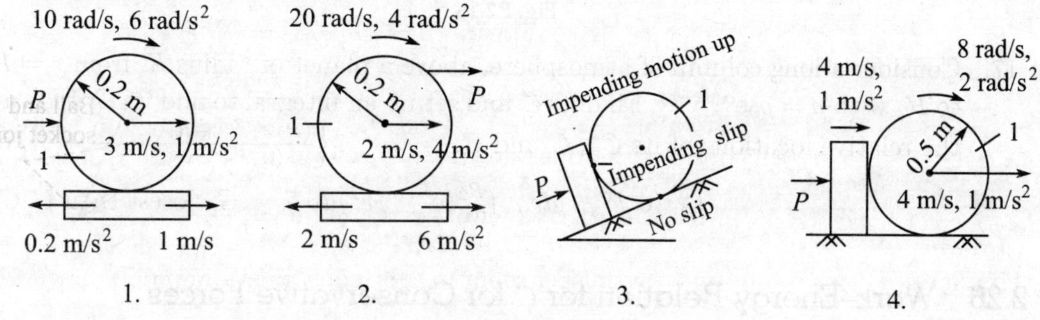

Fig. P2.14

15. (a) Find the work done by the force $\vec{F} = (x^2 + yt)\,\vec{\text{i}} + \cos t\,\vec{\text{j}} + t^2\,\vec{\text{k}}$ N acting on a material point which moves from $\vec{r}_1 = 2\,\vec{\text{j}} + 3\,\vec{\text{k}}$ m at $t = 1$ s to $\vec{r}_2 = 4\,\vec{\text{i}} - 5\,\vec{\text{k}}$ m at $t = 5$ s with uniform velocity. (b) Find the work done by the force $\vec{F} = (r + 8\phi + 2t)\,\vec{\text{e}}_\phi + 7t\,\vec{\text{e}}_z$ N acting on a material point which moves with uniform speed from $\vec{r}_1 = 0.1\,\vec{\text{i}}$ m at $t = 0$ to $\vec{r}_2 = 0.1\,\vec{\text{j}}$ m at $t = 5$ s along a circular arc of radius 0.1 m with centre at the origin. [(a) $-51.10\,J$ (b) $1.788\,J$]

16. (a) Is the force $\vec{F} = (-2xy + yz)\,\vec{\text{i}} + (-x^2 + xz - z)\,\vec{\text{j}} + (xy - y)\,\vec{\text{k}}$ conservative? Find its potential energy if conservative. Find the work done by it in a closed path C and in an open quarter circular path C_1 (Fig. P2.16a). (b) Is the force $\vec{F} = (5x^2 + xy)\,\vec{\text{i}} + (10xy + y^2)\,\vec{\text{j}}$ conservative? Find the work done by it in a closed path (Fig. P2.16b). (c) A plane force field has magnitude prx and is directed towards O. Is it conservative? Find the work done by it for the three paths ACB, ADB

and AB shown in Fig. P2.16c. In each case, the point of application of the force starts at point A and ends at point B. (d) Is the force $\vec{F} = c_1 r^2\,\vec{e}_r + c_2\,\vec{e}_\phi + c_3 z\,\vec{e}_z$ conservative? Find the work done by it when its material point of application moves over a conical helix: $\vec{r}(\phi) = (r_0 - \alpha\phi)\,\vec{e}_r + \beta\phi\,\vec{e}_z$ from $\phi = 0$ to $\phi = \pi$. (e) Rework part (d) for the force $\vec{F} = 2r\phi z^2\,\vec{e}_r + rz^2\,\vec{e}_\phi + 2r^2\phi z\,\vec{e}_z$.

[(a) Yes, $x^2y - (xy - y)z$, 0, -6
(b) No, 875 (c) No, $pa(a^2/3 - b^2/2)$, $p(a^3 - b^3)/3$, $pa(a^2/3 - b^2/6)$
(d) No, $\frac{1}{3}c_1[(r_0 - \alpha\pi)^3 - r_0^3] + c_2[r_0\pi - \frac{1}{2}\alpha\pi^2] + \frac{1}{2}c_3\beta^2\pi^2$ (e) Yes, $\beta^2\pi^3(r_0 - \alpha\pi)^2$]

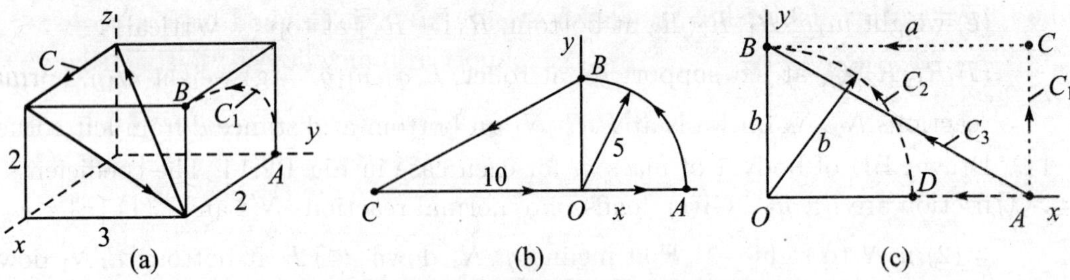

Fig. P2.16

17. Consider a long column of atmosphere, above a planet of radius R, from $r = R_1$ to R_2 with $\rho = \rho_0 e^{\alpha(R-r)}$. Find its C and set up an integral to find its G. Indicate the relative locations of its C^*, C and G. [$\left\{\frac{(R_2e^{-R_2\alpha} - R_1e^{-R_1\alpha})}{(e^{-R_2\alpha} - e^{-R_1\alpha})} + \frac{1}{\alpha}\right\}\vec{e}_r$, $\int_{R_1}^{R_2}(e^{-r\alpha}/r)dr \,/\, \int_{R_1}^{R_2}(e^{-r\alpha}/r^2)dr\,\vec{e}_r$, G closest then C, C^*]

2.26 Work–Energy Relation for C for Conservative Forces

In Sections 2.26 to 2.28, all statements are made with respect to an inertial frame I. For brevity this has not been mentioned explicitly in the equations used/derived.

Let $\dot{W}_c^*$ and $\dot{W}_{nc}^*$ be the rate of work done by the external conservative forces and the external non-conservative forces as if they were acting at the centre of mass C. Let V^* be the potential energy of the former with $\dot{W}_c^* = -\dot{V}^*$. Then, eq. (2.57): $\dot{T}_C = \dot{W}^* = \dot{W}_c^* + \dot{W}_{nc}^* = -\dot{V}^* + \dot{W}_{nc}$, yields

$$\dot{T}_C + \dot{V}^* = \dot{W}_{nc}^*, \qquad (T_{C_2} + V_2^*) - (T_{C_1} + V_1^*) = W_{nc_{1-2}}^* \tag{2.91}$$

with obvious notation in use. If all the forces, as if acting at C, are conservative, then $\dot{W}_{nc}^* = 0$ and eq. (2.91) reduces to conservation of mechanical energy (= kinetic energy + potential energy) associated with C:

$$\dot{T}_C + \dot{V}^* = 0, \qquad T_{C_2} + V_2^* = T_{C_1} + V_1^* \tag{2.92}$$

2.27 Dynamics of a Translating System

A system is *defined to translate* relative to I *if the displacements of its material points are the same for all time.* Hence, $\vec{d}_P(t) = \vec{d}_Q(t) \;\Rightarrow\; \vec{v}_P(t) = \vec{v}_Q(t),\ \vec{a}_P(t) = \vec{a}_Q(t) \;\Rightarrow$ $\dot{\vec{r}}_{PQ} = \vec{v}_{PQ} = \vec{v}_P - \vec{v}_Q = \vec{0} \;\Rightarrow\; \vec{r}_{PQ}(t) = \text{constant} \;\Rightarrow$ the distance between any two material points P and Q remains constant $\Rightarrow$ the system translates as a rigid system with no change in its orientation[7], irrespective of its constitution (rigid/deformable). $\vec{v}_{PA} = \vec{v}_P - \vec{v}_A = \vec{0} \;\Rightarrow\; \vec{v}_{PC} = \vec{0},\ \vec{v}_{iC} = \vec{0}$. Its $\vec{p}$, $\vec{H}_A$ about its material point A, $\vec{H}_C$, $\vec{H}_B$ about any point B, T and $\dot{W}$ of external forces are given by

$$\vec{p} = m\vec{v}_C, \qquad \vec{H}_A(t) \equiv \vec{0} \;\Rightarrow$$

$$\vec{H}_C(t) \equiv \vec{0}, \qquad \vec{H}_B = \vec{r}_{CB} \times m\vec{v}_{CB}$$

$$T = \tfrac{1}{2} m v_C^2 = T_C, \qquad \dot{W} = \vec{F} \cdot \vec{v}_C = \dot{W}^*$$

$\Rightarrow\; \vec{M}_C = \dot{\vec{H}}_C = \vec{0}$. The equations of motion of a translating system are:

$$\vec{F} = m\vec{a}_C, \qquad \vec{M}_C = \vec{0} \tag{2.93}$$

Thus, the resultant at C of the external force system is a single force $\vec{F}_R = \vec{F} = m\vec{a}_C$ since $\vec{C}_R = \vec{M}_C = \vec{0}$ (Fig. 2.73b). For any arbitrary point B, and any arbitrary line BD in direction $\vec{e}$, $\vec{M}_B = \vec{M}_C + \vec{r}_{CB} \times \vec{F} = \vec{r}_{CB} \times m\vec{a}_C$, i.e.,

$$\vec{M}_B = \vec{r}_{CB} \times m\vec{a}_C, \qquad M_{BD} = [\vec{r}_{CB} \times m\vec{a}_C] \cdot \vec{e} \tag{2.94}$$

It follows from eq. $(2.94)_1$ that

$$\vec{M}_B = \vec{0}, \qquad \text{if } \overrightarrow{BC} \text{ is in the direction of } \vec{a}_C \tag{2.95}$$

i.e., moment is zero *only* about any point on the line of acceleration vector $\vec{a}_C$ through C. To avoid solving many simultaneous equations, one should apply eq. (2.94) at a point where most unknown reactions act, or about a line through which they act.

Note that for a translating system (Fig. 2.73c) the moment is *not* zero about *any arbitrary* point, e.g., $\vec{M}_C = \vec{M}_D = \vec{M}_E = \vec{M}_H = \vec{M}_G = \vec{M}_K = \vec{0}$, though $\vec{M}_K \neq \dot{\vec{H}}_K$; and $\vec{M}_a \neq \vec{0},\ \vec{M}_b \neq \vec{0},\ \vec{M}_d \neq \vec{0},\ \vec{M}_e \neq \vec{0}$, though $\vec{M}_a = \dot{\vec{H}}_a,\ \vec{M}_b = \dot{\vec{H}}_b$ and $\vec{M}_d = \dot{\vec{H}}_d$.

For a translating system, the impulse–momentum relations are the same as given in Section 2.11. Further, since $\dot{W} = \dot{W}^*$, $V^* = V$ and $T = T_C$, the work–energy relations (2.57), (2.91), (2.92) become

[7] The converse is not true, e.g., for a thin rod with axial vibration superposed on its gross uniform axial motion, the orientation of all its material line elements remains the same (though their lengths may not), but it is not in translation as the displacements of all its material points are not the same.

$$\dot{T} = \dot{W}, \qquad T_2 - T_1 = W_{1-2}$$

$$\dot{T} + \dot{V} = \dot{W}_{\mathrm{nc}}, \qquad (T_2 + V_2) - (T_1 + V_1) = W_{\mathrm{nc}_{1-2}} \tag{2.96}$$

$$\text{conservative system}: \dot{T} + \dot{V} = 0, \qquad T_2 + V_2 = T_1 + V_1$$

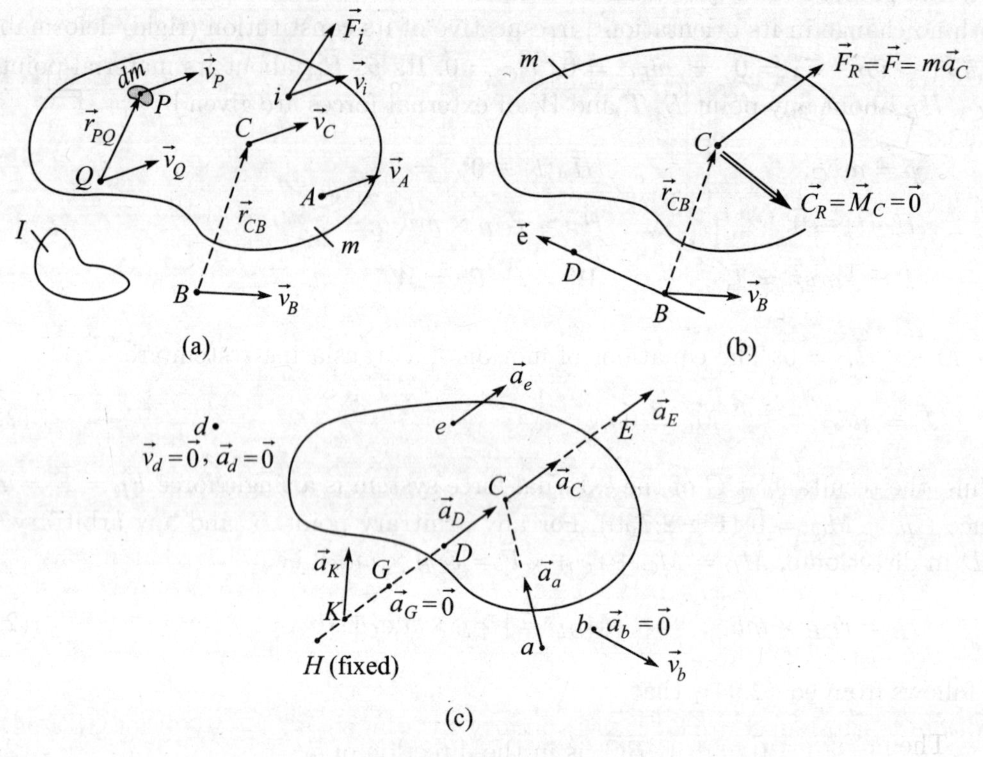

Fig. 2.73

For a system consisting of interconnected translating subsystems, eq. (2.96) is applied to each translating subsystem and added up to yield relations for the *whole system.* This may involve work done by the internal and external forces on the system since some of the forces which are external for an individual subsystem (Fig. 2.74) are in fact internal for the whole system. Denoting the contribution of internal and external forces by the subscript int + ext, the work–energy relations (2.96) become:

$$\begin{aligned}
&\dot{T} = \dot{W}_{\mathrm{int+ext}}, \qquad T_2 - T_1 = W_{\mathrm{int+ext}_{1-2}} \\
&\dot{T} + \dot{V}_{\mathrm{int+ext}} = \dot{W}^{\mathrm{nc}}_{\mathrm{int+ext}} \\
&T_2 + V_{\mathrm{int+ext}_2} - T_1 - V_{\mathrm{int+ext}_1} = W^{\mathrm{nc}}_{\mathrm{int+ext}_{1-2}} \\
&\text{conservative system}: \quad \dot{T} + \dot{V}_{\mathrm{int+ext}} = 0, \qquad T_2 + V_{\mathrm{int+ext}_2} = T_1 + V_{\mathrm{int+ext}_1}
\end{aligned} \tag{2.97}$$

If all internal forces of interaction are together workless, then eqs (2.97) reduce to eqs (2.96). The *work–energy relation is useful* if the interconnected system has *only one degree of freedom.* The *rate form is used to obtain acceleration,* which may then be integrated. If the work done can be computed independently of the actual motion, then the integrated form can be used directly to get the velocity in a given position.

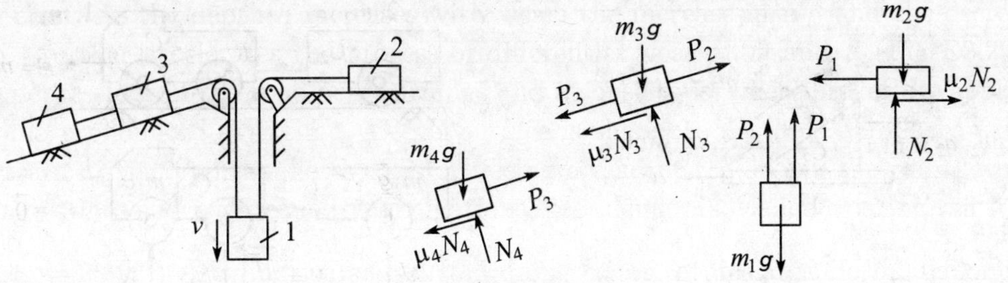

Fig. 2.74

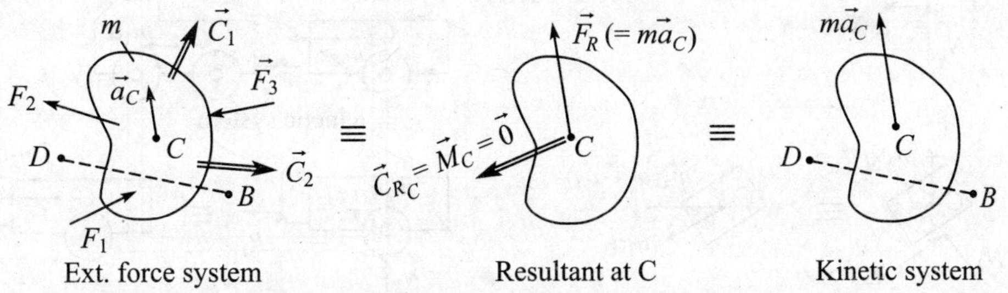

Ext. force system

Resultant at C

Kinetic system

Fig. 2.75

The problem of several connected *translating bodies* can be solved by three methods.

1. Using $\vec{F} = m\vec{a}_C$, $\vec{M}_C = \vec{0}$ for individual bodies which involve internal and external forces on the system.
2. Using $\vec{F} = m\vec{a}_C$ for the whole system and appropriate $\dot{\vec{H}}_O = \vec{M}_O$ for convenient fixed points for the whole system and some individual bodies such that only the external forces on the system appear in the equations.
3. The resultant of the external force system at C is a single force $\vec{F}_R = \vec{F} = m\vec{a}_C$ (Fig. 2.75). In this method, the moment $\vec{M}_B$ or moment M_{BD} of the external force system about a convenient point B (where most unknown reactions act) or line BD (through which most reactions act) is equated to the corresponding moment of the equivalent kinetic system of $m\vec{a}_C$ acting at C. Method 2 would require proper computation of $\dot{H}_{O_z}$, though the resulting moment equation would be the same as in method 3. Methods 2 and 3 are illustrated in Ex.2.47.

Note that using the equivalent kinetic systems $m_i\vec{a}_{C_i}$ at C_i for the individual translating systems and the fact that action and reaction are equal and opposite, the external force system on the whole system is equivalent to a kinetic system consisting of $m_i\vec{a}_{C_i}$ at C_i. Such equivalent systems for translating blocks on a conveyor and vehicles with light wheels, with driving moment M on rear wheels, are shown in Figs 2.76a,b.

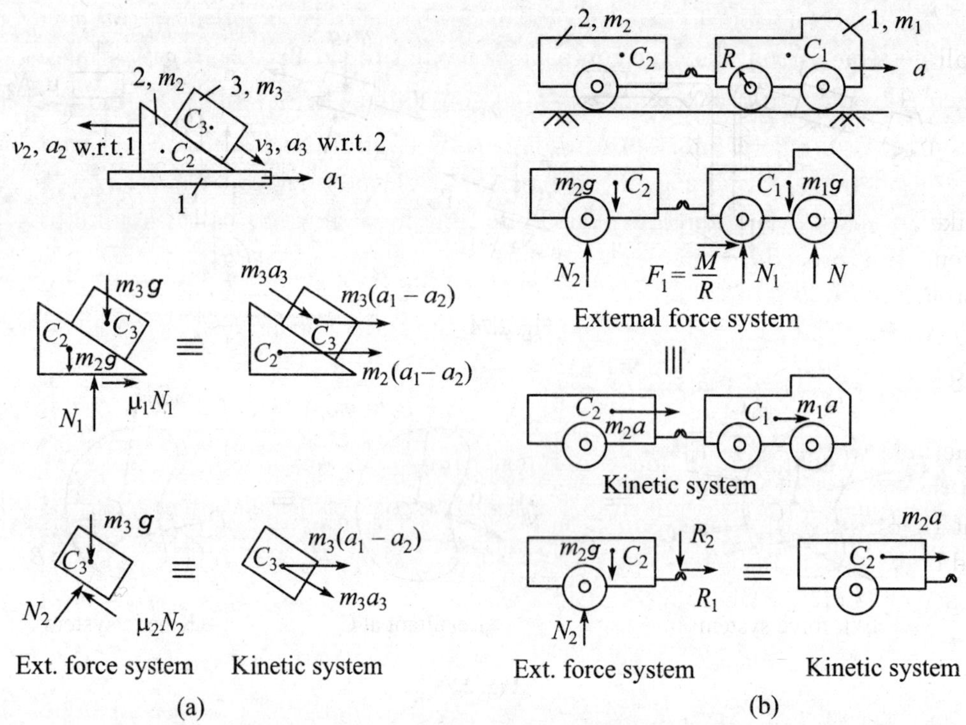

Fig. 2.76

2.27.1 Virtual work approach for equivalent force systems

The work done by equivalent force systems on a rigid body is the same. Therefore, for a translating rigid body, the virtual work, δW, done by the external force system on a rigid body in any infinitesimal virtual displacement to a neighbouring configuration in zero time, keeping the forces constant, equals the virtual work done by its equivalent system of $m\vec{a}_C$ acting at C. It is called virtual as this displacement is considered in zero time, whereas the actual displacement always occurs in a finite time interval. The virtual work in zero time is denoted by δW to distinguish it from the differential work dW in time dt. See Chapter 5 for a more detailed discussion of these concepts.

Hence, for a system of p connected rigid bodies with n dof q_i, $i = 1, \ldots, n$, with each one in translation, the virtual work of all the external and internal forces in any virtual displacement δq_i, $i = 1, \ldots, n$, equals the virtual work done by the equivalent system of $m_i \vec{a}_{C_i}$ acting at C_i. The internal forces are to be included since these are external forces for some individual bodies. The symbol δq_i for virtual displacement in zero time differs from the symbol of dq_i for differential displacement in time dt. Thus,

$$\sum_{j=1}^{p} m_j \vec{a}_{C_j} \cdot \delta \vec{r}_{C_j} = \delta W_{\text{ext+int}} = -\delta V_{\text{ext+int}} + \delta W^{\text{nc}}_{\text{ext+int}} \tag{2.98}$$

for all possible virtual displacements δq_i, where $\vec{r}_{C_j}(q_1, \ldots, q_n) \Rightarrow$

$$\delta \vec{r}_{C_j} = \sum_{i=1}^{n} \frac{d\vec{r}_{C_j}}{dq_i} \delta q_i$$

Unlike the usual work–energy relation which yields a single equation for the entire system, this procedure enables us to obtain n equations for an n dof system. This approach is explored further in Chapter 5.

2.28 Dynamics of a Rigid Body with Mass Concentrated at C

Sometimes, we model a stiff body with mass predominantly confined to a small volume as a rigid body with its mass m concentrated at its centre of mass C. For example, a light rigid rod with a heavy bob of mass m at one end (Fig. 2.77a) is modelled as a rigid body with mass m concentrated at C. However, the entire load on the body does not act at C. The FBD is shown in Fig. 2.77b. The spring is un-deformed for $\theta = 0$.

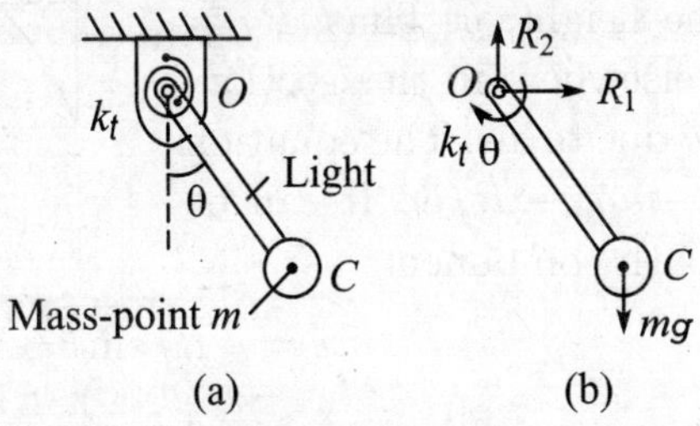

Fig. 2.77

For such a system: $\vec{H}_C(t) = \vec{r}_{CC} \times m\vec{v}_{CC}(t) \equiv \vec{0}$ and

$$\vec{M}_C = \dot{\vec{H}}_C = \vec{0}, \qquad \vec{F} = m\vec{a}_C \tag{a}$$

Hence, the resultant of the external force system at C is a single force $\vec{F}_R = \vec{F} = m\vec{a}_C$ and for an arbitrary point B: $\vec{M}_B = \vec{r}_{CB} \times \vec{F}_R \Rightarrow$

$$\vec{M}_B = \vec{r}_{CB} \times m\vec{a}_C \tag{b}$$

The kinetic energy T, $\dot{W}$ and work–energy relation for C yield:

$$T = \tfrac{1}{2}mv_C^2 = T_C, \qquad \dot{W} = \vec{F}\cdot\vec{v}_C + \vec{M}_C\cdot\vec{\omega} = \vec{F}\cdot\vec{v}_C = \dot{W}^* \qquad (c)$$

$$\dot{T}_C = \dot{W}^* \qquad \Rightarrow \qquad \dot{T} = \dot{W} \qquad (d)$$

Thus, the governing equations (a) to (d) of a 'rigid body mass-point' are exactly the same as those for a translating system. This is to be expected since, as for a translating system, for the present case also $\vec{H}_C \equiv \vec{0}$ and so rotation or $\vec{\omega}$ is irrelevant.

By eqs (3.2) and (3.37): $\vec{H}_C = I_{ij}^C \omega_j\, \vec{e}_i, \quad T = T_C + \frac{1}{2}I_{ij}^C\omega_i\omega_j,$ where I_{ij}^C are elements of inertia matrix at C. Consider the model of a **rigid body with negligible rotational inertia**, i.e., $I_{ij}^C \simeq 0 \;\forall\, i, j$. Thus,

$$\vec{H}_C(t) = I_{ij}^C\omega_j(t)\,\vec{e}_i \equiv \vec{0}\;\forall\, t, \qquad T = T_C + \tfrac{1}{2}I_{ij}^C\omega_i\omega_j = T_C$$

as for the 'rigid body point-mass' model. Hence, the governing equations of a rigid body with negligible rotational inertia are the same as those for a translating system.

In particular, eqs (a) to (d) are valid for a 'mass-point' model of a physical system.

2.29 Projectiles with Resistance

The effect of air resistance, $-R_f\,\vec{e}_t$, is important in determining the motion of an artillery shell (Fig. 2.78) (mass m and centre of mass C). R_f depends on the speed of the projectile, v, and the density of air, ρ, which in turn depends on the height, y. Thus, $R_f = R_f(y, v)$. Note that here, as elsewhere in this section, we neglect the buoyancy force due to air. The equation of motion for C is, $m\vec{a}_C = -mg\vec{j} - R_f\,\vec{e}_t$. It can be expressed in Cartesian and path components as:

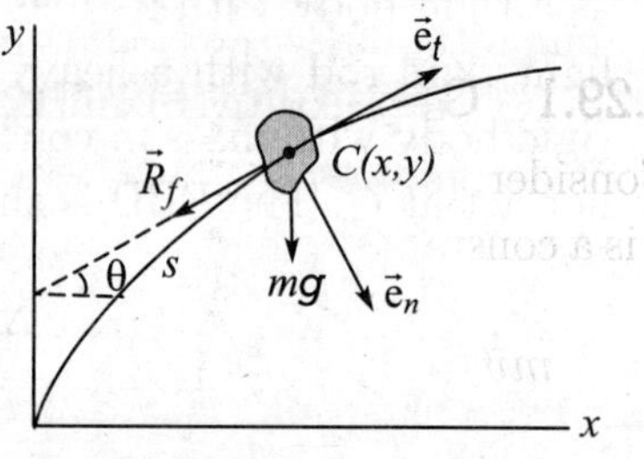

Fig. 2.78

$$m\ddot{x} = -R_f\cos\theta, \qquad m\ddot{y} = -R_f\sin\theta - mg,$$

$$m\dot{v} = mv\frac{dv}{ds} = -R_f - mg\sin\theta, \qquad \frac{mv^2}{\rho} = mg\cos\theta$$

with $v^2 = \dot{x}^2 + \dot{y}^2$, $\cos\theta = \dot{x}/v = \dot{x}/\sqrt{\dot{x}^2+\dot{y}^2}$, $\sin\theta = \dot{y}/\sqrt{\dot{x}^2+\dot{y}^2}$. These equations have to be solved numerically since the dependence of R_f on y and v is usually obtained from tables as analytical expressions are not always available.

To get a quantitative and qualitative idea of the effect of drag, we idealise the physical problem with $R_f(v) = B_1 v + B_2 v^2$, where B_1, B_2 are constants. For a sphere of radius r, $B_1 = c_1 r$, $B_2 = c_2 r^2$, where c_1, c_2 are constants and

$$R_f = c_1 r v + c_2 r^2 v^2 = c_1 r v(1 + v/v^*) = c_2 r^2 v^2(1 + v^*/v)$$

$$\Rightarrow \quad R_f \simeq c_1 r v \quad \text{if } v < 0.1v^*, \qquad R_f \simeq c_2 r^2 v^2 \quad \text{if } v > 10v^*$$

where $v^* = c_1/c_2 r$ is the velocity for which the two terms in R_f are equal. For air, $c_1 \simeq 3.1 \times 10^{-4}\,\text{kg.m}^{-1}.\text{s}^{-1}$, $c_2 \simeq 0.87\,\text{kg.m}^{-3}$, and

$$v^* = \frac{c_1}{c_2 r} = \frac{3.6 \times 10^{-4}}{r} \qquad \text{(SI units)} \tag{2.99}$$

For a macroscopic pebble-like object with $r \simeq 0.01\,\text{m}$, $v^* = 0.036\,\text{m/s}$. Starting from rest, in free fall under gravity, a speed of $10v^*$ occurs after a a fall of 0.8 cm in 0.04 s. Thus, for most practical problems with $r \sim 1\,\text{cm}$, the motion occurs essentially under a resistive force $R_f = B_2 v^2$. For a spherical rock of radius $r = 0.01\,\text{m}$, density $\rho = 2500\,\text{kg/m}^3$, mass $m \simeq 0.01\,\text{kg}$, the terminal speed v_t occurs for $mg = B_2 v_t^2$, i.e., $v_t = (mg/c_2)^{1/2}/r \simeq 30\,\text{m/s}$. In free fall, the speed v_t occurs after a fall of 45 m in 3 s. The effect of R_f becomes significant at times and distances much less than these. Thus, some idealised problems of free fall under gravity, are not physically valid in air.

By contrast, for microscopic objects, such as particles of dust with $r \simeq 1\,\mu\text{m} = 10^{-6}\,\text{m}$, eq. (2.99) yields $v^* = 360\,\text{m/s}$. Thus, for lower speeds ($0 < v < 0.1v^* = 36\,\text{m/s}$), R_f may be approximated as $B_1 v$, i.e., the resistance force is viscous.

Motion under force $R_f \propto$ (speed) and $R_f \propto$ (speed)^2 are considered next.

2.29.1 General motion under $\vec{R}_f = -mk\vec{v}$

Consider, motion of a body of mass m under linear resistance force $\vec{R}_f = -mk\vec{v}$, where k is a constant, with initial conditions: $\vec{v}(0) = \vec{v}_0$ and $\vec{r}(0) = \vec{0}$ (Fig. 2.79a). $\vec{F} = m\vec{a}_C \Rightarrow$

$$m\dot{\vec{v}} = m\vec{g} - mk\vec{v} \quad \Rightarrow \quad \dot{\vec{v}} + k\vec{v} = \vec{g} \tag{2.100}$$

Its complementary solution is $\vec{C}_1 \mathrm{e}^{-kt}$ and particular integral is $\vec{g}/k$:

$$\vec{v} = \vec{C}_1 \mathrm{e}^{-kt} + \frac{\vec{g}}{k} \quad \Rightarrow \quad \vec{v}(0) = \vec{C}_1 + \frac{\vec{g}}{k} = \vec{v}_0 \quad \Rightarrow \quad \vec{C}_1 = \vec{v}_0 - \frac{\vec{g}}{k} = \vec{v}_0 - \vec{v}_t$$

where, for $\dot{\vec{v}} = \vec{0}$, eq. (2.100) $\Rightarrow$ terminal velocity $\vec{v}_t = \vec{g}/k = -g\vec{\jmath}/k = -v_t\vec{\jmath}$, $v_t = g/k$.

$$\begin{aligned} \dot{\vec{r}} &= \vec{v} = (\vec{v}_0 - \vec{v}_t)\mathrm{e}^{-gt/v_t} + \vec{v}_t \\ \Rightarrow \quad \vec{r}(t) &= (\vec{v}_0 - \vec{v}_t)(1 - \mathrm{e}^{-gt/v_t})v_t/g + \vec{v}_t t \end{aligned} \tag{2.101}$$

The trajectory of C is in the plane of $\vec{v}_0$ and $\vec{v}_t = -g\vec{\jmath}/k$, i.e., in the vertical plane through $\vec{v}_0$, say the xy-plane (Fig. 2.79a). The component form of eq. (2.101) is

$$\begin{aligned} \dot{x} &= v_{x0}\mathrm{e}^{-gt/v_t}, & \dot{y} &= (v_{y0} + v_t)\mathrm{e}^{-gt/v_t} - v_t \\ x &= (1 - \mathrm{e}^{-gt/v_t})v_t v_{x0}/g, & y &= (v_{y0} + v_t)(1 - \mathrm{e}^{-gt/v_t})v_t/g - v_t t \end{aligned} \tag{2.102}$$

We non-dimensionalise eqs (2.102) with respect to the parameters associated with the maximum height in an un-resisted medium: $x_m = v_{x0}v_{y0}/g$, $y_m = v_{y0}^2/2g$, $t_m = v_{y0}/g$;

$$\xi = \frac{x}{x_m} = \frac{xg}{v_{x0}v_{y0}}, \qquad \eta = \frac{y}{2y_m} = \frac{yg}{v_{y0}^2}$$
$$\tau = \frac{t}{t_m} = \frac{gt}{v_{y0}}, \qquad \frac{v_{y0}}{v_t} = \frac{kv_{y0}}{g} = \alpha \tag{2.103}$$

Using $gt/v_t = v_{y0}\tau/v_t = \alpha\tau$, $d()/d\tau = [d()/dt][dt/d\tau] = (\dot{})v_{y0}/g$, eqs (2.102) reduce to

$$\frac{d\xi}{d\tau} = \mathrm{e}^{-\alpha\tau}, \qquad \xi = (1-\mathrm{e}^{-\alpha\tau})/\alpha$$
$$\frac{d\eta}{d\tau} = [(1+\alpha)\mathrm{e}^{-\alpha\tau} - 1]/\alpha, \qquad \eta = [(1+\alpha^{-1})(1-\mathrm{e}^{-\alpha\tau}) - \tau]/\alpha$$

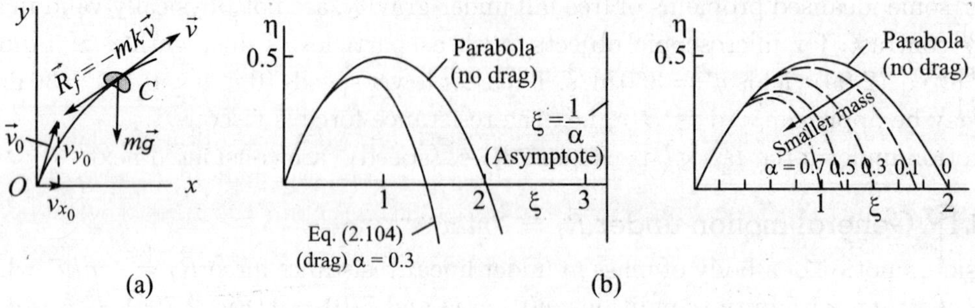

Fig. 2.79

Eliminating τ from the expressions of ξ, η and expanding the logarithmic term in Taylor series yields the equation of the trajectory:

$$\eta = (1+\alpha^{-1})\xi + \alpha^{-2}\ln(1-\alpha\xi)$$
$$\eta = \xi - \tfrac{1}{2}\xi^2 - \sum_{n=3}^{\infty}\frac{1}{n}\alpha^{n-2}\xi^n \tag{2.104}$$

The first two terms outside the series in eq. (2.104) are for free fall ($\alpha = 0$). Unlike free fall under gravity, the trajectory for the resisted motion depends on the mass of the object as $\alpha = kv_{y0}/g = c_1 r v_{y0}/mg = 3c_1 v_{y0}/4\pi r^2 \rho g$ depends on m. Equation (2.104) implies that the trajectory is not significantly affected in the ascending part (small ξ), but is affected considerably in the descending part, which becomes steeper and the fall tends to be vertical (Fig. 2.79b) for a larger time. This effect increases with α $\Rightarrow$

(1) for a given r, the effect is more for smaller ρ, i.e., for a body of smaller mass,

(2) for a given ρ, the effect is more for lesser r, i.e., for a smaller body of less mass.

The maximum height $\eta_{\max}$ is attained at $\tau = \tau_1$ when $\dfrac{d\eta}{d\tau} = [(1+\alpha)e^{-\alpha\tau_1} - 1]/\alpha = 0$,

$$\Rightarrow \quad \tau_1 = \alpha^{-1}\ln(1+\alpha), \qquad \eta_{\max} = \eta(\tau_1) = -\alpha^{-2}\ln(1+\alpha) + \alpha^{-1}$$

Letting $\eta = 0$ in eq. (2.104) yields range $\xi_r = 2 - 2\sum_{n=3}^{\infty}\frac{1}{n}\alpha^{n-2}\xi_r^{n-1}$. Its solution ξ_r is less than the value of 2 for free fall. As $t \to \infty$, $\tau \to \infty$:

$$\frac{d\xi}{d\tau} \to 0, \qquad \frac{d\eta}{d\tau} \to -\frac{1}{\alpha}, \qquad \xi \to \frac{1}{\alpha}, \qquad \eta \to -\frac{\tau}{\alpha} + \frac{(1+\alpha)}{\alpha^2}$$

i.e., $\eta = -\tau/\alpha + (1+\alpha)/\alpha^2$ is an asymptote to the trajectory in the η-τ plane (not shown) and $\xi = 1/\alpha$ is an asymptote in the ξ-η plane (Fig. 2.79b).

2.29.2 Body falling vertically under $\vec{R}_f = -mk\vec{v}$

The settling of particles under gravity in a viscous fluid has applications to settlement of solid, liquid and air pollutants such as ash and smoke from thermal power plant and sewer sediments in a river. Consider the vertical settlement of solid particles with initial velocity v_0 (Fig. 2.80a) under $\vec{R}_f = -B_1\vec{v} = -mk\vec{v}$, $k = B_1/m = c_1 r/m$.

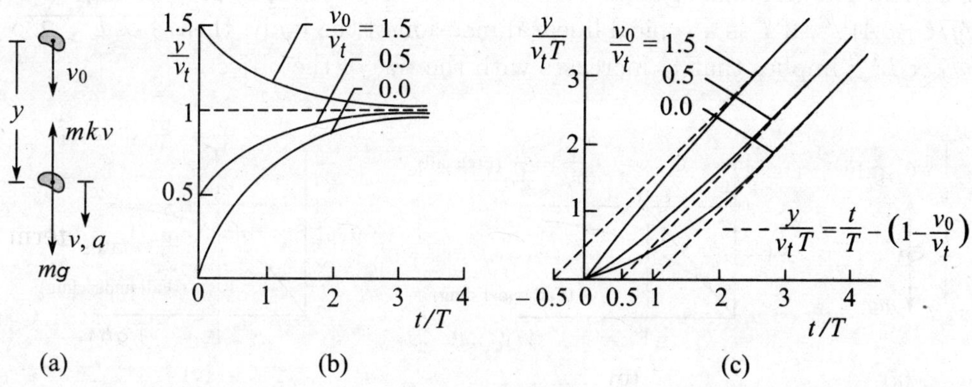

Fig. 2.80

The equation of motion: $\quad ma = mg - mkv \Rightarrow$

$$a = \frac{dv}{dt} = g - kv = g\left(1 - \frac{v}{v_t}\right)$$

$$\Rightarrow \quad \int_{v_0}^{v}\frac{1}{v_t - v}\,dv = \frac{g}{v_t}\int_0^t dt \quad \Rightarrow \quad -\ln\frac{v_t - v}{v_t - v_0} = \frac{gt}{v_t} \quad \Rightarrow$$

$$t = -T\ln\left[\frac{1 - v/v_t}{1 - v_0/v_t}\right], \qquad \dot{y} = v = v_t\left[1 - (1 - v_0/v_t)e^{-t/T}\right] \tag{2.105}$$

where $T = v_t/g$ is the time taken to attain v_t in free fall under gravity, starting from rest. Equation (2.105) implies that $v \to v_t$ as $t \to \infty$ and if $v_0 < v_t$, then $v(t) \le v_t$ and if $v_0 > v_t$, then $v(t) \ge v_t$. If L is a typical linear dimension of the body, then $m \propto L^3$, $k = B_1/m = c_1 r/m \propto 1/L^2$ and $v_t = g/k \propto L^2$. Hence, terminal velocity of larger-sized particles is more and they settle faster. Integration of eq. (2.105) yields:

$$y(t) = y(0) + \int_0^t v\,dt = v_t[t + T(1 - v_0/v_t)(\mathrm{e}^{-t/T} - 1)] \tag{2.106}$$

As $t \to \infty$: $y(t)/v_t T \to t/T - (1 - v_0/v_t)$, which is shown as dashed line in Fig. 2.80c. Relations (2.105) and (2.106) are illustrated in Fig. 2.80b and Fig. 2.80c.

Using $c_1 = 3.1 \times 10^{-4}\,\mathrm{kg.\,m^{-1}.s^{-1}}$, a particle of radius 10μm, with a density of $300\,\mathrm{kg/m^3}$ (soot), has $v_t \simeq 4\,\mathrm{mm/s} \Rightarrow$ it will take 7 hours to settle to the ground from a height of 100 m in still air, while a particle of radius 2.5μm would take 4.5 days.

2.29.3 Body falling vertically under $R_f = \frac{1}{2}C_D\rho A v^2$

The resistance force R_f on a body, moving in a fluid of density ρ at velocity v with cross-sectional area A normal to it, is given by $R_f = \frac{1}{2}C_D\rho A v^2$ for turbulent flow, where C_D is the drag coefficient. Consider a pebble-sized body falling from rest (Fig. 2.81a). The terminal speed v_t $(a = 0)$ is obtained from $\frac{1}{2}C_D\rho A v_t^2 = mg$, i.e., $v_t = (2mg/C_D\rho A)^{1/2}$. If L is a typical linear dimension of the body, then $A \propto L^2$, $m \propto L^3$ and $v_t \propto L^{1/2}$ implies that v_t increases with the size of the object.

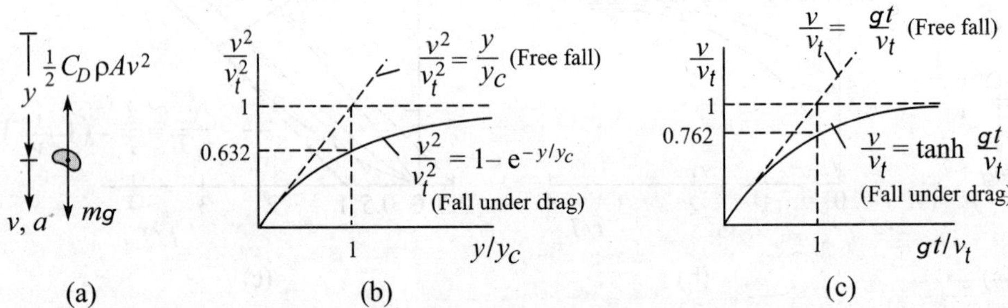

Fig. 2.81

The equation of motion

$$ma = mg - \tfrac{1}{2}C_D\rho A v^2 \qquad \Rightarrow \qquad a = \dot{v} = v\frac{dv}{dy} = \frac{g(v_t^2 - v^2)}{v_t^2}$$

$$\Rightarrow \quad \frac{v_t^2}{2g}\int_0^v \frac{2v}{v_t^2 - v^2}\,dv = \int_0^y dy \qquad \Rightarrow \qquad y = -y_c \ln(1 - v^2/v_t^2)$$

$$\text{and} \quad \int_0^v \frac{1}{v_t^2 - v^2}\,dv = \frac{g}{v_t^2}\int_0^t dt \qquad \Rightarrow \qquad \tanh^{-1}(v/v_t) = \frac{g}{v_t}t$$

$$v^2 = v_t^2(1 - \mathrm{e}^{-y/y_c}), \qquad v = v_t \tanh\frac{gt}{v_t} = v_t\,\frac{1-\mathrm{e}^{-2gt/v_t}}{1+\mathrm{e}^{-2gt/v_t}} \tag{2.107}$$

where $y_c = v_t^2/2g$ is the distance covered in free fall to attain a speed of v_t. The v^2–y and v–t relations are shown in Fig. 2.81b,c with $v \to v_t$ as $t \to \infty$. Equations (2.107) $\Rightarrow$ at $y = y_c$, $v = 0.795v_t$; at $t = v_t/g$, $v = 0.762v_t$; and $v = 0.95v_t$ at $y = 2.33y_c$, $t = 1.832v_t/g$; $v = 0.99v_t$ at $y = 3.92y_c$, $t = 2.647v_t/g$. The typical values of y_c (m) and v_t (m/s) for some objects are: (2.1, 6.5) for a rain drop of radius 0.1 cm, (21, 20) for hailstorm of radius 1 cm, (200, 63) for a human being, (149, 54) for a parachutist with closed canopy and (1.49, 5.4) for a parachutist with open canopy.

The parachutist falls for about 1500 m in 30 s with a closed canopy with most of this descent at nearly $v_t = 54$ m/s. Then the canopy is opened (about 150 m above the ground to account for the time of opening of the canopy) and attain a new $v_t = 5.4$ m/s in a very short descent of 20 m. This v_t equals the velocity after a free fall from 1.5 m height and is safe. If the canopy is opened earlier, the time of fall increases and the parachutist becomes an easy target. If it is not opened in time, the fall may be fatal.

2.29.4 Motion of a shell fired up under $R_f = \frac{1}{2}C_D\rho A v^2$

Consider, upward motion of a shell fired vertically up with velocity v_0 (Fig. 2.82). The equation of motion $ma = -mg - \frac{1}{2}C_D\rho A v^2 \Rightarrow$

$$a = \dot v = v\,\frac{dv}{dy} = -\frac{g(v_t^2+v^2)}{v_t^2}$$

$$\Rightarrow \quad \frac{v_t^2}{2g}\int_{v_0}^{v}\frac{2v}{v_t^2+v^2}\,dv = -\int_0^y dy, \qquad \frac{v_t}{g}\int_{v_0}^{v}\frac{v_t}{v_t^2+v^2}\,dv = -\int_0^t dt$$

$$t = \frac{v_t}{g}\left[\tan^{-1}\frac{v_0}{v_t} - \tan^{-1}\frac{v}{v_t}\right]$$

$$y = y_c \ln\frac{1+v_0^2/v_t^2}{1+v^2/v_t^2}, \qquad v = v_t \tan\left[\tan^{-1}\frac{v_0}{v_t} - \frac{gt}{v_t}\right] \tag{2.108}$$

The maximum height y_m is attained at time t_m for $v = 0$:

$$y_m = y_c \ln\left[1 + \frac{v_0^2}{v_t^2}\right], \qquad t_m = \frac{v_t}{g}\tan^{-1}\frac{v_0}{v_t} \le \tfrac{1}{2}\pi\left(\frac{v_t}{g}\right) \tag{2.109}$$

For small resistance, v_0/v_t is small. Expanding the logarithmic term by Taylor series:

$$y_m = y_c\left[v_0^2/v_t^2 - v_0^4/2v_t^4 + \ldots\right] \simeq \left(v_0^2/2g\right)\left[1 - v_0^2/2v_t^2\right]$$

The first term is the height for no resistance. The relations (2.108) are plotted in

Fig. 2.82b,c. Substituting y_m from eq. (2.109)$_1$ for y in eq. (2.107)$_1$ for down motion, yields the velocity with which the shell hits the ground as $v = v_0/[1 + v_0^2/v_t^2]^{1/2} < v_0$.

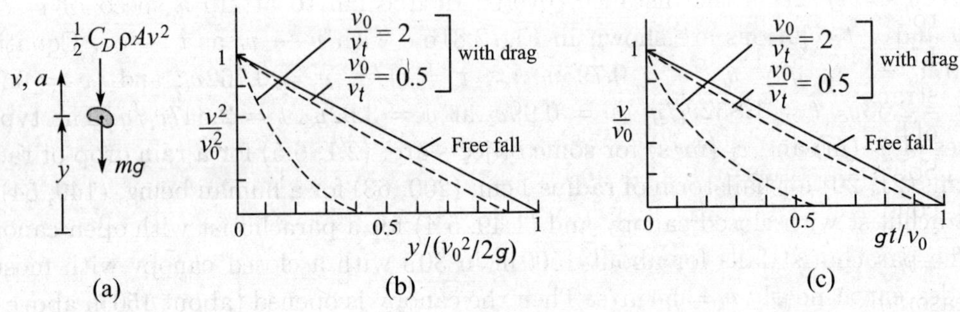

Fig. 2.82

2.30 Equation of Motion of C Relative to a Translating Frame

Consider a frame T translating with constant acceleration $\vec{a}_T$ relative to an inertial frame I. The equation of motion of the centre of mass C of any system, under external force sum $\vec{F} = \vec{F}_n + m\vec{g}$ where $\vec{F}_n$ is the sum of non-gravitational forces, is:

$$\vec{F} = \vec{F}_n + m\vec{g} = m\vec{a}_{C|I} = m(\vec{a}_{C|T} + \vec{a}_T)$$

$$\Rightarrow \quad \vec{F}_n + m\vec{g}_e = m\vec{a}_{C|T}, \qquad \text{where} \quad \vec{g}_e = \vec{g} - \vec{a}_T \tag{2.110}$$

$$\Rightarrow \quad (\vec{F}_n + m\vec{g}_e) \cdot \vec{v}_{C|T} = m\vec{a}_{C|T} \cdot \vec{v}_{C|T} = m\dot{\vec{v}}_{C|T} \cdot \vec{v}_{C|T} = \frac{d}{dt}\left(\tfrac{1}{2}mv_{C|T}^2\right)$$

$$\Rightarrow \quad \dot{T}_{C|T} = \dot{W}_{|T}^* \qquad \Rightarrow \qquad \dot{T}_{C|T} + \dot{V}_{|T}^* = \dot{W}_{\text{nc}|T}^* \tag{2.111}$$

where $\dot{W}_{|T}^*$ is the rate of work done relative to T by $\vec{F}_n + m\vec{g}_e$, as if acting at C; $\dot{W}_{\text{nc}|T}^*$ is the rate of work done by non-conservative part of forces, and $V_{|T}^*$ is the potential energy, relative to T, of the conservative part of forces. Thus, relative to T, the governing equation of motion of C and the work-energy relation are exactly of the same form as those relative to I, provided force $m\vec{g}$ and its potential energy $-m\vec{g} \cdot \vec{r}$ are replaced by force $m\vec{g}_e$ and its potential energy $-m\vec{g}_e \cdot \vec{r}$.

2.31 Examples on Motion of C and Translating Systems

In some cases, the motion of the centre of mass C can be obtained without studying the overall motion. For a translating system, this completes the study of dynamics. The following steps are useful for obtaining a solution.

1. List the given data and the entities to be computed.

2. Draw FBD of the system in general configuration. Select component directions to suit the description of kinematical quantities and the forces on the system.
3. Write the equations of motion and kinematic relations. List the unknowns. Go to step 5 if (number of unknowns) = (number of equations), else go to step 4.
4. List additional equations, if any, by using kinematic relationships for the constraints on the motion of various points, lines or bodies and using the conditions of friction for slip/no slip. If needed, draw FBDs of subsystems and repeat step 3.
5. Solve equations of motion and evaluate the constants of integration from initial conditions. Compute the required entities and analyse the results.

$\vec{F} = m\vec{a}_C$ or $\vec{F} = \sum m_i \vec{a}_{C_i}$ is used depending on whether the motion of C or C_is is reasonably simple and directly available. The force required for the given motion is obtained from $\vec{F} = m\vec{a}_C = m\ddot{\vec{r}}_C(t)$. Conversely, this equation is solved for $\vec{r}_C(t)$ for given $\vec{F}(t)$ and initial conditions $\vec{r}_C(0)$, $\dot{\vec{r}}_C(0)$. Closed form solution or a numerical solution is obtained by methods such as the fourth order Runge–Kutta method.

Generally, the energy relation is preferable if the force is a function of position and the system (even though interconnected) has a single dof, especially if the forces are conservative. The impulse–momentum relation is useful if the forces are known functions of time, or an impact occurs, or impulsive forces act. The moment of momentum equation is used if $F_\phi = 0$, or $M_{Oz} = 0$, or bodies rotate about a fixed axis, or the motion is under a central force $\vec{F} = F_r\,\hat{e}_r$.

§ **Exercise 2.10:** Prove that the tension on the two sides of a *mass-less belt* around a pulley is the same for the following three cases. (1) *The pulley is mass-less and has frictionless bearings.* This holds irrespective of the motion of the pulley and irrespective of whether the belt slips or not. (2) The pulley is heavy with its centre of mass on the axis of the frictionless bearings and is *in equilibrium.* (3) The pulley is *smooth*, i.e., $\mu = 0$. This holds irrespective of the mass, motion or the type of supports of the pulley.

Example 2.26: A system subjected to only 3 external forces and no couples is called a *three-force member.* Prove that these forces are necessarily coplanar and either parallel or concurrent for a mass-less 3-force member in motion and for a 3-force member with mass in equilibrium.

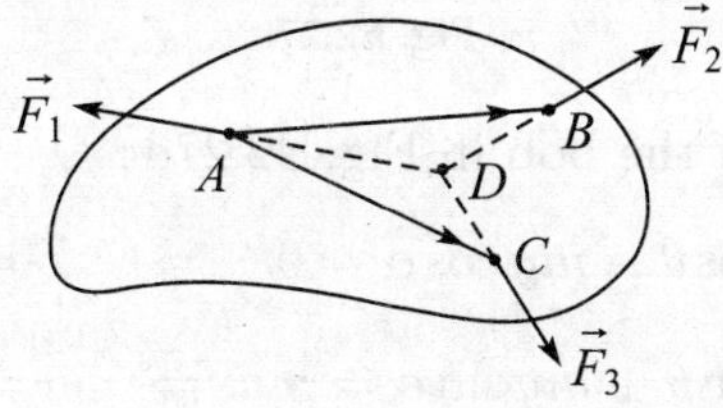

Fig. E2.26

Solution: Consider $\vec{F}_1$, $\vec{F}_2$, $\vec{F}_3$ acting at A, B, C (Fig. E2.26).

$$\vec{M}_A = \overrightarrow{AB} \times \vec{F}_2 + \overrightarrow{AC} \times \vec{F}_3 = \vec{0} \Rightarrow \quad \overrightarrow{AB} \times \vec{F}_2 = -\overrightarrow{AC} \times \vec{F}_3 = \vec{M} \quad \text{(say)}$$

$\Rightarrow \overrightarrow{AB}$ and $\vec{F}_2$ lie in a plane $\perp \vec{M}$, also $\overrightarrow{AC}$ and $\vec{F}_3$ lie in a plane $\perp \vec{M}$. These planes having the same normal are identical planes since A is common to them. Hence, $\vec{F}_2$, $\vec{F}_3$ lie in the plane of A, B, C. Similarly, $\vec{M}_B = \vec{0} \Rightarrow \vec{F}_1$ also lies in the same plane. $\Rightarrow \vec{F}_1$, $\vec{F}_2$, $\vec{F}_3$ are coplanar. The coplanar $\vec{F}_1$, $\vec{F}_2$ can be parallel or intersect at D. In the former case, $\vec{F} = \vec{0} \Rightarrow \vec{F}_3 = -\vec{F}_1 - \vec{F}_2 \Rightarrow \vec{F}_3$ is parallel to the parallel vectors $\vec{F}_1$ and $\vec{F}_2$. In the latter case, $\vec{M}_D = \overrightarrow{DC} \times \vec{F}_3 = \vec{0} \Rightarrow \vec{F}_3$ passes through the point of intersection, D. Hence, the forces are coplanar and either parallel or concurrent.

Example 2.27: (a) A simple device for measuring reasonably uniform acceleration a is a pendulum bob (Fig. E2.27a) connected to a post by a light inextensible string. Find the a–θ relation. (b) If the pendulum is released from rest with respect to a vehicle at $\theta = 0$, find $\dot{\theta}$ and tension T for any position θ.

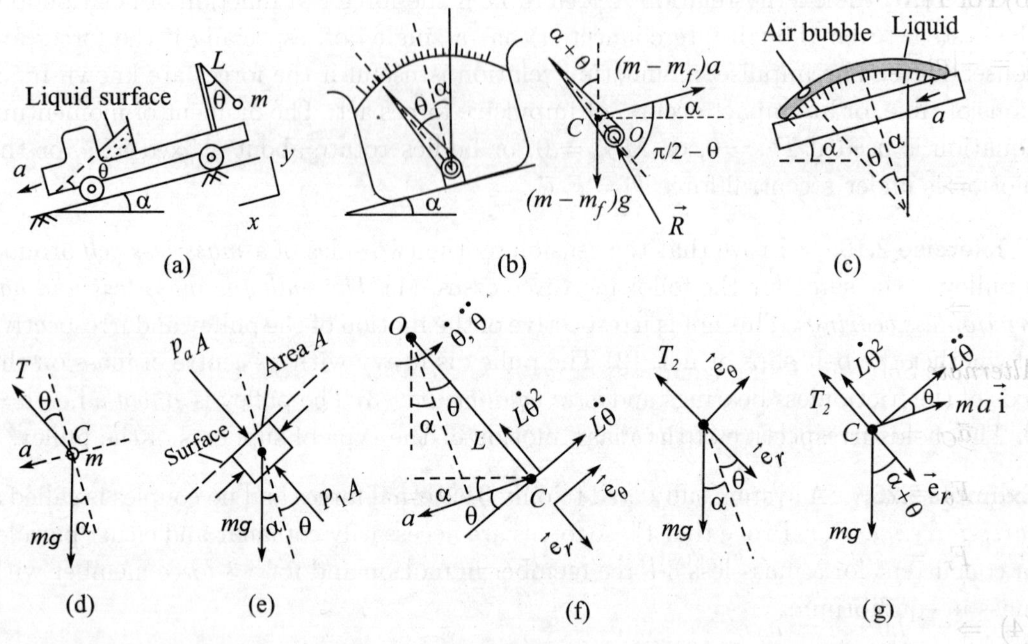

Fig. E2.27

Solution: From the FBD of the bob in Fig. E2.27d:

$$F_y = ma_{C_y}: \quad T\cos\theta - mg\cos\alpha = 0 \quad \Rightarrow \quad T = mg\cos\alpha/\cos\theta$$

$$F_x = ma_{C_x}: \quad T\sin\theta + mg\sin\alpha = ma \Rightarrow \quad a = g(\tan\theta\cos\alpha + \sin\alpha) \quad (1)$$

$$\Rightarrow a = g\tan\theta \quad \text{for } \alpha = 0$$

The inclination θ of the free surface of a liquid in a tanker moving down a plane at acceleration a is also given by eq. (1) since the FBD of an element of the liquid at the surface (Fig. E2.27e) is similar to that for the pendulum with T replaced by $(p_1 - p_a)A$, which is the net force due to pressure. The $a-\theta$ relation (1) also holds for an accelerometer consisting of a uniform pivoted float immersed in a liquid container (Fig. E2.27b). The float is in equilibrium relative to the translating frame T_1 of the truck moving at acceleration $\vec{a} = a\,\vec{\mathrm{i}}$. The augmented force system in terms of $\vec{g}_e = \vec{g} - \vec{a}$ consists of reaction $\vec{R}$ at O, $m\vec{g}_e$ and buoyancy force $-m_f\vec{g}_e$ at C. Let $OC = h$.

$$M_{Oz} = -(m - m_f)gh\sin(\alpha+\theta) + (m - mf)ah\sin(\pi/2 - \theta) = 0 \quad \Rightarrow \quad \text{eq. (1)}$$

A sensitive accelerometer, with quick response without much overshoot or oscillation, is a circular tube filled with acetone with a small air bubble in it (Fig. E2.27c). The bubble acts as a bob of an inverted pendulum with the resultant fluid thrust on it replacing tension T. a is found from the position θ of the bubble, using eq. (1).
(b) For T_1, $\vec{a}_{C|T_1}$ and the augmented force system (add $-ma\,\vec{\mathrm{i}}$) are shown in Fig. E2.27g.

$$V = -[m\vec{g} - m\vec{a}]\cdot\vec{r} = -m(\vec{g} - a\,\vec{\mathrm{i}})\cdot L\,\vec{\mathrm{e}}_r = -mL[g\cos(\alpha+\theta) + a\sin\theta],\ T = \tfrac{1}{2}m(L\dot{\theta})^2$$

$$T + V = T_0 + V_0 = 0 - mL[g\cos\alpha]$$

$$\Rightarrow \qquad \dot{\theta}^2 = 2[a\sin\theta + g\{\cos(\alpha+\theta) - \cos\alpha\}]/L \tag{2}$$

$$F_r = mg\cos(\alpha+\theta) + ma\sin\theta - T_2 = m(-L\dot{\theta}^2)$$

$$\Rightarrow \qquad T_2 = m[3g\cos(\alpha+\theta) - 2g\cos\alpha + 3a\sin\theta] \tag{3}$$

Alternate Solution: Using equations of motion relative to ground frame (Fig. E2.77f).

$$\vec{a}_C = a(-\sin\theta\,\vec{\mathrm{e}}_r - \cos\theta\,\vec{\mathrm{e}}_\theta) - \dot{\theta}^2 L\,\vec{\mathrm{e}}_r + \ddot{\theta}L\,\vec{\mathrm{e}}_\theta$$

$$F_\theta = ma_\theta : \qquad -mg\sin(\alpha+\theta) = m(L\ddot{\theta} - a\cos\theta) \tag{4}$$

$$F_r = ma_r : \qquad mg\cos(\alpha+\theta) - T_2 = -m(L\dot{\theta}^2 + a\sin\theta) \tag{5}$$

$$(4) \Rightarrow \quad L\dot{\theta}\frac{d\dot{\theta}}{d\theta} = a\cos\theta - g\sin(\alpha+\theta) \quad \Rightarrow \quad \tfrac{1}{2}L\dot{\theta}^2 = [a\sin\theta + g\cos(\alpha+\theta)]\Big|_0^\theta$$

$$\dot{\theta}^2 = 2[a\sin\theta + g\{\cos(\alpha+\theta) - \cos\alpha\}]/L$$

$$(5) \Rightarrow \quad T_2 = m[3g\cos(\alpha+\theta) - 2g\cos\alpha + 3a\sin\theta]$$

Example 2.28: A fly-ball governor (Fig. E2.28a), used to regulate the speed of a device such as a steam engine/turbine, is rotated by the device. The bars are light. Neglect the rotational inertia of the balls and the collar at A. The configuration θ

depends on ω and force P on the collar. The motion of the collar due to change in ω is used to open or close a valve to regulate the speed. Find P and the maximum bending moment in the top and bottom rods when the governor rotates at $\omega, \dot{\omega}$ with constant θ.

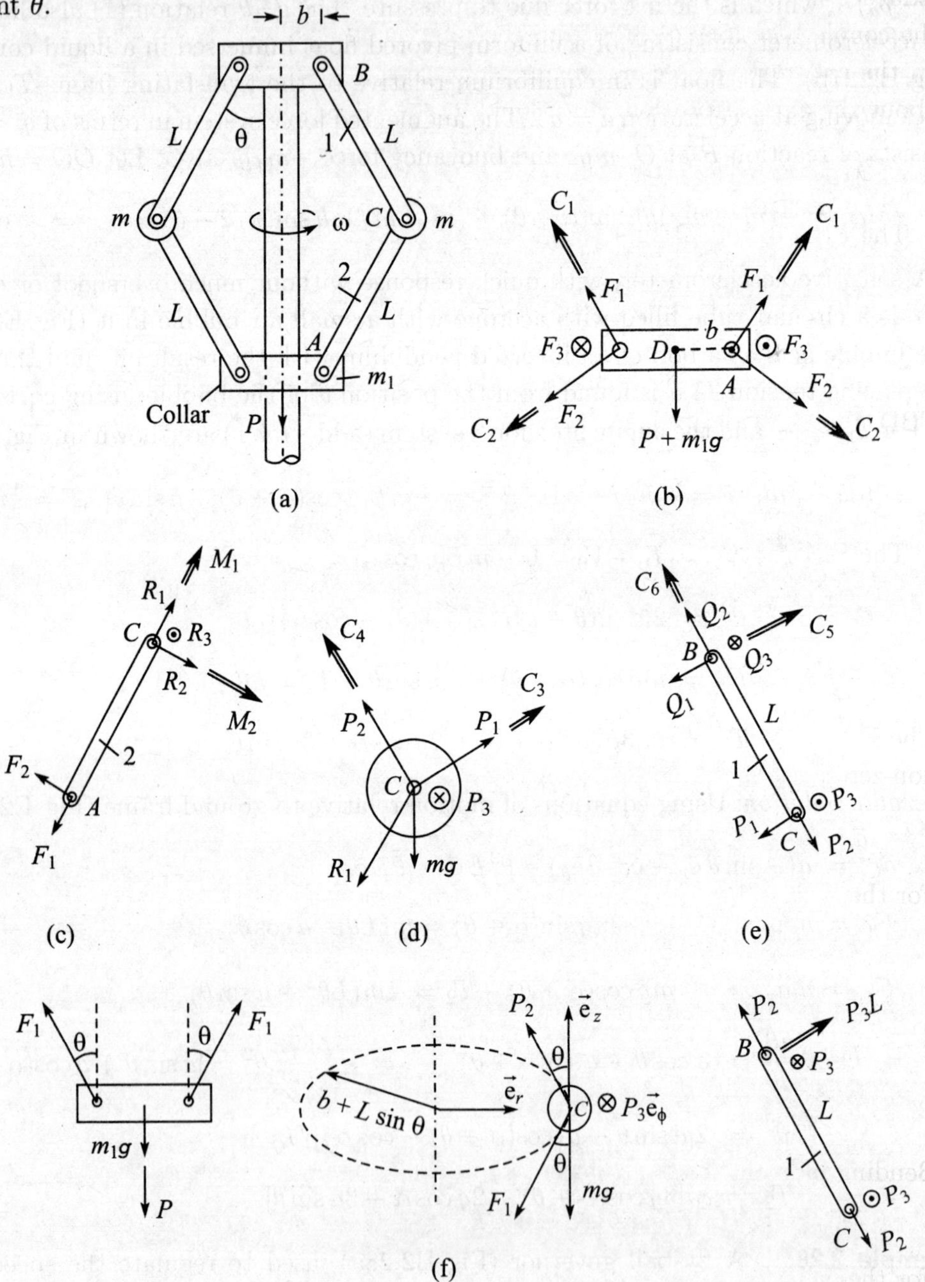

Fig. E2.28

Solution: For the transverse acceleration of the ball, the rod 2 must transmit a transverse force to C and hence there must be couple reactions on rod 1 and/or rod 2. The collar does not have any transverse acceleration and the pin reactions can provide the axial acceleration. Thus, for a determinate solution using rigid body models, we neglect the couple reactions C_1, C_2 from the pins on the collar (Fig. E2.28b). The reactions on the collar at the second pin are shown by rotating those on the first pin by 180° about the axis of rotation. Rotational inertia of the collar is neglected ⇒ for collar:

$$\vec{M}_D = 2rF_3\,\vec{e}_z = \vec{0} \qquad \Rightarrow \qquad F_3 = 0$$

The FBD of rod 2 (light rod) is drawn in Fig. E2.28c. $C_1 = C_2 = 0$, $F_3 = 0$ and

$$\vec{M}_C = \vec{0} \quad \Rightarrow \quad F_2 = 0, \qquad M_1 = M_2 = 0$$

$$\vec{F} = \vec{0} \quad \Rightarrow \quad R_2 = R_3 = 0, \qquad R_1 = F_1$$

FBD (Fig. E2.28d) of the ball (no rotational inertia) using $M_1 = M_2 = 0$, $R_2 = R_3 = 0$,

$$\Rightarrow \quad \vec{M}_C = \vec{0} \quad \Rightarrow \quad C_3 = C_4 = 0$$

The FBD of rod 1 (light rod) is drawn in Fig. E2.28e using $C_3 = C_4 = 0$ and

$$\vec{M}_B = \vec{0} \quad \Rightarrow \quad P_1 = 0, \qquad C_6 = 0, \qquad C_5 = P_3 L$$

$$\vec{F} = \vec{0} \quad \Rightarrow \quad Q_1 = 0, \qquad Q_2 = P_2, \qquad Q_3 = P_3$$

The FBDs of the collar, rod 1 and ball are presented in Fig. E2.28f showing only the non-zero actions. C moves in a circle of radius $r = b + L\sin\theta$ ⇒

$$\vec{a}_C = -\omega^2(b + L\sin\theta)\,\vec{e}_r + \dot{\omega}(b + L\sin\theta)\,\vec{e}_\phi$$

For the ball:

$$F_r = ma_{C_r}: \qquad -(P_2 + F_1)\sin\theta = -m\omega^2(b + L\sin\theta)$$

$$F_z = ma_{C_z}: \qquad (P_2 - F_1)\cos\theta - mg = 0$$

$$F_\phi = ma_{C_\phi}: \qquad P_3 = m\dot{\omega}(b + L\sin\theta)$$

$$\Rightarrow \qquad F_1 = \tfrac{1}{2}m[-g/\cos\theta + \omega^2(L + b/\sin\theta)]$$

Bending moment for bar 2 is zero and the maximum bending moment in bar 1 is:

$$M_{\max} = P_3 L = m\dot{\omega}(b + L\sin\theta)L$$

For the collar: $\quad F_z = m_1 a_z: \quad 2F_1\cos\theta - m_1 g - P = 0 \quad \Rightarrow$

$$P = -m_1 g + m[\omega^2(b\cot\theta + L\cos\theta) - g]$$

Example 2.29: Find the deceleration of the sled of a skier when its speed is v on a ground with elevation $y = h\sin(\pi x/L)$ (Fig. E2.29a). The coefficient of friction is μ.

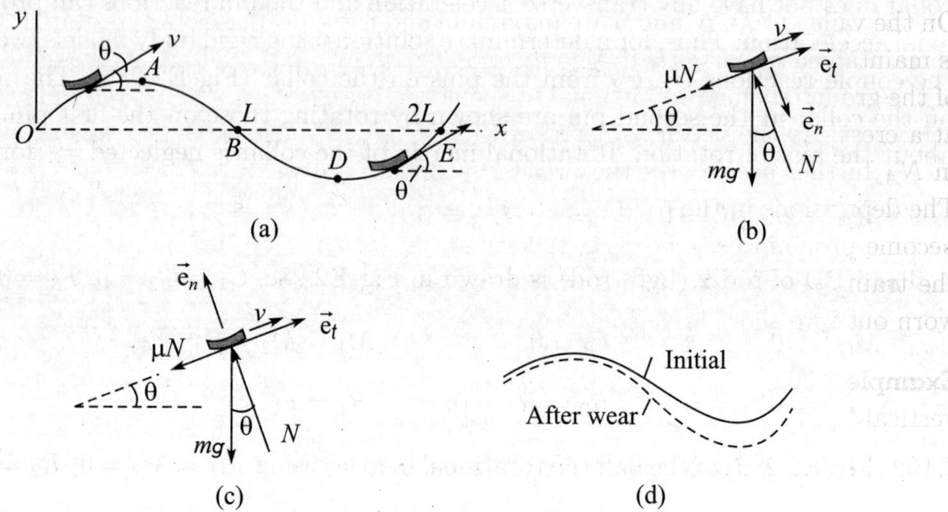

Fig. E2.29

Solution: We model the sled as a mass-point and describe its $\vec{a}$ in path coordinates:

$$y = h\sin\left(\frac{\pi x}{L}\right), \qquad y' = \frac{h\pi}{L}\cos\left(\frac{\pi x}{L}\right) = \tan\theta, \qquad y'' = -\frac{h\pi^2}{L^2}\sin\left(\frac{\pi x}{L}\right) \quad (1)$$

$$\rho = \frac{(1+y'^2)^{3/2}}{|y''|} = \frac{L^2[1+(h^2\pi^2/L^2)\cos^2(\pi x/L)]^{3/2}}{h\pi^2|\sin(\pi x/L)|} \quad (2)$$

In the FBD for $0 \le x \le L$, $\vec{e}_n$ is into the ground (Fig. E2.29b). Equations of motion:

$$F_n = ma_n: \quad mg\cos\theta - N = mv^2/\rho \quad \Rightarrow \quad N = m(g\cos\theta - v^2/\rho) \quad (3)$$

$$F_t = ma_t: \quad -\mu N - mg\sin\theta = m\dot{v} \quad \Rightarrow \quad \dot{v} = -\mu(g\cos\theta - v^2/\rho) - g\sin\theta \quad (4)$$

In the FBD for $L < x < 2L$, $\vec{e}_n$ is outwards (Fig. E2.29c). Equations of motion are:

$$-mg\cos\theta + N = mv^2/\rho, \qquad -\mu N - mg\sin\theta = m\dot{v}$$

$$\Rightarrow \quad N = m(g\cos\theta + v^2/\rho), \qquad \dot{v} = -\mu(g\cos\theta + v^2/\rho) - g\sin\theta \quad (5)$$

$$A: \; x = \frac{L}{2}, \quad \rho = \frac{L^2}{\pi^2 h}, \quad \theta = 0, \quad N = mg\left(1 - \frac{\pi^2 v^2 h}{L^2 g}\right), \quad \dot{v} = -\mu g\left(1 - \frac{\pi^2 v^2 h}{L^2 g}\right)$$

$$D: \; x = \frac{3L}{2}, \quad \rho = \frac{L^2}{\pi^2 h}, \quad \theta = 0, \quad N = mg\left(1 + \frac{\pi^2 v^2 h}{L^2 g}\right), \quad \dot{v} = -\mu g\left(1 + \frac{\pi^2 v^2 h}{L^2 g}\right)$$

$$B: \ x{=}L, \qquad \rho=\infty, \qquad \tan\theta=-\frac{\pi h}{L}, \qquad [N,\dot v]=\frac{[mg,-g(\mu-\pi h/L)]}{(1+\pi^2h^2/L^2)^{1/2}}$$

On the valley at D, N and $\dot v$ are maximum and these decrease on a crest at A. Contact is maintained in part AB if $N=m(g\cos\theta-v^2/\rho)>0$, i.e., if $v^2<\rho g\cos\theta$. The wear of the ground, being proportional to the normal reaction, is more at the valley and less at a crest. As the depth h increases with wear: the increase in N_D and the decrease in N_A further accelerates the process of differential wear at D and A (Fig. E2.29d). The depressions in the road near turnings and at bus-stops, where brakes are applied, become pronounced over time. A level surface is therefore unstable. Similarly, where the train driver often applies brakes on seeing a distant red signal, the rails become worn out into short wavelength corrugations, producing a so-called 'roaring rail'.

Example 2.30: A cylindrical vessel containing water rotates at uniform ω about its vertical axis (Fig. E2.30a). Show that the water level becomes paraboloid: $z=\omega^2r^2/2g$.

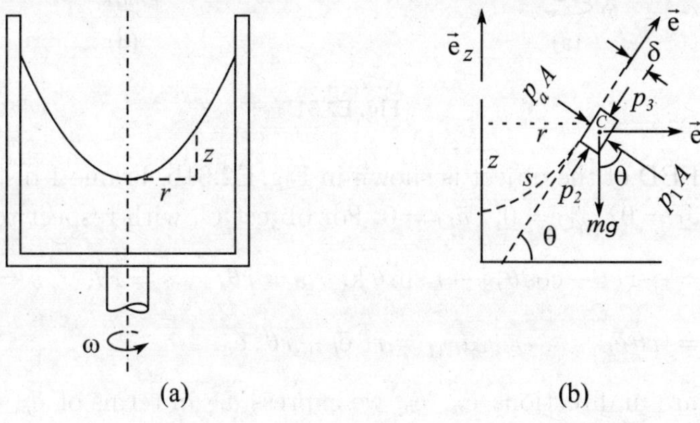

Fig. E2.30

Solution: At radius r, consider a volume element with outer face having small area $A=ds\times(r\,d\phi)$ and small thickness δ. Its mass $m=\rho A\delta$. In its FBD (Fig. E2.30b), $p_3=p_2+\dfrac{\partial p_2}{\partial s}ds$. In the limit as $\delta\to 0$, on the surface $p_2=p_a=\text{constant}\Rightarrow \dfrac{\partial p_2}{ds}=0$.

$$a_e \ = \vec a_C\cdot\vec e=(-\omega^2r\,\vec e_r)\cdot\vec e=-\omega^2r\cos\theta, \qquad\qquad F_e=ma_e$$

$$\Rightarrow \quad \Big(p_2 \ - \ p_2-\frac{\partial p_2}{\partial s}ds\Big)(r\,d\phi)\delta-\rho(r\,d\phi)ds\,\delta g\sin\theta=-\rho(r\,d\phi)ds\,\delta\omega^2r\cos\theta$$

$$\Rightarrow \quad \frac{\partial p_2}{\partial s} \ = -\rho(g\sin\theta-\omega^2r\cos\theta)=0 \qquad\qquad \Rightarrow \qquad \tan\theta=\frac{\omega^2r}{g}$$

$$\Rightarrow \quad \frac{dz}{dr} \ = \ \frac{\omega^2r}{g}: \qquad\qquad \int_0^z dz=\frac{\omega^2}{g}\int_0^r r\,dr \qquad\qquad \Rightarrow \qquad z=\frac{\omega^2r^2}{2g}$$

Example 2.31: In an automatic feeding system, small objects (modelled as mass-points) are released from rest at O at essentially zero radius and move out along a smooth circular path between radial vanes of the distributor which rotates at constant ω about a fixed vertical axis (Fig. E2.31a). Find the speed of the object and the reaction from the distributor on the object when it reaches the position θ.

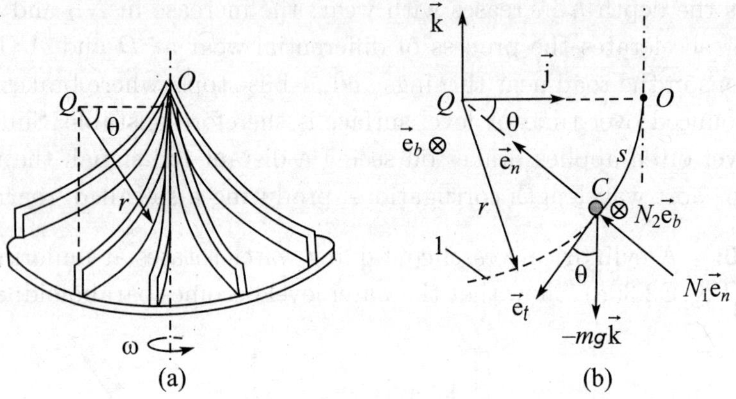

Fig. E2.31

Solution: The FBD of the object is shown in Fig. E2.31b. Frame 1 of the distributor has $\vec{\omega}_1 = \omega\,\vec{k}$, $\dot{\vec{\omega}}_1 = \vec{0}$, $\vec{v}_O = \vec{0}$, $\vec{a}_O = \vec{0}$. For object C, with respect to frame 1:

$$\overrightarrow{OC} = -r(1-\cos\theta)\,\vec{j} - r\sin\theta\,\vec{k}, \quad s = r\theta, \quad \dot{s} = r\dot{\theta}, \quad \ddot{s} = r\ddot{\theta}, \quad \rho = r$$

$$\Rightarrow \quad \vec{v}_{C|1} = r\dot{\theta}\,\vec{e}_t, \qquad \vec{a}_{C|1} = r\ddot{\theta}\,\vec{e}_t + r\dot{\theta}^2\,\vec{e}_n$$

Since N_1, N_2 are in directions $\vec{e}_n$, $\vec{e}_b$, we express $\vec{a}_C$ in terms of $\vec{e}_t$, $\vec{e}_n$, $\vec{e}_b$ using

$$\vec{i} = -\,\vec{e}_b, \qquad \vec{j} = -\sin\theta\,\vec{e}_t - \cos\theta\,\vec{e}_n, \qquad \vec{k} = -\cos\theta\,\vec{e}_t + \sin\theta\,\vec{e}_n$$

$$\begin{aligned}
\vec{a}_C &= \vec{a}_O + \dot{\vec{\omega}}_1 \times \overrightarrow{OC} + \vec{\omega}_1 \times (\vec{\omega}_1 \times \overrightarrow{OC}) + 2\,\vec{\omega}_1 \times \vec{v}_{C|1} + \vec{a}_{C|1} \\
&= \omega^2 r(1-\cos\theta)\,\vec{j} + 2\omega(-\cos\theta\,\vec{e}_t + \sin\theta\,\vec{e}_n) \times r\dot{\theta}\,\vec{e}_t + r\ddot{\theta}\,\vec{e}_t + r\dot{\theta}^2\,\vec{e}_n \\
&= \omega^2 r(1-\cos\theta)(-\sin\theta\,\vec{e}_t - \cos\theta\,\vec{e}_n) - 2\omega r\dot{\theta}\sin\theta\,\vec{e}_b + r\ddot{\theta}\,\vec{e}_t + r\dot{\theta}^2\,\vec{e}_n
\end{aligned}$$

$$F_t = ma_{C_t}: \qquad m[-\omega^2 r\sin\theta(1-\cos\theta) + r\ddot{\theta}] = mg\cos\theta$$

$$\Rightarrow \quad \ddot{\theta} = g\cos\theta/r + \omega^2(\sin\theta - \tfrac{1}{2}\sin 2\theta) \tag{1}$$

Since $\ddot{\theta}$ is a function of θ, eq. (1) is integrated using $\ddot{\theta} = \dfrac{d\dot{\theta}}{d\theta}\dot{\theta}$ to get

$$\int_0^{\dot{\theta}} \dot{\theta}\,d\dot{\theta} = \int_0^{\theta} [g\cos\theta/r + \omega^2(\sin\theta - \tfrac{1}{2}\sin 2\theta)]\,d\theta$$

$$\Rightarrow \quad \dot\theta^2 = 2g\sin\theta/r + \omega^2(-2\cos\theta + \tfrac{1}{2}\cos 2\theta + \tfrac{3}{2})$$

$$F_n = ma_{C_n}: \qquad m[-\omega^2 r\cos\theta(1-\cos\theta) + r\dot\theta^2] = N_1 - mg\sin\theta$$

$$F_b = ma_{C_b}: \qquad -2m\omega r\dot\theta\sin\theta = N_2$$

$$\Rightarrow \quad N_1 = m[3g\sin\theta + \omega^2 r(2 - 3\cos\theta + \cos 2\theta)]$$

$$N_2 = -2m\omega r\sin\theta[2g\sin\theta/r + \omega^2(\tfrac{3}{2} - 2\cos\theta + \tfrac{1}{2}\cos 2\theta)]^{1/2}$$

$$\vec v_C = \vec v_O + \vec\omega_1 \times \overrightarrow{OC} + \vec v_{C|1} = r\dot\theta\,\vec e_t + \omega r(1-\cos\theta)\,\vec i \qquad [\vec i = -\vec e_b]$$

$$v_C = r[\dot\theta^2 + \omega^2(1-\cos\theta)^2]^{1/2} = [2gr\sin\theta + \omega^2 r^2(3 - 4\cos\theta + \cos 2\theta)]^{1/2}$$

Example 2.32: (a) Find the steady settlement rate v of a small spherical body of radius R and density ρ in a liquid of density ρ_l assuming the fluid resistance force to be cRv. Find its settlement rate in water ($\rho_l = 1000\,\text{kg/m}^3$) if $\rho = 1.1\rho_l$, $R = 1\,\mu\text{m}$, $c = 0.02\,\text{kg.m}^{-1}.\text{s}^{-1}$. (b) Explain how this settlement rate is increased in the non-inertial rotating frame of a centrifuge.

Solution: (a) In the FBD of Fig. E2.32a, $m = \frac{4}{3}\pi R^3\rho$ and the buoyancy force $F_b = \frac{4}{3}\pi R^3\rho_l g$. For a steady settlement rate v, $a = 0$ and $F = ma$ yields

$$mg - F_b - cRv = \frac{4}{3}\pi R^3(\rho - \rho_l)g - cRv = 0 \quad \Rightarrow \quad v = \frac{4\pi}{3c}R^2 g(\rho - \rho_l) \qquad (1)$$

$\Rightarrow$ $v = 2.05 \times 10^{-7}$ m/s is so small that it is completely swamped by thermal motion. (b) For given size, eq. (1) $\Rightarrow$ v can only be changed by an effective g_e using a non-inertial frame. A centrifuge 1 rotates at a very high ω (Fig. 2.32b). For the motion of a particle of mass m with $\vec v_{|1}$ and $\vec a_{|1}$ under $mg\,\vec k$ and other forces $\vec P$,

$$\vec F = m\vec a_C: \quad \vec P + mg\,\vec k = m[\vec a_A + \dot{\vec\omega} \times r\,\vec e_r + \vec\omega \times (\vec\omega \times r\,\vec e_r) + 2\,\vec\omega \times \vec v_{|1} + \vec a_{|1}]$$
$$\simeq m\vec a_{|1} - m\omega^2 r\,\vec e_r$$

where $\vec a_A = \vec 0$, $\dot{\vec\omega} = \vec 0$; the Coriolis term $2\omega v_{|1} \ll \omega^2 r$ since ω is very large, and is neglected. As above, $\vec P$ is the resultant of the viscous drag force and buoyancy force,

$$m\vec a_{|1} \simeq \vec P + m(g\,\vec k + \omega^2 r\,\vec e_r) = \vec P + m\vec g_e$$

$$\vec g_e = g\,\vec k + \omega^2 r\,\vec e_r \simeq \omega^2 r\,\vec e_r$$

i.e., effective $g_e \simeq \omega^2 r$. The outward radial velocity v of settlement is thus given by eq. (1) with g_e replacing g: $v = 4\pi R^2 g_e(\rho - \rho_l)/3c$. For a speed of 1500 rpm and

$r = 0.15\,\text{m}$, $g_e \simeq \omega^2 r = 3700\,\text{m/s}^2 \simeq 377g$ and $v \simeq 0.77 \times 10^{-4}\,\text{m/s} = 28\,\text{cm/h}$. A centrifuge makes possible effective separation of small bodies of different density, suspended in a fluid, in reasonable time. Ultra-centrifuges exist wherein $g_e = 50\,000g$.

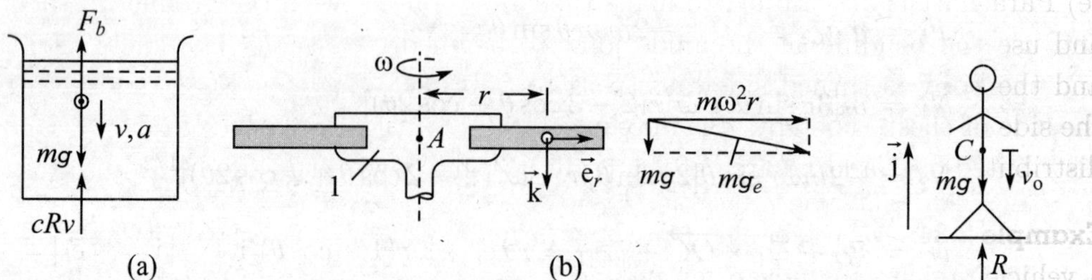

Fig. E2.32

Fig. E2.33

Example 2.33: A person of mass m hits the ground with speed $v_0 = \sqrt{2gh}$ after a fall from a height h. He controls his impact by crouching allowing his knees and back to bend so that he comes to rest after the centre of mass C of his body descends a distance d. (a) Assuming constant deceleration of C, find the reaction from the ground during collision. (b) Find the average force across a horizontal section of the body during collision. Where is bone fracture most likely to occur? (c) Find the maximum height of fall without fracture if the compressive fracture stress of tibia is $\sigma_0 = 150$ MPa and $m = 60\,\text{kg}$, $d = 1\,\text{cm}$, and minimum area A of the tibia is $3\,\text{cm}^2$. (d) Discuss the forces acting during jumps into soft sand or water. (e) How does a parachutist manage to land safely, even though v_0 is very high?

Solution: (a) Let R be the ground reaction, T be the duration of collision and a be the acceleration of C. The FBD of the body is shown in Fig. E2.33. $ma = mg - R$,

$$0^2 - v_0^2 = 2ad = 2d \times \frac{mg - R}{m} \quad \Rightarrow \quad R = m(g + v_0^2/2d) = mg(1 + h/d) \qquad (1)$$

$$I_y = (mg - R)T = \Delta p_y = m(0 - v_0) \quad \Rightarrow \quad R - mg = mv_0/T$$

(b) The average force F_{av}, across a horizontal section of the body, is given by eq. (1) with m replaced by the mass m^* of the body above that cross-section. The average stress σ_{av} across any cross-section of area A^* is $\sigma_{\text{av}} = F_{\text{av}}/A^* = m^*g(1 + h/d)/A^*$. Failure occurs where $(\sigma_{\text{av}})_{\text{max}}$ occurs. The likely section is not at the neck, since m^* above the neck is small, but it is in the tibia just above the ankle where $m^* \simeq m$ and A^* has the least value equal to $2A$ (for 2 tibiae).

(c) For no fracture: $(\sigma_{\text{av}})_{\text{max}} \simeq \dfrac{mg(1 + h/d)}{2A} < \sigma_0 \Rightarrow h < (2\sigma_0 A/mg - 1)d = 1.52\,\text{m}.$

Thus, fractures can occur if the person falls from a height of 2 m. The safe height can be increased to 2.25 m by more crouching so that $d = 1.5\,\text{cm}$.

(d) $(R - mg) \propto 1/T \Rightarrow R$ can be decreased by increasing T. T for landing into loose earth, sand or water is hundreds of times larger than that of landing on hard ground [discussed in part (a)]. Hence, the safe jumping height for such cases is much more.
(e) Parachutists are trained to make their first contact with the ground on their toes and use the bending of the ankle joint to begin deceleration. The knees then bend and the body is turned sideways so as to fall first on the leg, then thigh and lastly the side of chest. Thereby, d is maximized to about 1 m, thus decreasing R_{av} and it is distributed over a large area of the body in contact with the ground to reduce stress.

Example 2.34: An arrangement shown in Fig. E2.34a is used for hoisting a block by a vehicle which goes up an icy incline, operating its winch b. The winch b draws in rope at 0.15 m/s when the vehicle is moving up at 0.3 m/s. The winch a is operated such that the tension T in cable AB is $(3600 - 200e^{-t})$ N. The coefficient of friction for snow is 0.1. Find the speed of the block at $t = 60$ s when winch b is drawing in rope at 0.3 m/s. Find the work done by T on the vehicle.

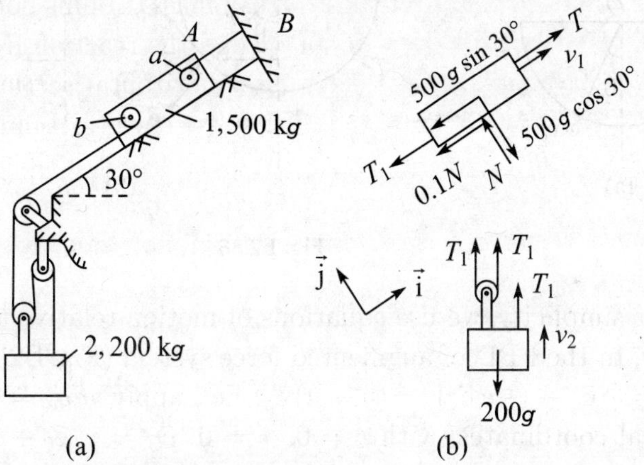

Fig. E2.34

Solution: The FBDs of the bodies are shown in Fig. E2.34b with $N = 500g\cos 30°$. Initially, $v_1(0) = 0.3$ m/s, $v_2(0) = [v_1(0) + 0.15]/3 = 0.15$ m/s and finally let $v_1(60) = v$ m/s. Then $v_2(60) = [v + 0.3]/3 = (0.1 + v/3)$ m/s. We apply the impulse–momentum relation in the direction of the motion of the bodies and eliminate T_1:

$$500(v - 0.3) = \int_0^{60} (T - T_1 - 500g\sin 30° - 0.1N)\, dt \tag{1}$$

$$200[(0.1 + v/3) - 0.15] = \int_0^{60} (3T_1 - 200g)\, dt \tag{2}$$

$$3(1) + (2): \quad \frac{4700v}{3} - 460 = \int_0^{60} [3(T - 500g\sin 30° - 0.1N) - 200g]\, dt$$

$$= \int_0^{60} (206.14 - 600e^{-t})\, dt = 206.14 \times 60 + 600(e^{-60} - 1)$$

$$\Rightarrow \quad v = 7.81\,\text{m/s}, \qquad v_2(60) = 0.1 + v/3 = 2.70\,\text{m/s}$$

T does not do any work on the vehicle as $v_A = 0$.
Alternate Solution: Write $F_i = m_i a_i$ for the 2 bodies, eliminate T_1 and integrate with respect to t.

Example 2.35: A small slider of mass m moves in a smooth hemispherical bowl (Fig. E2.35a) fixed in an elevator moving up with uniform acceleration a. It is imparted a horizontal circumferential velocity v_1 in the position shown. Find v_ϕ, v_θ, $\dot{\phi}$, $\dot{\theta}$, $\ddot{\phi}$, $\ddot{\theta}$ for the general position θ. Set up an equation for the maximum angle θ_2 made by OC with the vertical.

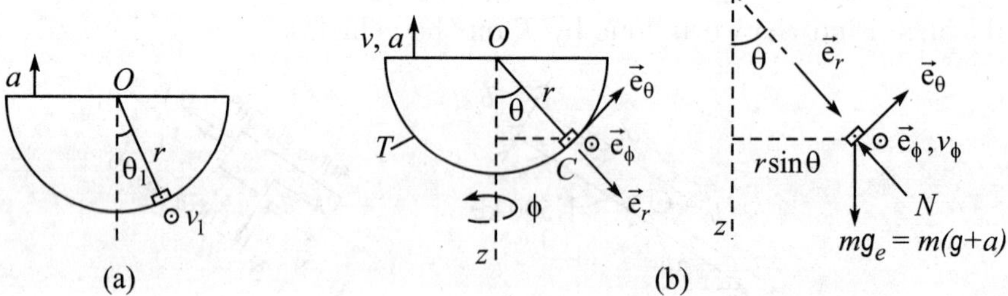

Fig. E2.35

Solution: For simplicity, we use equations of motion relative to translating frame T of the elevator. In the FBD of augmented force system (Fig. E2.35b), instead of $\vec{g}$, use $\vec{g}_e = \vec{g} - \vec{a}_T = g\,\vec{e}_z - (-a\,\vec{e}_z) = (g+a)\,\vec{e}_z$, i.e., apply $mg_e = m(g+a)$. Let r, θ, ϕ be the spherical coordinates, with $\dot{r} = 0$, $\ddot{r} = 0$. $\vec{v}_C = v_\theta\,\vec{e}_\theta + v_\phi\,\vec{e}_\phi$. From the FBD, $M_{Oz} = 0 \Rightarrow H_{Oz} = $ constant, since the z-axis through O is an axis fixed in T.

$$H_{Oz} = (mv_\phi) r \sin\theta = mv_1 r \sin\theta_1 \tag{1}$$

$$\Rightarrow \quad v_\phi = r\sin\theta\,\dot{\phi} = \frac{v_1 \sin\theta_1}{\sin\theta}, \qquad \dot{\phi} = \frac{v_1 \sin\theta_1}{r\sin^2\theta} \tag{2}$$

$v_\theta = r\dot{\theta}$. Relative to translating frame T, N does not do any work as $\vec{v}_C \perp \vec{N}$ and:

$$T + V = \tfrac{1}{2}m(r^2\dot{\theta}^2 + v_\phi^2) - m(g+a)rcos\theta = \tfrac{1}{2}mv_1^2 - m(g+a)r\cos\theta_1$$

Substitution of $\dot{\phi}$ from eq. (2), yields

$$\dot{\theta}^2 = \frac{2(g+a)}{r}(\cos\theta - \cos\theta_1) + \frac{v_1^2}{r^2}\left(1 - \frac{\sin^2\theta_1}{\sin^2\theta}\right) \tag{3}$$

$v_\theta = r\dot{\theta}$. $\ddot{\phi}$ is obtained using $\dot{H}_{Oz} = 0$, and $\ddot{\theta}$ is obtained by differentiating eq. (3):

$$\dot{H}_{Oz} = mr^2 \frac{d}{dt}(\sin^2\theta\,\dot{\phi}) = mr^2(\sin 2\theta\,\dot{\theta}\,\dot{\phi} + \sin^2\theta\,\ddot{\phi}) = 0$$

$$2\dot{\theta}\ddot{\theta} = 2\left[-\frac{(g+a)}{r}\sin\theta + \frac{v_1^2\cos\theta\sin^2\theta_1}{r^2\sin^3\theta}\right]\dot{\theta}$$

For $\theta_{\max} = \theta_2$, $\dot{\theta} = 0$ and eq. (3) yields

$$\frac{2(g+a)}{r}(\cos\theta_2 - \cos\theta_1) + \frac{v_1^2}{r^2}\left(1 - \frac{\sin^2\theta_1}{\sin^2\theta_2}\right) = 0$$

Example 2.36: (a) The system shown in Fig. E2.36a is released from rest. Specify the condition for downward displacement of block 2. Find its displacement s, velocity v and acceleration a at time t and $v(s)$. Assume light frictionless pulleys and a light inextensible string. The coefficients of friction for the inclined plane and block 1 are μ_s, μ_k. Assume $\theta < \tan^{-1}\mu_s$, so that all strings remain taut. (b) Rework part (a) if block 1 is subjected to additional resistance force $R = c\,v_1$, where v_1 is its speed.

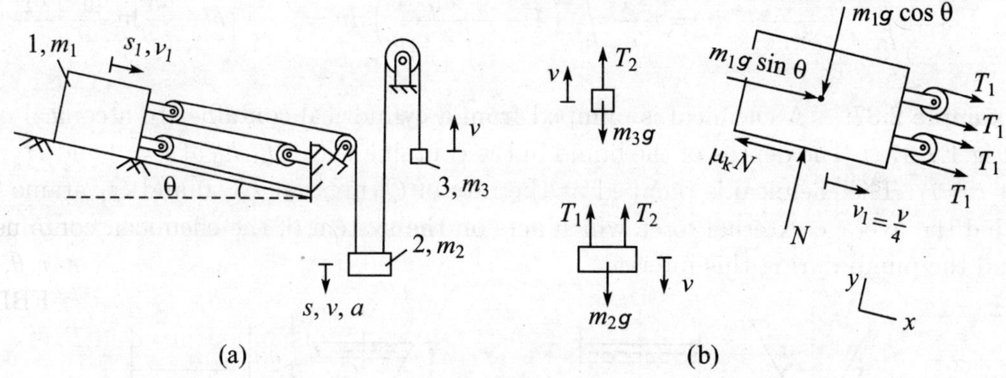

Fig. E2.36

Solution: (a) Inextensible string $\Rightarrow$ $v_1 = v/4$. The FBDs are shown in Fig. E2.36b with $N = m_1 g\cos\theta$, since $a_y = 0$ for m_1. The work–energy relation in the rate form for the whole system is the best to get acceleration since it is a single dof system in which the internal tensions are together workless and other forces are known.

$$T = \tfrac{1}{2}m_2v^2 + \tfrac{1}{2}m_1(v/4)^2 + \tfrac{1}{2}m_3v^2 = \tfrac{1}{2}(m_1/16 + m_2 + m_3)v^2$$

$$\dot{T} = \tfrac{1}{2}(m_1/16 + m_2 + m_3)2v\dot{v} = \dot{W}$$

$$= m_2gv - m_3gv + (m_1g\sin\theta - \mu_kN)v/4 \tag{1}$$

$$\Rightarrow \quad a = \dot{v} = \frac{g[m_2 - m_3 + m_1(\sin\theta - \mu_k \cos\theta)/4]}{m_1/16 + m_2 + m_3} = \text{const.} = a_0 \quad \text{(say)}$$

provided $m_2 + \frac{1}{4}m_1 \sin\theta > m_3 + \frac{1}{4}\mu_s m_1 \cos\theta$ to initiate motion. $v(0) = 0,\ s(0) = 0 \Rightarrow$

$$v(t) = v(0) + a_0 t = a_0 t, \qquad s(t) = s(0) + v(0)t + \tfrac{1}{2}a_0 t^2 = \tfrac{1}{2}a_0 t^2$$

$$v^2(s) - v^2(0) = 2a_0 s \qquad \Rightarrow \qquad v(s) = (2a_0 s)^{1/2}$$

(b) Add $-Rv_1 = -c(\frac{1}{4}v)(\frac{1}{4}v)$ in eq. (1) $\Rightarrow a = a_0 - c_1 v$ with $c_1 = \dfrac{c}{m_1 + 16(m_2 + m_3)}$.

$$\Rightarrow \quad \int_0^{v(t)} \frac{1}{a_0 - c_1 v}\, dv = \int_0^t dt \quad \Rightarrow \quad v = \frac{a_0}{c_1}(1 - e^{-c_1 t}) \tag{2}$$

$$\Rightarrow \quad s = \int_0^t \frac{a_0}{c_1}(1 - e^{-c_1 t})\, dt = \frac{a_0}{c_1}\left[t + \frac{1}{c_1}(e^{-c_1 t} - 1)\right] \tag{3}$$

Equation (2) implies that the terminal velocity is a_0/c_1 as $t \to \infty$. Separating the variables v and s in $\dfrac{dv}{dt} = v\,\dfrac{dv}{ds} = a_0 - c_1 v \Rightarrow$

$$s = \int_0^{v(s)} \frac{v}{a_0 - c_1 v}\, dv = -\frac{1}{c_1}\int_0^v \left(1 - \frac{a_0}{a_0 - c_1 v}\right) dv = -\frac{1}{c_1}\left[v + \frac{a_0}{c_1}\ln\frac{a_0 - c_1 v}{a_0}\right]$$

Example 2.37: A chemical is pumped from a cylindrical container to a conical one (Fig. E2.37a). The depths of the liquid in the containers are H_0, h_0 at $t = 0$ and H_1, h_1 at $t = t_1$. The chemical is pumped at the rate of $Q_0\,\text{m}^3/\text{s}$ at $t = 0$ and Q_1 at $t = t_1$. Find the average external force which acts on the system of the chemical, containers and the pump during this interval.

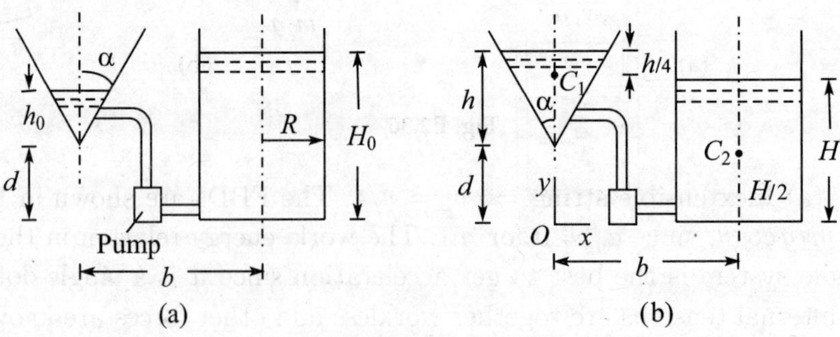

Fig. E2.37

Solution: The chemical has depths H and h in the two containers at time t and the flow rate is Q (Fig. E2.37b). The volumes V_1, V_2, mass m_1, m_2 and centres of mass C_1, C_2 of the chemical in the conical and cylindrical containers are

$$V_1 = \tfrac{1}{3}\pi (h\tan\alpha)^2 h, \qquad m_1 = \tfrac{1}{3}\pi\rho(\tan^2\alpha)h^3, \qquad \vec{r}_{C_1} = (d+\tfrac{3}{4}h)\,\vec{\jmath}$$

$$V_2 = \pi R^2 H, \qquad m_2 = \pi\rho R^2 H, \qquad \vec{r}_{C_2} = b\,\vec{\imath} + \tfrac{1}{2}H\,\vec{\jmath}$$

The mass m_3 of the rest of the system has centre of mass C_3 at a fixed location $\vec{r}_{C_3}$.

$$Q = \dot{V}_1 = \pi h^2 \dot{h}\tan^2\alpha \qquad \Rightarrow \qquad \dot{h} = Q/\pi h^2\tan^2\alpha$$

$$Q = -\dot{V}_2 = -\pi R^2 \dot{H} \qquad \Rightarrow \qquad \dot{H} = -Q/\pi R^2$$

$\vec{F}_{\mathrm{av}}$ on the system is obtained using the impulse–momentum relation:

$$m\vec{r}_C = \sum m_i \vec{r}_{C_i}$$

$$= \pi\rho\left[\tfrac{1}{3}\tan^2\alpha\left(dh^3 + \frac{3}{4}h^4\right)\vec{\jmath} + R^2\left(bH\vec{\imath} + \tfrac{1}{2}H^2\vec{\jmath}\right)\right] + m_3\vec{r}_{C_3}$$

$$m\vec{v}_C = \pi\rho\left[\tan^2\alpha\left(dh^2 + h^3\right)\dot{h}\,\vec{\jmath} + R^2\left(b\,\vec{\imath} + H\,\vec{\jmath}\right)\dot{H}\right] = \rho\left[(d+h-H)\,\vec{\jmath} - b\,\vec{\imath}\right]Q$$

$$\vec{F}_{\mathrm{av}} = \frac{\vec{I}}{t_1 - 0} = \frac{\Delta\vec{p}}{t_1} = \frac{m\vec{v}_C(t_1) - m\vec{v}_C(0)}{t_1}$$

$$= \rho\left[(bQ_0 - bQ_1)\,\vec{\imath} + \{(d+h_1-H_1)\,Q_1 - (d+h_0-H_0)\,Q_0\}\,\vec{\jmath}\right]/t_1$$

Example 2.38: A body of mass m_1 falls through a height h on a body of mass m_2 supported on a spring of stiffness k (Fig. E2.38a) and does not separate from it after impact. Find the energy lost on impact and the maximum compression of the spring.

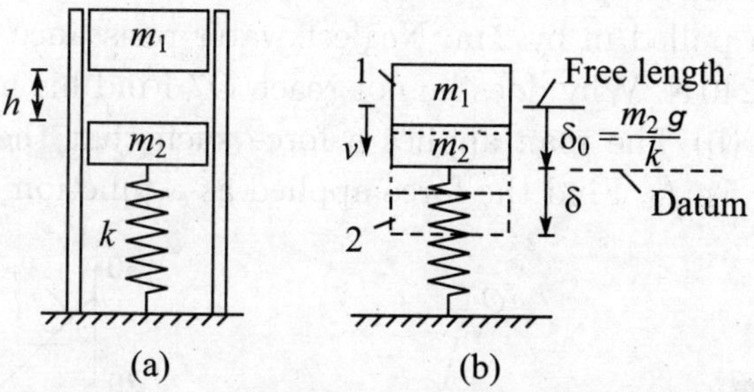

Fig. E2.38

Solution: The speed v_1 of m_1 just before impact is: $v_1 = (2gh)^{1/2}$. Let v be the common velocity of bodies just after impact. We assume that the duration of impact is very small compared to the period of the spring–mass system, i.e., the impact is modelled as instantaneous. The impulse of external forces on bodies 1 and 2 consisting of finite gravity and finite spring forces during the zero impact interval is zero. Hence, their momentum is conserved during impact:

$$(m_1 + m_2)v = m_1 v_1 \qquad \Rightarrow \qquad v = (2gh)^{1/2} m_1/(m_1 + m_2)$$

The relative kinetic energy lost on impact equals $[\frac{1}{2}m_1 v_1^2 - \frac{1}{2}(m_1 + m_2)v^2]/\frac{1}{2}m_1 v_1^2 = \frac{1}{1+m_1/m_2}$, which is less if $m_2 \ll m_1$. The initial compression δ_0 of the spring under gravity force $m_2 g$ is $\delta_0 = m_2 g/k$. Let the additional compression of the spring be δ when the velocities of the bodies are zero at their lowest position. After impact, the system is subjected to gravity and spring forces which are conservative (Fig. E2.38b):

$$T_1 + V_1 = \tfrac{1}{2}(m_1 + m_2)v^2 + \tfrac{1}{2}k\delta_0^2 + 0 = T_2 + V_2 = \tfrac{1}{2}k(\delta_0 + \delta)^2 - (m_1 + m_2)g\delta$$

$$\Rightarrow \qquad \delta^2 - \frac{2m_1 g}{k}\delta - \frac{2m_1^2 gh}{(m_1 + m_2)k} = 0$$

The downward deflection corresponds to the positive root of this quadratic equation:

$$\delta = \left[1 + \left\{1 + \frac{2hk}{(m_1 + m_2)g}\right\}^{1/2}\right] \frac{m_1 g}{k} \qquad (1)$$

The maximum spring compression is $\delta_0 + \delta$. Let $\delta_{st} = m_1 g/k$ be the static deflection under gravity load $m_1 g$. For $m_2 = 0$, eq.(1) yields $\delta/\delta_{st} = [1 + (1 + 2h/\delta_{st})^{1/2}] \gg 1$ if $h/\delta_{st} \gg 1$. If m_1 is released at the top of the spring, then $h = 0$ and $\delta/\delta_{st} = 2$.

Example 2.39: At the instant shown (Fig. E2.39a), a rope attached to O is thrown to the man in the rowboat at its centre of mass C. The man pulls on it to approach the deck. The mass of the person and the rowboat is 200 kg. Find v_r and v_ϕ of C after the rope has been pulled in by 2 m. Neglect water resistance. (a) The man applies a constant force of 240 N. Why does he not reach O? Find his minimum and maximum distance from O. (b) The man applies a force such that the rope is pulled in at a constant rate of 0.5 m/s. Find the force applied as a function of the radial distance.

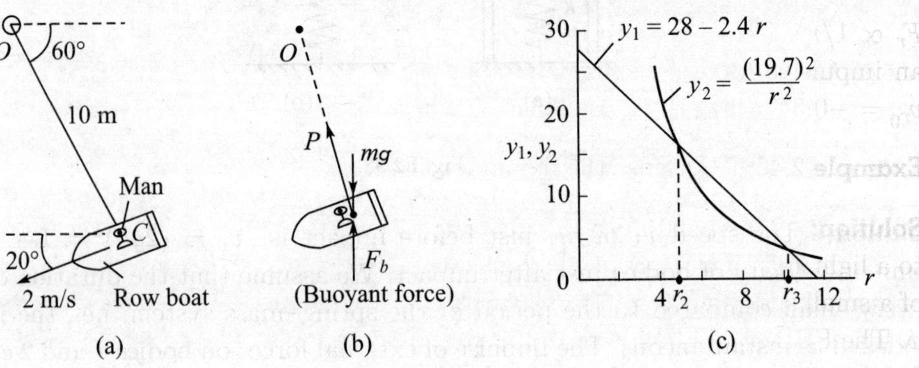

Fig. E2.39

Solution: The FBD of the man and the boat is shown in Fig. E2.39b. $F_b = mg \Rightarrow$ the resultant external force $\vec{F} = -P\,\vec{e}_r$ acts at C and passes through a fixed point O. Hence, the system moves under a central force and $F_\phi = 0 \Rightarrow$

$$F_\phi = ma_\phi = \frac{m}{r}\frac{d}{dt}(rv_\phi) = 0 \qquad \Rightarrow \qquad h_0 = rv_\phi = \text{constant}$$

Initial: $r_0 = 10\,\text{m}$, $v_0 = 2$, $v_{\phi_0} = v_0 \sin 80° = 1.970$, $v_{r_0} = -v_0 \cos 80° = -0.3473\,\text{m/s}$. Let v_1, v_{ϕ_1}, v_{r_1} be the values at $r_1 = 10 - 2 = 8\,\text{m}$.

$$h_0 = rv_\phi = r_0 v_{\phi_0} \quad \Rightarrow \quad v_\phi = \frac{r_0 v_{\phi_0}}{r} = \frac{19.7}{r} \quad \Rightarrow \quad v_{\phi_1} = \frac{19.7}{r_1} = 2.463\,\text{m/s}$$

(a) $F_r = -240\,\text{N}$. Work–energy relation for $m = 200\,\text{kg}$ for position r_0 and r_1:

$$T_1 - T_0 = W_{0-1}: \qquad \tfrac{1}{2}m(v_1^2 - v_0^2) = 240 \times 2 = 480$$

$$\Rightarrow \quad v_1 = 2.966\,\text{m/s}, \qquad v_r = -(v_1^2 - v_{\phi_1}^2)^{1/2} = -1.654\,\text{m/s}$$

The negative root is taken since r is decreasing.

The work–energy relation for positions r_0 and r yields

$$\tfrac{1}{2}(200)(v_\phi^2 + v_r^2 - v_0^2) = 240(10 - r) \qquad \Rightarrow \qquad v_r^2 = (28 - 2.4r) - 19.7^2/r^2$$

The terms $(28 - 2.4r)$ and $-19.7/r^2$ decrease to 28 and $-\infty$ as $r \to 0$. Hence, there exists $r_{\text{min}} = r_2$ where $v_r = 0$, then r increases to $r_{\text{max}} = r_3$. The boat does not reach O.

At the extremal values r_2 and r_3 of r: $v_r = \dot{r} = 0 \Rightarrow 28 - 2.4r = 19.7^2/r^2$. The smaller positive root of this cubic equation, obtained from the graph of Fig. 2.39c, is $r_2 = 4.882\,\text{m}$ and the larger positive root is $r_3 = 10.073\,\text{m}$.

(b) $v_r(t) = \dot{r} = -0.5\,\text{m/s} \Rightarrow \ddot{r} = 0$ and using $v_\phi = r_0 v_{\phi_0}/r$:

$$F_r = ma_r = m(\ddot{r} - r\dot{\phi}^2) = -mr\dot{\phi}^2 = -\frac{mv_\phi^2}{r} = -\frac{mr_0^2 v_{\phi_0}^2}{r^3} = -\frac{77620}{r^3}\,\text{N}$$

$F_r \propto 1/r^3$. It increases from 77.62 N at $r = 10\,\text{m}$ to 151.6 N at $r = 8\,\text{m}$. Moreover, an impulsive force $\hat{F}_r$ should be applied at $t = 0$ to change the radial speed from $v_{r_0} = -0.3473\,\text{m/s}$ to $v_r = -0.5\,\text{m/s}$, where $\hat{F}_r = m(v_r - v_{r_0}) = -30.54\,\text{N.s}$.

Example 2.40: Discuss the dynamics of a spherical pendulum.

Solution: This pendulum (Fig. E2.40a) consists of a small bob of mass m attached to a light rod with a smooth ball and socket joint at O. An equivalent problem is that of a small slider moving inside a smooth spherical bowl. Two dof of the bob are θ and ϕ. The FBD of the bob is shown in Fig. E2.40b. Using cylindrical coordinates (r, ϕ, z) for the centre of the bob:

$$r = L\sin\theta, \qquad \dot r = L\cos\theta\,\dot\theta, \qquad \ddot r = -L\sin\theta\,\dot\theta^2 + L\cos\theta\,\ddot\theta$$

$$z = L(1-\cos\theta), \qquad \dot z = L\sin\theta\,\dot\theta, \qquad \ddot z = L\cos\theta\,\dot\theta^2 + L\sin\theta\,\ddot\theta$$

$$\vec v = \dot r\,\vec e_r + r\dot\phi\,\vec e_\phi + \dot z\,\vec e_z = L\cos\theta\,\dot\theta\,\vec e_r + L\sin\theta\,\dot\phi\,\vec e_\phi + L\sin\theta\,\dot\theta\,\vec e_z \tag{1}$$

$$\Rightarrow \quad v^2 = L^2\dot\theta^2 + L^2\sin^2\theta\,\dot\phi^2$$

$$\begin{aligned}\vec a &= (\ddot r - r\dot\phi^2)\,\vec e_r + (2\dot r\dot\phi + r\ddot\phi)\,\vec e_\phi + \ddot z\,\vec e_z \\ &= [-\sin\theta\,(\dot\theta^2 + \dot\phi^2) + \cos\theta\,\ddot\theta]L\,\vec e_r \\ &\quad + L(2\cos\theta\,\dot\theta\dot\phi + \sin\theta\,\ddot\phi)\,\vec e_\phi + L(\cos\theta\,\dot\theta^2 + \sin\theta\,\ddot\theta)\,\vec e_z\end{aligned} \tag{2}$$

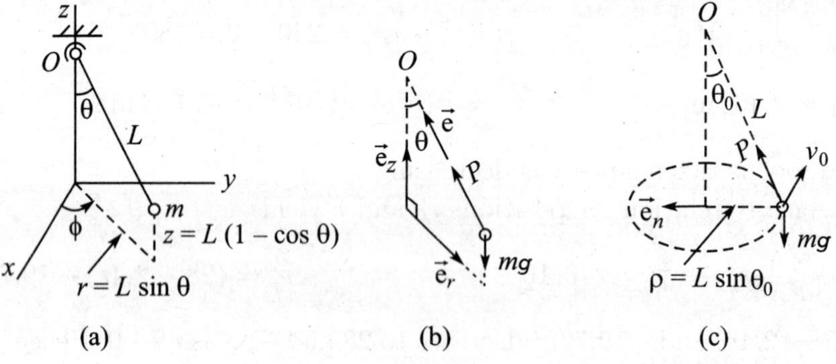

Fig. E2.40

For the bob, $-mg\,\vec e_z \parallel \vec e_z$ and $\vec P$ passes through $O \Rightarrow M_{Oz} = 0 \Rightarrow H_{Oz} = \text{constant}$. The 2nd equation is $T_C + V^* = \text{constant}$, as P is workless and mg is conservative.

$$H_{Oz} = mrv_\phi = mL^2\sin^2\theta\,\dot\phi \quad\Rightarrow\quad \sin^2\theta\,\dot\phi = c_1 \tag{3}$$

$$T_C + V^* = \tfrac{1}{2}m(L^2\dot\theta^2 + L^2\sin^2\theta\,\dot\phi^2) + mgL(1-\cos\theta)$$

$$\Rightarrow \quad \dot\theta^2 + \sin^2\theta\,\dot\phi^2 - 2g\cos\theta/L = c_2 \tag{4}$$

where c_1, c_2 are constants. Equations (3) and (4) are the first integrals of equations of motion and yield $\dot\phi$, $\dot\theta$ in any given position. Let the bob be projected circumferentially with speed v_0 at $\theta = \theta_0$. We get initial values $\dot\phi_0$, $\dot\theta_0$ of $\dot\phi$, $\dot\theta$ from eq. (1):

$$\vec v_0 = v_0\,\vec e_\phi = L\cos\theta_0\,\dot\theta_0\,\vec e_r + L\sin\theta_0\,\dot\phi_0\,\vec e_\phi + L\sin\theta_0\dot\theta_0\,\vec e_z$$

$$\Rightarrow \quad \dot\theta_0 = 0, \qquad \dot\phi_0 = v_0/L\sin\theta_0 \tag{5}$$

$$(3)\ \&\ (5) \Rightarrow \quad \sin^2\theta\,\dot\phi = \sin^2\theta_0\,\dot\phi_0 \quad\Rightarrow\quad \dot\phi = v_0\sin\theta_0/L\sin^2\theta \tag{6}$$

$$(4)-(6) \quad \Rightarrow \quad \dot\theta^2+\sin^2\theta\,\dot\phi^2-2g\cos\theta/L=\dot\theta_0^2+\sin^2\theta_0\,\dot\phi_0^2-2g\cos\theta_0/L$$

$$\Rightarrow \quad \dot\theta^2+v_0^2(\sin^2\theta_0-\sin^2\theta)/L^2\sin^2\theta+2g(\cos\theta_0-\cos\theta)/L=0 \qquad (7)$$

The maximum and minimum values of θ occur when $\dot\theta=0$. At these positions, using $\sin^2\theta_0-\sin^2\theta=\cos^2\theta-\cos^2\theta_0$, eq. (7) yields

$$(\cos\theta_0-\cos\theta)[2g-(\cos\theta_0+\cos\theta)v_0^2/L\sin^2\theta] = 0$$

$$\Rightarrow \quad \theta=\theta_0 \quad \text{or} \quad \sin^2\theta-(\cos\theta_0+\cos\theta)v_0^2/2gL = 0$$

$$\Rightarrow \quad \cos^2\theta+2\lambda^2\cos\theta-(1-2\lambda^2\cos\theta_0) = 0 \qquad (8)$$

with $\lambda^2=v_0^2/4gL$. $0<\theta<\pi \Rightarrow -1<\cos\theta<1 \Rightarrow$ acceptable root of eq. (8) is $\theta=\theta_1$, where $\cos\theta_1=-\lambda^2+(1-2\lambda^2\cos\theta_0+\lambda^4)^{1/2}$. The motion is between θ_0 to θ_1.

$$T_C+V^*=\tfrac{1}{2}mv^2+mgL(1-\cos\theta)=\tfrac{1}{2}mv_0^2+mgL(1-\cos\theta_0) \qquad (9)$$

$\vec P=P\,\vec e$ is obtained from the equation of motion in direction $\vec e=-\sin\theta\,\vec e_r+\cos\theta\,\vec e_z$ and using $\vec a$ from eq. (2) and v^2 from eq. (9):

$$F_e=P-mg\cos\theta=ma_e=m\vec a\cdot\vec e=mL(\dot\theta^2+\sin^2\theta\dot\phi^2)=mv^2/L$$

$$\Rightarrow \quad P=m(g\cos\theta+v^2/L)=m(3g\cos\theta-2g\cos\theta_0+v_0^2/L)$$

For a *conical pendulum* (Fig. E2.40c), $\theta(t)\equiv\theta_0 \Rightarrow \dot\theta=0 \Rightarrow \theta_0$ is a root of eq. (8) $\Rightarrow$

$$\cos^2\theta_0+4\lambda^2\cos\theta_0-1=0$$

$$\Rightarrow \quad v_0^2=gL\sin^2\theta_0/\cos\theta_0 \quad \& \quad \text{eq. (5)} \quad \Rightarrow \quad \dot\phi=v_0/L\sin\theta_0$$

Direct proof for conical pendulum: $\vec a=(v_0^2/\rho)\,\vec e_n=(v_0^2/L\sin\theta_0)\,\vec e_n$,

$$F_z=P\cos\theta_0-mg=0 \quad \Rightarrow \quad P=mg/\cos\theta_0$$

$$F_n=P\sin\theta_0=mv_0^2/L\sin\theta_0 \quad \Rightarrow \quad v_0^2=gL\sin^2\theta_0/\cos\theta_0$$

Example 2.41: A small packet descends a helix whose axis is vertical (Fig. E2.41a). Find its speed after descending a height h if (a) the coefficient of friction for the floor is μ and for the side walls is μ_1, (b) the helix is rotating about its axis at uniform ω. (c) Find the speed of a bead on a helical wire after descending a height h. The coefficient of friction is μ.

Solution: (a) The FBD of the package is shown in Fig. E2.41b. The helix has pitch $p=2\pi R\tan\alpha$. Cylindrical and path components are convenient for $\vec v$, $\vec a$:

$$r = R, \quad \dot{r} = 0, \quad \ddot{r} = 0, \quad z = p\phi/2\pi = R\phi\tan\alpha, \quad \dot{z} = R\dot{\phi}\tan\alpha$$

$$\vec{v} = \dot{r}\,\vec{e}_r + r\dot{\phi}\,\vec{e}_\phi + \dot{z}\,\vec{e}_z = R\dot{\phi}\,\vec{e}_\phi + R\dot{\phi}\tan\alpha\,\vec{e}_z = v\,\vec{e}_t$$

$$\Rightarrow \quad v = R\dot{\phi}(1+\tan^2\alpha)^{1/2}, \qquad \vec{e}_t = \cos\alpha\,\vec{e}_\phi + \sin\alpha\,\vec{e}_z$$

$$\dot{\phi} = v\cos\alpha/R, \qquad \ddot{\phi} = \dot{v}\cos\alpha/R$$

$$\Rightarrow \quad \dot{z} = R\dot{\phi}\tan\alpha = v\sin\alpha, \qquad \ddot{z} = \dot{v}\sin\alpha$$

$$\begin{aligned} \vec{a} &= (\ddot{r} - r\dot{\phi}^2)\,\vec{e}_r + (2\dot{r}\dot{\phi} + r\ddot{\phi})\,\vec{e}_\phi + \ddot{z}\,\vec{e}_z \\ &= -(v^2\cos^2\alpha/R)\,\vec{e}_r + \dot{v}(\cos\alpha\,\vec{e}_\phi + \sin\alpha\,\vec{e}_z) \\ &= \dot{v}\,\vec{e}_t - (v^2\cos^2\alpha/R)\,\vec{e}_r \end{aligned}$$

$\Rightarrow$ $\vec{e}_n = -\,\vec{e}_r$, $\rho = R/\cos^2\alpha$. Equations of motion of the packet are:

$$F_n = ma_n: \qquad N_1 = mv^2\cos^2\alpha/R$$

$$F_b = 0: \qquad mg\cos\alpha - N = 0$$

$$F_t = ma_t: \qquad m\dot{v} = mg\sin\alpha - \mu N - \mu_1 N_1$$

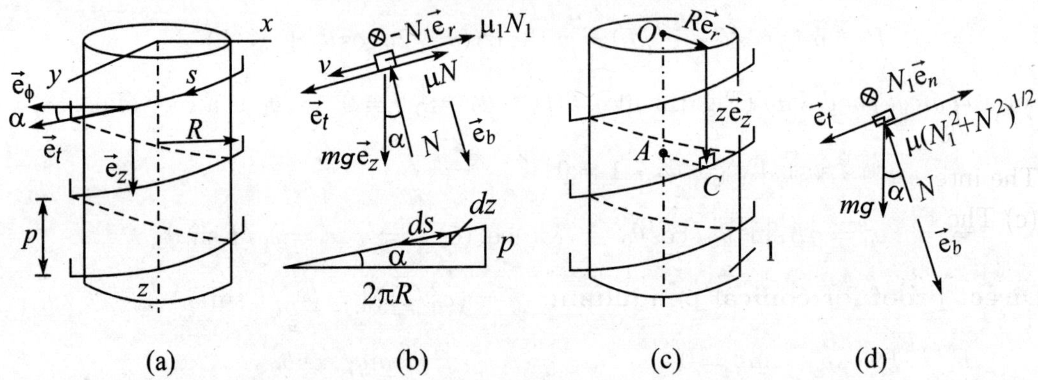

Fig. E2.41

Using expressions of N and N_1, the above equation yields:

$$\dot{v} = \frac{dv}{dz}\dot{z} = \frac{dv}{dz}v\sin\alpha = g(\sin\alpha - \mu\cos\alpha) - \mu_1 v^2\cos^2\alpha/R$$

$$\Rightarrow \quad \int_0^v \frac{v\sin\alpha}{g(\sin\alpha - \mu\cos\alpha) - \mu_1 v^2\cos^2\alpha/R}\,dv = \int_0^h dz$$

$$\Rightarrow \quad v^2 = gR(\sin\alpha - \mu\cos\alpha)[1 - \exp(-2\mu_1 h\cos\alpha\cot\alpha/R)]\,/\,\mu_1\cos^2\alpha$$

(b) $\vec{\omega}_1$ and $\dot{\vec{\omega}}_1$ of frame 1 of the helix are:

$$\vec{\omega}_1 = \omega\,\vec{e}_z = \omega(\sin\alpha\,\vec{e}_t + \cos\alpha\,\vec{e}_b), \qquad \dot{\vec{\omega}}_1 = \vec{0}$$

$$\vec{a}_{C|1} = \vec{a} = \dot{v}\,\vec{e}_t + \frac{v^2\cos^2\alpha}{R}\,\vec{e}_n, \qquad \vec{v}_{C|1} = \vec{v} = v\,\vec{e}_t$$

In Fig. E2.41c, $\vec{a}_A = \vec{0}$, $\overrightarrow{AC} = R\,\vec{e}_r$, and $\vec{a}_C$ with respect to the ground is

$$\begin{aligned}
\vec{a}_C &= \vec{a}_A + \dot{\vec{\omega}}_1 \times \overrightarrow{AC} + \vec{\omega}_1 \times (\vec{\omega}_1 \times \overrightarrow{AC}) + 2\,\vec{\omega}_1 \times \vec{v}_{C|1} + \vec{a}_{C|1} \\
&= -\omega^2 R\,\vec{e}_r + 2\omega(\sin\alpha\,\vec{e}_t + \cos\alpha\,\vec{e}_b) \times v\,\vec{e}_t + \vec{a} \\
&= (2\omega v\cos\alpha + \omega^2 R + v^2\cos^2\alpha/R)\,\vec{e}_n + \dot{v}\,\vec{e}_t
\end{aligned}$$

The equations of motion of the packet are:

$$\begin{aligned}
F_n &= ma_{C_n}: & N_1 &= m(2\omega v\cos\alpha + \omega^2 R + v^2\cos^2\alpha/R) \\
F_b &= 0: & mg\cos\alpha - N &= 0 \\
F_t &= ma_{C_t}: & m\dot{v} &= mg\sin\alpha - \mu N - \mu_1 N_1 \;\Rightarrow
\end{aligned}$$

$$\dot{v} = v\sin\alpha\frac{dv}{dz} = g(\sin\alpha - \mu\cos\alpha) - \mu_1(2\omega v\cos\alpha + \omega^2 R + v^2\cos^2\alpha/R)$$

$$\int_0^v v[g(1-\mu\cot\alpha) - \mu_1\omega^2 R\,\mathrm{cosec}\,\alpha - \mu_1\cot\alpha(2\omega v + v^2\cos\alpha/R)]^{-1}dv = h$$

The integral can be evaluated in closed form.

(c) The FBD of the bead is shown in Fig. 2.41d. Equations of motion:

$$\begin{aligned}
F_n &= ma_n: & N_1 &= mv^2\cos^2\alpha/R \\
F_b &= 0: & mg\cos\alpha - N &= 0 \\
F_t &= ma_t: & m\dot{v} &= mg\sin\alpha - \mu(N_1^2 + N^2)^{1/2}
\end{aligned}$$

$$\Rightarrow \quad \dot{v} = v\sin\alpha\frac{dv}{dz} = g\sin\alpha - \mu(v^4\cos^4\alpha/R^2 + g^2\cos^2\alpha)^{1/2}$$

which can be integrated numerically by the Runge–Kutta method.

Example 2.42: (a) Cylindrical cans are transported from one elevation to another by the translating horizontal arms (Fig. E2.42). The coefficient of friction is μ. Find the maximum upward acceleration a of the conveyor for which the can does not move relative to the arm. Rework for downward acceleration. (b) If relative motion starts with slip, find the relative acceleration and the time at which it starts to tilt.

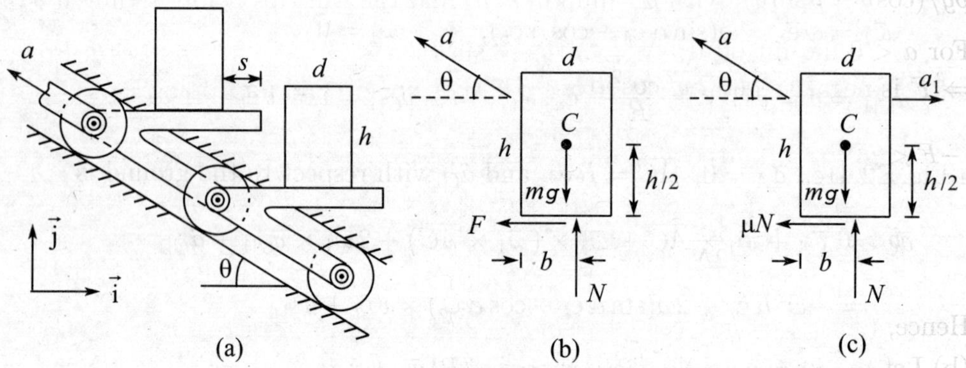

Fig. E2.42

Solution: (a) The FBD of the can is shown in Fig. E2.42b. The four unknowns a, N, F, b are solved using $F_x = ma_{Cx}$, $F_y = ma_{Cy}$, $M_{Cz} = 0$ and the slip condition ($|F| = \mu N$) or tipping condition ($b = 0$ or $b = d$).

$$F_x = ma_{Cx}: \qquad F = ma\cos\theta \tag{1}$$

$$F_y = ma_{Cy}: \qquad N - mg = ma\sin\theta \quad \Rightarrow \quad N = m(g + a\sin\theta) \tag{2}$$

$$M_{Cz} = 0: \qquad N(b - d/2) - Fh/2 = 0 \quad \Rightarrow \quad b = Fh/2N + d/2 \tag{3}$$

For $a > 0$: eq. (2) $\Rightarrow$ $N > 0$ (no lift-off), eq. (1) $\Rightarrow$ F is positive and eq. (3) $\Rightarrow$ $b > d/2$. For no slip $F < \mu N$ and for no tipping $b < d$.

$$F < \mu N: \qquad ma\cos\theta < \mu m(g + a\sin\theta)$$

$$\Rightarrow \qquad a(\cos\theta - \mu\sin\theta) < \mu g$$

$$\Rightarrow \quad \text{no slip } \forall a \text{ if} \quad \cos\theta \le \mu\sin\theta \quad \Rightarrow \quad \text{if } \theta \ge \theta_s = \tan^{-1}(1/\mu),$$

$$\text{else no slip if} \quad a < \frac{\mu g}{\cos\theta - \mu\sin\theta}$$

$$b < d: \qquad b = \frac{Fh}{2N} + \frac{d}{2} = \frac{ah\cos\theta}{2(g + a\sin\theta)} + \frac{d}{2} < d$$

$$\Rightarrow \qquad a(\cos\theta - d\sin\theta/h) < dg/h$$

$$\Rightarrow \text{ no tipping } \forall a \text{ if} \quad \cos\theta \le d\sin\theta/h \quad \Rightarrow \quad \text{if } \theta \ge \theta_t = \tan^{-1}(h/d),$$

$$\text{else no tipping if} \quad a < \frac{(d/h)g}{\cos\theta - (d/h)\sin\theta}$$

Hence, if $\theta \geq \max(\theta_s, \theta_t)$, then no relative motion of can $\forall a > 0$. Else, $a_{\max} = pg/(\cos\theta - p\sin\theta)$, with $p = \min(d/h, \mu)$ and the can tips before sliding if $d/h < \mu$. For $a < 0$, no lift-off if: $N \stackrel{(2)}{=} m(g + a\sin\theta) \geq 0$, i.e., if $-a \leq g\,\text{cosec}\,\theta$. Equation (1) $\Rightarrow F$ is negative and eq. (3) $\Rightarrow b < d/2$. For no slip and no tipping:

$$-F < \mu N: \quad -ma\cos\theta < \mu m(g + a\sin\theta) \quad \Rightarrow \quad -a < \frac{\mu g}{\cos\theta + \mu\sin\theta}$$

$$b > 0: \quad b = \frac{Fh}{2N} + \frac{d}{2} = \frac{ah\cos\theta}{2(g + a\sin\theta)} + \frac{d}{2} > 0 \quad \Rightarrow \quad -a < \frac{(d/h)g}{\cos\theta + (d/h)\sin\theta}$$

Hence, $(-a)_{\max} = pg/(\cos\theta + p\sin\theta)$, and the can tips before sliding if $d/h < \mu$.
(b) Let $a_1\vec{\imath}$ be the acceleration with respect to the translating arm (Fig. E2.42c).

$$\vec{a}_C = a_1\vec{\imath} + a(-\cos\theta\,\vec{\imath} + \sin\theta\,\vec{\jmath}) = (a_1 - a\cos\theta)\,\vec{\imath} + a\sin\theta\,\vec{\jmath}$$

$$F_y = ma_{C_y}: \quad N - mg = ma\sin\theta \quad \Rightarrow \quad N = m(g + a\sin\theta)$$

$$F_x = ma_{C_x}: \quad -\mu N = m(a_1 - a\cos\theta) \quad \Rightarrow \quad a_1 = a\cos\theta - \mu(g + a\sin\theta)$$

$$M_{C_z} = 0: \quad N(b - d/2) - \mu Nh/2 = 0 \quad \Rightarrow \quad b = \tfrac{1}{2}d + \tfrac{1}{2}\mu h$$

It certainly tilts at time t when the can has moved a distance $s + d - b$ so that the line of action of N is just at the edge of the arm: $\frac{1}{2}a_1t^2 = s + d - b$.

Example 2.43: The mass of a fork-lift truck (Fig. E2.43a), excluding the fork, is m_1 with centre of mass at C_1. Neglect the inertia of the wheels. The mass of the fork and the crate is m_2 with centre of mass at C_2. The fork is supported by a smooth roller at A and a connection at B which supports both horizontal and vertical reactions. (a) Find the maximum deceleration that the truck can have with all wheels braked so that the crate neither slips nor tips and the wheels neither slip nor lift off. The mass of the crate is m and its centre of mass is at height h_3 above the centre of its base. The coefficient of friction for the wheels and the ground is μ_1 and for the crate and the fork is μ_2. (b) Find the ground reactions for a rear-wheel driven fork-lift truck at the instant when the fork has an upward acceleration a_1 relative to the truck and the driving torque is M_1. Assume that there is no slip and no lift-off at all contacts.

Solution: (a) Let the backward acceleration of the truck be a. Since the wheels are braked, the friction forces on them are in the backward direction. The external force systems acting on the crate and truck are shown in Fig. E2.43b,c along with the equivalent kinetic systems: $m_i\vec{a}_{C_i}$ at C_i. For the truck:

$$M_{E_z} = N_2(b_2 + b_3) - m_1gb_2 + m_2gb_1 = -m_1ah_1 - m_2ah_2$$

$$\Rightarrow \quad N_2 = [m_1gb_2 - m_2gb_1 - (m_1ah_1 + m_2ah_2)]/(b_2 + b_3) \qquad (1)$$

$$F_y = N_1 + N_2 - (m_1 + m_2)g = 0 \quad \Rightarrow \quad N_1 + N_2 = (m_1 + m_2)g \qquad (2)$$

$$F_x = F_1 + F_2 = (m_1 + m_2)a \quad \Rightarrow \quad a = (F_1 + F_2)/(m_1 + m_2) \qquad (3)$$

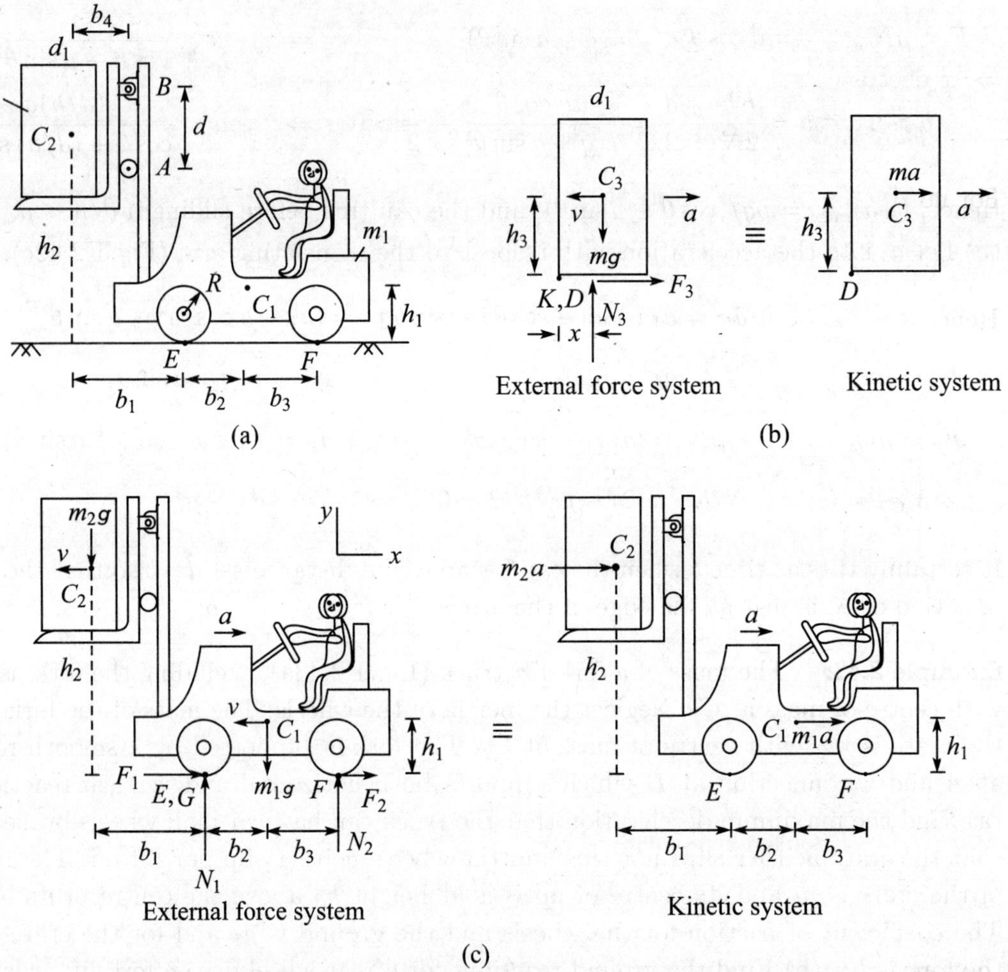

Fig. E2.43a–c

Equations (1) and (2) imply that N_2 decreases with increase in a and N_1 increases by the same amount. For no lift-off of rear wheels, using eq. (1) : $N_2 \geq 0 \Rightarrow$

$$a \leq a_1 = (m_1b_2 - m_2b_1)g\,/\,(m_1h_1 + m_2h_2) \qquad (4)$$

For no slip of wheels, using eqs (2), (3) :

$$a = (F_1 + F_2)/(m_1 + m_2) \leq \mu_1(N_1 + N_2)\,/\,(m_1 + m_2) = \mu_1 g \qquad (5)$$

For the crate:

$$F_x = F_3 = ma \tag{6}$$

$$F_y = N_3 - mg = 0 \quad \Rightarrow \quad N_3 = mg \tag{7}$$

$$M_{Dz} = N_3 x - \tfrac{1}{2} mgd_1 = -mah_3 \quad \Rightarrow \quad x = \tfrac{1}{2} d_1 - ah_3/g \tag{8}$$

$\Rightarrow$ x decreases with increase in a. For no slip of crate, using eqs (6), (7):

$$F_3 \leq \mu_2 N_3 \quad \Rightarrow \quad ma \leq \mu_2 mg \quad \Rightarrow \quad a \leq \mu_2 g \tag{9}$$

For no tipping of crate, using eq. (8):

$$x = \tfrac{1}{2} d_1 - ah_3/g \geq 0 \quad \Rightarrow \quad a \leq gd_1/2h_3 \tag{10}$$

Hence, the maximum retardation: $a = \min(a_1, \mu_1 g, \mu_2 g, gd_1/2h_3)$.

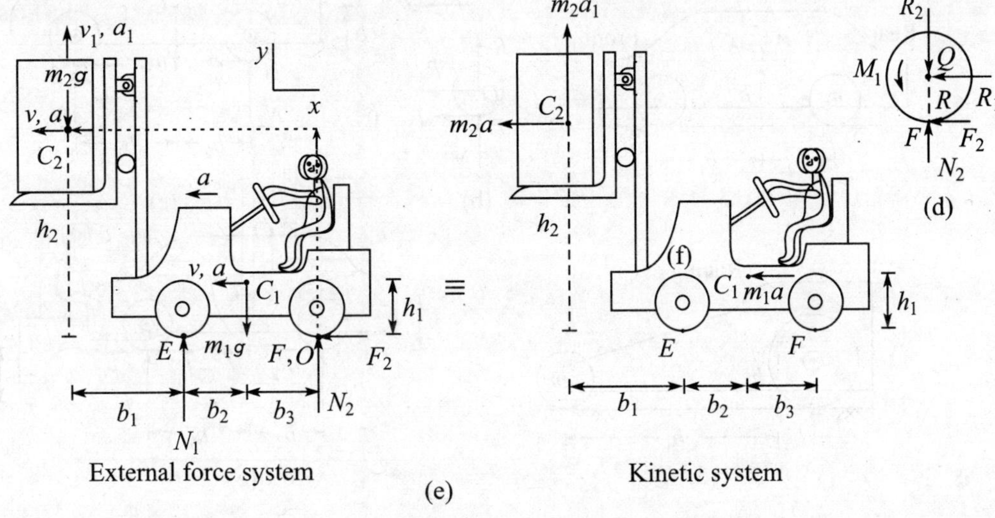

Fig. E2.43d,e

(b) The un-powered front wheels are two-force members with only normal ground reaction. For FBD of the driven rear wheels (Fig. E2.43d), $M_{Qz} = 0 \Rightarrow F_2 = M_1/R$. The external force system on the truck and its equivalent kinetic system of forces $m_i \vec{a}_{C_i}$ at C_i are shown in Fig. E2.43e:

$$F_x = -F_2 = -(m_1 + m_2)a \quad \Rightarrow \quad a = M_1/R(m_1 + m_2) \tag{11}$$

$$M_{Fz} = -N_1(b_2 + b_3) + m_1 g b_3 + m_2 g(b_1 + b_2 + b_3)$$

$$= -m_2 a_1(b_1 + b_2 + b_3) + m_1 a h_1 + m_2 a h_2$$

$$\Rightarrow \quad N_1 = \frac{m_1 g b_3 + m_2(g + a_1)(b_1 + b_2 + b_3) - m_1 a h_1 - m_2 a h_2}{b_2 + b_3} \tag{12}$$

$$F_y = N_1 + N_2 - m_1 g - m_2 g = m_2 a_1$$

$$\Rightarrow \quad N_2 = \frac{m_1 g b_2 - m_2(g + a_1) b_1 + m_1 a h_1 + m_2 a h_2}{b_2 + b_3} \tag{13}$$

where a in eqs (12) and (13) is given by eq. (11).

Example 2.44: (a) Find the maximum acceleration of a self-propelled vehicle moving on a straight level road (Fig. E2.44a). Neglect the inertia of the wheels. Consider rear-wheel, front-wheel and four-wheel drives. The coefficient of friction is μ. (b) Find the maximum deceleration of the vehicle if the brakes are applied to the rear wheels only, to the front wheels only and to all wheels.

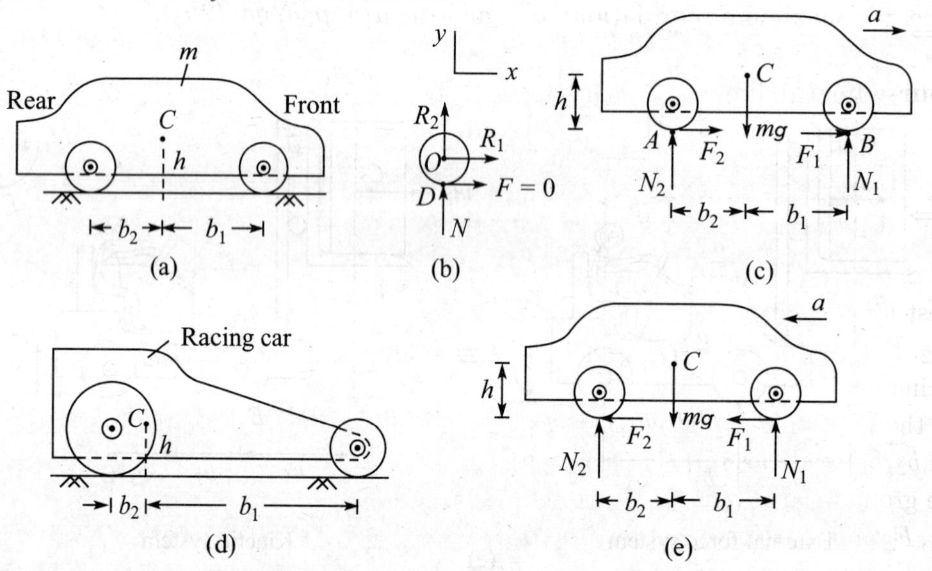

Fig. E2.44

Solution: For un-braked/un-powered light wheel (Fig. E2.44b), $M_{Oz} = 0 \Rightarrow F = 0$.
(a) The FBD of the vehicle is shown in Fig. E2.44c. The equations of motion of this translating body (since the inertia of rotating wheels is neglected) are:

$$F_x = ma_x: \qquad F_1 + F_2 = ma \tag{1}$$

$$F_y = ma_y: \qquad N_1 + N_2 - mg = 0 \tag{2}$$

$$M_{Cz} = 0 \quad : \qquad N_1 b_1 - N_2 b_2 + (F_1 + F_2)h = 0 \tag{3}$$

$$\Rightarrow \quad N_1 = m(gb_2 - ah)/(b_1 + b_2), \qquad N_2 = m(gb_1 + ah)/(b_1 + b_2) \tag{4}$$

N_1 decreases with a and for no lift-off: $N_1 \geq 0 \Rightarrow$

$$m(gb_2 - ah)/(b_1 + b_2) \geq 0 \qquad \Rightarrow \qquad a \leq gb_2/h \tag{5}$$

For **rear-wheel drive:** $F_1 = 0$ as there is no driving torque on the front wheels. Equations (1) and (4) imply that for no slip, $F_2 = ma \leq \mu N_2 \Rightarrow$

$$ma \leq \mu m(gb_1 + ah)/(b_1 + b_2) \qquad \Rightarrow \qquad a \leq \mu g/(1 + b_2/b_1 - \mu h/b_1) \tag{6}$$

$$\Rightarrow \quad a_{\max} = \min[\mu g/(1 + b_2/b_1 - \mu h/b_1),\ gb_2/h] \tag{7}$$

For most vehicles, $\mu g/(1+b_2/b_1-\mu h/b_1) < gb_2/h$ and so $a_{\max} = \mu g/(1+b_2/b_1-\mu h/b_1)$.

For **front-wheel drive**, $F_2 = 0$ for un-powered rear wheels. For no slip, eqs (1), (4)

$$\Rightarrow \quad F_1 = ma \leq \mu N_1 = \mu m(gb_2 - ah)/(b_1 + b_2)$$

$$\Rightarrow \quad a \leq a_1 = \mu g/(1 + b_1/b_2 + \mu h/b_2) \quad \Rightarrow \quad a_{\max} = \min(a_1,\ gb_2/h) \tag{8}$$

For **four-wheel drive**, $a_{\max}$ occurs for simultaneous slip, and eqs (1), (2) yield

$$a \leq (F_1 + F_2)/m = \mu(N_1 + N_2)/m = \mu mg/m = \mu g \tag{9}$$

$$\Rightarrow \quad a_{\max} = \min(\mu g,\ gb_2/h)$$

For $b_1 \simeq b_2$, eqs (7) to (9) $\Rightarrow$ $a_{\max}$ is the largest ($= \mu g$) for the four-wheel drive and smallest for the front-wheel drive, with $a_{\max} \simeq \frac{1}{2}\mu g$ for two-wheel drives. However, if $b_1 < b_2$, then the front-wheel drive would be better than the rear-wheel drive.

Racing cars, shown in Fig. E2.44d, are specially designed rear-wheel driven cars in which the limit on a given by eq. (6) is increased by having C close to the rear wheels (small b_2/b_1) and close to the ground (small h) and using wide tyres with good grip on the ground (large $\mu \simeq 1$). Hence, $a = \mu g$ if $\mu h = b_2$, i.e., if $b_2 \simeq h$.

(b) F_1, F_2 and a are reversed for braking (Fig. E2.44e). The equations of motion are

$$F_1 + F_2 = ma, \qquad\qquad N_1 + N_2 - mg = 0$$

$$M_{Cz} = N_1 b_1 - N_2 b_2 - (F_1 + F_2)h = 0$$

$$\Rightarrow \quad N_1 = m(gb_2 + ah)/(b_1 + b_2), \qquad N_2 = m(gb_1 - ah)/(b_1 + b_2) \tag{10}$$

N_2 decreases with a. For no lift-off: $N_2 \geq 0 \Rightarrow a \leq gb_1/h$ for all types of braking.

For only the rear wheels braked: $F_1 = 0$, $F_2 \leq \mu N_2 \Rightarrow a \leq \mu g/(1 + b_2/b_1 + \mu h/b_1)$.
For most vehicles, this provides a smaller bound than gb_1/h.

For only the front wheels braked: $F_2 = 0$, $F_1 \leq \mu N_1 \Rightarrow a \leq \mu g/(1 + b_1/b_2 - \mu h/b_2)$.

For all wheels braked, simultaneous slip $\Rightarrow a_{\max} = \dfrac{F_1 + F_2}{m} \leq \dfrac{\mu(N_1 + N_2)}{m} = \mu g$.

Example 2.45: A platform 3, used for transport over small distances (Fig. E2.45a), is hinged to two light bars. Find $\dot{\omega}$ and the reactions at A, B, D, E in the given position.

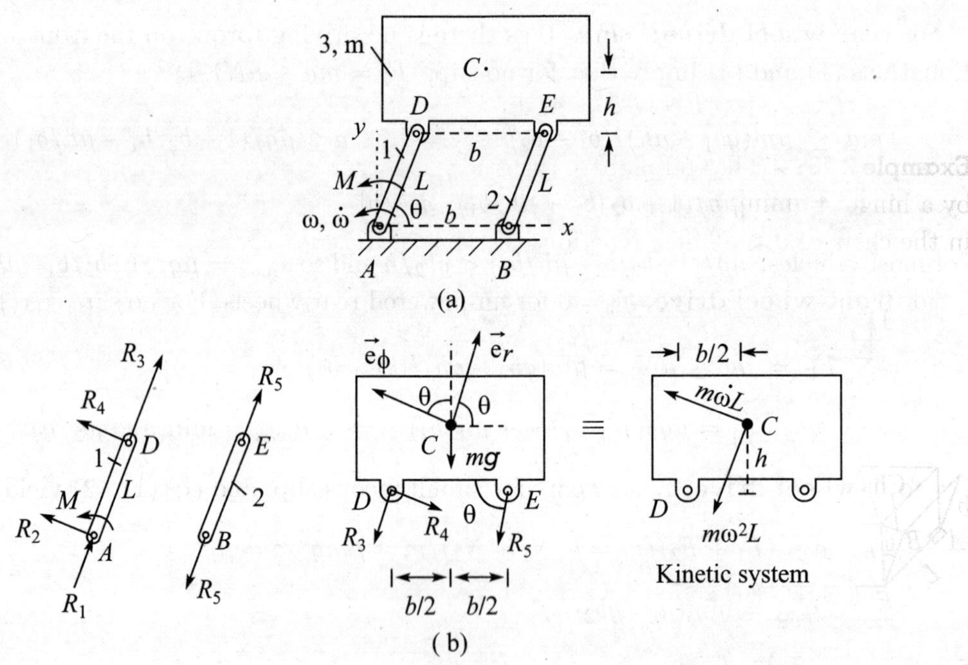

Fig. E2.45

Solution: Platform 3 translates. BE is a two-force light member. The reaction components are shown along $\vec{e}_r$, $\vec{e}_\phi$ in FBDs (Fig. 2.45b), since $\vec{a}_C = \vec{a}_D = -\omega^2 L\,\vec{e}_r + \dot{\omega} L\,\vec{e}_\phi$. The equivalent kinetic system for 3 is also shown. R_1 to R_5 and $\dot{\omega}$ are obtained from six equations of motion for 1 and 3.

$$1: \quad F_r = R_1 + R_3 = 0 \quad \Rightarrow \quad R_1 = -R_3 \tag{1}$$

$$F_\phi = R_2 + R_4 = 0$$

$$M_{A_z} = M + LR_4 = 0 \quad \Rightarrow \quad R_4 = -R_2 = -M/L \tag{2}$$

$$3: \quad F_\phi = ma_{C_\phi}: \qquad -R_4 - mg\cos\theta = m\dot{\omega}L$$

$$\Rightarrow \quad \dot{\omega} = (M - mgL\cos\theta)/mL^2 \tag{3}$$

$$F_r = ma_{C_r}: \qquad -R_3 - R_5 - mg\sin\theta = -m\omega^2 L \tag{4}$$

$$M_{D_z} = -\tfrac{1}{2}mgb - R_5\sin\theta\, b$$

$$= (m\dot{\omega}L\sin\theta + m\omega^2 L\cos\theta)h + (m\dot{\omega}L\cos\theta - m\omega^2 L\sin\theta)\tfrac{1}{2}b \tag{5}$$

R_5, R_3, R_1 are obtained in succession from eqs (5), (4), (1).
$\dot{\omega}$ could have been obtained directly from $\dot{T} + \dot{V} = \dot{W}_{\text{nc}}$:

$$\frac{d}{dt}(\tfrac{1}{2}mL^2\omega^2 + mgL\sin\theta) = M\omega$$

$$mL^2\omega\dot{\omega} + mgL\cos\theta\,\omega = M\omega \qquad \Rightarrow \qquad \dot{\omega} = (M - mgL\cos\theta)/mL^2$$

Example 2.46: The end plate of the truck (Fig. E2.46a) has mass m and is supported by a hinge A at the bottom edge and a chain BD at the top corner. Find the tension in the chain and the hinge reaction when the truck has acceleration a.

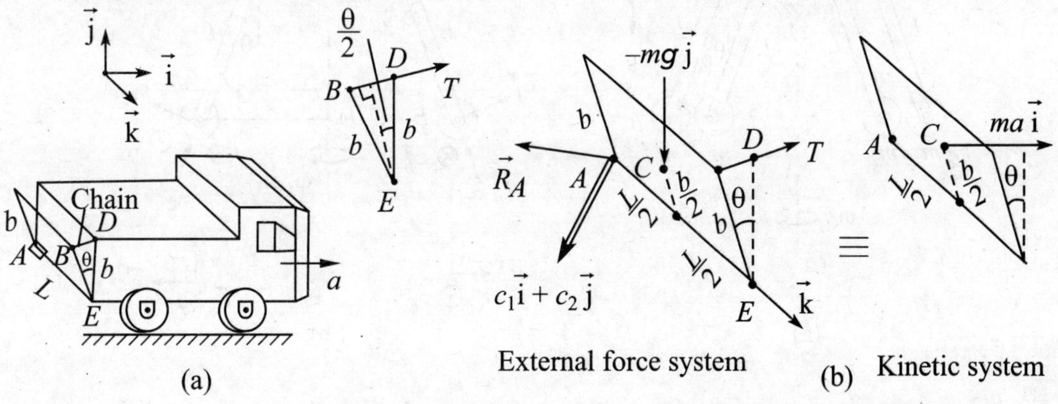

Fig. E2.46

Solution: The FBD of the end plate is shown in Fig. E2.46b, along with the equivalent kinetic system. $\vec{M}_A = \overrightarrow{AC} \times m\vec{a}_C$:

$$c_1\vec{\mathrm{i}} + c_2\vec{\mathrm{j}} + \left[\tfrac{1}{2}L\vec{\mathrm{k}} + \tfrac{1}{2}b\left(\cos\theta\,\vec{\mathrm{j}} - \sin\theta\,\vec{\mathrm{i}}\right)\right] \times (-mg\vec{\mathrm{j}}) + (L\vec{\mathrm{k}} + b\vec{\mathrm{j}})$$

$$\times T\left(\cos\tfrac{1}{2}\theta\,\vec{\mathrm{i}} + \sin\tfrac{1}{2}\theta\,\vec{\mathrm{j}}\right) = \left[\tfrac{1}{2}L\vec{\mathrm{k}} + \tfrac{1}{2}b(\cos\theta\,\vec{\mathrm{j}} - \sin\theta\,\vec{\mathrm{i}})\right] \times ma\vec{\mathrm{i}}$$

$$\vec{\mathrm{k}}: \qquad \tfrac{1}{2}mgb\sin\theta - Tb\cos\tfrac{1}{2}\theta = -\tfrac{1}{2}mab\cos\theta \quad \Rightarrow \quad T = \frac{m(a\cos\theta + g\sin\theta)}{2\cos\frac{1}{2}\theta}$$

$$\vec{\mathrm{i}}: \qquad c_1 + \tfrac{1}{2}mgL - TL\sin\tfrac{1}{2}\theta = 0 \qquad \Rightarrow \qquad c_1 = TL\sin\tfrac{1}{2}\theta - \tfrac{1}{2}mgL$$

$$\vec{\mathrm{j}}: \qquad c_2 + TL\cos\tfrac{1}{2}\theta = \tfrac{1}{2}maL \qquad \Rightarrow \qquad c_2 = \tfrac{1}{2}maL - TL\cos\tfrac{1}{2}\theta$$

$$\vec{F} = \vec{R}_A - mg\vec{\mathrm{j}} + T\left(\cos\tfrac{1}{2}\theta\,\vec{\mathrm{i}} + \sin\tfrac{1}{2}\theta\,\vec{\mathrm{j}}\right) = ma\vec{\mathrm{i}}$$

yields $\vec{R}_A$.

Example 2.47: The rotating system in an amusement park can be modelled as six light bars hinged at one end with a discrete mass m at the other end (Fig. E2.47a). The rotational springs are un-deformed for $\theta = 0$. The motor is shut-off when the system is in position θ_0 and has angular velocities ω_0 and $\dot{\theta}_0$. For the subsequent position θ, find ω, $\dot{\theta}$, $\dot{\omega}$, $\ddot{\theta}$ and the bearing reactions at A and O. Set up an equation for $\theta_{\text{max}} = \theta_1$ (say). Neglect the friction in the bearings.

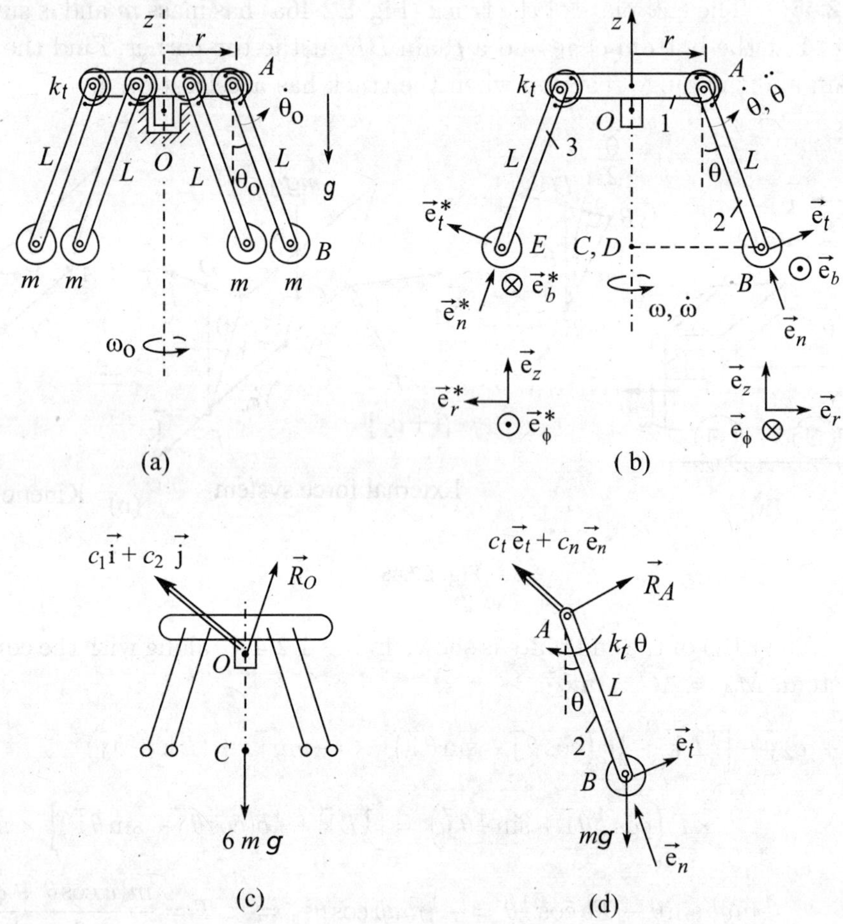

Fig. E2.47

Solution: From Fig. E2.47b, for point B: $\overrightarrow{DB} = (r + L\sin\theta)\,\vec{e}_r$,

$$\vec{r}_C = -L\cos\theta\,\vec{k}, \qquad \vec{v}_C = L\sin\theta\,\dot{\theta}\,\vec{k}, \qquad \vec{a}_C = L(\cos\theta\,\dot{\theta}^2 + \sin\theta\,\ddot{\theta})\,\vec{k}.$$

For point D on rigid extension of 1: $\vec{v}_D = \vec{0}, \qquad \vec{a}_D = \vec{0}.$

$$\vec{e}_r = \cos\theta\,\vec{e}_t - \sin\theta\,\vec{e}_n, \qquad \vec{e}_\phi = -\vec{e}_b, \qquad \vec{e}_z = \cos\theta\,\vec{e}_n + \sin\theta\,\vec{e}_t$$

$$\begin{aligned}
\vec{v}_{B|1} &= L\dot{\theta}\,\vec{e}_t, \qquad \vec{a}_{B|1} = L\ddot{\theta}\,\vec{e}_t + L\dot{\theta}^2\,\vec{e}_n \\
\vec{v}_B &= \vec{v}_D + \omega\,\vec{e}_z \times \overrightarrow{DB} + \vec{v}_{B|1} = \omega(r + L\sin\theta)\,\vec{e}_\phi + L\dot{\theta}\,\vec{e}_t \\
\vec{a}_B &= \vec{a}_D + \dot{\omega}\,\vec{e}_z \times \overrightarrow{DB} - \omega^2\overrightarrow{DB} + 2\omega\,\vec{e}_z \times \vec{v}_{B|1} + \vec{a}_{B|1} \\
&= \dot{\omega}(r + L\sin\theta)\,\vec{e}_\phi - \omega^2(r + L\sin\theta)\,\vec{e}_r \\
&\quad +2\omega(\cos\theta\,\vec{e}_n + \sin\theta\,\vec{e}_t) \times L\dot{\theta}\,\vec{e}_t + L\ddot{\theta}\,\vec{e}_t + L\dot{\theta}^2\,\vec{e}_n \\
&= [\ddot{\theta}L - \omega^2(r + L\sin\theta)\cos\theta]\,\vec{e}_t + [L\dot{\theta}^2 + \omega^2(r + L\sin\theta)\sin\theta]\,\vec{e}_n \\
&\quad -[\dot{\omega}(r + L\sin\theta) + 2\omega L\dot{\theta}\,\cos\theta]\,\vec{e}_b
\end{aligned}$$

Momenta $\vec{p}_2,\ \vec{p}_3$ of bars 2 and 3 (Fig. E2.47b) are given by

$$\vec{p}_2 = m\vec{v}_B = m\omega(r + L\sin\theta)\,\vec{e}_\phi + mL\dot{\theta}\,\vec{e}_t, \qquad \vec{p}_3 = m\omega(r + L\sin\theta)\,\vec{e}_\phi^{\,*} + mL\dot{\theta}\,\vec{e}_t^{\,*}$$

The contributions of $\vec{e}_t,\ \vec{e}_t^{\,*}$ components of $\vec{p}_1,\ \vec{p}_2$ to $\vec{H}_O$ cancel out as their resultant passes through O, whereas the contributions of their $\vec{e}_\phi,\ \vec{e}_\phi^{\,*}$ components to $\vec{H}_O$ add up yielding $2m\omega(r + L\sin\theta)^2\,\vec{e}_z$. FBD of whole system of 6 bars is shown in Fig. E2.47c. For this system (considering 3 pairs of bars):

$$\begin{aligned}
\vec{H}_O &= 6m(r + L\sin\theta)^2\omega\,\vec{k} = H_{Oz}\,\vec{k} \\
\dot{\vec{H}}_O &= \vec{M}_O: \qquad \dot{H}_{Oz}\,\vec{k} = c_1\,\vec{i} + c_2\,\vec{j} \quad \Rightarrow \quad c_1 = c_2 = \dot{H}_{Oz} = 0 \Rightarrow \\
H_{Oz} &= 6m(r + L\sin\theta)^2\omega = 6m(r + L\sin\theta_0)^2\omega_0 \qquad (1) \\
\Rightarrow \quad \omega &= (r + L\sin\theta_0)^2\omega_0/(r + L\sin\theta)^2 \\
T + V &= \tfrac{1}{2}m\left[\omega^2(r + L\sin\theta)^2 + L^2\dot{\theta}^2\right] \times 6 - 6mgL\cos\theta + \tfrac{1}{2}k_t\theta^2 \times 6 = E_0 \\
&= 3m\left[\omega_0^2(r + L\sin\theta_0)^2 + L^2\dot{\theta}_0^2\right] - 6mgL\cos\theta_0 + 3k_t\theta_0^2 \qquad (2) \\
\Rightarrow \quad \dot{\theta}^2 &= \dot{\theta}_0^2 + 2g(\cos\theta - \cos\theta_0)/L + k_t(\theta_0^2 - \theta^2)/mL^2 \\
&\quad + \omega_0^2\,(r/L + \sin\theta_0)^2\left\{1 - \frac{(r + L\sin\theta_0)^2}{(r + L\sin\theta)^2}\right\}
\end{aligned}$$

$\dot{\omega}, \ddot{\theta}$ are obtained by differentiating eqs(1) and (2): $\dot{H}_{Oz} = 0,\ \ \dot{T} + \dot{V} = 0,$

$$(r + L\sin\theta)^2\dot{\omega} + 2(r + L\sin\theta)L\cos\theta\,\dot{\theta}\,\omega = 0 \qquad (3)$$

$$m[\omega\dot{\omega}(r + L\sin\theta)^2 + \omega^2(r + L\sin\theta)L\cos\theta\,\dot{\theta} + L^2\dot{\theta}\,\ddot{\theta}] + mgL\sin\theta\,\dot{\theta} + k_t\theta\,\dot{\theta} = 0 \qquad (4)$$

Substituting $\dot{\omega}$ from eq.(3) in eq.(4) yields

$$mL[L\ddot{\theta} - \omega^2(r + L\sin\theta)\cos\theta] + mgL\sin\theta + k_t\theta = 0 \tag{5}$$

This equation can be obtained directly from $\vec{M}_A \cdot \vec{e}_b = (\overrightarrow{AB} \times m\vec{a}_B) \cdot \vec{e}_b$ for bar 2. For the system, $\vec{F} = 6m\vec{a}_C$ yields

$$\vec{R}_O = [6mg + 6mL(\cos\theta\,\dot{\theta}^2 + \sin\theta\,\ddot{\theta})]\ \vec{k}$$

FBD of bar 2 is shown in Fig. E2.47d and $\vec{M}_A = \overrightarrow{AB} \times m\vec{a}_B \Rightarrow$

$$c_t\,\vec{e}_t + c_n\,\vec{e}_n - (mgL\sin\theta + k_t\theta)\,\vec{e}_b = -L\,\vec{e}_n \times m\vec{a}_B$$

$$\vec{e}_n: \qquad c_n = 0$$

$$\vec{e}_t: \qquad c_t = -mL[\dot{\omega}(r + L\sin\theta) + 2\omega L\dot{\theta}\,\cos\theta]$$

$$\vec{e}_b: \qquad -mgL\sin\theta - k_t\theta = mL[\ddot{\theta}L - \omega^2(r + L\sin\theta)\cos\theta]$$

which is the same as eq. (5) derived earlier from $\dot{T} + \dot{V} = 0$, $\dot{H}_{O_z} = 0$. $\vec{R}_A$ is obtained using $\vec{F} = m\vec{a}_B$ for bar 2:

$$\vec{R}_A - mg(\cos\theta\,\vec{e}_n + \sin\theta\,\vec{e}_t) = m\vec{a}_B$$

At $\theta = \theta_1$: $\dot{\theta} = 0$, $\omega_1 = (r + L\sin\theta_0)^2\omega_0/(r + L\sin\theta_1)^2$, eq. (2) for $\theta = \theta_1$ yields:

$$\tfrac{1}{2}m\frac{(r + L\sin\theta_0)^4\omega_0^2}{(r + L\sin\theta_1)^2} \times 6 - 6mgL\cos\theta_1 + 3k_t\theta_1^2 = E_0$$

Example 2.48: Find the applied force P so that the system translates with the pendulum remaining vertical (Fig. E2.48a). The spring is un-deformed when the pendulum is at right angles to the inclined plane. Find the reactions at the pin and from the inclined plane. The centre of mass of the block is at O.

Solution: The FBD of the pendulum and its equivalent kinetic system (Fig. E2.48b)

$$\Rightarrow \quad M_{O_z} = k_t\theta = m_1 a\tfrac{1}{2}L\cos\theta \qquad \Rightarrow \qquad a = 2k_t\theta/m_1L\cos\theta$$

$$F_x = R_1 - m_1 g\sin\theta = m_1 a \qquad \Rightarrow \qquad R_1 = m_1 g\sin\theta + 2k_t\theta/L\cos\theta$$

$$F_y = R_2 - m_1 g\cos\theta = 0 \qquad \Rightarrow \qquad R_2 = m_1 g\cos\theta$$

The FBD for the system and its equivalent kinetic system are shown in Fig. E2.48c.

$$F_y = N - (m + m_1)g\cos\theta = 0 \qquad \Rightarrow \qquad N = (m + m_1)g\cos\theta$$

$$F_x = P - (m + m_1)g\sin\theta - \mu N = (m + m_1)a$$

$$\Rightarrow \quad P = (m+m_1)[g\sin\theta + \mu g\cos\theta + 2k_t\theta/m_1 L\cos\theta]$$

$$M_{Oz} = (\eta - b)N - \mu N d - P(h-d) = m_1 a(\tfrac{1}{2}L\cos\theta) \qquad \text{yields } \eta.$$

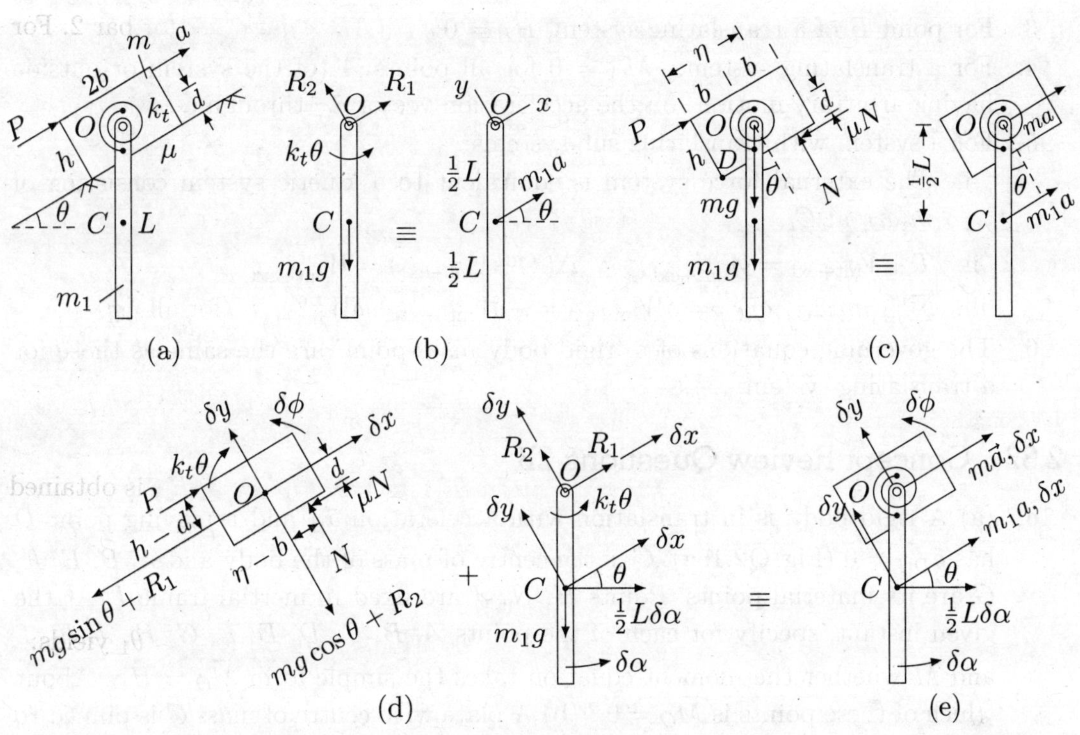

Fig. E2.48

Alternative solution using virtual work. Consider arbitrary virtual displacements $\delta x, \delta y$ of O, virtual rotation $\delta\phi$ of the block about O and virtual rotation $\delta\alpha$ of the pendulum about O. We equate the virtual work done by the internal and external loads mg, m_1g, P, N, $\mu N, \pm R_1, \pm R_2, \pm k_t\theta$ (Fig. E2.48d) to that of the equivalent kinetic system (Fig. E2.48e). The virtual work of $\pm R_1$, $\pm R_2$ is zero.

$$\begin{aligned}\delta W &= P[\delta x - (h-d)\delta\phi] - mg\sin\theta\,\delta x - mg\cos\theta\,\delta y - \mu N(\delta x + d\,\delta\phi)\\ &\quad + N[\delta y + (\eta - b)\delta\phi] - k_t\theta\,\delta\phi - m_1 g(\delta x\sin\theta + \delta y\cos\theta) + k_t\theta\,\delta\alpha\\ &= ma\delta x + m_1 a(\delta x + \tfrac{1}{2}L\,\delta\alpha\,\cos\theta)\end{aligned}$$

The coefficients of δx, δy, $\delta\alpha$, $\delta\phi$ yield the governing equations obtained earlier.

Review 2B

1. $\dot{T}_C + \dot{V}^* = \dot{W}^*_{\text{nc}}, \quad \Delta(T_C + V^*) = W^*_{\text{nc}}$
2. For a translating system: $\vec{F} = m\vec{a}_C$, $\vec{M}_C = \vec{0}$, $\dot{T} + \dot{V} = \dot{W}_{\text{nc}}$, $\Delta(T + V) = W_{\text{nc}}$
3. For point B of a translating system, $\vec{H}_B = \vec{0}$
4. For a translating system: $\vec{M}_A = \vec{0}$ for all points A (of the system or outside having arbitrary motion) on the acceleration vector $\vec{a}_C$ through C.
5. For a system with translating sub-systems:
   i. The external force system is equivalent to a kinetic system consisting of $m_i\vec{a}_{C_i}$ at C_i.
   ii. $\dot{T} + \dot{V}_{\text{int+ext}} = \dot{W}^{\text{nc}}_{\text{int+ext}}, \qquad \Delta(T + V_{\text{int+ext}}) = W^{\text{nc}}_{\text{int+ext}}$
   iii. $\sum_{j=1}^{p} m_j\vec{a}_{C_j} \cdot \delta\vec{r}_{C_j} = \delta W_{\text{int+ext}} = -\delta V_{\text{int+ext}} + \delta W^{\text{nc}}_{\text{int+ext}}$ for all δq_i.
6. The governing equations of a 'rigid body mass-point' are the same as those for a translating system.

2.32 Concept Review Questions 2B

16. (a) A rigid body is in translation with acceleration $\vec{a}_{|I}$ and a moving point D has $\vec{a}_{D|I} = \vec{0}$ (Fig. Q2.16a). C is the centre of mass of the body and A, B, E, F, G are its material points. Points H, N, J are fixed in inertial frame I. At the given instant, specify for each of the points A, B, C, D, E, F, G, H, J, K, L and M whether the moment equation takes the simple form $\vec{M}_O = \dot{\vec{H}}_O$. About which of these points is $\vec{M}_O = \vec{0}$? (b) A plate with centre of mass C is pinned to two rods of length L each, which rotate in a vertical plane at constant angular velocity ω (Fig. Q2.16b) with $AB = b$. About which of the points A, C, O, P, Q, R, S or T is the moment zero for the external forces acting on the plate?

[(a) C, D, E, F, H, J, L, M, N; C, E, F, L, K, N (b) C, P, S, T

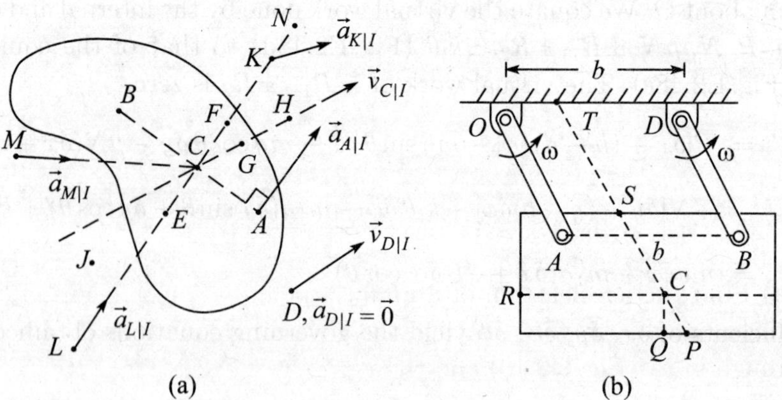

Fig. Q2.16

17. A mechanical shaker in a lift moving up with acceleration a_0 provides simple harmonic motion of amplitude A and circular frequency ω rad/s to a horizontal platform relative to the lift. A block of mass m rests on the platform with coefficient of friction μ. Find the minimum value of A for the block to separate from the platform or slip on it for (a) vertical shaking and (b) horizontal shaking.

[(a) $(g + a_0)/\omega^2$ (b) $(g + a_0)\mu/\omega^2$]

18. (a) A bead of mass m and coefficient of friction μ slides on a horizontal circular wire of radius R with speed v. The wire is translating up at acceleration a. Find the rate of decrease of v relative to the wire and the distance travelled on the wire before coming to rest. (b) A small coin is on a turntable (Fig. Q2.18) which is rotating at ω, $\dot{\omega}$ relative to an aeroplane translating with acceleration a. The coin is about to slip at the instant shown. Find the coefficient of friction between the turntable and the coin. (c) Find the total frictional force on an automobile of mass m which travels without slip in a horizontal plane around a bend. Its centre of mass is moving in a circle of radius r with speed v, which is increasing at $\dot{v}$. The coefficient of friction is μ. (d) In an amusement park, wooden horses are suspended from cables attached to periphery of a horizontal platform of radius R. The steady position of the cables is at an angle of θ to the vertical when the platform is rotating at ω about its vertical axis. The distance of the centre of mass C of a horse from the point of suspension is L. Find ω.

[(a) $\mu[(g + a)^2 + v^4/R^2]^{1/2}$, $(R/2\mu)\sinh^{-1}[v^2/(g + a)R]$

(b) $[\dot{\omega}^2 r^2 + (g \sin\theta - \omega^2 r - a\cos\theta)^2]^{1/2}/(g\cos\theta + a \sin\theta)$

(c) $m[\dot{v}^2 + (v^4/r^2)]^{1/2}$ (d) $\{g \tan\theta/(R + L\sin\theta)\}^{1/2}$]

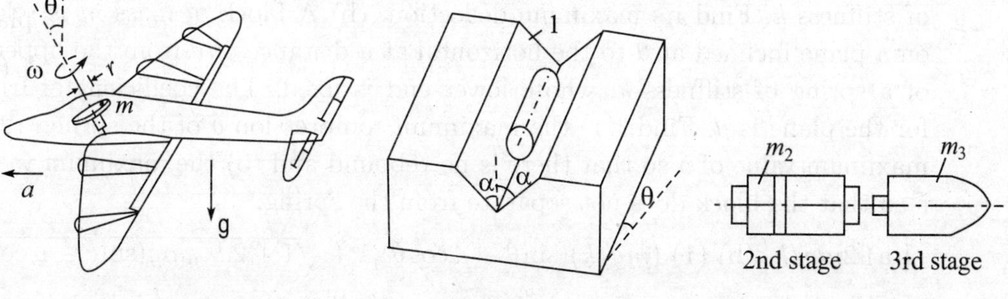

Fig. Q2.18 Fig. Q2.19 Fig. Q2.20

19. (a) Find the acceleration of a uniform cylinder which slides down a trough inclined at angle θ with the horizontal. The angle between the two flats of the trough is 2α (Fig. Q2.19) and these are symmetrical with respect to the vertical mid-plane. The coefficient of friction is μ. (b) Rework part (a) if face 1 is vertical.

[(a) $g \sin\theta - \mu g \cos\theta/\sin\alpha$ (b) $g\sin\theta - \mu g\cos\theta \cot\alpha$]

20. A rocket is moving at v_0 when the third stage engine is ignited by firing explosives, separating the 2nd and 3rd stages (Fig. Q2.20) in a time interval t_1. The stages separate at a velocity v_s. Assume that only η % of the chemical energy of explosives gets converted into mechanical energy. Neglect all forces other than that due to explosives. Find the final velocity of the 3rd stage, the average thrust on it during separation and the amount of chemical energy consumed.

$v_0 + m_2 v_s/(m_2+m_3),\ m_2 m_3 v_s/(m_2+m_3)t_1,\ m_2 m_3 v_s^2/2(m_2+m_3)\eta]$

21. A boy is riding a cycle at 10 m/s with the wheels rolling without slip when another boy running at 4 m/s in the same direction jumps on without a vertical component of velocity. The mass of each boy is 40 kg and that of the cycle is 20 kg. The radius of the wheel is 0.5 m. Find the speed of the cycle and angular speed of the wheels just after the boy gets on it. [7.6 m/s, 20 rad/s]

22. (a) A train has 3 cars, each of mass m and an engine of mass M. It is going up an incline, at elevation θ at speed v_0 when power is shut off and brakes are applied. The coefficient of friction is μ. The brakes are functioning only on the engine and the middle car and their wheels skid. Find the time t in which the train comes to rest. Assume that the couplings transmit only axial forces. (b) A driver moving at high speed v suddenly sees a blind T-crossing at distance d ahead. Which of the three options is the best, i.e., requires the least μ? (1) To try to stop by using brakes. (2) To try to turn in the maximum possible radius of d. (3) To try to stop for distance $\frac{1}{2}d$ and then take a turn of radius $\frac{1}{2}d$.

[(a) $(3m+M)v_0/g[(3m+M)\sin\theta + \mu(M+m)\cos\theta]$ (b) case (1) is the best]

23. (a) A block of mass m is released at the top of an un-deformed vertical spring of stiffness k. Find its maximum deflection. (b) A block of mass m is released on a plane inclined at θ to the horizontal at a distance of d from the upper end of a spring of stiffness k, whose lower end is fixed. The coefficient of friction for the plane is μ. Find: (1) the maximum compression δ of the spring, (2) the maximum value of δ so that there is no rebound and (3) the maximum value of δ so that the block does not separate from the spring.

[(a) $2mg/k$ (b) (1) $(mg/k)(\sin\theta - \mu\cos\theta)[1+\sqrt{1+2dk/mg(\sin\theta-\mu\cos\theta)}\,]$,

(2) $mg(\sin\theta+\mu\cos\theta)/k$, (3) $2mg(\sin\theta+\mu\cos\theta)/k$]

24. A gunner can exert an average force of 175 N against the gun. Find the maximum number n of bullets of 50 g each at 1000 m/s that he can fire per minute. [210]

25. A man sits at the front right end of a stationary barge. He rises up and moves to the back left end of the barge and sits down again. The water resistance on the barge is $c\,v$, where v is the speed of the barge. Where will the barge come to rest? [Hint: $\Delta\vec{p} = \vec{0} = \int \vec{F}\,dt$] [At initial position]

26. Find the path of the block on a smooth inclined plane (Fig. Q2.26). Find d and when it hits the bottom. $[y = g\sin\theta(x^2/2v_0^2),\ d = v_0\sqrt{2L/g\sin\theta},\ \sqrt{2L/g\sin\theta}]$

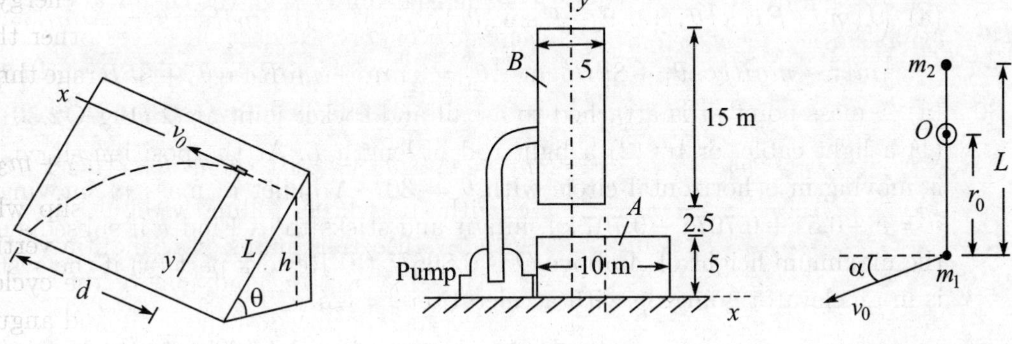

Fig. Q2.26 Fig. Q2.27 Fig. Q2.28

27. A cylindrical tank A is initially full of water and the cylindrical tank B is empty. Water is pumped from A to B (Fig. Q2.27). Initially the flow rate is 1 m^3/s, which increases at the rate of 0.1 m^3/s^2 for 30 s thereafter. What is the average force on the tanks and piping system from the water during this period aside from the weight of the water? $[-250\vec{i} + 886.7\vec{j}\ \text{N}]$

28. Two mass-points with mass m_1, m_2 rest on a smooth horizontal plane and are connected by a string which passes through a small fixed smooth ring at O (Fig. Q2.28). A horizontal velocity as shown is given to m_1. In their subsequent motion, find v_r, v_ϕ of m_1 when it is at distance r from O.

$$[[(m_1 + m_2\sin^2\alpha - m_1 r_0^2\cos^2\alpha/r^2)/(m_1+m_2)]^{1/2}v_0,\ r_0 v_0\cos\alpha/r]$$

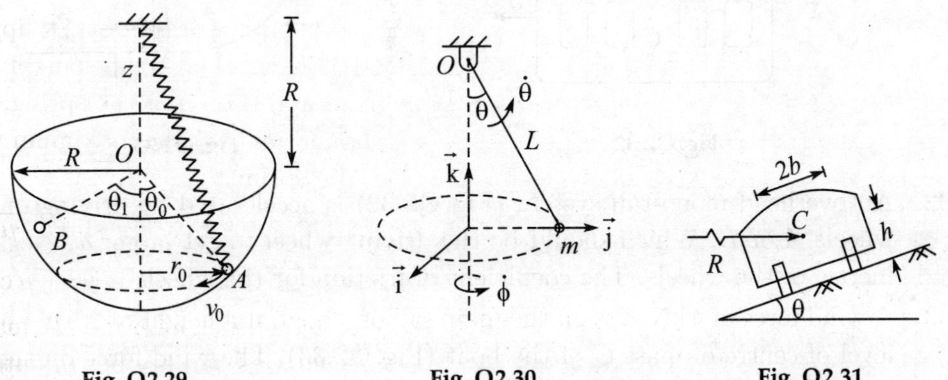

Fig. Q2.29 Fig. Q2.30 Fig. Q2.31

29. A mass-point is projected horizontally with speed v_0 at A on the inside of a smooth spherical surface of radius R (Fig. Q2.29). The spring stiffness is k and its free length is $2R$. (a) Find v_0 if it moves in a horizontal circle. (b) If it does

not move in a circle, set up two equations for finding the extreme angle θ_1 of its trajectory and the corresponding speed v_1. [with $P = 4kR\sin^2\frac{1}{4}\theta_0$

(a) $\left(\{(mg+P\cos\frac{1}{2}\theta_0)\tan\theta_0 - P\sin\frac{1}{2}\theta_0\}r_0/m\right)^{1/2}$ (b) $mR\sin\theta_1 v_1 = mr_0 v_0$ & $\frac{1}{2}mv_1^2 - mgR\cos\theta_1 + 8kR^2\sin^4\frac{1}{4}\theta_1 = \frac{1}{2}mv_0^2 - mgR\cos\theta_0 + 8kR^2\sin^4\frac{1}{4}\theta_0]$

30. (a) A mass-point m is attached to a ball and socket joint at O (Fig. Q2.30) by (1) a light cable, or by (2) a light rod of length L. At the position shown, m is moving in a horizontal circle with $\theta = 30°$. A bullet of mass m, moving at $\vec{v} = v(-0.5\vec{\imath} + 0.707\vec{\jmath} - 0.707\vec{k})$, hits it and sticks to it. Find v if subsequently the maximum height of A above O is $0.866L$. (b) Rework part (a) if the system is in an elevator going up with acceleration $a = \frac{1}{2}g$.

[For both (1) and (2): (a) $14.39\sqrt{gL}$ (b) $17.62\sqrt{gL}$]

31. A vehicle of mass m moves on a banked road (Fig. Q2.31), in a circle of radius R at speed v. The coefficient of friction for the ground is μ. Neglect the inertia of the wheels and the rotational inertia of the main body of the vehicle. (a) Find v for which the friction force is zero. (b) Find the range of v for which there is no slip. (c) Find the condition of tipping and show that tipping occurs before slip if $\mu > b/h$. [Let $s = \sin\theta$, $c = \cos\theta$ (a) $\sqrt{gR\tan\theta}$

(b) $\left[\dfrac{gR(s-\mu c)}{c+\mu s}\right]^{1/2} < v < \left[\dfrac{gR(\mu c + s)}{c - \mu s}\right]^{1/2}$ (c) $v^2 > gR(bc+sh)/(ch-sb)$]

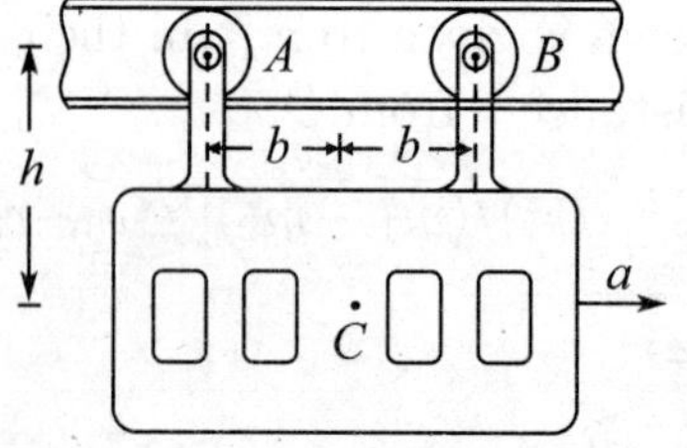

Fig. Q2.32

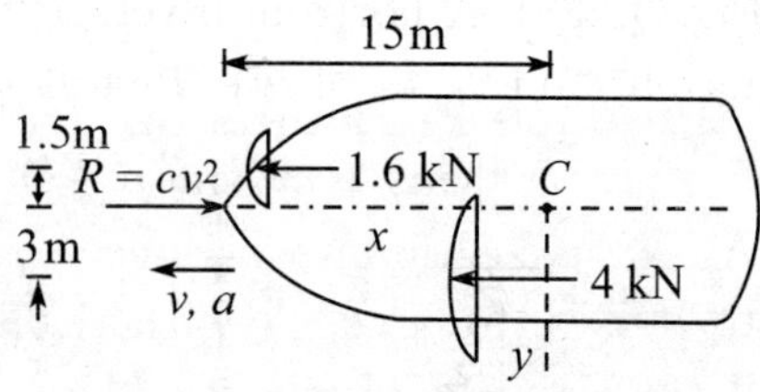

Fig. Q2.33

32. An overhead mono-rail system (Fig. Q2.32) is accelerated by driving one of its wheels A or B. Which should be the driving wheel to get larger a? Neglect the inertia of the wheels. The coefficient of friction for the wheels is μ. [B]

33. A wind force of 4 kN acts on the main sail of a boat at a height of 2.5 m above the level of centre of mass C of the boat (Fig. Q2.33). The wind force on the minor sail is 1.6 kN at a height of 2 m above the level of C. The water resistance $R = cv^2$ acts at 0.25 m above C. The weight of the boat is 800 kg and $a = 0.8\,\text{m/s}^2$. Find R, the buoyancy force B and the resultant of the lateral forces on the boat.

[$R = 4960$ N, $B = 7848$ N at 1.524 m ahead of C, $\vec{F}_R = \vec{0}$ & $\vec{C}_{R_C} = 9600\vec{k}$ N.m]

34. A jet-propelled sled with light wheels moves over an inclined plane (Fig. Q2.34) Find the maximum value of the thrust T so that there is no lift-off. [$2.165mg$]

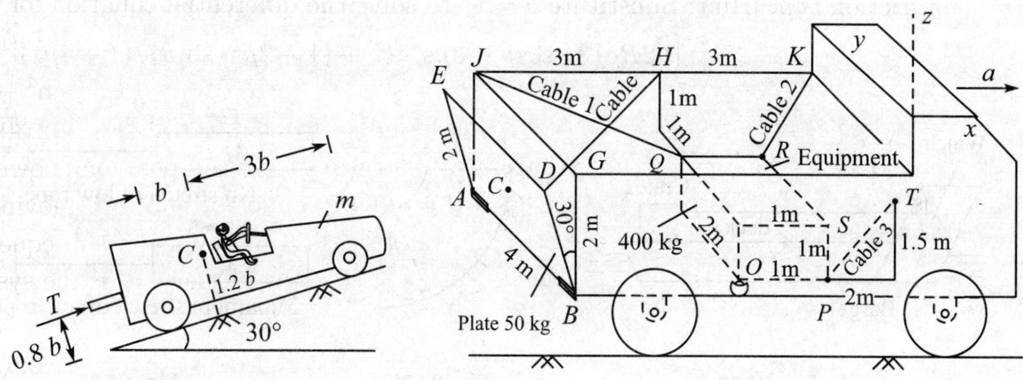

Fig. Q2.34

Fig. Q2.35

35. A truck is moving with acceleration $a = 0.6\,\text{m/s}^2$ (Fig. Q2.35). (a) A hinged plate of mass 50 kg at the back is restrained by cable DH. Find the cable tension and the hinge reactions. The hinge at A does not provide axial restraint. (b) An equipment of mass 400 kg carried by the truck is supported by a ball and socket joint at O and three cables (JQ, KR and PT). Find the reaction at O.

[(a) $213.4, -22.69\vec{i}+89.80\vec{k}, -98.08\vec{i}-150.8\vec{j}+390.6\vec{k}$ (b) $-2059\vec{i}-3924\vec{j}-642\vec{k}$ N]

2.33 Practice Problems 2B

18. An athlete is running at speed v_1 when he throws the javelin at speed v_0 relative to himself at angle α with the horizontal. Prove that the range is maximum if $\cos\alpha = -v_1/4v_0 + [(v_1/4v_0)^2 + \frac{1}{2}]^{1/2}$. Neglect the height of the athlete in comparison to the other dimensions.

19. A 1000 kg crate is placed on a 4000 kg barge, which is equipped with a winch for pulling the crate along its smooth deck (Fig. P2.19). Initially the barge and the crate are at rest relative to the water. (a) Find their velocities when the winch is drawing the load towards it at 5 m/s. (b) Find the position of the barge when 10 m of the rope has been drawn in. (c) Find the velocities of the crate and the barge after the winch stops operating. (d) Rework parts (a), (b), (c) if the coefficient of friction for the crate and the deck is 0.4. Neglect water resistance.

[(a) & (c) $\vec{i}$ m/s, $-4\vec{i}$ m/s (b) $2\vec{i}$ m (d_a) $\vec{i}$ m/s, $-4\vec{i}$ m/s (d_b) $2\vec{i}$ m (d_c) $(1-0.1gt)\vec{i}$, $(0.4gt-4)\vec{i}$ for $t \le 1.019$ s, no motion after 1.019 s]

20*. A small packet is released from rest at position 1 and slides down a circular path of radius R (Fig. P2.20). Find its velocity v at position 2. The coefficient of friction is μ. [Hint: Substitute $u = v^2$ to solve the differential equation for v.]

$[\,[2Rg\{3\mu\cos\theta - 3\mu e^{-2\mu\theta} + (1-2\mu^2)\sin\theta\}/(1+4\mu^2)]^{1/2}]$

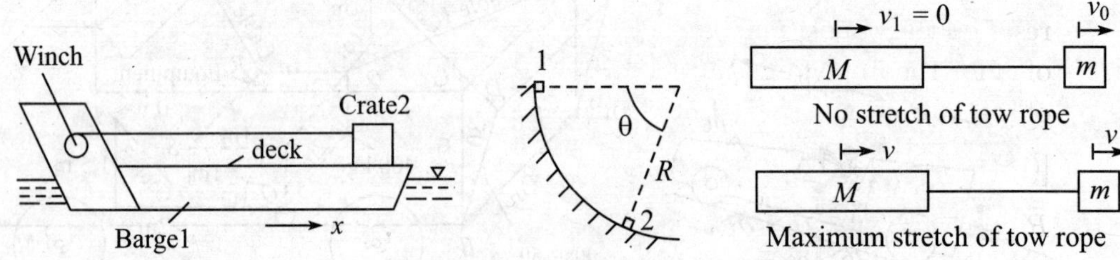

Fig. P2.19 **Fig. P2.20** **Fig. P2.21**

21. A tugboat of mass m moving at v_0 starts towing a stationary ship of mass M (Fig. P2.21). Neglect the impulse of the propeller thrust of the tugboat and the resistance of water during the period the tow-rope is stretched to its maximum when the boat and the ship have common velocity. The tow-rope extends e m/m length for 1 kN load and its allowable load is P_0. Prove that the minimum required length L of the rope is: $L = 1000(mv_0^2/eP_0^2)M/(M+m)$.

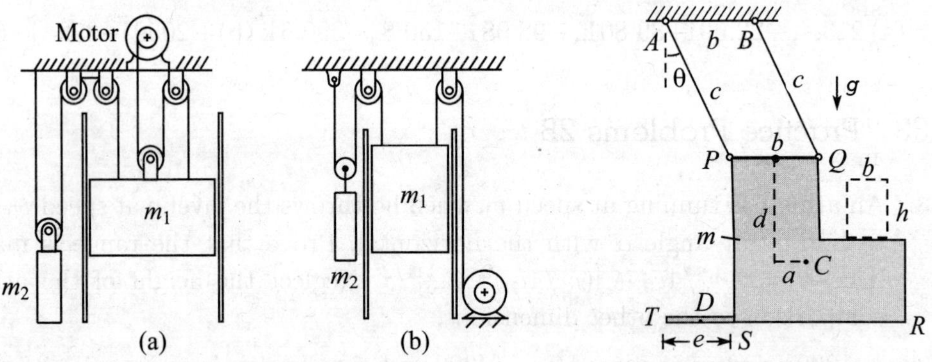

Fig. P2.22 **Fig. P2.23**

22. Find the power that must be supplied to the motor of an elevator if its efficiency is η and the elevator is going up with velocity v and acceleration a. Neglect the inertia of the pulleys and friction at their axles. Solve for the two cases shown in Fig. P2.22. [(a) & (b) $[(m_1 - \frac{1}{2}m_2)g + (m_1 + \frac{1}{4}m_2)a]v/\eta$]

23. (i) Model tests of a control panel $PQRS$ of mass m are conducted by using two light bars AP and BQ (Fig. P2.23). The support at A is frictionless and a frictional couple M_f acts at support at B. The system is released from rest at

$\theta = \theta_0$. Find (a) the support reactions at release and (b) velocity, acceleration and the support reactions at position θ. If $\dot\theta = \omega, \ddot\theta = \dot\omega$ in position θ, find the normal force, shear force and bending moment at the middle D of the bar ST of mass m_1. (c) Find the minimum speed to be imparted to the panel at $\theta = 0$ so that it makes a complete circle. (ii) A cuboidal block of height h and base b rests on the panel. The bars are rotated at constant ω. Find the minimum value of ω for the impending motion of the block relative to the panel. The coefficient of static friction is μ_s. [Hint: Consider sliding, tipping and separation.]
[(i) (b) $v = [2cg(\cos\theta - \cos\theta_0) - \frac{2M_f}{m}(\theta_0 - \theta)]^{1/2}$, $\vec{a} = -\frac{v^2}{c}\,\vec{e}_r + (\frac{M_f}{mc} - g\sin\theta)\,\vec{e}_\theta$,
$\vec{R}_A = -\frac{1}{2}(P + mg\cos\theta + \frac{mv^2}{c})\,\vec{e}_r$, $\vec{R}_B = -\frac{1}{2}(mg\cos\theta + \frac{mv^2}{c} - P)\,\vec{e}_r + \frac{M_f}{c})\,\vec{e}_\theta$
$P = [2m(g\cos\theta + v^2/c)(d\sin\theta - a\cos\theta) + M_f\{(b-2a)\sin\theta - 2d\cos\theta\}/c]/b\cos\theta$,
$N = \frac{1}{2}m_1(\dot\omega c\cos\theta - \omega^2 c\sin\theta)$, $S = \frac{1}{2}m_1(g + \dot\omega c\sin\theta + \omega^2 c\cos\theta)$, $M = -.25Se$
(a) as in (b) with $v = 0$, $\theta = \theta_0$ (c) $4gc + 2\pi M_f/m$
(ii) $[\min\{1,\ \mu^2/(1+\mu^2),\ b^2/(b^2+h^2)\}]^{1/4}\sqrt{g/c}$]

24. (a) A body of mass m and area of cross-section A falls from a height h above an absorbing medium and comes to rest after sinking a distance δ. Find the average pressure p_{av} from the medium on the body. (b) A person of mass 60 kg jumps from the window of a burning building into a life net 10 m below. How much should the net sink to avoid injury which occurs for $p > 20\,\text{N/cm}^2$? $A = 0.2\,\text{m}^2$ and assume that $p_{max} \simeq 3p_{av}$. (c) An elastic resistance band of stiffness k is stretched horizontally with a large initial tension T_0 (Fig. P2.24). It catches a descending parcel of mass m. Find h_{max} from which the parcel can fall if the maximum force in the band: $T_{max} \le F_m$. [(a) $(1 + h/\delta)mg/A$ (b) $> 0.4618\,\text{m}$
(c) $(F_m^2 - T_0^2)/2mgk - [\{(F_m - T_0)/2k + a\}^2 - a^2]^{1/2}$]

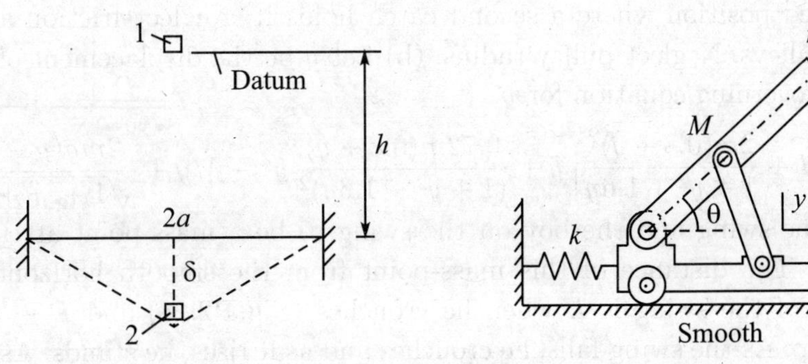

Fig. P2.24 Fig. P2.25

25. A gun of mass M with a recoil spring of stiffness k fires a shell of mass m with velocity v_0 with respect to the barrel (Fig. P2.25). Assume that the recoil velocity

of the gun is instantaneously achieved and the ground is smooth. Find (a) the maximum compression of the spring, (b) the impulsive reaction from the ground and (c) the ratio of the kinetic energy T_s of the shell to the total kinetic energy T just after firing. Show that $T_s/T \simeq 1$ if $m/M \ll 1$. [(a) $mv_0 \cos\theta\sqrt{M/k}/(M+m)$ (b) $mv_0 \sin\theta$ (c) $1/[1+m/M\{1+\tan^2\theta(1+m/M)^2\}]$]

26. A compressed air gun consists of a reservoir of volume V_0 containing compressed air at pressure p_0 (Fig. P2.26). When fired, the air expands according to $pV^n =$ constant, to propel a striker of mass m in a barrel of length L and cross-sectional area A. The coefficient of friction for the barrel is μ. The atmospheric pressure is p_a. Find the speed of the striker at exit.
[$[2\{p_0V_0\{(1+AL/V_0)^{1-n}-1\}/(1-n)-p_aAL-\mu mgL\}/m]^{1/2}$]

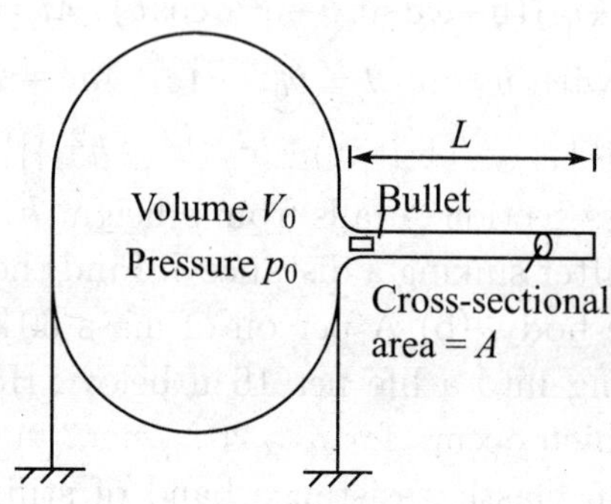

Fig. P2.26

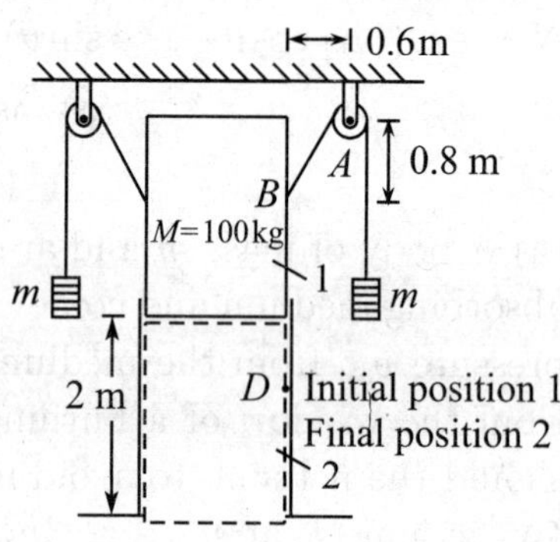

Fig. P2.27

27. (a) A 100 kg overhead door (Fig. P2.27) is held in open position by a catch. Design counterweights so that when the catch is released, the door gently reaches the closed position where a second catch holds it. Neglect friction and inertia of the pulleys. Neglect pulley radius. (b) Let y be the displacement of the door. Derive governing equation for y. [(a) 53.65 kg

(b) $\left[M+\dfrac{2m(0.8+y)^2}{1+y^2+1.6y}\right]\ddot{y}+\dfrac{0.72m(0.8+y)}{(1+y^2+1.6y)^2}\dot{y}^2-Mg+\dfrac{2mg(y+0.8)}{\sqrt{1+y^2+1.6y}}=0$]

28. Model the swing and the boy on the swing to be a mass-point at their centre of mass. The distance of this mass-point from the smooth horizontal axis of rotation of the swing is L when he crouches (Fig. P2.28) and $L-h$ when he stands up. As the swing falls, he crouches and as it rises, he stands. Assume that the changeover to standing position is instantaneous at the vertical position of the swing and changeover to crouching position at the extreme positions of the swing is also instantaneous. The maximum displacement at the start of a cycle is θ_1. Find position θ_3 at the end of one cycle. [$2\sin^{-1}\{\sin(\theta_1/2)/(1-h/L)^3\}$]

29. The system shown in Fig. P2.29 is rotating about a fixed vertical axis at ω_0. A catch keeps the collar of mass m_2 from moving relative to the shaft. The catch is removed. When the collars of mass m_1 have moved out to radius r, find their velocity v and acceleration a relative to the shaft and ω, $\dot{\omega}$ of the shaft. The contacts are smooth. Neglect inertia of the pulleys and the rods, and the rotational kinetic energy of m_2.

$[v^2 = 2[(r - r_0)(2m_1 \cos\theta - m_2)g\mathrm{cosec}\,\theta + m_1\omega_0^2 r_0^2(1 - r_0^2/r^2)]/(m_2 + 2m_1),$
$[(2m_1 \cos\theta - m_2)g\mathrm{cosec}\,\theta + 2m_1\omega_0^2 r_0^4/r^3] \sin\theta/(m_2 + 2m_1),\ \frac{\omega_0 r_0^2}{r^2},\ -\frac{2\omega_0 r_0^2 v \sin\theta}{r^3}]$

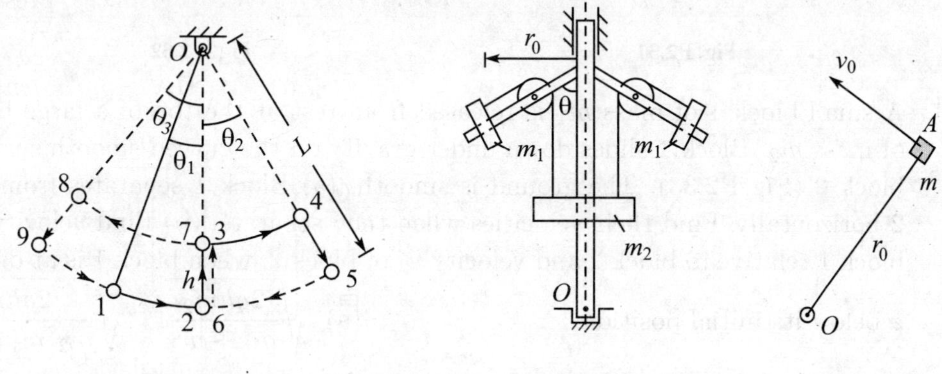

Fig. P2.28 Fig. P2.29 Fig. P2.30

30. A block of mass m rests at A on a smooth horizontal table and is tied to a fixed pin O by an elastic cord of stiffness k (Fig. P2.30). The free length of the cord is $L\,(< r_0)$. The block is imparted a horizontal speed v_0. (a) Find the radius of curvature of the trajectory of the block at location A (b) Find the speed and transverse component of velocity of the block when it is at a distance $r\,(> L)$ from O. (c) Find $(v_0)_{\min}$ so that the elastic cord remains taut at all times. (d) If $v_0 < (v_0)_{\min}$, find the speed of the block when the cord becomes slack and the subsequent closest distance of the block from O.

$[(a)\ mv_0^2/k(r_0 - L)\quad (b)\ [v_0^2 + k\{(r_0 - L)^2 - (r - L)^2\}/m]^{1/2},\ v_0 r_0/r$
$(c)\ [(r_0 - L)kL^2/(r_0 + L)m]^{1/2}\quad (d)\ v = [v_0^2 + k(r_0 - L)^2/m]^{1/2},\ v_0 r_0/v]$

31. Small sliders, each of mass m, move along smooth slots in a disc which rotates about the vertical axis at constant ω (Fig. P2.31). Each slider starts from rest at $\theta = 0$. Find the reaction of the slot on the slider at θ and the speed of the slider relative to the disc at exit. $[mg\vec{\mathrm{k}} + 2m\omega^2 r \sin\frac{\theta}{2}[3\sin\frac{\theta}{2} - 2]\,\vec{\mathrm{e}}_n,\ \sqrt{2}\,\omega r]$

32. Two blocks are connected by a cord. A spring between them, not connected to either, has compression x_0 (Fig. P2.32). The system moves on a smooth horizontal plane with speed v_0. Find the final velocities of the blocks if the cord is severed. [Hint: Choose the frame translating with C to simplify calculations]

$[\vec{v}_1 = v_0\,\vec{e} - [km_2/m_1(m_1+m_2)]^{1/2}x_0\,\vec{i},\ \vec{v}_2 = v_0\,\vec{e} + [km_1/m_2(m_1+m_2)]^{1/2}x_0\,\vec{i}]$

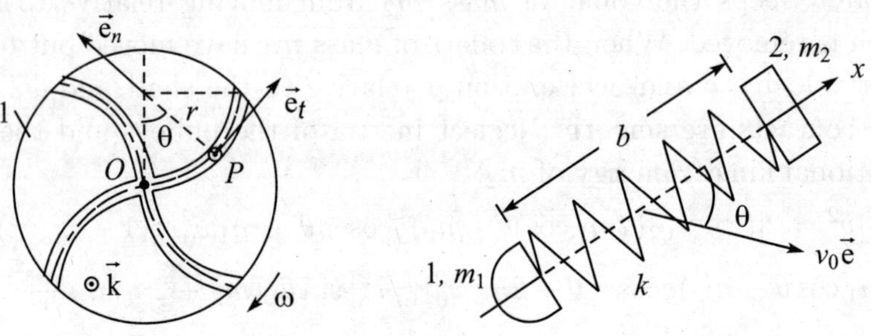

Fig. P2.31 Fig. P2.32

33. A small block 1 of mass m_1 is released from rest at the top of a large block 2 of mass m_2. Block 1 slides down under gravity on the curved smooth surface of block 2 (Fig. P2.33). The ground is smooth. (a) Block 1 separates from block 2 horizontally. Find their velocities when they separate. (b) Find velocity v_1 of block 1 relative to block 2 and velocity v_2 of block 2, when block 1 is at distance z below its initial position. $\left[\text{(a)} \sqrt{\dfrac{2ghm_2}{m_1+m_2}},\ -\sqrt{\dfrac{2ghm_1^2}{m_2(m_1+m_2)}}\right.$

$$\text{(b) } v_1 = \left[\frac{2gz(m_1+m_2)}{m_2+m_1\sin^2\alpha}\right]^{1/2},\quad v_2 = -m_1\cos\alpha\left[\frac{2gz}{(m_1+m_2)(m_2+m_1\sin^2\alpha)}\right]^{1/2}\Bigg]$$

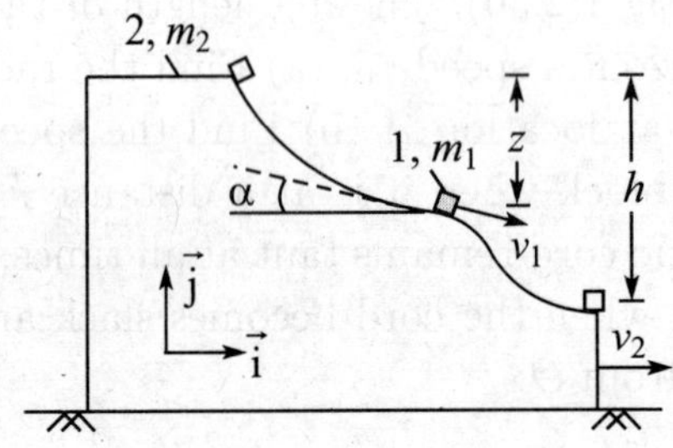

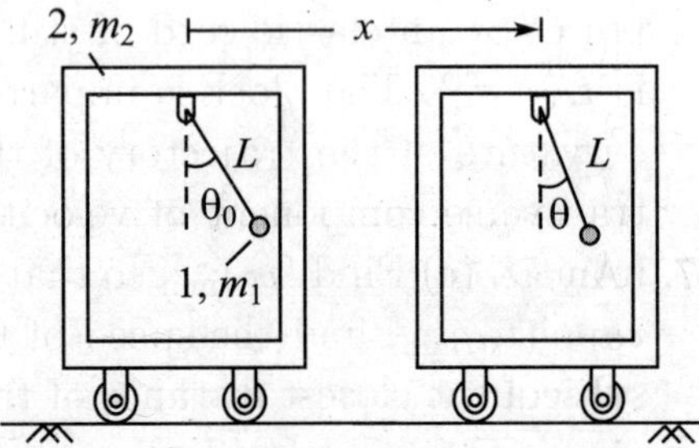

Fig. P2.33 Fig. P2.34

34. A simple pendulum 1 (Fig. P2.34) is suspended in a trolley 2 which has light wheels. It is released from rest at $\theta = \theta_0$. Find the displacement, velocity and acceleration of the trolley for position θ of the pendulum.

$$\left[\ x = \frac{m_1L(\sin\theta_0 - \sin\theta)}{m_1+m_2},\quad \dot{x} = -\text{sign}(\dot{\theta})\left[\frac{2m_1gL(\cos\theta - \cos\theta_0)}{(m_1+m_2)(\tan^2\theta + m_2\sec^2\theta/m_1)}\right]^{1/2},\right.$$

$$\left.\ddot{x} = \frac{m_1(g\cos\theta + L\dot{\theta}^2)\sin\theta}{m_1\sin^2\theta + m_2}\quad \text{with}\quad \dot{\theta}^2 = \frac{2(m_1+m_2)g(\cos\theta - \cos\theta_0)}{(m_1\sin^2\theta + m_2)L}\right]$$

35^d. A ramp is to be designed for a shipping warehouse so that the boxes come to rest within distance b after landing (Fig. P2.35). Assume a rounded transition at corner A. Find the optimum value of b for given $\mu_k = \mu$. However, for a lesser coefficient of friction, they will not come to rest at B. Design an appropriate value of stiffness k so that the boxes, having mass m each, do not rebound and the maximum compression is δ for $\mu_k = 0.5\mu$, $\mu_s = 0.6\mu$. Neglect m_1.

$[b = h(1 - \mu \cot\theta)/\mu, \quad mg(h - \mu\delta)/\delta^2 \quad \text{with} \quad h/\mu > \delta > h/1.6\mu]$

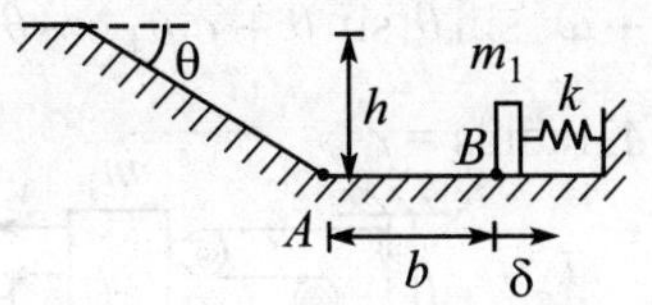

Fig. P2.35

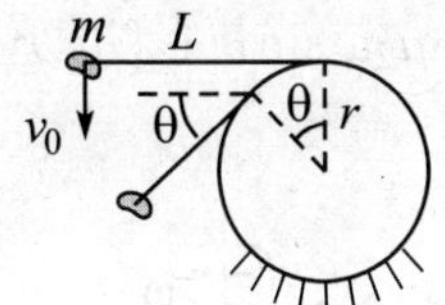

Fig. P2.36

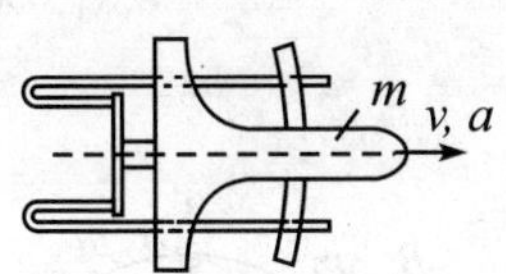

Fig. P2.37

36. (a) A taut light cable (Fig. P2.36) has a small bead of mass m tied to it at one end. Its other end is tied to a fixed disc with horizontal axis. The bead is imparted an initial velocity v_0. Find its speed and the cable tension after the cable has turned through angle θ. Find the force per unit length, p, on the disc from the cable if it is smooth. (b) Rework if the axis of the disc is vertical and the bead moves on a horizontal surface with: (1) Coulomb friction coefficient μ and (2) viscous drag cv. The cable is inextensible.

[(a) $v^2 = v_0^2 + 2g[r(1-\cos\theta) + (L - r\theta)\sin\theta]$, $\quad T = m[g\sin\theta + v^2/(L - r\theta)]$, T/r

(b) (1) $v = [v_0^2 - 2\mu g(L\theta - \frac{1}{2}r\theta^2)]^{1/2}$, $\quad T = mv^2/(L - r\theta)$, $\quad T/r$

(2) $v = v_0 - c(L\theta - \frac{1}{2}r\theta^2)/m$, $\quad T = mv^2/(L - r\theta)$, $\quad T/r$]

37. An aeroplane lands with speed v_0 on a runway of limited length. Brakes are applied and it gets hooked to two chains (mass λ kg/m) lying on the ground (Fig. P2.37). The coefficient of friction for the light wheels is μ and for the chain is μ_1. Find v, a for displacement x. $[a = -(\mu mg + 2\mu_1\lambda xg + 2\lambda v^2)/(m + 2\lambda x)$

$v = [m^2v_0^2 - 2g\{\mu m^2 x + \frac{4}{3}\mu_1\lambda^2x^3 + m\lambda(\mu + \mu_1)x^2\}]^{1/2}/(m + 2\lambda x)]$

38. (a) Two beads, connected by a light inextensible string, move over a smooth vertical ring (Fig. P2.38). These are released in the given position. Find their speed v and $\dot{v}$ when the bead A is in position A'. Find their initial acceleration. Does it matter whether the string is in contact with the ring? (b) A and B are connected by a light rigid rod. Find their maximum speed. If, instead, the bar AB is at rest in its lowest position, find the minimum velocity that has to be imparted to B so that it makes a complete revolution. [(b) $1.936\sqrt{gR}$, $1.936\sqrt{gR}$

(a) $[.5gR(1.5\sin\theta - 3.598\cos\theta + 3.598)]^{1/2}$, $.25g(1.5\cos\theta + 3.598\sin\theta)$, $.375g$, no]

39. A square tube attached to a vertical shaft is rotating at constant angular velocity ω (Fig. P2.39). A slider of mass m has coefficient of friction μ_1 for its slanting faces and μ_2 for its vertical faces. (a) Find the range of s: $s_1 \le s \le s_2$, for which the slider does not move with respect to the tube if $\mu_1 < \tan\theta < 1/\mu_1$. (b) Find the solution $s(t)$ if (1) $s(0) = s_1$, $\dot{s}(0) = v_0 < 0$ and (2) $s(0) = s_2$, $\dot{s}(0) = v_0 > 0$.

[(a) $s_1,\ s_2 = g(\cot\theta \mp \mu_1)/\omega^2(\sin\theta \pm \mu_1 \cos\theta)$

(b) (1): $i = 1$, $c_1 = 1$; (2): $i = 2$, $c_2 = -1$; $\quad s = s_i + \frac{v_0}{\lambda_1 - \lambda_2}(e^{\lambda_1 t} - e^{\lambda_2 t})$,

$\lambda_1,\ \lambda_2 = -\mu_2\omega\sin\theta \pm \{\mu_2^2\omega^2\sin^2\theta + \omega^2\sin\theta(\sin\theta + c_i\mu_1\cos\theta)\}^{1/2}$]

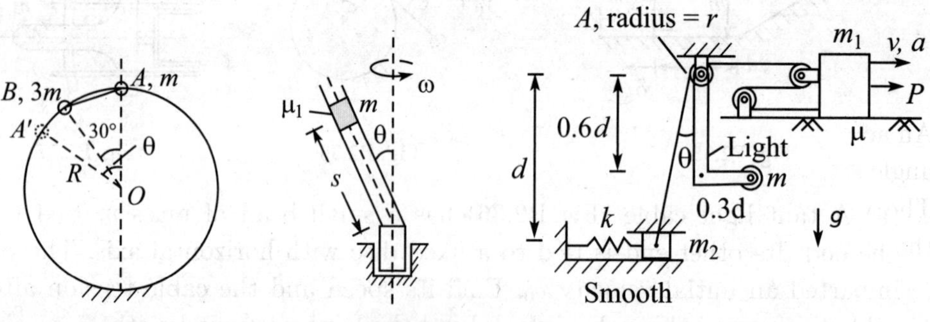

Fig. P2.38 Fig. P2.39 Fig. P2.40

40*. For the system shown in Fig. P2.40, the light bar rigidly attached to pulley, A, has a point-mass m at its end. Find the acceleration a in this position. The spring is undeformed for $\theta = 30°$ and the light cable does not slip on pulley A.

$$\left[\frac{P - \mu m_1 g + 0.9mgd/r + 3kd(\tan\theta - \tan 30°)/\sin\theta - 27v^2 m_2 \cot^3\theta/d\sin\theta}{m_1 + 4.05md^2/r^2 + 9m_2/\sin^2\theta}\right]$$

41. (a) Two blocks, of mass m each, are connected by an inextensible cable and rest on a smooth table (Fig. P2.41a). An arrow moving at v_0 hits the cable at its centre. Find the speed of the arrow when the cable segments make angle $2\theta = 60°$ between them. (b) Two blocks are at rest in the position shown in Fig. P2.41b. They are pulled by a force F such that there is no lift-off. Find the initial acceleration of the blocks. Find their velocity when θ has increased to α and also the velocity with which they make an impact. Find the maximum value of F for no lift-off. The coefficient of friction for the floor is μ.

[(a) $0.8088v_0$ (b) $[F\cot\theta - \mu(2mg - F)]/2m$, For impact: $v(\frac{1}{2}\pi)$.

$v(\alpha) = [\{F(\sin\alpha - \sin\theta) - \mu(2mg - F)(\cos\theta - \cos\alpha)\}b/m]^{1/2}$, $F_{\max} = 2mg$]

42. A jet plane of mass m kg (Fig. P2.42) reduces its speed on landing from $v_1 =$ 180 km/h to $v_2 = 54$ km/h by applying negative thrust from its jet thrust reversers in distance $d = 500$ m with constant deceleration. This operation occurs

before application of mechanical brakes. Find the reaction under the nose wheel and the reverse thrust R for $q = 14\,\text{m}$, $p = 2.5\,\text{m}$, $b = 2\,\text{m}$, $h = 3\,\text{m}$. Neglect aerodynamic forces and inertia of the wheels. [$1.914m\,\text{N}$, $2.275m\,\text{N}$]

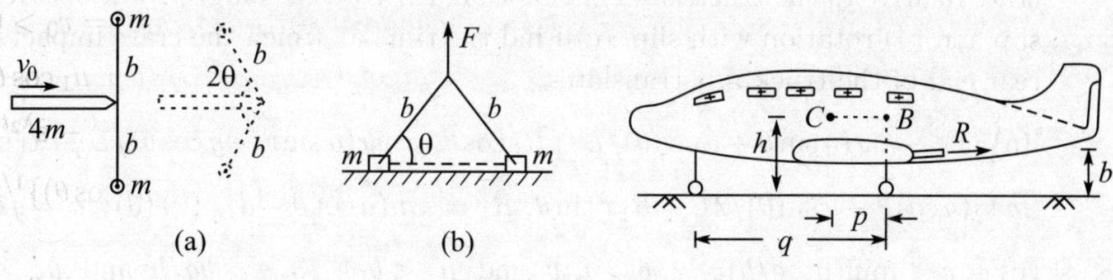

Fig. P2.41

Fig. P2.42

43[d]. An accelerometer is installed in a vehicle moving at constant acceleration a at an angle θ to the horizontal (Fig. P2.43). Its mass is m and C is its centre of mass. The moment from the torsional spring is M_0 to ensure larger contact force. Find the value of a at which the electric contact will open.

$[(M_0 + mgb)/(p\cos\theta + b\sin\theta)m]$

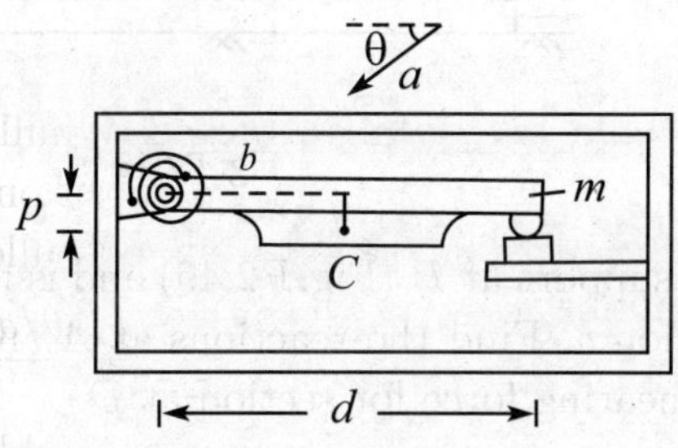

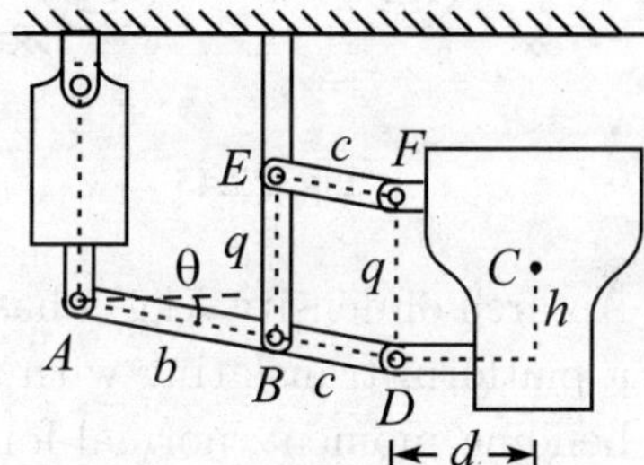

Fig. P2.43

Fig. P2.44

44. Find the initial acceleration of a container of mass m with centre of mass at C, which is lifted from rest at the given position (Fig. P2.44) by the application of force P at A by the piston of the hydraulic cylinder. Neglect inertia of the bars. Find also the supporting forces at pins B and E.

$$[(Pb/cm - g)\cos\theta,\ R_E = [mgd + (Pb/c - mg)\cos\theta(d\cos\theta - h\sin\theta)]/q\cos\theta,$$

$$\vec{R}_B:\ \ R_E - (P + mg)\sin\theta \ \text{ along } BD \ \&\ P\cos\theta(1 + b/c)\ \perp\ BD]$$

45. (a) A truck with rear-wheel drive has uniform acceleration a. A bar of mass m and length L is hinged to it at A and is supported on a smooth surface at B (Fig. P2.45). (1) Find the normal force, shear force and bending moment in the bar at a distance x from B. (2) What is the minimum value of a for which the

contact is lost at B? (3) Find the minimum value of a for which the bar topples over. (b) A crate at the back is at distance d from the rear end. The coefficients of friction are μ_s, μ_k. Find the value(s) of a for which the initial motion of the crate relative to the truck is: (1) no motion, (2) translation, (3) rotation without slip and (4) rotation with slip. (c) Find the time at which the crate impacts the rear end of the truck if it translates.

[(a) (1) $-mx(g\sin\theta + a\cos\theta)/L - R_1\cos\theta,\quad mx(a\sin\theta - g\cos\theta)/L + R_1\sin\theta,$

$mx^2(g\cos\theta - a\sin\theta)/2L - R_1 x\sin\theta,\ R_1 = \frac{1}{2}m(g\cot\theta - a),\ $ (2) (3): $a = g\cot\theta$

(b) 1. $a < \min(\mu_s, b/h)g$ 2. $a > \mu_s g$ and $\mu_s < b/h$ 3. $a > bg/h$ and $\mu_s > b/h$

4. $a > \max(\mu_s, b/h)g$ (c) $[2d/(a - \mu_k g)]^{1/2}$]

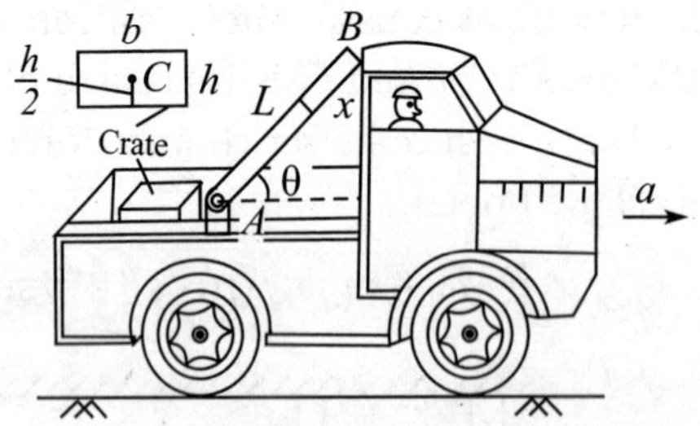

Fig. P2.45

Fig. P2.46

46. An arch of density λ kg/s has a roller support at B (Fig. P2.46) and is pinned to a platform translating with acceleration a. Find the reactions at A and B and bending moment, normal force and shearing force for section at D.

$$[r\lambda[\pi a\vec{i} + \tfrac{1}{2}(\pi g - 2a)\vec{j}],\quad \tfrac{1}{2}\lambda r(\pi g + 2a)\vec{j};$$

$$N = \lambda r\theta(g\cos\theta + a\sin\theta) - \tfrac{1}{2}\lambda r(\pi g + 2a)\cos\theta,$$

$$S = \lambda r\theta(g\sin\theta - a\cos\theta) - \tfrac{1}{2}\lambda r(\pi g + 2a)\sin\theta,$$

$$M = \lambda r^2[g\{\sin\theta - \theta\cos\theta - \pi\sin^2(\theta/2)\} - a\theta\sin\theta]\ ]$$

47. A rear-wheel driven pick-up truck has driving moment M and is connected to a trolley of mass m by a hitch (Fig. P2.47). Find its acceleration a. Find the reactions at A, B, D, E. Neglect inertia of the wheels.

[$a = M/r(m+m_1),\quad \vec{R}_A = m[gb-(p-h)a]\vec{j}/d,\quad \vec{R}_B = P_1\vec{i} + P_2\vec{j}$ with $P_1 = ma,$

$P_2 = m[(p-h)a + g(d-b)]/d,\quad N_E = [m_1 g b_2 - P_2 b_3 - P_1(p - h_1)]/(b_1 + b_2),$

$F_D = M/r,\quad N_D = m_1 g + P_2 - N_E$]

48. Find the force P for which the system (Fig. P2.48) translates. Find the corresponding reactions at A and B. [$0.7826mg$, $1.630mg$, $1.370mg$]

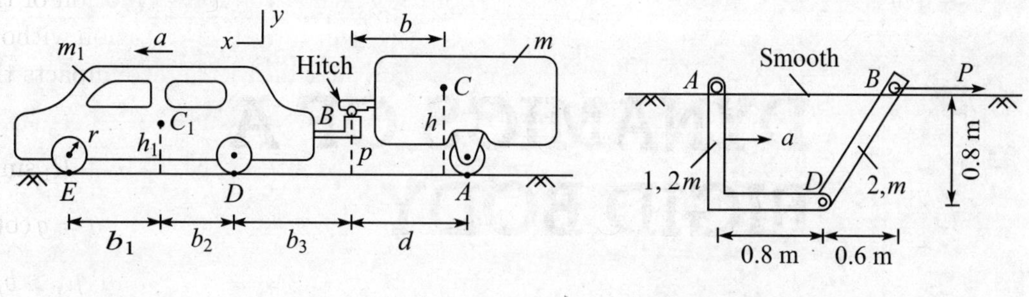

Fig. P2.47 **Fig. P2.48**

49 Consider a suspension cable supporting a suspension bridge (Fig. 4.25), under uniform load q_0 per unit projected horizontal length. From the FBD of part CD, show that $T\cos\theta = H =$ constant, $T\sin\theta = q_0 x \Rightarrow y' = \tan\theta = q_0 x/H$. Hence, show that the shape of the cable is a parabola: $y = q_0 x^2/2H$. (b) Consider transmission line cable subjected to uniform self-weight of q per unit length. As in part (a), show that $T\cos\theta = H =$ constant, $T\sin\theta = qs \Rightarrow y' = \tan\theta = qs/H \Rightarrow ds/dx = (1+y'^2)^{1/2} = (1+q^2s^2/H^2)^{1/2} \Rightarrow s = \frac{H}{q}\sinh\frac{qx}{H} = \frac{H}{q}y' \Rightarrow$ the shape of the cable is a common catenary: $y = \frac{H}{q}(\cosh\frac{qx}{H} - 1)$.

50. The force-deflection relation of the flexible base of a packing case of a delicate instrument of mass m is $P = cx^2$. Find the maximum height from which the base of the package can be dropped if the maximum force on the instrument from the base is not to exceed P_1.

$[\sqrt{mg/c} - \sqrt{P_1/c} + c[(P_1/c)^{3/2} - (mg/c)^{3/2}]/3mg]$

51. A bomb is dropped from an airplane flying horizontally at height h with speed v_0. Assume that a *small* resistance $R = mkv$ acts on it. Show that the time of its fall is approximately $(2h/g)^{1/2}[1 + (2h/g)^{1/2}k/6]$ and the horizontal distance covered in its fall is approximately $v_0(2h/g)^{1/2}[1 - (2h/g)^{1/2}k/3]$.

52. A small object is projected from origin O at speed v_0 at angle θ_0 to the horizontal direction $\vec{i}$ in a wind having horizontal velocity $-u_1\vec{i}$. The resistance force on the object is $-mk \times$ (velocity relative to wind). Find its position at any time. Show that (a) it returns to its point of projection if $\tan\theta_0 = g/ku_1$ and (b) the highest point of its trajectory is located above its point of projection if $(1 + g/kv_0\sin\theta_0)\ln(1 + kv_0\sin\theta_0/g) = 1 + v_0\cos\theta_0/u_1$.

$[x = (v_0\cos\theta_0 + u_1)(1 - e^{-kt})/k - u_1 t, \quad y = [v_0\sin\theta_0 + g/k](1 - e^{-kt})/k - gt/k]$

CHAPTER

3

DYNAMICS OF A RIGID BODY

THE equations of motion of a rigid body and their applications are covered in this chapter. Most of the materials used in engineering are stiff enough so that local deformation of bodies in motion can be neglected compared to their overall motion. The mathematically convenient rigid body model usually yields accurate results for their gross motion. In an equilibrium problem, a body may be modelled as rigid if its deformation is small so that the equilibrium equations in the deformed configuration can be approximated by the ones in the undeformed configuration. However, in problems where the reactions cannot be found from the equilibrium equations alone, the rigid body model has to be discarded even for small deformation.

In a rigid body model, the transmission of forces from one material point to another is instantaneous. This is not so for a deformable body. For the simplified model of a bar, elastic stress waves travel longitudinally at a speed of $c_0 = (E/\rho)^{1/2}$, where E is Young's modulus of elasticity and ρ is mass density. The time t required to traverse a bar of length L is $t = L/c_0$. For a steel bar, $t \sim 10^{-5}\,\mathrm{s}$ for $L \sim 0.1\,\mathrm{m}$ and $t \sim 10^{-4}\,\mathrm{s}$ for $L \sim 1\,\mathrm{m}$. The elastic stress-wave velocity for transverse vibration of a bar is smaller than c_0 and the transit time is more than 10 times that of longitudinal travel. If the time required for significant changes in the external forces is at least an order of magnitude larger than the transit time of internal waves, then the rigid body approximation is accurate. Typically, a time greater than 10^{-4} s or 10^{-5} s for significant changes in the external forces will permit a component to be modelled as rigid.

Euler's axioms yield $\vec{F} = m\vec{a}_{C|I}$, $\vec{M}_C = \dot{\vec{H}}_{C|I}$, which are six necessary and sufficient differential equations for the coordinates x_C, y_C, z_C of the centre of mass C and the Euler angles ϕ, θ, ψ (see Sec. 3.6.2) for orientation of a rigid body. In part A of this chapter, $\vec{H}_A$ about a material point A of the rigid body is expressed in terms of the inertia matrix at A and $\vec{\omega}$. The angular impulse–moment of momentum relation, Euler's equations of motion and the work–energy relation are presented in part B. These are applied to 3D problems and 2D applications such as centre of percussion, balancing of rotors, etc. The impact of rigid bodies is covered in part C. Part D deals with the application of Euler's equations to special 3D problems such as stability and torque-free motion of rigid bodies, steady precession and general motion of gyroscopes.

A. INERTIA MATRIX

3.1 $\vec{H}_A$ about a Point A of a Rigid Body

Let point A be a material point of a rigid body or a point on its hypothetical massless extension. In this section, all kinematical relations are with respect to a frame F and the explicit use of the subscript F has been omitted. For points P and A of a rigid body (Fig. 3.1a), $\vec{v}_{PA} = \vec{v}_P - \vec{v}_A = \vec{\omega} \times \vec{r}_{PA}$ and

$$\vec{H}_A = \int_m \vec{r}_{PA} \times \vec{v}_{PA}\, dm = \int_m \vec{r}_{PA} \times (\vec{\omega} \times \vec{r}_{PA})\, dm$$

It implies that each component of $\vec{H}_A$ depends linearly on the three components, $\omega_1, \omega_2, \omega_3$ of $\vec{\omega}$, through three coefficients each. These nine coefficients in all are the elements of the inertia matrix at A. Let $\vec{\omega} = \omega_i\, \vec{e}_i$, $\vec{r}_{PA} = x_i\, \vec{e}_i$. Then,

$$\vec{r}_{PA} \cdot \vec{r}_{PA} = r^2 = x_1^2 + x_2^2 + x_3^2, \qquad \vec{r}_{PA} \cdot \vec{\omega} = x_1\omega_1 + x_2\omega_2 + x_3\omega_3 = x_j\omega_j.$$

$$\begin{aligned}
\vec{H}_A &= \int_m \vec{r}_{PA} \times \vec{v}_{PA}\, dm = \int_m \vec{r}_{PA} \times (\vec{\omega} \times \vec{r}_{PA})\, dm \\
&= \int_m [(\vec{r}_{PA} \cdot \vec{r}_{PA})\vec{\omega} - (\vec{r}_{PA} \cdot \vec{\omega})\vec{r}_{PA}]\, dm \\
\Rightarrow \quad H_{A_i} &= \int_m [r^2\omega_i - (x_j\omega_j)x_i]\, dm = \int_m (r^2\delta_{ij} - x_i x_j)\, dm\, \omega_j
\end{aligned}$$

where, using $\omega_i = \vec{\omega} \cdot \vec{e}_i = (\omega_j\, \vec{e}_j) \cdot \vec{e}_i = \omega_j\delta_{ij}$, ω_i is expressed in terms of ω_j so that ω_j can be factored from the two terms on the right hand side. Defining

$$I_{ij}^A = \int_m (r^2\delta_{ij} - x_i x_j)\, dm \tag{3.1}$$

$$H_{A_i} = I_{ij}^A\omega_j = I_{i1}^A\omega_1 + I_{i2}^A\omega_2 + I_{i3}^A\omega_3, \qquad \vec{H}_A = I_{ij}^A\omega_j\, \vec{e}_i \tag{3.2}$$

i.e.,
$$
\begin{aligned}
H_{A_1} &= I_{11}^A\omega_1 + I_{12}^A\omega_2 + I_{13}^A\omega_3 \\
H_{A_2} &= I_{21}^A\omega_1 + I_{22}^A\omega_2 + I_{23}^A\omega_3 \\
H_{A_3} &= I_{31}^A\omega_1 + I_{32}^A\omega_2 + I_{33}^A\omega_3
\end{aligned}
$$

$$
\Rightarrow \begin{bmatrix} H_{A_1} \\ H_{A_2} \\ H_{A_3} \end{bmatrix} = \begin{bmatrix} I_{11}^A & I_{12}^A & I_{13}^A \\ I_{21}^A & I_{22}^A & I_{23}^A \\ I_{31}^A & I_{32}^A & I_{33}^A \end{bmatrix} \begin{bmatrix} \omega_1 \\ \omega_2 \\ \omega_3 \end{bmatrix}, \qquad [\vec{H}_A] = [I^A]\,[\vec{\omega}] \tag{3.3}
$$

where

$$
[\vec{H}_A] = \begin{bmatrix} H_{A_1} \\ H_{A_2} \\ H_{A_3} \end{bmatrix}, \quad [I^A] = \begin{bmatrix} I_{11}^A & I_{12}^A & I_{13}^A \\ I_{21}^A & I_{22}^A & I_{23}^A \\ I_{31}^A & I_{32}^A & I_{33}^A \end{bmatrix}, \quad [\vec{\omega}] = \begin{bmatrix} \omega_1 \\ \omega_2 \\ \omega_3 \end{bmatrix} \tag{3.4}
$$

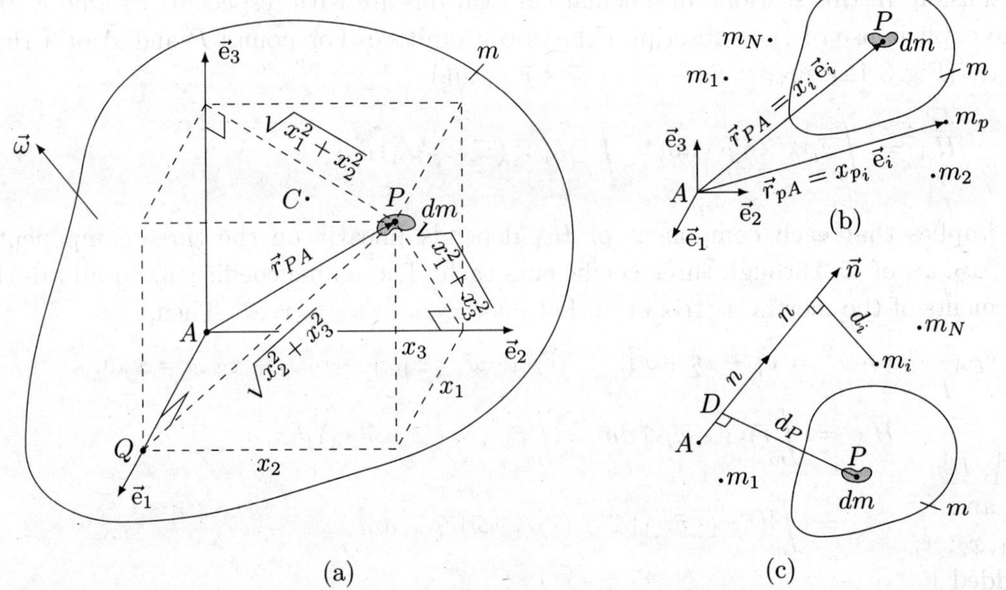

Fig. 3.1

Note that $I_{ij}^A = I_{ji}^A$. The real symmetric matrix $[I^A]$ is called the *inertia matrix at A relative to the axes* $x_i\,(\vec{e}_i)$ at A. In general, $\vec{H}_A$ is not in the direction of $\vec{\omega}$. If a point B is not a point of the rigid body, then

$$
\vec{H}_B = \vec{H}_C + \vec{r}_{CB} \times m\vec{v}_{CB} = I_{ij}^C\omega_j\,\vec{e}_i + \vec{r}_{CB} \times m\vec{v}_{CB}
$$

Equation (3.3) is very simple with respect to axes at A for which $[I^A]$ is a diagonal matrix. To establish the existence of such axes, we further study the inertia matrix.

3.2 Inertia Tensor at A

The integral appearing in eq. (3.1) is a Riemann integral. If the body includes N mass-points m_p at $\vec{r}_{pA} = x_{p_i}\,\vec{e}_i$ (Fig. 3.1b), then $\sum_{p=1}^{N}(r_{pA}^2\delta_{ij} - x_{p_i}x_{p_j})m_p$ is added to the right hand side of eq. (3.1). For mass distributed over a curve, a surface or a volume, dm is taken as $\lambda\,ds$, $\sigma\,dA$, or $\rho\,dV$ with ds, dA, dV as elements of arc length, surface area, volume and λ, σ, ρ as mass densities. The elements of $[I^A]$, given by eq. (3.1), are:

$$I_{11}^A = \int_m [r^2\delta_{11} - x_1^2]dm = \int_m (x_2^2 + x_3^2)dm = \int_m (PQ)^2 dm$$

$$I_{12}^A = \int_m (r^2\delta_{12} - x_1x_2)dm = -\int_m x_1x_2\,dm$$

$$\begin{aligned} I_{11}^A &= \int_m (x_2^2 + x_3^2)dm, & I_{12}^A = I_{21}^A &= -\int_m x_1x_2\,dm \\ I_{22}^A &= \int_m (x_3^2 + x_1^2)dm, & I_{23}^A = I_{32}^A &= -\int_m x_2x_3\,dm \\ I_{33}^A &= \int_m (x_1^2 + x_2^2)dm, & I_{31}^A = I_{13}^A &= -\int_m x_3x_1\,dm \end{aligned} \tag{3.5}$$

$$\begin{aligned} I_{xx}^A &= \int_m (y^2 + z^2)dm \equiv m(k_x^A)^2, & I_{xy}^A = I_{yx}^A &= -\int_m xy\,dm \\ I_{yy}^A &= \int_m (z^2 + x^2)dm \equiv m(k_y^A)^2, & I_{yz}^A = I_{zy}^A &= -\int_m yz\,dm \\ I_{zz}^A &= \int_m (x^2 + y^2)dm \equiv m(k_z^A)^2, & I_{zx}^A = I_{xz}^A &= -\int_m zx\,dm \end{aligned} \tag{3.6}$$

I_{11}^A, I_{22}^A, I_{33}^A are called the *moments of inertia* of the body about axes x_1, x_2, x_3 at A and I_{12}^A, I_{23}^A, I_{31}^A, I_{21}^A, I_{32}^A, I_{13}^A are called the *products of inertia* with respect to axes x_1, x_2; x_2, x_3; x_3, x_1; ...; at A. If N mass-points are present, then appropriate sums are added in eq. (3.5); e.g., $\sum_p (x_{p2}^2 + x_{p3}^2)m_p$ in I_{11}^A and $\sum_p -x_{p1}x_{p2}m_p$ in I_{12}^A. $PQ^2 = (x_2^2 + x_3^2)$ in the expression of I_{11}^A is the square of the distance of P from the axis x_1 through A. Unless specified otherwise, the summation convention will not be implied for the index n in I_{nn}^A and s in I_{ss}^A. In general, the mass moment of inertia I_{nn}^A about a line with unit vector $\vec{n}$ through A is defined by $I_{nn}^A = \int_m d_P^2\,dm$, where d_P is the distance of P from the line nn through A (Fig. 3.1c). $I_{nn}^A \geq 0$, with the equality holding iff all the mass is on the line nn. The product of inertia I_{12}^A involves the product x_1x_2 of the coordinates of P with respect to axes x_1, x_2 at A and can be of any sign. In general,

$$I_{xx}^A + I_{yy}^A = \int_m (y^2 + x^2 + 2z^2)\,dm \geq I_{zz}^A$$

and the equality holds only if $z \equiv 0$, i.e., for mass distribution in plane xy through A.

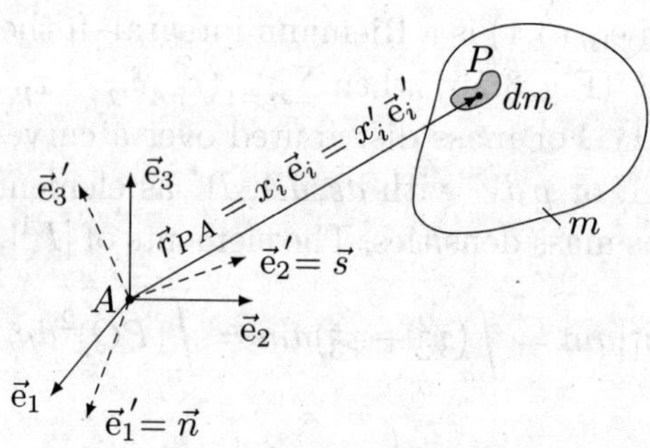

Fig. 3.2

The *radius of gyration* k_n^A of a body of mass m with respect to an axis $\vec{n}$ at A is defined as the distance from that axis at which a discrete mass m can be placed to have the same moment of inertia I_{nn}^A about that axis $\vec{n}$ as the given body, i.e., $m(k_n^A)^2 = I_{nn}^A \Rightarrow k_n^A = (I_{nn}^A/m)^{1/2}$. k_x^A, k_y^A, k_z^A are the radii of gyration about axes x, y, z at A.

For the case of *mass distribution on a plane area in the xy plane* with A in this plane, eq. (3.6) yields $I_{zz}^A = I_{xx}^A + I_{yy}^A$ as $z \equiv 0$, but for B not in the xy plane, $I_{zz}^B \neq I_{xx}^B + I_{yy}^B$.

Having computed the elements I_{pq}^A of the inertia matrix $[I^A]$ with respect to convenient axes $x_i\,(\vec{e}_i)$ at A, let us establish the *transformation relations* between them and the elements $I_{ij}^{A\prime}$ of the required inertia matrix $[I^{A\prime}]$ with respect to another set of axes $x_i'\,(\vec{e}_i')$ at A (Fig. 3.2), where

$$\vec{e}_i' = a_{i1}\,\vec{e}_1 + a_{i2}\,\vec{e}_2 + a_{i3}\,\vec{e}_3 = a_{ip}\,\vec{e}_p \tag{3.7}$$

$$\begin{aligned} a_{ip} &= (\text{component of } \vec{e}_i' \text{ along } \vec{e}_p) = \vec{e}_i' \cdot \vec{e}_p \\ &= \cos[\angle(\vec{e}_i', \vec{e}_p)] = (\text{direction cosine of } \vec{e}_i' \text{ with respect to } \vec{e}_p) \end{aligned} \tag{3.8}$$

Equation (3.7) $\Rightarrow$

$$\begin{aligned} \vec{e}_j' &= a_{jq}\,\vec{e}_q, \qquad \delta_{ij}' = \vec{e}_i' \cdot \vec{e}_j' = a_{ip}\,\vec{e}_p \cdot a_{jq}\,\vec{e}_q = a_{ip}\,a_{jq}\,\delta_{pq} \\ \vec{r}_{PA} &= x_i\,\vec{e}_i = x_i'\,\vec{e}_i' \qquad \Rightarrow \quad \vec{r}_{PA} \cdot \vec{r}_{PA} = r^2 = r'^2 \\ x_i' &= \vec{r}_{PA} \cdot \vec{e}_i' = x_p\,\vec{e}_p \cdot \vec{e}_i' = a_{ip}x_p \quad \Rightarrow \quad x_j' = a_{jq}\,x_q \end{aligned} \tag{a}$$

The dummy summation index p in $\vec{e}_i'$ is changed to index q in $\vec{e}_j'$ since in $\vec{e}_i' \cdot \vec{e}_j'$ each dummy index can only appear twice. Similarly index p in x_i' is changed to index q in x_j'. Using eqs (3.1) and (a),

$$I_{ij}^{A\prime} = \int_m (r'^2 \delta'_{ij} - x'_i x'_j) dm = a_{ip} a_{jq} \int_m (r^2 \delta_{pq} - x_p x_q) dm$$

$$\Rightarrow \quad I_{ij}^{A\prime} = a_{ip} a_{jq} I_{pq}^{A} \tag{3.9}$$

The summation convention implies that the right hand side of eq. (3.9) is the sum of 9 terms: $\sum_{p=1}^{3} \sum_{q=1}^{3} a_{ip} a_{jq} I_{pq}^{A}$. Let $[a] = [a_{ij}]$. This relation can be expressed in matrix form:

$$I_{ij}^{A\prime} = a_{ip} a_{jq} I_{pq}^{A} = a_{ip} I_{pq}^{A} ([a]^{\mathrm{T}})_{qj} \quad \Rightarrow \quad [I^{A\prime}] = [a]\,[I^{A}]\,[a]^{\mathrm{T}}$$

Denote the elements of an entity with respect to bases $\vec{e}_i$ and $\vec{e}'_i$ by ( ) and ( )$'$.

1. A scalar, like temperature T, with 3^0 (1) element, such that $T' = T$, is called a *zero-order tensor.*
2. A vector, like $\vec{r}_{PA}$, with 3^1 (3) elements, such that $x'_i = a_{ip} x_p$, is called a *first-order tensor.*
3. An entity, like $[I^A]$, with 3^2 (9) elements, such that $I_{ij}^{A\prime} = a_{ip} a_{jq} I_{pq}^{A}$, is called a *second-order tensor.*
4. An entity D with 3^3 (27) elements, such that $D'_{ijk} = a_{ip} a_{jq} a_{kr} D_{pqr}$, is called a *third-order tensor.*

The inertia matrix, with elements I_{ij}^{A} with respect to $\vec{e}_i$, is the representation of the symmetric second order inertia tensor at A.

The moment of inertia (moi) I_{nn}^{A} about an axis $\vec{n} = n_p \vec{e}_p$ at A (Fig. 3.2), is obtained using eq. (3.9) with $\vec{e}'_1 = a_{1p} \vec{e}_p = \vec{n}$, i.e., $a_{1p} = n_p$:

$$I_{nn}^{A} = I_{11}^{A\prime} = a_{1p} a_{1q} I_{pq}^{A} = n_p n_q I_{pq}^{A}$$

$$I_{nn}^{A} = I_{pq}^{A} n_p n_q = I_{11}^{A} n_1^2 + I_{22}^{A} n_2^2 + I_{33}^{A} n_3^2 + 2 I_{12}^{A} n_1 n_2 + 2 I_{23}^{A} n_2 n_3 + 2 I_{31}^{A} n_3 n_1 \tag{3.10}$$

The product of inertia I_{ns}^{A} about orthogonal axes $\vec{n}$ and $\vec{s} = s_q \vec{e}_q$ at A (Fig. 3.2), is obtained using eq. (3.9) with $\vec{e}'_2 = a_{2q} \vec{e}_q = \vec{s}$, i.e., $a_{2q} = s_q$:

$$I_{ns}^{A} = I_{12}^{A\prime} = a_{1p} a_{2q} I_{pq}^{A} = n_p s_q I_{pq}^{A}$$

$$I_{ns}^{A} = I_{pq}^{A} n_p s_q = I_{11}^{A} n_1 s_1 + I_{22}^{A} n_2 s_2 + I_{33}^{A} n_3 s_3 + I_{12}^{A} (n_1 s_2 + n_2 s_1)$$

$$+ I_{23}^{A} (n_2 s_3 + n_3 s_2) + I_{31}^{A} (n_3 s_1 + n_1 s_3) \tag{3.11}$$

Note that some expressions involving inertia elements I_{ij}^{A} are invariant (have the same value) for all triads, e.g., I_{ii}^{A} is an invariant:

$$I_{11}^{A} + I_{22}^{A} + I_{33}^{A} = 2 \int_m (x_1^2 + x_2^2 + x_3^2)\, dm = 2 \int_m r^2\, dm = I_{11}^{A\prime} + I_{22}^{A\prime} + I_{33}^{A\prime}$$

3.2.1 Principal axes of inertia at A

An axis $\vec{n}$ is called a *principal axis of inertia* at A if $I^A_{ns} = 0 \; \forall \, \vec{s} \perp \vec{n}$. The corresponding I^A_{nn} is called the *principal moment of inertia.* Let $v_q = I^A_{pq} n_p = I^A_{qp} n_p$ in eq. (3.11):

$$I^A_{ns} = I^A_{pq} n_p s_q = v_q s_q = \vec{v} \cdot \vec{s} = 0 \quad \forall \, \vec{s} \perp \vec{n} \quad \Rightarrow \quad \vec{v} = \lambda \vec{n}$$

$$\Rightarrow \quad v_q = I^A_{qp} n_p = \lambda n_q \quad \Rightarrow \quad I^A_{q1} n_1 + I^A_{q2} n_2 + I^A_{q3} n_3 = \lambda n_q \tag{3.12}$$

$$\text{i.e.,} \quad \begin{aligned} I^A_{11} n_1 + I^A_{12} n_2 + I^A_{13} n_3 &= \lambda n_1 \\ I^A_{21} n_1 + I^A_{22} n_2 + I^A_{23} n_3 &= \lambda n_2 \\ I^A_{31} n_1 + I^A_{32} n_2 + I^A_{33} n_3 &= \lambda n_3 \end{aligned}$$

$$\Rightarrow \quad \begin{bmatrix} I^A_{11} & I^A_{12} & I^A_{13} \\ I^A_{21} & I^A_{22} & I^A_{23} \\ I^A_{31} & I^A_{32} & I^A_{33} \end{bmatrix} \begin{bmatrix} n_1 \\ n_2 \\ n_3 \end{bmatrix} = \lambda \begin{bmatrix} n_1 \\ n_2 \\ n_3 \end{bmatrix}, \qquad [I^A]\,[\vec{n}] = \lambda\,[\vec{n}] \tag{3.13}$$

where $[\vec{n}] = [n_1 \; n_2 \; n_3]^{\mathrm{T}}$. Hence, $\vec{n}$ and λ are an *eigen-vector and an eigen-value* of $[I^A]$, since u and λ are defined as an eigen-vector and eigen-value pair of matrix B if $Bu = \lambda u$. Equations (3.10) and (3.12) yield

$$I^A_{nn} = I^A_{pq} n_p n_q = \lambda n_q n_q = \lambda \vec{n} \cdot \vec{n} = \lambda \tag{3.14}$$

Hence, the *principal axes of inertia and principal moments of inertia at A are, respectively, the eigen-vectors and eigen-values of the inertia matrix* $[I^A]$ *at* A. From linear algebra, we know that a real symmetric 3×3 matrix has at least 3 orthogonal eigen-vectors $\vec{e}^{\,*}_i$ with 3 eigen-values λ_i. Hence, *there always exist at least 3 orthogonal principal axes of inertia* $\vec{e}^{\,*}_i$ *at A (Fig. 3.3a) with respect to which the inertia matrix takes a diagonal form* with principal moments of inertia $I^{A*}_{11} = \lambda_1$, $I^{A*}_{22} = \lambda_2$, $I^{A*}_{33} = \lambda_3$:

$$[I^{A*}] = \begin{bmatrix} \lambda_1 & 0 & 0 \\ 0 & \lambda_2 & 0 \\ 0 & 0 & \lambda_3 \end{bmatrix} = \begin{bmatrix} I^{A*}_{11} & 0 & 0 \\ 0 & I^{A*}_{22} & 0 \\ 0 & 0 & I^{A*}_{33} \end{bmatrix} \tag{3.15}$$

§ **Exercise 3.1:** Show that the moment of inertia about any line through A cannot be greater than $\frac{1}{2}(I^A_{11} + I^A_{22} + I^A_{33})$.

§ **Exercise 3.2:** Prove that $\vec{n}$ is a principal axis of inertia at A if $I^A_{ns^{(1)}} = 0$, $I^A_{ns^{(2)}} = 0$, for two independent axes $\vec{s}^{\,(1)} \perp \vec{n}$ and $\vec{s}^{\,(2)} \perp \vec{n}$. [Hint: Use eq. (3.11) to show that $I^A_{ns} = 0 \; \forall \; \vec{s} \perp \vec{n}$, i.e., $\forall \; \vec{s} = c_1 \vec{s}^{\,(1)} + c_2 \vec{s}^{\,(2)}$.]

From Exercise 3.2 it is clear that if $I_{xz}^A = I_{yz}^A = 0$, then z is a principal axis at A.

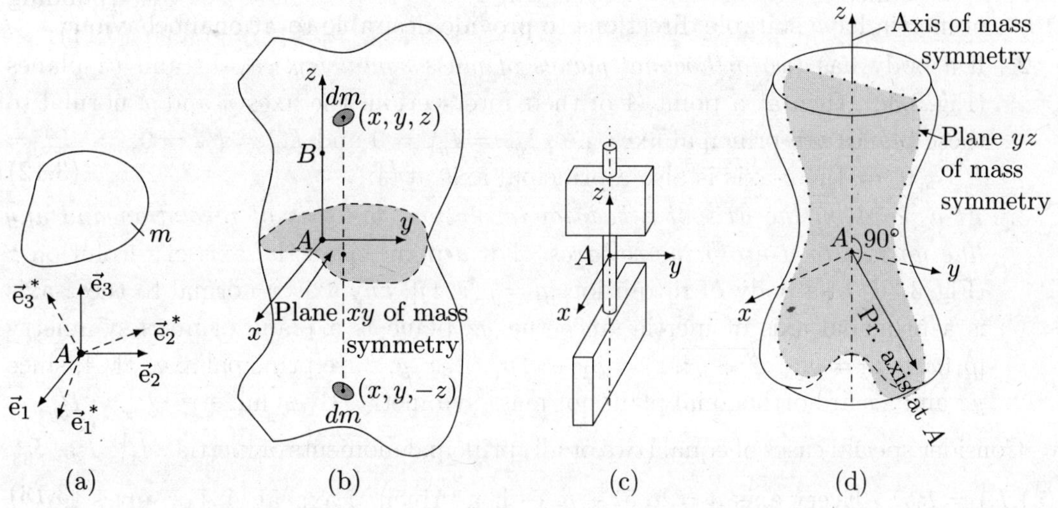

Fig. 3.3

Let $\vec{\omega} = \omega_i^* \,\vec{e}_i^*$. Using eqs (3.3) and (3.15), $\vec{H}_A$ in terms of the principal axes at A is

$$\vec{H}_A = I_{11}^{A*}\omega_1^*\,\vec{e}_1^{\,*} + I_{22}^{A*}\omega_2^*\,\vec{e}_2^{\,*} + I_{33}^{A*}\omega_3^*\,\vec{e}_3^{\,*} \tag{3.16}$$

In general, $\vec{H}_A$ is not in the direction of $\vec{\omega}$, not even for plane motion. It is in the direction of $\vec{\omega} = \omega\,\vec{n}$ if

$$[\vec{H}_A] = [I^A][\vec{\omega}] = \lambda[\vec{\omega}] \qquad \text{i.e., if} \qquad [I^A][\vec{n}] = \lambda[\vec{n}]$$

$\Rightarrow$ if $\vec{n}$ is a principal axis of inertia at A.

Hence, *$\vec{H}_A$ is in the direction of $\vec{\omega}$ iff $\vec{\omega}$ is in the direction of a principal axis of inertia at A,* for example, if $\vec{\omega} = \omega\,\vec{e}_1^{\,*}$, then $\vec{H}_A = I_{11}^{A*}\omega\,\vec{e}_1^{\,*} = I_{11}^{A*}\,\vec{\omega}$.

The principal axis of inertia can be found by inspection for 3 kinds of mass symmetry.

1. *A normal to a plane of mass symmetry is a principal axis of inertia at its point of intersection with it.* In Fig. 3.3b, the xy-plane is a plane of mass symmetry if

$$\rho(x,y,z) = \rho(x,y,-z) \quad \Rightarrow I_{zx}^A = -\int_m zx\,dm = 0, \quad I_{zy}^A = -\int_m zy\,dm = 0$$

since the total contribution to the integrals due to mass elements dm at (x,y,z) and dm at $(x,y,-z)$ is zero. Hence, the z-axis is a principal axis of inertia at A for any point A in the xy plane. However, the z-axis may not be a principal axis of inertia at B. Thus, for thin plates modelled as mass distribution in the xy-plane, the z-axis is a principal axis of inertia at any point A in the xy-plane.

This type of mass symmetry occurs in ships, automobiles, and aircraft, with the lay-out of engine, turbines, pumps, etc., so designed that the principal axes of inertia have suitable directions to provide desirable rotational behaviour.

2. If a body has *two orthogonal planes of mass symmetry,* say xz and yz planes (Fig. 3.3c), then at a point A of their intersection, the axes y and x normal to these planes are principal axes, i.e., $I^A_{yz} = I^A_{yx} = 0$ and $I^A_{xy} = I^A_{xz} = 0 \Rightarrow I^A_{xz} = I^A_{yz} = 0 \Rightarrow$ the z-axis is also a principal axis at A.
3. *At a point on the axis of a body of revolution, the axis of revolution and any line normal to it are principal axes.* For a point A on the axis of revolution z (Fig. 3.3d) of a body of revolution $[\rho = \rho(r, z)]$, *any* axis x normal to the z-axis is a principal axis of inertia since the yz plane is a plane of mass symmetry $[\rho_1(x, y, z) = \rho(\sqrt{x^2 + y^2}, z) = \rho_1(-x, y, z)]$. x, y, z are principal axes at A, since yz and zx are orthogonal planes of mass symmetry. By symmetry, $I^A_{xx} = I^A_{yy}$.

Consider special cases of equal (two or all) principal moments of inertia: I^A_{11}, I^A_{22}, I^A_{33}.

(a) $I^A_{11} = I^A_{22}$: Every axis $\vec{n} = n_1\,\vec{e}_1 + n_2\,\vec{e}_2$ is a principal axis at A. For any $\vec{s} \perp \vec{n}$:

$$\begin{aligned} I^A_{ns} &= I^A_{ij} n_i s_j = I^A_{11} n_1 s_1 + I^A_{22} n_2 s_2 = I^A_{11}(n_1 s_1 + n_2 s_2) = I^A_{11}\,\vec{n}\cdot\vec{s} = 0 \\ I^A_{nn} &= I^A_{ij} n_i n_j = I^A_{11} n_1^2 + I^A_{22} n_2^2 = I^A_{11}(n_1^2 + n_2^2) = I^A_{11} \end{aligned}$$

Hence, all directions $\vec{n}$ at A normal to $\vec{e}_3$ are principal axes of inertia at A with the same moment of inertia I^A_{11}. For example, for a cuboid with square cross-section, every axis through its centre of mass C in the square cross-section is a principal axis.

(b) $I^A_{11} = I^A_{22} = I^A_{33}$: For *any* $\vec{n}$ and any $\vec{s} \perp \vec{n}$:

$$\begin{aligned} I^A_{ns} &= I^A_{ij} n_i s_j = I^A_{11}(n_1 s_1 + n_2 s_2 + n_3 s_3) = I^A_{11}\,\vec{n}\cdot\vec{s} = 0 \\ I^A_{nn} &= I^A_{ij} n_i n_j = I^A_{11}(n_1^2 + n_2^2 + n_3^2) = I^A_{11} \end{aligned}$$

Hence, all directions at A are principal axes of inertia at A with the same moment of inertia I^A_{11}, e.g., any axis through the centre C of a cube is a principal axis of inertia at C.

In general, the directions of principal axes of inertia and the values of principal moments of inertia vary from point to point.

Example 3.1: A body has two non-orthogonal planes of mass symmetry. Show that at a point A on their line of intersection, every line normal to it is a principal axis of inertia at A with the same moment of inertia.

Solution: Let the planes of mass symmetry $BDEF$ (normal $\vec{e}_1$) and $PQRS$ (normal $\vec{n}$) intersect in the x_3-axis (Fig. E3.1a). Their normal directions $\vec{e}_1$ and $\vec{n}$ are principal directions of inertia at A, where $\vec{n} = \cos\theta\,\vec{e}_1 + \sin\theta\,\vec{e}_2$, $n_1 = \cos\theta$, $n_2 = \sin\theta$, $n_3 = 0$.

Hence, $I^A_{12} = I^A_{13} = I^A_{n3} = I^A_{nt} = 0$, where

$$\vec{t} = \vec{e}_3 \times \vec{n} = -\sin\theta\,\vec{e}_1 + \cos\theta\,\vec{e}_2, \qquad t_1 = -\sin\theta, \quad t_2 = \cos\theta, \quad t_3 = 0$$

Let $\vec{s} = \vec{e}_3$, $(s_1 = s_2 = 0,\ s_3 = 1)$,

$$I^A_{n3} = I^A_{ns} = I^A_{ij}\,n_i\,s_j = I^A_{i3}\,n_i\,s_3 = I^A_{13}\,n_1 + I^A_{23}\,n_2 = I^A_{23}\sin\theta = 0 \quad\Rightarrow\quad I^A_{23} = 0$$

since $\theta \neq 0$. $I^A_{12} = I^A_{23} = I^A_{31} = 0 \Rightarrow \vec{e}_1, \vec{e}_2, \vec{e}_3$ are the principal axes of inertia at A.

$$I^A_{nt} = I^A_{ij} n_i t_j = I^A_{11}\,n_1 t_1 + I^A_{22}\,n_2 t_2 = (I^A_{22} - I^A_{11})\sin\theta\cos\theta = 0$$

$\Rightarrow I^A_{11} = I^A_{22}$. Since these two principal moments of inertia are equal, all lines normal to the x_3-axis at A are principal axes of inertia at A with moments of inertia equal to I^A_{11}.

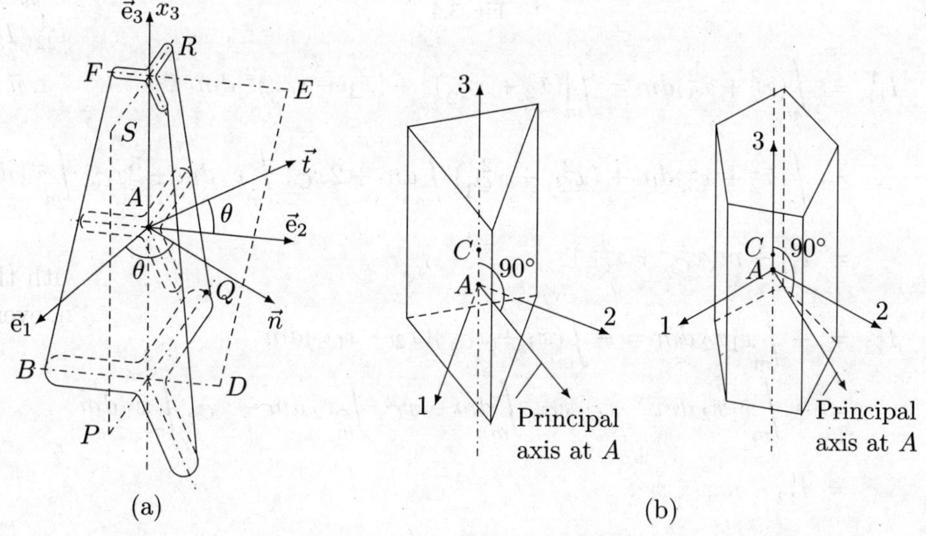

Fig. E3.1

Hence, for a uniform cylinder/pyramid whose cross-section is a regular polygon, at any point A on its axis, the axis and all transverse lines normal to it are principal axes of inertia with equal moment of inertia about these transverse axes (Fig. E3.1b).

3.2.2 Parallel axes theorem in terms of C

This theorem gives the relations to find $[I^A]$ at the point of interest A from the known $[I^C]$ at the centre of mass C. Consider two sets of parallel axes, x_i at A and $\bar{x}_i$ at centre of mass C ($\Rightarrow \bar{x}_{C_i} = 0$) (Fig. 3.4a). Let $\vec{r}_{PA} = x_i\,\vec{e}_i$, $\vec{r}_{PC} = \bar{x}_i\,\vec{e}_i$, $\vec{r}_{CA} = x_{C_i}\,\vec{e}_i$, then $\int_m \bar{x}_i dm = m\bar{x}_{C_i} = 0$ and $\vec{r}_{PA} = \vec{r}_{PC} + \vec{r}_{CA} \Rightarrow x_i = \bar{x}_i + x_{C_i}$.

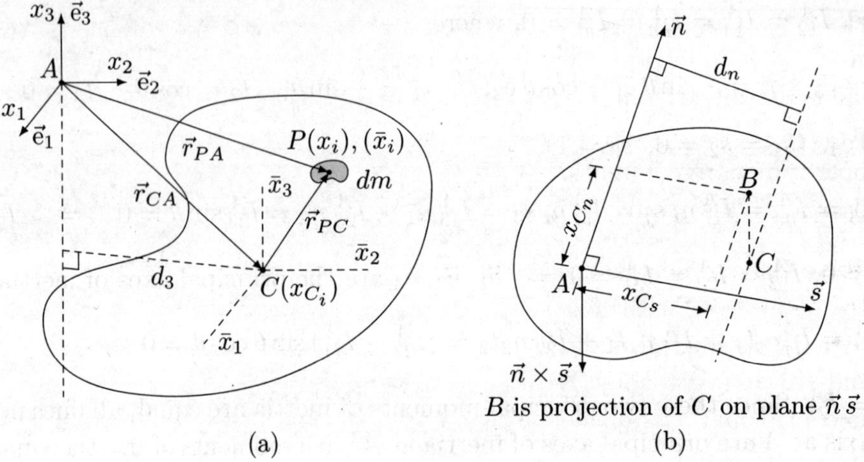

B is projection of *C* on plane $\vec{n}\,\vec{s}$

(a) (b)

Fig. 3.4

$$
\begin{aligned}
I_{11}^A &= \int_m (x_2^2 + x_3^2)dm = \int_m [(\bar{x}_2 + x_{C_2})^2 + (\bar{x}_3 + x_{C_3})^2]dm \\
&= \int_m (\bar{x}_2^2 + \bar{x}_3^2)dm + (x_{C_2}^2 + x_{C_3}^2)\int_m dm + 2x_{C_2}\int_m \bar{x}_2\, dm + 2x_{C_3}\int_m \bar{x}_3\, dm \\
&= I_{11}^C + m(x_{C_2}^2 + x_{C_3}^2)
\end{aligned}
$$

$$
\begin{aligned}
I_{12}^A &= -\int_m x_1 x_2\, dm = -\int_m (\bar{x}_1 + x_{C_1})(\bar{x}_2 + x_{C_2})dm \\
&= -\int_m \bar{x}_1\bar{x}_2\, dm - x_{C_1}x_{C_2}\int_m dm - x_{C_1}\int_m \bar{x}_2\, dm - x_{C_2}\int_m \bar{x}_1\, dm \\
&= I_{12}^C - m x_{C_1} x_{C_2}
\end{aligned}
$$

$$
\begin{aligned}
I_{11}^A &= I_{11}^C + m(x_{C_2}^2 + x_{C_3}^2), \qquad I_{12}^A = I_{12}^C - m x_{C_1} x_{C_2} \\
I_{22}^A &= I_{22}^C + m(x_{C_3}^2 + x_{C_1}^2), \qquad I_{23}^A = I_{23}^C - m x_{C_2} x_{C_3} \qquad (3.17) \\
I_{33}^A &= I_{33}^C + m(x_{C_1}^2 + x_{C_2}^2), \qquad I_{31}^A = I_{31}^C - m x_{C_3} x_{C_1} \\
[I^A] &= [I^C] + \text{ (inertia matrix at } A \text{ for discrete mass } m \text{ at } C)
\end{aligned}
$$

$$
I_{33}^A = I_{33}^C + m d_3^2, \quad I_{nn}^A = I_{nn}^C + m d_n^2, \quad I_{ns}^A = I_{ns}^C - m x_{Cn} x_{Cs} \qquad (3.18)
$$

where $d_3^2 = x_{C_1}^2 + x_{C_2}^2$ is the distance of C from the x_3-axis at A, i.e., the distance between lines through C and A parallel to $\vec{e}_3$ (Fig. 3.4a). In general, for a set of axes through A and C parallel to $\vec{n}$ and $\vec{s}$ (Fig. 3.4b), eq. (3.17) yields relations (3.18).

Remarks: **1.** The parallel axes theorem is not valid for any two points. To obtain I_{ij}^B from I_{ij}^A, it is applied first to get I_{ij}^C from I_{ij}^A and then again to get I_{ij}^B from I_{ij}^C. **2.** $I_{ij}^{A\prime}$ can be obtained from $[I^C]$ by first finding $I_{ij}^{C\prime}$ using transformation relation (3.9) for rotation of axes at C and then applying the parallel axes theorem (3.17) to get $I_{ij}^{A\prime}$. These operations are commutative, i.e., we can apply the parallel axes theorem first and subsequently use transformation relation for rotation of axes at A.

§ **Exercise 3.3:** Let the z-axis be a principal axis of inertia of a rigid body at its centre of mass C with $\vec{\omega} = \omega\,\vec{k}$. Prove using parallel axes theorem (3.17) that (a) the z-axis passing through C is a principal axis of inertia at every point A on itself, and (b) at every point B in the xy plane through C, the axis parallel to the z-axis is a principal axis of inertia. (c) Hence, show that for a point D of the rigid body, $\vec{H}_D = I_{zz}^D\,\vec{\omega}$ only if D lies on the z-axis through C or lies in a plane through C normal to the z-axis. (d) Show that of all the moments of inertia about a set of parallel axes, the moment of inertia about a line through C is the least.

The inertia matrix at A for a composite body, built up of standard shapes, is obtained by decomposing it and adding the contributions of its parts (since the integral over mass m equals the sum of the integrals over mass m_i of parts P_i, $i = 1, \ldots n$,) using parallel axes theorems and the transformation relations for rotation, if needed.

3.2.3 Inertia elements for some symmetric uniform bodies

The coordinates which best fit the boundary of the body should be used in the integral of I_{ij}^A. An element of mass dm of the lowest order should be used (as in Section 2.2 for C) and its inertia element with respect to the given axis (axes) be obtained using transformation relations (3.9) and / or (3.17), if needed. For a body of revolution, use a thin disc element normal to the axis. We consider some common bodies.

1. A thin rod of length L and mass m is modelled as a 1-D distribution of mass over its centreline. The principal axis/axes at A, B, C, D, based on symmetry, are shown in Fig. 3.5a. At A and C any axis normal to x-axis is a principal axis.

$$I_{xx}^C = 0, \qquad I_{zz}^C = I_{yy}^C = \int_m x^2\,dm = \int_{-L/2}^{L/2} x^2 \frac{m}{L}\,dx = \tfrac{1}{12}\,mL^2$$

$$I_{xx}^B = I_{xx}^C + md^2 = md^2, \qquad I_{yy}^B = I_{yy}^C = \tfrac{1}{12}\,mL^2$$

$$I_{zz}^B = I_{zz}^C + md^2 = m(L^2/12 + d^2)$$

$$I_{xx}^A = 0, \qquad I_{zz}^A = I_{yy}^A = I_{yy}^C + m(L/2)^2 = \tfrac{1}{3}\,mL^2$$

2. A thin rectangular plate of sides a, b and mass m is modelled as a 2-D distribution of mass over its mid-plane (Fig. 3.5b). The principal axis/axes at A, B, C, D, E, O, based on symmetry, are shown in Fig. 3.5b; e.g., planes xy and xz are planes of mass symmetry through A. We find I_{xx}^C by dividing the rectangle into thin strips of mass dm:

$$I_{xx}^C = \int_m \tfrac{1}{12}\, b^2\, dm = \tfrac{1}{12}\, mb^2, \qquad \text{similarly,} \quad I_{yy}^C = \tfrac{1}{12}\, ma^2$$

$$I_{zz}^C = I_{xx}^C + I_{yy}^C = \tfrac{1}{12}\, m(a^2 + b^2)$$

$$I_{xx}^O = I_{xx}^C + m(b/2)^2 = \tfrac{1}{3}\, mb^2, \qquad I_{yy}^O = \tfrac{1}{3}\, ma^2$$

$$I_{zz}^O = I_{xx}^O + I_{yy}^O = \tfrac{1}{3}\, m(a^2 + b^2), \qquad I_{zx}^O = I_{zy}^O = 0$$

$$I_{xy}^O = I_{xy}^C - m x_{CO} y_{CO} = -m(\tfrac{1}{2}a)(\tfrac{1}{2}b) \neq 0 \Rightarrow \; x,\, y \text{ are not principal axes at } O.$$

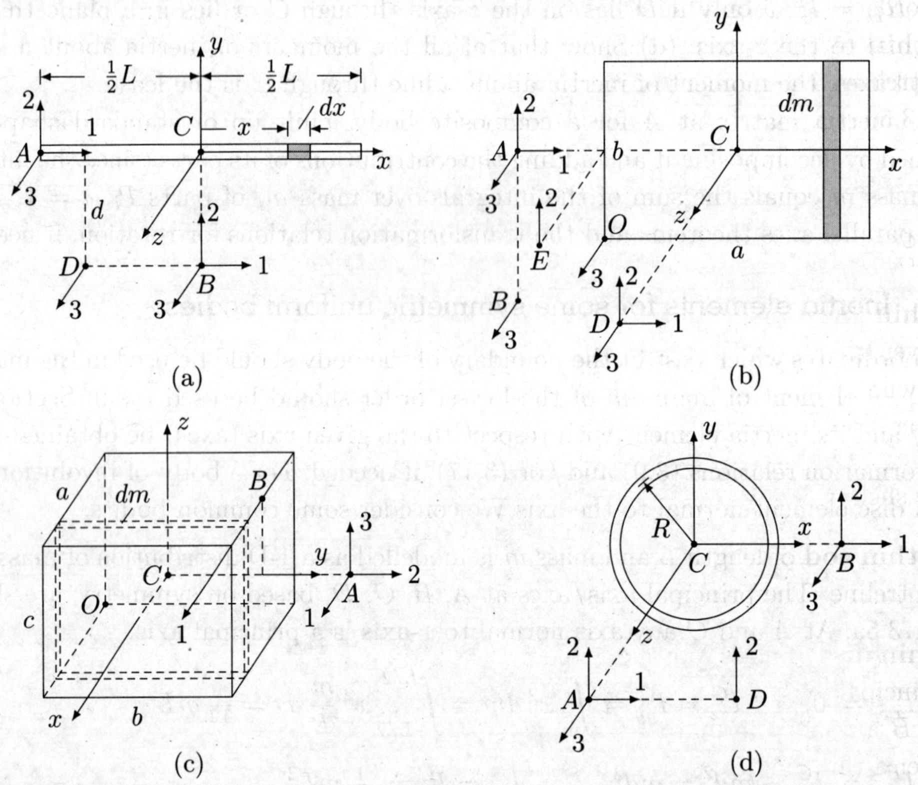

Fig. 3.5

3. A cuboid of mass m. The principal axis/axes at A, B, C, based on symmetry, are shown in Fig. 3.5c. We find I_{xx}^C by dividing the cuboid into thin plates normal to x-axis:

$$I_{xx}^C = \int_m \tfrac{1}{12}\, (b^2 + c^2)\, dm = \tfrac{1}{12}\, m(b^2 + c^2)$$

$$I^C_{yy} = \tfrac{1}{12} m(c^2 + a^2), \qquad I^C_{zz} = \tfrac{1}{12} m(a^2 + b^2)$$

with the sides normal to the concerned axis being squared.

$$I^O_{xx} = I^C_{xx} + m(y^2_{CO} + z^2_{CO}) = \frac{m(b^2 + c^2)}{12} + m\left[\left(\tfrac{1}{2}b\right)^2 + \left(\tfrac{1}{2}c\right)^2\right] = \tfrac{1}{3} m(b^2 + c^2)$$

$$I^O_{xy} = I^C_{xy} - m x_{CO} y_{CO} = 0 - m\left(\tfrac{1}{2}a\right)\left(\tfrac{1}{2}b\right) = -\tfrac{1}{4}mab$$

$$I^O_{yz} = I^C_{yz} - m y_{CO} z_{CO} = 0 - m\left(\tfrac{1}{2}b\right)\left(\tfrac{1}{2}c\right) = -\tfrac{1}{4}mbc$$

The x, y, z are are not principal at O. Comparison of the results of a thin plate and a cuboid shows that I^C_{zz} for a plate is exact but the expressions of I^C_{xx} and I^C_{yy} have errors of the order of c^2/b^2 and c^2/a^2, i.e., error less than 1% if $c/b < 0.1$ and $c/a < 0.1$.

4. A thin circular ring of mass m is modelled as a 1-D distribution of mass over its centreline. The principal axis/axes at A, B, C, D, based on symmetry, are shown in Fig. 3.5d. At C and A any axis normal to z-axis is a principal axis.

$$I^C_{zz} = \int_m (x^2 + y^2)\, dm = R^2 \int_m dm = mR^2$$

$$I^C_{xx} + I^C_{yy} = I^C_{zz} \quad \Rightarrow \quad I^C_{xx} = I^C_{yy} = \tfrac{1}{2}mR^2$$

5. A thin circular cylindrical shell of mass m is modelled as a 2-D distribution of mass over its mid-surface. The principal axis/axes at A, B, C, D, based on symmetry, are shown in Fig. 3.6a. At C and A any axis normal to z-axis is a principal axis. For the thin ring of width dz and mass $dm = (m/L)dz$, $I^C_{xx} = \tfrac{1}{2}R^2 dm + z^2 dm$, $I^C_{zz} = R^2 dm$.

For the shell: $$I^C_{xx} = \int_{-L/2}^{L/2} \left(\tfrac{1}{2}R^2 + z^2\right) \frac{m}{L} dz = m\left(\tfrac{1}{2}R^2 + \tfrac{1}{12}L^2\right)$$

$$I^C_{yy} = I^C_{xx}, \qquad I^C_{zz} = \int_m R^2 dm = mR^2$$

6. A thin disc of mass m is modelled as a 2-D distribution of mass over its mid-plane. The principal axis/axes at A, B, C, D, based on symmetry, are shown in Fig. 3.6b. At C and B any axis normal to z-axis is a principal axis. $I^C_{xx} = I^C_{yy}$, $I^C_{xx} + I^C_{yy} = I^C_{zz}$. Consider a thin ring element of mass $dm = (m/\pi R^2) 2\pi r\, dr$:

$$I^C_{zz} = \int_m (x^2 + y^2)\, dm = \int_m r^2\, dm = \int_0^R r^2 \left(\frac{m}{\pi R^2}\right) 2\pi r\, dr = \tfrac{1}{2}mR^2$$

$$I^C_{xx} = I^C_{yy} = \tfrac{1}{2}I^C_{zz} = \tfrac{1}{4}mR^2$$

The part up to radius R_1 has mass $m_1 = mR_1^2/R^2$ and contributes $m_1 R_1^2/2 = I^C_{zz}(R_1/R)^4$ to I^C_{zz}, which is only $0.4, 6.25, 13\%$ of I^C_{zz} for $R_1/R = 0.25, 0.5, 0.6$.

7. A solid circular cylinder of mass m: The principal axis/axes at A, B, C, D, based on symmetry, are shown in Fig. 3.6c. At C ana A any axis normal to z-axis is a principal axis. For the thin disc of mass $dm = (m/L)dz$, $I_{xx}^C = \frac{1}{4}R^2 dm + z^2 dm$, $I_{zz}^C = \frac{1}{2}R^2 dm$.

For the cylinder: $$I_{xx}^C = \int_{-L/2}^{L/2} \left(\tfrac{1}{4}R^2 + z^2\right)\frac{m}{L}\,dz = m\left(\tfrac{1}{4}R^2 + \tfrac{1}{12}L^2\right)$$

$$I_{yy}^C = I_{xx}^C, \qquad I_{zz}^C = \int_m \tfrac{1}{2}R^2\,dm = \tfrac{1}{2}mR^2$$

(a) (b) (c) (d)

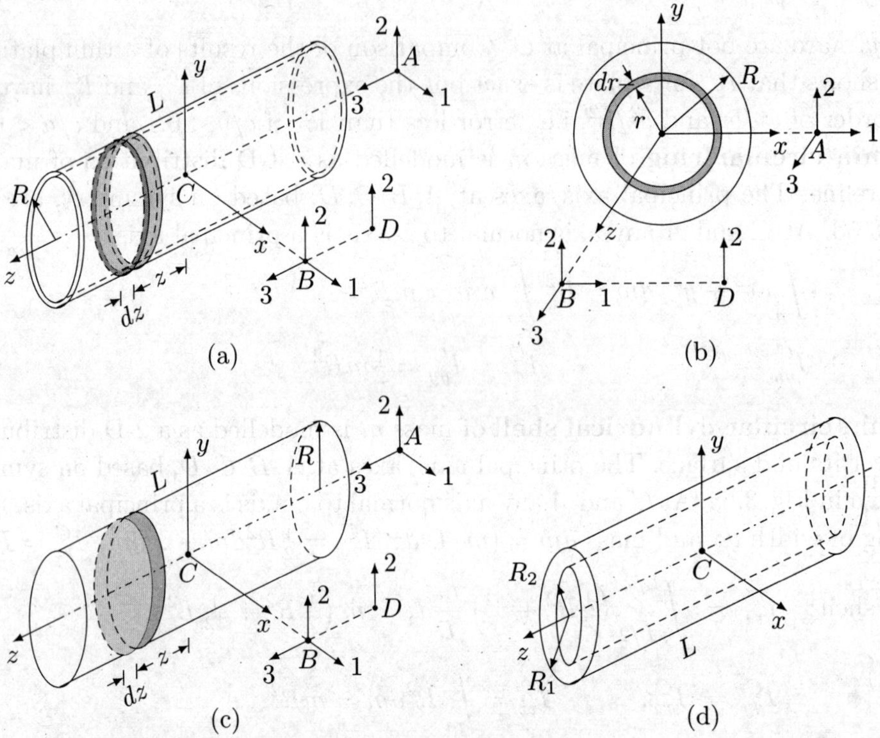

Fig. 3.6

Let us clarify the meaning of 'thin' in: (a) a thin bar, (b) a thin disc, (c) a thin shell.

(a) $I_{xx}^C \to mL^2/12$ as $R/L \to 0$. $I_{xx}^C = 1.0075(mL^2/12)$ for $R = L/20$
$\Rightarrow$ $I_{xx}^C = mL^2/12$ for a thin bar has an error $\leq 0.75\%$ for $2R/L \leq 0.1$.

(b) $I_{xx}^C \to mR^2/4$ as $L/R \to 0$. $I_{xx}^C = 1.0133(mR^2/4)$ for $L = R/5$
$\Rightarrow$ $I_{xx}^C = mR^2/4$ for a thin disc has an error $\leq 1.33\%$ for $L/2R \leq 0.1$.

(c) For a hollow cylinder (Fig. 3.6d) of mean radius R, thickness t and mass m,

$$I_{zz}^C = \tfrac{1}{2}(\rho\pi R_1^2 L)R_1^2 - \tfrac{1}{2}(\rho\pi R_2^2 L)R_2^2 = \tfrac{1}{2}m(R_1^2 + R_2^2) = mR^2[1 + t^2/4R^2]$$

$I_{zz}^C = mR^2$ for a thin circular cylindrical shell has an error of $\leq 1\%$ for $t/R \leq 0.2$.

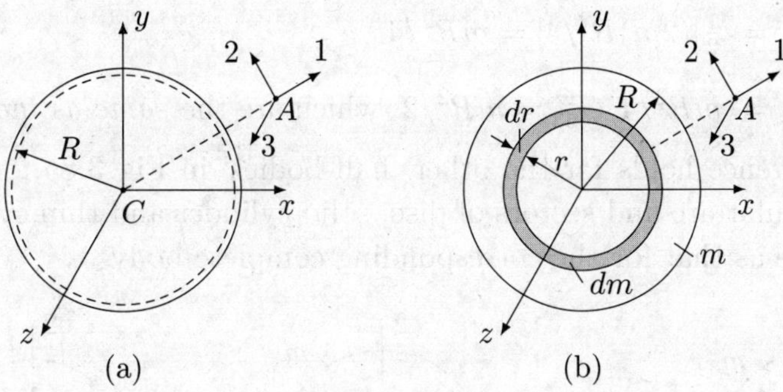

Fig. 3.7

8. A thin spherical shell of mass m is modelled as a 2-D distribution of mass over its mid-surface. The principal axes at A and C, based on symmetry, are shown in Fig. 3.7a. At C every axis is a principal axis. At A any axis normal to axis 1 is a principal axis. By symmetry $I_{xx}^C = I_{yy}^C = I_{zz}^C$ and:

$$I_{xx}^C + I_{yy}^C + I_{zz}^C = \int_m [(y^2+z^2)+(z^2+x^2)+(x^2+y^2)]dm = 2R^2 \int_m dm = 2mR^2$$

$$\Rightarrow \quad I_{xx}^C = I_{yy}^C = I_{zz}^C = \tfrac{2}{3}\, mR^2$$

9. Consider a sphere of mass m. The principal axes at A and C are shown in Fig. 3.7b. At C every axis is a principal axis. At A any axis normal to axis 1 is a principal axis. For thin spherical shell of radius r, thickness dr, mass $dm = \rho(4\pi r^2\, dr) = [m/(4\pi R^3/3)]4\pi r^2\, dr = 3mr^2 dr/R^3$, $I_{xx}^C = 2r^2\, dm/3$. For sphere,

$$I_{xx}^C = \int_m \frac{2}{3} r^2\, dm = \frac{2m}{R^3}\int_0^R r^4\, dr = \frac{2}{5} mR^2$$

The contribution to I_{xx}^C due to mass m_1 up to radius R_1 is $2m_1R_1^2/5 = I_{xx}^C R_1^5/R^5$, which is $0.1, 3.1, 17\%$ for $R_1/R = 0.25, 0.5, 0.7$ respectively.

§ Exercise 3.4: Prove that for a hollow sphere of mean radius R, thickness t and mass m, $I_{xx}^C = (2mR^2/3)(1+t^2/2R^2+t^4/80R^4)/(1+t^2/12R^2)$. Hence, show that for a thin spherical shell $I_{xx}^C = 2mR^2/3$ has an error within 1.7% for $t \leq R/5$.

10. 'Half-bodies' The inertia matrix for 'half-bodies', such as a semicircular ring, semicircular disc, semi-cylinder, hemisphere, half-cone, semi-circular cylindrical shell (Fig. 3.8a) at a point on the axis of revolution, can be obtained from the inertia matrix at that point for the full bodies. For example, axes 1, 2, 3 are the principal axes of inertia at point O of the semi-circular thin disc of radius R and mass m

$$I_{11}^O = \tfrac{1}{2}[I_{11}^O \text{ of full disc of mass } 2m \text{ obtained by adding dashed half-disc}]$$

$$= \tfrac{1}{2}[(2m)R^2/4] = mR^2/4$$

Similarly, $I_{22}^O = mR^2/4$, $I_{33}^O = mR^2/2$ which are the *same as those for a full disc.*

The same inference holds for the other 'half-bodies' in Fig. 3.8a for the axes shown. For a thin circular arc and sectors of disc, solid cylinder and thin cylinder (Fig. 3.8b), I_{33}^O is the same as that for the corresponding complete body.

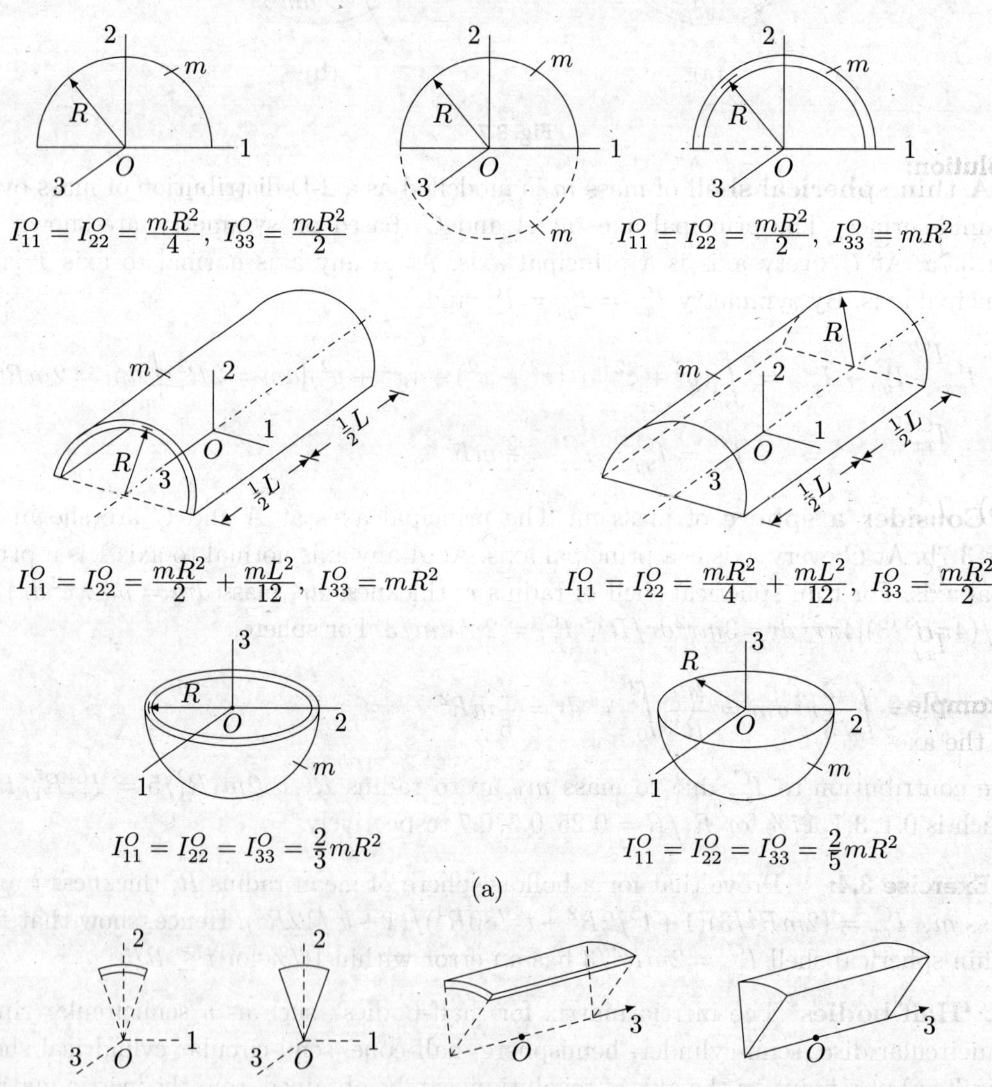

Fig. 3.8

Example 3.2: Find the principal moments of inertia at O and C (Fig. E3.2).

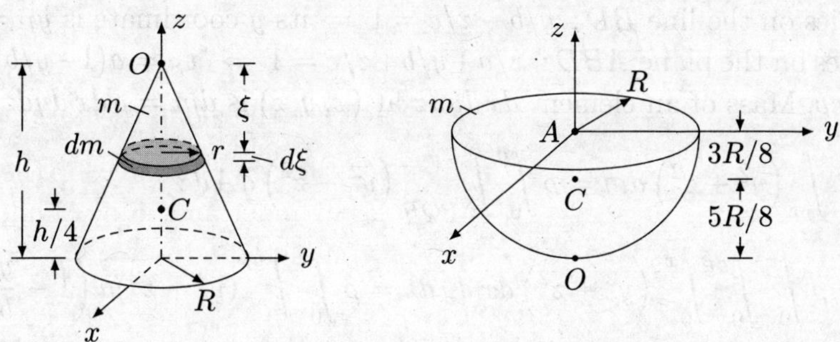

Fig. E3.2

Solution: (a) $m = \rho\pi R^2 h/3$. For disc: $r/\xi = R/h \Rightarrow r = R\xi/h,\ dm = \rho\pi r^2 d\xi$.

$$I_{xx}^O = \int_m \left(\frac{r^2}{4} + \xi^2\right) dm = \pi\rho\frac{R^2}{h^2}\left(\frac{R^2}{4h^2} + 1\right)\int_0^h \xi^4 d\xi = \frac{3}{5}m\left(\frac{R^2}{4} + h^2\right) = I_{yy}^O$$

$$I_{zz}^O = \int_m \frac{r^2}{2} dm = \pi\rho\frac{R^4}{2h^4}\int_0^h \xi^4 d\xi = \frac{\pi\rho R^4 h}{10} = \frac{3}{10}mR^2 = I_{zz}^C$$

$$I_{xx}^C = I_{xx}^O - m(3h/4)^2 = 3m(4R^2 + h^2)/80 = I_{yy}^C$$

(b) $$I_{xx}^A = I_{yy}^A = I_{zz}^A = I_{zz}^C = I_{zz}^O = 2mR^2/5$$

$$I_{xx}^C = I_{xx}^A - m(3R/8)^2 = 83mR^2/320, \qquad I_{yy}^C = I_{xx}^C$$

$$I_{xx}^O = I_{xx}^C + m(5R/8)^2 = 13mR^2/20, \qquad I_{yy}^O = I_{xx}^O$$

Example 3.3: Find the inertia matrix at O and at the centre of mass C with respect to the axes shown in Fig. E3.3a for a tetrahedron of mass m.

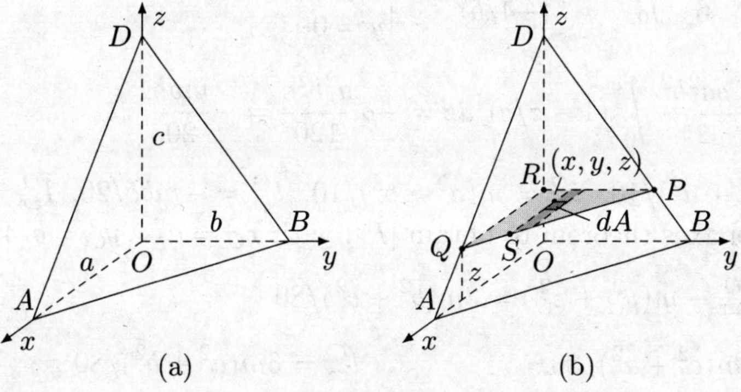

Fig. E3.3

Solution: Take ΔPQR parallel to xy plane at a distance z from O (Fig. E3.3b). Point P $(z_P = z)$ lies on the line BD: $y/b + z/c = 1 \Rightarrow$ its y coordinate is $y_P = (c-z)b/c$. S $(z_S = z)$ lies on the plane ABD: $x/a + y/b + z/c = 1 \Rightarrow x_S = a(1 - y/b - z/c)$. Mass $m = \frac{1}{3}(\frac{1}{2}ab)c\rho$. Mass of an element $dx\,dy\,dz$ at (x, y, z) is $dm = \rho\,dx\,dy\,dz = \rho\,dA\,dz$,

$$\begin{aligned}
I^O_{xx} &= \int_m \left(y^2+z^2\right) dm = \rho \int_0^c \int_{\Delta PQR} \left(y^2+z^2\right) dA\,dz \\
&= \rho \int_0^c \int_0^{y_P} \int_0^{x_S} \left(y^2+z^2\right) dx\,dy\,dz = \rho \int_0^c \int_0^{y_P} (y^2+z^2) a \left(1 - \frac{y}{b} - \frac{z}{c}\right) dy\,dz \\
&= a\rho \int_0^c \left[\left(\frac{y^3}{3} + z^2 y\right)\left(1 - \frac{y}{b} - \frac{z}{c}\right)\Big|_{y=0}^{y=y_P} + \int_0^{y_P} \left(\frac{y^3}{3} + z^2 y\right) \frac{1}{b}\,dy \right] dz \\
&= \frac{a}{b}\rho \int_0^c \left[\left(\frac{y^4}{12} + z^2 \frac{y^2}{2}\right)\Big|_{y=0}^{y=y_P} \right] dz \\
&= \frac{a}{b}\rho \int_0^c \left[\frac{(c-z)^4 b^4}{12c^4} + z^2 \frac{(c-z)^2 b^2}{2c^2} \right] dz = \frac{\rho abc}{60}(b^2+c^2) = \frac{m(b^2+c^2)}{10}
\end{aligned}$$

$$\begin{aligned}
I^O_{xy} &= -\int_m xy\,dm = -\rho \int_0^c \int_0^{y_P} \int_0^{x_S} xy\,dx\,dy\,dz \\
&= -\rho \int_0^c \int_0^{y_P} \frac{a^2}{2}\left(1 - \frac{y}{b} - \frac{z}{c}\right)^2 y\,dy\,dz \\
&= -\rho \frac{a^2}{2} \int_0^c \left[\frac{(1 - y/b - z/c)^3}{(-3/b)} y \Big|_{y=0}^{y=y_P} - \int_0^{y_P} \frac{(1 - y/b - z/c)^3}{(-3/b)} dy \right] dz \\
&= -\frac{\rho a^2 b}{6} \int_0^c \frac{(1 - y/b - z/c)^4}{(-4/b)} \Big|_{y=0}^{y=y_P} dz \\
&= -\frac{\rho a^2 b^2}{24} \int_0^c (1 - z/c)^4 dz = -\rho \frac{a^2 b^2 c}{120} = -\frac{mab}{20}
\end{aligned}$$

$\Rightarrow I^O_{yy} = m(c^2+a^2)/10$, $I^O_{zz} = m(a^2+b^2)/10$, $I^O_{yz} = -mbc/20$, $I^O_{zx} = -mca/20$. We use the parallel axes theorem to obtain $[I^C]$ with $x_C = a/4, y_C = b/4, z_C = c/4$:

$$I^C_{xx} = I^O_{xx} - m(y_C^2 + z_C^2) = 3m(b^2+c^2)/80$$

$$I^C_{yy} = 3m(c^2+a^2)/80, \qquad I^C_{zz} = 3m(a^2+b^2)/80$$

$$I^C_{xy} = I^O_{xy} + m x_C y_C = mab/80, \qquad I^C_{yz} = mbc/80, \qquad I^C_{zx} = mca/80$$

Example 3.4: Find the elements of the inertia matrix for point A with respect to x, y, z axes for the system shown in Fig. E3.4a consisting of a thin rectangular plate ① and a cylinder ②. The density of the body i is ρ_i. Neglect the mass of the shaft. List the principal axes of inertia which can be determined by inspection for various bodies at the following points: body ① at A, J, K and body ② at A, D, P, Q.

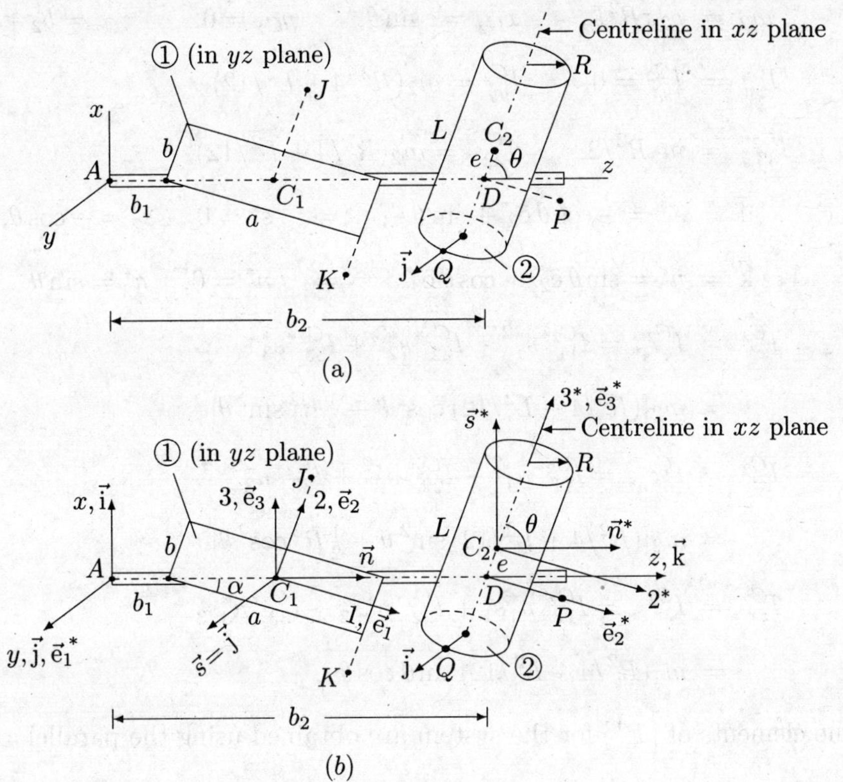

Fig. E3.4

Solution: The principal axes of inertia at C_i of the ith body are shown in Fig. E3.4b.
Body ①: Axes 1, 2, 3 are the principal axes of inertia at C_1. Plane yz is a plane of mass symmetry through C_1 $\Rightarrow$ axis 3 parallel to x-axis is a principal axis at C_1 with

$$m_1 = \rho_1 ab, \quad x_{C_1} = y_{C_1} = 0, \quad z_{C_1} = b_1 + \tfrac{1}{2}(a^2 + b^2)^{1/2}, \quad \alpha = \tan^{-1}\frac{b}{a}$$

$$I_{xy}^{C_1} = I_{zx}^{C_1} = 0, \quad I_{xx}^{C_1} = \tfrac{1}{12}\, m_1(a^2 + b^2), \quad I_{11}^{C_1} = \tfrac{1}{12}\, m_1 b^2, \quad I_{22}^{C_1} = \tfrac{1}{12}\, m_1 a^2$$

$$\vec{\text{j}} = \vec{s} = \sin\alpha\, \vec{\text{e}}_1 - \cos\alpha\, \vec{\text{e}}_2 \quad \Rightarrow \quad s_1 = \sin\alpha, \quad s_2 = -\cos\alpha, \quad s_3 = 0$$

$$\vec{\text{k}} = \vec{n} = \cos\alpha\, \vec{\text{e}}_1 + \sin\alpha\, \vec{\text{e}}_2 \quad \Rightarrow \quad n_1 = \cos\alpha, \quad n_2 = \sin\alpha, \quad n_3 = 0$$

$$I_{yy}^{C_1} = I_{ss}^{C_1} = I_{11}^{C_1} s_1^2 + I_{22}^{C_1} s_2^2 + I_{33}^{C_1} s_3^2 = \tfrac{1}{12}\, m_1(b^2 \sin^2\alpha + a^2 \cos^2\alpha)$$

$$I_{zz}^{C_1} = I_{nn}^{C_1} = I_{11}^{C_1} n_1^2 + I_{22}^{C_1} n_2^2 + I_{33}^{C_1} n_3^2 = \tfrac{1}{12} m_1 (b^2 \cos^2 \alpha + a^2 \sin^2 \alpha)$$

$$I_{yz}^{C_1} = I_{sn}^{C_1} = I_{11}^{C_1} s_1 n_1 + I_{22}^{C_1} s_2 n_2 + I_{33}^{C_1} s_3 n_3 = \tfrac{1}{12} m_1 (b^2 - a^2) \sin\alpha \cos\alpha$$

Body ②: Axes 1* $(= \vec{\mathrm{j}},\, y)$, 2*, 3* are the principal axes of inertia at C_2.

$$m_2 = \rho_2 \pi R^2 L, \qquad x_{C_2} = e \sin\theta, \qquad y_{C_2} = 0, \qquad z_{C_2} = b_2 + e\cos\theta$$

$$I_{xy}^{C_2} = I_{yz}^{C_2} = 0, \qquad I_{yy}^{C_2} = m_2 (R^2/4 + L^2/12)$$

$$I_{33}^{C_2 *} = m_2 R^2/2, \qquad I_{22}^{C_2 *} = m_2 (R^2/4 + L^2/12)$$

$$\vec{\mathrm{i}} = \vec{s}^{\,*} = -\cos\theta\, \vec{\mathrm{e}}_2^{\,*} + \sin\theta\, \vec{\mathrm{e}}_3^{\,*} \quad \Rightarrow \quad s_1^* = 0, \quad s_2^* = -\cos\theta, \quad s_3^* = \sin\theta$$

$$\vec{\mathrm{k}} = \vec{n}^{\,*} = \sin\theta\, \vec{\mathrm{e}}_2^{\,*} + \cos\theta\, \vec{\mathrm{e}}_3^{\,*} \quad \Rightarrow \quad n_1^* = 0, \quad n_2^* = \sin\theta, \quad n_3^* = \cos\theta$$

$$\begin{aligned} I_{xx}^{C_2} &= I_{s^*s^*}^{C_2} = I_{11}^{C_2 *} s_1^{*2} + I_{22}^{C_2 *} s_2^{*2} + I_{33}^{C_2 *} s_3^{*2} \\ &= m_2 [(R^2/4 + L^2/12) \cos^2\theta + \tfrac{1}{2} R^2 \sin^2\theta] \\ I_{zz}^{C_2} &= I_{n^*n^*}^{C_2} = I_{11}^{C_2 *} n_1^{*2} + I_{22}^{C_2 *} n_2^{*2} + I_{33}^{C_2 *} n_3^{*2} \\ &= m_2 [(R^2/4 + L^2/12) \sin^2\theta + \tfrac{1}{2} R^2 \cos^2\theta] \\ I_{zx}^{C_2} &= I_{n^*s^*}^{C_2} = I_{11}^{C_2 *} n_1^* s_1^* + I_{22}^{C_2 *} n_2^* s_2^* + I_{33}^{C_2 *} n_3^* s_3^* \\ &= m_2 (R^2/4 - L^2/12) \sin\theta \cos\theta \end{aligned}$$

The elements of $[\,I^A\,]$ for the system are obtained using the parallel axes theorems:

$$I_{xx}^A = \textstyle\sum_{i=1}^2 [I_{xx}^{C_i} + m_i (y_{C_i}^2 + z_{C_i}^2)], \qquad I_{xy}^A = \sum_{i=1}^2 [I_{xy}^{C_i} - m_i x_{C_i} y_{C_i}]$$

$$I_{yy}^A = \textstyle\sum_{i=1}^2 [I_{yy}^{C_i} + m_i (z_{C_i}^2 + x_{C_i}^2)], \qquad I_{yz}^A = \sum_{i=1}^2 [I_{yz}^{C_i} - m_i y_{C_i} z_{C_i}]$$

$$I_{zz}^A = \textstyle\sum_{i=1}^2 [I_{zz}^{C_i} + m_i (x_{C_i}^2 + y_{C_i}^2)], \qquad I_{zx}^A = \sum_{i=1}^2 [I_{zx}^{C_i} - m_i z_{C_i} x_{C_i}]$$

The principal axes at A can be obtained sometimes by looking for a plane of mass symmetry of the body passing through A. The axis normal to such a plane is a principal axis at A. If two principal axes are known at A by the above method, then the third principal axis follows from mutual orthogonality of these axes. The principal axes at various points are listed with the planes of mass symmetry given in the parenthesis:

Body ① $A:\ \vec{\mathrm{i}}\ (yz)$, $J:\ \vec{\mathrm{e}}_1, \vec{\mathrm{e}}_2, \vec{\mathrm{e}}_3\ (12, 23)$, $K:\ \vec{\mathrm{i}}\ (yz)$

Body ② $A:\ \vec{\mathrm{j}}\ (xz)$, $D:\ \vec{\mathrm{e}}_1^{\,*}, \vec{\mathrm{e}}_2^{\,*}, \vec{\mathrm{e}}_3^{\,*}\ (1^*3^*, 2^*3^*)$, $P:\ \vec{\mathrm{e}}_1^{\,*}\ (2^*3^*)$, $Q:\ \vec{\mathrm{e}}_2^{\,*}\ (1^*3^*)$

Review 3A

1. The product of inertia can be negative. $I^A_{yz} = I^C_{yz} - my_Cz_C$.
2. $I^A_{yy} = I^C_{yy} + m(z^2_C + x^2_C),\ \ I^A_{nn} = I^A_{ij}n_in_j,\ \ I^A_{ns} = I^A_{ij}n_is_j,\ \ I^{A\prime}_{ij} = a_{ip}a_{jq}I^A_{pq}$.
3. An axis through A for which the two products of inertia elements associated with it vanish is a principal axis of inertia at A.
4. If an axis is a principal axis of inertia at C, then it is so at every point on it.
5. If an axis is a principal axis of inertia at B, then it may not be a principal axis of inertia at other points on itself.
6. A normal to a plane of mass symmetry is a principal axis of inertia at the point where it intersects it.
7. At any point on the axis of a body of revolution, this axis and all lines normal to it are principal axes of inertia.
8. The principal axes of inertia at a material point of a rigid body do not change their orientation with respect to the body as it moves.
9. $I^A_{xx} + I^A_{yy} + I^A_{zz} = I^A_{x'x'} + I^A_{y'y'} + I^A_{z'z'},\ \ I^A_{zz} \leq I^A_{xx} + I^A_{yy}$.
10. $I^A_{zz} = I^A_{xx} + I^A_{yy}$ is *only* valid for mass distributed in xy plane, *and* only if A is in it.
11. Two equal principal moments of inertia at $A \Rightarrow$ every line through A in the plane of the corresponding principal axes is a principal axis for A, with the same moment of inertia.
12. Equal principal moments of inertia for $A \Rightarrow$ every axis is a principal axis at A.
13. For point A of rigid body: $\vec{H}_A = I^A_{ij}\omega_j\vec{e}_i$, for any B: $\vec{H}_B = I^C_{ij}\omega_j\vec{e}_i + \vec{r}_{CB} \times m\vec{v}_{CB}$
14. For point A of rigid body, $\vec{H}_A \parallel \vec{\omega}$ *only if* $\vec{\omega}$ is parallel to a principal axis at A.
15. Even for plane motion of a rigid body, for a point A on it, in general, $\vec{H}_A \nparallel \vec{\omega}$.

3.3 Concept Review Questions 3A

1. (a) Can I^A_{11}, I^A_{22}, I^A_{33} have values (i) $50, 20, 30$, (ii) $50, 10, 30$, (iii) $50, 22, 35$? (b) Of all the moments of inertia about a set of parallel lines, the moment of inertia about the line through the centre of mass is the least. (T/F) (c) The moment of inertia of a uniform thin plate with the shape of a regular polygon of n sides about any axis in its plane through its centre of mass is the same. (T/F) (d) For a body with its points on the z-axis being centres of symmetry, i.e., dm at $(x, y, z) \Rightarrow dm$ at $(-x, -y, z)$ (for example, a body with a parallelogram cross-section of variable size), z-axis is a principal axis of inertia at all points on itself. (T/F) [(a) (i) T (ii) F (iii) T (b) T (c) T (d) T]

2. (a) For a thin quarter-circular plate (Fig. Q3.2a) of mass m, find I^O_{xx}, I^O_{zz}, I^O_{xy}, I^A_{xx}, I^A_{zz}, I^A_{xy} and the principal axes of inertia at O. (b) Find I^A_{xx}, I^A_{xz}, I^A_{xy} for a thin plate with mass σ kg/m^2 (Fig. Q3.2b). (c) Given $I^O_{xy} = -30$, $I^O_{yy} = 140\,\text{kg.m}^2$ for a body of mass 10 kg (Fig. Q3.2c), find I^A_{xy}, I^A_{yy}.

[(a) $mR^2[1/4,\ 1/2,\ -1/2\pi,\ 1/4,\ 3/2 - 8/3\pi,\ 5/6\pi]$, $\vec{\text{k}}$, $(\vec{\text{j}} \pm \vec{\text{i}})/\sqrt{2}$
(b) $[h^2/3(n+1),\ 0,\ -bh/4(n+1)]\sigma bh$ (c) -60, $90\,\text{kg. m}^2$]

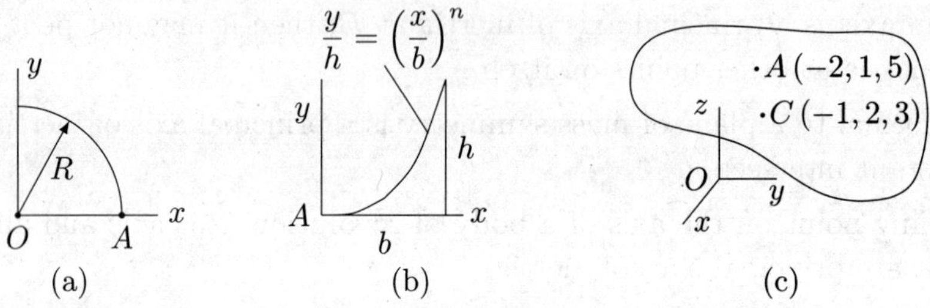

Fig. Q3.2

3. (a) At point A of a body: $I^A_{xx} = 5$, $I^A_{yy} = 8$, $I^A_{zz} = 4$, $I^A_{xy} = 0$, $I^A_{xz} = 2$, $I^A_{yz} = 0$. Which of the axes x, y, z is a principal axis of inertia at A? (b) Consider a set of axes parallel to the x, y, z-axes at A, B, C, D, O in Fig. Q3.3. Specify for each point whether any of these axes is a principal axis. [(a) y (b) **1.** A, D, O: z **2.** A: x, y, z, B: y, C: x, D none **3.** A: y, B: x, y, z, D none **4.** A: x, y, z]

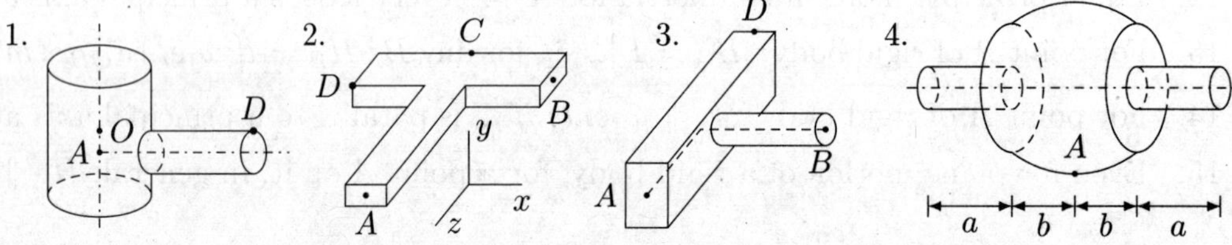

Fig. Q3.3

3.4 Practice Problems 3A

1. (a) For a rigid body with equal principal moments of inertia at C, $\vec{H}_C \parallel \vec{\omega}$ for all $\vec{\omega}$. (T/F). Find R/L of a cylinder so that $\vec{H}_C \parallel \vec{\omega},\ \forall\, \vec{\omega}$. (b) For a rigid body with $\vec{\omega}$ normal to the principal axis z at C, find the condition which the moments of inertia should satisfy so that $\vec{H}_C \parallel \vec{\omega}$. (c) A thin disc of radius R has a light axle fixed to its centre C whose other end is A. Find the distance CA so that $\vec{H}_A \parallel \vec{\omega},\ \forall\, \vec{\omega}$. [(a) T, $1/\sqrt{3}$ (b) $I^C_{xx} = I^C_{yy}$ (c) $\frac{1}{2}R$]

2. (a) Find I^A_{xx}, I^A_{xz}, I^A_{xy} for a plate with density σ kg/m^2 (Fig. P.3.2a). (b) Find I^A_{xx}, I^A_{xz} for a body with base ① of 20 kg and top ② of 60 kg (Fig. P3.2b). (c) Find I^O_{22}, I^O_{12} of a square plate of density σ kg/m^2 with a hole (Fig. P3.2c). Find the principal axes of inertia at its centre of mass. (d) Find I^C_{zz}, I^C_{xz}, I^C_{yz}, I^A_{zz}, I^A_{xz}, I^A_{yz} for a hemisphere of mass m (Fig. P3.2d).

[(a) $\frac{1}{3}\sigma a^3 R - \frac{1}{4}\sigma\pi R^2[a^2 + \frac{1}{4}R^2 - 8aR/3\pi]$, 0, $-\frac{1}{4}\sigma a^2 R^2 + \sigma R^3[(6\pi - 8)a - 5R]/24$

(c) $\frac{7}{12}\sigma a^4 - \pi R^2 \sigma(\frac{1}{4}R^2 + \frac{1}{2}a^2)$, $-\frac{1}{4}\pi\sigma a^2 R^2$, $\parallel$ diagonals & $\perp$ plane

(b) 2521.7, -1420 kg. m^2 (d) (19.84, 0, 4.32, 25.6, 0, -9.6) m]

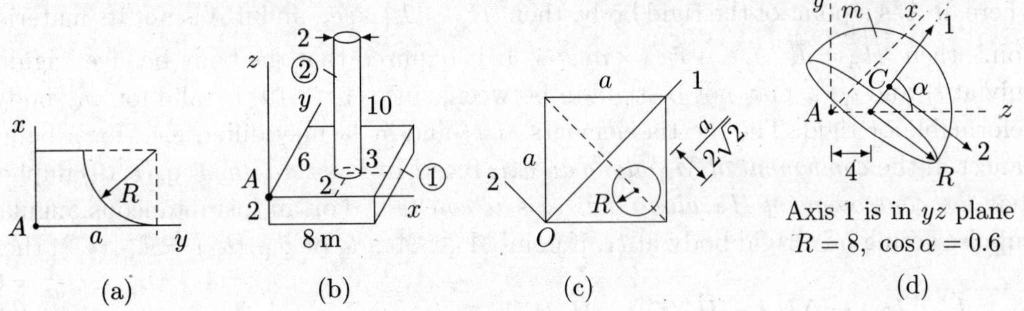

Fig. P3.2

3. Find $[I^A]$ for the bodies in Fig. P3.3 with reference to the axes x, y, z. List the principal axes of inertia, at A by inspection. [I^A_{xx}, I^A_{yy}, I^A_{zz}, I^A_{xy}, I^A_{yz}, I^A_{zx} =

1. $\frac{1}{4}m_1R^2 + \frac{1}{2}m_2L^2(1 + \sin^2\theta) + m_3[\frac{1}{12}b^2 + (\frac{1}{2}b + L\sin\theta)^2] + m_4[\frac{1}{4}r^2 + \frac{1}{12}a^2 + (L\sin\theta + b + \frac{1}{2}a)^2]$, I^A_{xx}, $\frac{1}{2}m_1R^2 + m_2L^2\cos^2\theta + \frac{1}{2}m_4r^2, 0, 0, 0$

2. $222.33\pi\rho r^5$, $222.33\pi\rho r^5$, $103\pi\rho r^5$, 0, 0, 0

3. $\rho abc[\frac{1}{30}(2b^2 + c^2)$, $\frac{1}{30}(2a^2 + c^2)$, $\frac{1}{15}(a^2 + b^2)$, $-ab/20$, $-bc/40$, $-ac/40]$,

principal axes: 1. z and all lines $\perp$ it 2. z and all lines $\perp$ it 3. none]

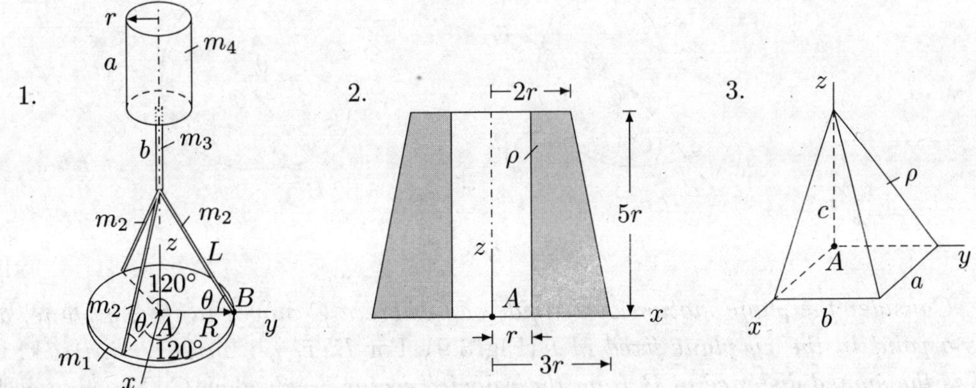

Fig. P3.3

B. EQUATIONS FOR DYNAMICS OF A RIGID BODY

3.5 Angular Impulse–Moment of Momentum Relation for a Rigid Body

The angular impulse–moment of momentum relations (2.53) and (2.54) for a rigid body: If point $A \ni A \equiv C$ or $\vec{a}_{A|I} = \vec{0}$ or $\vec{a}_{A|I}$ is along AC, $\forall t \in (t_1, t_2)$, then

$$\vec{I}_{\text{ang}_A}(t_1, t_2) = \Delta\vec{H}_A = \vec{H}_A(t_2) - \vec{H}_A(t_1) \tag{3.19}$$

where, if A is a point of the rigid body, then $\vec{H}_A = I^A_{ij}\,\omega_j\,\vec{e}_i$, and if A is not its material point, then $\vec{H}_A = I^C_{ij}\,\omega_j\,\vec{e}_i + \vec{r}_{CA} \times m\,\vec{v}_{CA}$. It is required that the body behaves rigidly only at t_1 and t_2, it may not be rigid in-between, since eq. (3.19) is valid for any body, deformable or rigid. The inertia elements at t_1 and at t_2 may differ, e.g., for a ballet dancer. If the *component of $\vec{M}_A$ along an axis* **fixed in** *I is zero, then* eq. (3.19) implies that *the component of $\vec{H}_A$ along that axis is constant.* For an instantaneous angular impulse acting on a rigid body at t_1, if point $A \ni A \equiv C$ or $\vec{v}_{A|I}(t_1^+) = \vec{v}_{A|I}(t_1^-)$, then

$$\vec{I}_{\text{ang}_A}(t_1) = \Delta\vec{H}_A = \vec{H}_A(t_1^+) - \vec{H}_A(t_1^-) \tag{3.20}$$

Let the axes x, y, z be attached to a rigid body and *let $\vec{k}$ remain constant relative to $I \Rightarrow$* $\vec{\omega} = \omega\,\vec{k}$. For *a material point A of the rigid body* $\ni A \equiv C$ or $\vec{a}_{A|I} = \vec{0}$ or $\vec{a}_{A|I}$ is along AC, the component of eq. (3.19) along $\vec{k}$ (fixed in I) yields

$$I_{\text{ang}_{A_z}}(t_1, t_2) = \int_{t_1}^{t_2} M_{A_z}(t)\,dt = [I^A_{zz}\omega]|_{t_2} - [I^A_{zz}\omega]|_{t_1} \tag{3.21}$$

For rotation of a rigid body about a fixed axis, eq. (3.21) is valid for any point O on it.

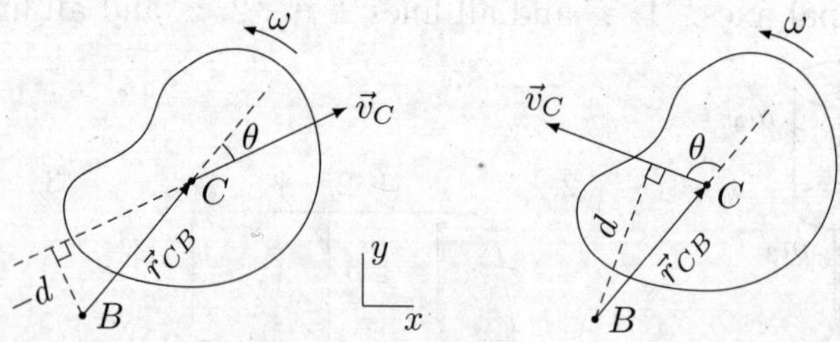

Fig. 3.9

Consider the plane motion of a rigid body wherein C moves in the xy plane, and B is a point in the xy plane fixed in I (Fig. 3.9). For B, $\vec{r}_{CB} \times m\vec{v}_{CB} = mv_C d\,\vec{k}$, where *$d$ is the signed distance of B from the velocity vector $\vec{v}_C$ through C*. The sign assigned is positive if the cross-product is in the direction $\vec{k}$ and negative otherwise. $\Rightarrow$ The

z-component of the angular impulse–moment of momentum relation for B is:

$$I_{\text{ang}B_z}(t_1, t_2) = \int_{t_1}^{t_2} M_{Bz}(t)\,dt = \Delta H_{Bz} = [I_{zz}^C \omega + m v_C d]|_{t_1}^{t_2} \tag{3.22}$$

The conservation relations corresponding to eqs (3.21) and (3.22) are:

(i) If $I_{\text{ang}A_z}(t_1, t_2) = 0,$ then $\left[I_{zz}^A \omega\right]\Big|_{t_2} = \left[I_{zz}^A \omega\right]\Big|_{t_1}.$ (3.23)

(ii) If $I_{\text{ang}B_z}(t_1, t_2) = 0,$ then $\left[I_{zz}^C \omega + m v_C d\right]\Big|_{t_2} = \left[I_{zz}^C \omega + m v_C d\right]\Big|_{t_1}.$ (3.24)

For an instantaneous angular impulse at t_1, $[I^A]$ is unaltered as the position remains the same, and for a *point A of the rigid body* $\ni A \equiv C$ or $\vec{v}_{A|I}(t_1^+) = \vec{v}_{A|I}(t_1^-)$, eq. (3.21)

$$\Rightarrow \quad I_{\text{ang}A_z}(t_1) = I_{zz}^A \left[\omega(t_1^+) - \omega(t_1^-)\right] \tag{3.25}$$

and for a point $B \ni \vec{v}_{B|I}(t_1^+) = \vec{v}_{B|I}(t_1^-)$, eq. (3.22)

$$\Rightarrow \quad I_{\text{ang}B_z}(t_1) = I_{zz}^C \left[\omega(t_1^+) - \omega(t_1^-)\right] + md\left[v_C(t_1^+) - v_C(t_1^-)\right] \tag{3.26}$$

Example 3.5: A thin plate of mass m with centre of mass C and axis of symmetry x hangs vertically from a ball-and-socket joint (Fig. E3.5a). The radii of gyration about principal axes x and y at O are k_x and k_y. A bullet of mass m_1 moving at velocity $v\vec{k}$ hits the plate at $B(a, b, 0)$ and gets embedded in it. Find the angular velocity of the plate just after the impact and the impulsive reaction at O during impact.

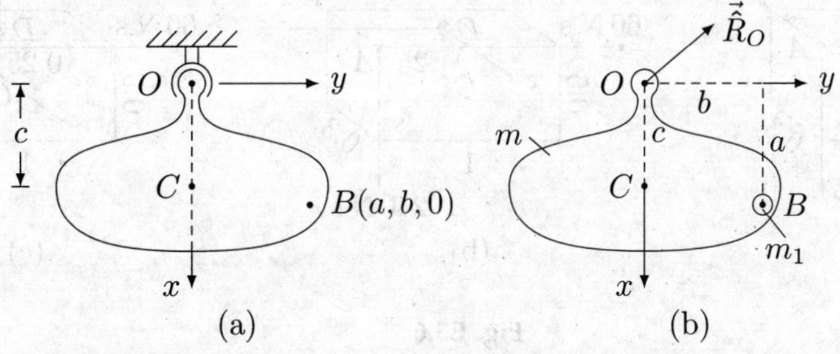

Fig. E3.5

Solution: Let $\vec{\omega}' = \omega_x' \vec{i} + \omega_y' \vec{j} + \omega_z' \vec{k}$ be the angular velocity just after impact, with

$$I_{xx}^O = mk_x^2 + m_1 b^2, \qquad I_{yy}^O = mk_y^2 + m_1 a^2, \qquad I_{xy}^O = -m_1 ab$$

$$I_{zz}^O = I_{xx}^O + I_{yy}^O = m(k_x^2 + k_y^2) + m_1(a^2 + b^2), \qquad I_{xz}^O = I_{yz}^O = 0$$

For bullet + plate (Fig. E3.5b), $\vec{I}_{\text{ang}O} = \vec{0} \Rightarrow \vec{H}_O' = \vec{H}_O$, which yields $\vec{\omega}'$. $\vec{\hat{R}}_O$ is obtained using $\vec{I} = \Delta\vec{p}$.

$$\vec{H}_O = \overrightarrow{OB} \times m_1 v \vec{k} = (a\vec{i} + b\vec{j}) \times m_1 v \vec{k} = m_1 v b \vec{i} - m_1 v a \vec{j}$$

$$[\vec{H}'_O] = \begin{bmatrix} I^O_{xx} & I^O_{xy} & 0 \\ I^O_{xy} & I^O_{yy} & 0 \\ O & 0 & I^O_{zz} \end{bmatrix} \begin{bmatrix} \omega'_x \\ \omega'_y \\ \omega'_z \end{bmatrix} = m_1 v \begin{bmatrix} b \\ -a \\ 0 \end{bmatrix}$$

$$\Rightarrow \quad \vec{\omega}' = \frac{m_1 v}{I^O_{xx} I^O_{yy} - I^{O\,2}_{xy}} \left[\left(b I^O_{yy} + a I^O_{xy} \right) \vec{i} - \left(b I^O_{xy} + a I^O_{xx} \right) \vec{j} \right]$$

$$= \frac{m_1 v}{m + m_1 (a^2/k_y^2 + b^2/k_x^2)} \left[\frac{b}{k_x^2} \vec{i} - \frac{a}{k_y^2} \vec{j} \right]$$

$$\vec{I} = \Delta\vec{p} \quad \Rightarrow \quad \hat{\vec{R}}_O = m\vec{\omega}' \times c\vec{i} + m_1 [\vec{\omega}' \times (a\vec{i} + b\vec{j}) - v\vec{k}]$$

Example 3.6: (a) A uniform block of mass $m = 120$ kg is supported at D (the centre of its top face) by a ball-and-socket joint (Fig. E3.6a). C is its centre of mass. It is subjected to 3 instantaneous impulses at $t = 0$ with $\hat{P} = 240$ N.s. Find the impulsive reaction at D. Find $\vec{v}_C(0^+)$ and $\vec{\omega}(0^+)$. (b) Rework part (a) if the block is supported at D by a vertical inextensible cable for two cases: (1) $\hat{P} = 240$ N.s, (2) $\hat{P} = -240$ N.s. (c) Rework part (b) if the block is supported at D by a spring of stiffness 120 kN/m.

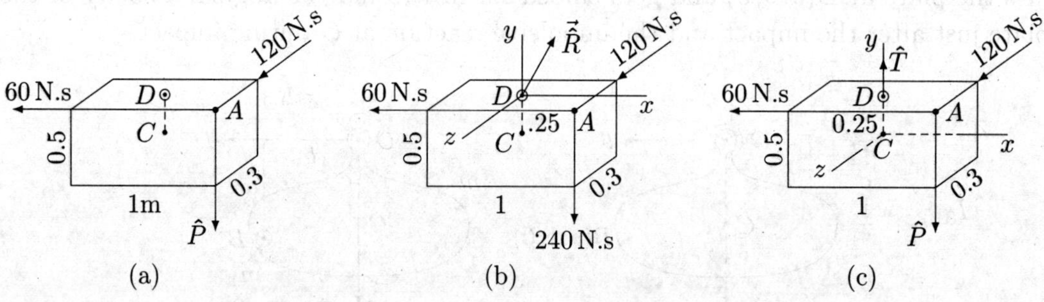

Fig. E3.6

Solution: (a) x, y and z are the principal axes of inertia at fixed point D with

$$I^D_{xx} = 120[(0.5^2 + 0.3^2)/12 + 0.25^2] = 10.9\,\text{kg.m}^2$$

$$I^D_{yy} = 120(0.3^2 + 1^2)/12 = 10.9\,\text{kg.m}^2$$

$$I^D_{zz} = 120[(1^2 + 0.5^2)/12 + 0.25^2] = 20\,\text{kg.m}^2, \qquad \overrightarrow{DA} = 0.5\vec{i} + 0.15\vec{k}\ \text{m}$$

$$\vec{I}_{\text{ang}_D} = \overrightarrow{DA} \times (-60\vec{i} - 240\vec{j} + 120\vec{k}) = 36\vec{i} - 69\vec{j} - 120\vec{k}\ \text{Nm.s}$$

Let $\vec{\omega}(0^+) = \omega_x \vec{\mathrm{i}} + \omega_y \vec{\mathrm{j}} + \omega_z \vec{\mathrm{k}}$. As $\hat{\vec{R}}$ passes through fixed point D (Fig. E3.6b), $\vec{\omega}(0^+)$ is obtained using $\Delta\vec{H}_D = \vec{I}_{\mathrm{ang}_D}$:

$$\begin{aligned}
I^D_{xx}\omega_x \vec{\mathrm{i}} + I^D_{yy}\omega_y \vec{\mathrm{j}} + I^D_{zz}\omega_z \vec{\mathrm{k}} &= 10.9\omega_x \vec{\mathrm{i}} + 10.9\omega_y \vec{\mathrm{j}} + 20\omega_z \vec{\mathrm{k}} \\
&= 36\vec{\mathrm{i}} - 69\vec{\mathrm{j}} - 120\vec{\mathrm{k}} \\
\Rightarrow \quad \omega_x &= 36/10.9 = 3.303, \quad \omega_y = -69/10.9 = -6.330, \quad \omega_z = -6 \\
\vec{\omega} &= 3.303\vec{\mathrm{i}} - 6.330\vec{\mathrm{j}} - 6\vec{\mathrm{k}} \text{ rad/s} \\
\vec{v}_C &= \vec{v}_D + \vec{\omega} \times \overrightarrow{DC} = (3.303\vec{\mathrm{i}} - 6.330\vec{\mathrm{j}} - 6\vec{\mathrm{k}}) \times (-0.25\vec{\mathrm{j}}) \\
&= -1.5\vec{\mathrm{i}} - 0.8258\vec{\mathrm{k}} \text{ m/s}
\end{aligned}$$

The impulse–momentum relation $m\Delta\vec{v}_C = \vec{I}$:

$$120\vec{v}_C = \hat{\vec{R}} - 60\vec{\mathrm{i}} - 240\vec{\mathrm{j}} + 120\vec{\mathrm{k}} \quad \Rightarrow \quad \hat{\vec{R}} = -120\vec{\mathrm{i}} + 240\vec{\mathrm{j}} - 219.1\vec{\mathrm{k}} \text{ N.s}$$

(b) Let the instantaneous impulse applied by the cable be $\hat{T}\vec{\mathrm{j}}$ ($\hat{T} \geq 0$) (Fig. E3.6c). The angular impulse–moment of momentum equation is applied at C and not at D, since D may not be a valid point for application of this equation as $\vec{v}_D$ may change. The principal axes at C are x, y, z axes with $\overrightarrow{CA} = 0.5\vec{\mathrm{i}} + 0.25\vec{\mathrm{j}} + 0.15\vec{\mathrm{k}}$ m,

$$\begin{aligned}
I^C_{xx} &= 120(0.5^2 + 0.3^2)/12 = 3.4 \text{ kg.m}^2 \\
I^C_{yy} &= 120(0.3^2 + 1^2)/12 = 10.9 \text{ kg.m}^2 \\
I^C_{zz} &= 120(1^2 + 0.5^2)/12 = 12.5 \text{ kg.m}^2 \\
\vec{I}_{\mathrm{ang}_C} &= \overrightarrow{CA} \times (-60\vec{\mathrm{i}} - \hat{P}\vec{\mathrm{j}} + 120\vec{\mathrm{k}}) = (30 + 0.15\hat{P})\vec{\mathrm{i}} - 69\vec{\mathrm{j}} + (15 - 0.5\hat{P})\vec{\mathrm{k}} \\
&= \Delta\vec{H}_C = I^C_{xx}\omega_x \vec{\mathrm{i}} + I^C_{yy}\omega_y \vec{\mathrm{j}} + I^C_{zz}\omega_z \vec{\mathrm{k}} = 3.4\omega_x \vec{\mathrm{i}} + 10.9\omega_y \vec{\mathrm{j}} + 12.5\omega_z \vec{\mathrm{k}} \\
\Rightarrow \quad \omega_x &= (30 + 0.15\hat{P})/3.4, \quad \omega_y = -6.330, \quad \omega_z = (15 - 0.5\hat{P})/12.5 \\
\vec{\omega} &= [(30 + 0.15\hat{P})/3.4]\vec{\mathrm{i}} - 6.330\vec{\mathrm{j}} + [(15 - 0.5\hat{P})/12.5]\vec{\mathrm{k}} \text{ rad/s} \qquad (1) \\
\vec{v}_C &= \vec{I}/m = [-60\vec{\mathrm{i}} + (\hat{T} - \hat{P})\vec{\mathrm{j}} + 120\vec{\mathrm{k}}]/120 \text{ m/s} \qquad (2) \\
\vec{v}_D &= \vec{v}_C + \vec{\omega} \times \overrightarrow{CD} = \vec{v}_C + \vec{\omega} \times 0.25\vec{\mathrm{j}} \\
\Rightarrow \quad & v_{D_y} = v_{C_y} = (\hat{T} - \hat{P})/120 \qquad (3)
\end{aligned}$$

1. $\hat{P} = 240$ N.s. *Assume* that the cable remains taut, then $v_{D_y} = 0$ and eq. (3)

yields $\hat{T} = \hat{P} = 240$ N.s $> 0 \Rightarrow$ assumption is valid. Equations (1), (2) $\Rightarrow$

$$\vec{\omega} = 19.41\,\vec{\imath} - 6.330\,\vec{\jmath} - 8.4\,\vec{k} \text{ rad/s}, \qquad \vec{v}_C = -0.5\,\vec{\imath} + \vec{k} \text{ m/s}.$$

2. $\hat{P} = -240$ N.s. *Assume* that the cable remains taut. Then $v_{Dy} = 0$ and eq. (3) yields $\hat{T} = \hat{P} = -240$ N.s $\not> 0 \Rightarrow$ assumption is not valid.

   Thus, the cable becomes slack with $\hat{T} = 0$. Equations (1), (2) $\Rightarrow$

$$\vec{\omega} = -1.765\,\vec{\imath} - 6.330\,\vec{\jmath} + 10.8\,\vec{k} \text{ rad/s}, \qquad \vec{v}_C = -0.5\,\vec{\imath} + 2\,\vec{\jmath} + \vec{k} \text{ m/s}.$$

(c) The spring is extensible and does not provide any impulsive reaction since its stiffness is finite. Hence, the solution is given by eqs (1), (2) with $\hat{T} = 0$.

Example 3.7: (a) A space probe, with solar panel open (Fig. E3.7a), has angular velocity $\omega_0\,\vec{k}$. There is no external moment on it. Find the angular velocity at the instant an identical solar panel is opened by an internal mechanism into the configuration of Fig. E3.7b. The main body ① has mass M and its principal radii of gyration at C_1 are: $k_x = k_y = 0.7R$, $k_z = 0.9R$. The panel ② has mass $m = 0.1M$ and two of its principal radii of gyration at C_2 with respect to axes $\vec{e}_i$ are: $k_1 = 0.4R$, $k_3 = 0.5R$. (b) Rework if this system is on the ground and rotates about the fixed vertical z-axis.

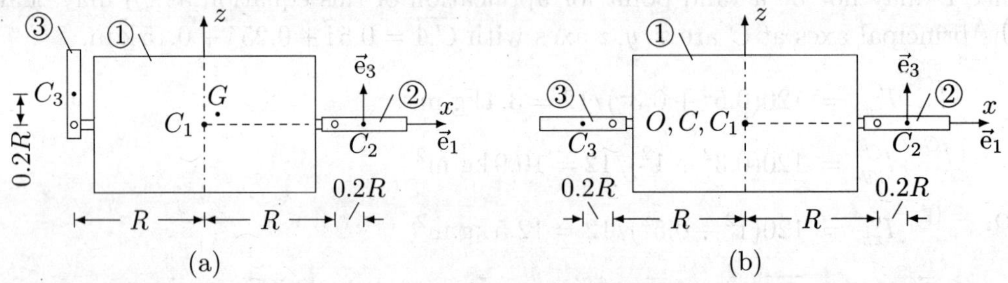

Fig. E3.7

Solution: (a) Let G be the centre of mass in the configuration (a). $C_1(0,\,0,\,0)$, $C_2(1.2R,\,0,\,0)$, $C_3(-R,\,0,\,0.2R)$, $y_G = 0$.

$$x_G = \frac{1.2Rm - Rm}{M + 2m} = 0.01667R, \qquad z_G = \frac{0.2Rm}{M + 2m} = 0.01667R$$

$$\Rightarrow \quad G(0.01667R,\,0,\,0.01667R), \qquad C_1G = (0.01667R,\,0,\,0.01667R)$$

$$C_2G = (-1.1833R,\,0,\,0.01667R), \qquad C_3G = (1.0167R,\,0,\,-0.1833R)$$

Since x, y, z are principal axes of inertia of bodies ①, ②, ③ at C_1, C_2, C_3, using the parallel axes theorems: $I^G_{yz} = 0$,

$$I_{xz}^G = -M(0.01667R)(0.01667R) - m(-1.1833R)(0.01667R)$$
$$-m(1.0167R)(-0.1833R) = 0.02033MR^2$$
$$I_{zz}^G = M[k_z^2 + (0.01667R)^2] + m[k_3^2 + (1.1833R)^2]$$
$$+m[k_1^2 + (1.0167R)^2] = 1.0947MR^2$$
$$[\vec{H}_G] = [I^G][0\ 0\ \omega_0]^{\mathrm{T}} \quad \Rightarrow \quad \vec{H}_G = I_{xz}^G\omega_0\,\vec{\mathrm{i}} + I_{yz}^G\omega_0\,\vec{\mathrm{j}} + I_{zz}^G\omega_0\,\vec{\mathrm{k}}$$

Let $\vec{\omega} = \omega_x\,\vec{\mathrm{i}} + \omega_y\,\vec{\mathrm{j}} + \omega_z\,\vec{\mathrm{k}}$ be the angular velocity just after opening the panel. In configuration (b), the centre of mass C is at O with x, y, z as principal axes at C and

$$I_{xx}^C = Mk_x^2 + 2mk_1^2 = 0.522MR^2$$
$$I_{zz}^C = Mk_z^2 + 2m[k_3^2 + (1.2R)^2] = 1.148MR^2$$

Since the external moment is zero, $\vec{H}_C = \vec{H}_G$:

$$\vec{H}_C = I_{xx}^C\omega_x\,\vec{\mathrm{i}} + I_{yy}^C\omega_y\,\vec{\mathrm{j}} + I_{zz}^C\omega_z\,\vec{\mathrm{k}} = I_{xz}^G\omega_0\,\vec{\mathrm{i}} + I_{yz}^G\omega_0\,\vec{\mathrm{j}} + I_{zz}^G\omega_0\,\vec{\mathrm{k}}$$
$$\Rightarrow \quad \omega_x = I_{xz}^G\omega_0/I_{xx}^C = 0.03895\omega_0, \qquad \omega_y = 0, \qquad \omega_z = I_{zz}^G\omega_0/I_{zz}^C = 0.9536\omega_0$$

(b) After opening of the panel, let $\vec{\omega} = \omega\,\vec{\mathrm{k}}$. Initial and final I_{zz}^O are:

$$(I_{zz}^O)_i = Mk_z^2 + m[k_3^2 + (1.2R)^2] + m[k_1^2 + R^2] = 1.095MR^2$$
$$(I_{zz}^O)_f = Mk_z^2 + 2m[k_3^2 + (1.2R)^2] = 1.148MR^2$$

$$M_{Oz} = 0 \Rightarrow H_{Oz} = \text{constant}: \quad (I_{zz}^O)_f\,\omega = (I_{zz}^O)_i\,\omega_0 \Rightarrow$$
$$\omega = (I_{zz}^O)_i\,\omega_0/(I_{zz}^O)_f = 0.9538\omega_0$$

Example 3.8: A satellite of mass M and centre of mass at C is in translation (Fig. E3.8a) when impacted by a meteoroid of mass m at $P(a, b, c)$ with velocity $\vec{v} = v_1\,\vec{\mathrm{i}} + v_2\,\vec{\mathrm{j}} + v_3\,\vec{\mathrm{k}}$ relative to it. The radii of gyration of the satellite at C about principal axes x, y, z are k_1, k_2, k_3. The meteoroid gets embedded in the satellite. (a) Find the angular velocity $\vec{\omega}'$ of the satellite and the change $\vec{v}_C'$ in the velocity of C just after impact. (b) Simplify the results for $m \ll M$. Note that for an axisymmetric satellite, $k_1 = k_2$ and without loss of generality, we can take $a = 0$. (c) Rework if instead, a bullet of mass m is fired from P with velocity $\vec{v}$ relative to P. In this case, the data M and k_i for the satellite exclude the contribution of m.

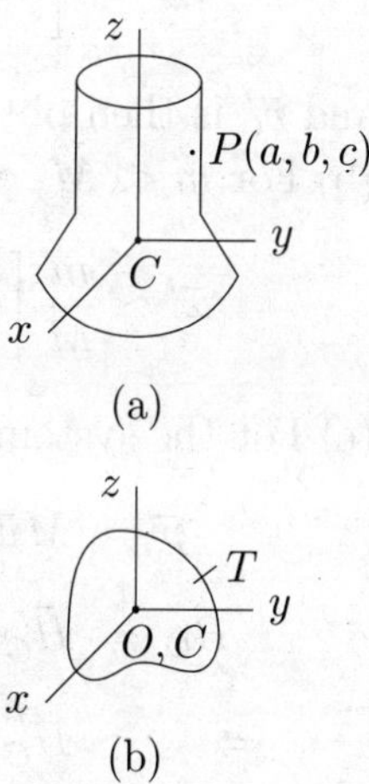

Fig. E3.8

Solution: (a) We analyse the dynamics with respect to an inertial frame T translating with the initial velocity of the satellite (Fig. E3.8b). For the satellite, $[\vec{H}'_C] = [I^C][\vec{\omega}']$ and $\vec{v}'_P = \vec{v}'_C + \vec{\omega}' \times \vec{r}_P$. For the system of satellite and meteoroid, $\vec{I} = \vec{0}$ and for the point O (coincident with C) fixed in frame T, $\vec{I}_{\text{ang}_O} = \vec{0} \Rightarrow \vec{p},\ \vec{H}_O$ are conserved:

$$\vec{p} = M\vec{v}'_C + m[\vec{v}'_C + \vec{\omega}' \times \vec{r}_P) = m\vec{v}$$

$$\Rightarrow \quad \vec{v}'_C = m(\vec{v} - \vec{\omega}' \times \vec{r}_P)/(M+m) \tag{1}$$

$$\vec{v}'_P = \vec{v}'_C + \vec{\omega}' \times \vec{r}_P = (m\vec{v} + M\vec{\omega}' \times \vec{r}_P)/(M+m)$$

$$\vec{H}_O = \vec{H}'_C + \vec{r}_P \times m\,\frac{m\vec{v} + M\vec{\omega}' \times \vec{r}_P}{M+m} = \vec{r}_P \times m\vec{v}$$

$$\Rightarrow \quad \left(1 + \frac{m}{M}\right)\vec{H}'_C + \vec{r}_P \times (m\vec{\omega}' \times \vec{r}_P) = \vec{r}_P \times m\vec{v}$$

$$\left[\left(1 + \frac{m}{M}\right) I^C + I_m\right][\vec{\omega}'] = m[\vec{r}_P \times \vec{v}]$$

$$\Rightarrow \quad [\vec{\omega}'] = \left[\left(1 + \frac{m}{M}\right) I^C + I_m\right]^{-1} m[\,bv_3 - cv_2 \quad cv_1 - av_3 \quad av_2 - bv_1\,]^{\mathrm{T}} \tag{2}$$

where

$$[I^C] = \begin{bmatrix} Mk_1^2 & 0 & 0 \\ 0 & Mk_2^2 & 0 \\ 0 & 0 & Mk_3^2 \end{bmatrix}$$

$$[I_m] = \begin{bmatrix} m(b^2+c^2) & -mab & -mac \\ -mab & m(c^2+a^2) & -mbc \\ -mac & -mbc & m(a^2+b^2) \end{bmatrix}$$

and $\vec{v}'_C$ is then obtained using eq. (1).
(b) For $m \ll M$, eqs (2), (1) yield $\quad [\vec{\omega}'] \simeq [I^C]^{-1} m[\vec{r}_P \times \vec{v}] \ \Rightarrow$

$$\vec{\omega}' \simeq \frac{m}{M}\left[\frac{bv_3 - cv_2}{k_1^2}\,\vec{\imath} + \frac{cv_1 - av_3}{k_2^2}\,\vec{\jmath} + \frac{av_2 - bv_1}{k_3^2}\,\vec{k}\right], \qquad \vec{v}'_C \simeq \frac{m}{M}\vec{v}$$

(c) For the system of satellite and the bullet

$$\vec{p} = M\vec{v}'_C + m(\vec{v}'_C + \vec{\omega}' \times \vec{r}_P + \vec{v}) = \vec{0} \ \Rightarrow \ M\vec{v}'_C + m(\vec{v}'_C + \vec{\omega}' \times \vec{r}_P) = -m\vec{v}$$

$$\vec{H}_O = \vec{H}'_C + \vec{r}_P \times m(\vec{v}'_C + \vec{\omega}' \times \vec{r}_P + \vec{v}) = \vec{0}$$

$$\Rightarrow \quad \vec{H}'_C + \vec{r}_P \times m(\vec{v}'_C + \vec{\omega}' \times \vec{r}_P) = -\vec{r}_P \times m\vec{v}$$

These are as in case (a) with $-\vec{v}$ replacing $\vec{v}$ $\Rightarrow$ Same solution with $-\vec{v}$ replacing $\vec{v}$.

Example 3.9: An axisymmetric body of mass m and axial radius of gyration k is at rest (Fig. E3.9a). An impulsive force $\hat{P}$ acts in the vertical plane containing the centre of mass C. (a) Find ω' and v'_C of C just after the impulsive action (Fig. E3.9b). (b) Find h so that there is no slip. (c) If there is initial slip, find the angular velocity when slip stops. The coefficients of friction are μ_s, μ_k.

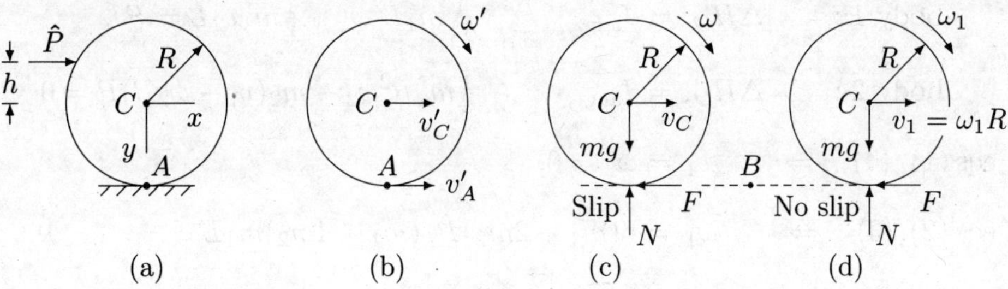

Fig. E3.9

Solution: (a) Normal $\hat{N} = 0$ since v_C is horizontal. Hence, there is no frictional impulsive reaction. $\vec{I} = \Delta\vec{p}$ and $I_{\text{ang}C_z} = I^C_{zz}\Delta\omega$ yield v'_C and ω'.

$$\hat{P} = mv'_C, \qquad \hat{P}h = mk^2\omega' \quad \Rightarrow \quad v'_C = \hat{P}/m, \qquad \omega' = \hat{P}h/mk^2$$

(b) For no slip, $v'_A = v'_C - \omega' R = \hat{P}/m - \hat{P}hR/mk^2 = 0 \quad \Rightarrow \quad h = k^2/R$. For a solid cylinder, solid sphere, cylindrical shell and spherical shell, $k^2 = R^2/2$, $2R^2/5$, R^2 and $2R^2/3$ respectively, and for no slip, $h = k^2/R = 0.5R$, $0.4R$, R and $2R/3$ respectively. (c) In subsequent motion, $N = mg$ (Fig. E3.9c). Hence, for a fixed point B, $I_{\text{ang}B_z} = 0$ $\Rightarrow H_{Bz}$ is conserved. Let $\omega_1 = v_1/R$ when the slip stops (Fig. E3.9d):

$$\begin{aligned} H_{Bz} &= mk^2\omega' + mv'_C R = mk^2\omega_1 + (m\omega_1 R)R \\ \Rightarrow \quad \omega_1 &= (k^2\omega' + v'_C R)/(k^2 + R^2) \end{aligned}$$

Example 3.10: A rod hinged to a disc is at rest on a smooth table (Fig. E3.10a). Find their angular velocities just after a blow is struck at A with impulsive force $\hat{P}$.

Solution: The FBDs and the kinematics just after the blow are shown in Fig. E3.10b.

$$\begin{aligned} \vec{v}_{C_2} &= \vec{v}_B + \vec{\omega}_2 \times \overrightarrow{BC}_2 = \vec{v}_{C_1} + \vec{\omega}_1 \times \overrightarrow{C_1B} + \vec{\omega}_2 \times \overrightarrow{BC}_2 \\ &= v_1\,\vec{\imath} + v_2\,\vec{\jmath} + \omega_1\,\vec{k} \times (-\tfrac{1}{2}L\,\vec{\imath}) + \omega_2\,\vec{k} \times R\,\vec{\jmath} \\ &= (v_1 - \omega_2 R)\,\vec{\imath} + (v_2 - \tfrac{1}{2}L\omega_1)\,\vec{\jmath} \end{aligned}$$

We set up 4 independent equations for $\omega_1, \omega_2, v_1, v_2$. Let O be a point (coincident with

B) fixed to the table. Using $\vec{H}_O = \vec{H}_C + \vec{r}_{CO} \times m\vec{v}_{CO}$:

$$\text{bodies } 1+2: \quad \Delta p_x = I_x, \qquad m_1 v_1 + m_2(v_1 - \omega_2 R) = 0 \tag{1}$$

$$\text{bodies } 1+2: \quad \Delta p_y = I_y, \qquad m_1 v_2 + m_2\left(v_2 - \tfrac{1}{2}\omega_1 L\right) = \hat{P} \tag{2}$$

$$\text{body } 1: \quad \Delta H_{Oz} = I_{\text{ang}_{Oz}}, \qquad \tfrac{1}{12} m_1 L^2 \omega_1 + \tfrac{1}{2} m_1 v_2 L = \hat{P}L \tag{3}$$

$$\text{body } 2: \quad \Delta H_{Oz} = I_{\text{ang}_{Oz}}, \qquad \tfrac{1}{2} m_2 R^2 \omega_2 - m_2(v_1 - \omega_2 R)R = 0 \tag{4}$$

$$\text{eqs (1), (4)} \quad \Rightarrow \quad v_1 = \omega_2 = 0,$$

$$\text{eqs (2), (3)} \quad \Rightarrow \quad \omega_1 = 6(m_1 + 2m_2)\hat{P}/(m_1 + 4m_2)m_1 L$$

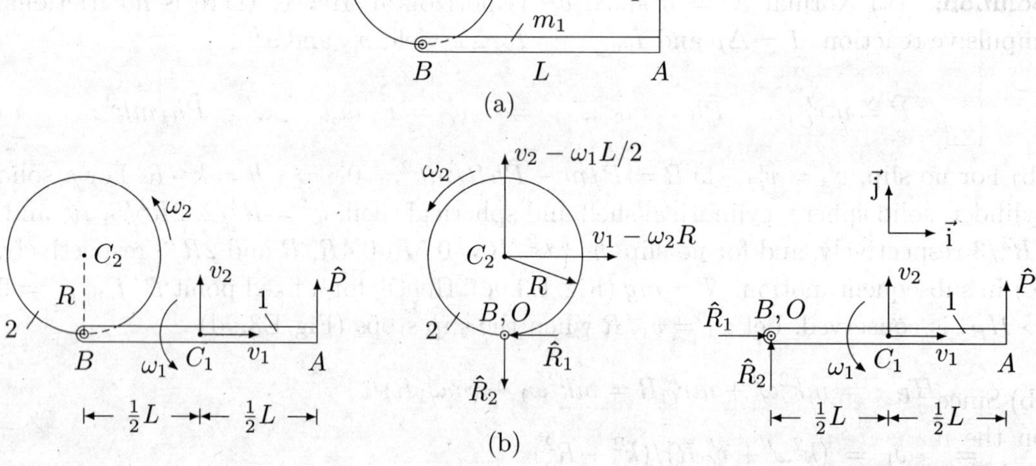

Fig. E3.10

Example 3.11: A head-on collision of two trolleys moving in opposite directions at speed v_0 occurs at O (Fig. E3.11a). Find the vertical velocity v of O at the instant the compression in the buffers is the maximum. Neglect the inertia of the wheels. Each trolley has mass m and radius of gyration k about the transverse horizontal axis through its centre of mass C. Consider the main frame to be (a) not spring-borne and (b) spring-borne.

Solution: (a) The instantaneous centre I of the main frame is at the intersection of the normals to $\vec{v}_D$ and $\vec{v}_O$ (Fig. E3.11b). $\vec{\omega} = (v/d)\,\vec{k}$. The FBD for lift off at B is shown in Fig. E3.11b. There is no frictional impulsive force at A since the wheels are inertia-less. The point G (coincident with I) fixed in the ground is a point of the body just after impact and $(I_{\mathrm{ang}_G})_z = 0 \Rightarrow$

$$H_{Gz} = I^G_{zz}\omega = m(k^2 + a^2 + h^2)v/d = mv_0h$$

$$\Rightarrow \quad v = dhv_0/(k^2 + a^2 + h^2)$$

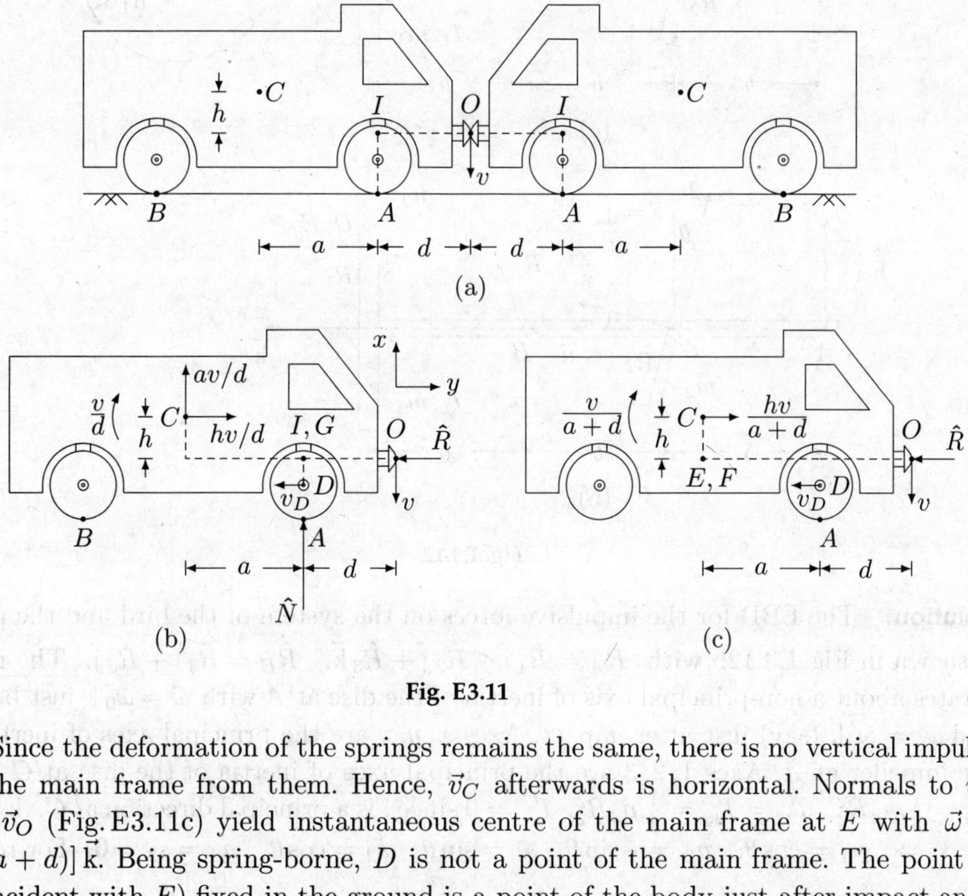

(a)

(b)

(c)

Fig. E3.11

(b) Since the deformation of the springs remains the same, there is no vertical impulse on the main frame from them. Hence, $\vec{v}_C$ afterwards is horizontal. Normals to $\vec{v}_C$ and $\vec{v}_O$ (Fig. E3.11c) yield instantaneous centre of the main frame at E with $\vec{\omega} = [v/(a+d)]\,\vec{k}$. Being spring-borne, D is not a point of the main frame. The point F (coincident with E) fixed in the ground is a point of the body just after impact and

$$(I_{\mathrm{ang}_F})_z = 0 \quad \Rightarrow \quad H_{Fz} = I^F_{zz}\omega = m(k^2 + h^2)v/(a+d) = mv_0h$$

$$\Rightarrow \quad v = (a+d)hv_0/(k^2 + h^2)$$

Example 3.12: The rotor shown in Fig. E3.12a consists of an impeller of axial moment of inertia I_1 and a thin disc with a slight misalignment of θ. It has angular velocity ω_0 just before a bird of mass m_0 hits the impeller at D. Model the bird as a mass-point and the impact as instantaneous. The support at B does not provide any axial thrust to the shaft. Find the impulsive reactions of the support if the bird gets embedded in the impeller.

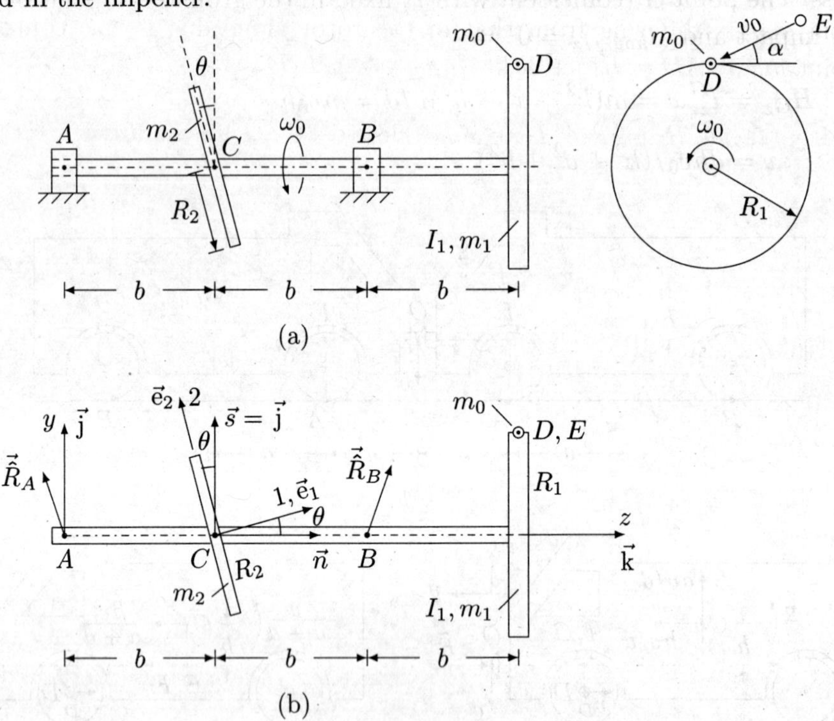

Fig. E3.12

Solution: The FBD for the impulsive forces on the system of the bird and the rotor is shown in Fig. E3.12b with $\hat{\vec{R}}_A = \hat{R}_1\vec{\imath} + \hat{R}_2\vec{\jmath} + \hat{R}_3\vec{k}$, $\hat{\vec{R}}_B = \hat{R}_4\vec{\imath} + \hat{R}_5\vec{\jmath}$. The rotor rotates about a non-principal axis of inertia of the disc at A with $\vec{\omega} = \omega_0\vec{k}$ just before and $\vec{\omega}' = \omega'\vec{k}$ (say) just after impact. Axes x, y, z are the principal axes of inertia of the impeller at A. Axes $1, 2, 3$ are the principal axes of inertia of the disc at C with $I_{11}^C = \frac{1}{2}m_2R_2^2$, $I_{22}^C = I_{33}^C = \frac{1}{4}m_2R_2^2$. $I_{xz}^C = 0$ since $\vec{\imath}$ is a principal direction at C. $\vec{k} = \vec{n}$, $\vec{\jmath} = \vec{s} \Rightarrow \quad n_1 = \cos\theta, \quad n_2 = -\sin\theta, \quad s_1 = \sin\theta, \quad s_2 = \cos\theta, \quad n_3 = s_3 = 0$. For rotor

$$\begin{aligned}
I_{xz}^A &= I_{xz}^C - m_2(0)b = 0 \\
I_{yz}^A &= I_{yz}^C - m_2(0)b = I_{yz}^C = I_{sn}^C \\
&= I_{11}^C s_1 n_1 + I_{22}^C s_2 n_2 + I_{33}^C s_3 n_3 = \tfrac{1}{4}m_2R_2^2 \sin\theta\cos\theta \\
I_{zz}^A &= I_1 + I_{11}^C n_1^2 + I_{22}^C n_2^2 + I_{33}^C n_3^2 = I_1 + \tfrac{1}{4}m_2R_2^2(\sin^2\theta + 2\cos^2\theta) \\
\vec{H}_A &= (I_{xz}^A\vec{\imath} + I_{yz}^A\vec{\jmath} + I_{zz}^A\vec{k})\omega
\end{aligned}$$

The unknowns, ω', $\hat{R}_1$, $\hat{R}_2$, $\hat{R}_3$, $\hat{R}_4$, $\hat{R}_5$, are obtained using $\vec{I}_{\text{ang}_A} = \Delta\vec{H}_A$, and

$\vec{I} = \Delta\vec{p}$ for the rotor + bird. After impact, the rotor + bird forms one rigid body with its inertia matrix differing from that of the rotor alone due to the contribution of mass-point m_0 at $\overrightarrow{AD} = R_1\vec{\jmath} + 3b\vec{k}$. $\vec{I}_{\mathrm{ang}_A} = \Delta\vec{H}_A$:

$$
\begin{aligned}
2b\vec{k} \times (\hat{R}_4\vec{\imath} + \hat{R}_5\vec{\jmath}) &= [\{I^A_{xz} - m_0(0)3b\}\vec{\imath} + \{I^A_{yz} - m_0R_1(3b)\}\vec{\jmath} \\
&\quad + \{I^A_{zz} + m_0(0^2 + R_1^2)\}\vec{k}]\omega' - [(I^A_{xz}\vec{\imath} + I^A_{yz}\vec{\jmath} + I^A_{zz}\vec{k})\omega_0 \\
&\quad + (R_1\vec{\jmath} + 3b\vec{k}) \times m_0v_0(-\cos\alpha\vec{\imath} - \sin\alpha\vec{\jmath})] \\
\Rightarrow \quad \vec{k}: \qquad 0 &= I^A_{zz}(\omega' - \omega_0) + m_0R_1^2\omega' - m_0v_0R_1\cos\alpha \\
\vec{\imath}: \qquad -2b\hat{R}_5 &= I^A_{xz}(\omega' - \omega_0) - 3m_0v_0b\sin\alpha \\
\vec{\jmath}: \qquad 2b\hat{R}_4 &= I^A_{yz}(\omega' - \omega_0) - 3m_0R_1b\omega' + 3m_0v_0b\cos\alpha
\end{aligned}
$$

which successively yield ω', $\hat{R}_5$, $\hat{R}_4$. $\vec{I} = \Delta\vec{p}$ for the rotor + bird yields

$$
\begin{aligned}
\hat{R}_1\vec{\imath} + \hat{R}_2\vec{\jmath} + \hat{R}_3\vec{k} + \hat{R}_4\vec{\imath} + \hat{R}_5\vec{\jmath} &= m_0(\vec{v}'_E - \vec{v}_E) \\
&= m_0[-\omega'R_1\vec{\imath} - v_0(-\cos\alpha\vec{\imath} - \sin\alpha\vec{\jmath})] \\
\Rightarrow \quad \vec{\imath}: \qquad \hat{R}_1 + \hat{R}_4 &= m_0(v_0\cos\alpha - \omega'R_1) \\
\vec{\jmath}: \qquad \hat{R}_2 + \hat{R}_5 &= m_0v_0\sin\alpha \\
\vec{k}: \qquad \hat{R}_3 &= 0
\end{aligned}
$$

$\hat{R}_1, \hat{R}_2$ are obtained from these equations.

3.6 Moment of Momentum Equations of Motion of a Rigid Body

Euler's axioms can be applied to a rigid body in two alternative ways:

1. Euler's axioms yield the equations of motion (Fig. 3.10b):

$$\vec{F} = m\vec{a}_{C|I}, \qquad \vec{M}_C = \dot{\vec{H}}_{C|I} \tag{3.27}$$

2. Equations (3.27) imply that the external force system (Fig. 3.10a) is equivalent to a kinetic system at C consisting of a 'force' $m\vec{a}_{C|I}$ and a couple of 'moment' $\dot{\vec{H}}_{C|I}$ (Fig. 3.10c). Hence, the moment of these equivalent systems about *any* point B of the body or outside, having arbitrary motion, can be equated:

$$\vec{M}_B = \dot{\vec{H}}_{C|I} + \vec{r}_{CB} \times m\vec{a}_{C|I} \tag{3.28}$$

where $\vec{M}_B$ is the moment of the external forces about B.

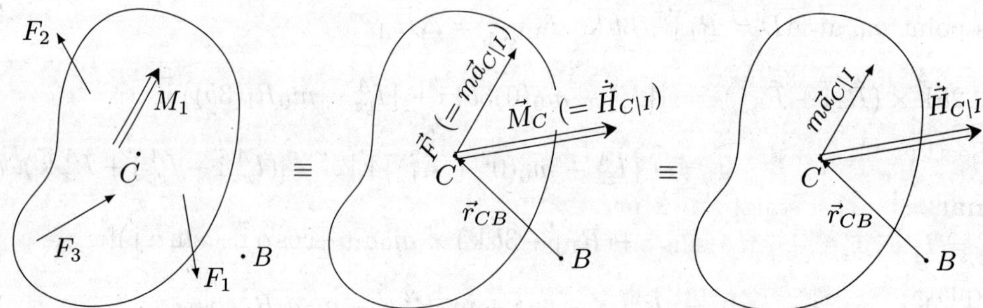

(a) External force system (b) Equivalent system at C (c) Equivalent kinetic system

Fig. 3.10

Most engineering components execute plane motion. Rotodynamic machines rotate about a fixed axis. Therefore, quite a lot of engineering dynamic problems concern plane motion. The moment of momentum equation (i) for the case in which $\vec{e}_3$ = constant, which includes plane motion, (ii) for the case of three-dimensional general rotation and (iii) an alternative equation for the case of general rotation of a body of revolution, are derived in Sections 3.6.1, 3.6.2 and 3.6.3.

3.6.1 Motion of a rigid body with $\vec{e}_3$ = constant, i.e., $\vec{\omega} = \omega\,\vec{e}_3$

If A *is a point of the rigid body* or its 3-dimensional massless extension, then eq. (3.3)

$$\Rightarrow \quad [\vec{H}_A] = \begin{bmatrix} I_{11}^A & I_{12}^A & I_{13}^A \\ I_{21}^A & I_{22}^A & I_{23}^A \\ I_{31}^A & I_{32}^A & I_{33}^A \end{bmatrix} \begin{bmatrix} 0 \\ 0 \\ \omega \end{bmatrix} = \begin{bmatrix} I_{13}^A\omega \\ I_{23}^A\omega \\ I_{33}^A\omega \end{bmatrix}$$

$$\Rightarrow \quad \vec{H}_A = I_{13}^A\omega\,\vec{e}_1 + I_{23}^A\omega\,\vec{e}_2 + I_{33}^A\omega\,\vec{e}_3 \tag{3.29}$$

In general, $\vec{H}_A$ is not parallel to $\vec{\omega}$, even for plane motion. For convenience, we *fix* $\vec{e}_i$ *to the rigid body* so that $\dot{I}_{ij}^A = 0$. If *in addition A coincides with C or* $\vec{a}_{A|I} = \vec{0}$ *or* $\vec{a}_{A|I}$ *is along AC*, then

$$\begin{aligned} \vec{M}_A &= \dot{\vec{H}}_{A|I} = \dot{\vec{H}}_{A|\,\vec{e}_i=\text{const.}} + \vec{\omega}\times\vec{H}_A \\ &= I_{13}^A\dot{\omega}\,\vec{e}_1 + I_{23}^A\dot{\omega}\,\vec{e}_2 + I_{33}^A\dot{\omega}\,\vec{e}_3 + \omega\,\vec{e}_3 \times (I_{13}^A\omega\,\vec{e}_1 + I_{23}^A\omega\,\vec{e}_2 + I_{33}^A\omega\,\vec{e}_3) \\ &= (I_{13}^A\dot{\omega} - I_{23}^A\omega^2)\,\vec{e}_1 + (I_{23}^A\dot{\omega} + I_{13}^A\omega^2)\,\vec{e}_2 + I_{33}^A\dot{\omega}\,\vec{e}_3 \end{aligned}$$

$$\text{i.e.,}\quad \begin{aligned} M_{A_1} &= I^A_{13}\dot\omega - I^A_{23}\omega^2, \\ M_{A_2} &= I^A_{23}\dot\omega + I^A_{13}\omega^2, \\ M_{A_3} &= I^A_{33}\dot\omega, \end{aligned} \quad \Rightarrow \quad \text{for } \vec{e}_3 = \vec{k}: \quad \begin{aligned} M_{Ax} &= I^A_{xz}\dot\omega - I^A_{yz}\omega^2 \\ M_{Ay} &= I^A_{yz}\dot\omega + I^A_{xz}\omega^2 \\ M_{Az} &= I^A_{zz}\dot\omega \end{aligned} \tag{3.30}$$

In Section 3.6, A will be a point of the body $\ni A \equiv C$ or $\vec{a}_{A|I} = \vec{0}$ or $\vec{a}_{A|I}$ is along AC.

Remarks: **1.** Though, in general, $\vec{M}_A \neq I^A_{33}\dot\omega\,\vec{e}_3$, the equation of motion $(3.30)_3$: $M_{A3} = I^A_{33}\dot\omega$, is the same whether or not axis 3 is a principal axis of inertia at A.

2. Equations (3.30) are nonlinear in ω. Often eq. $(3.30)_3$ is solved for $\dot\omega$ and eqs $(3.30)_{2,3}$ yield the moment or force required to constrain the body to ensure that $\vec{\omega} = \omega\,\vec{k}\ \ \forall\, t$.

3. Usually eqs (3.30) are applied at C for general plane motion. For fixed axis rotation, eqs (3.30) are applied for a point on the axis of rotation where the largest number of unknown reactions act so as to eliminate them from these equations.

4. If angular velocity $\vec{\omega} = \omega\,\vec{e}_3$ is constant, then $\dot\omega = 0$ and eq. (3.30) yields

$$\vec{M}_A = -I^A_{23}\omega^2\,\vec{e}_1 + I^A_{13}\omega^2\,\vec{e}_2 \tag{a}$$

Hence, a constant moment $\vec{M}_A$ with respect to the body and a variable moment $\vec{M}_A$ with respect to inertial frame I (since $\vec{e}_1$, $\vec{e}_2$ rotate with the body) is needed for motion at constant $\vec{\omega}$. However, if $\vec{e}_3$ is a principal axis of inertia at A, then $I^A_{13} = I^A_{23} = 0$ and eq. (a) implies $\vec{M}_A = \vec{0}$, i.e., *no moment $\vec{M}_A$ is needed to maintain motion at constant $\vec{\omega}$ in the direction of a principal axis of inertia at A.* Thus, if $\vec{e}_3$ is a principal axis at C and $\vec{\omega} = \omega\,\vec{e}_3$ is constant, then $\vec{M}_C = \vec{0}$. The converse of this statement is not true. This aspect will be discussed in Section 3.18.

5. The moment equation **for an arbitrary point** B is obtained as follows:

$$\begin{aligned} \vec{M}_B &= \vec{M}_C + \vec{r}_{CB} \times \vec{F} = \dot{\vec{H}}_{C|I} + \vec{r}_{CB} \times m\vec{a}_{C|I} \quad \Rightarrow \\ \vec{M}_B &= \vec{r}_{CB} \times m\vec{a}_{C|I} + (I^C_{xz}\dot\omega - I^C_{yz}\omega^2)\,\vec{i} + (I^C_{yz}\dot\omega + I^C_{xz}\omega^2)\,\vec{j} + I^C_{zz}\dot\omega\,\vec{k} \end{aligned} \tag{3.31}$$

B can be chosen where many unknown reaction components act, or using eq. (3.31) the moment can be taken about a line BD through which several reaction components act. If the z-axis is a principal axis of inertia at C, then eq. (3.31) reduces to

$$\vec{M}_B = I^C_{zz}\dot{\vec{\omega}} + \vec{r}_{CB} \times m\vec{a}_{C|I} \tag{3.32}$$

6. *If axis 3 is a principal axis of inertia at A*, then $I^A_{13} = I^A_{23} = 0$ and eqs (3.29), (3.30) $\Rightarrow$

$$\vec{H}_A = I^A_{33}\omega\,\vec{e}_3, \quad M_{A_1} = 0, \quad M_{A_2} = 0, \quad M_{A_3} = I^A_{33}\dot\omega \quad \Rightarrow \quad \vec{M}_A = I^A_{33}\dot{\vec{\omega}} \tag{3.33}$$

[$\Rightarrow$ If $\vec{e}_3 = \vec{k}$, then $\vec{M}_A = I^A_{zz}\dot{\vec{\omega}}$.] Equation (3.33) is definitely valid for the centre of mass C and any point A on $\vec{e}_3$ through C for which $\vec{a}_{A|I} = \vec{0}$ or $\vec{a}_{A|I}$ is along AC for the following bodies (Fig. 3.11):

1. Body with $\vec{e}_3$ along the intersection of its *two planes of mass symmetry.*
2. *Body of revolution* with $\vec{e}_3$ along the axis of revolution.
3. Slab like *cylindrical body* with $\vec{e}_3$ parallel to the generator.

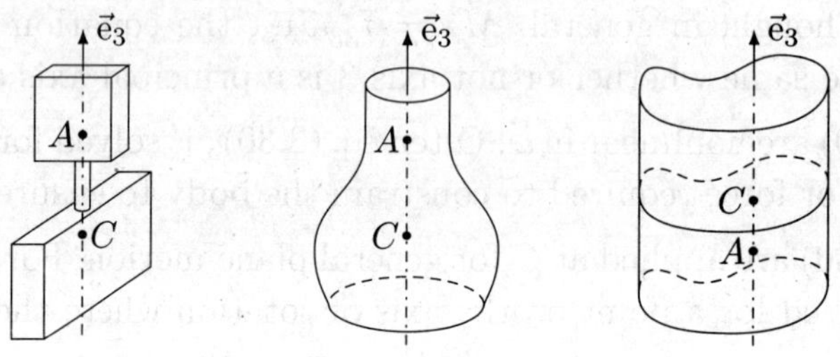

Fig. 3.11

For *this special case of rotation with $\vec{\omega}$ in the direction of a principal axis of inertia x_3 at A, $\vec{e}_1$, $\vec{e}_2$ need not be fixed to the rigid body* as they do not appear in eq. (3.33).

3.6.2 Euler's equations for three-dimensional motion of a rigid body

The expression of $\vec{H}_A$ in terms of $\vec{\omega}$ is simple in terms of the principal axes at A. Let *A be a point of the rigid body such that $A \equiv C$ or $\vec{a}_{A|I} = \vec{0}$ or $\vec{a}_{A|I}$ is along AC.* Then with respect to the principal axes of inertia $\vec{e}_i$ (fixed to the body) at A:

$$[\vec{H}_A] = \begin{bmatrix} I^A_{11} & 0 & 0 \\ 0 & I^A_{22} & 0 \\ 0 & 0 & I^A_{33} \end{bmatrix} \begin{bmatrix} \omega_1 \\ \omega_2 \\ \omega_3 \end{bmatrix} = \begin{bmatrix} I^A_{11}\omega_1 \\ I^A_{22}\omega_2 \\ I^A_{33}\omega_3 \end{bmatrix}$$

$$\Rightarrow \qquad \vec{H}_A = I^A_{11}\omega_1\,\vec{e}_1 + I^A_{22}\omega_2\,\vec{e}_2 + I^A_{33}\omega_3\,\vec{e}_3$$

$$\vec{M}_A = \dot{\vec{H}}_{A|I} = \dot{\vec{H}}_{A|\,\vec{e}_i=\text{const.}} + \vec{\omega}\times\vec{H}_A$$

$$= I^A_{11}\dot{\omega}_1\,\vec{e}_1 + I^A_{22}\dot{\omega}_2\,\vec{e}_2 + I^A_{33}\dot{\omega}_3\,\vec{e}_3 + \begin{vmatrix} \vec{e}_1 & \vec{e}_2 & \vec{e}_3 \\ \omega_1 & \omega_2 & \omega_3 \\ I^A_{11}\omega_1 & I^A_{22}\omega_2 & I^A_{33}\omega_3 \end{vmatrix}$$

$$= [I^A_{11}\dot{\omega}_1 - (I^A_{22} - I^A_{33})\omega_2\omega_3]\,\vec{e}_1 + \ldots$$

i.e.,

$$M_{A_1} = I^A_{11}\dot{\omega}_1 - (I^A_{22} - I^A_{33})\omega_2\omega_3$$

$$M_{A_2} = I^A_{22}\dot{\omega}_2 - (I^A_{33} - I^A_{11})\omega_3\omega_1 \qquad (3.34)$$

$$M_{A_3} = I^A_{33}\dot{\omega}_3 - (I^A_{11} - I^A_{22})\omega_1\omega_2$$

These are *Euler's equations of motion of a rigid body for principal axes of inertia at A.*

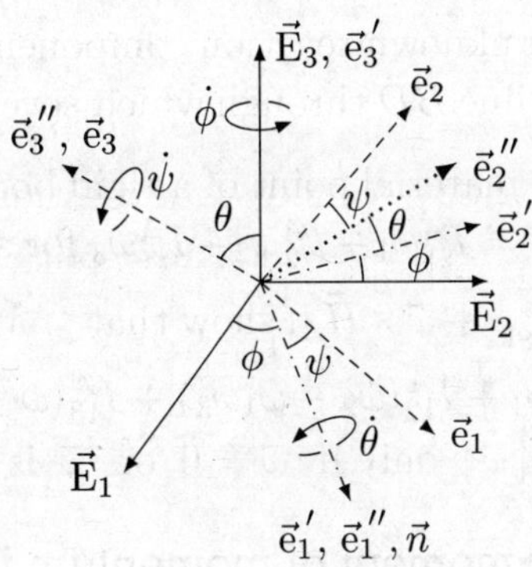

Fig. 3.12

Remarks:

1. Equations (3.34) are coupled non-linear ordinary differential equations in $\omega_1, \omega_2, \omega_3$. Their solution does not directly give the orientation of the body in I, unless $\vec{\omega}$ has a fixed direction. To get orientation, Euler expressed $\vec{\omega}$ in terms of Euler's angles ϕ, θ, ψ (Fig. 3.12). $\vec{n}$ is normal to $\vec{\mathrm{E}}_3$ and $\vec{\mathrm{e}}_3$ along $\vec{\mathrm{E}}_3 \times \vec{\mathrm{e}}_3$. $\phi = \angle(\vec{n}, \vec{\mathrm{E}}_1)$ is rotation about $\vec{\mathrm{E}}_3$. $\theta = \angle(\vec{\mathrm{e}}_3, \vec{\mathrm{E}}_3)$ is the rotation about $\vec{n}$. $\psi = \angle(\vec{n}, \vec{\mathrm{e}}_1)$ is rotation about $\vec{\mathrm{e}}_3$.

$$\begin{aligned}\vec{\omega} &= \dot{\phi}\,\vec{\mathrm{E}}_3 + \dot{\theta}\,\vec{n} + \dot{\psi}\,\vec{\mathrm{e}}_3 \\ &= \dot{\phi}[\cos\theta\,\vec{\mathrm{e}}_3 + \sin\theta(\cos\psi\,\vec{\mathrm{e}}_2 + \sin\psi\,\vec{\mathrm{e}}_1)] + \dot{\theta}(\cos\psi\,\vec{\mathrm{e}}_1 - \sin\psi\,\vec{\mathrm{e}}_2) + \dot{\psi}\,\vec{\mathrm{e}}_3\end{aligned}$$

$$\Rightarrow \quad \omega_1 = \dot{\theta}\cos\psi + \dot{\phi}\sin\theta\sin\psi, \quad \omega_2 = -\dot{\theta}\sin\psi + \dot{\phi}\sin\theta\cos\psi, \quad \omega_3 = \dot{\psi} + \dot{\phi}\cos\theta$$

$$\begin{aligned}\Rightarrow \quad \dot{\omega}_1 &= \ddot{\theta}\cos\psi + \ddot{\phi}\sin\theta\sin\psi - \dot{\theta}\dot{\psi}\sin\psi + \dot{\phi}\dot{\theta}\cos\theta\sin\psi + \dot{\phi}\dot{\psi}\sin\theta\cos\psi \\ \dot{\omega}_2 &= -\ddot{\theta}\sin\psi + \ddot{\phi}\sin\theta\cos\psi - \dot{\theta}\dot{\psi}\cos\psi + \dot{\phi}\dot{\theta}\cos\theta\cos\psi - \dot{\phi}\dot{\psi}\sin\theta\sin\psi \\ \dot{\omega}_3 &= \ddot{\psi} + \ddot{\phi}\cos\theta - \dot{\phi}\dot{\theta}\sin\theta\end{aligned}$$

2. However, eqs (3.34) are useful to find the loads for a given motion of the rigid body.

3. If $\vec{\omega}$ and $\dot{\vec{\omega}}$ are in the direction of the principal axis of inertia $\vec{\mathrm{e}}_3$, then eqs (3.34) reduce to $\vec{M}_A = I_{33}^A \dot{\omega}_3\,\vec{\mathrm{e}}_3 = I_{33}^A\,\dot{\vec{\omega}}$, derived earlier [eq. (3.33)].

4. If $M_{A_3}(t) \equiv 0$, then eq. $(3.34)_3 \Rightarrow \omega_3(t) =$ constant, only if $I_{11}^A = I_{22}^A$.

5. If $\vec{M}_A(t) \equiv \vec{0}$, then $\dot{\vec{H}}_{A|I}(t) \equiv \vec{0}$, i.e., $\vec{H}_A =$ constant. This *does* **not** *necessarily imply that $\vec{\omega}$ is constant* (see Section 3.18).

6. Moment equation **for an arbitrary point** B in terms of the principal axes at C:

$$\vec{M}_B = \vec{M}_C + \vec{r}_{CB} \times \vec{F} = \dot{\vec{H}}_{C|I} + \vec{r}_{CB} \times m\vec{a}_{C|I} \quad \Rightarrow$$

$$\vec{M}_B = \vec{r}_{CB} \times m\vec{a}_{C|I} + [I^C_{xx}\dot{\omega}_x - (I^C_{yy} - I^C_{zz})\omega_y\omega_z]\,\vec{i} + \ldots \tag{3.35}$$

B can be chosen where some unknown reaction components act, or using eq. (3.35) the moment can be taken about a line BD through which several reaction components act.

§ **Exercise 3.5:** Let A be a material point of a rigid body with $A \equiv C$ or $\vec{a}_{A|I} = \vec{0}$ or $\vec{a}_{A|I}$ along AC. (a) Using $H_{A_i} = I^A_{i1}\omega_1 + I^A_{i2}\omega_2 + I^A_{i3}\omega_3$ for an arbitrary set of body-fixed axes $\vec{e}_i$ and $\vec{M}_A = \dot{\vec{H}}_{A|\,\vec{e}_i=\text{const.}} + \vec{\omega} \times \vec{H}_A$, show that $M_{A_1} = \dot{H}_{A_1} + \omega_2 H_{A_3} - \omega_3 H_{A_2}$ $= I^A_{11}\dot{\omega}_1 - (I^A_{22} - I^A_{33})\omega_2\omega_3 + I^A_{12}(\dot{\omega}_2 - \omega_1\omega_3) + I^A_{13}(\dot{\omega}_3 + \omega_1\omega_2) + I^A_{23}(\omega_2^2 - \omega_3^2)$.
(b) Prove that $[\vec{M}_A] = [I^A][\dot{\vec{\omega}}]$ only if $\vec{\omega} = \vec{0}$ or $\vec{\omega}$ is along a principal axis at A.

3.6.3 Direct equations of moment of momentum if $I^A_{11} = I^A_{22}$

Let A be a *material point of a rigid body* with $\vec{e}_i$ as the principal axes at A, $I^A_{11} = I^A_{22}$ *and* $A \equiv C$ *or* $\vec{a}_{A|I} = \vec{0}$ *or* $\vec{a}_{A|I}$ *is along* AC. As shown in Section 3.2.1, every axis $\vec{n} \perp \vec{e}_3$ is a principal axis at A (Fig. 3.13) with $I^A_{nn} = I^A_{11}$ for all $\vec{n}$. Hence, the axes $\vec{e}_i$ are chosen so that axis $\vec{e}_3$ is aligned with the principal axis 3 at A (Fig. 3.13) and is fixed to the body, whereas axes $\vec{e}_1$, $\vec{e}_2$ could rotate relative to the rigid body about $\vec{e}_3$ at the rate $\dot{\phi}\,\vec{e}_3$. Thus, I^A_{11}, I^A_{22}, I^A_{33} remain constant for all positions of $\vec{e}_1$, $\vec{e}_2$.

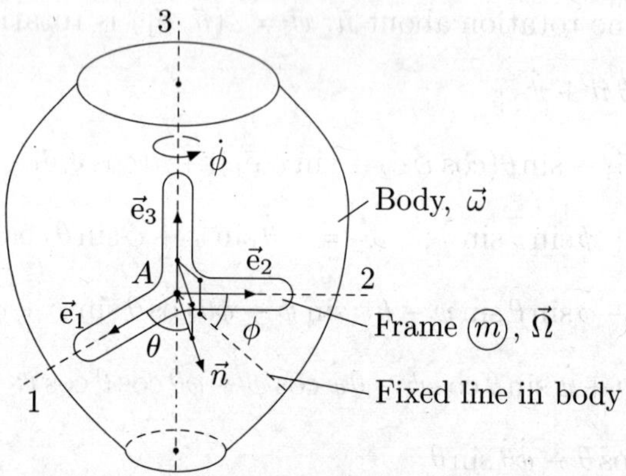

Fig. 3.13

Let (m) be the frame attached at A with $\vec{e}_i$ fixed to it. The *angular velocity* $\vec{\Omega}$ *of the frame* (m) *differs from the angular velocity* $\vec{\omega}$ *of the rigid body in its* $\vec{e}_3$ *component:*

$$\vec{\Omega} = \Omega_1\vec{e}_1 + \Omega_2\vec{e}_2 + \Omega_3\vec{e}_3 = \vec{\omega} + \dot{\phi}\,\vec{e}_3$$

$$\vec{\omega} = \omega_1\vec{e}_1 + \omega_2\vec{e}_2 + \omega_3\vec{e}_3$$

$$\Omega_1 = \omega_1, \qquad \Omega_2 = \omega_2, \qquad \Omega_3 = \omega_3 + \dot{\phi} \neq \omega_3$$

$$\vec{H}_A = I^A_{11}\,\omega_1\vec{e}_1 + I^A_{22}\,\omega_2\vec{e}_2 + I^A_{33}\,\omega_3\vec{e}_3 = I^A_{11}\,(\omega_1\vec{e}_1 + \omega_2\vec{e}_2) + I^A_{33}\,\omega_3\vec{e}_3$$

$$\begin{aligned}\vec{M}_A &= \dot{\vec{H}}_{A|I} = \dot{\vec{H}}_{A|\,\vec{e}_i=\text{const.}} + \vec{\Omega}\times\vec{H}_A \\ &= I^A_{11}(\dot{\omega}_1\,\vec{e}_1 + \dot{\omega}_2\,\vec{e}_2) + I^A_{33}\dot{\omega}_3\,\vec{e}_3 + \vec{\Omega}\times\vec{H}_A \qquad (3.36)\end{aligned}$$

We normally use eq. (3.36) directly rather than the component form of it listed in Exercise 3.6. Such a procedure can definitely be used for a body of revolution. Note that $\dot{\vec{\omega}}_{|I} \neq \dot{\omega}_1\,\vec{e}_1 + \dot{\omega}_2\,\vec{e}_2 + \dot{\omega}\,\vec{e}_3$, but $\dot{\vec{\omega}}_{|I} = \dot{\omega}_1\,\vec{e}_1 + \dot{\omega}_2\,\vec{e}_2 + \dot{\omega}_3\,\vec{e}_3 + \vec{\Omega}\times\vec{\omega}$.

§ **Exercise 3.6:** Show that the component form of eq. (3.36) is:
$M_{A_3} = I^A_{33}\dot{\omega}_3,\quad M_{A_1} = I^A_{11}\dot{\omega}_1 - I^A_{11}\omega_2\Omega_3 + I^A_{33}\omega_3\Omega_2,\quad M_{A_2} = I^A_{11}\dot{\omega}_2 - I^A_{33}\omega_3\Omega_1 + I^A_{11}\omega_1\Omega_3.$

Example 3.13: Find the bearing reactions when the shaft rotates at $\omega = 100\,\text{rad/s}$ and the driving torque $T = 0.2\,\text{N.m}$ (Fig. E3.13a). Bearing B does not provide any axial constraint.

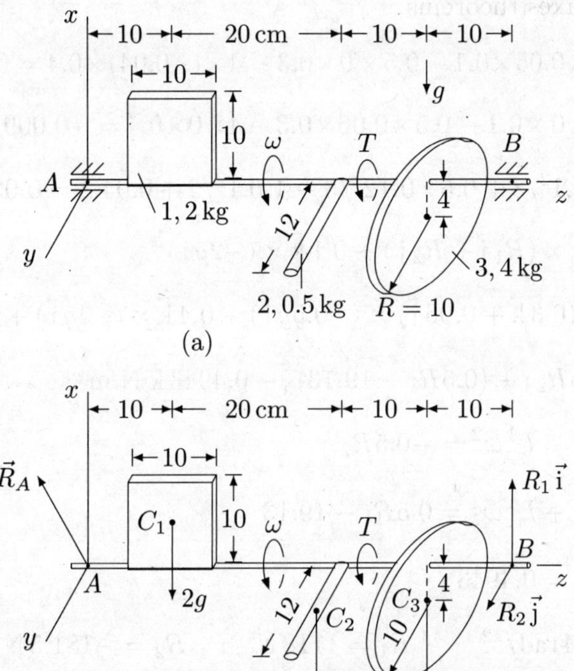

Fig. E3.13

Solution: For a rotor with two hinged supports, each bearing provides a force reaction $\vec{R} = R_x\,\vec{i} + R_y\,\vec{j} + R_z\,\vec{k}$ and a moment reaction $\vec{C} = C_x\,\vec{i} + C_y\,\vec{j}$, resulting in 11[5+5+1 ($\dot{\omega}$)] unknowns, whereas there are only 6 equations of motion for a rigid body. Hence, to enable a solution within the mathematically simple rigid body model, the bearings are modelled to provide only force reaction (length of bearing being small) with one bearing not providing any axial constraint force, reducing the number of unknowns to 6.[1]

Let $\vec{R}_B = R_1\vec{\imath}+R_2\vec{\jmath}$, $\vec{R}_A = R_3\vec{\imath}+R_4\vec{\jmath}+R_5\vec{k}$. The FBD is shown in Fig. E3.13b when the x-axis is vertical. The unknowns $\dot{\omega}$, R_1, R_2, R_3, R_4, R_5, are obtained from the moment equation for the fixed point A for plane motion with $\vec{\omega}$ in the direction of the non-principal axis $\vec{k}$ at A, and $\vec{F} = \sum m_i\vec{a}_{C_i}$. $C_i\,(x_i, y_i, z_i)$ is the centre of mass of part i of mass m_i at distance r_i from the axis and $\vec{a}_{C_i} = -\omega^2 r_i\,\vec{e}_{r_i} + \dot{\omega} r_i\,\vec{e}_{\phi_i}$ (no sum on i).

$$
\begin{aligned}
m_1 &= 2\,\text{kg}, \quad x_1 = r_1 = 0.05\,\text{m}, \quad y_1 = 0, \quad z_1 = 0.1\,\text{m}, \quad \omega = 100\,\text{rad/s}\\
m_2 &= 0.5\,\text{kg}, \quad x_2 = 0, \quad y_2 = r_2 = 0.06\,\text{m}, \quad z_2 = 0.3\,\text{m}\\
m_3 &= 4\,\text{kg}, \quad x_3 = -0.04\,\text{m}, \quad y_3 = 0, \quad z_3 = 0.4\,\text{m}, \quad r_3 = 0.04\,\text{m}
\end{aligned}
$$

Using the parallel axes theorems:

$$
\begin{aligned}
I^A_{xz} &= -2\times 0.05\times 0.1 - 0.5\times 0\times 0.3 - 4\times(-0.04)\times 0.4 = 0.054\,\text{kg.m}^2\\
I^A_{yz} &= -2\times 0\times 0.1 - 0.5\times 0.06\times 0.3 - 4\times 0\times 0.4 = -0.009\,\text{kg.m}^2\\
I^A_{zz} &= 2\times 0.1^2/3 + 0.5\times 0.12^2/3 + 4(0.1^2/2 + 0.04^2) = 0.035467\,\text{kg.m}^2\\
\vec{M}_A &= 0.5\vec{k}\times(R_1\vec{\imath} + R_2\vec{\jmath}) + 0.1\vec{k}\times(-2g\vec{\imath})\\
&\quad +(0.3\vec{k} + 0.06\vec{\jmath})\times(-0.5g\vec{\imath}) + 0.4\vec{k}\times(-4g\vec{\imath}) + 0.2\vec{k}\\
&= -0.5R_2\vec{\imath} + (0.5R_1 - 19.13)\vec{\jmath} + 0.4943\,\vec{k}\ \text{N.m}\\
M_{A_x} &= I^A_{xz}\dot{\omega} - I^A_{yz}\omega^2 = -0.5R_2\\
M_{A_y} &= I^A_{yz}\dot{\omega} + I^A_{xz}\omega^2 = 0.5R_1 - 19.13\\
M_{A_z} &= I^A_{zz}\dot{\omega} = 0.4943\\
\Rightarrow \quad \dot{\omega} &= 13.94\,\text{rad/s}^2, \quad R_1 = 1118\,\text{N}, \quad R_2 = -181.5\,\text{N}\\
\vec{F} &= \vec{R}_A + R_1\vec{\imath} + R_2\vec{\jmath} - (2 + 0.5 + 4)g\vec{\imath} = \sum m_i\vec{a}_{C_i}\\
&= 2[-\omega^2(.05)\vec{\imath} + \dot{\omega}(.05)\vec{\jmath}] + .5[-\omega^2(.06)\vec{\jmath} - \dot{\omega}(.06)\vec{\imath}] + 4[\omega^2(.04)\vec{\imath} - \dot{\omega}(.04)\vec{\jmath}]\\
\Rightarrow \quad \vec{R}_A &= -454.7\vec{\imath} - 119.3\vec{\jmath}\ \text{N}
\end{aligned}
$$

[1]Not having an axial constraint at one of the bearings is desirable, especially when ω is very large since if both ends are fixed axially, then an increase in temperature would cause the shaft to buckle. This could lead to catastrophic failure of the system since the centre of mass would no longer lie on the axis and the rotor would not be balanced (see Section 3.9).

Example 3.14: A body ②, consisting of a light shaft BD and a rod GE of length 0.2 m and mass $m = 0.3$ kg, is rotating relative to platform ① (Fig. E3.14a). The bearing at B does not provide any axial restraint. At the given instant, find (a) the reactions at B and D and the applied couple M, and (b) the twisting moment, bending moment, normal force and shear force at section A of the rod. $\omega_\alpha = 80$ rad/s, $\omega_\beta = 20$ rad/s, $\dot{\omega}_\alpha = 3$ rad/s^2, $\dot{\omega}_\beta = 1$ rad/s^2, $d = 0.05$ m, $b = 0.25$ m, $r = 0.1$ m.

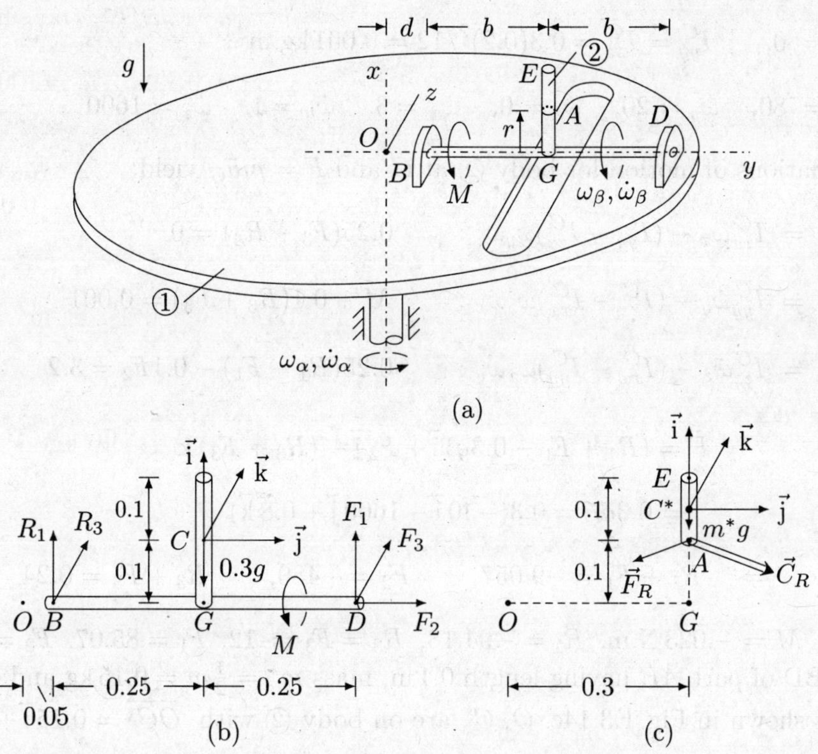

Fig. E3.14

Solution: (a) Let $\vec{\omega}_1$, $\vec{\omega}_2$ and $\dot{\vec{\omega}}_1, \dot{\vec{\omega}}_2$ be of bodies ① and ② relative to the ground:

$$\vec{\omega}_1 = \omega_\alpha \vec{i} = 80\vec{i}, \qquad \dot{\vec{\omega}}_1 = \dot{\omega}_\alpha \vec{i} = 3\vec{i}$$

$$\vec{\omega}_2 = \vec{\omega}_1 + \omega_\beta \vec{j} = 80\vec{i} + 20\vec{j}$$

$$\dot{\vec{\omega}}_2 = \dot{\vec{\omega}}_1 + \dot{\omega}_\beta \vec{j} + \vec{\omega}_1 \times \omega_\beta \vec{j} = 3\vec{i} + \vec{j} + 80\vec{i} \times 20\vec{j} = 3\vec{i} + \vec{j} + 1600\vec{k}$$

The 6 unknowns, R_1, R_3, F_1, F_2, F_3, M (Fig. E3.14b), are obtained from the Euler's equations at C and $\vec{F} = m\vec{a}_C$. Point O on the rigid extension of body ② remains fixed, $\vec{a}_O = \vec{0}$. Points O, C are on body ② with $\overrightarrow{OC} = 0.3\vec{j} + 0.1\vec{i}$:

$$\vec{a}_C = \vec{a}_O + \dot{\vec{\omega}}_2 \times \overrightarrow{OC} + \vec{\omega}_2 \times (\vec{\omega}_2 \times \overrightarrow{OC}) = -40\vec{i} - 1600\vec{j} + 0.8\vec{k}$$

$$\vec{M}_C = M\,\vec{j} + (-0.1\,\vec{i} - 0.25\,\vec{j}) \times (R_1\,\vec{i} + R_3\,\vec{k}) \\ +(-0.1\,\vec{i} + 0.25\,\vec{j}) \times (F_1\,\vec{i} + F_2\,\vec{j} + F_3\,\vec{k}) \\ = 0.25(F_3 - R_3)\,\vec{i} + (M + 0.1R_3 + 0.1F_3)\,\vec{j} + [0.25(R_1 - F_1) - 0.1F_2]\,\vec{k}$$

The principal axes of body ② at C are parallel to the x, y, z axes and

$$I^C_{xx} = 0, \quad I^C_{yy} = I^C_{zz} = 0.3(0.2)^2/12 = 0.001\,\text{kg.m}^2$$

$$\omega_x = 80, \quad \omega_y = 20, \quad \omega_z = 0, \quad \dot{\omega}_x = 3, \quad \dot{\omega}_y = 1, \quad \dot{\omega}_z = 1600$$

Euler's equations of motion for body ② at C and $F = m\vec{a}_C$ yield:

$$M_{Cx} = I^C_{xx}\dot{\omega}_x - (I^C_{yy} - I^C_{zz})\omega_y\omega_z : \qquad 0.25(F_3 - R_3) = 0 \qquad (1)$$

$$M_{Cy} = I^C_{yy}\dot{\omega}_y - (I^C_{zz} - I^C_{xx})\omega_z\omega_x : \qquad M + 0.1(R_3 + F_3) = 0.001 \qquad (2)$$

$$M_{Cz} = I^C_{zz}\dot{\omega}_z - (I^C_{xx} - I^C_{yy})\omega_x\omega_y : \qquad 0.25(R_1 - F_1) - 0.1F_2 = 3.2 \qquad (3)$$

$$\vec{F} = (R_1 + F_1 - 0.3g)\,\vec{i} + F_2\,\vec{j} + (R_3 + F_3)\,\vec{k}$$

$$= 0.3\vec{a}_C = 0.3(-40\,\vec{i} - 1600\,\vec{j} + 0.8\,\vec{k})$$

$$\Rightarrow \quad R_1 + F_1 = -9.057, \qquad F_2 = -480, \qquad R_3 + F_3 = 0.24 \qquad (4)$$

(1)–(4) $\Rightarrow$ $M = -.023\,\text{N.m}$, $R_1 = -94.13$, $R_3 = F_3 = .12$, $F_1 = 85.07$, $F_2 = -480\,\text{N}$.
(b) The FBD of part AE having length 0.1 m, mass $m^* = \frac{1}{2}m = 0.15\,\text{kg}$ and centre of mass C^* is shown in Fig. E3.14c. O, C^* are on body ② with $\overrightarrow{OC^*} = 0.15\,\vec{i} + 0.3\,\vec{j}$,

$$\vec{a}_{C^*} = \vec{a}_O + \dot{\vec{\omega}}_2 \times \overrightarrow{OC^*} + \vec{\omega}_2 \times (\vec{\omega}_2 \times \overrightarrow{OC^*}) = -60\,\vec{i} - 1440\,\vec{j} + 0.75\,\vec{k}$$

$$\vec{F} = -m^* g\,\vec{i} + \vec{F}_R = m^*\vec{a}_{C^*}$$

$$\Rightarrow \quad \vec{F}_R = -7.528\,\vec{i} - 216\,\vec{j} + 0.1125\,\vec{k}\,\text{N}$$

The principal axes of body AE at C^* are parallel to the x, y, z axes with $I^{C^*}_{xx} = 0$, $I^{C^*}_{yy} = I^{C^*}_{zz} = m^*(0.1)^2/12 = 1.25 \times 10^{-4}\,\text{kg.m}^2$. Euler's equations of motion at C^* :

$$M_{C^*_x} = I^{C^*}_{xx}\dot{\omega}_x - (I^{C^*}_{yy} - I^{C^*}_{zz})\omega_y\omega_z = 0$$

$$M_{C^*_y} = I^{C^*}_{yy}\dot{\omega}_y - (I^{C^*}_{zz} - I^{C^*}_{xx})\omega_z\omega_x = 1.25 \times 10^{-4}$$

$$M_{C^*_z} = I^{C^*}_{zz}\dot{\omega}_z - (I^{C^*}_{xx} - I^{C^*}_{yy})\omega_x\omega_y = 0.4$$

$$\Rightarrow \quad \vec{M}_{C*} = \vec{C}_R - 0.05\,\vec{\mathrm{i}} \times \vec{F}_R = 1.25 \times 10^{-4}\,\vec{\mathrm{j}} + 0.4\,\vec{\mathrm{k}}$$

$$\Rightarrow \quad \vec{C}_R = -0.0055\,\vec{\mathrm{j}} - 10.4\,\vec{\mathrm{k}}\ \mathrm{kN.m}$$

The normal to the cross-section at A is $\vec{\mathrm{n}} = -\vec{\mathrm{i}} \Rightarrow$ The $-\vec{\mathrm{i}}$ component of $\vec{F}_R$ is the normal force $N = 7.528\,\mathrm{N}$ and the shear force: $\vec{S} = -216\,\vec{\mathrm{j}} + 0.1125\,\vec{\mathrm{k}}\ \mathrm{N}$, $S = 216\,\mathrm{N}$. The twisting moment $M_t = -\vec{\mathrm{i}} \cdot \vec{C}_R = 0 \Rightarrow C_R$ itself is the bending moment.

3.7 Kinetic Energy Expression of a Rigid Body

Using eq. (2.31), the expression for the kinetic energy T in terms of C, and $\vec{v}_{PC} = \vec{\omega} \times \vec{r}_{PC}$ (Fig. 3.14a), we can express T of a rigid body as

$$\begin{aligned} T &= \tfrac{1}{2} m v_C^2 + \tfrac{1}{2} \int_m v_{PC}^2\, dm = \tfrac{1}{2} m v_C^2 + \tfrac{1}{2} \int_m (\vec{\omega} \times \vec{r}_{PC}) \cdot \vec{v}_{PC}\, dm \\ &= \tfrac{1}{2} m v_C^2 + \tfrac{1}{2} \int_m \vec{r}_{PC} \times \vec{v}_{PC}\, dm \cdot \vec{\omega} = \tfrac{1}{2} m v_C^2 + \tfrac{1}{2} \vec{H}_C \cdot \vec{\omega} \end{aligned}$$

Using eq. (3.2), $\vec{H}_C \cdot \vec{\omega} = H_{C_i} \omega_i = I_{ij}^C \omega_i \omega_j$ and hence,

$$T = \tfrac{1}{2} m v_C^2 + \tfrac{1}{2} \vec{H}_C \cdot \vec{\omega} = \tfrac{1}{2} m v_C^2 + \tfrac{1}{2} I_{ij}^C \omega_i \omega_j \tag{3.37}$$

$$T = \tfrac{1}{2} m v_C^2 + \tfrac{1}{2} \left[I_{11}^C \omega_1^2 + I_{22}^C \omega_2^2 + I_{33}^C \omega_3^2 + 2 I_{12}^C \omega_1 \omega_2 + 2 I_{23}^C \omega_2 \omega_3 + 2 I_{31}^C \omega_3 \omega_1 \right] \tag{3.38}$$

$$T = \tfrac{1}{2} m v_C^2 + \tfrac{1}{2} \left[I_{11}^{C*} \omega_1^{*2} + I_{22}^{C*} \omega_2^{*2} + I_{33}^{C*} \omega_3^{*2} \right] \tag{3.39}$$

where $*$ refers to the principal axes of inertia at C. The two terms in the expression (3.37) of T are called *translational kinetic energy* T_C (as if the body were translating at $\vec{v}_C$) and *rotational kinetic energy* T_R (as if the body were rotating about C):

$$T_C = \tfrac{1}{2} m v_C^2 = \tfrac{1}{2} \vec{p} \cdot \vec{v}_C, \qquad T_R = \tfrac{1}{2} \vec{H}_C \cdot \vec{\omega} = \tfrac{1}{2} I_{ij}^C \omega_i \omega_j \tag{3.40}$$

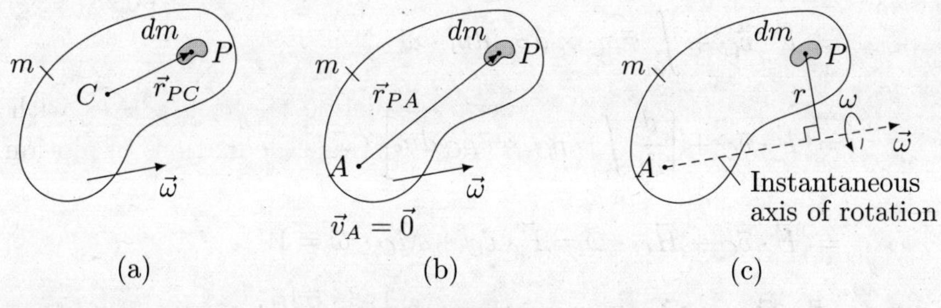

Fig. 3.14

If a point A of the rigid body has $\vec{v}_A = \vec{0}$, then $\vec{v}_P = \vec{v}_{PA} = \vec{\omega} \times \vec{r}_{PA}$ (Fig. 3.14b),

$$T = \tfrac{1}{2}\int_m v_P^2\,dm = \tfrac{1}{2}\int_m (\vec{\omega}\times\vec{r}_{PA})\cdot\vec{v}_{PA}\,dm$$

$$= \tfrac{1}{2}\int_m \vec{r}_{PA}\times\vec{v}_{PA}\,dm\cdot\vec{\omega} = \tfrac{1}{2}\vec{H}_A\cdot\vec{\omega} = \tfrac{1}{2}I_{ij}^A\omega_i\omega_j \tag{3.41}$$

$$T = \tfrac{1}{2}\left[I_{11}^A\omega_1^2 + I_{22}^A\omega_2^2 + I_{33}^A\omega_3^2 + 2I_{12}^A\omega_1\omega_2 + 2I_{23}^A\omega_2\omega_3 + 2I_{31}^A\omega_3\omega_1\right] \tag{3.42}$$

$$T = \tfrac{1}{2}\left[I_{11}^{A*}\omega_1^{*2} + I_{22}^{A*}\omega_2^{*2} + I_{33}^{A*}\omega_3^{*2}\right] \tag{3.43}$$

where $*$ refers to the principal axes of inertia at A. If instantaneous axis of rotation (Fig. 3.14c) of the rigid body exists through A, then T can be expressed compactly as

$$T = \tfrac{1}{2}\int_m v_P^2\,dm = \tfrac{1}{2}\int_m (\omega r)^2\,dm = \tfrac{1}{2}\int r^2\,dm\,\omega^2 = \tfrac{1}{2}I_{\omega\omega}^A\,\omega^2 \tag{3.44}$$

where $I_{\omega\omega}^A$ is the moment of inertia about the instantaneous axis of rotation.
For the case of $\vec{\omega} = \omega\,\vec{e}_3$, eqs (3.37) & (3.44) $\Rightarrow$

$$T = \tfrac{1}{2}mv_C^2 + \tfrac{1}{2}I_{33}^C\,\omega^2, \qquad T = \tfrac{1}{2}I_{33}^A\,\omega^2 \tag{3.45}$$

where $\vec{v}_A = \vec{0}$. Equations (3.45) are valid even if $\vec{e}_3$ is not a principal axis at C (or A).

3.8 Work–Energy Relation for a Rigid Body

For simplicity, we drop the suffix I for inertial frame. By eq. (2.33), the rate of work done by the external loads on a rigid body is: $\dot{W} = \vec{F}\cdot\vec{v}_C + \vec{M}_C\cdot\vec{\omega}$. For a rigid body

$$T = \tfrac{1}{2}mv_C^2 + \tfrac{1}{2}\int_m v_{PC}^2\,dm = \tfrac{1}{2}m\vec{v}_C\cdot\vec{v}_C + \tfrac{1}{2}\int_m \vec{v}_{PC}\cdot\vec{v}_{PC}\,dm$$

$$\Rightarrow\quad \dot{T} = m\vec{a}_C\cdot\vec{v}_C + \int_m \vec{v}_{PC}\cdot\vec{a}_{PC}\,dm = \vec{F}\cdot\vec{v}_C + \int_m \vec{\omega}\times\vec{r}_{PC}\cdot\vec{a}_{PC}\,dm$$

$$= \vec{F}\cdot\vec{v}_C + \int_m \vec{r}_{PC}\times\vec{a}_{PC}\,dm\cdot\vec{\omega}$$

$$= \vec{F}\cdot\vec{v}_C + \left[\frac{d}{dt}\int_m \vec{r}_{PC}\times\vec{v}_{PC}\,dm\right]\cdot\vec{\omega}$$

$$= \vec{F}\cdot\vec{v}_C + \dot{\vec{H}}_C\cdot\vec{\omega} = \vec{F}\cdot\vec{v}_C + \vec{M}_C\cdot\vec{\omega} = \dot{W}$$

using $m\vec{a}_C = \vec{F}$, $\dot{\vec{H}}_C = \vec{M}_C$. This is the *work–energy relations in rate form*:

$$\dot{T} = \dot{W}, \qquad \Rightarrow \qquad T_2 - T_1 = W_{1\text{-}2} \tag{3.46}$$

where T_1, T_2 are the values of T in configurations 1 and 2 at times t_1 and t_2 and $W_{1\text{-}2}$ is the work done by the external loads in the displacement of the body from configuration 1 to 2. The work–energy relation (3.46) can be split into two independent relations using eq. (2.57): $\dot{T}_C = \dot{W}^* = \vec{F}\cdot\vec{v}_C \Rightarrow \dot{T}_R = \dot{T} - \dot{T}_C = \dot{W} - \dot{W}^* = \vec{M}_C\cdot\vec{\omega} = \dot{W}_R$ (say), where $\dot{W}_R$ is the rate of work done by the external loads for motion of the points of application of loads relative to C, i.e., due to rotation about C. Hence,

$$\dot{T}_R = \dot{W}_R \qquad \Rightarrow \qquad T_{R_2} - T_{R_1} = W_{R_{1\text{-}2}} = \int_{t_1}^{t_2} \vec{M}_C\cdot\vec{\omega}\,dt \tag{3.47}$$

Denoting the work done by the conservative and the non-conservative forces by W^{c} and W^{nc}, we have $\dot{W} = \dot{W}^{\text{c}} + \dot{W}^{\text{nc}} = -\dot{V} + \dot{W}^{\text{nc}}$ and eqs (3.46) reduce to

$$\dot{T} + \dot{V} = \dot{W}^{\text{nc}}, \qquad (T_2 + V_2) - (T_1 + V_1) = W^{\text{nc}}_{1\text{-}2} \tag{3.48}$$

where V_1, V_2 are the values of V in configurations 1 and 2. If all forces are conservative, then $\dot{W}^{nc} = 0$ and eqs (3.48) reduce to conservation of mechanical energy $(T + V)$:

$$\dot{T} + \dot{V} = 0, \qquad T_2 + V_2 = T_1 + V_1 \tag{3.49}$$

For *torque-free motion* of a rigid body ($\vec{M}_C(t) \equiv \vec{0}$), two constants for the motion are:

$$\dot{\vec{H}}_C = \vec{M}_C \equiv \vec{0} \qquad \Rightarrow \qquad \vec{H}_C(t) = I^C_{11}\omega_1\,\vec{e}_1 + I^C_{22}\omega_2\,\vec{e}_2 + I^C_{33}\omega_3\,\vec{e}_3 = \vec{H}_C(0)$$

$$\dot{T}_R = \vec{M}_C\cdot\vec{\omega} \equiv 0 \quad \Rightarrow \quad T_R(t) = \tfrac{1}{2}\vec{H}_C\cdot\vec{\omega} = \tfrac{1}{2}(I^C_{11}\omega_1^2 + I^C_{22}\omega_2^2 + I^C_{33}\omega_3^2) = T_R(0)$$

$\Rightarrow$ $\vec{\omega}$ is not constant, but its component in the invariable direction of $\vec{H}_C$ is constant.

Fig. 3.15

3.8.1 Work–energy relation for interconnected system of rigid bodies

Work–energy relations are applied to each body and added to get relations for the *whole system* which involve work done by the internal (due to interaction between bodies and due to the links and cables connecting them) and external forces on the system, since some of the forces which are external for an individual body are internal forces for the whole system. Denoting the contribution of the internal and external forces by the subscript int+ext, the work–energy relations reduce to the following form

$$\begin{aligned}
&\dot{T} = \dot{W}_{\text{int+ext}}, \qquad T_2 - T_1 = W_{\text{int+ext}_{1\text{-}2}} \\
&\dot{T} + \dot{V}_{\text{int+ext}} = \dot{W}^{\text{nc}}_{\text{int+ext}} \\
&(T_2 + V_{\text{int+ext}_2}) - (T_1 + V_{\text{int+ext}_1}) = W^{\text{nc}}_{\text{int+ext}_{1\text{-}2}} \\
&\dot{T} + \dot{V}_{\text{int+ext}} = 0, \qquad T_2 + V_{\text{int+ext}_2} = T_1 + V_{\text{int+ext}_1}
\end{aligned} \tag{3.50}$$

where the last line is valid only for conservative systems. If all internal forces are together workless, e.g., for systems in Fig. 3.15, then in eqs (3.50) the work of the internal forces is zero. The rate form and the integrated form are convenient to obtain acceleration (or angular acceleration) and velocity (or angular velocity) respectively, as a function of position, if the system of connected bodies has one dof and $\dot{W}$ and W can be obtained a priori. Work–energy relations are not valid for impulsive loads that cause instantaneous infinite accelerations, since the proof of the work–energy relation is based on the acceleration being finite. The impulse–momentum and the angular-impulse–moment-of-momentum relations are useful for impact problems and for obtaining velocity and angular velocity as a function of time, if the forces are given functions of time.

3.8.2 Conditions of equilibrium of a rigid body

The necessary and sufficient conditions for the equilibrium of a rigid body are:

1. $\vec{F}(t) = \vec{0}$, 2. $\vec{M}_A(t) = \vec{0}$ *and* 3. *the body is initially in equilibrium at* $t = 0$.

Proof: Let $\vec{e}_i$ be the principal axes of inertia at C.

$$\vec{F} = \vec{0}, \qquad \vec{M}_A = \vec{0} \quad \Rightarrow \quad \vec{M}_C = \vec{0}$$

$$\Rightarrow \quad \dot{T} = \dot{W} = \vec{F} \cdot \vec{v}_C + \vec{M}_C \cdot \vec{\omega} = 0 \quad \forall\, t$$

$$\Rightarrow \quad T = \tfrac{1}{2} m v_C^2 + \tfrac{1}{2} I_{11}^C \omega_1^2 + \tfrac{1}{2} I_{22}^C \omega_2^2 + \tfrac{1}{2} I_{33}^C \omega_3^2 = \text{constant} = T(0) = 0 \tag{a}$$

As each of the four terms on the left hand side of eq. (a) is positive:

$$v_C(t) = 0, \quad \omega_i(t) = 0 \quad \Rightarrow \quad \vec{v}_C(t) = \vec{0}, \quad \vec{\omega}(t) = \vec{0}$$

i.e., the rigid body is in equilibrium for all time.

Example 3.15: A planetary gear system consists of a sun gear 1, a spider 2, carrying four planet gears 3, and a fixed annulus 4 (Fig. E3.15a). The driving torque on the input shaft of the sun gear 1 is M_1 and the load torque on the output shaft of the spider 2 is M_2. The frictional torques on these shafts are M_{f_1}, M_{f_2}, and on the axle of each planet is M_{f3}. Find the angular acceleration of the input shaft. The axial moments of inertia of the sun gear and the spider are I_1, I_2, and of a planet it is I_3.

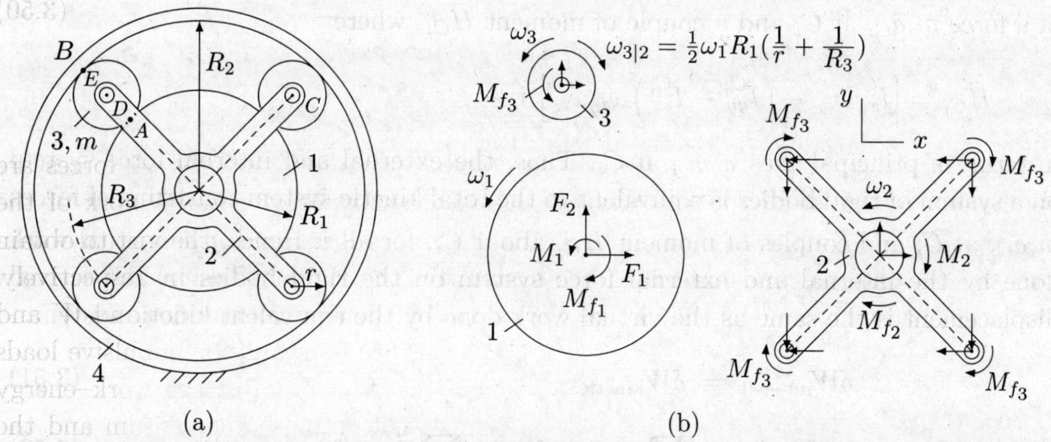

Fig. E3.15

Solution: The energy method is convenient as the system has one dof. Let the angular velocity of body i relative to the ground be ω_i $(\omega_4 = 0)$. Points A on 1, D on 3; and B on 4 (Fig. E3.15b), E on 3 have circular motion relative to the frame of spider 2 on which the axes of planet 3, sun gear 1 and annulus 4 are fixed. The gears mesh externally at A and internally at B. $\omega_{i|2} = \omega_i - \omega_2$. No slip conditions $\vec{v}_{B|2} = \vec{v}_{E|2}$, $\vec{v}_{A|2} = \vec{v}_{D|2}$:

$$(\omega_4 - \omega_2)(R_1 + 2r) = (\omega_3 - \omega_2)r \quad \Rightarrow \quad \omega_3 = -\omega_2 R_3/r \tag{1}$$

$$(\omega_1 - \omega_2)R_1 = -(\omega_3 - \omega_2)r \quad \Rightarrow \quad \omega_2 = \omega_1 R_1/2R_3 \tag{2}$$

Equations (1), (2) $\Rightarrow$ $\quad \omega_3 = -\omega_1 R_1/2r, \qquad v_C = \omega_2 R_3 = \frac{1}{2}\omega_1 R_1.$

$$T = \tfrac{1}{2}I_1\omega_1^2 + \tfrac{1}{2}I_2\omega_2^2 + 4\left(\tfrac{1}{2}I_3\omega_3^2 + \tfrac{1}{2}mv_C^2\right) = \tfrac{1}{2}\left[I_1 + \frac{I_2R_1^2}{4R_3^2} + \frac{I_3R_1^2}{r^2} + mR_1^2\right]\omega_1^2$$

The frictional M_{f_3} acting on the axle of planet 3 is opposite to $\vec{\omega}_{3|2} = \vec{\omega}_3 - \vec{\omega}_2 = -\omega_1 R_1(1/2r + 1/2R_3)\,\vec{\mathrm{k}}$, i.e., it is $M_{f_3}\,\vec{\mathrm{k}}$ and on spider 2 is $-M_{f_3}\,\vec{\mathrm{k}}$. For the system

$$\begin{aligned}
\dot{W}_{\text{int+ext}} &= (M_1 - M_{f_1})\omega_1 - (M_2 + M_{f_2})\omega_2 + 4M_{f_3}(\omega_3 - \omega_2)\\
&= \omega_1[M_1 - M_{f_1} - (M_2 + M_{f_2})R_1/2R_3 - 2M_{f_3}R_1(1/r + 1/R_3)]\\
&= \dot{T} = [I_1 + I_2R_1^2/4R_3^2 + I_3R_1^2/r^2 + mR_1^2]\omega_1\dot{\omega}_1
\end{aligned}$$

$$\Rightarrow \quad \dot{\omega}_1 = \frac{M_1 - M_{f_1} - (M_2 + M_{f_2})R_1/2R_3 - 2M_{f_3}R_1(1/r + 1/R_3)}{I_1 + I_2R_1^2/4R_3^2 + I_3R_1^2/r^2 + mR_1^2}$$

3.8.3 Virtual work relation for inter-connected rigid bodies in motion

The external force system on a rigid body i is equivalent to a kinetic system consisting of a force $m_i\vec{a}_{C_i}$ at C_i and a couple of moment $\dot{\vec{H}}_{C_i}$, where

$$\dot{\vec{H}}_{C_i} = \left[I_{xx}^{C_i}\dot{\omega}_{ix} - \left(I_{yy}^{C_i} - I_{zz}^{C_i}\right)\omega_{iy}\omega_{iz}\right]\vec{\imath} + \ldots$$

in terms of principal axes x, y, z at C_i. Thus, the external and internal force system on a system of rigid bodies is equivalent to the total kinetic system consisting of forces $m_i\vec{a}_{C_i}$ at C_i and couples of moment $\dot{\vec{H}}_{C_i}$ about C_i, for all i. Hence, the virtual work done by the internal and external force system on the rigid bodies in any virtual displacement is the same as the virtual work done by the equivalent kinetic system:

$$\delta W_{\text{int+ext}} = \delta W_{\text{kinetic}} \tag{3.51}$$

$$\delta W_{\text{int+ext}}^{\text{nc}} - \delta V_{\text{int+ext}} = \sum m_i\vec{a}_{C_i}\cdot\delta\vec{r}_{C_i} + \sum \dot{\vec{H}}_{C_i}\cdot\overrightarrow{\delta\theta_i} \tag{3.52}$$

for all possible virtual displacements, where $\delta\vec{r}_{C_i}$ is the virtual displacement of C_i and $\overrightarrow{\delta\theta_i}$ is the virtual rotation of body i. By taking virtual displacement corresponding to each dof of the system, the virtual work equation (3.52) yields the required number of governing equations. The pairs of internal forces,(e.g., pair of action and reaction), if any, which together are workless in a virtual displacement, do not contribute to δW.

Example 3.16: Derive the equations of motion of the system shown in Fig. E3.16a.

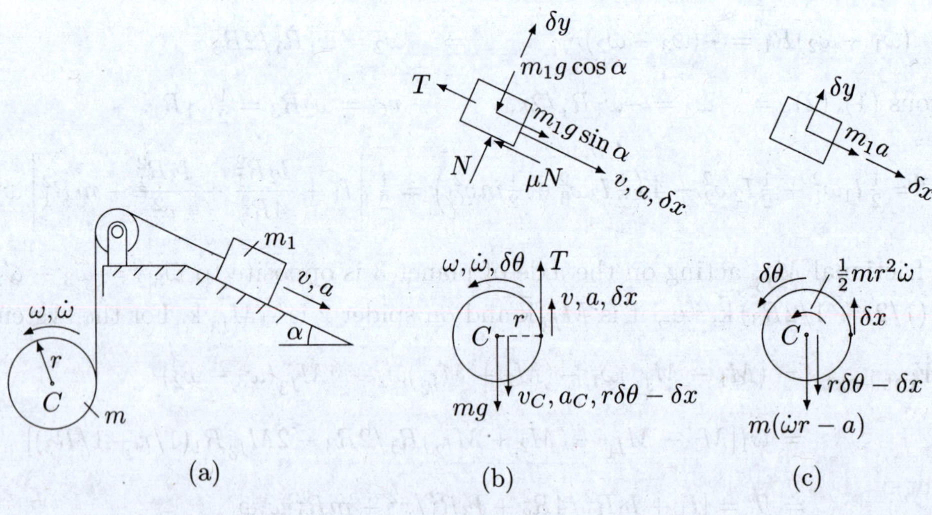

Fig. E3.16

Solution: We present two solutions. The first uses $\vec{F} = m\vec{a}_C$ and $\vec{M}_C = I_{zz}^C\dot{\vec{\omega}}$ and the second uses the virtual work relation. The FBDs are shown in Fig. E3.16b.

1. For the block: $N = m_1 g\cos\alpha$ and $m_1 a_x = F_x$ yields

$$m_1 a = m_1 g\sin\alpha - \mu N - T \tag{1}$$

For the disc: $v_C = \omega r - v$, $a_C = \dot{\omega} r - a$ and $ma_C = F$, $I^C_{zz}\dot{\omega} = M_{Cz}$ yield

$$m(\dot{\omega} r - a) = mg - T \tag{2}$$

$$\tfrac{1}{2}mr^2\dot{\omega} = Tr \tag{3}$$

$\Rightarrow T = \frac{1}{2}mr\dot{\omega} = m(g+a-\dot{\omega}r)$ and put in eqs (1), (2) to get the equations of motion:

$$m_1 a = m_1 g\sin\alpha - \mu m_1 g\cos\alpha - m(g + a - \dot{\omega}r) \tag{4}$$

$$\tfrac{3}{2}m\dot{\omega}r - ma = mg \tag{5}$$

2. The equivalent kinetic system is shown in Fig. E3.16c. Consider the virtual work equation of the system of block and disc for virtual displacement corresponding to δx, δy, $\delta\theta$ shown in Fig. E3.16c. The virtual work of the pair of internal tensions T on the block and the disc is $-T\,\delta x + T\,\delta x = 0$. Hence, $\delta W_{\text{int+ext}} = \delta W_{\text{kinetic}} \Rightarrow$

$$(m_1 g\sin\alpha - \mu N)\delta x + (N - m_1 g\cos\alpha)\delta y + mg(r\delta\theta - \delta x)$$
$$= m_1 a\delta x + m(\dot{\omega}r - a)(r\delta\theta - \delta x) + \tfrac{1}{2}mr^2\dot{\omega}\delta\theta$$

for all δx, δy, $\delta\theta$. Equating the coefficients of δx, δy, $\delta\theta$ yields

$\delta y:\quad N - m_1 g\cos\alpha = 0$

$\delta x:\quad m_1 g\sin\alpha - \mu N - mg = m_1 a - m(\dot{\omega}r - a) \quad\Rightarrow\quad (4)$

$\delta\theta:\quad mgr = m(\dot{\omega}r - a)r + \frac{1}{2}mr^2\dot{\omega} \quad\Rightarrow\quad (5)$

Note: In this approach the tension T is automatically eliminated.

3.8.4 Dynamics of a rigid body relative to a translating frame

We present the governing equations of motion of a rigid body B of mass m moving relative to a translating frame T moving at $\vec{v}_T(t)$, $\vec{a}_T(t)$ with respect to an inertial frame I. Angular velocity $\vec{\omega}$ of rigid body relative to I and T is the same. If A is a point of the rigid body, then $\vec{H}_{A|T} = I^A_{ij}\omega_j\,\hat{e}_i$. The expression of kinetic energy T' relative to frame T follows from eq. (3.37) as:

$$T' = \tfrac{1}{2}mv^2_{C|T} + \tfrac{1}{2}\vec{H}_{C|T}\cdot\vec{\omega} = \tfrac{1}{2}mv^2_{C|T} + \tfrac{1}{2}I^C_{ij}\omega_i\omega_j \tag{1}$$

Let the actual external force system be augmented by an 'inertia force' $-m\vec{a}_{A|T}$ acting at C. It has been shown in Section 2.8.2 that Euler's two axioms yield

$$m\vec{a}_{C|T} = \vec{F}', \qquad \dot{\vec{H}}_{A|T} = \vec{M}'_A \tag{2}$$

if point A is coincident with C / $\vec{a}_{A|T} = \vec{0}$ / $\vec{a}_{A|T}$ is along AC. $\vec{F}'$ and $\vec{M}'_A$ are the force sum and moment sum for the augmented force system. The relations of eq. (2) are identical to those relative to I, with the augmented force system replacing the external force system. Time integrals of eqs (2) yield augmented impulse–momentum and augmented angular impulse–moment of momentum relations:

$$\vec{\mathrm{I}}'(t_1, t_2) = m\Delta\vec{v}_{C|T}, \qquad \vec{\mathrm{I}}'_{\mathrm{ang}_A}(t_1, t_2) = \Delta\vec{H}_{A|T} \tag{3}$$

Equation (2) ⇒ Euler's equations of motion: eq. (3.34) for body-fixed principal axes of inertia $\vec{\mathrm{e}}_i$ at its material point A, or eq. (3.30) for case of $\vec{\omega} = \omega\,\vec{\mathrm{k}}$ for body-fixed axes x, y, z at A are in terms of components of moment $\vec{M}'_A$ of augmented force system:

$$\begin{aligned} M'_{A_1} &= I^A_{11}\dot{\omega}_1 - (I^A_{22} - I^A_{33})\omega_2\omega_3, & & & M'_{A_x} &= I^A_{xz}\dot{\omega} - I^A_{yz}\omega^2 \\ M'_{A_2} &= I^A_{22}\dot{\omega}_2 - (I^A_{33} - I^A_{11})\omega_3\omega_1, & &\text{or} & M'_{A_y} &= I^A_{yz}\dot{\omega} + I^A_{xz}\omega^2 \\ M'_{A_3} &= I^A_{33}\dot{\omega}_3 - (I^A_{11} - I^A_{22})\omega_1\omega_2, & & & M'_{A_z} &= I^A_{zz}\dot{\omega} \end{aligned} \tag{4}$$

For a rigid body, the work–energy relation (3.46) relative to I follows from $\vec{F} = m\vec{a}_{C|I}$ and $\vec{M}_C = \dot{\vec{H}}_{C|I}$. Similarly, from eqs (2) follows the work–energy relation relative to T:

$$\dot{T}' = \dot{W}', \qquad T'_2 - T'_1 = W'_{1\text{–}2} \tag{5}$$

$$\dot{T}' + \dot{V}' = \dot{W}^{\mathrm{nc}'}, \qquad T'_2 + V'_2 - (T'_1 + V'_1) = W^{\mathrm{nc}'}_{1\text{–}2} \tag{6}$$

T' is the kinetic energy relative to T, W' is the work done by the augmented force system and V' is the potential energy of conservative forces in the augmented system.

3.9 Balancing of Rotors

Consider a rotor (a body rotating about a fixed axis) rotating at constant $\vec{\omega}$ about an axis ($\vec{\mathrm{k}}$) fixed in the inertial frame I of the ground (Fig. 3.16). Its mass is m with centre of mass $C\,(x_C, y_C, z_C)$. Its inertia matrix elements at A involving z are $I^A_{xz}, I^A_{yz}, I^A_{zz}$. The body-fixed $\vec{\mathrm{i}}, \vec{\mathrm{j}}$ are related to ground-fixed $\vec{\mathrm{I}}, \vec{\mathrm{J}}$ by

$\vec{\mathrm{i}} = \cos\omega t\,\vec{\mathrm{I}} + \sin\omega t\,\vec{\mathrm{J}}, \quad \vec{\mathrm{j}} = -\sin\omega t\,\vec{\mathrm{I}} + \cos\omega t\,\vec{\mathrm{J}}$. For $\dot{\omega} = 0$, eqs (3.30) yield:

$$\begin{aligned} \vec{M}_A &= -I^A_{yz}\omega^2\,\vec{\mathrm{i}} + I^A_{xz}\omega^2\,\vec{\mathrm{j}} \\ &= -\omega^2(I^A_{xz}\sin\omega t + I^A_{yz}\cos\omega t)\,\vec{\mathrm{I}} + \omega^2(I^A_{xz}\cos\omega t - I^A_{yz}\sin\omega t)\,\vec{\mathrm{J}} \end{aligned} \tag{a}$$

Hence, a constant moment $\vec{M}_A$ relative to the body and a harmonic moment relative to I is needed for constant $\vec{\omega}$. The harmonic forces needed from the bearings to provide $\vec{M}_A$ will cause undesirable forced vibrations of the structure and the body causing wear. This effect, being proportional to ω^2, is more pronounced at high ω.

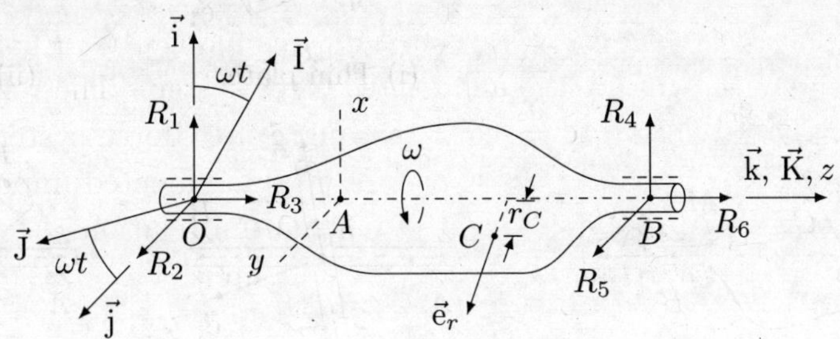

Fig. 3.16

A rotor is said to be *balanced if the dynamic bearing reactions are zero when it rotates at constant* ω, i.e., if

$$m\vec{a}_C = -m\omega^2 r_C\,\vec{e}_r = \vec{F} = \vec{0} \quad \Rightarrow \quad r_C = 0, \quad x_C = 0, \quad y_C = 0$$

$$-I^C_{xz}\omega^2\,\vec{\imath} + I^C_{yz}\omega^2\,\vec{\jmath} = \vec{M}_C = \vec{0} \quad \Rightarrow \quad I^C_{xz} = 0, \quad I^C_{yz} = 0$$

i.e., if C is on the axis of rotation and this axis is a principal axis at C. Further, $I^A_{xz} = I^C_{xz} - mx_Cz_C = 0$, $I^A_{yz} = I^C_{yz} - my_Cz_C = 0$, $\Rightarrow$ the z-axis is a principal axis at every point on it. Hence, *the necessary and sufficient conditions of balancing are:*

1. *Centre of mass C lies on the axis of rotation.*
2. *Axis of rotation is a principal axis at one point A of itself.*

Analytically, the four conditions of balancing are

$$x_C = 0, \qquad y_C = 0, \qquad I^A_{xz} = 0, \quad I^A_{yz} = 0 \tag{3.53}$$

For an unbalanced rotor with $r_C = 0.5\,\text{mm}$, $\omega = 6000\,\text{rpm}$ (typical speed for a jet engine), the harmonic force on the bearings has amplitude $m\omega^2 r_C \simeq 20mg$, i.e., twenty times the static load. The dynamic magnification factor for harmonic forced response (see Section 6.15) further enhances the effect on deflection, wear and noise. Balancing the rotor of a jet engine, the crankshaft of an internal combustion engine, the drum of a washing machine, etc., is extremely important to avoid forced oscillations.

Example 3.17: A rigid body of mass m with centre of mass C is rotating about a non-principal axis of inertia (Fig. E3.17a). The inertia matrix at point O for axes x, y, z attached to the body has elements I^O_{xx}, I^O_{xy}, ... Find the bearing reactions for the given position. The bearing at B does not provide any axial constraint. Consider the particular cases shown in Fig. E3.17b.

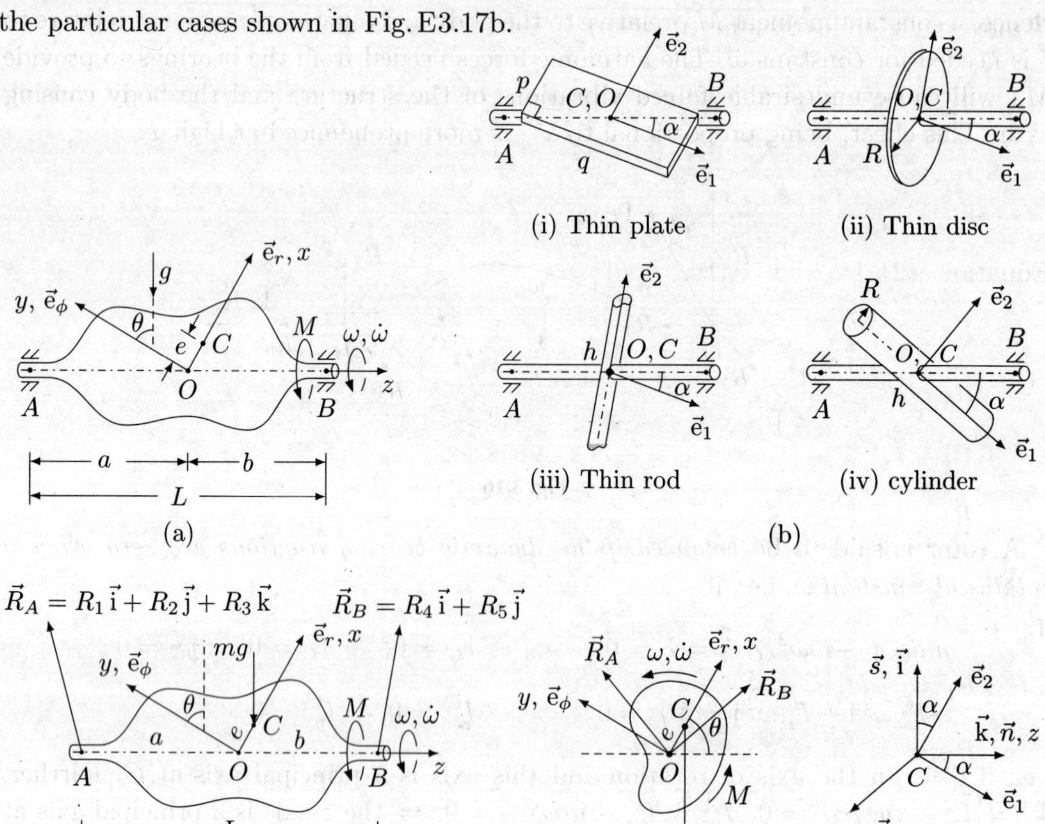

Fig. E3.17

Solution: $\vec{\omega} = \omega\,\vec{k}$, $\overrightarrow{OC} = e\,\vec{i}$. Let $\vec{\dot{\omega}} = \dot{\omega}\,\vec{k}$, $\vec{R}_A = R_1\,\vec{i} + R_2\,\vec{j} + R_3\,\vec{k}$, $\vec{R}_B = R_4\,\vec{i} + R_5\,\vec{j}$. The FBD of the body is shown in Fig. E3.17c. The 6 unknowns, $\dot{\omega}$, R_1, R_2, R_3, R_4, R_5, are obtained using the moment equation for the fixed point O (or A) and $\vec{F} = m\vec{a}_C$:

$$\begin{aligned}
\vec{M}_O &= -a\,\vec{k} \times (R_1\,\vec{i} + R_2\,\vec{j} + R_3\,\vec{k}) + b\,\vec{k} \times (R_4\,\vec{i} + R_5\,\vec{j}) \\
&\quad + M\,\vec{k} + e\,\vec{i} \times mg(-\sin\theta\,\vec{i} - \cos\theta\,\vec{j}) \\
&= (aR_2 - bR_5)\,\vec{i} + (bR_4 - aR_1)\,\vec{j} + (M - mge\cos\theta)\,\vec{k}
\end{aligned}$$

$$\Rightarrow \quad M_{Ox} = aR_2 - bR_5 = I^O_{xz}\dot{\omega} - I^O_{yz}\omega^2 \tag{1}$$

$$M_{Oy} = bR_4 - aR_1 = I^O_{yz}\dot{\omega} + I^O_{xz}\omega^2 \tag{2}$$

$$M_{Oz} = M - mge\cos\theta = I^O_{zz}\dot{\omega} \quad \Rightarrow \quad \dot{\omega} = (M - mge\cos\theta)/I^O_{zz} \tag{3}$$

C moves in a circle of radius e: $\Rightarrow \vec{a}_C = -\omega^2 e\,\vec{\mathrm{i}} + \dot{\omega} e\,\vec{\mathrm{j}}$. Component-wise $\vec{F} = m\vec{a}_C \Rightarrow$

$$\vec{\mathrm{i}}: \qquad R_1 + R_4 - mg\sin\theta = -m\omega^2 e \tag{4}$$

$$\vec{\mathrm{j}}: \qquad R_2 + R_5 - mg\cos\theta = m\dot{\omega} e \tag{5}$$

$$\vec{\mathrm{k}}: \qquad R_3 = 0 \tag{6}$$

Equations (2), (4) and (1), (5) yield

$$\begin{aligned} R_1 &= -(I^O_{yz}\dot{\omega} + I^O_{xz}\omega^2 + m\omega^2 eb - mgb\sin\theta)/L \\ R_4 &= (I^O_{yz}\dot{\omega} + I^O_{xz}\omega^2 - m\omega^2 ea + mga\sin\theta)/L \\ R_2 &= (I^O_{xz}\dot{\omega} - I^O_{yz}\omega^2 + m\dot{\omega} eb + mgb\cos\theta)/L \\ R_5 &= (-I^O_{xz}\dot{\omega} + I^O_{yz}\omega^2 + m\dot{\omega} ea + mga\cos\theta)/L \end{aligned} \tag{7}$$

We consider some particular cases, using eq. (7).

1. $e = 0$, i.e., the centre of mass C is on the axis of rotation.

$$R_1 = -(I^O_{yz}\dot{\omega} + I^O_{xz}\omega^2 - mgb\sin\theta)/L, \qquad R_2 = (I^O_{xz}\dot{\omega} - I^O_{yz}\omega^2 + mgb\cos\theta)/L$$

$$R_4 = (I^O_{yz}\dot{\omega} + I^O_{xz}\omega^2 + mga\sin\theta)/L, \qquad R_5 = (-I^O_{xz}\dot{\omega} + I^O_{yz}\omega^2 + mga\cos\theta)/L$$

2. $I^O_{xz} = I^O_{yz} = 0$, i.e., z-axis is a principal axis of inertia at O.

$$R_1 = (-m\omega^2 eb + mgb\sin\theta)/L, \qquad R_2 = (m\dot{\omega} eb + mgb\cos\theta)/L$$

$$R_4 = (-m\omega^2 ea + mga\sin\theta)/L, \qquad R_5 = (m\dot{\omega} ea + mga\cos\theta)/L$$

3. $e = 0$, $I^O_{xz} = I^O_{yz} = 0$, i.e., the axis of rotation is a principal axis of inertia at C.

$$R_1 = mgb\sin\theta/L, \qquad R_2 = mgb\cos\theta/L$$

$$R_4 = mga\sin\theta/L, \qquad R_5 = mga\cos\theta/L$$

i.e., only static reactions due to force mg act since the *rotor is balanced.*

4. $e = 0$. The axis of rotation $\vec{\mathrm{k}}$ is coplanar with the principal axes of inertia $\vec{\mathrm{e}}_1$, $\vec{\mathrm{e}}_2$ at O (Fig. E3.17d). Choose $\vec{\mathrm{j}} = \vec{\mathrm{e}}_3 \Rightarrow \vec{\mathrm{j}}$ is a principal axis at O, i.e., $I^O_{yz} = 0$. $\vec{\mathrm{i}} = \vec{\mathrm{j}} \times \vec{\mathrm{k}}$ is in the plane of $\vec{\mathrm{e}}_1$ and $\vec{\mathrm{e}}_2$. We compute I^O_{xz} using the transformation relations:

$$\vec{\mathrm{k}} = \vec{n} \quad \Rightarrow \quad n_1 = \cos\alpha, \qquad n_2 = \sin\alpha, \qquad n_3 = 0$$

$$\vec{\mathrm{i}} = \vec{s} \quad \Rightarrow \quad s_1 = -\sin\alpha, \qquad s_2 = \cos\alpha, \qquad s_3 = 0$$

$$I^O_{xz} = I^O_{sn} = I^O_{ij}s_i n_j = I^O_{11}s_1 n_1 + I^O_{22}s_2 n_2 + I^O_{33}s_3 n_3 = \tfrac{1}{2}(I^O_{22} - I^O_{11})\sin 2\alpha \tag{8}$$

$$R_1 = (mgb\sin\theta - I^O_{xz}\omega^2)/L, \qquad R_2 = (mgb\cos\theta + I^O_{xz}\dot{\omega})/L$$
$$R_4 = (mga\sin\theta + I^O_{xz}\omega^2)/L, \qquad R_5 = (mga\cos\theta - I^O_{xz}\dot{\omega})/L$$

where I^O_{xz} is given by eq. (8). For the bodies in Fig. E3.17b:

$$\text{(i)} \quad I^O_{xz} = \tfrac{1}{2}\left(\tfrac{1}{12}mq^2 - \tfrac{1}{12}mp^2\right)\sin 2\alpha = \tfrac{1}{24}m\left(q^2 - p^2\right)\sin 2\alpha$$
$$\text{(ii)} \quad I^O_{xz} = \tfrac{1}{2}\left(\tfrac{1}{4}mR^2 - \tfrac{1}{2}mR^2\right)\sin 2\alpha = -\tfrac{1}{8}mR^2\sin 2\alpha$$
$$\text{(iii)} \quad I^O_{xz} = \tfrac{1}{2}\left(0 - \tfrac{1}{12}mh^2\right)\sin 2\alpha = -\tfrac{1}{24}mh^2\sin 2\alpha$$
$$\text{(iv)} \quad I^O_{xz} = \tfrac{1}{2}\left[m\left(\tfrac{1}{4}R^2 + \tfrac{1}{12}h^2\right) - \tfrac{1}{2}mR^2\right]\sin 2\alpha = \tfrac{1}{2}m\left(\tfrac{1}{12}h^2 - \tfrac{1}{4}R^2\right)\sin 2\alpha$$

It is interesting to note that the dynamic reactions are zero for a square plate and a misaligned cylinder if $h = \sqrt{3}R$, since the axis of rotation is a principal axis of inertia.

3.10 Centre of Percussion

Consider a rigid body (Fig. 3.17a,b) with fixed axis of rotation and centre of mass $C\,(h, 0, 0)$, subjected to an impulsive force $\vec{\hat{F}} = \hat{F}_x\vec{\imath} + \hat{F}_y\vec{\jmath} + \hat{F}_z\vec{k}$ at $Q\,(x, y, z)$. Let the reaction at O be an impulsive force $\vec{\hat{R}}_O = \hat{R}_x\vec{\imath} + \hat{R}_y\vec{\jmath} + \hat{R}_z\vec{k}$ and an impulsive couple $\vec{\hat{C}}_O = \hat{C}_x\vec{\imath} + \hat{C}_y\vec{\jmath}$. $\vec{\hat{R}}_O$, $\vec{\hat{C}}_O$ and the instantaneous $\Delta\omega$ are obtained using $\vec{I}_{\text{ang}_O} = \Delta\vec{H}_O$:

$$\hat{C}_x\vec{\imath} + \hat{C}_y\vec{\jmath} + (x\vec{\imath} + y\vec{\jmath} + z\vec{k})\times(\hat{F}_x\vec{\imath} + \hat{F}_y\vec{\jmath} + \hat{F}_z\vec{k}) = (I^O_{xz}\vec{\imath} + I^O_{yz}\vec{\jmath} + I^O_{zz}\vec{k})\Delta\omega$$

$\vec{k}$: $$\Delta\omega = (x\hat{F}_y - y\hat{F}_x)/I^O_{zz} \tag{1}$$

$\vec{\jmath}$: $$\hat{C}_y = I^O_{yz}\Delta\omega + x\hat{F}_z - z\hat{F}_x \tag{2}$$

$\vec{\imath}$: $$\hat{C}_x = I^O_{xz}\Delta\omega + z\hat{F}_y - y\hat{F}_z \tag{3}$$

and $\vec{I} = m\Delta\vec{v}_C$: $\quad (\hat{R}_x + \hat{F}_x)\vec{\imath} + (\hat{R}_y + \hat{F}_y)\vec{\jmath} + (\hat{R}_z + \hat{F}_z)\vec{k} = mh\,\Delta\omega\,\vec{\jmath} \quad \Rightarrow$

$\vec{\imath}$: $$\hat{R}_x = -\hat{F}_x \tag{4}$$

$\vec{\jmath}$: $$\hat{R}_y = mh\,\Delta\omega - \hat{F}_y \tag{5}$$

$\vec{k}$: $$\hat{R}_z = -\hat{F}_z \tag{6}$$

Thus, in general, there is an impulsive reaction at O. Let us find the conditions to be satisfied by: $\hat{\vec{F}}$, its point of application Q and $[I^C]$, so that there is no impulsive reaction, i.e., $\hat{R}_x = \hat{R}_y = \hat{R}_z = 0$, $\hat{C}_x = \hat{C}_y = 0$. Equations (4), (6), (2) $\Rightarrow \hat{F}_x = \hat{F}_z = 0$, $I^O_{yz} = 0$. Substituting $\hat{F}_y = mh\,\Delta\omega$ from eq. (5) in eqs (3) and (1) yields

$$z = -I^O_{xz}/mh, \qquad x = I^O_{zz}/mh \tag{7}$$

As $z_C = 0$, using parallel axes theorem: $I^O_{xz} = I^C_{xz}$, $I^O_{yz} = I^C_{yz}$.

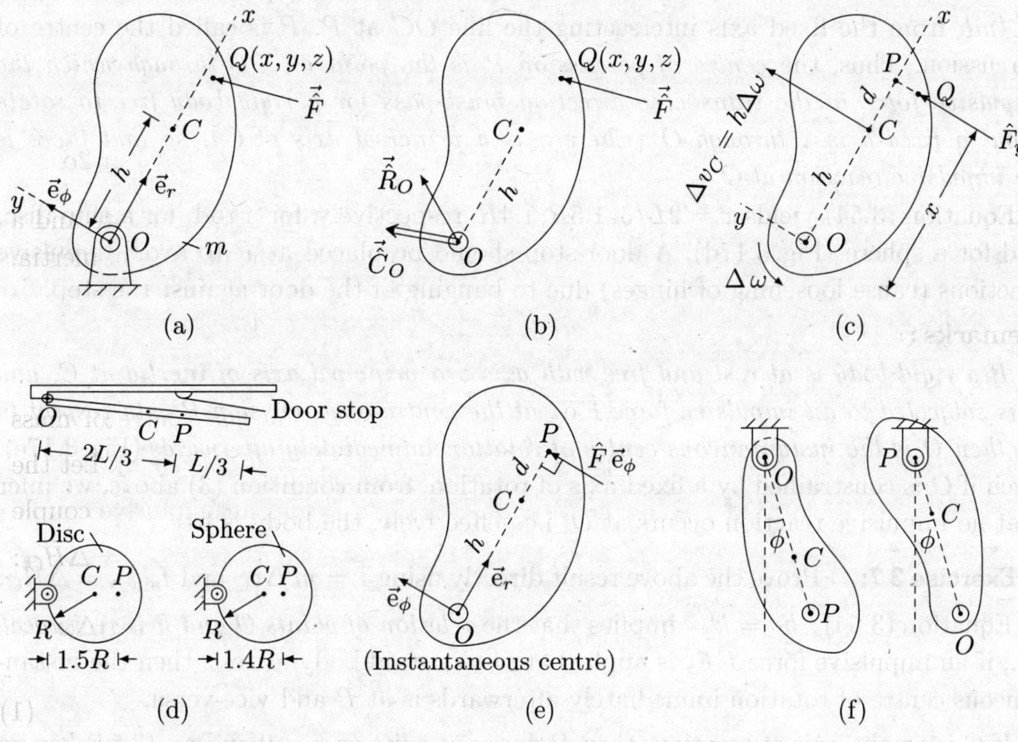

Fig. 3.17

Hence, the conditions for no impulsive reaction at O are:

1. The radial ($\vec{e}_r = \vec{i}$) and axial components of $\hat{\vec{F}}$ must be zero, i.e., it should act in the transverse direction ($\vec{e}_\phi = \vec{j}$) at Q. Hence, the y-coordinate of Q is immaterial.

2. $I^C_{yz} = 0$.

3. The radial distances x and $d = x - h$ (Fig. 3.17c) should satisfy eq. (7), i.e.,

$$mhx = I^O_{zz} = I^C_{zz} + mh^2 \;\Rightarrow\; mhd = I^C_{zz} \;\Rightarrow$$

$$hd = (k^C_z)^2 \;\Leftrightarrow\; hx = (k^O_z)^2 \tag{3.54}$$

This relation can be derived directly using FBD of Fig. 3.17c:

$$\hat{F}_y = \Delta p_y = m\,\Delta v_C = m(h\,\Delta\omega)$$

$$I_{\text{ang}_{C_z}} = d\hat{F}_y = \Delta H_{C_z} = I^C_{zz}\,\Delta\omega$$

$$\Rightarrow \quad d(mh\,\Delta\omega) = I^C_{zz}\,\Delta\omega \quad \Rightarrow \quad mhd = I^C_{zz} \quad \Rightarrow \quad hd = (k^C_z)^2$$

4. The z-coordinate of Q should satisfy $z = -I^C_{xz}/mh$. If $I^C_{xz} = 0$, then z is a principal axis of inertia at C and $z = 0$, i.e., $\vec{\hat{F}}$ should act in the xy plane at distance $x = I^O_{zz}/mh$ from the fixed axis intersecting the line OC at P. P is called the centre of percussion. Thus, the *centre of percussion P is the point on OC through which the impulsive force in the transverse direction must pass for a rigid body free to rotate about a fixed axis z through O (where z is a principal axis at C), so that there is no impulsive reaction at O.*

Equation (3.54) yields $x = 2L/3, 1.5R, 1.4R$, respectively, for a rod, for a thin disc, and for a sphere (Fig. 3.17d). A door-stop should be placed at P to avoid impulsive reactions (cause loosening of hinges) due to banging of the door against the stop.

Remarks:

1. *If a rigid body is at rest and free with axis z a principal axis of inertia at C, and it is subjected to an impulsive force $\hat{F}\,\vec{e}_\phi$ at the centre of percussion P with respect to O, then O is the instantaneous centre of rotation immediately afterwards* (Fig. 3.17e). Even if O is constrained by a fixed axis of rotation, from condition (3) above, we infer that no impulsive reaction occurs at O, i.e., effectively, the body is free.

§ **Exercise 3.7:** Prove the above result directly using $\vec{I} = m\,\Delta\vec{v}_C$ and $\vec{I}_{\text{ang}_C} = \Delta\vec{H}_C$.

2. Equation (3.54): $hd = k^{C^2}_z$ implies that the *relation of points O and P is reciprocal*, i.e., if an impulsive force $\hat{F}\,\vec{e}_\phi$ is applied at O to a rigid body at rest, then the instantaneous centre of rotation immediately afterwards is at P and vice-versa.

3. If C is on the axis of rotation, then P does not exist as $h = 0$ and eq. (3.54) has no solution for d. Every impulsive force applied would cause impulsive bearing reactions.

4. A ballistic pendulum and the pendulum hammer of an impact testing machine, should be so designed that the impulsive forces act in the circumferential direction at P. This will decrease the impulsive reactions at the bearings and decrease their wear.

5. While hitting with a hammer, or striking the ball with a cricket bat/tennis racquet, the impact point should be the centre of percussion with respect to the point where you hold it. This ensures no impulsive reaction (characteristic 'sting') on the hand. In the tennis racquet the 'sweet spot' is usually marked with strings of a distinctive colour.

6. It is shown in the next section that the simplest resultant of the external force system acting on a rigid body rotating about a fixed axis of rotation z, where the z-axis is a principal axis of inertia at C, is a single force through the centre of percussion P.

§ Exercise 3.8: (a) Prove that the motion of a rigid body suspended as a pendulum (Fig. 3.17f) with axis of rotation z, either from O or from the centre of percussion P is the same. (b) Show that the period T of small oscillations is $T = 2\pi[(OP)/g]^{1/2}$. (c)Prove that the period T of a compound pendulum is minimum if it is suspended from O at a distance $OC = k_z^C$. The error due to inaccurate length $h = OC$ of a pendulum clock is given by $\Delta T = (dT/dh)|_h \Delta h + \frac{1}{2}(d^2T/dh^2)|_h(\Delta h)^2 + \ldots$. Show that in the design of high precision pendulum clocks, OC is chosen equal to k_z^C so that the error ΔT for this case is of second order.

Example 3.18: An impact testing machine (Fig. E3.18a) is used to find the energy absorption capacity of a specimen during impact. A pendulum, with centre of mass C and radius of gyration k_A about the axis of rotation, is released from $\theta = \theta_0$ and strikes specimen at S when $\theta = 90°$. The frictional couple is M_f. Find the support reaction for the position θ. Find the position of S to avoid impulsive reactions.

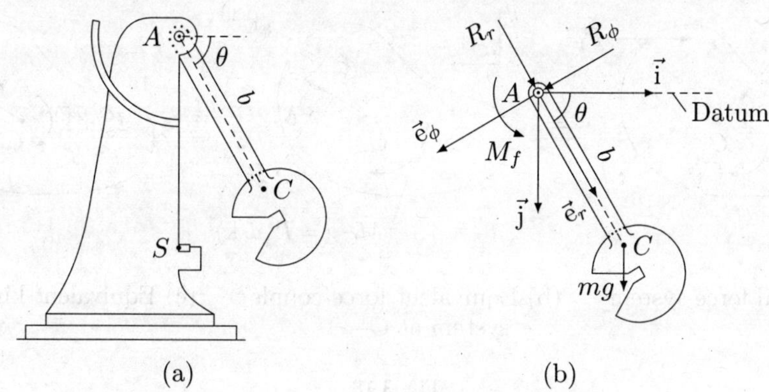

Fig. E3.18

Solution: The FBD of the pendulum is shown in Fig. E3.18b with M_f in a direction opposite to ω. $\dot{\omega}, R_r, R_\phi$ are found from the 3 equations of motion:

$$I_{zz}^A \dot{\omega} = M_{A_z}: \qquad mk_A^2 \dot{\omega} = mgb\cos\theta - M_f$$

$$F_r = ma_{C_r}: \qquad R_r + mg\sin\theta = -m\omega^2 b$$

$$F_\phi = ma_{C_\phi}: \qquad R_\phi + mg\cos\theta = m\dot{\omega}b$$

$$\Rightarrow \qquad \dot{\omega} = (mgb\cos\theta - M_f)/mk_A^2 \qquad (1)$$

$$R_r = -m(\omega^2 b + g\sin\theta) \qquad (2)$$

$$R_\phi = [(b^2 - k_A^2)mg\cos\theta - M_f b]/k_A^2 \qquad (3)$$

ω required for R_r can be found by integrating eq. (1) or by the work–energy relation:

$$T_2 + V_2 - T_1 - V_1 = \tfrac{1}{2} mk_A^2\omega^2 - mgb\sin\theta - (-mgb\sin\theta_0) = W_{1-2}^{\text{nc}} = -M_f(\theta - \theta_0)$$

$$\Rightarrow \quad \omega = [2\{mgb(\sin\theta - \sin\theta_0) - M_f(\theta - \theta_0)\}/mk_A^2]^{1/2}$$

For no impulsive reaction, S should be the centre of percussion with respect to A, i.e., $k_A^2 = (AC)(AS) \Rightarrow AS = k_A^2/b$.

3.11 Plane Motion with z-Axis as a Principal Axis of Inertia at C

There are two ways of applying Euler's axioms as given in eqs (3.27) and (3.28) for plane motion of a rigid body with $\vec{\omega} = \omega\,\mathrm{k}$, $\dot{\vec{\omega}} = \dot{\omega}\,\mathrm{k}$, where z-axis is a principal axis of inertia at C ($\Rightarrow \vec{H}_C = I_{zz}^C\omega\,\mathrm{k}$), under the external force system shown in Fig. 3.18a.

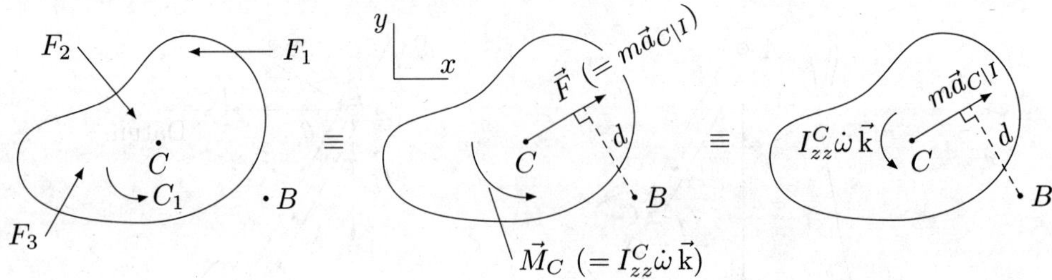

(a) External force system (b) Equivalent force-couple system at C (c) Equivalent kinetic system

Fig. 3.18

1. $$\vec{F} = m\vec{a}_{C|I}, \qquad \vec{M}_C = I_{zz}^C\dot{\omega}\,\mathrm{k} \tag{3.55}$$

In particular, for coplanar loads, the three non-trivial equations are:

$$M_{C_z} = I_{zz}^C\dot{\omega}$$

and $F_x = ma_{C_x}$, $F_y = ma_{C_y}$

or $F_r = ma_{C_r}$, $F_\phi = ma_{C_\phi}$

or $F_t = ma_{C_t}$, $F_n = ma_{C_n}$

2. The resultant of the external force system at C consists of a force $\vec{F}$ ($= m\vec{a}_{C|I}$) and a couple of moment $\vec{C}_{R_C} = \vec{M}_C$ ($= I_{zz}^C\dot{\omega}\,\mathrm{k}$) (Fig. 3.18b). Thus, eqs (3.55) imply that a kinetic system (Fig. 3.18c) consisting of a 'force' $m\vec{a}_{C|I}$ at C and a couple of 'moment' $I_{zz}^C \dot{\omega}\,\vec{k}$ is equivalent to the external force system. Hence, the moments of these equivalent force systems about *any* point B, having arbitrary motion, are equal:

$$\vec{M}_B = \vec{M}_C + \vec{r}_{CB} \times \vec{F} = I_{zz}^C \dot{\omega}\,\vec{k} + \vec{r}_{CB} \times m\vec{a}_{C|I} \tag{3.56}$$

where $\vec{M}_B$ is the moment of the external forces about B. For *any point B in the plane of motion of C*, in terms of the signed distance d of vector $\vec{a}_{C|I}$ from B,

$$\vec{M}_B = (I_{zz}^C \dot{\omega} + m a_C d)\,\vec{k} \tag{3.57}$$

where the sign assigned to d is positive if $\overrightarrow{BC} \times \vec{a}_{C|I}$ is along $+\vec{k}$ and negative otherwise. For the given position, the sign of d is negative.

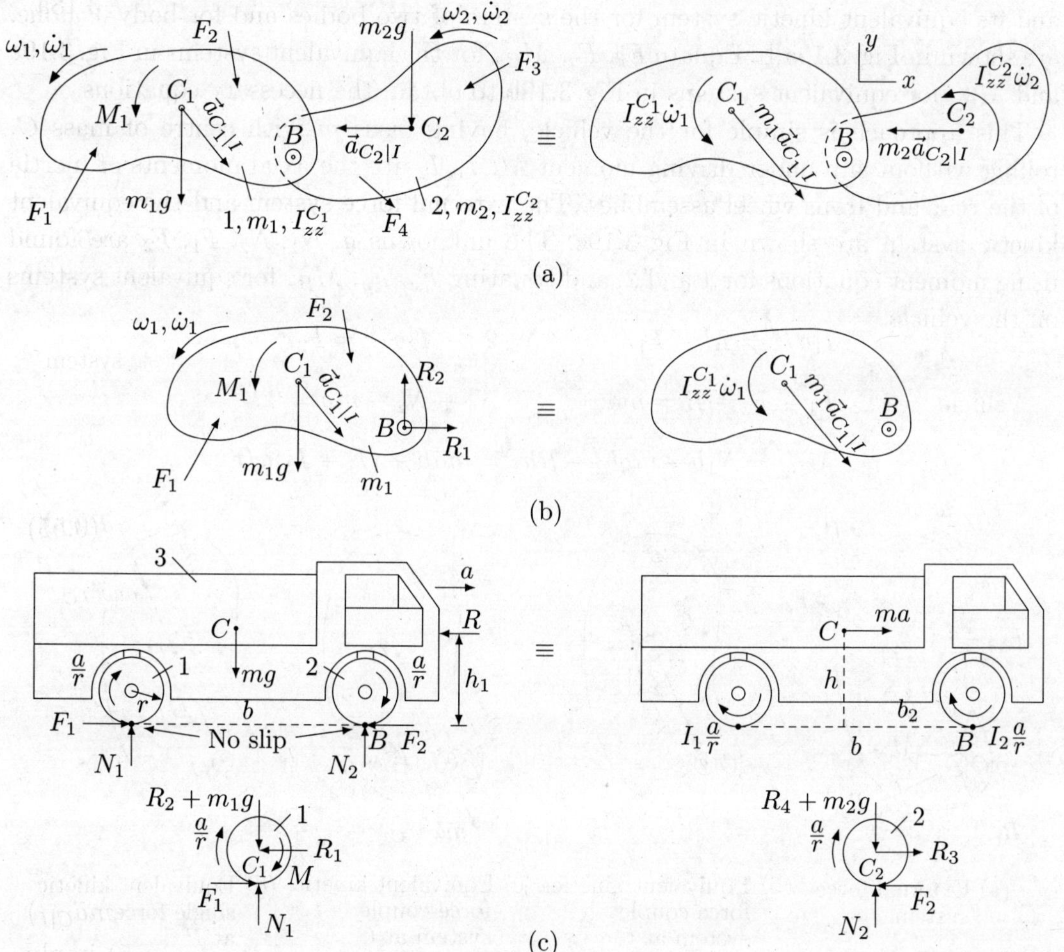

Fig. 3.19

This alternative perspective of equivalent kinetic system is sometimes useful in problems of connected bodies. The kinetic system consists of forces $m_i\vec{a}_{C_i|I}$ at C_i and moments $I_{zz}^{C_i}\dot{\omega}_i\,\vec{k}$ applied on each (ith) of its bodies. For the external force system and the equivalent kinetic system acting on the connected bodies (or any of its parts), we can equate the force sum and the moment sum $\vec{M}_B$ or the moment sum M_{BD}. This approach may completely or partially avoid finding the internal forces in the system.

Alternatively, we can equate the virtual work of the internal and external force system to the virtual work of the equivalent kinetic system for any arbitrary virtual displacement to obtain the governing equations of motion.

For example, consider the two rigid bodies in Fig. 3.19a hinged together with the z axis being a principal axis of inertia at C_1 and C_2. For the given loads, the unknowns are: $\vec{a}_B = a_1\,\vec{i} + a_2\,\vec{j}$, $\dot{\omega}_i$. Express $\vec{a}_{C_i}$ in terms of $\vec{a}_B$ and $\dot{\omega}_i$. The external force system and its equivalent kinetic system for the system of two bodies and for body 1 alone, are shown in Fig. 3.19a,b. Equate F_x, F_y, M_{B_z} for the equivalent systems in Fig. 3.19a and M_{B_z} for equivalent systems in Fig. 3.19b to obtain the necessary equations.

This approach is simple for the vehicle, having mass m with centre of mass C, rolling without slip under driving moment M. I_1, I_2 are the axial moments of inertia of the rear and front wheel assemblies. The external force system and the equivalent kinetic system are shown in Fig. 3.19c. The unknowns a, N_1, N_2, F_1, F_2 are found using moment equations for 1 and 2, and equating F_x, F_y, M_{B_z} for equivalent systems on the vehicle.

$$1: \quad I_1 a/r = M - F_1 r, \qquad 2: \quad I_2 a/r = F_2 r$$

$$\text{Vehicle}: \quad F_1 - F_2 - R = ma, \qquad N_1 + N_2 - mg = 0,$$

$$M_{B_z} = N_1 b - mgb_2 - Rh_1 = mah + (I_1 + I_2)a/r$$

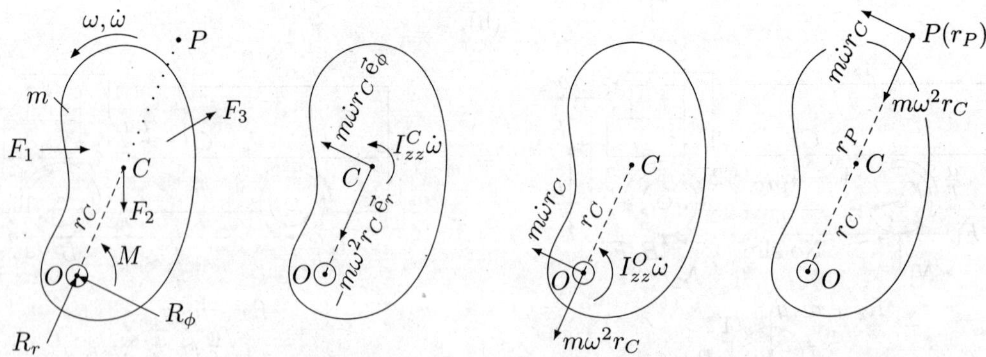

(a) External force system (b) Equivalent kinetic force-couple system at C (c) Equivalent kinetic force-couple system at O (d) Equivalent kinetic single force at P

Fig. 3.20

Consider rotation of a rigid body hinged at O (fixed in I) with the z-axis being a principal axis of inertia at C, under a coplanar force system. The external force system and its equivalent kinetic force–couple system $m\vec{a}_{C|I}$, $I^C_{zz}\dot{\omega}\,\vec{\mathrm{k}}$ at C are shown in Figs 3.20a,b. The *alternative equivalent kinetic force–couple system at O* (Fig. 3.20c) consists of a 'force' $m\vec{a}_{C|I}$ at O and a couple of 'moment'

$$\vec{C}_{R_O} = r_C\,\vec{\mathrm{e}}_r \times m(-\omega^2 r_C\,\vec{\mathrm{e}}_r + \dot{\omega} r_C\,\vec{\mathrm{e}}_\phi) + I^C_{zz}\dot{\omega}\,\vec{\mathrm{k}} = (I^C_{zz} + m r_C^2)\dot{\omega}\,\vec{\mathrm{k}} = I^O_{zz}\dot{\omega}\,\vec{\mathrm{k}}$$

The simplest resultant of the external force system is a force $m\vec{a}_{C|I}$ at P (Fig. 3.20d)

$$M_{Oz} = (m\dot{\omega} r_C) r_P = I^O_{zz}\dot{\omega}$$

$$\Rightarrow \quad r_C r_P = k_z^{O2} = k_z^{C2} + r_C^2 \quad \Rightarrow \quad r_C(r_P - r_C) = k_z^{C2}$$

which is independent of ω, $\dot{\omega}$. Comparison with eq. (3.54) reveals that the point P through which the simplest resultant acts is the centre of percussion for the given axis of rotation at O. Since the external force system is equivalent to a force $m\vec{a}_{C|I}$ at P, $\vec{M}_P = \vec{0}$. Hence, if the transverse component R_ϕ of the bearing reactions (Fig. 3.20a) has to be zero, then the applied external forces, other than the bearing reactions, should have a resultant passing through P since

$$M_{Pz} = -R_\phi(OP) + (\text{moment of external forces other than } R_r, R_\phi, \text{ about} P)$$

$$= -R_\phi(OP) + 0 = 0 \quad \Rightarrow \quad R_\phi = 0$$

In particular, an external force applied through P does not affect component R_ϕ.

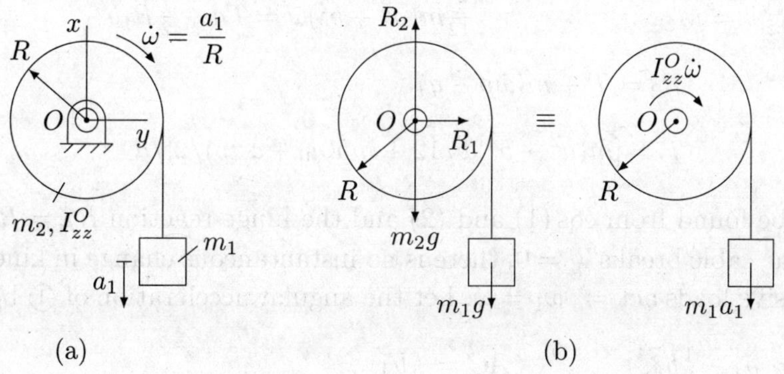

Fig. 3.21

The equation of motion, of connected system of winch (Fig. 3.21), consists of equating M_{Oz} for the external and kinetic systems, which eliminates the support reaction:

$$M_{Oz} = m_1 g R = m_1 a_1 R + I^O_{zz} a_1/R$$

Example 3.19: The instrument panels attached to a rotor are restrained by cables and have hinge supports (Fig. E3.19a). The rotor is mounted on a vehicle moving with acceleration a_0. The centre C of the panel has the same y coordinate as O. (a) Find the hinge reaction at A and the tension in cable BD for the given position. (b) If the cable BD snaps in this position, find the hinge reaction and the angular acceleration of the panel immediately afterwards. The panel is a thin rectangular plate of mass m.

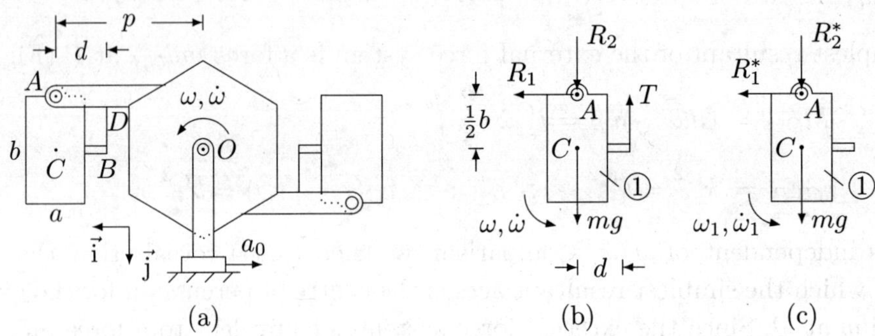

Fig. E3.19

Solution: (a) $\vec{a}_O = -a_0\,\vec{\imath}$, $\vec{a}_C = \vec{a}_O - \omega^2 p\,\vec{\imath} + \dot{\omega} p\,\vec{\jmath} = -(a_0 + \omega^2 p)\,\vec{\imath} + \dot{\omega} p\,\vec{\jmath}$. The FBD of ① is shown in Fig. E3.19b. T, R_1, R_2 are found using 3 equations of motion:

$$F_x = ma_{C_x}: \qquad R_1 = -m(a_0 + \omega^2 p) \tag{1}$$

$$F_y = ma_{C_y}: \qquad R_2 - T + mg = m\dot{\omega} p$$

$$I^C_{zz}\dot{\omega} = M_{C_z}: \qquad \tfrac{1}{12}m(a^2 + b^2)\dot{\omega} = Td + \tfrac{1}{2}R_1 b$$

$$\Rightarrow \qquad R_2 = T + m(\dot{\omega} p - g) \tag{2}$$

$$T = [m(a^2 + b^2)\dot{\omega}/12 + mb(a_0 + \omega^2 p)/2]/d \tag{3}$$

R_1, R_2 can be found from eqs (1) and (2) and the hinge reaction $\vec{R}_A = R_1\,\vec{\imath} + R_2\,\vec{\jmath}$.
(b) When the cable breaks $T = 0$, there is no instantaneous change in kinematics of ① as no impulsive loads act $\Rightarrow$ $\omega_1 = \omega$. Let the angular acceleration of ① be $\dot{\vec{\omega}}_1 = \dot{\omega}_1\,\vec{k}$.

$$\overrightarrow{OA} = p\,\vec{\imath} - \tfrac{1}{2}b\,\vec{\jmath}, \qquad \overrightarrow{AC} = \tfrac{1}{2}b\,\vec{\jmath}$$

$$\vec{a}_A = \vec{a}_O + \dot{\omega}\,\vec{k} \times \overrightarrow{OA} - \omega^2\overrightarrow{OA} = (\tfrac{1}{2}\dot{\omega} b - \omega^2 p - a_0)\,\vec{\imath} + (\dot{\omega} p + \tfrac{1}{2}\omega^2 b)\,\vec{\jmath}$$

$$\vec{a}_C = \vec{a}_A + \dot{\omega}_1\,\vec{k} \times \overrightarrow{AC} - \omega^2\overrightarrow{AC} = (\tfrac{1}{2}\dot{\omega} b - \omega^2 p - a_0 - \tfrac{1}{2}\dot{\omega}_1 b)\,\vec{\imath} + \dot{\omega} p\,\vec{\jmath}$$

FBD is shown in Fig. E3.19c. R_1^*, R_2^* and $\dot{\omega}_1$ are found from the equations of motion:

$$F_x = ma_{Cx}: \qquad R_1^* = m(\tfrac{1}{2}\dot{\omega}b - \omega^2 p - a_0 - \tfrac{1}{2}\dot{\omega}_1 b) \tag{4}$$

$$F_y = ma_{Cy}: \qquad R_2^* + mg = m\dot{\omega}p$$

$$I_{zz}^C\dot{\omega}_1 = M_{Cz}: \qquad \tfrac{1}{12}m(a^2+b^2)\dot{\omega}_1 = \tfrac{1}{2}R_1^* b$$

$$\Rightarrow \qquad R_2^* = m(\dot{\omega}p - g) \tag{5}$$

$$\dot{\omega}_1 = 6(\tfrac{1}{2}b\dot{\omega} - a_0 - \omega^2 p)b/(a^2+4b^2) \tag{6}$$

R_1^*, R_2^* can then be found from eqs (4) and (5) and the hinge reaction $\vec{R}_A^* = R_1^*\vec{\mathrm{i}} + R_2^*\vec{\mathrm{j}}$.

3.12 Illustrative Examples

The following steps are useful for solving problems of plane motion.

1. Locate the centre of mass of each body.
2. Draw the FBD of each body with external loads from within/outside the system.
3. Select A at C for general plane motion, and on the axis (for fixed axis rotation) where most unknown forces act. Select the triad $\vec{\mathrm{i}}, \vec{\mathrm{j}}, \vec{\mathrm{k}}$ for each body with $\vec{\mathrm{k}}$ normal to the plane of motion and x, $y \ni I_{xz}^A, I_{yz}^A$ are easy to compute.
4. Write any other given information in the form of equations.
5. Write the kinematical relations, if any, between $\vec{\omega}_i$, $\dot{\vec{\omega}}_i$, $\vec{a}_{C_i}$.
6. Compute I_{xz}^A, I_{yz}^A, I_{zz}^A for each body that has rotational motion.
7. For each body, set up equations of motion:

$$F_x = ma_{Cx}, \qquad F_y = ma_{Cy}, \qquad F_z = ma_{Cz}$$

$$M_{Ax} = I_{xz}^A\dot{\omega} - I_{yz}^A\omega^2, \qquad M_{Ay} = I_{yz}^A\dot{\omega} + I_{xz}^A\omega^2, \qquad M_{Az} = I_{zz}^A\dot{\omega}$$

   Depending on the given data, replace F_x, F_y, F_z by F_r, F_ϕ, F_z or F_t, F_n, F_b.
8. If the number of equations and the number of unknowns do not match, then look for some unused kinematical constraint or a friction condition.
9. Solve the equations of motion and check whether the initial assumptions, if any, regarding the nature of the motion or the forces, are satisfied. For example, if no slip is assumed, then check whether $|\vec{F}|/|\vec{N}| < \mu_s$. If not, rework with assumed slip with the friction force $\mu_k|\vec{N}|$ opposite to the actual slip. If slip was assumed initially, then check whether the friction force at P in contact with Q is opposite to the component of $\vec{a}_{PQ}$ in the tangent plane of contact.

If the body/system of bodies have one dof and if T and $\dot{W}$ can be written easily in the general configuration, then the best method is to use $\dot{T} = \dot{W}$. If impulsive loads act, then use impulse–momentum and angular-impulse–moment-of-momentum relations.

The following steps are useful for problems of three-dimensional motion.

1. Draw the FBD and write in equation form the given information that does not appear in the FBD. List all entities required in the solution.
2. Select a material point A of the rigid body, usually either C or a point fixed in I, for which the moment of momentum equation is valid.
3. Select a frame m $(\vec{\mathrm{e}}_i)$ with origin at $A \ni [I^A]$ remains constant and its computation is simple. Select axes $\vec{\mathrm{E}}_i$ in frame I that best fit the fixed axis of rotation.
4. Find $[I^A]$ using integration, rotational transformation and parallel axes relations.
5. Using the constraints, find expressions for $\vec{\omega}$, $\dot{\vec{\omega}}$, $\vec{a}_C$, $\vec{\Omega}$ (of frame m relative to I).
6. If $\vec{\mathrm{e}}_i$ are fixed to the body, use Euler's equations of motion of a rigid body:

   $$M_{Ax} = I^A_{xx}\dot{\omega}_x - (I^A_{yy} - I^A_{zz})\omega_y\omega_z, \quad \ldots$$

   In case $\vec{\mathrm{e}}_i$ are not fixed to the body but have angular velocity $\vec{\Omega}$ relative to frame I, use $H_{A_i} = I^A_{ij}\omega_j$ and $\vec{M}_A = \dot{\vec{H}}_{A|I} = \dot{\vec{H}}_{A|m} + \vec{\Omega} \times \vec{H}_A$.
7. Apply $\vec{F} = m\vec{a}_{C|I}$. If the number of equations do not match the number of unknowns, then look for some constraint/friction condition that has not been used.
8. Solve the equations for the unknowns.

Sometimes, using a combination of principles is more efficient, e.g., $\vec{F} = m\vec{a}_C$, $\dot{T}+\dot{V} = \dot{W}^{\mathrm{nc}}$, $\Delta(T+V) = \delta W^{\mathrm{nc}}$, Euler's equation for $M_{A\mathrm{e}}$ where $\vec{\mathrm{e}}$ is a body fixed axis and Euler's axiom $M_{O_{\mathrm{E}}} = \dot{H}_{O_{\mathrm{E}}}$, where O is either C or fixed in I and $\vec{\mathrm{E}}$ is fixed in I.

Sometimes, a moment equation applied at a point where a large number of unknown forces act is very useful. Such a moment equation in terms of *arbitrary point P* is:

$$\vec{M}_P = \dot{\vec{H}}_C + \vec{r}_{CP} \times m\vec{a}_C, \qquad \text{where, } \dot{\vec{H}}_C \text{ is obtained as follows.}$$

(a) For principal axes of inertia x, y, z at C: $\dot{\vec{H}}_C = [I^C_{xx}\dot{\omega}_x - (I^C_{yy} - I^C_{zz})\omega_y\omega_z]\,\vec{\imath} + \ldots$

(b) For $\vec{\omega} = \omega\,\vec{\mathrm{k}}$, $\dot{\vec{\omega}} = \dot{\omega}\,\vec{\mathrm{k}}$: $\dot{\vec{H}}_C = (I^C_{xz}\dot{\omega} - I^C_{yz}\omega^2)\,\vec{\imath} + (I^C_{yz}\dot{\omega} + I^C_{xz}\omega^2)\,\vec{\jmath} + I^C_{zz}\dot{\omega}\,\vec{\mathrm{k}}$

and $M_{P_z} = I^C_{zz}\dot{\omega} + (\vec{r}_{CP} \times m\vec{a}_C)\cdot\vec{\mathrm{k}}$.

If P is an *arbitrary point of the rigid body*, then

$$\vec{M}_P = \dot{\vec{H}}_P + \vec{r}_{CP} \times m\vec{a}_P, \qquad \text{where } \dot{\vec{H}}_P \text{ is obtained as follows.}$$

(a) For principal axes of inertia x, y, z at P: $\dot{\vec{H}}_P = [I^P_{xx}\dot{\omega}_x - (I^P_{yy} - I^P_{zz})\omega_y\omega_z]\,\vec{\imath} + \ldots$

(b) For $\vec{\omega} = \omega\,\vec{\mathrm{k}}$, $\dot{\vec{\omega}} = \dot{\omega}\,\vec{\mathrm{k}}$: $\dot{\vec{H}}_P = (I^P_{xz}\dot{\omega} - I^P_{yz}\omega^2)\,\vec{\imath} + (I^P_{yz}\dot{\omega} + I^P_{xz}\omega^2)\,\vec{\jmath} + I^P_{zz}\dot{\omega}\,\vec{\mathrm{k}}$

and $M_{P_z} = I^P_{zz}\dot{\omega} + (\vec{r}_{CP} \times m\vec{a}_P)\cdot\vec{\mathrm{k}}$.

Example 3.20: *Banking or super-elevation of highway and railroad curves.* An automobile is travelling over a circular road 'banked' at an angle θ. Find proper value of θ if its centre of mass C travels at constant speed v (Fig. E3.20a). Neglect the inertia of the wheels. The principal axes of inertia at C are the axes 1, 2, 3 with the principal moments of inertia I_{11}^C, I_{22}^C, I_{33}^C. Find also the condition of tipping or sliding sideways.

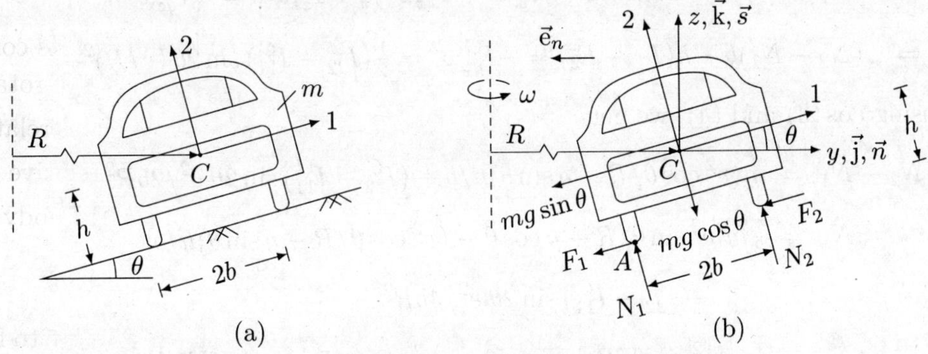

Fig. E3.20

Solution: Super-elevation/banking is the difference in the elevation of the outer and inner edges of a curved track to reduce the tendency of a vehicle to slide away from the centre or to overturn. The main body rotates about a fixed vertical axis. Using $\vec{v}_C:\ v = \omega R\ \Rightarrow\ \omega = v/R\ \Rightarrow\ \dot{\omega} = \dot{v}/R = 0$. C moves in a circle of radius R $\Rightarrow$

$$\vec{a}_C = \frac{v^2}{R}\,\vec{e}_n = \frac{v^2}{R}(-\cos\theta\,\vec{e}_1 + \sin\theta\,\vec{e}_2) \tag{1}$$

In the FBD (Fig. E3.20b), N_1, F_1 are the normal and frictional components of the sum of the forces on the front and rear inner wheels. N_2, F_2 correspond to the outer wheels. This assumes that the front and rear wheels point in the same direction, which is a simplification of the actual situation. No force is shown along $\vec{e}_3$, i.e., the forward frictional force exerted by the road is balanced by the air resistance. We use the moment equation about C. $I_{xz}^C = 0$, as $\vec{i} = \vec{e}_3$ is a principal axis of inertia.

$$\vec{j} = \vec{n}\quad\Rightarrow\quad n_1 = \cos\theta,\quad n_2 = -\sin\theta,\quad n_3 = 0$$

$$\vec{k} = \vec{s}\quad\Rightarrow\quad s_1 = \sin\theta,\quad s_2 = \cos\theta,\quad s_3 = 0$$

$$I_{yz}^C = I_{ns}^C = I_{ij}^C n_i s_j = I_{11}^C n_1 s_1 + I_{22}^C n_2 s_2 + I_{33}^C n_3 s_3 = \tfrac{1}{2}(I_{11}^C - I_{22}^C)\sin 2\theta \tag{2}$$

Using eq. (1), the components of $\vec{F} = m\vec{a}_C$ yield

$$\vec{e}_1:\qquad -F_1 - F_2 - mg\sin\theta = m(-v^2\cos\theta/R)$$

$$\vec{e}_2:\qquad N_1 + N_2 - mg\cos\theta = m(v^2\sin\theta/R)$$

$$\Rightarrow \quad F_1 + F_2 = m(v^2 \cos\theta/R - g\sin\theta) \tag{3}$$

$$N_1 + N_2 = m(v^2 \sin\theta/R + g\cos\theta) \tag{4}$$

Using eq. (2) and $\dot{\omega} = 0$, $I^C_{xz} = 0$,

$$\vec{M}_C = (I^C_{xz}\dot{\omega} - I^C_{yz}\omega^2)\,\vec{\mathrm{i}} + (I^C_{yz}\dot{\omega} + I^C_{xz}\omega^2)\,\vec{\mathrm{j}} + I^C_{zz}\dot{\omega}\,\vec{\mathrm{k}} = -I^C_{yz}\omega^2\,\vec{\mathrm{i}}$$

$$\Rightarrow \quad (N_2 - N_1)b - h(F_1 + F_2) = -I^C_{yz}\omega^2 = \tfrac{1}{2}(I^C_{22} - I^C_{11})\sin 2\theta\,(v/R)^2$$

and using eqs (3) and (4), we get

$$N_2 - N_1 = m(v^2\cos\theta/R - g\sin\theta)h/b + (I^C_{22} - I^C_{11})\sin 2\theta\, v^2/2bR^2 \tag{5}$$

$$N_1 = \tfrac{1}{2}m[v^2\sin\theta/R + g\cos\theta - (v^2\cos\theta/R - g\sin\theta)h/b] - (I^C_{22} - I^C_{11})\sin 2\theta\, v^2/4bR^2 \tag{6}$$

$$N_2 = \tfrac{1}{2}m[v^2\sin\theta/R + g\cos\theta + (v^2\cos\theta/R - g\sin\theta)h/b] + (I^C_{22} - I^C_{11})\sin 2\theta\, v^2/4bR^2 \tag{7}$$

We consider some special cases.

1. *For proper banking*, $F_1 + F_2 = 0$. Eq. (3) yields the *optimum angle* θ for banking

$$F_1 + F_2 = m(v^2\cos\theta/R - g\sin\theta) = 0 \quad \Rightarrow \quad \tan\theta = v^2/Rg \tag{8}$$

2. *For no outward slip*, $F_1 + F_2 \le \mu(N_1 + N_2)$, where $\mu = \tan\phi$ is the static coefficient of friction, eqs (3), (4) $\Rightarrow$

$$m(v^2\cos\theta/R - g\sin\theta) \le \mu m(v^2\sin\theta/R + g\cos\theta)$$

$$\Rightarrow \quad v^2 \le gR(\tan\theta + \mu)/(1 - \mu\tan\theta) = gR\tan(\theta + \phi) \tag{9}$$

3. *For no inward slip*:

$$-(F_1 + F_2) \le \mu(N_1 + N_2) \quad \Rightarrow \quad v^2 \ge gR\tan(\theta - \phi) \tag{10}$$

4. *For no tipping*, the smaller normal reaction $N_1 \ge 0$. Using eq. (6), $N_1 \ge 0 \Rightarrow$

$$v^2 \le gR(\tan\theta + b/h)\,/\,[1 - (b/h)\tan\theta + (I^C_{22} - I^C_{11})\sin\theta/mRh] \tag{11}$$

Equation (11) $\Rightarrow$ the car is more likely to overturn for smaller values of R, b, θ and for larger values of h, $(I^C_{22} - I^C_{11})$, v. If $(I^C_{22} - I^C_{11})/mRh \ll 1$, then eq. (11)$\Rightarrow$

$$v^2 \le gR(\tan\theta + b/h)\,/\,[1 - (b/h)\tan\theta] \tag{12}$$

Comparison of eqs (9) and (12) $\Rightarrow$ tipping would occur before slipping if $\mu > b/h$.

5. *For no banking*, $(\theta = 0)$. z-axis is a principal axis at C. Equations (6), (7) yield

$$N_1, N_2 = \tfrac{1}{2}mg(1 \mp hv^2/gbR)$$

eqs (9), (11) yield the speeds v_s and v_t for the initiation of slipping and tipping:

$$v_s^2 = \mu gR, \qquad v_t^2 = (b/h)gR, \qquad \text{smaller than those for the banked road.}$$

Tipping occurs before slipping if $\mu > b/h$.

6. *For a straight banked road* $(R = \infty)$, eqs (6), (7) $\Rightarrow$ $N_1, N_2 = \frac{1}{2}mg[\cos\theta \pm \frac{h}{b}\sin\theta]$ i.e., unlike the circular case, the inner wheels have greater normal reaction.

Example 3.21: A body with centre of mass at C (Fig. E3.21a) is hinged to a shaft at O. Its principal moments of inertia at O are $I_{xx}^O, I_{yy}^O, I_{zz}^O$. The shaft is rotated at a constant rate ω such that OC has a constant inclination θ to the vertical. Find θ.

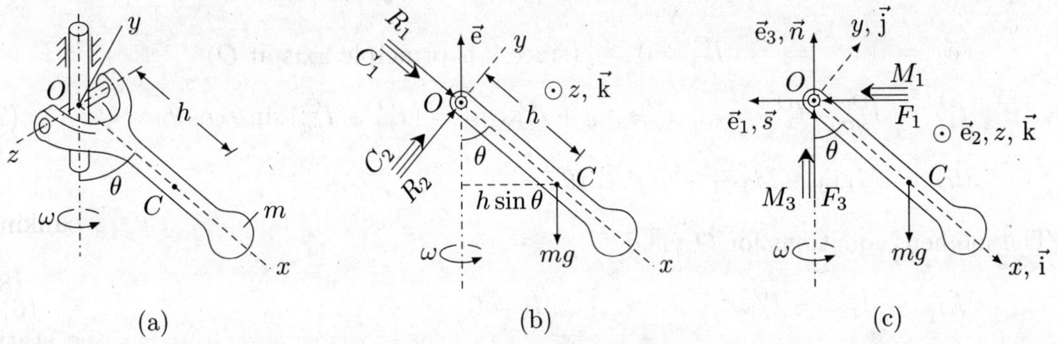

Fig. E3.21

Solution: The body has plane motion. We obtain the solution using (1) Euler's equations for principal axes, (2) moment equation for plane motion for non-principal axes.

1. The FBD is shown in Fig. E3.21b. The hinge joint does not exert any moment along the hinge axis. C_1, C_2, θ, are obtained from the Euler's equations for the fixed point O.

$$\vec{\omega} = \omega\,\vec{e} = -\omega\cos\theta\,\vec{i} + \omega\sin\theta\,\vec{j}, \qquad \dot{\vec{\omega}} = \vec{0}$$

$$\Rightarrow \quad \omega_x = -\omega\cos\theta, \quad \omega_y = \omega\sin\theta, \quad \omega_z = 0, \quad \dot{\omega}_x = \dot{\omega}_y = \dot{\omega}_z = 0 \tag{1}$$

$$\vec{M}_O = C_1\,\vec{i} + C_2\,\vec{j} - mgh\sin\theta\,\vec{k} \tag{2}$$

Euler's equations of motion for the body for O, using eqs (1) and (2) yield:

$$M_{Ox} = I_{xx}^O\dot{\omega}_x - (I_{yy}^O - I_{zz}^O)\omega_y\omega_z = C_1 \quad \Rightarrow \quad C_1 = 0 \tag{3}$$

$$M_{Oy} = I_{yy}^O\dot{\omega}_y - (I_{zz}^O - I_{xx}^O)\omega_z\omega_x = C_2 \quad \Rightarrow \quad C_2 = 0 \tag{4}$$

$$M_{Oz} = I_{zz}^O\dot{\omega}_z - (I_{xx}^O - I_{yy}^O)\omega_x\omega_y = -mgh\sin\theta$$

$$\Rightarrow \quad -mgh\sin\theta = (I_{xx}^O - I_{yy}^O)\omega^2\sin\theta\cos\theta \tag{5}$$

$$\Rightarrow \quad \sin\theta = 0\ (\theta = 0,\ \text{or}\ \theta = \pi), \quad \text{or} \quad \theta = \cos^{-1}[mgh/(I_{yy}^O - I_{xx}^O)\omega^2]$$

which exists if

$$-1 < [mgh/(I_{yy}^O - I_{xx}^O)\omega^2] < 1, \quad \text{i.e., if} \quad \omega^2 > |mgh/(I_{yy}^O - I_{xx}^O)|$$

2. The FBD of the body is shown in Fig. E3.21c. M_1, M_3, θ, are obtained from the moment equation for the fixed point O. The axis of rotation is a non-principal axis $\vec{n}$:

$$\vec{e}_3 = \vec{n} \quad \Rightarrow \quad n_x = -\cos\theta, \quad n_y = \sin\theta, \quad n_z = 0$$

$$\vec{e}_1 = \vec{s} \quad \Rightarrow \quad s_x = -\sin\theta, \quad s_y = -\cos\theta, \quad s_z = 0$$

$$\vec{e}_2 = \vec{k} \quad \Rightarrow \quad I_{23}^O = 0 \quad (\text{since } \vec{k} \text{ is principle axis at } O)$$

$$I_{13}^O = I_{sn}^O = I_{xx}^O s_x n_x + I_{yy}^O s_y n_y + I_{zz}^O s_z n_z = (I_{xx}^O - I_{yy}^O)\sin\theta\cos\theta$$

$$\vec{M}_O = M_1\vec{e}_1 + M_3\vec{e}_3 - mgh\sin\theta\,\vec{e}_2$$

The moment equations for O yield

$$M_{O_1} = I_{13}^O\dot{\omega} - I_{23}^O\omega^2 \quad \Rightarrow \quad M_1 = 0 \tag{6}$$

$$M_{O_2} = I_{23}^O\dot{\omega} + I_{13}^O\omega^2 \quad \Rightarrow \quad -mgh\sin\theta = (I_{xx}^O - I_{yy}^O)\omega^2\sin\theta\cos\theta \tag{7}$$

$$M_{O_3} = I_{33}^O\dot{\omega} \quad \Rightarrow \quad M_3 = 0 \tag{8}$$

Equation (7) for θ is the same as eq. (5). The rest of the solution is the same as before.

Example 3.22: A crane is rotating at ω, $\dot{\omega}$ about a fixed axis (Fig. E3.22a). The boom OB is supported by a cable and a hinge joint. Model it as a uniform slender rod of mass m. (a) Find the cable tension and the hinge reaction on the boom. (b) The crane rotates at uniform ω and the cable breaks. Find the initial $\ddot{\theta}$; and $\dot{\theta}$ when $\theta = \pi/2$.

Solution: (a) The principle axes of inertia x, y, z at O are chosen so that axis y is coplanar with the axis of the boom and the vertical axis. The FBD of the boom is shown in Fig. E3.22b, where C is its centre of mass. The 6 unknowns, $R_1, R_2, R_3, C_1, C_2, T$, are found from the 6 equations of motion of the boom: $\vec{F} = m\vec{a}_C$ and 3 Euler's equations for the fixed point O. With $\vec{e} = \cos\theta\,\vec{i} + \sin\theta\,\vec{j}$, $\vec{\omega} = \omega\,\vec{e}$, $\dot{\vec{\omega}} = \dot{\omega}\,\vec{e}$ are:

$$\vec{\omega} = \omega\cos\theta\,\vec{i} + \omega\sin\theta\,\vec{j} \quad \Rightarrow \quad \omega_x = \omega\cos\theta, \quad \omega_y = \omega\sin\theta, \quad \omega_z = 0$$

$$\dot{\vec{\omega}} = \dot{\omega}\cos\theta\,\vec{\mathrm{i}} + \dot{\omega}\sin\theta\,\vec{\mathrm{j}} \quad \Rightarrow \quad \dot{\omega}_x = \dot{\omega}\cos\theta, \quad \dot{\omega}_y = \dot{\omega}\sin\theta, \quad \dot{\omega}_z = 0$$

Points O, C are on body 2 with $\overrightarrow{OC} = \frac{1}{2}L\,\mathrm{i}$.

$$\vec{a}_C = \vec{a}_O + \dot{\vec{\omega}} \times \overrightarrow{OC} + \vec{\omega} \times (\vec{\omega} \times \overrightarrow{OC})$$

$$= \tfrac{1}{2}L\sin\theta(-\omega^2\sin\theta\,\vec{\mathrm{i}} + \omega^2\cos\theta\,\vec{\mathrm{j}} - \dot{\omega}\,\vec{\mathrm{k}})$$

$$\vec{M}_O = C_1\,\vec{\mathrm{i}} + C_2\,\vec{\mathrm{j}} + [TL\cos\theta - mg(\tfrac{1}{2}L\sin\theta)]\,\vec{\mathrm{k}}$$

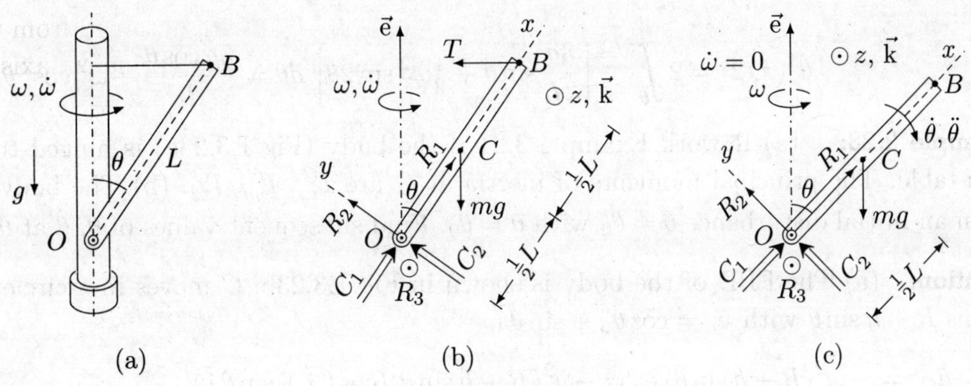

Fig. E3.22

With $I_{xx}^O = 0,\ I_{yy}^O = I_{zz}^O = mL^2/3$, Euler's equations of motion for the boom for O:

$$M_{Ox} = I_{xx}^O\dot{\omega}_x - (I_{yy}^O - I_{zz})\omega_y\omega_z = C_1 \quad \Rightarrow \quad C_1 = 0$$

$$M_{Oy} = I_{yy}^O\dot{\omega}_y - (I_{zz}^O - I_{xx})\omega_z\omega_x = C_2 \quad \Rightarrow \quad C_2 = \tfrac{1}{3}mL^2\dot{\omega}\sin\theta$$

$$M_{Oz} = I_{zz}^O\dot{\omega}_z - (I_{xx}^O - I_{yy})\omega_x\omega_y = TL\cos\theta - \tfrac{1}{2}mgL\sin\theta$$

$$\Rightarrow \quad T = \tfrac{1}{2}mg\tan\theta + \tfrac{1}{3}mL\omega^2\sin\theta$$

$\Rightarrow \vec{C}_O = C_1\,\vec{\mathrm{i}} + C_2\,\vec{\mathrm{j}} = \frac{1}{3}mL^2\dot{\omega}\sin\theta\,\vec{\mathrm{j}}$. $\vec{F} = m\vec{a}_C$ yields the force reaction $\vec{R}$:

$$\vec{R} + T(-\sin\theta\,\vec{\mathrm{i}} + \cos\theta\,\vec{\mathrm{j}}) + mg(-\cos\theta\,\vec{\mathrm{i}} - \sin\theta\,\vec{\mathrm{j}})$$

$$= \tfrac{1}{2}mL\sin\theta(-\omega^2\sin\theta\,\vec{\mathrm{i}} + \omega^2\cos\theta\,\vec{\mathrm{j}} - \dot{\omega}\,\vec{\mathrm{k}})$$

$$\Rightarrow \quad \vec{R} = [mg(1+\cos^2\theta)/2\cos\theta - \tfrac{1}{6}mL\omega^2\sin^2\theta]\,\vec{\mathrm{i}}$$

$$+\tfrac{1}{6}m\sin\theta(\omega^2L\cos\theta + 3g)\,\vec{\mathrm{j}} - \tfrac{1}{2}m\dot{\omega}L\sin\theta\,\vec{\mathrm{k}}$$

(b) After breakage, for general value of θ (Fig. E3.22c):

$$\vec{\omega} = \omega\,\vec{\mathrm{e}} - \dot{\theta}\,\vec{\mathrm{k}} \quad \Rightarrow \quad \omega_x = \omega\cos\theta, \quad \omega_y = \omega\sin\theta, \quad \omega_z = -\dot{\theta}$$

$$\dot{\vec{\omega}} = -\ddot{\theta}\,\vec{k} + \omega\vec{e}\times(-\dot{\theta}\,\vec{k}) = -\omega\dot{\theta}\sin\theta\,\vec{i} + \omega\dot{\theta}\cos\theta\,\vec{j} - \ddot{\theta}\,\vec{k}$$

$$\Rightarrow \qquad \dot{\omega}_x = -\omega\dot{\theta}\sin\theta, \qquad \dot{\omega}_y = \omega\dot{\theta}\cos\theta, \qquad \dot{\omega}_z = -\ddot{\theta}$$

Euler's eq.: $\quad M_{O_z} = I^O_{zz}\dot{\omega}_z - (I^O_{xx} - I_{yy})\omega_x\omega_y$

$$\Rightarrow \qquad \tfrac{1}{3}mL^2[-\ddot{\theta} + \omega^2\sin\theta\cos\theta] = -\tfrac{1}{2}mgL\sin\theta$$

$$\Rightarrow \qquad \ddot{\theta} = \frac{d\dot{\theta}}{d\theta}\dot{\theta} = \frac{3g}{2L}\sin\theta + \tfrac{1}{2}\omega^2\sin 2\theta$$

$$\dot{\theta}^2(\pi/2) = 2\int_\theta^{\pi/2}\left[\frac{3g}{2L}\sin\theta + \tfrac{1}{2}\omega^2\sin 2\theta\right]d\theta = \frac{3g\cos\theta}{L} + \omega^2\cos^2\theta$$

Example 3.23: (a) Rework Example 3.21 if the body (Fig. E3.23a) is hinged to a turn-table. The principal moments of inertia at C are I^C_{xx}, I^C_{yy}, I^C_{zz}. (b) The body is given an initial disturbance $\theta = \theta_0$ with $\dot{\theta} = \dot{\theta}_0$. Find subsequent values of $\ddot{\theta}$, $\dot{\theta}$ at θ.

Solution: (a) The FBD of the body is shown in Fig. E3.23b. C moves in a circle of radius $R + h\sin\theta$ with $\vec{e}_r = \cos\theta\,\vec{j} + \sin\theta\,\vec{i}$,

$$\vec{a}_C = -\omega^2(R + h\sin\theta)\,\vec{e}_r = -\omega^2(R + h\sin\theta)(\cos\theta\,\vec{j} + \sin\theta\,\vec{i})$$

$$\vec{\omega} = -\omega\cos\theta\,\vec{i} + \omega\sin\theta\,\vec{j}, \qquad \dot{\vec{\omega}} = \vec{0}$$

$$\omega_x = -\omega\cos\theta, \quad \omega_y = \omega\sin\theta, \quad \omega_z = 0, \quad \dot{\omega}_x = \dot{\omega}_y = \dot{\omega}_z = 0$$

Using these, Euler's equations for C yield

$$\vec{M}_C = -(I^C_{xx} - I^C_{yy})\omega_x\omega_y\,\vec{k} = (I^C_{xx} - I^C_{yy})\omega^2\sin\theta\cos\theta\,\vec{k}$$

The equivalent force-couple system at C consists of $\vec{F}$ and $\vec{M}_C$. We use the equation for moment component M_{A_z} about point A to eliminate the reactions:

$$M_{A_z} = \vec{M}_A\cdot\vec{k} = (\vec{M}_C + \overrightarrow{AC}\times\vec{F})\cdot\vec{k}$$

$$\Rightarrow \quad -mgh\sin\theta = M_{C_z} + h\,\vec{i}\times m\vec{a}_C\cdot\vec{k}$$

$$-mgh\sin\theta = (I^C_{xx} - I^C_{yy})\omega^2\sin\theta\cos\theta - hm\omega^2(R + h\sin\theta)\cos\theta$$

$$\Rightarrow \quad \tfrac{1}{2}\omega^2(I^C_{xx} - I^C_{yy} - mh^2)\sin 2\theta = mh(\omega^2R\cos\theta - g\sin\theta) \tag{1}$$

The transcendental equation (1) can be expressed as

$$B\sin 2\theta = D\sin(\theta - \alpha)$$

Its solution depends on the relative values of B and D. The three possible cases are shown in Fig. E3.23c. For steady θ, there are two solutions θ_1, θ_2 if $D > B$, and four solutions $\theta_1, \theta_2, \theta_3, \theta_4$ if $D \leq B$.

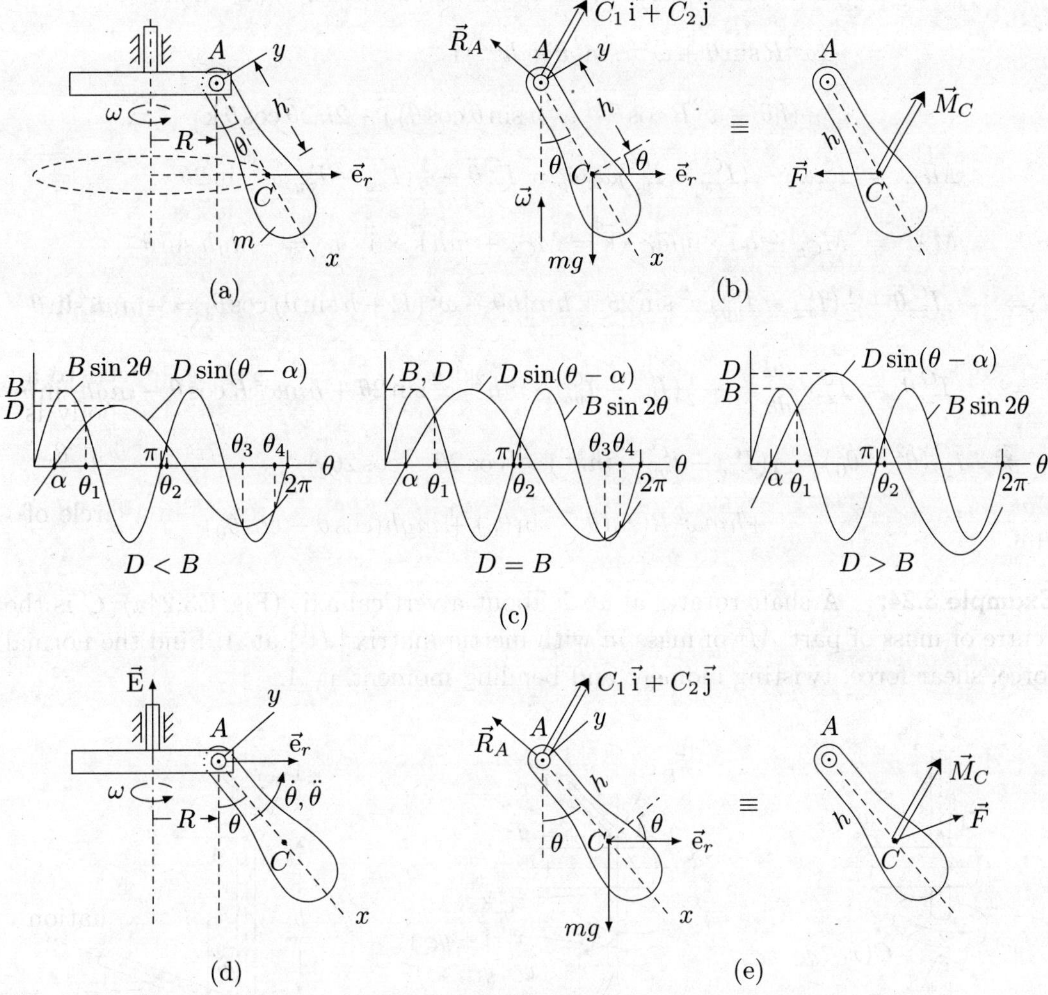

Fig. E3.23

(b) For the case of variable θ (Fig. E3.23d,e), $\vec{a}_A = -\omega^2 R\,\vec{e}_r = -\omega^2 R(\sin\theta\,\vec{i} + \cos\theta\,\vec{j})$

$$\vec{\omega} = \omega\,\vec{E} + \dot{\theta}\,\vec{k} = -\omega\cos\theta\,\vec{i} + \omega\sin\theta\,\vec{j} + \dot{\theta}\,\vec{k}$$

$$\dot{\vec{\omega}} = \ddot{\theta}\,\vec{k} + \omega\,\vec{E} \times \dot{\theta}\,\vec{k} = \ddot{\theta}\,\vec{k} + \omega(-\cos\theta\,\vec{i} + \sin\theta\,\vec{j}) \times \dot{\theta}\,\vec{k}$$

$$= \omega\dot{\theta}\sin\theta\,\vec{i} + \omega\dot{\theta}\cos\theta\,\vec{j} + \ddot{\theta}\,\vec{k}$$

$$\omega_x = -\omega\cos\theta, \quad \omega_y = \omega\sin\theta, \quad \omega_z = \dot{\theta}$$

$$\dot{\omega}_x = \omega\dot{\theta}\sin\theta, \quad \dot{\omega}_y = \omega\dot{\theta}\cos\theta, \quad \dot{\omega}_z = \ddot{\theta}$$

$$\begin{aligned}\vec{a}_C &= \vec{a}_A + \dot{\vec{\omega}} \times h\vec{\mathrm{i}} + \vec{\omega} \times (\vec{\omega} \times h\vec{\mathrm{i}}) \\ &= -\omega^2 R(\sin\theta\,\vec{\mathrm{i}} + \cos\theta\,\vec{\mathrm{j}}) + \dot{\vec{\omega}} \times h\vec{\mathrm{i}} + (\vec{\omega}\cdot h\vec{\mathrm{i}})\vec{\omega} - (\vec{\omega}\cdot\vec{\omega})h\vec{\mathrm{i}} \\ &= -(\omega^2 R\sin\theta + \omega^2 h\sin^2\theta + h\dot{\theta}^2)\vec{\mathrm{i}} \\ &\qquad +(h\ddot{\theta} - \omega^2 R\cos\theta - \omega^2 h\sin\theta\cos\theta)\vec{\mathrm{j}} - 2h\omega\dot{\theta}\cos\theta\,\vec{\mathrm{k}} \\ M_{Cz} &= I^C_{zz}\dot{\omega}_z - (I^C_{xx} - I^C_{yy})\omega_x\omega_y = I^C_{zz}\ddot{\theta} + \tfrac{1}{2}(I^C_{xx} - I^C_{yy})\omega^2\sin 2\theta \\ M_{Az} &= M_{Cz} + h\vec{\mathrm{i}} \times m\vec{a}_C \cdot \vec{\mathrm{k}} = M_{Cz} + mh\vec{\mathrm{k}} \times \vec{\mathrm{i}} \cdot \vec{a}_C = -mgh\sin\theta \\ \Rightarrow \quad & I^C_{zz}\ddot{\theta} + \tfrac{1}{2}(I^C_{xx} - I^C_{yy})\omega^2\sin 2\theta + hm[h\ddot{\theta} - \omega^2(R + h\sin\theta)\cos\theta] = -mgh\sin\theta \\ \Rightarrow \quad I^O_{zz}\ddot{\theta} &= I^O_{zz}\dot{\theta}\frac{d\dot{\theta}}{d\theta} = -\tfrac{1}{2}(I^C_{xx} - I^C_{yy} - mh^2)\omega^2\sin 2\theta + hm\omega^2 R\cos\theta - mgh\sin\theta \\ \Rightarrow \quad & I^O_{zz}(\dot{\theta}^2 - \dot{\theta}_0^2) = \tfrac{1}{4}(I^C_{xx} - I^C_{yy} - mh^2)\omega^2(\cos 2\theta - \cos 2\theta_0) \\ &\qquad + hm\omega^2 R(\sin\theta - \sin\theta_0) + mgh(\cos\theta - \cos\theta_0)\end{aligned}$$

Example 3.24: A shaft rotates at ω, $\dot{\omega}$ about a vertical axis (Fig. E3.24a). C is the centre of mass of part AB of mass m with inertia matrix $[I^A]$ at A. Find the normal force, shear force, twisting moment and bending moment at A.

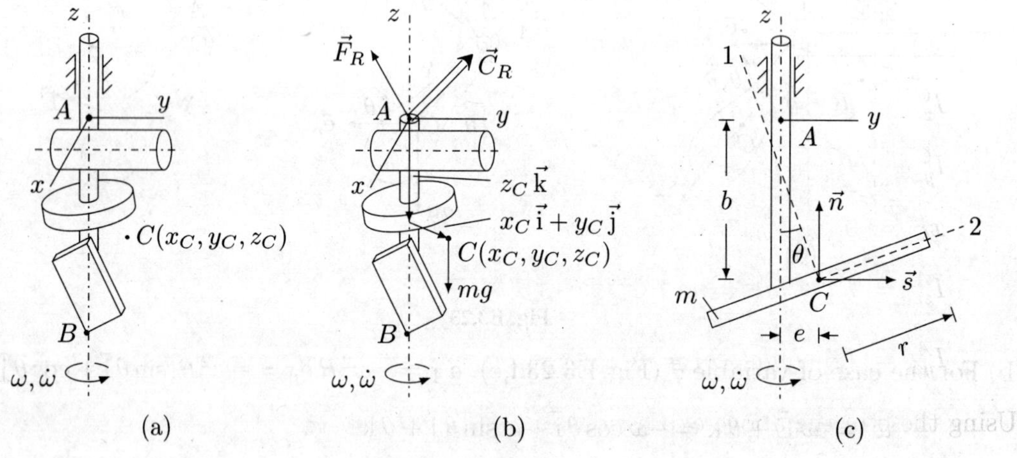

Fig. E3.24

Solution: The FBD is shown in Fig. E3.24b. The axial direction at A is $\vec{\mathrm{k}}$. So, $\vec{\mathrm{k}}$ components of $\vec{F}_R$ and $\vec{C}_R$ are normal force $\vec{N}$ and twisting moment $\vec{M}_t$, and the remaining components are the shear force $\vec{S}$ and the bending moment $\vec{M}_b$. $\vec{F} = m\vec{a}_C$

and Euler's equations for fixed point A yield $\vec{F}_R$ and $\vec{C}_R$.

$$
\begin{aligned}
\vec{F}_R - mg\,\vec{\mathrm{k}} &= m\vec{a}_C = m[\dot{\omega}\,\vec{\mathrm{k}} \times (x_C\,\vec{\mathrm{i}} + y_C\,\vec{\mathrm{j}}) - \omega^2(x_C\,\vec{\mathrm{i}} + y_C\,\vec{\mathrm{j}})] \\
\Rightarrow \qquad \vec{F}_R &= -(m\dot{\omega}y_C + m\omega^2 x_C)\,\vec{\mathrm{i}} + (m\dot{\omega}x_C - m\omega^2 y_C)\,\vec{\mathrm{j}} + mg\,\vec{\mathrm{k}} \\
\Rightarrow \qquad \vec{N} &= mg\,\vec{\mathrm{k}} \\
\vec{S} &= -(m\dot{\omega}y_C + m\omega^2 x_C)\,\vec{\mathrm{i}} + (m\dot{\omega}x_C - m\omega^2 y_C)\,\vec{\mathrm{j}} \\
\vec{M}_A &= \vec{C}_R + (x_C\,\vec{\mathrm{i}} + y_C\,\vec{\mathrm{j}}) \times (-mg\,\vec{\mathrm{k}}) \\
&= (I^A_{xz}\dot{\omega} - I^A_{yz}\omega^2)\,\vec{\mathrm{i}} + (I^A_{yz}\dot{\omega} + I^A_{xz}\omega^2)\,\vec{\mathrm{j}} + I^A_{zz}\dot{\omega}\,\vec{\mathrm{k}} \\
\Rightarrow \qquad \vec{C}_R &= (I^A_{xz}\dot{\omega} - I^A_{yz}\omega^2 + mgy_C)\,\vec{\mathrm{i}} + (I^A_{yz}\dot{\omega} + I^A_{xz}\omega^2 - mgx_C)\,\vec{\mathrm{j}} + I^A_{zz}\dot{\omega}\,\vec{\mathrm{k}} \\
\Rightarrow \qquad \vec{M}_t &= I^A_{zz}\dot{\omega}\,\vec{\mathrm{k}} \\
\vec{M}_b &= (I^A_{xz}\dot{\omega} - I^A_{yz}\omega^2 + mgy_C)\,\vec{\mathrm{i}} + (I^A_{yz}\dot{\omega} + I^A_{xz}\omega^2 - mgx_C)\,\vec{\mathrm{j}}
\end{aligned}
$$

We can apply these results to the tilted and eccentrically mounted disc shown in Fig. E3.24c. The yz-plane is a plane of mass symmetry $\Rightarrow$ the x-axis is a principal axis at $A \Rightarrow I^A_{xz} = 0$. Axes 1, 2 are principal axes at C.

$$
\begin{aligned}
\vec{\mathrm{k}} &= \vec{n}, \qquad n_1 = \cos\theta, \qquad n_2 = \sin\theta, \qquad n_3 = 0, \qquad I^C_{11} = \tfrac{1}{2}mr^2, \\
\vec{\mathrm{j}} &= \vec{s}, \qquad s_1 = -\sin\theta, \qquad s_2 = \cos\theta, \qquad s_3 = 0, \qquad I^C_{22} = \tfrac{1}{4}mr^2, \\
I^C_{zz} &= I^C_{nn} = I^C_{11}n_1^2 + I^C_{22}n_2^2 = \tfrac{1}{4}mr^2(2\cos^2\theta + \sin^2\theta) \\
I^C_{yz} &= I^C_{sn} = I^C_{11}s_1n_1 + I^C_{22}s_2n_2 = -\tfrac{1}{4}mr^2\sin\theta\cos\theta \\
x_C &= 0, \qquad y_C = e, \qquad z_C = -b \\
I^A_{zz} &= I^C_{zz} + me^2 = m\left[\tfrac{1}{4}r^2(2\cos^2\theta + \sin^2\theta) + e^2\right] \\
I^A_{yz} &= I^C_{yz} - me(-b) = m(eb - \tfrac{1}{4}r^2\sin\theta\cos\theta)
\end{aligned}
$$

Using these in the above expressions of internal force resultants, we get:

$$
\begin{aligned}
\vec{N} &= mg\,\vec{\mathrm{k}}, \qquad \vec{S} = -m\dot{\omega}e\,\vec{\mathrm{i}} - m\omega^2 e\,\vec{\mathrm{j}} \\
\vec{M}_t &= I^A_{zz}\dot{\omega}\,\vec{\mathrm{k}}, \qquad \vec{M}_b = (mge - I^A_{yz}\omega^2)\,\vec{\mathrm{i}} + I^A_{yz}\dot{\omega}\,\vec{\mathrm{j}}
\end{aligned}
$$

Example 3.25: *A gyroscopic grinder* (Fig. E3.25a) consists of a 'muller' or grinder in the form of a disc of radius r and mass m, mounted on rod OA which is rotated about a fixed vertical axis at a constant rate Ω. The disc rolls without slip and can rotate freely on the rod. Neglect radial frictional force on it. Find the force exerted by the grinder on the pan and the reaction of the pin at O on the light rod OA.

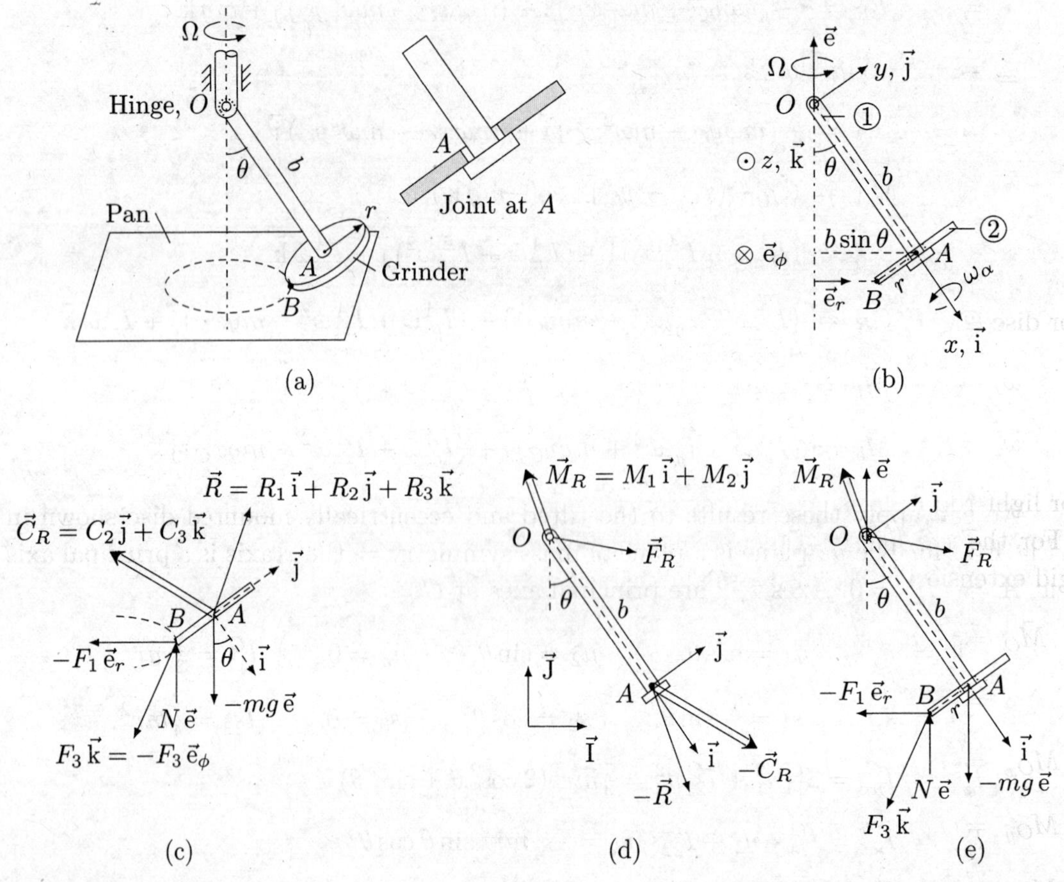

Fig. E3.25

Solution: The axis x of the disc is a principal axis at a point O on its rigid extension. O remains fixed in the ground reference. At the instant shown (Fig. E3.25b), the principal axis y is chosen in the vertical plane through the x-axis. The FBDs of the disc, the rod and the rod + disc, are shown in Figs E3.25c–e. The radial friction force $F_1 \simeq 0$ (given). $\vec{C}_R$ has no component along the axis of the disc. $\vec{M}_R$ has no component along the axis of the pin. The 12 unknowns, M_1, M_2, C_2, C_3, N, F_3, $\vec{F}_R\,(3)$, $\vec{R}\,(3)$, are determined from the 12 equations of motion of the rod (6) and rod + disc (6). The angular velocity of the disc ② relative to body ① is along $\vec{\mathrm{i}}$, say $\omega_\alpha \vec{\mathrm{i}}$. Let $\vec{\omega}_1$, $\vec{\omega}_2$ and $\dot{\vec{\omega}}_1$, $\dot{\vec{\omega}}_2$ be the angular velocities and angular accelerations of bodies ①, ② with respect to the ground. $\overrightarrow{OB} = b\,\vec{\mathrm{i}} - r\,\vec{\mathrm{j}}$, and $\vec{v}_B = \vec{0}$ (no slip).

$$
\begin{aligned}
\vec{\omega}_1 &= \Omega\,\vec{e} = \Omega(-\cos\theta\,\vec{\imath} + \sin\theta\,\vec{\jmath}), \qquad \dot{\vec{\omega}}_1 = \vec{0}, \qquad \vec{v}_O = \vec{0}\\
\vec{\omega}_2 &= \vec{\omega}_1 + \omega_\alpha\,\vec{\imath} = (\omega_\alpha - \Omega\cos\theta)\,\vec{\imath} + \Omega\sin\theta\,\vec{\jmath}\\
\vec{v}_B &= \vec{v}_O + \vec{\omega}_2 \times \overrightarrow{OB} = -[(\omega_\alpha - \Omega\cos\theta)r + b\Omega\sin\theta]\,\vec{k} = \vec{0}\\
\Rightarrow \quad \omega_\alpha &= \Omega(\cos\theta - b\sin\theta/r) \quad \Rightarrow \quad \dot{\omega}_\alpha = 0\\
\dot{\vec{\omega}}_2 &= \dot{\vec{\omega}}_1 + \dot{\omega}_\alpha\,\vec{\imath} + \vec{\omega}_1 \times \omega_\alpha\,\vec{\imath} = -\Omega\omega_\alpha\sin\theta\,\vec{k}\\
&= -\Omega^2\sin\theta(\cos\theta - b\sin\theta/r)\,\vec{k}\\
\vec{\omega}_2 &= (\omega_\alpha - \Omega\cos\theta)\,\vec{\imath} + \Omega\sin\theta\,\vec{\jmath} = -(b\Omega\sin\theta/r)\,\vec{\imath} + \Omega\sin\theta\,\vec{\jmath}
\end{aligned}
$$

For disc ②, $I^O_{xx} = \frac{1}{2}mr^2$, $I^O_{yy} = I^O_{zz} = m(\frac{1}{4}r^2 + b^2)$ and

$$
\begin{aligned}
\omega_x &= -b\Omega\sin\theta/r, \quad \omega_y = \Omega\sin\theta, \quad \omega_z = 0\\
\dot{\omega}_x &= 0, \quad \dot{\omega}_y = 0, \quad \dot{\omega}_z = -\Omega^2\sin\theta(\cos\theta - b\sin\theta/r)
\end{aligned}
$$

For light rod OA: since $\vec{R}$ intersects the axis, $M_{Ox} = M_1 = 0$.

For the rod + disc, Euler's equations for the fixed point O, on the 3-D mass-less rigid extension of the disc, are applicable since the rod is inertia-less:

$$
\begin{aligned}
\vec{M}_O &= -mgb\sin\theta\,\vec{k} + (b\sin\theta - r\cos\theta)N\,\vec{k} + M_2\,\vec{\jmath} + (b\,\vec{\imath} - r\,\vec{\jmath}) \times F_3\,\vec{k}\\
&= -rF_3\,\vec{\imath} + (M_2 - bF_3)\,\vec{\jmath} + [N(b\sin\theta - r\cos\theta) - mgb\sin\theta]\,\vec{k}\\
M_{Ox} &= I^O_{xx}\dot{\omega}_x - (I^O_{yy} - I^O_{zz})\omega_y\omega_z = -rF_3 \qquad \Rightarrow \quad F_3 = 0\\
M_{Oy} &= I^O_{yy}\dot{\omega}_y - (I^O_{zz} - I^O_{xx})\omega_z\omega_x = M_2 - bF_3 \qquad \Rightarrow \quad M_2 = 0\\
M_{Oz} &= I^O_{zz}\dot{\omega}_z - (I^O_{xx} - I^O_{yy})\omega_x\omega_y
\end{aligned}
$$

$$
\Rightarrow \quad N(b\sin\theta - r\cos\theta) - mgb\sin\theta = m\Omega^2\sin\theta[\tfrac{1}{2}br\sin\theta - (b^2 + \tfrac{1}{4}r^2)\cos\theta]
$$

$$
N = [mgb\sin\theta + m\Omega^2\sin\theta\{\tfrac{1}{2}br\sin\theta - (b^2 + \tfrac{1}{4}r^2)\cos\theta\}]/(b\sin\theta - r\cos\theta) \qquad (1)
$$

Let N for the static case ($\Omega = 0$) be $N_{\rm st}$. Equation (1) $\Rightarrow$

$$
N_{\rm st} = mgb\sin\theta/(b\sin\theta - r\cos\theta)
$$

$$
\Rightarrow \quad N/N_{\rm st} = 1 + [\sin\theta - (r/2b + 2b/r)\cos\theta]\Omega^2 r/2g \qquad (2)
$$

For $\theta = 90°$, eq. (2) $\Rightarrow$ $N/N_{\rm st} = 1 + \Omega^2 r/2g$. Thus, a large Ω $\Rightarrow$ very large N and the crushing action of the grinder is better. A moves in a circle of radius $b\sin\theta$ at rate Ω. Hence, $\vec{a}_A = -\Omega^2 b\sin\theta\,\vec{I}$. For the rod + disc, $\vec{F} = m\vec{a}_A$ $\Rightarrow$

$$\vec{F}_R + (N - mg)\,\vec{\mathrm{J}} = -\Omega^2 b \sin\theta\, \vec{\mathrm{I}}$$

$$\Rightarrow \quad \vec{F}_R = -\Omega^2 b \sin\theta\, \vec{\mathrm{I}} + (mg - N)\,\vec{\mathrm{J}}$$

where N is given by eq. (1). Note that $\vec{M}_R = M_1\,\vec{\imath} + M_2\,\vec{\jmath} = \vec{0}$.

The purpose of considering the FBD of rod OA was to show that $M_1 = 0$. This system is kinematically equivalent to one wherein rod OA is rigidly attached to the disc and the joint at O is now a ball-and-socket joint for which a priori $\vec{M}_R = \vec{0}$.

Example 3.26: A *flywheel* is a heavy axisymmetric body (Fig. E3.26a) with large axial moment of inertia I. Discuss its action as a reservoir of kinetic energy. Find the response $\omega(t)$ for net applied moment $M = M_0 + M_1 \cos pt - c\,\omega$.

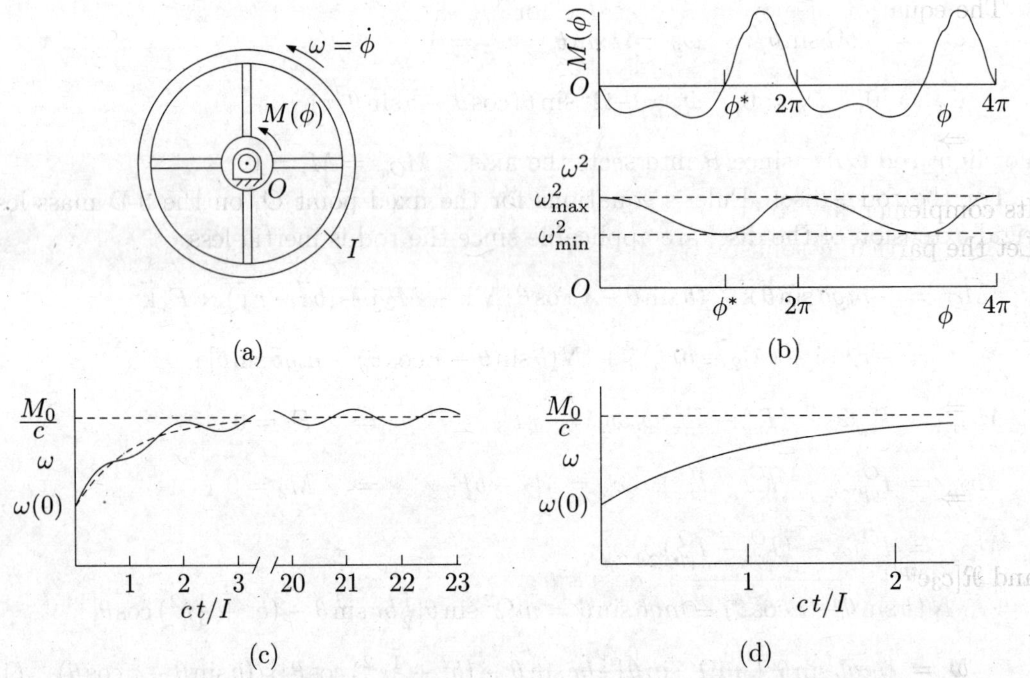

Fig. E3.26

Solution: In steam and internal combustion piston engines, the power input to the crankshaft varies periodically. If the load torque on it is constant and the net torque $M(\phi)$ is periodic with zero mean (Fig. E3.26b), then its angular velocity is periodic and will fluctuate about the mean value ω_m. The work–energy relation yields

$$\tfrac{1}{2} I(\omega_2^2 - \omega_1^2) \;=\; W_{1\text{-}2} = \int_{\phi_1}^{\phi_2} M(\phi)\, d\phi$$

$$\Rightarrow \quad \tfrac{1}{2}I(\omega_{\max}^2 - \omega_{\min}^2) = \int_{\phi^*}^{2\pi} M(\phi)\, d\phi = W \ \text{(say)}$$

$$(\omega_{\max} - \omega_{\min})/\omega_m = W/I\omega_m^2$$

For prescribed W and ω_m, the relative fluctuation of ω is less if flywheel has large I. For $0 < \phi < \phi^*$, $M(\phi) < 0$ and the flywheel slows down giving up its kinetic energy. For $\phi^* < \phi < 2\pi$, $M(\phi) > 0$ and the flywheel accelerates accumulating kinetic energy. Flywheels are also used when it is necessary to do work in a short interval of time which otherwise would require larger power, e.g., in heavy presses and giant shears for metal cutting. To increase I, the diameter of a flywheel is made large and its rim is made massive. By contrast, the rim of a pulley, used in power transmission, is made light to avoid wastage of power in accelerating it.

The equation of motion, $M_{O_z} = I\dot{\omega}$, for a given net torque yields

$$I\dot{\omega} = M_0 + M_1 \cos pt - c\,\omega$$

$$\Rightarrow \quad \dot{\omega} + \frac{c}{I}\omega = \frac{M_0}{I} + \frac{M_1}{I}\cos pt \tag{1}$$

Its complementary solution is $\omega_c = c_1 e^{\lambda t}$ with $\lambda + c/I = 0$, i.e., $\omega_c = c_1 e^{-ct/I}$. Let the particular solution be $\omega_p = c_2 + \Re[c_3 e^{ipt}]$. Substituting in eq. (1) yields

$$\frac{c_2 c}{I} + \Re\left[\left(ip + \frac{c}{I}\right) c_3 e^{ipt}\right] = \frac{M_0}{I} + \Re\left[\frac{M_1}{I} e^{ipt}\right]$$

$$\Rightarrow \quad \frac{c_2 c}{I} = \frac{M_0}{I}, \qquad \left(ip + \frac{c}{I}\right) c_3 = \frac{M_1}{I}$$

$$\Rightarrow \quad c_2 = \frac{M_0}{c}, \qquad c_3 = \frac{M_1(c/I - ip)}{I(c^2/I^2 + p^2)}$$

and $\Re[c_3 e^{ipt}] = \dfrac{M_1}{I(c^2/I^2 + p^2)}[\dfrac{c}{I}\cos pt + p \sin pt]$. The complete solution of eq. (1) is

$$\omega = c_1 e^{-ct/I} + \frac{M_0}{c} + \frac{M_1}{I(c^2/I^2 + p^2)}\left[\frac{c}{I}\cos pt + p\sin pt\right]$$

$$= c_1 e^{-ct/I} + \frac{M_0}{c} + \frac{M_1}{I(c^2/I^2 + p^2)^{1/2}}\cos(pt - \phi) \tag{2}$$

where $\phi = \tan^{-1}(pI/c)$. $\Rightarrow$ If $M_1 \neq 0$, then the steady state solution for large t is

$$\omega \simeq \frac{M_0}{c}\left[1 + \frac{M_1 c}{M_0 I(c^2/I^2 + p^2)^{1/2}}\cos(pt - \phi)\right]$$

$$\Rightarrow \quad \omega_m = M_0/c, \qquad (\omega_{\max} - \omega_{\min})/\omega_m = 2M_1 c/M_0(c^2 + I^2p^2)^{1/2}$$

The fluctuation of ω would be less if I is large, irrespective of M_1/M_0 (Fig. E3.26c).

If $M_1 = 0$, then $\omega(0) = c_1 + M_0/c = \omega_0$, $\Rightarrow$ $c_1 = \omega_0 - M_0/c$ and

$$\omega = (\omega_0 - M_0/c)e^{-ct/I} + M_0/c$$

$\Rightarrow$ as $t \to \infty$, $\omega \to M_0/c$ exponentially (Fig. E3.26d).

Example 3.27: Discuss the plane motion of an axisymmetric body, of the type of a circular cylinder or a sphere, down a plane inclined at angle α to the horizontal. Its mass is m and axial radius of gyration is k. Compare its acceleration with that of an un-powered vehicle with four light wheels shown in Fig. E3.27g.

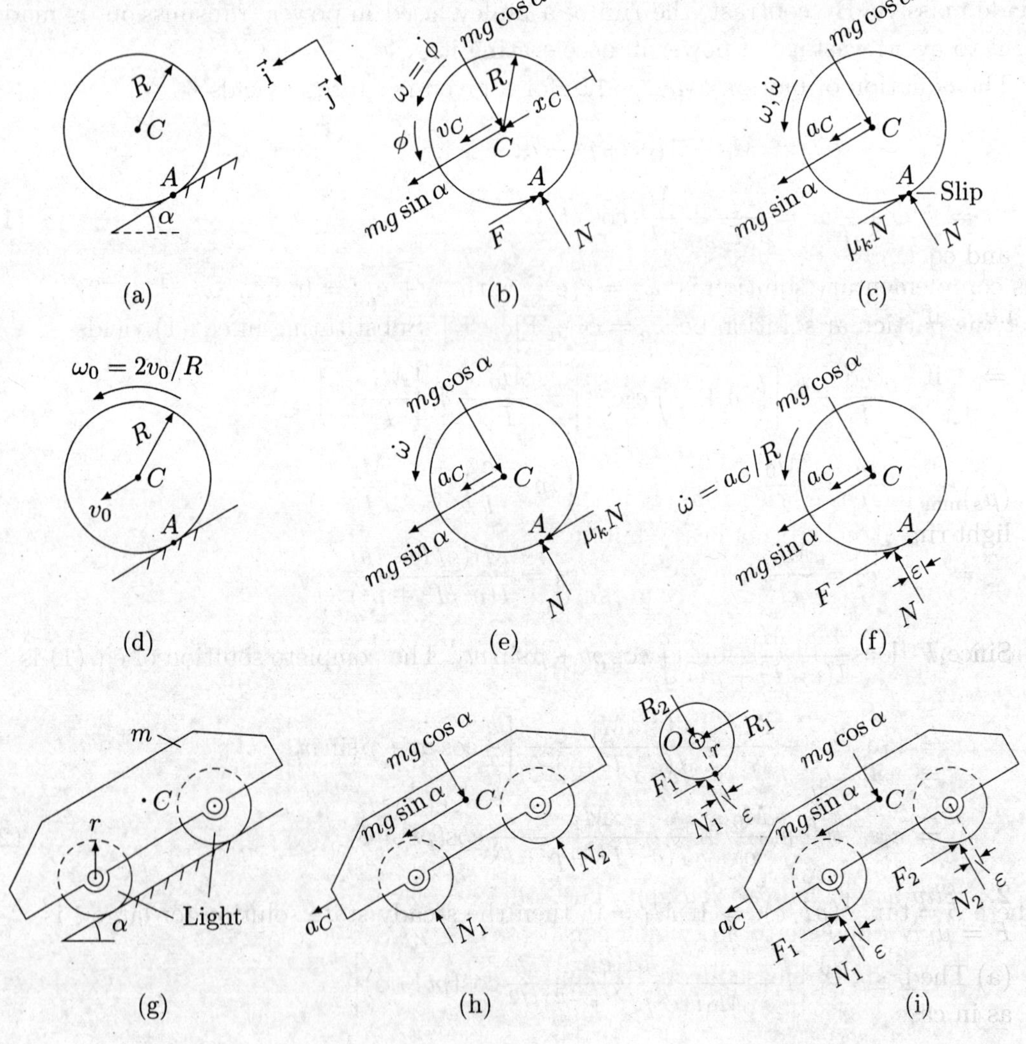

Fig. E3.27

Solution: The 2 degrees of freedom of the body (Fig. E3.27a,b) are ϕ and x_C. The FBD is shown in Fig. E3.27b. The 4 unknowns, N, F, x_C, ϕ, are obtained using 3 equations of motion: $F_x = ma_{Cx}$, $F_y = ma_{Cy}$, $I^C_{zz}\dot{\omega} = M_{Cz}$, and one condition at the contact point A. $\vec{v}_A = (v_C - \omega R)\,\vec{\mathrm{i}}$. We consider 4 cases: No slip with Coulomb friction, slip with Coulomb friction, no slip with rolling friction and the case of a smooth plane.

1. *No slip with Coulomb friction.* No slip $\Rightarrow$ $v_C = \omega R,\ a_C = \dot{\omega} R$: the 4th equation.

$$F_y = ma_{Cy}: \quad mg\cos\alpha - N = 0$$

$$F_x = ma_{Cx}: \quad mg\sin\alpha - F = ma_C$$

$$I^C_{zz}\dot{\omega} = M_{Cz}: \quad mk^2(a_C/R) = FR$$

$$\Rightarrow \quad N = mg\cos\alpha \tag{1}$$

$$F = m(g\sin\alpha - a_C) \tag{2}$$

$$a_C = g\sin\alpha/(1 + k^2/R^2) \tag{3}$$

and eq. (2) $\Rightarrow$ $F = mg\sin\alpha/(1 + R^2/k^2)$. The body rolls without slip if $|F| < \mu_s N$,

i.e., if $\quad mg\sin\alpha/(1 + R^2/k^2) < \mu_s mg\cos\alpha$

$\Rightarrow$ if $\quad \mu_s > (\mu_s)_{\min} = \tan\alpha/(1 + R^2/k^2)$

$$\alpha < \tan^{-1}[\mu_s(1 + R^2/k^2)]$$

$(\mu_s)_{\min}$ increases with k/R. It is the largest for a thin ring and very small for a large light rim attached to a heavy hub. $\vec{a}_A$ is along AC $\Rightarrow$ a_C can *also* be obtained using

$$M_{A_z} = I^A_{zz}\dot{\omega}: \quad mg\sin\alpha\, R = m(k^2 + R^2)(a_C/R)$$

Since F does no work for no slip, a_C can *also* be obtained using $\dot{T} = \dot{W}$:

$$T = \tfrac{1}{2} I^A_{zz}\omega^2 = \tfrac{1}{2} m(k^2 + R^2)v_C^2/R^2$$

$$\Rightarrow \quad \dot{T} = m(k^2 + R^2)v_C a_C/R^2 = \dot{W} = mg\sin\alpha\, v_C$$

$$\Rightarrow \quad a_C = g\sin\alpha/(1 + k^2/R^2)$$

2. *Slip with Coulomb friction.* The fourth equation is the kinetic condition of slip: $F = \mu_k N$ opposite to $\vec{v}_A$, which can be either way. We consider two cases.
(a) The body is released from rest and $\mu_s < (\mu_s)_{\min}$. $F = \mu_k N$ is in the same direction as in case 1 (Fig. E3.27c). $N, a_C, \dot{\omega}$ are found from the 3 equations of motion:

$$F_y = ma_{Cy}: \quad mg\cos\alpha - N = 0$$

$$F_x = ma_{C_x}: \qquad mg\sin\alpha - \mu_k N = ma_C$$

$$I_{zz}^C \dot{\omega} = M_{C_z}: \qquad mk^2\dot{\omega} = \mu_k NR$$

$$\Rightarrow \qquad N = mg\cos\alpha \tag{4}$$

$$a_C = g(\sin\alpha - \mu_k\cos\alpha) \tag{5}$$

$$\dot{\omega} = \mu_k gR\cos\alpha/k^2 \tag{6}$$

We check the assumed direction of slip. $\vec{a}_A = (a_C - \dot{\omega}R)\vec{\imath} - \omega^2 R\vec{\jmath}$:

$$a_{A_x} = a_C - \dot{\omega}R = [\tan\alpha - \mu_k(1 + R^2/k^2)]g\cos\alpha > 0$$

since $\mu_k < (\mu_s)_{\min} = \tan\alpha/(1 + R^2/k^2)$. The assumed $\mu_k N$ is opposite to a_{A_x} (o.k.).
(b) A spherical ball has initial $\omega_0 = 2v_0/R$ and v_0 (Fig. E3.27d). $\vec{v}_A(0) = (v_0 - \omega_0 R)\vec{\imath} = -v_0\vec{\imath} \Rightarrow$ an initial *slip phase* with $F = \mu_k N$ in direction $\vec{\imath}$ (Fig. E3.27e).

$$F_y = ma_{C_y}: \qquad mg\cos\alpha - N = 0$$

$$F_x = ma_{C_x}: \qquad mg\sin\alpha + \mu_k N = ma_C$$

$$I_{zz}^C \dot{\omega} = M_{C_z}: \qquad \frac{2}{5}mR^2\dot{\omega} = -\mu_k NR$$

$$\Rightarrow \qquad N = mg\cos\alpha \tag{7}$$

$$a_C = g(\sin\alpha + \mu_k\cos\alpha) \tag{8}$$

$$\dot{\omega} = -2.5\mu_k g\cos\alpha/R \tag{9}$$

$$\text{Eqs (8), (9)} \Rightarrow \quad v_C(t) = v_0 + g(\sin\alpha + \mu_k\cos\alpha)t$$

$$\omega(t) = 2v_0/R - 2.5\mu_k gt\cos\alpha/R$$

The slip phase continues up to $t = t_1$ when $v_C(t_1) = \omega(t_1)R$:

$$v_0 + g(\sin\alpha + \mu_k\cos\alpha)t_1 = 2v_0 - 2.5\mu_k gt_1\cos\alpha$$

$$\Rightarrow \qquad t_1 = v_0/g(\sin\alpha + 3.5\mu_k\cos\alpha)$$

For $t > t_1$, the direction of the friction force changes sign and
(i) if $\mu_s \geq (\mu_s)_{\min} = 2\tan\alpha/7$ (from case 1), then there is no slip and $a_C \overset{(3)}{=} 5g\sin\alpha/7$,
(ii) if $\mu_s < (\mu_s)_{\min} = 2\tan\alpha/7$, then there is slip and $a_C, \dot{\omega}$ are given by eqs (5), (6).
3. *No slip with rolling friction.* No slip $\Rightarrow v_C = \omega R,\ a_C = \dot{\omega}R$: the 4th equation. Let the coefficient of rolling friction be ϵ. N, F, a_c in Fig. E3.27f are obtained using

$$F_y = ma_{Cy}: \qquad mg\cos\alpha - N = 0$$

$$F_x = ma_{Cx}: \qquad mg\sin\alpha - F = ma_C$$

$$I_{zz}^C\dot{\omega} = M_{Cz}: \qquad mk^2(a_C/R) = FR - N\epsilon$$

$$\Rightarrow \qquad N = mg\cos\alpha \tag{10}$$

$$F = m(g\sin\alpha - a_C) \tag{11}$$

$$a_C = gR(R\sin\alpha - \epsilon\cos\alpha)/(k^2 + R^2) \tag{12}$$

Alternatively, we can use $\dot{T} = \dot{W}$ with $T = \frac{1}{2}I_{zz}^A\omega^2 = \frac{1}{2}m(k^2 + R^2)\omega^2$ and $\dot{W} = \vec{F}\cdot\vec{v}_C + \vec{M}_C\cdot\vec{\omega}$, since the material point of application of F and N keeps changing.

$$\dot{T} = m(k^2 + R^2)\omega\,\dot{\omega} = \dot{W} = (mg\sin\alpha - F)\omega R + (FR - N\epsilon)\omega$$

$$\Rightarrow \quad \dot{\omega} = g(R\sin\alpha - \epsilon\cos\alpha)/(k^2 + R^2)$$

4. *Smooth plane.* The fourth equation is $F = 0$. $\Rightarrow$

$$M_{Cz} = mk^2\dot{\omega} = 0 \qquad \Rightarrow \qquad \omega = \text{constant}$$

$$F_x = mg\sin\alpha = ma_C \qquad \Rightarrow \qquad a_C = g\sin\alpha$$

which equals that of a cuboid translating down the smooth incline.

5. Consider a vehicle (Fig. E3.27g). For Coulomb friction, the light wheels without axle friction are two-force members $\Rightarrow$ only normal ground reactions act (Fig. E3.27h).

$$ma_C = mg\sin\alpha \quad \Rightarrow \quad a_C = g\sin\alpha$$

For rolling friction, FBDs are shown in Fig. E3.27i. For front wheels,

$$M_{Oz} = F_1 r - N_1\varepsilon = 0 \quad \Rightarrow \quad F_1 = N_1\varepsilon/r$$

Similarly, $F_2 = N_2\varepsilon/r$. For the complete vehicle:

$$F_y = ma_{Cy} = 0 \qquad \Rightarrow \qquad N_1 + N_2 = mg\cos\alpha$$

$$ma_C = F_x = mg\sin\alpha - F_1 - F_2 = mg\sin\alpha - (N_1 + N_2)\varepsilon/r$$

$$\Rightarrow \quad a_C = g\sin\alpha - g\cos\alpha\,\varepsilon/r$$

Example 3.28: Each wheel in the mechanism of Fig. E3.28a has axial radius of gyration k_0. Each link (OB and O_1B) is modelled as a thin rod. The collar slides on the fixed vertical shaft with a constant friction force F. The system is released from rest at position $\theta = \theta_1$ when the torsional springs are undeformed. Assume no slip of the wheels. Find v and $\dot{v}$ of the collar when $\theta = \theta_2 > 0$, and $\dot{v}$ at the instant of release.

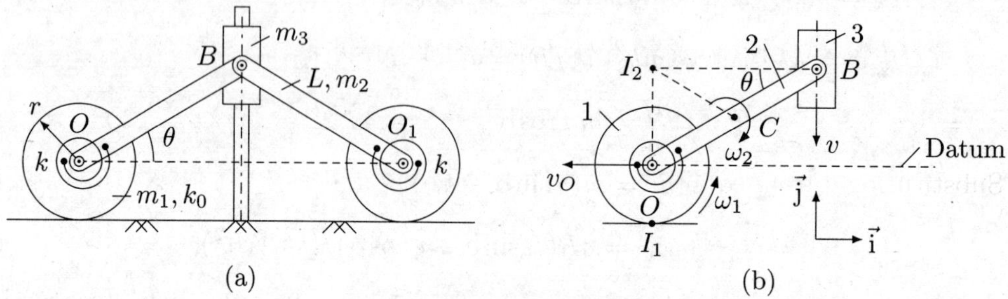

Fig. E3.28

Solution: The work–energy principle, applied for the complete system, is the best as it has only one dof and T, V and W^{nc} can be written easily for the general position.

The instantaneous centre of link OB is at the point of intersection I_2 of the normals to $\vec{v}_O = -v_O\vec{i}$ at O and $\vec{v}_B = -v\vec{j}$ at B (Fig. E3.28b). The instantaneous centre of wheel 1 is at I_1 as no slip $\Rightarrow v_{I_1} = 0$. $I_2B = L\cos\theta$, $I_2C = L/2$, $I_2O = L\sin\theta$.

$$|\omega_2| = \frac{v}{I_2B} = \frac{v_O}{I_2O} = \frac{v_C}{I_2C}$$

$$\Rightarrow \quad |\omega_2| = v\sec\theta/L, \qquad v_C = \tfrac{1}{2}v\sec\theta, \qquad v_O = v\tan\theta \tag{1}$$

$$\vec{\omega}_2 = -\frac{v\sec\theta}{L}\vec{k} \quad \Rightarrow \quad \dot{\theta} = -\frac{v\sec\theta}{L}, \qquad \omega_1 = \frac{v_O}{r} = \frac{v\tan\theta}{r} \tag{2}$$

As bar 2 rotates by $\phi_2\vec{k} = -(\theta_1 - \theta)\vec{k}$, O moves to left by $L(\cos\theta - \cos\theta_1)$ and wheel 1 rotates by $\phi_1\vec{k} = L(\cos\theta - \cos\theta_1)\vec{k}/r$. The twist ϕ of the spring is: $\phi = \phi_1 - \phi_2 = \theta_1 - \theta + L(\cos\theta - \cos\theta_1)/r$. T and V are expressed in terms of v and θ:

$$\begin{aligned} T &= \tfrac{1}{2}m_3v^2 + 2[\tfrac{1}{2}m_1v_O^2 + \tfrac{1}{2}m_1k_0^2\omega_1^2] + 2[\tfrac{1}{2}m_2v_C^2 + \tfrac{1}{2}(m_2L^2/12)\omega_2^2] \\ &= \tfrac{1}{2}[m_3 + 2m_1(1 + k_0^2/r^2)\tan^2\theta + \tfrac{2}{3}m_2\sec^2\theta]v^2 \end{aligned} \tag{3}$$

$$\begin{aligned} V &= m_3gL\sin\theta + 2m_2g(\tfrac{1}{2}L\sin\theta) + \tfrac{1}{2}k\phi^2 \times 2 \\ &= (m_2 + m_3)gL\sin\theta + k[\theta_1 - \theta + L(\cos\theta - \cos\theta_1)/r]^2 \end{aligned} \tag{4}$$

$\dot{W}^{nc} = -Fv$. We apply the work–energy relation from position 1 at $\theta = \theta_1$ to position 2 at $\theta = \theta_2$ to get v at position 2: $(T_2 + V_2) - (T_1 + V_1) = W^{nc}_{1-2} \Rightarrow$

$$\begin{aligned} -FL(\sin\theta_1 - \sin\theta_2) &= \tfrac{1}{2}[m_3 + 2m_1(1 + k_0^2/r^2)\tan^2\theta_2 + \tfrac{2}{3}m_2\sec^2\theta_2]v^2 \\ &\quad + (m_2 + m_3)gL\,(\sin\theta_2 - \sin\theta_1) + k[\theta_1 - \theta_2 + L(\cos\theta_2 - \cos\theta_1)/r]^2 \end{aligned} \tag{5}$$

which yields v. Using eqs(1)–(4) in $\dot{T}+\dot{V}=\dot{W}^{\rm nc}$ yields

$$[m_3+2m_1(1+k_0^2/r^2)\tan^2\theta+\tfrac{2}{3}m_2\sec^2\theta]v\dot{v}$$

$$+\tfrac{1}{2}[4m_1(1+k_0^2/r^2)+\tfrac{4}{3}m_2]\sec^2\theta\tan\theta\dot{\theta}v^2+(m_2+m_3)gL\cos\theta\,\dot{\theta}$$

$$-2k[\theta_1-\theta+L(\cos\theta-\cos\theta_1)/r](1+L\sin\theta/r)\dot{\theta}=-Fv \qquad (6)$$

Substituting $\dot{\theta}$ from eq. (2) in eq. (6) and cancelling factor v yields an equation for $\dot{v}$:

$$[m_3+2m_1(1+k_0^2/r^2)\tan^2\theta+\tfrac{2}{3}m_2\sec^2\theta]\dot{v}$$

$$=(m_2+m_3)g-F+\tfrac{1}{2}[4m_1(1+k_0^2/r^2)+\tfrac{4}{3}m_2]v^2\sec^3\theta\tan\theta/L$$

$$-2k[\theta_1-\theta+L(\cos\theta-\cos\theta_1)/r](1+L\sin\theta/r)/L\cos\theta \qquad (7)$$

Substituting $\theta=\theta_2$ and v from eq. (5) in eq. (7) yields $\dot{v}$. At the instant of release, $\theta=\theta_1$ and $v=0$. Hence, eq. (7) yields

$$\dot{v}=[(m_2+m_3)g-F]\,/\,[m_3+2m_1(1+k_0^2/r^2)\tan^2\theta_1+\tfrac{2}{3}m_2\sec^2\theta_1]$$

Example 3.29: (a) A tricycle rolls without slip due to a force P applied at A on the driving paddle (Fig. E3.29a) at a perpendicular distance d from the axis of the axle. The radii of the sprockets are r_1, r_2. The wind resistance force is cv. The front wheel, rear wheels and paddle assemblies have axial moments of inertia I_f, I_r and I_p, respectively. The total mass of the cycle including the rider is m. Neglect the mass of the legs of the rider. Find the acceleration of the cycle. (b) The cycle rests on a horizontal plane and an external force P is applied to the paddle arm in the backward direction at distance d from the axis of the axle (Fig. E3.29b). Find its acceleration.

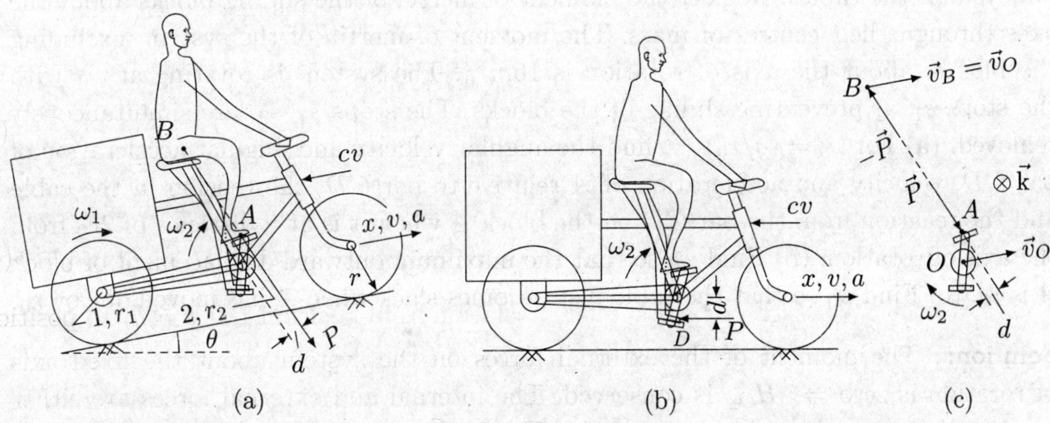

Fig. E3.29

Solution: (a) $\omega_1 = v/r$, $\omega_2 = \omega_1 r_1/r_2 = vr_1/rr_2$. The light leg is a 2-force member. $\Rightarrow$ The internal actions are collinear forces $\vec{P}$ at A and $-\vec{P}$ at B (Fig. E3.29c), with

$$\begin{aligned}
\dot{W}_{\text{int}} &= \vec{P}\cdot(\vec{v}_A - \vec{v}_B) = \vec{P}\cdot(\vec{v}_O + \vec{\omega}_2 \times \overrightarrow{OA} - \vec{v}_O) = \vec{\omega}_2 \cdot (\overrightarrow{OA} \times \vec{P}) \\
&= \vec{\omega}_2 \cdot \vec{M}_O = \omega_2 \vec{\mathrm{k}} \cdot Pd\vec{\mathrm{k}} = P\omega_2 d = Pdvr_1/rr_2 \\
T &= \tfrac{1}{2}mv^2 + \tfrac{1}{2}(I_f + I_r)v^2/r^2 + \tfrac{1}{2}I_p(vr_1/rr_2)^2 \\
V &= mgx\sin\theta \\
\dot{T} + \dot{V} &= \dot{W}^{\text{nc}}_{\text{int+ext}} = -(cv)v + Pdvr_1/rr_2 \quad \Rightarrow \\
&[m + (I_f + I_r + I_p r_1^2/r_2^2)/r^2]\dot{v}v + mgv\sin\theta = -cv^2 + Pdvr_1/rr_2 \\
a &= \dot{v} = [Pdr_1/rr_2 - cv - mg\sin\theta]/[m + (I_f + I_r + I_p r_1^2/r_2^2)/r^2]
\end{aligned}$$

(b) $\theta = 0$. The rate of work done by the single force $\vec{P}$ is

$$\begin{aligned}
\dot{W} &= \vec{P}\cdot\vec{v}_D = -P(v - \omega_2 d) = P(dvr_1/rr_2 - v) \\
\dot{T} &= \dot{W}_{\text{int+ext}} = \dot{W}_{\text{ext}} = P(dvr_1/rr_2 - v) - (cv)v \quad \Rightarrow \\
a &= \dot{v} = [P(dr_1/rr_2 - 1) - cv]/[m + (I_f + I_r + I_p r_1^2/r_2^2)/r^2]
\end{aligned}$$

with $v = 0$ initially.

Example 3.30: The system shown in Fig. E3.30a is free to rotate about the fixed vertical axis. Neglect friction everywhere. The springs are attached to O and the sliding blocks. The free length of the springs is r_0 and $k = 2.5mg/r_0$. Neglect the inertia of the pulleys and the cables. Neglect the moment of inertia of the sliding blocks about the axes through their centres of mass. The moment of inertia of the system, excluding the blocks, about the axis of rotation is $16mr_0^2$. The system is rotating at ω_0 with the stops s_1, s_2 preventing sliding of the blocks. The stops s_1, s_2, are simultaneously removed. (a) For $\omega_0 = (g/r_0)^{1/2}$, find the angular velocity and angular acceleration of bar CD, velocity and acceleration of A relative to bar CD, the tension in the cable and the reaction from the bar CD on the block A when it is at a distance of $2r_0$ from the axis of rotation. (b) Find ω_0 so that the maximum outward displacement of block A is r_0. (c) Find ω_0 so that the cable just becomes slack when A has moved out by r_0.

Solution: The moment of the external forces on the system about the fixed axis of rotation is zero $\Rightarrow$ H_{O_z} is conserved. The internal and external forces are either conservative or workless in pairs $\Rightarrow$ $E = T + V$ is conserved. Let CD rotate at ω, $\dot{\omega}$ when A is at radial position r from the axis of rotation with $\dot{r} = v$ (Fig. E3.30b), and the downward displacement of P, Q is $\frac{1}{2}(r-r_0) \Rightarrow v_{Pz} = -\frac{1}{2}\dot{r} = -\frac{1}{2}v,\ a_{Pz} = -\frac{1}{2}\ddot{r}$.

$$H_{Oz} = 16mr_0^2\omega + 2mr^2\omega + 2(2m)(3r_0)^2\omega = m(52r_0^2 + 2r^2)\omega$$

$$\Rightarrow \quad m(52r_0^2 + 2r^2)\omega = m(52r_0^2 + 2r_0^2)\omega_0 = 54mr_0^2\omega_0 \tag{1}$$

$$\Rightarrow \quad \dot{H}_{Oz} = m(52r_0^2 + 2r^2)\dot{\omega} + 4mr\dot{r}\omega = 0 \tag{2}$$

$$\begin{aligned} E &= \tfrac{1}{2}(16mr_0^2)\omega^2 + 2[\tfrac{1}{2}m\{(\omega r)^2 + \dot{r}^2\}] \\ &\quad + 2[\tfrac{1}{2}(2m)\{(3\omega r_0)^2 + (\tfrac{1}{2}\dot{r})^2\}] + 2[\tfrac{1}{2}k(r-r_0)^2] - 4mg(r-r_0)/2 \\ &= m(26r_0^2 + r^2)\omega^2 + 1.5m\dot{r}^2 + 2.5mg(r-r_0)^2/r_0 - 2mg(r-r_0) \qquad (3) \\ &= m(26r_0^2 + r_0^2)\omega_0^2 = 27mgr_0 \end{aligned}$$

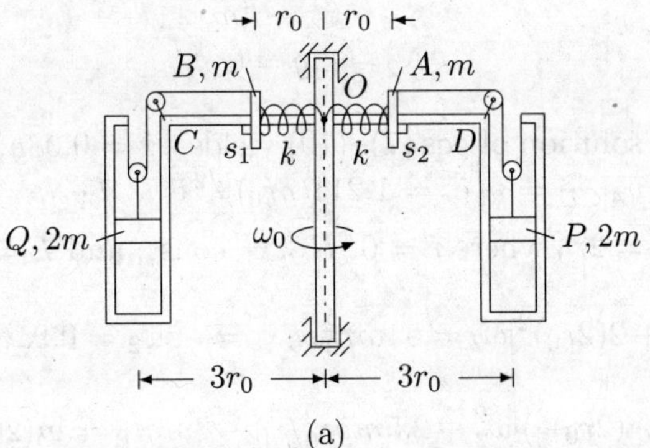

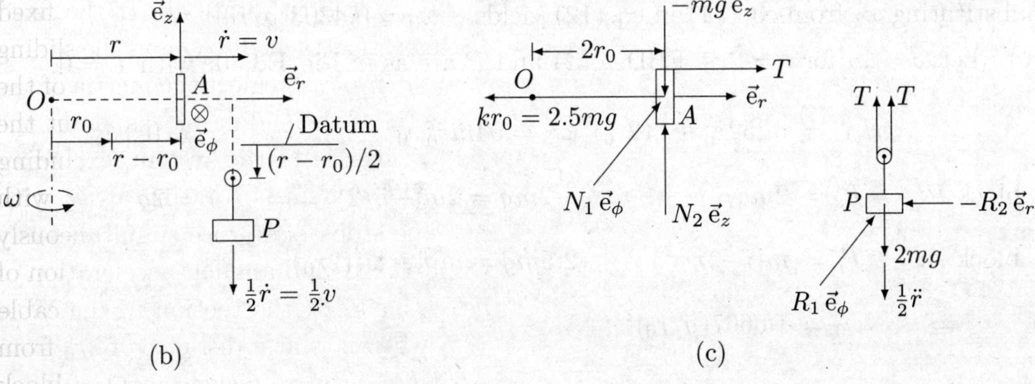

Fig. E3.30

(a) For $r = 2r_0$, let $v_1 = \dot{r}$ and CD rotate at $\omega_1, \dot{\omega}_1$. $\omega_1, v_1, \dot{\omega}_1$ are obtained from $H_{Oz} =$ constant [eq. (1)], $E =$constant [using eq. (3)], and $\dot{H}_{Oz} = 0$ [eq. (2)].

Eq. (1) $\Rightarrow \quad m[52r_0^2 + 2(2r_0)^2]\omega_1 = 54mr_0^2\omega_0 \quad \Rightarrow \quad \omega_1 = 0.9\omega_0 = 0.9(g/r_0)^{1/2} \qquad (4)$

Eq. (3) $\Rightarrow$ $\quad E = m[26r_0^2 + (2r_0)^2)]\omega_1^2 + 1.5mv_1^2 + 2.5mgr_0^2/r_0 - 2mgr_0 = 27mgr_0$

$$\Rightarrow \quad v_1 = 1.211(gr_0)^{1/2} \tag{5}$$

Eq. (2) $\Rightarrow$ $\quad m[52r_0^2 + 2(2r_0)^2]\dot{\omega}_1 + 4m(2r_0)v_1\omega_1 = 0$

$$\Rightarrow \quad \dot{\omega}_1 = -8v_1\omega_1/60r_0 = -0.1453(g/r_0) \tag{6}$$

The FBDs of blocks A and P for this position are shown in Fig. E3.30b.

$$\text{block } P: \quad F_z = 2ma_{P_z}: \qquad 2T - 2mg = 2m(-\ddot{r}/2) \tag{7}$$

$$\text{block } A: \quad F_r = m(\ddot{r} - r\dot{\phi}^2): \qquad T - 2.5mg = m[\ddot{r} - \omega_1^2(2r_0)] \tag{8}$$

$$F_\phi = m(r\ddot{\phi} + 2\dot{r}\dot{\phi}): \qquad N_1 = m(2r_0\dot{\omega}_1 + 2v_1\omega_1) \tag{9}$$

$$F_z = m\ddot{z} \qquad N_2 - mg = 0 \tag{10}$$

Using eqs (4)–(6), the solution of eqs (7)–(10) yields $\ddot{r} = 0.08g,\ T = 0.96mg,\ N_1 = 1.889mg,\ N_2 = mg.\ \vec{v}_{A|CD} = v_1\,\vec{e}_r = 1.211(gr_0)^{1/2}\,\vec{e}_r,\ \ \vec{a}_{A|CD} = \ddot{r}\,\vec{e}_r = 0.08g\,\vec{e}_r.$

(b) Let $\omega = \omega_2$ for $r = 2r_0$ where $\dot{r} = 0$. $H_{O_z} =$ const. and $E =$ const. yield ω_2, ω_0:

$$H_{Oz} \overset{(1)}{=} m[52r_0^2 + 2(2r_0)^2]\omega_2 = 54mr_0^2\omega_0 \quad \Rightarrow \quad \omega_2 = 0.9\omega_0 \tag{11}$$

$$E \overset{(3)}{=} m[26r_0^2 + (2r_0)^2)]\omega_2^2 + 2.5mgr_0^2/r_0 - 2mgr_0 = m(26r_0^2 + r_0^2)\omega_0^2 \tag{12}$$

Substituting ω_2 from eq. (11) in eq. (12) yields $\quad \omega_0 = 0.4303(g/r_0)^{1/2}$.

(c) Let $\omega = \omega_3$ for $r = 2r_0$. FBDs of A and P are as in Fig. E3.30c with $T = 0$.

$$H_{Oz} \overset{(1)}{=} m[52r_0^2 + 2(2r_0)^2]\omega_3 = 54mr_0^2\omega_0 \quad \Rightarrow \quad \omega_3 = 0.9\omega_0$$

$$\text{block } P: \quad F_z = 2ma_{P_z}: \qquad -2mg = 2m(-\ddot{r}/2) \quad \Rightarrow \quad \ddot{r} = 2g$$

$$\text{block } A: \quad F_r = m(\ddot{r} - r\dot{\phi}^2): \quad -2.5mg = m[\ddot{r} - \omega_3^2(2r_0)]$$

$$\Rightarrow \quad \omega_0 = 1.6667(g/r_0)^{1/2}$$

Example 3.31: In a crash, a car comes to stop with constant deceleration a. The motion of the driver is obtained by modelling the torso above the hip as a rigid body rotating about the hip joint modelled as a hinge joint. The mass of this part of the body is m with centre of mass C and radius of gyration k_O about the rotation axis at O (Fig. E3.31a). The mass of the arms is neglected. Find the velocity v with respect to the car with which the head strikes the dashboard. $OB = r$.

Solution: We present 3 solutions.

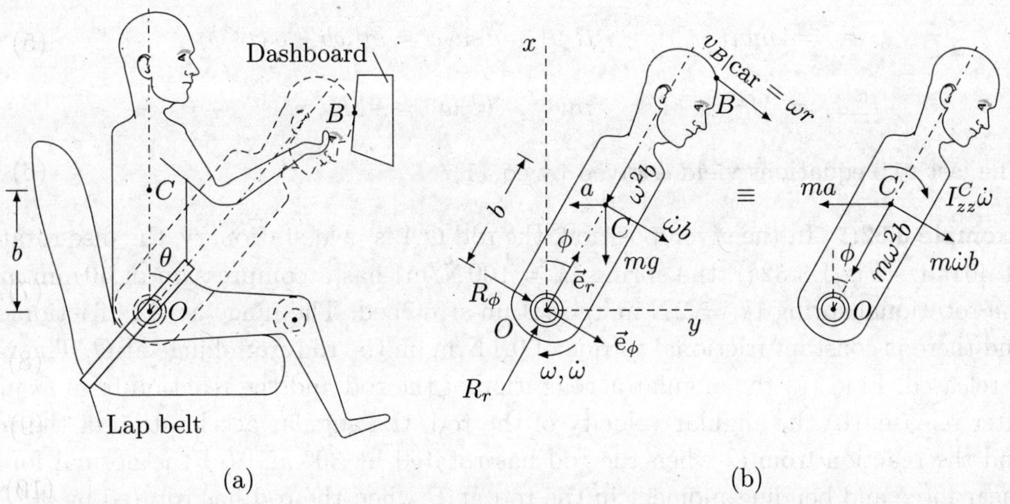

Fig. E3.31

(i) The augmented force system relative to translating frame T of the car: $\vec{R}_O$ at O and $-mg\,\vec{\mathrm{i}} - m(-a\,\vec{\mathrm{j}})$ at C with potential energy $V' = -(-mg)x_C - may_C = mgb\cos\phi - mab\sin\phi$. Work–energy relation $T' + V' = \text{const.}$ from $\phi = 0$ to θ:

$$\tfrac{1}{2}mk_0^2\omega^2 + mgb\cos\theta - mab\sin\theta = mgb \qquad \Rightarrow \qquad \omega^2 = 2b[a\sin\theta + g(1-\cos\theta)]/k_0^2$$

$$v_{B|\text{car}} = \omega r = r[2b\{a\sin\theta + g(1-\cos\theta)\}]^{1/2}/k_0$$

(ii) $\vec{a}_C = \vec{a}_O - \omega^2 b\,\vec{e}_r + \dot\omega b\,\vec{e}_\phi$. For the external force system and the equivalent kinetic system consisting of a force $m\vec{a}_C$ at C and a couple $I^C_{zz}\dot\omega\,\vec{\mathrm{k}}$, shown in Fig. E3.31b:

$$M_{O_z} = mg\sin\phi\, b = m(\dot\omega b - a\cos\phi)b + I^C_{zz}\dot\omega$$

$$\Rightarrow \qquad \dot\omega = (a\cos\phi + g\sin\phi)b/k_0^2 \tag{1}$$

using $I^C_{zz} + mb^2 = I^O_{zz} = mk_0^2$. Equation (1) is integrated using $\dot\omega = \dfrac{d\omega}{d\phi}\dot\phi = \omega\dfrac{d\omega}{d\phi}$:

$$\int_0^\omega \omega\, d\omega = \int_0^\theta \frac{b}{k_0^2}(a\cos\phi + g\sin\phi)\, d\phi$$

$$\Rightarrow \qquad \omega^2 = 2b[a\sin\theta + g(1-\cos\theta)]/k_0^2$$

(iii) $\quad \vec{a}_C = \vec{a}_O - \omega^2 b\,\vec{e}_r + \dot\omega b\,\vec{e}_\phi = -(\omega^2 b + a\sin\phi)\,\vec{e}_r + (\dot\omega b - a\cos\phi)\,\vec{e}_\phi.$

The 3 equations of motion are:

$$F_r = ma_{C_r}: \qquad R_r - mg\cos\phi = -m(\omega^2 b + a\sin\phi)$$

$$F_\phi = ma_{C_\phi}: \qquad R_\phi + mg\sin\phi = m(\dot{\omega} b - a\cos\phi)$$

$$I_{zz}^C\dot{\omega} = M_{Cz}: \qquad m(k_0^2 - b^2)\dot{\omega} = -bR_\phi$$

The last two equations yield $\dot{\omega}$ given by eq. (1).

Example 3.32: In the given position, the rod OA is held stationary, the disc rotates at 40 rad/s(Fig. E3.32a), the spring ($k = 100$ N/m) has a compression of 40 mm and the rotational spring ($k_t = 2$ N.m/rad) is un-stretched. The hinge at A is frictionless and there is constant frictional torque of 0.1 N.m on the rod from hinge at O. The rod is released. Find (a) the angular acceleration of the rod and the reaction from O just after release, (b) the angular velocity of the rod, the angular acceleration of the rod and the reaction from O when the rod has rotated by 30° and (c) the normal force, shear force and bending moment in the rod at B when the rod has rotated by 30°.

Solution: The FBDs of the rod + disc and the disc are shown in Fig. E3.32b. For rod 1, $m_1 = 0.6$ kg, $I_{zz}^O = 0.6(0.8)^2/3 = 0.128$ kg.m^2 and for disc 2, $m_2 = 0.1$ kg, $I_{zz}^A = 0.1(0.1)^2/2 = 0.0005$ kg.m^2. Let the angular velocities of the rod and the disc be ω_1, ω_2.

Disc 2: $\quad I_{zz}^A\dot{\omega}_2 = M_{A_z} = 0 \quad \Rightarrow \quad \dot{\omega}_2 = 0 \quad \Rightarrow \quad \omega_2 = 40\,\text{rad/s}$

The spring extension $\delta = 0.2\sin\theta - 0.04$ m. $\vec{v}_A = 0.8\omega_1\,\vec{e}_\phi$. For the rod + disc, H_{Oz}, T and V, for the general position θ ($\omega_1 = \dot{\theta}$) are given by

$$\begin{aligned}
H_{Oz} &= I_{zz}^O\omega_1 + (I_{zz}^A\omega_2 + 0.8m_2v_A) \\
&= 0.128\omega_1 + 0.0005(40) + 0.8(0.1)(0.8\omega_1) = 0.192\omega_1 + 0.02 \qquad (1) \\
T &= \tfrac{1}{2}I_{zz}^O\omega_1^2 + [\tfrac{1}{2}m_2v_A^2 + \tfrac{1}{2}I_{zz}^A\omega_2^2] \\
&= \tfrac{1}{2}(0.128)\omega_1^2 + \tfrac{1}{2}[0.1(0.8\omega_1)^2 + 0.0005(40)^2] = 0.096\omega_1^2 + 0.4 \\
V &= \tfrac{1}{2}k_t\theta^2 + \tfrac{1}{2}k(0.2\sin\theta - 0.04)^2 - m_1g(0.4\sin\theta) - m_2g(0.8\sin\theta) \\
&= \theta^2 + 50(0.2\sin\theta - 0.04)^2 - 0.32g\sin\theta
\end{aligned}$$

(a) FBD of rod + disc (in intial position) is shown in Fig. E3.32c. $\omega_1 = 0$. $\dot{H}_{Oz} = M_{Oz}$ yields $\dot{\omega}_1$. R_1, R_2 are obtained using $\vec{F} = \sum m_i\vec{a}_{C_i}$ and kinematics. $\dot{H}_{Oz} = M_{Oz}$:

$$0.192\dot{\omega}_1 = -0.1 + 4(0.2) + 0.6g(0.4) + 0.1g(0.8) \quad \Rightarrow \quad \dot{\omega}_1 = 20.00\,\text{rad/s}^2$$

$$\vec{a}_C = \dot{\omega}_1(0.4)\,\vec{j} = 8\,\vec{j}\ \text{m/s}^2, \qquad \vec{a}_A = \dot{\omega}_1(0.8)\,\vec{j} = 16\,\vec{j}\ \text{m/s}^2$$

$$\vec{F} = R_1\vec{\mathrm{i}} + R_2\vec{\mathrm{j}} + (4+0.6g+0.1g)\vec{\mathrm{j}} = \sum m_i\vec{a}_{C_i} = (0.6)8\vec{\mathrm{j}} + (0.1)16\vec{\mathrm{j}}$$

$$\Rightarrow \quad R_1 = 0, \qquad R_2 = -4.467\,\mathrm{N}$$

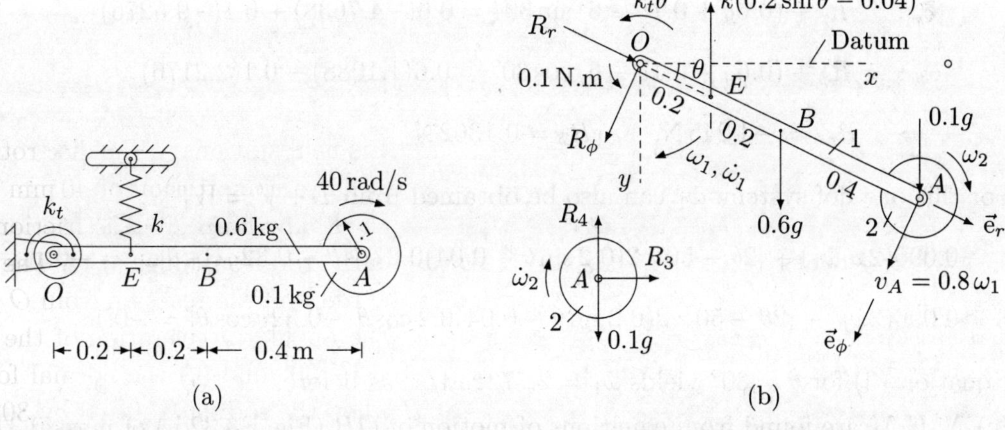
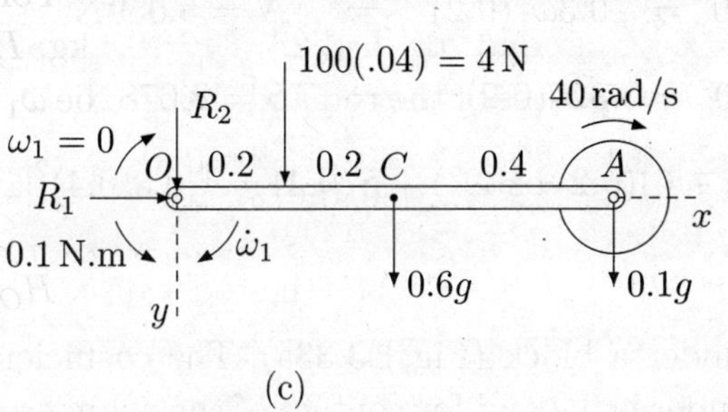
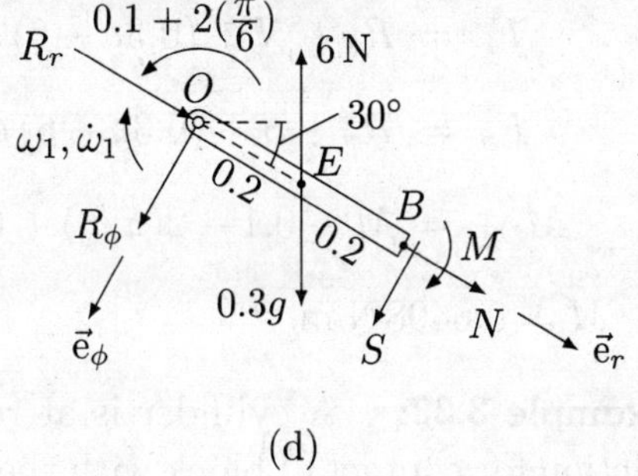

Fig. E3.32

(b) The work–energy relation from the position 1 for $\theta = 0$ to the position 2 for $\theta = \pi/6$ yields ω_1. Using eq. (1), $\dot{H}_{Oz} = M_{Oz}$ yields $\dot{\omega}_1$. R_r, R_ϕ are obtained using $\vec{F} = \sum m_i\vec{a}_{C_i}$ and kinematics. The pull F_s on spring for $\theta = 30°$ is $F_s = 100(0.2\sin 30° - 0.04) = 6\,\mathrm{N}$. $(T_2 + V_2) - (T_1 + V_1) = W_{1-2}^{\mathrm{nc}} \Rightarrow$

$$[(.096\omega_1^2 + .4) + \{(\frac{\pi}{6})^2 + 50(.2\sin 30° - .04)^2 - .32g\sin 30°\}] - [.4 + 50(.04)^2] = -\frac{.1\pi}{6}$$

$$\Rightarrow \qquad \omega_1 = 3.451\,\mathrm{rad/s}$$

$$\dot{H}_{Oz} = M_{Oz}: \quad 0.192\dot{\omega}_1 = -0.1 + (0.6g\times 0.4 + 0.1g\times 0.8 - 6\times 0.2)\cos 30° - \frac{2\pi}{6} \qquad (2)$$

$$\Rightarrow \qquad \dot{\omega}_1 = 2.772\,\mathrm{rad/s^2}$$

$$\vec{a}_C = -\omega_1^2(0.4)\,\vec{e}_r + \dot{\omega}_1(0.4)\,\vec{e}_\phi = -4.7638\,\vec{e}_r + 1.1088\,\vec{e}_\phi\ \mathrm{m/s^2}$$

$$\vec{a}_A = -\omega_1^2(0.8)\,\vec{e}_r + \dot{\omega}_1(0.8)\,\vec{e}_\phi = -9.5275\,\vec{e}_r + 2.2176\,\vec{e}_\phi \text{ m/s}^2$$

$$\vec{F} = \sum m_i \vec{a}_{C_i} \quad \Rightarrow$$

$$\vec{e}_r: \quad R_r + (0.6g + 0.1g - 6)\sin 30^\circ = 0.6(-4.7638) + 0.1(-9.5275)$$

$$\vec{e}_\phi: \quad R_\phi + (0.6g + 0.1g - 6)\cos 30^\circ = 0.6(1.1088) + 0.1(2.2176)$$

$$\Rightarrow \quad R_r = -4.245\,\text{N}, \qquad R_\phi = 0.1362\,\text{N}$$

For this one dof system, $\dot{\omega}_1$ can also be obtained from $\dot{T} + \dot{V} = \dot{W}_{1-2}^{\text{nc}}$:

$$0.096(2\omega_1\dot{\omega}_1) + [2\theta + 50\times 2(0.2\sin\theta - 0.04)0.2\cos\theta - 0.32g\cos\theta]\omega_1 = -0.1\omega_1$$

$$\Rightarrow \quad 0.096(2\dot{\omega}_1) + [2\theta + 50\times 2(0.2\sin\theta - 0.04)0.2\cos\theta - 0.32g\cos\theta] = -0.1 \qquad (3)$$

Equation (3) for $\theta = 30^\circ$ yields $\dot{\omega}_1 = 2.772\,\text{rad/s}^2$ as before.

(c) N, F, M are found from equations of motion of OB (Fig. E3.32d.) of mass 0.3 kg.

$$F_r = R_r + N + (0.3g - 6)\sin 30^\circ = -0.3\omega_1^2(0.2) \quad \Rightarrow \quad N = 5.059\,\text{N}$$

$$F_\phi = R_\phi + S + (0.3g - 6)\cos 30^\circ = 0.3\dot{\omega}_1(0.2) \quad \Rightarrow \quad S = 2.678\,\text{N}$$

$$M_{O_z} = M - 0.1 - 2(\pi/6) + (0.3g - 6)(0.2\cos 30^\circ) + S(0.4) = \tfrac{1}{3}[0.3(0.4)^2]\dot{\omega}_1$$

$\Rightarrow M = 0.6498\,$N.m.

Example 3.33: A cylinder is at rest under a block (Fig. E3.33a). The coefficient of friction for contact of block with the cylinder is 0.2 and for contact of the cylinder with the ground is 0.1. (a) Find the value P_1 of P for impending motion of the cylinder for two cases: (1) $m_1 = 3m$, (2) $m_1 = 9.6m$. (b) For $m_1 = 9.6m$, find the initial a_C and $\dot{\omega}$ if the applied force $P > P_1$. (c) If $P = 2.3mg$ for $m_1 = 9.6mg$, find v_C and the position of C when slip starts at both contacts. Find a_C and $\dot{\omega}$ during subsequent motion.

Solution: (a) Although it is not known a priori where the slip occurs first (A/B), the directions of F_1 and F_2 (Fig. E3.33b) are unambiguously obtained from $M_{A_z} = 0$, $M_{B_z} = 0$ for the cylinder. P, N_1, F_1, N_2, F_2 are found using $M_{O_z} = 0$ for the block, 3 equations of equilibrium for the cylinder and the condition of slip.

$$M_{O_z} = N_1(0.6b) - m_1 g(0.5b) = 0 \quad \Rightarrow \quad N_1 = 5m_1 g/6 \qquad (1)$$

We *assume* impending rolling without slip at B and with slip at A with $F_1 = 0.2N_1$ opposite to the direction of impending slip (Fig. E3.33c).

$$M_{B_z} = 0.2N_1(2R) - 1.2RP = 0 \Rightarrow \quad P = N_1/3 \qquad (2)$$

$$F_x = P - 0.2N_1 - F_2 = 0 \quad \Rightarrow \quad F_2 = 0.1333N_1 \tag{3}$$

$$F_y = N_2 - mg - N_1 = 0 \quad \Rightarrow \quad N_2 = mg + N_1 \tag{4}$$

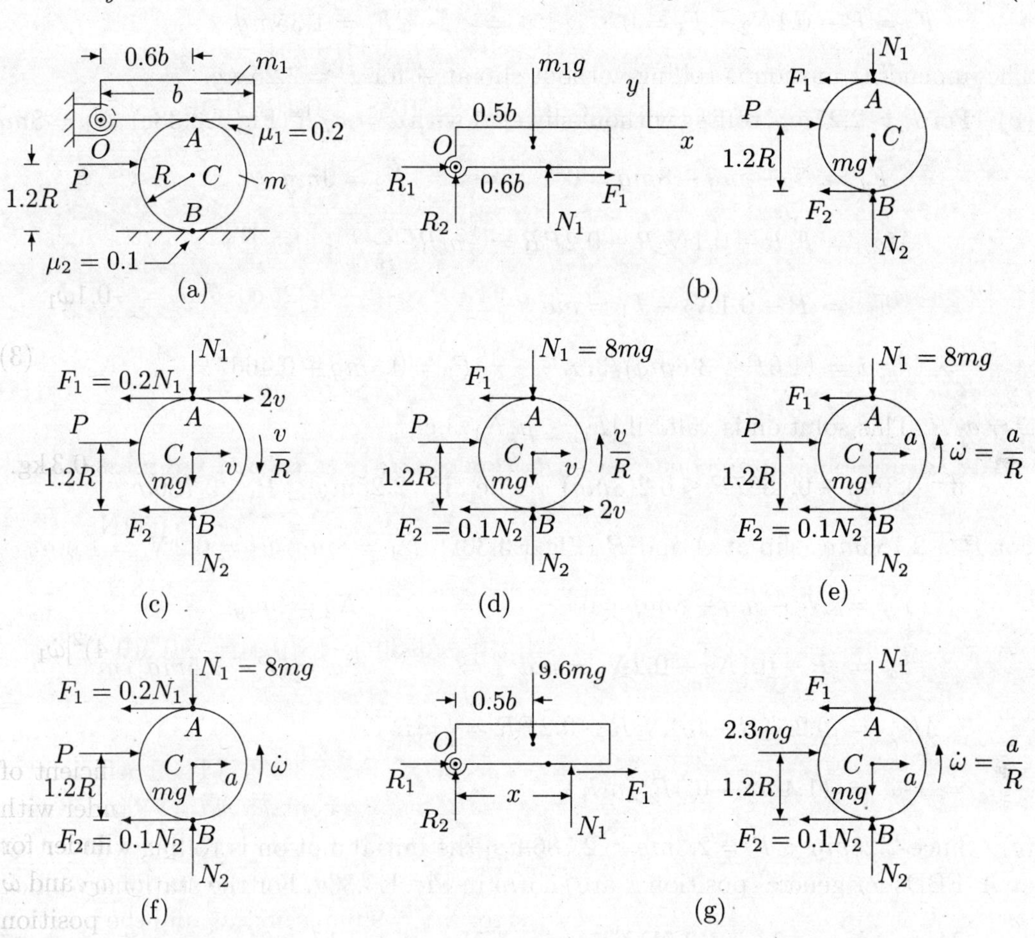

Fig. E3.33

(1) For $m_1 = 3m$, eqs (1)–(4) yield $(N_1,\ P,\ F_2,\ N_2) = (2.5,\ 0.8333,\ 0.3333,\ 3.5)mg$.

Check: $\quad |F_2|/N_2 = 0.3333mg/3.5mg = 0.0952 \leq \mu_2 = 0.1$

The assumption is correct. $\Rightarrow$ Rolling without slip at B impends for $P = 0.8333mg$. It should be noted that slip does not necessarily occur at the surface where μ is smaller.

(2) For $m_1 = 9.6m$, eqs (1)–(4) $\Rightarrow$ $(N_1,\ P,\ F_2,\ N_2) = (8,\ 2.667,\ 1.067,\ 9)mg$.

Check: $\quad |F_2|/N_2 = 1.067mg/9mg = 0.1186 \nleq \mu_2 = 0.1$

The assumption is false. $\Rightarrow$ The impending motion is rolling without slip at A and with slip at B with $F_2 = 0.1N_2$ opposite to the direction of impending slip (Fig. E3.33d).

$$F_y = N_2 - mg - 8mg = 0 \quad \Rightarrow \quad N_2 = 9mg$$

$$M_{A_z} = 0.8RP - 0.1N_2(2R) = 0 \quad \Rightarrow \quad P = 2.25mg$$

$$F_x = P - 0.1N_2 - F_1 = 0 \quad \Rightarrow \quad F_1 = 1.35mg$$

The impending motion is rolling without slip at A for $P = 2.25mg$.

(b) For $P > 2.25mg$, rolling without slip at A with $\dot{\omega} = a/R$ (Fig. E3.33e). $N_1 = 8mg$,

$$F_y = N_2 - mg - 8mg = 0 \quad \Rightarrow \quad N_2 = 9mg$$

$$M_{C_z} = F_1 R - 0.1N_2 R - 0.2PR = \tfrac{1}{2}mR^2 \frac{a}{R}$$

$$F_x = P - 0.1N_2 - F_1 = ma$$

$$\Rightarrow \quad a = (1.6P - 3.6mg)/3m, \qquad F_1 = 0.3mg + 0.4667P$$

$\dot{\omega} = a/R$. This solution is valid if $|F_1| \leq \mu_1 N_1$, i.e.,

$$\text{if} \quad 0.3mg + 0.4667P \leq 0.2(8mg) \qquad \text{i.e., if} \quad 2.25mg \leq P \leq 2.786mg$$

For $P > 2.786mg$: slip at A and B (Fig. E3.33f). $N_1 = 8mg$, $F_1 = 0.2N_1 = 1.6mg$.

$$F_y = N_2 - mg - 8mg = 0 \quad \Rightarrow \quad N_2 = 9mg$$

$$F_x = P - 0.1N_2 - 0.2N_1 = ma \quad \Rightarrow \quad a = (P - 2.5mg)/m$$

$$M_{C_z} = 0.2N_1 R - 0.1N_2 R - 0.2PR = \tfrac{1}{2}mR^2\dot{\omega}$$

$$\Rightarrow \quad \dot{\omega} = (1.4mg - 0.4P)/mR$$

(c) Since $2.25mg < P = 2.3mg < 2.786mg$, the initial motion is rolling without slip at A. FBDs for general position x are shown in Fig. E3.33g. For the stationary block:

$$M_{O_z} = N_1 x - 9.6mg(0.5b) = 0 \quad \Rightarrow \quad N_1 = 4.8mgb/x \tag{5}$$

The equations of motion of the cylinder for general position x yield

$$F_y = N_2 - mg - N_1 = 0$$

$$M_{C_z} = F_1 R - 0.1N_2 R - 0.2R(2.3mg) = \tfrac{1}{2}mR^2 \frac{a}{R}$$

$$F_x = 2.3mg - 0.1N_2 - F_1 = ma$$

$$\Rightarrow \quad N_2 = (1 + 4.8b/x)mg \tag{6}$$

$$a = \dot{\omega}R = (1.09333 - 0.64b/x)g$$

$$F_1 = (1.1067 + 0.16b/x)mg$$

$$|F_1| \le \mu_1 N_1 \quad \Rightarrow \quad (1.1067 + 0.16b/x)mg \le 0.2(4.8mgb/x) \quad \Rightarrow \quad x \le 0.7229b.$$

Let $v_C = v_1$ when slip starts at both contacts for $x = 0.7229b$. v_1 is obtained using:

$$a = v\frac{dv}{dx} = \left(1.09333 - 0.64\frac{b}{x}\right)g$$

$$\int_0^{v_1} v\,dv = \int_{0.6b}^{0.7229b} \left(1.09333 - 0.64\frac{b}{x}\right) g\,dx$$

$$v_1 = [2\{1.09333(0.7229b - 0.6b) - 0.64b\ln(0.7229/0.6)\}g]^{1/2} = 0.1738\sqrt{gb}$$

Subsequently, $F_1 = 0.2N_1$ and $F_x = ma,\ M_{C_z} = I_{zz}^C\dot{\omega} \Rightarrow$

$$F_x = 2.3mg - 0.1N_2 - 0.2N_1 = ma \tag{7}$$

$$M_{C_z} = 0.2N_1R - 0.1N_2R - 0.2R(2.3mg) = \tfrac{1}{2}mR^2\dot{\omega} \tag{8}$$

Using N_1, N_2 from eqs (5) and (6) in eqs (7) and (8) yields

$$a = (2.2 - 1.44b/x)g, \qquad \dot{\omega} = (0.96b/x - 1.12)g/R$$

Example 3.34: The mass of a self-propelled vehicle (Fig. E3.34a) including the wheels is m and the centre of mass is at C. The rear and front wheel assemblies have mass m_1 and m_2 and axial radii of gyration of k_1 and k_2 respectively. The coefficients of friction for the wheels are μ_s and μ_k. (a) A rear-wheel driven car, starting from rest, moves without slip under a driving torque M_1. The wind resistance is cv^2, where v is the velocity of the car. Find its acceleration a, $v(t)$ and $v(x)$ after travelling distance x. (b) Discuss the method of analysis for motion with possible slip of a rear-wheel driven car. (c) Find a_{max} for a rear-wheel driven car without slipping and the driving torque required to achieve it. Neglect wind resistance. (d) Find a_{max} for a four-wheel drive car without slipping and the driving torques M_1, M_2 required on the rear and front wheels to achieve it. Neglect wind resistance. (e) Find a and the angular accelerations of its wheels if the driving torques are M_1, M_2 and the wheels slip. Neglect wind resistance.

Solution: (a) We model the rear and front wheel assemblies and the chassis as rigid bodies. The car is a single dof system for the case of *rolling without slip* assuming no motion of the centre of the wheels relative to the chassis. Hence, $\dot{T} = \dot{W}$ is handy for the governing equation. The FBDs of the car, front wheels and rear wheels are shown in Fig. E3.34b. There is no input torque to the front wheels and F_2 is backwards to provide the desired angular acceleration. F_1 on the rear wheels is forwards to provide forward acceleration to the car. The work done by the external and internal forces is either zero (such as $F_1,\ F_2$ as $\vec{v}_A = \vec{v}_B = \vec{0}$) or pair-wise zero (such as R_1) except the work done by the wind force cv^2 and the driving torque M_1 on the rear wheels.

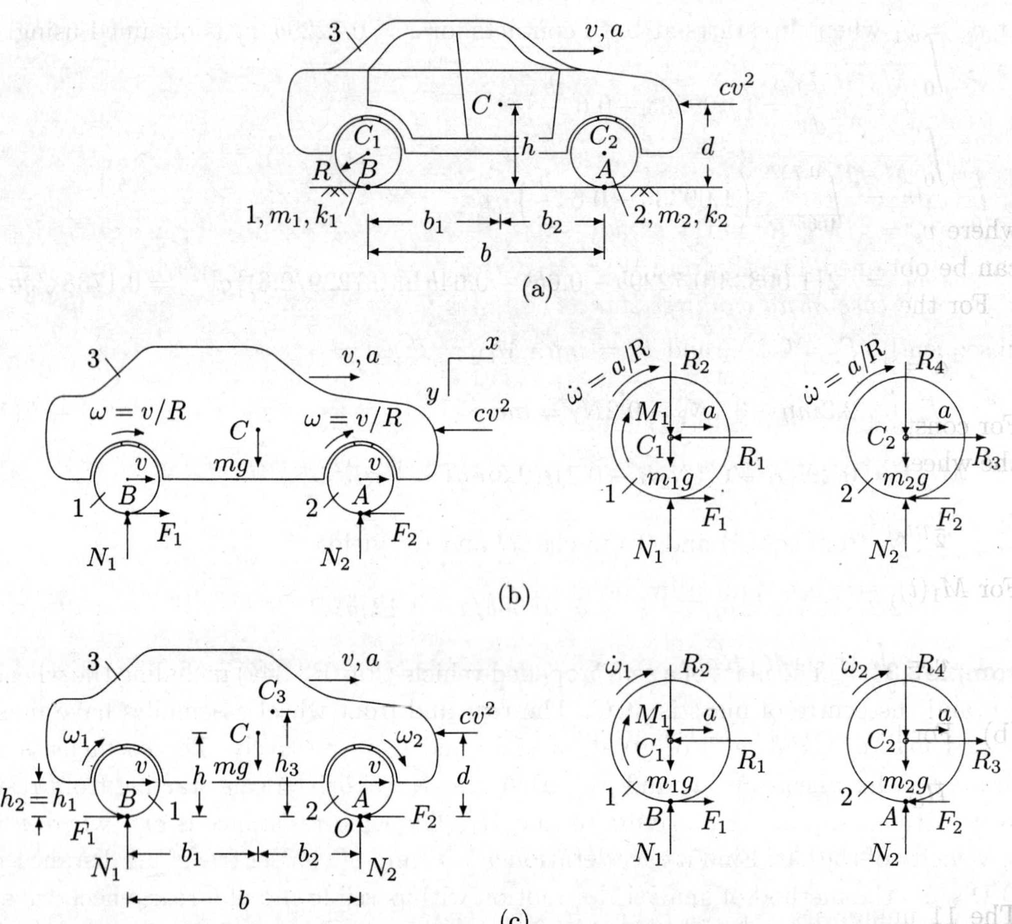

Fig. E3.34a, b, c

Let the wheels rotate at ω, $\dot{\omega}$. No slip $\Rightarrow$ $\omega = v/R$, $\dot{\omega} = a/R$. For the car :

$$T = \tfrac{1}{2}(m - m_1 - m_2)v^2 + \tfrac{1}{2}m_1 v^2 + \tfrac{1}{2}m_1 k_1^2 \frac{v^2}{R^2} + \tfrac{1}{2}m_2 v^2 + \tfrac{1}{2}m_2 k_2^2 \frac{v^2}{R^2} = \tfrac{1}{2}m_e v^2 \quad (1)$$

$$\dot{W} = M_1\omega + cv^2(-v) = (M_1/R - cv^2)v \quad (2)$$

where $m_e = m + (m_1k_1^2 + m_2k_2^2)/R^2$. The acceleration is obtained using $\dot{T} = \dot{W}$:

$$m_e v\dot{v} = (M_1/R - cv^2)v$$

$$\Rightarrow \quad a = (M_1/R - cv^2)\,/\,m_e \quad (3)$$

For constant M_1, $a = \dfrac{dv}{dt}$ and $a = v\,\dfrac{dv}{dx}$ yield $v(t)$ and $v(x)$:

$$\int_0^v \frac{1}{M_1/R - cv^2}\,dv = \frac{t}{m_e} \quad\Rightarrow\quad v = v_s\,\frac{e^{2cv_s t/m_e} - 1}{e^{2cv_s t/m_e} + 1}$$

$$\int_0^v \frac{v}{M_1/R - cv^2}\,dv = \frac{x}{m_e} \quad\Rightarrow\quad v = v_s(1 - e^{-2cx/m_e})^{1/2}$$

where $v_s = \sqrt{M_1/Rc}$ is the steady state uniform speed for driving moment M_1 and can be obtained directly from $\dot{W} = M_1 v/R - (cv^2)v = \dot{T} = 0$. Note that $v \to v_s$.

For the *case of no wind resistance*, eq. (3) yields

$$a = M_1/Rm_e \tag{4}$$

For constant M_1, the work done by M_1 in displacement x of the car (rotation x/R of the wheels) is $M_1(x/R)$, and $v(x)$ can also be directly obtained using $T_2 - T_1 = W_{1\text{-}2}$:

$$\tfrac{1}{2}m_e v^2 = M_1 x/R$$

For $M_1(t)$, $v(t)$ is obtained by integrating eq. (4) with respect to time:

$$v = \int_0^t M_1(t)\,dt/Rm_e$$

(b) For fixed point O of the ground coincident with A (Fig. E3.34c), for the car:

$$\begin{aligned} H_{O_z} &= (m_1k_1^2\omega_1 + m_1vh_1) + (m_2k_2^2\omega_2 + m_2vh_2) + m_3vh_3 \\ &= mvh + m_1k_1^2\omega_1 + m_2k_2^2\omega_2 \end{aligned} \tag{5}$$

The 11 unknowns, $N_1, N_2, F_1, F_2, R_1, R_2, R_3, R_4$, and a, $\dot{\omega}_1$, $\dot{\omega}_2$, can be determined using 9 (3×3) equations of motion of 3 bodies (main body, front and rear wheel assemblies of car) and 2 conditions of slip or no slip at A and B. However, we shall write 7 equations involving N_1, N_2, F_1, F_2 and a, $\dot{\omega}_1$, $\dot{\omega}_2$. For the whole car:

$$M_{O_z} = \dot{H}_{O_z}: \qquad N_1 b - mgb_2 - cv^2 d = mah + m_1k_1^2\dot{\omega}_1 + m_2k_2^2\dot{\omega}_2 \tag{6}$$

$$F_y = 0: \qquad mg - N_1 - N_2 = 0 \tag{7}$$

$$F_x = ma: \qquad F_1 - F_2 - cv^2 = ma \tag{8}$$

for rear and front wheels:

$$M_{C_{1_z}} = M_1 - F_1 R = m_1k_1^2\dot{\omega}_1 \tag{9}$$

$$M_{C_{2_z}} = F_2 R = m_2k_2^2\dot{\omega}_2 \tag{10}$$

Contact conditions at B and A:

$$\text{no slip at } B: \quad \dot{\omega}_1 = a/R, \qquad \text{or} \quad \text{slip at } B: \quad F_1 = \mu_k N_1 \tag{11}$$

$$\text{no slip at } A: \quad \dot{\omega}_2 = a/R, \qquad \text{or} \quad \text{slip at } A: \quad F_2 = \mu_k N_2 \tag{12}$$

For the case of impending slip, μ_s replaces μ_k in eqs (11) and (12). An assumption is made regarding the condition of slip or no slip at contacts A and B. N_1, N_2, F_1, F_2 are expressed in terms of a, $\dot{\omega}_1$, $\dot{\omega}_2$ using successively eqs (6), (7), (9), (10). These are substituted in eq. (8) and appropriate equations of (11) and (12) to yield 3 equations for a, $\dot{\omega}_1$, $\dot{\omega}_2$. These are solved and the forces are computed. A check is finally made about the proper directions of friction forces at the contact(s) with slip and whether the friction forces at contact(s) with no slip satisfy $|F_i| \leq \mu_s N_i$. If these checks are not satisfied, then the problem is reworked with other assumptions at points of contact.

(c) The FBDs are the same as in Fig. E3.34b without cv^2. For no slip, a given by eq. (4) has a bound since at some stage either the front wheels will lift-off or the rear wheels will slip. $a_{\max}$ can be obtained by first finding the forces as in part (b). For no slip, $\omega_1 = \omega_2 = v/R$, $\dot{\omega}_1 = \dot{\omega}_2 = a/R$ and eq. (5) yields

$$H_{Oz} = [mh + (m_1 k_1^2 + m_2 k_2^2)/R]v \tag{13}$$

To find $N_1, N_2, F_1, F_2, a, M_1$, we use the following 6 equations. For the whole car:

$$M_{Oz} = \dot{H}_{Oz}: \qquad N_1 b - mgb_2 = [mh + (m_1 k_1^2 + m_2 k_2^2)/R]a \tag{14}$$

$$F_y = 0: \qquad mg - N_1 - N_2 = 0 \tag{15}$$

$$F_x = ma: \qquad F_1 - F_2 = ma \tag{16}$$

For rear and front wheels:

$$M_{C_{1z}} = M_1 - F_1 R = m_1 k_1^2 a/R \tag{17}$$

$$M_{C_{2z}} = F_2 R = m_2 k_2^2 a/R \tag{18}$$

and conditions of no slip at B and no lift-off of front wheels:

$$F_1 \leq \mu_s N_1, \qquad N_2 \geq 0 \tag{19}$$

The forces N_1, N_2, F_2, F_1 and M_1 are expressed in terms of a using successively eqs (14), (15), (18), (16) and (17) [or using $\dot{T} = \dot{W}$, i.e., eq. (4)]:

$$N_1 = [mgb_2 + \{mh + (m_1 k_1^2 + m_2 k_2^2)/R\}a]/b \tag{20}$$

$$N_2 = [mgb_1 - \{mh + (m_1 k_1^2 + m_2 k_2^2)/R\}a]/b \tag{21}$$

$$F_2 = m_2k_2^2a/R^2, \qquad F_1 = (m + m_2k_2^2/R^2)a \tag{22}$$

$$M_1 = [m + (m_1k_1^2 + m_2k_2^2)/R^2]aR \tag{23}$$

Substituting these in eqs (19) yields no slip at rear wheels if

$$a \le a_1 = \mu_s g b_2 \,/\, [b(1 + m_2k_2^2/mR^2) - \mu_s\{h + (m_1k_1^2 + m_2k_2^2)/mR\}]$$

and no lift-off of front wheels if

$$a \le a_2 = gb_1/\,[h + (m_1k_1^2 + m_2k_2^2)/mR]$$

$a_{\max} = \min(a_1, a_2)$. M_1 needed to achieve a_{max} is obtained from eq. (23). The lift-off of front wheel of a two-wheel scooter when the clutch is suddenly engaged is due to $a > a_2$.

The analysis for a front-wheel driven car is similar.

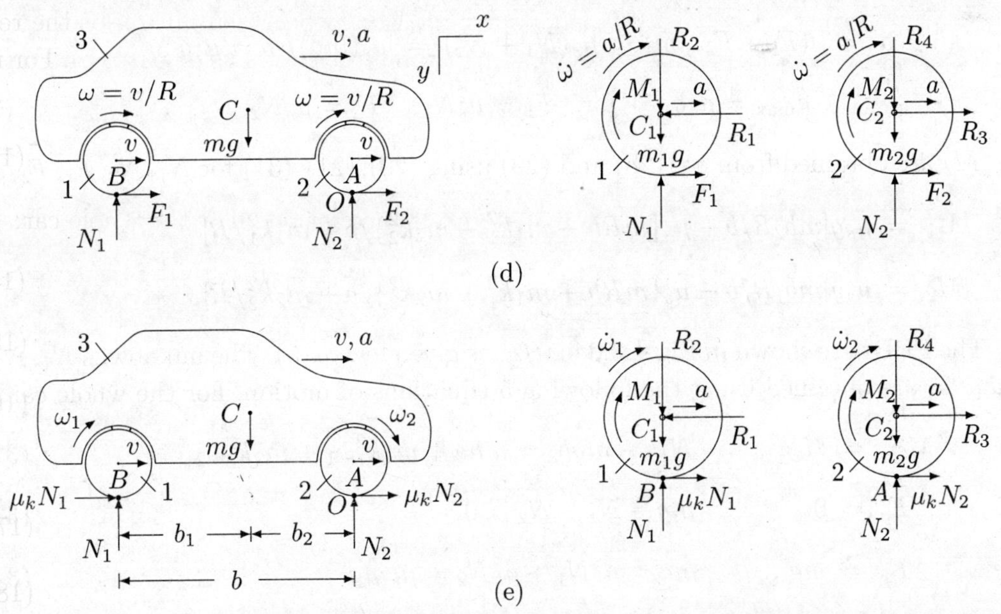

Fig. E3.34d, e

(d) In the FBDs for a four-wheel drive car (Fig. E3.34d), notice the direction of friction force on the front wheels to achieve greater acceleration. For no slip, $\dot{T} = \dot{W}$:

$$m_e v\dot{v} = (M_1 + M_2)v/R \qquad \Rightarrow \qquad a = (M_1 + M_2)\,/\,m_eR \tag{24}$$

Provided there is no lift-off of the wheels, we use the following equations for unknowns $N_1, N_2, F_1, F_2, a, M_1, M_2$. For the whole car:

$$M_{O_z} = \dot{H}_{O_z}: \quad N_1 b - mgb_2 = [mh + (m_1k_1^2 + m_2k_2^2)/R]a \tag{25}$$

$$F_y = 0: \quad mg - N_1 - N_2 = 0 \tag{26}$$

$$F_x = ma: \quad F_1 + F_2 = ma \tag{27}$$

For rear and front wheels:

$$M_{C_{1z}} = M_1 - F_1R = m_1k_1^2a/R \tag{28}$$

$$M_{C_{2z}} = M_2 - F_2R = m_2k_2^2a/R \tag{29}$$

and conditions of no slip at B and A:

$$F_1 \leq \mu_s N_1, \qquad F_2 \leq \mu_s N_2 \tag{30}$$

Equations (25) and (26) yield N_1, N_2 given by eqs (20) and (21). For $a_{\max}$:

$$a \overset{(27)}{=} (F_1 + F_2)/m \overset{(30)}{\leq} \mu_s(N_1 + N_2) \overset{(26)}{=} \mu_s(mg)/m = \mu_s g$$

$$\Rightarrow \quad a_{\max} = \mu_s g \quad \text{for} \quad F_1 = \mu_s N_1, \quad F_2 = \mu_s N_2 \tag{31}$$

M_1, M_2 are obtained from eqs (28) and (29) using (20), (21), (31) for N_1, N_2, F_1, F_2, a:

$$M_1 = \mu_s g[mb_2R/b + \mu_s\{mRh + m_1k_1^2 + m_2k_2^2\}/b + m_1k_1^2/R]$$

$$M_2 = \mu_s g[mb_1R/b - \mu_s\{mRh + m_1k_1^2 + m_2k_2^2\}/b + m_2k_2^2/R]$$

(e) The FBDs are shown in Fig. E3.34e. H_{O_z} is given by eq. (5). The unknowns N_1, N_2, a, $\dot{\omega}_1$, $\dot{\omega}_2$ are obtained using the following 5 equations of motion. For the whole car:

$$M_{O_z} = \dot{H}_{O_z}: \quad N_1 b - mgb_2 = mha + m_1k_1^2\dot{\omega}_1 + m_2k_2^2\dot{\omega}_2 \tag{32}$$

$$F_y = 0: \quad mg - N_1 - N_2 = 0 \tag{33}$$

$$F_x = ma: \quad ma = \mu_k N_1 + \mu_k N_2 = \mu_k mg \tag{34}$$

$$\Rightarrow \quad a = \mu_k g \tag{35}$$

For rear and front wheels:

$$M_{C_{1z}} = M_1 - \mu_k N_1 R = m_1k_1^2\dot{\omega}_1 \tag{36}$$

$$M_{C_{2z}} = M_2 - \mu_k N_2 R = m_2k_2^2\dot{\omega}_2 \tag{37}$$

Substituting a, $\dot{\omega}_1$, $\dot{\omega}_2$ from eqs (35) to (37) in eq. (32) and using $N_1 + N_2 = mg$ yields

$$N_1 = [mg\{b_2 + \mu_k(h - R)\} + M_1 + M_2]\,/\,b \tag{38}$$

$$N_2 = [mg\{b_1 - \mu_k(h-R)\} - M_1 - M_2]/b \tag{39}$$

$\dot{\omega}_1$, $\dot{\omega}_2$ are computed from eqs (36) and (37) using N_1, N_2 from eqs (38) and (39).

The analyses of the retardation of a vehicle with only the rear wheels braked, and with all the wheels braked, are similar to those presented for acceleration. Comparison of equations (31) and (35) reveals that a greater retardation of $\mu_s g$ is obtained if all the wheels are braked in such a manner that there is no slip. On the other hand, if all the wheels are locked, then there is actual slip, yielding a lesser retardation of $\mu_k g$.

Example 3.35: An arm of a robot consists of hinged joints at the elbow A and the shoulder O with torsional springs (Fig. E3.35a). The motors at the shoulder and the elbow provide driving torques M_s and M_e. Derive the governing equations for $\ddot{\theta}$ and $\ddot{\phi}$ for the given configuration. The springs are un-deformed for $\theta = \phi = 0$. $I_{zz}^{C_i} = I_i$.

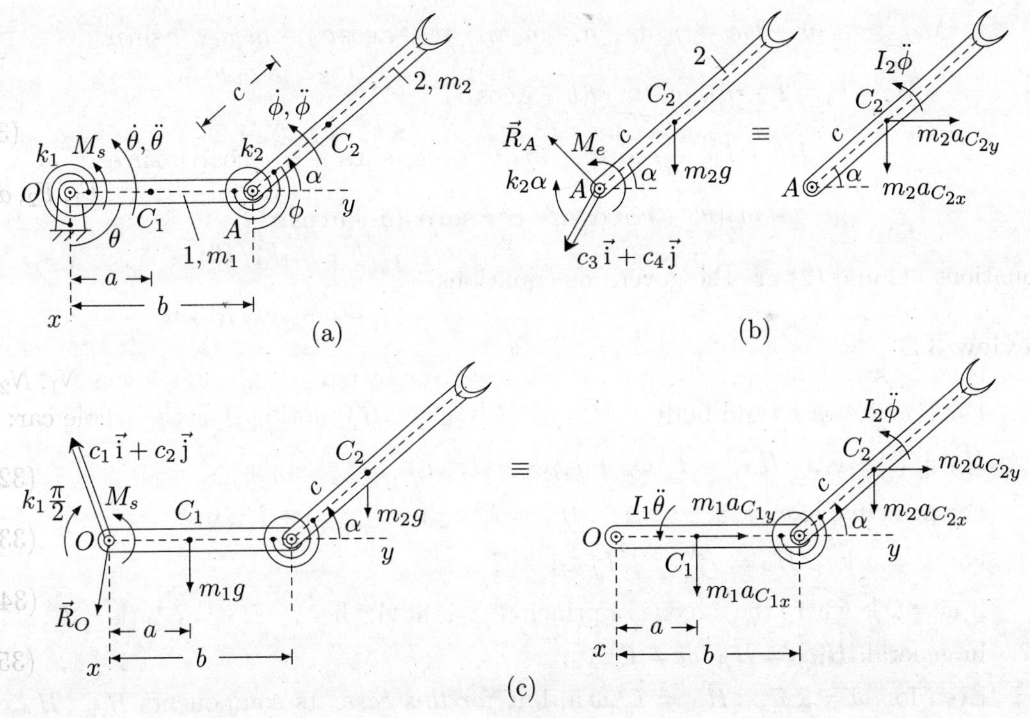

Fig. E3.35

Solution: It is convenient to work with the equivalent kinetic systems. The moments exerted by the springs on the bars are $k_1\theta = k_1\pi/2$ and $k_2(\phi - \theta) = k_2\alpha$. The external force system on link 2 and the equivalent kinetic system are shown in Fig. E3.35b.

$$\overrightarrow{OC}_1 = a\,\vec{\mathrm{j}}, \qquad \overrightarrow{OA} = b\,\vec{\mathrm{j}}, \qquad \alpha = \phi - \theta, \qquad \theta = \pi/2$$

$$\overrightarrow{AC}_2 = c(\cos\alpha\,\vec{\jmath} - \sin\alpha\,\vec{\imath}), \qquad \overrightarrow{OC}_2 = (b + c\cos\alpha)\,\vec{\jmath} - c\sin\alpha\,\vec{\imath}$$

$$\vec{a}_{C_1} = -\ddot{\theta}a\,\vec{\imath} - \dot{\theta}^2 a\,\vec{\jmath}, \qquad \vec{a}_A = -\ddot{\theta}b\,\vec{\imath} - \dot{\theta}^2 b\,\vec{\jmath}$$

$$\begin{aligned}\vec{a}_{C_2} &= \vec{a}_A + \ddot{\phi}\,\vec{k} \times \overrightarrow{AC}_2 - \dot{\phi}^2\overrightarrow{AC}_2 \\ &= \vec{a}_A + \ddot{\phi}\,\vec{k} \times c(\cos\alpha\,\vec{\jmath} - \sin\alpha\,\vec{\imath}) - \dot{\phi}^2 c(\cos\alpha\,\vec{\jmath} - \sin\alpha\,\vec{\imath}) \\ &= (-\ddot{\theta}b - c\ddot{\phi}\cos\alpha + \dot{\phi}^2 c\sin\alpha)\,\vec{\imath} - (\dot{\theta}^2 b + c\ddot{\phi}\sin\alpha + \dot{\phi}^2 c\cos\alpha)\,\vec{\jmath}\end{aligned}$$

$$\begin{aligned}M_{A_z} &= I_2\ddot{\phi} - m_2 a_{C_{2x}} c\cos\alpha - m_2 a_{C_{2y}} c\sin\alpha \\ \Rightarrow \quad M_e - k_2\alpha - m_2 gc\cos\alpha &= (I_2 + m_2c^2)\ddot{\phi} + m_2\ddot{\theta}bc\cos\alpha + m_2\dot{\theta}^2 bc\sin\alpha \qquad (1)\end{aligned}$$

The external and equivalent kinetic systems on links 1 and 2 are shown in Fig. E3.35c.

$$\begin{aligned}M_{O_z} &= I_1\ddot{\theta} + I_2\ddot{\phi} - m_1 a_{C_{1x}} a - m_2 a_{C_{2x}}(b + c\cos\alpha) - m_2 a_{C_{2y}} c\sin\alpha \\ \Rightarrow \quad M_s - k_1\pi/2 - m_1 ga - m_2 g(b + c\cos\alpha) & \\ &= I_1\ddot{\theta} + I_2\ddot{\phi} + m_1a^2\ddot{\theta} + m_2(\dot{\theta}^2 b + c\ddot{\phi}\sin\alpha + \dot{\phi}^2 c\cos\alpha)c\sin\alpha \\ &\quad + m_2(\ddot{\theta}b + c\ddot{\phi}\cos\alpha - \dot{\phi}^2 c\sin\alpha)(b + c\cos\alpha) \qquad (2)\end{aligned}$$

Equations (1) and (2) are the governing equations.

Review 3 B

1. For point A of a rigid body: $[\vec{H}_A] = [I^A]\,[\vec{\omega}]$, $\quad H_{A_i} = I^A_{ij}\omega_j$,
   $\vec{H}_A = I^A_{ij}\omega_j\,\vec{e}_i$, $\quad H_{A_x} = I^A_{xx}\omega_x + I^A_{xy}\omega_y + I^A_{xz}\omega_z$.
   For principal axes x, y, z at A: $\vec{H}_A = I^A_{xx}\omega_x\,\vec{\imath} + I^A_{yy}\omega_y\,\vec{\jmath} + I^A_{zz}\omega_z\,\vec{k}$.
   For $\vec{\omega} = \omega\,\vec{k}$: $\vec{H}_A = I^A_{xz}\omega\,\vec{\imath} + I^A_{yz}\omega\,\vec{\jmath} + I^A_{zz}\omega\,\vec{k}$.
   If $\vec{\omega} = \omega\,\vec{k}$ where the z-axis is a principal axis at A, then: $\vec{H}_A = I^A_{zz}\omega\,\vec{k} = I^A_{zz}\,\vec{\omega}$.
2. In general, $H_{A_n} = \vec{H}_A \cdot \vec{n} \neq I^A_{nn}\omega_n$.
   Even for $\vec{\omega} = \omega\,\vec{n}$, $\quad \vec{H}_A \neq I^A_{nn}\omega\,\vec{n}$; but *for this case*, its components H_{A_n}, H_{A_s} along $\vec{n}$ and along a direction $\vec{s} \perp \vec{n}$, are given by $H_{A_n} = I^A_{nn}\omega$, $\quad H_{A_s} = I^A_{ns}\omega$.
3. For $\vec{\omega} = \omega\,\vec{k}$, for a point A *of the rigid body* and given $\vec{a}_{A|I} = \vec{0}$ or $A \equiv C$ or $\vec{a}_{A|I}$ along CA; and *x, y, z axes attached to the rigid body*:
   $M_{A_x} = I^A_{xz}\dot{\omega} - I^A_{yz}\omega^2$, $\quad M_{A_y} = I^A_{yz}\dot{\omega} + I^A_{xz}\omega^2$, $\quad M_{A_z} = I^A_{zz}\dot{\omega}$.
   a. In general, $\vec{M}_A \neq I^A_{zz}\dot{\omega}\,\vec{k}$, i.e., directions of $\dot{\vec{\omega}}$ and $\vec{M}_A$ are different.
   b. *Only if z is a principal axis at A*, then $\vec{H}_A = I^A_{zz}\,\vec{\omega}$, $\vec{M}_A = I^A_{zz}\dot{\vec{\omega}} \Rightarrow \dot{\vec{\omega}} \parallel \vec{M}_A$.

c. For constant $\vec{\omega}$: $\vec{M}_A = -I^A_{yz}\omega^2\,\vec{\mathrm{i}} + I^A_{xz}\omega^2\,\vec{\mathrm{j}} \neq \vec{0}$, and $\vec{M}_A = \vec{0}$ *only if* z *is a principal axis of inertia at* A.

d. If $M_{A_z} \equiv 0$, then $\vec{\omega}$ is constant.

4. For general $\vec{\omega}$, for a point A *of the rigid body* and given $\vec{a}_{A|I} = \vec{0}$ or $A \equiv C$ or $\vec{a}_{A|I}$ along CA; and x, y, z principal axes of inertia at A and *axes* x, y, z *attached to the rigid body*: $M_{A_x} = I^A_{xx}\dot{\omega}_x - (I^A_{yy} - I^A_{zz})\omega_y\omega_z, \quad \ldots$

   a. For $\vec{\omega} =$ const.: $\vec{M}_A \neq \vec{0}$ *except when* $\vec{\omega}$ *is parallel to a principal axis at* A.

   b. If $\dot{\omega}_z = 0$, then $M_{A_z} \neq 0$, *except when* $I^A_{xx} = I^A_{yy}$ *or* $\omega_x = 0$ *or* $\omega_y = 0$.

   c. If $M_{A_z} = 0$, then $\dot{\omega}_z \neq 0$, *except when* $I^A_{xx} = I^A_{yy}$ *or* $\omega_x = 0$ *or* $\omega_y = 0$.

   d. If $\vec{M}_A = \vec{0}$, then $\dot{\vec{H}}_{A|I} = \vec{0} \Rightarrow \vec{H}_A =$ const. $\not\Rightarrow$ necessarily that $\vec{\omega} =$ const.

   e. In general, $\dot{\vec{\omega}}$ is not in the direction of $\vec{M}_C$, even for plane motion.

5. $T = \frac{1}{2}mv_C^2 + \frac{1}{2}\vec{H}_C \cdot \vec{\omega} = \frac{1}{2}mv_C^2 + \frac{1}{2}I^C_{ij}\omega_i\omega_j, \quad T_R = \frac{1}{2}I^C_{ij}\omega_i\omega_j.$
   If a point A on the body has $\vec{v}_A = \vec{0}$, then $T = \frac{1}{2}I^A_{ij}\omega_i\omega_j = \frac{1}{2}I^A_{\omega\omega}\omega^2$.
   If $\vec{\omega} = \omega\,\vec{\mathrm{k}}$, where $\vec{\mathrm{k}}$ may not be a principal axis at C, then
   $T = \frac{1}{2}mv_C^2 + \frac{1}{2}I^C_{zz}\omega^2$, and if its point A has $\vec{v}_A = \vec{0}$, then $T = \frac{1}{2}I^A_{zz}\omega^2$.

6. $\dot{T} + \dot{W} = \dot{W}^{\text{nc}}, \quad \dot{T}_C = \dot{W}^*, \quad \dot{T}_R = \vec{M}_C \cdot \vec{\omega} = \dot{W}_R. \quad (T_2 + V_2) - (T_1 + V_1) = W^{\text{nc}}_{1\text{-}2}.$
   The total work done by the internal forces on a rigid body is zero.

7. The *necessary and sufficient conditions of equilibrium for a rigid body are*: (1) $\vec{F}(t) \equiv \vec{0}$, (2) $\vec{M}_A(t) \equiv \vec{0}$ *and* (3) *the body should be initially in equilibrium.*

8. The necessary and sufficient conditions of a rigid rotor to be balanced: (a) C lies on the axis of rotation and (b) the axis of rotation is a principal axis at C.

9. For axis of rotation along z at O and centre of percussion P: $m(OC)(OP) = I^O_{zz}$.

10. For a one dof system if T and $\dot{W}$ can be written easily in the general configuration, then $\dot{T} = \dot{W}$ is the easiest means to get acceleration/angular acceleration.

11. If impulsive loads act, then $\dot{T} = \dot{W}$, $\vec{F} = m\vec{a}_{C|I}$, $\vec{M}_C = \dot{\vec{H}}_{C|I}$ are invalid. Use the impulse–momentum and angular impulse-moment of momentum relations.

12. For motion relative to a reference frame T translating at $\vec{a}_{|I}$, all equations of dynamics of a rigid body are exactly the same as for the inertial frame I if the external force system is augmented by an 'inertial force' $-m\vec{a}_{|I}$ acting at C.

3.13 Concept Review Questions 3B

4. A body of mass m with a ball-and-socket joint (Fig. Q3.4) has angular velocity $-\dot{\theta}\,\vec{\mathrm{i}}$. Its principal moments of inertia at O are $I^O_{xx} = I^O_{yy}$, I^O_{zz}. It is subjected to an impulsive force $\hat{P}\,\vec{\mathrm{i}}$ at point (a, b, c). Find the impulsive reaction at O and the angular velocity of the body just after the application of the impulsive force.
   $[\hat{P}[mhc/I^O_{yy} + mbh\tan\theta/I^O_{xx} - 1]\,\vec{\mathrm{i}}, \; -\dot{\theta}\,\vec{\mathrm{i}} + c\hat{P}\,\vec{\mathrm{j}}/I^O_{yy} - b\hat{P}\,\vec{\mathrm{k}}/I^O_{zz}]$

5. A rigid body with a ball-and-socket joint at O has angular velocity $\vec{\omega} = \omega_i \, \vec{e}_i$ just before it engages with an obstruction at A such that point A comes to rest (Fig. Q3.5). The direction cosines of $\overrightarrow{OA}$ with respect to $\vec{e}_i$ are n_i. Show that the angular velocity $\omega \, \vec{n}$ just after impact is given by $\omega = I^O_{ij} n_i \omega_j / I^O_{pq} n_p n_q$.

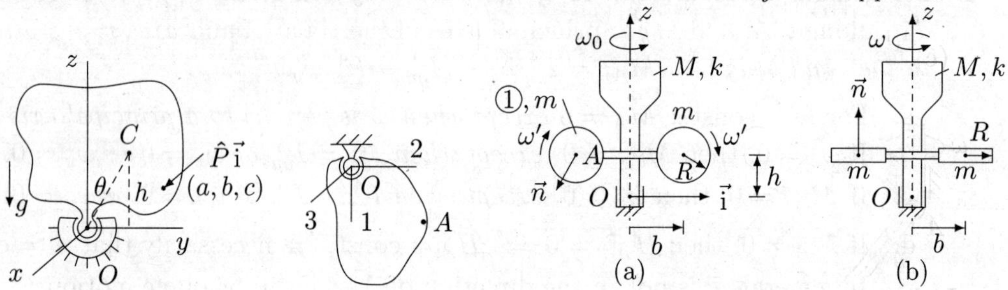

Fig. Q3.4 Fig. Q3.5 Fig. Q3.6

6. (a) The system shown in Fig. Q3.6a consists of two discs attached to a rotor with axial radius of gyration k. The rotor is free to rotate about a fixed vertical axis. Initially, the rotor rotates at ω_0 and the discs rotate at ω' relative to the rotor. Find $\vec{H}_A$ for disc ① and $\vec{H}_O$ of the system. Compute $\dot{\vec{H}}_A$ and $\dot{\vec{H}}_O$ to show that ω' and ω_0 are constant and find the moment reaction at A from the support on the disc ①, and moment reaction at O from the support on the rotor. (b) By an internal mechanism, the supports of the discs are rotated by 90° to the configuration shown in Fig. Q3.6b. Find the final ω of the rotor. Find the work done by the internal mechanism. (c) In the final configuration, another internal mechanism is operated to provide $\omega_1 \, \vec{k}$ to the discs relative to the rotor. Find ω^* of the rotor. Find the work done by this internal mechanism. Find the work done by this internal mechanism on disc ① and its angular impulse on disc ① about its centre A.
[(a) $\frac{1}{2} mR^2 \omega' \, \vec{j} + \frac{1}{4} mR^2 \omega_0 \, \vec{k}$, $mR^2 \omega' \, \vec{j} + (Mk^2 + \frac{1}{2} mR^2 + 2mb^2)\omega_0 \, \vec{k}$, $-\frac{1}{2} mR^2 \omega_0 \omega' \, \vec{i}$, $-mR^2 \omega_0 \omega' \, \vec{i}$ (b) $[(Mk^2 + \frac{1}{2} mR^2 + 2mb^2)\omega_0 + mR^2\omega']/(Mk^2 + 2mb^2)$, $\frac{1}{2}(Mk^2 + 2mb^2)(\omega^2 - \omega_0^2) - \frac{1}{4} mR^2 \omega_0^2$
(c) $[(Mk^2 + 2mb^2 + \frac{1}{2} mR^2)\omega_0 - mR^2\omega_1]/(Mk^2 + 2mb^2 + mR^2)$, $\frac{1}{2}(Mk^2 + 2mb^2)(\omega^{*2} - \omega^2) + \frac{1}{2} mR^2[(\omega^* + \omega_1)^2 - \omega'^2]$, $\frac{1}{4} mR^2[(\omega^* + \omega_1)^2 - \omega'^2]$, $\frac{1}{2} mR^2(\omega^* + \omega_1 - \omega') \, \vec{k}$]

7. (a) A cylinder (m, r, b) is welded to a light rod which is pinned to a vertical shaft (Fig. Q3.7). When the shaft rotates at constant ω, the rod makes a constant angle θ with the vertical. Axes y, z are in the vertical plane of AB at the given instant. Prove that $\theta = 0$ for all ω and also $\theta = \cos^{-1}[gd/\omega^2(d^2 + \frac{1}{12} b^2 - \frac{1}{4} r^2)]$ if $\omega^2 > gd/(d^2 + \frac{1}{12} b^2 - \frac{1}{4} r^2)$. Find the reaction from the pin A on the rod. (b) Find the governing equation for the steady angle θ for a similar rotating body if the

elements of its inertia matrix at A with respect to body-fixed axes $\vec{e}_i$ are I^A_{ij}. (c) The body in part (b) is imparted $\dot\theta = \dot\theta_0$ at $\theta = \theta_0$. Find its $\dot\theta$ at position θ.

[(a) $-m\omega^2 d \sin\theta\,\vec{j} + mg\,\vec{k}$

(b) $[I^A_{21}(c^2 - s^2) + (I^A_{11} - I^A_{22})sc]\omega^2 + mgds = 0$ with $s = \sin\theta,\ c = \cos\theta$

(c) $[\dot\theta_0^2 + \{\frac{1}{2}(I^A_{11} - I^A_{22})\omega^2(\cos 2\theta - \cos 2\theta_0) - I^A_{12}\omega^2(\sin 2\theta - \sin 2\theta_0)$
$+2mgd(c - \cos\theta_0)\}/I^A_{33}]^{1/2}$]

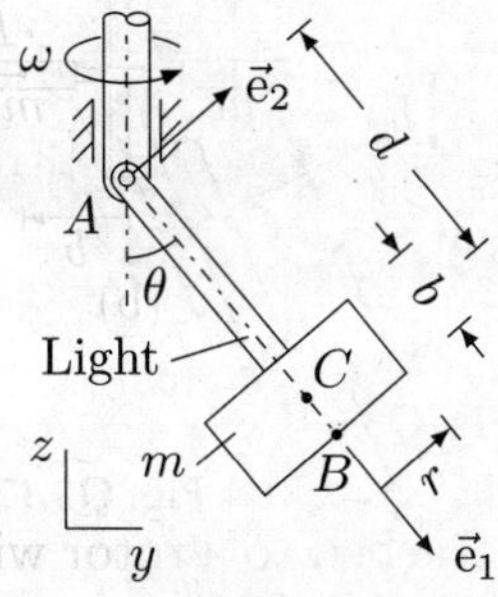

Fig. Q3.7

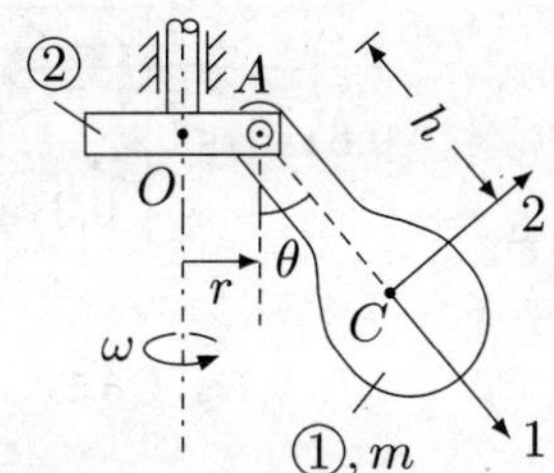

Fig. Q3.8

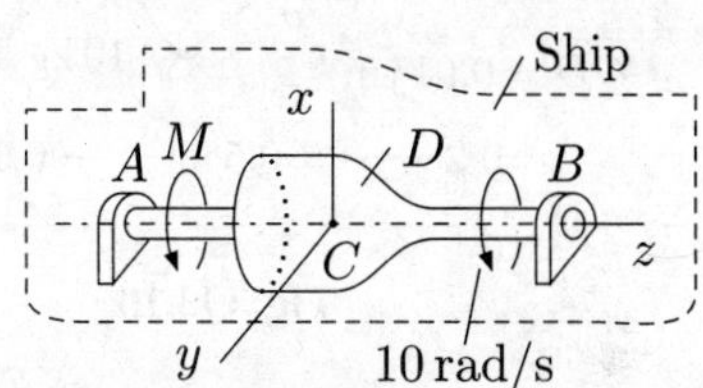

Fig. Q3.9

8. (a) The turn-table ② rotates at ω (Fig. Q3.8). The elements of the inertia matrix of body ① at C with respect to axes $\vec{e}_i$ are I^C_{ij}. Set up an equation for the steady value of θ. (b) Body ① is imparted $\dot\theta = \dot\theta_0$ at $\theta = \theta_0$. Find its $\dot\theta$ at position θ.
[(a) $\omega^2[\frac{1}{2}(I^C_{11} - I^C_{22})\sin 2\theta + I^C_{12}\cos 2\theta - m(r + h\sin\theta)h\cos\theta] + mgh\sin\theta = 0$
(b) $\frac{1}{2}(I^C_{33} + mh^2)(\dot\theta^2 - \dot\theta_0^2) = mgh(\cos\theta - \cos\theta_0) + m\omega^2 rh(\sin\theta - \sin\theta_0)$
$-\frac{1}{2}\omega^2 I^C_{12}(\sin 2\theta - \sin 2\theta_0) - \frac{1}{4}\omega^2(mh^2 - I^C_{11} + I^C_{22})(\cos 2\theta - \cos 2\theta_0)$]

9. The body D on a ship (Fig. Q3.9) is rotated by a motor with a constant torque $M = 800$ N.m. At the given instant, relative to the ground the ship has: $\vec\omega_1 = \vec{i} + 2\vec{j} + 3\vec{k}$ rad/s, $\dot{\vec\omega}_1 = 4\vec{i} + 5\vec{j} + 6\vec{k}$ rad/s^2. The angular velocity of body D relative to the ship is 10 rad/s. The x, y, z-axes are the principal axes of inertia of the body D at its centre of mass C with $I^C_{xx} = 120$, $I^C_{yy} = 90$, $I^C_{zz} = 100$ kg.m^2. Find the angular acceleration of the body D relative to the ship and the moment $\vec{M}_C$ of the bearing reactions on body D. [2.6 rad/s^2, $3140\vec{i} - 190\vec{j}$ N.m]

10. The shaft in Fig. Q3.10 rotates at $\omega = 200$ rad/s and a couple $M_0 = 20$ N.m is applied to it. Find its $\dot\omega$ and the bearing reactions at this instant. The bearing at E does not provide any axial constraint. $OD = 0.01$ m, $R = 0.2$ m.
[70.64 rad/s^2, $-1048\vec{i} - 798\vec{j}$ N, $1237\vec{i} - 3202\vec{j}$ N]

11. A swing-wing plane is moving at 1440 km/h. The pilot is rolling the plane at 0.6 rad/s and performing a loop (Fig. Q3.11). He is swinging his wings back at the rate of 0.3 rad/s. The wing has mass of 500 kg and its principal radii of

gyration at the centre of mass C are: $k_x = 0.7\,\text{m}$, $k_y = 3\,\text{m}$, $k_z = 3.05\,\text{m}$. Find the total external moment on the wing about C from the fuselage and from the aerodynamic forces at this instant. $[-71.28\vec{\imath} - 356.2\vec{\jmath} - 712.5\vec{k}\ \text{N.m}]$

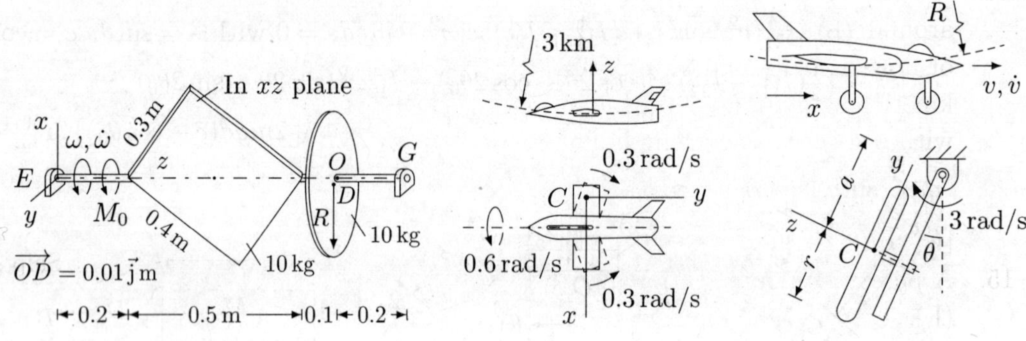

Fig. Q3.10 Fig. Q3.11 Fig. Q3.12

12. (a) A plane (Fig. Q3.12) just after take-off is flying at $v = 200\,\text{km/h}$, $\dot{v} = 50\,\text{km/h}/s$ in a vertical trajectory of radius $R = 1000\,\text{m}$. The back wheels are being retracted at constant rate of 3 rad/s. The wheel has $r = 300\,\text{mm}$ and weight of 900 N. The radius of gyration about its axis is 250 mm and about a transverse axis through C is 180 mm. Find the total moment coming on to the bearings of the wheel (excluding the effect of gravity force). Neglect wind and bearing friction. $\theta = 0°$. (b) Rework part (a) if the given data is some time after take-off when $\theta = 30°$. For this case, the plane was moving without slip at 183.6 km/h just before take-off. $[(a)\ 3187\vec{\jmath}\ \text{N.m}\quad (b)\ -27.12\vec{\imath} + 2929\vec{\jmath}\ \text{N.m}]$

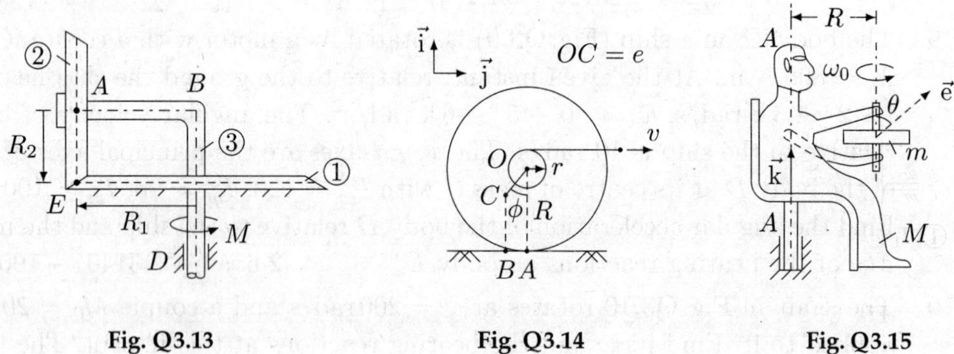

Fig. Q3.13 Fig. Q3.14 Fig. Q3.15

13. Fixed gear ① is in mesh with gear ② whose axis AB is rotated by applying a couple M to the vertical shaft BD (Fig. Q3.13). Initially the system is at rest. The moment of inertia of body ③ about its axis of rotation is I. The mass of gear ② is m and its radii of gyration about axis AB and about a transverse axis at A are k_1, k_2. Find the angular velocity ω of body ③ after a rotation of θ

(a) if M = constant and (b) if $M = ct$.

[(a) $(2M\theta/I_1)^{1/2}$ (b) $(4.5c\,\theta^2/I_1)^{1/3}$ with $I_1 = I + m(R_1^2 + k_2^2 + k_1^2 R_1^2/R_2^2)$]

14. A cylinder of mass m rolls without slip with the end of the light rope wrapped around its axle having velocity v at the given instant (Fig. Q3.14). Its centre of mass is at C and its axial principal moment of inertia at O is I_{zz}^O. Find its kinetic energy, momentum $\vec{p}$ and moments of momentum $\vec{H}_C, \vec{H}_O, \vec{H}_B$ and $\vec{H}_A$ with respect to the ground. $[[I_{zz}^O + m(R^2 - 2eR\cos\phi)]v^2/2(R+r)^2$, $mv[e\sin\phi\,\vec{\text{i}} + (R - e\cos\phi)\,\vec{\text{j}}\,]/(R+r)$, $(I_{zz}^O - me^2)v\,\vec{\text{k}}/(R+r)$, $I_{zz}^O v\,\vec{\text{k}}/(R+r)$, $[I_{zz}^O - me^2 + m(R - e\cos\phi)^2]v\,\vec{\text{k}}/(R+r)$, $[I_{zz}^O + m(R^2 - 2eR\cos\phi)]v\,\vec{\text{k}}/(R+r)]$

15. A person sits on a stationary stool which is free to rotate about a vertical axis (Fig. Q3.15). The moment of inertia of the person and the stool about the axis AA is I_1 and their total mass is M. Another person hands him a wheel rotating freely on its axle at $\omega_0\,\vec{\text{k}}$. The mass of the wheel and axle is m, with moment of inertia I_0 about its axis and I_2 about a transverse axis through its centre of mass. (a) The person inverts the wheel in his hands. Find the angular velocity ω_1 of the stool. (b) The person turns the axis of rotation of the wheel by 90° so that it now points radially outward from the axis of rotation of the stool. Find the angular velocity ω_2 of the stool. Neglect change of moment of inertia due to change in the position of the person's arms. (c) The person turns the axis of rotation of the wheel such that it is at an angle θ to the axis of stool and its spin rate is in direction $\vec{\text{e}}$. Find the angular velocity ω_3 of the stool. [(a) $2I_0\omega_0/(I_1 + mR^2)$ (b) $I_0\omega_0/(I_1 + I_2 + mR^2)$ (c) $(1 - \cos\theta)I_0\omega_0/(I_1 + mR^2 + I_2\sin^2\theta)$]

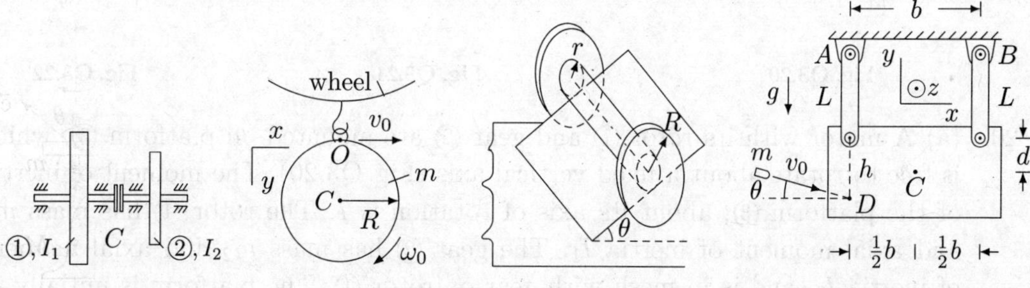

Fig. Q3.16 **Fig. Q3.17** **Fig. Q3.18** **Fig. Q3.19**

16. (a) Flywheel ① is spinning and flywheel ② is at rest. Their axial moments of inertia are I_1, I_2. They are engaged by the clutch C (Fig. Q3.16) and during the process to full engagement a quantity of heat H is developed. Find the initial kinetic energy of the system. (b) The axial force between the circular clutch plates of radius R is P and the coefficients of friction are μ_s, μ_k. Assume uniform wear. Find the time taken and the rotation of flywheel ② before full engagement.

[(a) $(I_1+I_2)H/I_2$ (b) $[8I_1I_2H/(I_1+I_2)]^{1/2}/\mu_k RP,\ 2I_1H/(I_1+I_2)\mu_k RP$]

17. In an automatic process (Fig. Q3.17), a stationary disc with axial radius of gyration k, is picked up by a rotating wheel. The wheel imparts velocity v_0 to point O. Find the instantaneous angular velocity ω_0 of the disc. [$v_0R/(k^2+R^2)$]

18. The axle of the rigid body rolls without slip (Fig. Q3.18) with angular velocity ω on the inclined plane just before the disc makes contact with the horizontal surface. The mass of the system is m and the axial moment of inertia is I. Assuming that the disc rolls without slip on the horizontal surface, find its angular velocity Ω just after impact. Find the minimum coefficient of friction to ensure this. [$(I+mrR\cos\theta)\omega/(I+mR^2),\ I(R-r\cos\theta)/(I+mR^2)r\sin\theta$]

19. A bullet impacts a body suspended by two light rods (Fig. Q3.19). This body has mass M and radius of gyration k_C about the principal axis z at its centre of mass C. The bullet gets embedded in the body at D. Find v_C just after impact, the impulsive reactions at A and B and the maximum rise of C.

[$mv_0\cos\theta/(M+m),\ \hat{T}_B = M(d-h)mv_0\cos\theta/(M+m)b,\ mv_0\sin\theta-\hat{T}_B,\ v_C^2/2g$]

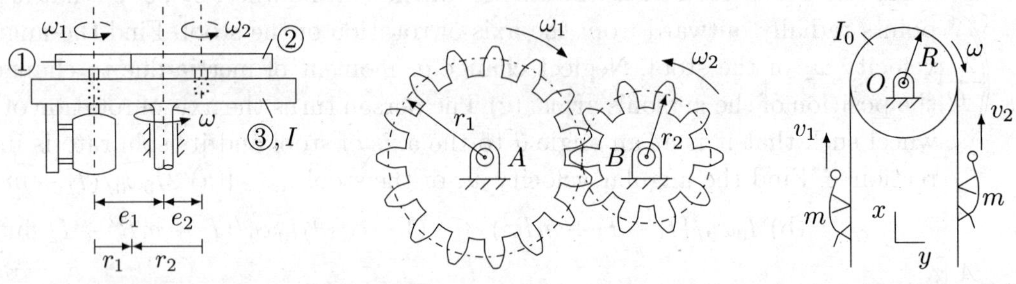

Fig. Q3.20 Fig. Q3.21 Fig. Q3.22

20. (a) A motor with its rotor ① and gear ② are mounted on platform ③, which is free to rotate about a fixed vertical axis (Fig. Q3.20). The moment of inertia of the platform ③, about its axis of rotation is I. The rotor ① has mass m_1 and axial moment of inertia I_1. The gear ② has mass m_2 and axial moment of inertia I_2, and is in mesh with gear on rotor ①. The platform is initially at rest. The motor is turned on. Find the angular velocity ω of the platform with respect to the ground when the angular velocity of the motor is ω_1 with respect to the platform. (b) The platform is restrained from rotation and the motor is started. Find the restraining couple C_1 which has to be applied about the axis of rotation of the platform if the motor has uniform angular acceleration $\dot{\omega}_1$.

[(a) $(I_2r_1/r_2-I_1)\omega_1/(I+I_1+I_2+m_1e_1^2+m_2e_2^2)$ (b) $(I_1-I_2r_1/r_2)\dot{\omega}_1$]

21. Two gears with axial moments of inertia I_1, I_2, are rotating freely at ω_1, ω_2

(Fig. Q3.21). One of them is pushed on its splined shaft to engage with the other. Find the tangential impulse on them at the contact line during engagement and their final angular velocities. If instead of gears, two discs are engaged together, find the duration of slip if the normal force between them is N and the coefficients of friction are μ_s, μ_k. $[\hat{F}_t = (\omega_2 r_2 - \omega_1 r_1)/(r_1^2/I_1 + r_2^2/I_2)$,

$\omega_1' r_1 = \omega_2' r_2 = r_1 r_2 (I_2 \omega_2 r_1 + I_1 \omega_1 r_2)/(I_2 r_1^2 + I_1 r_2^2),\ t = \hat{F}_t/\mu_k N]$

22. A rope of mass m and length $3\pi R$ goes around a pulley of mass m_0 and axial moment of inertia I_0 (Fig. Q3.22). Two monkeys hang from the ends of the rope. Initially, the system is at rest. The monkeys start moving with velocities v_1 and v_2 ($v_2 > v_1$) relative to the rope. The rope does not slip over the pulley. (a) Find the initial ω_0 of the pulley and the initial impulsive reaction at O. (b) Find $\dot{\omega}$, ω of the pulley and the reaction at O after the pulley has rotated by 90°.

[(a) $m(v_2 - v_1)R/(I_0 + 3mR^2)$, $m(v_1 + v_2)\vec{\imath} + \frac{2}{3\pi} mR\omega_0 \vec{\jmath}$ (b) $\dot{\omega} = mgR/3(I + 3mR^2)$,

$\omega^2 = \omega_0^2 + \pi mgR/6(I + 3mR^2)$, $(3m + m_0)g\vec{\imath} - \frac{1}{3} m\dot{\omega}R\vec{\imath} + \frac{2}{3\pi} mR(\dot{\omega}\vec{\jmath} - \omega^2 \vec{\imath})$]

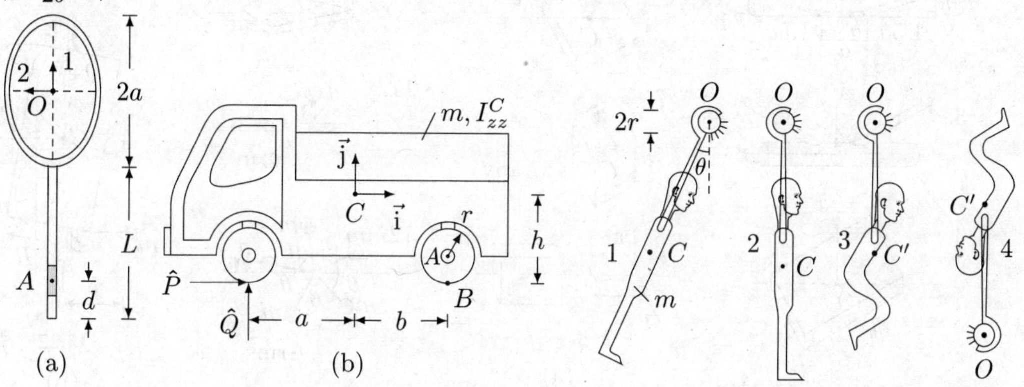

Fig. Q3.23

Fig. Q3.24

23[d]. (a) The handle and head of a tennis racket are modelled as: a rod of length L with mass m_1 and an oval of mass m_2 with $I_{22}^O = I_2$ (Fig. Q3.23a). The centre of the hand grip is at distance d from the end. Show that if a hard stroke to the ball at the centre of the racket head is not to jar the hand of the player, then: $m_1[L(a + L/3) - d(L + 2a)] = 2I_2$. (b) The design requirement of a vehicle (Fig. Q3.23b) is that if the front wheels go over a bump, the rear wheels are not subjected to any impulsive force and vice-versa. Find the condition for this requirement to be satisfied approximately. Work out for spring-borne and non spring-borne frame. Neglect inertia of the wheels.

[(b) Spring-borne: no condition, non spring-borne: $I_{zz}^C = mab$]

24. An acrobat translates down from a high platform at height h above a parallel bar and grasps it (Fig. P3.24) in position 1. Assume the bar as smooth. In position 1, the centre of mass C of the acrobat is at $OC = b$ and its $I_{zz}^O = I$. In the vertical position, he crouches from configuration 2 to configuration 3 and his centre of mass C shifts to C' with $OC' = b'$, $I_{zz}^O = I'$. Find h so that he is just able to reach the vertical position 4 at which point he straightens out.
$[\{2b'I' - Ib(1 - \cos\theta)\}/mb^2 \sin^2\theta]$

25. (a) Which of the rotors shown in Fig. Q3.25a are balanced for rotation about the z axis? (b) The rotor in Fig. Q3.25b is to be balanced by attaching mass-points, each of mass m, at A and B. Find m.
$[(a)\ 2, 5, 6, 7, 8, 9, 10, 12\ \ (b)\ (L^2/24 - R^2/8)M \sin 2\theta / b(2R\sin\theta + L\cos\theta)]$

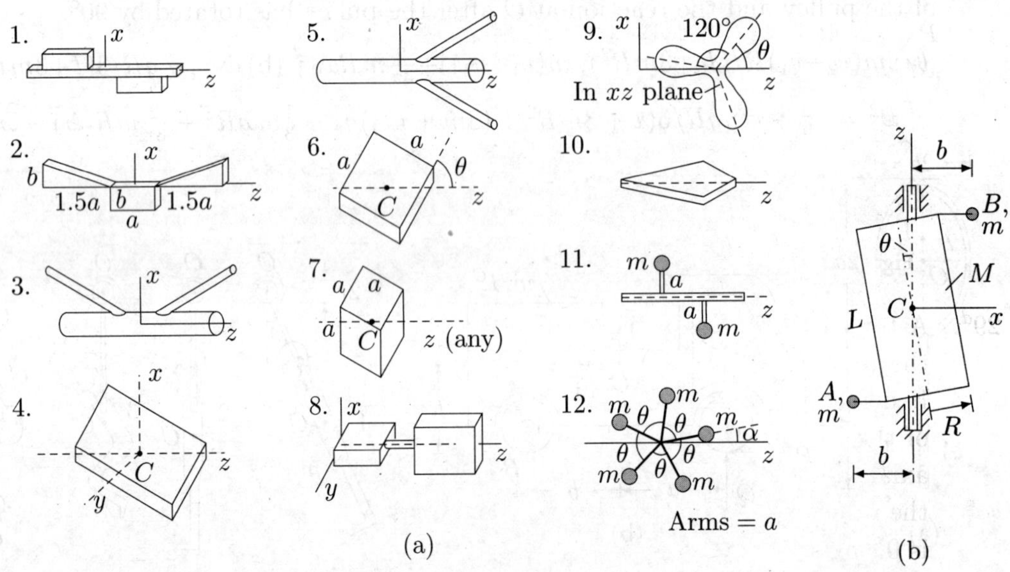

Fig. Q3.25

26. (a) An impulsive force $\hat{P}$ is applied on an axisymmetric body having axial moment of inertia I_{zz}^C (Fig. Q3.26). The body rolls without slip after impact. Find the angular velocity ω just after impact and $\mu_{\min}$. (b) Find ω just after impact if the body is slipping to the right on the ground and the coefficient of friction is μ. $[(a)\ \{\hat{P}(h+R)\cos\theta - \sqrt{R^2 - h^2}\sin\theta\}/I$, where $I = (I_{zz}^C + mR^2)$, $|(I_{zz}^C - mRh)\cot\theta + mR\sqrt{R^2 - h^2}|/I$ (b) $\hat{P}[h\cos\theta + (\mu R - \sqrt{R^2 - h^2})\sin\theta]/I_{zz}^C]$

27. A slab moving at speed v_0 comes in contact with two rollers with axial radius of gyration k_C (Fig. Q3.27). Find the velocity v_1 of the slab after slip has stopped everywhere. Find the time taken and the distance moved by the slab to achieve this. $[2mv_0r^2/[2mr^2 + M(k_C^2 + r^2)],\ (v_0 - v_1)/\mu g,\ (v_0^2 - v_1^2)/2\mu g]$

28. The four-wheeled vehicle in Fig. Q3.28 has endless treads on each side with mass λ per unit length. The mass of the vehicle without the treads and wheels is m_1, the mass of each wheel is m_2 and its axial radius of gyration is k. The engine torque is M_e, the angular velocity of the crank shaft is ω_e and the efficiency of transmission is η. The gear ratio n is the ratio of ω_e to the angular velocity ω of the wheels. The effective moment of inertia of the rotating parts of the engine about the crank shaft axis is I. The wind resistance is P. (a) Find the acceleration of the vehicle. (b) What is the minimum gear ratio necessary for the acceleration of the vehicle? (c) Find the gear ratio for maximum acceleration.
[(a) $a = [\eta n M_e/r - P]/(m_e + n^2 I/r^2)$ with $m_e = m_1 + 4m_2(1 + k^2/r^2) + 8\lambda(b + \pi r)$
(b) $Pr/\eta M_e$ (c) $(Pr/\eta M_e)[1 + (1 + m_e \eta^2 M_e^2/IP^2)^{1/2}]$

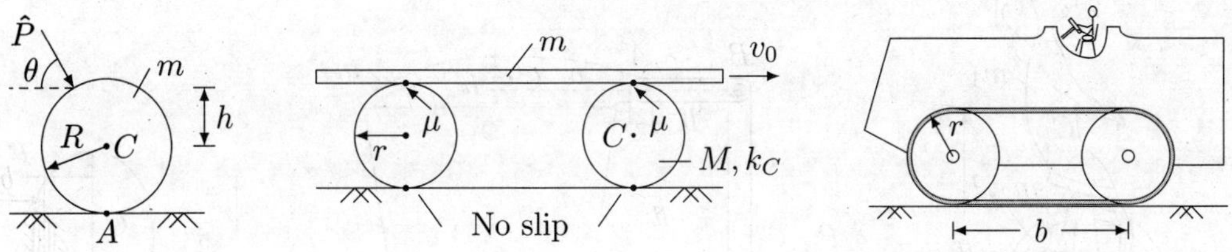

Fig. Q3.26 Fig. Q3.27 Fig. Q3.28

29[d]. A toy car is driven by a torsional spring of stiffness k_t mounted on the axle of the rear wheels. It is given eight turns before the car is released on the floor. It detaches from the axle after unwinding. Assume no slip. Neglect the inertia of the front wheels. The rear wheel assembly has mass 30 g, radius 50 mm and axial radius of gyration of 30 mm. The main body has mass of 70 g. Determine the value of k_t so that the speed of the car immediately after the spring unwinds is 0.8 m/s. [28.07×10^{-6} N.m/rad]

30. A man rides a unicycle (Fig. Q3.30). The rotating parts have mass m_2, centre of mass O and axial moment of inertia I. The man and the translating parts of the cycle have mass m_1 and centre of mass C. The coefficient of friction is μ. Assume no slip. Find the acceleration a of O. [$-m_1 gb/\{m_2 R + m_1 h + I/R\}$]

31. The lid of a pressure vessel has mass m and radius of gyration k_C about the z-axis through its centre of mass C (Fig. Q3.31). A vertical force P is applied to a foot-paddle to open it. Neglect the mass of links OE, AB and BD. (a) Find the angular velocity of the lid after rotating by angle θ. (b) Find the initial angular acceleration of the lid and the reaction of the hinge at A on the lid.
[(a) $\{2PL\sin\theta + 2mg(h - h\cos\theta - a\sin\theta)/m(k_C^2 + a^2 + h^2)\}^{1/2}$
(b) $\dot{\omega} = (PL - mga)/m(k_C^2 + a^2 + h^2)$, $\{(\dot{\omega}a + g)m + PL/b\}\vec{\jmath} - m\dot{\omega}h\vec{\imath}$]

32[d]. An inverted pendulum of mass m with centre of mass C (Fig. Q3.32). is subjected to a constant force F and a moment M. Its radius of gyration about the axis of rotation is k. The spring is undeformed for $\theta = 0$. (a) Derive the equation of motion. (b) The angular velocity is ω_0 at $\theta = 0$. At position θ, find $\dot{\theta}$, $\ddot{\theta}$ and the hinge reaction. Set up an equation for θ_{max}. (c) For $F = 0, M = 0$, set up the equation of motion for small values of θ. Using dynamic analysis, determine the required value of k_t so that the inverted pendulum is stable.

[(a) $mk^2\ddot{\theta} = mgh\sin\theta - Fb\cos\theta + M - k_t\theta$ (c) $mk^2\ddot{\theta} + (k_t - mgh)\theta \simeq 0,\ k_t > mgh$
(b) $\dot{\theta}^2(\theta) = \omega_0^2 + 2[mgh(1-\cos\theta) - \frac{1}{2}k_t\theta^2 + M\theta - Fb\sin\theta]/mk^2$, $\ddot{\theta}$ from (a),
$\vec{R}_O = [m(h\ddot{\theta} - g\sin\theta) + F\cos\theta]\,\vec{e}_\theta + [m(g\cos\theta - h\dot{\theta}^2) + F\sin\theta]\,\vec{e}_r$, θ_{max} for $\dot{\theta} = 0$]

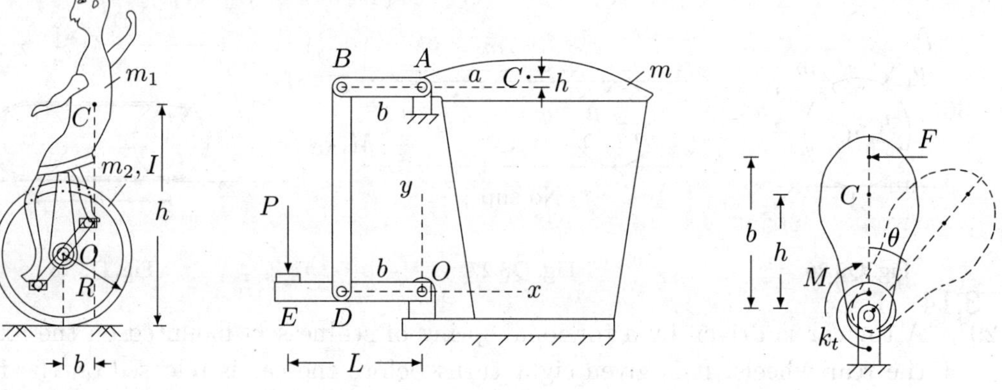

Fig. Q3.30 **Fig. Q3.31** **Fig. Q3.32**

33. A block, of mass m and I_{zz}^C at centre of mass C, is at rest on a railway wagon which starts with acceleration a_0 (Fig. Q3.33). The coefficient of friction is μ. (a) If μ is large enough, find the minimum value a_1 of a_0 for which the block tips about A. (b) If μ is small enough, find the minimum value a_2 of a_0 for which the block slips. (c) If $\mu > b/h$, find the initial angular acceleration of the block for different values of a_0. (d) If the block does not slip, find its angular velocity when it has rotated by θ. [(a) gb/h (b) μg (c) $a_0 < gb/h:\ 0,\ \ gb/h < a < \mu g:$
$m(a_0h - gb)/[I_{zz}^C + m(h^2 + b^2)],\ \ a > \mu g:\ (\mu h - b)mg/[I_{zz}^C + mb(b - \mu h)]$
(d) $[2m\{(a_0b + gh)(1-\cos\theta) + (a_0h - gb)\sin\theta\}/\{I_{zz}^C + m(b^2 + h^2)\}]^{1/2}$]

34. The coefficient of friction between the hoop and the cylinder is μ (Fig. Q3.34). There is no slip on the inclined plane. The moments of inertia about the central rotation axes are I_1, I_2. Prove that the velocity of the hoop centre, after moving a distance s starting from rest, is $[2gs\sin\alpha\{(m_1+m_2)(\cos\beta + \mu\sin\beta) + \mu(m_1 - m_2)\}/\{(I_1/r_1^2 + I_2/r_2^2)(\cos\beta + \mu\sin\beta) + \mu(I_1/r_1^2 - I_2/r_2^2)\}]^{1/2}$.

35. A spool, resting on a horizontal surface, has coefficients of friction μ_s, μ_k, mass m and axial moment of inertia $\frac{1}{4}mR^2$ (Fig. Q3.35). A cord wrapped around the bobbin is pulled out at end A by a force $P = \alpha mgt$. Find the distance moved by C and A before slip changes into no slip or vice-versa as the case may be. Find also the position of C and A at the instant the rotation of the spool changes its direction. $\mu_k = 0.8\mu_s$. $[[2/3,\ 5/12,\ 10/3,\ 2.738]\mu_s^3 g/\alpha^2]$

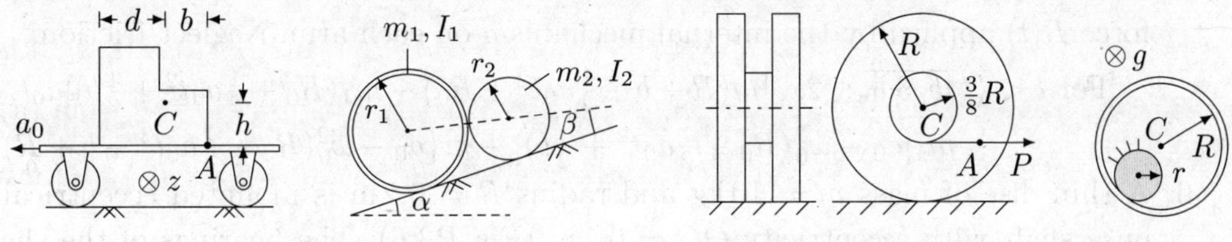

Fig. Q3.33 **Fig. Q3.34** **Fig. Q3.35** **Fig. Q3.36**

36. A ring, of internal radius R, axial width b and mass m, rotates without slip around a fixed vertical cylinder of radius r (Fig. Q3.36). The axial moment of inertia of the ring is I_C. The coefficient of friction is μ. For what value of v_C would the ring not fall down? $[v_C > [g(R-r)\max\{1/\mu,\ 2R/b\}]^{1/2}]$

3.14 Practice Problems 3B

4. A gear A is running at ω_A and is in mesh with gears B_1 and B_2 (Fig. P3.4). A tooth on A breaks, bringing B_1 and B_2 suddenly to rest relative to gear box G, which is free to rotate about axis $\vec{j}$. Show that the angular velocity $\omega\vec{j}$ of A and $\omega_G\vec{j}$ of G just afterwards are given by $\omega = (1 + 2r_A^2 I_B/r_B^2 I_A)\omega_A$ and $\omega_G = -2r_A^2 I_B\omega_A/r_B^2[I_G + 2(I_2 + m_B r_A^2)]$, where I_A, I_B are moments of inertia of the gears about their axes and m_B is mass of B_1. I_2 is the moment of inertia of B_1 about the transverse central axis and I_G is the axial moment of inertia of G.

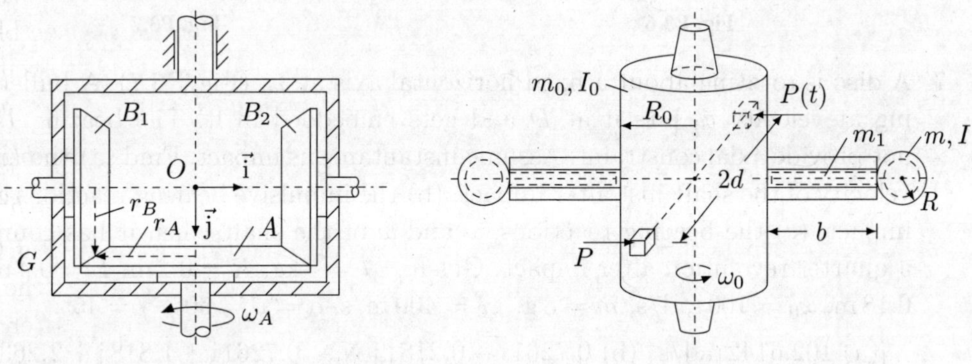

Fig. P3.4 **Fig. P3.5**

5. A satellite has two telescopic arms with thick spherical shells at their ends. Its mass, without the movable parts, is m_0 with axial moment of inertia I_0. Each shell has mass m and moment of inertia I about the shell axis. Initially, the satellite is rotating at ω_0 (Fig. P3.5). By an internal mechanism the arms are extended at a constant rate $a_0\,\mathrm{m/s^2}$ by moving out by distance $2b/3$ the inner segments of length b and mass m_1 each. Two jets on the satellite are activated to provide thrust $P(t)$ so that its angular velocity remains constant. Find $P(t)$ and the radial force $R(t)$ applied by the internal mechanism on each arm. Neglect friction.

$$[\text{For } t < \sqrt{4b/3a_0}: \; 2\omega_0[m(R+b+\tfrac{1}{2}a_0t^2+R_0)+m_1(R_0+\tfrac{1}{2}a_0t^2+\tfrac{1}{2}b)]a_0t/d,$$
$$m_1[a_0-\omega_0^2(R_0+\tfrac{1}{2}a_0t^2+\tfrac{1}{2}b)]+m\{a_0-\omega_0^2(R_0+\tfrac{1}{2}a_0t^2+b+R)\}]$$

6 A thin disc of mass $m = 10\,\mathrm{kg}$ and radius $R = 0.2\,\mathrm{m}$ is mounted eccentrically on a shaft with eccentricity $OC = 0.1\,\mathrm{m}$ (Fig. P3.6). The bearings of the shaft are mounted on a platform which rotates about a fixed vertical axis. Find the moment M applied to the shaft BD and the bearing reactions at the given instant when OC is vertical. $OA = 0.5$ m and $AO \perp BD$. The bearing at B does not provide any axial restraint.

$$[51.8\,\mathrm{N.m},\; -251.8\vec{j}+7.3\vec{k}\,\mathrm{N},\; 55\vec{i}-254.3\vec{j}+74.8\vec{k}\,\mathrm{N}]$$

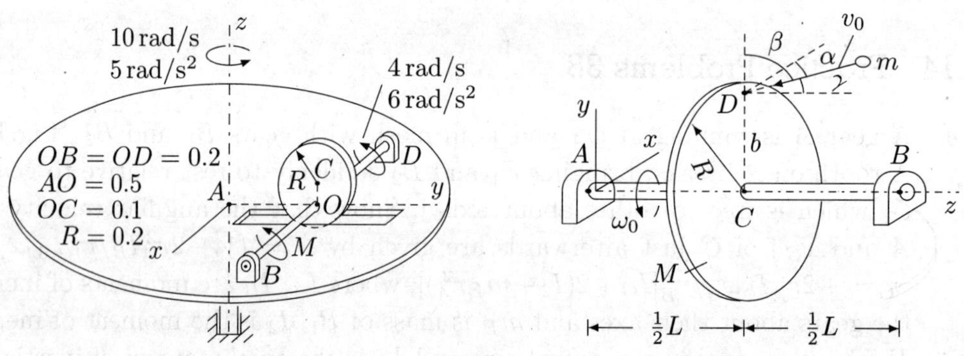

Fig. P3.6

Fig. P3.7

7 A disc is rotating about a fixed horizontal axis at ω_0 (Fig. P3.7). A bullet moving at velocity v_0 hits it at D and gets embedded in it. The bearing B does not provide axial constraint. Assume instantaneous impact. Find (a) the angular velocity of the shaft just after impact, (b) the impulsive bearing reactions during impact, (c) the bearing reactions, ω and $\dot{\omega}$ of the shaft when it has completed a quarter revolution after impact. Given: $M = 5\,\mathrm{kg}$, $R = 0.2\,\mathrm{m}$, $L = 0.4\,\mathrm{m}$, $b = 0.18\,\mathrm{m}$, $\omega_0 = 100\,\mathrm{rad/s}$, $m = 8$ g, $v_0 = 400\,\mathrm{m/s}$, $\alpha = \beta = 60°$, $\gamma = 45°$.

[(a) $102.6142\,\mathrm{rad/s}$ (b) $0.7261\vec{i}-0.2181\vec{j}\,\mathrm{N.s}$, $0.7261\vec{i}+1.818\vec{j}+2.263\vec{k}\,\mathrm{N.s}$ (c) $\omega = 102.6156\,\mathrm{rad/s}$, $\dot{\omega} = .1409\,\mathrm{rad/s^2}$, $24.564\vec{i}-7.582\vec{j}\,\mathrm{N}$, $24.56\vec{i}-7.582\vec{j}\,\mathrm{N}$]

8. (a) A disc is free to turn about O (Fig. P3.8). It spins at ω about the vertical axis. A mass-point hits the disc and adheres to it. Prove that just after impact it moves at angle θ to the horizontal where $\tan\theta = 4\alpha mv(M+2\alpha^2 m)/[(M+4\alpha^2 m)RM\omega]$. (b) Find the final angular velocity of the disc if it has this motion in outer space without any support. (c) Find the final angular velocity of the disc in parts (a) and (b) if instead of a mass-point an arrow of mass m and length L parallel to z-axis, translating at v, hits the disc at D and gets embedded in it.

$[$(b) $4\alpha mv\vec{\jmath}/(M+m+4\alpha^2 m)R + \omega\vec{k}/[1+2\alpha^2 m/(m+M)]$

(c$_a$) $\omega'_x = -I^O_{xz}\omega'_z/I^O_{xx} = \frac{1}{2}mR^2 I^O_{xz}\omega/({I^O_{xz}}^2 - I^O_{xx}I^O_{zz})$, $\quad \omega'_y = mv\alpha R/I^O_{yy}$

with $I^O_{xx} = \frac{1}{4}MR^2 + \frac{1}{3}mL^2$, $\quad I^O_{xz} = -\frac{1}{2}m\alpha RL$, $\quad I^O_{zz} = \frac{1}{2}MR^2 + m\alpha^2 R^2$

$I^O_{yy} = \frac{1}{4}MR^2 + \frac{1}{3}mL^2 + m(\alpha^2 R^2 + \frac{1}{4}L^2)$

(c$_b$) $\omega'_x = -I^C_{xz}\omega'_z/I^C_{xx} = \frac{1}{2}mR^2 I^C_{xz}\omega/({I^C_{xz}}^2 - I^C_{xx}I^C_{zz})$, $\omega'_y = v\alpha RMm/(M+m)I^C_{yy}$

with $I^C_{xx} = \frac{1}{4}MR^2 + \frac{1}{12}mL^2 + \frac{1}{4}L^2 Mm/(M+m)$, $\quad I^C_{xz} = -\frac{1}{2}mM\alpha RL/(M+m)$,

$I^C_{yy} = \frac{1}{4}MR^2 + \frac{1}{12}mL^2 + (\alpha^2 R^2 + \frac{1}{4}L^2)\frac{Mm}{M+m}$, $I^C_{zz} = \frac{1}{2}MR^2 + mM\alpha^2 R^2/(M+m)]$

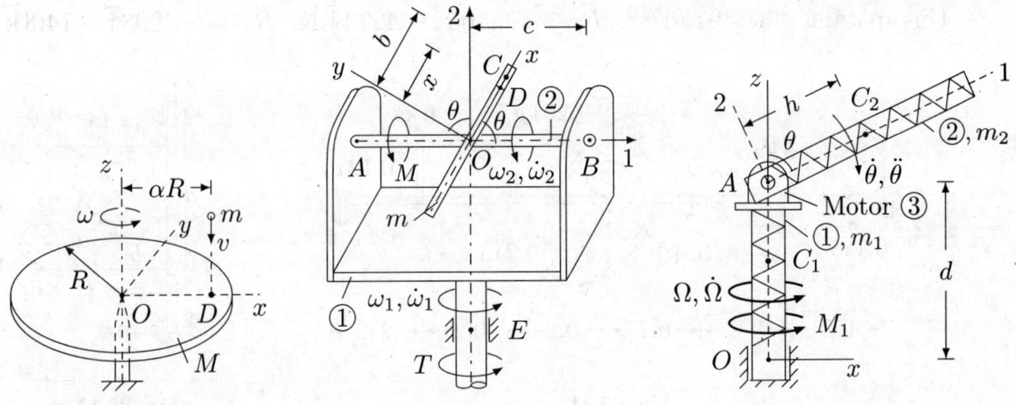

Fig. P3.8 Fig. P3.9 Fig. P3.10

9. Platform ① rotates about a vertical axis (Fig. P3.9). A thin rod is welded to shaft ② at angle θ. Shaft ② is rotating relative to platform ① at the rates shown under a torque M applied to it. The bearing at A does not provide axial constraint. (a) Find the bearing reactions, M and the internal force resultants for the section of the rod at D. (b) Find the driving torque T on platform ① if its $I^O_{zz} = I_2$. Given: $m = 3\,\text{kg}$, $b = 0.4\,\text{m}$, $\cos\theta = 0.6$, $\omega_1 = 150\,\text{rad/s}$, $\dot{\omega}_1 = 40\,\text{rad/s}^2$, $\omega_2 = 100\,\text{rad/s}$, $\dot{\omega}_2 = 20\,\text{rad/s}^2$, $c = 0.5\,\text{m}$, $x = 0.2\,\text{m}$, $I_2 = 0.8\,\text{kg.m}^2$.

[(a) $\vec{R}_A = 2127\vec{e}_2 + 0.768\vec{e}_3$, $\vec{R}_B = -2097\vec{e}_2 - 0.768\vec{e}_3$ N, $M = -1.024$ N.m, $\vec{F}_{R_D} = -16.61\vec{\imath} - 2966\vec{\jmath} - 1.8\vec{k}$ N, $\vec{M}_{R_D} = 0.2\vec{\jmath} - 329.6\vec{k}$ N.m (b) 32.77 N.m]

10. The vertical column ① of a tower crane (Fig. P3.10), including the stator of motor ③, has mass m_1 and inertia elements $I^O_{xx}, I^O_{yy}, \ldots$ at O. The boom ②, including the rotor of motor ③, has mass m_2, centre of mass C_2 and principal moments of inertia I_1, I_2, I_3 about axes $1, 2, 3$ at C_2. The motors apply driving moments $M_1\vec{\mathrm{k}}$ on ① and $-M_2\vec{\mathrm{e}}_3$ on ②. For the given position, find M_1, M_2 and the reactions at A and O.

$$[I^O_{xz}\dot{\Omega} + c_1 c + c_2 s, \ -m_2 ghs + (I_3 + m_2 h^2)\ddot{\theta} + (I_1 - I_2 - m_2 h^2)\Omega^2 sc,$$
$$\vec{R}_A = m_2 h[(\ddot{\theta}c - \dot{\theta}^2 s - \Omega^2 s)\,\vec{\mathrm{i}} + (\dot{\Omega}s + 2\Omega\dot{\theta}c)\,\vec{\mathrm{j}} - (\ddot{\theta}s + \dot{\theta}^2 c - g/h)\,\vec{\mathrm{k}}],$$
$$\vec{R}_O = \vec{R}_A + m_1 g\,\vec{\mathrm{k}}, \ \ \vec{C}_A = c_1\vec{\mathrm{e}}_1 + c_2\vec{\mathrm{e}}_2, \ \text{ with } \ c_1 = I_1(\dot{\Omega}c - \Omega\dot{\theta}s) + (I_1 - I_3)\Omega\dot{\theta}s,$$
$$c = \cos\theta, \ \ s = \sin\theta, \ \ c_2 = (I_2 + m_2 h^2)(\dot{\Omega}s + \Omega\dot{\theta}c) + (I_3 + m_2 h^2 - I_1)\Omega\dot{\theta}c,$$
$$\vec{C}_O = d\,\vec{\mathrm{k}} \times \vec{R}_A + (I^O_{xz}\dot{\Omega} - I^O_{yz}\Omega^2)\,\vec{\mathrm{i}} + (I^O_{yz}\dot{\Omega} + I^O_{xz}\Omega^2 + M_2)\,\vec{\mathrm{j}} + (c_1 s - c_2 c)\,\vec{\mathrm{i}}\,]$$

11. The shaft in Fig. P3.11 carries two rectangular plates and a circular disc welded to it. It rotates at $150\,\vec{\mathrm{k}}$ rad/s and a couple $M_0 = 10$ N.m is applied to it. Find its angular acceleration and the bearing reactions at this instant. The bearing at D does not provide axial restraint. The axis 1 of the disc lies in the xz plane at this instant. [$31.49\,\text{rad/s}^2$, $\vec{R}_O = -3505\,\vec{\mathrm{i}} - 4214\,\vec{\mathrm{j}}$ N, $\vec{R}_E = -4298\,\vec{\mathrm{i}} - 1400\,\vec{\mathrm{j}}$ N]

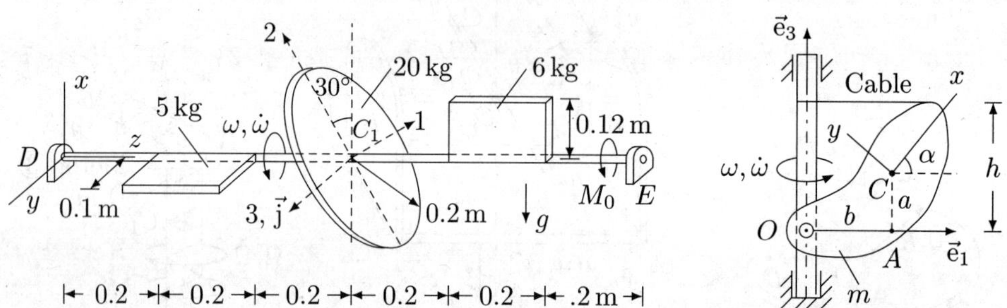

Fig. P3.11

Fig. P3.12

12. A body is hinged to a vertical shaft at O (Fig. P3.12) and restrained by a cable. The shaft rotates at ω, $\dot{\omega}$. The principal moments of inertia at C are $I^C_{xx}, I^C_{yy}, I^C_{zz}$. (a) Find the tension in the cable and the reaction at O. (b) For the case of $\dot{\omega} = 0$, the cable breaks. Find $\ddot{\theta}$, $\dot{\theta}$ when OA has rotated by angle θ.

[(a) $T = (mgb - I^O_{31}\omega^2)/h, \quad \vec{C}_O = I^O_{31}\dot{\omega}\,\vec{\mathrm{e}}_1 + I^O_{33}\dot{\omega}\,\vec{\mathrm{e}}_3,$

$\vec{R}_O = T\,\vec{\mathrm{e}}_1 + mg\,\vec{\mathrm{e}}_3 + m(\dot{\omega}b\,\vec{\mathrm{e}}_2 - \omega^2 b\,\vec{\mathrm{e}}_1)$ with

$I^O_{31} = (I^C_{xx} - I^C_{yy})\cos\alpha\sin\alpha - mab, \quad I^O_{33} = I^C_{xx}\sin^2\alpha + I^C_{yy}\cos^2\alpha + mb^2$

(b) $\ddot{\theta} = [mg(b\cos\theta + a\sin\theta) + \frac{1}{2}(I^O_{11} - I^O_{33})\sin 2\theta\omega^2 - I^O_{31}\cos 2\theta\omega^2]/I^O_{22},$

$\dot{\theta} = [[2mg\{b\sin\theta + a(1 - \cos\theta)\} + \frac{1}{2}\omega^2(I^O_{11} - I^O_{33})(1 - \cos 2\theta) - \omega^2 I^O_{31}\sin 2\theta]/I^O_{22}]^{1/2}$

with $I_{22}^O = I_{zz}^C + m(a^2+b^2)$, $\quad I_{11}^O = I_{xx}^C \cos^2\alpha + I_{yy}^C \sin^2\alpha + ma^2$]

13. Wagon ②, moving at v_0, hits a stationary wagon ① (Fig. P3.13). At the instant of maximum compression of the buffer springs, let u be the horizontal velocity of their points of contact O_1, O_2, and assume that they have a common downward velocity v. At this instant, let ω_3, ω_4 be the angular velocities of the main-frames ③ and ④. The wagon bodies are spring mounted. Main-frames ③ and ④ have mass m_3 and m_4, and radii of gyration k_3 and k_4 about the horizontal transverse axis through their centres of mass C_3 and C_4 at heights h_3 and h_4 above O. The wagons have total mass m_1 and m_2 and centres of mass at C_1 and C_2. Set up four equations for finding u, v, ω_3 and ω_4.

$$[m_3(v+\omega_3 b_1) + m_4(v-\omega_4 b_2) = 0, \quad m_3(u+\omega_3 h_1) + m_4(u+\omega_4 h_2) = m_4 v_0$$
$$m_4(u+\omega_4 h_2)h_2 - m_4(v-\omega_4 b_2)b_2 + m_4 k_4^2 \omega_4 = m_4 v_0 h_2,$$
$$m_3(u+\omega_3 h_1)h_1 + m_3(v+\omega_3 b_1)b_1 + m_3 k_3^2 \omega_3 = 0]$$

14. Body ① of mass $m = 10\,\text{kg}$ with centre of mass at C has radius of gyration $k_O = 0.2\,\text{m}$ about its horizontal axis of rotation through O. A disc ② of mass $0.5m$ and axial radius of gyration $0.5k_O$ is mounted on body ①. Initially, body ① is held stationary and the disc rotates as shown (Fig. P3.14) with the springs undeformed. Body ① is released. Find the angular velocity and angular acceleration of body ①, and the reactions from its hinged support after it has rotated by 30°. $k = 200\,\text{N/m}$, $k_t = 2\,\text{N.m/rad}$.

$$[5.743\,\text{rad/s},\ 25.67\,\text{rad/s}^2,\ -211.3\,\vec{e}_r + 11.03\,\vec{e}_\phi\,\text{N} \ \&\ 0.5743\,\vec{e}_r\,\text{N.m}]$$

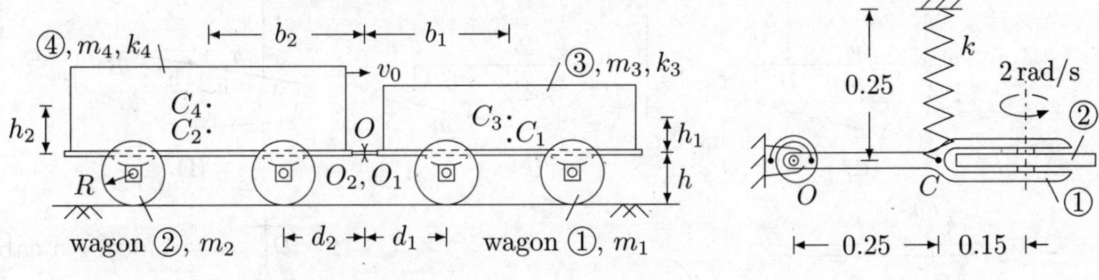

Fig. P3.13 **Fig. P3.14**

15. (a) A plate of mass m and moment of inertia I_{zz}^C about axis z through its centre of mass C is being lifted up by a crane with acceleration ng when the cable at B breaks (Fig. P3.15a). Find the angular acceleration of the plate immediately after the cable breaks. Assume inextensible cable. (b) Rework part (a) if the cable at A is replaced by (1) a spring (Fig. P3.15b), (2) a torsional spring (Fig. P3.15c), (3) a hinge joint (Fig. P3.15d), (4) a cable passing over a light pulley and supporting a counterweight of $mg/2$ (Fig. P3.15e). (c) Rework part (a) if the plate is initially at angle θ to the horizontal (Fig. P3.15f). (d) Rework part

(a) if the supporting cables are initially at angle θ to the vertical (Fig. P3.15g). (e) The problem of a thin plate or a cylinder released on the smooth floor of the elevator (Fig. P3.15h) is at the initial instant identical to the one in part (a) with the normal reaction replacing the tension T. The arrangement in Fig. P3.15i is used to find I_{zz}^C. Find I_{zz}^C if the reaction measured by a piezoelectric sensor at A is P_0 before release and P_1 just after the system is released from rest by cutting cable BD. (f) Rework part (a) if the plate is swinging with respect to the crane at velocity $v\vec{\imath}$. (a) $(1+n)mgb/(I_{zz}^C+b^2)$ (b) (1) $(1+n)mgb/2I_{zz}^C$ (2) $(1+n)mgb/I_{zz}^C$ (3) $(1+n)mgb/[I_{zz}^C+m(b^2+h^2)]$ (4) $2(1+n)mgb/(3I_{zz}^C+mb^2)$ (c) $(1+n)mg(b\cos\theta+h\sin\theta)/[I_{zz}^C+m(b\cos\theta+h\sin\theta)^2]$ (d) $(1+n)mg\cos\theta(b\cos\theta-h\sin\theta)/[I_{zz}^C+m(b\cos\theta-h\sin\theta)^2]$ (e) $P_1mL^2(1-P_0/mg)^2/(mg-P_1)$ (f) $[g(1+n)+v^2/L]mb/(I_{zz}^C+mb^2)]$

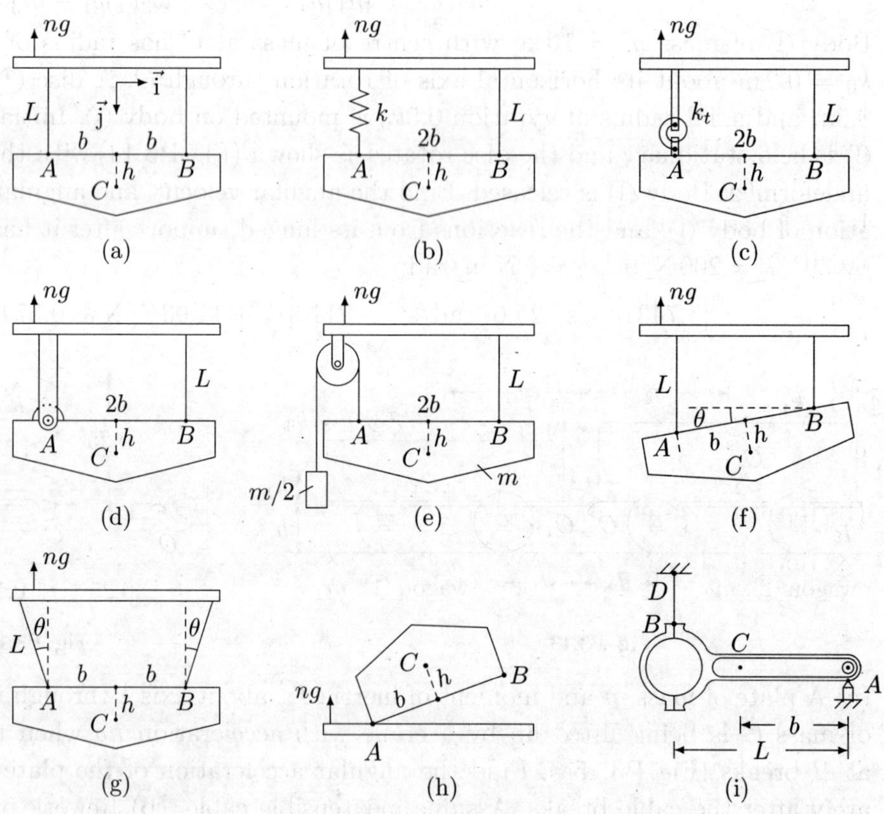

Fig. P3.15

16. (a) A cylindrical spool, of radius r and axial radius of gyration k, is placed on a horizontal mechanical shaker which oscillates horizontally with $x = x_0(1 -$

$\cos\omega t$). The coefficient of friction is μ. Find the condition for no slip and the amplitude of oscillation of the spool. (b) The spool rests on a truck. The coefficient of friction is μ. The truck starts moving with uniform acceleration a. The spool slips. Find the distance moved by the spool on the truck in time t. Find $\mu_{\min}$ for no slip. (c) The spool is at rest on a stationary horizontal conveyor. By application of impulsive forces, the conveyor gets velocity v_0 which remains constant. Find the distance moved by the spool with respect to the conveyor before the slip stops. [(a) $\omega^2 x_0 < \mu g(1+r^2/k^2)$, $x_0/(1+r^2/k^2)$ (b) $\frac{1}{2}(a-\mu g)t^2$, $a/g(1+r^2/k^2)$ (c) $v_0^2(1+2r^2/k^2)/2\mu g(1+r^2/k^2)^2$]

17. A solid sphere rolls without slip down a channel with included angle 2β, which is at an angle α to the horizontal. Find the minimum coefficient of friction to ensure no slip for the following cases. (a) The plane of symmetry of the channel is vertical. (b) One face of the channel is vertical and $\beta < 45°$.

[(a) $2\tan\alpha\sin\beta/(2+5\sin^2\beta)$ (b) $\tan\alpha\tan 2\beta/(2+5\sin^2\beta)$]

18. (a) A wheel of mass m and axial radius of gyration k has axle of radius r. C is imparted velocity v_0 and the wheel is given a back spin ω_0 just before placing its axle on the two rails fixed to the ground (Fig. P3.18). Find the minimum value of ω_0 so that the wheel returns. Find the time taken and the distance moved by C before returning to the initial position. The coefficient of friction is μ. (b) Rework part (a) for the case of landing of the wheel between two conveyors moving at constant velocity $v_1\vec{\mathrm{i}}$. The wheel lands in such a manner that the axle of radius r is horizontal and makes contact with both conveyors.

[(a) $\omega_0 > v_0 r/k^2$. For $\omega_0 > (1+2r^2/k^2)v_0/r$: $t = 2v_0/\mu g$, distance $d = v_0^2/\mu g$,

else $t = \dfrac{(v_0+\omega_0 r)^2}{2\mu g(1+r^2/k^2)(\omega_0 r - v_0 r^2/k^2)}$, $d = \dfrac{[v_0(1+2r^2/k^2)-\omega_0 r](v_0+\omega_0 r)}{2\mu g(1+r^2/k^2)^2}$

(b) Return, distance in conveyor frame: Same as (a), v_0 replaced by v_0+v_1.

Return, distance in ground frame: Same as (a), ω_0 replaced by $\omega_0 + v_1/r$]

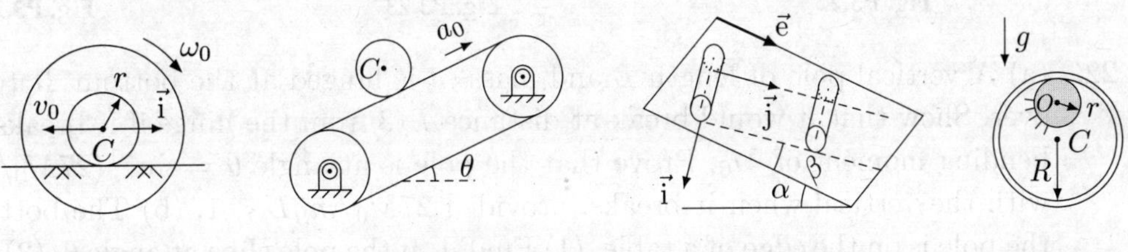

Fig. P3.18 Fig. P3.19 Fig. P3.20 Fig. P3.21

19. A conveyor belt in Fig. P3.19 is moving with acceleration a_0 and a cylinder of mass m, radius R and axial radius of gyration k lies on it. (a) It rolls without slip. Show that $a_0 < [(1+R^2/k^2)\mu g\cos\theta - g\sin\theta]$ where μ is the coefficient of static

friction and the upward acceleration a_C of C is $[a_0 - gR^2 \sin\theta/k^2]/(1 + R^2/k^2)$. Hence, show that (1) if $a_0 = gR^2 \sin\theta/k^2$ and $v_C(0) = 0$, then the angular acceleration of the cylinder is $\dot{\omega} = gR \sin\theta/k^2$ and C remains stationary relative to the ground; (2) if the belt moves down with acceleration $g \sin\theta$ and $\omega(0) = 0$, then the cylinder translates with C remaining stationary relative to the belt. (b) If there is a downward slip at the contact point and the kinetic coefficient of friction is μ_k, find $\dot{\omega}$ and a_C. [(b) $\mu_k gR \cos\theta/k^2$, $(\mu_k \cos\theta - \sin\theta)g$]

20. Find the acceleration of the axis of a cylinder, of radius r and mass m, rolling without slip down an inclined plane (Fig. P3.20). Find the resultant frictional force and the resultant normal reaction and their location. [$\frac{2}{3} g \sin\alpha \cos\beta$,

$\vec{F} = -mg \sin\alpha(\sin\beta\,\vec{\mathrm{i}} + \frac{1}{3}\cos\beta\,\vec{\mathrm{j}}\,)$ at mid-point of contact line

$\vec{N} = mg \cos\alpha\,\vec{\mathrm{k}}$ at distance $r \sin\beta \tan\alpha$ from mid-point of contact line]

21. A vertical ring ($R = 4r$) rests on a horizontal cylinder (Fig. Q3.21). (1) Find the angular frequency ω_n for small oscillations without slip. (2) Find the initial velocity v_0 for C so that the ring makes a complete turn without separation and without slip. The coefficient of friction is μ. (3) In case of slip, set up governing equations for subsequent motion in terms of ω and angle θ of OC with the vertical. [(1) $(g/6r)^{1/2}$ (2) $v_0^2 > 3[1 + (4 + 1/4\mu^2)^{1/2}]rg$

(3) $4r\dot{\omega} + \mu(g\cos\theta + 3r\dot{\theta}^2) = 0$, $3r\ddot{\theta} + g(\sin\theta - \mu\cos\theta) - 3\mu r\dot{\theta}^2 = 0$]

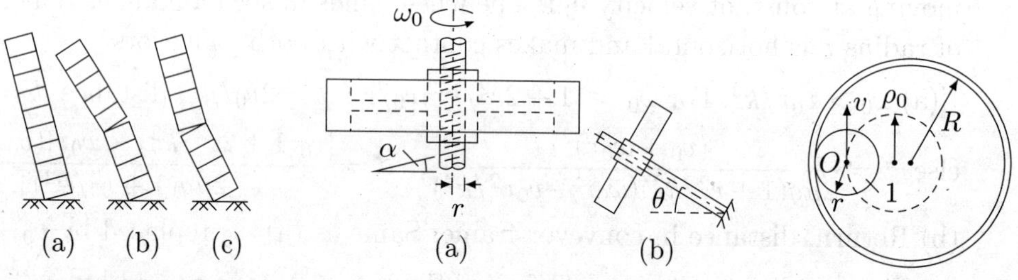

Fig. P3.22 Fig. P3.23 Fig. P3.24

22. (a) A vertical pole of length L and mass m is hinged at the bottom. It topples over. Show that it would break at distance $L/3$ from the hinge if it breaks for a bending moment of M_0. Prove that the pole is at angle $\theta = \sin^{-1}(27M_0/mgL)$ with the vertical when it breaks, provided $27M_0/mgL < 1$. (b) The bottom of the pole is on the edge of a table. (1) Find μ_s if the pole slips at angle θ. (2) If the pole is notched at the bottom so that it cannot slip, find θ for lift-off. (c) Solve part $(b)_1$ if the pole is at the centre of the table. (d) A tower is built of blocks. When slightly disturbed, it topples over initially (Fig. P3.22a) and then breaks up. Does it do so as in Fig. P3.22b/Fig. 3.22c? Justify by analysis.

[(b) $\dfrac{\sin\theta}{10\cos\theta-6}$, 53.13° (c) $(9\cos\theta-6)\sin\theta/(1-6\cos\theta+9\cos^2\theta)$ (d) Fig. P3.22c]

23. (a) A wheel, of mass m and axial moment of inertia I, rotates initially at ω_0 on a fixed threaded vertical spindle (Fig. P3.23). Find its axial advance in time t. The mean radius of the thread is r, its helix angle is α and the coefficient of friction is $\tan\psi$. (b) The wheel instead moves down a spindle inclined at an angle θ to the horizontal under a frictional moment M_f. Find the velocity of the wheel's centre after it has moved a distance s starting from rest. [(a) Let $t_1 = c\omega_0 r\tan\alpha/g$, $c = 1 + I/mr^2\tan\alpha\tan(\alpha+\psi)$. $t \le t_1$: $d = \omega_0 rt\tan\alpha - gt^2/2c$, for $t > t_1$: $d = \frac{1}{2}gt_1^2/c$ if $\alpha \le \psi$, else $d = \frac{1}{2}gt_1^2/c - g(t-t_1)^2/2[1+I/mr^2\tan\alpha\tan(\alpha-\psi)]$
(b) $[2(mg\sin\theta - M_f/r\tan\alpha)s/(m + I/r^2\tan^2\alpha)]^{1/2}$]

24. A hula-hoop of radius R, width b and mass m moves without slip around a cylindrical body 1 which translates with uniform speed v in a horizontal circular path of radius ρ_0 (Fig. P3.24). The coefficient of friction is μ. Find v so that the hoop does not fall. [$v^2 > \max[1/\mu,\ 2R/b]\rho_0^2 g/(R-\rho_0-r)$]

25. (a) A disc is pinned to a vertical rod (Fig. P3.25). (1) Find their angular accelerations just after a force P and an axial torque $T = PR$ are applied to the disc. (2) Find their angular velocities just after an instantaneous impulse $\hat{P}$ and an instantaneous angular axial impulse $\hat{T} = \hat{P}R$ are applied to the disc. (b) Rework part (a) if the rod and the disc are on a smooth horizontal surface and the rod is not hinged to the ground. [(a) $-P/8mR$, $3P/2mR$; $-\hat{P}/8mR$, $3\hat{P}/2mR$
(b) $-3P/13mR$, $20P/13mR$; $-3\hat{P}/13mR$, $20\hat{P}/13mR$]

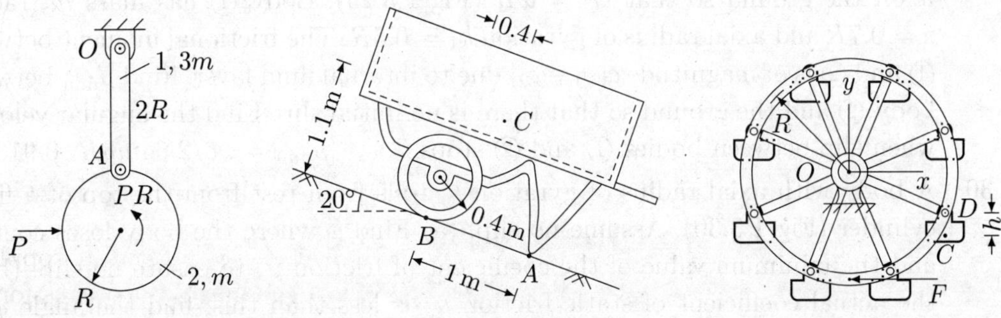

Fig. P3.25 **Fig. P3.26** **Fig. P3.27**

26. The hand-cart has total mass 30 kg with centre of mass C (Fig. P3.26). The mass of the rear wheel assembly is 10 kg with radius of gyration $k = 0.3$ m about the axis of symmetry. (a) Find the acceleration a_C if the wheels roll without slip, and the minimum value of the coefficient of friction at B to ensure this. The

coefficient of friction at A is $\mu = 0.3$. (b) Find a_C if $\mu_1 = 0.05$ at B is not sufficient to ensure no slip. [(a) $1.367\,\mathrm{m/s^2}$, 0.07437 (b) $1.472\,\mathrm{m/s^2}$]

27. A Ferris wheel (Fig. P3.27) of mass m_0 has axial moment of inertia I_0 and n swings. Each swing has mass m, centre of mass C and radius of gyration k about the axis of rotation at the point of suspension D. Initially the wheel is at rest. Force F is applied to impart initial $\dot{\omega}$ to the wheel. Prove that the reaction $\vec{R}_O$ at O is given by $\vec{R}_O = [mRn(1 - h^2/2k^2) + I_O/R]\dot{\omega}\,\vec{\jmath} + (m_0 + nm)g\,\vec{\imath}$.

28. Find the acceleration of the centre of mass C of a vehicle of mass m with respect to the ground if the conveyor has acceleration a_0 (Fig. P3.28). The axial moment of inertia of the rear and front wheel assemblies is I_0. Assume no slip. The coefficients of rolling friction for the front and rear wheels are e_1 and e_2.
$[\{2I_0a_0 + mgR(e_1b_2 + e_2b_1)/(b_1 + b_2)\}/\{2I_0 + mR^2 + (e_1 - e_2)mR(h - R)/(b_1 + b_2)\}]$

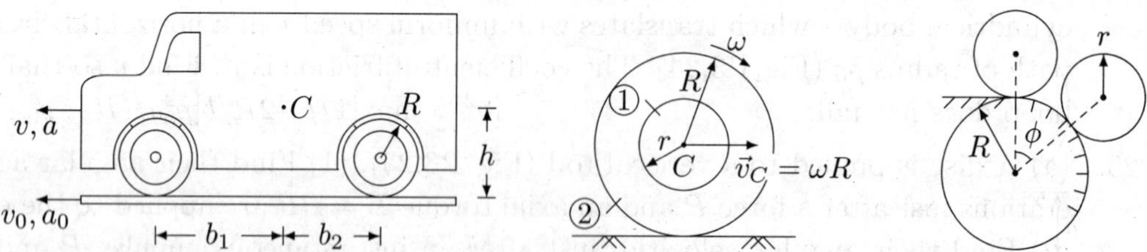

Fig. P3.28 Fig. P3.29 Fig. P3.30

29. Body ② of mass m and axial radius of gyration $k_2 = 0.8R$, is spun at ω, with stationary body ① inside. It is given a horizontal impulse just before releasing it on the ground so that $v_C = \omega R$ (Fig. P3.29). Body ① has mass m, radius $r = 0.7R$ and axial radius of gyration $k_1 = 0.5R$. The frictional moment between ① and ② is of magnitude $c|\omega_1 - \omega_2|$ due to internal fluid layer. Find $\mu_{\min}$ between body ② and the ground so that there is no initial slip. Find the angular velocity when slip between bodies ① and ② stops. [$\mu_{\min} = \omega c/2.64mgR$, 0.9135ω]

30. A body with axial radius of gyration k, rolls from rest from the top of a fixed cylinder (Fig. P3.30). Assume no slip. (a) Find ϕ where the body loses contact and the minimum value of the coefficient of friction μ_s to ensure no slip. (b) If the actual coefficient of static friction μ_s is less than this, find the angle ϕ at which slip starts. Analyse velocity v_C of C for the subsequent motion during slip. The kinetic coefficient of friction is μ_k. (c) Show that if the body is a disc, then slip occurs before flying off. Find angle ϕ for flying off in this case for $\mu_s = 0.4$, $\mu_k = 0.3$. [(a) $\cos^{-1}[2r^2/(k^2 + 3r^2)]$, ∞
(b) $\phi_1 = \cos^{-1}[2r^2/\{(k^2 + 3r^2)^2 + k^4/\mu_s^2\}^{1/2}] - \tan^{-1}[k^2/\mu_s(k^2 + 3r^2)]$,
$v_C^2 = Ae^{2\mu_k\phi} - 2g(R + r)[3\mu_k \sin\phi + (1 - 2\mu_k^2)\cos\phi]/(1 + 4\mu_k^2)$

where A is obtained from condition at $\phi = \phi_1$ (c) 52.9°]

31. A wheel of radius R and axial radius of gyration k rolls down a staircase without slipping or bouncing (Fig. P3.31). The angular velocity on the first step is ω_1. (a) Prove that the angular velocity ω^* just before impact with the second step is $[\omega_1^2 + 2gh/(R^2 + k^2)]^{1/2}$. Find $\mu_{\min}$ to ensure no slip. Find the maximum ω_1 so that the wheel does not jump off at A. (b) Prove that the angular velocity ω_2 just after this impact is $\omega_2 = \omega^*(R^2 + k^2 - Rh)/(R^2 + k^2)$. Find $\mu_{\min}$ to ensure no slip. (c) Show that when steady state has been achieved, the angular velocity ω_∞ for each step is given by $\omega_\infty^2 = (R^2+k^2-Rh)^2 g/(R^2+k^2-Rh/2)(R^2+k^2)R$.

[(a) $k^2(2Rh - h^2)^{1/2}/[(R-h)(k^2 + 3R^2) - 2R^3 - v_1^2(k^2 + R^2)/g]$, $[g\{(R-h)(k^2 + 3R^2) - 2R^3\}/(k^2 + R^2)]^{1/2}$ (b) $1/(1 + R^2/k^2)\sqrt{2R/h - 1}$]

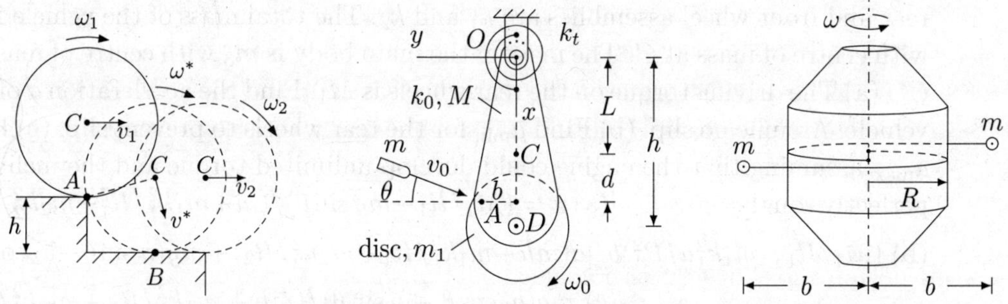

Fig. P3.31 Fig. P3.32 Fig. P3.33

32. A bullet impacts a pendulum whose radius of gyration is k_0 about the axis of rotation (a principal axis at O) and whose centre of mass is at C (Fig. P3.32). A disc of radius R is pinned to its end. Just before impact, the pendulum is at rest and the disc has angular velocity ω_0. (a) Find the impulsive reaction at O if the bullet gets embedded in the pendulum at A and lodges there. Assume the impact to be instantaneous. (b) When the pendulum has rotated by ϕ, find the bearing reactions at O and D. (c) Set up an equation to find the maximum angle of swing α of the pendulum. (d) For $\theta = 0$, find b and d so that there is no impulsive reaction on the support at O. Initially, the spring is undeformed.

[(a) $[c\omega + mv_0\cos\theta]\vec{\jmath} - m(\omega b + v_0\sin\theta)\vec{\imath}$ with $\omega = -mv_0[(L+d)\cos\theta + b\sin\theta]/I$,

$$I = Mk_0^2 + mb^2 + m(L+d)^2 + m_1h^2,\ c = ML + m(L+d) + m_1h$$

(b) $[\dot\omega_1 bm - (M+m+m_1)g\cos\phi - \omega_1^2 c]\,\vec{e}_r + [(M+m+m_1)g\sin\phi + \dot\omega_1 c + \omega_1^2 bm]\,\vec{e}_\phi$

$m_1[-(g\cos\phi + \omega_1^2 h)\,\vec{e}_r + (\dot\omega_1 h + g\sin\phi)\,\vec{e}_\phi]$, with

$\omega_1^2 = \omega^2 + [2cg(\cos\phi - 1) + 2mgb\sin\phi - k_t\phi^2]/I$, $\dot\omega_1 = [mgb\cos\phi - cg\sin\phi - k_t\phi]/I$

(c) $\omega_1^2(\alpha) = 0$ (d) $b = 0$, $d = (Mk_0^2 + m_1h^2)/(ML + m_1h) - L$]

33^d. The model of a mechanical device for reducing the spin of a satellite consists of two small objects, attached to the ends of thin metal tapes wound around it in a sense opposite to its rotation (Fig. P3.33) and ends strapped to it. The moment of inertia of the satellite, excluding the device, about the axis of rotation is I. The satellite is spinning at ω_0 before the ends of the tapes are released. (a) Find ω when length b of the tape is unwound and the time taken for this. Neglect gravity. (b) Find the length of the tape unwound when the spin of the satellite is reduced to zero. At this instant, the tape gets detached. (c) Rework parts (a) and (b) if the tape is wound in the same direction as the initial spin.

$$[\text{(a)}\ \frac{\omega_0[I+2m(R^2-b^2)]}{[I+2m(R^2+b^2)]},\ \frac{b}{\omega_0 R}\quad \text{(b)}\ (I/2m+R^2)^{1/2}\quad \text{(c) no unwinding}]$$

34. A road-roller is going up an incline (Fig. P3.34). The axial radii of gyration of the rear and front wheel assemblies are k_1 and k_2. The total mass of the vehicle is m with centre of mass at C. The mass of the main body is m_3 with centre of mass at C^*. (a) The driving torque on the rear wheels is M_1. Find the acceleration a of the vehicle. Assume no slip. (b) Find $\mu_{\min}$ for the rear wheels to prevent slip. (c) Find $a_{\max}$ assuming that the engine could develop unlimited torque and the incline is perfectly rough. [(a) $a = (M_1/R_1 - mg\sin\theta)/(m + m_1k_1^2/R_1^2 + m_2k_2^2/R_2^2)$ (b) $(M_1/R_1 - m_1k_1^2a/R_1^2)b/[a(mh + m_1k_1^2/R_1 + m_2k_2^2/R_2) + mg(h\sin\theta + b_2\cos\theta)]$ (c) $mg(b_1\cos\theta - h\sin\theta)/(mh + m_1k_1^2/R_1 + m_2k_2^2/R_2)$

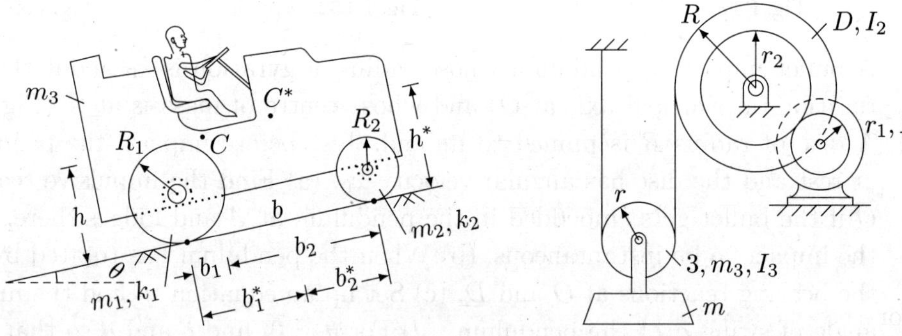

Fig. P3.34 Fig. P3.35

35. The winding drum D of a hoist is driven by an electric motor by a gear drive (Fig. P3.35). The pinion on the motor shaft has N_1 teeth and the gear on D has N_2 teeth. The rotating parts on the motor and the drum shafts have axial moments of inertia I_1 and I_2. The pulley 3 has axial moment of inertia I_3. The frictional couples at the bearings of the motor and drum shafts are M_{f_1}, M_{f_2}. The torque applied by the motor is M_1. The viscous damping torques on the motor and drum shafts are $c_1\omega_1$, $c_2\omega_2$, where ω_1, ω_2 are their angular speeds. Assume

the rope to be inertia-less and inextensible. Assume no slip of the rope. (a) Find the angular acceleration of the motor shaft. (b) Find the angular velocity of the motor shaft at time t if $M_1 = ct$. [(b) $Ic[e^{-(ct-M_0)c_0/cI} - 1]/c_0^2 + (ct - M_0)/c_0$

[(a) $(M_1 - M_0 - c_0\omega_1)/I$ with $M_0 = M_{f_1} + [M_{f_2} + \frac{1}{2}(m + m_3)gR]r_1/r_2$, $c_0 = c_1 + c_2 r_1^2/r_2^2$, $I = I_1 + [I_2 + \frac{1}{4}(m + m_3 + I_3/r^2)R^2]r_1^2/r_2^2]$

36. The crank OA (Fig. P3.36) has radius of gyration k_0 about its axis of rotation. The moment of inertia of the connecting rod AB about an axis through its centre of mass C normal to its plane of motion is $m_2 k_C^2$. (a) Find M so that the crank has angular velocity ω after a rotation of $\pi/2$ starting from rest. (b) Find the initial angular acceleration of the crank and the reactions at the pin joints.

[(a) $M = [\{m_1 k_0^2 + m_2(k_C^2 + (L-b)^2)R^2/L^2\}\omega^2 + m_1 gR$
$+2m_2 g(R + b - b\sqrt{L^2 - R^2}/L) + 2m_3 g(R + L - \sqrt{L^2 - R^2})]/\pi$
(b) $\dot\omega_1 = [M - gR(0.5m_1 + m_2 + m_3)]/[m_1 k_0^2 + (m_2 + m_3)R^2]$
$\vec{R}_O = R_1\vec{i} + R_2\vec{j}$, $\vec{R}_B = R_3\vec{i} + R_4\vec{j}$, $\vec{R}_A = R_5\vec{i} + R_6\vec{j}$,
$R_1 = R_3 = R_5 = -R(m_3 + m_2 b/L)(g + \dot\omega_1 R)/\sqrt{L^2 - R^2}$, $R_4 = m_3(g + \dot\omega_1 R)$,
$R_2 = m_1(g + 0.5\dot\omega_1 R) + (m_2 + m_3)(g + \dot\omega_1 R)$, $R_6 = (m_2 + m_3)(g + \dot\omega_1 R)$]

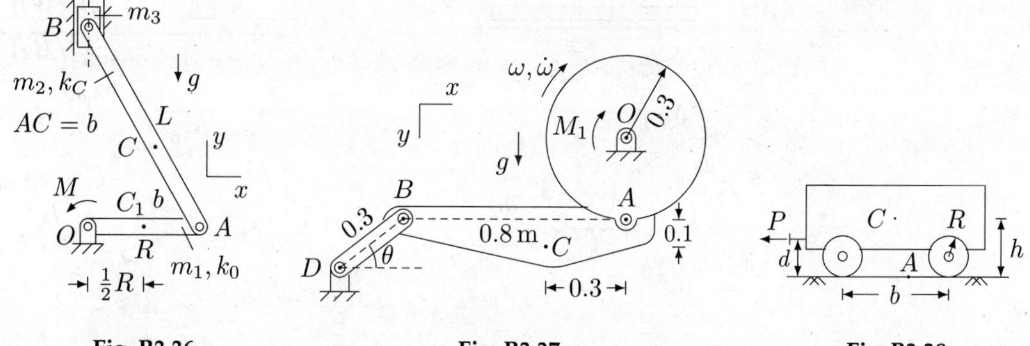

Fig. P3.36 Fig. P3.37 Fig. P3.38

37. For the plane 4-bar mechanism (Fig. P3.37), torque $M_1 = 10$ N.m is applied to the disc of mass 1.8 kg and radius 0.3 m at the given position. The mass of member AB is 0.8 kg and its centre of mass is at C with $I_{zz}^C = 0.06$ kg.m^2. Rod BD has mass 0.3 kg. Find $\dot\omega$ if $\omega = 20$ rad/s and $\theta = \cos^{-1} 0.8$. [514 rad/s^2]

38[d]. A wagon has total mass M and centre of mass C (Fig. P3.38). The mass of all wheels is m. Each wheel has axial radius of gyration k. Show that for acceleration without slip, the axles carry equal load if $d = (Mh + mk^2/R)/(M + mk^2/R^2)$.

39. Wheel ③ is pinned to ② and rolls without slip (Fig. P3.39). The springs are undeformed for $\theta = 30°$. The driving force applied by the hydraulic cylinder is F. The system is at rest for $\theta = \theta_0$. Find $\dot\theta$ and $\ddot\theta$ at position θ.

$[\dot{\theta}^2 = 2[V_e(\theta_0) - V_e(\theta)]/I_e,$

$I_e\ddot{\theta} + \frac{1}{2}[(7.375 + \frac{4k_B^2}{r^2})m - (4M + 1.175m)]\sin 2\theta\, b^2\dot{\theta}^2 + k_1(\theta - \frac{1}{6}\pi) + 2k_2(2\theta - \frac{1}{3}\pi)$
$+b(2M + 2.75m)g\cos\theta - 1.6bF\cos\theta - Pb(2 - \cos 2\theta) = 0$, with
$I_e = [\frac{59m}{24} + \frac{m}{25} + (4M + 1.175m)\cos^2\theta + (7.375 + 4k_B^2/r^2)m\sin^2\theta]b^2, \quad V_e(\theta) =$
$\frac{1}{2}k_1(\theta - \frac{\pi}{6})^2 + \frac{1}{2}k_2(2\theta - \frac{\pi}{3})^2 + (2M + 1.95m)gb\sin\theta - 1.6bF\sin\theta - Pb(2\theta - \frac{1}{2}\sin 2\theta)]$

40. A toy car has mass m_1 (= 20 m) for the main body (Fig. P3.40). Each of the rear and front wheel assemblies have mass m and axial radius of gyration $k = 0.9r$. (a) The front and rear wheels are imparted spin ω_0 before placing the car on the ground ($\mu_k = 0.3$). Find the velocity of the car and the distance moved (1) when one set of wheels stop slipping and (2) when slip stops. (b) Rework part (a), if only the front wheels are imparted ω_0. (c) Rework parts (a) and (b) if the car is placed on a conveyor moving at velocity $v_1\,\vec{i}$. [(a)(1) $.06527\omega_0 r$, $7.101\omega_0^2 r^2 \times 10^{-3}/g$ (2) $.06858\omega_0 r$, $8.672\omega_0^2 r^2 \times 10^{-3}/g$ (b) (1) 0, 0 (2) $.03430\omega_0 r$, $4.172\omega_0^2 r^2 \times 10^{-3}/g$

(c) Replace ω_0 by $\omega_0 + v_1/r$ in solutions (a) and (b)]

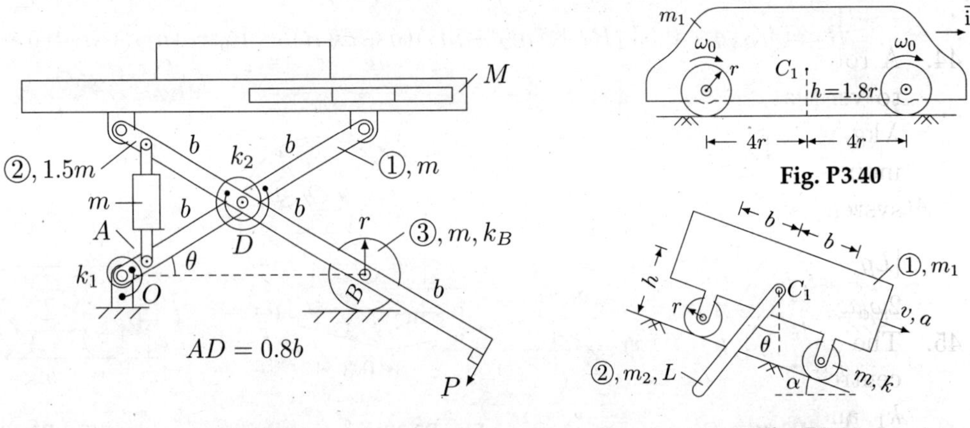

Fig. P3.39

Fig. P3.40

Fig. P3.41

41. (a) A pendulum is suspended from a cart of mass m_1 (without wheels) which rolls without slip (Fig. P3.41). Each of the front and rear wheel assemblies has axial radius of gyration k. Derive equations of motion for a and $\ddot{\theta}$. (b) Find the initial acceleration of the cart when released from rest with $\theta = 0$.

[(a) $L\ddot{\theta} + 1.5g\sin\theta - 1.5\cos(\theta - \alpha)a = 0$,
$[m_1 + m_2 + 2m(1 + \frac{k^2}{r^2})]a + \frac{1}{2}m_2 L[\dot{\theta}^2\sin(\theta - \alpha) - \ddot{\theta}\cos(\theta - \alpha)] = (m_1 + 2m + m_2)g\sin\alpha$
(b) $(m_1 + m_2 + 2m)g\sin\alpha/\{m_1 + m_2(1 - 0.75\cos^2\alpha) + 2m(1 + k^2/r^2)\}]$

42. The principal moments of inertia at C of a rigid body (Fig. P3.42) of mass 100 kg are: $I_{xx}^C = 50$, $I_{yy}^C = 60$, $I_{zz}^C = 100\,\text{kg.m}^2$. Given $\vec{\omega} = \vec{i} + 2\vec{j} + 3\vec{k}$ rad/s, find $\dot{\vec{\omega}}$

and the reaction at O. $[-1.6\vec{\mathrm{i}}+0.8824\vec{\mathrm{j}}-0.5\vec{\mathrm{k}}\ \mathrm{rad/s^2},\ 194.1\vec{\mathrm{i}}+320\vec{\mathrm{j}}-731\vec{\mathrm{k}}\ \mathrm{N}]$

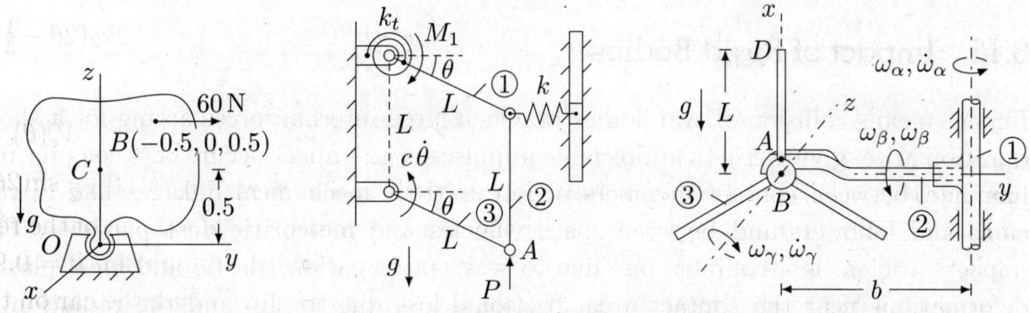

Fig. P3.42 **Fig. P3.43** **Fig. P3.44**

43. Each rod has mass m and length L (Fig. P3.43). The springs are undeformed for $\theta = 0$. The driving torque M_1 acts on rod ① and the load torque $c\,\dot\theta$ and a force P at A act on rod ③. (a) Find $\dot\omega$ of rod ① at θ for given $\dot\theta$. (b) The system is released from rest at $\theta = 0$. Find ω of rod ① in position θ if $c = 0$.

 [(a) $0.6[M_1 - PL\cos\theta - k_t\theta - kL^2(1-\cos\theta)\sin\theta + 2mgL\cos\theta - c\,\dot\theta]/mL^2$,

 (b) $[1.2\{M_1\theta - PL\sin\theta - \frac{1}{2}k_t\theta^2 - \frac{1}{2}kL^2(1-\cos\theta)^2 + 2mgL\sin\theta\}/mL^2]^{1/2}$]

44. A rotor ③ rotates at ω_γ relative to an arm ② which rotates at ω_β, $\dot\omega_\beta$ relative to vertical shaft ① rotating at $\omega_\alpha, \dot\omega_\alpha$ (Fig. P3.44). The mass of thin blades is λ kg/m. The driving torque of a motor on ③ is T. The axial and transverse moments of inertia of the rotor are I_{33} and I_{11}. Find $\dot\omega_\gamma$ and the force-couple system acting at A from the hub on the blade. Neglect hub radius. [$\frac{T}{I_{33}} - \omega_\alpha\omega_\beta$ $[Lg - \frac{1}{2}L^2(\omega_\beta^2 + \omega_\gamma^2)]\lambda\vec{\mathrm{i}} + [L\omega_\alpha^2 b + \frac{1}{2}L^2(\dot\omega_\gamma + 2\omega_\alpha\omega_\beta)]\lambda\vec{\mathrm{j}} - [L\dot\omega_\alpha b + \frac{1}{2}L^2(\dot\omega_\beta - 2\omega_\alpha\omega_\gamma)]\lambda\vec{\mathrm{k}}$, $L^2\lambda[\{\frac{1}{2}\dot\omega_\alpha b + \frac{1}{3}L(\dot\omega_\beta - 2\omega_\alpha\omega_\gamma)\}\vec{\mathrm{j}} + \{\frac{1}{2}\omega_\alpha^2 b + \frac{1}{3}L(\dot\omega_\gamma + 2\omega_\alpha\omega_\beta)\}\lambda\vec{\mathrm{k}}]$]

45. The radii of gyration about z-axis at centres of mass of bodies ① and ② are k_1 and k_2 (Fig. P3.45). Body ① rolls without slip from given rest position. Set up equations to find ω_1, ω_2, v_C after CC_1 has rotated by θ in case (a) and C_1 has moved a distance d relative to body ② in case (b). Assume smooth ground.

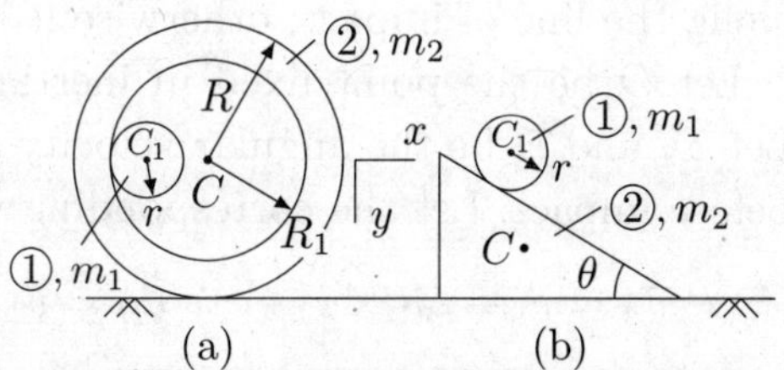

Fig. P3.45

[(a) $m_1k_1^2\omega_1/r = -m_2k_2^2\omega_2/R_1$, $m_2v_C + m_1[v_C - \omega_2R_1\sin\theta + \omega_1 r\sin\theta] = 0$,

$\frac{1}{2}m_2v_C^2 + \frac{1}{2}m_2k_2^2\omega_2^2 + \frac{1}{2}m_1k_1^2\omega_1^2 + \frac{1}{2}m_1[v_C^2 + (\omega_1r - \omega_2R_1)^2 + 2v_C(\omega_1r - \omega_2R_1)\sin\theta]$

$-m_1g(R_1 - r)\sin\theta = 0$ (b) $m_1(v_C + \omega_1r\cos\theta) + m_2v_C = 0$, $\omega_2 = 0$

$\frac{1}{2}m_2v_C^2 + \frac{1}{2}m_1(\omega_1^2r_1^2 + v_C^2 + 2v_C\omega_1r_1\cos\theta) + \frac{1}{2}m_1k_1^2\omega_1^2 - m_1gd\sin\theta = 0$]

C. IMPACT

3.15 Impact of Rigid Bodies

Impact means collision of two bodies in which large internal forces acting for a short duration arise, giving rise to appreciable impulses, e.g., impact occurs between pile and hammer, between contact members of intermittent mechanical linkages like valves, cams and followers and between space vehicles and meteoritic dust particles. The impact problem is a complex one due to wave propagation, elastic and local plastic deformations near the contact area, frictional loss due to slip and the radiation of energy as light, heat and sound. The techniques to deal with these interrelated events are complex and so, simplifications are postulated to model the impact situation. The usefulness of the results derived depends on the degree of validity of the idealisations.

For a simple analysis of impact of rigid bodies, the *idealisations* made are: bodies are modelled as rigid just before and just after impact, zero contact time, impulsive contact forces and moments, and the energy dissipation is modelled in terms of an empirical parameter called the coefficient of restitution. The impulse of all finite forces is zero since the contact time is modelled as zero. The impulsive forces cause instantaneous change in the velocities of the bodies, with their configurations remaining the same.

Consider the impact of two unconstrained rigid bodies (Fig. 3.22a), so that point A of body ① of mass m_1 collides with point B of body ② of mass m_2. The common tangent plane at A is called the *plane of impact.* Let $\vec{e}_t$ and $\vec{e}_b$ be orthogonal unit vectors in this plane. The direction $\vec{e}_n$ at A normal to the plane of impact is called the *line of impact.* The impact is called *central impact if the centres of mass C_i of the two bodies lie on the line of impact*, otherwise it is called *non-central impact.* Central impact is called *direct central impact* if the velocities of C_i just before the impact are along the line of impact; otherwise it is called *oblique central impact.*

Let O be the point fixed in inertial space which is coincident with A at impact. Let $\vec{\omega}_i$ and $\vec{v}_i$ be the angular velocity and velocity of centre of mass C_i of body i just before impact. Let the corresponding values just after be $\vec{\omega}_i'$ and $\vec{v}_i'$. For A and B:

$$\vec{v}_A = \vec{v}_1 + \vec{\omega}_1 \times \vec{r}_1, \qquad \vec{v}_B = \vec{v}_2 + \vec{\omega}_2 \times \vec{r}_2$$

$$\vec{v}_A' = \vec{v}_1' + \vec{\omega}_1' \times \vec{r}_1, \qquad \vec{v}_B' = \vec{v}_2' + \vec{\omega}_2' \times \vec{r}_2$$

The impulsive forces $\pm\vec{\hat{R}}$ act during impact (Fig. 3.22b). We can set up 6 equations $\vec{I} = \Delta\vec{p}$, $\vec{I}_{\text{ang}C_i} = \Delta\vec{H}_{C_i}$ for each body for $\vec{\omega}_1'$, $\vec{\omega}_2'$, $\vec{v}_1'$, $\vec{v}_2'$, $\vec{\hat{R}}$: $(5 \times 3 = 15$ unknowns).

$$-\vec{\hat{R}} = m_1(\vec{v}_1' - \vec{v}_1), \qquad \vec{\hat{R}} = m_2(\vec{v}_2' - \vec{v}_2) \tag{3.58}$$

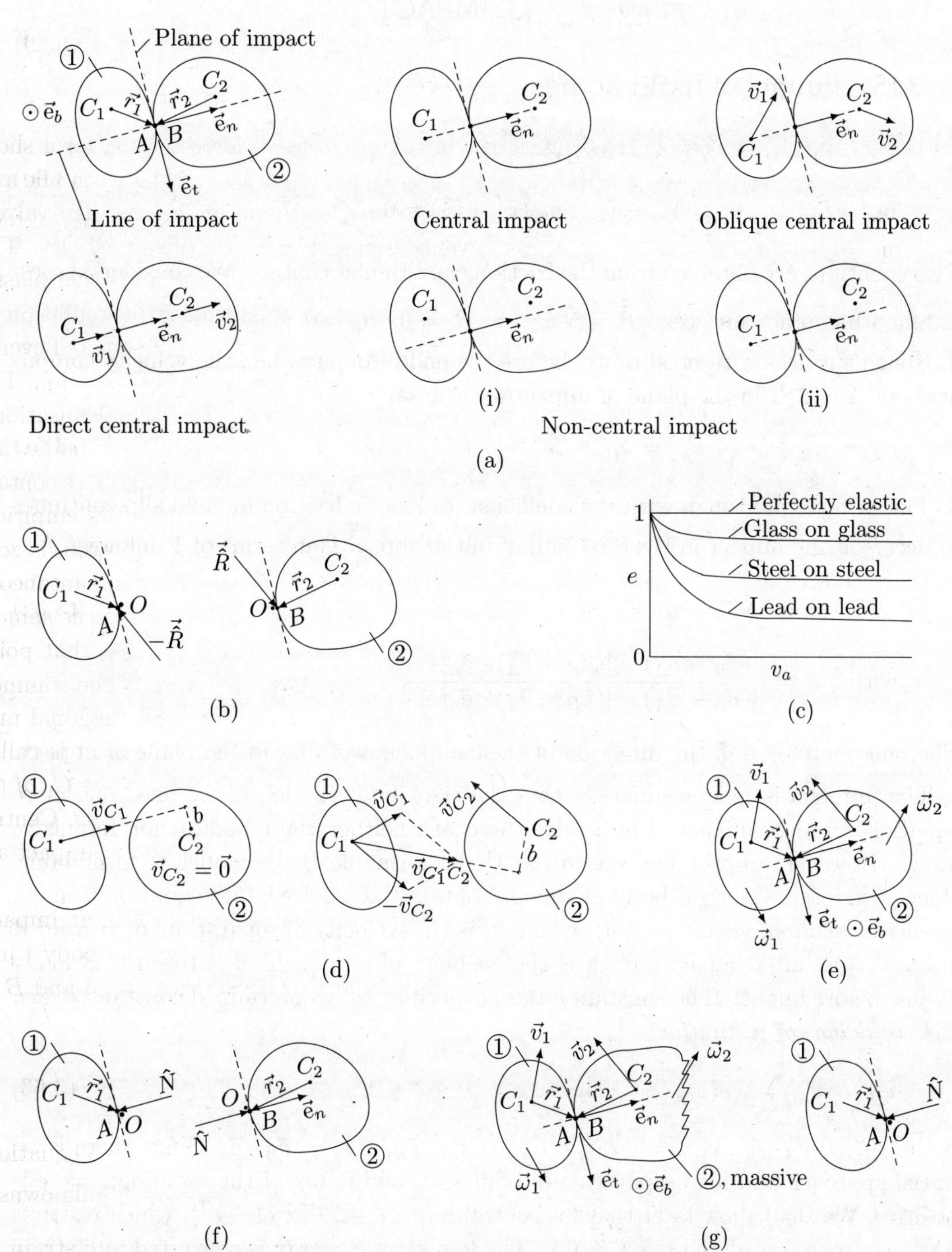

Plane of impact
Line of impact
Central impact
Oblique central impact
Direct central impact
(i)
(ii)
Non-central impact
(a)
Perfectly elastic
Glass on glass
Steel on steel
Lead on lead
(b)
(c)
(d)
(e)
(f)
②, massive
(g)

Fig. 3.22

$$\vec{r}_1 \times (-\hat{\vec{R}}) = I_{ij}^{C_1}(\omega'_{1_j} - \omega_{1_j})\,\vec{e}_i$$
$$\text{or} \quad I_{ij}^{C_1}(\omega'_{1_j} - \omega_{1_j})\,\vec{e}_i - \vec{r}_1 \times m_1(\vec{v}'_1 - \vec{v}_1) = \vec{0} \tag{3.59}$$

$$\vec{r}_2 \times \hat{\vec{R}} = I_{ij}^{C_2}(\omega'_{2_j} - \omega_{2_j})\,\vec{e}_i$$
$$\text{or} \quad I_{ij}^{C_2}(\omega'_{2_j} - \omega_{2_j})\,\vec{e}_i - \vec{r}_2 \times m_2(\vec{v}'_2 - \vec{v}_2) = \vec{0} \tag{3.60}$$

Two equations are obtained from the friction condition at contact. We consider 3 cases.

1. Smooth contact surfaces: $\hat{\vec{R}} = \hat{N}\,\vec{e}_n \;\Rightarrow\; \hat{R}_t = 0,\; \hat{R}_b = 0.$

2. Rough surfaces with no slip just before the end of impact, i.e., the velocity components of A and B in the plane of impact are the same:

$$v'_{A_t} = v'_{B_t}, \qquad v'_{A_b} = v'_{B_b} \tag{3.61}$$

3. The surfaces are rough with the coefficient of kinetic friction μ_k, and slip continues to occur during impact in the direction of initial slip. $\hat{\vec{R}}$ is in terms of 1 unknown $\hat{N}$:

$$\hat{\vec{R}} = \hat{N}\,\vec{e}_n - \mu_k \hat{N}\,\vec{e} \tag{3.62}$$

$$\text{with} \quad \vec{e} = \frac{(\vec{v}_B - \vec{v}_A) - [(\vec{v}_B - \vec{v}_A)\cdot\vec{e}_n]\,\vec{e}_n}{|(\vec{v}_B - \vec{v}_A) - [(\vec{v}_B - \vec{v}_A)\cdot\vec{e}_n]\,\vec{e}_n|}$$

The unit vector $\vec{e}$ is in the direction of the component of $\vec{v}_{BA}$ in the plane of impact.

The problem is indeterminate within the rigid body model, as we need one more equation for determinacy. Physically, there are neither rigid bodies nor impulsive forces. However, impact analysis using the deformable body model is formidable. Hence, we retain the rigid-body model and obtain the required 15th equation from an *empirical relation*, viz., $e = v_s/v_a$, where v_s is the 'velocity of separation' of B from A along $\vec{e}_n$ just after impact and v_a is the 'velocity of approach' of A towards B along $\vec{e}_n$ just before impact. The constant e is an experimentally determined constant, called the *coefficient of restitution*:

$$v'_{Bn} - v'_{An} = -e(v_{Bn} - v_{An}), \qquad 0 \le e \le 1 \tag{3.63}$$

with $v'_{An} = \vec{v}'_A \cdot \vec{e}_n$, $v'_{Bn} = \vec{v}'_B \cdot \vec{e}_n$, $v_{An} = \vec{v}_A \cdot \vec{e}_n$, $v_{Bn} = \vec{v}_B \cdot \vec{e}_n$. Since there is an initial approach and final separation of bodies, v_a and v_s are of the same sign. $\Rightarrow$ e is positive. We shall show later that for central impact $-\Delta T \propto (1 - e^2)$, which restricts e to the range postulated in eq. (3.63). The lost kinetic energy is converted into strain energy, thermal, sound and light energies, etc. The case $e = 1$ is called a *perfectly elastic* impact, for which $\Delta T = 0$ and $v_s = v_a$. A better word for it is an *ideal impact* (i.e., T is conserved), since for impact of deformable elastic bodies the presence of an elastic strain energy during vibrations after impact means that the kinetic energy is not conserved. Ideal impact is necessarily a noiseless impact. The case $e = 0$ is called *perfectly plastic* impact, for which the loss of kinetic energy is maximum and $v_s = 0$.

The coefficient of restitution e depends on the materials, geometry and size of the bodies, the type of impact (central/non-central, direct/oblique, with/without spin) and the speed of approach, since the plastic deformation depends on the nature of the stress distributions and also on the rate of loading. Hence, the above analysis is reliable if e is determined under conditions which closely approximate those for which it is to be used. The velocity dependent nature of e is dramatic for a silicon compound called 'silly putty'. A ball made out of it bounces from the floor if struck at high speed but at low speed, its rebound is just like that of an ordinary putty. In general, $e \to 1$ as $v_a \to 0$, but e decreases with v_a as plastic deformations and hence, loss of kinetic energy increases with v_a. The variation of e with v_a is shown in Fig. 3.22c. For steel, cast-iron and brass, $e \in (0.5, 1)$. For rubber, wood and aluminium, $e \in (0, 0.5)$.

If $\mu_k \neq 0$, we *assume* no slip and solve for $\hat{\vec{R}}$. If the no slip check, $|\hat{\vec{R}} \times \vec{e}_n| < \mu_s(\hat{\vec{R}} \cdot \vec{e}_n)$, is not satisfied, then the analysis is repeated using eq. (3.62) for the case of slip.

If body ② is massive, i.e., $m_2 \gg m_1$, then we approximate that its kinematics remains unchanged: $\vec{\omega}_2' = \vec{\omega}_2$, $\vec{v}_2' = \vec{v}_2$. Hence, eqs $(3.58)_2$ and (3.60) are not applied.

The *impact parameter* b of a collision is the minimum distance between the relative velocity vector $\vec{v}_{C_1C_2}$ at C_1 from C_2 (Fig. 3.22d). The difference in the initial and final kinetic energies in a reaction (collision/explosion) is called the *Q-value* of the reaction. A reaction is said to be *endoergic, exoergic or elastic* according to whether Q is negative (absorption of energy), positive (release of energy) or zero (neither).

We consider some special cases of impact.

1. General Impact of Unconstrained Smooth Rigid Bodies

Consider the case of smooth rigid bodies (Fig. 3.22e). Let $\vec{v}_i = v_{i_t}\,\vec{e}_t + v_{in}\,\vec{e}_n + v_{i_b}\,\vec{e}_b$, $\vec{v}_i' = v_{i_t}'\,\vec{e}_t + v_{in}'\,\vec{e}_n + v_{i_b}'\,\vec{e}_b$ be the velocities of C_i just before and just after the impact. An impulsive force $\hat{N}$ (Fig. 3.22f) acts during impact. We can set up 13 equations for the 13 unknowns: $\vec{\omega}_i'$, $\vec{v}_i'$ and $\hat{N}$. The momenta of each body in the $\vec{e}_t$ and $\vec{e}_b$ directions are conserved, since the external impulse on each is along $\vec{e}_n$. $\Rightarrow$

$$v_{1_t}' = v_{1_t}, \qquad v_{1_b}' = v_{1_b}, \qquad v_{2_t}' = v_{2_t}, \qquad v_{2_b}' = v_{2_b} \tag{3.64}$$

The external impulse $\vec{I} = \vec{0}$ on the system of two bodies $\Rightarrow$ their total p_n is conserved:

$$m_1 v_{1n}' + m_2 v_{2n}' = m_1 v_{1n} + m_2 v_{2n} \tag{3.65}$$

$\vec{H}_O$ of each body is conserved [eqs (3.59), (3.60)], since the external angular impulse on each about the fixed point O is zero. Using eqs (3.64) and eqs (3.60), eqs (3.59) yield

$$I_{ij}^{C_1}(\omega'_{1_j} - \omega_{1_j})\,\vec{e}_i - \vec{r}_1 \times m_1(v'_{1n} - v_{1n})\,\vec{e}_n = \vec{0} \tag{3.66}$$

$$I_{ij}^{C_2}(\omega'_{2_j} - \omega_{2_j})\,\vec{e}_i - \vec{r}_2 \times m_2(v'_{2n} - v_{2n})\,\vec{e}_n = \vec{0} \tag{3.67}$$

Equations (3.64)–(3.67) yield 11 equations. The 12th equation is the e relation (3.63):

$$v'_{Bn} - v'_{An} = -e(v_{Bn} - v_{An}) \tag{3.68}$$

$\hat{N}$ is found from the impulse–momentum relation applied to body ①:

$$-\hat{N} = m_1(v'_{1n} - v_{1n}) \tag{3.69}$$

If body ② is massive (Fig. 3.22g), then we approximate $\vec{\omega}'_2 = \vec{\omega}_2,\ \vec{v}'_2 = \vec{v}_2,\ v'_{Bn} = v_{Bn}$. Hence, eqs (3.65) and (3.67) are not applied. The 6 unknowns in $\vec{\omega}'_1,\ \vec{v}'_1$ are found from eq. $(3.64)_{1,2}$, (3.66) for body ① and (3.68):

$$\begin{gathered} v'_{1_t} = v_{1_t}, \qquad v'_{1_b} = v_{1_b}, \qquad v_{Bn} - v'_{An} = -e(v_{Bn} - v_{An}) \\ I_{ij}^{C_1}(\omega'_{1_j} - \omega_{1_j})\,\vec{e}_i - \vec{r}_1 \times m_1(v'_{1n} - v_{1n})\,\vec{e}_n = \vec{0} \end{gathered} \tag{3.70}$$

If body ② is a massive body at rest, then eq. $(3.70)_3$ reduces to $v'_{An} = -ev_{An}$.

2. Impact of Unconstrained Rigid Bodies in Plane Motion

Consider the impact of two unconstrained rigid bodies (Fig. 3.23a) with C_1 and C_2 coplanar with the line of impact and let the axis $\vec{e}_b$ normal to this xy-plane be a principal axis of inertia at C_i for body i. The motion before impact is plane motion with points of the bodies moving parallel to xy-plane. Choose $\vec{e}_n$ and $\vec{e}_t$ in xy-plane, with $\vec{e}_n$ along the line of impact. The bodies continue in plane motion after impact.

Proof: Let $\vec{e}_1 = \vec{e}_b = \vec{k}$ and $\vec{e}_2,\ \vec{e}_3$ be the principal axes at C_1 and $\vec{\omega}_1 = \omega_1\,\vec{e}_1,\ \vec{\omega}'_1 = \omega'_1\,\vec{e}_1 + \Omega'_2\,\vec{e}_2 + \Omega'_3\,\vec{e}_3$. Since $\hat{\vec{R}} = \hat{R}_n\,\vec{e}_n + \hat{R}_t\,\vec{e}_t$ and $\vec{r}_1$ has only $\vec{e}_t,\ \vec{e}_n$ components:

$$\vec{I}_{\text{ang}_{C_1}} = \vec{r}_1 \times (-\hat{\vec{R}}) = I_{\text{ang}_{C_1}}\vec{e}_1 = I_{11}^{C_1}(\omega'_1 - \omega_1)\,\vec{e}_1 + I_{22}^{C_1}(\Omega'_2 - 0)\,\vec{e}_2 + I_{33}^{C_1}(\Omega'_3 - 0)\,\vec{e}_3$$

$$\Rightarrow\ \Omega'_2 = \Omega'_3 = 0 \quad \Rightarrow \quad \vec{\omega}'_1 = \omega'_1\,\vec{e}_1 = \omega'_1\,\vec{k} \quad \Rightarrow \quad \text{plane motion.}$$

Hence, the 8 unknowns are: $\omega'_1\,\vec{k},\ \omega'_2\,\vec{k},\ v'_{1n}\,\vec{e}_n + v'_{1_t}\,\vec{e}_t,\ v'_{2n}\,\vec{e}_n + v'_{2_t}\,\vec{e}_t$, and $\hat{\vec{R}} = \hat{R}_n\,\vec{e}_n + \hat{R}_t\,\vec{e}_t$. The governing equations are: $\vec{I} = \Delta\vec{p},\ \Delta H_{Oz} = 0$ i.e., 3 equations each for ① and ②, the e relation and an equation from the nature of the friction condition at the contact.

If body ② is massive, then approximate: $\omega_2' = \omega_2$, $v_{2n}' = v_{2n}$, $v_{2t}' = v_{2t}$. Hence, exclude 3 equations $\vec{I} = \Delta\vec{p}$, $\Delta H_{Oz} = 0$ for body ②.

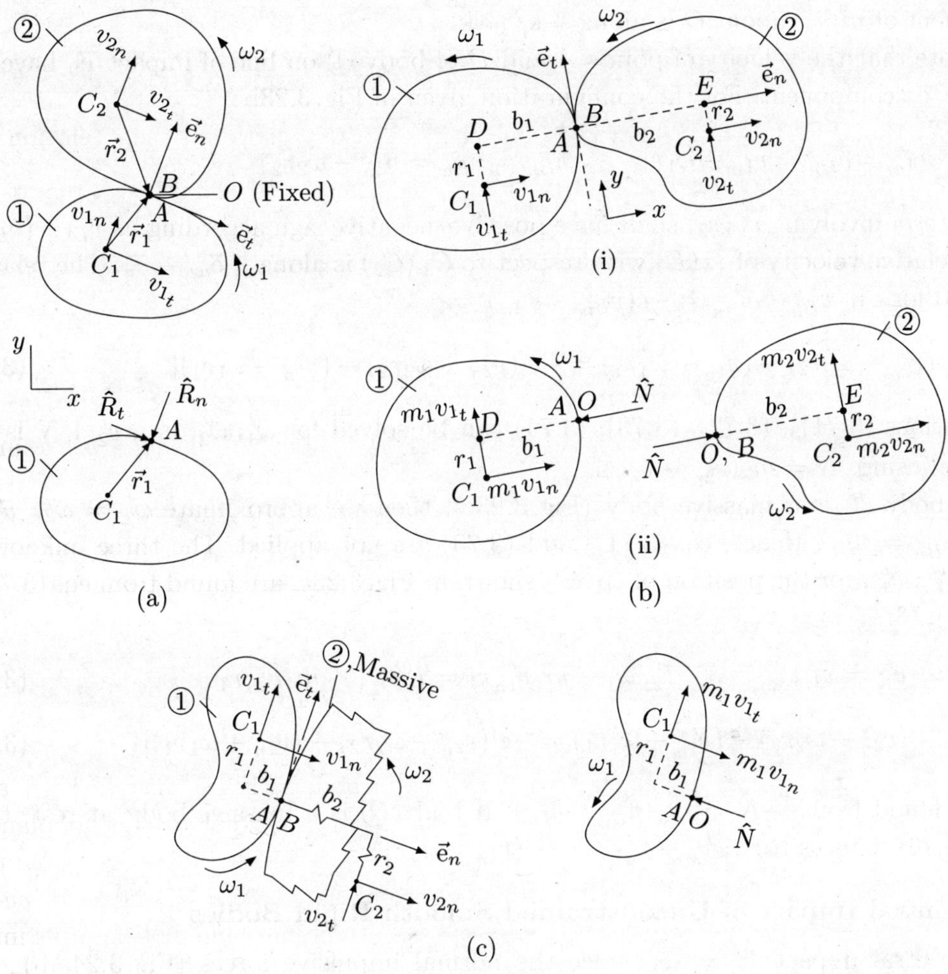

Fig. 3.23

Special Case of Smooth Rigid Bodies in Plane Motion

Consider the general impact of unconstrained smooth rigid bodies (Fig. 3.23b) in plane motion. v_{1n}', v_{1t}', v_{2t}', v_{2n}', ω_1', ω_2', $\hat{N}$ (7 unknowns) are obtained using eqs (3.64), (3.65):

$$v_{1_t}' = v_{1_t}, \qquad v_{2_t}' = v_{2_t}, \qquad m_1 v_{1n}' + m_2 v_{2n}' = m_1 v_{1n} + m_2 v_{2n} \tag{3.71}$$

and conservation of H_{Oz} for each of the bodies ① and ②, which for the configuration of C_1, C_2, O shown in Fig. 3.23b yields:

$$I_{zz}^{C_1}\omega_1' + m_1 v_{1n}' r_1 = I_{zz}^{C_1}\omega_1 + m_1 v_{1n} r_1 \tag{3.72}$$

$$I_{zz}^{C_2}\omega_2' + m_2 v_{2n}' r_2 = I_{zz}^{C_2}\omega_2 + m_2 v_{2n} r_2 \tag{3.73}$$

The terms involving r_i shall have positive/negative sign according to whether the moment of $m_i v_{in}$ about O is along $+\vec{k}/-\vec{k}$.

Note that the velocity of points A and D of body ① on line of impact $\vec{e}_n$ have the same $\vec{e}_n$ component. For the configuration given in Fig. 3.23b:

$$v_{An} = v_{Dn} = v_{1n} - \omega_1 r_1, \qquad v_{Bn} = v_{En} = v_{2n} - \omega_2 r_2$$

The term involving r_1 (r_2) shall have positive/negative sign according as $\vec{v}_{DC_1}$ ($\vec{v}_{EC_2}$) the relative velocity of D (E) with respect to C_1 (C_2) is along $+\vec{e}_n/-\vec{e}_n$. The relation (3.63) for e: $v_{Bn}' - v_{An}' = -e(v_{Bn} - v_{An}) \Rightarrow$

$$(v_{2n}' - \omega_2' r_2) - (v_{1n}' - \omega_1' r_1) = -e[(v_{2n} - \omega_2 r_2) - (v_{1n} - \omega_1 r_1)] \tag{3.74}$$

Equations $(3.71)_3$, (3.72), (3.73), (3.74) can be solved for ω_1', ω_2', v_{1n}', v_{2n}'. $\hat{N}$ is obtained using $\hat{N} = m_2(v_{2n}' - v_{2n})$.

If body ② is a massive body (Fig. 3.23c), then we approximate $\omega_2' = \omega_2$, $\vec{v}_2' = \vec{v}_2$, $v_{Bn}' = v_{Bn}$. Hence, eqs $(3.71)_3$ and (3.73) are not applied. The three unknowns, ω_1', v_{1n}', v_{1t}', for the position of C_1, C_2 shown in Fig. 3.23c, are found from eqs $(3.71)_1$, (3.72), (3.74):

$$v_{1t}' = v_{1t}, \qquad I_{zz}^{C_1}\omega_1' - m_1 v_{1n}' r_1 = I_{zz}^{C_1}\omega_1 - m_1 v_{1n} r_1 \tag{3.75}$$

$$(v_{2n} - \omega_2 r_2) - (v_{1n}' + \omega_1' r_1) = -e[(v_{2n} - \omega_2 r_2) - (v_{1n} + \omega_1 r_1)] \tag{3.76}$$

$\hat{N}$ is found from $-\hat{N} = m_1(v_{1n}' - v_{1n})$. If body ② is a massive body at rest, then eq. (3.76) reduces to: $v_{1n}' + \omega_1' r_1 = -e(v_{1n} + \omega_1 r_1)$.

3. Central Impact of Unconstrained Smooth Rigid Bodies

For central impact $\vec{\omega}_i' = \vec{\omega}_i$, since the normal impulsive forces (Fig. 3.24a,b) pass through C_i. $\vec{v}_i'$ are obtained using eqs (3.64), (3.65), (3.68):

$$v_{1t}' = v_{1t}, \quad v_{1b}' = v_{1b}, \quad v_{2t}' = v_{2t}, \quad v_{2b}' = v_{2b}$$

$$m_1 v_{1n}' + m_2 v_{2n}' = m_1 v_{1n} + m_2 v_{2n} \tag{3.77}$$

$$v_{2n}' - v_{1n}' = -e(v_{2n} - v_{1n})$$

since $v_{An} = (v_{C_1})_n = v_{1n}$, $v_{Bn} = (v_{C_2})_n = v_{2n}$.

For the impact of two unconstrained bodies, rigid or not, the velocity of the centre of mass C of the system remains constant since the external impulse on the system is zero. The governing algebraic equations are much simpler in the inertial frame I_C translating relative to the ground, in which C is a fixed point. Denoting velocities relative to I_C by $\vec{v}^*$ (Fig. 3.24c): $v_{1_t}^{*\prime} = v_{1_t}^*$, $v_{2_t}^{*\prime} = v_{2_t}^*$,

$$m_1 v_{1n}^{*\prime} + m_2 v_{2n}^{*\prime} = m_1 v_{1n}^* + m_2 v_{2n}^* = 0, \quad v_{2n}^{*\prime} - v_{1n}^{*\prime} = -e(v_{2n}^* - v_{1n}^*)$$

$$\Rightarrow \quad v_{1n}^{*\prime} = -e v_{1n}^*, \qquad v_{2n}^{*\prime} = -e v_{2n}^* \tag{a}$$

i.e., the $\vec{e}_n$ components of $\vec{v}^*$ get reversed and their magnitudes get multiplied by e.

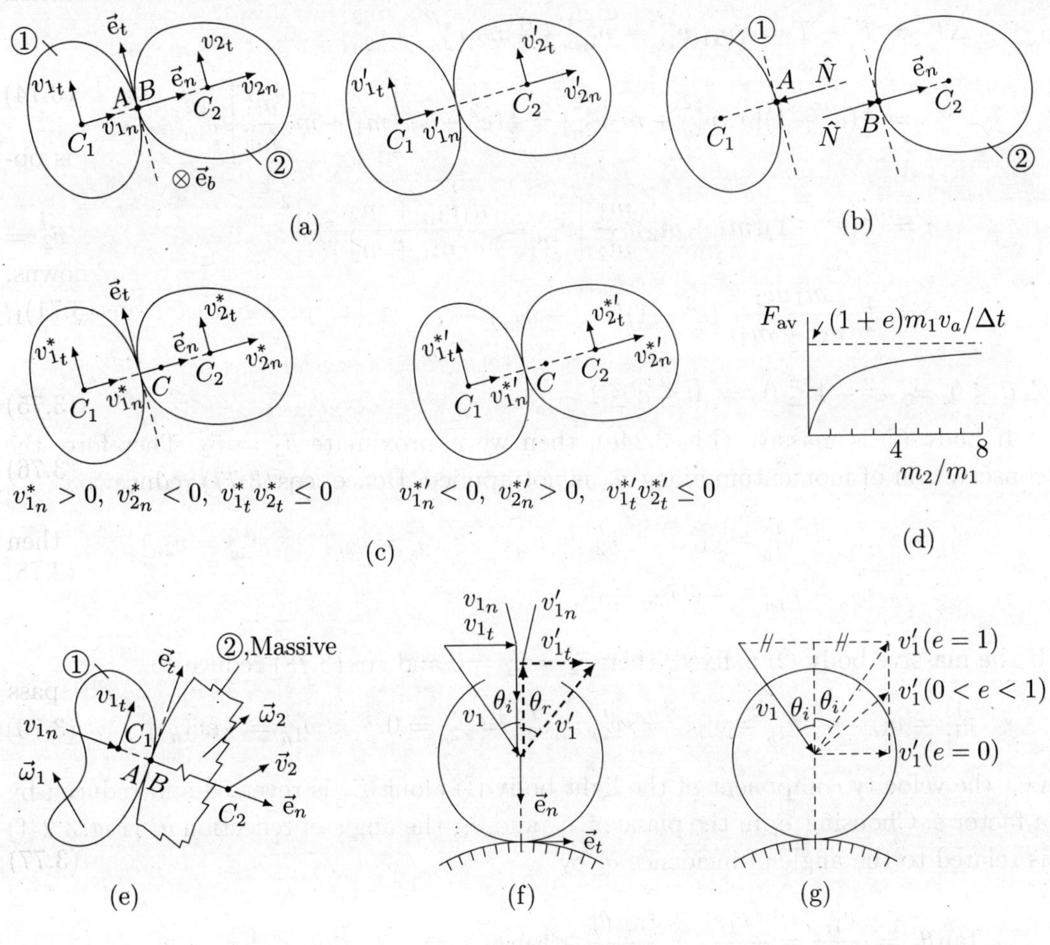

Fig. 3.24

The impulse $\hat{N}$ and the average force F_{av} for an actual impact duration Δt are

$$\hat{N} = m_2(v_{2n}^{*\prime} - v_{2n}^*) = -(1+e)m_2 v_{2n}^*$$

$$= -(1+e)m_2[v_{2n} - (m_1 v_{1n} + m_2 v_{2n})/(m_1+m_2)]$$
$$= (1+e)m_1 m_2 (v_{1n} - v_{2n})/(m_1+m_2)$$
$$F_{\text{av}} = \hat{N}/\Delta t = (1+e)m_1 v_a/\Delta t(1+m_1/m_2)$$

$F_{\text{av}} \propto v_a$ and it is inversely proportional to Δt, and increases for greater e. For a given m_1, it also increases with m_2 and as $m_1/m_2 \to 0$, $F_{\text{av}} \to (1+e)m_1 v_a/\Delta t$ (Fig. 3.24d).

Since $\vec{\omega}_i$, v_{i_t}, v_{i_b} and T_C remain the same, ΔT relative to the ground during impact is only because of change in $\vec{e}_n$ component of $\vec{v}_{C_iC}$, i.e., v^*_{in} :

$$\Delta T = T' - T = \tfrac{1}{2}m_1[v^{*\prime 2}_{1n} - v^{*2}_{1n}] + \tfrac{1}{2}m_2[v^{*\prime 2}_{2n} - v^{*2}_{2n}]$$
$$= \tfrac{1}{2}(e^2-1)[m_1 v^{*2}_{1n} + m_2 v^{*2}_{2n}] = \tfrac{1}{2}(e^2-1)\left[m_1 + m_2\frac{m_1^2}{m_2^2}\right] v^{*2}_{1n}$$
$$= \tfrac{1}{2}(e^2-1)(m_1+m_2)\frac{m_1}{m_2}\left[v_{1n} - \frac{m_1 v_{1n} + m_2 v_{2n}}{m_1+m_2}\right]^2$$
$$= \frac{1}{2}\frac{m_1 m_2}{(m_1+m_2)}(e^2-1)(v_{1n} - v_{2n})^2$$

$\Delta T \le 0 \Rightarrow e^2 - 1 \le 0 \Rightarrow 0 \le e \le 1.$

If body ② is massive (Fig. 3.24e), then we approximate $\vec{v}_2' = \vec{v}_2$. Therefore, the conservation of momentum along $\vec{e}_n$ is not applied. Hence, eqs (3.77) reduce to

$$\begin{gathered} v'_{1_t} = v_{1_t}, \quad v'_{1_b} = v_{1_b}, \quad v'_{2_t} = v_{2_t}, \quad v'_{2_b} = v_{2_b}, \quad v'_{2n} = v_{2n} \\ v_{2n} - v'_{1n} = -e(v_{2n} - v_{1n}) \end{gathered} \tag{3.78}$$

If the massive body ② is fixed, then $\vec{v}_2' = \vec{v}_2 = \vec{0}$ and eqs (3.78) reduce to

$$v'_{1_t} = v_{1_t}, \quad v'_{1_b} = v_{1_b}, \quad v'_{2_t} = v'_{2n} = v'_{2_b} = 0, \quad v'_{1n} = -e v_{1n} \tag{3.79}$$

i.e., the velocity component of the light body ① along $\vec{e}_n$ is reversed and reduced by a factor e. Choosing $\vec{e}_t$ in the plane of $\vec{e}_n$ and $\vec{v}_1$, the angle of reflection θ_r (Fig. 3.24f) is related to the angle of incidence θ_i by

$$\tan\theta_r = \frac{v'_{1_t}}{|v'_{1n}|} = \frac{v_{1t}}{e|v_{1n}|} = \frac{\tan\theta_i}{e} \ge \tan\theta_i \quad \Rightarrow \quad \theta_i \le \theta_r \le \tfrac{1}{2}\pi$$

$\theta_r = \theta_i, \pi/2$ correspond respectively to elastic and perfectly plastic impact (Fig. 3.24g).

Direct central impact of two unconstrained smooth rigid bodies without rotation (Fig. 3.25a), with body ② at rest before impact, is employed extensively in industry.

v'_{1n}, v'_{2n} are obtained from

$$m_1 v'_{1n} + m_2 v'_{2n} = m_1 v_{1n}, \qquad v'_{2n} - v'_{1n} = -e(0 - v_{1n})$$

$$\Rightarrow \quad v'_{1n} = \frac{1 - em_2/m_1}{1 + m_2/m_1} v_{1n}, \qquad v'_{2n} = \frac{1+e}{1 + m_2/m_1} v_{1n} \tag{1}$$

$$\eta = \frac{T - T'}{T} = \frac{1 - e^2}{1 + m_1/m_2} \tag{2}$$

The fractional decrease of T is large if e and m_1/m_2 are small. For $e = 0.4$ and $m_2/m_1 = 15$, $\eta = 0.79$. This 'loss' of kinetic energy may be used in deforming a workpiece attached to body ②, for example, in a drop hammer (Fig. 3.25b) for forging operation. Body ① is the hammer, body ② is the workpiece along with the anvil and generally $m_2 > 15m_1$. In fact, the anvil is constrained by the foundation in this case.

For $m_1 \ll m_2$, eq. (1) $\Rightarrow$ $v'_{1n} = -ev_{1n}$, $v'_{2n} = 0$, $\eta = 1 - e^2$. For $m_1 \gg m_2$, eq. (1) $\Rightarrow$ $v'_{1n} = v_{1n}$, $v'_{2n} = (1+e)v_{1n}$, $\eta \simeq 0$, i.e., there is no 'loss' of kinetic energy. Such an arrangement is used in pile driving, nailing, etc., to impart velocity v'_{2n} to a pile by a massive hammer. It follows from eq. $(1)_2$ that to impart a large velocity to the pile, e should be large and m_2/m_1 should be small.

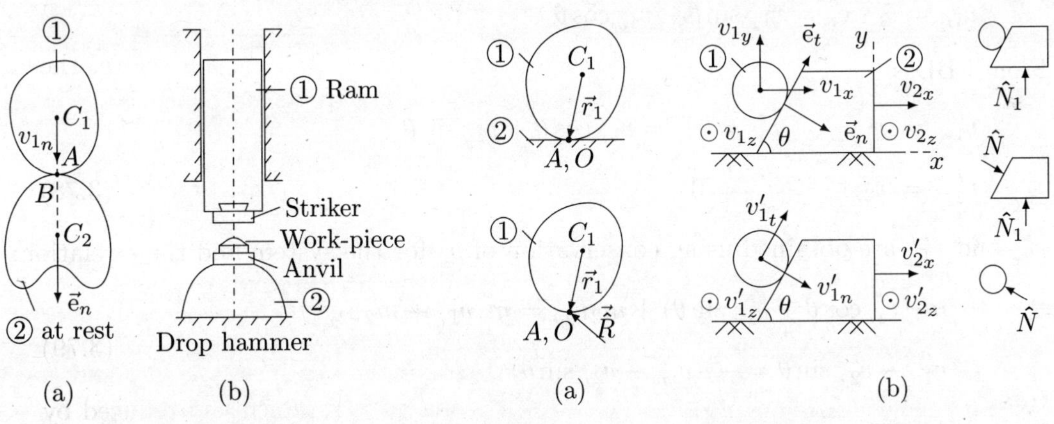

Fig. 3.25 **Fig. 3.26**

4. Other Cases

If body ① impacts a perfectly rough fixed body ② (Fig. 3.26a), then $\vec{v}'_A = 0$. $\vec{\omega}'_1$ is obtained from the conservation of $\vec{H}_O$ of body ①:

$$I^O_{ij}\omega'_{1_j}\,\vec{e}_i = I^{C_1}_{ij}\omega_{1_j}\,\vec{e}_i - \vec{r}_1 \times m_1\vec{v}_1$$

since just after the impact, $\vec{v}'_A = \vec{v}_O = \vec{0}$, i.e., O becomes a point of the rigid body ①.

Consider two unconstrained rigid bodies which stick together as a single rigid body after impact. An instantaneous angular impulse acts at the point of contact to constrain relative rotation. The post-impact $\vec{\omega}'$ and velocity of one point, say centre of mass C of the system are obtained using $\vec{v}'_C = \vec{v}_C$ and $\vec{H}'_O = \vec{H}_O$ for the system of two bodies, where O is a fixed point coincident with C. Alternatively $\vec{\omega}'$ and $\vec{v}'_{C_1}$ of C_1 are obtained using $\vec{p}' = \vec{p}$ and $\vec{H}'_D = \vec{H}_D$ for the system of two bodies, where D is a fixed point coincident with C_1.

If one or both bodies are constrained at some point, then FBD of each body is drawn showing all the impulsive forces at the point of impact and the constrained points. The post-impact data is obtained by writing impulse–momentum equations, angular impulse–moment of momentum equations about the centre of mass (or about a fixed point, if any) for each body, and the coefficient of restitution equation. For impact of constrained bodies, often the solution for kinematics can only be obtained by solving them in conjunction with the constraining impulses and the impulsive action between them.

For example, consider the impact of a smooth sphere ① of mass m_1 (Fig. 3.26b) translating at $v_{1_x}\vec{\imath} + v_{1_y}\vec{\jmath} + v_{1_z}\vec{k}$, with a block ② of mass m_2 translating at $v_{2_x}\vec{\imath} + v_{2_z}\vec{k}$ on smooth ground. The line of impact $\vec{e}_n = \sin\theta\,\vec{\imath} - \cos\theta\,\vec{\jmath}$ and $\vec{e}_t = \cos\theta\,\vec{\imath} + \sin\theta\,\vec{\jmath}$.

$$v_{1n} = \vec{v}_1 \cdot \vec{e}_n = v_{1_x}\sin\theta - v_{1_y}\cos\theta$$

From FBDs:

$$v'_{1_z} = v_{1_z}, \qquad v'_{1_t} = v_{1_t} = v_{1_x}\cos\theta + v_{1_y}\sin\theta$$

$$v'_{2_z} = v_{2_z}, \qquad v'_{2_y} = 0$$

v'_{2_x} and v'_{1_n} are obtained using conservation of p_x for the system and the e relation:

$$m_1(v'_{1_t}\cos\theta + v'_{1_n}\sin\theta) + m_2 v'_{2_x} = m_1 v_{1_x} + m_2 v_{2_x}$$

$$v'_{1_n} - v'_{2_x}\sin\theta = -e(v_{1_n} - v_{2_x}\sin\theta)$$

Example 3.36: A cylinder rolling without slip on a horizontal plane (Fig. E3.36a) hits an obstruction. Its radius of gyration about the axis of symmetry is k. The coefficient of restitution is e, $h = 0.5r$ and $k^2 = r^2/2$. (a) Find the angular velocity of the cylinder just after impact if its surface is (1) perfectly rough, (2) smooth and (3) rough with $\mu_k = 0.1, e = 0.2$. (b) For $e = 0$, find the minimum v_0 for the cylinder to climb the obstruction for cases (1) and (2). (c) Find the angular velocity after impact with an inclined plane (Fig. E3.36b).

Solution: The FBD is shown in Fig. E3.36c. $\cos\theta = (r - h)/r$. Just before and just after impact:

$$\omega = v_0/r, \qquad v_{1_t} = v_0 \cos\theta, \qquad v_{1n} = -v_0 \sin\theta$$

$$v_{An} = v_{1n}, \qquad v_{Bn} = 0, \qquad v_{A_t} = v_{1_t} - \omega r = -v_0(1-\cos\theta) < 0$$

$$v'_{An} = v'_{1n}, \qquad v'_{Bn} = 0, \qquad v'_{A_t} = v'_{1_t} - \omega' r$$

The coefficient of restitution equation: $\quad v'_{An} - v'_{Bn} = -e(v_{An} - v_{Bn}) \Rightarrow$

$$v'_{1n} = -e v_{1n} = e v_0 \sin\theta \tag{1}$$

For fixed point O, coincident with A, $\vec{I}_{\text{ang}_O} = \vec{0} \;\Rightarrow\; \vec{H}'_O = \vec{H}_O$:

$$\begin{aligned} mk^2\omega' + mv'_{1_t} r &= mk^2\omega + mv_{1_t} r = mk^2 v_0/r + rmv_0\cos\theta \\ &= m(k^2 + r^2 - rh)v_0/r \end{aligned} \tag{2}$$

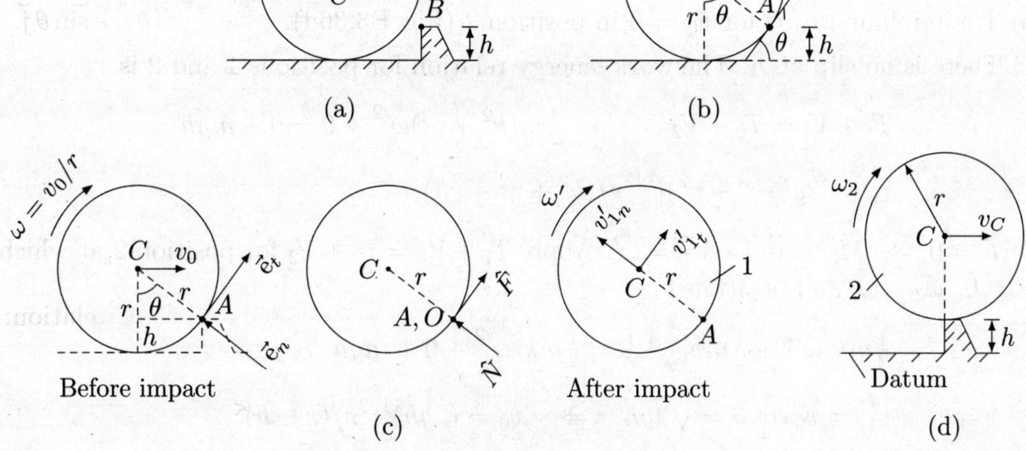

Fig. E3.36

(a) (1) For a perfectly rough surface, there is no slip at A:

$$v'_{A_t} = v'_{1_t} - \omega' r = 0 \quad\Rightarrow\quad v'_{1_t} = \omega' r \quad \text{and eq. (2)} \;\Rightarrow$$

$$\omega' = \frac{(k^2 + r^2 - rh)v_0}{(k^2 + r^2)r}, \qquad v'_{1_t} = \frac{(k^2 + r^2 - rh)v_0}{k^2 + r^2} \tag{3}$$

(2) For a smooth surface, $\hat{F} = 0 \Rightarrow v'_{1_t} = v_{1_t} = v_0\cos\theta$, eq. (2) $\Rightarrow \omega' = \omega$.

(3) $\cos\theta = (r-h)/r = 0.5 \;\Rightarrow\; \sin\theta = 0.866$. There can be slip/no slip at A just before the end of impact. *Assume* no slip, i.e., it is the same as case (1), but check $\hat{F}/\hat{N} < \mu_k$. Using eqs (1) and (3):

$$\hat{N} = m(v'_{1_n} - v_{1_n}) = (1+e)mv_0 \sin\theta$$

$$\hat{F} = m(v'_{1_t} - v_{1_t}) = mv_0 hk^2/r(k^2+r^2)$$

$$\frac{\hat{F}}{\hat{N}} = \frac{hk^2}{r(k^2+r^2)(1+e)\sin\theta} = \frac{0.5(0.5)}{(1.5)(1.2)0.866} = 0.16 > \mu_k = 0.1$$

$\Rightarrow$ The assumption is false. $\Rightarrow$ Slip just before the end of impact with $\hat{F}\vec{e}_t$ as $v_{A_t} < 0$:

$$\hat{F} = \mu_k \hat{N} = 0.1(1+e)mv_0 \sin\theta = m(v'_{1_t} - v_{1_t})$$

$$\Rightarrow \quad v'_{1_t} = v_{1_t} + 0.1(1+e)v_0 \sin\theta$$

$$\text{eq. (2)} \Rightarrow \quad mk^2\omega' + m[v_{1_t} + 0.1(1+e)v_0 \sin\theta]r = mk^2 v_0/r + mv_{1_t} r$$

$$\Rightarrow \quad \omega' = [1 - 0.1(1+0.2)0.866/0.5]v_0/r = 0.792 v_0/r$$

(b) The minimum v_0 is for $\vec{v}_C = 0$ in position 2 (Fig. E3.36d).

(1) There is no slip at A. The work–energy relation for positions 1 and 2 is:

$$T_1 + V_1 = T_2 + V_2: \qquad \tfrac{1}{2}m(k^2+r^2)\omega'^2 + 0 = 0 + mgh$$

$$\Rightarrow \quad v_0 = [2gh(k^2+r^2)]^{1/2} r/(k^2+r^2-rh)$$

(2) $F = 0 \;\Rightarrow\; \vec{M}_C = \vec{0} \;\Rightarrow\; \omega_2 = \omega'$. Apply $T_1 + V_1 = T_2 + V_2$ for position 2 at which $\vec{v}_C = \vec{0}$, $\omega_2 = \omega'$ and position 1:

$$\tfrac{1}{2}mk^2\omega'^2 + \tfrac{1}{2}mv'^2_{1_t} + 0 = \tfrac{1}{2}mk^2\omega'^2 + 0 + mgh$$

$$\Rightarrow \quad v'_{1_t} = v_0\cos\theta = \sqrt{2gh} \quad \Rightarrow \quad v_0 = (2gh)^{1/2} r/(r-h)$$

(c) The solution is the same as in part (a), since the velocity of C is normal to AC and the impulsive forces are of the same type.

Example 3.37: A uniform thin door is rotating at ω_0 when it bangs against a stop D at the ground level (Fig. E3.37a). The coefficient of restitution is 0.2. Find the impulsive forces on the door during the impact.

Solution: The FBD for impulsive forces is shown in Fig. E3.37b. The line of impact is along $\vec{\mathrm{i}}$. Before impact, $\vec{\omega} = \omega_0 \vec{\mathrm{k}}$. Let the angular velocity just after impact be $\vec{\omega}' = \omega' \vec{\mathrm{k}}$, and the impulsive bearing reactions be $\hat{\vec{R}}_A = \hat{R}_1 \vec{\mathrm{i}} + \hat{R}_2 \vec{\mathrm{j}} + \hat{R}_3 \vec{\mathrm{k}}$, $\hat{\vec{R}}_B =$

$\hat{R}_4\,\vec{\mathrm{i}} + \hat{R}_5\,\vec{\mathrm{j}} + \hat{R}_6\,\vec{\mathrm{k}}$. The velocities of point E of the door and point D of the stop are

$$\vec{v}_E = -\omega_0 p\,\vec{\mathrm{i}}, \qquad \vec{v}_D = 0, \qquad \vec{v}\,'_E = -\omega' p\,\vec{\mathrm{i}}, \qquad \vec{v}\,'_D = 0$$

$$v_{Ex} = -\omega_0 p, \qquad v_{Dx} = 0, \qquad v'_{Ex} = -\omega' p, \qquad v'_{Dx} = 0$$

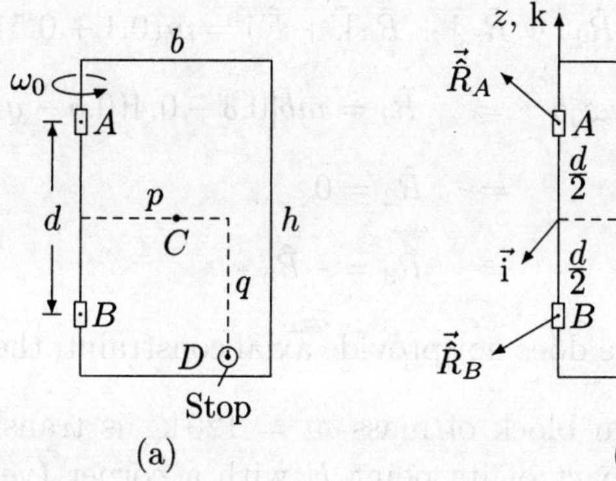

Fig. E3.37

The e relation: $\;v'_{Ex} - v'_{Dx} = -e(v_{Ex} - v_{Dx}) \;\Rightarrow$

$$-\omega' p = -0.2(-\omega_0 p) \qquad \Rightarrow \;\omega' = -0.2\omega_0$$

Axis z is not a principal axis of inertia at point A of the door. $\overrightarrow{AC} = \frac{1}{2}b\,\vec{\mathrm{j}} - \frac{1}{2}d\,\vec{\mathrm{k}}$,

$$I^A_{xz} = -m(0)\left(-\tfrac{1}{2}d\right) = 0$$

$$I^A_{yz} = -m\left(\tfrac{1}{2}b\right)\left(-\tfrac{1}{2}d\right) = \tfrac{1}{4}mbd$$

$$I^A_{zz} = m\left[\tfrac{1}{12}b^2 + \left(\tfrac{1}{2}b\right)^2\right] = \tfrac{1}{3}mb^2$$

$$\vec{H}_A = \left(I^A_{xz}\,\vec{\mathrm{i}} + I^A_{yz}\,\vec{\mathrm{j}} + I^A_{zz}\,\vec{\mathrm{k}}\right)\omega \;\Rightarrow$$

$$\Delta\vec{H}_A = \left(I^A_{yz}\,\vec{\mathrm{j}} + I^A_{zz}\,\vec{\mathrm{k}}\right)(\omega' - \omega_0) = -1.2\left(\tfrac{1}{4}mbd\,\vec{\mathrm{j}} + \tfrac{1}{3}mb^2\,\vec{\mathrm{k}}\right)\omega_0$$

For fixed point A: $\;\vec{I}_{\mathrm{ang}_A} = \Delta\vec{H}_A$ yields

$$[-(\tfrac{1}{2}d + q)\,\vec{\mathrm{k}} + p\,\vec{\mathrm{j}}] \times \hat{F}\,\vec{\mathrm{i}} - d\,\vec{\mathrm{k}} \times (\hat{R}_4\,\vec{\mathrm{i}} + \hat{R}_5\,\vec{\mathrm{j}} + \hat{R}_6\,\vec{\mathrm{k}}) = -1.2(\tfrac{1}{4}mbd\,\vec{\mathrm{j}} + \tfrac{1}{3}mb^2\,\vec{\mathrm{k}})\omega_0$$

$$\Rightarrow \quad \vec{\mathrm{i}}: \quad d\hat{R}_5 = 0 \qquad \Rightarrow \qquad \hat{R}_5 = 0$$

$$\vec{\mathrm{k}}: \quad -p\hat{F} = -0.4mb^2\omega_0 \qquad \Rightarrow \qquad \hat{F} = 0.4mb^2\omega_0/p$$

$$\vec{j}: \quad -(\tfrac{1}{2}d+q)\hat{F} - d\hat{R}_4 = -0.3mbd\omega_0$$

$$\Rightarrow \quad \hat{R}_4 = mb[0.3 - 0.4(0.5+q/d)b/p]\omega_0$$

$$\vec{v}_C = -0.5\omega_0 b\,\vec{i}, \quad \vec{v}_C' = -0.5\omega_0' b\,\vec{i} = 0.1\omega_0 b\,\vec{i}. \quad \vec{I} = m\Delta\vec{v}_C:$$

$$\hat{R}_1\vec{i} + \hat{R}_2\vec{j} + \hat{R}_3\vec{k} + \hat{R}_4\vec{i} + \hat{R}_5\vec{j} + \hat{R}_6\vec{k} + \hat{F}\vec{i} = m(0.1+0.5)\omega_0 b\,\vec{i}$$

$$\vec{i}: \quad \hat{R}_1 + \hat{R}_4 + \hat{F} = 0.6m\omega_0 b \quad \Rightarrow \quad \hat{R}_1 = mb[0.3 - 0.4(0.5-q/d)b/p]\omega_0$$

$$\vec{j}: \quad \hat{R}_2 + \hat{R}_5 = 0 \quad \Rightarrow \quad \hat{R}_2 = 0$$

$$\vec{k}: \quad \hat{R}_3 + \hat{R}_6 = 0 \quad \Rightarrow \quad \hat{R}_3 = -\hat{R}_6$$

$\hat{R}_3$, $\hat{R}_6$ can be found if one hinge does not provide axial constraint, then $\hat{R}_3 = \hat{R}_6 = 0$.

Example 3.38: (a) A uniform block of mass $m = 120\,\text{kg}$ is translating vertically down at 2 m/s just before impact of its point E with a corner (very near it) of a smooth horizontal table (Fig. E3.38a). Find its angular velocity just after impact. The coefficient of restitution $e = 0.6$. (b) Rework part (a) if the table is rough and point E has zero velocity just after impact. (c) The stationary block is supported at D by a ball-and-socket joint and it is hit at B by a mass-point of mass $m_1 = 2\,\text{kg}$ moving at $-2\vec{i} + \vec{j} - \vec{k}\,\text{m/s}$. Find the angular velocity just after impact for the following cases: (1) the impact is smooth and $e = 0.6$, (2) the mass-point sticks to the block after impact.

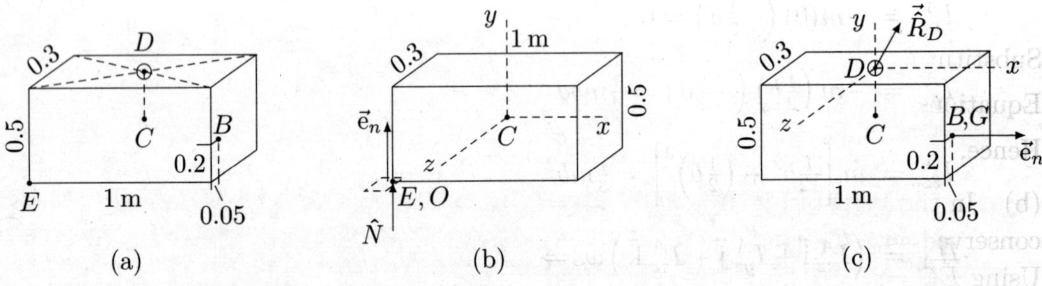

Fig. E3.38

Solution: (a) The line of impact is normal to the common horizontal tangent surface in direction $\vec{e}_n = \vec{j}$. Let $\vec{v}_C$, $\vec{v}_E$, $\vec{\omega}$ be the values just before the impact, the values just after being $\vec{v}_C'$, $\vec{v}_E'$, $\vec{\omega}'$. The FBD is shown in Fig. E3.38b.

$$\overrightarrow{CE} = -0.5\vec{i} - 0.25\vec{j} + 0.15\vec{k}\text{ m}, \qquad \overrightarrow{OC} = 0.5\vec{i} + 0.25\vec{j} - 0.15\vec{k}\text{ m}$$

$$I_{xx}^C = 120\frac{(0.3^2+0.5^2)}{12} = 3.4\,\text{kg.m}^2, \qquad I_{yy}^C = 120\frac{(1^2+0.3^2)}{12} = 10.9\,\text{kg.m}^2$$

$$I_{zz}^C = 120\,\frac{(1^2+0.5^2)}{12} = 12.5\,\text{kg.m}^2, \qquad \vec{v}_C = \vec{v}_E = -2\,\vec{\jmath}\,\text{m/s}, \quad \vec{\omega} = \vec{0}$$

$$\vec{\omega}' = \omega_x'\,\vec{\imath} + \omega_y'\,\vec{\jmath} + \omega_z'\,\vec{k}, \qquad \vec{v}_C' = v_y'\,\vec{\jmath} \tag{1}$$

$$\vec{v}_E' = \vec{v}_C' + \vec{\omega}' \times \overrightarrow{CE}$$

$$= (0.15\omega_y' + 0.25\omega_z')\,\vec{\imath} + (v_y' - 0.15\omega_x' - 0.5\omega_z')\,\vec{\jmath} + (0.5\omega_y' - 0.25\omega_x')\,\vec{k} \tag{2}$$

ω_x', ω_y', ω_z', v_y', are obtained using conservation of $\vec{H}_O$ $(= \vec{H}_C + \overrightarrow{OC} \times m\vec{v}_C)$ about fixed point O coincident with E, and the e relation. $\vec{H}_O' = \vec{H}_O \Rightarrow$

$$I_{xx}^C\omega_x'\,\vec{\imath} + I_{yy}^C\omega_y'\,\vec{\jmath} + I_{zz}^C\omega_z'\,\vec{k} + \overrightarrow{OC} \times m\vec{v}_C' = \vec{0} + \overrightarrow{OC} \times m\vec{v}_C$$

$$\Rightarrow \quad 3.4\omega_x'\,\vec{\imath} + 10.9\omega_y'\,\vec{\jmath} + 12.5\omega_z'\,\vec{k} + (0.5\,\vec{\imath} + 0.25\,\vec{\jmath} - 0.15\,\vec{k}) \times 120v_y'\,\vec{\jmath}$$

$$= (0.5\,\vec{\imath} + 0.25\,\vec{\jmath} - 0.15\,\vec{k}) \times 120(-2\,\vec{\jmath})$$

$$\vec{\imath}: \quad 3.4\omega_x' + 18v_y' = -36 \quad \Rightarrow \quad \omega_x' = -10.59 - 5.294v_y' \tag{3}$$

$$\vec{\jmath}: \quad 10.9\omega_y' = 0 \quad \Rightarrow \quad \omega_y' = 0 \tag{4}$$

$$\vec{k}: \quad 12.5\omega_z' + 60v_y' = -120 \quad \Rightarrow \quad \omega_z' = -9.6 - 4.8v_y' \tag{5}$$

The e relation: $\quad v_{En}' - 0 = -e(v_{En} - 0) \quad \Rightarrow \quad v_{Ey}' = -0.6v_{Ey} \Rightarrow$

$$v_y' - 0.15\omega_x' - 0.5\omega_z' = -0.6(-2) \tag{6}$$

Substituting ω_x', ω_z' from eqs (3), (5) in eq. (6) yields $v_y' = -1.237\,\text{m/s}$.

Equations (3), (5) $\Rightarrow$ $\omega_x' = -4.041\,\text{rad/s}$, $\qquad \omega_z' = -3.662\,\text{rad/s}$.

Hence, $\qquad \vec{v}_C' = -1.237\,\vec{\jmath}\,\text{m/s}, \qquad \vec{\omega}' = -4.041\,\vec{\imath} - 3.662\,\vec{k}\,\text{rad/s}.$

(b) In the FBD of Fig. E3.38b replace $\hat{N}$ by arbitrary $\hat{\vec{R}}$. $\vec{H}_O$ about fixed point O is conserved with O being a point of the rigid body just after impact, since $\vec{v}_E' = \vec{0} = \vec{v}_O'$. Using $\overrightarrow{EC} = (0.5, 0.25, -0.15)$ and parallel axes theorems,

$$I_{xx}^E = 120[(0.5^2 + 0.3^2)/12 + 0.25^2 + 0.15^2] = 13.6\,\text{kg.m}^2$$

$$I_{yy}^E = 120[(0.3^2 + 1^2)/12 + 0.15^2 + 0.5^2] = 43.6\,\text{kg.m}^2$$

$$I_{zz}^E = 120[(1^2 + 0.5^2)/12 + 0.5^2 + 0.25^2] = 50\,\text{kg.m}^2$$

$$I_{xy}^E = 0 - 120(0.5)(0.25) = -15\,\text{kg.m}^2$$

$$I_{yz}^E = 0 - 120(0.25)(-0.15) = 4.5\,\text{kg.m}^2$$

$$I^E_{zx} = 0 - 120(-0.15)(0.5) = 9\,\text{kg.m}^2$$

$$\vec{H}_O = \vec{0} + \overrightarrow{OC} \times m\vec{v}_C = (0.5\,\vec{i} + 0.25\,\vec{j} - 0.15\,\vec{k}) \times 120(-2\,\vec{j}) = -36\,\vec{i} - 120\,\vec{k}$$

$[\vec{H}'_O] = [I^E][\vec{\omega}']$ and $\vec{H}'_O = \vec{H}_O$

$$\Rightarrow \begin{bmatrix} 13.6 & -15 & 9 \\ -15 & 43.6 & 4.5 \\ 9 & 4.5 & 50 \end{bmatrix} \begin{bmatrix} \omega'_x \\ \omega'_y \\ \omega'_z \end{bmatrix} = \begin{bmatrix} -36 \\ 0 \\ -120 \end{bmatrix} \Rightarrow \begin{bmatrix} \omega'_x \\ \omega'_y \\ \omega'_z \end{bmatrix} = \begin{bmatrix} -1.687 \\ -0.3674 \\ -2.063 \end{bmatrix}$$

$$\Rightarrow \quad \vec{\omega}' = -1.687\,\vec{i} - 0.3674\,\vec{j} - 2.063\,\vec{k}\ \text{rad/s}$$

Note that this case of $\vec{v}'_E = \vec{0}$ differs from the case of $e = 0$ for which only $v'_{E_y} = 0$.

(c) The FBD of the system of mass-point G and block is shown in Fig. E3.38c. $\overrightarrow{DB} = 0.5\,\vec{i} - 0.3\,\vec{j} + 0.1\,\vec{k}$. x, y, z are principal axes at D.

$$I^D_{xx} = I^C_{xx} + 120(0.25)^2 = 10.9, \qquad I^D_{yy} = I^C_{yy} = 10.9$$

$$I^D_{zz} = I^C_{zz} + 120(0.25)^2 = 20\,\text{kg.m}^2$$

$$\vec{v}_B = \vec{0}, \qquad \vec{v}_G = -2\,\vec{i} + \vec{j} - \vec{k}\ \text{m/s}, \qquad \vec{v}'_G = v'_x\,\vec{i} + v'_y\,\vec{j} + v'_z\,\vec{k} \tag{7}$$

$$\vec{v}'_B = \vec{v}'_D + \vec{\omega}' \times \overrightarrow{DB} = (\omega'_x\,\vec{i} + \omega'_y\,\vec{j} + \omega'_z\,\vec{k}) \times (0.5\,\vec{i} - 0.3\,\vec{j} + 0.1\vec{k})$$

$$= (0.1\omega'_y + 0.3\omega'_z)\,\vec{i} + (0.5\omega'_z - 0.1\omega'_x)\,\vec{j} - (0.3\omega'_x + 0.5\omega'_y)\,\vec{k} \tag{8}$$

$\vec{H}_D$ of the system of mass-point G and the block is conserved: $\vec{H}'_D = \vec{H}_D \Rightarrow$

$$I^D_{xx}\omega'_x\,\vec{i} + I^D_{yy}\omega'_y\,\vec{j} + I^D_{zz}\omega'_z\,\vec{k} + \overrightarrow{DB} \times m_1\vec{v}'_G = \vec{0} + \overrightarrow{DB} \times m_1\vec{v}_G$$

$$\Rightarrow 10.9\omega'_x\,\vec{i} + 10.9\omega'_y\,\vec{k} + 20\omega'_z\,\vec{k} + (0.5\,\vec{i} - 0.3\,\vec{j} + 0.1\,\vec{k}) \times 2(v'_x\,\vec{i} + v'_y\,\vec{j} + v'_z\,\vec{k})$$

$$= (0.5\,\vec{i} - 0.3\,\vec{j} + 0.1\,\vec{k}) \times 2(-2\,\vec{i} + \vec{j} - \vec{k})$$

$$\vec{i}: \quad 10.9\omega'_x - 0.2v'_y - 0.6v'_z = 0.4$$

$$\vec{j}: \quad 10.9\omega'_y + 0.2v'_x - v'_z = 0.6 \tag{9a}$$

$$\vec{k}: \quad 20\omega'_z + 0.6v'_x + v'_y = -0.2$$

$$\Rightarrow \quad \omega'_x = 0.036697 + 0.018349v'_y + 0.055046v'_z \tag{9}$$

$$\omega'_y = 0.055046 - 0.018349v'_x + 0.091743v'_z \tag{10}$$

$$\omega'_z = -0.01 - 0.03v'_x - 0.05v'_y \tag{11}$$

(1) $\vec{e}_n = \vec{i}$, $v'_y = v_{Gy} = 1$, $v'_z = v_{Gz} = -1 \Rightarrow \vec{v}'_G = v'_x \vec{i} + \vec{j} - \vec{k}$.

The e relation: $v'_{Bx} - v'_{Gx} = -e(v_{Bx} - v_{Gx}) \Rightarrow$

$$0.1\omega'_y + 0.3\omega'_z - v'_x = -0.6(0+2) \tag{12}$$

Substituting $v'_y = 1$, $v'_z = -1$ in eqs (9) to (11) yields

$$\omega'_x = 0, \quad \omega'_y = -0.036697 - 0.018349 v'_x, \quad \omega'_z = -0.06 - 0.03 v'_x \tag{13}$$

Using eq. (13) in eq. (12) yields: $v'_x = 1.1657\,\text{m/s}$ and eq. (13) $\Rightarrow$

$$\omega'_y = -0.05809, \quad \omega'_z = -0.09497, \quad \vec{\omega}' = -0.05809\,\vec{j} - 0.09497\,\vec{k}\ \text{rad/s}$$

(2) For mass-point + block:

$$I^D_{xx} = 10.9 + 2(0.3^2 + 0.1^2) = 11.1, \qquad I^D_{xy} = -2(0.5)(-0.3) = 0.3$$

$$I^D_{yy} = 10.9 + 2(0.5^2 + 0.1^2) = 11.42, \qquad I^D_{yz} = -2(-0.3)(0.1) = 0.06 \tag{14}$$

$$I^D_{zz} = 20 + 2(0.5^2 + 0.3^2) = 20.68, \qquad I^D_{zx} = -2(0.1)(0.5) = -0.1\,\text{kg.m}^2$$

$$\vec{H}'_D = \vec{H}_D: \begin{bmatrix} 11.1 & 0.3 & -0.1 \\ 0.3 & 11.42 & 0.06 \\ -0.1 & 0.06 & 20.68 \end{bmatrix} \begin{bmatrix} \omega'_x \\ \omega'_y \\ \omega'_z \end{bmatrix} = \begin{bmatrix} 0.4 \\ 0.6 \\ -0.2 \end{bmatrix} \Rightarrow \begin{bmatrix} \omega'_x \\ \omega'_y \\ \omega'_z \end{bmatrix} = \begin{bmatrix} 0.03455 \\ 0.05168 \\ -0.009654 \end{bmatrix}$$

$$\Rightarrow \quad \vec{\omega}' = 0.03455\,\vec{i} + 0.05168\,\vec{j} - 0.009654\,\vec{k}\ \text{rad/s}$$

Example 3.39: A spherical satellite translates with velocity $v_0\vec{i}$ in a rarefied atmosphere having n molecules per unit volume, each of mass m. Find the drag force on the satellite. Assume frictionless collisions with coefficient of restitution e. Assume that the speed of the satellite is much larger than the r.m.s. speed of the molecules.

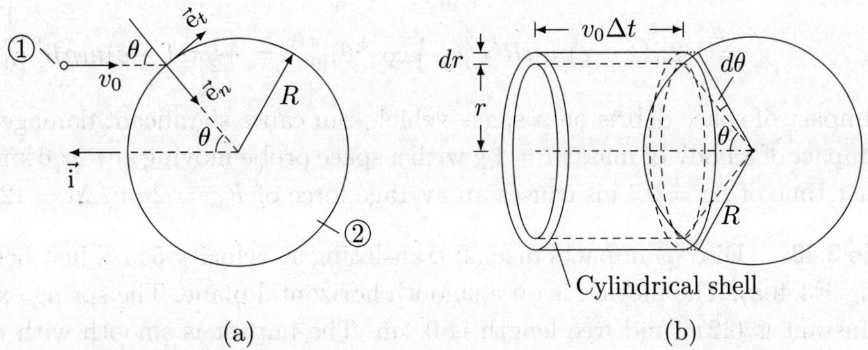

Fig. E3.39

Solution: We neglect change in speed of the satellite due to impact. Attach an inertial frame to the sphere and consider velocities relative to this frame. The drag force equals the impulse of the collisions per unit time. Molecule ① moving with v_0 makes impact with the line of impact along $\vec{e}_n$. The impulse $\vec{I}_1$ on it is expressed in terms of its velocity components just before and just after impact:

$$v_{1_n} = v_0\cos\theta, \qquad v'_{1_n} = -ev_0\cos\theta, \qquad v'_{1_t} = v_{1_t} = ev_0\sin\theta$$

$$\vec{I}_1 = m(v'_{1_n} - v_{1_n})\,\vec{e}_n = -(1+e)mv_0\cos\theta\,\vec{e}_n$$

The opposite impulse $\vec{I}_2$ on the sphere is $\vec{I}_2 = (1+e)mv_0\cos\theta\,\vec{e}_n$. The components of these impulses on the sphere in directions normal to $\vec{i}$ due to all impacts cancel out, whereas the $\vec{i}$ components add up. The component I_x of $\vec{I}_2$ in direction $\vec{i}$ is

$$I_x = [(1+e)mv_0\cos\theta](-\cos\theta) = -(1+e)mv_0\cos^2\theta$$

Let N be the number of molecules hitting the sphere in time Δt between angles θ and $\theta + d\theta$ (Fig. E3.39).

$$\begin{aligned}N &= n\times(\text{volume swept by the arc between } \theta \text{ and } \theta+d\theta \text{ in time } \Delta t)\\ &= n\times(\text{volume of cylindrical shell: } r = R\sin\theta,\ \ dr = R\cos\theta\,d\theta,\ \ L = v_0\Delta t)\\ &= n[2\pi(R\sin\theta)(R\cos\theta\,d\theta)v_0\Delta t] = 2\pi nR^2v_o\sin\theta\cos\theta\,d\theta\,\Delta t\end{aligned}$$

The impulse per unit time due to impact of N molecules is

$$NI_x/\Delta t = -2\pi(1+e)mnR^2v_0^2\sin\theta\cos^3\theta\,d\theta$$

$$\begin{aligned}\text{Drag force:}\quad D_x &= [\text{Total impulse per unit time for } \theta\in(0,\ \pi/2)]\\ D_x &= -2\pi(1+e)mnR^2v_0^2\int_0^{\pi/2}\sin\theta\cos^3\theta\,d\theta\\ &= -2\pi(1+e)mnR^2v_0^2(-\tfrac{1}{4}\cos^4\theta)\big|_0^{\pi/2} = -\tfrac{1}{2}\pi(1+e)mnR^2v_0^2\end{aligned}$$

The impact of space debris on a space vehicle can cause significant damage. Direct elastic impact of a body of mass $m = 1$ g with a space probe moving at $v = 6$ km/s with an impact time of $\Delta t = 0.1$ ms causes an average force of $F_{\text{av}} = 2mv/\Delta t = 120$ kN!

Example 3.40: Disc ① impacts disc ② translating at velocity 5 m/s just before impact (Fig. E3.40a). The motion is on a smooth horizontal plane. The spring extension at this instant is 0.2 m and free length is 0.4 m. The impact is smooth with $e = 0.7$. The subsequent maximum spring compression is 0.1 m. Find the velocities of the discs just after impact, the loss in kinetic energy in impact, impulsive force on disc ② during impact, the stiffness of the spring and its maximum extension in subsequent motion.

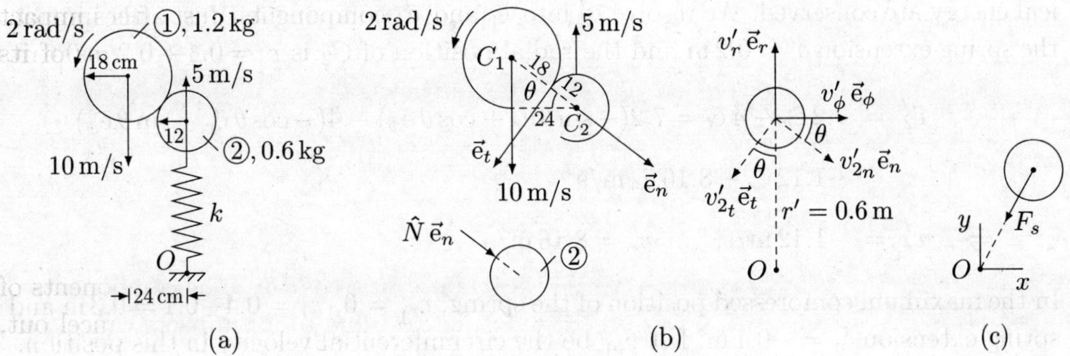

Fig. E3.40

Solution: Since the spring force is finite, it is not relevant during impact. The line of impact is along $\vec{e}_n$ (Fig. E3.40b) and $\cos\theta = 24/30 = 0.8 \;\Rightarrow\; \sin\theta = 0.6$. The velocities of the centres and angular velocities just before and just after impact are:

$$m_1 = 1.2\,\text{kg}, \qquad v'_{1_t} = v_{1_t} = 10\cos\theta = 8\,\text{m/s}$$

$$v_{1n} = 10\sin\theta = 6\,\text{m/s}, \qquad \omega'_1 = \omega_1 = 2\,\text{rad/s}$$

$$m_2 = 0.6\,\text{kg}, \qquad v'_{2_t} = v_{2_t} = -5\cos\theta = -4\,\text{m/s}$$

$$v_{2n} = -5\sin\theta = -3\,\text{m/s}, \qquad \omega'_2 = \omega_2 = 0$$

v'_{1n}, v'_{2n} are determined from the e relation and conservation of total momentum:

$$v'_{2n} - v'_{1n} = -e(v_{2n} - v_{1n}), \qquad m_1 v'_{1n} + m_2 v'_{2n} = m_1 v_{1n} + m_2 v_{2n}$$

$$\Rightarrow \quad v'_{2n} - v'_{1n} = -0.7(-3-6) = 6.3 \tag{1}$$

$$1.2v'_{1n} + 0.6v'_{2n} = 1.2(6) + 0.6(-3) = 5.4 \tag{2}$$

Equations (1) and (2) yield

$$v'_{1n} = 0.9\,\text{m/s}, \qquad v'_{2n} = 7.2\,\text{m/s}, \qquad \vec{v}'_1 = 0.9\,\vec{e}_n + 8\,\vec{e}_t, \qquad \vec{v}'_2 = 7.2\,\vec{e}_n - 4\,\vec{e}_t$$

The loss of kinetic energy is $-\Delta T$:

$$-\Delta T = -[\tfrac{1}{2}m_1(v'^2_{1n} - v^2_{1n}) + \tfrac{1}{2}m_2(v'^2_{2n} - v^2_{2n})] = 8.262\ \text{J}$$

The instantaneous impulse on disc 2 is $\hat{N}\,\vec{e}_n$ with

$$\hat{N} = m_2(v'_{2n} - v_{2n}) = 0.6(7.2+3) = 6.12\,\text{N.s}$$

The motion of disc ② after impact takes place under a spring force F_s (Fig. E3.40c), which is conservative and passes through $O \Rightarrow M_{Oz} = 0$. Hence, its H_{Oz} and mechanical energy are conserved. We resolve $\vec{v}_2'$ into $\vec{e}_r$ and $\vec{e}_\phi$ components. Just after impact, the spring extension $\delta' = 0.2\,\text{m}$ and the radial position of C_2 is $r' = 0.4 + 0.2 = 0.6\,\text{m}$,

$$\vec{v}_2' = 7.2\,\vec{e}_n - 4\,\vec{e}_t = 7.2(-\sin\theta\,\vec{e}_r + \cos\theta\,\vec{e}_\phi) - 4(-\cos\theta\,\vec{e}_r - \sin\theta\,\vec{e}_\phi)$$

$$= -1.12\,\vec{e}_r + 8.16\,\vec{e}_\phi\,\text{m/s}$$

$$\Rightarrow \quad v_r' = -1.12\,\text{m/s}, \qquad v_\phi' = 8.16\,\text{m/s}$$

In the maximum compressed position of the spring, $v_{r_1} = 0$, $r_1 = 0.4 - 0.1 = 0.3\,\text{m}$ and spring extension $\delta_1 = -0.1\,\text{m}$. Let v_{ϕ_1} be the circumferential velocity in this position.

$$H_{Oz} = m_2 r_1 v_{\phi_1} = m_2 r' v_\phi' \quad \Rightarrow \quad v_{\phi_1} = \frac{0.6(8.16)}{0.3} = 16.32\,\text{m/s}$$

$T_1 + V_1 = T' + V' \Rightarrow$

$$\tfrac{1}{2} m_2 (0^2 + v_{\phi_1}^2) + \tfrac{1}{2} k \delta_1^2 = \tfrac{1}{2} m_2 (v_r'^2 + v_\phi'^2) + \tfrac{1}{2} k \delta'^2$$

$$\Rightarrow \quad \tfrac{1}{2}(0.6)(16.32)^2 + \tfrac{1}{2} k (0.1)^2 = \tfrac{1}{2}(0.6)[8.16^2 + 1.12^2] + \tfrac{1}{2} k (0.2)^2$$

$$\Rightarrow \quad k = 3970\,\text{N/m} = 3.970\,\text{kN/m}$$

Let the maximum extension be δ for $r = 0.4 + \delta$ with $v_r = 0$ and circumferential velocity v_ϕ. $H_{Oz} = m_2 r v_\phi = m_2 r' v_\phi'$ and $T + V = T' + V' \Rightarrow$

$$v_\phi = 0.6(8.16)/(0.4 + \delta) = 4.896/(0.4 + \delta)$$

$$\tfrac{1}{2}(0.6)[0^2 + 4.896^2/(0.4 + \delta)^2] + \tfrac{1}{2}(3970)\delta^2 = \tfrac{1}{2}(0.6)[8.16^2 + 1.12^2] + \tfrac{1}{2}(3970)(0.2)^2$$

$$\Rightarrow \quad 3970\delta^4 + 3176\delta^3 + 435.7\delta^2 - 159.6\delta - 17.538 = 0 \tag{3}$$

$$\Rightarrow \quad (\delta + 0.1)[3970\delta^3 + 2779\delta^2 + 157.8\delta - 175.38] = 0$$

$$\Rightarrow \quad 3970\delta^3 + 2779\delta^2 + 157.8\delta - 175.38 = 0 \tag{4}$$

where $(\delta + 0.1)$ has been factored out from eq. (3) as $\delta_{\text{min}} = -0.1$ is its root. The only real root of the cubic equation (4), obtained by Cardan's method, is $\delta = 0.2005$.

Example 3.41: (a) A thin steel rod ① with sharp ends and a thin wooden plate ② have the given motion (Fig. E3.41a) just before their impact. Their mass $m_1 = 0.1\,\text{kg}$, $m_2 = 0.2\,\text{kg}$ and $\cos\theta = 0.8$. Find ω_2', (1) if the impact is smooth with $e = 0.4$ and (2) if the rod gets embedded in the plate at angle θ. (b) Rework part (a) for the case of free plate with the same motion just before impact. (c) Find the motion of the rod just after impact in part (a) if $m_2 \gg m_1$ and the impact is smooth with $e = 0.4$.

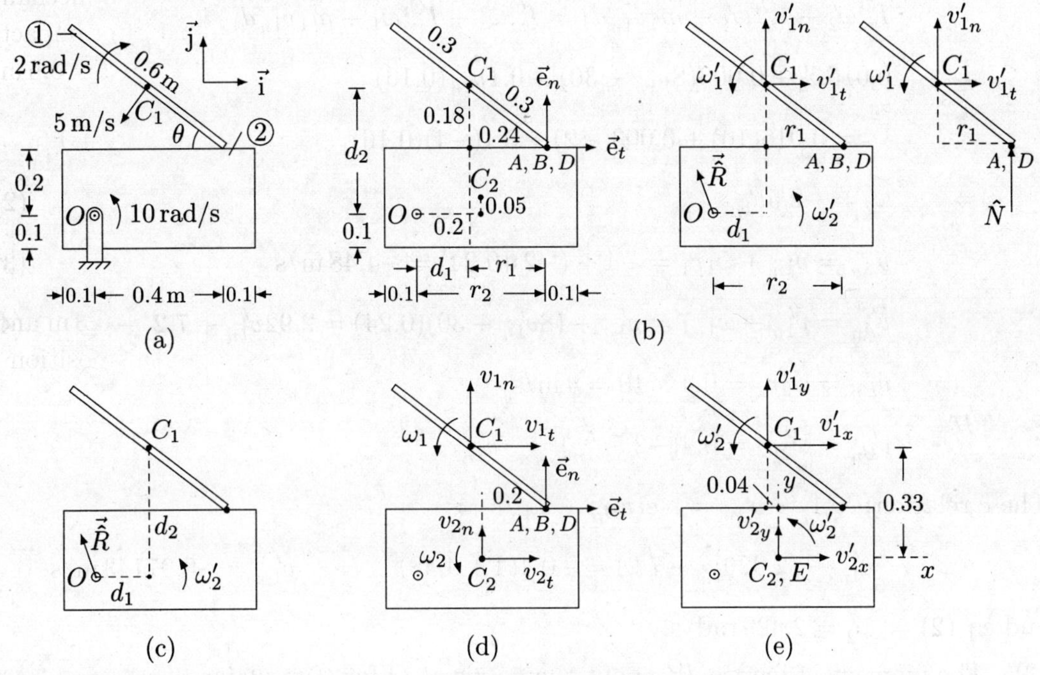

Fig. E3.41

Solution: (a) (1) The line of impact is along $\vec{e}_n = \vec{j}$ (Fig. E3.41b).

$$r_1 = 0.3\cos\theta = 0.24\,\text{m}, \quad d_1 = 0.16\,\text{m}, \quad r_2 = 0.4\,\text{m}, \quad d_2 = 0.38$$

$$v'_{1_t} = v_{1_t} = -5\sin\theta = -3\,\text{m/s}, \qquad v_{1_n} = -5\cos\theta = -4\,\text{m/s}$$

$$\omega_1 = -2\,\text{rad/s}, \qquad \omega_2 = 10\,\text{rad/s}$$

① : $I^{C_1}_{zz} = 0.1(0.6)^2/12 = 0.003\,\text{kg.m}^2$

② : $I^{O}_{zz} = 0.2[(0.6^2 + 0.3^2)/12 + 0.2^2 + 0.05^2] = 0.016\,\text{kg.m}^2$

ω'_1, ω'_2, v'_{1_n} are obtained using $H'_{D_z} = H_{D_z}$ for ①, $H'_{O_z} = H_{O_z}$ for ① + ② and the e relation, where D is a fixed point of the ground coincident with point A of ①.

For ①: $H'_{D_z} = H_{D_z} \Rightarrow$

$$I^{C_1}_{zz}\omega'_1 - m_1 v'_{1_n} r_1 = I^{C_1}_{zz}\omega_1 - m_1 v_{1_n} r_1$$

$$\Rightarrow \quad 0.003\omega'_1 - 0.1v'_{1_n}(0.24) = 0.003(-2) - 0.1(-4)(0.24)$$

$$\Rightarrow \omega'_1 = 8v'_{1_n} + 30 \tag{1}$$

B is a point of plate ②. For ① + ②: $H'_{Oz} = H_{Oz} \Rightarrow$

$$I^O_{zz}\omega'_2 + I^{C_1}_{zz}\omega'_1 + m_1 v'_{1n} d_1 = I^O_{zz}\omega_2 + I^{C_1}_{zz}\omega_1 + m_1 v_{1n} d_1$$

$$\Rightarrow \quad 0.016\omega'_2 + 0.003(8v'_{1n} + 30) + 0.1v'_{1n}(0.16)$$

$$= 0.016(10) + 0.003(-2) + 0.1(-4)(0.16)$$

$$\Rightarrow \quad \omega'_2 = -2.5v'_{1n} \tag{2}$$

$$v_{An} = v_{1n} + \omega_1 r_1 = -4 + (-2)(0.24) = -4.48\,\text{m/s} \tag{3}$$

$$v'_{An} = v'_{1n} + \omega'_1 r_1 = v'_{1n} + (8v'_{1n} + 30)(0.24) = 2.92v'_{1n} + 7.2 \tag{4}$$

$$v_{Bn} = \omega_2 r_2 = 0.4 \times 10 = 4\,\text{m/s}$$

$$v'_{Bn} = \omega'_2 r_2 = 0.4(-2.5v'_{1n}) = -v'_{1n}$$

The e relation: $v'_{Bn} - v'_{An} = -e(v_{Bn} - v_{An}) \Rightarrow$

$$-v'_{1n} - (2.92v'_{1n} + 7.2) = -0.4(4 + 4.48) \quad \Rightarrow \quad v'_{1n} = -0.97143\,\text{m/s}$$

and eq. (2) $\Rightarrow \omega'_2 = 2.429\,\text{rad/s}$.

(2) The moment of inertia I^O about the z-axis at O for ① + ② is:

$$I^O = I^O_{zz} + [I^{C_1}_{zz} + m_1(d_1^2 + d_2^2)]$$

$$= 0.016 + 0.003 + 0.1(0.16^2 + 0.38^2) = 0.036\,\text{kg.m}^2$$

ω'_2 (Fig. E3.41c) is obtained from $H'_{Oz} = H_{Oz}$ for ① + ②:

$$I^O\omega'_2 = I^O_{zz}\omega_2 + I^{C_1}_{zz}\omega_1 + m_1 v_{1n} d_1 - m_1 v_{1_t} d_2$$

$$0.036\omega'_2 = 0.016(10) + 0.003(-2) + 0.1(-4)(0.16) - 0.1(-3)(0.38)$$

$$\omega'_2 = 5.667\,\text{rad/s}$$

(b) (1) The FBD and kinematics are shown in Fig. E3.41d.

$$\vec{v}_{C_2} = \vec{v}_O + \omega_2 \vec{k} \times \overrightarrow{OC}_2 = 10\vec{k} \times (0.2\vec{i} + 0.05\vec{j}) = -0.5\vec{i} + 2\vec{j}\,\text{m/s}$$

$$\Rightarrow \quad v'_{2_t} = v_{2_t} = -0.5\,\text{m/s}, \qquad v_{2n} = 2\,\text{m/s}$$

$$I^{C_2}_{zz} = 0.2(0.6^2 + 0.3^2)/12 = 0.0075\,\text{kg.m}^2$$

ω'_1, v'_{1n}, ω'_2, v'_{2n}, are determined using $H'_{Dz} = H_{Dz}$ for rod ①, $H'_{Dz} = H_{Dz}$ for plate ②, $p'_n = p_n$ for ① + ② and the e relation, where D is a fixed point of the ground coincident with A. As in eq. (1), $H'_{Dz} = H_{Dz}$ for ① $\Rightarrow$

$$\omega'_1 = 8v'_{1n} + 30 \tag{5}$$

For plate ②, $H'_{Dz} = H_{Dz} \Rightarrow$

$$I_{zz}^{C_2}\omega'_2 - m_2 v'_{2n}(0.2) = I_{zz}^{C_2}\omega_2 - m_2 v_{2n}(0.2)$$

$$\Rightarrow \quad 0.0075\omega'_2 - 0.2v'_{2n}(0.2) = 0.0075(10) - 0.2(2)(0.2)$$

$$\Rightarrow \quad \omega'_2 = -0.66667 + 5.3333v'_{2n} \tag{6}$$

$$v_{Bn} = v_{2n} + \omega_2(0.2) = 2 + 10(0.2) = 4\,\text{m/s}$$

$$v'_{Bn} = v'_{2n} + \omega'_2(0.2) = v'_{2n} + (-0.66667 + 5.3333v'_{2n})(0.2)$$

$$= 2.0667v'_{2n} - 0.13334$$

For rod+plate, $p'_n = p_n \Rightarrow$

$$m_1 v'_{1n} + m_2 v'_{2n} = m_1 v_{1n} + m_2 v_{2n}$$

$$0.1v'_{1n} + 0.2v'_{2n} = 0.1(-4) + 0.2(2) \quad \Rightarrow \quad v'_{2n} = -0.5v'_{1n} \tag{7}$$

Using eqs (3), (4) and the e relation, $v'_{Bn} - v'_{An} = -0.4(v_{Bn} - v_{An}) \Rightarrow$

$$(2.0667v'_{2n} - 0.13334) - (2.92v'_{1n} + 7.2) = -0.4(4 + 4.48)$$

$$\Rightarrow \quad 2.0667v'_{2n} - 2.92v'_{1n} = 3.9413 \tag{8}$$

Using eqs (7) and (8) yields

$$[2.0667(-0.5) - 2.92]v'_{1n} = 3.9413 \quad \Rightarrow \quad v'_{1n} = -0.9970\,\text{m/s}$$

Equations (7) and (6) yield $v'_{2n} = 0.4985\,\text{m/s}$, $\omega'_2 = 1.992\,\text{rad/s}$.

(2) We present two solutions for angular velocity ω'_2 (Fig. E3.41e).

(i) Use $\vec{H}'_P = \vec{H}_P$ for ① + ②, where P is a fixed point coincident with their centre of mass C. $\vec{r}_{C_2} = \vec{0}$,

$$\vec{r}_{C_1} = -0.04\vec{\imath} + 0.33\vec{\jmath}, \qquad \vec{r}_C = \frac{0.1(-0.04\vec{\imath} + 0.33\vec{\jmath})}{0.1 + 0.2} = -0.013333\vec{\imath} + 0.11\vec{\jmath}$$

$$\vec{r}_{C_1C} = \vec{r}_{C_1} - \vec{r}_C = -0.02667\vec{\imath} + 0.22\vec{\jmath}, \qquad \vec{r}_{C_2C} = \vec{r}_{C_2} - \vec{r}_C = 0.01333\vec{\imath} - 0.11\vec{\jmath}$$

$$I_{zz}^C = 0.1[\tfrac{1}{12}(0.6)^2 + 0.02667^2 + 0.22^2] + 0.2[\tfrac{1}{12}(0.3^2 + 0.6^2) + 0.01333^2 + 0.11^2]$$

$$= 0.017867$$

$$\vec{v}_{C_1} = -3\vec{\imath} - 4\vec{\jmath}\,\text{m/s}, \qquad \vec{v}_{C_2} = -0.5\vec{\imath} + 2\vec{\jmath}\,\text{m/s}$$

$$\vec{H}'_P = I_{zz}^{C}\omega'_2\,\vec{k} \;=\; \vec{H}_P = I_{zz}^{C_1}\omega_1\,\vec{k} + \vec{r}_{C_1C}\times m_1\vec{v}_{C_1} + I_{zz}^{C_2}\omega_2\,\vec{k} + \vec{r}_{C_2C}\times m_2\vec{v}_{C_2}$$

$$= \frac{0.1}{12}(0.6)^2(-2\,\vec{k}) + (-0.02667\,\vec{i} + 0.22\,\vec{j}) \times 0.1(-3\,\vec{i} - 4\,\vec{j})$$

$$+\frac{0.2}{12}(0.3^2 + 0.6^2)10\,\vec{k} + (0.013333\,\vec{i} - 0.11\,\vec{j}) \times 0.2(-0.5\,\vec{i} + 2\,\vec{j})$$

$$\Rightarrow \quad 0.017867\omega'_2 \;=\; 0.14 \quad\Rightarrow\quad \omega'_2 = 7.836\,\text{rad/s}$$

(ii) ω'_2, v'_{2x}, v'_{2y} (Fig. E3.41e), are obtained using $p'_x = p_x$, $p'_y = p_y$, $H'_{Ez} = H_{Ez}$ for ① + ②, where E is a point fixed in the ground coincident with C_2.

$$\vec{v}_{C_1} \;=\; -3\,\vec{i} - 4\,\vec{j}\,\text{m/s}, \qquad \vec{v}_{C_2} = -0.5\,\vec{i} + 2\,\vec{j}\,\text{m/s}, \qquad \vec{v}'_{C_2} = v'_{2x}\,\vec{i} + v'_{2y}\,\vec{j}$$

$$v_{1x} \;=\; -3\,\text{m/s}, \qquad v_{1y} = -4\,\text{m/s}, \qquad v_{2x} = -0.5\,\text{m/s}, \qquad v_{2y} = 2\,\text{m/s}$$

$$\vec{v}'_{C_1} \;=\; \vec{v}'_{C_2} + \omega'_2\,\vec{k} \times \overrightarrow{C_2C_1} = v'_{2x}\,\vec{i} + v'_{2y}\,\vec{j} + \omega'_2\,\vec{k} \times (-0.04\,\vec{i} + 0.33\,\vec{j})$$

$$= (v'_{2x} - 0.33\omega'_2)\,\vec{i} + (v'_{2y} - 0.04\omega'_2)\,\vec{j}$$

$$v'_{1x} \;=\; v'_{2x} - 0.33\omega'_2, \qquad v'_{1y} = v'_{2y} - 0.04\omega'_2 \tag{9}$$

$p'_x = p_x$ and $p'_y = p_y$ $\Rightarrow$

$$m_1 v'_{1x} + m_2 v'_{2x} = m_1 v_{1x} + m_2 v_{2x}$$

$$0.1(v'_{2x} - 0.33\omega'_2) + 0.2v'_{2x} = 0.1(-3) + 0.2(-0.5)$$

$$\Rightarrow \quad v'_{2x} = -1.3333 + 0.11\omega'_2 \tag{10}$$

$$m_1 v'_{1y} + m_2 v'_{2y} = m_1 v_{1y} + m_2 v_{2y}$$

$$0.1(v'_{2y} - 0.04\omega'_2) + 0.2v'_{2y} = 0.1(-4) + 0.2(2)$$

$$\Rightarrow \quad v'_{2y} = 0.013333\omega'_2 \tag{11}$$

Equations (9) to (11) yield

$$\vec{v}'_{C_1} = (-1.3333 - 0.22\omega'_2)\,\vec{i} - 0.026667\omega'_2\,\vec{j}$$

For ① + ②, $H'_{Ez} = H_{Ez}$ $\Rightarrow$

$$I_{zz}^{C_2}\omega'_2\,\vec{k} + [I_{zz}^{C_1}\omega'_2\,\vec{k} + \overrightarrow{C_2C_1} \times m_2\vec{v}'_{C_1}] = I_{zz}^{C_2}\omega_2\,\vec{k} + [I_{zz}^{C_1}\omega_1\,\vec{k} + \overrightarrow{C_2C_1} \times m_2\vec{v}_{C_1}]$$

$$\Rightarrow \quad 0.0075\omega'_2 + 0.003\omega'_2 + 0.1[-0.04(-0.026667\omega'_2) - 0.33(-1.3333 - 0.22\omega'_2)]$$

$$= 0.0075(10) + 0.003(-2) + 0.1[-0.04(-4) - 0.33(-3)]$$

$$\Rightarrow \quad \omega_2' = 7.836\,\text{rad/s}$$

Obviously solution method (i) is direct and shorter.

(c) Since the plate is massive, $\omega_2' = \omega_2 \Rightarrow v_{Bn}' = v_{Bn} = 0.4(10) = 4\,\text{m/s}$. ω_1', v_{1n}', are determined from $H_{D_z}' = H_{D_z}$ for ① and the e relation, where D is a point fixed on the ground and coincident with A. As in eq. (1), $H_{D_z}' = H_{D_z}$ for ① $\Rightarrow$

$$\omega_1' = 8v_{1n}' + 30 \tag{12}$$

Using eqs (3), (4) and the e relation: $v_{Bn}' - v_{An}' = -0.4(v_{Bn} - v_{An}) \Rightarrow$

$$4 - (2.92v_{1n}' + 7.2) = -0.4(4 + 4.48) \quad \Rightarrow \quad v_{1n}' = 0.06575\,\text{m/s}$$

and eq. (12) $\Rightarrow \quad \omega_1' = 30.526\,\text{rad/s}$.

Note: I_{zz}^C can be obtained by the following general result. Given two bodies with mass m_i, centres of mass $C_i(x_i, y_i, 0)$ with principal moments of inertia $I_{zz}^{C_i}$, then

$$I_{zz}^C = I_{zz}^{C_1} + I_{zz}^{C_2} + (C_1C_2)^2 m_1 m_2/(m_1 + m_2)$$

$$C: \quad (m_1x_1 + m_2x_2,\; m_1y_1 + m_2y_2,\; 0)/(m_1 + m_2)$$

$$CC_1: \quad (x_1 - x_2,\; y_1 - y_2,\; 0)m_2/(m_1 + m_2)$$

$$CC_2: \quad (x_2 - x_1,\; y_2 - y_1,\; 0)m_1/(m_1 + m_2)$$

$$I_{zz}^C = I_{zz}^{C_1} + I_{zz}^{C_2} + [(x_1 - x_2)^2 + (y_1 - y_2)^2](m_1m_2^2 + m_2m_1^2)/(m_1 + m_2)^2$$

$$= I_{zz}^{C_1} + I_{zz}^{C_2} + (C_1C_2)^2 m_1 m_2/(m_1 + m_2)$$

Example 3.42: Rod ① rotating at ω hits rod ② at rest (Fig. E3.42a). Find ω_1' if the impact is smooth with coefficient of restitution e.

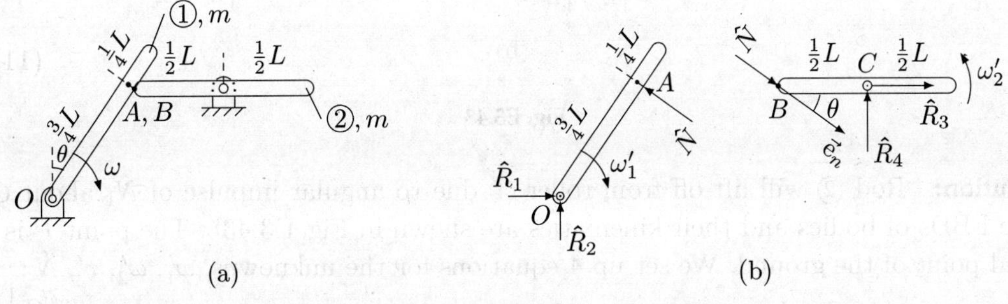

Fig. E3.42

Solution: The FBDs are drawn in Fig. E3.42b. Let ω_1', ω_2' be the angular velocities just after impact. The governing equations for ω_1', ω_2', $\hat{N}$ are:

$$
\begin{aligned}
①: \quad I_{\text{ang}_{O_z}} &= \Delta H_{Oz}, & -\hat{N}\left(\tfrac{3}{4}L\right) &= \tfrac{1}{3}mL^2(\omega_1' - \omega) \\
②: \quad I_{\text{ang}_{C_z}} &= \Delta H_{Cz}, & \hat{N}\sin\theta\left(\tfrac{1}{2}L\right) &= \tfrac{1}{12}mL^2\omega_2' \\
v'_{An} - v'_{Bn} &= -e(v_{An} - v_{Bn}): & \tfrac{3}{4}L\omega_1' - \omega_2'\left(\tfrac{1}{2}L\right)\sin\theta &= -\tfrac{3}{4}eL\omega \\
\Rightarrow \quad \omega_1' &= (16\sin^2\theta - 9e)\omega/(9 + 16\sin^2\theta)
\end{aligned}
$$

Example 3.43: Rework Example 3.42 if rod ② rests on two rollers (Fig. E3.43a). Assume all contacts (except between ① and ②) are sufficiently rough to prevent slip.

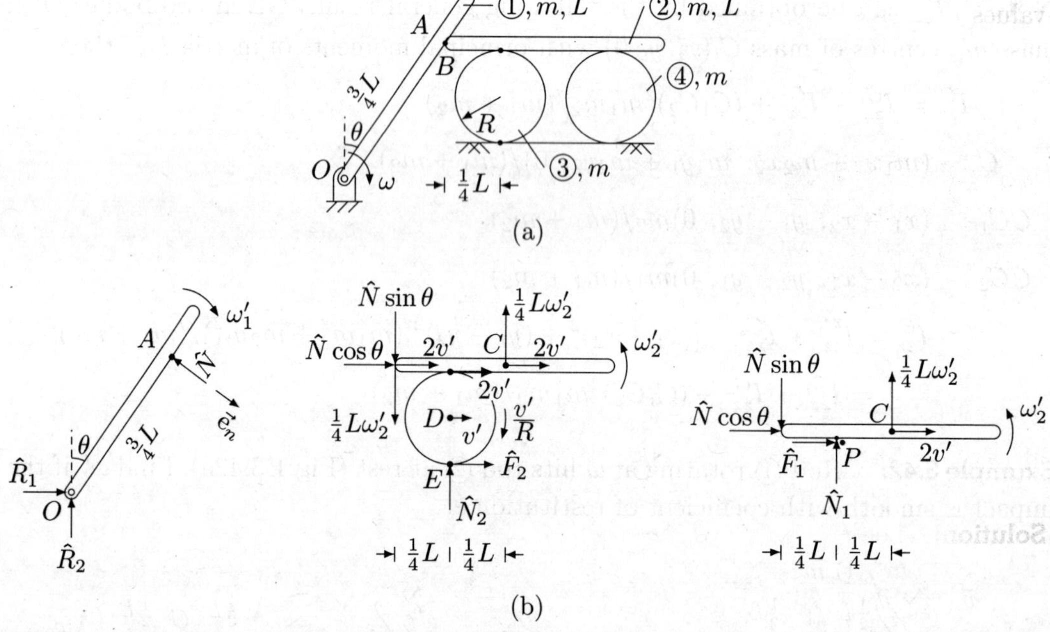

Fig. E3.43

Solution: Rod ② will lift-off from roller ④ due to angular impulse of $\hat{N}_1$ about C. The FBDs of bodies and their kinematics are shown in Fig. E3.43b. The point P is a fixed point of the ground. We set up 4 equations for the unknowns, ω_1', ω_2', v', $\hat{N}$:

$$①: \quad I_{\text{ang}_{O_z}} = \Delta H_{Oz}: \qquad -\hat{N}\left(\tfrac{3}{4}L\right) = \tfrac{1}{3}mL^2(\omega_1' - \omega) \tag{1}$$

$$②: \quad I_{\text{ang}_{P_z}} = \Delta H_{Pz}: \qquad \hat{N}\sin\theta\left(\tfrac{1}{4}L\right) = \tfrac{1}{12}mL^2\omega_2' + m\left(\tfrac{1}{4}L\omega_2'\right)\tfrac{1}{4}L \tag{2}$$

② + ③ : $I_{\text{ang}_{Ez}} = \Delta H_{Ez}$:

$$\hat{N}\sin\theta(\tfrac{1}{4}L) - \hat{N}\cos\theta(2R) = -\tfrac{3}{2}mR^2\cdot\frac{v'}{R} + \tfrac{1}{12}mL^2\omega_2' + m(\tfrac{1}{4}L\omega_2')\tfrac{1}{4}L - m(2v')2R \quad (3)$$

$$v_{An}' - v_{Bn}' = -e(v_{An} - v_{Bn}) : \qquad \omega_1'(\tfrac{3}{4}L) - \omega_2'(\tfrac{1}{4}L\sin\theta) - 2v'\cos\theta = -e(\tfrac{3}{4}L\omega) \quad (4)$$

Equations (1)–(4) can be solved for ω_1', ω_2', v', $\hat{N}$ to yield

$$\omega_1' = (16\sin^2\theta/21 + 128\cos^2\theta/99 - 3e)\omega/(16\sin^2\theta/21 + 128\cos^2\theta/99 + 3)$$

Example 3.44: A *super-ball* is a hard spherical rubber ball whose impact on a hard surface is almost elastic (idealised as conservation of kinetic energy) and essentially non-smooth at the contact point. Let such a ball bounce on the xy plane with $\vec{v}_C = v_1\,\vec{\mathrm{i}} + v_2\,\vec{\mathrm{j}} + v_3\,\vec{\mathrm{k}}$, $\vec{\omega} = \omega_1\,\vec{\mathrm{i}} + \omega_2\,\vec{\mathrm{j}} + \omega_3\,\vec{\mathrm{k}}$ (Fig. E3.44a). Set up equations to find the values v_i', ω_i' just after impact.

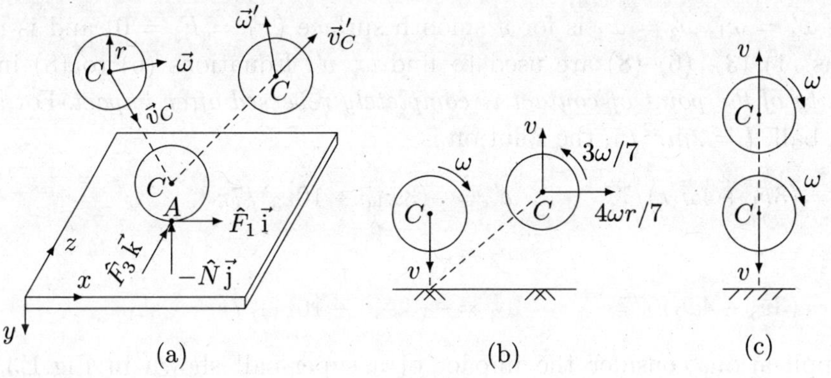

Fig. E3.44

Solution: Let $I_{xx}^C = I_{yy}^C = I_{zz}^C = I$.

$$\vec{v}_A = \vec{v}_C + \vec{\omega}\times r\,\vec{\mathrm{j}} = (v_1 - \omega_3 r)\,\vec{\mathrm{i}} + v_2\,\vec{\mathrm{j}} + (v_3 + \omega_1 r)\,\vec{\mathrm{k}}$$

$$\vec{v}_A' = (v_1' - \omega_3' r)\,\vec{\mathrm{i}} + v_2'\,\vec{\mathrm{j}} + (v_3' + \omega_1' r)\,\vec{\mathrm{k}}$$

$\vec{I} = \Delta\vec{p}$ and $\vec{I}_{\text{ang}_C} = r\,\vec{\mathrm{j}}\times(\hat{F}_1\,\vec{\mathrm{i}} + \hat{F}_3\,\vec{\mathrm{k}}) = r\hat{F}_3\,\vec{\mathrm{i}} - r\hat{F}_1\,\vec{\mathrm{k}} = \Delta\vec{H}_C \;\Rightarrow$

$$m(v_1' - v_1) = \hat{F}_1, \qquad m(v_2' - v_2) = -\hat{N}, \qquad m(v_3' - v_3) = \hat{F}_3$$

$$I(\omega_1' - \omega_1) = r\hat{F}_3 \quad\Rightarrow\quad I(\omega_1' - \omega_1) = mr(v_3' - v_3) \quad (1)$$

$$I(\omega_2' - \omega_2) = 0 \quad\Rightarrow\quad \omega_2' = \omega_2 \quad (2)$$

$$I(\omega_3' - \omega_3) = -r\hat{F}_1 \quad\Rightarrow\quad I(\omega_3' - \omega_3) = -mr(v_1' - v_1) \quad (3)$$

$\hat{F}_1$ alters v_1, ω_3; $\hat{F}_3$ alters v_3, ω_1; $\hat{N}$ alters v_2. *Since $\hat{F}_1$, $\hat{F}_2$, $\hat{N}$ are independent and the kinetic energy is conserved, it follows that the contributions to T due to v_1 and ω_3, v_3 and ω_1, v_2 are separately conserved*:

$$mv_1'^2 + \tfrac{1}{2}I\omega_3'^2 = \tfrac{1}{2}mv_1^2 + \tfrac{1}{2}I\omega_3^2 \quad \Rightarrow \quad I(\omega_3'^2 - \omega_3^2) = m(v_1^2 - v_1'^2) \tag{4}$$

$$\tfrac{1}{2}mv_3'^2 + \tfrac{1}{2}I\omega_1'^2 = \tfrac{1}{2}mv_3^2 + \tfrac{1}{2}I\omega_1^2 \quad \Rightarrow \quad I(\omega_1'^2 - \omega_1^2) = m(v_3^2 - v_3'^2) \tag{5}$$

$$\tfrac{1}{2}mv_2'^2 = \tfrac{1}{2}mv_2^2 \quad \Rightarrow \quad v_2' = -v_2 \tag{6}$$

Equations (1) and (5) $\Rightarrow$

$$(\omega_1' + \omega_1)r = -(v_3' + v_3) \qquad \text{provided} \quad \omega_1' \neq \omega_1 \tag{7}$$

Equations (2) and (4) $\Rightarrow$

$$(\omega_3' + \omega_3)r = v_1' + v_1 \qquad \text{provided} \quad \omega_3' \neq \omega_3 \tag{8}$$

The case $\omega_1' = \omega_1$, $\omega_3' = \omega_3$ is for a smooth surface ($\hat{F}_1 = \hat{F}_3 = 0$) and is ruled out. Equations (1)–(3), (6)–(8) are used to find ω_i', v_i'. Equations (6) to (8) imply that the *velocity of the point of contact is completely reversed after impact.* For a uniform spherical ball, $I = 2mr^2/5$, the solution is

$$\begin{aligned} v_1' &= (3v_1 + 4\omega_3 r)/7, & \omega_1' &= -(3\omega_1 r + 10v_3)/7r \\ v_2' &= -v_2, & \omega_2' &= \omega_2 \\ v_3' &= (3v_3 - 4\omega_1 r)/7, & \omega_3' &= -(3\omega_3 r - 10v_1)/7r \end{aligned} \tag{9}$$

As an application, consider the impact of a super-ball shown in Fig. E3.44b with $v_1 = v_3 = 0$, $v_2 = v$, $\omega_1 = \omega_2 = 0$, $\omega_3 = \omega$. Equations (9) yield $\omega_1' = \omega_2' = 0$, $v_3' = 0$, $v_2' = -v_2$, $v_1' = 4\omega r/7$, $\omega_3' = -3\omega/7$. Unlike the case of a smooth ball (Fig. E3.44c), the super-ball does not bounce back vertically and there is increase in the speed of C and decrease in the angular velocity.

Example 3.45: (a) The vehicle ① is released from rest at the position shown, on another vehicle ② at rest (Fig. E3.45a). They have mass m_1, m_2 with $m_2 = 4m_1$. Model the wheels to be rigid and light. Find the position and velocity of vehicle ② just before impact at B. Find the velocity of vehicle ② just after impact and the impulsive reaction from the ground during impact. The coefficient of restitution is 0.6. Assume that just after impact, main frame of vehicle ① translates with respect to main frame of ② and main frame of ② translates with reference to ground. $\cos\theta = 0.8$. (b) Rework part (a) if vehicle ① is replaced by a spherical ball of mass m_1 and radius r. The coefficient of friction for all surfaces is 0.3. Show that before impact, the ball rolls without slip.

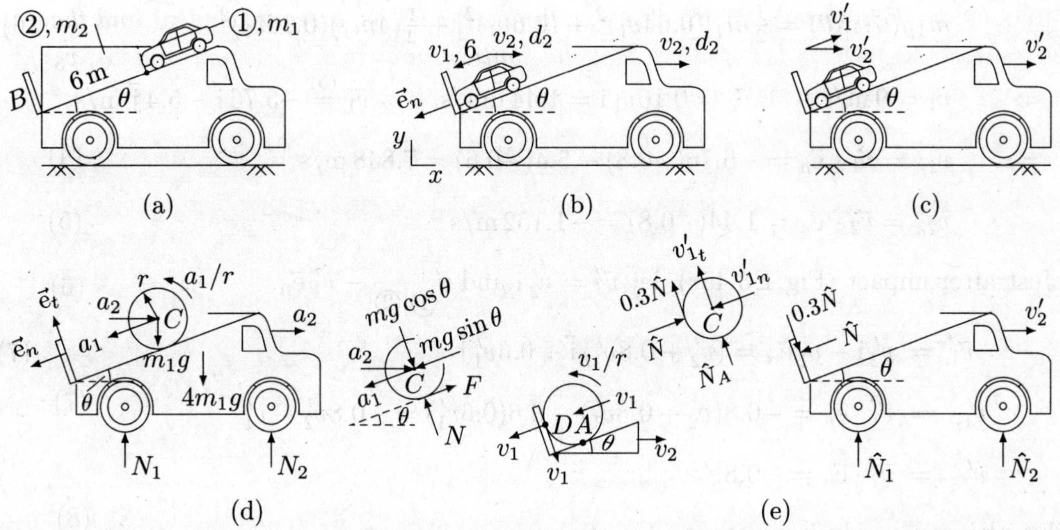

Fig. E3.45

Solution: $\sin\theta = 0.6$. The friction force on the light wheels is zero. Line of impact is

$$\vec{e}_n = -\cos\theta\,\vec{i} - \sin\theta\,\vec{j} = -0.8\,\vec{i} - 0.6\,\vec{j}$$

Just before impact, let $\vec{v}_{①|②} = v_1\,\vec{e}_n$, velocity of ② be $\vec{v}_2 = v_2\,\vec{i}$ (Fig. E3.45b), and displacement of ② be $\vec{d}_2 = d_2\,\vec{i}$. The velocity $\vec{v}_1$ and displacement $\vec{d}_1$ of ① are:

$$\vec{v}_1 = v_2\,\vec{i} + v_1(-0.8\,\vec{i} - 0.6\,\vec{j}) = (v_2 - 0.8v_1)\,\vec{i} - 0.6v_1\,\vec{j} \tag{1}$$

$$\vec{d}_1 = d_2\,\vec{i} + 6\,\vec{e}_n = (d_2 - 4.8)\,\vec{i} - 3.6\,\vec{j}\,\text{m}$$

For the system of two bodies, since $F_x \equiv 0$:

$$p_x = m_1 v_{1x} + m_2 v_{2x} \equiv \text{constant} \equiv 0 \tag{2}$$

$$\Rightarrow \quad m_1 d_{1x} + m_2 d_{2x} = 0 \tag{3}$$

Equations (3) and (2) $\Rightarrow$

$$m_1(d_2 - 4.8) + 4m_1 d_2 = 0 \quad \Rightarrow \quad d_2 = 0.96\,\text{m} \tag{4}$$

$$m_1(v_2 - 0.8v_1) + 4m_1 v_2 = 0 \quad \Rightarrow \quad v_2 = 0.16v_1$$

$$\vec{v}_1 = -0.64v_1\,\vec{i} - 0.6v_1\,\vec{j} \tag{5}$$

Conservation of mechanical energy $T_i + V_i = T_f + V_f$ for the system yields:

$$m_1 g(6\sin\theta) = \tfrac{1}{2}m_1[(0.64v_1)^2 + (0.6v_1)^2] + \tfrac{1}{2}(4m_1)(0.16v_1)^2 \tag{6}$$

$$\Rightarrow \quad v_1 = 9\,\text{m/s}, \qquad \vec{v}_2 = 0.16v_1\,\vec{\imath} = 1.44\,\vec{\imath}\,\text{m/s}, \qquad \vec{v}_1 \overset{(5)}{=} -5.76\,\vec{\imath} - 5.4\,\vec{\jmath}\,\text{m/s}$$

$$\Rightarrow \quad v_{1n} = \vec{v}_1 \cdot \vec{e}_n = -5.76(-0.8) - 5.4(-0.6) = 7.848\,\text{m/s}$$

$$v_{2n} = \vec{v}_2 \cdot \vec{e}_n = 1.44(-0.8) = -1.152\,\text{m/s}$$

Just after impact (Fig. E3.45c), let $\vec{v}_2' = v_2'\,\vec{\imath}$ and $\vec{v}_{\text{①|②}} \; - v_1'\,\vec{e}_n$.

$$\begin{aligned} \vec{v}_1' &= v_2'\,\vec{\imath} - v_1'\,\vec{e}_n = (v_2' + 0.8v_1')\,\vec{\imath} + 0.6v_1'\,\vec{\jmath} && (7) \\ v_{1n}' &= \vec{v}_1' \cdot \vec{e}_n = -0.8(v_2' + 0.8v_1') - 0.6(0.6v_1') = -0.8v_2' - v_1' \\ v_{2n}' &= \vec{v}_2' \cdot \vec{e}_n = -0.8v_2' \end{aligned}$$

For the system, during impact: $I_x = 0 \;\Rightarrow$

$$p_x = m_1(v_2' + 0.8v_1') + 4m_1 v_2' = 0 \qquad \Rightarrow \qquad v_2' = -0.16v_1'$$

The coefficient of restitution equation: $v_{2n}' - v_{1n}' = -e(v_{2n} - v_{1n})$ yields

$$-0.8v_2' - (-0.8v_2' - v_1') = -0.6(-1.152 - 7.848) \quad \Rightarrow \quad v_1' = 5.4\,\text{m/s}$$

$$v_2' = -0.16v_1' = -0.864\,\text{m/s}$$

$$\vec{v}_1' \overset{(7)}{=} 3.456\,\vec{\imath} + 3.24\,\vec{\jmath}\,\text{m/s}, \qquad \vec{v}_2' = -0.864\,\vec{\imath}\,\text{m/s}$$

The impulsive force on the system is the impulsive reaction $\hat{N}_y$ from the ground:

$$\begin{aligned} \hat{N}_y &= \Delta p_y = m_1(v_{1y}' - v_{1y}) + 4m_1(v_{2y}' - v_{2y}) \\ &= m_1[3.24 - (-5.4)] = 8.64m_1 \end{aligned}$$

(b) FBDs for sphere rolling without slip are shown in Fig. E3.45d. For the system:

$$F_x = 0 \quad \Rightarrow \quad m_1(a_2 - a_1\cos\theta) + 4m_1 a_2 = 0 \quad \Rightarrow \quad a_2 = 0.16a_1$$

For the sphere:

$$\begin{aligned} m_1 g\sin\theta - F &= m_1(a_1 - a_2\cos\theta) = m_1(1 - 0.16 \times 0.8)a_1 \\ \Rightarrow \quad F &= m_1(0.6g - 0.872a_1) \\ Fr &= \frac{2}{5}m_1 r^2\left(\frac{a_1}{r}\right) \quad \Rightarrow \quad F = 0.4m_1 a_1 = m_1(0.6g - 0.872a_1) \\ \Rightarrow \quad a_1 &= 4.627\,\text{m/s}^2, \qquad F = 0.4m_1 a_1 = 1.851m_1 \end{aligned}$$

$$N - m_1 g\cos\theta \;=\; m_1(-a_2\sin\theta) = -m_1(0.16a_1)0.6 \quad\Rightarrow\quad N = 7.404m_1$$

$$\frac{F}{N} = \frac{1.851m_1}{7.404m_1} = 0.2500 < \mu = 0.3 \qquad \text{(o.k.)}$$

We obtain $v_1 = (2a_1 \times 6)^{1/2} = 7.451\,\text{m/s}, \quad v_2 = 0.16v_1 = 1.192\,\text{m/s}$.

Equations (1) to (5) are valid and $\omega_1 = v_1/r$. v_1 can also be obtained from $T+V=$ const., which replaces eq. (6):

$$m_1 g(6\sin\theta) = \tfrac{1}{2}m_1[(0.64v_1)^2 + (0.6v_1)^2] + \tfrac{1}{2}\left(\frac{2}{5}m_1 r^2\right)\left(\frac{v_1}{r}\right)^2 + \tfrac{1}{2}(4m_1)(0.16v_1)^2$$

$$\Rightarrow \quad v_1 = 7.451\,\text{m/s}, \quad \vec{v}_2 = 1.192\,\vec{\text{i}}\,\text{m/s}$$

$$\vec{v}_1 \overset{(1)}{=} -4.769\,\vec{\text{i}} - 4.471\,\vec{\text{j}}\,\text{m/s}$$

$$v_{1n} = \vec{v}_1 \cdot \vec{\text{e}}_n = -4.769(-0.8) - 4.471(-0.6) = 6.498\,\text{m/s}$$

$$v_{1_t} = \vec{v}_1 \cdot (-\sin\theta\,\vec{\text{i}} + \cos\theta\,\vec{\text{j}})$$

$$= -4.769(-0.6) - 4.471 \times 0.8 = -0.7154\,\text{m/s}$$

$$v_{2n} = \vec{v}_2 \cdot \vec{\text{e}}_n = 1.192(-0.8) = -0.9536\,\text{m/s}$$

There is initial slip at impact point D (Fig. E3.45e). Assuming slip at the end of impact too, the impulsive reaction at D is $-\hat{N}\,\vec{\text{e}}_n + 0.3\hat{N}\,\vec{\text{e}}_t$. Impulse $0.3\hat{N}\,\vec{\text{e}}_t$ imparts velocity component to C in direction $\vec{\text{e}}_t$ and hence we assume there is separation at A so that $\hat{N}_A = 0$. Let $\vec{v}_2' = v_2'\,\vec{\text{i}}, \quad \vec{v}_1' = v_{1_n}'\,\vec{\text{e}}_n + v_{1_t}'\,\vec{\text{e}}_t, \quad \vec{\omega}_1' = \omega'\,\vec{\text{k}}$. For the sphere:

$$m_1(v_{1_n}' - v_{1n}) = -\hat{N} \qquad\Rightarrow\qquad v_{1_n}' = 6.498 - \frac{\hat{N}}{m_1} \tag{8}$$

$$m_1(v_{1_t}' - v_{1_t}) = 0.3\hat{N} \qquad\Rightarrow\qquad v_{1_t}' = \frac{0.3\hat{N}}{m_1} - 0.7154 \tag{9}$$

For vehicle 2: $\quad I_x = -\hat{N}\cos\theta + 0.3\hat{N}\sin\theta = 4m_1(v_2' - v_2)$

$$\Rightarrow \qquad v_2' = 1.192 - \frac{0.155\hat{N}}{m_1} \tag{10}$$

With $v_{2n}' = \vec{v}_2' \cdot \vec{\text{e}}_n = -0.8v_2'$, the e relation: $v_{2n}' - v_{1n}' = -e(v_{2n} - v_{1n})$ yields

$$-0.8v_2' - v_{1_n}' = -0.6(-0.9536 - 6.498) = 4.471 \tag{11}$$

Substituting from (8) to (10) in (11) yields:

$$-0.8\left(1.192 - \frac{0.155\hat{N}}{m_1}\right) + \frac{\hat{N}}{m_1} - 6.498 = 4.471$$

$$\Rightarrow \quad \hat{N} = 10.607m_1$$

$$(8), (9), (10) \Rightarrow \quad v'_{1_n} = -4.109, \quad v'_{1_t} = 2.467, \quad v'_2 = -0.4521\,\text{m/s}$$

For sphere, $(I_{\text{ang}_C})_z = I^C_{zz}\Delta\omega$:

$$\frac{2}{5}m_1r^2\left(\omega' - \frac{v_1}{r}\right) = -r \times 0.3\hat{N} = -0.3r\,(10.607m_1)$$

$$\Rightarrow \quad \omega' = -0.5043/r$$

$$v'_{2_t} = -v'_2\sin\theta = 0.2713\,\text{m/s} \neq v'_{1_t} \quad \Rightarrow \quad \text{separation at } A \quad \text{(o.k.)}$$

$$\frac{v'_{1_t} - v'_{2_t}}{r} = \frac{2.196}{r} \neq \omega' = -\frac{0.5043}{r} \quad \Rightarrow \quad \text{slip at } D \quad \text{(o.k.)}$$

Hence, the assumptions are correct. The impulsive ground reaction is:

$$\hat{N}_y = \hat{N}\sin\theta + 0.3\hat{N}\cos\theta = 0.84\hat{N} = 0.84(10.607m_1) = 8.901m_1$$

Review 3 C

1. Impact of unconstrained rigid bodies with point A of body ① making impact with point B of body ②. Consider a fixed point O coincident with A and B. The fifteen unknowns, $\vec{\omega}'_i$, $\vec{v}'_i$ and $\hat{\vec{R}}$, are obtained using:

   for ①: $m_1(\vec{v}'_1 - \vec{v}_1) = -\hat{\vec{R}}$, $\quad I^{C_1}_{ij}(\omega'_{1_j} - \omega_{1_j})\,\vec{e}_i - \vec{r}_1 \times m_1(\vec{v}'_1 - \vec{v}_1) = \vec{0}$

   or $\quad \vec{r}_1 \times (-\hat{\vec{R}}) = I^{C_1}_{ij}(\omega'_{1_j} - \omega_{1_j})\,\vec{e}_i$

   for ②: $m_2(\vec{v}'_2 - \vec{v}_2) = \hat{\vec{R}}$, $\quad I^{C_2}_{ij}(\omega'_{2_j} - \omega_{2_j})\,\vec{e}_i - \vec{r}_2 \times m_2(\vec{v}'_2 - \vec{v}_2) = \vec{0}$

   or $\quad \vec{r}_2 \times \hat{\vec{R}} = I^{C_2}_{ij}(\omega'_{2_j} - \omega_{2_j})\,\vec{e}_i$

   e relation: $\quad v'_{Bn} - v'_{An} = -e(v_{Bn} - v_{An})$

   a. smooth surfaces: $\hat{R}_t = 0$, $\hat{R}_b = 0$.
   b. rough surfaces and no slip just before end of impact: $v'_{A_t} = v'_{B_t}$, $v'_{A_b} = v'_{B_b}$
   c. rough surfaces and slip continues during impact: $\quad \hat{\vec{R}} = \hat{N}\,\vec{e}_n - \mu_k\hat{N}\,\vec{e}$
      with $\quad \vec{e} = [(\vec{v}_B - \vec{v}_A) - (\vec{v}_B - \vec{v}_A)\cdot\vec{e}_n]\,/\,|(\vec{v}_B - \vec{v}_A) - (\vec{v}_B - \vec{v}_A)\cdot\vec{e}_n|$

2. $e = 1$: elastic, noiseless, $v_s = v_a$, $T' = T$.
3. $e = 0$: plastic, $v_s = 0$, T'/T is the least, $v'_{An} = v'_{Bn}$ but $\vec{v}'_A$ may not equal $\vec{v}'_B$.
4. For smooth impact of unconstrained rigid bodies: $v'_{i_t} = v_{i_t}$, $v'_{i_b} = v_{i_b}$, and the eight unknowns, $\vec{\omega}'_i$, v'_{in}, are obtained from: $\quad \vec{H}'_O = \vec{H}_O$ for ①,

   $\vec{H}'_O = \vec{H}_O$ for ②, $\quad m_1v'_{1n} + m_2v'_{2n} = m_1v_{1n} + m_2v_{2n}$ and the e relation.

5. For smooth impact of unconstrained rigid bodies in plane motion: $v'_{it} = v_{it}$ and the four unknowns, ω'_i, v'_{in}, are obtained from: the e relation,
$$H'_{Oz} = H'_{Oz} \text{ for ①}, \quad H'_{Oz} = H'_{Oz} \text{ for ②}, \quad m_1 v'_{1n} + m_2 v'_{2n} = m_1 v_{1n} + m_2 v_{2n}$$
6. For central smooth impact of unconstrained rigid bodies in plane motion: $\omega'_i = \omega_i$, $v'_{it} = v_{it}$ and the two unknowns, v'_{in}, are obtained from
$$m_1 v'_{1n} + m_2 v'_{2n} = m_1 v_{1n} + m_2 v_{2n} \text{ and the } e \text{ relation.}$$
7. If body ② is massive, its kinematics remains the same; $\vec{p}$–$\vec{I}$ and $\vec{H}$–$\vec{I}_{\text{ang}}$ relations for ② alone and for the system ①+②, do not yield useful information. The $\vec{p}$–$\vec{I}$ and $\vec{H}$–$\vec{I}_{\text{ang}}$ relations for ① and the e relation yield the required equations.

3.16 Concept Review Questions 3C

37. A bent rod translates down with speed v and hits the edge of a table at A. Find its angular velocity just after smooth impact (Fig. Q3.37) with coefficient of restitution e. $[-0.4972(0.6\vec{\imath} + \vec{\jmath})(1+e)v/a]$

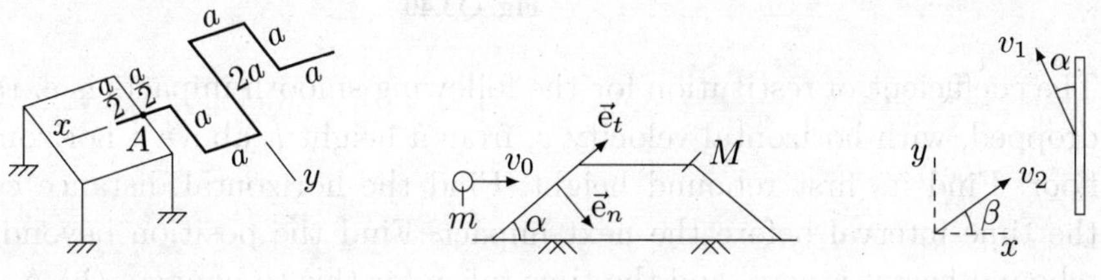

Fig. Q3.37 Fig. Q3.38 Fig. Q3.39

38. (a) A ball hits a stationary block with smooth surfaces (Fig. Q3.38). The coefficient of restitution is e. Find the velocity of the block just after impact if it translates. (b) Rework if block M is not on a rigid surface but on foam rubber. $[(\text{a})\ (1+e)v_0/(1+M/m \sin^2\alpha) \quad (\text{b})\ [(1+e)mv_0 \sin\alpha/(M+m)]\,\vec{e}_n]$
39. A thin plate of area A is moving in a rarefied atmosphere where the wind velocity is v_2 (Fig. Q3.39). The number of molecules (each of mass m) per unit volume is n. The coefficient of restitution is e. Find the drag force on the plate. Disregard the thermal motion of the molecules. $[(1+e)mnA(v_2\cos\beta + v_1 \sin\alpha)^2\,\vec{\imath}\,]$
40. (a) A ball moving at $v_0 = 10v_1$ hits an escalator (Fig. Q3.40a). The coefficient of restitution is e. (1) For smooth impact, find the rebound velocity component along the normal to the step. (2) If $\alpha = \beta = \tan^{-1} 0.75$ and $e = 0.6$, find ω of the ball just after impact for (i) $\mu = 0.1$ and (ii) $\mu = 0.3$. (b) A simple pendulum is released from the given position (Fig. Q3.40b). Find the velocity of the bob just after the thread becomes taut (i) if subsequently the thread remains taut, and (ii) if the kinetic energy is conserved. (c) A smooth billiard ball collides two adjacent sides of a billiard table once. Find e if its final velocity is parallel to the

initial velocity. (d) In which direction θ should a cue ball A be hit so that after its impact on ball B, B goes to the centre of the pocket C (Fig. Q3.40c) for $\alpha = 110°$? What is the least value of α for which the desired result can be achieved? (e) A hemispherical bowl attached to a string is released in the given position (Fig. Q3.40d). Find its angular velocity just after the thread becomes taut (i) if subsequently the thread remains taut, and (ii) if the kinetic energy is conserved.

[(a) (1) $v'_{1_n} = v_1 \sin\alpha + e(v_0 \sin\beta + v_1 \sin\alpha)$ (2) (i) $2.64\mu v_1/r$ (ii) $5.143v_1/r$
(b) (i), (ii) $(0.5,\ 1)(\sqrt{3gL})^{1/2}$ at 30° downward and upward to horizontal
(c) $\forall e$ (d) 21.41°, 101.54° (e) (i) $0.6702(\sqrt{3gL})^{1/2}/R$ (ii) $1.340(\sqrt{3gL})^{1/2}/R$]

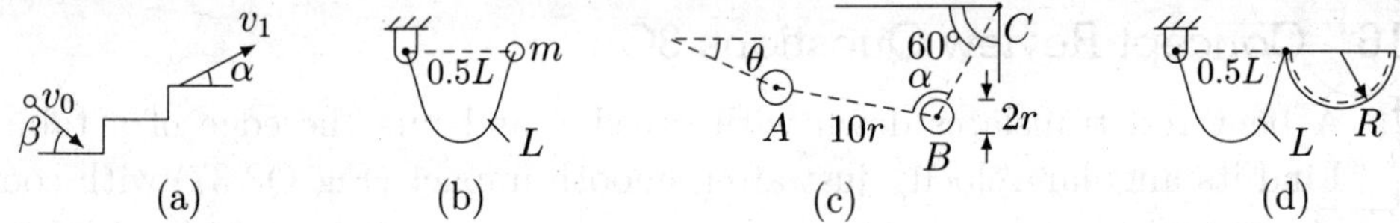

Fig. Q3.40

41. The coefficient of restitution for the following smooth impacts is e. (a) A ball is dropped, with horizontal velocity v, from a height h above a horizontal smooth floor. Find its first rebound height. Find the horizontal distance covered and the time interval before the next impact. Find the position beyond which the rebound height is zero, and the time taken for this to happen. (b) A pile driving hammer of mass m_1 impacts a pile of mass m_2 with speed v_0. Assume $e = 0$. The resistance force of the ground on the pile for displacement x is $a + bx$. Find the vertical distance through which the pile penetrates the ground. (c) A bumper protects a car from damage when it hits a rigid wall at speeds up to v_0. Find the speed at which it can hit another similar car moving at n times its speed in opposite direction without either incurring damage.

[(a) $e^2h,\ 2ev\sqrt{2h/g},\ 2e\sqrt{2h/g},\ [(1+e)/(1-e)]\sqrt{2h/g}(v,\ 1)$
(b) $[(m_1+m_2)g - a + [\{(m_1+m_2)g - a\}^2 + bm_1^2v_0^2/(m_1+m_2)]^{1/2}]/b$ (c) $\leq \frac{2v_0}{n+1}$]

42. A small ball lies above a bigger ball (Fig. Q3.42) ($M \gg m$). These fall one after the other on a smooth floor. Find the rebound height of m for smooth impact with coefficient of restitution e. . [$e^2(2+e)^2h$]

43. Three identical smooth discs are on a smooth horizontal plane. Before impact, ②, ③ are at rest (Fig. Q3.43) and ① is translating at v_0. (a) Discs ② and ③ are connected by an inextensible rope. The coefficient of restitution is e. Find the velocity of disc ③ after impact. (b) Rework for an extensible cable.

[(a) $(1+e)v_0\cos^2\theta/(3+\sin^2\theta)$ (b) 0]

44. Vehicle ① hits a stationary trolley ② which has a vehicle ③ resting on top (Fig. Q3.44). Subsequently trolley ② hits vehicle ④ at rest. All wheels are light. The coefficient of friction μ for all surfaces is 0.4 and the coefficient of restitution $e = 0.8$. Find the final velocities of all after impacts for (a) $d = 6$ m, and (b) $d = 10$ m. Consider two cases of wheels of vehicle 3 as (1) locked, and (2) unlocked. Assume that the trolley is long enough that vehicle ③ does not roll/slide off.
[(a) (1) (−8.8, 3.937, 3.937, 9.115) m/s (2) (−8.8, 4.608, 0, 10.37) m/s
(b) (1) (−8.8, 4.101, 4.101, 8.294) m/s (2) (−8.8, 4.608, 0, 10.37) m/s]

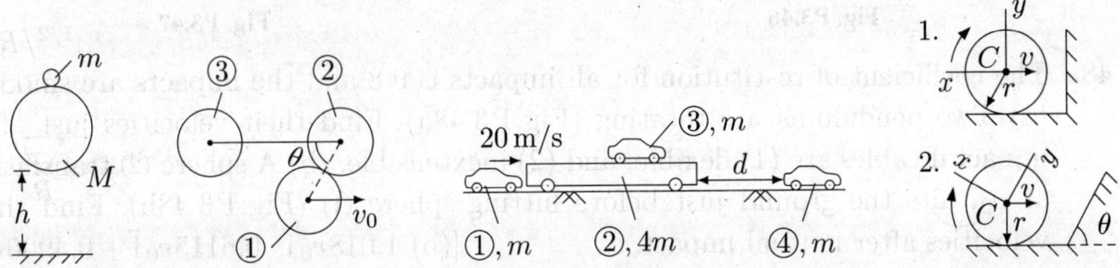

Fig. Q3.42 Fig. Q3.43 Fig. Q3.44 Fig. Q3.45

45. A sphere rolls without slip (Fig. Q3.45) with C moving at speed v just before impact. Find the angular velocity of the sphere and the velocity of C just after impact for the two cases. The coefficients of friction and restitution are μ and e.
[(1) If $\mu < 2/7(1+e)$, then $[1-2.5\mu(1+e)]v/r$, $ev\vec{\imath}+(1+e)\mu v\vec{\jmath}$, else: $\frac{2v}{7r}$, $ev\vec{\imath}+\frac{2v}{7}\vec{\jmath}$
(2) If $\mu \geq 2(1-\cos\theta)/7(1+e)\sin\theta$, then $(2+5\cos\theta)v/7r$, $ev\sin\theta\,\vec{\imath}+\frac{1}{7}(2+5\cos\theta)v\vec{\jmath}$
else: $[1-2.5\mu(1+e)\sin\theta]v/r$, $ev\sin\theta\,\vec{\imath}+[(1+e)\mu\sin\theta+\cos\theta]v\vec{\jmath}$]

3.17 Practice Problems 3C

46. The motion of 2 discs on a smooth horizontal table just before impact is shown in Fig. P3.46. Find their final velocities, loss in kinetic energy and the impulsive action between them if (1) the impact is smooth with coefficient of restitution $e = 0.5$, (2) $e = 0$, (3) the discs get locked together and become a single body after impact. [(1)$4\vec{e}_t - 2.5\vec{e}_n$ m/s, $6\vec{e}_t + 3\vec{e}_n$ m/s, 3.025 J, 1.1 N.s (2) $4\vec{e}_t - 0.6667\vec{e}_n$, $6\vec{e}_t - 0.6667\vec{e}_n$ m/s, 4.033 J, 0.7333 N.s (3) $3.024\vec{\imath}+3.101\vec{\jmath}$, $3.953\vec{\imath}+3.798\vec{\jmath}$ m/s, 4.092 J, $-0.3952\vec{\imath}+0.6202\vec{\jmath}$ N.s & 0.0005919 N.ms]

47. A hopper drops n small cylinders per second (Fig. P3.47) from height h on a conveyor belt. At any time, there are n_1 cylinders on the belt. The weight of the belt on the conveyor bed is W. The coefficient of friction for the bed and the belt is μ. Neglect weight of the belt around the pulleys. Assume that end tensions in slack side of the belt differ by $W\sin\theta$. Find the average value of driving torque M_1.

$[R\{nvw/g + \mu W\cos\theta + (n_1 w + n\sqrt{2gh}\, w/g)(\sin\theta + \mu\cos\theta)\}]$

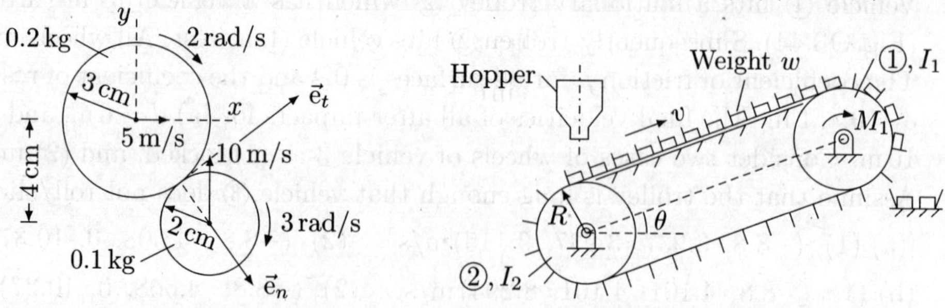

Fig. P3.46 **Fig. P3.47**

48. The coefficient of restitution for all impacts is 0.8 and the impacts are smooth. (a) Two pendulums are rotating (Fig. P3.48a). Find their velocities just after impact if cables are (1) flexible, and (2) inextensible. (b) A sphere ② translating at v_0, hits the ground just before hitting sphere ① (Fig. P3.48b). Find their velocities after mutual impact. $[(b)\ 1.018v_0\vec{\imath},\ 0.6115v_0\vec{\imath} + 0.4900v_0\vec{\jmath}$

$(a)\ (1)\ \omega r(7.758\vec{\imath}+3.620\vec{\jmath}),\ \omega r(12.56\vec{\imath}-0.904\vec{\jmath})\ (2)\ 7.528\omega r\vec{\imath}+3.703\omega r\vec{\jmath},\ 12.62\omega r\vec{\imath}]$

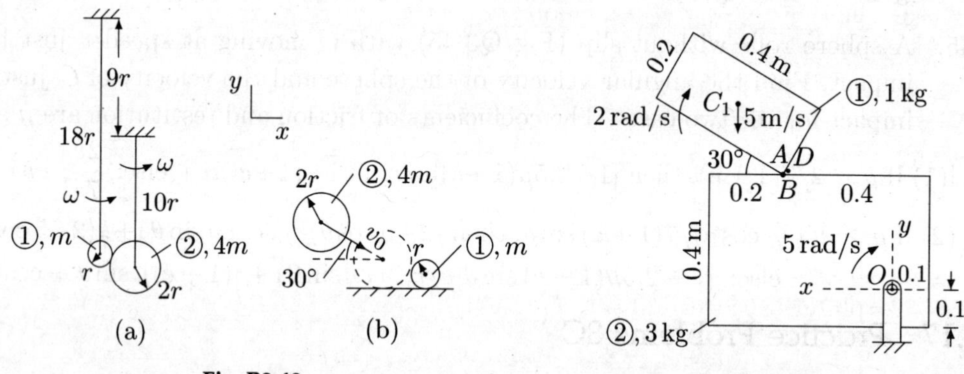

Fig. P3.48 **Fig. P3.49**

49. (a) Two thin rectangular plates have the motion shown in Fig. P3.49 just before impact. Find their angular velocities just after impact for (1) smooth impact with coefficient of restitution $e = 0.6$ and (2) plate ① getting embedded into plate ② in the given orientation. (b) Rework part (a) if both the plates are free. (c) Rework part (a) if plate ② is massive and the impact is smooth with $e = 0.6$.

[(a) (1) −33.75 rad/s, −0.1794 rad/s (2) −0.9581 rad/s

(b) (1) −32.38 rad/s, 1.423 rad/s (2) −1.574 rad/s (c) −39.76 rad/s, 5 rad/s]

50. (a) A uniformly loaded cuboidal crate is released from rest in the position shown in Fig. P3.50. The impact at A is perfectly plastic and there is no slip. $k_z^B = k_z^A = k$. Show that for $k/a < 1/\sqrt{2}$, the edge B will remain in contact with the floor. Find the normal component of the impulses from the ground. (b) For

$k/a > 1/\sqrt{2}$, find the angle through which the crate will rotate after A strikes the floor. Find also $\mu_{\min}$ to ensure no slip at A. [(a) $mk^2\omega/a,\ m\omega(a^2 - 2k^2)/2a$ (b) $\sin^{-1}[\{(1 - a^2/2k^2)^2(a \sin\theta + b \cos\theta - b) + b\}/\sqrt{a^2 + b^2}] - \tan^{-1}\frac{b}{a},\ \frac{ab}{4k^2-a^2}$]

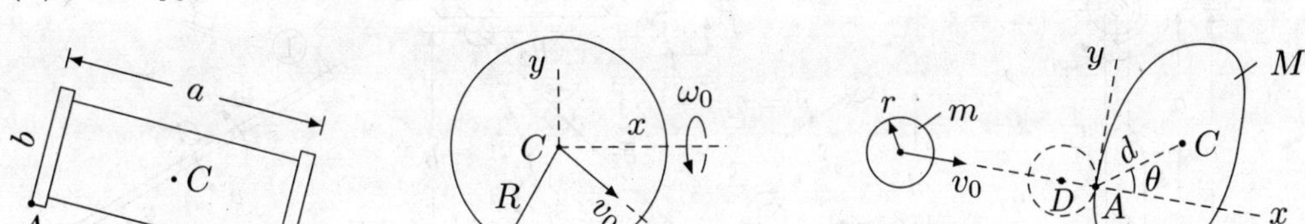

Fig. P3.50 Fig. P3.51 Fig. P3.52

51. A ball, of mass m and radius of gyration k about an axis through its centre (Fig. P3.51), strikes the ground with velocity v_0 and spin ω_0. (a) The coefficient of restitution is e. The ground is sufficiently rough to reduce slip at the point of contact to zero. Find $\vec{v}_C$ and the angular velocity of the ball just after impact. Find the required $\mu_{\min}$ to ensure this. (b) Rework part (a) if the impact point has zero velocity just before the end of impact.

[(a) $ev_0 \sin\theta\,\vec{\text{j}} + (k^2\omega_0\,\vec{\text{k}} + Rv_0 \cos\theta\,\vec{\text{i}})R/(k^2+R^2)$, $[k^2\omega_0\,\vec{\text{i}} - Rv_0 \cos\theta\,\vec{\text{k}}]/(k^2+R^2)$
$(\omega_0^2 R^2 + v_0^2 \cos^2\theta)^{1/2} k^2/(1+e)v_0(k^2 + R^2)\sin\theta$ (b) As in (a) with $e = 0$]

52. (a) A thin plate lies on a smooth surface and rotates at ω_0 about C. Its radius of gyration about the axis perpendicular to the plate at C is k. A disc translating at v_0 hits it and moves back along the same line (Fig. P3.52). The impact is frictionless with coefficient of restitution e. Find the angular velocity of the plate just after impact. (b) Rework part (a) if the disc sticks to the plate.

[(a) $\omega' = \omega_0 + d \sin\theta(1+e)(v_0 - \omega_0 d \sin\theta)/(Mk^2/m + d^2 \sin^2\theta + k^2)$

(b) $\omega' = \dfrac{[Mk^2\omega_0 + mv_0 d \sin\theta M/(M+m)]}{[Mk^2 + \frac{1}{2}mr^2 + Mm(r^2 + d^2 + 2rd\cos\theta)/(M+m)]}$]

53. (a) A cricket bat, of mass m and principal radius of gyration $k^O_{zz} = k$, is modelled to be pinned at O. It is swinging with angular velocity ω_0 just before making impact with a smooth ball (Fig. P3.53). The coefficient of restitution is e. Find the velocity of the ball just after impact. (b) Rework part (a) if the ball sticks to the bat. (c) Determine b in part (a) to have the largest rebound speed.

[(a) $v_0 \cos\theta\,\vec{\text{i}} + [(1+e)mk^2\omega_0 b + (mk^2 e - m_2 b^2)v_0 \sin\theta]\,\vec{\text{j}}/(m_2 b^2 + mk^2)$
(b) $\frac{(mk^2\omega_0 - m_2 b v_0 \sin\theta)b}{(mk^2 + m_2 b^2)}\,\vec{\text{j}}$ (c) $[(v_0^2 \sin^2\theta + mk^2\omega_0^2/m_2)^{1/2} - v_0 \sin\theta]/\omega_0$]

54. A super-ball (Fig. P3.54) moving on a smooth horizontal surface has $\vec{v} = v_1\,\vec{\text{i}} + v_2\,\vec{\text{j}}$, $\vec{\omega} = \vec{0}$ just before striking at A. Neglect gravity. (a) Find θ_1, θ_2, θ_3 after

rebounding from A, B and C. (b) Show that the ball returns to the same spot if $d = 7b$. [(a) $\tan^{-1}(7v_2/3v_1)$, $\tan^{-1}(49v_2/31v_1)$, $\tan^{-1}(343v_2/333v_1)$]

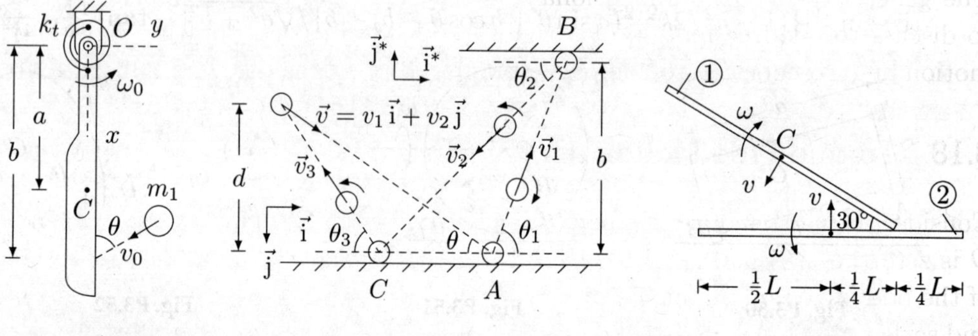

Fig. P3.53 **Fig. P3.54** **Fig. P3.55**

55. Two identical rods have the motion shown in Fig. P3.55 just before impact with $v = \omega L$. The impact is smooth with $e = 0.6$. Find their final angular velocities. [3.238ω, -1.447ω]

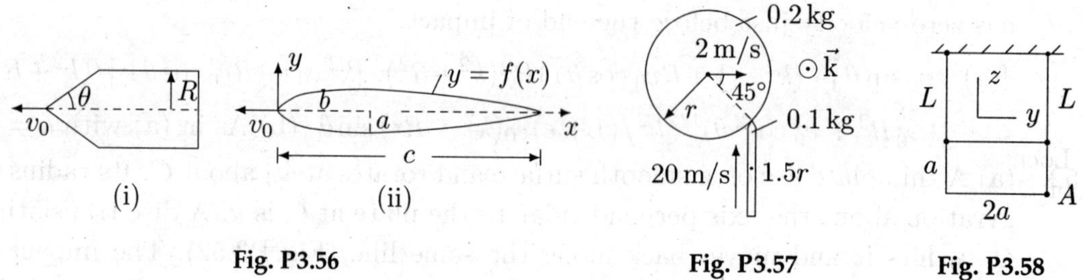

Fig. P3.56 **Fig. P3.57** **Fig. P3.58**

56. (i) A cylindrical satellite with a conical head (Fig. P3.56) moves in a rarefied atmosphere having n molecules per unit volume, each of mass m. The impact is smooth with coefficient of restitution e. v_0 is much greater than the speed of the molecules. Express the drag force in terms of the density $\rho = mn$ of the atmosphere. (ii) Rework part (i) if the shape is a cylinder of length L with an aerofoil section $y = f(x)$ (Fig. P3.56).
[(i) $(1+e)\rho v_0^2 \pi R^2 \sin^2\theta$ (ii) $2(1+e)L\rho v_0^2 \int_0^b \{f'^3/(1+f'^2)\}dx$]

57. An arrow moving at 20 m/s hits a disc translating at 2 m/s (Fig. P3.57). Find $\vec{\omega}$ of the arrow just after impact if: (a) The impact is smooth with $e = 0.3$. (b) The arrow gets embedded in the disc. [(a) $-19.06\vec{k}$ rad/s (b) $2.549\vec{k}$ rad/s]

58. A plate of mass m (Fig. P3.58) is suspended by two light rigid rods with ball and socket joints at the ends. It is hit on its face at A by a bullet of mass m, moving at $v_0(0.5\vec{i} - 0.5\vec{j} + 0.707\vec{k})$. Assume smooth impact with $e = 0.3$. Find the angular velocity of the plate just after impact. [$-(0.4875\vec{j} + 0.2438\vec{k})v_0/a$]

59. Rework Problem 3.58 if the bullet gets embedded in the plate. [$-(6\vec{i}+3\vec{k})v_0/16a$]

D. SPECIAL TOPICS IN 3-DIMENSIONAL MOTION

The general theory of 3-dimensional motion of a rigid body is applied in this part to discuss torque-free motion of a rigid body, stability of a rotating rigid body and motion of gyroscopes, gyro-compass, spinning top, rolling disc and billiard ball.

3.18 Torque-Free Motion of a Rigid Body

Consider the motion of a rigid body under a force system such that the moment about O is zero (i.e. $\vec{M}_O = \vec{0}$), where O is either the centre of mass, C, or a material point of the body fixed in an inertial frame I (as in Cardan's suspension or in a smooth ball and socket joint). All derivatives in this section are with respect to I and subscript I is omitted for simplicity. Let $\vec{e}_i$ be the principal axes of inertia at O. $\dot{\vec{H}}_O = \vec{M}_O = \vec{0} \Rightarrow$

$$\vec{H}_O = I_{11}^O \omega_1 \vec{e}_1 + I_{22}^O \omega_2 \vec{e}_2 + I_{33}^O \omega_3 \vec{e}_3 = \vec{H}_O(0) = \text{const.} \quad (1)$$

$$\Rightarrow \quad {I_{11}^O}^2 \omega_1^2 + {I_{22}^O}^2 \omega_2^2 + {I_{33}^O}^2 \omega_3^2 = H_O^2(0) = \text{const.} \quad (2)$$

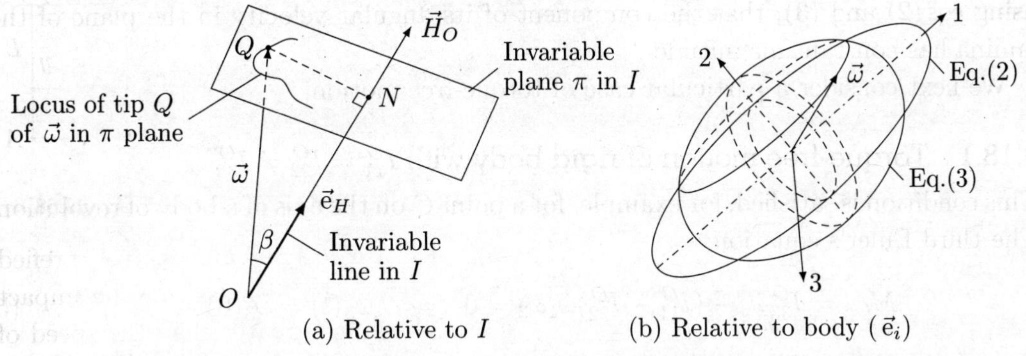

Fig. 3.27

The constant vector $\vec{H}_O$ provides an *invariable direction* $\vec{e}_H = \vec{H}_O/H_O$ in I, which is called an *invariable line* in I for the case of fixed O (Fig. 3.27a).

For O fixed: $\quad \dot{T} = \vec{F} \cdot \vec{v}_O + \vec{M}_O \cdot \vec{\omega} = \vec{F} \cdot \vec{v}_O = 0 \quad \Rightarrow \quad T = T(0) = C_0 \quad$ (say)

For $O \equiv C$: $\quad \dot{T}_R = \vec{M}_C \cdot \vec{\omega} = 0 \quad \Rightarrow \quad T_R = T_R(0) = C_0 \quad$ (say)

Thus, for O fixed or $O \equiv C$:

$$\tfrac{1}{2}(I_{11}^O \omega_1^2 + I_{22}^O \omega_2^2 + I_{33}^O \omega_3^2) = C_0 \quad (3)$$

$$\Rightarrow \quad \tfrac{1}{2}\vec{H}_O \cdot \vec{\omega} = \tfrac{1}{2} H_O \omega \cos\beta = C_0 > 0$$

$$\Rightarrow \quad \omega\cos\beta = ON = 2C_0/H_O = \text{const.} > 0 \quad \Rightarrow \quad (0 \le \beta < \pi/2) \tag{4}$$

i.e., the kinetic energy T for fixed O or 'rotational' kinetic energy T_R for $O \equiv C$, is constant. Relative to an observer moving with the rigid body, the vector $\vec{\omega}$ drawn from O moves in such a way that its extremity describes a curve which is the intersection of two ellipsoids fixed to the rigid body (Fig. 3.27b), described by eqs (2) and (3). Equation (4) implies that the tip of the vector $\vec{\omega}$ drawn from O lies in the fixed plane called the *invariable plane* π normal to $\vec{H}_O$ through N (Fig. 3.27a). If O is fixed in I, then O, N and plane π are fixed in I. If $O \equiv C$, then line ON (CN) and plane π normal to it have fixed orientation with respect to I, i.e., the rigid frame formed by them translates with respect to I with C. The vector $\vec{\omega}$ drawn from O changes with respect to I such that its tip Q always lies in the invariable plane π. In general, $\vec{\omega}$ is variable in magnitude as well as direction.

The only possible rotations with constant $\vec{\omega}$ under torque-free motion about O are the ones for which $\vec{\omega}$ is in the direction of the invariable direction of $\vec{H}_O$, i.e., rotations about any one of the principal axes of inertia at O.

§ **Exercise 3.9:** A rigid lamina with point O fixed in I moves under $\vec{M}_O = \vec{0}$. Show, using eqs (2) and (3), that the component of its angular velocity in the plane of the lamina has constant magnitude.

We next consider a particular case of torque-free motion.

3.18.1 Torque-free motion of rigid body with $I_{11}^O = I_{22}^O \neq I_{33}^O$

This condition is satisfied, for example, for a point O on the axis of a body of revolution. The third Euler's equation:

$$M_{O3} = I_{33}^O\dot{\omega}_3 - (I_{11}^O - I_{22}^O)\omega_1\omega_2 = 0 \quad \Rightarrow \quad \omega_3(t) = \omega_3(0) \tag{5}$$

Equations (2) and (5) $\Rightarrow$

$$H_O^2 = {I_{11}^O}^2(\omega_1^2 + \omega_2^2) + {I_{33}^O}^2\omega_3^2 = \text{const.}$$

$$\Rightarrow \quad \omega_1^2 + \omega_2^2 = \omega_1^2(0) + \omega_2^2(0) = \text{const.} \tag{6}$$

Equations (5) and (6) $\Rightarrow$

$$\omega^2 = \omega_1^2 + \omega_2^2 + \omega_3^2 = \omega_1^2(0) + \omega_2^2(0) + \omega_3^2(0) = \text{const.} \tag{7}$$

$$\alpha = \angle(\vec{\omega}, \vec{e}_3) = \tan^{-1}[(\omega_1^2 + \omega_2^2)^{1/2}/\omega_3] = \text{const.} \tag{8}$$

Equations (4) and (7) $\Rightarrow$

$$ON = \omega\cos\beta = \text{const.} \quad \Rightarrow \quad \beta = \angle(\vec{\omega}, \vec{e}_H) = \text{const.} \tag{9}$$

Using $\vec{H}_O$ from eq. (1), $\vec{\omega}$ is expressed in terms of $\vec{H}_O$ as follows:

$$\vec{H}_O = I_{11}^O(\omega_1\,\vec{e}_1 + \omega_2\,\vec{e}_2) + I_{33}^O\omega_3\,\vec{e}_3 = \vec{H}_O(0) = \text{const.} \tag{10}$$

$$\Rightarrow \quad \vec{\omega} = (\omega_1\,\vec{e}_1 + \omega_2\,\vec{e}_2) + \omega_3\,\vec{e}_3 = \frac{\vec{H}_O - I_{33}^O\omega_3\,\vec{e}_3}{I_{11}^O} + \omega_3\,\vec{e}_3$$

$$= \frac{\vec{H}_O}{I_{11}^O} + \frac{(I_{11}^O - I_{33}^O)}{I_{11}^O}\,\omega_3\,\vec{e}_3$$

$$\Rightarrow \quad \vec{\omega} = p\,\vec{e}_H + s\,\vec{e}_3 \tag{11}$$

with
$$p = \frac{H_O}{I_{11}^O}, \qquad s = \frac{(I_{11}^O - I_{33}^O)\omega_3}{I_{11}^O} \tag{12}$$

From eqs (5) and (10), it follows that

$$\vec{e}_H \cdot \vec{e}_3 = \frac{\vec{H}_O}{H_O} \cdot \vec{e}_3 = \frac{I_{33}^O\omega_3}{H_O} = \text{const.}$$

$$\Rightarrow \qquad \theta = \angle(\vec{e}_H, \vec{e}_3) = \text{const.} \tag{13}$$

Substituting $\omega_3 = s + p\cos\theta$ in eq. $(12)_2$ yields the precession rate

$$p = \frac{I_{33}^O s}{(I_{11}^O - I_{33}^O)\cos\theta} \tag{14}$$

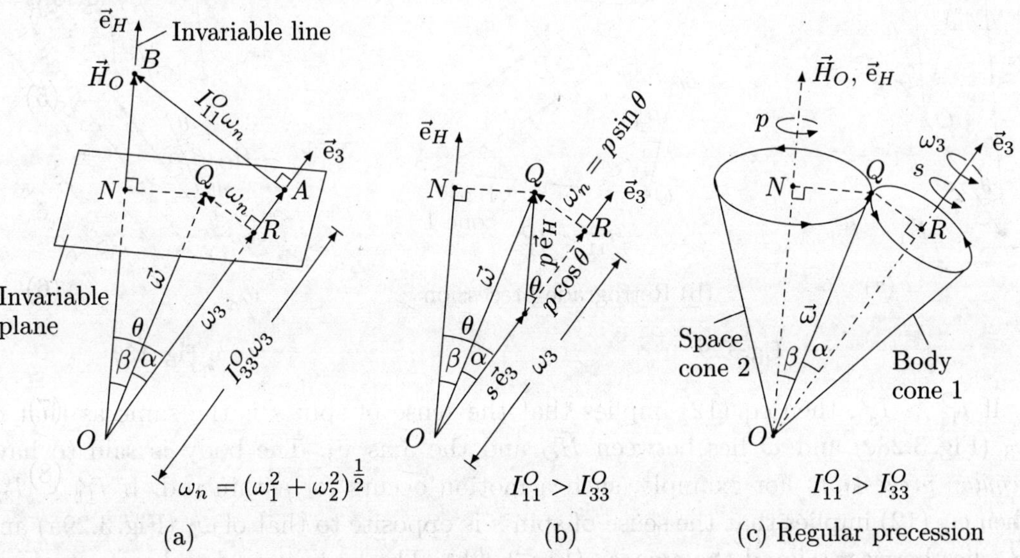

Fig. 3.28

Equation (12) implies that the components p and s are constant. The angular velocity $\vec{\omega}$ is a vector of constant magnitude (OQ) [Fig. 3.28a], whose inclination α with the body-fixed axis $\vec{e}_3$ and inclination β with the invariable line $\vec{e}_H$ along $\vec{H}_O$ fixed in the inertial frame I, are constant (Fig. 3.28b). The component ω_3 of $\vec{\omega}$ along $\vec{e}_3$ is constant and the component of $\vec{\omega}$ normal to $\vec{e}_3$ has constant magnitude $\omega_n = (\omega_1^2+\omega_2^2)^{1/2} = RQ$. $NQ = [(\mathrm{OQ})^2 - (\mathrm{ON})^2]^{1/2} =$ constant. Equation (11) shows that $\vec{\omega}$, $\vec{H}_O\,(\vec{e}_H)$ and $\vec{e}_3$ are always coplanar (Fig. 3.28a) and the inclination θ of the body-fixed axis $\vec{e}_3$ with the inertial space-fixed axis $\vec{e}_H$ is constant [for the case shown in Fig. 3.28a, $\theta = \alpha+\beta$]. Since ON, OQ, OR and angles α, β, θ are constants, we conclude that the plane formed by $\vec{\omega}$ and $\vec{e}_3$ originating from O, rotates rigidly about the invariable line $\vec{H}_O$ with fixed angles α and θ. The resolution of $\vec{\omega}$ as $\vec{\omega} = p\,\vec{e}_H + s\,\vec{e}_3$ is shown in Fig. 3.28b. It follows from eq. (11) that the motion consists of a constant precession rate p of axis $\vec{e}_3$ about $\vec{e}_H$ along with a constant spin rate s of the body about axis $\vec{e}_3$. The vector $\vec{\omega}$ being of constant magnitude and at constant angles β and α to $\vec{H}_O$ and $\vec{e}_3$, continuously sweeps out a space cone of semi-vertex angle β in I and a body cone of semi-vertex angle α in the body frame (Fig. 3.28c). Hence, the motion is as if the body-fixed cone 1 of angle α with axis $\vec{e}_3$, rolls without slip on an inertial space-fixed cone 2 of angle β with axis $\vec{e}_H$.

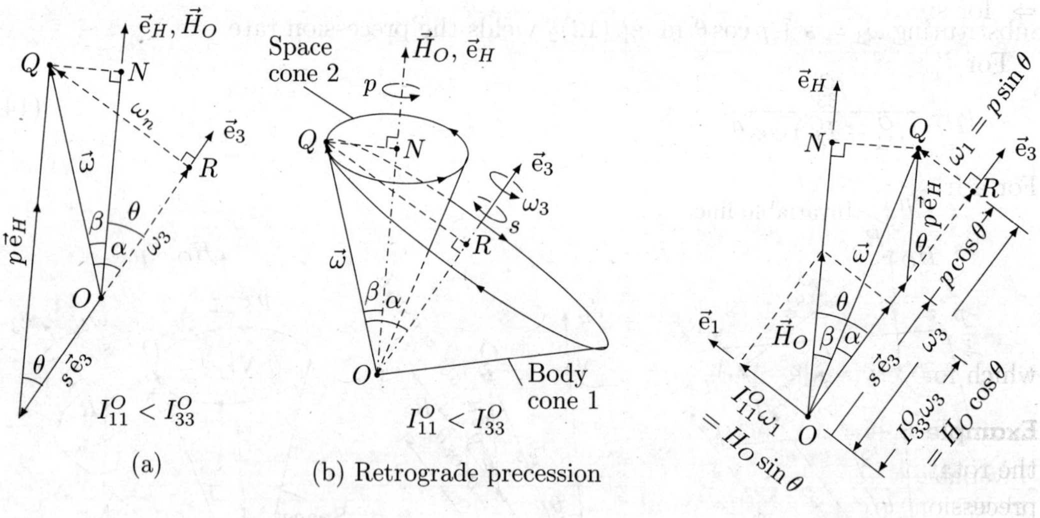

Fig. 3.29

Fig. 3.30

If $I_{11}^O > I_{33}^O$, then eq. (12) implies that the sense of spin s is the same as that of ω_3 (Fig. 3.28c) and $\vec{\omega}$ lies between $\vec{H}_O$ and the axis $\vec{e}_3$. The body is said to have *regular precession*. For example, such a motion occurs for a thin rod. If $I_{11}^O < I_{33}^O$, then eq. (12) implies that the sense of spin s is opposite to that of ω_3 (Fig. 3.29a) and $\vec{H}_O$ lies between $\vec{\omega}$ and the axis $\vec{e}_3$ (Fig. 3.29b). The body is said to have *retrograde precession*. For example, such a motion occurs for a thin disc.

The various entities for torque-free motion can be obtained abinitio from the initial data for $\vec{\omega}$ as follows. Choose $\vec{e}_1$ in the initial plane of $\vec{e}_3$ and $\vec{\omega}$ (Fig. 3.30) and resolve

$$\begin{aligned}
\vec{\omega} &= \omega_1\,\vec{e}_1 + \omega_3\,\vec{e}_3 = p\,\vec{e}_H + s\,\vec{e}_3 = p\sin\theta\,\vec{e}_1 + (s + p\cos\theta)\,\vec{e}_3 \\
\vec{H}_O &= I^O_{11}\omega_1\,\vec{e}_1 + I^O_{33}\omega_3\,\vec{e}_3 = H_O(\sin\theta\,\vec{e}_1 + \cos\theta\,\vec{e}_3) \\
\Rightarrow\ H_O\sin\theta &= I^O_{11}\omega_1, \qquad \omega_1 = p\sin\theta = H_O\sin\theta/I^O_{11}, \qquad p = H_O/I^O_{11} \\
H_O\cos\theta &= I^O_{33}\omega_3, \qquad \omega_3 = s + p\cos\theta = H_O\cos\theta/I^O_{33} \\
\tan\theta &= I^O_{11}\omega_1/I^O_{33}\omega_3 \\
s &= \omega_3 - p\cos\theta = \omega_3 - H_O\cos\theta/I^O_{11} = [I^O_{11} - I^O_{33}]\omega_3/I^O_{11} \\
\omega_3 &= \frac{sI^O_{11}}{I^O_{11} - I^O_{33}} \\
p &= \frac{\omega_3 - s}{\cos\theta} = \frac{I^O_{33}s}{(I^O_{11} - I^O_{33})\cos\theta}
\end{aligned}$$

$\Rightarrow$ for small wobble angle θ: $\ p \simeq I^O_{33}s/(I^O_{11} - I^O_{33})$.

For $I^O_{11} = I^O_{22} = I^O_{33}$:

$$\vec{H}_O = I^O_{11}(\omega_1\,\vec{e}_1 + \omega_2\,\vec{e}_2 + \omega_3\,\vec{e}_3) = I^O_{11}\,\vec{\omega} = \text{const.} \quad \Rightarrow \quad \vec{\omega} = \text{const.}$$

For a frisbee with $I^C_{33} = 2I^C_{11}$ and $\vec{\omega}$ at an angle α to its axis $\vec{e}_3$,

$$\begin{aligned}
s &= (I^C_{11} - I^C_{33})\omega_3/I^C_{11} = -\omega_3 = -\omega\cos\alpha \\
p &= H_C/I^C_{11} = [(I^C_{11}\omega\sin\alpha)^2 + (I^C_{33}\omega\cos\alpha)^2]^{1/2}/I^C_{11} = \omega(1 + 3\cos^2\alpha)^{1/2}
\end{aligned}$$

which for very small values of α reduce to $s \simeq -\omega,\ p \simeq 2\omega$.

Example 3.46: A meteorite imparts an impulse $2mR\omega/3$ normal to the plane of the rotating disc (a space station) at the edge (Fig. E3.46a). Find the precession axis, precession rate p, spin rate s and the angle of nutation θ between the axis of symmetry and the precession axis. Find the angular velocity when the axis of symmetry makes a 60° angle with its initial orientation.

Solution: Let $\vec{\omega} = \omega_x\,\vec{\imath} + \omega_y\,\vec{\jmath} + \omega_z\,\vec{k}$ be the angular velocity and $\vec{H}_C$ be the moment of momentum just after impact, where $\vec{\imath}, \vec{\jmath}, \vec{k}$ are shown in the initial position. $I^C_{xx} = I^C_{yy} = \frac{1}{4}mR^2$, $I^C_{zz} = \frac{1}{2}mR^2$. $\ \Delta\vec{H}_C = \vec{I}_{\text{ang}_C} \Rightarrow$

$$I^C_{xx}\omega_x\,\vec{\imath} + I^C_{yy}\omega_y\,\vec{\jmath} + I^C_{zz}\omega_z\,\vec{k} - I^C_{zz}\omega\,\vec{k} = R\,\vec{\imath} \times \left(-\frac{2}{3}mR\omega\,\vec{k}\right) = \frac{2}{3}mR^2\omega\,\vec{\jmath}$$

$$\Rightarrow \quad \vec{\mathrm{i}}: \quad \frac{1}{4}mR^2\omega_x = 0 \qquad \Rightarrow \qquad \omega_x = 0$$

$$\vec{\mathrm{j}}: \quad \frac{1}{4}mR^2\omega_y = \frac{2}{3}mR^2\omega \qquad \Rightarrow \qquad \omega_y = \frac{8}{3}\omega$$

$$\vec{\mathrm{k}}: \quad \frac{1}{2}mR^2(\omega_z - \omega) = 0 \qquad \Rightarrow \qquad \omega_z = \omega$$

Hence, $\vec{H}_C$ and the precession axis $\vec{\mathrm{e}}_H$ are given by

$$\vec{H}_C = \frac{1}{4}mR^2\left(\frac{8}{3}\omega\vec{\mathrm{j}}\right) + \frac{1}{2}mR^2\omega\vec{\mathrm{k}} = \frac{1}{6}mR^2\omega(4\vec{\mathrm{j}} + 3\vec{\mathrm{k}})$$

$$\Rightarrow \quad \vec{\mathrm{e}}_H = \frac{\vec{H}_C}{|\vec{H}_C|} = \frac{4\vec{\mathrm{j}} + 3\vec{\mathrm{k}}}{5} = 0.8\vec{\mathrm{j}} + 0.6\vec{\mathrm{k}}$$

$$\cos\theta = \vec{\mathrm{e}}_H \cdot \vec{\mathrm{k}} = 0.6 \qquad \Rightarrow \qquad \theta = 53.13^\circ$$

where θ is the angle between the precession and symmetry axes. Let $\vec{\omega} = p\,\vec{\mathrm{e}}_H + s\,\vec{\mathrm{k}}$:

$$\vec{\omega} = \frac{8}{3}\omega\vec{\mathrm{j}} + \omega\vec{\mathrm{k}} = p\,\vec{\mathrm{e}}_H + s\,\vec{\mathrm{k}} = p(0.8\vec{\mathrm{j}} + 0.6\vec{\mathrm{k}}) + s\,\vec{\mathrm{k}}$$

$$\Rightarrow \quad \vec{\mathrm{j}}: \quad \frac{8}{3}\omega = 0.8p \qquad \Rightarrow \qquad p = \frac{10}{3}\omega$$

$$\vec{\mathrm{k}}: \quad \omega = 0.6p + s \qquad \Rightarrow \qquad s = -\omega$$

The spin rate, precession rate and the precession axis $\vec{\mathrm{e}}_H$ are shown in Fig. E3.46b.

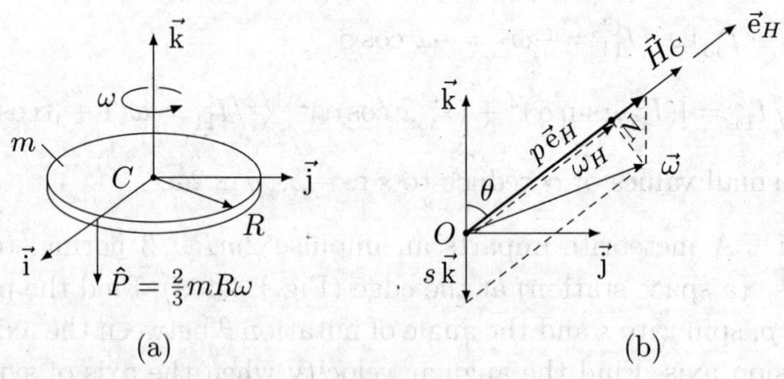

Fig. E3.46

Let the axis of the disc be along $\vec{\mathrm{e}}_3 = l\,\vec{\mathrm{i}} + m\,\vec{\mathrm{j}} + n\,\vec{\mathrm{k}}$, when it is at 60° to $\vec{\mathrm{k}}$, i.e., $\vec{\mathrm{e}}_3 \cdot \vec{\mathrm{k}} = n = \cos 60^\circ = 0.5$. In this position, let the angular velocity be $\vec{\omega}'$. m is obtained using the constant component ω_H of $\vec{\omega}$ along $\vec{\mathrm{e}}_H$:

$$\vec{\omega}' = p\,\vec{\mathrm{e}}_H + s\,\vec{\mathrm{e}}_3 = \frac{10}{3}\omega\,\vec{\mathrm{e}}_H - \omega(l\,\vec{\mathrm{i}} + m\,\vec{\mathrm{j}} + 0.5\,\vec{\mathrm{k}})$$

$$\begin{aligned}
\omega_H &= \vec{\omega}' \cdot \vec{e}_H = \vec{\omega}' \cdot (0.8\,\vec{j} + 0.6\,\vec{k}) = \frac{10}{3}\,\omega - 0.8m\omega - 0.3\omega \\
&= \vec{\omega} \cdot \vec{e}_H = \frac{8}{3}\,\omega(0.8) + \omega(0.6) \\
\Rightarrow \quad m &= 0.375, \qquad l = \pm\,(1 - m^2 - n^2)^{1/2} = \pm\,0.7806 \\
\vec{\omega}' &= \frac{10}{3}\,\omega\,\vec{e}_H - \omega(l\,\vec{i} + m\,\vec{j} + 0.5\,\vec{k}) \\
&= \frac{10}{3}\,\omega(0.8\,\vec{j} + 0.6\,\vec{k}) - \omega(\pm 0.7806\,\vec{i} + 0.375\,\vec{j} + 0.5\,\vec{k}) \\
&= \mp 0.7806\omega\,\vec{i} + 2.292\omega\,\vec{j} + 1.5\omega\,\vec{k}
\end{aligned}$$

3.19 Stability of Uniform Rotation of a Rigid Body

Consider the stability of rigid body motion with constant $\vec{\omega}$ in the direction of a principal axis of inertia at O under a system of forces such that $\vec{M}_O = \vec{0}$, where O is either the centre of mass, C, or a material point of the body fixed in an inertial frame I. An example of such motion with $\vec{\omega} = \omega_0\,\vec{e}_1$, where $\vec{e}_1$ is principal axis of inertia at O, is shown in Fig. 3.31a. We want to investigate whether the change in $\vec{\omega}$ remains bounded if a small disturbance is given to the body.

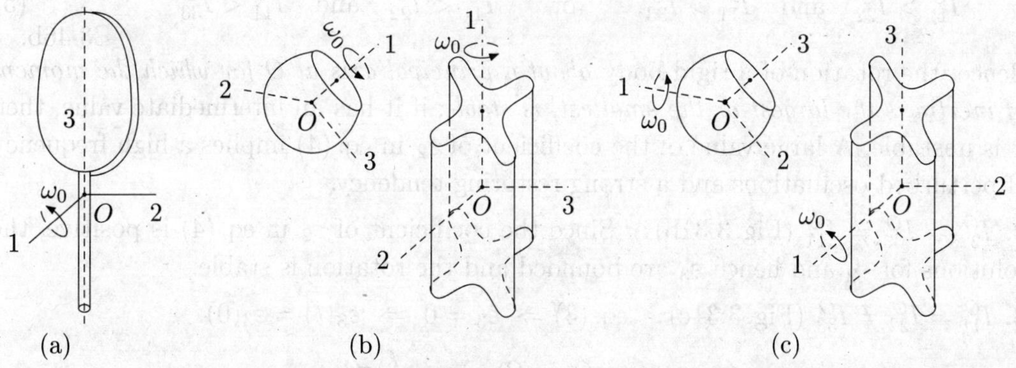

Fig. 3.31

Just after the disturbance is applied, let $\vec{\omega}(0) = \omega_0\,\vec{e}_1 + \varepsilon_i(0)\,\vec{e}_i$ with $\varepsilon_i \ll \omega_0$. Let $\vec{\omega}(t) = \omega_0\,\vec{e}_1 + \varepsilon_i(t)\,\vec{e}_i$ during the subsequent motion. Hence, $\dot{\vec{\omega}} = \dot{\varepsilon}_i\,\vec{e}_i$:

$$\omega_1 = \omega_0 + \varepsilon_1, \quad \omega_2 = \varepsilon_2, \quad \omega_3 = \varepsilon_3, \quad \dot{\omega}_1 = \dot{\varepsilon}_1, \quad \dot{\omega}_2 = \dot{\varepsilon}_2, \quad \dot{\omega}_3 = \dot{\varepsilon}_3$$

Euler's equations of motion at O yield:

$$M_{O_1} = I^O_{11}\dot{\omega}_1 - (I^O_{22} - I^O_{33})\omega_2\omega_3 : \qquad I^O_{11}\dot{\varepsilon}_1 - (I^O_{22} - I^O_{33})\varepsilon_2\varepsilon_3 = 0$$

$$M_{O_2} = I^O_{22}\dot{\omega}_2 - (I^O_{33} - I^O_{11})\omega_3\omega_1 : \qquad I^O_{22}\dot{\varepsilon}_2 - (I^O_{33} - I^O_{11})\varepsilon_3(\omega_0 + \varepsilon_1) = 0$$

$$M_{O_3} = I^O_{33}\dot{\omega}_3 - (I^O_{11} - I^O_{22})\omega_1\omega_2 : \qquad I^O_{33}\dot{\varepsilon}_3 - (I^O_{11} - I^O_{22})(\omega_0 + \varepsilon_1)\varepsilon_2 = 0$$

Neglecting terms of second order we have:

$$I^O_{11}\dot{\varepsilon}_1 = 0 \tag{1}$$

$$I^O_{22}\dot{\varepsilon}_2 - (I^O_{33} - I^O_{11})\omega_0\varepsilon_3 = 0 \tag{2}$$

$$I^O_{33}\dot{\varepsilon}_3 - (I^O_{11} - I^O_{22})\omega_0\varepsilon_2 = 0 \tag{3}$$

Equation (1) $\Rightarrow$ $\varepsilon_1(t) = \varepsilon_1(0)$. We eliminate ε_3 from the coupled eqs (2) and (3) by forming $\dot{\overline{(2)}}$ and using (3): $\quad I^O_{22}\ddot{\varepsilon}_2 - (I^O_{33} - I^O_{11})\omega_0\dot{\varepsilon}_3 = 0$

$$\Rightarrow \quad \ddot{\varepsilon}_2 + \frac{(I^O_{11} - I^O_{22})(I^O_{11} - I^O_{33})\omega_0^2}{I^O_{22}I^O_{33}}\varepsilon_2 = 0 \tag{4}$$

We consider 4 possible cases.

1. $I^O_{11} \neq I^O_{22}$, $I^O_{11} \neq I^O_{33}$: The solution of eq. (4) for ε_2 is bounded (harmonic) [and eq. (2) $\Rightarrow$ ε_3 is also bounded] iff the coefficient of ε_2 in eq. (4) is positive, i.e., iff

$$I^O_{11} > I^O_{22} \quad \text{and} \quad I^O_{11} > I^O_{33}, \qquad \text{or} \qquad I^O_{11} < I^O_{22} \quad \text{and} \quad I^O_{11} < I^O_{33} \tag{5}$$

Hence, the rotation of a rigid body *about a principal axis at O for which the moment of inertia is the largest or the smallest, is stable*; if it has an intermediate value, then it is unstable. A large value of the coefficient of ε_2 in eq. (4) implies a high frequency of perturbed oscillations and a strong restoring tendency.

2. $I^O_{22} = I^O_{33} \neq I^O_{11}$ (Fig. 3.31b): Since the coefficient of ε_2 in eq. (4) is positive, the solutions for ε_2 and hence ε_3 are bounded and the rotation is stable.

3. $I^O_{11} = I^O_{22} \neq I^O_{33}$ (Fig. 3.31c): eq. (3) $\Rightarrow$ $\dot{\varepsilon}_3 = 0$ $\Rightarrow$ $\varepsilon_3(t) = \varepsilon_3(0)$.

$$\text{eq. (2)} \Rightarrow \qquad \varepsilon_2 = \varepsilon_2(0) + (I^O_{33} - I^O_{11})\omega_0\varepsilon_3(0)t/I^O_{22}$$

Thus, ε_2 builds up with time and the rotation about axis $\vec{e}_1$ is unstable.

4. $I^O_{11} = I^O_{22} = I^O_{33}$: Equations (1)–(3) $\Rightarrow$ $\dot{\varepsilon}_i = 0$ $\Rightarrow$ $\varepsilon_i(t) = \varepsilon_i(0)$ $\Rightarrow$ the rotation about any axis is stable.

For a rigid body of revolution with point O on its axis $\vec{e}_1$, $\vec{\omega}$ in the direction of the axis is always stable since $I^O_{22} = I^O_{33} \neq I^O_{11}$ (case 2) or $I^O_{11} = I^O_{22} = I^O_{33}$ (case 4).

Similarly, for a rigid body of revolution with point O on its axis $\vec{e}_3$, the angular velocity in the direction $\vec{e}_1$ normal to the axis is unstable since $I_{11}^O = I_{22}^O \neq I_{33}^O$ (case 3) except when $I_{11}^O = I_{22}^O = I_{33}^O$ (case 4).

§ **Exercise 3.10:** A lamina in the 1-2 plane attached to it has a point O fixed in I with $\vec{M}_O = \vec{0}$ and principal axes of inertia $\vec{e}_i$ at O. Show that $\vec{\omega} = \omega\,\vec{e}_3$ is always stable and $\vec{\omega} = \omega\,\vec{e}_1$ is stable if $I_{11}^O < I_{22}^O$ and unstable if $I_{11}^O = I_{22}^O$.

3.19.1 Flying saucers are more stable than flying cigars

An early cylindrical satellite was put into orbit spinning about its longitudinal axis 3 (Fig. 3.32a). To the designer's surprise, it began to wobble more and more until finally it was rotating about a transverse axis 1 (Fig. 3.32b). This dramatically contradicted the above well-established theory that the angular velocity of a rigid body of revolution about its axis of symmetry is stable. The explanation of the actual observation lies in the fact that the *modelling of the object as a rigid body was wrong.* The body is deformable and though $\vec{H}_C$ is conserved since $\vec{M}_C = \vec{0}$, the 'rotational' kinetic energy T_R is not conserved but decreases because of the dissipation of energy within the body due to internal damping:

$$T_R = \tfrac{1}{2}(I_{11}^C\omega_1^2 + I_{22}^C\omega_2^2 + I_{33}^C\omega_3^2) = \tfrac{1}{2}\left(\frac{H_{C_1}^2}{I_{11}^C} + \frac{H_{C_2}^2}{I_{22}^C} + \frac{H_{C_3}^2}{I_{33}^C}\right) \tag{6}$$

$$\vec{H}_C = H_C\,\vec{e} = I_{33}^C\omega\,\vec{e} = I_{11}^C\omega'\,\vec{e}$$

$$T_{R_i} = \tfrac{1}{2}I_{33}^C\omega^2 = \frac{H_C^2}{2I_{33}^C}, \qquad T_{R_f} = \tfrac{1}{2}I_{11}^C\omega'^2 = \frac{H_C^2}{2I_{11}^C} < T_{R_i} \tag{7}$$

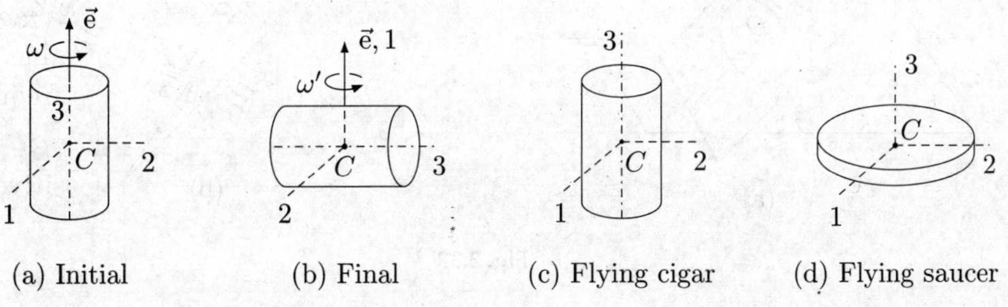

(a) Initial (b) Final (c) Flying cigar (d) Flying saucer

Fig. 3.32

T_R keeps decreasing due to dissipation and since H_C is constant, the body ends up turning about an axis for which T_R is minimum for the given H_C. Hence, it follows from the denominator of the terms in the expression (6) for T_R that eventually the body would *end up rotating about an axis for which the principal moment of inertia is the maximum.* Hence, the cylindrical satellite ends up rotating about a transverse axis 1 since $I_{11}^C > I_{33}^C$. By the same reason, cigar-shaped bodies with $I_{33}^C < I_{11}^C$ are not stable in axial rotation (Fig. 3.32c,d) whereas saucer-shaped bodies with $I_{33}^C > I_{11}^C$ are stable in axial rotation in the presence of internal damping whose effect for the latter case is to eliminate any initial nutation and wobbling in addition to axial rotation. Thus, saucer-shaped spacecraft are better than cigar-shaped spacecraft.

3.20 Steady Precession of Gyroscope

A *gyroscope is a rigid body with two equal principal moments of inertia* $I_{11}^O = I_{22}^O$ at its material point O which is either the centre of mass C or a point fixed in I, for example, a body of revolution is a gyroscope. Originally, the word 'gyroscope' referred to an instrument in which the earth's rotation produced an effect which could be observed. Gyroscopes have wide applications. The behaviour of a gyroscope is not in tune with our intuition. Consider a gyroscope spinning at high angular velocity $s\,\vec{e}_3$ (Fig. 3.33a). When an instantaneous angular impulse $\vec{I}_{\text{ang}_O}$ is applied to it, the instantaneous change $\Delta\vec{H}_O = \vec{I}_{\text{ang}_O}$. The angle θ between $(\vec{H}_O + \Delta\vec{H}_O)$ and $\vec{H}_O$ for a given $\Delta\vec{H}_O$ depends on H_O, being smaller for larger values of H_O (Fig. 3.33b). Since $\vec{H}_O = I_{33}^O s\,\vec{e}_3$ has a larger magnitude for a higher s, we infer that an angular impulse acting on a rapidly spinning gyroscope causes only a small change in the direction of $\vec{H}_O$ and hence, the change in the direction of $\vec{\omega}$ is also small, i.e., there is little change in the direction of its axis of rotation. Thus, spin imparts stability to the gyroscope.

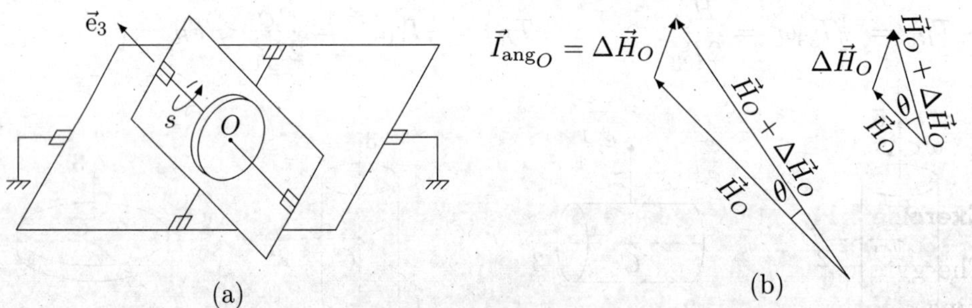

Fig. 3.33

Gyroscopes are used in automatic pilots, stabilized gun platforms, inertial guidance systems of aircraft, missiles and submarines. Spinning bullets and shells remain tangent to their trajectory because of gyroscopic action. It is easier to keep a bicycle balanced at high speeds because of the stabilizing effect of the spinning wheels.

We consider 3 specific motions of gyroscopes.

1. Consider a gyroscope spinning at constant rate $s\,\vec{e}_3$ about the body-fixed axis $\vec{e}_3$ through O with $\vec{e}_3$ itself precessing at constant rate $p\,\vec{E}$ about an orthogonal axis $\vec{E}$ fixed in I (Fig. 3.34a). The moment $\vec{C}_O$ about O required to maintain this motion is called *gyroscopic couple.* The frame formed by $\vec{e}_3$, $\vec{e}_2\,(=\vec{E})$ and $\vec{e}_1\,(=\vec{e}_2\times\vec{e}_3)$ rotates with respect to I at rate $\vec{\Omega}=p\,\vec{e}_2$. The angular velocity $\vec{\omega}$, $\vec{H}_O$ and $\vec{C}_O$ for the body using Euler's 2nd axiom are:

$$\begin{aligned}
\vec{\omega} &= p\,\vec{e}_2 + s\,\vec{e}_3 \\
\vec{H}_O &= I_{11}^O p\,\vec{e}_2 + I_{33}^O s\,\vec{e}_3 \\
\vec{C}_O &= \dot{\vec{H}}_{O|I} = \vec{\Omega}\times\vec{H}_O = p\,\vec{e}_2\times I_{33}^O s\,\vec{e}_3 \\
\vec{C}_O &= psI_{33}^O\,\vec{e}_1 = \vec{p}\times I_{33}^O\,\vec{s} \qquad (3.80)
\end{aligned}$$

where $\vec{p}=p\,\vec{E}$, $\vec{s}=s\,\vec{e}_3$. The gyroscopic couple is, surprisingly, normal to both precession and spin directions.

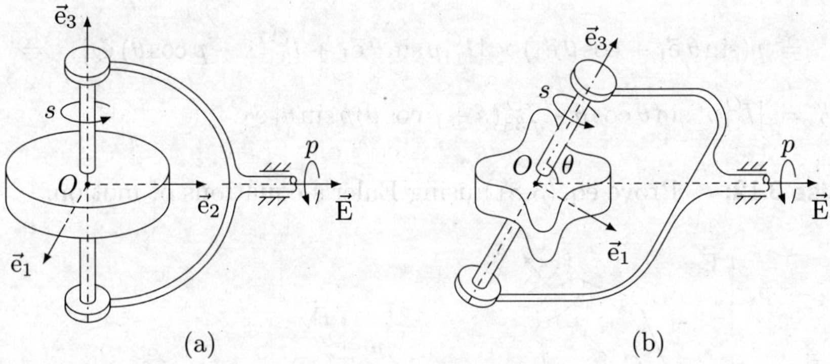

Fig. 3.34

§ **Exercise 3.11:** Prove eq. (3.80) using Euler's equations of motion.

The gyroscopic couple must be taken into account in the design of bearings supporting rotating axles carrying heavy rotors which are subjected to forced precession, e.g., the turbines in aeroplanes taking a turn. In a railway wagon, the axle may tilt in the vertical plane about a transverse horizontal axis due to imperfect ballasting of sleepers or localized weakness at a rail joint. This tilt, though small, is quick in action and the magnitude of angular velocity of tilt (precession) may be considerable. Thus, a large gyroscopic couple about the vertical axis acts on the axle and its reaction acts on the wagon causing it to swerve in the horizontal plane. If both wheels descend equally, no such swerve is induced—this is a sound reason for placing rail joints side by side, instead of staggering them. The same happens to a car if one wheel descends into a pot-hole. A swerving couple is induced and unless the steering is sufficiently rigid and irreversible, a succession of such impulses may give rise to an unpleasant and dangerous condition of front wheel wobble.

Rolling about the longitudinal axis is the most undesirable motion for ships. To avoid amplification of rolling due to gyroscopic action of the motors and turbines, their spin axis should be parallel to the longitudinal axis of the ship.

2. Consider the case of precession axis $\vec{\mathrm{E}}$ at an angle θ to the spin axis $\vec{e}_3$ (Fig. 3.34b). Choose a frame $\vec{e}_i$ such that $\vec{e}_1$ is in the plane of $\vec{\mathrm{E}}$ and $\vec{e}_3$. The angular velocity $\vec{\Omega}$ of this frame is $\vec{\Omega} = p\vec{\mathrm{E}} = p(\sin\theta\,\vec{e}_1 + \cos\theta\,\vec{e}_3)$. The angular velocity $\vec{\omega}$, $\vec{H}_O$ and the moment $\vec{M}_O$ for the body using Euler's 2nd axiom, are

$$\begin{aligned}
\vec{\omega} &= p\vec{\mathrm{E}} + s\,\vec{e}_3 = p(\sin\theta\,\vec{e}_1 + \cos\theta\,\vec{e}_3) + s\,\vec{e}_3 \\
\vec{H}_O &= I_{11}^O p\sin\theta\,\vec{e}_1 + I_{33}^O(s + p\cos\theta)\,\vec{e}_3 \\
\vec{M}_O &= \dot{\vec{H}}_{O|I} = \vec{\Omega}\times\vec{H}_O \\
&= p(\sin\theta\,\vec{e}_1 + \cos\theta\,\vec{e}_3)\times[I_{11}^O p\sin\theta\,\vec{e}_1 + I_{33}^O(s + p\cos\theta)\,\vec{e}_3] \quad \Rightarrow \\
\vec{M}_O &= [I_{11}^O p^2\sin\theta\cos\theta - I_{33}^O(s + p\cos\theta)p\sin\theta]\,\vec{e}_2
\end{aligned} \tag{3.81}$$

§ **Exercise 3.12:** Prove eq. (3.81) using Euler's equations of motion.

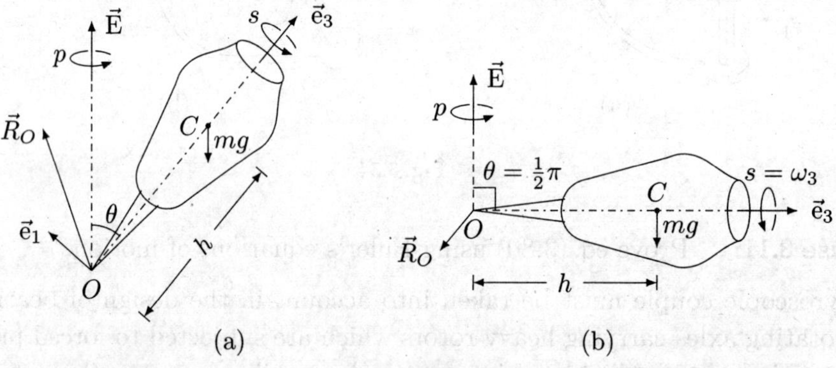

Fig. 3.35

3. Let a body with $I_{11}^O = I_{22}^O$ (a gyroscope) rotate about fixed point O (Fig. 3.35a). The *minimum value of s so that the precession at constant θ is possible* is obtained from eq. (3.81) using $\vec{M}_O = -mgh\sin\theta\,\vec{e}_2$:

$$[(I_{11}^O - I_{33}^O)\cos\theta\,p^2 - I_{33}^O sp + mgh]\sin\theta = 0 \;\Rightarrow$$

$$\sin\theta = 0 \quad (\theta = 0, \pi)$$

$$\text{or} \quad (I_{11}^O - I_{33}^O)\cos\theta\,p^2 - I_{33}^O sp + mgh = 0 \tag{1}$$

$$\Rightarrow \quad p_1, p_2 = \frac{I_{33}^O s}{2(I_{11}^O - I_{33}^O)\cos\theta}\left[1 \pm \left\{1 - \frac{4(I_{11}^O - I_{33}^O)}{{I_{33}^O}^2 s^2} mgh\cos\theta\right\}^{1/2}\right] \tag{2}$$

as the two precession rates p_1, p_2; provided the discriminant is positive, i.e., if

$$s^2 \geq 4(I_{11}^O - I_{33}^O)mgh\cos\theta/{I_{33}^O}^2 \tag{3.82}$$

which will definitely be satisfied if $s^2 \geq 4(I_{11}^O - I_{33}^O)mgh/{I_{33}^O}^2$. For large s, expanding by binomial theorem, eq. (2) yields fast and slow precession rates p_1, p_2 as

$$p_1 \simeq I_{33}^O s/(I_{11}^O - I_{33}^O)\cos\theta, \qquad p_2 = mgh/I_{33}^O s \tag{3}$$

For a spinning top, the initial precession at constant θ is followed by wobbling motion with nutation because as ω_3 decreases due to inevitable friction, at some stage, condition (3.82) gets violated and precession at constant θ is not possible (unless $I_{33}^O > I_{11}^O$).

Remarks:

1. We express eq. (1) in terms of ω_3 by replacing s with $\omega_3 - p\cos\theta$:

$$I_{11}^O \cos\theta\, p^2 - I_{33}^O\omega_3 p + mgh = 0 \tag{4}$$

$$\Rightarrow \quad p_1, p_2 = \frac{I_{33}^O\omega_3}{2I_{11}^O\cos\theta}\left[1 \pm \left(1 - \frac{4I_{11}^O mgh\cos\theta}{(I_{33}^O\omega_3)^2}\right)^{1/2}\right] \tag{5}$$

$$\text{if} \quad \omega_3^2 \geq 4I_{11}^O mgh\cos\theta/{I_{33}^O}^2 \tag{3.83}$$

The condition (3.83), for the existence of uniform precession rates p_1, p_2, is definitely satisfied for all θ if $\omega_3^2 \geq 4mghI_{11}^O/{I_{33}^O}^2$. It follows from eq. (3.83) that the *sleeping top* $[\theta(t) \equiv 0]$ motion exists if $\omega_3^2 \geq 4mghI_{11}^O/{I_{33}^O}^2$. For the case of $\omega_3^2 \gg 4I_{11}^O mgh\cos\theta/{I_{33}^O}^2$, the square-root term is expanded by binomial theorem to yield

$$p_1 \simeq I_{33}^O\omega_3/I_{11}^O\cos\theta, \qquad p_2 \simeq mgh/I_{33}^O\omega_3 \tag{6}$$

For C below O: $\theta > \pi/2$, $\cos\theta < 0$, eq. (3.83) holds and stable precessions exist $\forall\ \omega_3$.

2. If p is small or $\theta \simeq \pi/2$, then eq. (4) $\Rightarrow$

$$\omega_3 = mgh/I_{33}^O p + I_{11}^O p\cos\theta/I_{33}^O \simeq mgh/I_{33}^O p \simeq s$$

since $\omega_3 = s + p\cos\theta \simeq s$. In fact, it is the gyroscopic couple expression (3.80) with $C_O = mgh$ (Fig. 3.35b). For $\theta = \frac{1}{2}\pi$, eq. (4) does yield the exact relation $I_{33}^O\omega_3 p = mgh$.

3. If $\vec{M}_O = \vec{0}$, then $mgh = 0$ in eq. (1) $\Rightarrow$ $p = 0$, $I_{33}^O s/(I_{11}^O - I_{33}^O)\cos\theta$.

4. If the gyroscope has no spin: $s = 0$, for given p, eq. (1) $\Rightarrow$ $\cos\theta = mgh/p^2(I_{33}^O - I_{11}^O)$

We consider some applications of gyroscopes.

3.20.1 Spinning projectile

In case 3 above, the gyroscope's axis of spin $\vec{e}_3$ sweeps a conical surface about the axis of precession $\vec{E}$ (parallel to the gravitational force $-mg\,\vec{E}$). The motion of a *spinning projectile* (a bullet, a shell or a missile) under aerodynamic force $\vec{F}$ acting opposite to $\vec{v}_C$ at the centre of pressure B at distance L from C (Fig. 3.36a) is similar with O and $\vec{M}_O = -mgh\sin\theta\,\vec{e}_2$ replaced by C and $\vec{M}_C = -FL\sin\theta\,\vec{e}_2$. Stable steady precession of axis $\vec{e}_3$ about the axis along $\vec{v}_C$ through C occurs, provided eq. (3.82) is satisfied where O and mgh are replaced by C and FL. Hence, the high initial spin provided to a projectile stabilises its motion.

3.20.2 Navigational gyroscopes

Let the triad $\vec{e}_i$ be fixed to the gimbal of the gyroscope spinning about axis $\vec{e}_3$ at $\dot{\phi}\,\vec{e}_3$ in Fig. 3.36b. The gimbal can rotate about $\vec{e}_2$ axis fixed on the frame which is fixed on a vehicle. A restrained gyroscope has torsional spring or torsional dampers attached between the gimbal and the frame to provide a couple about axis $\vec{e}_2$ when the gimbal moves relative to the frame.

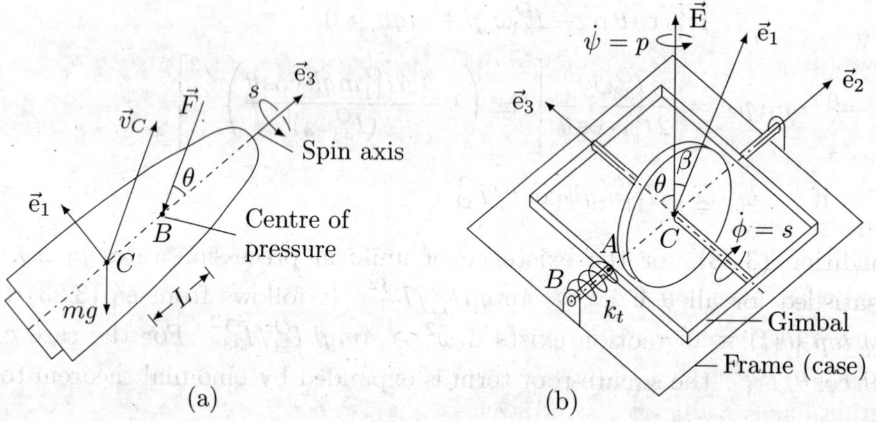

Fig. 3.36

For the gyroscope in Fig. 3.36b, a torsional spring of torsional stiffness k_t is undeformed for $\theta = \pi/2$ when the vehicle translates with $\vec{e}_3$ remaining normal to the axis $\vec{E}$ fixed in I. If the vehicle and hence, the frame gets rotated at a constant rate $\dot{\psi}$ about axis $\vec{E}$ (Fig. 3.36b), then the gimbal rotates by angle $\beta\,\vec{e}_2$ with respect to the frame to turn the spin axis $\vec{e}_3$ to an inclination $\theta = \pi/2 - \beta$ to $\vec{E}$. The spin axis $\vec{e}_3$ of the gyroscope precesses at rate $\dot{\psi}\,\vec{E}$ about the axis $\vec{E}$. The torsional spring is subjected to a moment $k_t\beta\,\vec{e}_2$ at A and it applies couple $\vec{M}_C = -k_t\beta\,\vec{e}_2$ to the gimbal. The angular velocities $\vec{\Omega}$ and $\vec{\omega}$ of the frame $\vec{e}_i$ and the gyroscope are

$$\vec{\Omega} = \dot{\psi}\,\vec{E} = \dot{\psi}(\cos\beta\,\vec{e}_1 + \sin\beta\,\vec{e}_3)$$

$$\vec{\omega} = \vec{\Omega} + \dot{\phi}\,\vec{e}_3 = \dot{\psi}\cos\beta\,\vec{e}_1 + (\dot{\phi} + \dot{\psi}\sin\beta)\,\vec{e}_3$$

$$\Rightarrow \quad \vec{H}_C = I_{11}^C\dot{\psi}\cos\beta\,\vec{e}_1 + I_{33}^C(\dot{\phi} + \dot{\psi}\sin\beta)\,\vec{e}_3$$

$M_{C_3} = 0$, $I_{11}^C = I_{22}^C$ and Euler's 3rd equation of motion $\Rightarrow$ $\omega_3 = \dot{\phi} + \dot{\psi}\sin\beta = \text{const.}$ Hence, $\dot{\vec{H}}_{C|\vec{e}_i} = I_{11}^C \frac{d}{dt}(\dot{\psi}\cos\beta)\,\vec{e}_1$. By Euler's second axiom:

$$\vec{M}_C = -k_t\beta\,\vec{e}_2 = \dot{\vec{H}}_{C|I} = \dot{\vec{H}}_{C|\vec{e}_i} + \dot{\psi}\,\vec{\mathrm{E}} \times \vec{H}_C$$

$$= I_{11}^C \frac{d}{dt}(\dot{\psi}\cos\beta)\,\vec{e}_1 + \dot{\psi}(\cos\beta\,\vec{e}_1 + \sin\beta\,\vec{e}_3) \times [I_{11}^C\dot{\psi}\cos\beta\,\vec{e}_1 + I_{33}^C(\dot{\phi} + \dot{\psi}\sin\beta)\,\vec{e}_3]$$

$$= I_{11}^C \frac{d}{dt}(\dot{\psi}\cos\beta)\,\vec{e}_1 + [(I_{11}^C - I_{33}^C)\dot{\psi}^2\sin\beta\cos\beta - I_{33}^C\dot{\phi}\dot{\psi}\cos\beta]\,\vec{e}_2$$

$$\Rightarrow \qquad 0 = \frac{d}{dt}(\dot{\psi}\cos\beta) = 0 \quad \Rightarrow \quad \dot{\psi}\cos\beta = \text{const.} \tag{1}$$

$$k_t\beta = (I_{33}^C - I_{11}^C)\dot{\psi}^2\sin\beta\cos\beta + I_{33}^C\dot{\phi}\dot{\psi}\cos\beta \tag{2}$$

It follows from eq. (1) that for constant rate $\dot{\psi}$, β is constant. Since the spin rate $\dot{\phi} \gg \dot{\psi}$ and the value of k_t is designed to ensure small β, eq. (2) $\Rightarrow$

$$k_t\beta \simeq I_{33}^C\dot{\phi}\dot{\psi}\cos\beta \qquad \Rightarrow \qquad \beta \simeq I_{33}^C\dot{\phi}\dot{\psi}/k_t \tag{3}$$

This result follows directly from $\beta \ll 1$. $\Rightarrow$ The precession axis $\vec{\mathrm{E}}$ is almost $\perp$ to the spin axis $\vec{e}_3$ $\Rightarrow$ (applied couple) $\simeq$ (gyroscopic couple) $= I_{33}^C ps = I_{33}^C\dot{\phi}\dot{\psi} = k_t\beta$. This gyroscope is called a *rate gyro* since the rotation β of the gimbal with respect to the frame directly measures *the rate of rotation* $\dot{\psi}$ *of the frame* (vehicle) along $\vec{\mathrm{E}}$. An electrical pick up is used to measure the rotation β. If $\dot{\psi}$ does not vary too fast, then the instantaneous value of $\dot{\psi}$ is obtained from eq. (2) or from the less accurate eq. (3).

If the spring is replaced by a viscous damper with an effective torsional damping constant c_t, then $\vec{M}_C = -c_t\dot{\beta}\,\vec{e}_2$ $\Rightarrow$ $k_t\beta$ in eq. (3), is replaced by $c_t\dot{\beta}$:

$$c_t\dot{\beta} \simeq I_{33}^C\dot{\phi}\dot{\psi}\cos\beta \qquad \Rightarrow \qquad \beta \simeq I_{33}^C\dot{\phi}\psi/c_t$$

Such a gyro is called a *rate integrating gyro* since the rotation β of the gimbal relative to the frame is itself proportional to the rotation ψ of the frame (vehicle).

In the guidance of space vehicles, a stable platform is oriented such that the 1st axis is east–west, the 2nd is north–south and the 3rd is vertical. Three integrating rate gyros are mounted on it in these directions. A deviation from its reference orientation is detected by the electrical pick up of a rate integrating gyro. A motor applies torque to realign the platform until the gyro indicates that it is back in its reference position.

3.20.3 Gyrocompass

A gyrocompass is an apparatus used to find the geographical north. It is a gyroscope with mass centre C lying at O whose latitude is λ (Fig. 3.37). Its spin axis $\vec{e}_3$ is mounted in a frame which can rotate only about the local vertical direction $\vec{e}_2$. The triad $\vec{e}_i$ is attached to this frame. Since the only rotation not allowed for the gyroscope is about $\vec{e}_1$, the constraining moment on it is $\vec{M}_C = M_{C_1}\,\vec{e}_1$. To ensure this, the frictional effects at the bearings have to be reduced as much as possible by using air bearings.

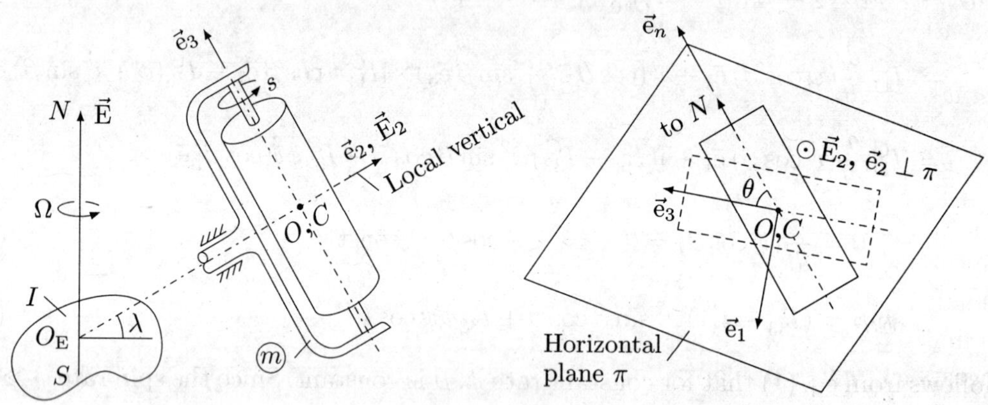

Fig. 3.37

Let $\vec{e}_3$ make angle $\theta(t)$ with ON, directed towards north of the earth's axis, in the horizontal plane π. The angular velocity $\Omega\,\vec{E}$ of the earth has a local vertical component $\Omega\sin\lambda\,\vec{e}_2$ and a horizontal component $\Omega\cos\lambda\,\vec{e}_n$ along ON. The angular velocities $\vec{\Omega}$ of the earth, $\vec{\Omega}^*$ of the frame ⓜ ($\vec{e}_i$), $\vec{\omega}$ of the gyroscope and its $\vec{H}_C$ with respect to an inertial frame I attached to earth's centre with orientation fixed relative to distant stars:

$$\vec{\Omega} = \Omega\vec{E} = \Omega[\sin\lambda\,\vec{e}_2 + \cos\lambda(\cos\theta\,\vec{e}_3 - \sin\theta\,\vec{e}_1)]$$

$$\vec{\Omega}^* = \vec{\Omega} + \dot{\theta}\,\vec{e}_2 = -\Omega\cos\lambda\sin\theta\,\vec{e}_1 + (\Omega\sin\lambda + \dot{\theta})\,\vec{e}_2 + \Omega\cos\lambda\cos\theta\,\vec{e}_3$$

$$\vec{\omega} = \vec{\Omega}^* + s\,\vec{e}_3 = -\Omega\cos\lambda\sin\theta\,\vec{e}_1 + (\Omega\sin\lambda + \dot{\theta})\,\vec{e}_2 + \omega_3\,\vec{e}_3$$

$$\vec{H}_C = -I_{11}^C\Omega\cos\lambda\sin\theta\,\vec{e}_1 + I_{11}^C(\Omega\sin\lambda + \dot{\theta})\,\vec{e}_2 + I_{33}^C\omega_3\,\vec{e}_3$$

where $\omega_3 = \Omega\cos\lambda\cos\theta + s$.

$$\dot{\vec{H}}_{C|\vec{e}_i} = -I_{11}^C\Omega\cos\lambda\cos\theta\dot{\theta}\,\vec{e}_1 + I_{11}^C\ddot{\theta}\,\vec{e}_2 + I_{33}^C\dot{\omega}_3\,\vec{e}_3$$

$$\vec{\Omega}^* \times \vec{H}_C = (\Omega\sin\lambda + \dot{\theta})(I_{33}^C\omega_3 - I_{11}^C\Omega\cos\lambda\cos\theta)\,\vec{e}_1$$
$$-[-I_{33}^C\Omega\omega_3\cos\lambda\sin\theta + I_{11}^C\Omega^2\cos^2\lambda\sin\theta\cos\theta]\,\vec{e}_2$$

By Euler's second axiom, $\vec{M}_C = \dot{\vec{H}}_{C|I} = \dot{\vec{H}}_{C|\vec{e}_i} + \vec{\Omega}^* \times \vec{H}_C \Rightarrow$

$$M_{C_1}\,\vec{e}_1 = [(\Omega\sin\lambda + \dot{\theta})(I_{33}^C\omega_3 - I_{11}^C\Omega\cos\lambda\cos\theta) - I_{11}^C\Omega\cos\lambda\cos\theta\dot{\theta}]\,\vec{e}_1$$

$$+[I_{11}^C\ddot{\theta} + I_{33}^C\,\Omega\,\omega_3\cos\lambda\sin\theta - I_{11}^C\Omega^2\cos^2\lambda\sin\theta\cos\theta]\,\vec{e}_2 + I_{33}^C\dot{\omega}_3\,\vec{e}_3$$

$$\vec{e}_3:\quad I_{33}^C\dot{\omega}_3 = 0 \quad\Rightarrow\quad \omega_3 = \text{const.}$$

$$\vec{e}_2:\quad I_{11}^C\ddot{\theta} + I_{33}^C\Omega\omega_3\cos\lambda\sin\theta - I_{11}^C\Omega^2\cos^2\lambda\sin\theta\cos\theta = 0 \tag{1}$$

$\Omega \ll \omega_3 \Rightarrow \Omega^2$ term in eq. (1) is neglected. Denote $n = (I_{33}^C\omega_3\Omega\cos\lambda/I_{11}^C)^{1/2}$, eq. (1) $\Rightarrow$

$$\ddot{\theta} + n^2\sin\theta = 0 \tag{2}$$

Equation (2) is identical to the equation of motion of a simple pendulum. A particular solution of eq. (2) is $\theta \equiv 0$. If the initial values of $\theta, \dot{\theta}$ are small, then the motion of the axis is oscillatory about the direction $\theta = 0$. By bisecting the angle of swing, we can find the geographical north. The gyroscope acts as a gyro-compass, indicating the geographical north unlike the magnetic compass which indicates the magnetic north. By geographical north, we mean the tip of the spin axis of the earth at the north pole. For small oscillations, the periodic time T of the swing of the axis of the gyroscope is $T = 2\pi/n = 2\pi(I_{11}^C/I_{33}^C\omega_3\Omega\cos\lambda)^{1/2}$. The latitude λ can be obtained by using this expression with measured values of I_{11}^C, I_{33}^C, Ω, ω_3, T. Since Ω is very small, the spin ω_3 must be large to make T reasonably small. For example, for $I_{33}^C\cos\lambda/I_{11}^C = 1, \omega_3 = 12000\,\text{rpm} = 1256.6\,\text{rad/s}, T = 20.76\,\text{s}$.

Example 3.47: The turbine of a ship, with axial moment of inertia I, rotates at ω about ship's longitudinal axis (Fig. E3.47a). The bearing B is axially free. Find the additional bearing reactions when the ship takes a right turn of radius R at speed v.

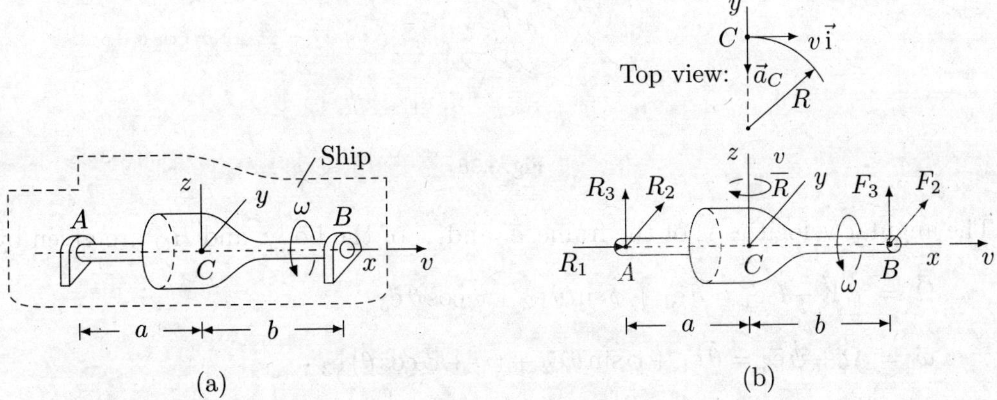

Fig. E3.47

Solution: Let $\vec{R}_A = R_1\,\vec{\mathrm{i}} + R_2\,\vec{\mathrm{j}} + R_3\,\vec{\mathrm{k}}$, $\vec{R}_B = F_2\,\vec{\mathrm{j}} + F_3\,\vec{\mathrm{k}}$ be the additional reactions. The FBD for the additional forces is shown in Fig. E3.47b. The spin rate $\vec{s}$, precession rate $\vec{p}$ and $\vec{a}_C$ are given by

$$\vec{s} = \omega\,\vec{\mathrm{i}}, \quad \vec{p} = -\frac{v}{R}\,\vec{\mathrm{k}}, \quad \vec{a}_C = -\frac{v^2}{R}\,\vec{\mathrm{j}}$$

$$\vec{F} = m\vec{a}_C \Rightarrow \quad R_1 = 0, \quad R_2 + F_2 = -mv^2/R, \quad R_3 + F_3 = 0 \qquad (1)$$

$$\vec{M}_C = \vec{p} \times I\vec{s} = -a\,\vec{\mathrm{i}} \times (R_1\,\vec{\mathrm{i}} + R_2\,\vec{\mathrm{j}} + R_3\,\vec{\mathrm{k}}) + b\,\vec{\mathrm{i}} \times (F_2\,\vec{\mathrm{j}} + F_3\,\vec{\mathrm{k}})$$

$$\Rightarrow \quad aR_3 - bF_3 = -I\omega v/R, \quad -aR_2 + bF_2 = 0 \qquad (2)$$

Equations (1) and (2) yield

$$R_1 = 0, \quad R_2 = \frac{b}{a}\,F_2 = -\frac{bmv^2}{R(a+b)}, \quad R_3 = -F_3 = -\frac{I\omega v}{R(a+b)}$$

3.21 General Motion of a Gyroscope

Consider a gyroscope with O being its centre of mass or its fixed point with principal moments of inertia $I_{11}^O = I_{22}^O$. The orientation of $\vec{\mathrm{e}}_3$ is described in terms of two angles θ, ϕ (Fig. 3.38), where θ is the angle between $\vec{\mathrm{e}}_3$ and $\vec{\mathrm{K}}$, and ϕ is the angle between the normal $\vec{n}\,(\vec{\mathrm{e}}_1)$ to the plane of $\vec{\mathrm{e}}_3$, $\vec{\mathrm{K}}$ and the X-axis. $\vec{\mathrm{e}}_2$ is coplanar with $\vec{\mathrm{e}}_3$ and $\vec{\mathrm{K}}$. The spin rate is the rate of rotation $\dot{\psi}$ about axis $\vec{\mathrm{e}}_3$. Only $\vec{\mathrm{e}}_3$ is attached to the body.

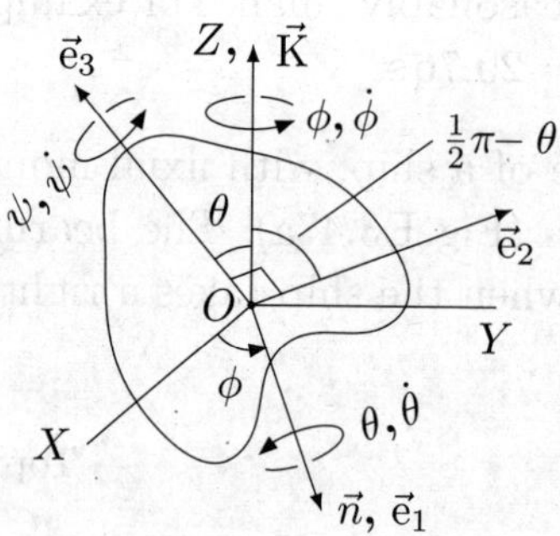

Fig. 3.38

The angular velocities $\vec{\Omega}$ of the frame $\vec{\mathrm{e}}_i$ and $\vec{\omega}$ of the body, and $\vec{H}_O$ are given by:

$$\vec{\Omega} = \dot{\phi}\,\vec{\mathrm{K}} + \dot{\theta}\,\vec{\mathrm{e}}_1 = \dot{\theta}\,\vec{\mathrm{e}}_1 + \dot{\phi}\sin\theta\,\vec{\mathrm{e}}_2 + \dot{\phi}\cos\theta\,\vec{\mathrm{e}}_3$$

$$\vec{\omega} = \vec{\Omega} + \dot{\psi}\,\vec{\mathrm{e}}_3 = \dot{\theta}\,\vec{\mathrm{e}}_1 + \dot{\phi}\sin\theta\,\vec{\mathrm{e}}_2 + (\dot{\psi} + \dot{\phi}\cos\theta)\,\vec{\mathrm{e}}_3$$

$$\vec{H}_O = I_{11}^O(\dot{\theta}\,\vec{\mathrm{e}}_1 + \dot{\phi}\sin\theta\,\vec{\mathrm{e}}_2) + I_{33}^O(\dot{\psi} + \dot{\phi}\cos\theta)\,\vec{\mathrm{e}}_3$$

The equations of motion are $\vec{F} = m\vec{a}_C$ and

$$\vec{M}_O = \dot{\vec{H}}_{O|I} = \dot{\vec{H}}_{O|\vec{e}_i} + \vec{\Omega} \times \vec{H}_O \quad \Rightarrow$$

$$M_{O_1} = I_{11}^O(\ddot{\theta} - \dot{\phi}^2 \sin\theta\cos\theta) + I_{33}^O\dot{\phi}\sin\theta(\dot{\psi} + \dot{\phi}\cos\theta)$$

$$M_{O_2} = I_{11}^O(\ddot{\phi}\sin\theta + 2\dot{\phi}\dot{\theta}\cos\theta) - I_{33}^O\dot{\theta}(\dot{\psi} + \dot{\phi}\cos\theta)$$

$$M_{O_3} = I_{33}^O\frac{d}{dt}(\dot{\psi} + \dot{\phi}\cos\theta)$$

3.22 Spinning Top

The triad $\vec{E}_i$ is fixed in inertial frame with $\vec{E}_3$ vertical. The FBD of a spinning top with fixed point O is shown in Fig. 3.39a with its axis $\vec{e}_3$ at angle θ to the vertical axis $\vec{E}_3$. Choose a triad $\vec{e}_i$ with $\vec{e}_1$ in the direction of $\vec{E}_3 \times \vec{e}_3$, i.e., normal to the vertical plane through $\vec{E}_3$ and $\vec{e}_3$ and making an angle ϕ with the horizontal axis $\vec{E}_1$. The 3rd vector $\vec{e}_2 = \vec{e}_3 \times \vec{e}_1$, is normal to $\vec{e}_3$ in the vertical plane of $\vec{e}_3$ and $\vec{E}_3$.

The axis $\vec{e}_3$ precesses at the rate $\dot{\phi}$ about $\vec{E}_3$ and nutates at rate $\dot{\theta}$ about $\vec{e}_1$ with reference to inertial frame I. Let the top have spin rate $s\,\vec{e}_3$ with respect to the triad $\vec{e}_i$. The angular velocity $\vec{\omega}$ and $\vec{H}_O$ of the top with respect to I are given by

$$\vec{\omega} = \dot{\phi}\,\vec{E}_3 + \dot{\theta}\,\vec{e}_1 + s\,\vec{e}_3 = \dot{\theta}\,\vec{e}_1 + \dot{\phi}\sin\theta\,\vec{e}_2 + (s + \dot{\phi}\cos\theta)\,\vec{e}_3$$

$$\Rightarrow \quad \omega_3 = s + \dot{\phi}\cos\theta$$

$$\vec{H}_O = I_{11}^O(\dot{\theta}\,\vec{e}_1 + \dot{\phi}\sin\theta\,\vec{e}_2) + I_{33}^O\omega_3\,\vec{e}_3$$

The moment about O is $\vec{M}_O = mgh\sin\theta\,\vec{e}_1$. The 3 integrals of the equations of motion for $\dot{\theta}$, $\dot{\phi}$, ω_3 (or s) can be obtained directly:

1. from $M_{O3} = \vec{M}_O \cdot \vec{e}_3 = 0$ using Euler's equation of motion,
2. from $\vec{M}_O \cdot \vec{E}_3 = 0 \;\Rightarrow\; \vec{H}_O \cdot \vec{E}_3 = \text{constant}$, and
3. from $T + V = T_0 + V_0 = E_0$ since $\vec{R}_O$ is workless and mg is conservative.

Since $I_{11}^O = I_{22}^O$ and $M_{O_3} = 0$, the third Euler's equation yields ω_3 as constant:

$$M_{O3} = I_{33}^O\dot{\omega}_3 - (I_{11}^O - I_{22}^O)\omega_1\omega_2 = 0 \quad \Rightarrow \quad \dot{\omega}_3 = 0$$

$$\Rightarrow \quad \omega_3 = s + \dot{\phi}\cos\theta = \text{constant} = \omega_3(0) \tag{1}$$

$$\vec{H}_O \cdot \vec{E}_3 = [I_{11}^O(\dot{\theta}\,\vec{e}_1 + \dot{\phi}\sin\theta\,\vec{e}_2) + I_{33}^O\omega_3\,\vec{e}_3] \cdot (\sin\theta\,\vec{e}_2 + \cos\theta\,\vec{e}_3) = \text{const.}$$

$$\Rightarrow \quad I_{11}^O\dot{\phi}\sin^2\theta + I_{33}^O\omega_3\cos\theta = H_V \text{ (say)} \tag{2}$$

$$T + V = \tfrac{1}{2}[I^O_{11}(\dot\theta^2 + \dot\phi^2 \sin^2\theta) + I^O_{33}\omega_3^2] + mgh\cos\theta = E_0 \tag{3}$$

$$\text{eq. (2)} \Rightarrow \quad \dot\phi = (H_V - I^O_{33}\omega_3\cos\theta)/I^O_{11}\sin^2\theta \tag{4}$$

$$\text{eq. (3)} \Rightarrow \quad I^O_{11}\left[\dot\theta^2 + \frac{(H_V - I^O_{33}\omega_3\cos\theta)^2}{{I^O_{11}}^2\sin^2\theta}\right] + I^O_{33}\omega_3^2 = 2(E_0 - mgh\cos\theta) \tag{5}$$

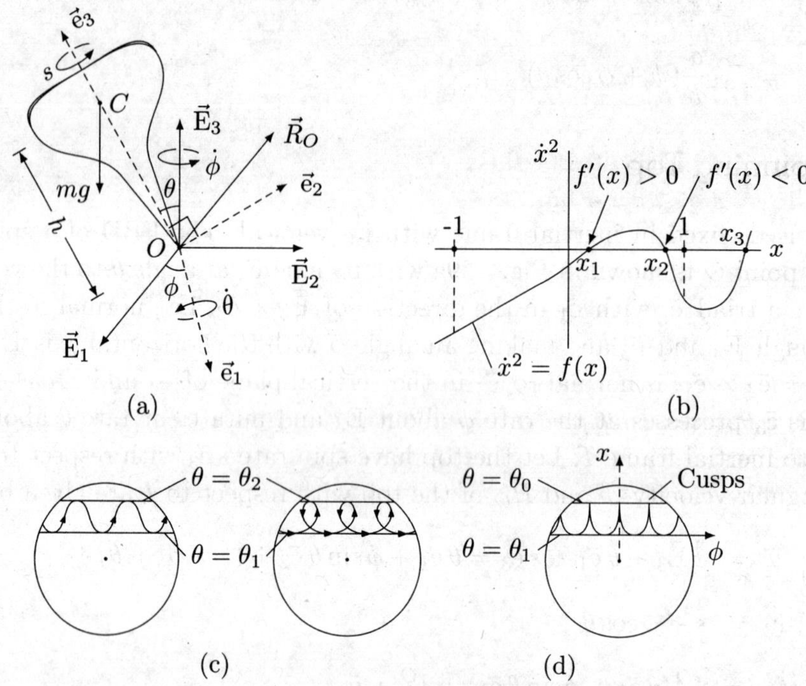

Fig. 3.39

From the initial conditions, $\omega_3(0)$, H_V, E_0 are computed. For a given θ, the values of $\dot\phi$, $\dot\theta$ and s are obtained from eqs (4), (5), (1). The solution of the un-coupled eq. (5) for θ can be expressed in terms of elliptic functions. We study the variation in the height $h\cos\theta$ of C, i.e., in the variable $x = \cos\theta$. With $\dot x = -\sin\theta\dot\theta$, eq. (5) transforms to

$$\dot x^2 = f(x) = (2E_0 - I^O_{33}\omega_3^2 - 2mghx)\frac{1 - x^2}{I^O_{11}} - \frac{(H_V - I^O_{33}\omega_3 x)^2}{{I^O_{11}}^2} \tag{6}$$

where $f(x)$ is a cubic function in x and the coefficient of x^3 is positive. $\Rightarrow$

$$f(x) > 0 \text{ for } x \gg 1, \qquad f(x) < 0 \text{ for } x \ll -1, \qquad f(\pm 1) \le 0$$

$$f(x) = \dot x^2 \ge 0 \text{ for actual range } \quad -1 \le x_1 \le x_2 \le 1$$

$\Rightarrow$ the graph of $f(x)$ has the general features shown in Fig. 3.39b, since a cubic function cannot have more than 3 changes in sign. $f(x)$ has 3 real zeros x_1, x_2, x_3, such that

$$-1 < x_1 < x_2 < 1 < x_3 \tag{7}$$

one or more inequality signs in it may be an equality for special cases. The motion occurs between x_1 and x_2. Equations (6) and (4) can be expressed as

$$\begin{aligned} f(x) &= 2mgh(x - x_1)(x - x_2)(x - x_3)/I_{11}^O \\ \dot{\phi} &= (H_V - I_{33}^O \omega_3 x)/I_{11}^O(1 - x^2) \end{aligned} \tag{8}$$

To show the essential features of the motion, we plot the locus of the point of intersection of the $\vec{e}_3$ axis with a unit sphere having its centre at O. The point oscillates between the two levels $\theta = \theta_1 = \cos^{-1} x_1$ and $\theta = \theta_2 = \cos^{-1} x_2$ (Fig. 3.39c). Loops will appear on the curve if $\dot{\phi}$ changes sign during motion, i.e., $\dot{\phi}$ becomes zero. By eq. (8), $\dot{\phi} = 0$ at $x = H_V / I_{33}^O \omega_3$, and loops will occur only if this value of x lies between x_1 and x_2. This depends on the values of ω_3, H_V, E_0, i.e., on how the motion is started.

Consider the case of motion of a top spinning at rate ω_3 about its axis $\vec{e}_3$ in the fixed position $\theta(0) = \theta_0$ ($x_0 = \cos\theta_0$) in I after its axis is released. Equations (2), (3), (6) $\Rightarrow$

$$H_V = I_{33}^O \omega_3 x_0, \qquad E_0 = mghx_0 + \tfrac{1}{2} I_{33}^O \omega_3^2 \tag{9}$$

$$f(x) = (x_0 - x)[2mgh(1 - x^2) - I_{33}^{O\,2} \omega_3^2 (x_0 - x)/I_{11}^O]/I_{11}^O \tag{10}$$

$$\Rightarrow \quad f'(x_0) = -2mgh(1 - x_0^2)/I_{11}^O < 0 \tag{11}$$

Hence, the zero x_0 of $f(x)$ corresponds to x_2 of Fig. 3.39b, where $f(x_2) = 0$, $f'(x_2) < 0$. The other roots of $f(x)$ are:

$$x_1, x_3 = \alpha \mp (\alpha^2 - 2\alpha x_0 + 1)^{1/2}$$

where $\alpha = I_{33}^{O\,2} \omega_3^2 / 4 I_{11}^O mgh$ and $x_1 \leq x \leq x_0$. The axis $\vec{e}_3$ of the top falls down from the initial inclination θ_0 to $\theta_1 = \cos^{-1} x_1$ and rises again up to $\theta = \theta_0$, where the axis is again instantaneously at rest. This cycle is repeated indefinitely. From eqs (8)–(10)

$$\dot{\phi} = I_{33}^O \omega_3 (x_0 - x)/I_{11}^O (1 - x^2) \geq 0$$

$$\frac{d\phi}{dx} = \frac{\dot{\phi}}{\dot{x}} = \pm \frac{I_{33}^O \omega_3 (x_0 - x)}{I_{11}^O (1 - x^2)[f(x)]^{1/2}} = \pm (x_0 - x)^{1/2} g(x) \tag{12}$$

where $\quad g(x) = I_{33}^O \omega_3 / (1 - x^2)[2 I_{11}^O mgh(1 - x^2) - I_{33}^{O\,2} \omega_3^2 (x_0 - x)]^{1/2}$

Equation (12) $\Rightarrow$ $\frac{d\phi}{dx}(x_0) = 0$. Thus, at the highest position of the axis at $\theta = \theta_0$, the path of the representative point on the unit sphere meets the circle $\theta = \theta_0$ at right angles, i.e., the path has cusps at these points directed upwards (Fig. 3.39d).

For fast spin rate ω_3, α is large and binomial expansion of x_1 yields $\theta_1 - \theta_0$ as small:

$$x_1 = \alpha - \alpha\left(1 + \frac{1}{\alpha^2} - \frac{2x_0}{\alpha}\right)^{1/2}$$

$$= -\tfrac{1}{2}\left(\frac{1}{\alpha^2} - \frac{2x_0}{\alpha}\right)\alpha - \tfrac{1}{2}(-\tfrac{1}{2})(\tfrac{1}{2})\left(\frac{1}{\alpha^2} - \frac{2x_0}{\alpha}\right)^2\alpha + \ldots \simeq x_0 - \frac{1 - x_0^2}{2\alpha}$$

$$\Rightarrow \qquad \cos\theta_0 - \cos\theta_1 = (1 - x_0^2)/2\alpha \ll 1 \quad \Rightarrow \quad \theta_1 - \theta_0 \ll 1$$

The case of uniform precession of axis $\vec{e}_3$ (θ =constant) has been discussed in Sec. 3.20.

§ **Exercise 3.13:** Differentiating eq. (6): $\ddot{x} - \frac{1}{2}f'(x) = 0$. Hence, for a sleeping top $\theta \equiv 0\,(x \equiv 1)$ and $f'(1) = 0$. For its perturbed motion, $x = 1 + \eta$, show that $\ddot{\eta} - \frac{1}{2}f''(1)\eta = 0$ and hence prove stability if $\alpha > 1$.

Example 3.48: A spinning top (Fig. E3.48a) consists of a thin disc joined to a thin rod . (a) Find its minimum angular velocity ω_0 for stable motion with its axis remaining vertical. (b) For (1) $\omega_3 = 2\omega_0$ and (2) $\omega_3 = \frac{1}{2}\omega_0$, find the range of θ for oscillatory motion if the top is initially disturbed to $\theta = 4°$. (c) If $\omega_3 = 2\omega_0$ and $\theta_0 = 30°$, find $\dot{\phi}_{\max}$ and the value of $\dot{\theta}$ for a given θ. (d) Initially $\theta = \theta_0 = 30°$ and the axis precesses at rate $\Omega\vec{E}$ with $\Omega = 0.5\omega_0$. Find $\dot{\phi}$, $\dot{\theta}$ for a given θ and the range of θ.

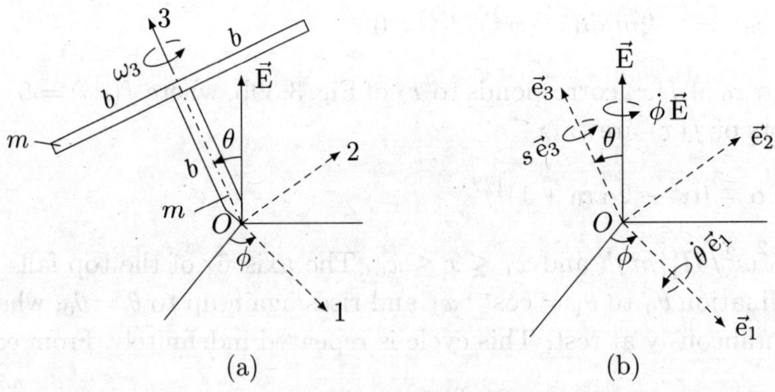

Fig. E3.48

Solution: $h = OC = (mb/2 + mb)/2m = 3b/4, \qquad M = 2m,$

$$Mgh = (2m)g(3b/4) = 3mgb/2, \qquad I_{33}^O = \tfrac{1}{2}mb^2,$$

$$I_{11}^O = (\tfrac{1}{4}mb^2 + mb^2) + \tfrac{1}{3}mb^2 = 19mb^2/12$$

(a) Equation (3.83) $\Rightarrow \quad \omega_0 = (4MghI_{11}^O/I_{33}^{O\,2})^{1/2} = (38g/b)^{1/2}$

which corresponds to the case $\alpha = 1$ as per Exercise 3.13.

(b) $\quad \alpha = I_{33}^{O\,2}\omega_3^2/4I_{11}^O Mgh, \qquad x_0 = \cos\theta_0 = \cos 4^\circ = 0.9976.$

(1) $\omega_3 = 2\omega_0 = 2(38g/b)^{1/2} \quad \Rightarrow \quad \alpha = 4,$

$$\Rightarrow \qquad \cos\theta_1 = \alpha - (\alpha^2 - 2\alpha x_0 + 1)^{1/2} = 0.99675 \quad \Rightarrow \quad \theta_1 = 4.62^\circ$$

The axis oscillates between $\theta = 4^\circ$ and $\theta = 4.62^\circ$ and so the motion is stable as it should be since $\omega_3 > \omega_0$.

(2) $\omega_3 = \frac{1}{2}\omega_0 = \frac{1}{2}(38g/b)^{1/2} \quad \Rightarrow \quad \alpha = 0.25,$

$$\Rightarrow \qquad \cos\theta_1 = \alpha - (\alpha^2 - 2\alpha x_0 + 1)^{1/2} = -0.5008 \quad \Rightarrow \quad \theta_1 = 120.1^\circ$$

The axis oscillates between $\theta = 4^\circ$ and $\theta = 120.1^\circ$ and so the motion is unstable as it should be since $\omega_3 < \omega_0$.

(c) $\omega_3 = 2\omega_0 = 2(38g/b)^{1/2}, \qquad \theta_0 = 30^\circ, \qquad x_0 = \cos\theta_0 = 0.866 \Rightarrow$

$$\alpha = 4 \quad \Rightarrow \quad \cos\theta_1 = \alpha - (\alpha^2 - 2\alpha x_0 + 1)^{1/2} = 0.8264 \quad \Rightarrow \quad \theta_1 = 34.27^\circ.$$

The top nutates between 30° and 34.27°.

$$\begin{aligned} H_V &= I_{33}^O\omega_3\cos\theta_0 = \dot\phi I_{11}^O\sin^2\theta + I_{33}^O\omega_3\cos\theta \\ \Rightarrow \quad \dot\phi &= 6\omega_3(0.866 - \cos\theta)/19\sin^2\theta \qquad (1) \\ E &= \tfrac{1}{2}I_{11}^O(\dot\theta^2 + \dot\phi^2\sin^2\theta) + \tfrac{1}{2}I_{33}^O\omega_3^2 + Mgh\cos\theta \\ &= E_0 = \tfrac{1}{2}I_{33}^O\omega_3^2 + Mgh\cos\theta_0 \qquad (2) \end{aligned}$$

Equations (1) and (2) $\Rightarrow$

$$\dot\theta^2 = -(36g/19b)(\cos^2\theta - 8\cos\theta + 5.928)(0.866 - \cos\theta)/\sin^2\theta \qquad (3)$$

$\dot\phi$ is locally maximum for $\frac{d\dot\phi}{d\theta} = 0$, which from eq. (1) yields

$$\cos^2\theta - 1.732\cos\theta + 1 = 0$$

with no admissible root. $\Rightarrow \dot\phi_{max}$ occurs at the end of the range of θ at $\theta = \theta_1 = 34.27^\circ$:

$$\dot\phi_{max} = \frac{6\omega_3(0.866 - \cos\theta_1)}{19\sin^2\theta_1} = 0.03944\omega_3$$

(d) Initially at $t=0$, $\vec{\omega}(0)=\Omega\,\vec{\mathrm{E}}=\Omega(\sin\theta_0\,\vec{\mathrm{e}}_2+\cos\theta_0\,\vec{\mathrm{e}}_3)$

$$\begin{aligned}
\Rightarrow \quad \vec{H}_O(0) &= I_{11}^O\Omega\sin\theta_0\,\vec{\mathrm{e}}_2+I_{33}^O\Omega\cos\theta_0\,\vec{\mathrm{e}}_3\\
\omega_3 &= \Omega\cos\theta_0\\
H_V &= \vec{H}_O\cdot\vec{\mathrm{E}}=(I_{11}^O\sin^2\theta_0+I_{33}^O\cos^2\theta_0)\Omega\\
E_0 &= \tfrac{1}{2}I_{11}^O\Omega^2\sin^2\theta_0+\tfrac{1}{2}I_{33}^O\omega_3^2+Mgh\cos\theta_0 \qquad (4)
\end{aligned}$$

For the general position θ (Fig. E3.48b),

$$\begin{aligned}
\vec{\omega} &= \dot{\phi}\,\vec{\mathrm{E}}+\dot{\theta}\,\vec{\mathrm{e}}_1+s\,\vec{\mathrm{e}}_3=\dot{\theta}\,\vec{\mathrm{e}}_1+\dot{\phi}\sin\theta\,\vec{\mathrm{e}}_2+(\dot{\phi}\cos\theta+s)\,\vec{\mathrm{e}}_3\\
&= \dot{\theta}\,\vec{\mathrm{e}}_1+\dot{\phi}\sin\theta\,\vec{\mathrm{e}}_2+\omega_3\,\vec{\mathrm{e}}_3\\
\vec{H}_O &= I_{11}^O(\dot{\theta}\,\vec{\mathrm{e}}_1+\dot{\phi}\sin\theta\,\vec{\mathrm{e}}_2)+I_{33}^O\omega_3\,\vec{\mathrm{e}}_3\\
H_V &= \vec{H}_O\cdot\vec{\mathrm{E}}=I_{11}^O\dot{\phi}\sin^2\theta+I_{33}^O\omega_3\cos\theta\\
&= I_{11}^O\dot{\phi}\sin^2\theta+I_{33}^O\Omega\cos\theta\cos\theta_0\\
\Rightarrow \quad \dot{\phi} &= [I_{11}^O\sin^2\theta_0+I_{33}^O\cos\theta_0(\cos\theta_0-\cos\theta)]\Omega/I_{11}^O\sin^2\theta \qquad (5)\\
E &= \tfrac{1}{2}I_{11}^O(\dot{\theta}^2+\dot{\phi}^2\sin^2\theta)+\tfrac{1}{2}I_{33}^O\omega_3^2+Mgh\cos\theta\\
&= E_0=\tfrac{1}{2}I_{11}^O\Omega^2\sin^2\theta_0+\tfrac{1}{2}I_{33}^O\omega_3^2+Mgh\cos\theta_0 \qquad (6)
\end{aligned}$$

Equations (5) and (6) $\Rightarrow$

$$\begin{aligned}
\dot{\theta}^2 &= 2Mgh(\cos\theta_0-\cos\theta)/I_{11}^O+\Omega^2\sin^2\theta_0\\
&\quad -[I_{11}^O\sin^2\theta_0+I_{33}^O\cos\theta_0(\cos\theta_0-\cos\theta)]^2\Omega^2/I_{11}^{O\,2}\sin^2\theta\\
&= \frac{(\cos\theta_0-\cos\theta)}{I_{11}^{O\,2}\sin^2\theta}[2I_{11}^OMgh(1-\cos^2\theta)+\Omega^2\{I_{11}^{O\,2}\sin^2\theta_0(\cos\theta_0+\cos\theta)\\
&\quad -2I_{11}^OI_{33}^O\sin^2\theta_0\cos\theta_0-I_{33}^{O\,2}\cos^2\theta_0(\cos\theta_0-\cos\theta)\}] \qquad (7)
\end{aligned}$$

$\dot{\theta}=0$ for $\theta=\theta_0$ and for $\theta=\theta_1$, where θ_1 is the admissible root of the quadratic factor $[\ldots]$ in eq. (7). With $\theta_0=30°$, $\Omega=0.5\omega_0=0.5(38g/b)^{1/2}$, eq. (7) becomes

$$\dot{\theta}^2=\left(\frac{144g}{361b}\right)(-4.75\cos^2\theta+7.735\cos\theta+5.1071)\frac{0.866-\cos\theta}{\sin^2\theta} \qquad (8)$$

For extremal values of θ, $\dot{\theta} = 0$, i.e., one root is $\theta = \theta_0$ and eq. (8) yields

$$-4.75\cos^2\theta + 7.735\cos\theta + 5.1071 = 0 \quad \Rightarrow \quad \cos\theta = 2.133, -0.5042$$

The root is $\theta_1 = \cos^{-1}(-0.5042) = 120.3°$. The top nutates between 30° and 120.3°.

3.23 Rolling Without Slip of a 'Disc'

For a rigid body, the governing equations of motion of its centre of mass C do not always decouple from the governing equations of its angular motion. For example, the equations of motion are coupled for a 'disc' rolling without slip. A child's hoop acquires stability from its spin and can pass over small obstacles without being upset.

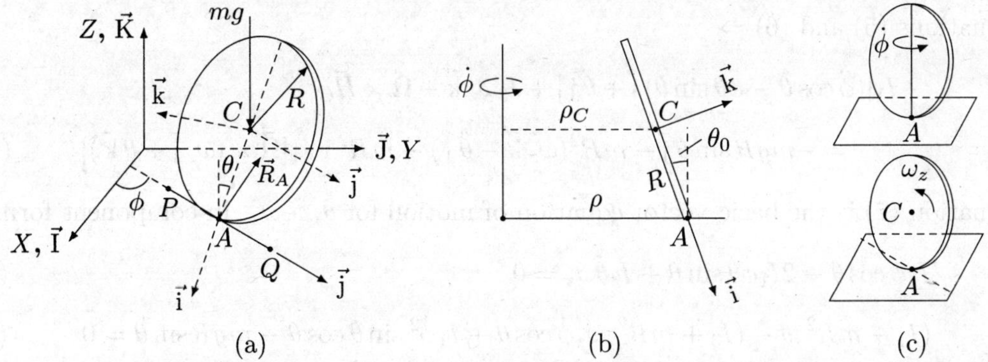

Fig. 3.40

Consider the case of such a body with a plane of symmetry xy having a sharp circular edge ('disc'), of radius R in it, which rolls without slip ($\vec{v}_A = \vec{0}$) on a horizontal plane (Fig. 3.40a). Its mass is m and the centre of mass C is at the centre of the 'disc'. The principal axes of inertia at C are the axis $\vec{k}$ of the 'disc' and $\vec{i}, \vec{j}$ with $I^C_{xx} = I^C_{yy} = I_1 \neq I^C_{zz} = I_3$. Hence, every line in the plane of symmetry is a principal axis of inertia at C. Let ϕ be the angle which the trace PQ of the plane of the 'disc' on the fixed XY plane makes with $\vec{I}$ and let θ be the inclination of the plane of the 'disc' to the vertical. Choose a frame 1 ($\vec{i}, \vec{j}, \vec{k}$) with $\vec{k}$ along the axis of the 'disc' and $\vec{i}$ directed from C to the point of contact A. Since $\vec{k}, \vec{i}$ are in a vertical plane, $\vec{j} = \vec{k} \times \vec{i}$ is a horizontal axis parallel to PQ. The angular velocity $\vec{\omega}$ of the rigid body differs from the angular velocity $\vec{\Omega}$ of frame 1 by its spin rate $s\,\vec{k}$ with respect to frame 1:

$$\begin{aligned}\vec{\Omega} &= -\dot{\theta}\vec{j} + \dot{\phi}\vec{K} = -\dot{\theta}\vec{j} + \dot{\phi}(-\cos\theta\,\vec{i} + \sin\theta\,\vec{k}) \\ &= -\cos\theta\dot{\phi}\,\vec{i} - \dot{\theta}\,\vec{j} + \dot{\phi}\sin\theta\,\vec{k} \end{aligned} \tag{1}$$

$$\vec{\omega} = \vec{\Omega} + s\vec{k} = -\cos\theta\dot{\phi}\vec{i} - \dot{\theta}\vec{j} + \omega_z\vec{k}, \qquad \omega_z = s + \dot{\phi}\sin\theta \tag{2}$$

$$\vec{v}_C = \vec{v}_A + \vec{\omega}\times\vec{r}_{CA} = \vec{\omega}\times(-R\vec{i}) = -R(\omega_z\vec{j} + \dot{\theta}\vec{k}) \tag{3}$$

$$\vec{H}_C = -I_1(\dot{\phi}\cos\theta\vec{i} + \dot{\theta}\vec{j}) + I_3\omega_z\vec{k} \tag{4}$$

$\vec{R}_A$, θ, ϕ, $s\,(=\dot{\psi})$ are governed by $\vec{F} = m\dot{\vec{v}}_{C|I}$, $\vec{M}_C = \dot{\vec{H}}_{C|I}$:

$$\vec{R}_A - mg\vec{K} = m\frac{d}{dt}[-R(\omega_z\vec{j} + \dot{\theta}\vec{k})]_{|I}$$

$$\Rightarrow \quad \vec{R}_A = mg(\sin\theta\vec{k} - \cos\theta\vec{i}) - mR(\dot{\omega}_z\vec{j} + \ddot{\theta}\vec{k}) - mR\vec{\Omega}\times(\omega_z\vec{j} + \dot{\theta}\vec{k}) \tag{5}$$

$$\dot{\vec{H}}_{C|I} = \frac{d}{dt}[-I_1(\dot{\phi}\cos\theta\vec{i} + \dot{\theta}\vec{j}) + I_3\omega_z\vec{k}]_{|I} = \vec{M}_C = R\vec{i}\times\vec{R}_A \tag{6}$$

Equations (5) and (6) $\Rightarrow$

$$-I_1[(\ddot{\phi}\cos\theta - \dot{\phi}\dot{\theta}\sin\theta)\vec{i} + \ddot{\theta}\vec{j}] + I_3\dot{\omega}_z\vec{k} + \vec{\Omega}\times\vec{H}_C$$
$$= -mgR\sin\theta\vec{j} - mR^2(\dot{\omega}_z\vec{k} - \ddot{\theta}\vec{j}) - mR^2\vec{i}\times[\vec{\Omega}\times(\omega_z\vec{j} + \dot{\theta}\vec{k})] \tag{7}$$

Equation (7) is the basic vector equation of motion for θ, ϕ, ω_z. In component form:

$$I_1\ddot{\phi}\cos\theta - 2I_1\dot{\phi}\dot{\theta}\sin\theta + I_3\dot{\theta}\omega_z = 0 \tag{8}$$

$$(I_1 + mR^2)\ddot{\theta} - (I_3 + mR^2)\omega_z\dot{\phi}\cos\theta + I_1\dot{\phi}^2\sin\theta\cos\theta - mgR\sin\theta = 0 \tag{9}$$

$$(I_3 + mR^2)\dot{\omega}_z + mR^2\dot{\phi}\dot{\theta}\cos\theta = 0 \tag{10}$$

We consider a special motion of rolling without slip with the disc inclined at constant angle $\theta = \theta_0$ (Fig. 3.40b) to the vertical and the trace PQ having constant $\dot{\phi}$ and always tangent to the locus of the contact points on the ground. Hence, $\dot{\theta} = 0$, $\ddot{\theta} = \ddot{\phi} = 0$ and eqs (10) and (9) become

$$\dot{\omega}_z = 0 \tag{11}$$

$$(I_3 + mR^2)\dot{\phi}\,\omega_z\cos\theta_0 - I_1\dot{\phi}^2\sin\theta_0\cos\theta_0 = -mgR\sin\theta_0 \tag{12}$$

$$\Rightarrow \quad \omega_z = (I_1\cos\theta_0 - mgR/\dot{\phi}^2)\dot{\phi}\tan\theta_0/(I_3 + mR^2) \tag{13}$$

Let ρ_C be the radius of curvature of the path of C (Fig. 3.40b). The radius of curvature ρ of the locus of points of contact on the ground is, $\rho = \rho_C + R\sin\theta_0$. Equation (3)

$$\Rightarrow \quad \vec{v}_C = -R\omega_z\vec{j} = \rho_C\dot{\phi}\vec{j} = v_C\vec{j}$$

$$\Rightarrow \quad \dot{\phi} = v_C/\rho_C = -\omega_z R/\rho_C, \qquad \omega_z = -\dot{\phi}\rho_C/R \tag{14}$$

As $\dot{\phi}$ is constant, eqs (13), (14) imply that ω_z, ρ_C and v_C are constant. Substituting ω_z from eq. (14) into eq. (12) :

$$\dot{\phi}^2 = \frac{v_C^2}{\rho_C^2} = \frac{\omega_z^2 R^2}{\rho_C^2} = \frac{mgR\tan\theta_0}{I_1\sin\theta_0 + (I_3 + mR^2)\rho_C/R} \tag{15}$$

C moves in a horizontal circle of radius ρ_C at constant speed v_C given by eq. (15). Of the 4 variables ρ_C, θ_0, ω_z, $\dot{\phi}$, any two can be chosen at will and the other two are obtained from eqs (15). For $\theta_0 \simeq 0$, eq. (15) $\Rightarrow$ $\rho_C \simeq v_C^2(I_3 + mR^2)/mR^2 g\theta_0$.

For a vertical 'disc', $\theta_0 = 0$ (Fig. 3.40c) and for constant $\dot{\phi}$, eq. (8) yields $\dot{\phi}\omega_z = 0 \Rightarrow$

(i) $\omega_z = 0$, $\dot{\phi}$ arbitrary, eq. (14) $\Rightarrow$ $v_C = 0$, $\rho_C = 0$, i.e., the 'disc' spins about a vertical diameter at the rate $\dot{\phi}$.

(ii) $\dot{\phi} = 0$, ω_z arbitrary, eq. (14) $\Rightarrow$ $v_C = -R\omega_z$, $\rho_C = \infty$, i.e., the 'disc' rolls with C moving in a straight line.

We discuss the *stability of these uniform motions for* $\theta_0 = 0$. For nearly vertical motion, using $\sin\theta \simeq \theta$, $\cos\theta \simeq 1$, eqs (8)–(10) yield

$$I_1\ddot{\phi} + I_3\dot{\theta}\omega_z - 2I_1\theta\dot{\theta}\dot{\phi} = 0 \tag{16}$$

$$(I_1 + mR^2)\ddot{\theta} - (I_3 + mR^2)\omega_z\dot{\phi} + I_1\dot{\phi}^2\theta - mgR\theta = 0 \tag{17}$$

$$(I_3 + mR^2)\dot{\omega}_z + mR^2\dot{\phi}\dot{\theta} = 0 \tag{18}$$

For case (i) : For unperturbed motion, $\theta = 0$, $\dot{\phi} \neq 0$, $\dot{\theta} = \omega_z = 0$ $\Rightarrow$ for perturbed motion ω_z, θ, $\dot{\theta}$ are small. Neglecting the 2nd order terms $\dot{\theta}\omega_z$, $\theta\dot{\theta}$ in eq. (16),

$$I_1\ddot{\phi} = 0 \quad \Rightarrow \quad \dot{\phi}(t) = \dot{\phi}(0)$$

$$\text{eq. (18)} \Rightarrow \quad (I_3 + mR^2)\omega_z = c_1 - mR^2\dot{\phi}\theta \tag{19}$$

where c_1 is a constant. Substituting ω_z from eq. (19) in eq. (17) yields

$$(I_1 + mR^2)\ddot{\theta} + [(I_1 + mR^2)\dot{\phi}^2 - mgR]\theta = c_1\dot{\phi} \tag{20}$$

The perturbed motion is bounded and hence position $\theta = 0$ is stable if the coefficient of θ in eq. (20) is positive, i.e., if

$$|\dot{\phi}| > (mgR/I_{yy}^A)^{1/2} \tag{21}$$

For case (ii) : For unperturbed motion, $\theta = 0$, $\omega_z \neq 0$, $\dot{\theta} = \dot{\phi} = 0$ $\Rightarrow$ for perturbed motion $\dot{\phi}, \theta, \dot{\theta}$ are small. So, the 2nd order term $\dot{\phi}\dot{\theta}$ in eqs (16) and (18), and 3rd order term $\theta\dot{\phi}^2$ in eq. (17) are neglected. Equation (18) $\Rightarrow$ $\dot{\omega}_z = 0$ $\Rightarrow$ $\omega_z(t) = \omega_z(0)$.

$$\text{eq. (16)} \Rightarrow \qquad \frac{d}{dt}(I_1\dot{\phi} + I_3\omega_z\theta) = 0$$

$$\Rightarrow \qquad I_1\dot{\phi} + I_3\omega_z\theta = \text{constant} = c_2 \quad \text{(say)} \tag{22}$$

Substituting $\dot{\phi}$ from eq. (22) into eq. (17) yields

$$(I_1 + mR^2)\ddot{\theta} + \left[\frac{I_3 + mR^2}{I_1} I_3\omega_z^2 - mgR\right]\theta = \frac{I_3 + mR^2}{I_1}\omega_z c_2 \tag{23}$$

The perturbed motion is bounded and hence position $\theta = 0$ is stable if the coefficient of θ in eq. (23) is positive, i.e., if

$$\frac{v_C}{R} = |\omega_z| > [mgRI_1/I_3(I_3 + mR^2)]^{1/2} \tag{24}$$

§ **Exercise 3.14:** For a lamina, a circular hoop, and a circular disc: (a) Using eq. (15), find required v_C for finite θ_0 and for $\theta_0 \simeq 0$, (b) using eq. (21), find the minimum $\dot{\phi}$ about vertical diameter for stability and (c) using eq. (24), find the minimum v_C for stable plane motion with C moving in a straight line.

[(a) $\rho_C\left[\dfrac{mgR\tan\theta_0}{I\sin\theta_0 + (2I + mR^2)\rho_C/R}\right]^{1/2}$, $\rho_C\left[\dfrac{mgR^2\theta_0}{(2I + mR^2)\rho_C}\right]^{1/2}$, (b) $\left(\dfrac{mgR}{I + mR^2}\right)^{1/2}$

(c) $\left[\dfrac{mgR}{2(2I + mR^2)}\right]^{1/2}$ with $I = I^C_{xx}$, $\frac{1}{2}mR^2$, $\frac{1}{4}mR^2$ for lamina, hoop and disc.]

3.24 Motion of a Billiard Ball

Consider the motion of a radially symmetric billiard ball of mass m, radius R and radius of gyration k_C about any axis through its centre of mass C on a rough horizontal table with coefficient of kinetic friction μ. After being stuck by a cue at $t = 0$, let its angular velocity be $\vec{\omega}(0)$ and the velocity of C be $\vec{v}_C(0)$.

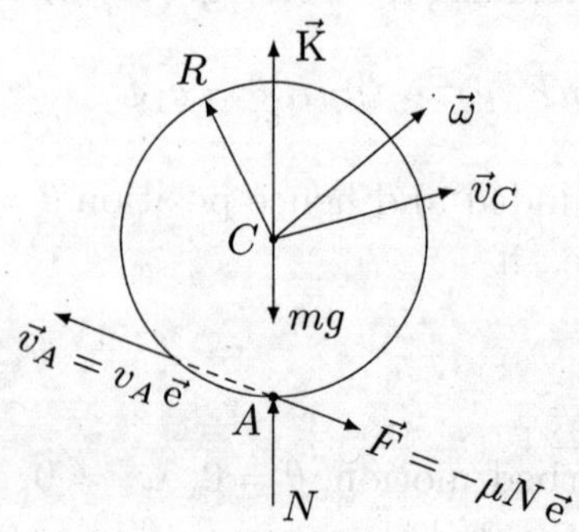

Fig. 3.41

The FBD of the sphere is shown in Fig. 3.41. The axes X, Y, Z are fixed to the table.

$$\vec{v}_{A(t)} = \vec{v}_C + \vec{\omega} \times (-R\vec{\mathrm{K}}) = \vec{v}_C - R\vec{\omega} \times \vec{\mathrm{K}} = v\,\vec{\mathrm{e}} \qquad \text{(say)} \tag{1}$$

$\vec{H}_C = mk_C^2\,\vec{\omega}$ and C moves in a horizontal plane. $\vec{F} = m\vec{a}_C$ and $\dot{\vec{H}}_{C|I} = \vec{M}_C$ yield:

$$-\mu N\,\vec{\mathrm{e}} + (N - mg)\,\vec{\mathrm{K}} = m\vec{a}_C$$

$$\Rightarrow \quad N = mg, \qquad \vec{a}_C = -\mu g\,\vec{\mathrm{e}} \tag{2}$$

$$\dot{\vec{H}}_{C|I} = mk_C^2\,\dot{\vec{\omega}} = \vec{M}_C = -R\,\vec{\mathrm{K}} \times (-\mu mg\,\vec{\mathrm{e}})$$

$$\Rightarrow \quad \dot{\vec{\omega}} = \frac{\mu Rg}{k_C^2}\,\vec{\mathrm{K}} \times \vec{\mathrm{e}} \tag{3}$$

Taking time derivative of eq. (1) yields

$$\begin{aligned}\dot{\vec{v}}_{A(t)} &= \vec{a}_C - R\dot{\vec{\omega}} \times \vec{\mathrm{K}} = -\mu g\,\vec{\mathrm{e}} - (\mu g R^2/k_C^2)(\vec{\mathrm{K}} \times \vec{\mathrm{e}}) \times \vec{\mathrm{K}} \\ &= -\mu g(1 + R^2/k_C^2)\,\vec{\mathrm{e}} \end{aligned} \tag{4}$$

As $\dot{\vec{v}}_{A(t)}$ is in the direction of $\vec{v}_{A(t)}$, the direction of $\vec{v}_{A(t)}$ remains fixed, i.e., $\vec{\mathrm{e}}$ is constant during the slip phase. Equations (2), (3) imply that $\vec{a}_C$ and $\dot{\vec{\omega}}$ are constant in the slip phase. Hence, the trajectory of C is, in general, a parabola with

$$\vec{v}_C(t) = \vec{v}_C(0) - \mu g t\,\vec{\mathrm{e}}, \qquad \vec{r}_C(t) = \vec{r}_C(0) + \vec{v}_C(0)t - \tfrac{1}{2}\mu g t^2\,\vec{\mathrm{e}}$$

Using $\dot{\vec{v}}_A(t) = \dot{v}\,\vec{\mathrm{e}}$ in eq. (4) yields

$$\dot{v} = -\mu g(1 + R^2/k_C^2), \qquad v = v(0) - \mu g(1 + R^2/k_C^2)t$$

$\Rightarrow$ slip stops for $v = 0$ at $t = t_1 = v(0)k_C^2/\mu g(R^2 + k_C^2)$. The subsequent motion is rolling without slip in a straight line with zero friction force, at constant velocity $\vec{v}_C(t)$ and at constant angular velocity $\vec{\omega}(t)$:

$$\vec{v}_C(t) = \vec{v}_C(t_1) = \vec{v}_C(0) - \mu g t_1\,\vec{\mathrm{e}}$$

$$\vec{\omega}(t) = \vec{\omega}(t_1) = \vec{\omega}(0) + (\mu R g t_1/k_C^2)\,\vec{\mathrm{K}} \times \vec{\mathrm{e}}$$

Review D

1. For a rigid body with $\vec{M}_C(t) \equiv \vec{0}$: $\quad \vec{H}_C(t) = \vec{H}_C(0)$ (invariable line),

   $T_R(t) = T_R(0), \qquad \vec{H}_C(t) \cdot \vec{\omega}(t) = \vec{H}_C(0) \cdot \vec{\omega}(0)$ (invariable plane).

   If $I_{11}^C = I_{22}^C$, then: $\quad \omega_3(t),\ \omega_1^2(t) + \omega_2^2(t),\ |\vec{\omega}|$ are constant.

   $\Rightarrow \angle(\vec{\omega}, \vec{\mathrm{e}}_3),\ \angle(\vec{\omega}, \vec{\mathrm{e}}_H),\ \angle(\vec{\mathrm{e}}_3, \vec{\mathrm{e}}_H)$ are constants and $\vec{\omega} = p\,\vec{\mathrm{e}}_H + s\,\vec{\mathrm{e}}_3$.

2. $\vec{\omega}$ about a principal axis at C with the largest/smallest I^C is stable.
3. Gyroscopic couple $\vec{M}_C = I^C_{ss}\vec{p} \times \vec{s}$ for $\vec{\omega} = \vec{s} + \vec{p}$ with $\vec{s} \perp \vec{p}$.
4. For an axisymmetric top with a fixed point O, spin axis z and vertical axis Z:
ω_z, $T + V$, $H_{O_Z|I}$ $(= \vec{H}_{O|I} \cdot \vec{K})$ are constant.

3.25 Concept Review Questions 3D

46. The system in Fig. Q3.46 is free to rotate about a fixed vertical axis. The principal moments of inertia of body ③ at O are $I_1 = I_2, I_3$. Body ① has moment of inertia I_{AA} about axis AA. Using conservation principles, write 3 governing equations of the system. $\dot{\psi}$ is angular velocity of body ③ with reference to body ②.
$[\dot{\theta} - \dot{\psi} = \text{const.},\ (I_{AA} + I_1 + mL^2 \sin^2\theta)\dot{\phi} = \text{const.},$
$\frac{1}{2}I_{AA}\dot{\phi}^2 + \frac{1}{2}m[(\dot{\phi}L\sin\theta)^2 + L^2\dot{\theta}^2] + \frac{1}{2}I_3(\dot{\theta} - \dot{\psi})^2 + \frac{1}{2}I_1\dot{\phi}^2 - mgL\cos\theta = \text{const.}]$

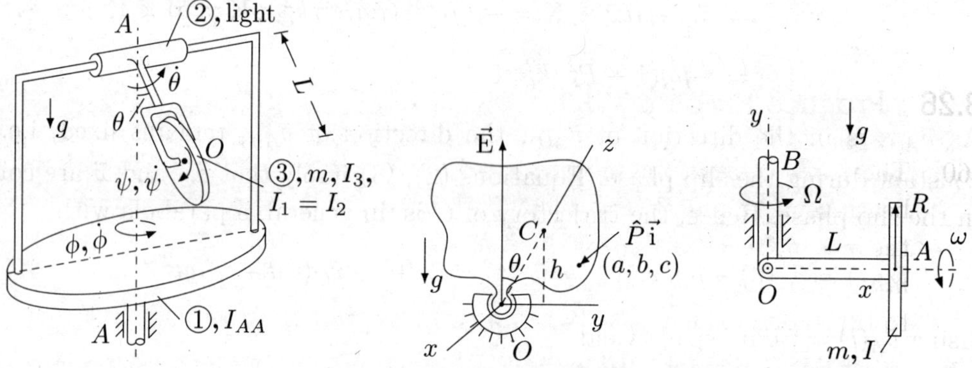

Fig. Q3.46 Fig. Q3.47 Fig. Q3.48

47. A body is rotating at $-\dot{\theta}\vec{i}$ when an impulse $\hat{P}\vec{i}$ acts on it (Fig. Q3.47). Set up an equation to find the maximum inclination θ_{max} of OC to the vertical in subsequent motion. Use $k^O_x = k^O_y = 0.4\,\text{m}$, $k^O_z = 0.5\,\text{m}$, $m = 5\,\text{kg}$, $a = 0$, $h = c = 0.2\,\text{m}$, $b = 0.25\,\text{m}$, $\theta = 30°$, $\dot{\theta} = 1\,\text{rad/s}$, $\hat{P} = 8\,\text{N.s}$.
$[\cos\theta_{max} = d,\ d^3 - 1.3247d^2 - 0.35474d + 0.66138 = 0, d = -0.671, \theta_{max} = 132.2°]$
48. (a) A cycle wheel with axial moment of inertia I rotates at ω relative to a light shaft OA pinned to a vertical shaft OB (Fig. Q3.48). The shaft OA rotates at a constant rate Ω about the fixed vertical axis and remains horizontal. Set up a relation between the variables. (b) Rework if the system is mounted in an elevator which itself is mounted on a vehicle. The elevator moves up with acceleration $a\vec{j}$ relative to the vehicle which translates down an incline with acceleration $a_1\vec{i} - a_2\vec{j}$ relative to the ground. $[(a)\ mgL = I\Omega\omega\ \ (b)\ m(g + a - a_2)L = I\Omega\omega]$

49. (a) A gyro is used to stabilize an overhead monorail car. Find the direction of its spin rate $\vec{s}$ to counteract the tendency of the car to lean outwards while going around a curve in either direction. (b) A gyro-stabiliser of mass m and axial radius of gyration k, is mounted on a car to completely counteract its tendency to tip when rounding a level curve of radius R at speed v. The total mass of the rotor and the car is M with the centre of mass at a height h above the ground. Neglect the inertia of the wheels. Find the desired spin rate of the gyro. Also, find the bearing reactions on the gyro shaft. The distance between its bearings is L and one of the bearings does not provide axial restraint. [(a) Towards left (b) Mvh/mk^2 to right, vertical: $\frac{1}{2}mg \pm Mv^2h/RL$, horizontal mv^2/R inwards]

50. A thin disc of mass m and radius R has its centre C tied to a fixed point O by a cable. C moves in a horizontal circle of radius b at constant speed v. There is no slip at the contact point A on the horizontal plane. The plane of the disc remains vertical. Find the tension in the cable and the ground reaction.

[$mv^2(1/R + 1/2b)$, $mg\vec{k} - mv^2\vec{i}/2b$ where $\vec{k}$ upward and $\vec{i}$ inward]

3.26 Practice Problems 3D

60. The rotor of a jet engine of an aeroplane has axial moment of inertia I and bearings at A and B (Fig. P3.60). The aeroplane makes a horizontal turn. Find the bearing reactions. $v_0 = 1000\,\text{km/h}$, $R = 4\,\text{km}$, $b = 1\,\text{m}$, $\omega_0 = 10000\,\text{rpm}$, $I = mk^2$, $m = 100\,\text{kg}$, $k = 0.4\,\text{m}$. In which direction is the tendency of the aeroplane to tilt? Is it better to use two oppositely turning jet engines rather than one?

[$mv^2\vec{j}/2R + (\frac{1}{2}mg \pm I\omega_0 v_0/bR)\vec{k} = 964.5\vec{j} + (490.5 \pm 1164)\vec{k}$ N, to pitch with tail going down, yes]

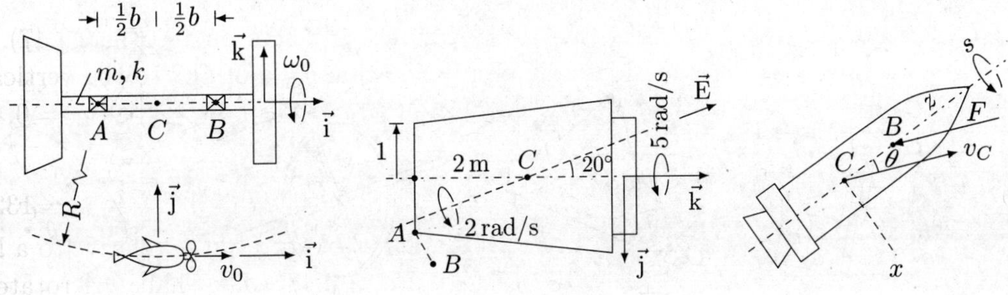

Fig. P3.60 Fig. P3.61 Fig. P3.62

61. A space capsule of mass 1000 kg has radii of gyration 0.6 m and 0.8 m about the symmetry axis and the transverse axis through the centre of mass C. It spins at 5 rad/s about its axis of symmetry and the spin axis precesses about axis $\vec{E}$ at

20° to it (Fig. P3.61). Due to impact with a meteorite, an impulse of 200 N.s acts on it at A in the direction B to A at an angle of 60° to $\vec{k}$ in the $\vec{k}$-$\vec{E}$ plane. Find the precession axis, the precession rate, the spin rate and the angle between the precession and spin axes for motion after the impact.

[$-0.1748\vec{\imath} - 0.1714\vec{\jmath} + 0.9696\vec{k}$, 3.991 rad/s, 3.008 rad/s, 14.16°]

62. A projectile spins at $s = 6000$ rpm at the instant of release, its axis being at an angle $\theta = 4°$ with respect to the velocity $\vec{v}_C$ of its centre of mass C (Fig. P3.62). $v_C = 600$ m/s and a resultant aerodynamic drag force F of 120 N acts at the centre of buoyancy B (80 mm from C as shown). The projectile has a mass of 20 kg and radii of gyration 50 mm and 200 mm about the axis of symmetry and transverse axis at C. Find the possible precession rates. Also find the minimum rate of spin required for stable motion. [41.68 & 0.3078 rad/s, 107.2 rad/s]

63. Show that a thin rigid lamina, thrown into air with spin ω_0 about the axis normal to its plane, has stable rotational motion irrespective of its shape and the frequency of the precessional wobble of the axis is $\omega_0/2\pi$.

64. Air friction exerts a torque $-c_1\vec{\omega}$ on a frisbee. Show that the axial component ω_3 decreases exponentially with time. Also show that the angle between the symmetry axis 3 and $\vec{\omega}$ decreases with time if $I_{33}^C > I_{11}^C$.

65. The total mass of the motorcyclist and the bike (Fig. P3.65) is m and the centre of mass is at C. The bike travels at fixed inclination θ to the vertical with C moving in a circle of radius R at constant speed v. The principal moments of inertia of the entire system of the bike and rider about axis 1 and 2 at C are I_1 and I_2. The axial moment of inertia of each wheel is I. Show that the minimum coefficient of friction for no slip is v^2/Rg. Show that for no slip, $\tan\theta \simeq (v^2/Rg)[1 + \{(I_2 - I_1)\sin\theta/R + 2I/r\}/hm] \simeq (v^2/Rg)(1 + 2I/mrh)$.

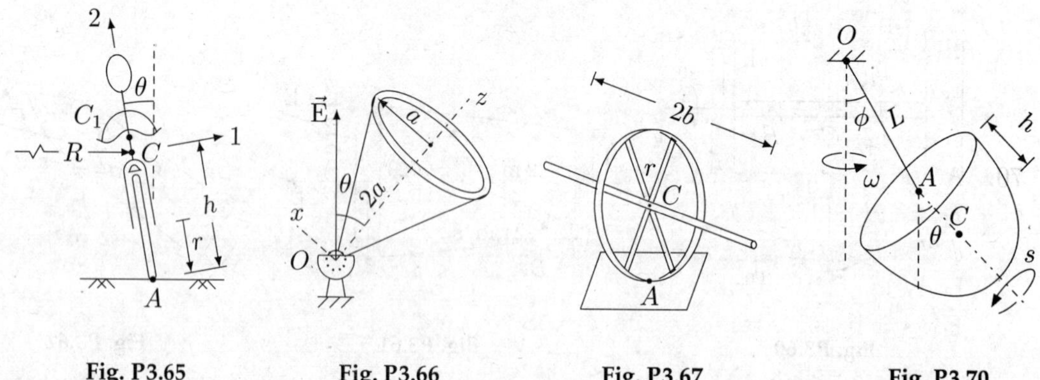

Fig. P3.65 Fig. P3.66 Fig. P3.67 Fig. P3.70

66. A conical thin shell of mass m, base radius a and height $2a$ is held by a fixed ball and socket joint at its vertex O (Fig. P3.66). (a) Find the angular velocity about

its vertical axis of symmetry ($\theta = 0°$) below which it is unstable. (b) The cone is set spinning about the vertical axis at $\omega_0\,\vec{\mathrm{E}}$. A slight disturbance causes the cone to fall and precess, the largest value of θ being 60°. Find ω_0 and the precession and spin rates at $\theta = 60°, 30°$. (c) The cone is made to precess at $(36g/a)^{1/2}\vec{\mathrm{E}}$ at $\theta = 30°$ without any spin or nutation rate. It is released. Find the extremal values of the nutation angle θ in subsequent motion. Find the precession and spin rates at these extreme positions. (d) What should be the minimum initial precession rate for $\theta = 120°$ before release, without any spin or nutation rate, so that the inclination θ decreases in subsequent motion? (e) What should be the initial precession rate in part (d) at $\theta = 120°$ so that the subsequent motion is between $\theta = 120°$ and $\theta = 90°$? [(a) $\sqrt{48g/a}$ (b) (6; 0.8889, 5.556; 0.7145, 5.381)$\sqrt{g/a}$ (c) 113.8°, $3.543\sqrt{g/a}$, $6.625\sqrt{g/a}$ (d) $\sqrt{32g/21a}$ (e) $2.421\sqrt{g/a}$]

67. (a) For the body shown in Fig. P3.67, find the minimum angular speed about the vertical diameter through A for stable rotation. Neglect the inertia of the ring and spokes. (b) Find the minimum speed of C in a straight line for stable rolling without slip. [(a) $[g/r(1 + b^2/3r^2)]^{1/2}$ (b) unstable $\forall\, v$]

68. A sphere rolls without slip on a horizontal plane which rotates about a fixed vertical axis at ω_0. Its centre moves in a circle of radius r at constant speed v about the axis of rotation of the plane. Show that $v = 2\omega_0 r/7$.

69. A satellite moves at speed v in a circular orbit of radius r. To control its rotational motion, three identical rotors are installed. At their centres of mass, the axial and transverse moments of inertia of the rotors are I_1 and I_2. The motors can apply variable driving torques about the axes of the rotors along x, y, z, respectively. The satellite is to have uniform angular velocity $\omega\,\vec{\mathrm{j}} = (v/r)\,\vec{\mathrm{j}}$ so that it continues to face the earth. At $t = 0$, the motors have angular velocities $\omega_x(0)$, $\omega_y(0)$ and $\omega_z(0)$ with respect to the satellite. Find $\omega_x(t)$, $\omega_y(t)$ and $\omega_z(t)$ and hence, find the torques applied by the motors on the satellite.
[$\omega_x(0)\cos\omega t - \omega_z(0)\sin\omega t$, $\omega_y(0)$, $\omega_z(0)\cos\omega t + \omega_x(0)\sin\omega t$, $I_1\omega(\omega_z\,\vec{\mathrm{i}} + \omega_x\,\vec{\mathrm{k}})$, $\vec{0}$, $-I_1\omega(\omega_z\,\vec{\mathrm{i}} + \omega_x\,\vec{\mathrm{k}})$]

70. A body of mass m and centroidal axial and transverse moments of inertia I_3 and I_1, is suspended at A from a cable (Fig. P3.70). OA and AC form fixed angles ϕ and θ with the vertical for sufficiently large rate of rotation ω of OAC. The body spins about its axis at constant rate s. (a) Given θ, ϕ, find ω and s. (b) For $s = 0$, for given ϕ, set up equations to find θ and ω. [(a) ω as in (b)

$$s = \frac{g[(I_3 - I_1)\sin\phi\sin\theta\cos\theta + mh(L\sin\phi + h\sin\theta)\sin(\theta - \phi)]}{I_3\omega\cos\phi\sin\theta(L\sin\phi + h\sin\theta)}$$

(b) $\omega^2 = \dfrac{g\tan\phi}{L\sin\phi + h\sin\theta}$, $\frac{1}{2}(I_3 - I_1)\sin 2\theta\sin\phi + mh(L\sin\phi + h\sin\theta)\sin(\theta - \phi) = 0$]

CHAPTER

4

STATICS

THE equilibrium problems involving smooth rigid bodies and frames, systems with friction, and bars (trusses, cables, beams), are covered in Parts A,B,C, respectively.

A. EQUATIONS OF EQUILIBRIUM

4.1 Remarks on the Equations of Equilibrium

Recall the following results.

1. For a system in equilibrium, the external loads on any of its parts P must satisfy,
$$\vec{F}(P) = \vec{0}, \qquad \vec{M}_A(P) = \vec{0} \tag{4.1}$$
2. If $\vec{F}(P) = \vec{0}$, $\vec{M}_A(P) = \vec{0}$ for all parts P of a system B, then $\vec{F}(B) = \vec{0}$, $\vec{M}_A(B) = \vec{0}$. The converse is not true even for a rigid body, e.g., for a disc B of mass m rotating at constant angular velocity about a fixed axis: $\vec{F}(B) = \vec{0}$, $\vec{M}_A(B) = \vec{0}$, but for its semi-circular part P with centre of mass C: $\vec{F}(P) = \frac{1}{2}m\vec{a}_C \neq \vec{0}$.
3. For a rigid body B, if for the whole body $\vec{F}(t) \equiv \vec{0}$, $\vec{M}_A(t) \equiv \vec{0}$, then it remains in equilibrium if it is initially in equilibrium.
4. For a deformable body initially in equilibrium, $\vec{F}(P) = \vec{0}$, $\vec{M}_A(P) = \vec{0}$ for all parts P, are not sufficient to ensure equilibrium. In addition, the *deformations* of different parts of the system *should be compatible with one another*. More than one distribution of internal forces may satisfy equilibrium equations, but only one/a few of these may correspond to compatible deformations.
5. It is difficult to verify eqs (4.1) for infinite number of parts of a deformable body. A set of partial differential equations of equilibrium in terms of stresses are derived. Satisfying these equations ensures that eqs (4.1) are satisfied for every part.

We can show that for a system in equilibrium under three concurrent forces, the forces F_1, F_2, F_3, can be represented by the sides of a triangle taken in order (Fig. 4.1).

Lame's theorem : $$\frac{F_1}{\sin\alpha_1} = \frac{F_2}{\sin\alpha_2} = \frac{F_3}{\sin\alpha_3} \tag{4.2}$$

follows from the sine formula for the triangle of forces, where $\alpha_1, \alpha_2, \alpha_3$ are the angles between F_2 and F_3, F_3 and F_1, F_1 and F_2, respectively.

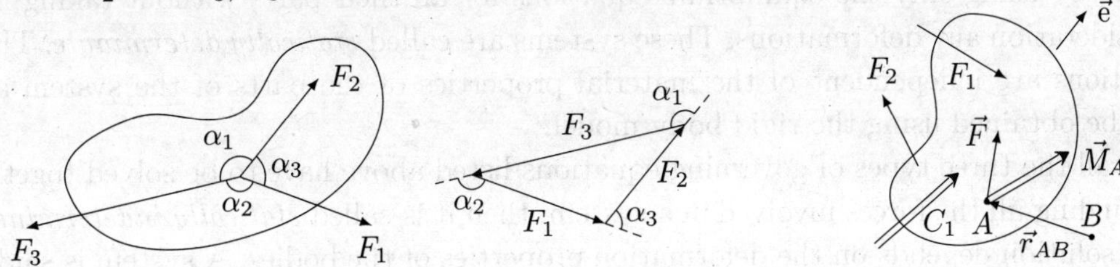

Fig. 4.1 **Fig. 4.2**

We can make the following observations about a system in equilibrium.

1. For its part P, $\vec{F} = \vec{0}$, $\vec{M}_A = \vec{0}$ yield 6 equations:
   $F_x = 0, \quad F_y = 0, \quad F_z = 0, \quad M_{Ax} = 0, \quad M_{Ay} = 0, \quad M_{Az} = 0.$
2. We do not get any more equations independent of $\vec{F} = \vec{0}$ and $\vec{M}_A = \vec{0}$ by writing $\vec{M}_B = \vec{0}$ (Fig. 4.2), since using the equivalent force–couple system $\vec{F}$ and $\vec{M}_A$ at A, $\vec{M}_B = \vec{M}_A + \vec{r}_{AB} \times \vec{F} = \vec{0}$ is automatically satisfied since $\vec{F} = \vec{0}$, $\vec{M}_A = \vec{0}$.
3. We do not get an independent equation by taking the component of $\vec{F}$ along another direction $\vec{e}$: $F_e = \vec{F} \cdot \vec{e} = 0$ (Fig. 4.2), as it holds automatically if $\vec{F} = \vec{0}$.[1]
4. For a system of n interconnected rigid bodies, we can write $6n$ scalar equations of equilibrium, 6 for each body. The equilibrium equations of the total system $\vec{F} = \vec{0}$, $\vec{M}_A = \vec{0}$ are not independent of these, being the sum of $3n$ force and $3n$ moment equations of individual bodies—the contribution of the internal loads of interaction pair-wise cancel, being equal in magnitude and opposite in direction.

The independent equilibrium equations for various force systems are listed below:

1. General: for $i = 1, 2, 3$ or x, y, z — (6): $F_i, M_{O_i} = 0$
2. Concurrent (spatial): — (3): $F_x, F_y, F_z = 0$
3. Parallel to z-axis (spatial): — (3): $F_z, M_{Ox}, M_{Oy} = 0$
4. Coplanar with forces in xy-plane: — (3): $F_x, F_y, M_{Oz} = 0$
5. Coplanar with forces parallel to x-axis in xy-plane: — (2): $F_x, M_{Oz} = 0$
6. Coplanar and concurrent in xy-plane: — (2): $F_x, F_y = 0$
7. All forces intersect a line (say x-axis): — (5): $F_x, F_y, F_z, M_{Oy}, M_{Oz} = 0$

4.1.1 Statically determinate and statically indeterminate systems

In a problem of equilibrium, the support reactions, internal force resultants and deformation can in general be solved using (i) equilibrium equations and boundary conditions on forces, (ii) deformation equations, conditions of geometric fit and boundary conditions on deformations, and (iii) force–deformation relations.

In some systems all the unknown reactions and internal force resultants can be found by using only the equilibrium equations for all their parts without taking into consideration any deformations. These systems are called *statically determinate.* Their solutions are independent of the material properties of the parts of the system and can be obtained using the rigid body model.

If all the three types of governing equations listed above have to be solved together for finding all the forces involved in a system, then it is called *statically indeterminate.* The solution depends on the deformation properties of the bodies. A system is said to be *statically indeterminate externally or externally redundant* if the support reactions cannot be found from the equilibrium equations alone. A system is called *internally statically indeterminate* if the internal force resultants cannot be found using only the equilibrium equations. A system is certainly indeterminate if the *number of unknown forces and moments exceeds the number of independent equations of equilibrium.* It is a *sufficient condition* of static indeterminacy. But *it may not be a necessary condition.* A single rigid body is statically indeterminate if the number of unknown reactions exceeds six and three for three-dimensional and coplanar force systems respectively.

If the number of equations of equilibrium, m, is more than the number of unknowns, n, and these equations are linear in the unknowns, then they can be written in a matrix form as $Au = b$, where A is a $m\times n$ matrix, u is an $n\times 1$ column matrix of the unknowns and b is an $m \times 1$ column matrix. From linear algebra we know that a unique solution exists if the rank of matrix A and that of the $m \times (n+1)$ augmented matrix $[A\,b]$ is each equal to n. Such a system is statically determinate. If $m = n$, then there is a unique solution, provided the rank of A is n, i.e., A is non-singular.

The cantilever and the three-pin arch in Fig. 4.3a,b are statically determinate. The reactions $R_1, \ldots, R_6$ at the three pins of the arch can be found from the six equilibrium equations of the system (3 for each body). The exception is when the three pins are collinear (Fig. 4.3c). In this case, for members

$$①+②: \quad M_{C_z} = P(b+c) - Qh - R_2(a+b) = 0, \qquad ①: \quad M_{B_z} = Pc - R_2 a = 0$$

$\Rightarrow R_2 = [P(b+c) - Qh]/(a+b)$ which contradicts $R_2 = Pc/a$. Thus, these equilibrium equations cannot be satisfied in the undeformed configuration. Hence, these have to be written in the deformed configuration and so the problem is statically indeterminate.

The propped cantilever and two-pin arch in Fig. 4.4 are statically indeterminate as the 4 reactions $R_1, R_2, R_3, C_1/R_4$ cannot be found from the 3 equations of equilibrium.

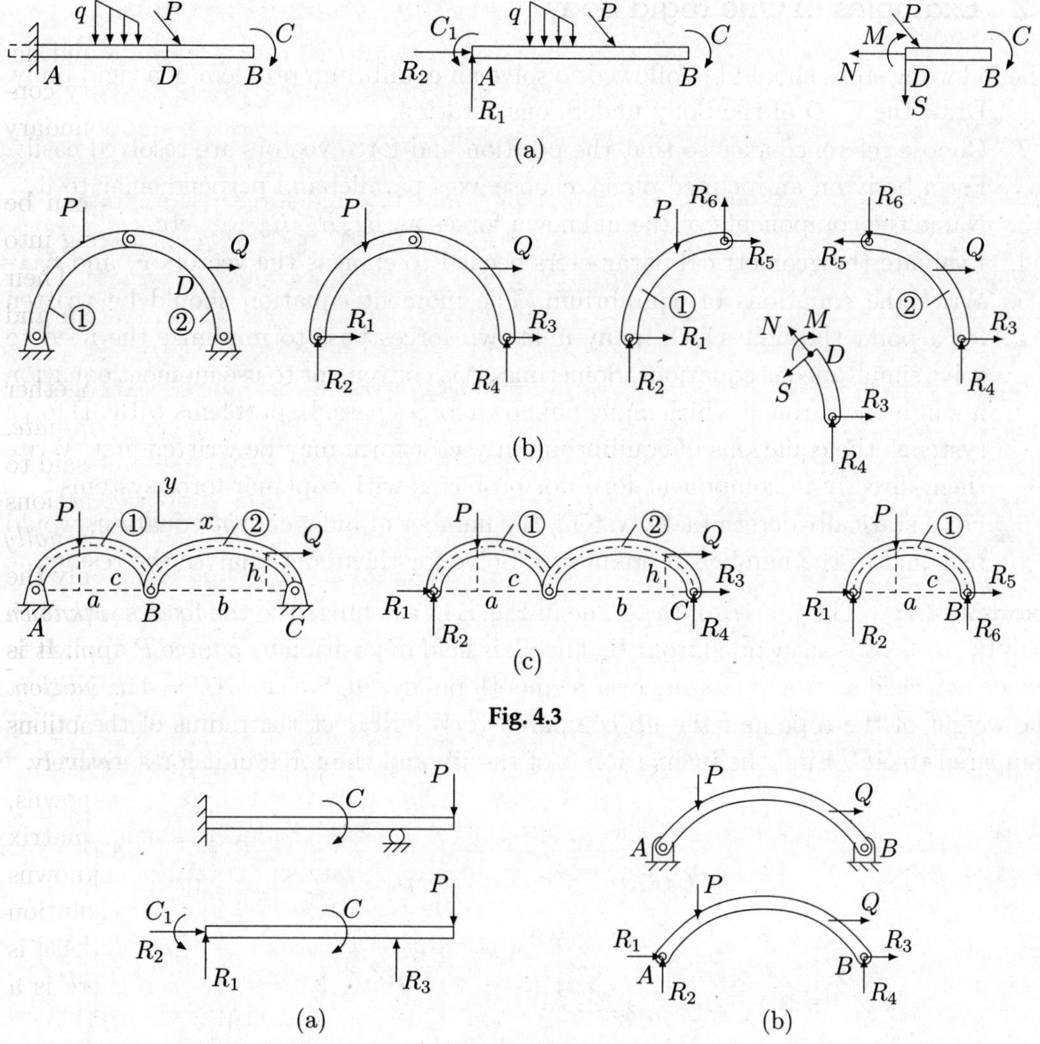

Fig. 4.3

Fig. 4.4

A deformable body model is needed for statically indeterminate problems. For example, the bending moment M and shear force S in the cantilever of Fig. 4.3a can be found by modelling it as a rigid body, whereas M and S in the propped cantilever of Fig. 4.4a can be found only by modelling it as a deformable body *even though the order of deformations is the same in both cases.* Hence, the *amount of deformation of a body does not necessarily indicate whether it should be modelled as a rigid body.*

As the solution of a statically indeterminate structure depends on the deformations, it can have initial stresses due to mismatch during assembly, and stresses due to temperature change. Statically determinate structures are not subjected to stresses due to these causes. We will confine this study to statically determinate problems.

4.2 Examples of One Rigid Body

The following steps should be followed to solve an equlibrium problem of a rigid body.

1. Draw the FBD of the body under consideration.
2. Choose reference axes so that the position and force vectors are resolved easily. For a body on an inclined plane, choose axes parallel and perpendicular to it.
3. Name the components of the unknown forces as $R_1, R_2, R_3, \ldots$, etc.
4. Evaluate the geometrical parameters needed to express the vectors $\vec{F_i}$ and $\vec{r_i}$.
5. Write the equations of equilibrium. The moment equation should be written for a point through which many unknown forces pass to minimize the need to solve simultaneous equations. Sometimes it is convenient to use moment equation about a line through which many unknown forces pass. In problems with 3D force systems, the equations of equilibrium in vector form may be written first. Write them directly in component form for problems with coplanar force systems.
6. For a statically determinate system, the number of independent equations would be equal to the number of unknowns. Solve for them and analyse the results.

Example 4.1: The jib AB of the crane in Fig. E4.1a is hinged to the fixed support. A load $W = 50\,\text{kN}$ is suspended from B. The jib is held in position by a force P applied to a rope attached at B and passing over a smooth pulley. $AC = 5\,\text{m}$, $AB = 4\,\text{m}$. Neglect the weight of the rope and the jib compared to W. Neglect the radius of the pulley compared to AC. Find the inclination α of the jib and the force in it for a given P.

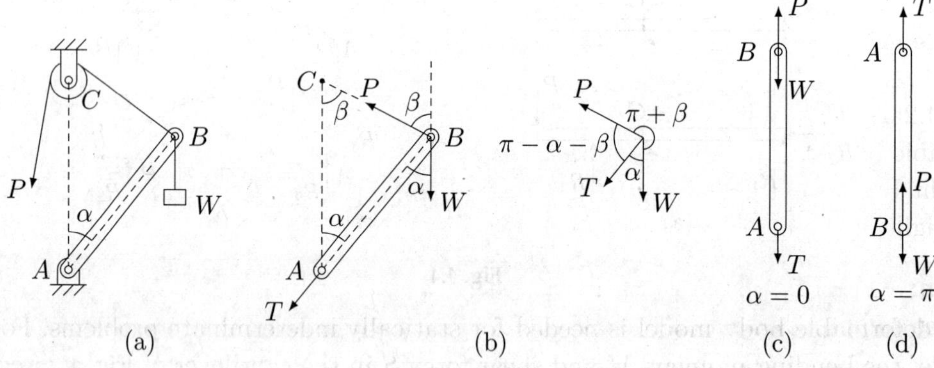

Fig. E4.1

Solution: The tensions in the light rope on the two sides of the smooth pulley are equal. Thus, the tension is P throughout, since the rope is massless. AB is a two-force member subjected to forces at A and B. Hence, tensions T at its ends are collinear with the line AB joining their points of application. For $\alpha = 0$ or $\alpha = \pi$, AB is in equilibrium under collinear forces P, W and T, $\forall P$ (Fig. E4.1c,d). The FBD of the jib ($\alpha \neq 0, \pi$) is shown in Fig. E4.1b. It is in equilibrium under 3 coplanar concurrent forces.

Lame's theorem: $$\frac{T}{\sin(\pi+\beta)} = \frac{P}{\sin\alpha} = \frac{W}{\sin(\pi-\alpha-\beta)}$$

and $$\frac{(AB)}{\sin\beta} = \frac{(BC)}{\sin\alpha} = \frac{(AC)}{\sin(\pi-\alpha-\beta)}$$

$$\Rightarrow \quad T = -W\sin\beta/\sin(\alpha+\beta) = -W(AB)/(AC) = -40\,\text{kN} \qquad (1)$$

$$P = W\sin\alpha/\sin(\alpha+\beta) = W(BC)/(AC)$$
$$= 10[5^2+4^2-2(5)(4)\cos\alpha]^{1/2} = 10[41-40\cos\alpha]^{1/2}\,\text{kN} \qquad (2)$$

Note: $T \neq -W\cos\alpha$; W is not being resolved into rectangular components. The 40 kN compressive force in jib is independent of α as long as $\alpha \neq 0$ or π. Equation (2) $\Rightarrow$

$$\text{for } \alpha \simeq 0: \; P = 10, \qquad \text{for } \alpha = \tfrac{1}{2}\pi: \; P = 64.03, \qquad \text{for } \alpha = \pi: \; P = 90\,\text{kN}$$

$$\Rightarrow \quad \text{if } 10 < P < 90, \text{ then} \qquad \alpha = \cos^{-1}[(41-0.01P^2)/40],\, 0,\, \pi$$

$$\text{if } P < 10\,\text{kN} \quad \text{or} \quad P > 90\,\text{kN}, \text{ then} \quad \alpha = 0, \pi$$

The FBDs of the jib for $\alpha = 0$ and $\alpha = \pi$ are given in Fig. E4.1c,d.

$$\text{For } \alpha = 0: \qquad T + W = P \quad \Rightarrow \quad T = P - W$$

$$\text{for } \alpha = \pi: \qquad T + P = W \quad \Rightarrow \quad T = W - P$$

Example 4.2: A widely used chain hoist called differential pulley block consists of two pulleys of slightly different radii fixed together with their axle free to rotate at O (Fig. E4.2a). The pulleys form two sprockets for an endless chain, making two loops. A movable pulley of weight W_1 is mounted on one of the loops and it carries a crate of weight W. A force P applied to the free loop hanging down from the larger pulley. The weight of the upper pulley block is W_2. Find P for equilibrium. Neglect friction.

Solution: The FBD is shown in Fig. E4.2b. P and $\vec{R}_O = R_1\vec{\imath} + R_2\vec{\jmath}$ are found from the three equations of equilibrium of the system subjected to coplanar force system

$$M_{Oz} = PR - (W+W_1)[\tfrac{1}{2}(R+r) - r] = 0$$

$$\Rightarrow \quad P = \tfrac{1}{2}(W+W_1)(1-r/R) \ll (W+W_1) \qquad \text{as } (1-r/R) \ll 1$$

$$\vec{F} = \vec{0} \quad \Rightarrow \quad \vec{R} = (W+W_1+W_2)\vec{\jmath} - \vec{P} \simeq (W+W_1+W_2)\vec{\jmath} \qquad \text{as } P \ll (W+W_1)$$

Let D be pulled at speed $v_D = v \Rightarrow \omega = v/R$, $\vec{v}_B = \vec{v}_E = \omega R\vec{\jmath} = v\,\vec{\jmath}$, $\vec{v}_A = \vec{v}_F = -\omega r\vec{\jmath} = -(vr/R)\vec{\jmath}$, $\vec{v}_C = \tfrac{1}{2}(\vec{v}_B + \vec{v}_A) = \tfrac{1}{2}v(1-r/R)\vec{\jmath}$. Mechanical advantage m_a is defined as the ratio of load $L = W + W_1$ to the effort P and velocity ratio v_r is defined as the ratio of the speeds of points of application D of the effort and C of the load. Let s_D and s_C be the distances moved by D and C in time t. Hence,

$$m_a = \frac{L}{P} = \frac{2}{1 - r/R}, \qquad v_r = \frac{v_D}{v_C} = \frac{2}{1 - r/R} = m_a, \qquad \frac{s_D}{s_C} = \frac{v_D t}{v_C t} = v_r$$

In general, for all simple machines with effort P to move load L, in the absence of any power loss, $m_a = v_r$, $P = L/m_a$, $s_C = s_D/m_a$. We infer that the desirable large m_a for small effort P to lift large load L causes L to move by a small distance s_C for very large displacement s_D of D. In the presence of power loss

$$(\text{Input power } Pv_D) = (\text{Output power } Lv_C) + (\text{Power loss } P_l)$$

$$\Rightarrow \quad m_a = \frac{L}{P} = \frac{v_D}{v_C} - \frac{P_l}{Pv_C} < v_r$$

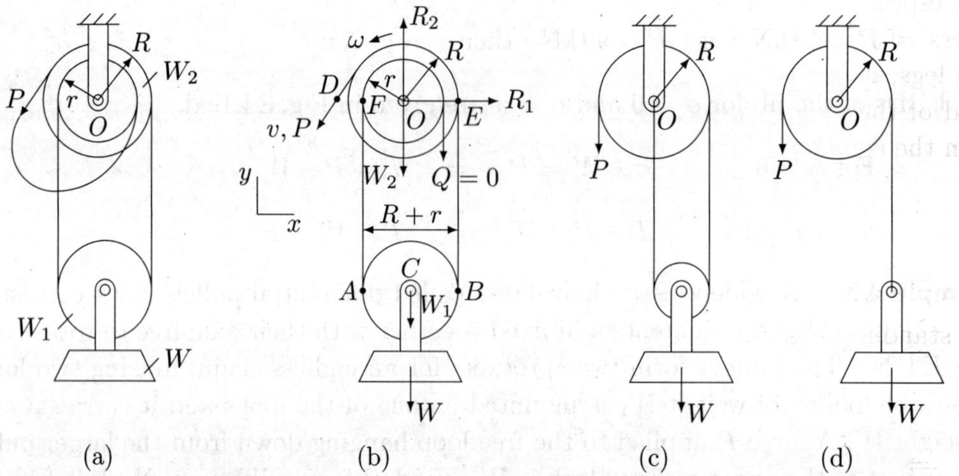

Fig. E4.2

$(1 - r/R)$ is made very small to have very large m_a, e.g., for vertically applied P and light pulleys, for $r/R = 24/25$: $P = W/50$, $R_O = 1.02W$ and $s_C = 2\,\text{cm}$ for $s_D = 1\,\text{m}$.

For $r = 0$ (Fig. E4.2c), $m_a = 2$, $P = \frac{1}{2}W$, $R_O = 1.5W$. For the simple pulley in Fig. E4.2d, $P = W$, $m_a = 1$, $R_O = 2W$. The differential pulley is thus the best to lift heavy load with a small force but without subjecting the support to enhanced load.

Example 4.3: The FBD of the leg of a man of weight W, standing on one leg, is shown in Fig. E4.3a. $N = W$ is the ground reaction, $W_1 = W/7$ is the weight of the leg, F_1 is the force exerted by the hip abductor muscles and $\vec{R}$ is the force exerted on the head of the femur. (a) Find F_1 and $\vec{R}$ if $a = 10\,\text{cm}$, $b = 8\,\text{cm}$, $c = 7\,\text{cm}$. (b) The

FBD of the leg when a cane carried under the other arm supports a load of $0.2W$ ($N = 0.8W$), is shown in Fig. E4.3b. Find F_1 and $\vec{R}$ if $a^* = 6.11\,\text{cm}$, $b^* = 4.89\,\text{cm}$, $c^* = c = 7\,\text{cm}$.

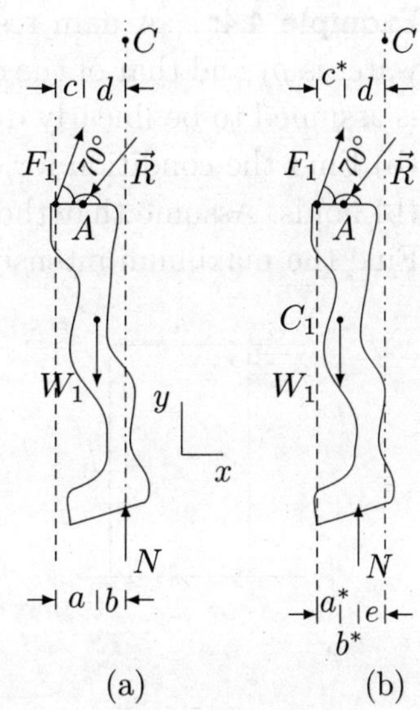

Fig. E4.3

Solution: (a) The three unknowns, F_1 and the two components of $\vec{R}$, can be found from the three equilibrium equations of the leg subjected to a coplanar force system. Point A is convenient for the moment equation because $\vec{R}$ passes through it.

$$M_{A_z} = -(F_1 \sin 70°)c - W_1(a-c) + N(a+b-c) = 0$$

$$\Rightarrow \quad F_1 = [N(a+b-c) - W_1(a-c)]/c \sin 70° \qquad (1)$$

We conclude from eq. (1) that if the man stands on two feet, especially with legs apart, then $N = \frac{1}{2}W$ and $a+b-c$ is less $\Rightarrow$ the muscle force F_1 is less. While standing on two legs, it is best to place the foot directly below the head of the femur so that $a + b - c = 0$. $\vec{R}$ is obtained from the equilibrium equation:

$$\vec{F} = (N - W_1)\vec{j} + F_1(\cos 70°\,\vec{i} + \sin 70°\,\vec{j}) + \vec{R} = \vec{0}$$

$$\Rightarrow \quad \vec{R} = -F_1 \cos 70°\,\vec{i} + (W_1 - N - F_1 \sin 70°)\,\vec{j} \qquad (2)$$

For standing on one leg, $N = W$, $W_1 = W/7$ and eqs (1) and (2) yield

$$F_1 = 1.607W, \qquad \vec{R} = -0.550W\,\vec{i} - 2.367W\,\vec{j}$$

$$\Rightarrow \quad R = (0.55^2 + 2.367^2)^{1/2} W = 2.43W$$

(b) The distance of N from A is $d - e = 11 - 7 = 4\,\text{cm}$ which is about 1/3rd of the distance of $d = 11\,\text{cm}$ for case (a). With $N = 0.8W$, $W_1 = W/7$, eqs (1) and (2) yield

$$F_1 = [0.8W(a^* + b^* - c^*) - (a^* - c^*)W/7]/c^* \sin 70° = 0.506W$$

$$\vec{R} = -F_1 \cos 70°\,\vec{i} + (W/7 - 0.8W - F_1 \sin 70°)\,\vec{j} = -0.173W\,\vec{i} - 1.133W\,\vec{j}$$

$$\Rightarrow \quad R = 1.146W$$

The values of F_1 and R for standing with a cane are much less than those for standing without it. Hence, the use of a cane following a surgery of hip or knee can play an important role in reducing the strain on the overworked good leg.

Example 4.4: A dam rests on a horizontal foundation (Fig. E4.4a). The density of water is ρ_1 and that of the dam is ρ. The normal reaction of the foundation on the dam is assumed to be linearly distributed. (a) Find its minimum and maximum intensities. (b) Find the condition for the reaction to become negative. (c) The condition of part (b) holds. Assume that there is no tensile force between the dam and the foundation. Find the maximum intensity of the normal reaction of the foundation on the dam.

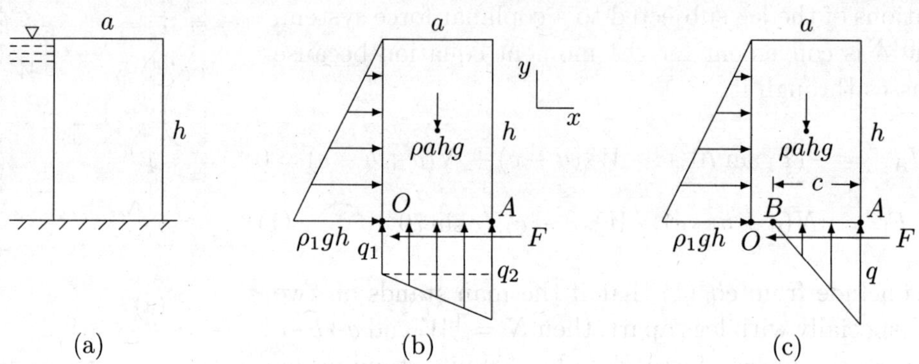

Fig. E4.4

Solution: (a) The FBD of a unit length of the dam is shown in Fig. E4.4b. The shear force exerted by the foundation on the dam is F and the ground pressure varies linearly from q_1 at O to q_2 at A. We find F, q_1, q_2 from the three equilibrium equations of the dam subjected to a coplanar force system:

$$F_x = (\rho_1 gh)\tfrac{1}{2}h - F = 0 \qquad \Rightarrow \qquad F = \tfrac{1}{2}\rho_1 gh^2 \tag{1}$$

$$F_y = q_1 a + (q_2 - q_1)\tfrac{1}{2}a - \rho ahg = 0 \qquad \Rightarrow \qquad q_1 + q_2 = 2\rho gh \tag{2}$$

$$M_{O_z} = (q_1 a)\tfrac{1}{2}a + [(q_2 - q_1)\tfrac{1}{2}a]\tfrac{2}{3}a - (\rho ahg)\tfrac{1}{2}a - [(\rho_1 gh)\tfrac{1}{2}h]\tfrac{1}{3}h = 0$$

$$\Rightarrow \qquad q_1 + 2q_2 = 3\rho gh + \rho_1 gh^3/a^2 \tag{3}$$

$$\text{eqs (2), (3)} \Rightarrow \quad q_1 = gh[\rho - \rho_1 h^2/a^2], \qquad q_2 = gh[\rho + \rho_1 h^2/a^2] \tag{4}$$

(b) Equations (4) $\Rightarrow$ $q_1 < q_2$ and there is tension between the dam and the foundation if $q_1 < 0$, i.e., if $h/a > (\rho/\rho_1)^{1/2}$. If such a situation arises, then a portion OB of the dam along the base does not press on the foundation and creates the danger of seepage.

(c) The pressure is distributed over width c (Fig. E4.4c). F, c, q are obtained from:

$$F_y = \tfrac{1}{2}qc - \rho ahg = 0 \qquad \Rightarrow \qquad q = 2\rho ahg/c$$

$$M_{A_z} = (\rho ahg)\tfrac{1}{2}a - (\tfrac{1}{2}qc)\tfrac{1}{3}c - [(\rho_1 gh)\tfrac{1}{2}h]\tfrac{1}{3}h = 0$$

$$\Rightarrow \qquad c = [1.5 - \rho_1 h^2/2\rho a^2]a, \qquad q = 4\rho gh/[3 - \rho_1 h^2/\rho a^2]$$

Example 4.5: Find the reactions at the three landing wheels of an aeroplane of weight 200 kN with centre of mass at C (Fig. E4.5a). The fuel tanks in the wings carry fuel of weight 50 kN and 25 kN.

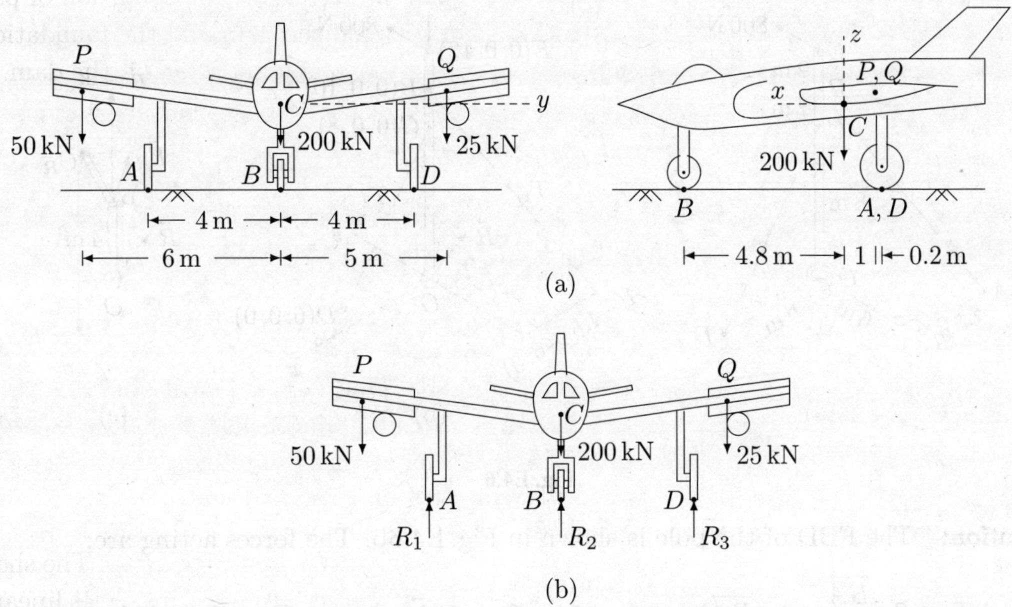

Fig. E4.5

Solution: The FBD is shown in Fig. E4.5b.

$$\overrightarrow{DA} = -8\vec{j}, \qquad \overrightarrow{DB} = 6\vec{i} - 4\vec{j}, \qquad \overrightarrow{DC} = 1.2\vec{i} - 4\vec{j} + h_C\,\vec{k}$$

$$\overrightarrow{DP} = 0.2\vec{i} - 10\vec{j} + h_P\,\vec{k}, \qquad \overrightarrow{DQ} = 0.2\vec{i} + \vec{j} + h_Q\,\vec{k}$$

where h_C, h_P, h_Q are the heights of C, P, Q above the ground. The reactions R_1, R_2, R_3 at A, B, D are obtained from the three equations of equilibrium for the 3D parallel force system:

$$\begin{aligned} \vec{M}_D = {} & -8\vec{j} \times R_1\vec{k} + (6\vec{i} - 4\vec{j}) \times R_2\vec{k} + (1.2\vec{i} - 4\vec{j} + h_C\,\vec{k}) \times (-200\,\vec{k}) \\ & +(0.2\vec{i} - 10\vec{j} + h_P\,\vec{k}) \times (-50\vec{k}) + (0.2\vec{i} + \vec{j} + h_Q\,\vec{k}) \times (-25\,\vec{k}) = \vec{0} \end{aligned}$$

$$\vec{j}: \qquad -6R_2 + 240 + 10 + 5 = 0 \qquad \Rightarrow \qquad R_2 = 42.5\,\text{kN}$$

$$\vec{i}: \qquad -8R_1 - 4R_2 + 800 + 500 - 25 = 0 \qquad \Rightarrow \qquad R_1 = 138.1\,\text{kN}$$

$$F_z = R_1 + R_2 + R_3 - 200 - 50 - 25 = 0 \qquad \Rightarrow \qquad R_3 = 94.4\,\text{kN}$$

Example 4.6: A pole is supported as shown in Fig. E4.6a. Find the forces in the light flexible cables AB, CD and the reaction at the ball and socket joint at O. Find the axial force, shear force and bending moment in the pole at a section 4 m from the base.

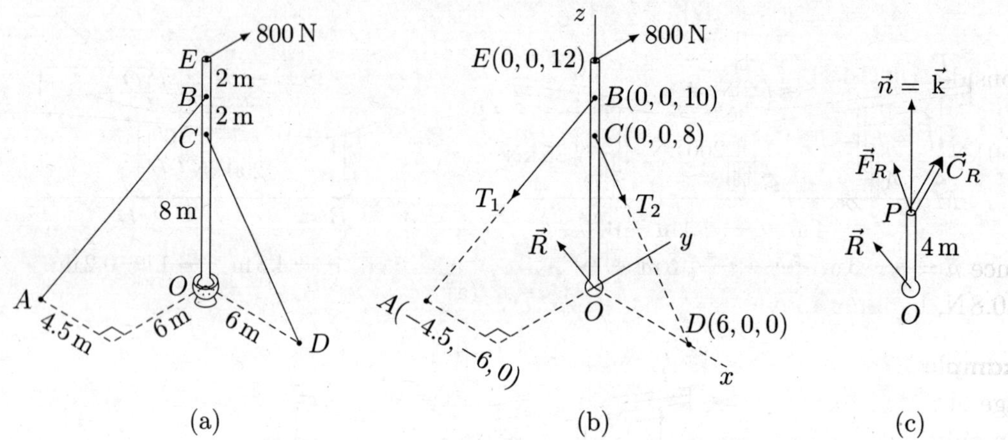

Fig. E4.6

Solution: The FBD of the pole is shown in Fig. E4.6b. The forces acting are:

$$\vec{T_1} = T_1 \frac{\overrightarrow{BA}}{BA} = t_1(\overrightarrow{BA}) = t_1(-4.5\vec{\imath} - 6\vec{\jmath} - 10\vec{k}) \quad \text{at } B: \ \vec{r}_B = 10\vec{k}\,\text{m}$$

$$\vec{T_2} = T_2 \frac{\overrightarrow{CD}}{CD} = t_2(\overrightarrow{CD}) = t_2(6\vec{\imath} - 8\vec{k}) \quad \text{at } C: \ \vec{r}_C = 8\vec{k}\,\text{m}$$

$$800\vec{\jmath}\,\text{N} \ \text{at } E: \ \vec{r}_E = 12\vec{k}\,\text{m}, \qquad \vec{R} \ \text{at } O: \ \vec{r}_O = \vec{0}$$

All forces intersect line OE. $\vec{M}_O = \vec{0}$ yields t_1 and t_2 and then $\vec{F} = \vec{0}$ yields $\vec{R}$.

$$\begin{aligned}
\vec{M}_O &= \vec{r}_B \times \vec{T_1} + \vec{r}_C \times \vec{T_2} + \vec{r}_E \times 800\vec{\jmath} \\
&= 10\vec{k} \times t_1(-4.5\vec{\imath} - 6\vec{\jmath} - 10\vec{k}) + 8\vec{k} \times t_2(6\vec{\imath} - 8\vec{k}) + 12\vec{k} \times 800\vec{\jmath} = \vec{0}
\end{aligned}$$

$$\Rightarrow \quad \vec{\imath}: \quad 60t_1 - 9600 = 0 \quad \Rightarrow \quad t_1 = 160$$

$$\vec{\jmath}: \quad -45t_1 + 48t_2 = 0 \quad \Rightarrow \quad t_2 = 150$$

$$\begin{aligned}
\vec{T_1} &= t_1(-4.5\vec{\imath} - 6\vec{\jmath} - 10\vec{k}) = -720\vec{\imath} - 960\vec{\jmath} - 1600\vec{k}\ \text{N} \\
T_1 &= (720^2 + 960^2 + 1600^2)^{1/2} = 2000\,\text{N} \\
\vec{T_2} &= t_2(6\vec{\imath} - 8\vec{k}) = 900\vec{\imath} - 1200\vec{k}\ \text{N} \\
T_2 &= (900^2 + 1200^2)^{1/2} = 1500\,\text{N}
\end{aligned}$$

$$\vec{F} = \vec{T}_1 + \vec{T}_2 + \vec{R} + 800\vec{j} = \vec{0}$$

$$\Rightarrow \quad \vec{R} = -\vec{T}_1 - \vec{T}_2 - 800\vec{j} = -180\vec{i} + 160\vec{j} + 2800\vec{k}\ \text{N}$$

$$\Rightarrow \quad R = (180^2 + 160^2 + 2800^2)^{1/2} = 2810\,\text{N}$$

Consider the FBD of OP (Fig. E4.6c). $\vec{C}_R$ and $\vec{F}_R$ are internal force resultants at P.

$$\vec{F} = \vec{R} + \vec{F}_R = \vec{0} \qquad \Rightarrow \qquad \vec{F}_R = -\vec{R} = 180\vec{i} - 160\vec{j} - 2800\vec{k}\ \text{N}$$

$$\vec{M}_P = -4\vec{k} \times \vec{R} + \vec{C}_R = \vec{0} \qquad \Rightarrow \qquad \vec{C}_R = 4\vec{k} \times \vec{R} = -640\vec{i} - 720\vec{j}\ \text{N.m}$$

Since $\vec{n} = \vec{k}$, axial force $= -2800\vec{k}$ N, shear force $= 180\vec{i} - 160\vec{j}$ N with magnitude of 240.8 N, bending moment $= -640\vec{i} - 720\vec{j}$ N.m with magnitude of 963.3 N.m.

Example 4.7: A uniform rectangular plate $ABCE$ of weight 6 kN is hinged about its edge AB and is supported by a light flexible cable CO (Fig. E4.7a). The hinge at B can only support a force normal to the hinge axis but the hinge at A can also support an axial load. Find the reactions at the hinges A and B and the tension in the cable.

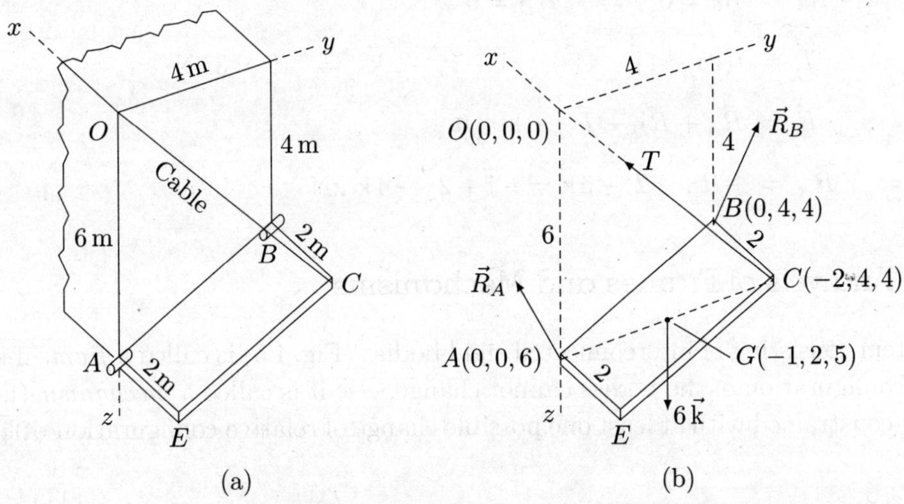

Fig. E4.7

Solution: The FBD is shown in Fig. E4.7b. The coordinates of the midpoint G of the diagonal AC are $\frac{1}{2}[(0,0,6)+(-2,4,4)] = (-1,2,5)$. The reactions and the tension are:

$$\vec{R}_A = R_1\vec{i} + R_2\vec{j} + R_3\vec{k}, \qquad \vec{R}_B = R_4\vec{i} + R_5\vec{j} + R_6\vec{k}$$

$$\vec{T} = T\frac{\overrightarrow{CO}}{CO} = t(2\vec{i} - 4\vec{j} - 4\vec{k})$$

where $t = T/(CO)$ and $\vec{R}_B \cdot \overrightarrow{AB} = 0$ since $\vec{R}_B \perp \overrightarrow{AB}$. The 7 unknowns, $R_1, \ldots, R_6, t$, are obtained from $\vec{M}_A = \vec{0},\ \vec{F} = \vec{0},\ \vec{R}_B \cdot \overrightarrow{AB} = 0$. Position vectors are obtained by subtracting the coordinates of their tails from those of the tips,

$$\overrightarrow{AG} = -\vec{\imath} + 2\vec{\jmath} - \vec{k}\,\text{m}, \qquad \overrightarrow{AO} = -6\vec{k}\,\text{m}, \qquad \overrightarrow{AB} = 4\vec{\jmath} - 2\vec{k}\,\text{m}$$

$$\begin{aligned}\vec{M}_A &= \overrightarrow{AG} \times 6\vec{k} + \overrightarrow{AO} \times \vec{T} + \overrightarrow{AB} \times \vec{R}_B \\ &= (-\vec{\imath} + 2\vec{\jmath} - \vec{k}) \times 6\vec{k} - 6\vec{k} \times t(2\vec{\imath} - 4\vec{\jmath} - 4\vec{k}) \\ &\quad + (4\vec{\jmath} - 2\vec{k}) \times (R_4\vec{\imath} + R_5\vec{\jmath} + R_6\vec{k}) = \vec{0}\end{aligned}$$

$$\begin{aligned}\vec{k}:&\quad -4R_4 = 0 &&\Rightarrow\quad R_4 = 0 \\ \vec{\jmath}:&\quad 6 - 12t - 2R_4 = 0 &&\Rightarrow\quad t = 0.5 \\ \vec{\imath}:&\quad 12 - 24t + 4R_6 + 2R_5 = 0 &&\Rightarrow\quad 2R_6 + R_5 = 0 \qquad (1)\end{aligned}$$

$$\vec{R}_B \cdot \overrightarrow{AB} = (R_4\vec{\imath} + R_5\vec{\jmath} + R_6\vec{k}) \cdot (4\vec{\jmath} - 2\vec{k}) = 4R_5 - 2R_6 = 0 \qquad (2)$$

$$\begin{aligned}\Rightarrow\quad R_5 &= R_6 = 0 \quad\Rightarrow\quad \vec{R}_B = \vec{0} \\ \vec{T} &= t(2\vec{\imath} - 4\vec{\jmath} - 4\vec{k}) = \vec{\imath} - 2\vec{\jmath} - 2\vec{k} \quad\Rightarrow\quad T = 3\,\text{kN} \\ \vec{F} &= \vec{R}_A + \vec{R}_B + \vec{T} + 6\vec{k} = \vec{0} \\ \Rightarrow\quad \vec{R}_A &= -\vec{R}_B - \vec{T} - 6\vec{k} = -\vec{\imath} + 2\vec{\jmath} - 4\vec{k}\,\text{kN}\end{aligned}$$

4.3 Analysis of Frames and Mechanisms

A system consisting of interconnected rigid bodies (Fig. 4.5) is called a *frame* if the relative configuration of the bodies cannot change, else it is called a *mechanism* (i.e., it is under-constrained with at least one possible change of relative configuration of bodies).

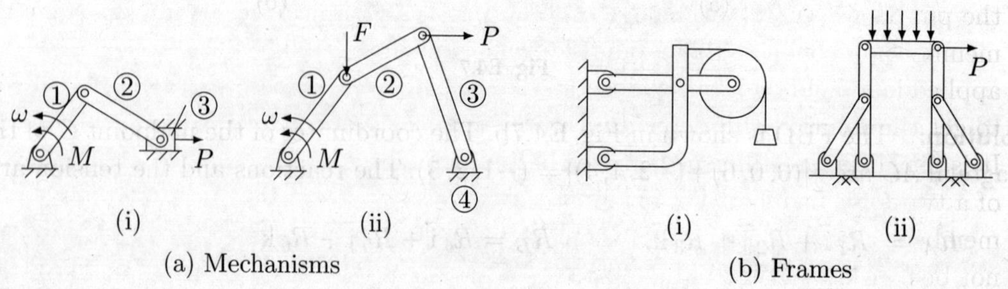

Fig. 4.5

A mechanism/machine is a mechanical device which transmits and modifies applied forces/moments. The basic mechanical devices used in mechanisms are: *lever, inclined plane and wedge, pulley, screw* (helical form of inclined plane), and the *wheel and axle.* These five devices are called 'simple machines'. For a given input load (force/moment), these help to obtain an output load having different magnitude and/or direction. Since a mechanism has possible change in configuration, only some of its configurations are equilibrium configurations under given loads. The unknowns in a mechanism may be some unknown forces/moments and some unknown coordinates specifying its equilibrium configuration. The unknowns in a frame are forces or moments.

A frame or a mechanism may include members that are not two-force members. Such members, even if straight, may be subjected to bending moment, shear and axial forces. A complex system is decomposed into a number of constituent members (subsystems). The theorem of action and reaction being equal in magnitude and opposite in direction is used to show the reactions at the inter-connections.

There is no universal guide to the best choice of subsystems for writing the equations of equilibrium. The solution of simultaneous equations can be avoided or minimised by a careful selection of the sequence of subsystems and by a careful choice of the base point for writing the moment equation of equilibrium. Generally, one should take subsystems with a large number of elements. If the frame is a rigid unit by itself, when removed from its external supports, and the supports provide 6 (3 for the 2D case) independent constraints, then the support reactions are found first from 6 (3 for the 2D case) equilibrium equations. Subsystems are considered next. By contrast, if the structure is non-rigid without the external supports, then the constraint reactions are greater than 6 (greater than 3 for the 2D case) to make it a rigid frame. Hence, the support reactions for this case cannot be determined from the equilibrium equations of the whole structure. In such a case, it is usually necessary to first consider the equilibrium of some portion of the system that itself is inherently rigid. A lot of labour can be saved by *first identifying the two-force members*, if any.

In drawing the FBDs of either of the two individual members connected by a pin with no external load on the pin, it does not matter which one is assumed to have the pin. However, it can make a difference for the following cases: (1) more than two members are connected by a pin, (2) two members are pinned to a support, and (3) an applied load acts on a pin connecting two members. In such cases, the effort required to get the solution will depend on where the pin is assumed to be in the subsystems. It is usually more efficient to take the pin as part of a multi-force member than that of a two-force member. For example, in the system of four members shown in Fig. 4.6, members ① and ② are two-force members for the given loading. Therefore, pins should not be considered part of members ① and ②. Pins C, D should form part of member ④ and pins A, B should form part of member ③ for consideration of their equilibrium.

The FBDs of the members are shown in Fig. 4.6. $F_1, F_2, R_1, R_2, R_3, R_4$ can be obtained by writing 6 equations of equilibrium for members ③ and ④ in the following order: $\vec{M}_C = \vec{0}$ for member ④ yields F_1, $\vec{M}_A = \vec{0}$ for member ③ yields F_2, $\vec{F} = \vec{0}$ for member ④ yields R_1, R_2 and $\vec{F} = \vec{0}$ for member ③ yields R_3, R_4.

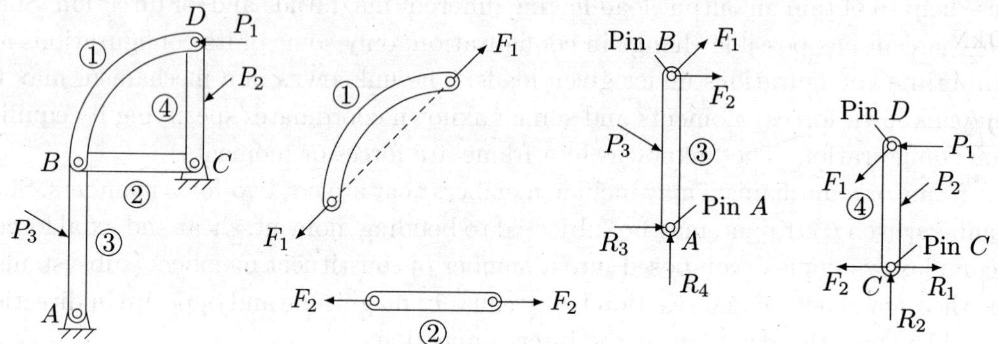

Fig. 4.6

Example 4.8: Find the force P required for impending motion of a block on rollers (Fig. E4.8a). The coefficient of friction of the rollers with the ground is μ_1 and with the block is μ_2.

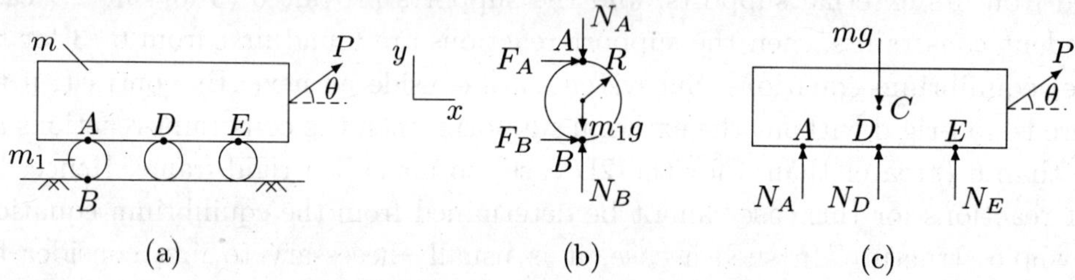

Fig. E4.8

Solution: From the FBD of a roller (Fig. E4.8b),

$$M_{A_z} = F_B(2R) = 0 \quad \Rightarrow \quad F_B = 0, \qquad F_x = F_A + F_B = 0 \quad \Rightarrow \quad F_A = 0$$

$\Rightarrow$ Only normal reactions N_A, N_B occur. In fact, roller is a two-force member with reaction at A and m_1g & ground reaction through B. Hence, the reactions are along AB. The FBD of the block is shown in Fig. E4.8c. For impending motion of block

$$F_x = P\cos\theta = ma_x = m(0^+) \quad \Rightarrow \quad P = 0^+$$

As friction force $F = 0$ on the block, the effective coefficient of friction $\mu_{\text{eff}} = F/N = 0$ for transport over rollers, but rollers have to be repeatedly shifted from back to front.

Example 4.9: Find the support reactions for the 3-pin arch in Fig. E4.9a.

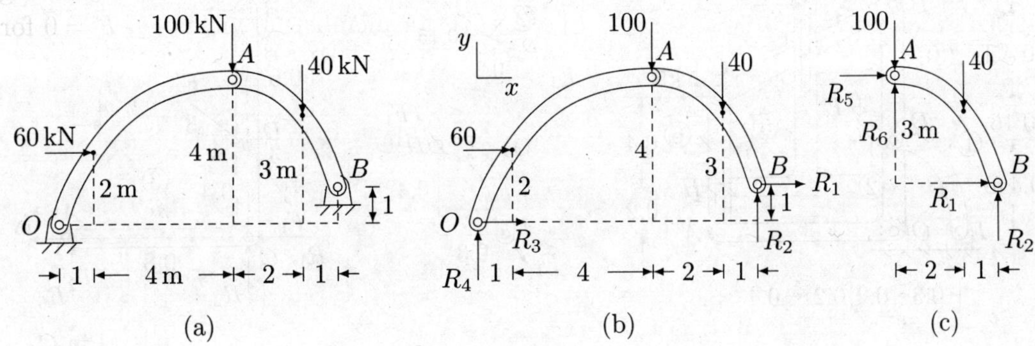

Fig. E4.9

Solution: The FBDs of the arch and part AB are shown in Fig. E4.9b,c. R_1, R_2, R_3, R_4 are found from $F_x = 0,\ F_y = 0,\ M_{O_z} = 0$ for the arch and $M_{A_z} = 0$ for part AB:

$$AB: \quad M_{A_z} = 3R_2 + 3R_1 - 40 \times 2 = 0 \tag{1}$$

$$\text{arch}: \quad M_{O_z} = 8R_2 - R_1 \times 1 - 60 \times 2 - 100 \times 5 - 40 \times 7 = 0 \tag{2}$$

$$F_x = R_1 + R_3 + 60 = 0 \tag{3}$$

$$F_y = R_2 + R_4 - 100 - 40 = 0 \tag{4}$$

$$(1), (2) \Rightarrow \quad R_1 = -76.296\,\text{kN}, \qquad R_2 = 102.96\,\text{kN}$$

$$\text{and } (3), (4) \Rightarrow \quad R_3 = 16.296\,\text{kN}, \qquad R_4 = 37.04\,\text{kN}$$

Example 4.10: For the structure shown in Fig. E4.10a, find the forces acting on member ③ at O, A, B, D. The slots are smooth.

Solution: Members ① and ② are two-force members with forces F_1 and F_2 along ID and DE, respectively. The FBDs are shown in Figs E4.10b–g. The 12 unknowns for the four bodies ③, ④, ⑤, ⑥ are: F_1, F_2; N_1 at G; N_2 at H; R_1, R_2 at O; R_3, R_4 at B; R_5, R_6 at A; R_7, R_8 at C. It is a statically determinate problem.

Analysing the number of reactions on bodies ③, ④, ⑤, ⑥, it is noted that only body ⑤ has three reactions N_1, N_2, F_2 acting on it (Fig. E4.10d). These are obtained using

$$F_y = -10 + 0.6F_2 = 0 \qquad \Rightarrow \quad F_2 = 16.67\,\text{kN}$$

$$M_{P_z} = 8(0.8) - 0.8F_2(0.35) - 0.6F_2(0.2) - 0.6N_1 = 0 \quad \Rightarrow \quad N_1 = -0.4444\,\text{kN}$$

$$F_x = N_1 - N_2 - 8 + 0.8F_2 = 0 \qquad \Rightarrow \quad N_2 = 4.889\,\text{kN}$$

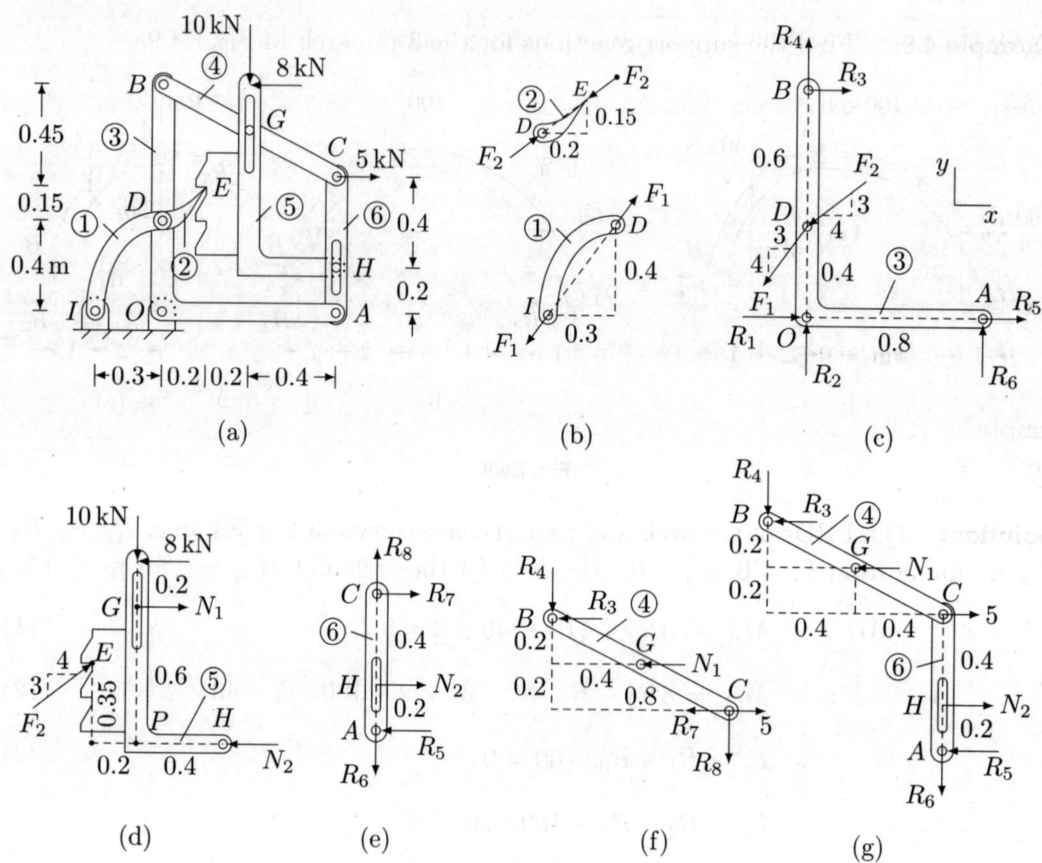

Fig. E4.10

Members ④, ⑥ are hinged together and to the remaining structure. This subsystem can be analysed for R_3 to R_8 as for a 3-hinged arch. For ④ + ⑥ (Fig. E4.10g):

$$M_{Bz} = 5 \times 0.4 - 1 \times R_5 - 0.8R_6 + 0.8N_2 - 0.2N_1 = 0$$

$$\Rightarrow \quad R_5 + 0.8R_6 = 6 \tag{1}$$

$$F_x = N_2 + 5 - N_1 - R_3 - R_5 = 0 \tag{2}$$

$$F_y = -R_4 - R_6 = 0 \tag{3}$$

We use another equation involving R_5, R_6 using $M_{Cz} = 0$ for body ⑥ (Fig. E4.10e):

$$M_{Cz} = 0.4N_2 - 0.6R_5 = 0 \quad \Rightarrow \quad R_5 = 3.259\,\text{kN}$$

Equations (1) to (3) yield $R_6 = 3.426$, $R_4 = -3.426$, $R_3 = 7.074\,\text{kN}$. Lastly, consider body ③ (Fig. E4.10c) to find F_1, R_1, R_2:

$$M_{Oz} = 0.8R_6 - R_3 \times 1 + (0.6F_1 + 0.8F_2) \times 0.4 = 0 \quad \Rightarrow \quad F_1 = -4.167\,\text{kN}$$

$$F_x = R_1 + R_3 + R_5 - 0.8F_2 - 0.6F_1 = 0 \quad \Rightarrow \quad R_1 = 0.5\,\text{kN}$$

$$F_y = R_2 + R_4 + R_6 - 0.6F_2 - 0.8F_1 = 0 \quad \Rightarrow \quad R_2 = 6.666\,\text{kN}$$

$$\vec{R}_O = 0.5\vec{\imath} + 6.666\vec{\jmath}, \qquad \vec{R}_B = 7.074\vec{\imath} - 3.426\vec{\jmath}, \qquad \vec{R}_A = 3.259\vec{\imath} + 3.426\vec{\jmath}\ \text{kN}$$

$$\vec{R}_D = (-0.6F_1 - 0.8F_2)\,\vec{\imath} - (0.8F_1 + 0.6F_2)\,\vec{\jmath} = -10.83\vec{\imath} - 6.666\vec{\jmath}\ \text{kN}$$

Example 4.11: Find the force on the pin at A (Fig. E4.11a) for the hoisting rig. The force P is just enough to support the crate of weight W. Neglect weight of the members.

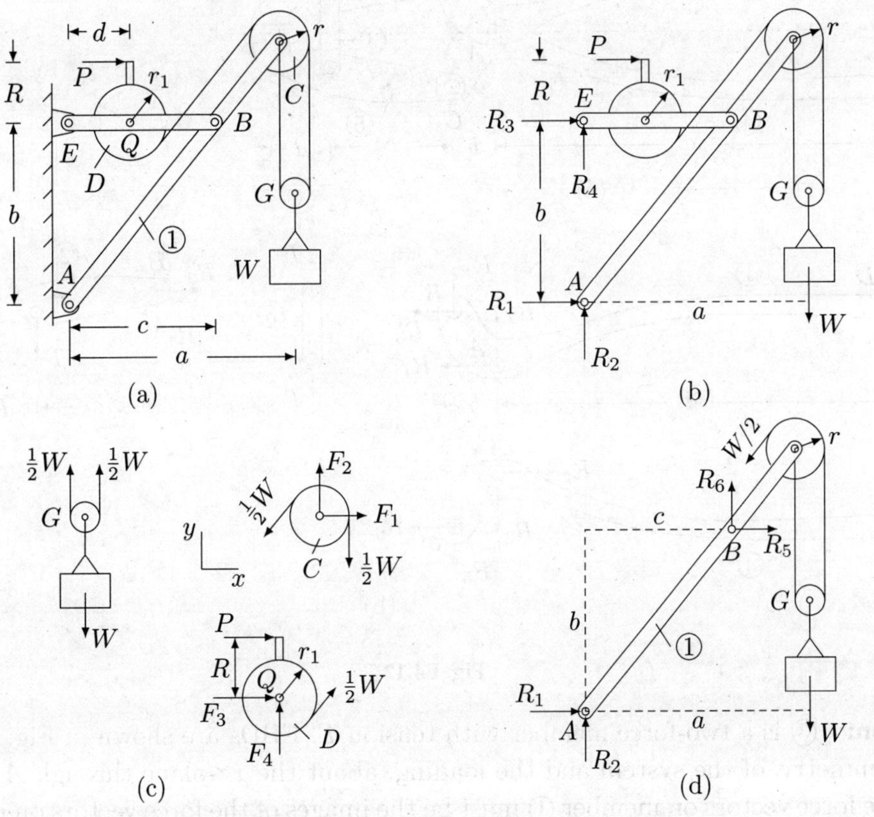

Fig. E4.11

Solution: The FBDs of the whole system and subsystems are shown in Fig. E4.11b,c,d. We use moment equilibrium equation about a point where most unknown forces act.

For pulley D (Fig. E4.11c) : $M_{Qz} = (\frac{1}{2}W)r_1 - PR = 0 \quad \Rightarrow \quad P = Wr_1/2R$

For rig (Fig. E4.11b) : $M_{Ez} = R_1 b - Wa - PR = 0 \Rightarrow R_1 = W(a + \frac{1}{2}r_1)/b$

For system of Fig. E4.11d:

$$M_{Bz} = \tfrac{1}{2}Wr - W(a - c) + R_1 b - R_2 c = 0 \quad \Rightarrow \quad R_2 = W[1 + (r + r_1)/2c]$$

Example 4.12: Find the forces exerted on the wire by the wire cutter (Fig. E4.12a) for applied forces P. Discuss the order of b/a, c/d to obtain large cutting force N.

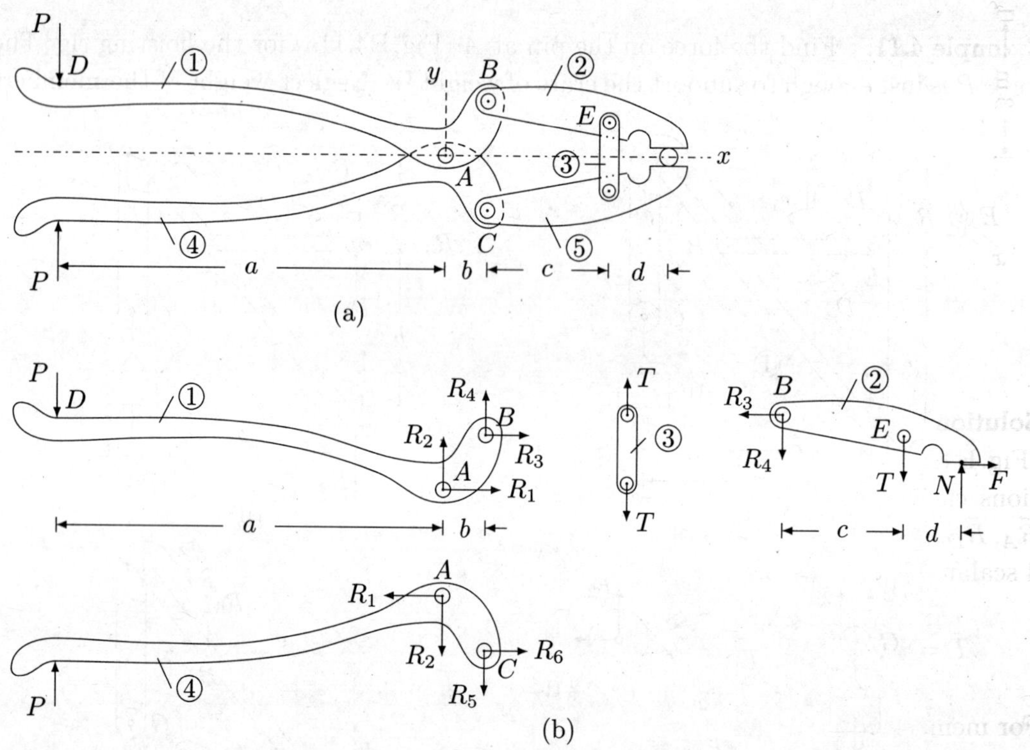

Fig. E4.12

Solution: ③ is a two-force member with tension T. FBDs are shown in Fig. E4.12b. The symmetry, of the system and the loading, about the xz-plane through A implies that the force vectors on member ① must be the images of the force vectors on member ④ in a mirror placed on xz-plane. $\Rightarrow \quad R_3 = R_6,\ R_4 = R_5,\ R_1 = -R_1,\ \Rightarrow\ R_1 = 0.$

①: $F_x = R_3 = 0, \quad M_{Az} = Pa + R_4 b = 0 \quad \Rightarrow \quad R_4 = -P\dfrac{a}{b}$

②: $F_x = F = 0, \quad M_{Ez} = R_4 c + Nd = 0 \quad \Rightarrow \quad N = -R_4\dfrac{c}{d} = P\dfrac{a}{b}\dfrac{c}{d}$

N can be increased by having $a/b \gg 1$, $c/d \gg 1$, e.g., if $P = 200\,\text{N}, a = 240\,\text{mm}, b = 8\,\text{mm}, c = 45\,\text{mm}, d = 15\,\text{mm}$, then $N = 90P = 18\,\text{kN}$. P gets magnified 90 times.

Example 4.13: Each connection of the space frame shown in Fig. E4.13a is idealised as a ball and socket joint. Find the reaction on the member AG at the support A.

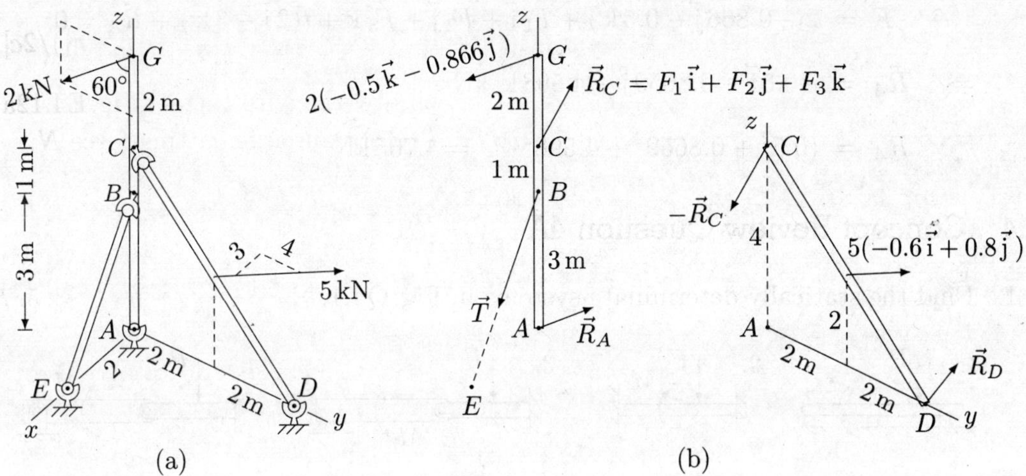

Fig. E4.13

Solution: EB is a two-force member. The force systems on members AG and CD (Fig. E4.13b) intersect lines AG and CD respectively. Hence, only 5 equilibrium equations can be written for each. The 10 unknowns are: T and 3 components each of $\vec{R}_A, \vec{R}_D, \vec{R}_C$. The equations of equilibrium $\vec{M}_D = 0$ for CD and $\vec{M}_A = 0$ for AG yield 4 scalar equations for T and F_1, F_2, F_3 ($\vec{R}_C$). Then, $\vec{F} = \vec{0}$ for AG yields $\vec{R}_A$.

$$\vec{T} = T\,\frac{\overrightarrow{BE}}{BE} = T\,\frac{-3\vec{k} + 2\vec{i}}{\sqrt{13}} = t(2\vec{i} - 3\vec{k}), \qquad t = \frac{T}{\sqrt{13}}$$

For member CD: $\vec{M}_D = \vec{0} \Rightarrow$

$$(-2\vec{j} + 2\vec{k}) \times 5(-0.6\vec{i} + 0.8\vec{j}) + (-4\vec{j} + 4\vec{k}) \times (-F_1\vec{i} - F_2\vec{j} - F_3\vec{k}) = \vec{0}$$

$$\vec{i}: \quad -8 + 4F_3 + 4F_2 = 0 \quad \Rightarrow \quad F_2 + F_3 = 2 \qquad (1)$$

$$\vec{j}: \quad -6 - 4F_1 = 0 \quad \Rightarrow \quad F_1 = -1.5$$

$$\vec{k}: \quad -6 - 4F_1 = 0 \quad \Rightarrow \quad F_1 = -1.5$$

For member AG: $\vec{M}_A = \vec{0} \Rightarrow$

$$6\vec{k} \times 2(-0.866\vec{j} - 0.5\vec{k}) + 4\vec{k} \times (F_1\vec{i} + F_2\vec{j} + F_3\vec{k}) + 3\vec{k} \times t(2\vec{i} - 3\vec{k}) = \vec{0}$$

$\vec{\imath}$: $10.39 - 4F_2 = 0 \quad \Rightarrow \quad F_2 = 2.598$

$\vec{\jmath}$: $4F_1 + 6t = 0 \quad \Rightarrow \quad t = -4(-1.5)/6 = 1$

and eq. (1) $\Rightarrow F_3 = 2 - F_2 = -0.598$. For member AG:

$$\vec{F} = 2(-0.866\vec{\jmath} - 0.5\vec{k}) + F_1\vec{\imath} + F_2\vec{\jmath} + F_3\vec{k} + t(2\vec{\imath} - 3\vec{k}) + \vec{R}_A = 0$$

$$\Rightarrow \quad \vec{R}_A = -0.5\vec{\imath} - 0.8659\vec{\jmath} + 4.598\vec{k} \text{ kN}$$

$$R_A = (0.5^2 + 0.8659^2 + 4.598^2)^{1/2} = 4.705\,\text{kN}$$

4.4 Concept Review Question 4A

1. Find the statically determinate systems in Fig. Q4.1. [3, 5, 7]

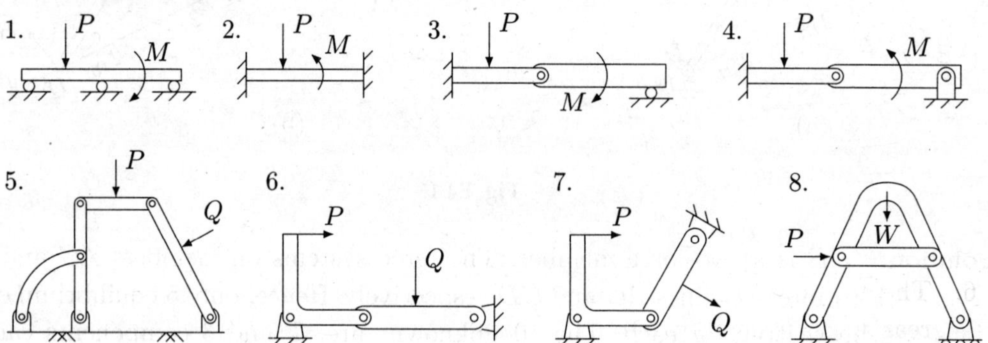

Fig. Q4.1

2. Find θ for equilibrium of the mechanism formed by light bars with smooth pins (Fig. Q4.2). [75.96°]

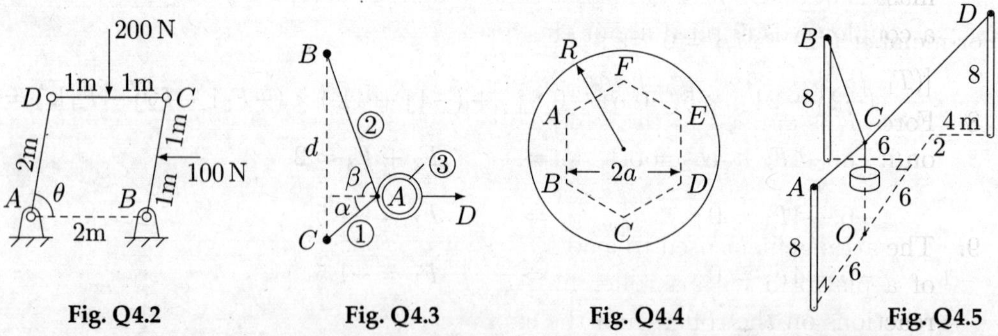

Fig. Q4.2 **Fig. Q4.3** **Fig. Q4.4** **Fig. Q4.5**

3. A buoy ③ is to be kept afloat in a river (Fig. Q4.3) by tying cables ① and ② to it. The drag on the buoy is D. (a) For $\beta = 70°$, find α (1) for minimum tension

in cable ① and (2) for minimum tension in cable ②. (b) If a single cable is to be used through a smooth ring at A, find the length L of the cable that can be used if the allowable tension is T_0. [(1) 20°, (2) 0° (b) $L \geq d/\sqrt{1 - D^2/4T_0^2}$]

4. A sphere of mass m rests on a hexagonal network of light rods fixed at A, B, C, D, E, F with ball and socket joints (Fig. Q4.4). Find the horizontal and vertical forces on pin A from the rods.

$$[H_A = \frac{0.1443mga}{\sqrt{R^2 - a^2}},\ V_A = \frac{mg}{6},\ R_A = \sqrt{H_A^2 + V_A^2} \text{ at } \theta = \tan^{-1}\tfrac{V_A}{H_A} \text{ to diagonal}]$$

5. Find the cable tensions which support a weight of 220 N (Fig. Q4.5). $OC = 6$ m. [379.5, 174.4, 275.0 N]

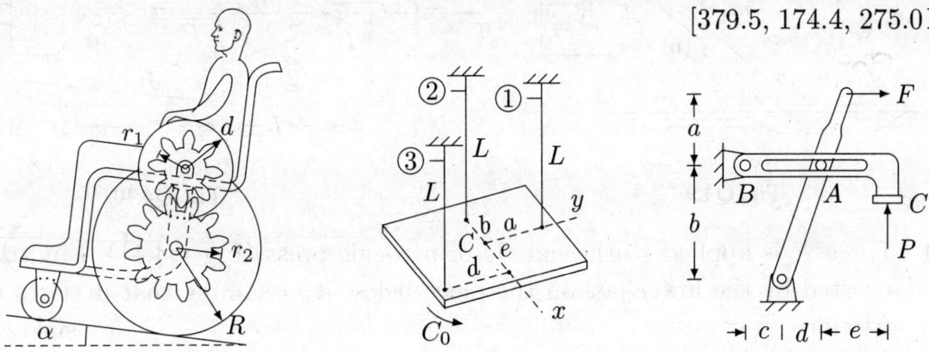

Fig. Q4.6 **Fig. Q4.7** **Fig. Q4.8**

6. The wheel-chair in Fig. Q4.6 has mass M including the rider. The mass of the rear wheel assembly is m. It is slowly moved up by the rider by applying a tangential force P to a hand wheel at distance d from its axis. A pinion of radius r_1 on the hand-wheel is in mesh with a gear of radius r_2 fixed to the wheel. Find P. [$Mg \sin\alpha\, Rr_1/r_2 d$]

7. Find the tensions in the supporting cables ①, ②, ③ of a rectangular plate whose mass is m and centre of mass is at C (Fig. Q4.7). Find its *small* rotation θ when a couple C_0 is applied about the vertical axis.

$$[(T_1, T_2, T_3) = (bd, ae, ab)mg/(ab + ae + bd),\ \ C_0 L/\{T_1 a^2 + T_2 b^2 + T_3(e^2 + d^2)\}]$$

8. Force F is applied to the handle of a stamping machine (Fig. Q4.8). The pin A on it can slide in a smooth slot in arm BC. Find the stamping force P at C.

$$[(c + d)(a + b)F/(c + d + e)d]$$

9. The arrangement used to load a barge, where the water level is variable, consists of a platform with a roller hinged to it at the bottom (Fig. Q4.9). Find the reactions on the rollers and the cable tension.

$$[N_B = [0.5mgL\cos\theta + m_1 g(b\cos\theta - h\sin\theta)]/L(0.7 - \sin^2\theta),$$
$$N_A = m_1 g + mg - N_B\cos\theta,\quad T = N_B\sin\theta]$$

10. An overload protection device (Fig. Q4.10) consists of the toggle operated jaw which shears the pin P when the cable tension exceeds preset value T. Find the shear load Q for which the pin must shear. Find the load on the pin at hinge A when the cable carries tension T.
[$T(a - e + ac/b)/2f$, $T\{a(f - c) - b(a - e)\}/2bf$]

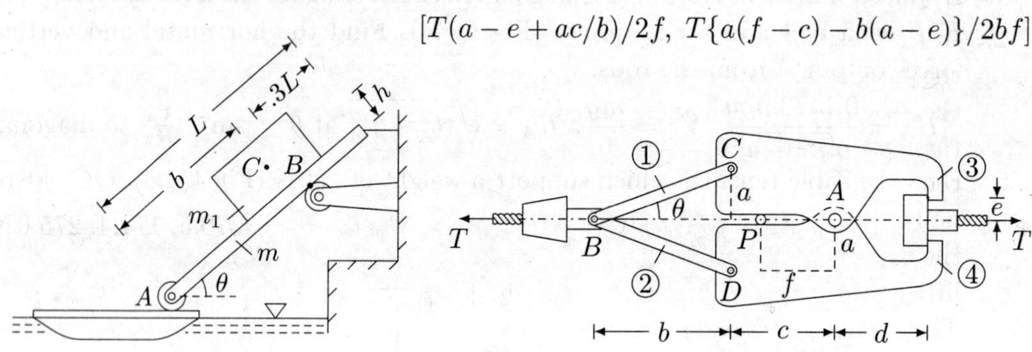

Fig. Q4.9 Fig. Q4.10

11. Force F is applied to the handle of a toggle press (Fig. Q4.11). Find the force exerted by the lower jaw on the piece below it, assuming that it slides without friction.
[$bF\sin\alpha/2a\sin\theta$]

12. A heavy pre-fabricated structural component is subjected to given loads during installation (Fig. Q4.12). It is supported by three cables and a ball and socket joint at O. Find the cable tensions and reaction at O.
[34, 21.88, 10; $24\vec{i} + 46.87\vec{j} + 2.5\vec{k}$ kN]

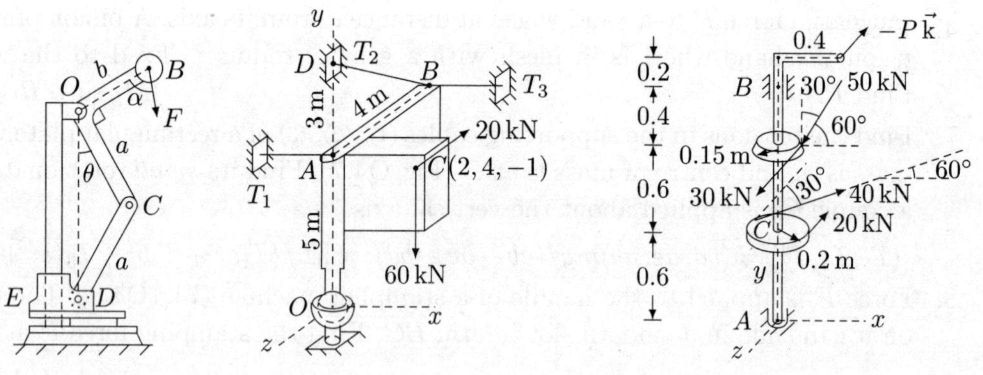

Fig. Q4.11 Fig. Q4.12 Fig. Q4.13

13. A shaft with a gear and a pulley has bearings at A and B (Fig. Q4.13). The tensions in the belt and the force on the gear are shown. Find the applied force P for equilibrium and the bearing reactions. The bearing at B does not provide axial restraint. [21.25, $-2.191\vec{i} + 43.30\vec{j} + 24.00\vec{k}$, $7.191\vec{i} + 21.89\vec{k}$ kN]

4.5 Practice Problems 4A

1[d]. A painter's scaffold is 7 m long and weighs 0.8 kN. It is supported in a horizontal position by vertical ropes attached at equal distance d from its ends. Find $d_{\max}$ so that a 0.6 kN man can stand safely anywhere on the scaffold. [2 m]

2. In the FBD of the vertebral column (Fig. P4.2), F_e is the force exerted by the erector spinal muscles. $\vec{R}$ is the reaction from the top of the sacrum pushing on the base of the lumbo-sacral disc. W_1 is the weight of the trunk. $W_1 = 0.4W$, where W is the weight of the body. W_2 is the sum of the weight of the head and the arms $(= 0.2W)$. W_0 is the weight of the box to be lifted. Find F_e and $\vec{R}$ for $\theta = 30°$ and $W_0 = 0.3W$. These forces are smaller if the distance of W_2 from O is smaller. Hence, lifting a load on the head is easier than on the back. Similarly, it is advisable to also bend the knees rather than just the back while lifting a load from the ground.
[$F_e = 4.374W$, $R = 4.730W$ at 28.43° to horizontal]

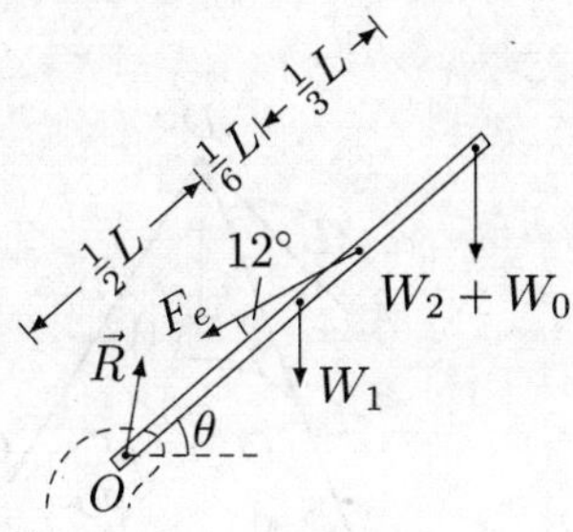

Fig. P4.2

Fig. P4.3

3. The distributed forces acting on the faces AB and CD of the embedded portion of the cantilever are assumed to vary linearly with zero intensities at B and C (Fig. P4.3). Find the maximum intensities of loading on the beam in the embedded portion. Comment on the desirable portion b that should be embedded.

[$(a + b,\ 2b + a)q_0 a/b^2$, large b]

4. Find the reactions at the fixed end of the curved beams in Fig. P4.4. In part(b): $q = q_0 \sin\phi$.
[(a) $\vec{R}_O = 3.571 q_0 r\,\vec{k}$ & $\vec{C}_O = q_0 r^2(6.142\vec{i} - 0.5708\vec{j})$
(b) $\vec{R}_A = -q_0 r\,\vec{k}$ & $\vec{C}_A = -(0.7854\vec{i} + 0.5\vec{j})q_0 r^2$]

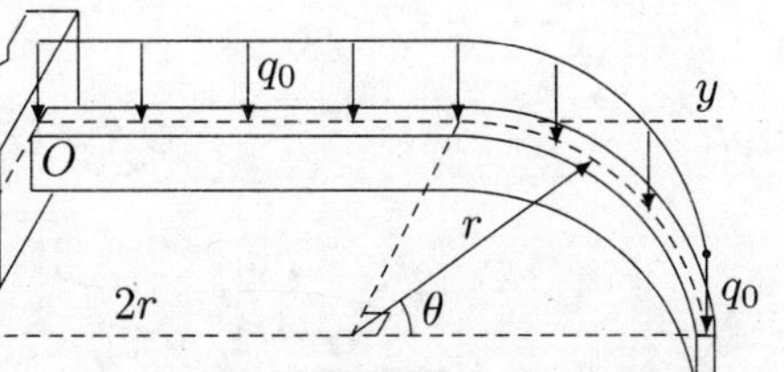

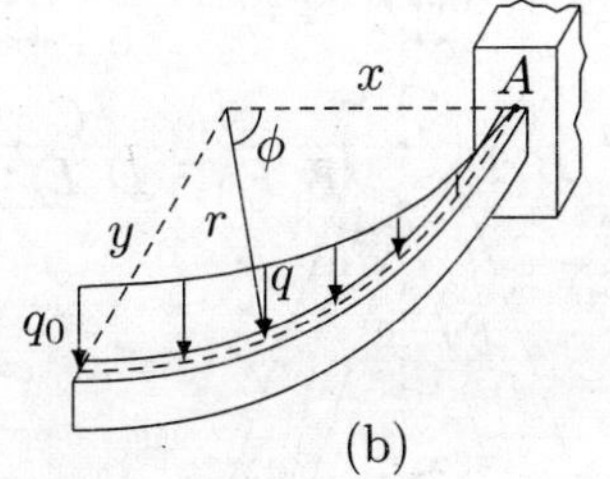

Fig. P4.4

5. Find the reactions from a smooth floor on the tripod (Fig. P4.5) supporting a load W. Each leg weighs w_1. Find also the force and couple exerted on each leg at D. What would happen if legs were connected at D by a ball and socket joint?
[$1.049w_1 + 0.3660W$, $0.817w_1 + 0.2113W$, $1.134w_1 + 0.4227W$;

$-(0.049w_1 + 0.3660W)\vec{k} \quad \& \quad (0.549w_1 + 0.366W)r\vec{i},$

$-(0.2113W - 0.183w_1)\vec{k} \quad \& \quad (0.317w_1 + 0.2113W)r\vec{j},$

$-(0.134w_1 + 0.4227W)\vec{k} \quad \& \quad (0.634w_1 + 0.4227W)r(0.866\vec{i} + 0.5\vec{j})$, collapse]

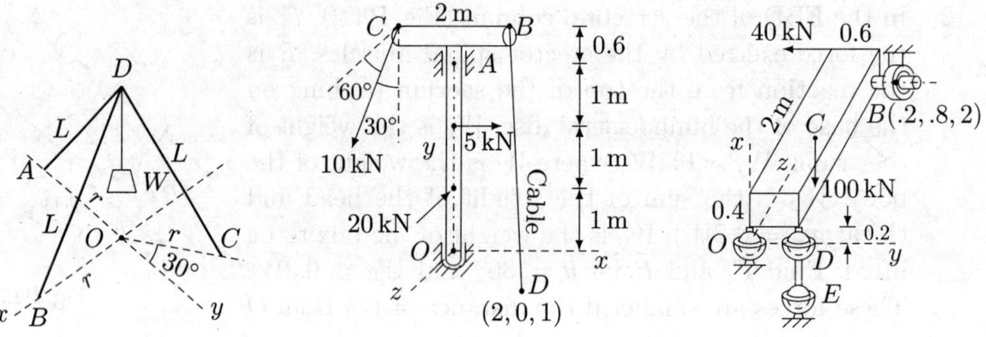

Fig. P4.5 **Fig. P4.6** **Fig. P4.7**

6. Find the cable tension and the bearing reactions at A and O (Fig. P4.6).

   [19.34, $-12.45\vec{i} - 5.333\vec{k}$, $2.450\vec{i} + 26.66\vec{j} + 15.33\vec{k}$ kN]

7. A box is supported by a ball and socket joint at O, an eye-bolt at B and a light rod DE with ball and socket joints at its ends (Fig. P4.7). Find the reactions.

   [$\vec{R}_D = -43.33\vec{i}$, $\vec{R}_B = 40\vec{i} - 100\vec{k}$, $\vec{R}_O = 103.3\vec{i} + 40\vec{j} + 100\vec{k}$ kN]

8. An antenna of weight 20 kN is supported by six bars placed symmetrically with respect to the xz-plane with ball and socket joints at the ends (Fig. P4.8). The centre of mass is at C. Find the forces in the bars.

   [F_i = 0.4225, -2.912, 13.06, 0.4225, -2.912, 13.06 kN]

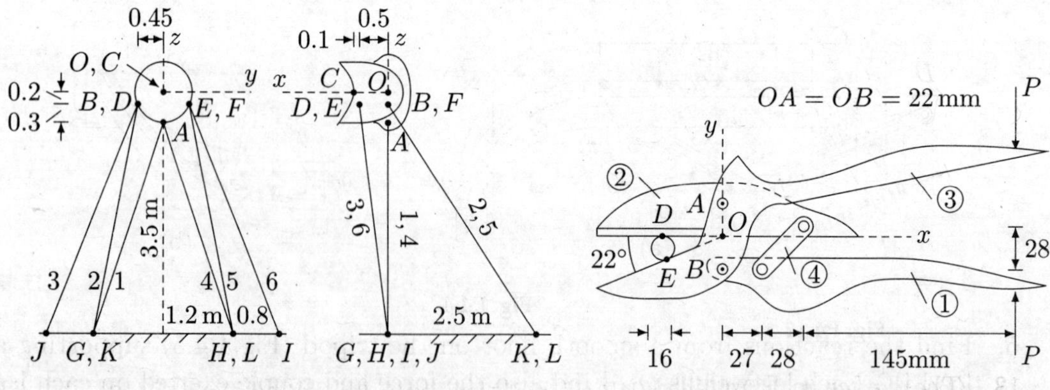

Fig. P4.8 **Fig. P4.9**

9*. Find the cutting force when forces $P = 200\,\text{N}$ are applied to a pruning shear (Fig. P4.9). $[449.9\vec{\text{i}} - 2315\vec{\text{j}}\ \text{N}]$

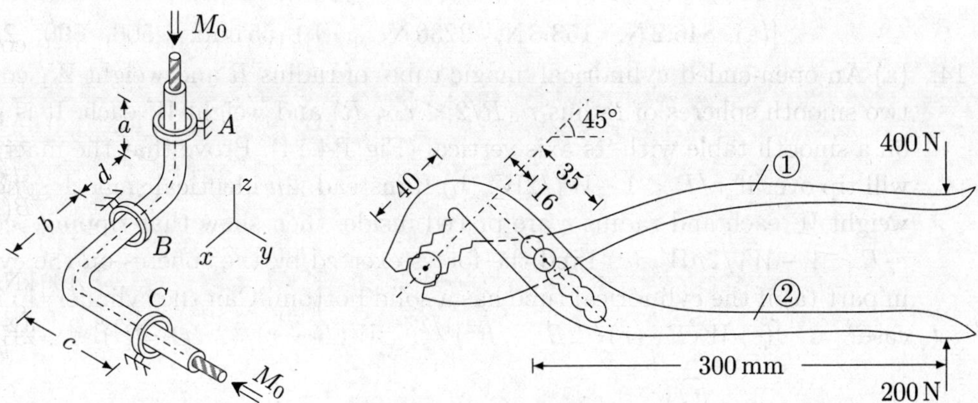

Fig. P4.10 **Fig. P4.11**

10. A rigid bent tube, supported by smooth eye-bolts, guides a flexible shaft transmitting a torque M_0 (Fig. P4.10). Find their reactions. Is it properly constrained? $[R_1\vec{\text{i}} + R_2\vec{\text{j}},\ \ R_3\vec{\text{j}} + R_4\vec{\text{k}},\ \ R_5\vec{\text{i}} + R_6\vec{\text{k}},$ with $R_1 = -R_5 = (ab - cd)M_0/ac(b-d),$ $R_2 = -R_3 = (a-c)M_0/a(b-d),\ \ R_4 = -R_6 = (c-a)M_0/c(b-d);$ yes if $b \neq d]$

11. Find the forces exerted by the pliers (Fig. P4.11) on a fixed bolt. How do these change when ① is shifted down? $[(F, N)$: $(4342, 6168)$, $(4484, 6026)$ N; decrease$]$

12. The nose-wheel assembly of an aeroplane (Fig. P4.12) is raised by applying a torque M to the link OA. The nose-wheel and the arm DE have a mass of 50 kg with the centre of mass at C. Neglect mass of links OA and AB. Find the value of M for equilibrium in the given position. [370.1 N.m]

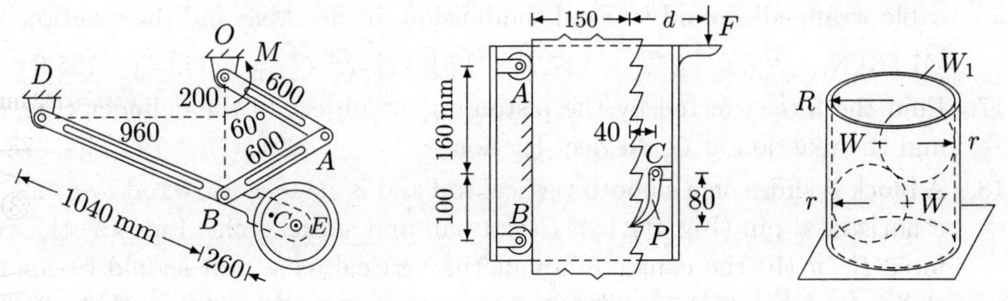

Fig. P4.12 **Fig. P4.13** **Fig. P4.14**

13. The device in Fig. P4.13 can support a load F at various heights by re-setting the pawl P in a tooth of the fixed ratchet at the desired height. Neglect weight of the device compared to F. (a) Find the reactions at A, B and C for $F =$

2000 N, $d = 100$ mm. (b) Find the distance d of the load F so that the rollers at A and B support optimum forces. Find the supporting forces for $F = 2000$ N.

[(a) 846.2 N, 153.8 N, 2236 N (b) 55 mm; 500, 500, 2236 N]

14. (a) An open-ended cylindrical 'magic tube' of radius R and weight W_1 contains two smooth spheres of radius r $(R/2 < r < R)$ and weight W each. It is placed on a smooth table with its axis vertical (Fig. P4.14). Prove that the magic tube will tip over if $r/R < 1 - W_1/2W$. (b) If instead, $2n$ identical smooth spheres of weight W each and radius r are placed inside, then show that tipping occurs if $r/R < 1 - W_1/2nW$. (c) Find the forces exerted by the spheres on the cylinder in part (a) if the cylindrical tube has a solid bottom. Can the cylinder tip in this case? [(c) $W(R-r)/(2Rr - R^2)^{1/2}$, $W(R-r)/(2Rr - R^2)^{1/2}$, $2W$, no]

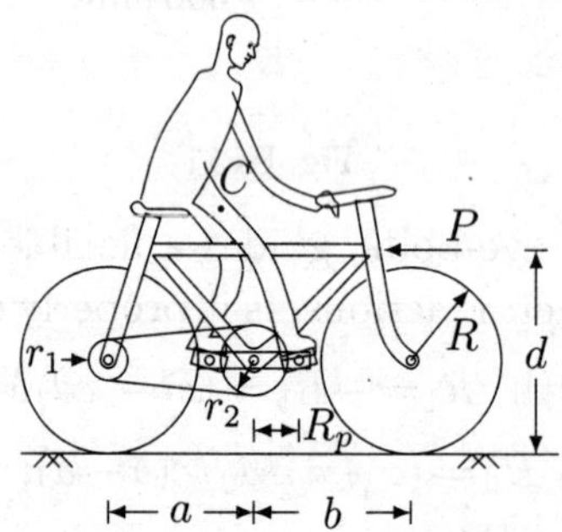

Fig. P4.15

z, D, E, 3 m, 1.2, A, 3 m, B, y, 1, C, 30°, x, 10 kN, 60°, $BC = 0.8$ m, 50 kN

Fig. P4.16

15. A cyclist drives a bike uniformly without slip (Fig. P4.15). The weight W of the bike and rider acts at C. Find the force applied to the paddle at radius R_p. Why are the spokes not radial? [PRr_2/r_1R_p, to transmit applied torque]

16. A bar ABC is hinged at A and has a small free pulley at B (Fig. P4.16) with a cable wrapped around it. Find the tension in the cable and the reaction at A.

[97.56 kN, $\vec{R}_A = 44.42\vec{\imath} + 147.7\vec{\jmath} + 2.5\vec{k}$ kN & $\vec{C}_A = -34.64\vec{\jmath} - 195.0\vec{k}$ kN.m]

17. Find the force exerted by the piston on the object in the cylinder (Fig. P4.17) and the reaction at O. Neglect friction. [921.6 N, $793.1\vec{\imath} - 378.9\vec{\jmath}$ N]

18. A block A slides on a smooth vertical rod and is attached to a rod by a clevis with a horizontal pin (Fig. P4.18). B is a ball and socket joint. Find (a) the vertical force P, or (b) the couple C about the vertical axis, that should be applied to slider A to maintain equilibrium. [(a) 60 N (b) −2400 N.m]

19*. A folding chair is subjected to force W (Fig. P4.19). Find the force on the pin at A for: (a) Smooth floor. Find also the location c of W to ensure equilibrium. (b) Rough floor and $c = (a+b)\cos\theta$. [(b) $-\frac{1}{2}W(\cot\theta\,\vec{\imath} + \vec{\jmath})$

(a) $\vec{R}_A = Wa\cos\theta\,\vec{\imath}/2b\sin\theta - W(b + a\cos^2\theta)\,\vec{\jmath}/2b\sin^2\theta$, $c = (a+b)\cos\theta/\sin^2\theta$]

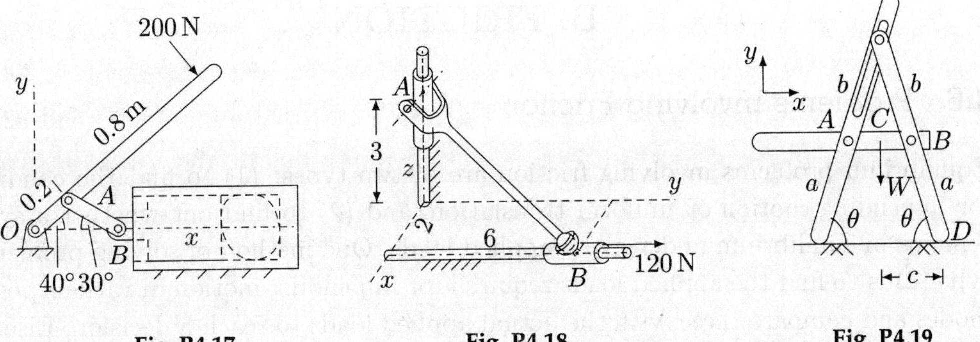

Fig. P4.17 **Fig. P4.18** **Fig. P4.19**

20. A back hoe is subjected to loads shown in Fig. P4.20. Find the forces at the pins A and C. $[10\vec{i} - 10.25\vec{j}$ kN, $21.80\vec{i} - 1.072\vec{j}$ kN]

21. A clamp (Fig. P4.21) is operated by applying moment M to the handle of a threaded shaft. The lead of the screw is l. Neglect friction. Find a relation between the clamping force P and M. $[M = 2lLP/\pi(Lc/b - c + a)]$

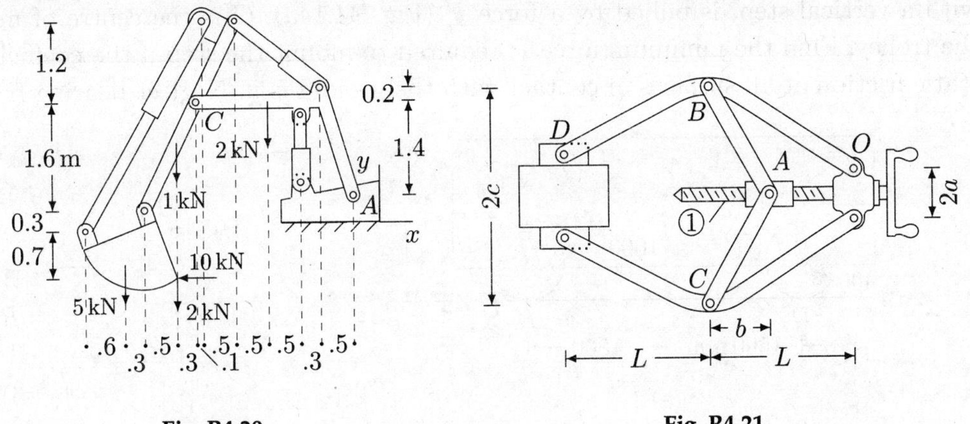

Fig. P4.20 **Fig. P4.21**

22. (a) A compound-lever snips (Fig. P4.22a) is gripped with a force of 180 N. Find the cutting force. (b) Find the punching force for the paper punch (Fig. P4.22b). The pins fixed to ① and ② slide freely in the slots. Show that clamping jaws remain parallel for objects of different thickness if $a = d$. [(a) 2304 N (b) Pb/d]

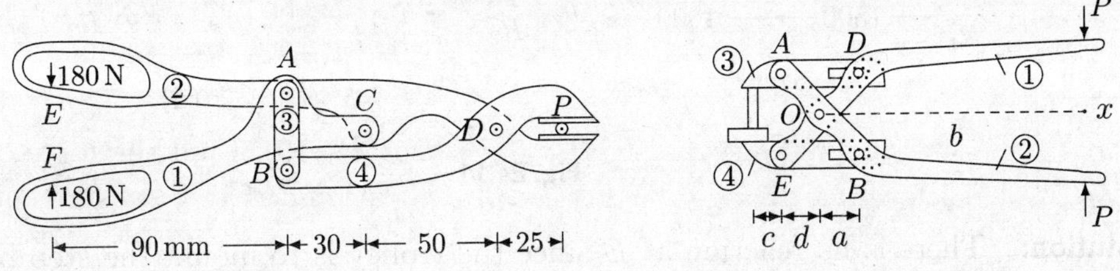

Fig. P4.22a **Fig. P4.22b**

B. FRICTION

4.6 Problems Involving Friction

Equilibrium problems involving friction are of two types: (1) to find the conditions for impending motion or uniform translation, and (2) to find out whether a system remains in equilibrium under given applied loads. One method of solving problems of type (2) is to find the applied loads required for impending motion in various possible modes and compare these with the actual applied loads to reach a decision. Else, find the value of a parameter such as μ or the position of a load (instead of the given value) for impending motion and compare it with its given value to reach a decision.

For the no slip case, the no slip condition should be checked in the form $|\tau| < \mu_s|p|$, or $|F| < \mu_s|N|$ for a two-sided constraint and in the form $|\tau| < \mu_s p$, or $|F| < \mu_s N$ if we know that N is positive. The methods of solution are illustrated by examples.

Example 4.14: A four wheel trolley of weight W with its front wheels in contact with a vertical step, is pulled by a force P (Fig. E4.14a). C is the centre of mass of the trolley. Find the minimum force P required to mount the step if the coefficient of static friction at all surfaces of contact with the ground is μ. Neglect bearing friction.

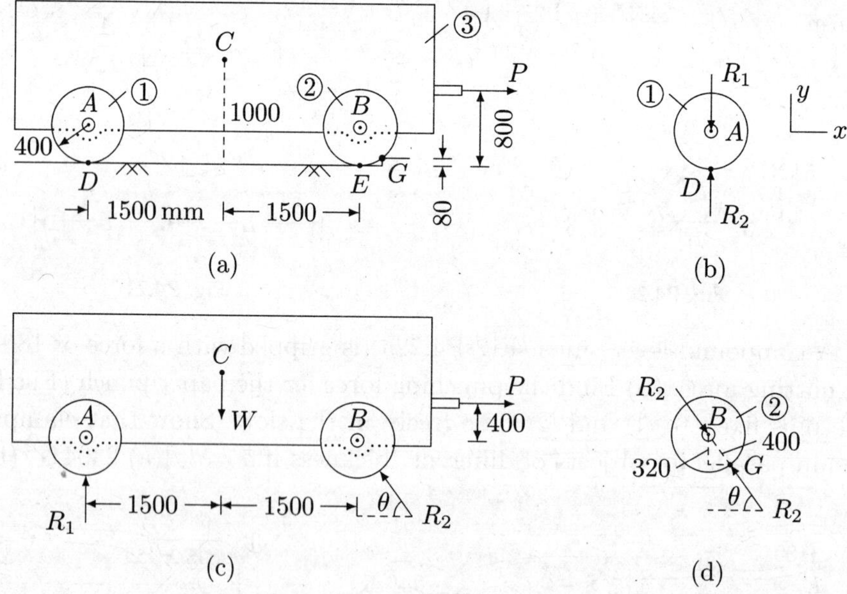

Fig. E4.14

Solution: There is no reaction at E since the trolley is to mount the step by losing contact at E. Wheel ① is subjected to only two forces through A (axle reaction and weight) and D (ground reaction). It is a 2-force member with the forces acting along AD (Fig. E4.14b). Similarly, wheel ② is a 2-force member with the forces at B and G acting along BG (Fig. E4.14d) with $\sin\theta = 320/400 = 0.8$, $\cos\theta = 0.6$. The FBD of the trolley is shown in Fig. E4.14c. Notice the interesting fact that the friction forces are zero. The 3 equations of equilibrium of this coplanar system yield R_1, R_2, P:

$$F_x = P - R_2\cos\theta = 0 \qquad \Rightarrow \quad R_2 = 5P/3$$

$$F_y = R_1 + R_2\sin\theta - W = 0 \qquad \Rightarrow \quad R_1 = W - 4P/3$$

$$M_{Bz} = 1500W - 3000R_1 - 400P = 0$$

$$\Rightarrow \quad 1500W - 3000(W - 4P/3) - 400P = 0 \Rightarrow \quad P = 5W/12$$

Example 4.15: A pipe wrench is engaged with a fixed pipe of radius R. Find the forces P_1 and P_2 applied to the handles (Fig. E4.15a) if the wrench develops a moment of M_0 about the pipe axis and its jaws are about to slip. $P_1 > P_2$ and the coefficient of friction is μ.

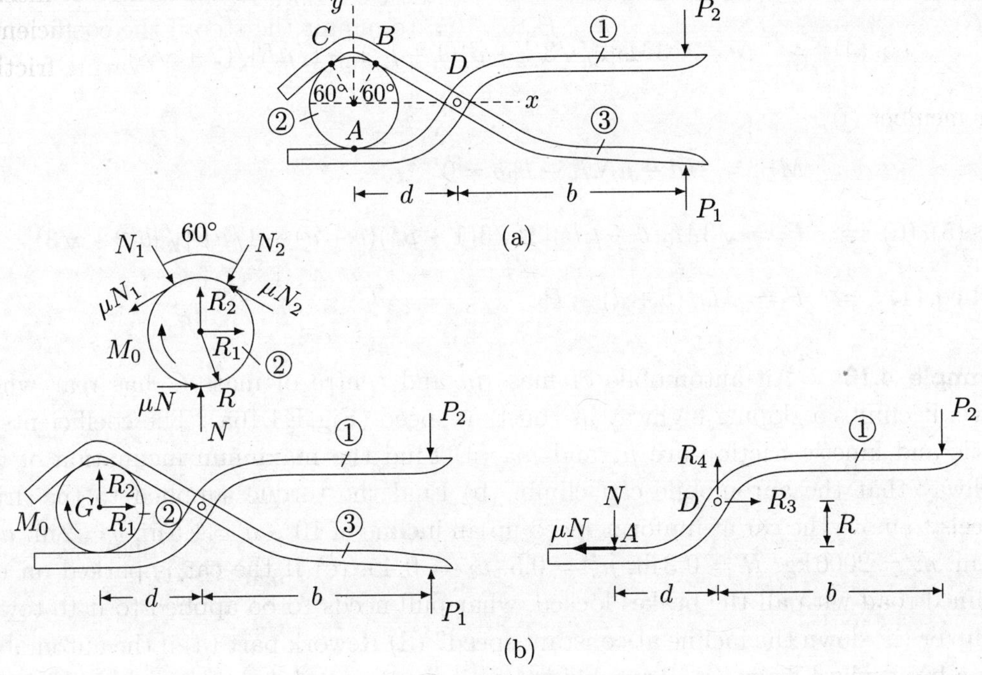

Fig. E4.15

Solution: The FBDs of members ①, ②, ③ together and members ① and ② individually are shown in Fig. E4.15b. For P_1, P_2: $\vec{M}_G = (P_1 - P_2)(d + b)\,\vec{k}$. Hence, M_0 in Fig. E4.15b is shown in direction $-\vec{k}$ and the jaw is about to rotate about the z-axis $\Rightarrow$ A, B, C have impending motion in direction $\vec{e}_\phi$ relative to pipe $\Rightarrow$ $\mu N, \mu N_1, \mu N_2$ on jaw are in $-\vec{e}_\phi$ direction and on pipe are in $\vec{e}_\phi$ direction. The 9 unknowns, $R_1, R_2, R_3, R_4, N, N_1, N_2, P_1, P_2$, can be found from the 9 equations of equilibrium (3 for each member). For members ①, ② and ③ together:

$$F_x = R_1 = 0$$

$$F_y = R_2 + P_1 - P_2 = 0$$

$$M_{Gz} = (P_1 - P_2)(d + b) - M_0 = 0 \quad \Rightarrow \quad R_2 = P_2 - P_1 = -M_0/(b + d) \qquad (1)$$

For member ②:

$$F_x = \mu N + R_1 + (N_1 - N_2)\sin 30° - \mu(N_1 + N_2)\cos 30° = 0 \qquad (2)$$

$$F_y = N + R_2 - (N_1 + N_2)\cos 30° - \mu(N_1 - N_2)\sin 30° = 0 \qquad (3)$$

$$M_{Gz} = \mu(N_1 + N_2 + N)R - M_0 = 0 \qquad (4)$$

$$\text{eqs (2), (3), (1)} \Rightarrow \quad N_1 + N_2 = [(1 + \mu^2)N - M_0/(b + d)]/(1 + \mu^2)\cos 30°$$

$$\text{eq. (4)} \Rightarrow \quad N = \sqrt{3}M_0[2/\sqrt{3}(b + d)(1 + \mu^2) + 1/\mu R]/(2 + \sqrt{3}) \qquad (5)$$

For member ①:

$$M_{Dz} = Nd - \mu NR - P_2 b = 0 \qquad (6)$$

$$\text{eqs (5), (6)} \Rightarrow \quad P_2 = \sqrt{3}M_0(d - \mu R)[2/\sqrt{3}(1 + \mu^2)(b + d) + 1/\mu R] \,/\, [b(2 + \sqrt{3})]$$

$$\text{and eq. (1)} \Rightarrow \quad P_1 = M_0/(b + d) + P_2.$$

Example 4.16: An automobile of mass m and centre of mass C has rear wheel drive. It climbs a sloping highway at constant speed (Fig. E4.16a). The coefficients of static and kinetic friction are μ_s and μ_k. (a) Find the maximum inclination of the highway that the automobile can climb. (b) Find the torque supplied to the drive wheels to move the car at uniform speed up an incline of 10°. $a = 1.5\,\text{m}$, $b = 2\,\text{m}$, $h = 0.6\,\text{m}$, $m = 2000\,\text{kg}$, $R = 0.3\,\text{m}$, $\mu_s = 0.5$, $\mu_k = 0.45$. (c) If the car is parked on the inclined road with all the brakes locked, what pull needs to be applied to it to tow it (i) up or (ii) down the incline at constant speed? (d) Rework part (a) if the automobile has a front wheel drive. (e) Rework part (a) if the automobile has four wheel drive.

Solution: (a) The FBDs of the wheels and the car are shown in Fig. E4.16b for the case of no slip. The front wheel ② is a two-force member subjected to force at A (weight and axle reaction) and B (ground reaction). Hence, these act along AB. The rear wheel ① is under the driving torque M_0 and forces at D and E. An internal load

M_0, does not appear in the FBD of the car. The equilibrium equations of the car:

$$
\begin{aligned}
F_x &= F - mg\sin\theta = 0 && \Rightarrow \quad F = mg\sin\theta \\
M_{Bz} &= mg\cos\theta\, a + mg\sin\theta\, h - N_1(a+b) = 0 && \Rightarrow \quad N_1 = \frac{mg(a\cos\theta + h\sin\theta)}{a+b} > 0 \\
F_y &= N_1 + N_2 - mg\cos\theta = 0 && \Rightarrow \quad N_2 = \frac{mg(b\cos\theta - h\sin\theta)}{a+b}
\end{aligned}
$$

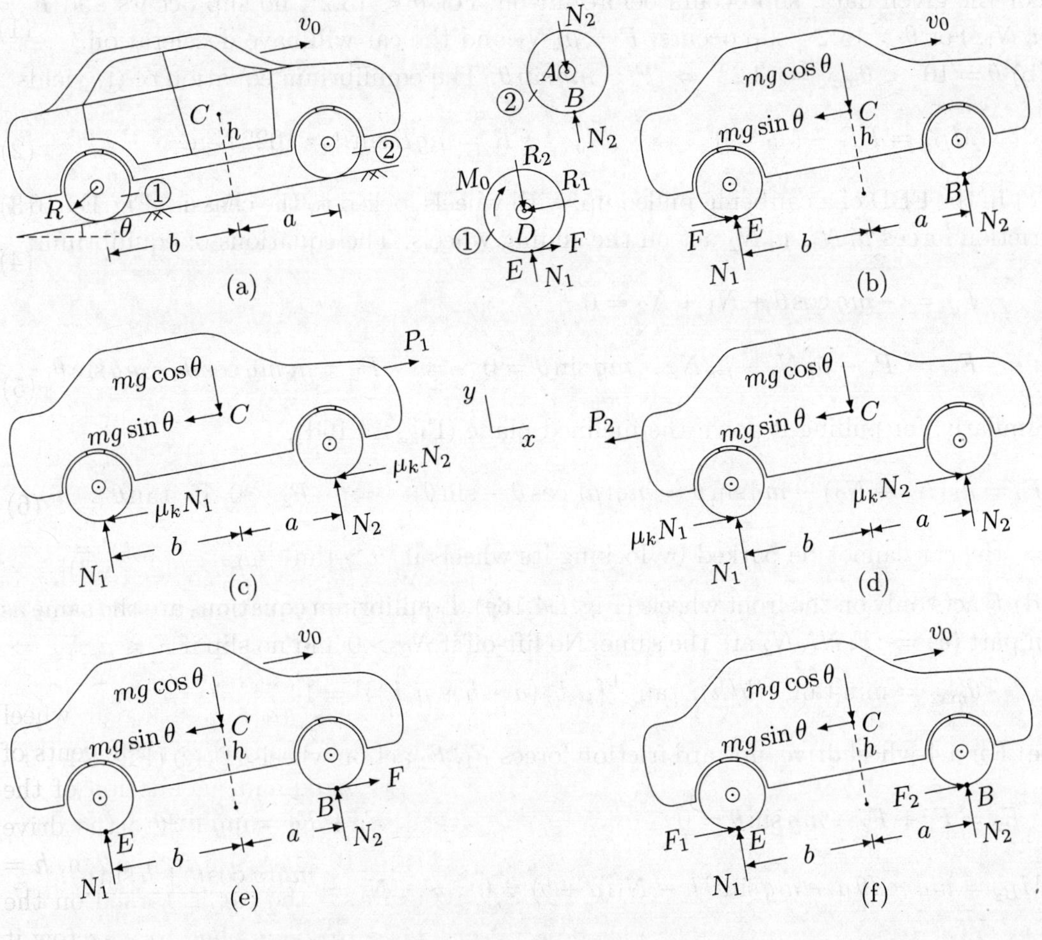

Fig. E4.16

For no lift-off, $N_2 > 0 \quad \Rightarrow \quad b\cos\theta - h\sin\theta > 0$ i.e., $\tan\theta < b/h$.

For no slip, $F < \mu_s N_1$, i.e.,

$$mg\sin\theta < \mu_s mg(a\cos\theta + h\sin\theta)/(a+b) \quad \Rightarrow \quad \tan\theta < \mu_s a/(a+b-\mu_s h)$$

If actual slip takes place, then $F = \mu_k N_1 \;\Rightarrow\; \tan\theta = \mu_k a/(a+b-\mu_k h)$. Thus, the maximum possible slope without slip or without lift-off of front wheels is given by

$$\begin{aligned}\theta_{\max} &= \min[\tan^{-1}(b/h),\ \tan^{-1}\{\mu_s a/(a+b-\mu_s h)\}] \\ &= \min[\tan^{-1}(2/0.6),\ \tan^{-1}\{0.5(1.5)/(1.5+2-0.5\times 0.6)\}] \\ &= \min[73.3°,\ 13.2°] = 13.2°\end{aligned}$$

For the given data, slip occurs before lift-off. For $\theta < 13.2°$, no slip occurs and $F \neq \mu_s N_1$. For $\theta > 13.2°$, slip occurs, $F = \mu_k N_1$ and the car will have deceleration.
(b) $\theta = 10° < \theta_{\max} = 13.2° \;\Rightarrow\; F = mg\sin\theta$. The equilibrium equation of ① yields

$$M_{D_z} = FR - M_0 = 0 \quad\Rightarrow\quad M_0 = FR = mgR\sin\theta = 1022\,\text{N.m}$$

(c) In the FBD of a car being pulled up, with wheels locked to the chassis (Fig. E4.16c), friction forces $\mu_k N_1$, $\mu_k N_2$ act on the sliding wheels. The equations of equilibrium:

$$\begin{aligned}F_y &= -mg\cos\theta + N_1 + N_2 = 0 \\ F_x &= P_1 - \mu_k N_1 - \mu_k N_2 - mg\sin\theta = 0 \quad\Rightarrow\quad P_1 = \mu_k mg\cos\theta + mg\sin\theta\end{aligned}$$

Similarly, for pulling it down the inclined plane (Fig. E4.16d),

$$P_2 = \mu_k(N_1+N_2) - mg\sin\theta = mg(\mu_k\cos\theta - \sin\theta) \quad\Rightarrow\quad P_2 < 0 \text{ if } \tan\theta > \mu_k$$

$\Rightarrow$ the car cannot be parked by locking its wheels if $\theta > \tan^{-1}\mu_k$.
(d) F acts only on the front wheels (Fig. E4.16e). Equilibrium equations are the same as in part (a) $\Rightarrow$ F, N_1, N_2 are the same. No lift-off if $N_2 > 0$ and no slip if $F < \mu_s N_2 \Rightarrow$

$$\theta_{\max} = \min[\tan^{-1}(b/h),\ \tan^{-1}\{\mu_s b/(a+b+\mu_s h)\}] = 14.7°$$

(e) For a 4-wheel drive, upward friction forces F_1, F_2 act on wheels ①, ② (Fig. E4.16f).

$$\begin{aligned}F_x &= F_1 + F_2 - mg\sin\theta = 0 &&\Rightarrow\quad F_1 + F_2 = mg\sin\theta \\ M_{B_z} &= mg\cos\theta\, a + mg\sin\theta\, h - N_1(a+b) = 0 &&\Rightarrow\quad N_1 = \frac{mg(a\cos\theta + h\sin\theta)}{a+b} > 0 \\ F_y &= N_1 + N_2 - mg\cos\theta = 0 &&\Rightarrow\quad N_2 = \frac{mg(b\cos\theta - h\sin\theta)}{a+b}\end{aligned}$$

For no lift-off: $\quad N_2 > 0 \quad\Rightarrow\quad \tan\theta < b/h.$

For impending slip, $F_1 + F_2 = mg\sin\theta = \mu_s(N_1+N_2) = \mu_s mg\cos\theta \;\Rightarrow\; \tan\theta = \mu_s$
$\Rightarrow \theta_{\max} = \min[\tan^{-1}(b/h),\ \tan^{-1}\mu_s] = 26.6° \Rightarrow$ largest $\theta_{\max}$ is for a 4-wheel drive.

Hence, jeeps with 4-wheel drive are the best for going uphill.

4.6.1 Impending motion with tipping or slipping

Sometimes, in trying to make a body slip, we observe that it tips. Here, we want to investigate whether a body initially in equilibrium and subjected to a gradually increasing load, loses its equilibrium by impending slipping or impending tipping. We assume impending slip and find the line of action of normal force N or values of N_1, N_2 for the two types of problems shown in Figs 4.7a and 4.7b. If the line of action of N in Fig. 4.7a is within the convex envelope of the area of contact, then impending slip is physically realisable. If $N_1 > 0$, $N_2 > 0$ in Fig. 4.7b, then impending slip is physically valid. If these conditions are not met, say $x < 0$ and $N_2 < 0$, then there is impending tipping. The FBDs for the tipping case are shown in Figs 4.7c,d with F being less than the limiting value and N acting at the boundary of the convex envelope of the area of contact. We illustrate the procedure by an example.

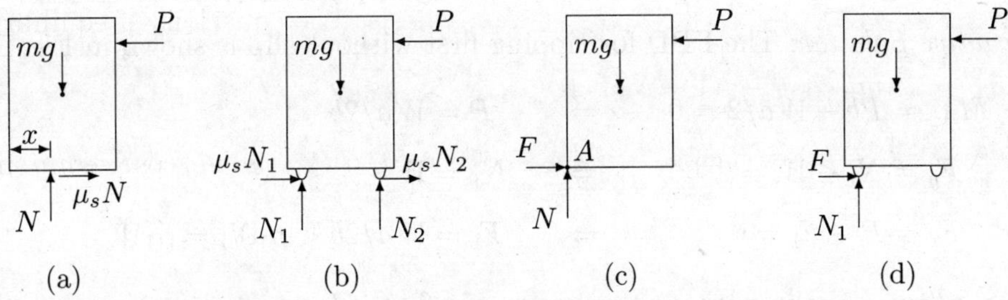

Fig. 4.7

Example 4.17: A door of weight W and centre of mass C is mounted on a horizontal rail on runners 1 and 2 with coefficients of friction μ_1 and μ_2 (Fig. E4.17a). The handle 3 is pulled to open the door. (a) Find the maximum value $b_{\max}$ of b for which the door will not tip when opened. (b) Find the force required to open the door if $b = \frac{2}{3}\, b_{\max}$.

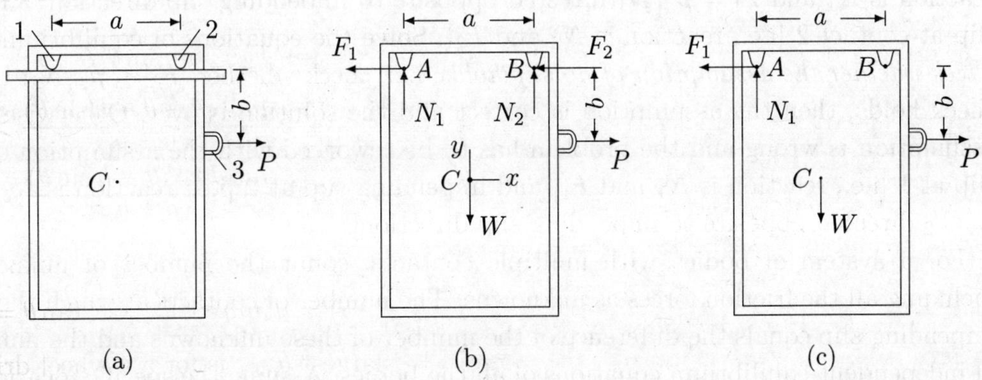

Fig. E4.17

Solution: The FBD of the door is shown in Fig. E4.17b. The equilibrium equations:

$$M_{A_z} = Pb - Wa/2 + N_2 a = 0 \quad \Rightarrow \quad N_2 = W/2 - Pb/a \qquad (1)$$

$$F_x = P - F_1 - F_2 = 0 \quad \Rightarrow \quad P = F_1 + F_2 \qquad (2)$$

$$F_y = N_1 + N_2 - W = 0 \quad \Rightarrow \quad N_1 = W/2 + Pb/a \qquad (3)$$

As N_2 decreases with increase in P (or b), lift-off at B and tipping at A would occur.
(a) If the door slides but does not tip, then $F_1 = \mu_1 N_1$, $F_2 = \mu_2 N_2$ and eq. (2) yields

$$P = \mu_1 N_1 + \mu_2 N_2 = (\mu_1 + \mu_2)W/2 + (\mu_1 - \mu_2)Pb/a$$

$$\Rightarrow \quad P = (\mu_1 + \mu_2)W/[2 - 2(\mu_1 - \mu_2)b/a] \qquad (4)$$

For no tipping, $N_2 = W/2 - Pb/a \geq 0 \Rightarrow$ [using eq.(4)] $b \leq a/2\mu_1$ i.e., $b_{\max} = a/2\mu_1$.

Alternative Solution: The FBD for tipping first without slip is shown in Fig. E4.17c.

$$M_{A_z} = Pb - Wa/2 = 0 \quad \Rightarrow \quad P = Wa/2b$$

$$F_y = N_1 - W = 0 \quad \Rightarrow \quad N_1 = W$$

$$F_x = P - F_1 = 0 \quad \Rightarrow \quad F_1 = Wa/2b < \mu_1 N_1 = \mu_1 W$$

$\Rightarrow \; b > a/2\mu_1 \;\Rightarrow$ slip without tipping occurs first if $b \leq a/2\mu_1$.
(b) $b = \frac{2}{3}[a/2\mu_1] = a/3\mu_1$, eq. (4) $\Rightarrow$ door slides at $P = 3\mu_1 W(\mu_1 + \mu_2)/(4\mu_1 + 2\mu_2)$.

4.6.2 Impending rolling of a cylinder with contact at two or more surfaces

It is only in exceptional cases that slip will take place simultaneously at line contacts with all surfaces. Consider a body with two contact surfaces with the possibility of slip at one and no slip at the other. *Assume* impending slip at contact 1 (say) [i.e., reaction is N_1 and $F_1 = \mu_{s_1} N_1$ directed opposite to impending slip direction] and no slip at contact 2 [i.e., reaction is N_2 and F_2]. Solve the equations of equilibrium and *check whether the assumption of no slip holds*, i.e., check whether $|F_2| \leq \mu_{s_2} N_2$. If the check holds, then the assumption is correct and the solution is over. Otherwise, the assumption is wrong and the problem has to be reworked with the assumption of no slip at 1 [i.e., reaction is N_1 and F_1] and impending slip at 2 [i.e., reaction is N_2 and $\mu_{s_2} N_2$ directed opposite to impending slip direction].

For a system of bodies with multiple contacts, count the number of unknowns including all the friction forces as unknowns. The number of contacts at which there is impending slip equals the difference of the number of these unknowns and the number of independent equilibrium equations of all the bodies. Assume the specific location of these contacts with impending slip, and employ at these contacts, $F_i = \mu_{s_i} N_i$ directed opposite to the direction of impending slip as obtained from assumed kinematics. Solve the equations of equilibrium and these kinetic conditions, and then check whether $|F_j| \leq \mu_{s_j} N_j$ holds at the contacts where no slip is assumed at the beginning. If the check holds then the solution has been obtained, else rework with different assumption.

Example 4.18: A wooden log of weight W is to be just pushed up a hard rock by the blade of a bulldozer (Fig. E4.18a). The coefficients of friction of wood with the rock and the blade are 0.3 and 0.4 respectively. Find the reaction on the log from the rock.

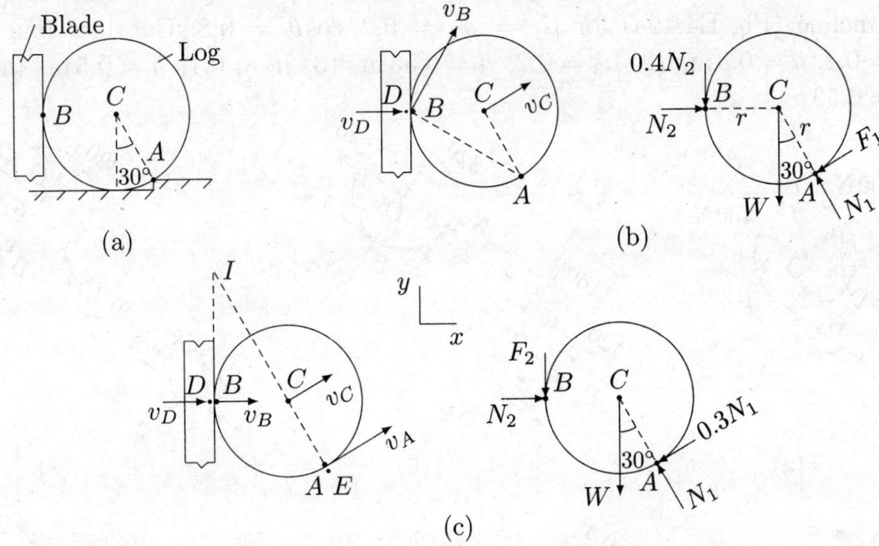

Fig. E4.18

Solution: The normal reactions and friction forces N_1, F_1, N_2, F_2 at A and B satisfy 3 equations of equilibrium. Hence, the impending motion is with impending slip at A or B which provides the fourth equation for the solution. *Assume* no slip at A and impending slip at B. With the instantaneous centre of rotation at A, the impending velocities of B and D are shown in Fig. E4.18b. F_1 at A (no slip) is arbitrary and $F_2 = 0.4N_2$ at B is downwards since $\vec{v}_{BD}$ is upwards. The equilibrium equations:

$$M_{Cz} = 0.4N_2 r - F_1 r = 0 \quad \Rightarrow \quad F_1 = 0.4N_2$$

$$F_x = N_2 - F_1 \cos 30° - N_1 \sin 30° = 0$$

$$\Rightarrow \quad N_2 = N_1 \sin 30°/(1 - 0.4 \cos 30°) = 0.7650 N_1$$

We check: $F_1/N_1 = 0.4N_2/N_1 = 0.4(0.765) = 0.306 \not\leq 0.3 \Rightarrow$ the assumption is wrong.

Hence, the motion impends with no slip at B and with impending slip at A. The instantaneous centre I is at the intersection of the normals to $\vec{v}_B$ and $\vec{v}_C$ (Fig. E4.18c). $F_1 = 0.3N_1$ is opposite to $\vec{v}_{AE}$ and F_2 (no slip) is arbitrary. The equilibrium equations:

$$M_{Cz} = F_2 r - 0.3N_1 r = 0 \qquad \Rightarrow \quad F_2 = 0.3N_1$$

$$F_x = N_2 - 0.3N_1 \cos 30° - N_1 \sin 30° = 0 \qquad \Rightarrow \quad N_2 = 0.7598N_1$$

$$F_y = N_1 \cos 30° - 0.3N_1 \sin 30° - W - F_2 = 0 \Rightarrow \quad N_1 = 2.404W$$

Example 4.19: Find the force P required to cause impending motion of the cylinder up the incline (Fig. E4.19a) for $\mu_1 = \mu_2 = 0.2$, $\cos\theta = 0.8$. Consider four cases: (1) $\mu_3 = 0.2$, $h = 0.5\,\text{m}$, (2) $\mu_3 = 0.2$, $h = 0.53\,\text{m}$, (3) $\mu_3 = 0.1$, $h = 0.5\,\text{m}$, (4) $\mu_3 = 0.1$, $h = 0.53\,\text{m}$.

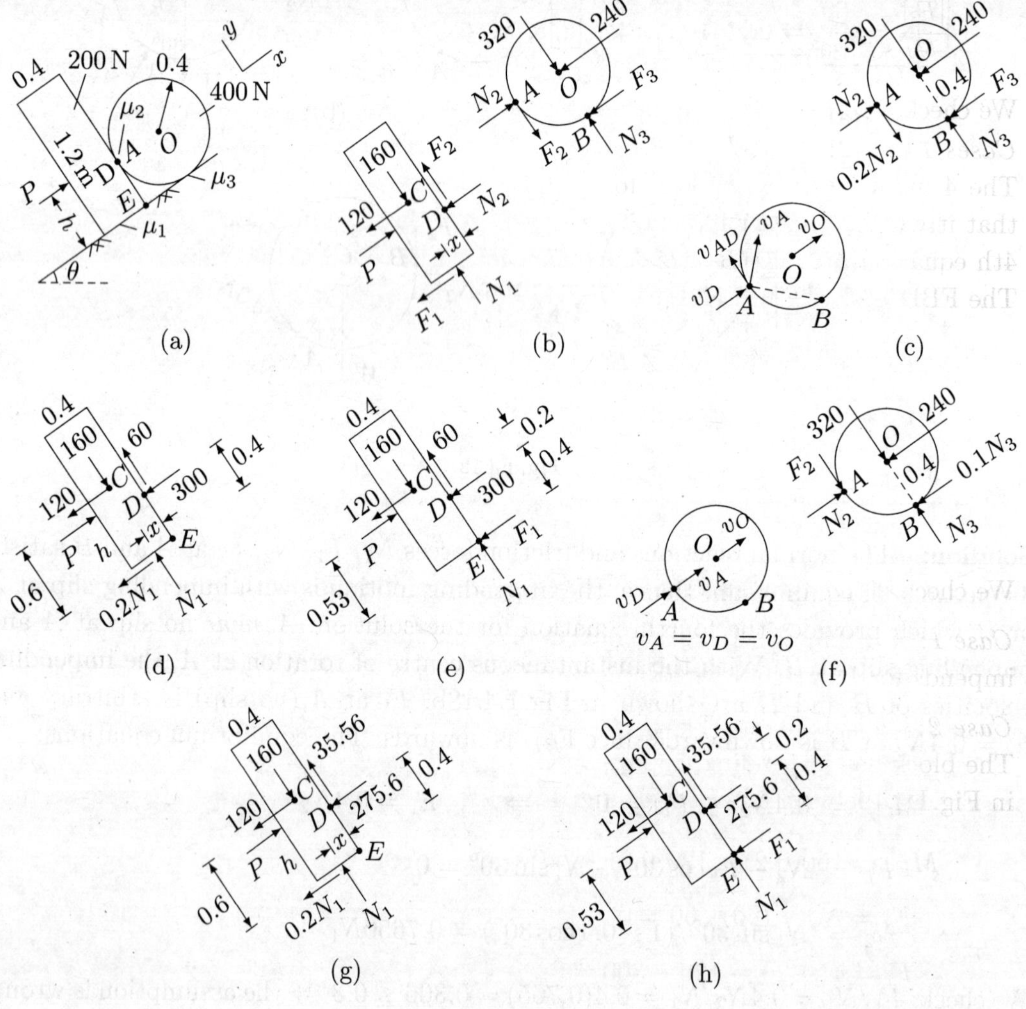

Fig. E4.19

Solution: There are 8 unknowns: N_1, F_1, N_2, F_2, N_3, F_3, x, P in the FBDs of the block and the cylinder (Fig. E4.19b), but only 6 equilibrium equations for them. N_2, F_2, N_3, F_3 for the cylinder and 3 equilibrium equations imply that it will have impending slip at either A or B, with the 4th equation provided by this condition.

Assume slip at A and no slip at B $\Rightarrow$ B is instantaneous centre $\Rightarrow$ $v_{AD}\,\vec{j}$ $\Rightarrow$ $\vec{F_2} = -0.2N_2\,\vec{j}$ in the FBD of the cylinder (Fig. E4.19c). Its equations of equilibrium are:

$$M_{Oz} = 0.2N_2(0.4) - F_3(0.4) = 0 \quad \Rightarrow \quad F_3 = 0.2N_2$$

$$F_x = N_2 - 240 - F_3 = 0 \quad \Rightarrow \quad N_2 - 240 - 0.2N_2 = 0 \;\Rightarrow\; N_2 = 300\,\text{N}$$

$$F_y = N_3 - 320 - 0.2N_2 = 0 \quad \Rightarrow \quad N_3 = 380\,\text{N}, \qquad F_3 = 60\,\text{N}$$

$$\frac{|F_3|}{N_3} = \frac{60}{380} = 0.1579 \tag{1}$$

We check whether the assumption of no slip at B is correct.

Cases 1 and 2: $|F_3|/N_3 = 0.1579 < \mu_3 = 0.2$ and hence the assumption is correct. The 4 unknowns, N_1, F_1, x, P for the block and only 3 equilibrium equations imply that it will have either impending slip or impending tipping about point E, with the 4th equation provided by this condition. *We assume impending slip without tipping.* The FBD is shown in Fig. E4.19d with $F_1 = \mu_1 N_1 = 0.2N_1$ downward and

$$F_y = N_1 - 160 + 60 = 0 \quad \Rightarrow \quad N_1 = 100\,\text{N}$$

$$F_x = P - 0.2N_1 - 120 - 300 = 0 \quad \Rightarrow \quad P = 440\,\text{N}$$

$$M_{Ez} = 160(0.2) + 120(0.6) + 300(0.4) - Ph - N_1 x = 0$$

$$x = 2.24 - 4.4h \tag{2}$$

We check if the assumption of no tipping is correct, i.e., whether $x > 0$.

Case 1: $h = 0.5$, eq. (2) $\Rightarrow x = 0.04 > 0 \Rightarrow$ no tipping assumption is correct. Motion impends for $P = 440\,\text{N}$ with sliding of block and rolling of cylinder without slip at B.

Case 2: $h = 0.53$, eq. (2) $\Rightarrow$ $x = -0.092 < 0$ $\Rightarrow$ no tipping assumption is wrong. The block has impending tipping about point E without slipping. The FBD is shown in Fig. E4.19e with N_1, F_1 at E and F_1 is arbitrary. The equations of equilibrium:

$$M_{Ez} = 160(0.2) + 120(0.6) + 300(0.4) - 0.53P = 0 \quad \Rightarrow \quad P = 422.6\,\text{N}$$

$$F_y = N_1 - 160 + 60 = 0 \quad \Rightarrow \quad N_1 = 100\,\text{N}$$

$$F_x = P - F_1 - 120 - 300 = 0 \quad \Rightarrow \quad F_1 = 2.64\,\text{N}$$

No slip assumption is o.k., as $|F_1|/N_1 = .0264 < \mu_1 = .2$. Motion impends for $P =$

422.6 N with tipping of the block about E and rolling of cylinder without slip at B.

Cases 3 and 4: $|F_3|/N_3 = 0.1579 > \mu_3 = 0.1 \Rightarrow$ assumption of no slip at B is wrong $\Rightarrow$ impending slip at B and no slip at A $\Rightarrow$ $\vec{v}_O = \vec{v}_D = \vec{v}_A$ $\Rightarrow$ impending translation up. In the FBD (Fig. E4.19f) $F_3 = \mu_3 N_3 = 0.1N_3$ is downwards and F_2 is arbitrary.

$$M_{O_z} = 0.4F_2 - 0.1N_3(0.4) = 0 \quad \Rightarrow \quad F_2 = 0.1N_3$$

$$F_y = N_3 - 320 - F_2 = 0 \quad \Rightarrow \quad N_3 - 320 - 0.1N_3 = 0 \Rightarrow N_3 = 355.6\,\text{N}$$

$$F_x = N_2 - 240 - 0.1N_3 = 0 \quad \Rightarrow \quad N_2 = 275.6\,\text{N}, \qquad F_2 = 35.56\,\text{N}$$

The 4 unknowns, N_1, F_1, x, P for the block and only 3 equilibrium equations imply that it will have either impending slip or impending tipping about point E, with the 4th equation provided by this condition. *We assume impending slip without tipping.* The FBD is shown in Fig. E4.19g with $F_1 = \mu_1 N_1 = 0.2N_1$ downward and

$$F_y = N_1 - 160 + 35.56 = 0 \quad \Rightarrow \quad N_1 = 124.4\,\text{N}$$

$$F_x = P - 0.2N_1 - 120 - 275.6 = 0 \quad \Rightarrow \quad P = 420.5\,\text{N}$$

$$M_{E_z} = 160(0.2) + 120(0.6) + 275.6(0.4) - Ph - N_1 x = 0$$

$$x = 1.722 - 3.380h \tag{3}$$

We check if the assumption of no tipping is correct, i.e., whether $x > 0$.

Case 3: $h = 0.5$, eq. (3) $\Rightarrow$ $x = 0.032 > 0$ $\Rightarrow$ assumption is correct. The motion impends for $P = 420.5\,\text{N}$ with sliding of the block and translation of the cylinder.

Case 4: $h = 0.53$, eq. (3) $\Rightarrow$ $x = -0.0694 < 0 \Rightarrow$ assumption of no tipping is wrong. The block has impending tipping about point E without slipping. The FBD is shown in Fig. E4.19h with N_1, F_1 at E and F_1 is arbitrary. The equations of equilibrium:

$$M_{E_z} = 160(0.2) + 120(0.6) + 275.6(0.4) - 0.53P = 0 \quad \Rightarrow \quad P = 404.23\,\text{N}$$

$$F_y = N_1 - 160 + 35.56 = 0 \quad \Rightarrow \quad N_1 = 124.4\,\text{N}$$

$$F_x = P - F_1 - 120 - 275.6 = 0 \quad \Rightarrow \quad F_1 = 8.63\,\text{N}$$

Motion impends for $P = 404.23\,\text{N}$ with block tipping about E and cylinder translating.

4.6.3 Rigid body in plane / line contact under 3D loads

If a body with planar contact in xy-plane has an impending general plane motion, then there exists an axis through the instantaneous centre $I(x, y)$ about which the body is about to rotate. The frictional force at any point of support is at right angles to the line connecting it to I and its magnitude is μ_s times the normal reaction there. The applied load for impending motion and x, y are found using $F_x = 0, F_y = 0, M_{A_z} = 0$.

The above procedure is applicable to impending plane motion of a rigid body supported on a plane at several points. An exceptional case arises when I coincides with one of its points of support. The two components F_1, F_2 of the frictional reaction at I, *assumed at one support*, and the applied load are found from the 3 equilibrium equations and the assumption is checked by verifying the no slip condition, $(F_1^2 + F_2^2)^{1/2} \le \mu_s N$ at I where normal force N acts.

These remarks are illustrated in the following examples.

Example 4.20: A small block of mass m rests on a plane inclined at an angle θ to the horizontal (Fig. E4.20a). Find the pull P (parallel to the inclined plane) applied at the end of a cord for which the block starts slipping. Find also the direction of initial slip of the block. The coefficient of friction is μ.

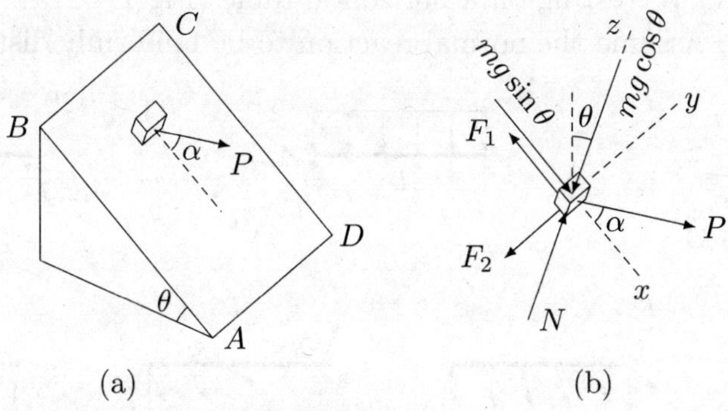

Fig. E4.20

Solution: We choose x and y axes in the plane $ABCD$ in the directions of BA and $BC \Rightarrow$ z-axis is inclined at angle θ to the vertical. Since we do not know the direction of impending slip of the block on the xy-plane, we have shown two components F_1 and F_2 of the frictional force $F = (F_1^2 + F_2^2)^{1/2}$ in the FBD of the block (Fig. E4.20b).

$$P(\cos\alpha\vec{\mathrm{i}} + \sin\alpha\vec{\mathrm{j}}) + N\vec{\mathrm{k}} - F_1\vec{\mathrm{i}} - F_2\vec{\mathrm{j}} + mg(-\cos\theta\vec{\mathrm{k}} + \sin\theta\vec{\mathrm{i}}) = \vec{0}$$

$$\vec{\mathrm{i}}: \quad P\cos\alpha - F_1 + mg\sin\theta = 0 \quad \Rightarrow \quad F_1 = P\cos\alpha + mg\sin\theta \qquad (1)$$

$$\vec{\mathrm{j}}: \quad P\sin\alpha - F_2 = 0 \quad \Rightarrow \quad F_2 = P\sin\alpha \qquad (2)$$

$$\vec{\mathrm{k}}: \quad N - mg\cos\theta = 0 \quad \Rightarrow \quad N = mg\cos\theta \qquad (3)$$

For impending slip, $F = \mu N \;\Rightarrow\; F_1^2 + F_2^2 = \mu^2 N^2 \;\Rightarrow$

$$P^2 + 2Pmg\cos\alpha\sin\theta + m^2g^2\sin^2\theta = \mu^2 m^2 g^2 \cos^2\theta$$

The physically meaningful positive root of this quadratic equation is

$$P = mg[(\mu^2 \cos^2\theta - \sin^2\theta \sin^2\alpha)^{1/2} - \sin\theta\cos\alpha] \tag{4}$$

The direction $\vec{e}$ of impending slip is opposite to the direction of frictional force $\vec{F}$.

$$\vec{e} = -\frac{\vec{F}}{F} = \frac{(P\cos\alpha + mg\sin\theta)\,\vec{i} + P\sin\alpha\,\vec{j}}{[P^2 + 2Pmg\cos\alpha\sin\theta + m^2g^2\sin^2\theta]^{1/2}} \tag{5}$$

where P is given by eq. (4). In general, $\vec{e}$ is neither in the direction of P nor in the direction $\vec{i}$ down the plane. If P is horizontal, then $\alpha = \pi/2$ and eqs (4), (5) $\Rightarrow$

$$P = mg(\mu^2\cos^2\theta - \sin^2\theta)^{1/2}, \qquad \vec{e} = [\sin\theta\,\vec{i} + (\mu^2\cos^2\theta - \sin^2\theta)^{1/2}\vec{j}\,]\,/\,\mu\cos\theta$$

Example 4.21: Find the horizontal force P for which there is impending motion of a thin bar of weight W resting on a horizontal table (Fig. E4.21a). The coefficient of static friction is μ. Assume the normal reaction to be uniformly distributed.

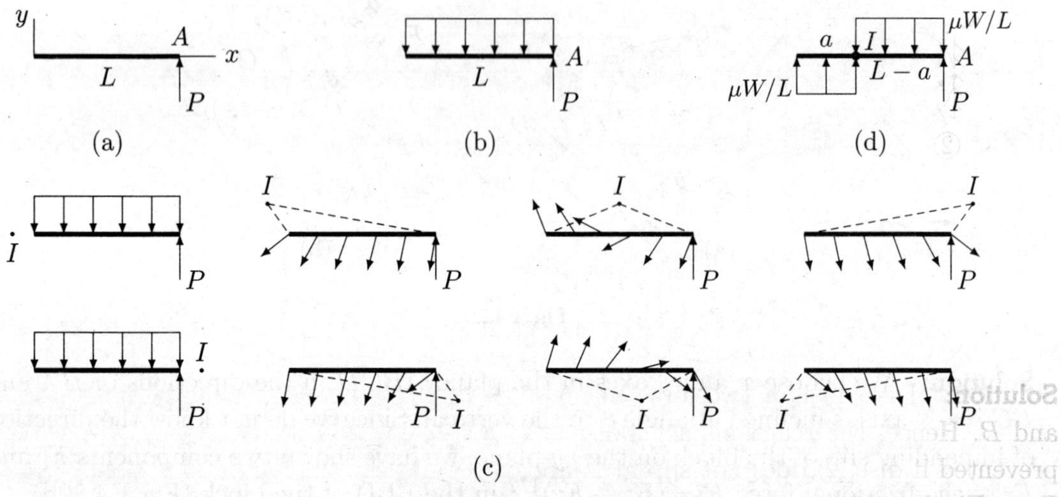

Fig. E4.21

Solution: The normal reaction per unit length is W/L. Assume impending translation in the y direction $\Rightarrow$ opposite distributed friction force in the FBD (Fig. E4.21b). This assumption is not correct since $M_{A_z} \neq 0$.

Assume that the instantaneous centre I of impending rotation is not on AB. From the FBDs for the 8 possible cases (Fig. E4.21c), it is clear that this assumption is also wrong since one of the equations $F_x = 0$ or $M_{A_z} = 0$ is not satisfied. Hence, I lies on AB and the FBD (Fig. E4.21d) has 2 unknowns P and a. Equilibrium equations:

$$M_{A_z} = (\mu W/L)(L-a)^2/2 - (\mu W/L)a(L - a + \tfrac{1}{2}a) = 0$$

$$\Rightarrow \quad 2a^2 - 4La + L^2 = 0 \quad \Rightarrow \quad a = (1 - 1/\sqrt{2})L$$

$$F_y = P - (\mu W/L)(L - a) + (\mu W/L)a = 0$$

$$\Rightarrow \quad P = (L - 2a)\mu W/L = (\sqrt{2} - 1)\mu W$$

where only the admissible root $a < L$ is considered.

4.6.4 Friction grips and friction locking

In some situations, the friction grip is designed to loosen, i.e., begin to slip at a specified load. In other situations, the friction grip is designed to lock itself, i.e., slip cannot occur irrespective of the magnitude of the applied load, by ensuring that the reaction is inclined to the normal at an angle less than $\phi_s = \tan^{-1}\mu_s$. We illustrate below.

Example 4.22: A disc ①, mounted on a shaft (Fig. E4.22a), is prevented from rotation irrespective of the value of torque M applied in the direction shown, by placing a light disc ② in contact with it. Find the condition that r, R and d must satisfy.

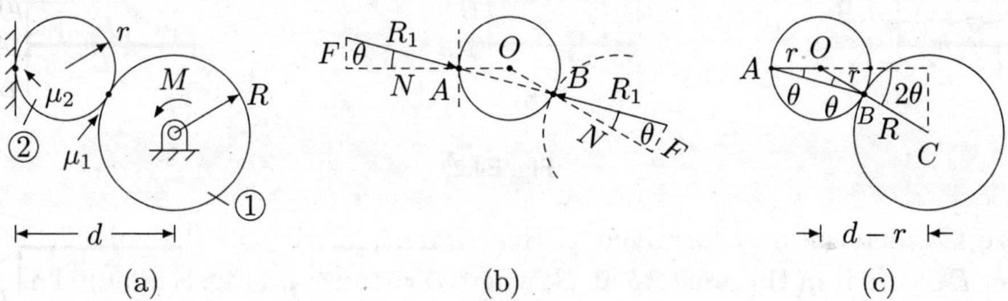

Fig. E4.22

Solution: Disc ② is a two-force member as it is subjected to two reactions at A and B. Hence, the reactions act along AB (Fig. E4.22b). The rotation of disc ① is prevented if disc ② does not slip at A and B, i.e., if $\theta < \tan^{-1}\mu_1$ and $\theta < \tan^{-1}\mu_2$. From Fig. E4.22c, $\cos 2\theta = (d - r)/(R + r)$. Hence, the required condition is

$$\theta = \tfrac{1}{2}\cos^{-1}[(d - r)/(R + r)] < \min(\tan^{-1}\mu_1, \ \tan^{-1}\mu_2)$$

$$\Rightarrow \quad (d - r)/(R + r) > \cos[2\min(\tan^{-1}\mu_1, \ \tan^{-1}\mu_2)]$$

Example 4.23: Prove that the light collar of a bracket (Fig. E4.23a) cannot slip down irrespective of the load P, if $x > a/2\mu$, where μ is the coefficient of friction.

Solution: Contact of collar occurs at A and B (Fig. E4.23b), due to its clockwise rotation under load P, because of inevitable play. The equations of equilibrium are:

$$F_x = N_2 - N_1 = 0 \quad \Rightarrow \quad N_1 = N_2 \qquad (1)$$

$$F_y = F_1 + F_2 - P = 0 \tag{2}$$

$$M_{A_z} = bF_2 + aN_2 - (x + \tfrac{1}{2}b)P = 0 \tag{3}$$

The problem is statically indeterminate for 4 unknown forces. But, the problem of finding the location $x = x_1$ for impending slip is determinate as $F_1 = \mu N_1, F_2 = \mu N_2$.

$$\text{eqs (1), (2)} \Rightarrow \quad F_1 + F_2 = \mu N_1 + \mu N_2 = P \qquad \Rightarrow \qquad N_1 = N_2 = P/2\mu$$

$$\text{eq. (3)} \Rightarrow \quad \mu b(P/2\mu) + aP/2\mu - (x_1 + \tfrac{1}{2}b)P = 0 \qquad \Rightarrow \qquad x_1 = a/2\mu$$

Hence, the bracket is self-locking and does not slip down if $x > a/2\mu$.

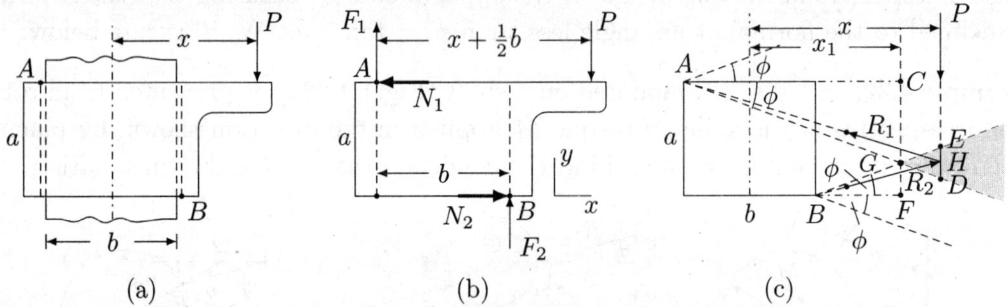

Fig. E4.23

There is indeterminacy for given x. Reaction R_1 at A is in the zone 2ϕ at A and reaction R_2 at B is in the zone 2ϕ at B, where $\phi = \tan^{-1}\mu$ (Fig. E4.23c). The point of concurrence of R_1, R_2, P can be at any (indeterminate) point on DE. There is no slip if the line of action of P is through the shaded zone. For the limiting case of impending slip, the line of action of P passes through G. x_1 can also be obtained from geometry:

$$a = FC = FG + GC = (x_1 - \tfrac{1}{2}b)\tan\phi + (x_1 + \tfrac{1}{2}b)\tan\phi = 2x_1\mu \Rightarrow x_1 = a/2\mu$$

4.6.5 Wedge

A wedge is a small block having two flat surfaces with a small angle ($\sim 5°$) between them. It transforms an applied force into a much larger force approximately at right angles to it. It is used to make small adjustments in the position of a body or to raise heavy loads. Consider equilibrium of a smooth wedge (Fig. 4.8a) with a *small angle* α:

$$F_x = P - (N_1 + N_2)\sin\frac{\alpha}{2} = 0$$

$$F_y = (N_2 - N_1)\cos\frac{\alpha}{2} = 0$$

$$\Rightarrow \quad N_1 = N_2 = P/2\sin\frac{\alpha}{2} \simeq P/\alpha$$

$\Rightarrow$ for $\alpha = 0.1\,\text{rad} \simeq 5.7°$, $N_1 = 10P$ $\Rightarrow$ A small applied force P can support a large load N_1. This is the basis of the use of an axe to enlarge a crack in a wooden block.

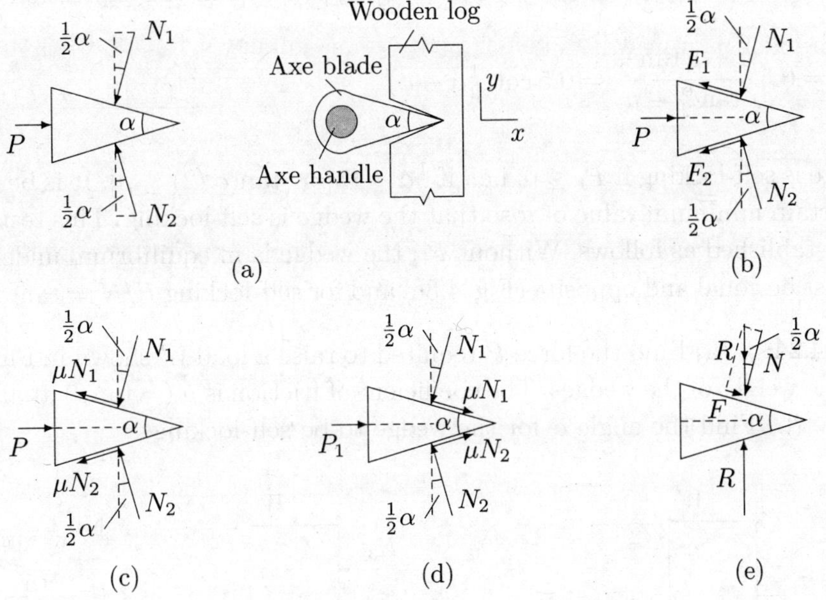

Fig. 4.8

Consider a wedge with friction (Fig. 4.8b). This equilibrium problem is statically indeterminate since 2 equilibrium equations ($F_x = 0$, $F_y = 0$) are insufficient to find N_1, F_1, N_2, F_2. The problem of impending motion ($F_1 = \mu N_1$, $F_2 = \mu N_2$) of the wedge to the right (Fig. 4.8c), though, is statically determinate:

$$F_y = 0 \quad \text{(or symmetry)} \qquad \Rightarrow \qquad N_1 = N_2$$

$$F_x = 0 \qquad \Rightarrow \qquad P = (N_1 + N_2)\sin\frac{\alpha}{2} + \mu(N_1 + N_2)\cos\frac{\alpha}{2}$$

$$\Rightarrow \quad N_1 = N_2 = 0.5P\Big/\left(\sin\frac{\alpha}{2} + \mu\cos\frac{\alpha}{2}\right)$$

The force F in the y direction on a face of the wedge is given by

$$F = N_2\cos\frac{\alpha}{2} - \mu N_2\sin\frac{\alpha}{2} = 0.5P\,\frac{1 - \mu\tan\frac{\alpha}{2}}{\tan\frac{\alpha}{2} + \mu} = 0.5P\cot(\tfrac{1}{2}\alpha + \phi)$$

with $\phi = \tan^{-1}\mu$. The force-multiplying ratio F/P of a wedge is very sensitive to the value of μ, for example, for $\alpha = 5.7°$, $F/P = 10, 3.32, 1.98, 1.41, 1.09, 0.89$ for $\mu = 0, 0.1, 0.2, 0.3, 0.4, 0.5$ $\Rightarrow$ F/P decreases rapidly with μ.

For a wedge of a given angle α, depending on μ, an inward force P_1 may be required to keep it in place without being squeezed out (Fig. 4.8d). The direction of F gets reversed, the ratio F/P_1 is obtained by changing μ to $-\mu$ in the expression of F/P:

$$\frac{F}{P_1} = 0.5\,\frac{1+\mu\tan\frac{\alpha}{2}}{\tan\frac{\alpha}{2}-\mu} = 0.5\cot(\tfrac{1}{2}\alpha - \phi)$$

The wedge is self-locking if $P_1 \le 0$, i.e., if $\frac{1}{2}\alpha \le \phi$ [$\Rightarrow \tan(\alpha/2) \le \mu$]. It is beneficial to have a certain minimum value of μ so that the wedge is self-locking. This result can be directly established as follows. Without P_1, the wedge is in equilibrium under 2 forces which must be equal and opposite (Fig. 4.8e) and for self-locking $F/N = \tan(\alpha/2) \le \mu$.

Example 4.24: (a) Find the force P required to raise a load W shown in Fig. E4.24a. Neglect the weight of the wedges. The coefficient of friction is μ ($= \tan\theta$) at all surfaces of contact. (b) Find the angle α for the wedge to be self-locking.

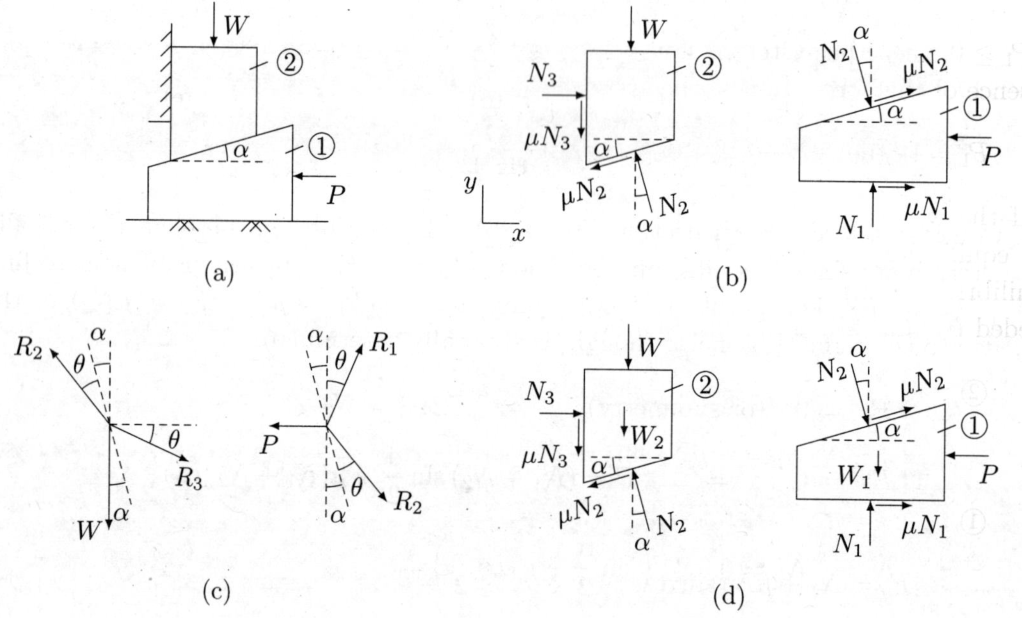

Fig. E4.24

Solution: (a) When block ② is about to move up, then block ① is about to move to the left and ② is be about to translate up relative to ①. The direction of the friction forces (Fig. E4.24b) have been shown on the basis of these impending relative motions. As the location of the forces is not known, there are only two relevant equations of equilibrium ($F_x = 0$, $F_y = 0$) for each body subjected to a coplanar force system. These

4 equations can be solved for 4 unknowns, N_1, N_2, N_3 and P. We can also directly apply Lame's theorem since each block is in equilibrium under 3 forces (Fig. E4.24c).

$$\text{Block ②}: \quad \frac{R_2}{\sin(\pi/2-\theta)} = \frac{W}{\sin(\pi/2+2\theta+\alpha)} \quad \Rightarrow \quad R_2 = \frac{W\cos\theta}{\cos(2\theta+\alpha)} \tag{1}$$

$$\text{Block ①}: \quad \frac{R_2}{\sin(\pi/2+\theta)} = \frac{P}{\sin(\pi-2\theta-\alpha)} \quad \Rightarrow \quad P = R_2\frac{\sin(2\theta+\alpha)}{\cos\theta} \tag{2}$$

$$\text{eqs (1), (2)} \Rightarrow \quad P = W\tan(2\theta+\alpha) \tag{3}$$

(b) When P is removed, there may not be equilibrium. If there is equilibrium, then the motion may not be impending. The frictional forces in this case are F_1, F_2, F_3. The problem is statically indeterminate for 6 unknowns, (F_i, N_i) from 4 equations of equilibrium. Instead, we find the force P_1 required for impending motion downwards. As friction forces get reversed, P_1 is obtained by replacing μ by $-\mu$ (θ by $-\theta$) in eq. (3):

$$P_1 = W\tan(-2\theta+\alpha) \tag{4}$$

If $P_1 \geq 0$, then the system is not self-locking because block ② will move down in the absence of this force. Hence, the system is self-locking if

$$P_1 = W\tan(-2\theta+\alpha) < 0 \qquad \text{i.e., if} \qquad \alpha < 2\theta = 2\tan^{-1}\mu \tag{5}$$

If the wedges are heavy, then the blocks are not three-force members and so we use equilibrium equations $F_x = 0,\ F_y = 0$. To investigate whether the system is in equilibrium under a given value of force, we proceed as follows. We find the force P_1 needed for impending upward motion of ②. The FBDs are shown in Fig. E4.24d.

$$②: \quad F_x = N_3 - N_2\sin\alpha - \mu N_2\cos\alpha = 0 \tag{6}$$

$$F_y = N_2\cos\alpha - \mu N_2\sin\alpha - W - W_2 - \mu N_3 = 0 \tag{7}$$

$$①: \quad F_x = \mu N_1 + \mu N_2\cos\alpha + N_2\sin\alpha - P_1 = 0 \tag{8}$$

$$F_y = N_1 + \mu N_2\sin\alpha - N_2\cos\alpha - W_1 = 0 \tag{9}$$

Solution of eqs (6), (7) yields N_2 and N_3. Then eqs (8), (9) are solved for N_1 and P_1.

Similarly, find the force P_2 for impending motion downward by repacing μ by $-\mu$ in the expression of P_1. If $P_2 < 0$, then the system remains in equilibrium even without the external force. Hence, such a system is self-locking. Three cases can arise:

1. $P_2 \leq P \leq P_1$: the system remains in equilibrium for given P.
2. $P > P_1$: the block ② moves up for given force P.
3. $P < P_2$: the block ② moves down for given force P.

4.7 Screw

A screw is a machine element used in jacks, presses, etc., for transforming a couple applied about its axis to a coaxial force that can be used to move objects or to fasten objects (bolt and nut). It is essentially an inclined plane wound around a cylinder. The thread employed in presses, screw-jacks and testing machines for transmission of heavy loads has a cross-section in the form of a square, buttress or acme (Fig. 4.9). The V-shaped thread is used in fasteners.

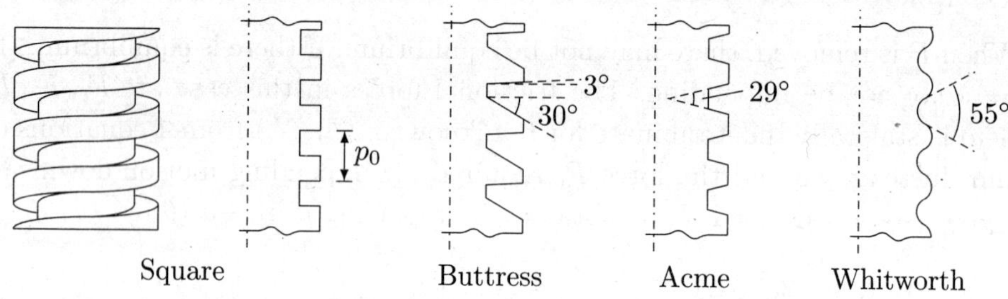

Fig. 4.9

4.7.1 Square thread

Consider a square thread of mean radius $r = \frac{1}{2}(r_1 + r_2)$ with pitch p_0 and n starts. The distance l advanced by the screw relative to the nut in a single turn is called *lead* and $l = np_0$. The pitch of the helix of the screw is l and the helix angle α is given by $\tan\alpha = l/2\pi r = np_0/2\pi r$. For single-start threads $n = 1$, $l = p_0$, $\tan\alpha = p_0/2\pi r$. Let the coefficients of static and kinetic friction be μ_s $(= \tan\theta)$ and μ_k. The axial moment M required for impending motion of the screw against an axial load P (Fig. 4.10) is given by (see Section 2.20):

$$\frac{M}{P} = r\,\frac{\tan\alpha + \mu_s}{1 - \mu_s\tan\alpha} = r\tan(\alpha + \theta) = \frac{r(np_0 + 2\pi r\mu_s)}{2\pi r - \mu_s np_0} \tag{4.3}$$

For the rotation of the screw with constant angular velocity ω, $M_{O_z} = I_{zz}^{O}\dot{\omega} = 0$, and $\vec{F} = m\vec{a}_C = \vec{0}$. Hence, eq. (4.3) is valid even when μ_s is replaced with μ_k. This replacement is also valid for arbitrary angular acceleration of the screw if its inertia can be neglected. In a torque-wrench used for tightening screws to a predetermined clamping force P, by applying a specified torque M, the relation of eq. (4.3) applies with μ_k replacing μ_s. To initiate further tightening (further impending movement), a greater torque would be needed as $\mu_s > \mu_k$.

If the moment M about the axis of the screw-jack is due to the application of transverse force F to a handle at distance d from the axis of the screw, then $M = Fd = rP\tan(\alpha + \theta)$. Thus, the *mechanical advantage* m_a for the screw-jack, defined as the ratio of the output force P to the input applied force F, is given by

$$m_a = \frac{P}{F} = \frac{d}{r\tan(\alpha+\theta)}$$

The *velocity ratio* v_r is defined as the ratio of the velocities of the points of application of F and P. Let the angular velocity of the screw be ω. v_r is given by

$$v_r = \frac{\omega d}{l\omega/2\pi} = \frac{2\pi d}{l}$$

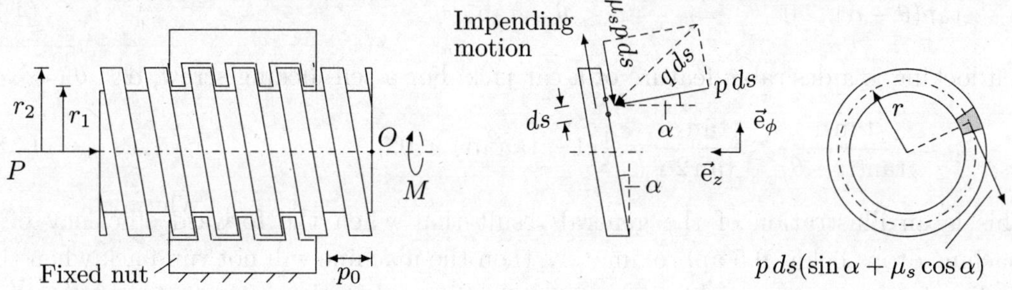

Fig. 4.10

The *efficiency* η of a screw is defined as the ratio of the moments M_0 and M required to cause impending motion against the force P for the ideal case of $\mu_s = 0$ ($\theta = 0$) and for the actual case of given μ_s. Equivalently, η is defined as the ratio of the output work to input work.[2] For an axial advance z, the rotation of the screw is $2\pi z/l$. Hence,

$$\eta \equiv \frac{M_0}{M} = \frac{rP\tan\alpha}{rP\tan(\alpha+\theta)} = \frac{\tan\alpha}{\tan(\alpha+\theta)}$$

$$\eta = \frac{Pz}{M(2\pi z/l)} = \frac{l}{2\pi r} \times \frac{P}{P\tan(\alpha+\theta)} = \frac{\tan\alpha}{\tan(\alpha+\theta)} \tag{4.4}$$

For η to be maximum, α should be so chosen that

$$\frac{d\eta}{d\alpha} = \frac{\tan(\alpha+\theta)\sec^2\alpha - \tan\alpha\sec^2(\alpha+\theta)}{\tan^2(\alpha+\theta)} = 0 \quad\Rightarrow\quad \sin 2(\alpha+\theta) = \sin 2\alpha$$

Since $\theta \neq 0$, the first root of this equation is given by $2(\alpha+\theta) = \pi - 2\alpha \Rightarrow$

$$\alpha = \frac{\pi}{4} - \frac{\theta}{2} \quad \text{and eq. (4.4)} \Rightarrow \quad \eta_{\max} = \frac{\tan(\pi/4-\theta/2)}{\tan(\pi/4+\theta/2)} = \frac{1-\sin\theta}{1+\sin\theta} \tag{4.5}$$

[2] Another definition is $\eta \equiv m_a/v_r$. It is easy see that one obtains the same value for η in each case.

If $\mu_s = 0.2$, then for maximum η: $\alpha = \pi/4 - \theta/2 = \pi/4 - (\tan^{-1}\mu_s)/2 = 39.3° \Rightarrow l/r = 2\pi \tan\alpha = 5.15$ which being very high is not used. Most screw designs are not intended to achieve maximum efficiency. A more important criterion is self-locking.

The moment required in opposite direction to retract/loosen a screw fastening, i.e., with impending motion in the direction of P, is given by (see Section 2.20):

$$M_1 = rP\tan(\theta - \alpha) = \frac{rP(2\pi r\mu_s - np_0)}{2\pi r + \mu_s np_0} \tag{4.6}$$

The screw is self-locking if it is in equilibrium without M_1, i.e., if $M_1 > 0$, i.e.,

$$\tan(\theta - \alpha) > 0 \qquad \Rightarrow \qquad \alpha < \theta = \tan^{-1}\mu_s \tag{4.7}$$

Self-locking is a desirable feature of a car jack. For a self-locking screw, $\alpha < \theta \Rightarrow$

$$\eta = \frac{\tan\alpha}{\tan(\alpha + \theta)} \le \frac{\tan\alpha}{\tan 2\alpha} = \tfrac{1}{2}(1 - \tan^2\alpha) < 0.5 \tag{4.8}$$

This is an illustration of the general result that when the forward efficiency of a machine drops below 0.5 approximately, then the machine will not run back when the applied force is removed. Its general proof follows. Let the output energy stored in the machine be O for input energy I during forward operation and L be the loss of energy: $\eta = O/I = O/(O+L) \Rightarrow L = O(1-\eta)/\eta$. The available energy for complete reversed motion is O. Assume losses in reversed motion to be the same as L. Hence, there is no automatic run back on removal of the driving force/torque if $O < L$:

$$O < L = O(1-\eta)/\eta \quad \Rightarrow \quad \eta < 1 - \eta \qquad \text{i.e.,} \qquad \text{if } \eta < 0.5.$$

4.7.2 V-thread

Let α be the helix angle of the V-thread and 2γ the angle of the V-groove in the view along $\vec{e}_\phi$ (Fig. 4.11a). Let 2β be the angle of the V-thread while looking along the helix (Fig. 4.11b). We relate angle 2γ and angle 2β in plane at an angle α to the plane of γ. In the latter view, the middle line $CD = h$ of V along the line of sight remains unchanged ($C'D' = CD = h$), but points A, B are observed at their projected positions A', B' on line normal to helix line along CD. The observed length $A'D' = (AD)\cos\alpha$ and

$$\tan\gamma = (A'D')/(C'D') = (AD)\cos\alpha\,/\,(CD) = \tan\beta\cos\alpha$$

$$\Rightarrow \quad \cos\gamma = (1 + \tan^2\beta\cos^2\alpha)^{-1/2} \tag{4.9}$$

If helix angle $\alpha \ll 1$, then $\cos\gamma \simeq \cos\beta \Rightarrow \gamma \simeq \beta$.

Let us find the axial torque M needed to impend motion of the screw against axial force P. Let the normal force per unit length be p at the centre-line of the surface of contact, i.e., normal to a side of the V-groove on which contact occurs (Fig. 4.11c). The frictional force per unit length is $\mu_s p$ parallel to the helix line. The component of $p\,ds$ along the normal to the helix line in the plane containing the screw axis is $pds\cos\beta$ (Fig. 4.11c). Its other radial component $pds\sin\beta$ does not contribute to the equilibrium equations: $F_z = 0,\ M_{Oz} = 0$. We eliminate p between the equilibrium equations:

$$\int_0^{s_1} (\cos\beta\cos\alpha - \mu_s \sin\alpha) p\,ds = P$$

$$r\int_0^{s_1} (\cos\beta\sin\alpha + \mu_s \cos\alpha) p\,ds = M$$

$$\Rightarrow \quad M = rP\,\frac{[\tan\alpha + \mu_s/\cos\beta]}{1 - \tan\alpha(\mu_s/\cos\beta)} \tag{4.10}$$

which is same as eq. (4.3) for the square thread with μ_s replaced by a greater effective coefficient of friction $\mu = \mu_s/\cos\beta > \mu_s$. Note that this μ is the ratio of the contact force per unit length $\mu_s p$ along and $p\cos\beta$ normal to the helix line. Hence, V-threads are preferred for fasteners since torque M needed for un-fastening is greater than that for a square thread. Square threads are preferred for power screws since they require less torque M for a given force P. In most commercial V-screws, $\beta \simeq 29°$, so that effective $\mu = 1.14\mu_s$.

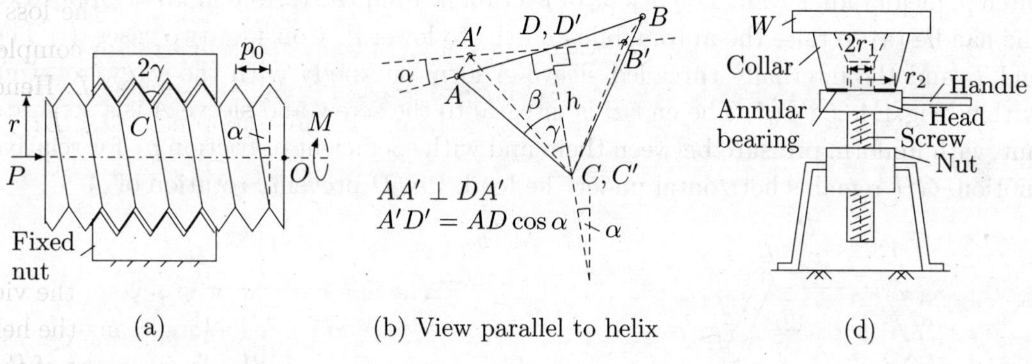

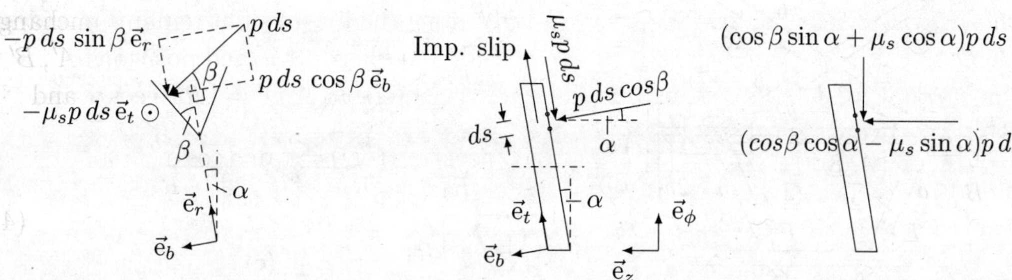

(c)

Fig. 4.11

Consider a V-threaded screw (μ_s) with a collar (μ) having annular contact area from r_1 to r_2 on the screw-head (Fig. 4.11d). The collar and the load W resting on it do not rotate, but translate up as the screw is rotated. Including the moment M_b required to overcome bearing friction between collar and screw-head, the axial moments $M\vec{e}_z$ and $-M_1\vec{e}_z$ (in opposite direction) for raising and lowering the load are obtained from eq. (4.10) and eq. (5) of Sec. 2.19 with μ_s replaced by $-\mu_s$ and sign reversed for M_1:

$$M = M_b + \frac{r(\tan\alpha + \mu_s/\cos\beta)}{1 - \mu_s\tan\alpha/\cos\beta}W, \qquad M_1 = M_b - \frac{r(\tan\alpha - \mu_s/\cos\beta)}{1 + \mu_s\tan\alpha/\cos\beta}W$$

$$M_b = \tfrac{1}{2}\mu W(r_1 + r_2) \text{ for uniform wear}, \quad M_b = \tfrac{2}{3}\mu W\frac{r_2^3 - r_1^3}{r_2^2 - r_1^2} \text{ for uniform pressure}$$

The contribution of bearing friction to M_1 is M_b as it is also in opposite direction. Self-locking occurs if $M_1 > 0$, e.g., for the case of uniform wear it occurs if

$$\tfrac{1}{2}\mu(1 + \mu_s\tan\alpha/\cos\beta)(r_1 + r_2) + (\mu_s/\cos\beta - \tan\alpha)r > 0$$

Example 4.25: A light double scissor-jack (Fig. E4.25a) supports an end of an automobile which exerts a force P on it. The screw has a double start square thread with pitch p, mean radius r and coefficient of friction μ. Find the torque M to be applied to the handle (a) to raise the automobile and (b) to lower it. Consider two cases. (i) The end B and the internally threaded sleeve A form one body with the screw rotating within (Fig. E4.25a). (ii) The end B is integral to the screw and sleeve A is a separate nut, with uniform pressure between them and with coefficient of friction μ_1 for relative motion. GH remains horizontal under the load. Pin D prevents rotation of A.

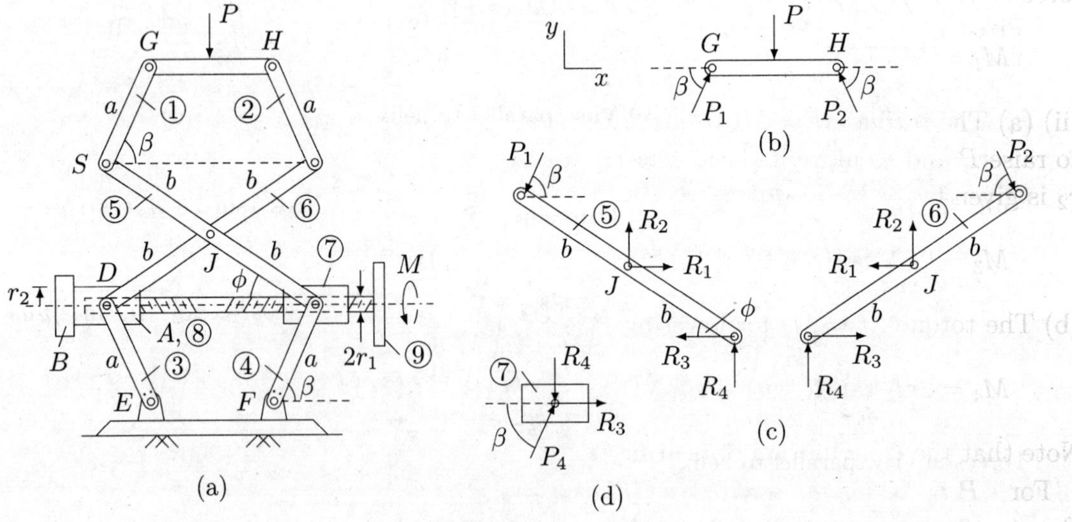

Fig. E4.25

Solution: Members ①, ②, ③, ④ are two-force members. The FBD of platform GH is shown in Fig. E4.25b. By symmetry, the compressive forces in members ① and ② are equal: $P_1 = P_2$. The equation of equilibrium of the platform GH yields

$$P = (P_1 + P_2)\sin\beta \qquad \Rightarrow \qquad P_1 = P_2 = P/2\sin\beta$$

Similarly, the compressive forces in members ③ and ④ are: $P_3 = P_4 = P/2\sin\beta$. The FBD of member ⑤ is shown in Fig. E4.25c along with its mirror image for member ⑥ due to symmetry. Since the reactions at J have to be $\vec{R}$ and $-\vec{R}$, we infer $R_2 = 0$.

$$\begin{aligned}\text{For ⑤:}\quad F_y &= R_4 - P_1\sin\beta = 0 \quad\Rightarrow\quad R_4 = \tfrac{1}{2}P\\ M_{J_z} &= R_4 b\cos\phi - R_3 b\sin\phi + P_1\sin\beta(b\cos\phi) + P_1\cos\beta(b\sin\phi) = 0\\ \Rightarrow\quad R_3 &= P\cot\phi + \tfrac{1}{2}P\cot\beta\end{aligned}$$

From the FBD of nut ⑦ (Fig. E4.25d), the axial force N on the screw is given by

$$N = R_3 + P_4\cos\beta = P\cot\phi + P\cot\beta \tag{1}$$

The helix angle of the thread and angle of friction are given by

$$\tan\alpha = 2p/(2\pi r), \qquad \tan\theta = \mu \tag{2}$$

We use the results for annular bearings, i.e., eq. $(5)_2$ of Section 2.19 and eqs (4.3), (4.6).
(i) Let $M = M_1$ be the torque applied to move the screw in the two nuts against axial force N to raise P, and $M = M_2$ be the torque applied to lower it.

$$M_1 = 2rN\tan(\alpha+\theta), \qquad M_2 = 2rN\tan(\alpha-\theta)$$

(ii) (a) The torque $M = M_3$ to move the screw in the two nuts against axial force N to raise P and to move its head relative to annular bearing (on the sleeve) from r_1 to r_2 is given by

$$M_3 = 2rN\tan(\alpha+\theta) + 2\mu_1 N(r_2^3 - r_1^3)/3(r_2^2 - r_1^2)$$

(b) The torque $M = M_4$ for lowering P is given by

$$M_4 = 2rN\tan(\alpha-\theta) - 2\mu_1 N(r_2^3 - r_1^3)/3(r_2^2 - r_1^2)$$

Note that the contribution of bearing friction to M_4 is negative.

For $P = 10\,\text{kN}$, $\mu = 0.3$, $\mu_1 = 0.35$, $(p, r, r_1, r_2) = (4, 12, 15, 25)\,\text{mm}$, $\beta = \phi = 30°$, we get $N = 34.64\,\text{kN}$, $\alpha = 6.057°$, $\theta = 16.7°$, $M_1 = 348.7\,\text{N.m}$, $M_2 = -156.23\,\text{N.m}$, $M_3 = 596.2\,\text{N.m}$, $M_4 = -403.8\,\text{N.m}$.

4.8 Flexible Cables on Rough Surfaces

Flexible belts, cables and ropes in contact with rough surfaces are used for moorings, belt drives, band brakes, hoisting rigs, belt conveyors, etc. For a flat belt drive, the rim of the pulley may be slightly crowned (given a convex shape) to produce a self-centering action. For a V-belt drive and circular rope drive, the rim of the pulley has a V-groove. The *condition of their impending slip is an important criterion for their design.* The basic results of a belt moving over a surface have been derived in Section 2.18. For $v = 0$: the relation $p = T/\rho$ for normal reaction p per unit length ('pressure') implies that p is proportional to T and $1/\rho$ and would be larger, where T is greater and ρ is lesser. For example, for a parcel tied by a string around it, the sharper the edge of the parcel, the smaller is ρ and greater is the tendency of the string to bite into it due to greater p. T and p vary exponentially with θ. Notice the remarkable fact that eq. (2.76): $T_1/T_2 = e^{\mu_s\theta}$ for impending slip implies that for contact on a plane curve, the ratio T_1/T_2 is independent of the size or the shape of the curve and depends only on the total change in the direction of the belt and the coefficient of friction. For a circular post, θ equals the angle of wrap of the belt.

The exponential increase of T_1 with θ is of great practical utility. One turn around a post makes $\theta = 2\pi$, irrespective of its shape. Even if μ_s is small, several turns around a post will make $e^{\mu_s\theta}$ a large number and a much smaller force T_2, on one end would hold much larger tension T_1 on the other end. For example, for $\mu_s = 0.25$ and 4 turns around a post, $T_2/T_1 = e^{-\mu_s(8\pi)} = 0.0019$. Hence, a load T_1 of 20 kN can be held by applying a very small force, $T_2 = 38$ N. The same 38 N force can hold a load of 10900 kN with 8 turns around the post. This effect is used to hold ships by ropes passed around mooring posts — with sufficient number of turns. The slack end may even be left on the ground. The tension T_2 generated due to the friction of the ground on the length of the rope lying on it suffices to hold a large tension T_1 on the end of the rope attached to the ship. The same phenomenon is used in hoists by passing a rope several turns around a drum. The tight side of the rope carries the heavy load which is held by applying a small tension by hand to the slack side. The relation $T_1/T_2 < e^{\mu_s\theta}$ for no slip is fundamental to the mechanics of cloth and the strength of knots.

§ **Exercise 4.1:** Consider a light cable in the form of a curve C (with $\vec{e}_t$, $\vec{e}_n$, $\vec{e}_b$) resting on a smooth convex surface S under the distributed normal reaction $\vec{q} = p\,\vec{e}_N$ from S (in the direction of the normal $\vec{e}_N$ to S) and the tensions at its ends. Prove that $d(T\,\vec{e}_t)/ds + p\,\vec{e}_N = \vec{0}$, $T = T_0 =$ constant, $\vec{e}_n = -\,\vec{e}_N$, $p = T_0/\rho$. Hence, C is a geodesic on S. If S is a sphere, then C is an arc of a great circle. If S is a cylinder, then C is its generating curve on the cylinder.

Example 4.26: Establish the condition so that the band brake shown in Fig. E4.26a is equally effective for either direction of rotation of the shaft.

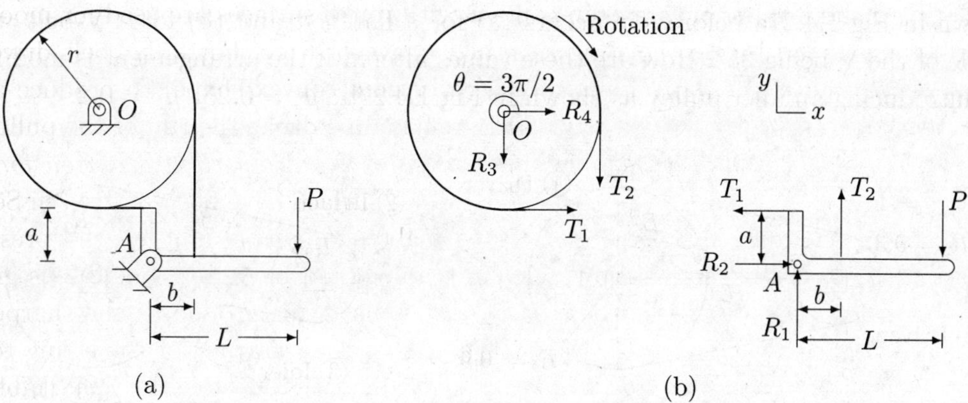

Fig. E4.26

Solution: The FBDs of the rotor and the lever are shown in Fig. E4.26b for one direction of rotation. There is actual slip between the band brake and the brake drum. The equilibrium equation of the lever is:

$$M_{A_z} = T_1 a + T_2 b - PL = 0 \tag{1}$$

The braking torque M provided by the band brake is opposite to the direction of rotation of the rotor $\Rightarrow \quad M = (T_1 - T_2)r > 0 \quad \Rightarrow \quad T_1 > T_2$. For angle of wrap $3\pi/2$, $T_1 = T_2 \mathrm{e}^{3\pi\mu_k/2}$ and using eq. (1), we get the solution as

$$T_1 = \frac{PL\mathrm{e}^{3\pi\mu_k/2}}{b + a\mathrm{e}^{3\pi\mu_k/2}}, \qquad T_2 = \frac{PL}{b + a\mathrm{e}^{3\pi\mu_k/2}}$$

$$\Rightarrow \quad M = (T_1 - T_2)r = \frac{(\mathrm{e}^{3\pi\mu_k/2} - 1)}{b + a\mathrm{e}^{3\pi\mu_k/2}} rPL$$

If the direction of rotation is reversed, then $T_1 < T_2$, $T_2 = T_1 \mathrm{e}^{3\pi\mu_k/2}$ and eq. (1) yields

$$T_1 = \frac{PL}{a + b\mathrm{e}^{3\pi\mu_k/2}}, \qquad T_2 = \frac{PL\mathrm{e}^{3\pi\mu_k/2}}{a + b\mathrm{e}^{3\pi\mu_k/2}}$$

$$\Rightarrow \quad M^* = (T_2 - T_1)r = \frac{(\mathrm{e}^{3\pi\mu_k/2} - 1)}{a + b\mathrm{e}^{3\pi\mu_k/2}} rPL$$

where M^* is the braking torque. The band brake is equally effective in either direction

$$\text{if} \quad M^* = M \quad \Rightarrow \quad \text{if} \quad a + b\mathrm{e}^{3\pi\mu_k/2} = b + a\mathrm{e}^{3\pi\mu_k/2} \quad \Rightarrow \quad \text{if} \quad a = b$$

Example 4.27: A light belt is used to transmit power. The allowable tension in the belt is 2.5 kN. Find the maximum torques that can be transmitted by the two shafts shown in Fig. E4.27a before slip occurs (a) for a flat belt and (b) for a V-belt with angle of the V being 35°. How are these values altered if the arrangement is modified by introducing an idler pulley as shown in Fig. E4.27b? $\mu_s = 0.25$, $\mu_k = 0.2$.

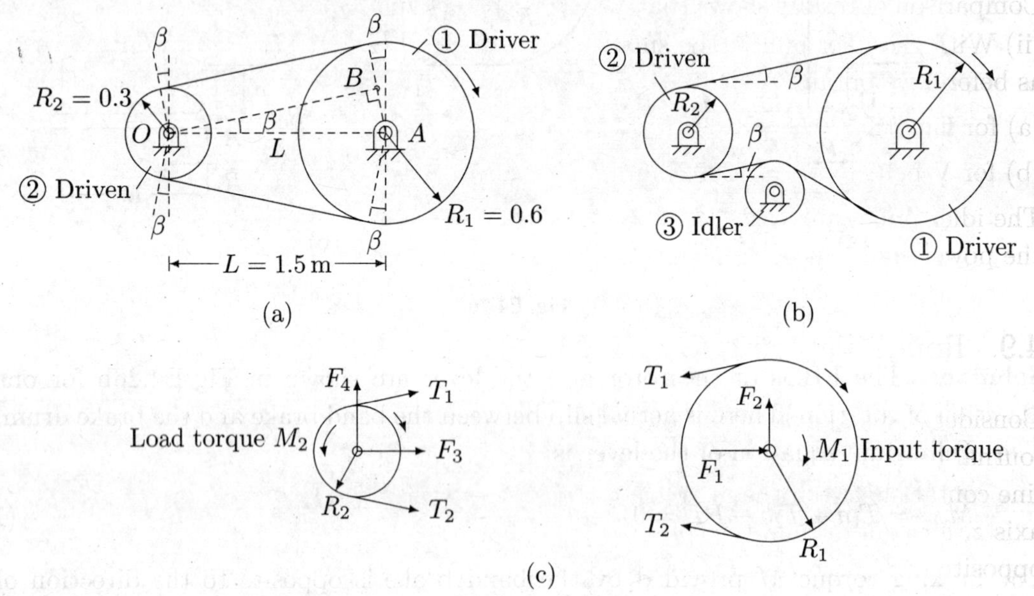

Fig. E4.27

Solution: The tight (T_1) and slack (T_2) sides for the two pulleys are shown in Fig. E4.27c since the driver pulley has input torque in the direction of rotation and the driven pulley has output torque opposite to the direction of its rotation. The angles of wrap on the two pulleys are $\pi + 2\beta$ and $\pi - 2\beta$, where

$$\sin\beta = \frac{(AB)}{L} = \frac{R_1 - R_2}{L} = \frac{0.3}{1.5} = 0.2 \quad \Rightarrow \quad \beta = 11.54^\circ = 0.2014\,\text{rad}$$

For impending slip, $T_1/T_2 = e^{\mu_s \theta_1}$, with θ_1 being the smaller angle of wrap, since μ_s and the two tensions are the same. Let M_1, M_2 be the torques on pulleys ① and ②.

(i) Without the idler pulley: $\theta_1 = \pi - 2\beta = 2.739\,\text{rad}$.

(a) Flat belt: $T_1 = 2.5\,\text{kN} = 2500\,\text{N}$,

$$\frac{T_1}{T_2} = e^{\mu_s \theta_1} = 1.983 \quad \Rightarrow \quad T_2 = \frac{T_1}{1.983} = 1260.6\,\text{N}$$

$$M_1 = (T_1 - T_2)R_1 = 743.6\,\text{N.m}, \qquad M_2 = (T_1 - T_2)R_2 = 371.8\,\text{N.m}$$

(b) V-belt: $T_1 = 2500\,\text{N}, \quad \alpha = 35/2 = 17.5°,$

$$\frac{T_1}{T_2} = e^{\mu_s \theta_1 / \sin\alpha} = 9.749 \qquad \Rightarrow \qquad T_2 = \frac{2500}{9.749} = 256.4\,\text{N}$$

$$M_1 = (T_1 - T_2)R_1 = 1346\,\text{N.m}, \qquad M_2 = (T_1 - T_2)R_2 = 673\,\text{N.m}$$

Comparison of results shows that V-belt is better than flat belt in power transmission.
(ii) With the idler pulley, the smaller angle of wrap is $\theta_1 = 180° = \pi\,\text{rad}$. Proceeding as before, we obtain

(a) for flat belt: $M_1 = 816\,\text{N.m}, \quad M_2 = 408\,\text{N.m},$

(b) for V-belt: $M_1 = 1390\,\text{N.m}, \quad M_2 = 695\,\text{N.m}.$

The idler pulley increases the angle of wrap on the smaller pulley and thus increases the power that can be transmitted.

4.9 Rough Pin Joint / Journal Bearing

Consider a shaft, axle or pin of radius r at rest in a short cylindrical cavity, called a journal bearing, of the same radius (Fig. 4.12a). The inevitable clearance results in a line contact between them. When a transverse force $P\vec{\mathrm{i}}$ acts on the axle normal to its axis z, for equilibrium, the collinear bearing reaction R will act on the line of contact opposite to P with $R = P$ (Fig. 4.12a).

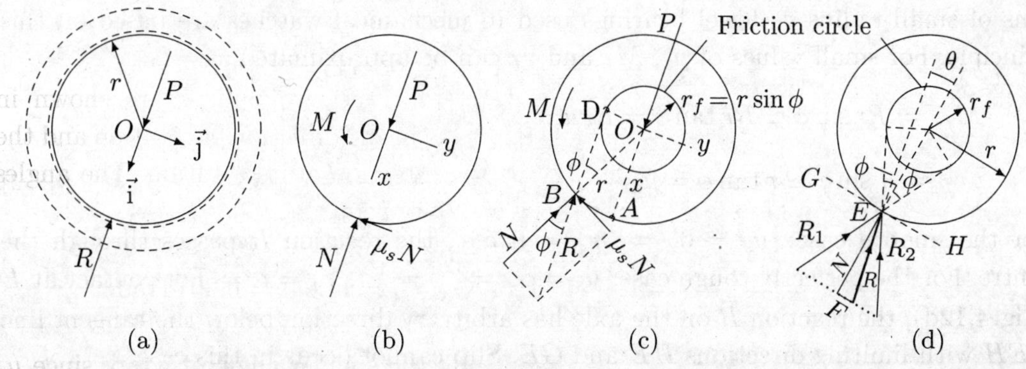

Fig. 4.12

Consider an external moment M applied to the axle about its axis to cause impending slip. The line of contact cannot be opposite to P (Fig. 4.12b) since $F_y = 0$ is violated. The axle is in equilibrium under applied force $P\vec{\mathrm{i}}$, applied couple $M\vec{\mathrm{k}}$ and the reaction $\vec{R}$ of the bearing. $\vec{F} = \vec{R} + P\vec{\mathrm{i}} = \vec{0} \Rightarrow \vec{R} = -P\vec{\mathrm{i}}$ is in a direction opposite to $P\vec{\mathrm{i}}$. It should be so positioned (Fig. 4.12c) on the axle that it causes a moment about the axis opposite to $M\vec{k}$ so that the equation $\vec{M}_O = \vec{0}$ is satisfied. The normal and tangential components N and $\mu_s N$ of reaction R at impending slip, satisfy

$$N = R\cos\phi, \qquad \mu_s N = R\sin\phi \quad \Rightarrow \quad \mu_s = \tan\phi$$

R is at an angle ϕ to N, where $\phi = \tan^{-1}\mu_s$ is the angle of friction. The moment equation, $M_{O_z} = M - \mu_s N r = 0 \Rightarrow$

$$M = \mu_s N r = R(r\sin\phi) = P(r\sin\phi) \tag{4.11}$$

Thus, for impending motion, R does not act radially but at a distance $r_f = r\sin\phi$ from the axis, i.e., along a tangent to a circle of radius $r_f = r\sin\phi$. This circle is called the *friction circle* for the axle and the bearing and is independent of the loading conditions on the axle. Contact occurs at B riding some distance away from A (located opposite to P), in a direction opposite to M. Physically, when the applied moment M is gradually increased on the axle, the axle rolls without slip on the bearing, so 'climbing' in the *direction opposite* to M takes place until slip occurs. For a rotating shaft, $\mu_k = \tan\phi_k$ and the radius of the friction circle is $r\sin\phi_k < r\sin\phi$ since $\mu_k < \mu_s$.

For a given line of contact at E, for no slip, R makes an angle $\theta \leq \phi$ (Fig. 4.12d) with the normal (radial line), i.e., the line of action of R lies between tangents to the friction circle from contact E with R_1, R_2 being the reactions for impending slip at E.

The equivalent reaction at the axis of the axle (at pin-axis of a pin-joint with friction) is a force $\vec{R}$ and a frictional moment $M_f = r\mu_s N = Rr\sin\phi$ about the pin-axis in a direction opposing impending rotation. The effect of friction can be decreased by using pins of small radius r. Jewel bearings used in mechanical watches are based on this principle. For small values of μ_s, M_f and r_f can be approximated as

$$\begin{aligned} M_f &= Rr\sin\phi \simeq Rr\tan\phi = Rr\mu_s \\ r_f &= r\sin\phi \simeq r\tan\phi = \mu_s r \end{aligned} \tag{4.12}$$

For the smooth case: $\mu_s = 0 \Rightarrow r_f = 0$, i.e., the reaction R passes through the centre. For the perfectly rough case: $\mu_s = \infty \Rightarrow \phi = \frac{1}{2}\pi$, $r_f = r$. $\Rightarrow$ For contact at E (Fig. 4.12d), the reaction $\vec{R}$ on the axle has arbitrary direction below the tangent line GEH with limiting directions HE and GE. Slip cannot occur in this case.

Example 4.28: *Bearing friction on the axis of the wheels of a cart.* Consider the cart shown in Fig. E4.28a which is pulled with a force P. The weights of the front wheel and the rear wheel assemblies are W_1 each. The weight of the main body (chassis) of the trolley including the load carried is W. What is the force P for impending motion of the trolley? The wheels have radius R and the radius of the axles is r. The coefficient of friction at the contact between axles and wheels is μ_s.

Solution: The direction of impending rotation of the wheels relative to the translating axle is shown in Fig. E4.28b. The radius r_f of the friction circle of the axle of radius r is $r_f = r\sin\phi$ ($\phi = \tan^{-1}\mu_s$). The reaction R_2 of the axle on the wheel has to be tangent to this circle in an appropriate direction so that it provides a moment to oppose the impending rotation of the wheel relative to the axle.

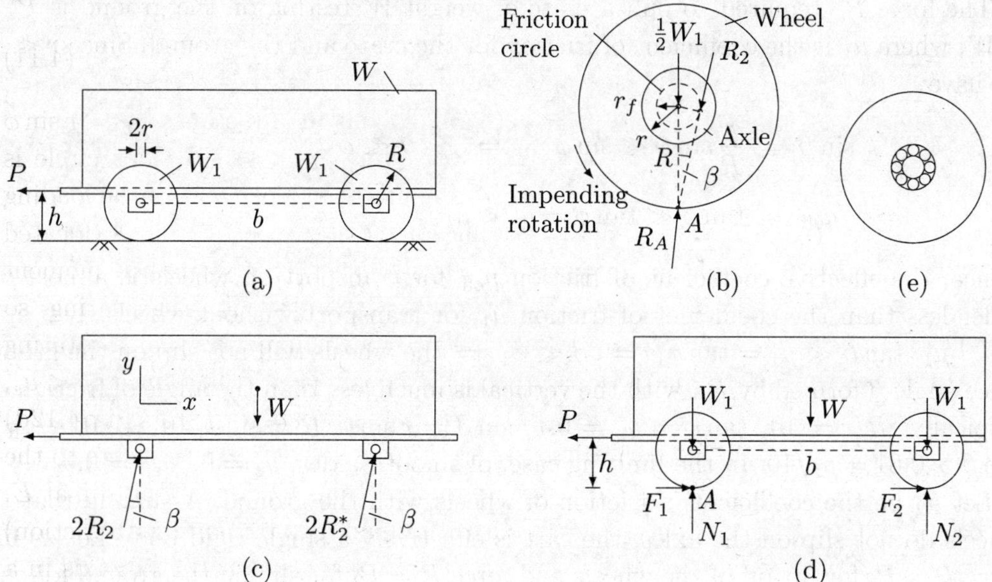

Fig. E4.28

The front wheel (FBD in Fig. E4.28b) is subjected to weight $\frac{1}{2}W_1$, ground reaction R_A and the reaction R_2 of the axle. Forces $\frac{1}{2}W_1$ and R_A intersect at A. R_2 should also pass through A as $M_{A_z} = 0$. The angle β formed by R_2 with the vertical is given by

$$\sin\beta = r_f/R = r\sin\phi/R \tag{1}$$

Similarly, the reaction R_2^* of the rear axle on a rear wheel is at an angle β with the vertical. The FBD of the main body of the trolley is shown in Fig. E4.28c. The equilibrium equation $\vec{F} = \vec{0}$ yields

$$F_y = 2(R_2 + R_2^*)\cos\beta - W = 0 \quad\Rightarrow\quad (R_2 + R_2^*) = W/2\cos\beta$$

$$F_x = 2(R_2 + R_2^*)\sin\beta - P = 0 \quad\Rightarrow\quad P = W\tan\beta \tag{2}$$

For equilibrium $P_{\max} = W\tan\beta$. The effective coefficient of friction μ_{eff} is given by

$$\mu_{\text{eff}} = P/W = \tan\beta \tag{3}$$

$$\sin\beta = \frac{r_f}{R} = \frac{r\sin\phi}{R} = \frac{r\mu_s}{R\sqrt{1+\mu_s^2}}$$

$$\tan\beta = \frac{\sin\beta}{\sqrt{1-\sin^2\beta}} = \frac{\mu_s r/R}{\sqrt{1+\mu_s^2(1-r^2/R^2)}} \simeq \frac{\mu_s r}{R}$$

The force P^* required to pull a crate of weight W resting on the ground is $P^* = \mu_1 W$, where μ_1 is the coefficient of friction for the crate and the ground. Since $r \ll R$, we have

$$\sin\beta = \frac{r}{R}\sin\phi \ll \sin\phi \quad \Rightarrow \quad \beta \ll \phi$$

$$\Rightarrow \quad \mu_{\text{eff}} = \tan\beta \ll \tan\phi = \mu_s < \mu_1$$

Hence, the effective coefficient of friction μ_{eff} for transport on wheels is almost one order less than the coefficient of friction μ_1 for transport without wheels. Let $\phi_1 = \tan^{-1}\mu_1$. $\tan\beta \ll \mu_1 = \tan\phi_1 \Rightarrow \beta \ll \phi_1 \Rightarrow$ the wheels will not slip on the ground since angle β formed by R_A with the vertical is much less than the angle of friction ϕ_1. Typically, $R/r = 10$, $\tan\phi = \mu_s = 0.1$, $\sin\beta = r\sin\phi/R = \sin\phi/10 \simeq 0.01$, $\mu_{\text{eff}} = \tan\beta \simeq 0.01 = \mu_s/10$. In the limiting case, of smooth axles, $\mu_s = 0$, $\mu_{\text{eff}} = 0$.

Let μ_2 be the coefficient of friction of wheels with the ground. Assuming that the wheels do not slip on the axles, the cart is effectively a single rigid body Let us find force $P = P_1$ for lift-off of the wheels and force $P = P_2$ for slip on the ground for such a rigid cart (Fig. E4.28d).

$$M_{A_z} = 0 \quad \Rightarrow \quad N_2 = \tfrac{1}{2}(W + 2W_1) - Ph/b \geq 0 \quad \Rightarrow \quad P \leq (W + 2W_1)b/2h = P_1$$

$$F_y = 0 \quad \Rightarrow \quad N_1 + N_2 = W + 2W_1$$

$$F_x = 0 \quad \Rightarrow \quad P = F_1 + F_2 \leq \mu_2(N_1 + N_2) = \mu_2(W + 2W_1) = P_2$$

There will be lift-off at rear wheels if $P_1 < P_2$ and $P_1 < P_{\text{max}}$, and slip on the ground if $P_2 < P_1$ and $P_2 < P_{\text{max}}$. If $P_1 > P_{\text{max}}$ and $P_2 > P_{\text{max}}$, then for case of $\mu_s \neq \infty$, the cart will roll without slip on the ground and with slip between the wheels and axles at $P = P_{\text{max}} = W\tan\beta$.

The arrangement of pin-joint has been improved in modern vehicles by introducing ball/roller bearing between the axle and the wheel (Fig. E4.28e). Neglecting the inertia of the balls, the balls are two-force members with forces acting along the line joining the points of contact, i.e., along the radial line. Hence, the total force from the wheel on the axle is radial, i.e., $r_f = 0$ ($\Rightarrow$ $\phi_{\text{eff}} = 0$, $\beta = 0$, $\mu_{\text{eff}} = \tan\beta = 0$). This is as good as transport on rollers without the disadvantage of repeatedly putting the rollers from the back of the block to its front.

Example 4.29: What is the relation between the forces P and Q for a lever with a pin joint as pivot (Fig. E4.29a) so that the lever remains in equilibrium? The coefficient of friction at the pin is $\mu\,(=\tan\phi)$.

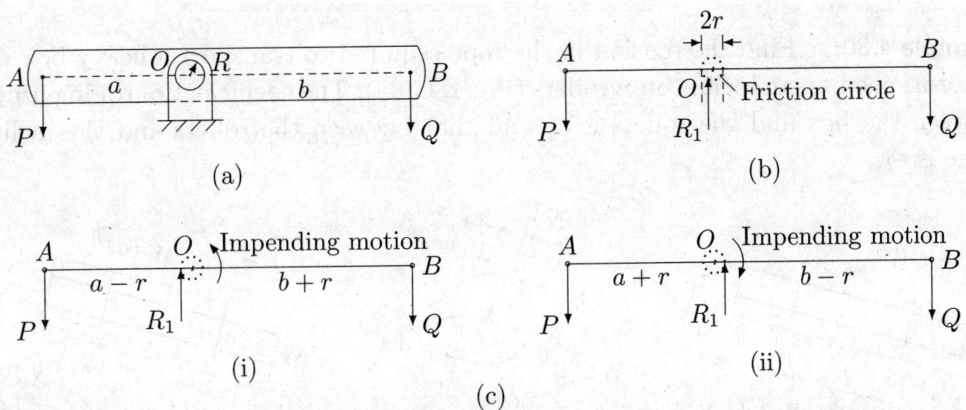

Fig. E4.29

Solution: For equilibrium, the reaction R_1 of the pin on the lever has to be $P+Q$ in a direction opposite to P and its line of action has to pass through the friction circle of radius $r = R\sin\phi$ as shown in Fig. E4.29b. The two limiting cases for impending rotations of the lever in the two directions are shown in Fig. E4.29c.

The equation of equilibrium for these limiting cases yields

$$\text{case (i)}: \qquad P(a-r) = Q(b+r) \qquad \Rightarrow \qquad P = Q\,\frac{b}{a}\cdot\frac{1+r/b}{1-r/a} > Q\frac{b}{a}$$

$$\text{case (ii)}: \qquad P(a+r) = Q(b-r) \qquad \Rightarrow \qquad P = Q\,\frac{b}{a}\cdot\frac{1-r/b}{1+r/a} < Q\frac{b}{a}$$

For the given load Q, the lever is in equilibrium if the balance load P satisfies the following inequality:

$$Q\,\frac{b}{a}\cdot\frac{1-r/b}{1+r/a} < P < Q\,\frac{b}{a}\cdot\frac{1+r/b}{1-r/a} \tag{1}$$

Thus, in a balance ($b = a$), used to measure weight of a commodity Q, depending on the procedure used, the balance weight P could be any value satisfying eq. (1). Therefore, under-weighing or over-weighing may occur. The culprit is the pivot friction and not necessarily the person doing the weighing. Note that as $R \to 0$, then $r = R\sin\phi \to 0$ and the balance becomes more accurate. Chemical balances are pivoted on knife edges for this reason.

4.10 Rolling Friction

The method of solving problems involving rolling friction (see Section 2.14.2) is illustrated by two examples.

Example 4.30: Find the tension in the rope required to transport a heavy box with uniform speed up an incline on n rollers (Fig. E4.30a). The coefficient of rolling friction between the box and the rollers is e_1 and that between the rollers and the inclined plane is e_2.

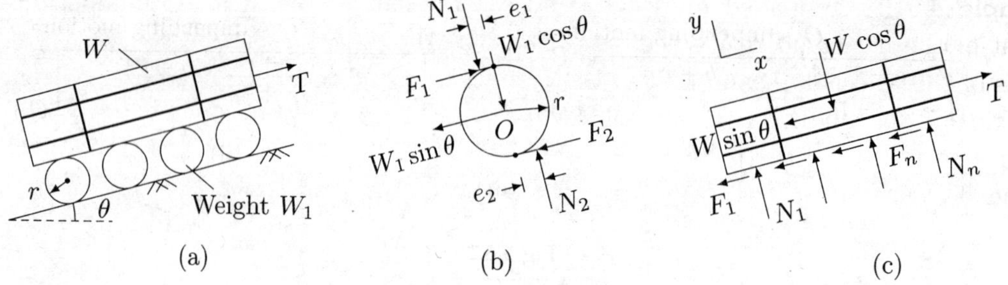

Fig. E4.30

Solution: In the FBD of the roller (Fig. E4.30b), the reactions are shifted towards the side of the roller which is to make contact subsequently. Its equations of equilibrium:

$$F_x = F_1 - W_1 \sin\theta - F_2 = 0 \qquad \Rightarrow \qquad F_2 = F_1 - W_1 \sin\theta$$

$$F_y = N_2 - N_1 - W_1 \cos\theta \qquad \Rightarrow \qquad N_2 = N_1 + W_1 \cos\theta$$

$$M_{0z} = e_2 N_2 + e_1 N_1 - r(F_1 + F_2) = 0$$

$$\Rightarrow \quad e_2(N_1 + W_1 \cos\theta) + e_1 N_1 - r(2F_1 - W_1 \sin\theta) = 0$$

$$\Rightarrow \quad F_1 = [(e_1 + e_2)N_1 + (e_2 \cos\theta + r \sin\theta)W_1]/2r \tag{1}$$

The FBD of the box is shown in Fig. E4.30c. The equations of equilibrium:

$$F_x = T - W \sin\theta - \sum F_i = 0 \qquad \Rightarrow \qquad T = W \sin\theta + \sum F_i \tag{2}$$

$$F_y = \sum N_i - W \cos\theta = 0 \qquad \Rightarrow \qquad \sum N_i = W \cos\theta \tag{3}$$

Equations(1) and (2) $\Rightarrow$

$$T = W \sin\theta + [(e_1 + e_2) \sum N_i + (e_2 \cos\theta + r \sin\theta) W_1 n]/2r$$

$$\stackrel{(3)}{=} (W + \tfrac{1}{2} n W_1) \sin\theta + [(e_1 + e_2) W + e_2 n W_1] \cos\theta / 2r$$

$$\simeq W \sin\theta + W \cos\theta\,(e_1 + e_2)/2r \qquad \text{since,}\quad nW_1 \ll W$$

$$\Rightarrow \quad T \simeq W \sin\theta + \mu_{\text{eff}} W \cos\theta \qquad \text{where} \quad \mu_{\text{eff}} = (e_1 + e_2)/2r$$

The tension T^* to slide the box directly on the incline with coefficient of friction μ is $T^* = W \sin\theta + \mu W \cos\theta$. Comparison of the expressions of T and T^* reveals that the effective coefficient of friction μ_{eff} for transporting on rollers: $\mu_{\text{eff}} = (e_1 + e_2)/2r \ll \mu$ $\Rightarrow T \ll T^*$. Hence, transport over rollers is advantageous since for typical values: $e_1 = 0.04\,\text{cm},\ e_2 = 0.16\,\text{cm},\ r = 5\,\text{cm}: \ \mu_{\text{eff}} = (e_1 + e_2)/2r = 0.02$ is indeed small.

Example 4.31: A loaded mine-car (Fig. E4.31a) has a weight W, excluding the weight of the front and rear wheel assemblies, each of which has a weight W_1. The radius of the wheels is R and that of the axle is r. The coefficient of friction between the axle and the wheel is $\mu = \tan\phi$. The coefficient of rolling friction of the wheels on the ground is e. Find the force P in the rope of the winch to haul the trolley up the incline at a uniform speed.

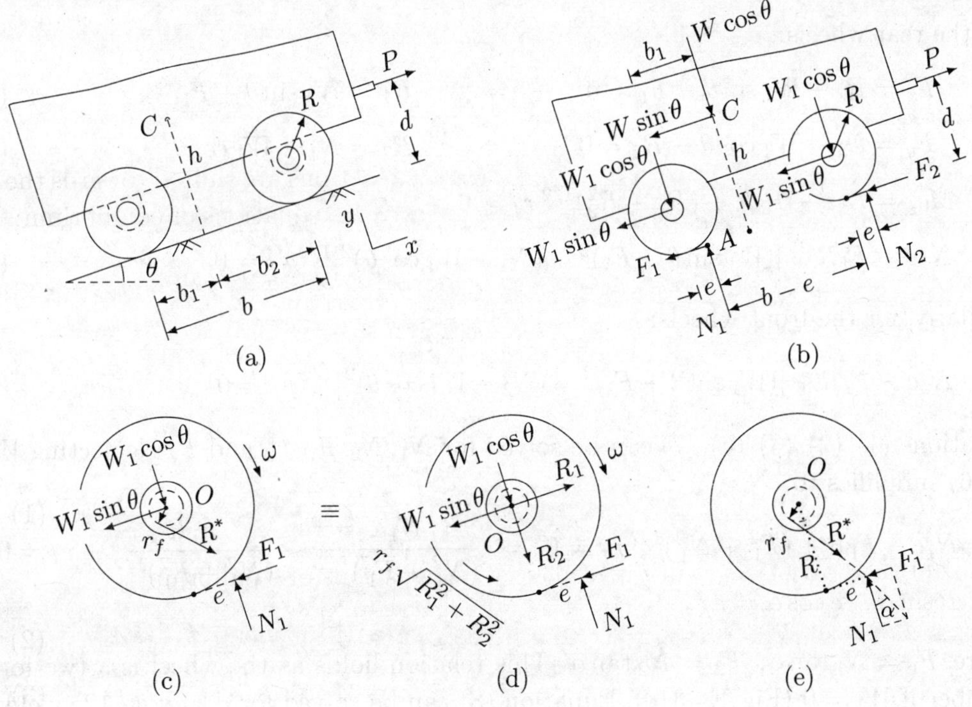

Fig. E4.31

Solution: The FBDs of the car and the rear wheels are shown in Figs E4.31b,c,d. The reaction R^* from the axle on rear wheels is tangent to the friction circle of radius $r_f = r \sin\phi$ (Fig. E4.31c) and is equivalent to components R_1, R_2 at centre O

and a moment $r_f R^* = r_f\sqrt{R_1^2 + R_2^2}$ (Fig. E4.31d) opposite to ω. Similarly, on front wheels reaction components R_3, R_4 act at centre and a moment $r_f\sqrt{R_3^2 + R_4^2}$. The 9 unknowns, $R_1, R_2, R_3, R_4, N_1, F_1, N_2, F_2, P$ are obtained from 3 equilibrium equations each for the car, and the front and rear wheel assemblies. For the car:

$$M_{A_z} = -W\cos\theta(b_1 - e) + W\sin\theta\, h - Pd$$

$$-W_1\cos\theta(b - e) + W_1\cos\theta\, e + 2W_1\sin\theta\, R + N_2 b = 0$$

$$\Rightarrow \quad N_2 = [W\cos\theta(b_1 - e) + Pd + W_1\cos\theta(b - 2e) - W\sin\theta\, h - 2W_1\sin\theta\, R]/b \qquad (1)$$

$$F_x = P - F_1 - F_2 - W\sin\theta - 2W_1\sin\theta = 0$$

$$\Rightarrow \quad P = F_1 + F_2 + (W + 2W_1)\sin\theta \qquad (2)$$

$$F_y = N_1 + N_2 - (W + 2W_1)\cos\theta = 0$$

$$\Rightarrow \quad N_1 = [W\cos\theta(b_2 + e) - Pd + W_1\cos\theta(b + 2e) + W\sin\theta\, h + 2W_1\sin\theta\, R]/b \qquad (3)$$

For the rear wheels:

$$F_x = R_1 - W_1\sin\theta - F_1 = 0 \quad \Rightarrow \quad R_1 = W_1\sin\theta + F_1 \qquad (4)$$

$$F_y = N_1 - W_1\cos\theta - R_2 = 0 \quad \Rightarrow \quad R_2 = N_1 - W_1\cos\theta \qquad (5)$$

$$M_{O_z} = N_1 e - F_1 R + (R_1^2 + R_2^2)^{1/2} r_f = 0$$

$$\Rightarrow \quad N_1 e - F_1 R + [(W_1\sin\theta + F_1)^2 + (N_1 - W_1\cos\theta)^2]^{1/2} r_f = 0 \qquad (6)$$

Similarly, for the front wheels:

$$N_2 e - F_2 R + [(W_1\sin\theta + F_2)^2 + (N_2 - W_1\cos\theta)^2]^{1/2} r_f = 0 \qquad (7)$$

Equations (1), (2), (3), (6), (7) can be solved for N_1, N_2, F_1, F_2 and P. Neglecting W_1, eq. (6) simplifies to

$$N_1 e - F_1 R + (F_1^2 + N_1^2)^{1/2} r_f = 0 \Rightarrow \frac{N_1 e}{(N_1^2 + F_1^2)^{1/2}} - \frac{F_1 R}{(N_1^2 + F_1^2)^{1/2}} + r_f = 0$$

$$\Rightarrow R\sin\alpha - e\cos\alpha = r_f \qquad (8)$$

where $F_1 = N_1\tan\alpha$, $F_2 = N_2\tan\alpha$. This relation holds as the wheel is a two-force member if $W_1 = 0$ (Fig. E4.31e). Equation (8) can be solved for α and eq. (2) yields

$$P = (N_1 + N_2)\tan\alpha + W\sin\theta = (W\cos\theta)\tan\alpha + W\sin\theta \qquad (9)$$

$$\text{eq. (8)} \Rightarrow \quad \tan\alpha = e/R + (r_f/R)\sec\alpha = e/R + r\sin\phi\sec\alpha/R$$

$$\Rightarrow \quad \tan\alpha \ll 1 \quad \text{as } e \ll R \quad \text{and} \quad r \ll R \qquad (10)$$

$$\Rightarrow \quad \tan\alpha \simeq e/R + (r/R)\sin\phi \simeq e/R + (r/R)\tan\phi = (e+\mu r)/R \qquad (11)$$

$$\Rightarrow \quad P \overset{(9)}{\simeq} W\cos\theta(e+\mu r)/R + W\sin\theta \qquad (12)$$

Thus, the effective coefficient of friction $\mu_{\text{eff}} = (e+\mu r)/R \ll \mu$. If the journal friction can be neglected, then eq. (12) yields $P \simeq W\cos\theta(e/R) + W\sin\theta$.

4.11 Concept Review Questions 4B

14. Find θ for which the force P needed to slide the block up the incline (Fig. Q4.14) is the least. The coefficient of friction is μ. [$\tan^{-1}\mu$]

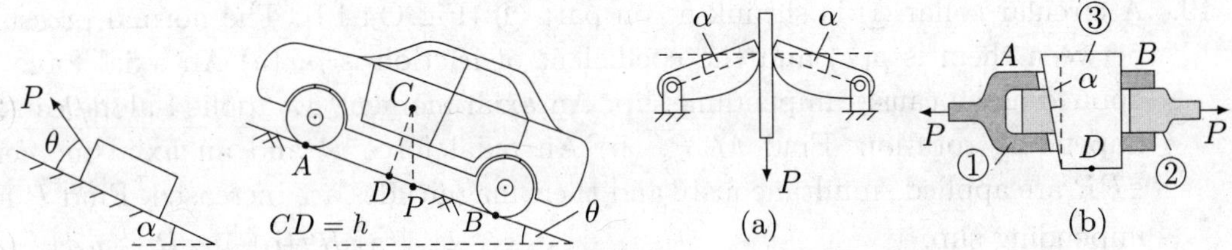

Fig. Q4.14 **Fig. Q4.15** **Fig. Q4.16**

15. (i) A car with light wheels and centre of mass at C (Fig. Q4.15) is parked with only the front wheels braked (locked to the chassis). The coefficient of friction of the ground is μ. Find the maximum θ for which the car remains at rest. (ii) Is it preferable to lock wheels nearer to P or farther from P? $AD = a$, $BD = b$.

[(i) $\tan^{-1}[\mu(b - h\tan\theta)/(a+b)]$ (ii) Nearer to P]

16^d. (a) Find the maximum value of α for which the systems in Fig. Q4.16 are self-locking. The coefficient of friction is μ. (b) Find the upward force Q that has to be applied at D to remove the cotter pin ③, supporting the forked part at A and B. Neglect horizontal friction between parts ① and ②.

[(a) (i) $\tan^{-1}\mu$ (ii) $2\tan^{-1}\mu$ (b) $P[2\mu\cos\alpha + (\mu^2-1)\sin\alpha]/(\cos\alpha + \mu\sin\alpha)$]

17^d. The system shown in Fig. Q4.17 is to be designed so that block ① can slip to the right for appropriate value of P. The coefficient of friction between block ① and the floor is μ_1. The coefficient of friction μ_2 between bodies ① and ② has a range of 0.2 to 0.6. However, the system should be self-locking under application of force P in the reversed direction. Design an appropriate value of θ. Does it depend on the range of μ_2? [$\theta < 11.31°$, yes]

18. (a) A cylindrical tube rests on a body of revolution (Fig. Q4.18) under axial force P. The coefficient of friction is μ. Find the axial moment M to cause impending

motion for $Q = 0$. (b) For $M = 0$, find Q for impending tipping without slip and the minimum value of μ to ensure this.

$[(a)\ \mu Pr/\sin\alpha \quad (b)\ Q = PR/h,\ \mu_{\min} = (h\cos\alpha + r\sin\alpha)/(h\sin\alpha - r\cos\alpha)]$

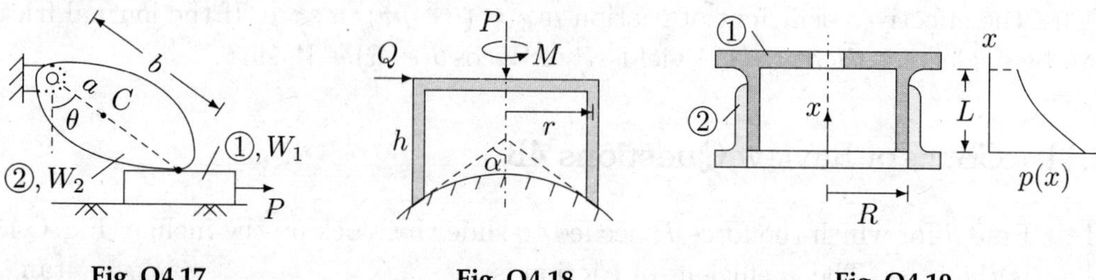

Fig. Q4.17 Fig. Q4.18 Fig. Q4.19

19. A circular collar ① is shrunk-fit on part ② (Fig. Q4.19). The normal pressure between them is $p(x)$ and the coefficient of friction is μ. (a) An axial force P applied alone causes impending slip. An axial moment M applied alone causes impending rotation. Find M/P. (b) An axial force P and an axial moment $\frac{1}{2}PR$ are applied simultaneously and their magnitudes are increased. Find P for impending slip. $[(a)\ R \quad (b)\ 4\mu\pi R\int_0^L p\,dx/\sqrt{5}]$

20. A rope ending in ring ① is looped around a pulley (Fig. Q4.20) and subjected to a pull. The coefficient of friction is μ for the rope and its contacts with the ring and the pulley. Set up an equation for finding the maximum value of α for no slip between the ring and the rope. Rework for the minimum value of α.

$[\sec\alpha = 2e^{\mu\alpha},\ 2\cos\alpha = e^{\mu\alpha}]$

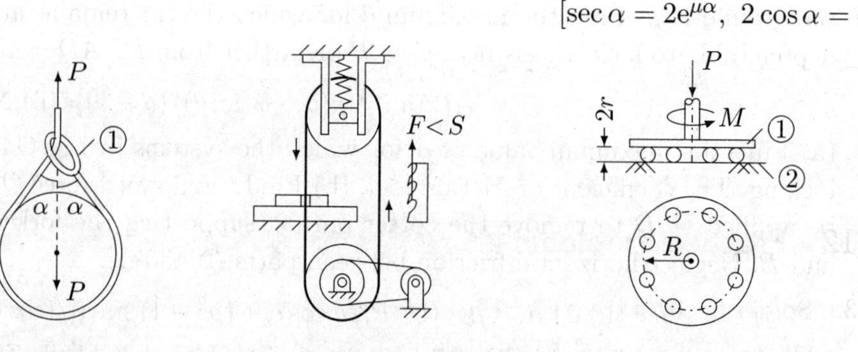

Fig. Q4.20 Fig. Q4.21 Fig. Q4.22

21[d]. The tension in the band-saw blade (Fig. Q4.21) is maintained by applying force P to the axis of the idler pulley (light) by a spring. The coefficient of friction between the band and the pulley is μ. Each tooth on the blade can withstand a maximum force S before being sheared off. Find the condition which should be satisfied so that the blade slips on the pulley before a tooth is sheared. Assume that only one tooth is in contact with the plank being sawed. $[P(e^{\mu\pi} - 1) < 2S]$

22. An axial force P on a shaft is supported by a ball bearing (Fig. Q4.22) with the centres of the light balls on a pitch circle of radius R. The coefficients of rolling resistance of the balls with the flange ① and the support ② are e_1 and e_2 with $e_1 > e_2$. Find the torque M for impending rotation of the shaft. $[PR(e_1+e_2)/2r]$

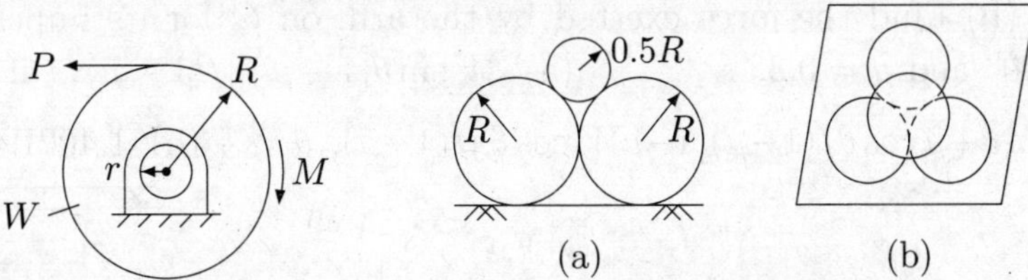

Fig. P4.23 **Fig. Q4.24**

23. Find the moment M required to slowly rotate a winch of weight W to which a force P is applied (Fig. Q4.23). The coefficient of friction for the axle of radius r is $\mu = 0.1$. $[PR + 0.0995(P^2 + W^2)^{1/2}r]$

24*. (a) A cylinder of radius $0.5R$ is placed on two cylinders of radius R (Fig. Q4.24a). The density of all cylinders is the same. Find the minimum coefficient of friction for all contacts so that they remain at rest. (b) A sphere is placed on the top of three spheres (Fig. Q4.24b) resting on a plane. All spheres are identical. Find the minimum coefficient of friction so that they remain at rest. (c) If the coefficient of friction μ between cylinders/spheres differs from the coefficient of friction μ_1 of the ground, find the condition so that the impending motion is (1) with no slip on the ground and (2) with slip on the ground. [(a) 0.3820 (b) 0.3178

(c) (1) (a) $\mu_1 > 0.04244$, $\mu < 0.3820$ (b) $\mu_1 > 0.07946$, $\mu < 0.3178$

(2) (a) $\mu_1 < 0.04244$, $\mu > 0.3820$ (b) $\mu_1 < 0.07946$, $\mu > 0.3178$]

4.12 Practice Problems 4B

23. Spiders, with three light wheels free to rotate on them, are provided on the two ends of the axle of a trolley (Fig. P4.23). These are free to rotate on the axle. Find the force components F_1 and F_2 of the pull exerted on the trolley to just move it up the staircase. The mass of the trolley is m and its centre of mass is at C. The coefficient of friction of the wheels with the step is μ.

$$[F_1 = mg[2a + (c\cos\alpha - d\sin\alpha)(\sqrt{3}\sin\alpha - 2\cos\alpha)]/\Delta,$$

$$F_2 = mg[2b + (c\cos\alpha - d\sin\alpha)(\sqrt{3}\cos\alpha + 2\sin\alpha)]/\Delta$$

$$\text{with } \Delta = (a\sqrt{3} + 2b)\cos\alpha + (2a - \sqrt{3}b)\sin\alpha]$$

24. Identical cylindrical cans, each of weight W, are to be raised up an incline by a series of arms (Fig. P4.24). (a) Find the force exerted by the arm on can A to move it. What is the condition which ensures that the impending motion of can A is rolling without slip on the incline? The coefficient of friction for all surfaces is μ. (b) Find the force exerted by the arm on C for its impending motion if $\theta = 30°$ and $\mu = 0.3$. [(a) $W \sin\theta[\vec{i} - \mu\vec{j}\,]/(1-\mu)$ if $\theta < \pi/4$, else $W[\sin\theta + \mu\cos\theta/(1-\mu)]\vec{i} - \mu W\cos\theta\,\vec{j}/(1-\mu)$; $\theta < \frac{\pi}{4}$ (b) $1.402W\vec{i} + 0.02659W\vec{j}$]

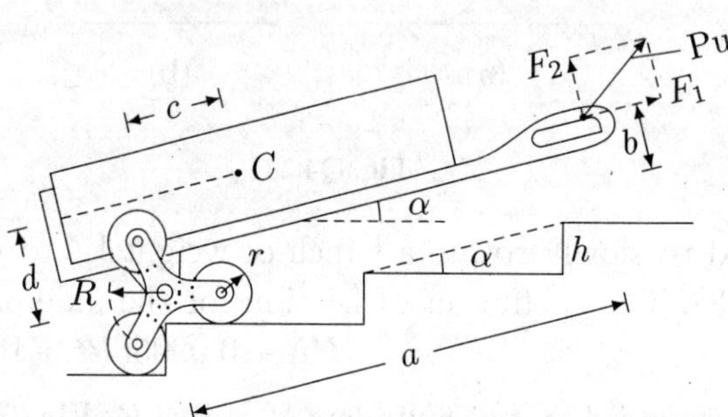

Fig. P4.23

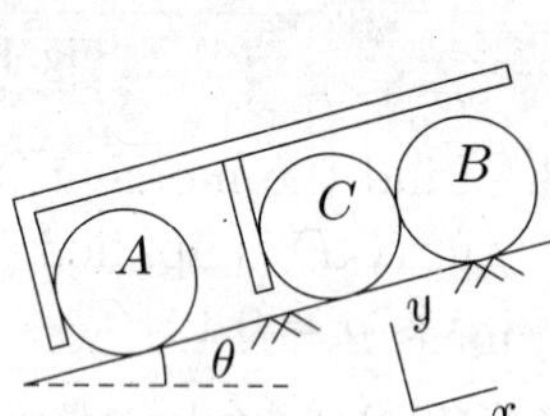

Fig. P4.24

25. The reel of a telephone cable has mass m. Its shaft, of radius r, is supported in 'V' blocks at its ends (Fig. P4.25) and the coefficient of friction between them is μ. Find P for impending rotation of the reel. [$\mu mgr/[(1+\mu^2)R\sin\alpha - \mu^2 r]$]

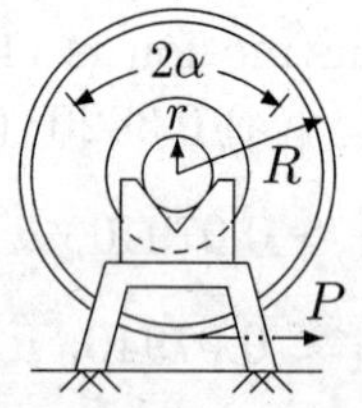

Fig. P4.25

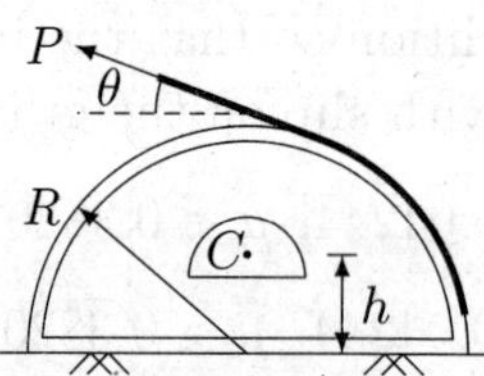

Fig. P4.26

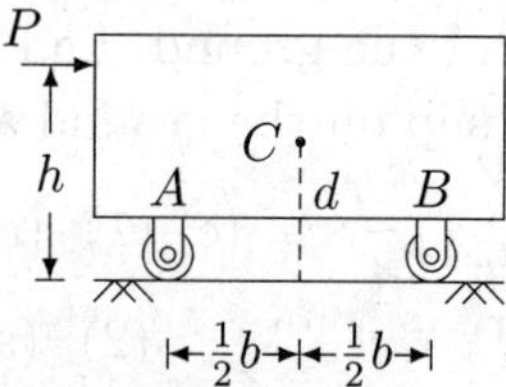

Fig. P4.27

26. Find the largest angle θ at which the cable can tow the cylindrical block, with kinetic coefficient of friction μ_k (Fig. P4.26), at uniform speed without tipping. The block has mass m and centre of mass C. [$\cos^{-1}\mu_k$]

27. (a) A trolley of mass m and centre of mass at C, is mounted on four casters (Fig. P4.27). The coefficient of friction between a locked caster and the floor is μ. Find the range of h for which the trolley will not tip when pushed with force P for the following cases: (i) all casters are locked, (ii) only casters B are locked, (iii) only casters A are locked. (b) Find P for impending motion for the three cases if (1) $h = b/3\mu$ and (2) $h = 2b/3\mu$.

[(a) (i), (ii) $h < b/2\mu$ (iii) $\forall h$ (b) (1) $(1,\ \frac{3}{4},\ \frac{3}{8})\mu mg$ (2) $(\frac{3}{4},\ \frac{3}{4},\ 0.3)\mu mg$]

28. A roller of weight W is at rest under a plate OCD, of weight $\frac{1}{2}W$ with centre of mass C, which is pinned at O (Fig. P4.28). A force P is applied such that there is impending motion of the roller with no slip at B and impending slip at A. The coefficient of friction is μ. Find the restriction on d so that the impending motion is of the given kind. $[d < 2R/3]$

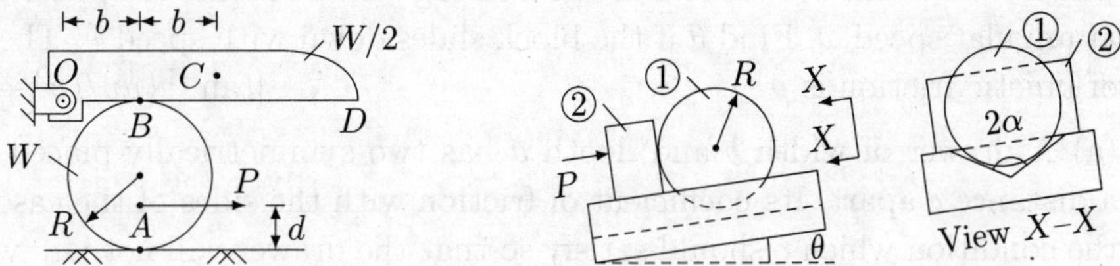

Fig. P4.28

Fig. P4.29

29. A force P is applied to a block of mass $2m$ in contact with a spherical ball of mass m resting in an inclined 'V' block (Fig. P4.29). (a) Find the least force P required for impending motion upwards. The coefficient of friction for all surfaces is μ. (b) Find the minimum θ for which the ball has impending translation.

$[(a)\ mg[(1-\mu^2)\sin\theta/(1-\mu/\sin\alpha)+2(\mu\cos\theta+\sin\theta)]$ if $\tan\theta < \dfrac{1-\mu/\sin\alpha}{1-\mu}$, else $mg[3\sin\theta+2\mu\cos\theta+\mu\cos\theta(1-\mu\sin\alpha)/(1-\mu)\sin\alpha]$ (b) $\tan^{-1}\dfrac{1-\mu/\sin\alpha}{1-\mu}]$

30. A tripod consists of uniform rigid legs, each of length L and mass m pinned together at the top to a disc of radius $r \ll L$, so that $r \simeq 0$. It supports a weight W hanging from the centre of the disc. The coefficient of friction with the floor is μ. The feet form an equilateral triangle. Find the greatest length of the side of this triangle for no motion of the tripod. $[2\sqrt{3}\mu L[4\mu^2+(3mg+2W)^2/(3mg+W)^2]^{-1/2}]$

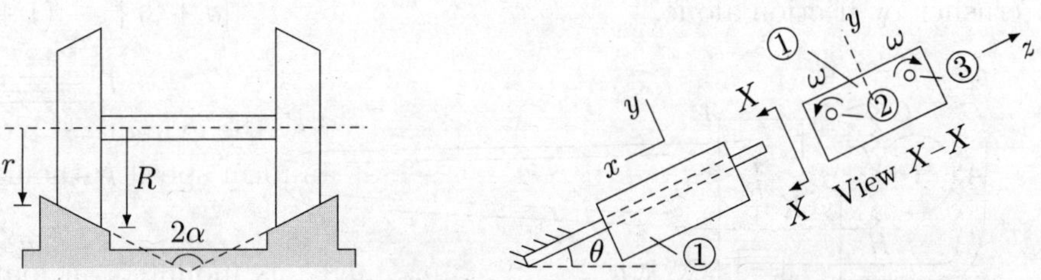

Fig. P4.31

Fig. P4.32

31*. A system consisting of a pair of wheels, welded to an axle, and of weight W, stands on longitudinal rails shaped to fit the conical treads of the wheels (Fig. P4.31). The coefficient of friction is μ. Assume uniform distribution of the normal force on the contact line. (a) Find the least force applied horizontally and longitudinally through the centre of the axle to cause impending motion. (b) Find the least axial couple applied to the axle to cause impending motion.

$[(a)\ \mu W[\sqrt{2(R^2+r^2)}-R-r]/(R-r)\sin\alpha \quad (b)\ \mu W(R-r)/4\sin\alpha]$

32*. Block ① of mass m slides down two parallel inclined shafts ② and ③ (Fig. P4.32) which have radius r each and are geared together to rotate in opposite directions at angular speed ω. Find θ if the block slides down with speed v. The coefficient of kinetic friction is μ. $[\tan^{-1}\{\mu v/(v^2+\omega^2r^2)^{1/2}\}]$

33^d. (a) A drawer of width b and depth a has two symmetrically placed handles at a distance c apart. Its coefficient of friction with the sides of the case is μ. Find the condition which c should satisfy so that the drawer will not jam when pulled by one handle (Fig. P4.33a). (b) A lift cage supported by two counter-weights (Fig. P4.33b) is to be designed so that when one of the supporting ropes breaks, the window jams. The coefficient of friction is μ. Find the design restriction on distance d. $[(a)\ c < a/\mu \quad (b)\ d > a/\mu]$

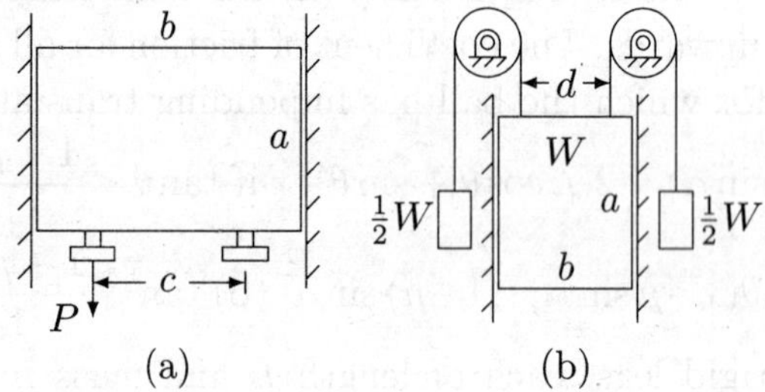

Fig. P4.33

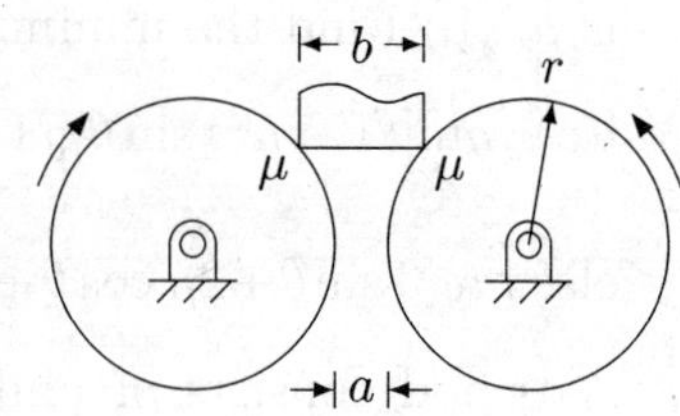

Fig. P4.34

34. The maximum size of the crushed aggregate in a stone crusher (Fig. P4.34) is a. Find the maximum thickness b of the stones which may be pulled through the crusher by friction alone. $[a+2r\{1-(1+\mu^2)^{-1/2}\}]$

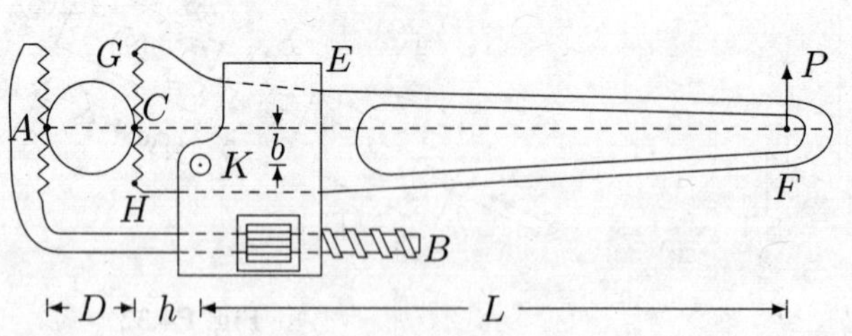

Fig. P4.35

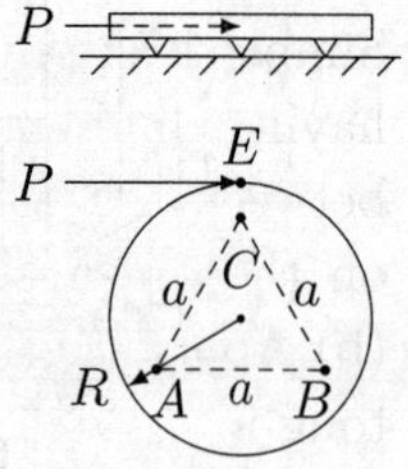

Fig. P4.36

35^d. (a) Show that the Stilson wrench in Fig. P4.35 would be self-locking (no slip) if the coefficient of friction μ with the pipe satisfies $\mu \geq b(D+h+L)/(h+L)(D+h)$, where D is the maximum size of the pipe. (b) Rework part (a) for the usual case in which the face GH is at an angle $\pi/2 + \alpha$ with AC.

$$[\text{(b)}\ \mu \geq \tan(\theta - \alpha) \ \text{ with } \ \tan\theta = \frac{(L + h + D)b/(h + D) - 0.5D \sin\alpha}{L + h + 0.5D(1 - \cos\alpha)}]$$

36. A thin plate of weight W (Fig. P4.36) is supported on the floor at three points A, B, C. Find the force P applied for impending motion. The coefficient of friction is μ. Assume equal normal reactions. In particular, find P if A, B, C are on outer circle of plate with C at E. $[\frac{1}{3}\mu W(1 - 2\sin\theta)$ with

$$\sin\left[\theta + \tan^{-1}\left\{\frac{a}{2R + a/\sqrt{3}}\right\}\right] = \frac{R - a/\sqrt{3}}{[a^2 + (2R + a/\sqrt{3})^2]^{1/2}},\ \frac{2}{3}\mu W]$$

37. Find the moment that should be applied by the engine to the rear wheels alone so that the bus can just ride over the kerb (Fig. P4.37). Its centre of mass is C. Find the minimum coefficient of friction for no slip at the rear wheels. The weight of the bus is 48 kN. [9 kN.m, 0.75]

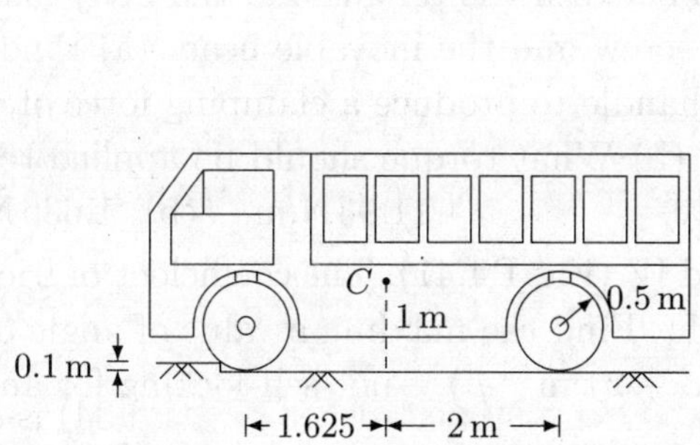

Fig. P4.37

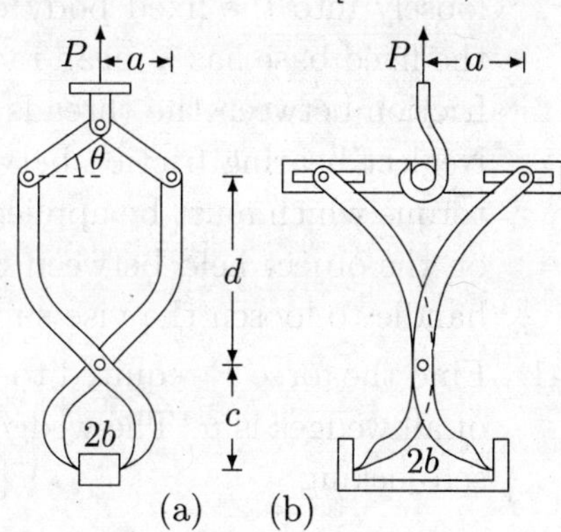

Fig. P4.38

38[d]. Simple friction grips have to be designed to lift anodised blocks of width $2b$ having different weights from the bath. The range of coefficient of friction μ between the blocks and the clamps is $\mu_1 \leq \mu \leq \mu_2$. (a) Find the restrictions on the dimensions of the clamp (Fig. P4.38a) so that the blocks do not slip. (b) Apply the results of part (a) to find the corresponding restriction for the tong in Fig. 4.54b. [(a) $c/(b + a + d\cot\theta) < \mu_1$ (b) $c/(a - b) < \mu_1$]

39. A panel of weight W is hinged about axis Ox by two small hinges A and B and is supported on an inclined plane at D (Fig. P4.39). Axis Ox is normal to the inclined plane. The centre of mass of the rectangular panel is at C. The coefficient of friction is μ. A force P is applied at E in a plane through E and D perpendicular to the panel. Find the value of P at which impending upward

motion of the panel occurs. Also, find the hinge reactions for $W = 100\,\text{N}, a = 0.5\,\text{m}, c = 0.25\,\text{m}, d = 0.5\,\text{m}, e = 0.1\,\text{m}, \mu = 0.3, \theta = \alpha = 30°, \beta = 60°$.

[$37.87\,\text{N}$, $-43.30\vec{j} + 18.43\vec{k}\,\text{N}$, $0.0022\vec{j} - 5.93\vec{k}\,\text{N}$]

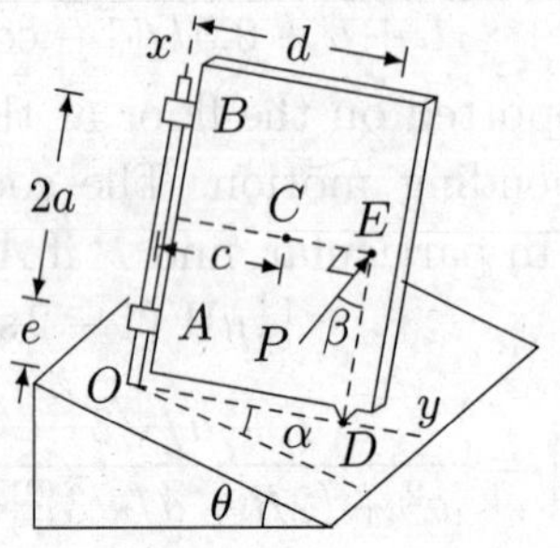

Fig. P4.39

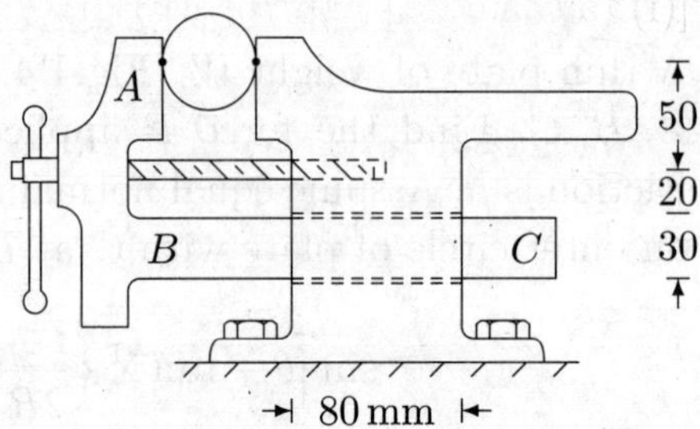

Fig. P4.40

40. A movable jaw A of a machinist's vise (Fig. P4.40) has a tongue BC which fits loosely into the fixed body of the vise. The single-threaded screw moving into the fixed base has a mean radius of 6 mm and a pitch of 5 mm. The coefficient of friction between the threads and also between the tongue and the body is 0.25. Neglect bearing friction between the screw and the movable head. (a) Find the torque which must be applied to the handle to produce a clamping force of 4 kN on the object held between the jaws. (b) What torque should be applied to the handle to loosen the vise subsequently? [11.93 N.m (b) 1.636 N.m]

41. Find the force P required to raise load W (Fig. P4.41). The coefficient of friction of all wedges is μ. The wedges are light. Find the maximum value of angle α for self-locking. [$W \tan^2(\alpha + 2\tan^{-1}\mu)$, not self-locking for any α]

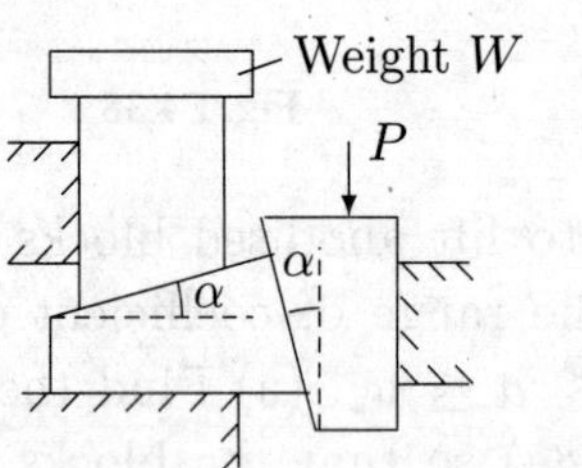

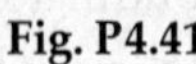

Fig. P4.41

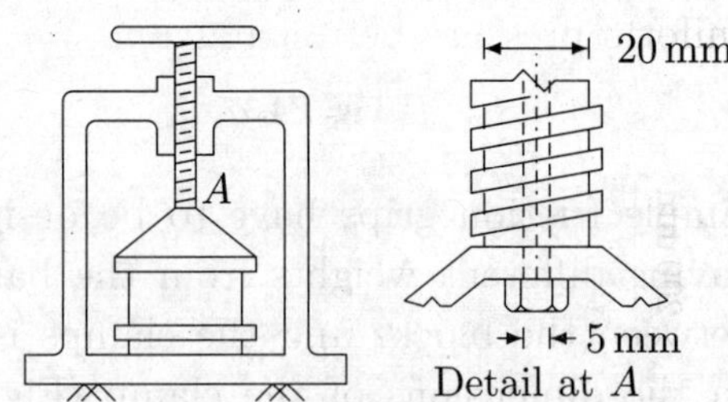

Fig. P4.42

42. The screw of a small press (Fig. P4.42) has a mean diameter of 30 mm and a double square thread with a pitch of 5 mm. The coefficient of friction μ between the nut and the screw is 0.25. The coefficient of friction μ_1 between the flat annular thrust bearing at A and the end of the screw is 0.2 and their surfaces are well-worn. (i) Find the torque M which should be applied to the hand wheel

(a) to produce a compressive force of 5 kN, and (b) to loosen the press from the 5 kN compression. (ii) Rework part (i) if the threads and the bearing are frictionless. (iii) Rework part (i) if $\mu = 0.25$, $\mu_1 = 0$.

[(i) (a) 33.69 (b) 16.76 (ii) (a) 7.958 (b) 7.958 (iii) (a) 27.44 (b) 10.51 N.m]

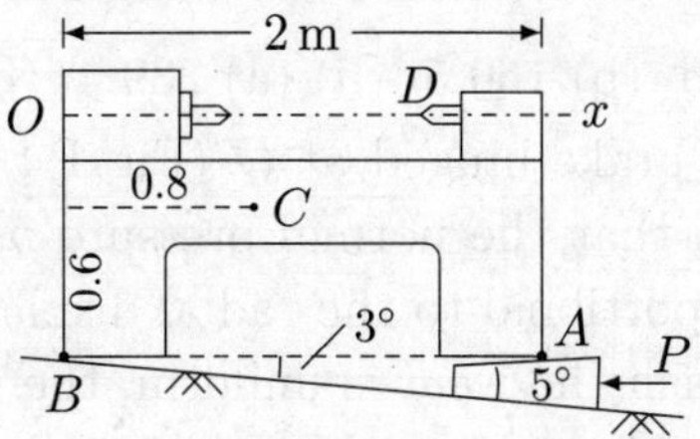

Fig. P4.43

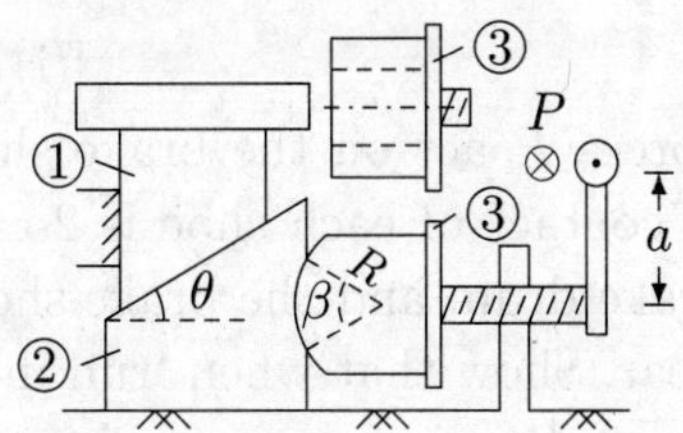

Fig. P4.44

43. A 2000 kg lathe (Fig. P4.43) is to be levelled by a steel wedge at A. The coefficient of friction between the wedge and its contacts is 0.2. (a) Find the force P to just raise A. The centre of mass of the lathe is at C. (b) Find the minimum coefficient of friction at B to ensure no sliding at B in the absence of a ledge at B. (c) Is the system with the wedge self-locking? Axis OD is horizontal.

[(a) 3.859 kN (b) 0.1044 (c) Yes]

44. Block ① of weight W (Fig. P.4.44) is to be raised by applying circumferential force P to the handle. The coefficient of friction is 0.3 at all surfaces of various blocks and it is 0.2 for the screw. The lead of the screw is 1 cm and its mean radius is 2 cm. Neglect weights other than W. Find P for slow upward movement of block ① for $\theta = 30°$. End ③ of the screw in contact with block ② is (a) spherical with radius 5 cm and $\beta = 60°$ and (b) annular with radii 5 cm and 2 cm. Assume uniform pressure between ② and ③. [(a) $0.02220W/a$ (b) $0.03359W/a$]

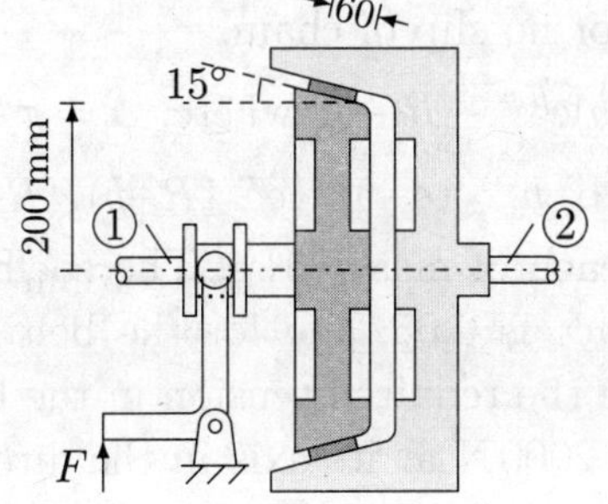

Fig. P4.45

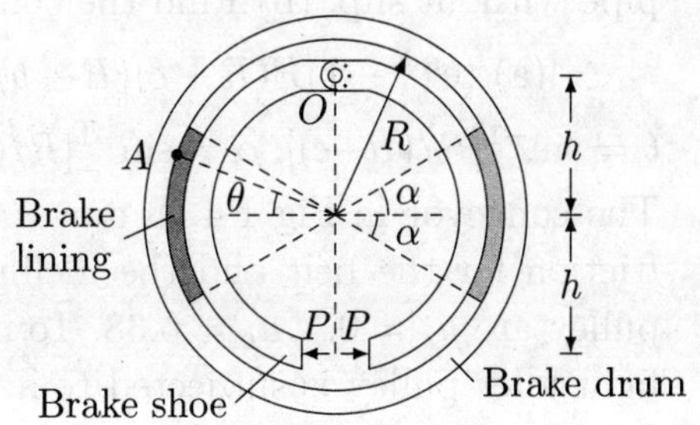

Fig. P4.46

45. A friction cone clutch connects two co-axial shafts (Fig. P4.45). The coefficient of friction is 0.25. The conical surface of the plate on the splined shaft ① is engaged with that of the plate fixed to the shaft by operating a bell-crank lever.

The lever exerts an axial force of 0.8 kN on the plate. (i) Find the torque that can be transmitted (a) if the contact pressure is uniform and (b) if the wear is uniform. Comment on the effect of the pattern of distribution of contact pressure on the torque. (ii) Rework part (i) for a semi-cone angle of 10° with the same radii of the contact surface. Comment on the effect of this angle on the torque.

[(i) (a) 160.6 (b) 160.5 (ii) (a) 239.4 (b) 239.3 N.m]

46*. Forces P act on the brake shoes of a brake hinged at O (Fig. P4.46). The arc of contact of each shoe is 2α. Assume that the normal pressure p between the brake drum and the brake shoe is proportional to the radial displacement after wear. Show that when 'run in', i.e., when the wear is uniform, the pressure p at point A varies as $\cos\theta$. Show that breaking torque M_1 provided by one shoe is given by $2hP/M_1 = h(\alpha + \sin\alpha\cos\alpha)/2\mu R\sin\alpha \pm 1$, where μ is the coefficient of friction. The sign in $\pm$ depends on the direction of rotation. Does the total braking torque from both shoes depend on the direction of rotation? [No]

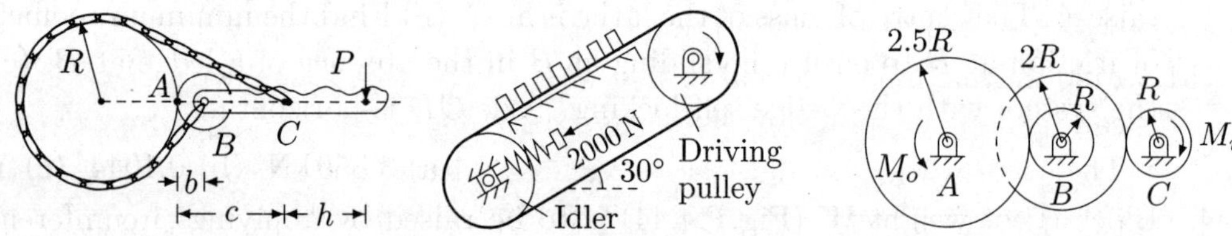

Fig. P4.47 Fig. P4.48 Fig. P4.49

47*. (a) A roller chain is connected to the handle at B and C (Fig. P4.47). The handle is pivoted to a small roller which rests against the pipe. A force P is applied to the handle. The coefficient of friction between the chain and the pipe is μ. (a) Find the minimum torque that can be transmitted by the grip to the pipe without slip. (b) Find the condition for no slip of chain.

[(a) $(e^{\mu\psi} - 1)P(R+c)(R+b)/[(R+b)e^{\mu\psi} - R - c]$ where, $\psi = \pi + \theta + \alpha$, $\theta = \sin^{-1}[R/(R+c)]$, $\alpha = \sin^{-1}[R/(R+b)]$ (b) $h > (c-b)/\{e^{\mu\psi}(R+b)/(R+c)-1\}$]

48. The conveyor in Fig. P4.48 moves boxes, each of mass 25 kg. The coefficient of friction for the belt and the inclined surface is 0.05, and for the belt and the pulley are $\mu_s = 0.4$, $\mu_k = 0.38$. To maintain the required tension in the belt, the light idler pulley is subjected to a force of 2000 N at its axis in the direction of the conveyor. Find the maximum tension in the belt and determine whether the boxes move up if their number on the conveyor is (a) 12, (b) 20, and (c) 24.

[(a) 2599 N, move up (b) 3300 N, in equilibrium (c) 3300 N, move down]

49. (a) For the compound gear train (Fig. P4.49), find the relation between the input and output torques M_i and M_o. The coefficient of friction for the axles of radius

r is μ. Neglect radial component of the force on the teeth in mesh. (b) Rework part (a) for $AB \perp BC$. [(a) $M_o = M_i(2.5R - r_f)(2R - r_f)/(R + r_f)^2$

(b) $\dfrac{2M_i R}{R + r_f} - \dfrac{M_o R}{2.5R - r_f} = r_f \left[\dfrac{M_i^2}{(R + r_f)^2} + \dfrac{M_o^2}{(2.5R - r_f)^2} \right]^{1/2}$]

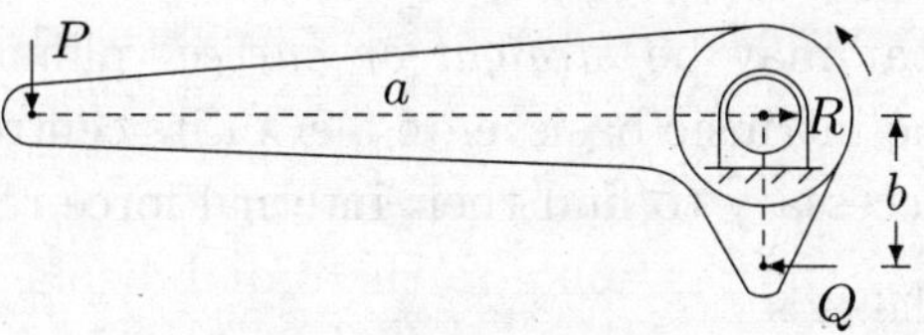

Fig. P4.50

Fig. P4.51

50. Forces P and Q are applied to a bell-crank lever (Fig. P4.50). The radius of the journal bearing is R and its coefficient of friction is μ. Find the force P needed to cause impending motion in the given direction for $b/a = 0.2$, $R/b = 0.25$, $\mu = 0.15$. Compare with the frictionless case. [$0.2076Q$, 3.8% more]
51. Find the maximum inclination θ at which a car (Fig. P4.51) with centre of mass C can remain in equilibrium under braking moment M_0 applied on the rear wheels. The coefficients of rolling resistance for the rear and front wheels are e_1 and e_2. [$\sin^{-1}[(c_1c_2+c_3\sqrt{c_1^2+c_3^2-c_2^2})/(c_1^2+c_3^2)]$ with $c_1 = 1-h(e_1-e_2)/R(b_1+b_2+e_1-e_2)$, $c_2 = M_0/mgR$, $c_3 = (e_2b_1 + e_1b_2)/R(b_1 + b_2 + e_1 - e_2)$]

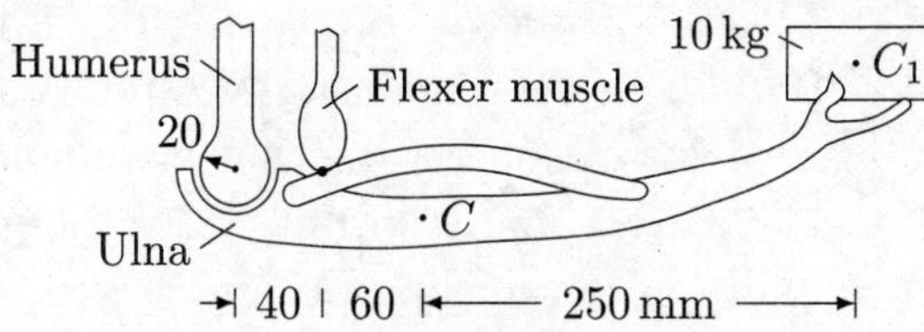

Fig. P4.52

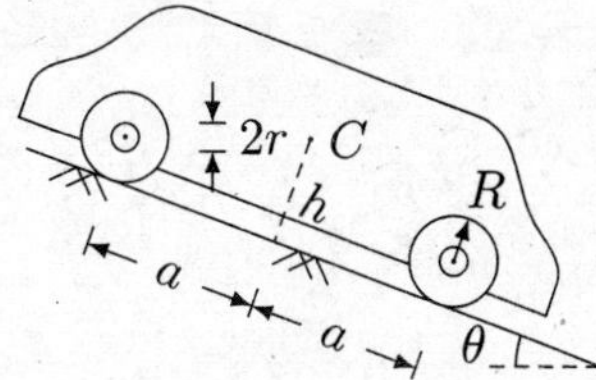

Fig. P4.53

52. A man holds a 10 kg block with the elbow at right angles (Fig. P4.52). The mass of forearm and the hand is 2 kg, with centre of mass at C. (a) Find the forces in the flexor muscle and in the humerus against the ulna to support the load. Neglect friction. (b) If the radius of curvature of the joint is 20 mm and its coefficient of friction is 0.016, find the force in the flexor muscle (i) to just raise the load, and (ii) to just hold it. [(a) 907.4 N, 789.7 N (b) (i) 913.8 N (ii) 901.2 N]
53. A toy car with light wheels has axles of radius r (Fig. P4.53). C is the centre of mass. The coefficient of friction between the axle and the wheels is μ. The car is designed to roll down a slope of θ. Find the diameter of wheels if their coefficient of rolling friction is (a) zero, and (b) e. [(b) $2(\mu r/\sqrt{1+\mu^2} + e\cos\theta)/\sin\theta$]

C. TRUSSES, CABLES AND BEAMS

4.13 Internal Force Resultants in Bars

A *bar* is a member whose transverse cross-sectional dimensions are much smaller ($< 0.1L$) than the axial length L (Fig. 4.13a), e.g., truss member, beam, column, shaft, belt, rope, cable and arch. A bar may be *straight or curved* (planar or spatial). A convenient basis for a bar is the intrinsic basis $\vec{e}_t$, $\vec{e}_n$, $\vec{e}_b$ of its centroidal axis (Fig. 4.13b). For designing bars, it is necessary to find their internal force resultants.

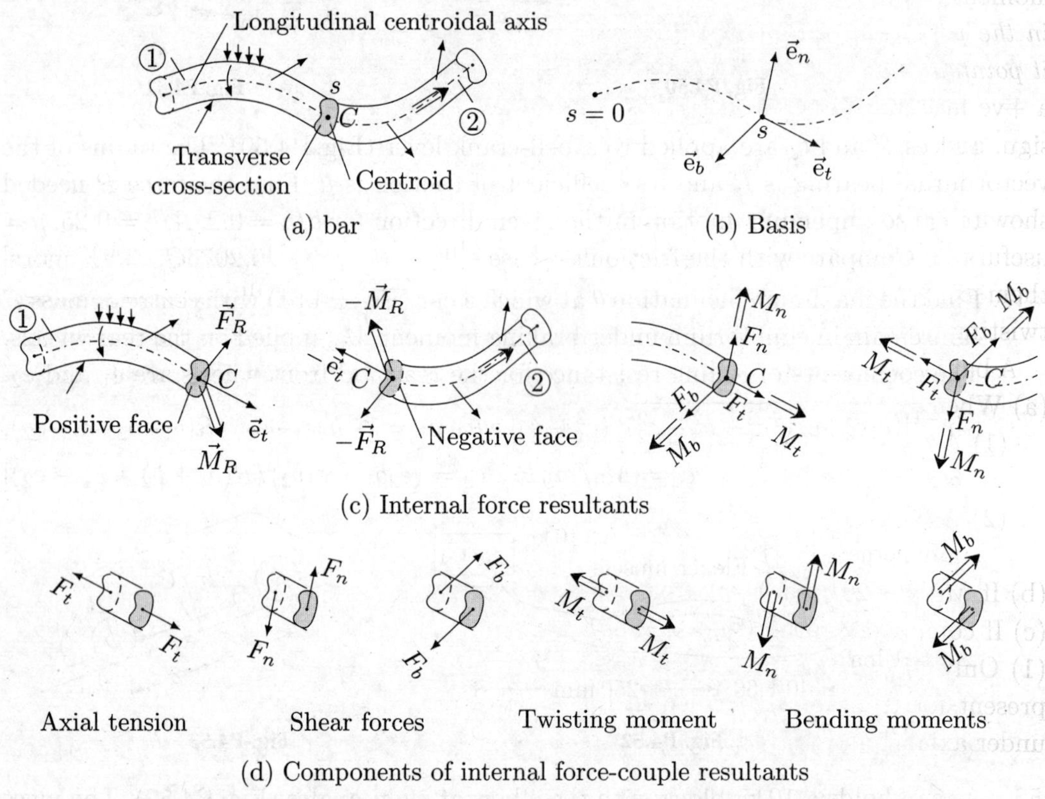

Fig. 4.13

Consider an internal cross-section of a bar, shown as shaded in Fig. 4.13a, which divides it into parts ① and ②. Choose the centroidal axis coordinate s. The FBDs of the two parts are shown in Fig. 4.13c. The outward normal to the face of the cut cross-section of part ① is in the $+\vec{e}_t$ direction; it is called a *positive face.* The outward normal to the face of the cut cross-section of part ② is in the $-\vec{e}_t$ direction; it is called a *negative face.* Let the internal force resultant at the centroid C of the positive face be a resultant force $\vec{F}_R$ and an associated resultant moment $\vec{M}_R$ (also denoted by $\vec{C}_R$). The internal force resultant at the centroid C of the negative face will be a resultant force $-\vec{F}_R$ and an associated resultant moment $-\vec{M}_R$:

$$\begin{aligned}\vec{F}_{R|\,\vec{e}_t} &= F_t\,\vec{e}_t + F_n\,\vec{e}_n + F_b\,\vec{e}_b = -\vec{F}_{R|-\,\vec{e}_t}\\ \vec{M}_{R|\,\vec{e}_t} &= M_t\,\vec{e}_t + M_n\,\vec{e}_n + M_b\,\vec{e}_b = -\vec{M}_{R|-\,\vec{e}_t}\end{aligned} \tag{4.13}$$

Force components $F_t\,\vec{e}_t$ of $\vec{F}_{R|\,\vec{e}_t}$ and $-F_t\,\vec{e}_t$ of $\vec{F}_{R|-\,\vec{e}_t}$ are both pulls on the two faces of the cross-section. Hence, we adopt the following *sign convention* which is physically meaningful and consistent with eq. (4.13): a force component F_i $(i = t, n, b)$ or a moment component M_i is taken as *positive if while acting on a positive face, it points in the positive direction of the basis vectors and if while acting on the negative face, it points in the negative direction of the basis vectors.* Thus, the component acting on a +ve face and directed in the −ve direction of a basis vector is assigned a negative sign, and the one acting on a −ve face and directed in the +ve direction of the basis vector is also assigned a negative sign. The positive directions of these components are shown in Fig. 4.13d. For the design of bars, diagrams showing their variation with s are useful to locate sections where the maximum value of each or a critical combination of them exists. These diagrams are called normal force $F_t(s)$, shearing force $F_n(s), F_b(s)$, twisting moment $M_t(s)$ and bending moment $M_n(s), M_b(s)$ diagrams.

A bar is named according to the predominant components of internal force resultant:

(a) When F_t is predominant:

(1) An axially loaded straight member under tensile F_t is called a *tie* and under compressive F_t, it is called a *strut* or a *column.*

(2) A transversely loaded curved member under compressive F_t is called an *arch* and under only tensile F_t is called a flexible *cable.*

(b) If M_t is predominant in a bar, then it is called a *shaft* or a *torque bar.*

(c) If components other than F_t and M_t are predominant, then it is called a *beam:* (1) Only M_n, M_b are present in a *beam* for pure bending, (2) M_n, F_b and M_b, F_n are present for transversely loaded beam, and (3) F_t, M_b, M_n are predominant in a bar under axial compressive load and transverse load. Such a bar is called a *beam-column.*

4.14 Equations of Equilibrium of a Straight Bar

We replace the external loads on an element of length ds of the bar by statically equivalent loads distributed over the length ds of its centroidal axis (Fig. 4.14a): A distributed external force $\vec{q}\,(s)$ N/m and a distributed external moment $\vec{m}(s)$ N.m/m:

$$\vec{q} = q_t\,\vec{e}_t + q_n\,\vec{e}_n + q_b\,\vec{e}_b, \qquad \vec{m} = m_t\,\vec{e}_t + m_n\,\vec{e}_n + m_b\,\vec{e}_b \tag{4.14}$$

It may also be subjected to discrete external forces and couples such as force $\vec{F}_0$ and moment $\vec{M}_0$ at $s = s_0$. Let $\vec{F}_R(s)$, $\vec{M}_R(s)$ be the internal force and couple resultant for the positive face of the cross-section at point s of the centroidal axis. For an exact description, we should refer all quantities: s, $\vec{e}_t$, $\vec{e}_n$, $\vec{e}_b$, $\vec{q}$, $\vec{m}$, $\vec{F}_R$, $\vec{M}_R$ to the *deformed loaded bar*. If the displacements and deformations are small, the deformed configuration can be approximated to the undeformed one for the equations of equilibrium.

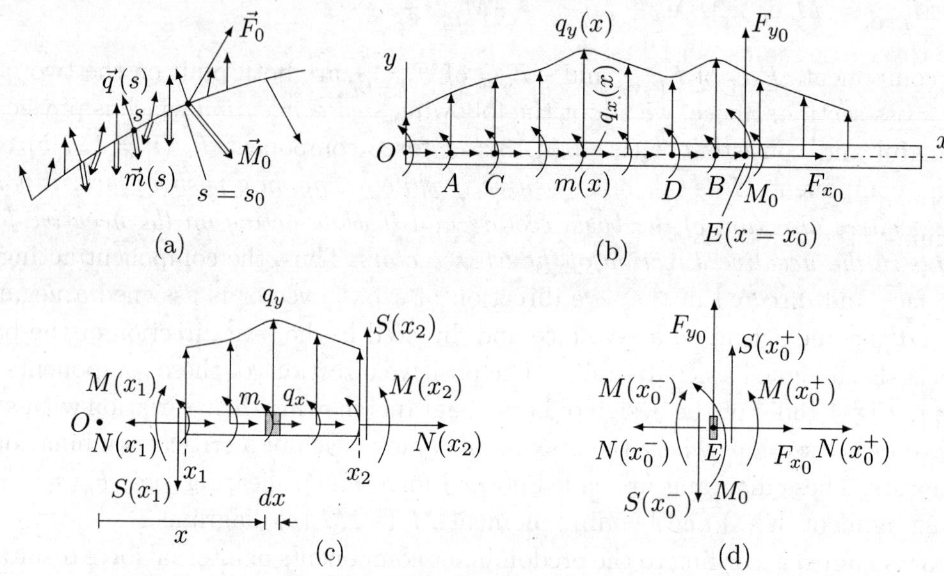

Fig. 4.14

Consider a straight bar, with the x-axis chosen along the centroidal axis, under a coplanar load (Fig. 4.14b) consisting of a distributed force and moment $\vec{q}(x)$, $\vec{m}(x)$:

$$\vec{q} = q_x\,\vec{\imath} + q_y\,\vec{\jmath}, \qquad \vec{m} = m\,\vec{k} \tag{4.15}$$

and a few discrete forces and moments such as force $F_{x0}\,\vec{\imath} + F_{y0}\,\vec{\jmath}$ and moment $M_0\,\vec{k}$ at $x = x_0$. On a positive face (with outward normal $\vec{n} = \vec{\imath}$), the internal force resultant $\vec{F}_{R|\vec{\imath}}$, $\vec{M}_{R|\vec{\imath}}$ and on a negative face (with outward normal $\vec{n} = -\vec{\imath}$), the internal force resultant $\vec{F}_{R|-\vec{\imath}}$, $\vec{M}_{R|-\vec{\imath}}$ are given by

$$\vec{F}_{R|\vec{\imath}} = N\vec{\imath} + S\vec{\jmath}, \qquad \vec{M}_{R|\vec{\imath}} = M\,\vec{k}, \qquad \vec{F}_{R|-\vec{\imath}} = -N\vec{\imath} - S\vec{\jmath}, \qquad \vec{M}_{R|-\vec{\imath}} = -M\,\vec{k}$$

where N, S, M are the normal force, shear force and bending moment. Consider the segment AB of the bar on which no external discrete force or moment acts. The FBD of its part CD from $x = x_1$ to $x = x_2$ is shown in Fig. 4.14c. Its equilibrium equations:

$$F_x = N(x_2) - N(x_1) + \int_{x_1}^{x_2} q_x\,dx = 0 \quad \Rightarrow \quad \int_{x_1}^{x_2} \left(\frac{dN}{dx} + q_x\right) dx = 0$$

$$F_y = S(x_2) - S(x_1) + \int_{x_1}^{x_2} q_y\,dx = 0 \quad \Rightarrow \quad \int_{x_1}^{x_2} \left(\frac{dS}{dx} + q_y\right) dx = 0$$

$$M_{Oz} = M(x_2) - M(x_1) + x_2 S(x_2) - x_1 S(x_1) + \int_{x_1}^{x_2} (x q_y + m)\, dx = 0$$
$$\Rightarrow \int_{x_1}^{x_2} \left(\frac{dM}{dx} + \frac{d(xS)}{dx} + x q_y + m \right) dx = 0$$

The three integrands must be zero, as the integrals are zero for arbitrary x_1, x_2:

$$\frac{dN}{dx} = -q_x, \qquad \frac{dS}{dx} = -q_y, \qquad \frac{dM}{dx} = -S - m \tag{4.16}$$

where eq. $(4.16)_2$ has been used to obtain eq. $(4.16)_3$. These are the equations of equilibrium of the bar where no external discrete force/moment acts. Eliminating S yields

$$\frac{d^2 M}{dx^2} = q_y - \frac{dm}{dx} \tag{4.17}$$

Consider an infinitesimal element of the bar from $x_0 - \varepsilon_1$ to $x_0 + \varepsilon_2$. Its equilibrium equations, in the limit $\varepsilon_1 \to 0,\ \varepsilon_2 \to 0$ when the contributions of distributed loads become zero (Fig. 4.14d), yield the *jump conditions of equilibrium* at x_0:

$$\begin{aligned} N(x_0^+) - N(x_0^-) + F_{x0} &= 0 \quad \Rightarrow \quad N(x_0^+) = N(x_0^-) - F_{x0} \\ S(x_0^+) - S(x_0^-) + F_{y0} &= 0 \quad \Rightarrow \quad S(x_0^+) = S(x_0^-) - F_{y0} \\ M_{Ez} = M(x_0^+) - M(x_0^-) + M_0 &= 0 \quad \Rightarrow \quad M(x_0^+) = M(x_0^-) - M_0 \end{aligned} \tag{4.18}$$

The jumps in N, S, M at x_0 are equal to the additive inverses of F_{x0}, F_{y0} and M_0.

For bars under axial loads only, the equilibrium equation is eq. $(4.16)_1$ and jump condition is eq. $(4.18)_1$. If the distributed axial force $q_x = 0$, then eq. $(4.16)_1 \Rightarrow N =$ constant. Such a situation exists in the two-force straight members used in trusses.

4.15 Trusses

Trusses are used in bridges, cranes, roof supports; cell-phone, radio, television and electric transmission line towers, etc. Truss design is based on forces in their members.

A bridge truss is shown in Fig. 4.15. Its members are built up of channels, angles and wide-flange I-sections, with or without cover plates. Their ends are connected to other members or to the foundation by bolting, riveting or welding to a gusset plate (Fig. 4.16a). Since the members are long compared to their lateral dimensions they bend easily and the gusset plate is able to exert very little rotational constraint on them. Thus, the action of the gusset plate on a member is modelled as only a force. This approximation is good if the centroidal axes of the connected members are concurrent. Hence, the

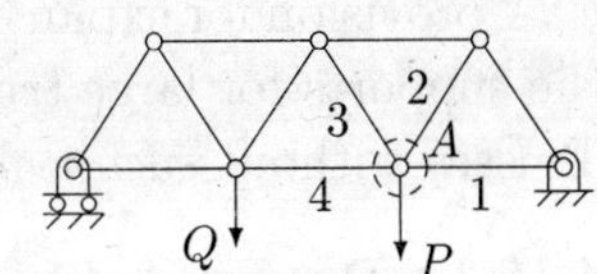

Fig. 4.15

joint is modelled as a smooth ball and socket joint at the point of concurrence for a three-dimensional truss (*space truss*) and as a smooth pin joint for a coplanar truss (*plane truss*) (Fig. 4.16b). The external loads on the truss, other than its self-weight, are transmitted to the joints by a system of floor beams and stringers (cables), such that the transmitted loads have very small moments about the joints. A truss is designed to support loads much greater than its own weight. Hence, the weight of members is usually neglected. For more accurate analysis, half the weight of each member is assigned to the joints connected to it and the member itself is treated as weightless.

Summarising, a framework formed by joining bars (members), so that their centroidal axes are *concurrent at the joints* and their *weight is negligible* compared to the *applied forces which act only at the joints*, is called a *truss*. Each member is joined to other members or to the foundation at only two joints which are idealised as *smooth hinged* joints for a plane truss and as *smooth ball and socket* joints for the space truss. Thus, all members are two-force members with the forces acting along the line joining their points of connection to the joints. Let F_i be the tensile force on the ith member at its ends. Its every cross-section is subjected to axial tensile force F_i. In an FBD of a joint, the forces due to members are shown as pulls on the joints as shown in Fig. 4.16c for joint A of the truss in Fig. 4.15. If the solution yields a negative value for F_i, then the actual force is compressive. A member is called a *tension member (a tie) or a compression member (a strut)* according as the forces at its ends are pulls or pushes.

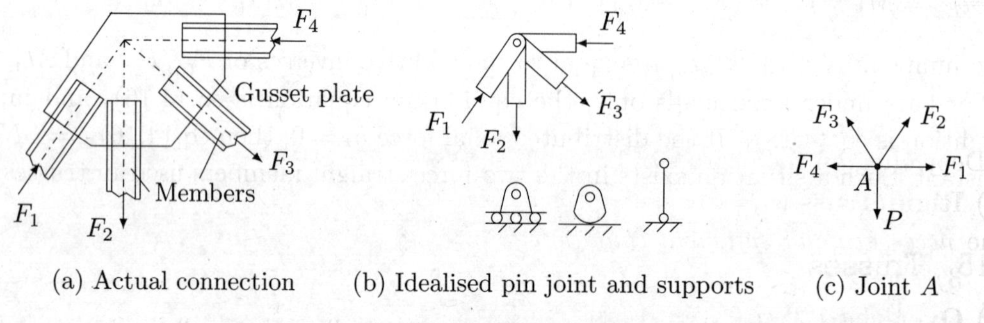

(a) Actual connection (b) Idealised pin joint and supports (c) Joint A

Fig. 4.16

A provision for expansion/contraction due to temperature change is usually made at the supports for large trusses by a roller, rocker or some kind of slip joint (Fig. 4.16b). Trusses without such a provision are statically indeterminate.

4.15.1 Degrees of freedom of a truss

Consider a truss made of b rigid bars, connected at j joints and with c foundation constraints. The number of coordinates needed to specify the positions of j joints is $3j$ ($2j$ for a plane truss). In the subsequent discussion in this section, simply replace $3j$ by $2j$ for coplanar trusses. The coordinates have to satisfy b constraint equations due to fixed lengths of b bars, e.g., $(x_A - x_B)^2 + (y_A - y_B)^2 + (z_A - z_B)^2 = L_{AB}^2$ for bar AB, and c constraint equations due to foundation constraints, e.g., $x_A = 0,\ y_C = 0$. Let n be the dof of the truss. If all the constraint equations are independent, then $n = 3j - b - c$, else $n > 3j - b - c$. Hence, $n \geq 3j - b - c$.

The $3j$ equilibrium equations for concurrent forces at j joints involve b member forces and c foundation reactions. The equilibrium equations for the whole truss do not furnish any additional independent equations. The *necessary condition* for a truss to be *statically determinate* is that the number of unknowns $b+c \leq 3j$. The equilibrium equations can be arranged as a matrix equation: $[A][F] = [P]$, where $[A]_{(3j)\times(b+c)}$ is the coefficient matrix, $[F]_{(b+c)}$ is the vector of unknown forces and $[P]_{(3j)}$ is the load vector. From linear algebra we know that a solution exists if $\text{rank}[A] = \text{rank}[A, P] = b + c$, which is the *necessary and sufficient condition for the truss to be statically determinate.*

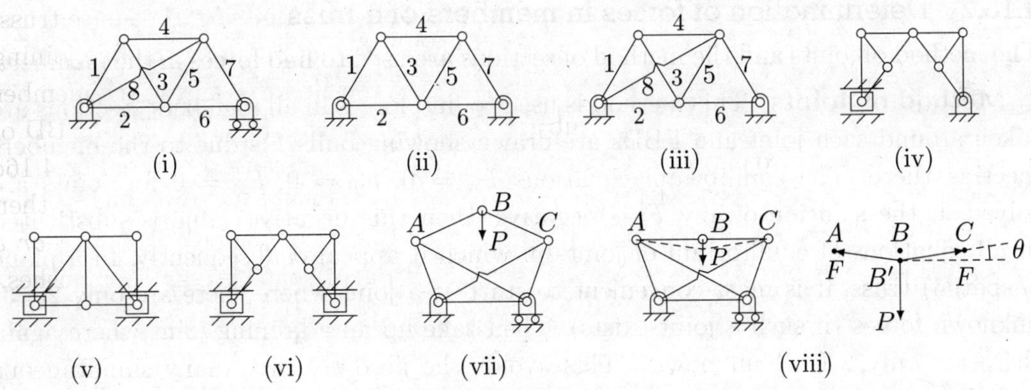

Fig. 4.17

Depending on the degrees of freedom, there are two types of trusses.

(1) Rigid Truss ($n = 0$)**:**

The *necessary and sufficient condition for a fully constrained rigid truss is*: $n = 0$.
$\Rightarrow\ 3j - b - c \leq n = 0$, i.e., $3j \leq b + c$ is a *necessary* but *not a sufficient condition.*

(a) Over-rigid Truss ($n = 0$ **and** $b + c > 3j$)**:**

A truss is called *over-rigid* if the removal of some member(s) or of some foundation constraint(s) still leaves it a fully constrained rigid truss $\Rightarrow$ $\exists$ at least one dependent constraint equation $\Rightarrow\ 0 = n > 3j - b - c\ \Rightarrow\ b + c > 3j\ \Rightarrow$ statically indeterminate. Such a member or foundation constraint is called *redundant* and the corresponding truss is called *internally redundant* or *externally redundant*, respectively. An over-rigid truss is safe-guarded against failure of a redundant member. The over-rigid trusses (i), (ii), and (iii) shown in Fig. 4.17 are respectively redundant internally (member 3 or 8), externally ($c = 4$) or both (member 3 or 8, and $c = 4$).

Redundancy of the type shown in truss (i) in Fig. 4.17 allows tensile bracing using cables for members 3 and 8, instead of bars. Since a cable cannot sustain compressive load, the problem becomes statically determinate, as either $F_3 = 0$ or $F_8 = 0$.

(b) Just-rigid Truss ($n = 0$ and $b + c = 3j$):

A truss is called *just-rigid* if the removal of any member or foundation constraint makes it a mechanism. The trusses (vii) and (viii) in Fig. 4.17 are just-rigid trusses, of which only (vii) is statically determinate. The equilibrium equations of joint B of truss (viii) can be satisfied only in the deformed configuration with large value of $F = P/2\sin\theta$.

(2) Mechanism ($n > 0$):

If $n > 0$ for a truss, then it is a mechanism in which some member(s) or the whole truss can move as a rigid body [if $c < 6$ for a space truss and $c < 3$ for a plane truss]. The trusses (iv) (v), and (vi) of Fig. 4.17 are mechanisms. A mechanism may be in equilibrium for given loads only in some specific configurations.

4.15.2 Determination of forces in members of a truss

The method of joints and the method of sections are used to find forces in the members.

1. Method of Joints This method is used to find forces in all members. Sections are taken around each joint and FBDs are drawn showing pulls F_is due to the members meeting there. The equilibrium equations $F_x = 0$, $F_y = 0$, $F_z = 0$ for joints are solved. If the solution of any F_i is negative, then this negative value is substituted in all equations of equilibrium of joints in which it appears subsequently. In a plane (/ spatial) truss, it is most convenient to start at a joint where there are only 2(/ 3) unknown forces (if such a joint exists). Then take up an adjoining joint where again there are only 2 (/ 3) unknowns. This avoids the need to solve many simultaneous equations. If the support reactions have been found from equilibrium equations of the whole truss, then at the end, 3 (/ 6) equations of equilibrium are redundant and provide a check on the calculations. For a cantilever type truss, the analysis may start from a joint at the free end without first finding the support reactions.

By writing the component of forces normal to the direction of two collinear members (if any) meeting at a joint (Fig. 4.18), the following useful results can be proved:

(a) If two pairs of collinear members meet at a joint where there is no load applied, then the forces in the collinear members are the same.

(b) If two collinear members form a joint with a third and the load P on the joint is in the direction of the third, then the forces in the collinear members are equal and the magnitude of the force in the third member is $|P|$.

(c) Same as the case (b) with $P = 0$. The third member is a zero force member.

(d) The forces in collinear members at a joint, with no other force on it, are equal.

(e) The forces in two non-collinear members meeting at a joint, with no external load applied on it, are each zero.

(f) The forces in 3 non-coplanar members meeting at a joint, with no external load applied on it, are each zero.

(g) If two collinear members and two more members meet at a joint without any external load acting on it and the members are non-coplanar, then the non-collinear members have zero force and the forces in collinear members are equal.

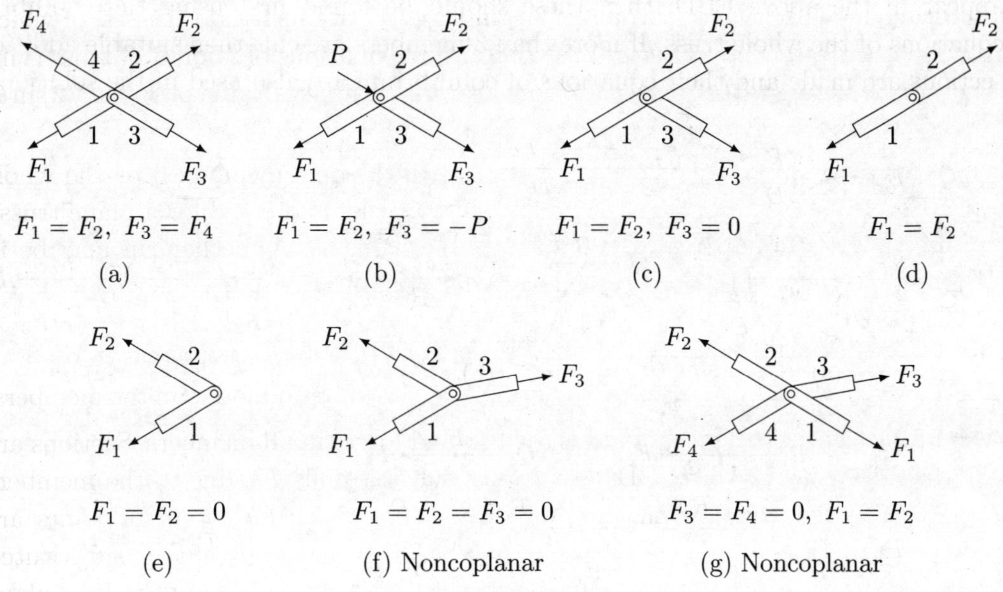

Fig. 4.18

For the curved member 4 in Fig. 4.19a, the simplest resultant of N, S, M across any section is the single force F_4 along BD (Fig. 4.19b). Thus, while drawing the FBD for a subsystem in which member 4 is partly included, the resultant of the force distribution across the section of member 4 should be shown as a pull F_4 along the line BD.

For the truss in Fig. 4.19a, R_3, R_1, R_2 (Fig. 4.19c) are found from $M_{A_z} = 0$, $F_x = 0$, $F_y = 0$ for the whole truss. $F_x = 0$, $F_y = 0$ for joints A, B, C, D, E in Fig. 4.19d, yield F_1 & F_2; F_3 & F_4; F_5 & F_6; F_7 & one check equation; and 2 check equations.

2. Method of Sections Very often in truss design it is known a priori that a few members are critical and hence the forces in these members need to be determined first. In such cases, the method of joints is laborious especially if the critical members are close to the centre of a large truss. Further, round-off errors accumulate with each calculation. In such cases, the method of sections is preferred.

The forces in just a few members of a truss can be computed by the method of 'sections', by taking a section across the truss so that it divides the truss into two parts by cutting the member of interest and a few more members. For a plane truss, the section should preferably not cut more than 3 bars with unknown forces and which are neither concurrent nor parallel. The FBD of a part of the truss to one side of this section (the side which involves smaller number of forces and simpler geometry) is drawn showing pulls on all members sectioned. The forces in the 3 members are found from the equilibrium equations: $M_{A_z} = 0$, $F_x = 0$, $F_y = 0$. It is often convenient to select A at the point of concurrence of many unknown member forces. If foundation reactions appear in the above FBD, then these should be found first using the equilibrium equations of the whole truss. If more than 3 members are cut, then suitable additional sections are made and their equations of equilibrium are also used in the solution.

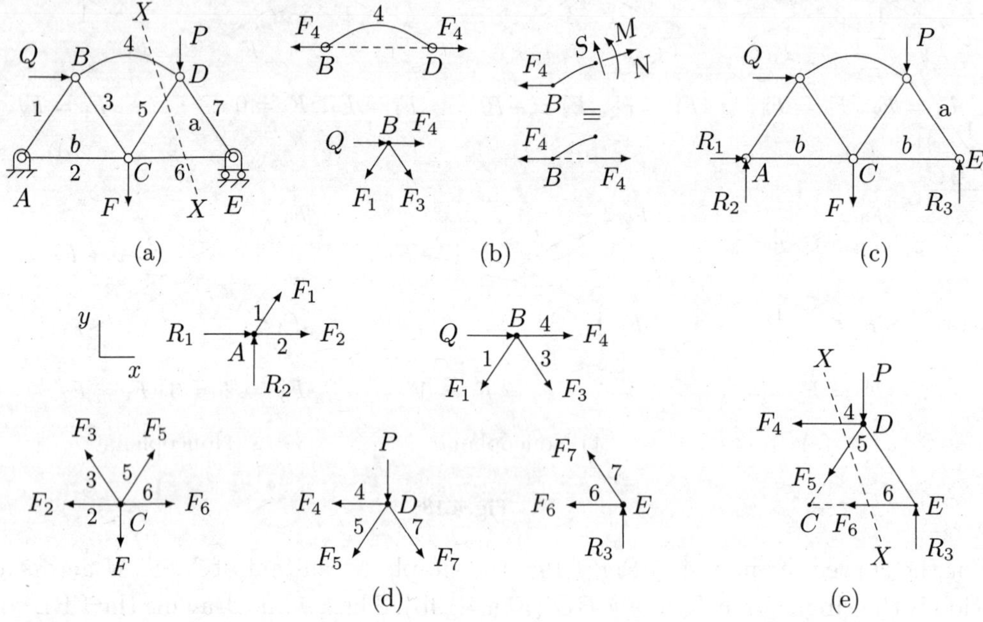

Fig. 4.19

For a space truss, the section should preferably not cut more than 6 bars with unknown forces, of which not more than 3 should be concurrent. These forces are found from the 6 equilibrium equations of the part of truss to one side of this section.

The forces in members 4 and 5 of the truss in Fig. 4.19a can be found by taking a section XX across it, drawing the FBD of the part to its right (Fig. 4.19e) and using $M_{C_z} = 0$ and $F_y = 0$ to yield F_4 and F_5.

4.15.3 Simple and complex trusses

A *simple space (or plane) truss* (Fig. 4.20a,b,c) is formed by a basic tetrahedron (or triangle) with 6 (3) bars and 4 (3) joints which is just rigid and successive addition of 3 non-coplanar (2 non-collinear) members to successively form the other $j-4$ $(j-3)$ joints and having $c = 6$ (3) independent foundation constraints. The number of members for a space truss is $b = 6 + 3(j-4) = 3j - c$ and for a plane truss it is $b = 3 + 2(j-3) = 2j - c$. The framework so formed is just-rigid at every stage of its formation and the foundation constraints prevent its motion as a rigid body. Hence, the b constraint equations provided by the fixed length of members are independent and the number of dof n of the truss is: $n = 3j - b - c = 0$ ($n = 2j - b - c = 0$). The simple plane truss in Fig. 4.20a is formed by triangle ABC with joints A, B, C followed by the other joints in the order D, E, G, F, H. The simple space truss in Fig. 4.20b is formed by tetrahedron $ABCD$ with joints A, B, C, D followed successively by joints E and F. A simple truss is not necessarily made of triangles (Fig. 4.20c).

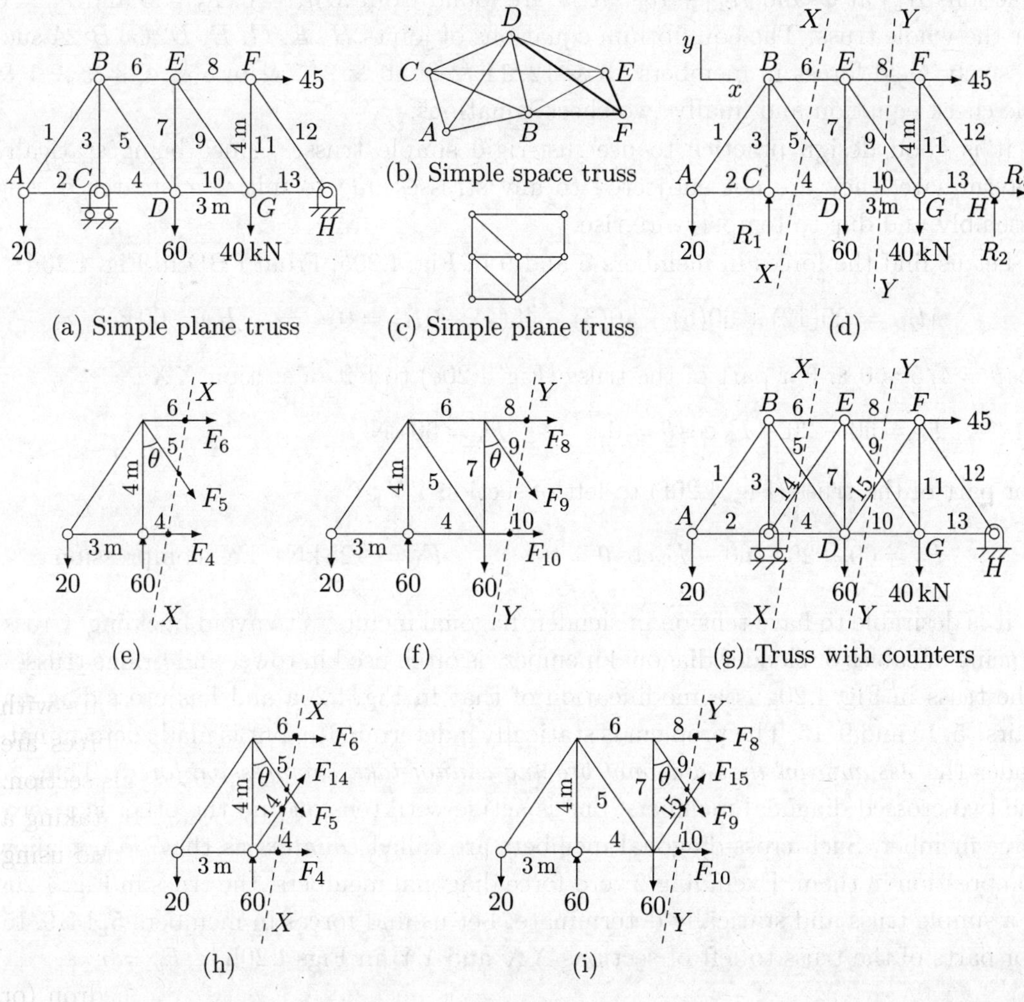

Fig. 4.20

The 6 (3) foundation reactions are determined from the 6 (3) equations of equilibrium of the whole truss. For the analysis of a simple truss by the method of joints, the joint which is the last to be formed during the construction of the truss is considered next. The forces in the 3 (2) members meeting at this joint are determined from its 3 (2) equations of equilibrium for the concurrent force system. The solution then proceeds in reverse-chronological order: the last but one joint that was formed is considered next and its 3 (2) equations of equilibrium are solved for the unknown forces in the 3 (2) members that created that joint and so on. This sequential procedure leads to the determination of the forces in all members. Hence, *a simple truss with* 6 (3) *independent foundation constraints is statically determinate.* For example, for the simple truss in Fig. 4.20a, the joints are formed in the order A, B, C, D, E, G, F, H. The reactions $R_1\vec{\jmath}$ at C and $R_2\vec{\jmath}+R_3\vec{\imath}$ at H are found from $M_{Hz}=0$, $F_x=0$ and $F_y=0$ for the whole truss. The equilibrium equations of joints H, F, G, E, D, C, B, A successively yield forces in members 13 & 12; 11 & 8; 10 & 9; 7 & 6; 5 & 4; 3 & 2; 1 & one check equation and finally two check equations.

It is good design practice to use just-rigid simple trusses, since being statically determinate, they are not subjected to any stresses due to initial mismatch during assembly and due to temperature rise.

Let us find the forces in members 5 and 9 in Fig. 4.20a. From FBD in Fig. 4.20d:

$$M_{Hz}=20(12)+60(6)+40(3)-45(4)-9R_1=0 \quad\Rightarrow\quad R_1=60\,\text{kN}$$

$\cos\theta=4/5=0.8$. For part of the truss (Fig. 4.20e) to left of section XX:

$$F_y=60-20-F_5\cos\theta=0 \quad\Rightarrow\quad F_5=50\,\text{kN}$$

For part of the truss (Fig. 4.20f) to left of section YY:

$$F_y=60-20-60-F_9\cos\theta=0 \quad\Rightarrow\quad F_9=-25\,\text{kN} \quad \text{i.e,. compression}$$

It is desirable to have tension in slender diagonal members to avoid buckling. Cross-bracing of cables or slender diagonal members is often used in tower and bridge trusses. The truss in Fig. 4.20g is a modification of that in Fig. 4.20a and has cross-diagonal pairs: 5, 14 and 9, 15. The problem is statically indeterminate, but is made determinate under the *assumption that diagonal bracing cannot take compressive forces.* Thus, of the two crossed diagonal members, one is active with tension and the other is a zero-force member. Such cross-diagonal members are called *counters* as they do not allow compression in them. Excluding 2 zero-force diagonal members, the truss in Fig. 4.20g is a simple truss and statically determinate. Let us find forces in members 5, 14, 9, 15. For parts of the truss to left of sections XX and YY in Figs 4.20h,i: $F_y=0\Rightarrow$

$$XX: \quad 60-20+(F_{14}-F_5)\cos\theta=0 \;\Rightarrow\; F_5-F_{14}=50 \;\Rightarrow\; F_{14}=0,\; F_5=50$$

$$YY: \quad 60-20-60+(F_{15}-F_9)\cos\theta=0 \;\Rightarrow\; F_{15}-F_9=25 \;\Rightarrow\; F_9=0,\; F_{15}=25$$

A just-rigid truss which is not a simple truss is called a *complex truss*, e.g., trusses in Fig. 4.21. We analyse it in terms of an unknown force in a member of our choice. The solution is carried out joint-wise for all members in terms of this unknown. At the last joint, this unknown can be evaluated and thereby all other member forces are found.

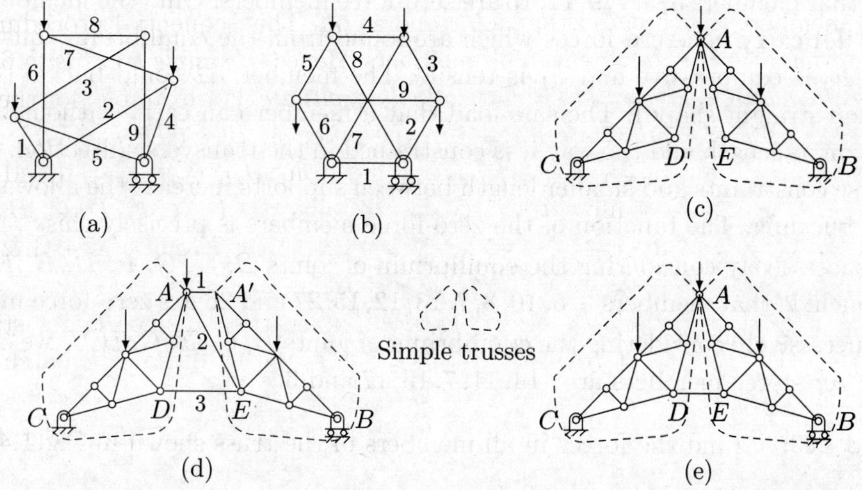

Fig. 4.21

A truss made up of several simple trusses connected together is called a *compound truss*, e.g., trusses in Fig. 4.21c,d,e. The truss in Fig. 4.21c consists of 2 simple trusses joined at A with 4 foundation constraints. Reactions R_1, R_2 at B are found from $M_{Cz} = 0$ for whole truss and $M_{A_z} = 0$ for truss ABE. The truss in Fig. 4.21d consists of 2 simple trusses joined by members 1, 2, 3 amounting to 3 additional internal forces. Reactions R_1, R_2 at C and R_3 at B are found from 3 equilibrium equations of the whole truss. Forces F_1, F_2, F_3 in members 1, 2, 3 are obtained from 3 equilibrium equations of truss ACD. The truss in Fig. 4.21e consists of 2 simple trusses joined by pin A and member DE yielding 3 more internal forces: two reaction components at pin A and force in bar DE. These are found from 3 equilibrium equations of truss ACD.

4.15.4 Examples on trusses

Example 4.32: Find zero-force members in the trusses in Fig. E4.32.

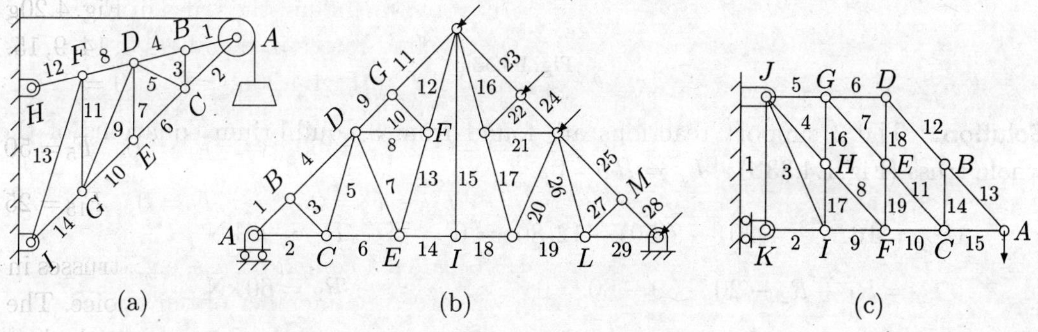

Fig. E4.32

Solution: (a) By successively considering the equilibrium of joints B, C, E, D, G, F, we infer that members $3, 5, 7, 9, 11, 13$ are zero-force members. Only the members along HA and IA carry non-zero forces which are found from the equilibrium equations of joint A. F_2 is compressive and F_1 is tensile. The member AI could buckle (bend to take a non-straight shape). The safe load that a member can carry without buckling depends on its length and the way it is constrained in the transverse direction. Greater transverse constraints and smaller length between supports increase the allowable load without buckling. The function of the zero-force members is precisely this.
(b) By successively considering the equilibrium of joints B, C, G, F, D, E, F, I, M, L, we conclude that members $3, 5, 10, 8, 7, 13, 12, 15, 27$ and 26 are zero-force members.
(c) By successively considering the equilibrium of joints B, C, E, G, H, I, we conclude that the zero-force members are: $14, 11, 7, 16, 17$ and 3.

Example 4.33: Find the forces in all members of the truss shown in Fig. E4.33a.

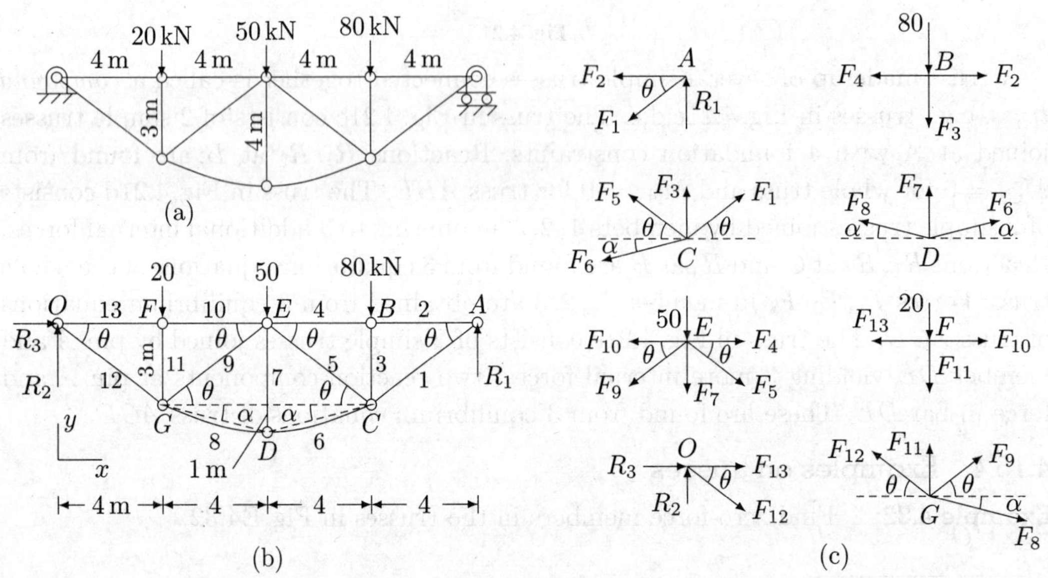

Fig. E4.33

Solution: The 3 support reactions are found from 3 equilibrium equations of the whole truss (Fig. E4.33b): $F_x = R_3 = 0$,

$$M_{Oz} = 16R_1 - 4(20) - 8(50) - 12(80) = 0 \quad \Rightarrow \quad R_1 = 90\,\text{kN}$$

$$F_y = R_1 + R_2 - 20 - 50 - 80 = 0 \quad \Rightarrow \quad R_2 = 60\,\text{kN}$$

The forces in the members are found by considering the equilibrium equations of joints in the sequence A, B, C, D, E, F, O, G so that at every stage, only two unknown member forces have to be determined.

$$\cos\theta = 4/(3^2+4^2)^{1/2} = 0.8 \qquad \Rightarrow \qquad \sin\theta = 0.6$$

$$\cos\alpha = 4/(4^2+1^2)^{1/2} = 0.9701 \qquad \Rightarrow \qquad \sin\alpha = 0.2425$$

FBDs of joints are given in Fig. E4.33c. The equilibrium equations of the joints are:

$$A: \quad F_y = R_1 - F_1\sin\theta = 0 \qquad \Rightarrow \qquad F_1 = 150\,\text{kN}$$

$$F_x = -F_2 - F_1\cos\theta = 0 \qquad \Rightarrow \qquad F_2 = -120\,\text{kN}$$

$$B: \quad F_x = F_2 - F_4 = 0 \qquad \Rightarrow \qquad F_4 = -120\,\text{kN}$$

$$F_y = -80 - F_3 = 0 \qquad \Rightarrow \qquad F_3 = -80\,\text{kN}$$

$$C: \quad F_x = F_1\cos\theta - F_5\cos\theta - F_6\cos\alpha = 0$$

$$F_y = F_3 + F_1\sin\theta + F_5\sin\theta - F_6\sin\alpha = 0$$

$$\Rightarrow \qquad 0.8F_5 + 0.9701F_6 = 120, \qquad 0.6F_5 - 0.2425F_6 = -10$$

$$F_5 = 25\,\text{kN}, \qquad F_6 = 103.1\,\text{kN}$$

$$D: \quad F_x = (F_6 - F_8)\cos\alpha = 0 \qquad \Rightarrow \qquad F_8 = 103.1\,\text{kN}$$

$$F_y = F_7 + (F_6 + F_8)\sin\alpha = 0 \qquad \Rightarrow \qquad F_7 = -50\,\text{kN}$$

$$E: \quad F_y = -50 - F_7 - (F_5 + F_9)\sin\theta = 0 \qquad \Rightarrow \qquad F_9 = -25\,\text{kN}$$

$$F_x = F_4 - F_{10} + (F_5 - F_9)\cos\theta = 0 \qquad \Rightarrow \qquad F_{10} = -80\,\text{kN}$$

$$F: \quad F_x = F_{10} - F_{13} = 0 \qquad \Rightarrow \qquad F_{13} = -80\,\text{kN}$$

$$F_y = -20 - F_{11} = 0 \qquad \Rightarrow \qquad F_{11} = -20\,\text{kN}$$

$$O: \quad F_x = R_3 + F_{13} + F_{12}\cos\theta = 0 \qquad \Rightarrow \qquad F_{12} = 100\,\text{kN}$$

$$F_y = R_2 - F_{12}\sin\theta = 0 \qquad \Rightarrow \qquad 60 - 100(0.6) = 0$$

The first check equation is satisfied. Joint G yields two more check equations:

$$F_x = F_8\cos\alpha + (F_9 - F_{12})\cos\theta = 0 \qquad \Rightarrow \qquad 0.0173 = 0$$

$$F_y = -F_8\sin\alpha + (F_9 + F_{12})\sin\theta + F_{11} = 0 \qquad \Rightarrow \qquad 0.0003 = 0$$

The check equations would be exactly satisfied if exact arithmetic is used instead of rounding off. Members $1, 5, 6, 8, 12$ are in tension as the forces in them are positive. Members $2, 3, 4, 7, 9, 10, 11, 13$ are in compression as the forces in them are negative.

Example 4.34: Find forces in members 5, 9 of truss (a) and in 7, 8, 12 of K-truss (b).

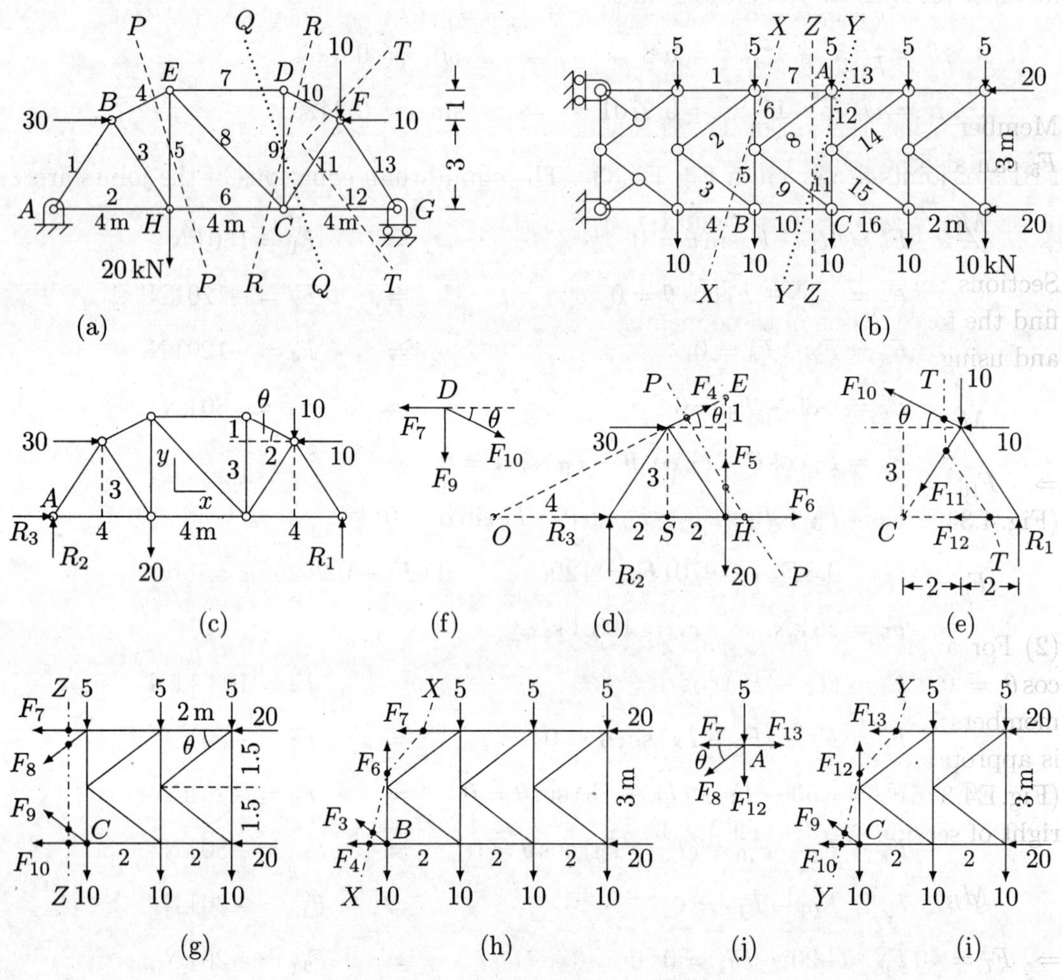

Fig. E4.34

Solution: (1) For θ in Fig. E4.34c: $\cos\theta = 2/(2^2+1^2)^{1/2} = 0.8944$, $\sin\theta = 0.4472$.
The reactions are found from the equilibrium equations of the truss (Fig. E4.34c):

$$M_{Az} = 12R_1 - 20(4) - 10(10) + 10(3) - 30(3) = 0 \quad \Rightarrow \quad R_1 = 20\,\text{kN}$$

$$F_x = R_3 + 30 - 10 = 0 \quad \Rightarrow \quad R_3 = -20\,\text{kN}$$

$$F_y = R_1 + R_2 - 20 - 10 = 0 \quad \Rightarrow \quad R_2 = 10\,\text{kN}$$

F_5 is found from equilibrium equations of the part on left of section PP (Fig. E4.34d):

$$M_{E_z} = 4F_6 + 30(1) + 4R_3 - 4R_2 = 0 \quad \Rightarrow \quad F_6 = 22.5\,\text{kN}$$

$$F_x = F_4 \cos\theta + R_3 + F_6 + 30 = 0 \quad \Rightarrow \quad F_4 = -36.34\,\text{kN}$$

$$F_y = F_4 \sin\theta + R_2 - 20 + F_5 = 0 \quad \Rightarrow \quad F_5 = 26.25\,\text{kN}$$

Member 5 has tension of 26.25 kN. Let F_4, F_6 intersect at O with $OS = 3\cot\theta = 6$. F_5 can also be found using

$$M_{O_z} = 8F_5 - 20(8) - 30(3) + 4R_2 = 0 \quad \Rightarrow \quad F_5 = 26.25\,\text{kN}$$

Sections through member 9, such as QQ and RR, cut 4 members. We have to first find the force in one of these members, separately. We find F_{10} by taking a section TT and using $M_{C_z} = 0$ for part of the truss on its right (Fig. E4.34e):

$$M_{C_z} = 4R_1 + 10(3) - 10(2) + F_{10}(3\cos\theta + 2\sin\theta) = 0$$

$\Rightarrow \quad F_{10} = -25.16\,\text{kN}$. F_9 is obtained by considering the equilibrium of joint D (Fig. 4.34f), where only two unknown forces act:

$$F_y = -F_{10}\sin\theta - F_9 = 0 \quad \Rightarrow \quad F_9 = 11.25\,\text{kN}$$

(2) For angle θ of the K-truss (Fig. E4.34g), $\tan\theta = 1.5/2 = 3/4 \Rightarrow \sin\theta = 0.6$, $\cos\theta = 0.8$. A section like ZZ (Fig. E4.34g) is not convenient as it intersects four members $7, 8, 9$ and 10 and no three of them are concurrent. The slanting section XX is appropriate as it cuts four members $7, 6, 3, 4$ with members $6, 3, 4$ concurrent at B (Fig. E4.34h). Consider the equilibrium equation $M_{B_z} = 0$ of part of the truss to the right of section XX (Fig. E4.34h):

$$M_{B_z} = 3F_7 + 20(3) - 15(2) - 15(4) - 15(6) = 0$$

$\Rightarrow F_7 = 40\,\text{kN}$. This is the procedure for the outer members of the K-truss. To find forces in the inner members, we first find forces in adjacent outer members and then consider the equilibrium of the joint where these meet. A section YY cuts members $13, 12, 9, 10$ of which $12, 9, 10$ are concurrent at C (Fig. E4.34i). The equilibrium equation $M_{C_z} = 0$ of part of the truss to the right of section YY (Fig. E4.34i):

$$M_{C_z} = 3F_{13} + 20(3) - 15(2) - 15(4) = 0 \quad \Rightarrow \quad F_{13} = 10\,\text{kN}$$

Consider the equilibrium of joint A (Fig. E4.34j):

$$F_x = F_{13} - F_7 - F_8\cos\theta = 0 \quad \Rightarrow \quad F_8 = (10 - 40)/0.8 = -37.5\,\text{kN}$$

$$F_y = -F_{12} - F_8\sin\theta - 5 = 0 \quad \Rightarrow \quad F_{12} = 37.5 \times 0.6 - 5 = 17.5\,\text{kN}$$

F_8 can also be found using $M_{C_z} = 0$ for system in Fig. E4.34g, after obtaining F_7.

Example 4.35: Find the forces in all members and the ground reactions for the space truss of Fig. E4.35a.

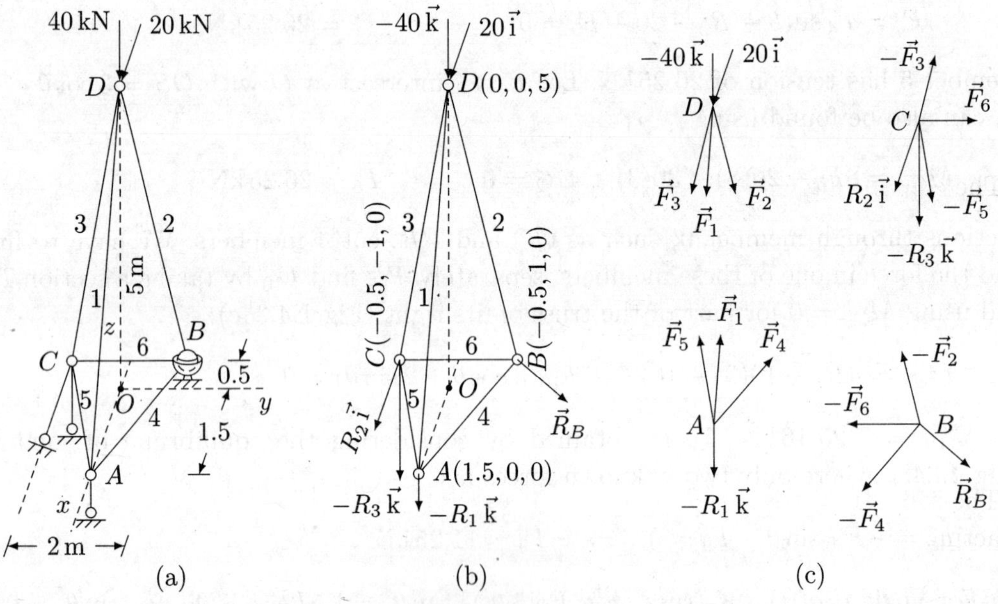

Fig. E4.35

Solution: The coordinates of points A, B, C, D are shown in Fig. E4.35b. We use the method of joints proceeding in the order D, A, C, B so that at every stage, there are only three unknowns to be found. The FBDs of various joints are shown in Fig. E4.35c. Consider the forces acting at joint D.

$$\vec{F}_1 = F_1\frac{\overrightarrow{DA}}{DA} = f_1\overrightarrow{DA} = f_1(1.5\vec{\imath} - 5\vec{k})\text{ kN}$$

$$\vec{F}_2 = f_2(\overrightarrow{DB}) = f_2(-0.5\vec{\imath} + \vec{\jmath} - 5\vec{k})\text{ kN}$$

$$\vec{F}_3 = f_3(\overrightarrow{DC}) = f_3(-0.5\vec{\imath} - \vec{\jmath} - 5\vec{k})\text{ kN}$$

The equilibrium equation of joint D yields

$$f_1(1.5\vec{\imath} - 5\vec{k}) + f_2(-0.5\vec{\imath} + \vec{\jmath} - 5\vec{k}) + f_3(-0.5\vec{\imath} - \vec{\jmath} - 5\vec{k}) + 20\vec{\imath} - 40\vec{k} = \vec{0}$$

$$\vec{\imath}: \quad 1.5f_1 - 0.5f_2 - 0.5f_3 + 20 = 0$$

$$\vec{\jmath}: \quad f_2 - f_3 = 0$$

$$\vec{k}: \quad -5f_1 - 5f_2 - 5f_3 - 40 = 0$$

The solution of these equations is $f_1 = -12$, $f_2 = 2$, $f_3 = 2$.

Consider the forces acting at joint A: force due to member 1 is

$$-\vec{F}_1 = -f_1(1.5\vec{i} - 5\vec{k})\ \text{kN}, \qquad \text{and} \qquad \vec{R}_A = -R_1\,\vec{k}$$

$$\vec{F}_4 = f_4(\overrightarrow{AB}) = f_4(-2\vec{i} + \vec{j})\ \text{kN}$$

$$\vec{F}_5 = f_5(\overrightarrow{AC}) = f_5(-2\vec{i} - \vec{j})\ \text{kN}$$

The equilibrium equation of joint A yields

$$f_4(-2\vec{i} + \vec{j}) + f_5(-2\vec{i} - \vec{j}) - R_1\vec{k} - f_1(1.5\vec{i} - 5\vec{k}) = \vec{0}$$

$$\vec{i}: \qquad -2f_4 - 2f_5 - 1.5f_1 = 0$$

$$\vec{j}: \qquad f_4 - f_5 = 0$$

$$\vec{k}: \qquad -R_1 + 5f_1 = 0$$

The solution of these equations is $R_1 = -60$, $f_4 = 4.5$, $f_5 = 4.5$. Consider the forces acting at joint C:

$$\vec{R}_C = R_2\vec{i} - R_3\vec{k}, \qquad \vec{F}_6 = f_6(\overrightarrow{CB}) = f_6(2\vec{j})$$

$$-\vec{F}_3 = -f_3(-0.5\vec{i} - \vec{j} - 5\vec{k}), \qquad -\vec{F}_5 = -f_5(-2\vec{i} - \vec{j})$$

The equilibrium equation of joint C yields

$$R_2\vec{i} - R_3\vec{k} + 2f_6\vec{j} - f_3(-0.5\vec{i} - \vec{j} - 5\vec{k}) - f_5(-2\vec{i} - \vec{j}) = \vec{0}$$

$$\vec{i}: \qquad R_2 + 0.5f_3 + 2f_5 = 0$$

$$\vec{j}: \qquad 2f_6 + f_3 + f_5 = 0$$

$$\vec{k}: \qquad -R_3 + 5f_3 = 0$$

The solution of these equations is $R_2 = -10$, $R_3 = 10$, $f_6 = -3.25$. The forces acting at point B are $\vec{R}_B$ and

$$-\vec{F}_6 = -2f_6\vec{j}, \qquad -\vec{F}_4 = -f_4(-2\vec{i} + \vec{j}), \qquad -\vec{F}_2 = -f_2(-0.5\vec{i} + \vec{j} - 5\vec{k})$$

The equilibrium equation of joint B:

$$\vec{R}_B - 2f_6\vec{j} - f_4(-2\vec{i} + \vec{j}) - f_2(-0.5\vec{i} + \vec{j} - 5\vec{k}) = \vec{0}$$

$\Rightarrow$ $\vec{R}_B = -10\vec{i} - 10\vec{k}$ kN. The solution can be checked by considering the equilibrium of the complete truss (Fig. E4.35b):

$$\vec{F} = -40\vec{k} + 20\vec{i} - R_1\vec{k} + R_2\vec{i} - R_3\vec{k} + \vec{R}_B$$
$$= (20 - 10 - 10)\vec{i} + (-40 + 60 - 10 - 10)\vec{k} = \vec{0} \quad \text{is o.k.}$$

$$\vec{M}_B = \overrightarrow{BD} \times (20\vec{i} - 40\vec{k}) + \overrightarrow{BA} \times (-R_1\vec{k}) + \overrightarrow{BC} \times (R_2\vec{i} - R_3\vec{k})$$
$$= (0.5\vec{i} - \vec{j} + 5\vec{k}) \times (20\vec{i} - 40\vec{k})$$
$$+(2\vec{i} - \vec{j}) \times 60\vec{k} - 2\vec{j} \times (-10\vec{i} - 10\vec{k}) = \vec{0} \quad \text{is o.k}$$

$$F_1 = f_1(DA) = -12(1.5^2 + 5^2)^{1/2} = -62.64\,\text{kN}$$

$$F_2 = f_2(DB) = 2(0.5^2 + 1^2 + 5^2)^{1/2} = 10.25\,\text{kN}$$

$$F_3 = f_3(DC) = 2(0.5^2 + 1^2 + 5^2)^{1/2} = 10.25\,\text{kN}$$

$$F_4 = f_4(AB) = 4.5(2^2 + 1^2)^{1/2} = 10.06\,\text{kN}$$

$$F_5 = f_5(AC) = 4.5(2^2 + 1^2)^{1/2} = 10.06\,\text{kN}$$

$$F_6 = f_6(CB) = -3.25(2) = -6.5\,\text{kN}$$

$$\vec{R}_A = -R_1\vec{k} = 60\vec{k}\ \text{kN}, \qquad \vec{R}_B = -10\vec{i} - 10\vec{k}\ \text{kN}$$

$$\vec{R}_C = R_2\vec{i} - R_3\vec{k} = -10\vec{i} - 10\vec{k}\ \text{kN}$$

Members 2, 3, 4, 5 are under tension and 1, 6 are under compression.

Example 4.36: Find the forces in the members of the compound truss shown in Fig. E4.36a.

Solution: Let the reaction at A be $R_1\vec{i} + R_2\vec{j}$. From the symmetry of the truss: $R_2 = \frac{1}{2}(6P) = 3P$. The FBD of half of the truss is shown in Fig. E4.36b. R_1 is obtained using

$$M_{Ez} = R_2(3a) - P(2a) - Pa + R_1(\sqrt{3}a) = 0 \quad \Rightarrow \quad R_1 = -3.464P$$

The equilibrium equations of the joints (Fig. E3.36c) are:

$$A: \quad F_y = R_2 + F_1 \sin 30^\circ = 0 \qquad\qquad \Rightarrow \quad F_1 = -6P$$

$$F_x = R_1 - F_1 \cos 30^\circ - F_2 = 0 \qquad\qquad \Rightarrow \quad F_2 = 1.732P$$

$$B: \quad F_x = (F_1 - F_3 - F_4)\cos 30^\circ = 0$$

$$F_y = (F_3 - F_4 - F_1)\sin 30° - P = 0$$

$$\Rightarrow \quad F_3 = -5P, \qquad F_4 = -P$$

$$C: \quad F_5 = -P, \qquad F_6 = F_3 = -5P$$

$$D: \quad F_y = F_5 + F_4 \sin 30° + F_7 \sin 60° = 0 \quad \Rightarrow \quad F_7 = 1.732P$$

$$F_x = F_2 + F_4 \cos 30° - F_7 \cos 60° = 0 \quad \Rightarrow \quad 0 = 0 \quad \text{(o.k.)}$$

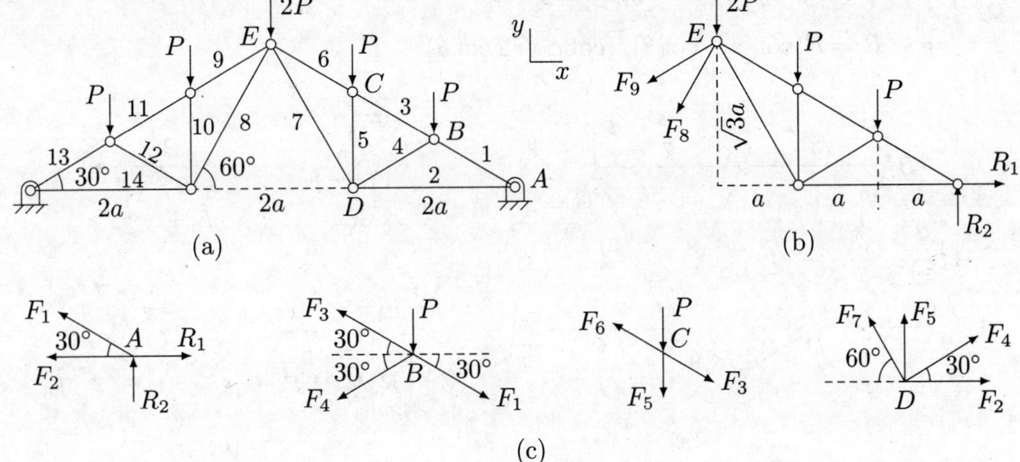

Fig. E4.36

The forces in the other half of the truss are obtained by symmetry:

$$F_8 = F_7 = 1.732P, \quad F_9 = F_6 = -5P, \quad F_{10} = F_5 = -P, \quad F_{11} = F_3 = -5P,$$

$$F_{12} = F_4 = -P, \quad F_{13} = F_1 = -6P, \quad F_{14} = F_2 = 1.732P$$

Example 4.37: Find the forces in the members of the complex truss shown in Fig. E4.37a.

Solution: Let $c = \cos\theta$, $s = \sin\theta$, $c_1 = \cos\alpha$, $s_1 = \sin\alpha$. The reaction at A is $R\vec{\jmath}$. By symmetry,

$$F_7 = F_4, \quad F_8 = F_5, \quad F_9 = F_6, \quad F_{10} = F_3, \quad F_{11} = F_1, \quad F_{12} = F_2$$

We successively solve for the member forces in terms of R using equilibrium equations of the joints A, B, C, D (Fig. E4.37b), till we are able to obtain R itself.

$$
\begin{aligned}
A:\quad & F_y = F_1 s + R = 0 && \Rightarrow \quad F_1 = -R/s \\
& F_x = F_2 + F_1 c = 0 && \Rightarrow \quad F_2 = cR/s \\
B:\quad & F_y = -P - (F_1 + F_3)s = 0 && \Rightarrow \quad F_3 = (R - P)/s \\
& F_x = F_4 + (F_3 - F_1)c = 0 && \Rightarrow \quad F_4 = c(P - 2R)/s \\
C:\quad & F_y = -(F_5 + F_8)s = 0 && \Rightarrow \quad F_5 = 0 \\
D:\quad & F_y = (F_3 + F_5)s - F_6 s_1 = 0 && \Rightarrow \quad F_6 = (R - P)/s_1 \\
& F_x = F_6 c_1 + (F_5 - F_3)c - F_2 = 0 && \\
\Rightarrow \quad & R = P(\cot\alpha - \cot\theta)/(\cot\alpha - 2\cot\theta) &&
\end{aligned}
$$

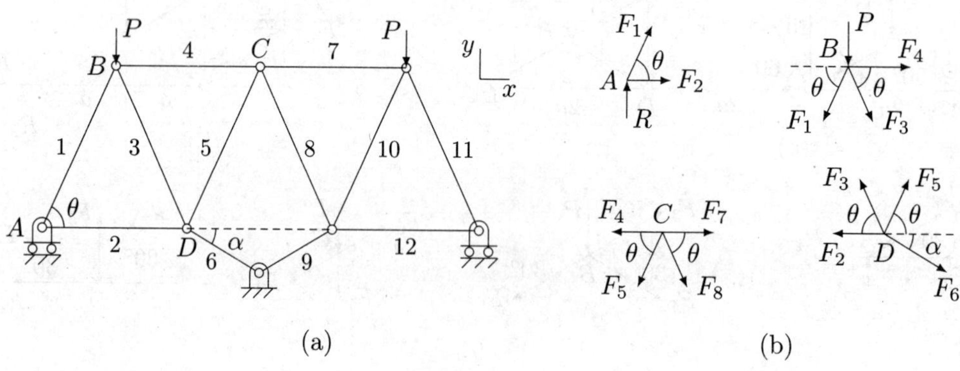

Fig. E4.37

Using this value of R, all the member forces are obtained. Note that R becomes indeterminate when the denominator in the expression above becomes zero, i.e, the truss is a critical form for $\cot\alpha = 2\cot\theta$.

4.16 Flexible Cables

A perfectly flexible cable (Fig. 4.22a) is a bar whose transverse dimensions r are very small compared to its length L ($r/L < 10^{-3}$). Hence, it offers negligible resistance to bending moments, twisting moment and shear forces, i.e., its constitutive equation is

$$\vec{M}_R = \vec{0}, \qquad \vec{F}_R = T\,\vec{e}_t, \qquad T \geq 0 \tag{4.19}$$

since it will bow out under a small compressive axial load. Thus, it cannot sustain any distributed or discrete moment load ($\vec{m} = \vec{0}$, $\vec{C} = \vec{0}$).

This model of a perfectly flexible cable is meaningful for chains, ropes, belts, strings and threads. Cables find use in suspension bridges, electric transmission lines, aerial tram-ways, guy wires for high transmission towers, etc. Flexible tubular marine risers behave as cables and link the sub-sea bore hole with the equipment on drilling rigs.

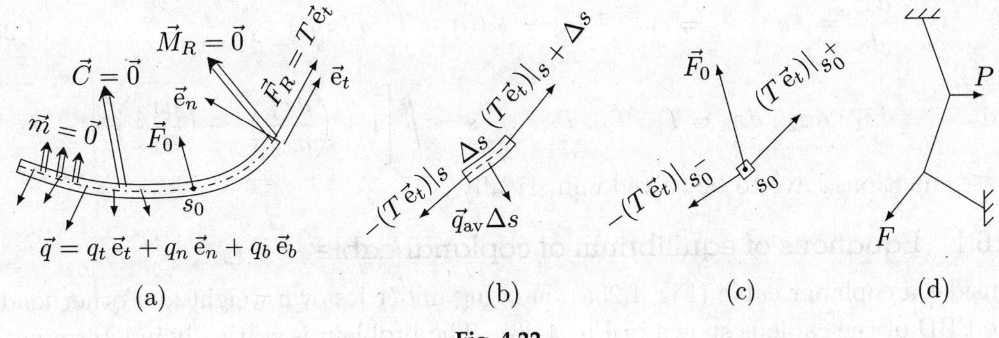

Fig. 4.22

We usually assume small strains in the cable, i.e., its extension causes negligible change in its geometry. Hence, the equilibrium equations can be applied in the undeformed/deformed geometry of the cable. For an element Δs (Fig. 4.22b), $\vec{F} = \vec{0}$:

$$(T\,\vec{e}_t)|_{s+\Delta s} - (T\,\vec{e}_t)|_s + \vec{q}_{av}\Delta s = \vec{0} \tag{1}$$

$$\lim_{\Delta s\to 0}\frac{\text{eq. (1)}}{\Delta s} \quad\Rightarrow\quad \frac{d}{ds}(T\,\vec{e}_t) + \vec{q} = \vec{0} \tag{4.20}$$

$$\Rightarrow \quad \frac{dT}{ds}\,\vec{e}_t + T\,\frac{d\vec{e}_t}{ds} + \vec{q} = \frac{dT}{ds}\,\vec{e}_t + \frac{T}{\rho}\,\vec{e}_n + \vec{q} = \vec{0} \quad\Rightarrow\quad q_b = 0$$

$$\frac{dT}{ds} + q_t = 0, \qquad \frac{T}{\rho} + q_n = 0 \tag{4.21}$$

These are differential equations of equilibrium of a cable at points where no discrete external force acts. Thus, $\vec{q} = q_t\,\vec{e}_t + q_n\,\vec{e}_n$ is in the osculating plane. The solution of eqs (4.21) yields the form of the cable and $T(s)$. At a location s_0 of discrete external force $\vec{F}_0$ (Fig. 4.22c), the cable has a corner. The equilibrium equation at s_0 is

$$T(s_0^+)\,\vec{e}_t(s_0^+) - T(s_0^-)\,\vec{e}_t(s_0^-) = -\vec{F}_0 \tag{4.22}$$

For a part of the cable on which no external forces act, $q_t = q_n = 0$ and eq. (4.21) $\Rightarrow$ $dT/ds = 0$, $T/\rho = 0$ $\Rightarrow$ $T =$ constant. $\Rightarrow$ (1) If the cable is slack, then $T = 0$ and ρ is arbitrary i.e., it can take any shape. (2) If the cable is taut, then $T \neq 0$ and $\rho = \infty$, i.e., its shape is a straight line. (3) If a cable is only subjected to discrete external forces, then its shape consists of a set of line segments with corners at the points of application of discrete forces and each line segment has constant value of T (Fig. 4.22d).

Equations (4.20) and (4.21) are also applicable for an *arch carrying only a thrust along its centre line* with $\vec{M}_R = \vec{0}$, $\vec{F}_R = F_t\,\vec{e}_t$, but F_t is negative (compressive).

For spatial cables, in rectangular Cartesian components, $\vec{T} = T\,\vec{e}_t = T\,d\vec{r}/ds = (T dx_i/ds)\,\vec{e}_i = T_i\,\vec{e}_i$. With $T_i = T(dx_i/ds)$, $\vec{q} = q_i\,\vec{e}_i$, eq. (4.20) yields

$$\frac{dT_i}{ds} = -q_i \qquad \Rightarrow \qquad T_i(s) = T_i(0) - \int_0^s q_i(s)ds \tag{4.23}$$

and $\qquad dx_1 : dx_2 : dx_3 = T_1 : T_2 : T_3, \qquad s = \int \left[1 + \left(\frac{dx_1}{dx_3}\right)^2 + \left(\frac{dx_2}{dx_3}\right)^2\right]^{1/2} dx_3$

These equations have to be solved numerically.

4.16.1 Equations of equilibrium of coplanar cable

Consider a coplanar cable (Fig. 4.23a), hanging under its own weight and other loads. The FBD of the cable is shown in Fig. 4.23b. The problem is statically indeterminate: 4 unknowns, T_A, T_B, θ_A, θ_B and 3 equilibrium equations. To enable solution without having to determine deformations, one additional piece of information needs to be specified; the length/ the maximum dip/ the horizontal component of cable tension/ the maximum tension.

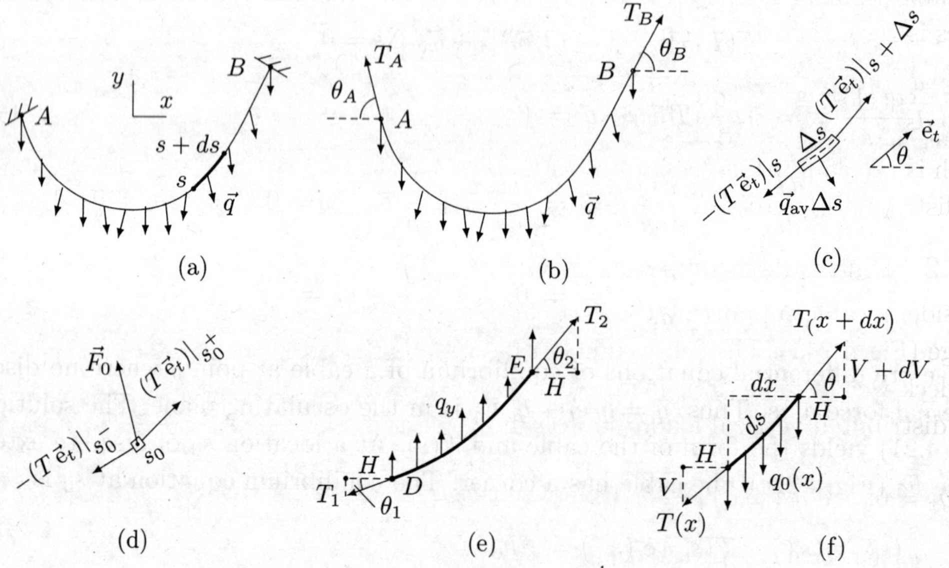

Fig. 4.23

At s: $\vec{e}_t = \cos\theta\,\vec{i} + \sin\theta\,\vec{j}$, and for element Δs (Fig. 4.23c):

$$\vec{F} = T\,\vec{e}_t|_{s+\Delta s} - T\,\vec{e}_t|_s + \vec{q}_{\text{av}}\Delta s = \vec{0} \tag{1}$$

$$\lim_{\Delta s \to 0} \frac{\text{eq. (1)}}{\Delta s} \;\Rightarrow\; \frac{d}{ds}(T\,\vec{e}_t) + \vec{q} = \vec{0} \tag{2}$$

$$\frac{d}{ds}[T(\cos\theta\,\vec{i} + T\sin\theta\,\vec{j}\,)] + q_x\,\vec{i} + q_y\,\vec{j} + q_z\,\vec{k} = \vec{0}$$

$$\Rightarrow \quad q_z = 0, \qquad \vec{q} = q_x \vec{\imath} + q_y \vec{\jmath}$$

$$\frac{d}{ds}(T\cos\theta) + q_x = 0, \qquad \frac{d}{ds}(T\sin\theta) + q_y = 0 \tag{4.24}$$

These are differential equations of equilibrium at points where no discrete force acts. At s_0 where discrete external force $\vec{F}_0$ acts (Fig. 4.23d), the cable has a corner and $\vec{F} = \vec{0} \Rightarrow$

$$T(s_0^+)\,\vec{e}_t(s_0^+) - T(s_0^-)\,\vec{e}_t(s_0^-) = -\vec{F}_0 \tag{3}$$

Equations (2) and (3) imply that the load on a coplanar cable is necessarily in its plane. For a part of the cable on which no external forces act, eq. (2) $\Rightarrow$ $T\vec{e}_t = \text{const.}$ $\Rightarrow$ the cable segment is straight with constant tension T.

If $q_x = 0$, i.e. the distributed load is along y-axis alone, then eq. $(4.24)_1 \Rightarrow$

$$T\cos\theta = H \text{ (horizontal component of } T) = \text{constant} \tag{4.25}$$

in a cable segment in which no discrete forces act. It follows directly from $F_x = 0$ for the cable segment DE (Fig. 4.23e). If the horizontal component of all external discrete forces is zero, then H is constant throughout. Using $T = H/\cos\theta$ in eq. $(4.24)_2$,

$$\frac{d}{ds}(\tan\theta) = -\frac{q_y}{H} \qquad \text{i.e.,} \qquad \frac{d}{ds}\left(\frac{dy}{dx}\right) = -\frac{q_y}{H} \tag{4.26}$$

which is the equilibrium equation of a cable loaded in the y-direction. We apply it for (1) distributed force $q_0(x)$ per unit projected length of the cable, and (2) constant q_y.

4.16.2 Distributed force $q_0(x)$ per unit projected length

Consider a cable supporting the vertical stringers attached to the deck of a suspension bridge (Fig. 4.24a). The self-weight of the cable is neglected, being a small fraction of the deck weight. The deck load is transmitted to it by the stringers and is modelled as a distributed vertical load $q_0(x)$ per unit projected length of the cable on x-axis:

$$q_x = 0, \qquad q_0(x)\,dx = -q_y\,ds \quad \Rightarrow \quad q_y = -q_0(x)\,\frac{dx}{ds}, \qquad \frac{d(\;)}{ds} = \frac{d(\;)}{dx}\frac{dx}{ds},$$

$$\text{eq. (4.26)} \quad \Rightarrow \quad \frac{d}{dx}\left(\frac{dy}{dx}\right) = \frac{q_0(x)}{H} \tag{4.27}$$

Direct proof of this equation of equilibrium from Fig. 4.23f: $F_x = 0 \Rightarrow H = \text{constant}$,

$$V = H\tan\theta = Hy' \quad \Rightarrow \quad dV = d(Hy') = q_0\,dx \quad \Rightarrow \quad \frac{dy'}{dx} = \frac{q_0}{H} \quad \Rightarrow \quad \text{eq. (4.27)}$$

The solution of eq. (4.27) is

$$\frac{dy}{dx} = \frac{1}{H}\int q_0(x)\,dx + c_1, \qquad y = \frac{1}{H}\int\left[\int q_0(x)\,dx\right]dx + c_1 x + c_2 \tag{4.28}$$

The constants c_1, c_2 can be found from the boundary conditions at A and B.

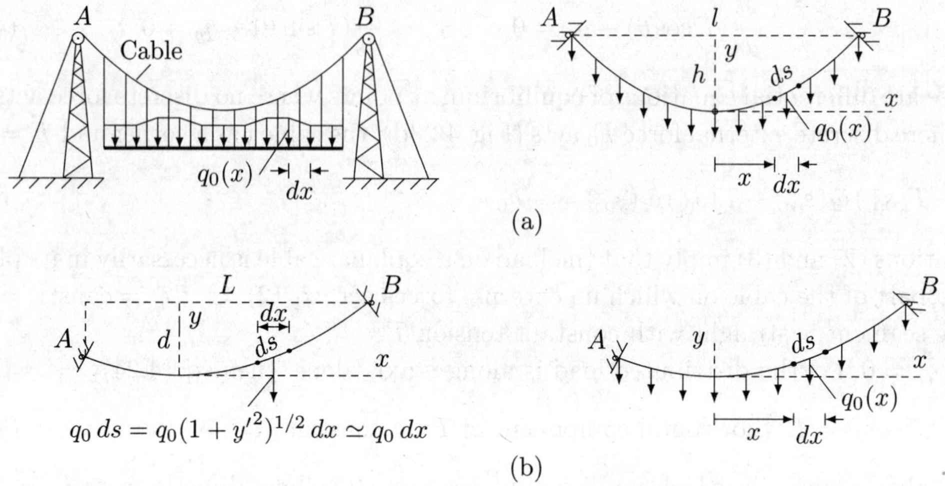

Fig. 4.24

1. Uniformly Distributed Load $q_0(x)$

Let $q_0(x) = q_0$ (a constant). The solution of eq. (4.27) is

$$\frac{dy}{dx} = \frac{q_0 x}{H} + c_1, \qquad y = \frac{q_0 x^2}{2H} + c_1 x + c_2 \tag{4.29}$$

Choose the origin $(0,0)$ at the lowest point $(dy/dx = 0)$ of the cable (Fig. 4.25a). The conditions $y = 0,\ dy/dx = 0$ at $x = 0 \ \Rightarrow\ c_2 = 0,\ c_1 = 0$. Hence,

$$y = q_0 x^2 / 2H \tag{4.30}$$

The shape of the cable is a parabola. For sag h in a span of l (Fig. 4.25a), $B(\frac{1}{2}l, h) \Rightarrow$

$$H = q_0 l^2/8h \tag{4.31}$$

From the FBD of CD (Fig. 4.25c), $V(x) = q_0 x$,

$$\Rightarrow \quad T(x) = [H^2 + (q_0 x)^2]^{1/2} \tag{4.32}$$

$$T_{\max} = [H^2 + q_0^2 l^2/4]^{1/2} = (q_0 l/2)[1 + (l/4h)^2]^{1/2} \tag{4.33}$$

If the sag to span ratio $d/L < 0.1$ for a cable (Fig. 4.24b) under self-weight q_0, then

$$q_0\, ds = q_0(1 + y'^2)^{1/2} dx = q_0(1 + \tfrac{1}{2}y'^2 + \ldots)\, dx \simeq q_0\, dx$$

as $y'^2 \sim (d/L)^2 \ll 1$. Hence, for this case, the analysis of a suspension cable is valid.

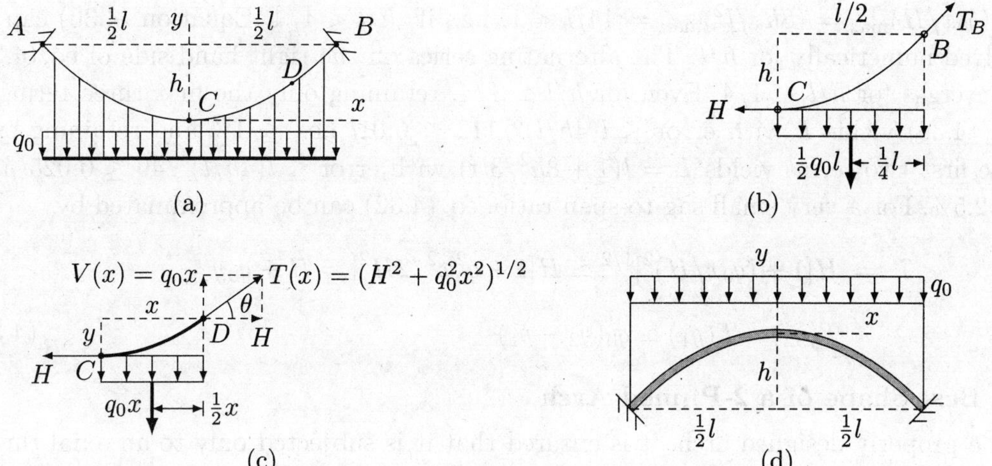

Fig. 4.25

§ **Exercise 4.2:** From the FBDs of the parts CB and CD of the cable (Figs 4.25b,c), derive eqs (4.30) and (4.31) directly, using $M_{Bz} = 0$ for CB and $M_{Dz} = 0$ for CD.

The sag h is related to the length L of the loaded cable:

$$L = \int ds = \int \left[(dx)^2 + (dy)^2\right]^{1/2} = 2\int_0^{l/2} \left[1 + y'^2\right]^{1/2} dx$$

$$= 2\int_0^{l/2} \left[1 + \frac{q_0^2 x^2}{H^2}\right]^{1/2} dx \tag{4.34}$$

Integrating by substituting $q_0 x/H = 8hx/l^2 = \sinh u$, we get

$$\frac{L}{l} = \tfrac{1}{2}\left[1 + \left(\frac{4h}{l}\right)^2\right]^{1/2} + \tfrac{1}{2}\,\frac{\sinh^{-1}(4h/l)}{(4h/l)} \tag{4.35}$$

Given length L and span l, eq. (4.35) can be solved numerically for the ratio h/l.

If the span l and the length L are of the same order of magnitude, then the sag-to-span ratio h/l is small and an approximate expression of L can be obtained from eq. (4.34) by using the binomial theorem:

$$L = 2\int_0^{l/2} \left[1 + \frac{1}{2}\left(\frac{8hx}{l^2}\right)^2 - \frac{1}{8}\left(\frac{8hx}{l^2}\right)^4 + \frac{1}{16}\left(\frac{8hx}{l^2}\right)^6 - \ldots\right] dx$$

$$= l\left[1 + \frac{1}{6}\left(\frac{4h}{l}\right)^2 - \frac{1}{40}\left(\frac{4h}{l}\right)^4 + \frac{1}{112}\left(\frac{4h}{l}\right)^6 - \ldots\right] \tag{4.36}$$

if $(q_0x/H)_{\max} = (8hx/l^2)_{\max} = 4h/l < 1$, i.e., if $h/l < 1/4$. Equation (4.36) can be solved numerically for h/l. The alternating series on the right hand side of eq. (4.36) converges for $h/l < 1/4$. Even for $h/l \simeq 1/4$, retaining only the first three terms in eq. (4.36) yields L with error $\leq l(4h/l)^6/112 \leq 0.01l$ i.e., $\leq 1\%$ and retaining only the first two terms yields $L = l(1 + 8h^2/3l^2)$ with error $\leq l(4h/l)^4/40 \leq 0.025l$ i.e., $\leq 2.5\%$. For a very small sag-to-span ratio, eq. (4.32) can be approximated by

$$T = H[1 + (q_0x/H)^2]^{1/2} \simeq H[1 + q_0^2x^2/2H^2] = H + q_0y$$

$$\Rightarrow \quad T(y_2) - T(y_1) \simeq q_0(y_2 - y_1) \tag{4.37}$$

2. Best Shape of a 2-Pinned Arch

In a properly designed arch, it is ensured that it is subjected only to an axial thrust without shear force or bending moment, with optimum use of its cross-section through uniform stress distribution. Thus, the optimal arch is like an inverted suspension bridge, with compression replacing tension. The equilibrium equation (4.27) becomes $d^2y/dx^2 = -q_0(x)\,/\,H$, where H is the constant horizontal thrust in the arch. The best shape of the arch (Fig. 4.25d) for a uniformly distributed load q_0 is obtained by integration as $y = -q_0x^2/2H$, i.e., a parabolic arch.

3. Illustrative Examples: Rather than use general results, we illustrate how to solve problems using first principles when both discrete and distributed loads act.

Example 4.38: A pipe line of mass 250 kg/m is supported by a cable as it passes over a gorge (Fig. E4.38a). The anchor cables are at 30° to the horizontal. The cable passes over a pulley at A and is attached to a saddle on rollers at B. (a) Find the minimum and maximum tensions in the cable. (b) Find the length of the cable. (c) Calculate the vertical and horizontal loads on the top of the towers.

Solution: The load on an element dx of the cable in Fig. E4.38b is $250g\,dx$ sec $20° = q_0\,dx$. Hence, $q_0 = 2.610$ kN/m. The FBD of part CB of the cable is shown in Fig. E4.38c. Its equations of equilibrium are:

$$M_{B_z} = (q_0d)(\tfrac{1}{2}d) - 10H = 0 \qquad \Rightarrow \qquad H = q_0d^2/20 \tag{1}$$

$$\vec{F} = -H\vec{\imath} - q_0d\vec{\jmath} + \vec{T}_B = \vec{0} \qquad \Rightarrow \qquad \vec{T}_B = H\vec{\imath} + q_0d\vec{\jmath}$$

$$\Rightarrow \quad T_B = (H^2 + q_0^2d^2)^{1/2} \tag{2}$$

The equations of equilibrium of part CA (Fig. E4.38d) are:

$$M_{A_z} = 25H - q_0(200 - d)\tfrac{1}{2}(200 - d) = 0$$

$$\Rightarrow \quad H = q_0(200 - d)^2/50 \tag{3}$$

$$\vec{F} = H\,\vec{\imath} - q_0(200-d)\,\vec{\jmath} + \vec{T}_A = \vec{0}$$

$$\Rightarrow \vec{T}_A = -H\,\vec{\imath} + q_0(200-d)\,\vec{\jmath}$$

$$T_A = [H^2 + q_0^2(200-d)^2]^{1/2} \tag{4}$$

Equations (1) and (3) $\Rightarrow$ $q_0 d^2/20 = q_0(200-d)^2/50 \quad \Rightarrow$

$$200 - d = \pm\sqrt{2.5}\,d \quad \Rightarrow \quad d = 77.49\,\text{m},\ -344.2\,\text{m}$$

The acceptable root is $d = 77.49\,\text{m}$. Equations (1) to (4) yield $H = 783.6\,\text{kN}$ and

$$\vec{T}_B = 783.6\vec{\imath} + 202.2\vec{\jmath}\ \text{kN}, \qquad T_B = 809.3\,\text{kN}$$
$$\vec{T}_A = -783.6\vec{\imath} + 319.8\vec{\jmath}\ \text{kN}, \qquad T_A = 846.3\,\text{kN}$$

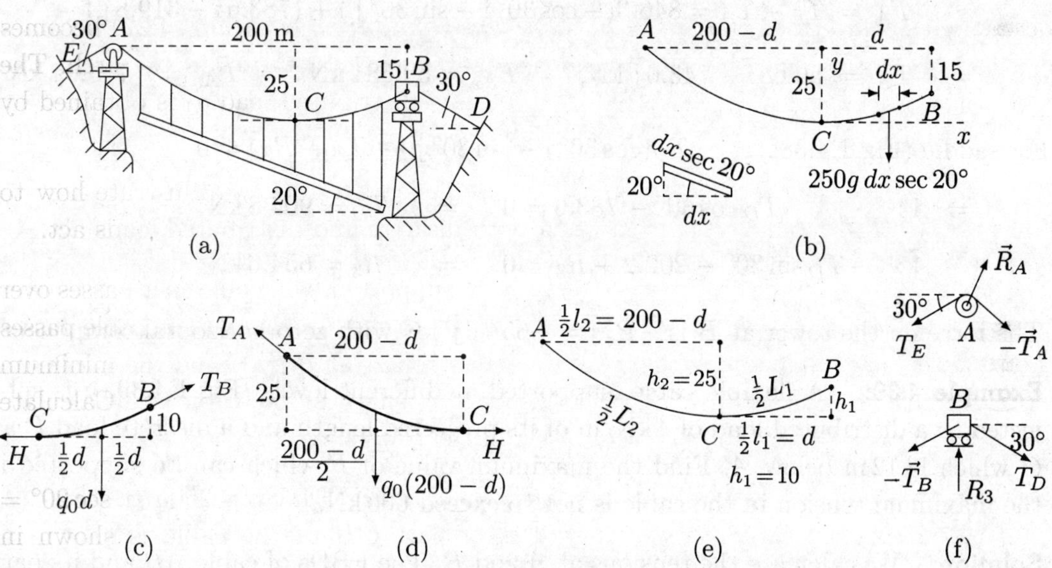

Fig. E4.38

(a) $T_{\min}$ is at C, $T_{\min} = H = 783.6\,\text{kN}$. $T(x) = (H^2 + q_0^2 x^2)^{1/2}$ is maximum for $x_{\max}$, i.e., at the highest point if there are no discrete loads, $\Rightarrow$ $T_{\max} = T_A = 846.3\,\text{kN}$.

If discrete vertical downward forces act on the cable in addition to q_0, then also $T_{\max}$ occurs at one of the supports but not necessarily the higher one.

(b) The total length of the cable is obtained by using eq. (4.35):

$$\tfrac{1}{2}L = \tfrac{1}{4}l\left[\left\{1 + \left(\frac{4h}{l}\right)^2\right\}^{1/2} + \frac{\sinh^{-1}(4h/l)}{(4h/l)}\right] \tag{5}$$

Referring to Fig. E4.38e, using eq. (5) with

$$\tfrac{1}{2}l = \tfrac{1}{2}l_1 = d = 77.49\,\text{m}, \qquad h = h_1 = 10\,\text{m} \quad \Rightarrow \quad \tfrac{1}{2}L_1 = 78.34\,\text{m}$$

$$\tfrac{1}{2}l = \tfrac{1}{2}l_2 = 200 - d = 122.51\,\text{m}, \qquad h = h_2 = 25\,\text{m} \quad \Rightarrow \quad \tfrac{1}{2}L_2 = 125.83\,\text{m}$$

The required total length $L = \frac{1}{2}L_1 + \frac{1}{2}L_2 = 204.17\,\text{m}$. Since $\frac{h_1}{l_1} = \frac{10}{2(77.49)} = 0.0645 < \frac{1}{4}$, $\frac{h_2}{l_2} = \frac{25}{2(122.51)} = 0.102 < \frac{1}{4}$, we can use approximate relation (4.36) with first 2 terms:

$$L \simeq \tfrac{1}{2}l_1\left(1 + \frac{8h_1^2}{3l_1^2}\right) + \tfrac{1}{2}l_2\left(1 + \frac{8h_2^2}{3l_2^2}\right) = 78.35 + 125.91 = 204.26\,\text{m}$$

(c) Let $\vec{R}_A$ be the reaction at A on the pulley (Fig. E4.38f). The force on the tower at A is $\vec{F}_A = -\vec{R}_A$. $T_E = T_A = 846.3\,\text{kN}$. For pulley: $\vec{R}_A - \vec{T}_A + \vec{T}_E = \vec{0} \Rightarrow$

$$\vec{F}_A = \vec{T}_E - \vec{T}_A = 846.3(-\cos 30°\,\vec{\imath} - \sin 30°\,\vec{\jmath}) + (783.6\,\vec{\imath} - 319.8\,\vec{\jmath})$$

$$\Rightarrow \quad \vec{F}_A = 50.68\,\vec{\imath} - 743.0\,\vec{\jmath}\,\text{kN}, \qquad \vec{F}_{Ax} = 50.68\,\vec{\imath}\,\text{kN}, \qquad \vec{F}_{Ay} = -743.0\,\vec{\jmath}\,\text{kN}$$

For saddle (Fig. E4.38f): $\quad T_D(\cos 30°\,\vec{\imath} - \sin 30°\,\vec{\jmath}) - \vec{T}_B + R_3\vec{\jmath} = \vec{0}$

$$\Rightarrow \vec{\imath}: \qquad T_D \cos 30° - 783.6 = 0 \quad \Rightarrow \quad T_D = 904.8\,\text{kN}$$

$$\vec{\jmath}: \quad -T_D \sin 30° - 202.2 + R_3 = 0 \quad \Rightarrow \quad R_3 = 654.6\,\text{kN}$$

The force on the tower at B is $-R_3\vec{\jmath} = -654.6\vec{\jmath}$ kN with zero horizontal component.

Example 4.39: A flexible cable, supported at different levels (Fig. E4.39a), is subjected to a distributed load of 4 kN/m of its projected length and a discrete load P at C which is 12 m below A. Find the maximum value of P which can be supported if the maximum tension in the cable is not to exceed 600 kN.

Solution: We calculate the tensions at A and B. The FBDs of cable AB and its part CB are shown in Fig. E4.39b,c. The equations of equilibrium of AB are:

$$F_x = H - H_1 = 0 \qquad \Rightarrow \qquad H_1 = H \tag{1}$$

$$F_y = V + V_1 - 400 - P = 0 \qquad \Rightarrow \qquad V_1 = 400 + P - V \tag{2}$$

$$M_{Az} = 100V - 4H - 50(400) - 30P = 0 \tag{3}$$

$M_{Cz} = 0$ for cable CB yields another equation for H and V:

$$M_{Cz} = 70V - 16H - 35(280) = 0 \qquad \Rightarrow \qquad V = 8H/35 + 140 \tag{4}$$

Eliminating P from eqs (2) and (3) (or directly using $M_{Cz} = 0$ for cable AC) yields

$$V_1 = 400 - V + (100V - 4H - 20000)/30 \overset{(4)}{=} 0.4H + 60 \tag{5}$$

$$T_B = (V^2 + H^2)^{1/2}$$

$$T_A = (V_1^2 + H_1^2)^{1/2} = (V_1^2 + H^2)^{1/2} \tag{6}$$

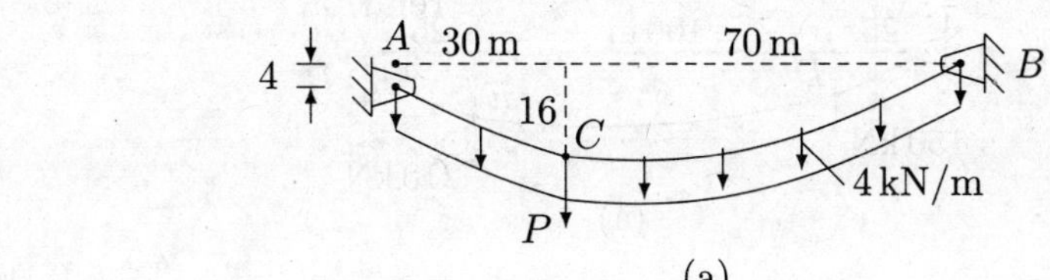

(a)

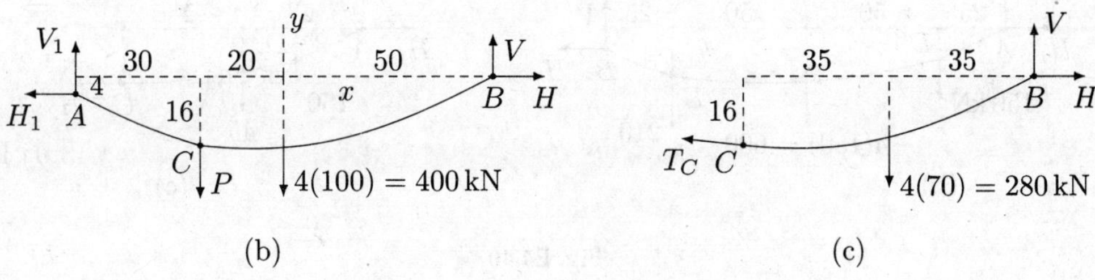

(b) (c)

Fig. E4.39

Assume that $T_{\max}$ occurs at the higher support B:

$$T_B^2 = V^2 + H^2 = (8H/35 + 140)^2 + H^2 = 600^2$$

$$\Rightarrow \quad 1.05224H^2 + 64H - 3.404 \times 10^5 = 0$$

$$H = \frac{-64 + \{64^2 + 4(1.05224)\times 3.404\times 10^5\}^{1/2}}{2 \times 1.05224} = 539.17\,\text{kN}$$

Equations (5), (6) $\Rightarrow$

$$V_1 = 0.4H + 60 = 275.67\,\text{kN}$$

$$T_A = (V_1^2 + H^2)^{1/2} = 605.6\,\text{kN} > 600\,\text{kN}$$

Hence, the assumption is wrong. We infer that $T_{\max}$ occurs at the lower support A:

$$T_A^2 = V_1^2 + H^2 = (0.4H + 60)^2 + H^2 = 600^2$$

$$\Rightarrow \quad 1.16H^2 + 48H - 35.64 \times 10^4 = 0$$

$$H = \frac{-48 + (48^2 + 4\times 1.16\times 35.64\times 10^4)^{1/2}}{2\times 1.16} = 533.99\,\text{kN}$$

eqs (4), (6) $\Rightarrow$ $\quad V = 8H/35 + 140 = 262.05\,\text{kN}, \quad V_1 = 0.4H + 60 = 273.6\,\text{kN}$

$$P = V_1 + V - 400 = 135.6\,\text{kN}$$

Example 4.40: A flexible cable weighing 4 kN/m (Fig. E4.40a) supports two vertical loads. Its maximum sag below its lower support point is 5 m. Find $T_{\max}$.

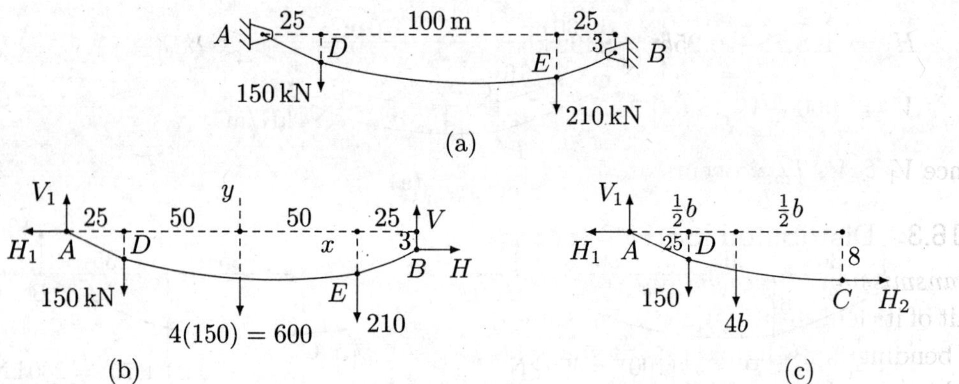

Fig. E4.40

Solution: Since the sag-to-span ratio $8/150 < 0.1$, we model the distributed load on the cable as 4 kN/m of its horizontal projected length. Its shape is a piecewise parabola between discrete loads. The position C of the maximum sag is not given. The FBD of cable AB is shown in Fig. E4.40b and the equations of equilibrium are:

$$F_x = H - H_1 = 0 \quad \Rightarrow \quad H_1 = H \tag{1}$$

$$F_y = V + V_1 - 150 - 210 - 600 = 0 \quad \Rightarrow \quad V + V_1 = 960 \tag{2}$$

$$M_{Bz} = 3H_1 + 25(210) + 75(600) + 125(150) - 150V_1 = 0$$

$$\Rightarrow \quad H_1 - 50V_1 = -23000 \tag{3}$$

Assume that C is within DE at horizontal distance b from A, i.e., $25 < b < 125$. The FBD of part AC of the cable is shown in Fig. E4.40c. Its equilibrium equations are:

$$F_x = H_2 - H_1 = 0 \quad \Rightarrow \quad H_2 = H_1 \tag{4}$$

$$F_y = V_1 - 150 - 4b = 0 \quad \Rightarrow \quad V_1 = 150 + 4b \tag{5}$$

$$M_{Az} = 8H_2 - \tfrac{1}{2}b \times 4b - 25 \times 150 = 0 \quad \Rightarrow \quad H_2 = 468.75 + 0.25b^2 \tag{6}$$

$$\Rightarrow \quad H = H_1 = H_2 = 468.75 + 0.25b^2 \tag{7}$$

Substituting expressions for H_1 and V_1 from eqs (7) and (5) into eq. (3), we obtain:

$$0.25b^2 - 200b + 15969 = 0 \quad \Rightarrow \quad b = 710\,\mathrm{m},\ 89.96\,\mathrm{m}$$

Assumption $b \in (25,\ 125)$ $\Rightarrow$ root $b = 89.96$ is acceptable. [If instead neither root satisfied the assumed condition, then the assumption is wrong. The assumption on the location of C is modified until an acceptable position is found.]. Equations (7), (5), (2) $\Rightarrow$

$$\begin{aligned} H &= 468.75 + 0.25b^2 = 2492\,\mathrm{kN}, \qquad V_1 = 150 + 4b = 509.8\,\mathrm{kN} \\ V &= 960 - V_1 = 450.2\,\mathrm{kN} \end{aligned}$$

Since $V_1 > V$, $T_{\max}$ occurs at A: $T_A = (H_1^2 + V_1^2)^{1/2} = 2544\,\mathrm{kN}$.

4.16.3 Distributed load $\vec{q}(s)$ per unit length

Transmission line cables are subjected to self-weight which is a distributed load per unit of its length (Fig. 4.26a). The same holds for *flexible marine risers* whose resistance to bending is negligible (Fig. 4.26b). These risers are flexible piping connected at one end to a semi-submersible or fixed marine structure and at the other end to some fixed structure on the sea-bed. An intermediate buoy B may be attached to some risers. Risers are normally subjected to time-varying waves and current loads. A dynamic analysis is required for complete design. The static solution of risers under self-weight helps at the conceptual design stage to decide a system layout and it is the initial approximation for the nonlinear equations of dynamics.

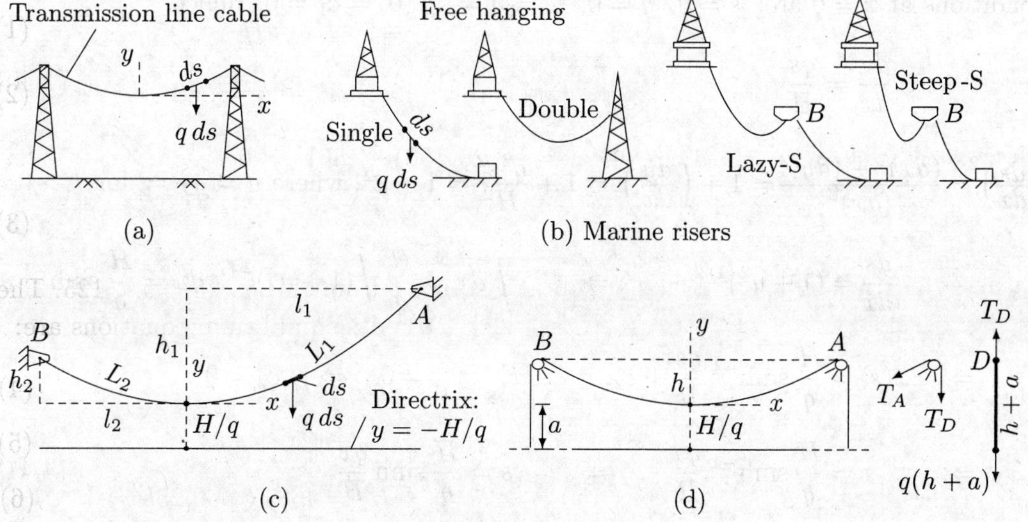

Fig. 4.26

Let the distributed self-weight of a transmission line cable or the self-weight in water of a marine riser be $\vec{q} = -q\,\vec{\jmath} \;\Rightarrow\; q_x = 0,\; q_y = -q$. Equations (4.25) and (4.26) $\Rightarrow$

$$T\cos\theta = H = \text{constant}, \qquad \frac{d}{ds}\left(\frac{dy}{dx}\right) = \frac{q}{H}$$

$$\Rightarrow \frac{dy}{dx} = \frac{1}{H}\int q(s)\,ds + c_1 \tag{4.38}$$

$\left(\dfrac{ds}{dx}\right)^2 = \dfrac{(dx)^2 + (dy)^2}{(dx)^2} = 1 + \left(\dfrac{dy}{dx}\right)^2$ and substitute $\dfrac{dy}{dx}$ from eq. (4.38) to eliminate y:

$$\frac{ds}{dx} = \left[1 + \left\{\frac{1}{H}\int q(s)\,ds + c_1\right\}^2\right]^{1/2}$$

$$\Rightarrow \quad x = \int\left[1 + \left\{\frac{1}{H}\int q(s)\,ds + c_1\right\}^2\right]^{-1/2} ds + c_2 \tag{4.39}$$

Equation (4.39) is solved for $s(x)$ which is substituted in eq. (4.38) to obtain $y(x)$. Boundary conditions are used to find c_1, c_2 and H.

These steps are illustrated for uniform load q. Equation (4.38) $\Rightarrow$

$$y'(s) = \frac{dy}{dx} = \frac{qs}{H} + c_1$$

Choosing the origin $(0,0)$ at the lowest point of the cable (Fig. 4.26c), the boundary conditions at $x = 0$ are: $s = 0,\; y = 0,\; y' = 0 \Rightarrow y'(0) = c_1 = 0$. Hence,

$$\frac{dy}{dx} = \frac{qs}{H} \tag{4.40}$$

$\left(\dfrac{ds}{dx}\right)^2 = \dfrac{(dx)^2 + (dy)^2}{(dx)^2} = 1 + \left(\dfrac{dy}{dx}\right)^2 = 1 + \dfrac{q^2 s^2}{H^2} = 1 + u^2$, where $u = \frac{qs}{H} \Rightarrow$ for $x > 0$:

$$\frac{ds}{dx} = (1+u^2)^{1/2} \qquad \Rightarrow \qquad \int dx = \frac{H}{q}\int (1+u^2)^{-1/2} du = \frac{H}{q}\sinh^{-1} u$$

$$\Rightarrow \quad x(s) = \frac{H}{q}\sinh^{-1}\frac{qs}{H} + c_2, \qquad x\Big|_{s\,=\,0} = 0 \quad \Rightarrow \quad c_2 = 0$$

$$\Rightarrow \quad x = \frac{H}{q}\sinh^{-1}\frac{qs}{H} \qquad \Rightarrow \qquad s = \frac{H}{q}\sinh\frac{qx}{H} \tag{4.41}$$

Eq. (4.40) $\Rightarrow \frac{dy}{dx} = \sinh\frac{qx}{H} \;\Rightarrow\; y = \frac{H}{q}\cosh\frac{qx}{H} + c_3$, and $y\Big|_{x\,=\,0} = 0 \;\Rightarrow\; c_3 = -\frac{H}{q}$

$$\Rightarrow \quad y = \frac{H}{q}\left[\cosh\frac{qx}{H} - 1\right], \quad \text{i.e.,} \quad \frac{qy}{H} = \cosh\frac{qx}{H} - 1 \tag{4.42}$$

The curve of eq. (4.42) is called a *common catenary* since a hanging uniform chain (*catena* in Latin) has this shape. Its intrinsic eq. (4.40) is $\tan\theta = qs/H$. Using eq. (4.42):

$$\sec\theta = (1 + \tan^2\theta)^{1/2} = \left[1 + \sinh^2\frac{qx}{H}\right]^{1/2} = \cosh\frac{qx}{H} = \frac{qy}{H} + 1$$

$$\Rightarrow \quad T = H\sec\theta = H\cosh\frac{qx}{H} = H + qy \tag{4.43}$$

$$T(y_2) - T(y_1) = (H + qy_2) - (H + qy_1) = q(y_2 - y_1)$$

$$T_{\max} = H\cosh\frac{qx_{\max}}{H} = H + qy_{\max} \tag{4.44}$$

The sag h_1 below the support at $x = l_1$ is obtained from eq. (4.42) as

$$h_1 = \frac{H}{q}\left[\cosh\frac{ql_1}{H} - 1\right] \tag{4.45}$$

For a given sag, eq. (4.45) can be solved numerically for H. The length L_1 of the cable between the lowest point and the support at $x = l_1$, is obtained from eq. (4.41) as

$$L_1 = \frac{H}{q}\sinh\frac{ql_1}{H} \quad \Rightarrow \quad \frac{L_1}{l_1} = \frac{H}{ql_1}\sinh\frac{ql_1}{H} \tag{4.46}$$

Let V be the vertical component of tension T. Each of the 5 variables x, y, T, s, θ are expressible in terms of others, for example, <u>in terms of x</u>:

$$y = \frac{H}{q}\left[\cosh\frac{qx}{H} - 1\right], \qquad \tan\theta = \frac{dy}{dx} = \sinh\frac{qx}{H} \tag{4.47}$$

$$V = T\sin\theta = H\tan\theta = H\sinh\frac{qx}{H} \tag{4.48}$$

$$T = H\cosh\frac{qx}{H} = H + qy, \qquad s = \frac{H}{q}\sinh\frac{qx}{H} \tag{4.49}$$

and <u>in terms of s</u>:

$$x = \frac{H}{q}\sinh^{-1}\frac{qs}{H}, \qquad \tan\theta = \frac{dy}{dx} = \frac{qs}{H} \tag{4.50}$$

$$T = H\sec\theta = H(1 + \tan^2\theta)^{1/2} = (H^2 + q^2s^2)^{1/2} \tag{4.51}$$

$$y = \frac{H}{q}\left[\cosh\left\{\sinh^{-1}\frac{qs}{H}\right\} - 1\right] = \frac{H}{q}\left[\left(1 + \frac{q^2s^2}{H^2}\right)^{1/2} - 1\right] \tag{4.52}$$

Remarks:

1. Solution of (4.45) for H: Newton-Raphson method, 1st approximation

$$h_1 = \frac{H}{q}\left[\cosh\frac{ql_1}{H} - 1\right] = \frac{H}{q}\left[1 + \frac{q^2 l_1^2}{2H^2} + \ldots - 1\right] \simeq \frac{ql_1^2}{2H} \quad \text{i.e.,} \quad H \simeq \frac{ql_1^2}{2h_1}$$

which is H for a cable under uniform q per unit horizontal projected length.

2. Solution of eq. (4.46) for ql_1/H: Newton–Raphson method, 1st approximation:

$$\frac{L_1}{l_1} = \frac{\sinh(ql_1/H)}{ql_1/H} = 1 + \frac{1}{6}\left(\frac{ql_1}{H}\right)^2 + \frac{1}{120}\left(\frac{ql_1}{H}\right)^4 + \ldots \simeq 1 + \frac{1}{6}\left(\frac{ql_1}{H}\right)^2$$

$$\text{i.e.,} \quad ql_1/H \simeq \sqrt{6}\,[L_1/l_1 - 1]^{1/2}$$

This also follows from the approximate length of a parabolic cable under uniform q per unit horizontal projected length: $\dfrac{L_1}{l_1} = 1 + \dfrac{8h_1^2}{3(2l_1)^2} = 1 + \dfrac{1}{6}\left(\dfrac{ql_1}{H}\right)^2$, using $H = \dfrac{ql_1^2}{2h_1}$.

3. Equation (4.44) implies that the cable can hang in equilibrium over two smooth pegs (Fig. 4.26d), provided it overhangs on either side with $a = H/q$ below the minimum point. $T_D = q(h + a) = q(h + H/q) = H + qh = T_A$.

4. The optimum shape of an arch, under uniform self-weight, is an inverted catenary.

§ **Exercise 4.3:** Show using eq. (4.45) that if $h/l \ll 1$, then $h/l = (ql/H)/8 + (ql/H)^3/384 + \ldots$, i.e., $ql/H \ll 1$ and eq. (4.42) is approximated by $y \simeq qx^2/2H$. Thus, the approximation of replacing a distributed load per unit length of the cable by a distributed load per unit projected horizontal length is quite good for $h/l < 0.1$. Show that for $h/l = 0.15$, error of this approximation for L and $T_{\max}$ is less than 5%.

§ **Exercise 4.4:** (a) Consider a cable with a horizontal span l and sag h. If h is small, then the tensions at the ends are large as these forces supporting the vertical load are almost horizontal. If h is large, then also these are large because the weight of the cable is large due to greater length. Show that the optimum h/l for which $T_{\max} = H\cosh(ql/2H)$ is minimum occurs for $\tanh(ql/2H) = 2H/ql$. Prove that its solution by Newton–Raphson method is $ql/2H = 1.2$ (Fig. Ex4.4). Using eq. (4.45), show that $h_{\text{opt}}/l = 0.3378 \simeq \frac{1}{3}$ with $T_{\max} = 0.7544ql \simeq 3ql/4$. (b) Let A be the cross-sectional area, ρ the mass density and σ the allowable stress of a cable. Prove that under self-weight $q = \rho Ag$, $l_{\max} = \sigma/0.7544\rho g$, if there is no restriction on h, e.g., for a steel cable with $\rho = 8000\,\text{kg/m}^3$, $\sigma = 700\,\text{MPa}$; $l_{\max} = 11.8\,\text{km}$ with $h = h_{\text{opt}} = 4\,\text{km}$, which is impractical. (c) Show using eq. (4.43) that for a given h, $l_{\max} = 2(\sigma/\rho g - h)\cosh^{-1}[\sigma/(\sigma - \rho gh)]$. For the steel cable, $l_{\max} = 4.15\,\text{km}$ for $h = 250\,\text{m}$ and $l_{\max} = 1.333\,\text{km}$ for $h = 25\,\text{m}$.

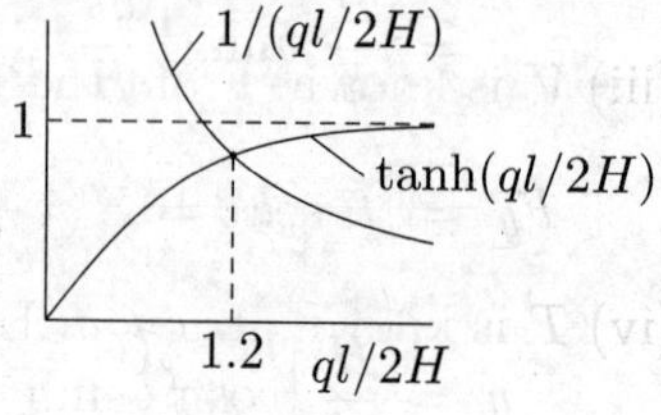

Fig. Ex4.4

Example 4.41: A *free hanging* marine riser (Fig. E4.41a) has length L from the touch down point to the barge where it is hooked at (b, d). The self-weight of the riser in water is q N/m of its length. The tension at the barge end is T making an angle θ_1 with the horizontal. Its horizontal and vertical components are H and V. For a given depth d and a given value of $H\,/\,L\,/\,V\,/\,T\,/\,\theta_1$, set up relations to obtain b and the shape of the curve formed by the marine riser.

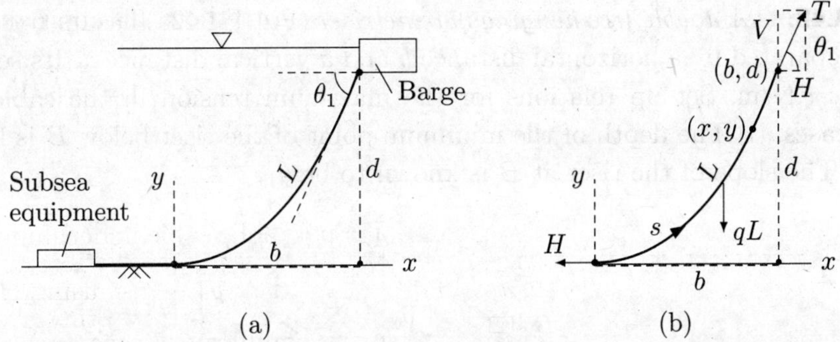

Fig. E4.41

Solution: (i) H is known. We use the basic relations (4.47) to (4.52) to obtain the shape of the curve:

$$y = \frac{H}{q}\left[\cosh\frac{qx}{H} - 1\right], \qquad s = \frac{H}{q}\sinh\frac{qx}{H} \tag{1}$$

$y = d$ for $x = b \;\Rightarrow$

$$d = \frac{H}{q}\left[\cosh\frac{qb}{H} - 1\right] \quad\Rightarrow\quad b = \frac{H}{q}\cosh^{-1}\left(\frac{qd}{H} + 1\right) \tag{2}$$

H required in the solution (1) and (2) for the other cases is obtained as follows.

(ii) L is known. Using eq. (1) for $x = b:\; y = d,\; s = L \;\Rightarrow$

$$d = \frac{H}{q}\left[\cosh\frac{qb}{H} - 1\right], \qquad L = \frac{H}{q}\sinh\frac{qb}{H}$$

Eliminating b: $1 = \cosh^2\dfrac{qb}{H} - \sinh^2\dfrac{qb}{H} = \left(\dfrac{qd}{H} + 1\right)^2 - \left(\dfrac{qL}{H}\right)^2 \;\Rightarrow\; H = \dfrac{q(L^2 - d^2)}{2d}$

(iii) V is known. From the FBD in Fig. E4.41b,

$$F_y = V - qL = 0 \quad\Rightarrow\quad L = V/q \quad \text{and then as in case (ii).}$$

(iv) T is known at $y = d$. Equation $(4.49)_1 \;\Rightarrow\quad T = H + qd \quad\Rightarrow\quad H = T - qd.$

(v) θ_1 is known at $x = b,\; y = d$. Equation (4.47) $\Rightarrow$

$$d = \frac{H}{q}\left[\cosh\frac{qb}{H} - 1\right], \qquad \tan\theta_1 = \sinh\frac{qb}{H}$$

Eliminating b: $1 = \cosh^2\left(\frac{qb}{H}\right) - \sinh^2\left(\frac{qb}{h}\right) = \left(\frac{qd}{H} + 1\right)^2 - \tan^2\theta_1$

$\Rightarrow \quad H = qd/(\sec\theta_1 - 1).$

Example 4.42: *A double free hanging marine riser* (Fig. E4.42a) has its two supports *A* and *B* separated by a horizontal distance *b* and a vertical distance *d*. Its self-weight in water is q N/m. Set up relations for the maximum tension in the cable for the following cases. (i) The depth of the minimum point of the riser below *B* is known to be d_1. (ii) The slope of the riser at *B* is known to be θ_1.

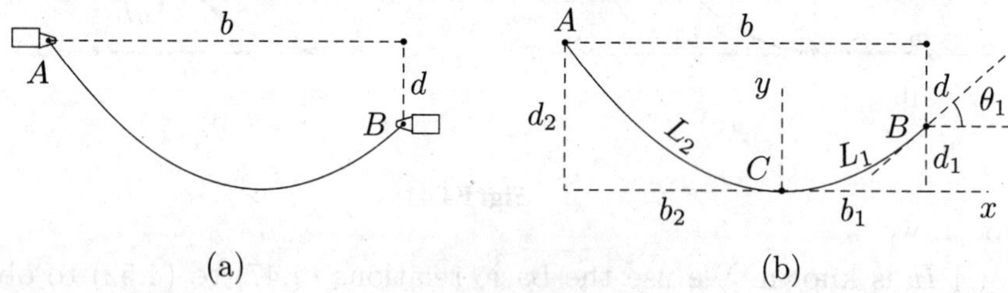

Fig. E4.42

Solution: Choose origin of the coordinate system at the lowest point *C* (Fig. E4.42b) of the riser. We use equations (4.47) and (4.49):

$$y = \frac{H}{q}\left[\cosh\frac{qx}{H} - 1\right], \qquad s = \frac{H}{q}\sinh\frac{qx}{H} \tag{1}$$

Using the data from Fig. E4.42b: $y = d_1$ at $x = b_1$, $y = d_2$ at $x = -b_2$, $s = L_1$ at $x = b_1$, $s = -L_2$ at $x = -b_2$ in eqs (1), we get

$$b_1 = (H/q)\cosh^{-1}(d_1 q/H + 1), \qquad L_1 = (H/q)\sinh(qb_1/H) \tag{2}$$

$$b_2 = (H/q)\cosh^{-1}(d_2 q/H + 1), \qquad L_2 = (H/q)\sinh(qb_2/H) \tag{3}$$

$$b = b_1 + b_2 = (H/q)[\cosh^{-1}(d_1 q/H + 1) + \cosh^{-1}(d_2 q/H + 1)] \tag{4}$$

$$L = L_1 + L_2 = (H/q)[\sinh(qb_1/H) + \sinh(qb_2/H)] \tag{5}$$

$$T_{\max} = T_A = H + qy_A = H + q(d + d_1) \qquad [\text{eq. } (4.49)_1] \tag{6}$$

Hence, to obtain $T_{\max}$, we need to know H and d_1.
(i) d_1 is given. With $d_2 = d_1 + d$, and known values of b, q, d_1, eq. (4) is solved numerically for H using the method of bisection. Then, $T_{\max}$ is obtained from eq. (6).

(ii) θ_1 is given. Equation $(2)_1$ and $(4.47)_2$ yield for $B(b_1, d_1)$:

$$d_1 q/H + 1 = \cosh(qb_1/H), \qquad \tan\theta_1 = \sinh(qb_1/H)$$

Eliminating b_1: $\quad 1 = \cosh^2(qb_1/H) - \sinh^2(qb_1/H) = (d_1 q/H + 1)^2 - \tan^2\theta_1 \;\Rightarrow$

$$d_1 q/H + 1 = \sec\theta_1 \tag{7}$$

Substituting this and $d_2 = d + d_1$ in eq. (4) $\Rightarrow$

$$b = (H/q)[\cosh^{-1}(\sec\theta_1) + \cosh^{-1}(dq/H + \sec\theta_1)]$$

which is solved numerically for H/q. Then eq. (7) yields d_1 and eq. (6) yields $T_{\max}$.

4.16.4 Cable subjected to discrete transverse loads

Consider a flexible cable (Fig. 4.27a) subjected to n discrete loads which are much larger than the weight of the cable, which is neglected. The cable segments between successive loads are straight with constant tension in each. Given the location h_{n+1}, b_{n+1} of the support, we have to find the tensions T_i, $i = 1, \ldots, n+1$, in the cable segments.

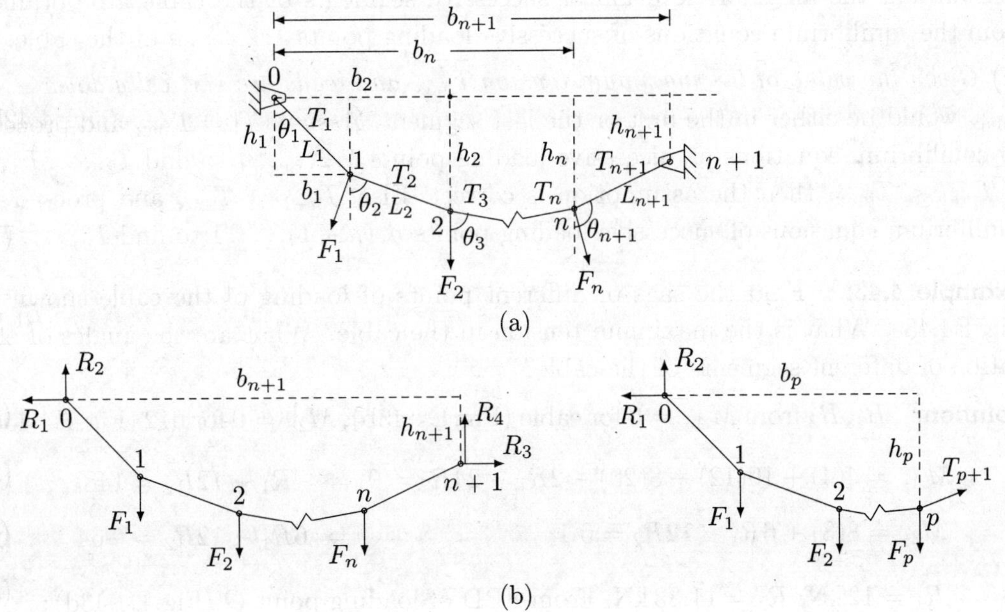

Fig. 4.27

We consider two types of problems.

1. Lengths $L_i, i = 1, \ldots, n+1$ of various segments and all loads are given The $(2n+2)$ unknowns: the inclinations θ_i of the segments and T_i are solved from the $2n$

equilibrium equations for n loading points and two geometrical relations:

$$\sum_{i=1}^{n+1} L_i \cos\theta_i = h_{n+1}, \qquad \sum_{i=1}^{n+1} L_i \sin\theta_i = b_{n+1}$$

The solution of these non-linear algebraic equations is quite complex.

2. The x coordinates $b_1, \ldots, b_n$ of all loading points and all loads are given
The n sags $h_1, \ldots, h_n$ and $(n+1)$ tensions $T_1, \ldots, T_{n+1}$ are governed by $2n$ equilibrium equations of n loading points. To make it statically determinate, an additional parameter has to be specified as follows.

(a) *Given the length L of the cable*

$$L = \sum_{i=1}^{n+1} [(h_i - h_{i-1})^2 + (b_i - b_{i-1})^2]^{1/2} \qquad \text{with} \quad h_0 = b_0 = 0$$

The solution of the nonlinear governing equations is complicated.

(b) *Given the sag of one loading point (say pth one) and loads are vertical*
Reactions R_1, R_2 are solved using the equilibrium equation $M_{n+1} = 0$ for the whole cable and the equilibrium equation $M_p = 0$ for the cable $012\ldots p$ (Fig. 4.27b). The tension and the angle of elevation of successive segments of the cable are obtained from the equilibrium equations of successive loading points $1, 2, \ldots, n$ of the cable.

(c) *Given the value of the maximum tension $T_{\max}$ and loads are vertically down*
$T_{\max}$ would be either in the first or the last segment. *Assume* $T_1 = T_{\max}$ and proceed to equilibrium equations of successive loading points $1, 2, \ldots, n$ to find $T_2, \ldots, T_{n+1}$. If $T_{n+1} < T_{\max}$, then the assumption is correct. Else, $T_{n+1} = T_{\max}$ and proceed to equilibrium equations of successive loading points $n, (n-1), \ldots, 1$ to find $T_n, \ldots, T_1$.

Example 4.43: Find the sags of different points of loading of the cable shown in Fig. E4.43a. What is the maximum tension in the cable? What are the angles of elevation of different segments of the cable?

Solution: R_1, R_2 from $M_{A_z} = 0$ for cable (Fig. E4.43b), $M_{2_z} = 0$ for 012 (Fig. E4.43c):

$$M_{A_z} = 4(4) + 10(12) + 8(20) - 2R_1 - 24R_2 = 0 \Rightarrow \quad R_1 + 12R_2 = 148 \qquad (1)$$

$$M_{2_z} = 8(8) + 6R_1 - 12R_2 = 0 \qquad \Rightarrow 6R_1 - 12R_2 = -64 \qquad (2)$$

$\Rightarrow \quad R_1 = 12\,\text{kN}$, $R_2 = 11.33\,\text{kN}$. From FBD of loading point O (Fig. E4.43d):

$$F_x = T_1 \cos\theta_1 - R_1 = 0 \qquad \Rightarrow \quad T_1 \cos\theta_1 = 12$$

$$F_y = R_2 - T_1 \sin\theta_1 = 0 \qquad \Rightarrow \quad T_1 \sin\theta_1 = 11.33$$

$$T_1 = (12^2 + 11.33^2)^{1/2} = 16.50\,\text{kN}$$

$$\tan\theta_1 = 11.33/12, \qquad \theta_1 = 43.36°, \qquad h_1 = 4\tan\theta_1 = 3.777\,\text{m}$$

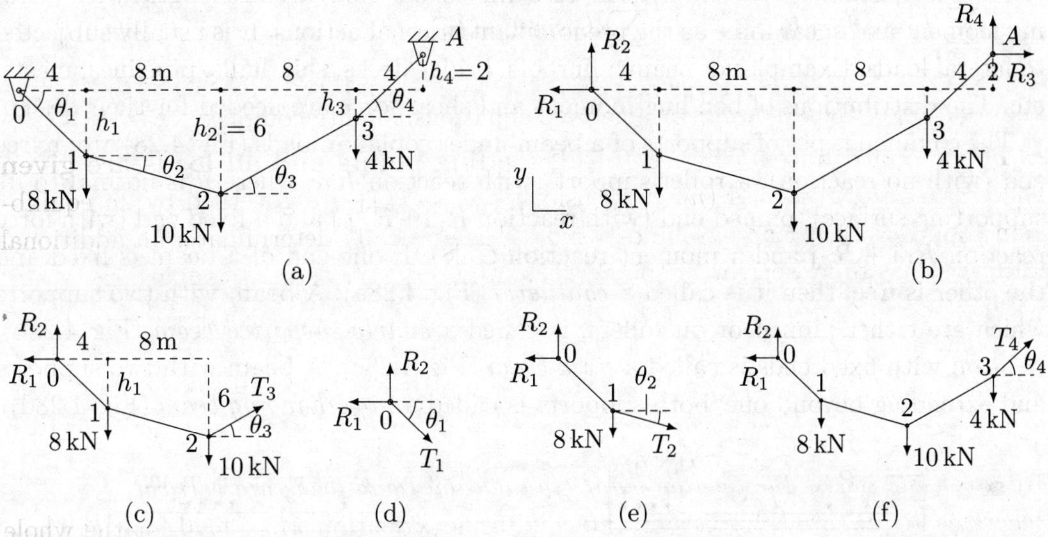

Fig. E4.43

The FBD of part 01 of the cable is shown in Fig. E4.43e.

$$T_2\cos\theta_2 = R_1 = 12, \qquad T_2\sin\theta_2 = R_2 - 8 = 3.333$$

$$T_2 = (12^2 + 3.333^2)^{1/2} = 12.45\,\text{kN}, \qquad \tan\theta_2 = 3.333/12, \quad \theta_2 = 15.52°$$

$$h_2 = h_1 + 8\tan\theta_2 = 3.777 + 8\tan 15.52° = 5.999 \simeq 6\,\text{m} \qquad \text{(it is ok)}$$

The difference is due to round-off error. It provides a check on the calculations. The FBD of part 012 is shown in Fig. E4.43c.

$$T_3\cos\theta_3 = R_1 = 12, \qquad T_3\sin\theta_3 = 8 + 10 - R_2 = 6.667$$

$$T_3 = (12^2 + 6.667^2)^{1/2} = 13.73\,\text{kN}, \quad \tan\theta_3 = 6.667/12, \quad \theta_3 = 29.06°$$

$h_3 = h_2 - 8\tan\theta_3 = 1.555\,\text{m}$. The FBD of part 0123 of cable is shown in Fig. E4.43f.

$$T_4\cos\theta_4 = R_1 = 12, \qquad T_4\sin\theta_4 = 8 + 10 + 4 - R_2 = 10.667$$

$$T_4 = (12^2 + 10.667^2)^{1/2} = 16.06\,\text{kN}, \quad \tan\theta_4 = 10.667/12, \quad \theta_4 = 41.63°$$

$h_4 = 4\tan\theta_4 - h_3 = 2.001\,\text{m}$ is another check which holds.

$$T_{\max} = \max(T_1, T_2, T_3, T_4) = 16.50\,\text{kN}.$$

4.17 Beams

A beam is a member with small transverse dimensions compared to length, with bending moment and shear force as the predominant internal actions. It is usually subjected to lateral loads. Examples of beams: in roofs, bridge decks, ship hulls; pipelines, masts, etc. The distributions of bending moment and shear force are needed for their design.

The common types of supports of a beam under coplanar loads (Fig. 4.28) are: a free end (with no reaction), a roller support (with reaction $R\vec{n}$, where $\vec{n}$ is normal to its supporting surface), pinned end (with reaction $R_1\vec{\mathrm{i}}+R_2\vec{\mathrm{j}}$) and a fixed end (with force reaction $R_1\vec{\mathrm{i}}+R_2\vec{\mathrm{j}}$ and a moment reaction $C_1\vec{\mathrm{k}}$). If one end of a beam is fixed and the other is free, then it is called a *cantilever* (Fig. 4.28a). A beam with two supports which are either pinned or on rollers, is called a *simply-supported beam* (Fig. 4.28b). A beam with fixed ends is called a *fixed beam* (Fig. 4.28c). A beam with two supports and extending beyond one/both supports is called an *overhanging beam* (Fig. 4.28d).

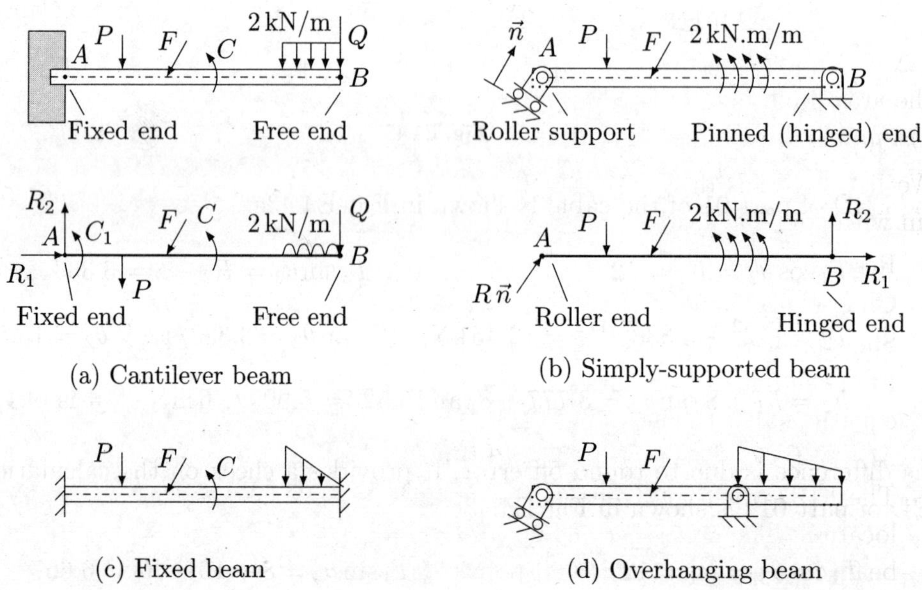

Fig. 4.28

The plots of bending moment $M(x)$, shear force $S(x)$ and axial force $N(x)$ versus x are called bending moment (BM), shear force (SF) and axial force (AF) diagrams. The geometrical interpretation of the differential equations of equilibrium (4.16), viz.,

$$\begin{aligned}(\text{slope of } N\text{-}x \text{ curve}) &= -q_x \\ (\text{slope of } S\text{-}x \text{ curve}) &= -q_y \\ (\text{slope of } M\text{-}x \text{ curve}) &= -S - m, \quad [= -S \quad \text{if } m = 0]\end{aligned} \tag{4.53}$$

is very useful in drawing these diagrams. Integration of eqs (4.16) within a segment in which no discrete load acts, yields

$$\begin{aligned} N(x_2) - N(x_1) &= -\int_{x_1}^{x_2} q_x\,dx = -(\text{area under } q_x\text{-}x \text{ diagram}) \\ &= -(\text{load along } x \text{ over } x_1 \text{ to } x_2) \qquad (4.54) \end{aligned}$$

$$\begin{aligned} S(x_2) - S(x_1) &= -\int_{x_1}^{x_2} q_y\,dx = -(\text{area under } q_y\text{-}x \text{ diagram}) \\ &= -(\text{load along } y \text{ over } x_1 \text{ to } x_2) \qquad (4.55) \end{aligned}$$

$$\begin{aligned} M(x_2) - M(x_1) &= -\int_{x_1}^{x_2} S\,dx - \int_{x_1}^{x_2} m\,dx \\ &= -(\text{area under } S\text{-}x \text{ and } m\text{-}x \text{ diagrams}) \qquad (4.56) \end{aligned}$$

i.e., $\Delta N, \Delta S, \Delta M$ between two cross-sections of a beam equal respectively the *negative* of the area under axial distributed force $q_x(x)$, transverse distributed force $q_y(x)$ and $S(x)+m(x)$ (shear force+ distributed moment) diagrams between these cross-sections.

We list steps for drawing BM, SF and AF diagrams for a statically determinate beam with coplanar loads.

1. Resolve forces in directions transverse and parallel to the beam axis.
2. Choose x-axis along the beam axis and the y-axis normal to it.
3. Show positive directions of S, M, N on the +ve and −ve faces of a beam element.
4. Find the reactions by drawing the FBD of the beam and writing its equilibrium equations. For the beam in Fig. 4.29a R_3 at B and R_2, R_1 at A are obtained from $M_{A_z} = 0$, $F_x = 0$, $F_y = 0$. This step is not strictly necessary for a cantilever.
5. The FBD of the part of a beam on one side of its cross-section differs with its location. The cross-sections at which these FBDs differ *qualitatively* divide the beam into segments. The end-points of these segments are the points where the nature of external loading changes, e.g., at $x = 0, a_1, a_2, a_3, a_4, a_5$ for the beam in Fig. 4.29a.
6. Obtain N, S, M at any section at x within each beam segment using method (a) or (b) described below.

(a) By integrating the differential equations of equilibrium, i.e., by using eqs (4.54) to (4.56) with the values of N, S, M at the beginning of the segment obtained from the values at the end of the previous segment using jump conditions (4.18), if necessary. The right hand side of eqs (4.54) to (4.56) is evaluated geometrically if the area is a rectangle or a trapezium and analytically if $q_y \neq c_1 + c_2 x$, where c_i are constants. This is the simpler method if uniformly distributed and discrete loads act.

(b) By considering equilibrium of a part of the beam to one side of the section at x (A). The part selected should generally be the one which has lesser number of applied loads. In the FBD of this part, N, S, M are drawn in +ve or −ve coordinate directions according as the cut cross-section is a +ve or a −ve face. N, S, M are obtained from $F_x = 0$, $F_y = 0$, $M_{A_z} = 0$. For example, for the beam in Fig. 4.29a, the FBDs for the convenient selected parts are shown in Fig. 4.29b for x in 5 different segments.

7. Calculate N, S, M at the end points of each segment.
8. An extremum (a maximum/minimum) of $S(x)$ occurs where $dS/dx = -q_y$ changes sign. We first locate the segment(s) of the beam within which q_y becomes zero, find the value of x at which this occurs and find S there. The greatest magnitude of S will be either one of these computed values or be the value where a discrete force acts in the y direction/where q_y changes sign. The extrema of N are obtained similarly.

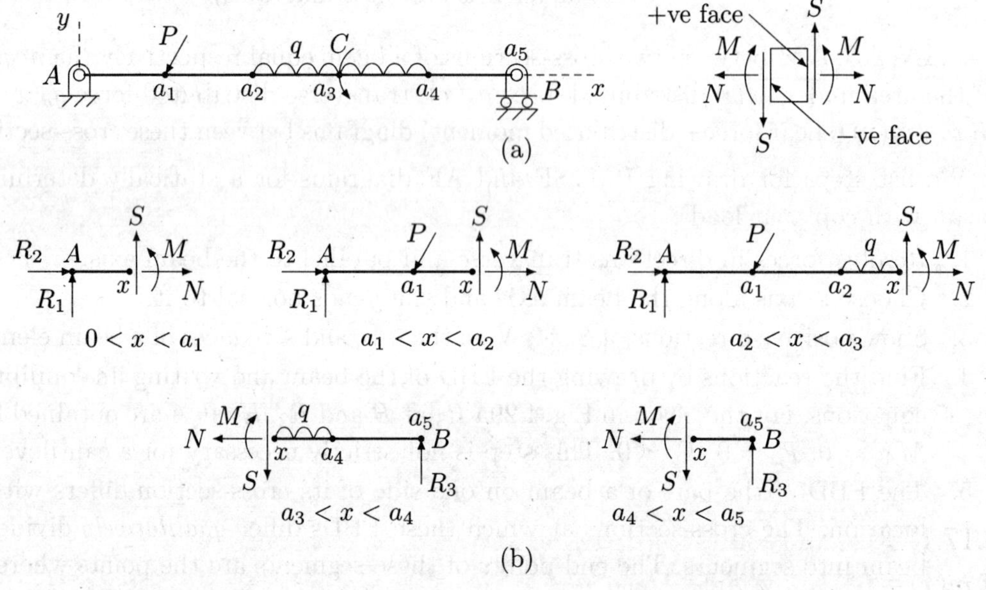

Fig. 4.29

9. Sketch the axial force and shear force diagrams. There will be jumps in S (N) at locations of discrete forces in y (x) direction equal to negative of these forces.
10. An extremum of $M(x)$ occurs where $dM/dx = -S - m$ changes sign. Find M at these sections. The greatest magnitude of M will be either one of these computed values or be the value of M where a discrete moment acts on the beam. For example, $M_{\max}$ in a simply-supported beam subjected to a discrete couple occurs at the location of this couple where S is neither zero nor changes sign (Fig. 4.30a).

11. Sketch the bending moment diagram. There will be jumps in M at locations of discrete applied moments equal to negative of these moments. The points where M changes sign are called *points of contraflexure* since the curvature of the deflected beam changes sign at such points. For the beam in Fig. 4.30a, C is the point of contraflexure where M changes sign. For the beam in Fig. 4.30b, M changes sign if $b < \frac{1}{2}a$ with two points of contraflexure at C_1 and C_2.

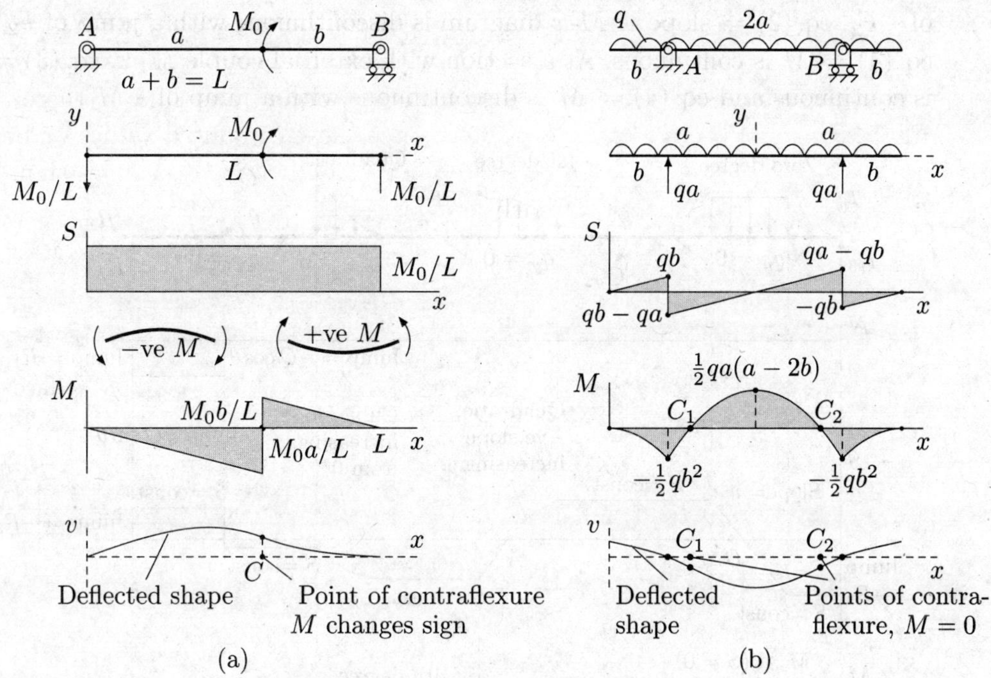

Fig. 4.30

4.17.1 Useful results for drawing BM and SF diagrams

If $m(x) \equiv 0$, then we infer some general results from eqs (4.16) and (4.18):

$$\frac{dS}{dx} = -q_y \quad (1) \qquad \frac{dM}{dx} = -S \quad (2)$$

$$S(x_0^+) = S(x_0^-) - F_{y0} \quad (3) \qquad M(x_0^+) = M(x_0^-) - M_0 \quad (4)$$

1. For a segment without any distributed transverse load on it, i.e., $q_y = 0$, eq. (1) $\Rightarrow$ $S =$ constant and eq. (2) $\Rightarrow$ M-x curve is a straight line with a slope of $-S$.
2. For a segment with a uniformly distributed transverse load $q_y = q_0 \neq 0$, eq. (1) $\Rightarrow$ S-x curve is a straight line with a slope of $-q_0$ and eq. (2) $\Rightarrow$ M-x curve is a parabola with its curvature obtained from the variation of its slope ($= -S$).

3. For a segment with q_y varying linearly with x, eq. (1) $\Rightarrow$ S is quadratic in x and eq. (2) $\Rightarrow$ M is cubic in x. The curvatures of S-x and M-x curves are obtained by studying the variation of their slopes which are respectively $-q_y$ and $-S$.
4. For a segment with a distributed load q_y which varies quadratically with x, eq. (1) $\Rightarrow$ S is cubic in x and eq. (2) $\Rightarrow$ M is of 4th degree in x with the curvatures of S-x and M-x curves obtained as in case 3.
5. At a section with external force $F_2\,\vec{j}$, eq. (3) $\Rightarrow$ S is discontinuous with a jump of $-F_2$, eq. (2) $\Rightarrow$ slope of M-x diagram is discontinuous with a jump of F_2 and eq. (4) $\Rightarrow$ M is continuous. At a section with external couple $M_1\,\vec{k}$, eq. (3) $\Rightarrow$ S is continuous and eq. (4) $\Rightarrow$ M is discontinuous with a jump of $-M_1$.

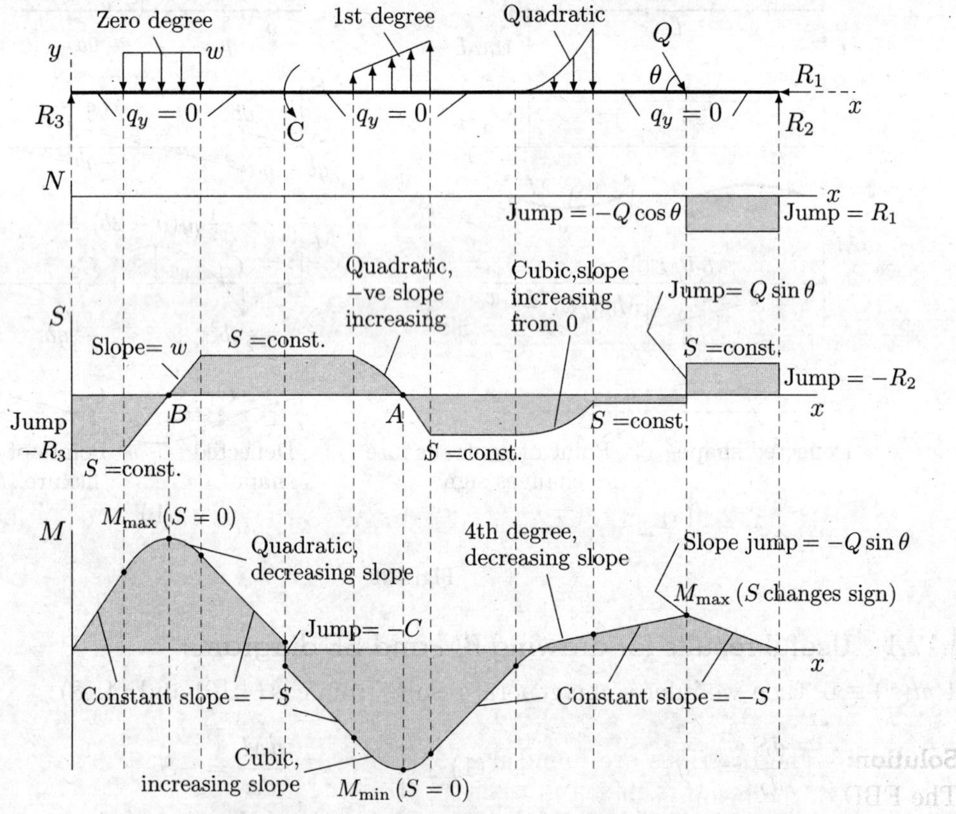

Fig. 4.31

6. *Location of the greatest magnitude of* S is at a section where q_y is zero/changes sign, or is at a location of an external transverse discrete force. *Location of the greatest magnitude of* M is at a section where S is zero/changes sign, or is at a location of an external discrete moment.

All these basic results have been illustrated in Fig. 4.31.

Example 4.44: Draw the AF, SF and BM diagrams for the beam in Fig. E4.44a. $\cos\theta = 0.6$.

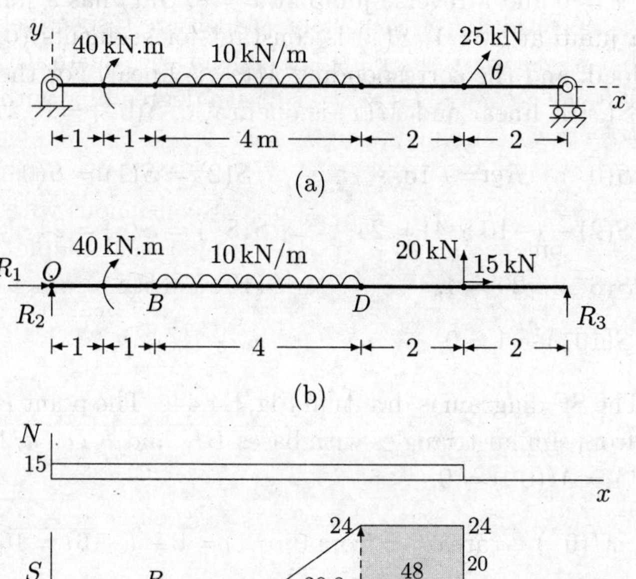

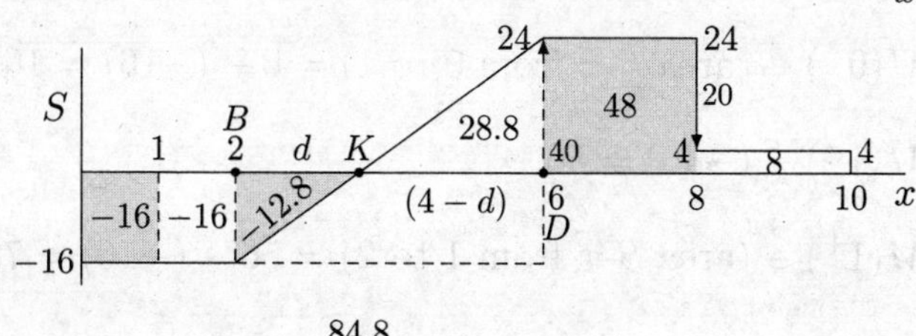

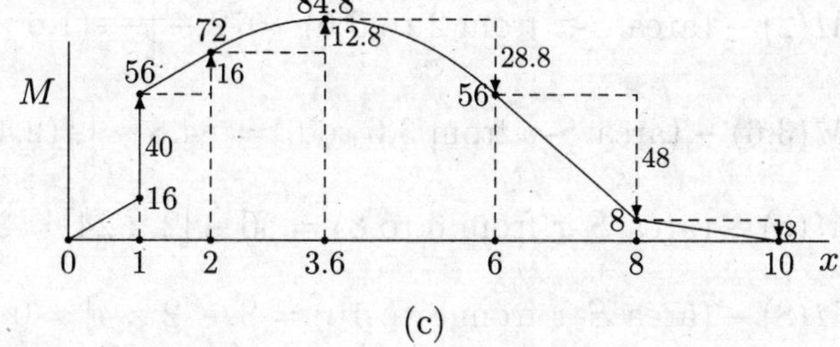

Fig. E4.44

Solution: The reactions are found first since the beam is supported on both sides. The FBD of the beam is shown in Fig. E4.44b. Its equations of equilibrium:

$$M_{Oz} = 10R_3 - 40 - (10 \times 4)4 + 20(8) = 0 \quad \Rightarrow \quad R_3 = 4\,\text{kN}$$

$$F_y = R_2 + R_3 + 20 - 10 \times 4 = 0 \quad \Rightarrow \quad R_2 = 16\,\text{kN}$$

$$F_x = R_1 + 15 = 0 \quad \Rightarrow \quad R_1 = -15\,\text{kN}$$

Since only discrete loads and uniformly distributed load act, we find the N, S and M distributions directly using jump conditions at discrete loads and finding segment-wise $\Delta S = -\Delta F_y$ and $\Delta M = -\Delta$(area under the S-x curve). $N(x)$ has a jump of $-R_1 = 15\,\text{kN}$ at $x = 0$ and a reverse jump at $x = 8$. $S(x)$ has a jumps at $x = 0, 8, 10$ and $M(x)$ has a jump at $x = 1$. $S(x)$ is constant for segments $(0, 2)$, $(6, 8)$, $(8, 10)$ which carry no load, and the corresponding $M(x)$ is linear. For the uniformly loaded segment $(2, 6)$: $S(x)$ is linear and $M(x)$ is quadratic. $S(0^-) = 0$, $M(0^-) = 0$.

$$S(0^+) = S(0^-) - R_2 = -16, \qquad S(2) = S(1) = S(0^+) = -16$$

$$S(6) = S(2) - (-10 \times 4) = 24, \qquad S(8^-) = S(6) = 24$$

$$S(8^+) = S(8^-) - 20 = 4, \qquad S(10^-) = S(8^+) = 4$$

$$S(10^+) = S(10^-) - 4 = 0$$

as it should be. The SF diagram is drawn in Fig. E4.44c. The point K where V changes sign is obtained from similar triangles with bases BK and KD: $d/16 = (4-d)/24 \Rightarrow d = 1.6\,\text{m}$. $M(0^+) = M(0^-) = 0$,

$$\begin{aligned}
M(1^-) &= M(0^+) - (\text{area } S\text{-}x \text{ from 0 to 1}) = 0 - (-16) = 16 \\
M(1^+) &= M(1^-) - (-40) = 56 \\
M(2) &= M(1^+) - (\text{area } S\text{-}x \text{ from 1 to 2}) = 56 - (-16) = 72 \\
M(3.6) &= M(2) - (\text{area } S\text{-}x \text{ from 2 to 3.6}) = 72 - \left[-\tfrac{1}{2}(1.6 \times 16)\right] = 84.8 \\
M(6) &= M(3.6) - (\text{area } S\text{-}x \text{ from 3.6 to 6}) = 84.8 - \left[\tfrac{1}{2}(2.4 \times 2.4)\right] = 56 \\
M(8) &= M(6) - (\text{area } S\text{-}x \text{ from 6 to 8}) = 56 - [2 \times 24] = 8 \\
M(10) &= M(8) - (\text{area } S\text{-}x \text{ from 8 to 10}) = 8 - [2 \times 4] = 0
\end{aligned}$$

The BM diagram is shown in Fig. E4.44c. M varies quadratically in BKD with slope $(= -S)$ decreasing from 16 at B to 0 at K and to -24 at D.

Example 4.45: Draw the axial force, shear force and bending moment diagrams for the beam shown in Fig. E4.45a. $\cos\theta = 0.6$.

Solution: The force system acting on the bracket DKL is replaced (Fig. E4.45b) by its equivalent load at point D consisting of a force $-90\vec{\imath} + 10\vec{\jmath}$ kN and a couple of moment $C_z = 10 \times 0.5 - 90 \times 0.5 = -40\,\text{kN.m}$. The reactions need not be found as the beam is supported on one side. Since the load is discrete or uniformly distributed, we obtain the N (kN), S (kN) and M (kN.m) distributions directly using jump conditions at discrete loads and finding segment-wise $\Delta S = -\Delta F_y$ and

$\Delta M = -\Delta$(area under the S-x curve).

$N(x)$ has a jump of -7.5 at $x = 2$, where force $7.5\vec{i}$ acts and of 90 at $x = 7$ where force $-90\vec{i}$ acts. The N-x diagram is drawn in Fig. E4.45c.

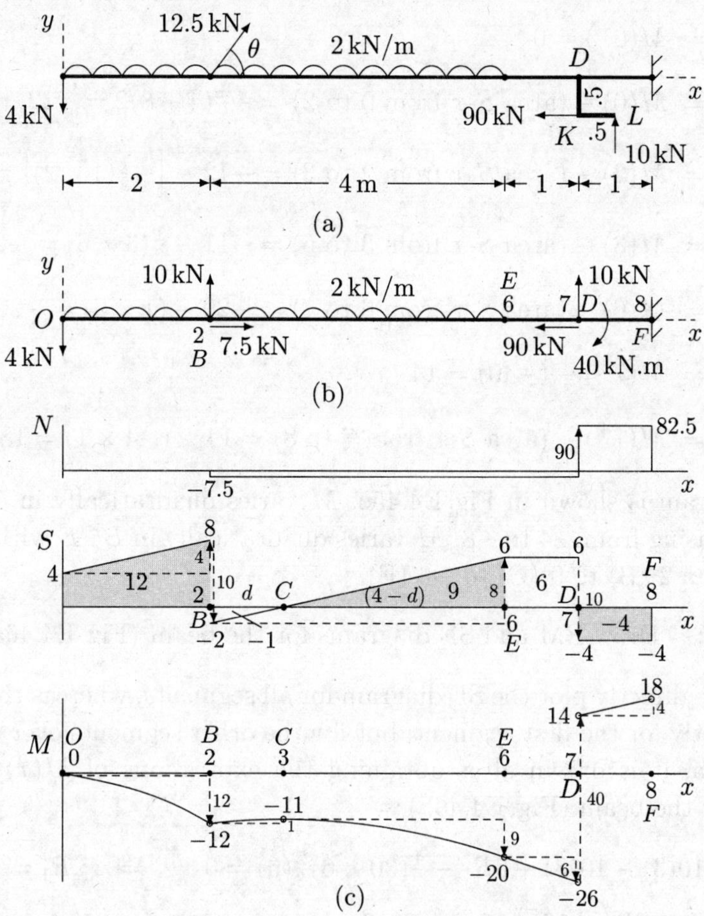

Fig. E4.45

$S(x)$ has jumps at $x = 0, 2, 7$ and $M(x)$ has a jump at $x = 7$. For segments $(6, 7)$ and $(7, 8)$ which are without load: $S(x)$ is constant and $M(x)$ is linear. For the uniformly loaded segments $(0, 2)$ and $(2, 6)$: $S(x)$ is linear and $M(x)$ is quadratic. $S(0^-) = 0$, $M(0^-) = 0$.

$$S(0^+) = S(0^-) - (-4) = 4, \qquad S(2^-) = S(0^+) - (-2 \times 2) = 8$$

$$S(2^+) = S(2^-) - 10 = -2, \qquad S(6) = S(2^+) - (-4 \times 2) = 6$$

$$S(7^-) = S(6) = 6, \qquad S(7^+) = S(7^-) - 10 = -4$$

$$S(8) = S(7^+) = -4$$

The SF diagram is drawn in Fig. E4.45c. The point C where S changes sign is obtained from similar triangles with bases BC and CE: $d/2 = (4-d)/6 \Rightarrow d = 1\,\text{m}$.

$$\begin{aligned}
M(0^+) &= M(0^-) = 0 \\
M(2) &= M(0) - (\text{area } S\text{-}x \text{ from 0 to 2}) = -\tfrac{1}{2}(4+8)2 = -12 \\
M(3) &= M(2) - (\text{area } S\text{-}x \text{ from 2 to 3}) = -12 - \left[-\tfrac{1}{2}(1\times 2)\right] = -11 \\
M(6) &= M(3) - (\text{area } S\text{-}x \text{ from 3 to 6}) = -11 - \tfrac{1}{2}(3\times 6) = -20 \\
M(7^-) &= M(6) - (\text{area } S\text{-}x \text{ from 6 to 7}) = -20 - (1\times 6) = -26 \\
M(7^+) &= M(7^-) - (-40) = 14 \\
M(8) &= M(7^+) - (\text{area } S\text{-}x \text{ from 7 to 8}) = 14 - (-4\times 1) = 18
\end{aligned}$$

The BM diagram is shown in Fig. E4.45c. M varies quadratically in OB with slope $(= -S)$ decreasing from -4 to -8. M varies quadratically in BCE with slope $(= -S)$ decreasing from 2 (B) to 0 (C) to -6 (E).

Example 4.46: Draw BM and SF diagrams for the beam (Fig. E4.46a).

Solution: We directly plot the SF diagram for all segments, whereas the BM diagram is drawn directly for the first segment, but for the other segments over which the load is not uniform, it is drawn after obtaining the expressions of $M(x)$. Equations of equilibrium of the beam (Fig. E4.46b):

$$\begin{aligned}
M_{B_z} &= 10(3) - 10(3) + 6R_1 - \tfrac{1}{2}(30\times 6)\tfrac{2}{3}(6) = 0 \qquad \Rightarrow \qquad R_1 = 60 \\
F_y &= R_2 + R_1 - 10 - 10 - \tfrac{1}{2}(30)6 = 0 \qquad \Rightarrow \qquad R_2 = 50
\end{aligned}$$

$S(x)$ has jumps at $x = 0, 3, 6$ and 9. $M(x)$ has no jump. For the segment $(0, 3)$ which is without load: $S(x)$ is constant and $M(x)$ is linear. For the linearly loaded segments $(3, 6)$ and $(6, 9)$: $S(x)$ is quadratic and $M(x)$ is cubic. $S(0^-) = 0$, $M(0^-) = 0$.

$$\begin{aligned}
&S(0^+) = S(0^-) - (-10) = 10, && S(3^-) = S(0^+) = 10 \\
&S(3^+) = S(3^-) - 50 = -40, && S(6^-) = S(3^+) - \left[-\tfrac{1}{2}(3\times 15)\right] = -17.5 \\
&S(6^+) = S(6^-) - (-10) = -7.5, && S(9^-) = S(6^+) - \left[-\tfrac{1}{2}(15+30)3\right] = 60 \\
&S(9^+) = S(9^-) - 60 = 0
\end{aligned}$$

The SF diagram is drawn in Fig. E4.46c. To find x where S changes sign in the last segment $(6, 9)$ requires an expression for $S(x)$ in that segment. $M(0^+) = M(0^-) = 0$,

$M(3) = M(0) - (\text{area } S\text{-}x \text{ from 0 to 3}) = -10 \times 3 = -30$

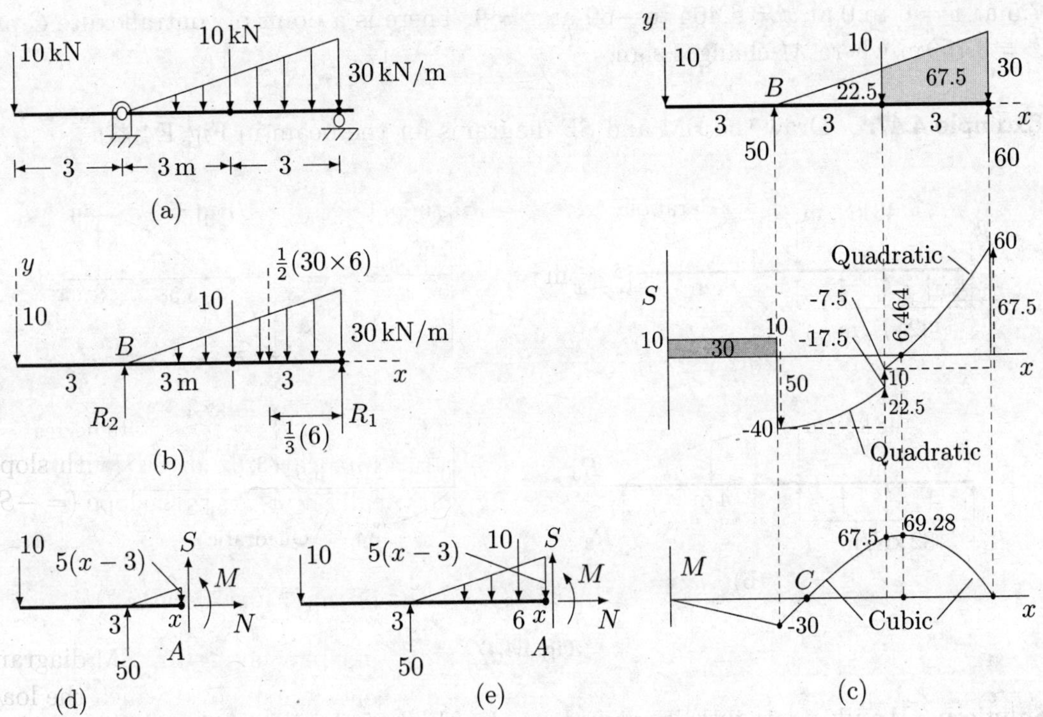

Fig. E4.46

For x in the segment $3 < x < 6$, FBD is drawn in Fig. E4.46d and

$$M_{A_z} = M + 10x + 5(x-3)\tfrac{1}{2}(x-3)\tfrac{1}{3}(x-3) - 50(x-3) = 0$$

$$\Rightarrow \quad M = 40x - 150 - 5(x-3)^3/6 \qquad \Rightarrow \qquad M(6^-) = 67.5$$

and $M = 0$ at $x = 3.759$.

For x in the segment $6 < x < 9$, the FBD is drawn in Fig. E4.46e and

$$F_y = S + 50 - 20 - 5(x-3)\tfrac{1}{2}(x-3) = 0 \quad \Rightarrow \quad S = 2.5(x-3)^2 - 30$$

and $S = 0$ for $x = 3 + \sqrt{12} = 6.464$.

$$M_{A_z} = M + 10x + 5(x-3)\tfrac{1}{2}(x-3)\tfrac{1}{3}(x-3) - 50(x-3) + 10(x-6) = 0$$

$$\Rightarrow\ M = 30x - 90 - 5(x-3)^3/6, \qquad M(6.464) = 69.28$$

$S_{\max} = 60\,\text{kN}$, $M_{\max} = 69.28\,\text{kN.m}$. The BM diagram is shown in Fig. E4.46c. M varies cubically in the segment $(3, 6)$ with the slope $(= -S)$ decreasing from 40 to 17.5. M varies cubically in the segment $(6, 9)$ with the slope $(= -S)$ decreasing from 7.5 at $x = 6$ to 0 at $x = 6.464$ to -60 at $x = 9$. There is a point of contraflexure C at $x = 3.759\,\text{m}$ where M changes sign.

Example 4.47: Draw the BM and SF diagrams for the beam in Fig. E4.47a.

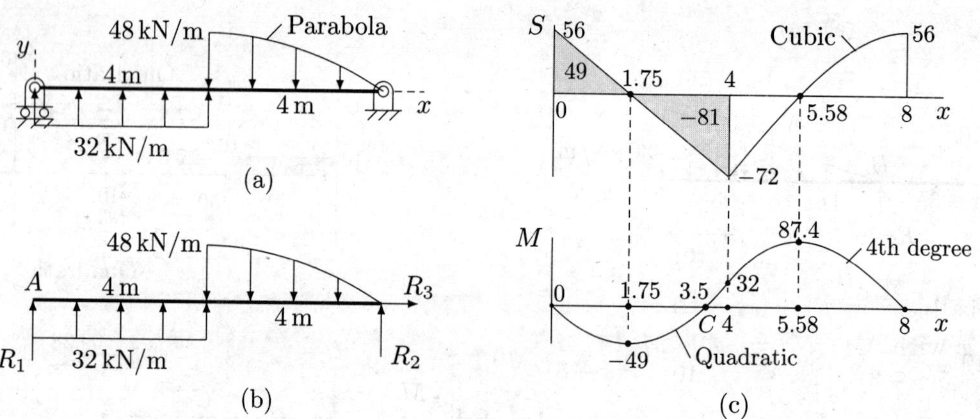

Fig. E4.47

Solution: We illustrate in this example the method of obtaining M and S from the differential equations of equilibrium and the jump conditions. For $0 < x < 4:\ q = 32$. For $4 < x < 8:$ let

$$q = ax^2 + bx + c \qquad \Rightarrow \qquad q(4) = 16a + 4b + c = -48$$

$$q'(4) = 8a + b = 0, \qquad q(8) = 64a + 8b + c = 0$$

$$\Rightarrow\ a = 3, \quad b = -24, \quad c = 0 \qquad \Rightarrow \qquad q = 3x^2 - 24x$$

The FBD of the beam is shown in Fig. E4.47b. The equations of equilibrium yield, $F_x = R_3 = 0$,

$$M_{A_z} = 8R_2 + \int_0^4 x(32\,dx) + \int_4^8 x(3x^2 - 24x)\,dx = 0$$

$$\Rightarrow\ 8R_2 + 32\left(\frac{4^2}{2}\right) + 3\left(\frac{8^4 - 4^4}{4}\right) - 24\left(\frac{8^3 - 4^3}{3}\right) = 0 \quad \Rightarrow \quad R_2 = 56\,\text{kN}$$

$$F_y = R_1 + R_2 + \int_0^4 32\,dx + \int_4^8 (3x^2 - 24x)\,dx = 0$$

$$\Rightarrow \quad R_1 + R_2 + 32(4) + 3\left(\frac{8^3 - 4^3}{3}\right) - 24\left(\frac{8^2 - 4^2}{2}\right) = 0 \quad \Rightarrow \quad R_1 = -56\,\text{kN}$$

At $x = 0$: $S(0^-) = 0, \quad M(0^-) = M(0^+) = 0$ and $S(0^+) = S(0^-) - R_1 = 56\,\text{kN}$.

For $0 < x < 4$: $q_y = 32$ and

$$\frac{dS}{dx} = -q_y = -32$$

$$\Rightarrow \quad S = S(0^+) - 32x = 56 - 32x$$

$$\frac{dM}{dx} = -S = -56 + 32x$$

$$\Rightarrow \quad M = M(0^+) - 56x + 16x^2 = -56x + 16x^2$$

S is linear in x with $S = 0$ at $x = 56/32 = 1.75\,\text{m}$ and $S(4^-) = -72$. M is quadratic in x with $M(1.75) = -49$, $M(4^-) = 32$.

$$S(4^+) = S(4^-) = -72\,\text{kN}, \qquad M(4^+) = M(4^-) = 32\,\text{kN.m}$$

For $4 < x < 8$: $q_y = 3x^2 - 24x$ and

$$\frac{dS}{dx} = -q_y = -3x^2 + 24x$$

$$\Rightarrow \quad S = S(4^+) - 3(x^3 - 4^3)/3 + 24(x^2 - 4^2)/2 = -200 + 12x^2 - x^3$$

$$\frac{dM}{dx} = -S = 200 - 12x^2 + x^3$$

$$\Rightarrow \quad M = M(4^+) + 200(x - 4) - 12(x^3 - 4^3)/3 + (x^4 - 4^4)/4$$

$$= -576 + 200x - 4x^3 + x^4/4$$

$S(8^-) = 56\,\text{kN}$, $S(4^+) = -72\,\text{kN}$ $\Rightarrow$ S is zero at x given by $-200 + 12x^2 - x^3 = 0$. The roots of this equation obtained by Cardan's method are $-3.58, 5.58, 10$, of which 5.58 is the acceptable root in the range $4 < x < 8$. $M(5.58) = 87.4\,\text{kN.m}$, $M(8) = 0$. The SF and BM diagrams are shown in Fig. E4.47c. The point of contraflexure C is in the range $0 < x < 4$, where $M = -56x + 16x^2 = 0 \Rightarrow C$ is at $x = 3.5\,\text{m}$.

Example 4.48: A flagpole, pivoted at its one end, is to be erected by a winch (Fig. E4.48a). Where should the rope be attached so that the greatest bending moment in the beam is as small as possible?

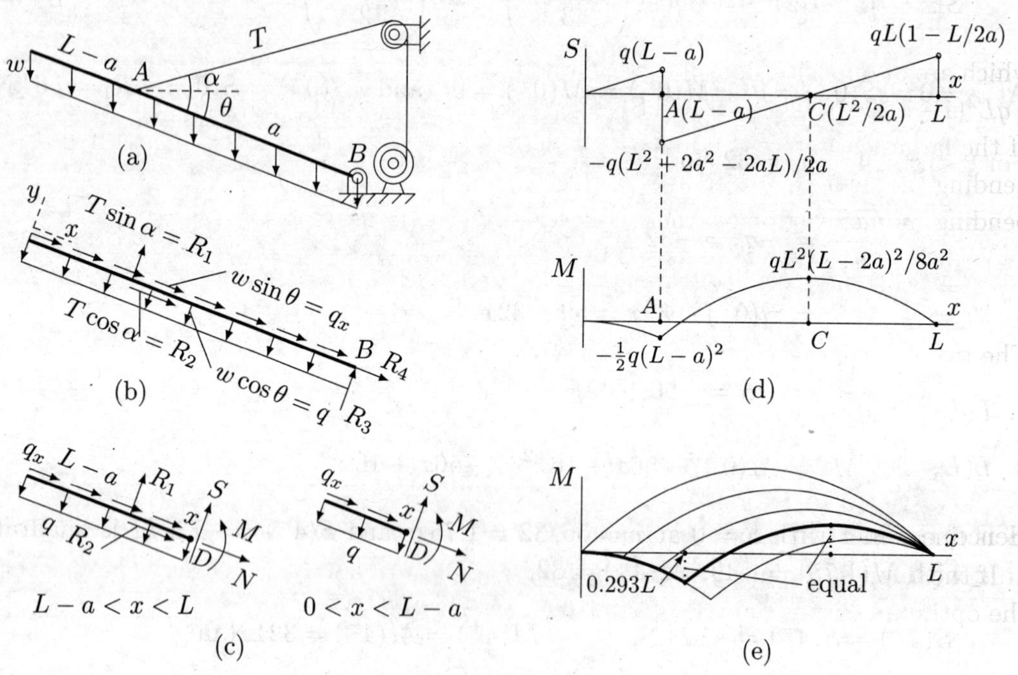

Fig. E4.48

Solution: The FBD of the pole is shown in Fig. E4.48b with the loads resolved in convenient directions. The equilibrium equation

$$M_{Bz} = (qL)\left(\tfrac{1}{2}L\right) - R_1 a = 0 \quad \Rightarrow \quad R_1 = qL^2/2a$$

FBDs of parts of the beam are shown in Fig. E4.48c for two ranges of x. Their equilibrium equations yield S, M. For $0 < x < (L-a)$:

$$F_y = S - qx = 0 \qquad \Rightarrow \qquad S = qx$$

$$M_{Dz} = M + \tfrac{1}{2}x(qx) = 0 \qquad \Rightarrow \qquad M = -\tfrac{1}{2}qx^2$$

$$\Rightarrow \quad S[x = (L-a)^-] = q(L-a)$$

$$M_A = M[x = (L-a)] = -\tfrac{1}{2}q(L-a)^2$$

For $L - a < x < L$:

$$F_y = S - qx + R_1 = 0 \quad \Rightarrow \quad S = qx - qL^2/2a$$

$$M_{Dx} = M + \tfrac{1}{2}x(qx) - R_1[x - (L-a)] = 0$$

$$\Rightarrow \quad M = -\tfrac{1}{2}qx^2 + qL^2(x - L + a)/2a$$

$$S[x = (L - a)^+] = -q(L^2 + 2a^2 - 2aL)/2a < 0, \qquad S(L) = qL(2a - L)/2a$$

which are of opposite signs if $a > \frac{1}{2}L$ and $S = 0$ at $x = x_C = L^2/2a$, $M_C = M(x_C) = \frac{1}{2}qL^2(1 - L/2a)^2$. The BM and SF diagrams are drawn in Fig. E4.48d. The magnitude of the largest negative bending moment M_A increases and that of the largest positive bending moment M_C decreases with decrease in a for $a > L/2$. Hence, the greatest bending moment will have the smallest value if $M_C = |M_A| = -M_A = \frac{1}{2}q(L-a)^2$:

$$\Rightarrow \quad \tfrac{1}{2}q(L-a)^2 = \tfrac{1}{2}qL^2(1 - L/2a)^2 \quad \Rightarrow \quad L(L - 2a) = \pm 2a(L-a)$$

The two solutions are:

1. $L(L-2a) = 2a(L-a) \quad \Rightarrow \quad a = (1 \pm 0.707)L \Rightarrow$ not admissible as $a \notin (\frac{1}{2}L, L)$
2. $L(L-2a) = -2a(L-a) \Rightarrow \quad a = 0.707L \quad \Rightarrow \quad M_A = 0.0429qL^2$

Hence, the greatest bending moment has the least value $0.0429qL^2$ when $a = 0.707L$.

If the flagpole is pulled at the end, then $M_{\max} = qL^2/8$. Hence, the greatest BM in the optimum case is only 34.3% of $M_{\max}$ when the pull is at the end.

4.17.2 Non-straight beams and three-dimensional loads

Consider a non-straight beam, whose axis is a plane curve in xy-plane, subjected to coplanar loads in xy-plane. The axial force N, shear force S and bending moment M at point A in it are found using $F_x = 0$, $F_y = 0$ (or better $F_n = 0$, $F_t = 0$) and $M_{A_z} = 0$ for a part of the beam on one side of the section. $\vec{e}_t$, $\vec{e}_n$ are in the direction of the tangent and the principal normal to the beam axis. N, S, M are expressed as functions of the coordinate (such as the arc length s) representing the position of a section.

For a beam under 3D load, the internal force resultant $\vec{F}_R$ and moment resultant $\vec{M}_R$ at a section at A are obtained from the equilibrium equations (in vector form) $\vec{F} = \vec{0}$, $\vec{M}_A = \vec{0}$ of a part of the beam to one side of the section. Their components along the tangent $\vec{e}_t$ to the beam axis and normal to it are the axial force $N = \vec{F}_R \cdot \vec{e}_t$, twisting moment $M_t = \vec{M}_R \cdot \vec{e}_t$; and shear force $\vec{S}$ and bending moment $\vec{M}_b$:

$$\vec{S} = \vec{F}_R - N\,\vec{e}_t = (\vec{e}_t \cdot \vec{e}_t)\vec{F}_R - (\vec{F}_R \cdot \vec{e}_t)\,\vec{e}_t = (\vec{e}_t \times \vec{F}_R) \times \vec{e}_t$$

$$\vec{M}_b = \vec{M}_R - M_t\,\vec{e}_t = (\vec{e}_t \times \vec{M}_R) \times \vec{e}_t$$

We illustrate these procedures through solutions of some problems.

Example 4.49: Draw the BM, SF and AF diagrams for the structure in Fig. E4.49a.

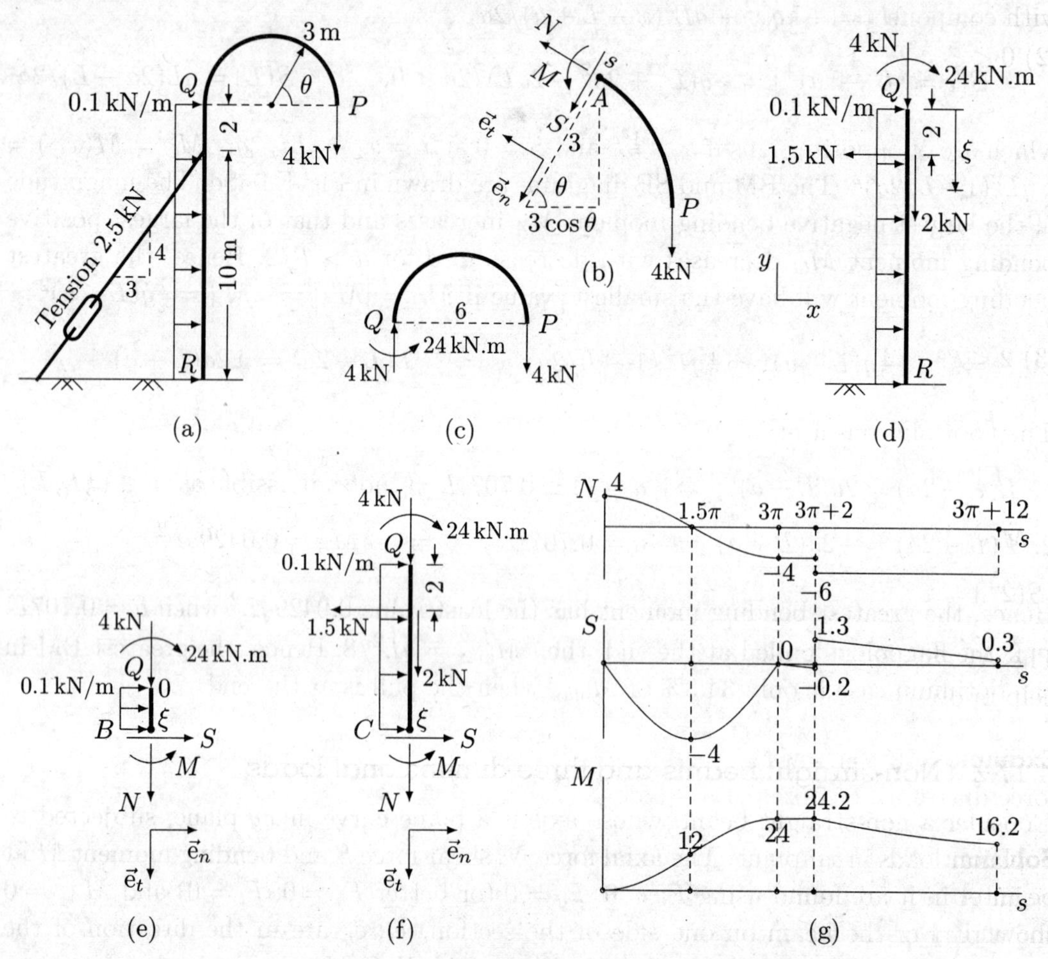

Fig. E4.49

Solution: We find expressions of N (kN), S (kN), M (kN.m) by dividing the beam in 3 segments: (1) the curved segment, (2) the middle segment of 2 m length. and (3) the bottom segment of 10 m length.

(1) $0 < s = 3\theta < 3\pi$: The equations of equilibrium of part PA (Fig. E4.49b) are:

$$F_t = N - 4\cos\theta = 0 \quad \Rightarrow \quad N = 4\cos\theta = 4\cos(s/3)$$

$$F_n = S + 4\sin\theta = 0 \quad \Rightarrow \quad S = -4\sin\theta = -4\sin(s/3)$$

$$M_{A_z} = M - 4 \times 3(1 - \cos\theta) = 0 \quad \Rightarrow \quad M = 12\,[1 - \cos(s/3)]$$

The values at $s = 3\pi$ are: $N = -4\,\text{kN}$, $S = 0$, $M = 24\,\text{kN.m}$. The FBD of part PQ is shown in Fig. E4.49c. The loading on part QR of the beam is shown in Fig. E4.49d with components 1.5 kN and 2 kN of the 2.5 kN tension.

(2) $0 < \xi < 2$: The equations of equilibrium of part QB (Fig. E4.49e) are:

$$F_t = N + 4 = 0 \quad \Rightarrow \quad N = -4$$

$$F_n = 0.1\xi + S = 0 \quad \Rightarrow \quad S = -0.1\xi$$

$$M_{Bz} = M - 24 - 0.1\xi\left(\tfrac{1}{2}\xi\right) = 0 \quad \Rightarrow \quad M = 24 + 0.05\xi^2$$

$$\Rightarrow \quad S(2^-) = -0.2\,\text{kN}, \quad M(2) = 24.2\,\text{kN.m}, \quad M(1) = 24.05\,\text{kN.m}$$

(3) $2 < \xi < 12$: The equations of equilibrium of the part QC (Fig. E4.49f) are:

$$F_t = N + 2 + 4 = 0 \quad \Rightarrow \quad N = -6$$

$$F_n = 0.1\xi - 1.5 + S = 0 \quad \Rightarrow \quad S = -0.1\xi + 1.5$$

$$M_{Cz} = M - 24 - 0.1\xi\left(\tfrac{1}{2}\xi\right) + 1.5(\xi - 2) = 0 \quad \Rightarrow \quad M = 27 - 1.5\xi + 0.05\xi^2$$

$$S(2^+) = 1.3\,\text{kN}, \quad S(12) = 0.3\,\text{kN}, \quad M(12) = 16.2\,\text{kN.m}, \quad M(7) = 18.95\,\text{kN.m}$$

The AF, SF and BM diagrams are shown in Fig. E4.49g. The values of $M(7), M(1)$ help in drawing a proper curvature of the BM diagram.

Example 4.50: Draw the bending moment, twisting moment, axial force and shearing force diagrams for different segments of the beam $ABCD$ shown in Fig. E4.50a.

Solution: The beam is composed of segments AB, BC and CD. Let $P(\xi)$, $Q(\eta)$, $V(\rho)$ be intermediate points within these segments. The FBDs of parts AP, BQ, CV are shown in Figs E4.50b,c,d.

Segment AB : $\vec{e}_t = \vec{k}$. The equilibrium equations of AP are:

$$\vec{F} = \vec{F}_R - 4\vec{k} + 2\vec{j} + \xi^2\vec{j} = \vec{0} \quad \Rightarrow \quad \vec{F}_R = -(2 + \xi^2)\vec{j} + 4\vec{k}$$

$$\vec{M}_P = \vec{M}_R - \xi\vec{k} \times 2\vec{j} - (\tfrac{1}{3}\xi)\vec{k} \times \xi^2\vec{j} = \vec{0} \quad \Rightarrow \quad \vec{M}_R = -(2\xi + \tfrac{1}{3}\xi^3)\vec{i}$$

$$N = \vec{F}_R \cdot \vec{e}_t = 4\,\text{kN}, \quad \vec{S} = -(2 + \xi^2)\vec{j}, \quad S = 2 + \xi^2$$

$$M_t = \vec{M}_R \cdot \vec{e}_t = 0, \quad \vec{M}_b = -(2\xi + \tfrac{1}{3}\xi^3)\vec{i}, \quad M_b = 2\xi + \tfrac{1}{3}\xi^3$$

$\Rightarrow$ $S(0) = 2$, $S(2) = 6\,\text{kN}$, $M_b(0) = 0$, $M_b(2) = 6.667\,\text{kN.m}$. N, S, M_b diagrams are shown in Fig. E4.50e.

Segment BC : $\vec{e}_t = \vec{i}$: It is subjected to $-\vec{F}_R(\xi = 2) = 6\vec{j} - 4\vec{k}$ and $-\vec{M}_R(\xi = 2) =$

$6.667\vec{i}$ at B. The equilibrium equations of BQ are:

$$\vec{F} = 6\vec{j} - 4\vec{k} - 2\eta\vec{k} + \vec{F}_R = \vec{0} \;\Rightarrow\; \vec{F}_R = -6\vec{j} + (4+2\eta)\vec{k}$$

$$\vec{M}_Q = -\tfrac{1}{2}\eta\vec{i} \times (-2\eta\vec{k}) + 6.667\vec{i} - \eta\vec{i} \times (6\vec{j} - 4\vec{k}) + \vec{M}_R = \vec{0} \;\Rightarrow$$

$$\vec{M}_R = -6.667\vec{i} + (4\eta + \eta^2)\vec{j} + 6\eta\vec{k}$$

$$N = \vec{F}_R \cdot \vec{e}_t = 0, \qquad \vec{S} = \vec{F}_R - N\vec{e}_t = -6\vec{j} + (4+2\eta)\vec{k}$$

$$S_y = -6\,\text{kN}, \qquad S_z = 4 + 2\eta, \qquad S = [36 + (4+2\eta)^2]^{1/2}$$

$$M_t = \vec{M}_R \cdot \vec{e}_t = -6.667\,\text{kN.m}, \qquad \vec{M}_b = \vec{M}_R - M_t\vec{e}_t = (4\eta + \eta^2)\vec{j} + 6\eta\vec{k}$$

$$M_{by} = 4\eta + \eta^2, \qquad M_{bz} = 6\eta, \qquad M_b = \eta[6^2 + (4+\eta)^2]^{1/2}$$

$$\Rightarrow S_z(0) = 4, \qquad S_z(2) = 8, \qquad S(0) = 7.211, \qquad S(2) = 10\,\text{kN},$$

$$M_b(0) = M_{by}(0) = M_{bz}(0) = 0, \quad M_{by}(2) = M_{bz}(2) = 12, \quad M_b(2) = 16.97\,\text{kN.m}$$

$N, M_t, S_y, S_z, S, M_{by}, M_{bz}, M_b$ diagrams are shown in Fig. E4.50f.

$$\vec{F}_R(\eta = 2) = -6\vec{j} + 8\vec{k}\ \text{kN}, \qquad \vec{M}_R(\eta = 2) = -6.667\vec{i} + 12\vec{j} + 12\vec{k}\ \text{kN.m}$$

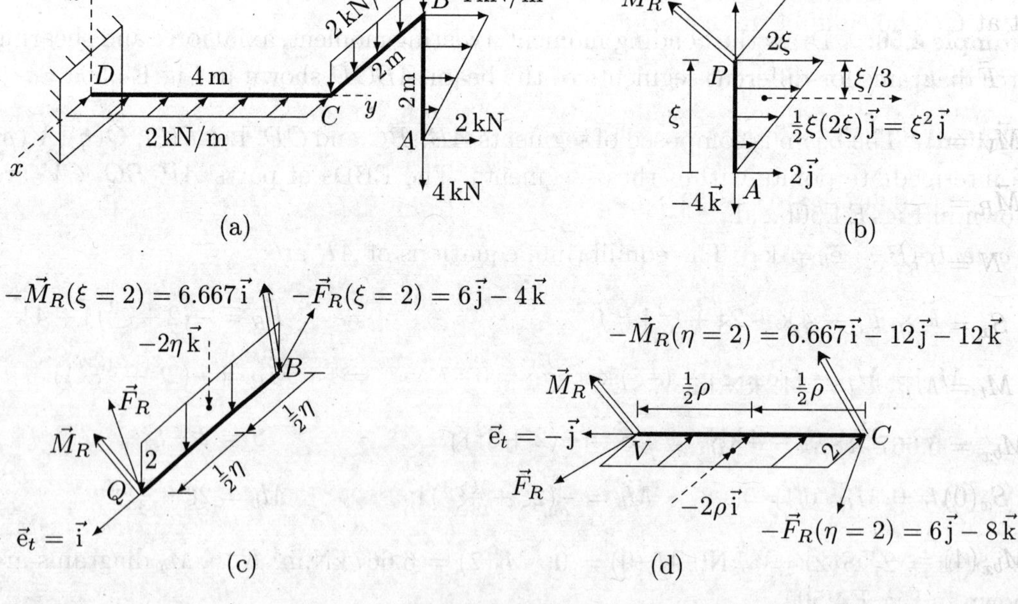

Fig. E4.50a-d

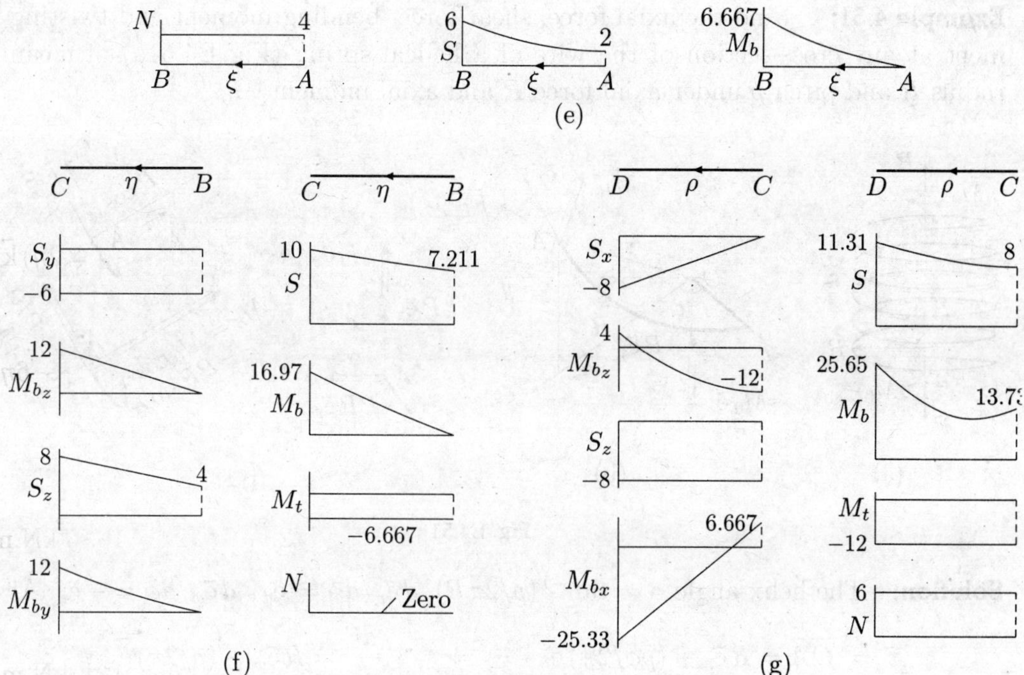

Fig. E4.50e-g

Segment CD: $\vec{e}_t = -\vec{j}$. $-\vec{F}_R(\eta = 2) = 6\vec{j} - 8\vec{k}$ and $-\vec{M}_R(\eta = 2) = 6.667\vec{i} - 12\vec{j} - 12\vec{k}$ act at C. The equilibrium equations of CV are:

$$\vec{F} = \vec{F}_R - 2\rho\vec{i} + 6\vec{j} - 8\vec{k} = \vec{0} \qquad \Rightarrow \qquad \vec{F}_R = 2\rho\vec{i} - 6\vec{j} + 8\vec{k}$$

$$\vec{M}_V = \vec{M}_R + \tfrac{1}{2}\rho\vec{j} \times (-2\rho\vec{i}) + \rho\vec{j} \times (6\vec{j} - 8\vec{k}) + 6.667\vec{i} - 12\vec{j} - 12\vec{k} = \vec{0}$$

$$\vec{M}_R = (-6.667 + 8\rho)\vec{i} + 12\vec{j} + (-\rho^2 + 12)\vec{k}$$

$$N = \vec{F}_R \cdot \vec{e}_t = 6\,\text{kN}, \qquad \Rightarrow \qquad \vec{S} = 2\rho\vec{i} + 8\vec{k}$$

$$S_x = -2\rho, \qquad S_z = -8, \qquad S = (4\rho^2 + 64)^{1/2}$$

$$M_t = \vec{M}_R \cdot \vec{e}_t = -12\,\text{kN.m}, \qquad \Rightarrow \qquad \vec{M}_b = (-6.667 + 8\rho)\vec{i} + (12 - \rho^2)\vec{k}$$

$$M_{bx} = 6.667 - 8\rho, \qquad M_{bz} = \rho^2 - 12 \qquad M_b = [(\rho^2 - 12)^2 + (8\rho - 6.667)^2]^{1/2}$$

$$S_x(0) = 0, \quad S_x(4) = -8, \quad S(0) = 8, \quad S(4) = 11.31\,\text{kN}, \quad M_{bx}(0) = 6.667$$

$$M_{bx}(4) = -25.33, \quad M_{bz}(0) = -12, \quad M_{bz}(4) = 4, \quad M_b(0) = 13.73, \quad M_b(4) = 25.65$$

Since normal to face at V in Fig. E4.50d points in $-\vec{j}$ direction, it is a −ve face. Hence, components of $\vec{F}_R$ and $\vec{M}_R$ in −ve coordinate directions are positive. The diagrams in Fig. E4.50g show the variation of $N, M_t, S_x, S_z, S, M_{bx}, M_{bz}$ and M_b with ρ.

Example 4.51: Find the axial force, shear force, bending moment and twisting moment at any cross-section of the wire of a helical spring (Fig. E4.51a) of mean coil radius R and pitch p under axial force P and axial moment M_0.

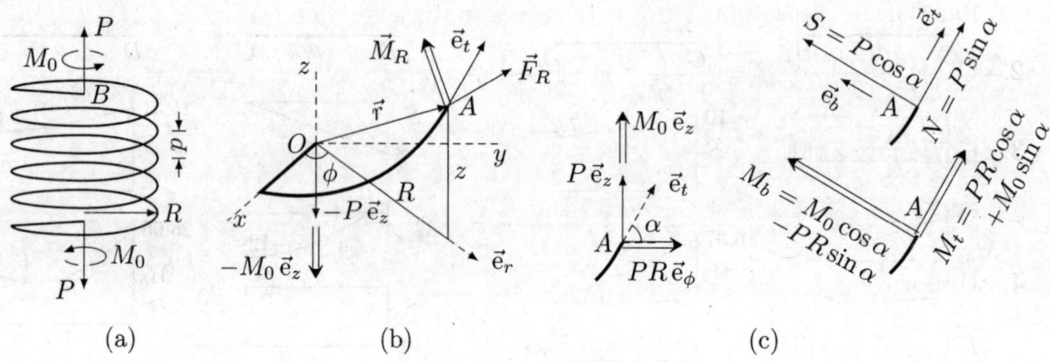

Fig. E4.51

Solution: The helix angle $\alpha = \tan^{-1}(p/2\pi R)$, $d\vec{e}_r/d\phi = \vec{e}_\phi$, $d\vec{e}_\phi/d\phi = -\vec{e}_r$. At A

$$\vec{r} = \overrightarrow{OA} = R\,\vec{e}_r + (p\phi/2\pi)\,\vec{e}_z \quad \Rightarrow \quad \frac{d\vec{r}}{d\phi} = R\,\vec{e}_\phi + (p/2\pi)\,\vec{e}_z$$

$$\Rightarrow \quad \vec{e}_t = \frac{d\vec{r}}{ds} = \frac{d\vec{r}/d\phi}{|d\vec{r}/d\phi|} = \frac{R\,\vec{e}_\phi + (p/2\pi)\,\vec{e}_z}{[R^2 + p^2/4\pi^2]^{1/2}} = \cos\alpha\,\vec{e}_\phi + \sin\alpha\,\vec{e}_z$$

$$\Rightarrow \quad \frac{d\vec{e}_t}{ds} = -\cos\alpha\,\vec{e}_r\,\frac{d\phi}{ds} \quad \Rightarrow \quad \vec{e}_n = -\,\vec{e}_r, \quad \vec{e}_b = \vec{e}_t \times \vec{e}_n = \cos\alpha\,\vec{e}_z - \sin\alpha\,\vec{e}_\phi$$

The equilibrium equations of element OA in Fig. E4.51b are:

$$\vec{F} = \vec{F}_R - P\,\vec{e}_z = \vec{0} \quad \Rightarrow \quad \vec{F}_R = P\,\vec{e}_z$$

$$\vec{M}_A = \vec{M}_R - \vec{r} \times (-P\,\vec{e}_z) - M_0\,\vec{e}_z = \vec{0} \quad \Rightarrow \quad \vec{M}_R = PR\,\vec{e}_\phi + M_0\,\vec{e}_z$$

$$N = \vec{F}_R \cdot \vec{e}_t = P\sin\alpha, \quad \vec{S} = \vec{F}_R - N\,\vec{e}_t \quad \Rightarrow \quad S = (F_R^2 - N^2)^{1/2} = P\cos\alpha$$

$$M_t = \vec{M}_R \cdot \vec{e}_t = PR\cos\alpha + M_0 \sin\alpha, \qquad \vec{M}_b = \vec{M}_R - M_t\,\vec{e}_t$$

$$M_b = (M_R^2 - M_t^2)^{1/2} = (P^2R^2\sin^2\alpha + M_0^2\cos^2\alpha - PRM_0\sin 2\alpha)^{1/2}$$

$$= M_0\cos\alpha - PR\sin\alpha$$

Or directly in Fig. E4.51c as $\vec{e}_t$, $\vec{e}_b$ components of $\vec{F}_R = P\,\vec{e}_z$, $\vec{M}_R = PR\,\vec{e}_\phi + M_0\,\vec{e}_z$.

For a close-coiled spring $p \ll R \Rightarrow \alpha = \tan^{-1}(p/2\pi R) \ll 1 \Rightarrow \sin\alpha \simeq 0$, $\cos\alpha \simeq 1$ $\Rightarrow$ $N \simeq 0$, $S \simeq P$, $M_t \simeq PR$, $M_b \simeq M_0$. The axial force P causes shear force and twisting moment on the wire, while the axial moment M_0 causes a bending moment.

Review 4 C

1. A simple plane (space) truss is built from a triangle (tetrahedron) by successively adding two (three) members to creat new joints. With three (six) properly placed foundation constraints, simple trusses are statically determinate.

2. Cable: $\dfrac{d(T\vec{e}_t)}{ds} + \vec{q} = \vec{0}, \quad \dfrac{dT}{ds} + q_t = 0, \quad \dfrac{T}{\rho} + q_n = 0, \ q_b = 0,$ jump: $\Delta(T\vec{e}_t) = -\vec{F}_0$

3. Coplanar cable: $\dfrac{d(T\cos\theta)}{ds} + q_x = 0, \quad \dfrac{d(T\sin\theta)}{ds} + q_y = 0,$
   and if $q_x = 0$, then: $T\cos\theta = H =$ constant, $\dfrac{d}{ds}\left(\dfrac{dy}{dx}\right) = -\dfrac{q_y}{H}.$

4. Uniform load $-q_0\,dx\,\vec{\jmath}$ on ds: $y = \dfrac{q_0 x^2}{2H}, \quad H = \dfrac{q_0 l^2}{8h}, \quad T = [H^2 + (q_0 x)^2]^{1/2},$
   $L\,/\,l = \frac{1}{2}[1 + (4h/l)^2]^{1/2} + \frac{1}{2}\sinh^{-1}(4h/l)\,/\,(4h/l) \simeq 1 + \dfrac{8h^2}{3l^2}.$

5. Uniform load $-q\,ds\,\vec{\jmath}$ on ds:

$$y = \frac{H}{q}\left[\cosh\frac{qx}{H} - 1\right] = \left[(H^2 + q^2 s^2)^{1/2} - H\right]/q, \qquad \tan\theta = \sinh\frac{qx}{H} = \frac{qs}{H},$$

$$s = \frac{H}{q}\sinh\frac{qx}{H}, \qquad V = T\sin\theta = H\tan\theta = H\sinh\frac{qx}{H},$$

$$T = H\cosh\frac{qx}{H} = H + qy = (H^2 + q^2 s^2)^{1/2}.$$

6. $\dfrac{dN}{dx} = -q_x, \ \dfrac{dS}{dx} = -q_y, \ \dfrac{dM}{dx} = -S - m; \ \Delta N = -F_{x0}, \Delta S = -F_{y0}, \Delta M = -M_0$

   $\Delta N = -$(area under q_x-x curve), $\Delta S = -$(area under q_y-x) curve,

   $\Delta M = -$(area under S-x and m-x curves).

   For no load segment: S is constant and M varies linearly. For uniform load segment: S varies linearly and M varies quadratically.

   An extreme value of S occurs where q_y changes sign, or where a transverse discrete force acts. An extremum of M occurs where $S+m$ changes sign, or where a discrete moment acts. The point of contraflexure occurs where M changes sign.

4.18 Concept Review Questions 4C

25. (i) Find the zero-force members in the trusses in Fig. Q4.25 and specify the simple trusses. (ii) Find the forces in members 1 and 27 of the truss in Fig. Q4.25a by taking a zig-zag section across the truss. [(i) (a) 2, 8, 18, 20, 25, 26 (b) 3, 5, 7 (c) 3, 5, 6, 7, 8, 9, 11, 12. Only (b) is simple truss (ii) −13.33 kN, −26.67 kN]

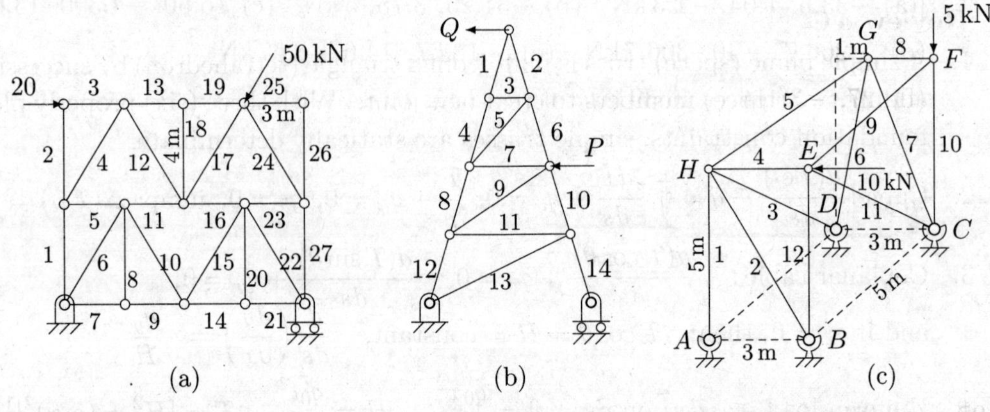

Fig. Q4.25

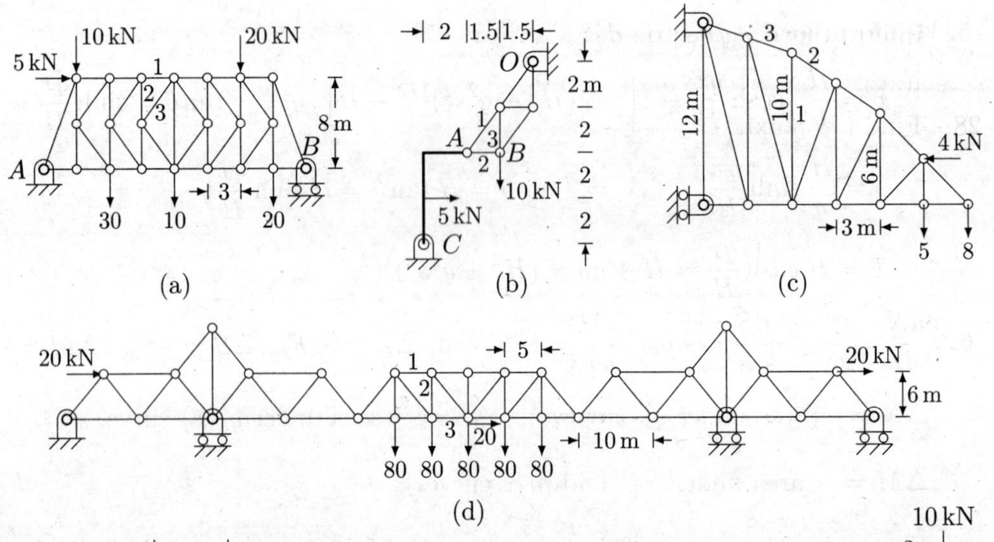

Fig. Q4.26

26. Find, if possible, the forces in members 1, 2, 3 in Fig. Q4.26. In case (g), the diagonal members are slender and so carry either zero force or only tension.

[(a) −33.6, 1.04, −1.3 kN (b) −31.25, 3.75, 5 kN (c) 15.50, −4.300, 13.60 kN (d) −266.7, −40, 306.7 kN (e) −18.67, 14.67, 8.67 kN (f) 27, −39.66, −46.39 kN (g) 72.15, 0, 14.14 kN]

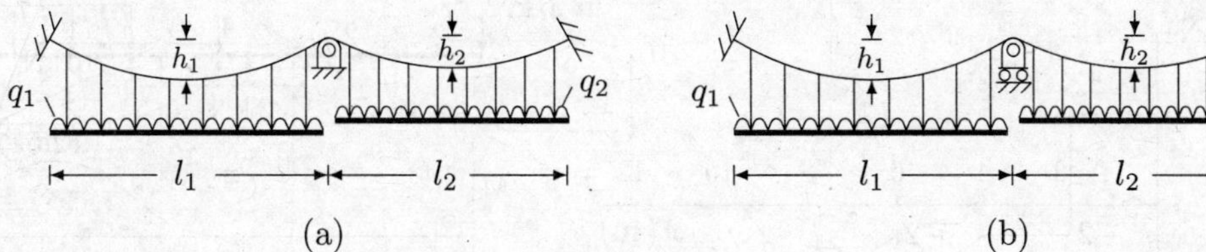

Fig. Q4.27

27. (a) A cable supports uniform vertical load q_1, q_2 per unit of its projected length in the spans l_1, l_2 and passes over an intermediate support of a free pulley (Fig. Q4.27a). Find the relation between h_1 and h_2. (b) Rework part (a) if the intermediate support is on rollers (Fig. Q4.27b).

[(a) $(\frac{1}{2}q_1l_1)^2+(q_1l_1^2/8h_1)^2 = (\frac{1}{2}q_2l_2)^2+(q_2l_2^2/8h_2)^2$ (b) $q_1l_1 = q_2l_2$ & $l_1h_2 = l_2h_1$]

28. Find the maximum BM in the beams shown in Fig. Q4.28.

[(a) 225 kN.m (b) 1.5aP]

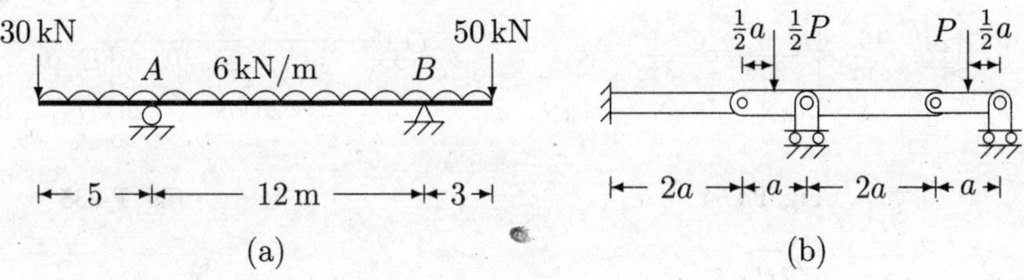

Fig. Q4.28

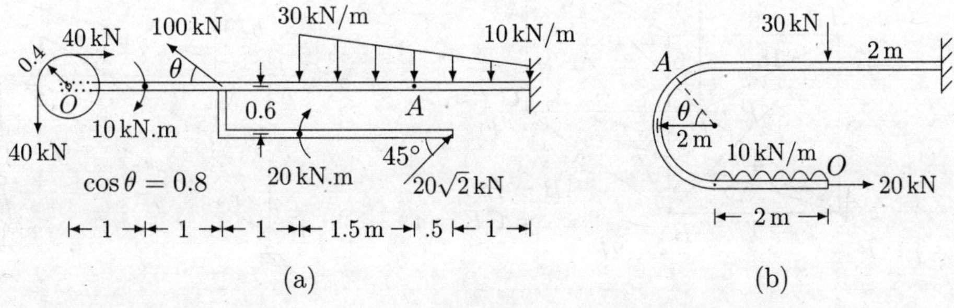

Fig. Q4.29

29. Draw the FBD of part OA of the beams in Fig. Q4.29 and find the AF, SF and BM for the section at A. [(a) 20 kN, −2.5 kN, −52 kN.m

(b) $20(\cos\theta - \sin\theta)$ kN, $20(\sin\theta + \cos\theta)$ kN, $20(2\cos\theta - 2\sin\theta - 1)$ kN.m]

30. Draw the loading and BM diagrams of a beam subjected to transverse coplanar forces, whose SF diagram is shown in Fig. P4.30.

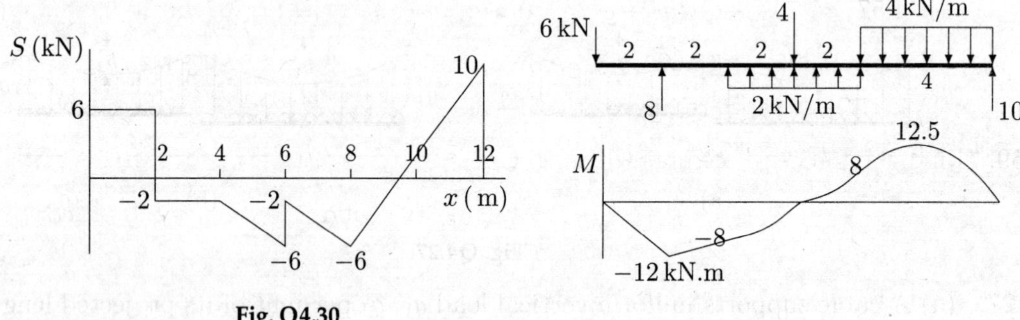

Fig. Q4.30

4.19 Practice Problems 4C

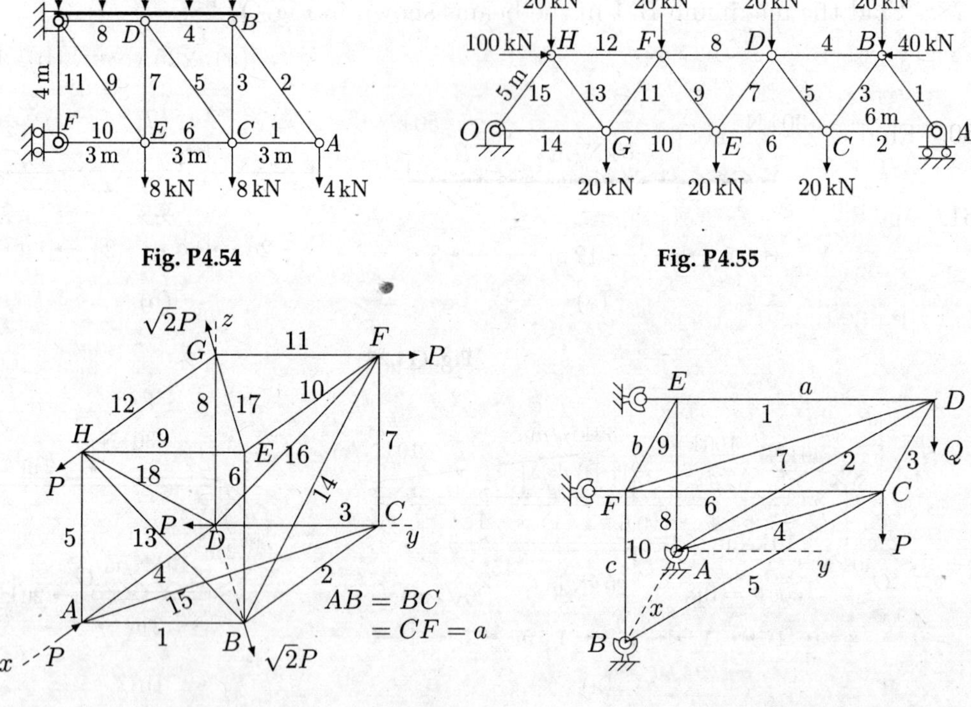

Fig. P4.54

Fig. P4.55

Fig. P4.56

Fig. P4.57

54. to 58. Find the forces in members of the trusses given in Figs P4.54 to P4.58.

[P4.54 −3, 5, −7, 3, 18.75, −14.25, −21, 14.25, 36.25, −36, 0 kN]

[P4.55 −100, 60, 75, −145, 135, 25, −190, 0, 150, −25, −175, 50, 105, −75 kN]

[P4.56 non-zero: $F_1 = F_2 = F_3 = F_{11} = F_{12} = P$, $F_{15} = -\sqrt{2}P$]

[P4.57 aQ/c, $-Q\sqrt{a^2+c^2}/c$, 0, 0, $-P\sqrt{a^2+c^2}/c$, aP/c, 0, 0, 0, 0]

[P4.58 9.375, −15.625, 5, 9.375, 4.375, −12.65, 1.677, 11.25, 5, 11.25, 1.677, −12.65, 4.375, 9.375, 5, 9.375, −15.625 kN]

59. Find the forces in members 1 and 2 of the truss shown in Fig. P4.59.

[$mg/12\sin\alpha_k$, $(\cot\alpha_{k+1} - \cot\alpha_k)mg/24\cos 75°$]

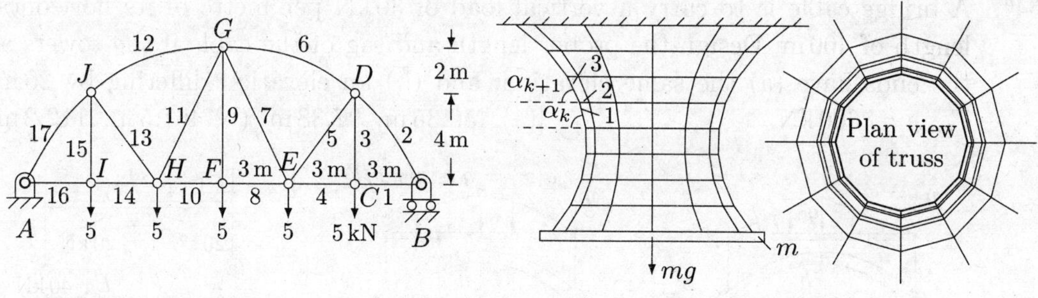

Fig. P4.58

Fig. P4.59

60. Find forces in members 1, 2, 3, 4 of the truss given in Fig. P4.60. What value of a/b makes it a critical form? [$\{-3c,\ 0,\ 3\sqrt{b^2+c^2},\ (2a-b)c/a\}P/(a-2b)$; 2]

61. and 62. Find the forces in members of the non-simple trusses in Figs P4.61 and P4.

[P4.61 1.5, 1.5, 1.5, 1.5, −5.408, 0, 0, 0, 4, 0, −1.803, −1.803, −5.408 kN]

[P4.62 −6.591, 10.28, −5.682, −5.682, −6.591, 10.28, 4.545, −2.811, −2.811 kN]

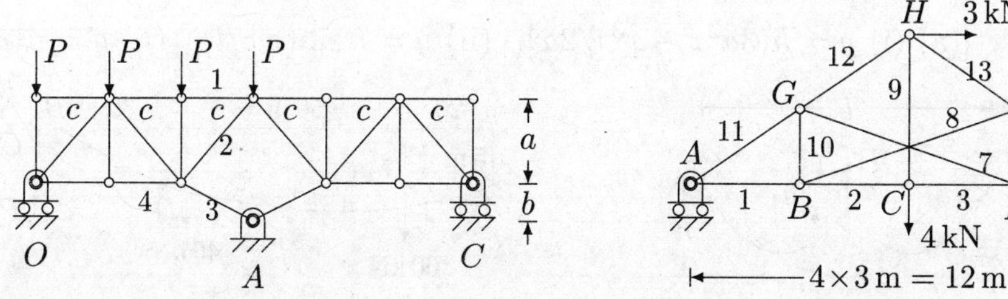

Fig. P4.60

Fig. P4.61

63. Find the tensions at the ends of a cable (Fig. P4.63) and its maximum deviation from AB, for uniform load q_0 N/m of its horizontal projected length.

[$\{H^2+(H\cot\alpha+q_0l)^2\}^{1/2}$, $H/\sin\alpha$ with $H = q_0l^2/2(d-l\cot\alpha)$; $\dfrac{l(d-l\cot\alpha)}{4(l^2+d^2)^{1/2}}$]

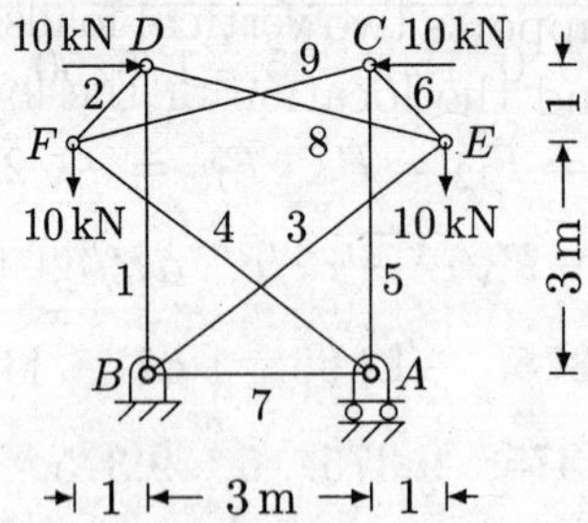

Fig. P4.62

A
l
d
α
q_0
B

Fig. P4.63

64[d]. A bridge cable is to carry a vertical load of 30 kN per metre of its horizontal length of 400 m. Design the proper length and sag of the cable if the towers at the ends have (a) the same elevation, and (b) an elevation differing by 20 m. $T_{\max} = 8600$ kN. [(a) 456.35 m, 97.38 m (b) 461.5 m, 112.2 m]

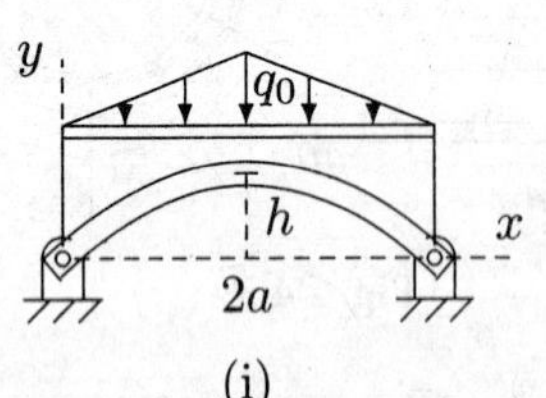

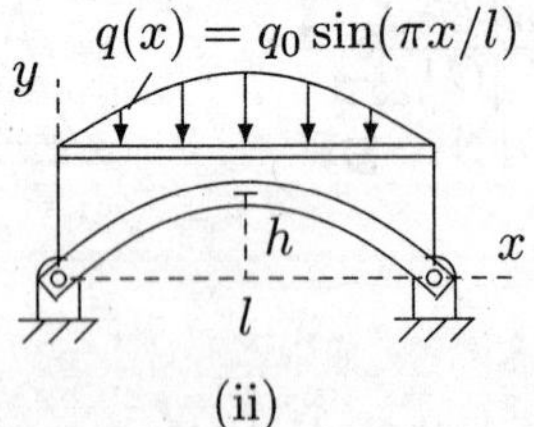

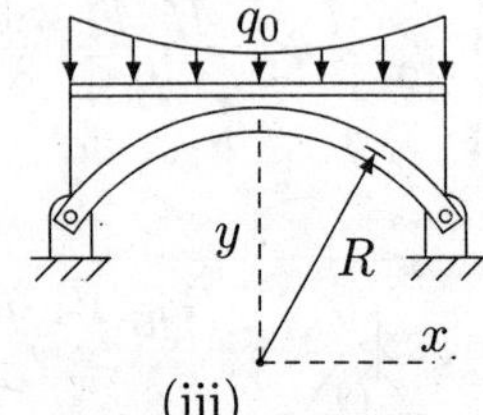

Fig. P4.65

65[d]. (a) What should be the shapes of the centre-line of the arches (i) and (ii) in Fig. P4.65 so that they carry only the axial thrust? (b) What vertical loading on a circular arch (iii) in Fig. P4.65 gives the line of thrust coinciding with its centre-line? The load intensity at the peak of the arch is q_0.

[(a) (i) $y = h(3a^2x - x^3)/2a^3$, (ii) $y = h\sin(\pi x/l)$ (b) $q = q_0R^3/y^3$]

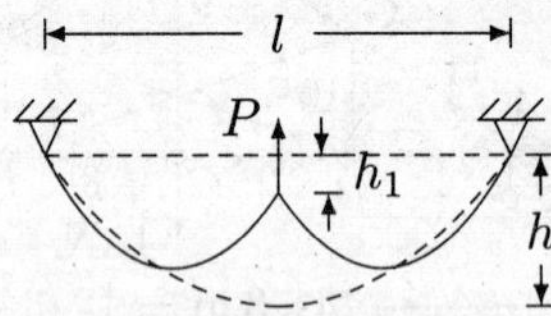

Fig. P4.66

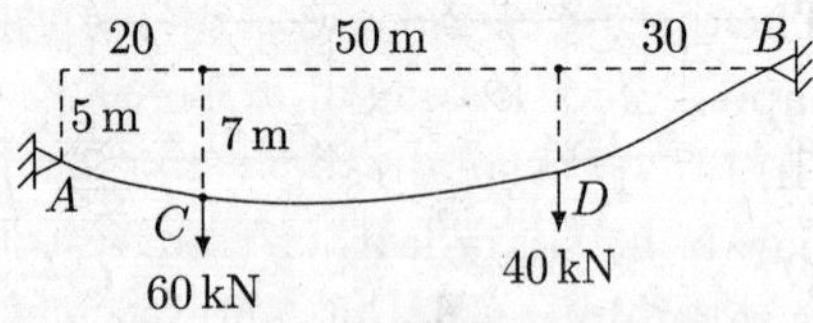

Fig. P4.67

66. A uniform inextensible cable of weight W and length L is suspended between fixed points at the same level with a small value of sag-to-span ratio h/l. An upward force P is applied at the centre to reduce the sag to h_1 (Fig. P4.66). Use $L \simeq l + 8h^2/3l$ to show that $h_1 = (1 - 2P/W)h/[1 + 3(P/W - 1)P/W]^{1/2}$.

67. A cable has a self-weight of 1 kN/m and supports two vertical loads of 40 kN and 60 kN (Fig. P4.67). C is 2 m below A. Find the location of the lowest point of the cable and the maximum tension in it. [C, 678 kN]

68. (a) A cable of length L and self-weight q N/m is translated at constant speed by force F (Fig. P4.68a) up an inclined plane. The coefficient of friction between the cable and the ground is μ. Find F. (b) A boat drags a chain of length 150 m and weight 150 N/m in water, at constant speed (Fig. P4.68b). A 15 kN thrust is developed by the boat as it moves against a water resistance of 10 kN. The lift exerted by the boat on the cable is 6 kN. Find the height h of point A above B, the length b of the chain being dragged at the bottom and the coefficient of friction μ between the cable and the ground.

[(a) $\sqrt{H^2+V^2}$ with $H = qbc(s+\mu c)$, $V = qbs(s+\mu c)+q(L-b)$, $s = \sin\alpha$, $c = \cos\alpha$, $b = [c_2 - \sqrt{c_2^2 - c_1c_3}]/c_1$, $c_1 = 1-2s^2-2\mu sc$, $c_2 = Lc(c-\mu s)+h(s+\mu c)$, $c_3 = L^2-h^2$

(b) 18.73 m, 110 m, 0.3030]

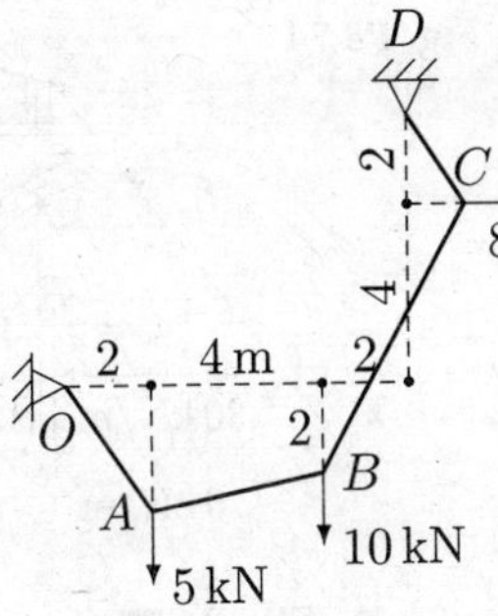

Fig. P4.69

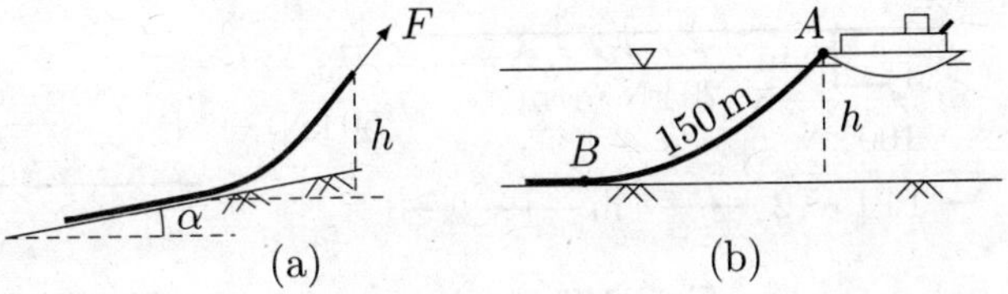

Fig. P4.68

69. Find the location of points A and C and the maximum tension in the cable in Fig. P4.69. Neglect the self-weight of the cable.

[A is 2.136 m below O, C is 0.6818 m right of D, 11.12 kN]

70. The distance l between two points at the same height measured with a measuring tape is L with a sag of h. (a) Show that $l = [(L^2/4-h^2)/h]\ln[(L/2+h)/(L/2-h)]$. (b) If h/L is small, show that $l \simeq [1-8h^2/3L^2]L$. The last result corresponds to a parabolic shape rather than a catenary shape. (c) If $h/L \ll 1$ and the extension is taken into account, show that $l \simeq L-8h^2/3L+\rho gL^3/8hE$, where ρ is density and E is Young's modulus.

71[d]. A transmission line cable of weight 150 N/m can sustain maximum tension of 25 kN. It is suspended from towers at equal elevation and 200 m apart. Find the minimum length of the cable that can be used and its sag. [222.3 m, 42.49 m]

72. to 77 Draw the AF, SF and BM diagrams for the beams in Figs P4.72 to P4.77. Find $N_{\max}$, $M_{\max}$, S_{max}. Locate points of contraflexure, if any. $P =$

$5q_0L/6, M_0 = q_0L^2/6$.

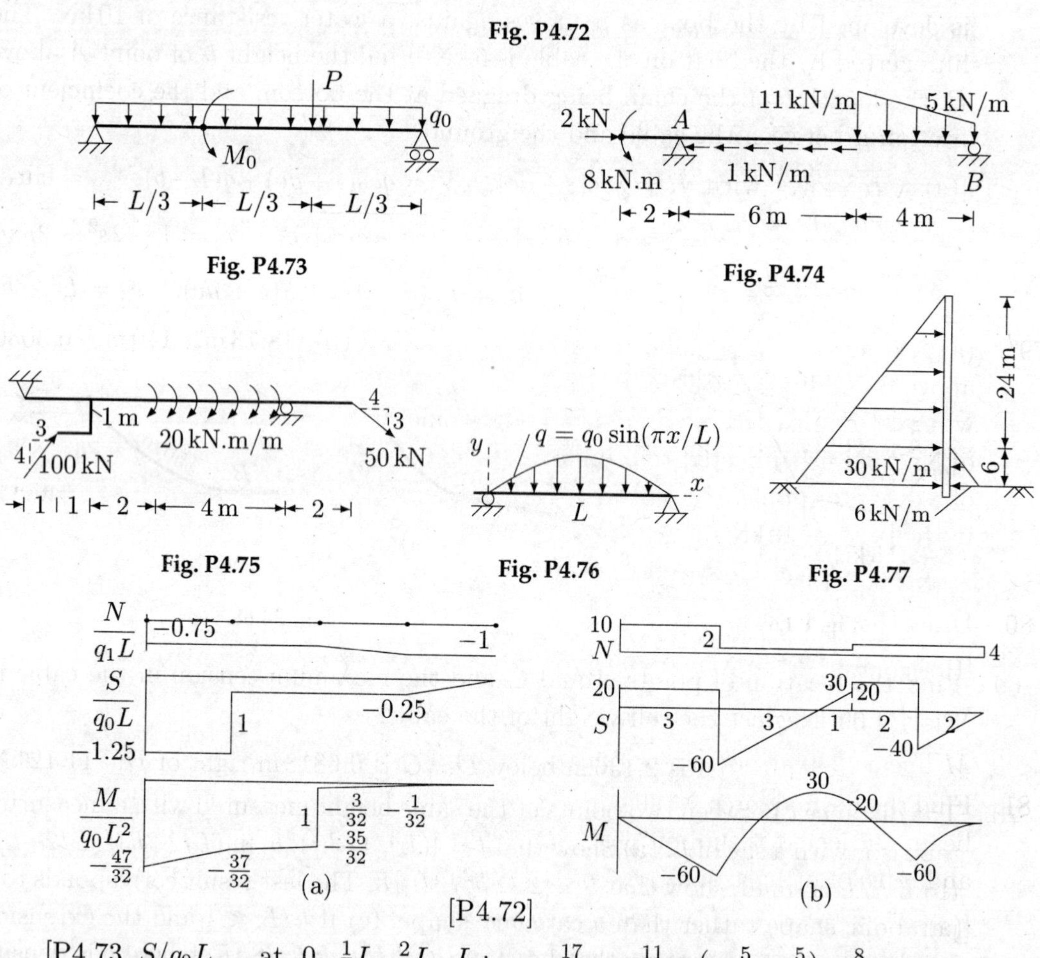

Fig. P4.72

Fig. P4.73

Fig. P4.74

Fig. P4.75

Fig. P4.76

Fig. P4.77

[P4.72]

[P4.73. S/q_0L at $0, \frac{1}{3}L, \frac{2}{3}L, L$: $-\frac{17}{18}, -\frac{11}{18}, (-\frac{5}{18}, \frac{5}{9}), \frac{8}{9}$;

M/q_0L^2 at $0, \frac{1}{3}L, \frac{2}{3}L, L$: $0, (\frac{7}{27}, \frac{5}{54}), \frac{13}{54}, 0$; $\frac{17}{18}q_0L, \frac{7}{27}q_0L^2$]

[P4.74 N at 0, 2, 8, 12: −2, (−2, −6), 0, 0; S at 0, 2, 8, 8.767, 10: 0, (0, −8), −8, 0, 24
M at 0, 2, 8, 8.767, 12: −8, −8, 40, 43.01, 0 kN.m; 6 kN, 24 kN, 43.01 kN.m]

[P4.75 N at 0, 2, 10: 100, (100, 40), 40; S at 0, 2, 8, 10: 80, (80, 0), (0, −30), 30;
M at 0, 2, 4, 8, 10: 0, (−160, −140), −140, −60, 0; 100 kN, 80 kN, 160 kN.m]

[P4.76 $-(q_0L/\pi)\cos(\pi x/L)$, $(q_0L^2/\pi^2)\sin(\pi x/L)$; q_0L/π, q_0L^2/π^2]

[P4.77 S at 24, 30: $-288, -432$; M at 24, 30: 2304, 4464; 432 kN, 4464 kN.m]

78[d]. A uniform heavy beam (Fig. P4.78) is to be lifted by two ropes at distance a apart. Find a/L so that the maximum bending moment is as small as possible (This is also the optimum ratio of the distance between the rails to the length of the sleepers supporting them.). Show that the maximum bending moment for the case of optimum location of the ropes is only 17% of the maximum bending moment if the ropes were attached at the ends of the beam. [0.5858]

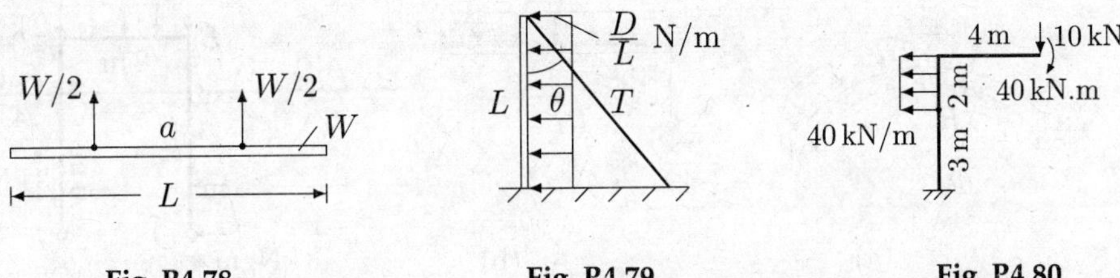

Fig. P4.78 Fig. P4.79 Fig. P4.80

79[d]. (a) A drag force D is uniformly distributed on a radio antenna which protrudes above a building (Fig. P4.79). Find the tension that should be given to the guy wire to minimize the maximum bending moment in the antenna. (b) A cantilever of length L is subjected to a uniformly distributed load q_0. Find the magnitude of a discrete couple M_0 applied at its end so that the maximum bending moment in the beam is as small as possible. Draw the BM diagram for the optimum case.

[(a) $0.414D/\sin\theta$ (b) $\frac{1}{4}q_0L^2$. M varies quadratically from $\frac{1}{4}q_0L^2$ to $-\frac{1}{4}q_0L^2$]

80. Draw the AF, SF and BM diagrams for the beam shown in Fig. P4.80.

[Horizontal part: $N = 0$, $S = 10$ kN, M varies linearly from -40 to -80 kN.m; Vertical part: $N = -10$ kN, $S = -80$ up to 3 m height then linear to 0 at top, M linear: 240 at bottom to 0 at 3m height, then quadratic to -80 kN.m at top]

81. Find the normal force N, shear force S and bending moment M at θ for a circular beam subjected to uniformly distributed load q per unit length in Fig. P4.81a and q per unit horizontal projected length in Fig. P4.81b.

[(a) $-qR$, 0, 0 (b) $qR[\frac{1}{2}(\cos\theta - \sin\theta)(1 - \cos\theta + \sin\theta) - \cos\theta]$, $qR[\frac{1}{2}(\cos\theta + \sin\theta)(1 - \cos\theta + \sin\theta) - \sin\theta]$, $\frac{1}{4}qR^2[4 - 4\cos\theta - (1 - \cos\theta + \sin\theta)^2]$]

82. to 84 Draw the AF, SF, BM and twisting moment diagrams for the beams in Figs P4.82 to P4.84. The system in Fig. P4.83 is in equilibrium.

[P4.82 OB: $N = -1.25$, $S = -.5$ to 4 kN, $M = 0$ to -7.5 cubic, $M_{\max} = .0425$ kN.m

BC: $N = -4$, $S = -1.25$ to 2.75 linear, $M = -7.5$ to -9 quad., $M_{\max} = -7.109$ kN.m

CE: $N = -2.75$ kN, in top half: $S = -4$ kN, $M = 3$ to 9 kN.m linear]

[P4.83 S in N and M_b, M_t in N.m, $x < 0.4: \quad S = 522.4,\ M_b = 522.4x,\ M_t = 0$
$0.4 < x < 0.9: \quad S = 305.9, \quad M_b = 293.54(1 + 1.086x^2 - 1.667x)^{1/2}, \quad M_t = 45$
$x > 0.9$: $S = 452,\ M_b = (1.3 - x)452,\ M_t = 0;\ S_{\max} = 522.4, \quad M_b^{\max} = 209.0$]

[P4.84 OA: $N = 2,\ S = 6.083,\ M_t = 5,\ M_b = [(13 - 6x)^2 + x^2]^{1/2}$: 13 to 2.24
BA: $N = 6, \quad S = 3.606, \quad M_t = -2, \quad M_b = [(2 + 3\xi)^2 + 4\xi^2]^{1/2}$: 2 to 5.38
BC: $N = -3, \quad S = 2, \quad M_t = 0, \quad M_b = 2 - 2\eta$: 2 to 0
CD: $N = -3, \quad S = 0, \quad M_t = 0, \quad M_b = 0$. N, S kN and M_t, M_b kN.m]

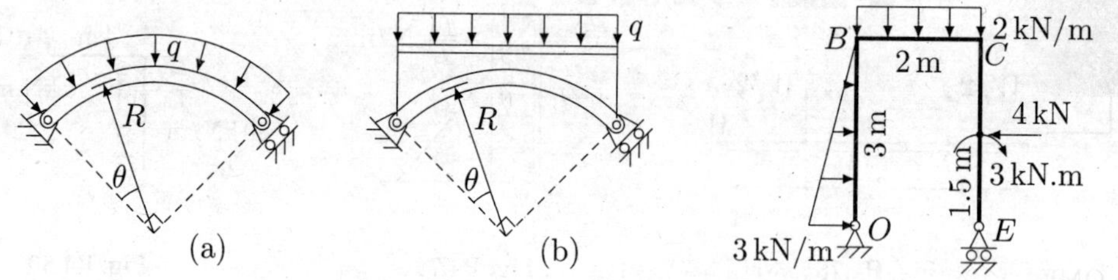

Fig. P4.81 **Fig. P4.82**

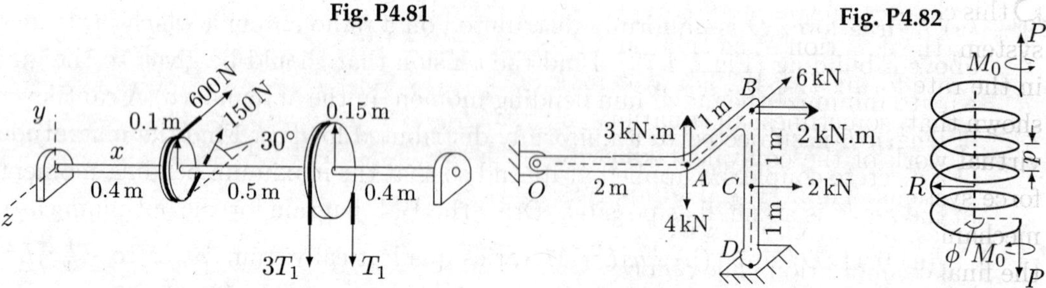

Fig. P4.83 **Fig. P4.84** **Fig. P4.85**

85. Find the AF, SF, BM and twisting moment at angle ϕ in a spring of helix angle α, subjected to the loads shown in Fig. P4.85. In particular, discuss the results for a close coiled spring. [$P \sin\alpha,\ P \cos\alpha,\ PR(\cos\phi - 1)\cos\alpha + M_0 \sin\alpha$, $[P^2R^2\{(\cos\phi - 1)^2 \sin^2\alpha + \sin^2\phi\} + M_0^2 \cos^2\alpha - PRM_0 \sin 2\alpha(\cos\phi - 1)]^{1/2}$; for close coiled spring: 0, P, $PR(\cos\phi - 1)$, $(P^2R^2 \sin^2\phi + M_0^2)^{1/2}$]

86. For a circular beam with centroidal radius R, the coplanar load consists of distributed force per unit length $\vec{q} = q_r\,\vec{e}_r + q_\phi\,\vec{e}_\phi$ and distributed moment per unit length $\vec{m} = m\,\vec{e}_z$. The internal force resultant for the positive face of the cross-section at ϕ with outward normal $\vec{e}_\phi$ is: $\vec{F}_R = N\,\vec{e}_\phi + S\,\vec{e}_r,\ \vec{C}_R = M\,\vec{e}_z$. Use $\vec{F} = \vec{0},\ \vec{M}_O = \vec{0}$ for a beam element from $\phi = \phi_1$ to $\phi = \phi_2$ to prove that $\dfrac{d\vec{F}_R}{d\phi} + \vec{q}\,R = \vec{0}, \quad \dfrac{dM}{d\phi}\,\vec{e}_z + \dfrac{d(\vec{r} \times \vec{F}_R)}{d\phi} + \vec{r} \times \vec{q}\,R + mR\,\vec{e}_z = \vec{0}$. Hence, derive the equilibrium equations: $\dfrac{1}{R}\dfrac{dN}{d\phi} + \dfrac{S}{R} + q_\phi = 0, \ \dfrac{1}{R}\dfrac{dS}{d\phi} - \dfrac{N}{R} + q_r = 0, \ \dfrac{1}{R}\dfrac{dM}{d\phi} - S + m = 0.$

CHAPTER 5

VARIATIONAL MECHANICS

SOME elements of the variational formulation of classical mechanics are presented in this chapter. In Chapters 2 and 3, we have seen that often for a one degree of freedom system, the equation of motion can be obtained easily using the work–energy relation in the rate form: $\dot{T} = \dot{W}$. For systems with more degrees of freedom, it has also been shown that, sometimes, all equations of motion can be obtained easily by equating the virtual work of the equivalent kinetic system with that of the external and internal force systems. This variational approach is generalised in this chapter. In variational mechanics, we consider geometrically admissible paths of the system from the initial to the final configuration. The difference in any physical entity from one admissible path to a varied path in its immediate neighbourhood is called its variation. Hamilton's principle is presented as a basic axiom and the Lagrange's equations of motion for a system with finite degrees of freedom are derived from it. Lagrange's equations for impulsive forces and the principle of virtual work are also derived. The concept of stability of equilibrium and the conditions required for this are discussed.

5.1 Hamilton's Principle

A configuration of a system at a given time means the specified position of its every material point with respect to a reference frame. The parameters which specify the configuration, unambiguously, are called *generalised coordinates* and are denoted as q_i. Their time rates of change $\dot{q}_i$ are called *generalised velocities*. The generalised coordinates of a system are not unique and a *general* parameter like a length, an angle or an area may serve as a generalised coordinate. Their number is finite for a system of rigid bodies and infinite for a deformable system. For a system with N

generalised coordinates, the position vector of a material point P can be expressed as $\vec{r}_P(t) = \vec{f}_P(q_1, q_2, \ldots, q_N, t)$. Sometimes, the generalised coordinates and velocities are subject to some constraint equations, e.g., a boat moving with orientation tangent to the trajectory of C (Fig. 5.1a). Its generalised coordinates x, y, θ are constrained by $\dot{y} = \dot{x}\tan\theta$.

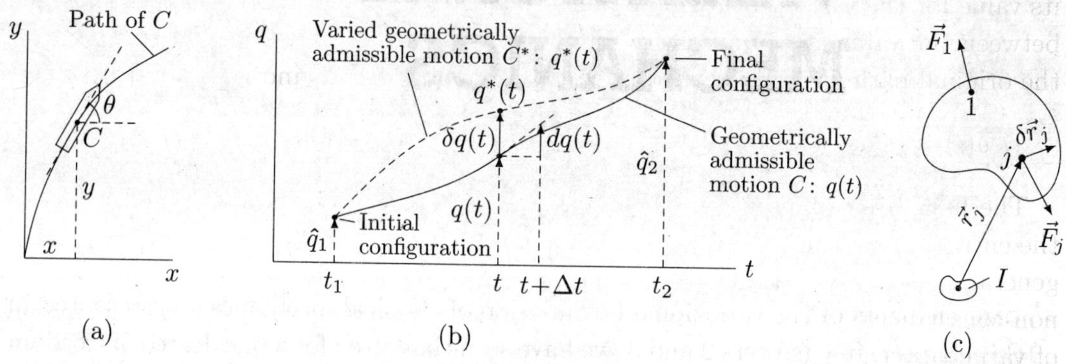

Fig. 5.1

We restrict attention herein to those systems for which there exist generalised coordinates without any constraints on them. Consider an n-dof system and let the set of n independent generalised coordinates required to describe its configuration be denoted as q. Let the system move from the configuration $\hat{q}_1$ at time t_1 to the configuration $\hat{q}_2$ at time t_2. Consider a *geometrically admissible motion* $q(t)$ of the system between these configurations such that $q(t_1) = \hat{q}_1,\ q(t_2) = \hat{q}_2$ (Fig. 5.1b), which may not satisfy the equations of motion at all times in the interval (t_1, t_2). Now, consider a *varied motion* $q^*(t)$ of the system between these configurations such that $q^*(t_1) = \hat{q}_1,\ q^*(t_2) = \hat{q}_2$, which is an *arbitrary* varied motion in the *small* neighbourhood of $q(t)$, i.e., $q^*(t)$ is obtained from $q(t)$ by adding arbitrary first order term denoted as $\delta q(t):\ q^*(t) = q(t) + \delta q(t)$. The *infinitesimal* 'displacement' from an admissible configuration $q(t)$ to a neighbouring configuration $q^*(t)$ at a given time t (i.e., the displacement takes place in zero time) is called a *virtual displacement* $\delta q(t)$:

$$\delta q(t) = q^*(t) - q(t), \qquad \ni\ \delta q(t_1) = \delta q(t_2) = 0 \tag{5.1}$$

This is called a 'virtual' displacement since an infinitesimal differential displacement $dq(t) = \dot{q}(t)dt$ is the first order term in the displacement $q(t+dt) - q(t)$ in infinitesimal time dt along a given admissible path of motion (Fig. 5.1b), whereas the virtual displacement $\delta q(t)$ takes place in zero time and is not along any path of motion. Thus, the notation '$\delta(\ )$' differs from the notation '$d(\ )$'. Let a force $\vec{F}_j(t)$ act at the material point j, where $\vec{r}_j(t) = \vec{r}_j(q_1, q_2, \ldots, q_N, t)$. Let the virtual displacement of the material point j due to the virtual displacement $\delta q(t)$ of the system be $\delta\vec{r}_j$ (Fig. 5.1c). The work done by $\vec{F}_j(t)$ in this virtual displacement in zero time, *keeping the value of the force constant at its value at time t*, is $\vec{F}_j(t)\cdot\delta\vec{r}_j$. It is called the *virtual work* done by $\vec{F}_j$. The virtual work done by all the external and internal forces acting on the system for the virtual displacement $\delta q(t)$ of the system is denoted by $\delta W(t)$.

If the value of an entity for the original admissible path C is denoted as ( ), then its value for the varied path will be denoted with a superscript as $(\ )^*$. The difference between the value $(\ )^*$ of an entity for the varied path $C^*\,[q^*(t)]$ and its value ( ) for the original path $C\,[q(t)]$ is called the *variation* of the entity and is denoted by $\delta(\ )$:

$$\delta(\ ) = (\ )^* - (\ )$$

The first order difference in the values of the entity is called the *first variation* of the entity and is denoted by $\delta^{(1)}(\ )$. The notation δW for virtual work is historical. In general, it is not the variation of W, since no such work function for a path exists for non-conservative forces. We now state *Hamilton's principle*, which is the basic axiom of variational mechanics, for any system (rigid or deformable):

'A geometrically admissible motion $q(t)$ of a system between the prescribed configurations $\hat{q}_1$ at t_1 and $\hat{q}_2$ at t_2 is the actual motion of the system if the following variational indicator ($V.I.$),

$$V.I. \equiv \int_{t_1}^{t_2} [\,\delta^{(1)}T + \delta W\,]\,dt = 0 \tag{5.2}$$

for any *arbitrary* varied motion of the system: $q^*(t) = q(t) + \delta q(t)$ in the small neighbourhood of $q(t)$ with $\delta q(t_1) = \delta q(t_2) = 0$'. The reference frame in which Hamilton's principle is valid is called an inertial frame.

It can be shown that Euler's axioms and Hamilton's principle are equivalent formulations of classical mechanics in differential and variational forms.

Remark: For a free rigid body, the differential displacements $d\vec{r}_A$ and $d\vec{r}_B$ of points A and B are related by

$$d\vec{r}_B = \vec{v}_B\,dt = \vec{v}_A\,dt + \vec{\omega}\times\vec{r}_{BA}\,dt = d\vec{r}_A + \overrightarrow{d\phi}\times\vec{r}_{BA}$$

where $\overrightarrow{d\phi} = \vec{\omega}\,dt$ is the differential rotation of the rigid body. Since this is a relation between first order terms, it follows that the relation between the *virtual displacements* $\delta\vec{r}_A, \delta\vec{r}_B$ and *virtual rotation* $\overrightarrow{\delta\phi}$ is given *exactly* by

$$\delta\vec{r}_B = \delta\vec{r}_A + \overrightarrow{\delta\phi}\times\vec{r}_{BA}$$

Note that the variations in q_i, $\vec{r}_A$, i.e., virtual displacements δq_i, $\delta\vec{r}_A$, being to a neighbouring varied path, are first order variations $\Rightarrow$ $\delta^{(1)}q_i = \delta q_i$, $\delta^{(1)}\vec{r}_A = \delta\vec{r}_A$.

5.2 Lagrange's Equations for a System with Finite DOF

Consider a system of n degrees of freedom with generalised coordinates $q_1, \ldots, q_n$, which are all independent of each other. Let $q = [\,q_1 \quad q_2 \quad \ldots \quad q_{n-1} \quad q_n\,]^{\mathrm{T}}$ be the the column vector of the generalised coordinates. The virtual work done by the internal and external forces $\vec{F}_j$, in zero time, acting at $\vec{r}_j = \vec{r}_j(q_1, \ldots, q_n, t)$, can be expressed in terms of δq using $\delta\vec{r}_j = \sum_i \frac{\partial \vec{r}_j}{\partial q_i}\,\delta q_i$:

$$\delta W \;=\; \sum_j \vec{F}_j \cdot \delta\vec{r}_j = \sum_j \vec{F}_j \cdot \left(\sum_i \frac{\partial \vec{r}_j}{\partial q_i}\,\delta q_i\right) = \sum_i \left(\sum_j \vec{F}_j \cdot \frac{\partial \vec{r}_j}{\partial q_i}\right)\delta q_i$$

$$\delta W \;=\; \sum_i Q_i\,\delta q_i = Q_i\,\delta q_i, \qquad \text{where} \qquad Q_i = \sum_j \vec{F}_j \cdot \frac{\partial \vec{r}_j}{\partial q_i} \tag{5.3}$$

and the summation convention has been used. Q_i is called the *generalised force* for the ith generalised coordinate q_i. The dimension of the product $Q_i\delta q_i$ is that of work. Hence, if q_i is a length coordinate like r, then Q_i has dimensions of force and if q_i is an angle coordinate like ϕ, then Q_i has dimensions of a moment. In practice, we will see that rather than compute Q_i from eq. (5.3), it is much easier to write the complete expression for δW and read off the Q_is as the coefficients of δq_is in this expression.

The first variation of T is obtained as follows. Let $\delta q(t) = \varepsilon\eta(t)$, i.e.,

$$q^*(t) = q(t) + \varepsilon\eta(t), \qquad \text{with} \quad \eta(t_1) = \eta(t_2) = 0 \quad \text{and} \quad \varepsilon \ll 1 \tag{a}$$

Since $\vec{v}_P = \frac{d}{dt}\vec{r}_P[q(t), t] = (\partial\vec{r}_P/\partial q_i)\dot{q}_i + \partial\vec{r}_P/\partial t$, the kinetic energy is a function of $q, \dot{q}, t$. The kinetic energy function T^* for the varied path $[\,q^*(t),\ \dot{q}^*(t) = \dot{q}(t) + \varepsilon\dot{\eta}(t)]$ is expressed as a function of ε, and expanded in a Taylor's series about $\varepsilon = 0$, i.e., about the original path $q(t)$:

$$\begin{aligned}
T^*(t) &= T[\dot{q}^*(t), q^*(t), t] = T[\dot{q} + \varepsilon\dot{\eta}, q + \varepsilon\eta, t] = T(\varepsilon) \\
&= T|_{\varepsilon\,=\,0} + \left.\frac{dT}{d\varepsilon}\right|_{\varepsilon\,=\,0}\varepsilon + \tfrac{1}{2}\left.\frac{d^2T}{d\varepsilon^2}\right|_{\varepsilon\,=\,0}\varepsilon^2 + \ldots \\
\delta T &= T[\dot{q}^*(t), q^*(t), t] - T[\dot{q}(t), q(t), t] = T(\varepsilon) - T|_{\varepsilon\,=\,0} \\
&= \left.\frac{dT}{d\varepsilon}\right|_{\varepsilon\,=\,0}\varepsilon + \tfrac{1}{2}\left.\frac{d^2T}{d\varepsilon^2}\right|_{\varepsilon\,=\,0}\varepsilon^2 + \ldots
\end{aligned}$$

The first order term on the right hand side is $\delta^{(1)}T$. Using $q_i^*(t) = q_i(t) + \varepsilon\eta_i(t)$, $\dot{q}_i^* = \dot{q}_i + \varepsilon\dot{\eta}_i$, $\delta q_i(t) = \varepsilon\eta_i(t)$, $\delta\dot{q}_i = \varepsilon\dot{\eta}_i$ and the chain rule of differentiation, $\delta^{(1)}T$ is expressed as:

$$\begin{aligned}\delta^{(1)}T &= \left.\frac{dT}{d\varepsilon}\right|_{\varepsilon=0} \varepsilon = \sum_{i=1}^{n}\left[\frac{\partial T}{\partial \dot{q}_i^*}\frac{\partial \dot{q}_i^*}{\partial \varepsilon} + \frac{\partial T}{\partial q_i^*}\frac{\partial q_i^*}{\partial \varepsilon}\right]\Bigg|_{\varepsilon=0} \varepsilon \\ &= \left[\frac{\partial T}{\partial \dot{q}_i}\dot{\eta}_i + \frac{\partial T}{\partial q_i}\eta_i\right]\varepsilon = \frac{\partial T}{\partial \dot{q}_i}\delta\dot{q}_i + \frac{\partial T}{\partial q_i}\delta q_i \end{aligned} \tag{5.4}$$

Note that eq. (5.4) relates the first variation of an entity with the first order derivatives. Substituting δW and $\delta^{(1)}T$ from eqs (5.3) and (5.4) in the expression for the variational indicator, eq. (5.2), and using $\delta q_i = \varepsilon\eta_i$ yields

$$\int_{t_1}^{t_2}\left[\frac{\partial T}{\partial \dot{q}_i}\dot{\eta}_i + \frac{\partial T}{\partial q_i}\eta_i + Q_i\eta_i\right]dt\,\varepsilon = 0, \quad \forall\, \eta_i(t) \;\ni\; \eta_i(t_1) = \eta_i(t_2) = 0 \tag{b}$$

The first term in the integrand of eq. (b) is integrated by parts to get a factor of η_i as in the other two terms:

$$\left[\frac{\partial T}{\partial \dot{q}_i}\eta_i(t)\right]\Bigg|_{t_1}^{t_2} - \int_{t_1}^{t_2}\left[\frac{d}{dt}\left(\frac{\partial T}{\partial \dot{q}_i}\right) - \frac{\partial T}{\partial q_i} - Q_i\right]\eta_i\,dt = 0, \quad \forall\, \eta_i(t) \tag{c}$$

The first term in eq. (c) is zero since $\eta_i(t_1) = \eta_i(t_2) = 0$ by eq. (a). Hence, the integrand in eq. (c) should be zero as eq. (c) has to be valid for arbitrary $\eta_i(t)$. Thus,

$$\frac{d}{dt}\left(\frac{\partial T}{\partial \dot{q}_i}\right) - \frac{\partial T}{\partial q_i} = Q_i. \qquad i = 1, \ldots, n \tag{5.5}$$

These are the *Lagrange's equations of motion for a finite dof system with all generalised coordinates independent of each other.* If some forces are conservative, then δW can be split into δW^{c} and δW^{nc}, the work done by conservative and non-conservative forces respectively, with $V(q)$ being the potential energy of the conservative forces. The work function W^{c} for the admissible path does exist for the conservative forces. Correspondingly, Q_i is split into conservative and non-conservative parts $Q_i^{\mathrm{c}}, Q_i^{\mathrm{nc}}$:

$$\begin{aligned}\delta W &= Q_i\,\delta q_i = \delta W^{\mathrm{c}} + \delta W^{\mathrm{nc}} = (Q_i^{\mathrm{c}} + Q_i^{\mathrm{nc}})\delta q_i \\ \delta W^{\mathrm{c}} &= \delta^{(1)}W^{\mathrm{c}} = -\delta^{(1)}V = -\frac{\partial V}{\partial q_i}\delta q_i = Q_i^{\mathrm{c}}\,\delta q_i\end{aligned}$$

and $\delta W^{\mathrm{nc}} = Q_i^{\mathrm{nc}}\delta q_i$, i.e.,

$$Q_i^{\mathrm{c}} = -\frac{\partial V}{\partial q_i}, \qquad Q_i = Q_i^{\mathrm{nc}} - \frac{\partial V}{\partial q_i} \tag{d}$$

Substituting for Q_i from eq. (d) in eq. (5.5) yields

$$\frac{d}{dt}\left(\frac{\partial T}{\partial \dot{q}_i}\right) - \frac{\partial T}{\partial q_i} + \frac{\partial V}{\partial q_i} = Q_i^{\text{nc}}, \qquad i = 1, \ldots, n \tag{5.6}$$

We define the *Lagrangian function* L of the system by $L(\dot{q}, q, t) \equiv T(\dot{q}, q, t) - V(q)$,

$$\Rightarrow \qquad \frac{\partial L}{\partial \dot{q}_i} = \frac{\partial T}{\partial \dot{q}_i}, \qquad \frac{\partial L}{\partial q_i} = \frac{\partial T}{\partial q_i} - \frac{\partial V}{\partial q_i}$$

and eq. (5.6) $\Rightarrow$

$$\frac{d}{dt}\left(\frac{\partial L}{\partial \dot{q}_i}\right) - \frac{\partial L}{\partial q_i} = Q_i^{\text{nc}}, \qquad i = 1, \ldots, n \tag{5.7}$$

The great advantage of Lagrange's equations is that they are scalar equations involving scalars like T, V, δW. The forces of constraint due to ideal constraints do not appear in them. Their first integrals are often easily obtained. For example, if q_3 does not appear explicitly in L and $Q_3^{\text{nc}} = 0$, then $\partial L/\partial q_3 = 0$ and the corresponding Lagrange's equation yields $\frac{d}{dt}(\partial L/\partial \dot{q}_3) = 0 \Rightarrow \partial L/\partial \dot{q}_3 = \text{constant}$.

5.2.1 Illustrative examples

We use the following steps to apply Lagrange's equations in illustrated examples.

1. Choose generalized coordinates q_i.
2. Find the virtual work δW^{nc} done by the non-conservative forces in virtual displacement δq and read off Q_i^{nc}.
3. Express $\vec{v}_{C_j}$ and $\vec{\omega}_j$ for body j in terms of q, $\dot{q}$.
4. Express V and T in terms of q, $\dot{q}$, t.
5. Set up Lagrange's equations and solve them.

Example 5.1: The system shown in Fig. E5.1 consists of a thin disc with a thin bar pinned to it. The disc rolls without slip on a horizontal track and is restrained by a spring. A torsional spring is connected between the rod and the disc. The springs are undeformed when $\theta = 0, x = 0$. Derive the equations of motion.

Solution: The frictional force on the rotor does no work since the velocity of the point of contact is zero. The displacement x of the disc centre and no slip imply that the rotation ϕ of the disc equals x/R. The angular velocities ω_1 of the disc and ω_2 of the bar are given by $\omega_1 = \dot{x}/R$, $\omega_2 = \dot{\theta}$. The ends of the torsional spring rotate by θ and ϕ in opposite directions, causing a relative twist of $\theta + \phi = (\theta + x/R)$. Since, $x_A = x + L\sin\theta$, the virtual displacements δx and $\delta\theta \Rightarrow \delta x_A = \delta x + L\cos\theta\,\delta\theta$ and

$$\delta W^{\text{nc}} = P\delta x_A = P[\delta x + L\cos\theta\,\delta\theta] = Q_x^{\text{nc}}\delta x + Q_\theta^{\text{nc}}\delta\theta$$

$$\Rightarrow \quad Q_x^{\rm nc} = P, \qquad Q_\theta^{\rm nc} = PL\cos\theta$$

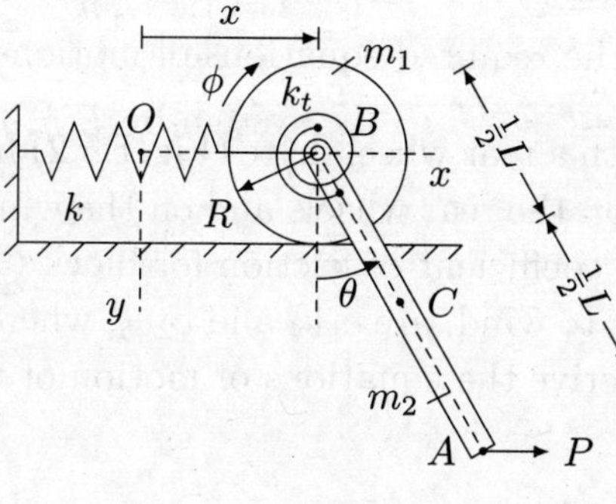

Fig. E5.1

The velocity $\vec{v}_C$ of $C\,(x_C, y_C)$, T and V are given by

$$x_C = x + \tfrac{1}{2}L\sin\theta, \qquad y_C = \tfrac{1}{2}L\cos\theta$$

$$\Rightarrow \quad \dot{x}_C = \dot{x} + \tfrac{1}{2}L\dot{\theta}\cos\theta, \qquad \dot{y}_C = -\tfrac{1}{2}L\dot{\theta}\sin\theta$$

$$\begin{aligned} T &= \tfrac{1}{2}m_1\dot{x}^2 + \tfrac{1}{2}(\tfrac{1}{2}m_1R^2)\left(\frac{\dot{x}}{R}\right)^2 \\ &\quad + \tfrac{1}{2}m_2[(\dot{x} + \tfrac{1}{2}L\dot{\theta}\cos\theta)^2 + (\tfrac{1}{2}L\dot{\theta}\sin\theta)^2] + \tfrac{1}{2}(\tfrac{1}{12}m_2L^2)\dot{\theta}^2 \\ &= \tfrac{1}{2}(\tfrac{3}{2}m_1 + m_2)\dot{x}^2 + \tfrac{1}{6}m_2L^2\dot{\theta}^2 + \tfrac{1}{2}m_2L\dot{x}\dot{\theta}\cos\theta \\ V &= \tfrac{1}{2}kx^2 + \tfrac{1}{2}k_t(\theta + x/R)^2 - \tfrac{1}{2}m_2gL\cos\theta \end{aligned}$$

$$\frac{\partial T}{\partial \dot{x}} = (\tfrac{3}{2}m_1 + m_2)\dot{x} + \tfrac{1}{2}m_2L\dot{\theta}\cos\theta, \qquad \frac{\partial T}{\partial x} = 0$$

$$\frac{\partial T}{\partial \dot{\theta}} = \tfrac{1}{3}m_2L^2\dot{\theta} + \tfrac{1}{2}m_2L\dot{x}\cos\theta, \qquad \frac{\partial T}{\partial \theta} = -\tfrac{1}{2}m_2L\dot{x}\dot{\theta}\sin\theta$$

$$\frac{\partial V}{\partial \theta} = k_t(\theta + x/R) + \tfrac{1}{2}m_2gL\sin\theta, \qquad \frac{\partial V}{\partial x} = kx + k_t(\theta + x/R)/R$$

Thus, the Lagrange's equations of motion are:

$$\frac{d}{dt}\left(\frac{\partial T}{\partial \dot{x}}\right) - \frac{\partial T}{\partial x} + \frac{\partial V}{\partial x} = Q_x^{\rm nc}, \qquad \frac{d}{dt}\left(\frac{\partial T}{\partial \dot{\theta}}\right) - \frac{\partial T}{\partial \theta} + \frac{\partial V}{\partial \theta} = Q_\theta^{\rm nc}:$$

$$\left(\tfrac{3}{2}m_1 + m_2\right)\ddot{x} + \tfrac{1}{2}m_2L(\ddot{\theta}\cos\theta - \dot{\theta}^2\sin\theta) + kx + k_t(\theta + x/R)/R = P \qquad (1)$$

$$\begin{aligned} &\tfrac{1}{3}m_2L^2\ddot{\theta} + \tfrac{1}{2}m_2L(\ddot{x}\cos\theta - \dot{x}\dot{\theta}\sin\theta) + \tfrac{1}{2}m_2L\dot{x}\dot{\theta}\sin\theta \\ &\qquad + k_t(\theta + x/R) + \tfrac{1}{2}m_2gL\sin\theta = PL\cos\theta \end{aligned}$$

$$\Rightarrow \qquad \tfrac{1}{3}m_2L^2\ddot{\theta} + \tfrac{1}{2}m_2L\ddot{x}\cos\theta + k_t(\theta + x/R) + \tfrac{1}{2}m_2gL\sin\theta = PL\cos\theta \qquad (2)$$

Equations (1) and (2) are the required equations of motion of the system.

Example 5.2: A truck with a rear wheel drive (Fig. E5.2) has a winch mounted at its back. The driving torques on the rear wheels and on the winch are T_1 and T_2. Assume no slip for the wheels. The coefficient of friction for block ⑤ is μ. The drag forces on the truck and block ⑤, due to wind, are c_1v_4 and c_2v_5, where v_4, v_5 are their velocities and c_1, c_2 are constants. Derive the equations of motion of this system using x and ϕ as generalised coordinates.

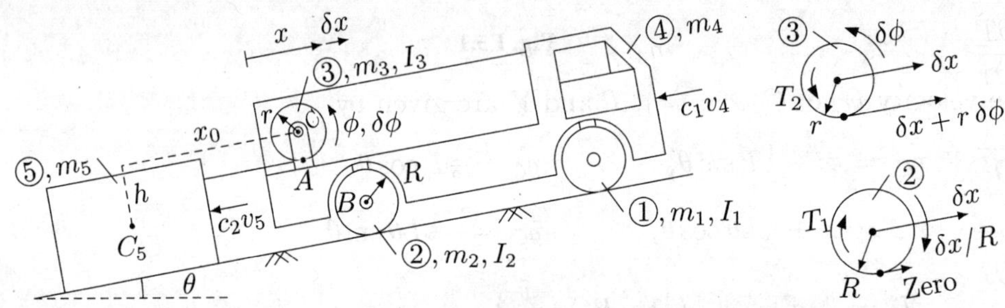

Fig. E5.2

Solution: $v_4 = \dot{x}$. For no slip, the angular velocity of the wheels is $v_B/R = \dot{x}/R$, and the upward velocity of block ⑤ is $v_5 = v_A = v_C + r\dot{\phi} = \dot{x} + r\dot{\phi}$. Hence, for displacement x of the truck and rotation ϕ of the winch, the wheels rotate by x/R and ⑤ displaces up by $(x + r\phi)$. We take the initial configuration of the system as the datum for potential energy V. The potential energy and kinetic energy are given by

$$\begin{aligned} T &= \tfrac{1}{2}[m_1\dot{x}^2 + I_1(\dot{x}/R)^2] + \tfrac{1}{2}[m_2\dot{x}^2 + I_2(\dot{x}/R)^2] \\ &\quad + \tfrac{1}{2}[m_3\dot{x}^2 + I_3\dot{\phi}^2] + \tfrac{1}{2}m_4\dot{x}^2 + \tfrac{1}{2}m_5(\dot{x} + r\dot{\phi})^2 \\ &= \tfrac{1}{2}[m_1 + m_2 + m_3 + m_4 + m_5 + (I_1 + I_2)/R^2]\dot{x}^2 + \tfrac{1}{2}(I_3 + m_5r^2)\dot{\phi}^2 + m_5r\dot{x}\dot{\phi} \\ V &= (m_1 + m_2 + m_3 + m_4)gx\sin\theta + m_5g(x + r\phi)\sin\theta \end{aligned}$$

Note that the locations of the datums for m_3 and m_5 are separated by dashed segments of lengths x_0 and h as shown between initial positions of C and C_5. In a virtual displacement involving δx, $\delta\phi$, the wheels rotate by $\delta x/R$ and block ⑤ moves up by $(\delta x + r\delta\phi)$. The normal force N on block ⑤ from the ground is $N = m_5g\cos\theta$ and the frictional force on it is $\mu N = \mu m_5 g\cos\theta$. The driving torques T_1 and T_2 do virtual work on the rear wheels and winch respectively. The opposite torques acting on chassis ④ do no virtual work since it translates. The drag forces on the truck and block ⑤ are

$c_1\dot{x}$ and $c_2(\dot{x}+r\dot{\phi})$. The frictional forces on the wheels do no work since the velocities of the points of contact are zero for no slip. Hence,

$$\begin{aligned}
\delta W^{\text{nc}} &= T_1(\delta x/R)+T_2\delta\phi-\mu m_5 g\cos\theta(\delta x+r\delta\phi)-c_1\dot{x}\delta x-c_2(\dot{x}+r\dot{\phi})(\delta x+r\delta\phi)\\
&= [T_1/R-\mu m_5 g\cos\theta-c_1\dot{x}-c_2(\dot{x}+r\dot{\phi})]\delta x\\
&\qquad +[T_2-\mu m_5 gr\cos\theta-c_2(\dot{x}+r\dot{\phi})r]\delta\phi = Q_x^{\text{nc}}\delta x+Q_\phi^{\text{nc}}\delta\phi\\
\Rightarrow Q_x^{\text{nc}} &= T_1/R-\mu m_5 g\cos\theta-c_1\dot{x}-c_2(\dot{x}+r\dot{\phi})\\
Q_\phi^{\text{nc}} &= T_2-\mu m_5 gr\cos\theta-c_2(\dot{x}+r\dot{\phi})r\\
\frac{\partial T}{\partial\dot{x}} &= [m_1+m_2+m_3+m_4+m_5+(I_1+I_2)/R^2]\dot{x}+m_5 r\dot{\phi}\\
\frac{\partial T}{\partial\dot{\phi}} &= (I_3+m_5r^2)\dot{\phi}+m_5 r\dot{x}\\
\frac{\partial V}{\partial x} &= (m_1+m_2+m_3+m_4+m_5)g\sin\theta\\
\frac{\partial V}{\partial\phi} &= m_5 gr\sin\theta, \qquad \frac{\partial T}{\partial x}=0, \qquad \frac{\partial T}{\partial\phi}=0
\end{aligned}$$

Hence, the Lagrange's equations of motion are:

$$\frac{d}{dt}\left(\frac{\partial T}{\partial\dot{x}}\right)-\frac{\partial T}{\partial x}+\frac{\partial V}{\partial x}=Q_x^{\text{nc}}, \qquad \frac{d}{dt}\left(\frac{\partial T}{\partial\dot{\phi}}\right)-\frac{\partial T}{\partial\phi}+\frac{\partial V}{\partial\phi}=Q_\phi^{\text{nc}}:$$

$$\begin{aligned}
&[m_1+m_2+m_3+m_4+m_5+(I_1+I_2)/R^2]\ddot{x}+m_5r\ddot{\phi}\\
&\quad +(m_1+m_2+m_3+m_4+m_5)g\sin\theta = T_1/R-\mu m_5 g\cos\theta-c_1\dot{x}-c_2(\dot{x}+r\dot{\phi})\\
&(I_3+m_5r^2)\ddot{\phi}+m_5r\ddot{x}+m_5gr\sin\theta = T_2-\mu m_5 gr\cos\theta-c_2(\dot{x}+r\dot{\phi})r
\end{aligned}$$

Example 5.3: The system shown in Fig. E5.3 consists of a rotor with axial moment of inertia I and two bars pinned to it. Each bar has mass m and the moments of inertia about the principal axes $\vec{e}_i$ at the centre of mass C are: I_{11}, I_{22}, I_{33}. The springs are undeformed for $\theta=0$. The torque applied to the rotor is T_1 and an internal mechanism applies torque T_2 to each bar to raise them. (a) Derive the equations of motion and show that these are related to the principles of moment of momentum and work–energy. (b) Obtain the first integrals of these equations. (c) Find $\dot{\phi}$ for which the bars make constant angle θ_0 to the vertical for $T_1=T_2=0$. (d) Consider the case of free rotation of the rotor, i.e., $T_1=0$. Initially, $\dot{\phi}=\omega_1, \dot{\theta}=0$, when $\theta=\pi/2$. The internal mechanism applies torque to each bar in such a way that the bars are lowered to $\theta = 0$ with $\dot\theta = 0$ at $\theta = 0$. Find the angular velocity of the rotor when $\theta = 0$, and the work done by the internal mechanism to change the configuration of the bars.

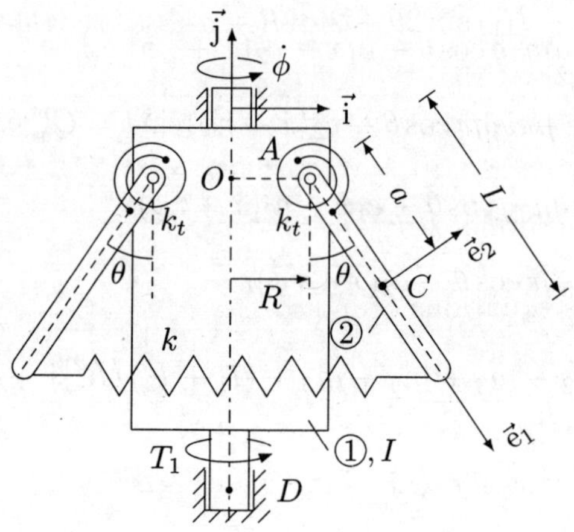

Fig. E5.3

Solution: The angular velocities of rotor ① and bar ② are $\vec\omega_1 = \dot\phi\,\vec{\rm j}$, $\vec\omega_2 = \dot\phi\,\vec{\rm j} + \dot\theta\,\vec{\rm k}$.

$$\begin{aligned} \vec v_C &= \vec v_A + \vec\omega_2 \times \overrightarrow{AC} = (\vec v_O + \vec\omega_1 \times \overrightarrow{OA}) + \vec\omega_2 \times \overrightarrow{AC} \\ &= \dot\phi\,\vec{\rm j} \times R\,\vec{\rm i} + (\dot\phi\,\vec{\rm j} + \dot\theta\,\vec{\rm k}) \times a(\sin\theta\,\vec{\rm i} - \cos\theta\,\vec{\rm j}) \\ &= -(R + a\sin\theta)\dot\phi\,\vec{\rm k} + a\dot\theta\cos\theta\,\vec{\rm i} + a\dot\theta\sin\theta\,\vec{\rm j} \\ \Rightarrow \quad v_C^2 &= (R + a\sin\theta)^2\dot\phi^2 + a^2\dot\theta^2 \end{aligned}$$

The spring extension is $2L\sin\theta$ and $\vec\omega_2 = \dot\phi(-\cos\theta\,\vec{\rm e}_1 + \sin\theta\,\vec{\rm e}_2) + \dot\theta\,\vec{\rm e}_3$. The kinetic energy and potential energy (with gravitational potential energy datum at O) are:

$$\begin{aligned} T &= \tfrac{1}{2}I\dot\phi^2 + 2(\tfrac{1}{2})[m\{(R + a\sin\theta)^2\dot\phi^2 + a^2\dot\theta^2\} + I_{11}\dot\phi^2\cos^2\theta + I_{22}\dot\phi^2\sin^2\theta + I_{33}\dot\theta^2] \\ &= [\tfrac{1}{2}I + I_{11}\cos^2\theta + I_{22}\sin^2\theta + m(R + a\sin\theta)^2]\dot\phi^2 + [I_{33} + ma^2]\dot\theta^2 \\ V &= -2mga\cos\theta + \tfrac{1}{2}k(2L\sin\theta)^2 + 2(\tfrac{1}{2}k_t\theta^2) \\ &= -2mga\cos\theta + 2kL^2\sin^2\theta + k_t\theta^2 \end{aligned}$$

In a virtual displacement $\delta\phi, \delta\theta$, virtual work is done by the torques T_2 on the bars but the reversed torques T_2 acting on the rotor do no virtual work since the rotor has no virtual rotation in direction $\vec{\rm k}$. Hence,

$$\delta W^{\rm nc} = T_1\,\delta\phi + 2T_2\,\delta\theta = Q_\phi^{\rm nc}\delta\phi + Q_\theta^{\rm nc}\delta\theta \quad \Rightarrow \quad Q_\phi^{\rm nc} = T_1, \qquad Q_\theta^{\rm nc} = 2T_2$$

$$\frac{\partial T}{\partial \dot{\phi}} = 2[\tfrac{1}{2}I + I_{11}\cos^2\theta + I_{22}\sin^2\theta + m(R + a\sin\theta)^2]\dot{\phi}, \qquad \frac{\partial T}{\partial \phi} = 0$$

$$\frac{\partial T}{\partial \theta} = [(I_{22} - I_{11})\sin 2\theta + 2m(R + a\sin\theta)a\cos\theta]\dot{\phi}^2$$

$$\frac{\partial T}{\partial \dot{\theta}} = 2[I_{33} + ma^2]\dot{\theta}$$

$$\frac{\partial V}{\partial \theta} = 2mga\sin\theta + 2kL^2\sin 2\theta + 2k_t\theta, \qquad \frac{\partial V}{\partial \phi} = 0$$

(a) The Lagrange's equations of motion are:

$$\frac{d}{dt}\left(\frac{\partial T}{\partial \dot{\phi}}\right) - \frac{\partial T}{\partial \phi} + \frac{\partial V}{\partial \phi} = Q_{\phi}^{\text{nc}}, \qquad \frac{d}{dt}\left(\frac{\partial T}{\partial \dot{\theta}}\right) - \frac{\partial T}{\partial \theta} + \frac{\partial V}{\partial \theta} = Q_{\theta}^{\text{nc}}:$$

$$\frac{d}{dt}[2\{\tfrac{1}{2}I + I_{11}\cos^2\theta + I_{22}\sin^2\theta + m(R + a\sin\theta)^2\}\dot{\phi}] = T_1 \tag{1}$$

$$2[I_{33} + ma^2]\ddot{\theta} - [(I_{22} - I_{11})\sin 2\theta + 2m(R + a\sin\theta)a\cos\theta]\dot{\phi}^2$$
$$+2mga\sin\theta + 2kL^2\sin 2\theta + 2k_t\theta = 2T_2 \tag{2}$$

Equation (1) is $\dot{H}_{Oy} = M_{Oy}$ for the system about the fixed axis DO. Equation (2) is $\vec{M}_A = \dot{\vec{H}}_{A|I} + \vec{r}_{CA} \times m\vec{a}_{A|I}$ for one of the bars about the z-axis at A.

(b) We can combine eqs (1) and (2) as: eq. (1)$\dot{\phi}$ + eq. (2)$\dot{\theta}$:

$$[2\{\tfrac{1}{2}I + I_{11}\cos^2\theta + I_{22}\sin^2\theta + m(R + a\sin\theta)^2\}\ddot{\phi}$$
$$+2\{(I_{22} - I_{11})\sin 2\theta + 2m(R + a\sin\theta)a\cos\theta\}\dot{\theta}\,\dot{\phi}]\dot{\phi}$$
$$+2[I_{33} + ma^2]\dot{\theta}\,\ddot{\theta} - [(I_{22} - I_{11})\sin 2\theta + 2m(R + a\sin\theta)a\cos\theta]\dot{\theta}\,\dot{\phi}^2$$
$$+[2mga\sin\theta + 2kL^2\sin 2\theta + 2k_t\theta]\dot{\theta} = T_1\dot{\phi} + 2T_2\,\dot{\theta}$$

$$\Rightarrow \quad [2\{\tfrac{1}{2}I + I_{11}\cos^2\theta + I_{22}\sin^2\theta + m(R + a\sin\theta)^2\}\dot{\phi}\,\ddot{\phi}$$
$$+\{(I_{22} - I_{11})\sin 2\theta + 2m(R + a\sin\theta)a\cos\theta\}\dot{\theta}\,\dot{\phi}^2$$
$$+2[I_{33} + ma^2]\dot{\theta}\,\ddot{\theta} + [2mga\sin\theta + 2kL^2\sin 2\theta + 2k_t\theta]\dot{\theta} = T_1\dot{\phi} + 2T_2\,\dot{\theta}$$

$$\Rightarrow \quad \frac{d}{dt}[\{\tfrac{1}{2}I + I_{11}\cos^2\theta + I_{22}\sin^2\theta + m(R + a\sin\theta)^2\}\dot{\phi}^2$$
$$+(I_{33} + ma^2)\dot{\theta}^2 - 2mga\cos\theta + 2kL^2\sin^2\theta + k_t\theta^2] = T_1\dot{\phi} + 2T_2\,\dot{\theta}$$

$$\Rightarrow \quad \dot{T} + \dot{V} = \dot{W}^{\text{nc}} \tag{3}$$

which is the work–energy relation for the whole system and just like eq. (1), it could have been written directly. Integrating eqs (1) and (3), we get

$$\Delta[2\{\tfrac{1}{2}I + I_{11}\cos^2\theta + I_{22}\sin^2\theta + m(R + a\sin\theta)^2\}\dot{\phi}] = \int T_1 dt \tag{4}$$

$$\Delta(T + V) = W^{\text{nc}} \tag{5}$$

(c) $T_1 = T_2 = 0,\ \theta = \theta_0\ \Rightarrow\ \dot{\theta} = 0,\ \ddot{\theta} = 0$, eq. (2) yields $\dot{\phi} = \text{constant} = \dot{\phi}_0$ (say):

$$\dot{\phi}_0^2 = \frac{2mga\sin\theta_0 + 2kL^2\sin 2\theta_0 + 2k_t\theta_0}{(I_{22} - I_{11})\sin 2\theta_0 + 2m(R + a\sin\theta_0)a\cos\theta_0}$$

(d) Let ω_2 be the angular velocity of the rotor when the bars are vertical. We apply eqs (4) and (5) for the displacement from $\theta = \pi/2$ to $\theta = 0$ to get

$$2(\tfrac{1}{2}I + I_{11} + mR^2)\omega_2 = 2[\tfrac{1}{2}I + I_{22} + m(R + a)^2]\omega_1$$

$$\begin{aligned}
\Rightarrow \qquad \omega_2 &= [\tfrac{1}{2}I + I_{22} + m(R + a)^2]\omega_1/(\tfrac{1}{2}I + I_{11} + mR^2) \\
W^{\text{nc}} &= (T + V)_f - (T + V)_i = [(\tfrac{1}{2}I + I_{11} + mR^2)\omega_2^2 - 2mga] \\
&\qquad - [\{\tfrac{1}{2}I + I_{22} + m(R + a)^2\}\omega_1^2 + 2kL^2 + k_t(\pi/2)^2]
\end{aligned}$$

which is the work done by the internal mechanism.

5.3 Lagrange's Equations for Impulsive Forces

For an impulsive force $\hat{\vec{F}}_j$ acting on a system at a material point $\vec{r}_j$, we define the *impulsive work* $d\hat{W}$ done by it in displacement $d\vec{r}_j$ as $d\hat{W} = \hat{\vec{F}}_j \cdot d\vec{r}_j$. The S.I. unit of impulsive work is J.s (N.s.m). The prefix *impulsive* changes the nature of the entity involved; its physical dimensions being altered with multiplication by time. Similar to the expression for δW, the impulsive virtual work, $\delta\hat{W}$ due to impulsive forces $\hat{\vec{F}}_j$ acting at time t_1, in a virtual displacement $\delta\vec{r}_j$, can be expressed as

$$\delta\hat{W} = \sum_j \hat{\vec{F}}_j \cdot \delta\vec{r}_j = \hat{Q}_i\delta q_i, \qquad \hat{Q}_i(t_1) = \lim_{t_2\to t_1}\int_{t_1}^{t_2} Q_i(t)dt \tag{5.8}$$

where $\hat{Q}_i$ is the *impulsive generalised force.* For an n-dof system with independent generalised coordinates, the Lagrange's equations for $i = 1, \ldots, n$ are:

$$\frac{d}{dt}\left(\frac{\partial T}{\partial \dot{q}_i}\right) - \frac{\partial T}{\partial q_i} = Q_i \qquad \Rightarrow \qquad \left.\frac{\partial T}{\partial \dot{q}_i}\right|_{t_1}^{t_2} = \int_{t_1}^{t_2}\left(\frac{\partial T}{\partial q_i} + Q_i\right)dt \tag{a}$$

For such a system subjected to impulsive forces at time t_1 (as in an impact modelled to occur in zero time), there is no change in the configuration of the system, but there is a jump in the generalised velocities at t_1 and eq. (a) $\Rightarrow$

$$\left.\frac{\partial T}{\partial \dot{q}_i}\right|_{t_1^-}^{t_1^+} = \lim_{t_2 \to t_1} \int_{t_1}^{t_2} \left(\frac{\partial T}{\partial q_i} + Q_i \right) dt \tag{b}$$

The first term in the integrand on the right hand side is finite and the second term is finite for non-impulsive forces. Hence, the only contribution to the right hand side is from the impulsive forces and eq. (b) simplifies to

$$\left.\frac{\partial T}{\partial \dot{q}_i}\right|_{t_1^-}^{t_1^+} = \left.\frac{\partial T}{\partial \dot{q}_i}\right|_{t_1^+} - \left.\frac{\partial T}{\partial \dot{q}_i}\right|_{t_1^-} = \hat{Q}_i, \qquad i = 1, \ldots, n \tag{5.9}$$

These Lagrange's equations for impulsive forces are algebraic equations. The kinetic energy expressions for the pre-collision and post-collision regimes are written for the given configuration in a perfectly general way, as the partial derivatives are evaluated with respect to $\dot{q}_i$ only at the instant of impact.

Example 5.4: A disc of mass m and radius R is at rest on a smooth horizontal table. A rod of mass m and length $2R$ is hinged to it (Fig. E5.4). Find the angular velocities of the disc and the rod just after the application of an impulsive force $\hat{P}$ to the rod.

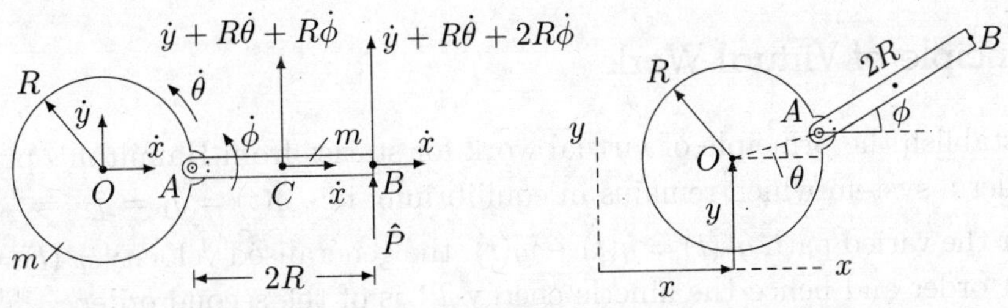

Fig. E5.4

Solution: Choose generalised coordinates x, y, θ, ϕ. $\overrightarrow{OA} = R\vec{\imath}$, $\overrightarrow{AC} = R\vec{\imath}$, $\overrightarrow{AB} = 2R\vec{\imath}$,

$$\begin{aligned}
\vec{v}_A &= \vec{v}_O + \dot{\theta}\vec{k} \times \overrightarrow{OA} = \dot{x}\,\vec{\imath} + (\dot{y} + \dot{\theta}R)\,\vec{\jmath} \\
\vec{v}_C &= \vec{v}_A + \dot{\phi}\vec{k} \times \overrightarrow{AC} = \dot{x}\,\vec{\imath} + (\dot{y} + \dot{\theta}R + \dot{\phi}R)\,\vec{\jmath} \\
\vec{v}_B &= \vec{v}_A + \dot{\phi}\vec{k} \times \overrightarrow{AB} = \dot{x}\,\vec{\imath} + (\dot{y} + \dot{\theta}R + 2\dot{\phi}R)\,\vec{\jmath} \\
\Rightarrow \quad \delta\hat{W} &= \hat{P}\vec{\jmath} \cdot \delta\vec{r}_B = \hat{P}(\delta y + R\delta\theta + 2R\delta\phi)
\end{aligned}$$

$$\Rightarrow \quad \hat{Q}_x = 0, \quad \hat{Q}_y = \hat{P}, \quad \hat{Q}_\theta = \hat{P}R, \quad \hat{Q}_\phi = 2\hat{P}R$$

$$T = \tfrac{1}{2}m(\dot{x}^2 + \dot{y}^2) + \tfrac{1}{2}(\tfrac{1}{2}mR^2)\dot{\theta}^2 + \tfrac{1}{2}m[\dot{x}^2 + (\dot{y} + R\dot{\theta} + R\dot{\phi})^2] + \tfrac{1}{2}[\tfrac{1}{12}m(2R)^2]\dot{\phi}^2$$

$$\frac{\partial T}{\partial \dot{x}} = 2m\dot{x}, \qquad \frac{\partial T}{\partial \dot{y}} = 2m\dot{y} + mR(\dot{\theta} + \dot{\phi})$$

$$\frac{\partial T}{\partial \dot{\theta}} = \frac{3}{2}mR^2\dot{\theta} + mR(\dot{y} + R\dot{\phi}), \qquad \frac{\partial T}{\partial \dot{\phi}} = \frac{4}{3}mR^2\dot{\phi} + mR(\dot{y} + R\dot{\theta})$$

Let $\dot{x}, \dot{y}, \dot{\theta}, \dot{\phi}$ be the values just after impact. Lagrange's equations for impulsive actions:

$$\Delta\left(\frac{\partial T}{\partial \dot{x}}\right) = \hat{Q}_x \quad \Rightarrow \quad 2m\dot{x} - 0 = 0 \tag{1}$$

$$\Delta\left(\frac{\partial T}{\partial \dot{y}}\right) = \hat{Q}_y \quad \Rightarrow \quad 2m\dot{y} + mR(\dot{\theta} + \dot{\phi}) - 0 = \hat{P} \tag{2}$$

$$\Delta\left(\frac{\partial T}{\partial \dot{\theta}}\right) = \hat{Q}_\theta \quad \Rightarrow \quad \frac{3}{2}mR^2\dot{\theta} + mR(\dot{y} + R\dot{\phi}) - 0 = \hat{P}R \tag{3}$$

$$\Delta\left(\frac{\partial T}{\partial \dot{\phi}}\right) = \hat{Q}_\phi \quad \Rightarrow \quad \frac{4}{3}mR^2\dot{\phi} + mR(\dot{y} + R\dot{\theta}) - 0 = 2\hat{P}R \tag{4}$$

Solution of eqs (1) to (4) is: $\dot{x} = 0, \;\; \dot{y} = -2\hat{P}/7m, \;\; \dot{\theta} = -4\hat{P}/7mR, \;\; \dot{\phi} = 15\hat{P}/7mR.$

5.4 Principle of Virtual Work

We can establish the principle of virtual work for statics from Hamilton's principle. (a) Consider a system which remains in equilibrium, i.e., $q(t) = \hat{q}_1 = \hat{q}_2 \Rightarrow \dot{q} = 0 \Rightarrow T = 0$. For the varied path $q^*(t) = q(t) + \delta q(t)$, the generalised velocity $\dot{q}^*(t) = \overline{\delta q(t)}$ is of the first order and hence the kinetic energy T^* is of the second order $\Rightarrow \delta^{(1)}T = 0$. Hamilton's principle (5.2) reduces to

$$\int_{t_1}^{t_2} \delta W \, dt = 0 \;\forall\; \delta q(t) \ni \delta q(t_1) = \delta q(t_2) = 0 \text{ for any } t_1, t_2 \Rightarrow \delta W = 0 \;\forall\; \delta q$$

(b) Conversely, if the system is initially in equilibrium and $\delta W = 0 \;\forall\; \delta q$, then the system can be proved to remain in equilibrium by contradiction. Assume that the system does not remain in equilibrium. Apply additional forces to maintain equilibrium.

Given condition: (δW of the original forces) $= 0 \;\;\forall\; \delta q$

By part (a): (δW of original and additional forces) $= 0 \;\;\forall\; \delta q$

$$\Rightarrow \qquad (\delta W \text{ of additional forces}) = 0 \ \forall \ \delta q$$

since δq is arbitrary $\Rightarrow$ all additional forces are zero. Hence, the system remains in equilibrium without any need for these additional forces.

Thus, we have proved the **principle of virtual work**: *If a system is in equilibrium, then the virtual work done by the internal and external forces in an arbitrary virtual displacement is zero. Conversely, a geometrically admissible configuration of a system is an equilibrium configuration if the virtual work done by the external and internal forces is zero for an arbitrary virtual displacement, provided the system is initially in equilibrium*:

$$\delta W = 0 \ \ \forall \ \delta q \tag{5.10}$$

The converse statement is true only if the system is initially in equilibrium. For example, for a rigid bar initially rotating (without friction) about an axis passing through its centre of mass C, at angular velocity ω (Fig. 5.2), $\delta W = 0 \ \forall \ \delta\theta$, yet it continues to move since it was not in equilibrium initially.

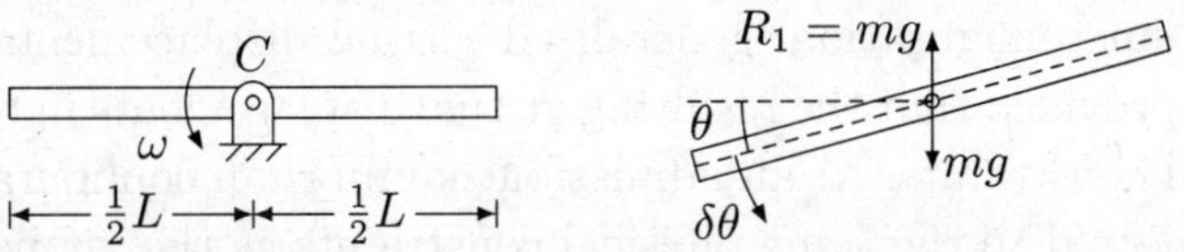

Fig. 5.2

Three alternative forms of the principle of virtual work are presented below.

1. Null Generalised Forces: For an n-dof system in equilibrium, with independent q_i, eq. (5.10) $\Rightarrow$

$$\begin{aligned} \delta W &= Q_i \delta q_i = 0 \quad \forall \ \delta q_i \\ \Rightarrow \quad Q_i &= 0, \qquad i = 1, \ldots, n \end{aligned} \tag{5.11}$$

i.e., *all the generalised forces are zero.*

2. Principle of Stationary Potential Energy: Let δW^{c} and δW^{nc} be the virtual work done by conservative and non-conservative external and internal forces with potential energy V of the conservative forces, then

$$\begin{aligned} \delta W &= \delta W^{\mathrm{c}} + \delta W^{\mathrm{nc}} = -\delta^{(1)} V + \delta W^{\mathrm{nc}} = 0 \\ \Rightarrow \quad \delta W^{\mathrm{nc}} &= \delta^{(1)} V \quad \forall \ \delta q \end{aligned} \tag{5.12}$$

If all internal and external forces are conservative, then eq. (5.12) $\Rightarrow$

$$\delta^{(1)}V = 0 \quad \forall\, \delta q \tag{5.13}$$

i.e., the first variation of potential energy is zero. Thus, the potential energy has a stationary value (an extremum) at an equilibrium configuration. The **principle of potential energy** follows:

If a system, subjected only to conservative internal and external forces, is in equilibrium, then $\delta^{(1)}V = 0 \;\forall\; \delta q$. Conversely, if $\delta^{(1)}V = 0 \;\forall\; \delta q$ and if the system is initially in equilibrium, then it remains in equilibrium.

For an n-dof conservative system with independent generalised coordinates q_i :

$$\text{eq. (5.13)} \Rightarrow \quad \delta^{(1)}V = \frac{\partial V}{\partial q_i}\,\delta q_i = 0 \quad \forall\, \delta q_i$$

$$\Rightarrow \quad \frac{\partial V}{\partial q_i} = 0, \qquad i = 1, \ldots, n \tag{5.14}$$

Equations (5.14) are the *equations of equilibrium of an n-dof conservative system* and provide the most systematic way to find *all* its positions of equilibrium.

3. Principle of Virtual Rate of Work: For a constrained system, at an admissible equilibrium configuration, generalised virtual displacements δq_j can be any algebraic entities provided that the resulting virtual displacements in the system satisfy the given physical constraints. At an admissible equilibrium configuration, the virtual velocities are subjected to the same physical constraints as the virtual displacements. Hence, the virtual generalised velocities $\delta\dot{q}_i$ at an admissible configuration provide algebraic entities to which we relate the virtual displacements in the principle of virtual work by $\delta q_i = \delta\dot{q}_i c$ where c is a constant having the dimension of time. Thus, the contribution $Q_i\delta q_i$ to δW is given by $c(Q_i\delta\dot{q}_j)$, where $Q_i\delta\dot{q}_i$ is the contribution of Q_i to the virtual rate of work $\delta\dot{W}$ under virtual velocities, i.e., $\delta W = c(\delta\dot{W})$. Hence, the principle of virtual work $\delta W = 0$ yields the principle of virtual rate of work,

$$\delta\dot{W} = 0 \quad \forall\, \delta\dot{q} \tag{5.15}$$

which for a discrete system yields

$$\delta\dot{W} = Q_i\,\delta\dot{q}_i = 0 \quad \forall\, \delta\dot{q}_i \tag{5.16}$$

i.e., *a system initially in equilibrium, will remain in equilibrium, if the virtual rate of work of all internal and external forces is zero for arbitrary virtual velocities.* This **principle of virtual rate of work** is very convenient for solving equilibrium problems of many mechanisms wherein finding the virtual velocities of various points in the equilibrium configuration is much simpler than determining the positions of various points in the general configuration in terms of generalised coordinates (for the purpose of finding their virtual displacements).

Note: If the system is constrained and the forces of constraint (internal or external) together do zero work in virtual displacement consistent with the constraints, then in eqs (5.10)–(5.16), the virtual work is only due to the active internal and external forces. If all the internal forces together are workless, then in eqs (5.10)–(5.16), the virtual work is only due to the external forces on the system. Thus, in general, the principle of virtual work is preferred for a system of interconnected rigid bodies if the interconnections and external constraints are workless, provided we are not interested in the internal and constraint forces. Whenever friction forces, which do net work, follower forces (forces dependent on the displacement or deformation) and velocity-dependent forces which do work, are present, the principle of virtual work does not offer any computational advantage.

Direct Proof of Principle of Virtual Work for a Rigid Body:

$\dot{W}$ for a rigid body can be expressed as $\dot{W} = \vec{F} \cdot \vec{v}_C + \vec{M}_C \cdot \vec{\omega} \Rightarrow$ with $\overrightarrow{d\theta} = \vec{\omega} dt$:

$$
\begin{aligned}
dW &= \vec{F} \cdot \vec{v}_C \, dt + \vec{M}_C \cdot \vec{\omega} \, dt = \vec{F} \cdot d\vec{r}_C + \vec{M}_C \cdot \overrightarrow{d\theta} \\
\delta W &= \vec{F} \cdot \delta\vec{r}_C + \vec{M}_C \cdot \overrightarrow{\delta\theta}
\end{aligned}
\tag{5.17}
$$

A rigid body is in equilibrium iff $\vec{F} = \vec{0}$, $\vec{M}_C = \vec{0}$, provided it is initially in equilibrium. Hence, eq. (5.17) $\Rightarrow$ a rigid body is in equilibrium iff $\delta W = 0$, for all virtual displacements $\delta\vec{r}_C$, $\overrightarrow{\delta\theta}$, provided it is initially in equilibrium.

5.4.1 Illustrative examples

The following steps yield the solution in a systematic manner

1. Decide on the type of virtual displacement desired. For a fully constrained system, the desired virtual displacement is the one which violates the constraint corresponding to the reaction of interest. For partially constrained systems, the virtual displacements should correspond to the degrees of freedom of motion of the system, had it not been in equilibrium.
2. Draw the system in its general configuration. Choose x, y, z-axes whose origin is at a point that does not move during virtual displacement and whose orientation suits the geometry or loading.
3. Choose generalized coordinates to locate the configuration of the system.
4. Find the virtual displacements of the points of application of the forces. In general, we prefer the analytical approach. Express the Cartesian coordinates of the points of application of the loads in terms of the generalised coordinates. Their virtual displacements are the differential changes in these due to an infinitesimal increase of the generalised coordinates. If the generalised coordinate has a specific value in the equilibrium configuration, substitute it *after* finding the derivatives.

Occasionally, when the body is given a planar virtual rotation $\delta\theta$, it is easier to get the virtual displacement $\delta\vec{r}$ of a point as the product of $\delta\theta$ and its radial distance from the instantaneous axis of virtual rotation. This $\delta\vec{r}$ is at right angles to the radial line in the sense of rotation.

When it is difficult to specify position vectors in terms of generalised coordinates, we employ the principle of virtual rate of work and express the velocities of all points of application of the forces in terms of the virtual generalised velocities.

5. Express the forces doing work / rate of work in terms of their components. Find δW or $\delta\dot{W}$ as the sum of the dot product of the forces with the virtual displacement or virtual velocity of their points of application.
6. Solve $\delta W = 0$ or $\delta\dot{W} = 0$ for the unknowns.

The simplest approach is to treat all forces as constant and use $\partial V/\partial q_i = 0\ \forall\, i$.

Example 5.5: (a) The equilibrium position of the telescopic link OA about the z-axis is controlled by the torque M applied to the shaft OB (Fig. E5.5a). The end A slides on fixed rod CD. The link has mass m with centre of mass G which is always at one-third of the distance from O to A. Neglect friction. Find M. (b) The double scissor-jack (Fig. E5.5b) supports an end of automobile which exerts a force P on it. The screw has a double start square thread of mean radius r and pitch p. Neglect friction and weight of the members. Find the torque M to be applied to raise the automobile.

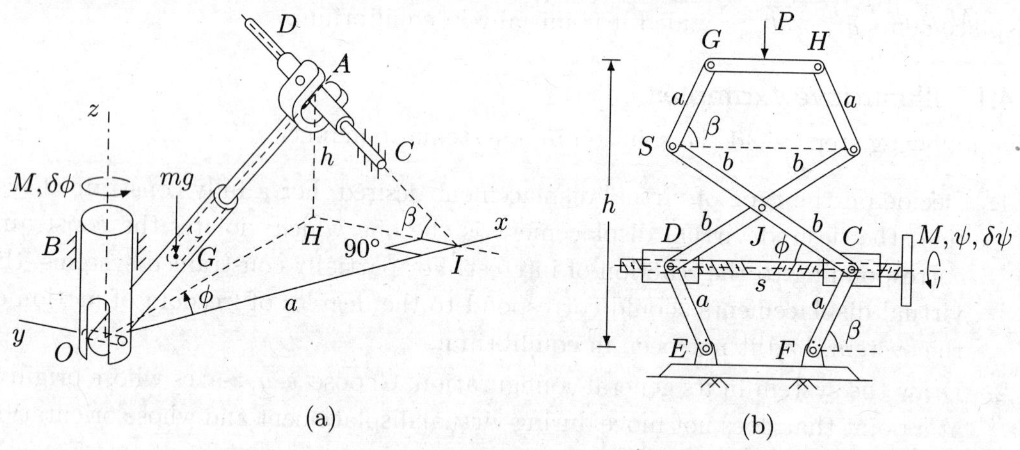

Fig. E5.5

Solution: (a) It is a 1-dof system with generalised coordinate ϕ related to h by

$$HI = a\tan\phi = h\cot\beta \qquad \Rightarrow \qquad a\sec^2\phi\,\delta\phi = \delta h\,\cot\beta$$

$$\delta W = M\vec{k}\cdot\delta\phi\vec{k} + (-mg)\delta(\tfrac{1}{3}h) = (M - \tfrac{1}{3}mga\sec^2\phi\tan\beta)\delta\phi = 0 \quad \forall\,\delta\phi$$

$$\Rightarrow \quad M = \tfrac{1}{3}mga\sec^2\phi\tan\beta = \tfrac{1}{3}mga[1 + h^2\cot^2\beta/a^2]\tan\beta$$

(b) Distance s decreases by 2(lead) $= 4p$ for rotation $\psi = 2\pi$ of the handle. It is a 1-dof system, since β and ϕ are related to ψ by

$$s = (CD) = 2b\cos\phi = (GH) + 2a\cos\beta$$

$$\Rightarrow \quad \delta s = -2b\sin\phi\,\delta\phi = -2a\sin\beta\,\delta\beta = -\delta\psi(4p/2\pi)$$

$$\Rightarrow \quad \delta\phi = p\,\delta\psi/\pi b\sin\phi, \qquad \delta\beta = p\,\delta\psi/\pi a\sin\beta$$

$$h = 2a\sin\beta + 2b\sin\phi$$

$$\Rightarrow \quad \delta h = 2a\cos\beta\,\delta\beta + 2b\cos\phi\,\delta\phi = 2p(\cot\beta + \cot\phi)\,\delta\psi/\pi$$

$$\delta W = M\,\delta\psi - P\,\delta h = [M - 2Pp(\cot\beta + \cot\phi)/\pi]\,\delta\psi \quad \forall\,\delta\psi$$

$$\Rightarrow \quad M = 2Pp(\cot\beta + \cot\phi)/\pi$$

Example 5.6: An instrument meant to monitor conditions in a hazardous environment is positioned using the lazy-tong type device (Fig. E5.6a), by applying a torque M to the handle. The mass of the instrument is m and the linear density of the rods is λ kg/m. The pitch of the screw is p. Neglect friction. Find M for equilibrium.

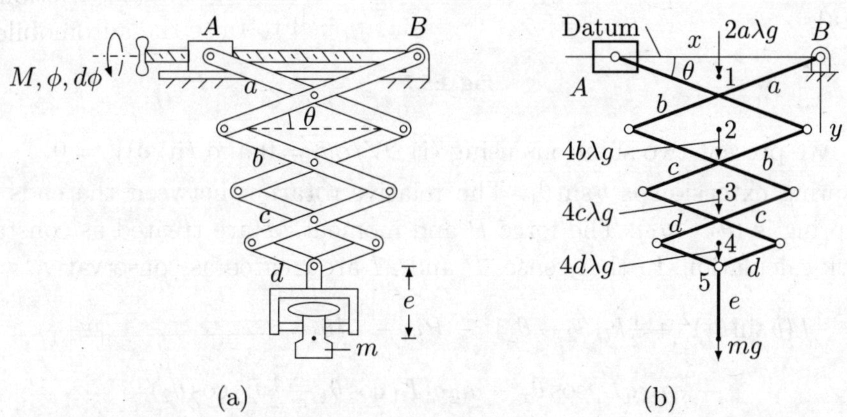

Fig. E5.6

Solution: Let θ be the generalised coordinate (Fig. E5.6b). We use $dV/d\theta = 0$. A rotation $d\phi$ of the handle causes an increase in length AB by $p\,d\phi/2\pi$:

$$d(\text{AB}) \;=\; d(2a\cos\theta) = -2a\sin\theta\,d\theta = \frac{p\,d\phi}{2\pi} \quad \Rightarrow \quad \frac{d\phi}{d\theta} = -\frac{4\pi a\sin\theta}{p}$$

$$V(\theta) \;=\; -\lambda g[2a(\tfrac{1}{2}a) + 4b(a+b) + 4c(a+2b+c) + 4d(a+2b+2c+d)]\sin\theta$$
$$-mg[(a+2b+2c+2d)\sin\theta + e] - M\phi$$

$$\frac{dV}{d\theta} = -g[\lambda\{a^2 + 4b(a+b) + 4c(a+2b+c) + 4d(a+2b+2c+d)\}$$
$$+m(a+2b+2c+2d)]\cos\theta - M\frac{d\phi}{d\theta} = 0$$
$$\Rightarrow \quad M = pg\cot\theta[\lambda\{a^2 + 4b(a+b) + 4c(a+2b+c) + 4d(a+2b+2c+d)\}$$
$$+m(a+2b+2c+2d)]\,/\,4\pi a$$

Example 5.7: Find the force P and the moment M for equilibrium of the pendulums (Fig. E5.7a), modelled as uniform bars. The springs are undeformed for $\theta_1 = \theta_2 = 0$.

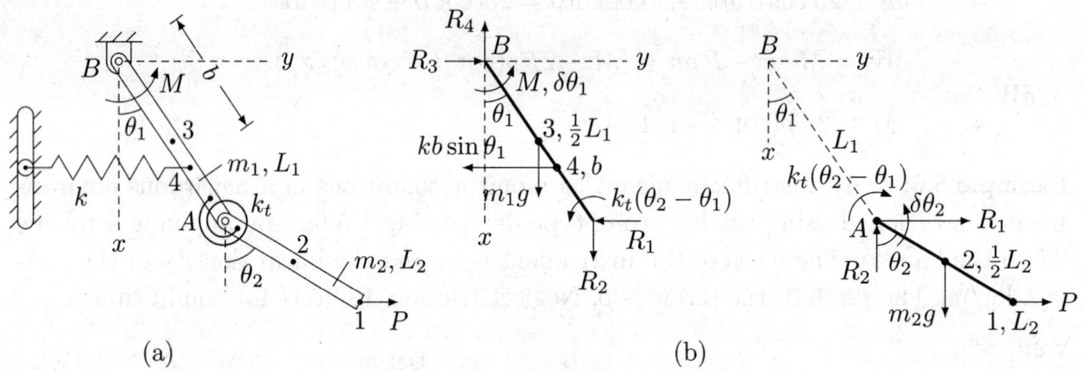

Fig. E5.7

Solution: We present two solutions using (i) $\partial V/\partial\theta_i = 0$ and (ii) $\delta W = 0$.

(i) The spring extension is $b\sin\theta_1$. The relative rotation between the ends of the torsional spring is $(\theta_2 - \theta_1)$. The force P and moment M are treated as constant for virtual work calculation. In that sense, P and M are treated as conservative with

$$V = \tfrac{1}{2}k(b\sin\theta_1)^2 + \tfrac{1}{2}k_t(\theta_2 - \theta_1)^2 - Py_1 - M\theta_1$$
$$-\tfrac{1}{2}m_1 gL_1\cos\theta_1 - m_2 g(L_1\cos\theta_1 + \tfrac{1}{2}L_2\cos\theta_2)$$
$$= \tfrac{1}{2}kb^2\sin^2\theta_1 + \tfrac{1}{2}k_t(\theta_2 - \theta_1)^2 - P(L_1\sin\theta_1 + L_2\sin\theta_2)$$
$$-M\theta_1 - (\tfrac{1}{2}m_1 + m_2)gL_1\cos\theta_1 - \tfrac{1}{2}m_2 gL_2\cos\theta_2$$

$$\frac{\partial V}{\partial\theta_1} = kb^2\sin\theta_1\cos\theta_1 - k_t(\theta_2 - \theta_1) - PL_1\cos\theta_1 - M + (\tfrac{1}{2}m_1 + m_2)gL_1\sin\theta_1 = 0 \quad (1)$$

$$\frac{\partial V}{\partial\theta_2} = k_t(\theta_2 - \theta_1) - PL_2\cos\theta_2 + \tfrac{1}{2}m_2 gL_2\sin\theta_2 = 0 \quad (2)$$

Equation (2) yields P and then eq. (1) yields M. Note that eq. (2) is $M_{A_z} = 0$ for the bar 2 and eq. (1) + eq. (2) is $M_{B_z} = 0$ for the system of two bars.

(ii) FBDs of the bars are shown in Fig. E5.7b. The pair of moments due to the torsional spring does net virtual work since the virtual rotations $\delta\theta_1, \delta\theta_2$ of the bars are different. Forces in the y direction act at points 1 and 4, whereas forces in the x-direction act at points 2 and 3. Hence, we need to calculate $\delta y_1, \delta y_4$ and $\delta x_2, \delta x_3$:

$$y_1 = L_1 \sin\theta_1 + L_2 \sin\theta_2, \qquad y_4 = b\sin\theta_1$$

$$\delta y_1 = L_1 \cos\theta_1\, \delta\theta_1 + L_2 \cos\theta_2\, \delta\theta_2, \qquad \delta y_4 = b\cos\theta_1\, \delta\theta_1$$

$$x_2 = L_1 \cos\theta_1 + \tfrac{1}{2}L_2 \cos\theta_2, \qquad x_3 = \tfrac{1}{2}L_1 \cos\theta_1$$

$$\delta x_2 = -L_1 \sin\theta_1\, \delta\theta_1 - \tfrac{1}{2}L_2 \sin\theta_2\, \delta\theta_2, \qquad \delta x_3 = -\tfrac{1}{2}L_1 \sin\theta_1\, \delta\theta_1$$

$$\begin{aligned}\delta W &= P\,\delta y_1 + m_2 g\,\delta x_2 + m_1 g\,\delta x_3 + (-kb\sin\theta_1)\,\delta y_4 \\ &\quad + M\,\delta\theta_1 + [k_t(\theta_2-\theta_1)]\,\delta\theta_1 + [-k_t(\theta_2-\theta_1)]\,\delta\theta_2 \\ &= [PL_1\cos\theta_1 - (\tfrac{1}{2}m_1 + m_2)gL_1\sin\theta_1 - kb^2\sin\theta_1\cos\theta_1 + M + k_t(\theta_2-\theta_1)]\delta\theta_1 \\ &\quad + [PL_2\cos\theta_2 - \tfrac{1}{2}m_2 g L_2 \sin\theta_2 - k_t(\theta_2-\theta_1)]\delta\theta_2 = 0\end{aligned}$$

$\forall\, \delta\theta_1$ and $\forall\, \delta\theta_2$. Hence, the coefficients of $\delta\theta_1$ and $\delta\theta_2$ should separately be zero, i.e.,

$$PL_1\cos\theta_1 - (\tfrac{1}{2}m_1 + m_2)gL_1\sin\theta_1 - kb^2\sin\theta_1\cos\theta_1 + M + k_t(\theta_2-\theta_1) = 0$$

$$PL_2\cos\theta_2 - \tfrac{1}{2}m_2 g L_2\sin\theta_2 - k_t(\theta_2-\theta_1) = 0$$

Example 5.8: Force F is applied to the handle of a toggle press (Fig. E5.8a). Find the force exerted by the lower jaw on the piece below it, assuming it slides without friction.

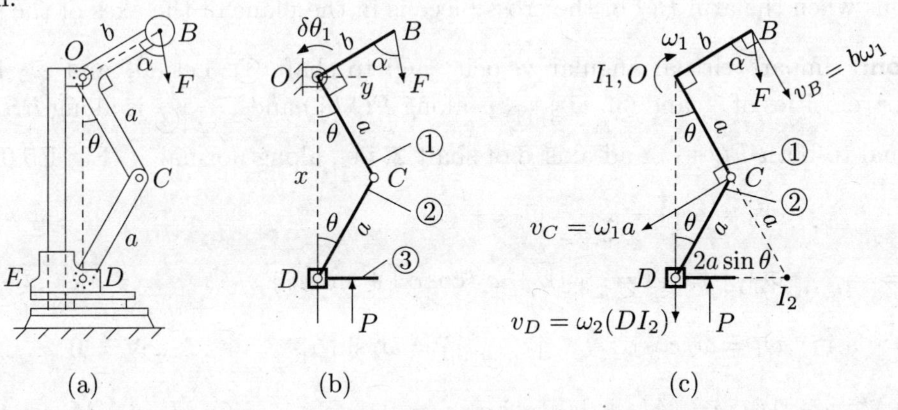

Fig. E5.8

Solution: We present 3 solutions for one-dof system with generalised coordinate θ.

1. $\delta\dot{W} = 0$: Give a virtual angular velocity ω_1 to the lever BOC. $v_B = b\omega_1$, $v_C = a\omega_1$ (Fig. E5.8c). D moves vertically. The instantaneous centre I_2 of link ② is located at the intersection of the normals to the velocities of points C and D.

$$\frac{v_D}{v_C} = \frac{I_2D}{I_2C} = \frac{2a\sin\theta}{a} = 2\sin\theta \quad \Rightarrow \quad v_D = 2a\omega_1\sin\theta$$

$$\delta\dot{W} = \vec{P}\cdot\vec{v}_D + \vec{F}\cdot\vec{v}_B = -P(2a\omega_1\sin\theta) + Fb\omega_1\cos(\pi/2-\alpha) = 0 \quad \forall\,\omega_1$$

$$\Rightarrow \quad P = bF\sin\alpha/2a\sin\theta \tag{1}$$

P is very large if θ is a small angle and P is the largest for $\alpha = \pi/2$.
2. $dV/d\theta = 0$: The moment $\vec{M}_O$ of F is $\vec{M}_O = -bF\sin\alpha\,\vec{k}$. Force F at B is equivalent to force F at O along with couple $C = M_{O_z} = -bF\sin\alpha$ whose potential energy is $-C\theta$. The potential energy of P is $-F_x x_D = -(-P)x_D$:

$$\begin{aligned} V(\theta) &= -(-P)2a\cos\theta - (-bF\sin\alpha)\theta = 2Pa\cos\theta + bF\sin\alpha\,\theta \\ V'(\theta) &= -2Pa\sin\theta + bF\sin\alpha = 0 \quad \Rightarrow \quad \text{eq. (1)} \end{aligned}$$

3. $\delta W = 0$: The virtual work of F on ① can be obtained using its $\vec{M}_O$ as $\vec{M}_O\cdot\delta\vec{\theta}_1$.

$$\begin{aligned} \delta W &= \vec{M}_O\cdot\delta\vec{\theta}_1 - P\delta x_D = -bF\sin\alpha\,\vec{k}\cdot\delta\theta\,\vec{k} - P\delta(2a\cos\theta) \\ &= (2aP\sin\theta - bF\sin\alpha)\delta\theta = 0 \quad \forall\,\delta\theta \quad \Rightarrow \quad \text{eq. (1)} \end{aligned}$$

Example 5.9: Shafts ① and ② are connected by a universal joint at A (Fig. E5.9a). The bearings at B and C do not exert axial force. A moment M_1 is applied to shaft ①. Find the moment M_2 to be applied to shaft ② for equilibrium and the bearing reactions when the arm PQ of the cross piece is in the plane of the axes of the shafts.

Solution: Impart virtual angular velocity $\omega_1\vec{i}$ to shaft ①. Let $\vec{\omega}_2$ and $\vec{\omega}_3$ be the angular velocities of ② and ③. $\vec{\omega}_{③|①}$ is along PQ ($\vec{j}$) and $\vec{\omega}_{③|②}$ is along RS which is normal to both PQ ($\vec{j}$) and axis $\vec{e}$ of shaft 2, i.e., along normal $\vec{k}$ (Fig. E5.9b).

$$\vec{\omega}_3 = \vec{\omega}_{3|1} + \vec{\omega}_1 = \vec{\omega}_{3|2} + \vec{\omega}_2$$

$$\Rightarrow \quad \omega_{3|1}\vec{j} + \omega_1\vec{i} = \omega_{3|2}\vec{k} + \omega_2(\cos\alpha\,\vec{i} + \sin\alpha\,\vec{j})$$

$$\Rightarrow \quad \vec{i}:\ \omega_1 = \omega_2\cos\alpha, \qquad \vec{j}:\ \omega_{3|1} = \omega_2\sin\alpha, \qquad \vec{k}:\ \omega_{3|2} = 0$$

$$\delta\dot{W} = M_1\omega_1 + M_2\omega_2 = (M_1 + M_2/\cos\alpha)\omega_1 = 0 \quad \forall\,\omega_1 \quad \Rightarrow \quad M_2 = -M_1\cos\alpha \tag{1}$$

The FBDs of shaft ② and the whole system are shown in Fig. E5.9c,d. $\vec{C}_R$ and $\vec{C}_S$ are normal to arm RS ($\vec{k}$) of ③. $R_6\,\vec{e}_1$ at C in xy-plane is normal to the axis $\vec{e}$ of ②.

For ②: $M_{RS} = cR_6 = 0 \Rightarrow R_6 = 0$. Use eq. (1) and $\vec{M}_D = \vec{0}$, $\vec{F} = \vec{0}$ for whole system:

$$\vec{M}_D = M_1\,\vec{\mathrm{i}} + M_2(\cos\alpha\,\vec{\mathrm{i}} + \sin\alpha\,\vec{\mathrm{j}}) + b\,\vec{\mathrm{i}} \times (R_4\,\vec{\mathrm{j}} + R_5\,\vec{\mathrm{k}})$$

$$+[(b + a + c\cos\alpha)\,\vec{\mathrm{i}} + c\sin\alpha\,\vec{\mathrm{j}}] \times R_7\,\vec{\mathrm{k}} = \vec{0}$$

$\vec{\mathrm{i}}$: $\quad M_1 + M_2\cos\alpha + c\sin\alpha R_7 = 0 \quad \Rightarrow \quad R_7 = -M_1\sin\alpha\,/c$

$\vec{\mathrm{j}}$: $\quad M_2\sin\alpha - bR_5 - (b + a + c\cos\alpha)R_7 = 0 \quad \Rightarrow \quad R_5 = M_1\sin\alpha(b + a)/bc$

$\vec{\mathrm{k}}$: $\quad bR_4 = 0 \quad \Rightarrow \quad R_4 = 0$

$$\vec{F} = \vec{R}_D + R_4\,\vec{\mathrm{j}} + R_5\,\vec{\mathrm{k}} + R_7\,\vec{\mathrm{k}} = \vec{0} \quad \Rightarrow \quad \vec{R}_D = -(M_1 a\sin\alpha\,/bc)\,\vec{\mathrm{k}}$$

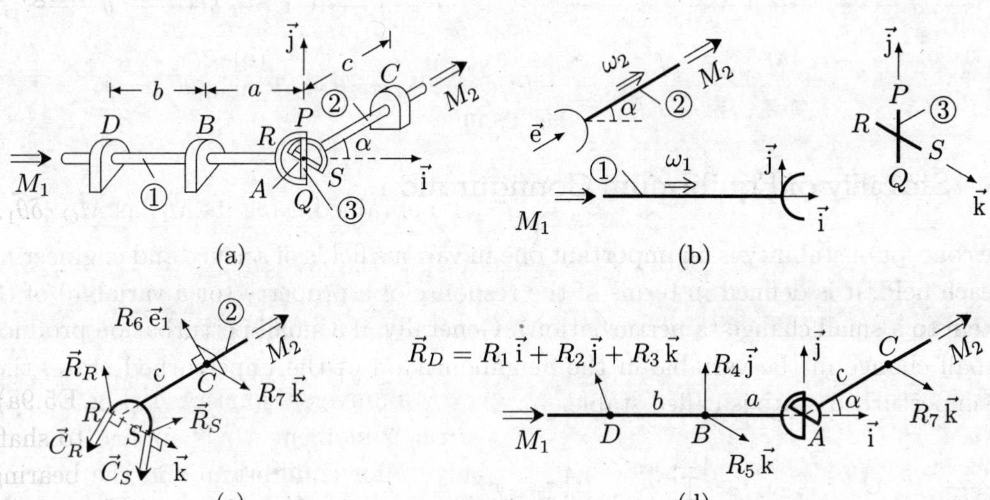

Fig. E5.9

Example 5.10: Inaccessible objects are held by a mechanism (Fig. E5.10a). The unstretched length of the spring is L_0. Find the clamping force P for applied forces F.

Solution: We apply principle of virtual rate of work to the system in Fig. E5.10b. Keeping O as fixed, a virtual angular velocity ω_1 is given to link ①. For point Q of a rigid body rotating at $\omega\,\vec{\mathrm{k}}$ about the instantaneous centre I, with $\overrightarrow{IQ} = x\,\vec{\mathrm{i}} + y\,\vec{\mathrm{j}}$:

$$\vec{v}_Q = \vec{v}_I + \omega\,\vec{\mathrm{k}} \times \overrightarrow{IQ} = \omega\,\vec{\mathrm{k}} \times (x\,\vec{\mathrm{i}} + y\,\vec{\mathrm{j}}) = \omega x\,\vec{\mathrm{j}} - \omega y\,\vec{\mathrm{i}} \quad \Rightarrow \quad v_{Q_y} = \omega x \qquad (1)$$

Symmetry $\Rightarrow$ v_B is along OB. The instantaneous centre I_2 of ② is at the point of intersection of the normals to $\vec{v}_A$ and $\vec{v}_B$. Using eq. (1), v_{A_y} and v_{C_y} are shown in Fig. E5.10b. The virtual angular velocity ω_2 of ② is found by writing v_{A_y} in 2 ways:

$$v_{Ay} = \omega_1 c = \omega_2 b \quad \Rightarrow \quad \omega_2 = \omega_1 c/b, \qquad v_{Cy} = \omega_2 a = \omega_1 ca/b$$

$$\delta \dot{W} = F\omega_1 d + k(L - L_0)\omega_1 c - P\omega_1 ca/b = 0 \quad \forall\ \omega_1$$

$$\Rightarrow \quad P = [Fd + ck(L - L_0)]b/ac$$

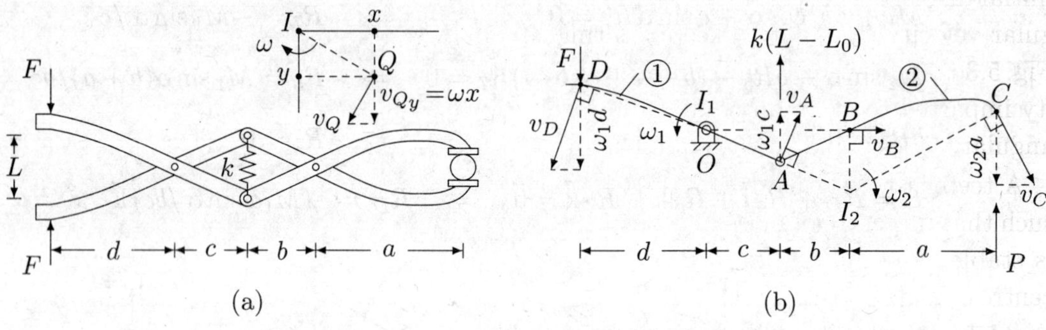

Fig. E5.10

5.5 Stability of Equilibrium Configuration

The concept of stability is an important one in various fields of science and engineering. In each field, it is defined in terms of the response of a property (or a variable) of the system to a small change (a perturbation). Generally, if a small perturbation produces a small change in the variable in the neighbourhood of the unperturbed state, then the unperturbed state is called stable.

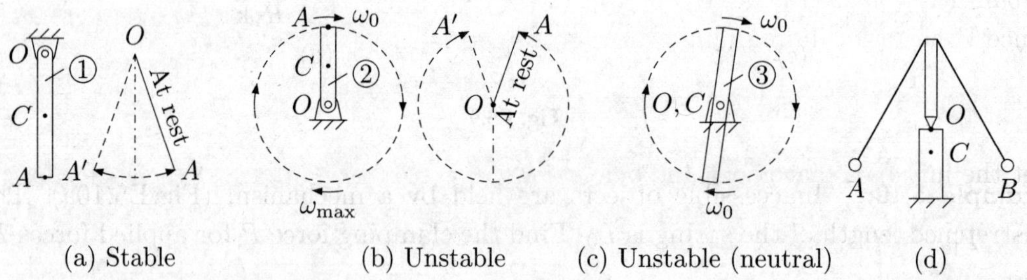

Fig. 5.3

An equilibrium configuration of a system is said to be **stable** if *an infinitesimal and once for all initial disturbance* (displacement and / velocity) from the equilibrium state, results in a motion which is limited to a *small neighbourhood* of the equilibrium configuration. Else, it is **unstable**, i.e., there exists *some* arbitrary small disturbance which causes the system to depart away from the position of equilibrium. If the neighbouring configurations of an equilibrium configuration are also equilibrium configurations, then this kind of *unstable equilibrium* [since a velocity disturbance causes the system to move well away from the equilibrium configuration] is sometimes called *neutral.*

The equilibrium position of rod ① in Fig. 5.3a is stable, since a small initial angular velocity and / initial small displacement causes rod ① to vibrate with a small displacement. Equilibrium position of rod ② in Fig. 5.3b is unstable, since a small initial displacement causes rod ② to vibrate with large displacement and a small angular velocity causes it to keep rotating around in the same direction. For rod ③ in Fig. 5.3c, all positions are equilibrium configurations. Any small initial angular velocity imparted to ③ in an equilibrium position, causes it to keep rotating at the same angular velocity and so the equilibrium positions of rod ③ are unstable.

A teeter toy (Fig. 5.3d) has A, B of equal mass attached to two arms. It is designed such that its centre of mass C is below its point of suspension O and so this 'pendulum' is stable. The human body in its erect position is an 'inverted pendulum' with its centre of mass above the support. It is mechanically unstable. An elaborate sensing and control system constantly exerts torque on the legs to maintain the vertical position. If a person has a tumor in the cerebrum, the control system can fail and he/she can fall.

5.5.1 Stability of a conservative system of rigid bodies

The conditions of stability are as follows.

(a) *If the potential energy of a conservative system of connected rigid bodies has a local minimum at a position of equilibrium, then that equilibrium configuration is stable.*

(b) *Conversely, if the equilibrium configuration of a conservative system of connected rigid bodies is stable, then its potential energy has a* **local minimum** *at that position.*

Proof: **(a)** Choose the datum of V at the equilibrium configuration q_0, i.e., $V(q_0) = 0$. Since V has a local minimum at q_0, for any position q in the neighbourhood of q_0,

$$V(q) > V(q_0) = 0 \tag{1}$$

Let the initial displacement and velocity imparted to the system be q^* and $\dot{q}^*$. Then,

$$T|_q + V(q) = T|_{q^*} + V(q^*) = E(q^*) > 0 \quad \Rightarrow \quad T|_q = E(q^*) - V(q) \tag{2}$$

$E(q^*)$ has a small positive value since $T|_{q^*} > 0$ and $V(q^*) > 0$ by eq. (1). Equation (1) $\Rightarrow$ there exists a set of q' in the neighbourhood of q_0 $\ni$ $V(q') = E(q^*)$ and eq. (2) $\Rightarrow$

$$T|_{q'} = E(q^*) - V(q') = 0 \qquad \Rightarrow \qquad \dot{q}' = 0$$

Thus, q lies in the neighbourhood of q_0 bounded by set of q' $\Rightarrow$ the position is stable.

(b) Let the system be given small initial velocity $\dot{q}_0$ at the equilibrium configuration q_0 and the maximum deviation be q' at which $\dot{q}' = 0$, since it is stable. Hence,

$$T|_{q'} + V(q') = T|_{q_0} + V(q_0) \quad \Rightarrow \quad V(q') - V(q_0) = T|_{q_0} > 0$$

$$\Rightarrow \quad V(q') > V(q_0) \tag{3}$$

where q' is in the neighbourhood of q_0 and depends on $\dot{q}_0$. Any q in the neighbourhood of q_0 can be reached as an extremal position, by an appropriate choice of $\dot{q}_0$. Hence, eq. (3) is valid $\forall\ q'$ in the neighbourhood of q_0 and so V has a local minimum at q_0.

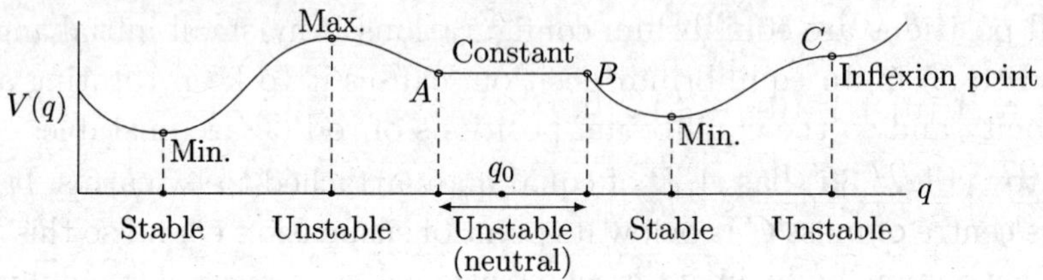

Fig. 5.4

If $V(q)$ is constant in some interval of q (Fig. 5.4), then a small velocity $\dot{q}(0) = \dot{q}^*$ imparted at equilibrium position q_0 causes motion away from q_0 since velocity is constant in this interval of q. Therefore, q_0 is an unstable (neutral) equilibrium.

The nature of equilibrium configurations is shown in Fig. 5.4 for a 1-dof conservative system. Between any two positions of stable equilibrium, there must be at least one position of unstable equilibrium. The positions of stable and unstable equilibrium alternate, except when the instability is at an inflexion point (C) or at a plateau AB.

5.5.2 Conditions of stability of 1-dof conservative system

We use the notation $V' = dV/dq$, $V^{(n)} = d^nV/dq^n$. Consider a 1-dof conservative system with generalised coordinate q and mass m.

1. The equilibrium configurations are given by values q_0 for which $V'(q_0) = 0$.

2. The equilibrium configuration q_0 is stable iff V has a local minimum at q_0, i.e., if

$$\begin{aligned} &V''(q_0) > 0, \qquad \text{or} \\ &V^{(2)}(q_0) = V^{(3)}(q_0) = \ldots = V^{(2n-1)}(q_0) = 0 \quad \text{and} \quad V^{(2n)}(q_0) > 0 \end{aligned} \tag{5.18}$$

Else, the equilibrium configuration q_0 is unstable, i.e., V has a local maximum or an inflection point or a plateau at q_0. The last is a case of unstable neutral equilibrium.

3. Consider *small* oscillations around a position of 'stable equilibrium', q_0, of a 1-dof conservative system for which $V''(0) > 0$. We choose the origin of generalised coordinate at this position of equilibrium $\Rightarrow q_0 = 0$. Choose the datum for V at $q = 0$. Hence, $V(0) = 0$, $V'(0) = 0$. Let P be a point within mass element dm:

$$T = \int_m v_P^2\, dm = \int_m \dot{\vec{r}}_P^{\,2}\, dm = \int_m \frac{d\vec{r}_P}{dq}\, \dot{q} \cdot \frac{d\vec{r}_P}{dq}\, \dot{q}\, dm$$

$$\Rightarrow \quad T = \tfrac{1}{2}m_e(q)\dot{q}^2, \quad \text{where} \quad m_e(q) = \int_m \frac{d\vec{r}_P}{dq} \cdot \frac{d\vec{r}_P}{dq}\, dm \tag{1}$$

is the *effective mass* of the system. For small values of q and $\dot{q}$, we need to retain only linear terms in q and $\dot{q}$ in Lagrange's equation of motion involving $\partial T/\partial \dot{q}$, $\partial T/\partial q$, $\partial V/\partial q \Rightarrow$ we retain up to quadratic terms in T and V. We expand $m_e(q)$ and $V(q)$ around $q = 0$ by Taylor's series and retain up to quadratic terms (provided $V''(0) \neq 0$):

$$T = \tfrac{1}{2}[m_e(0) + m_e'(0)q + \ldots]\dot{q}^2 \simeq \tfrac{1}{2}m_e(0)\dot{q}^2 > 0 \quad \Rightarrow \quad m_e(0) > 0 \tag{5.19}$$

$$V(q) = V(0) + V'(0)q + \tfrac{1}{2}V''(0)q^2 + \ldots \simeq \tfrac{1}{2}V''(0)q^2 \tag{5.20}$$

$$\frac{d}{dt}\left(\frac{\partial T}{\partial \dot{q}}\right) - \frac{\partial T}{\partial q} + \frac{\partial V}{\partial q} = m_e(0)\ddot{q} + V''(0)q = 0$$

$$\Rightarrow \quad \ddot{q} + \omega^2 q = 0 \quad \text{with} \quad \omega = [V''(0)/m_e(0)]^{1/2} \tag{5.21}$$

Its solution is simple harmonic $\Rightarrow$ motion is bounded near 'stable equilibrium' position.

4. For eq. $(5.18)_2$: $V(q) \simeq \dfrac{V^{(2n)}(0)}{(2n)!}q^{2n}, \quad T + V = \tfrac{1}{2}m_0\dot{q}^2 + \dfrac{V^{(2n)}(0)}{(2n)!}q^{2n} = E_0.$

$$\Rightarrow \quad \dot{q} = 0 \quad \text{for} \quad q = \pm q_m, \quad \text{where} \quad q_m = \left[\frac{(2n)!E_0}{V^{(2n)}(0)}\right]^{1/2} \quad \Rightarrow$$

The system vibrates within small bounds $-q_m$ and q_m $\Rightarrow$ stable equilibrium position.

We apply the conditions (5.18) to study various kinds of instabilities. The stability of a cylindrical body resting on another is explored in Ex.5.11. *Stable symmetric point of bifurcation* is illustrated by the model of a column in Ex.5.14. The post-buckling path is shown to be stable. The point of intersection of the trivial equilibrium path with the post-buckling path is called the *point of bifurcation.* The corresponding load is called the *critical load.* The effect of *initial imperfections* is shown to be insignificant. The case of an *unstable symmetric point of bifurcation* in which the post-buckling path is unstable is illustrated in Ex.5.15. The critical load is shown to be very sensitive to initial imperfection. An example of *asymmetric point of bifurcation* in which the post-buckling path is not symmetric about the pre-buckling path is shown in Ex.5.16. *Limit point* and *snap-through* buckling of shallow arches is discussed by the model in Ex.5.17.

Example 5.11: Find the condition of stability of a cylinder resting on a fixed cylinder (Fig. E5.11a). The curvature of their perimeters is continuous and there is no slip.

Solution: (a) Let C be the centre of mass of ② (Fig. E5.11a1). Mark the collinear points A, O_2, C on ②. In the displaced position (Fig. E5.11a2), the point of contact is D with $\angle BO_1D = \theta$. The no slip condition implies *for the case of constant* R_1, R_2

$$\widehat{BD} = \widehat{DA} \quad \Rightarrow \quad R_2\phi = R_1\theta \quad \Rightarrow \quad \phi = R_1\theta \,/\, R_2$$

In the displaced position, lines O_1DO_2 and AO_2C are at angles θ and $\theta + \phi$ with the vertical. The potential energy V of body ② of mass m with datum at O_1 is

$$
\begin{aligned}
V(\theta) &= mg\,[(O_1O_2)\cos\theta + (O_2C)\cos(\theta+\phi)] \\
&= mg\left[(R_1+R_2)\cos\theta + (h-R_2)\cos\left(1+\frac{R_1}{R_2}\right)\theta\right] \qquad (1)
\end{aligned}
$$

$$
V'(\theta) = -mg\left[(R_1+R_2)\sin\theta + (h-R_2)\left(1+\frac{R_1}{R_2}\right)\sin\left(1+\frac{R_1}{R_2}\right)\theta\right] \qquad (2)
$$

$$
V''(\theta) = -mg\left[(R_1+R_2)\cos\theta + (h-R_2)\left(1+\frac{R_1}{R_2}\right)^2\cos\left(1+\frac{R_1}{R_2}\right)\theta\right] \qquad (3)
$$

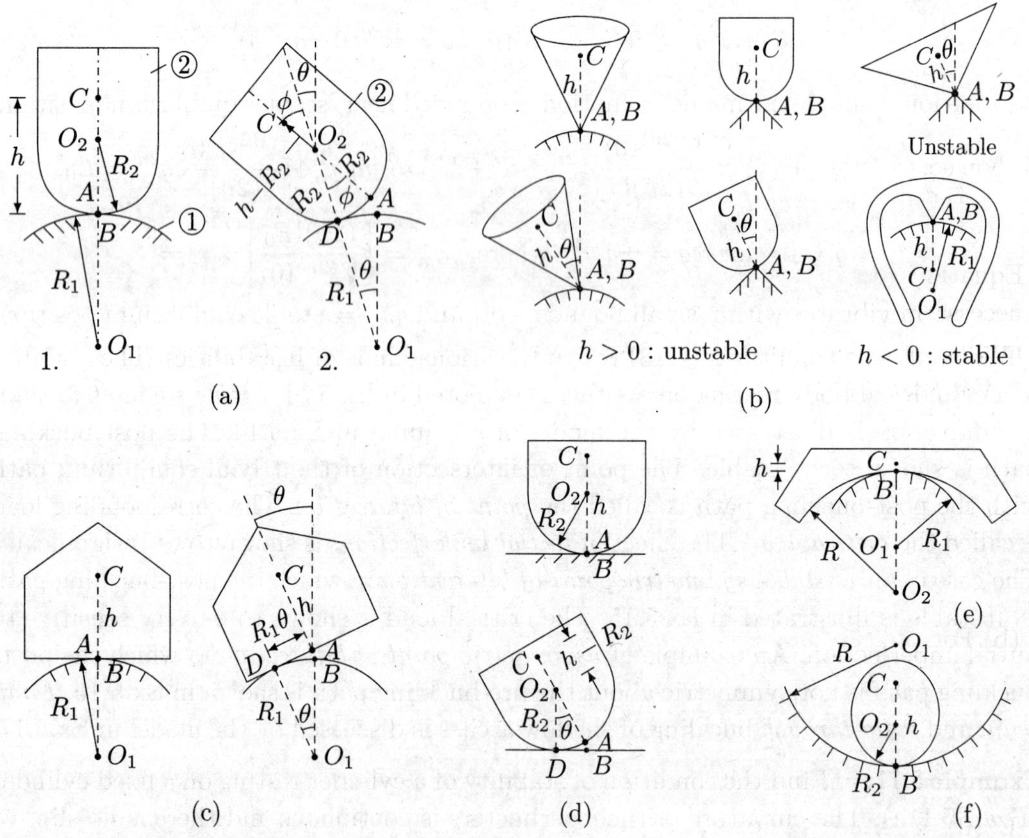

Fig. E5.11

Using eq. (2), the equilibrium condition $V' = 0$, yields $\theta = 0$ as the equilibrium configuration, which is otherwise obvious. This equilibrium position is stable if $V''(0) > 0$,

$$\text{i.e., if} \quad V''(0) \overset{(3)}{=} -mg\left[(R_1+R_2)+(h-R_2)\left(1+\frac{R_1}{R_2}\right)^2\right]$$

$$= mg\left(\frac{R_1}{R_2}+1\right)^2\left[\frac{R_1R_2}{R_1+R_2}-h\right] > 0$$

$$\text{i.e., if} \qquad h < \frac{R_1R_2}{R_1+R_2} = \frac{1}{1/R_1+1/R_2} \tag{4}$$

It is unstable for $V''(0) < 0$, i.e., for $h > R_1R_2/(R_1+R_2)$. For $h = R_1R_2/(R_1+R_2)$, the stability is decided from higher derivatives as $V''(0) = 0$. For this case, eq. (3) $\Rightarrow$

$$V''(\theta) = -mg(R_1+R_2)[\cos\theta - \cos\{(1+R_1/R_2)\theta\}] \tag{5}$$

$$V'''(\theta) = mg(R_1+R_2)[\sin\theta - (1+R_1/R_2)\sin\{(1+R_1/R_2)\theta\}] \tag{6}$$

$$V'''(0) = 0 \tag{7}$$

$$V''''(\theta) = mg(R_1+R_2)[\cos\theta - (1+R_1/R_2)^2\cos\{(1+R_1/R_2)\theta\}] \tag{8}$$

$$V''''(0) = mg(R_1+R_2)[1-(1+R_1/R_2)^2] < 0 \tag{9}$$

Equation 9 holds for: (i) R_1, R_2 are finite positive values, and (ii) R_2 is negative and necessarily $|R_2| > R_1$ (Fig. E5.11e). Equations (7), (9) $\Rightarrow$ V does not have a local minimum at $\theta = 0$ and therefore, this position is unstable for $h = R_1R_2/(R_1+R_2)$.

For the stable case, the frequency ω of small oscillation is obtained as follows:

$$T = \tfrac{1}{2}I^D_{zz}(\dot\theta+\dot\phi)^2 \simeq \tfrac{1}{2}I^A_{zz}(\dot\theta+\dot\phi)^2 = \tfrac{1}{2}m(k_C^2+h^2)(1+R_1/R_2)^2\dot\theta^2 = \tfrac{1}{2}m_e\dot\theta^2$$

$$\Rightarrow \quad m_e = m(k_C^2+h^2)(1+R_1/R_2)^2$$

$$\omega^2 = V''(0)/m_e = g[R_1R_2/(R_1+R_2)-h]/(k_C^2+h^2)$$

(b) For the case of R_1 and / or $R_2 = 0$ (Fig. E5.11b), $D = B$ and with datum at B,

$$V(\theta) = mgh\cos\theta \quad \Rightarrow \quad V'(\theta) = -mgh\sin\theta$$

$$V''(\theta) = -mgh\cos\theta \tag{10}$$

$V''(0) = -mgh > 0$ for $h < 0$ $\Rightarrow$ The position $\theta = 0$ is stable if $h < 0$ and unstable if $h > 0$ and also unstable for $h = 0$ as in this case $V(\theta) = 0$ (a constant).

$$T = \tfrac{1}{2}I^B_{zz}\dot\theta^2 = \tfrac{1}{2}m(k_C^2+h^2)\dot\theta^2 = \tfrac{1}{2}m_e\dot\theta^2 \quad \Rightarrow \quad m_e = m(k_C^2+h^2)$$

$$\omega^2 = V''(0)/m_e = -gh/(k_C^2+h^2)$$

(c) For the case of $R_2 = \infty$ (Fig. E5.11c):

$$\begin{aligned} V(\theta) &= mg[(h+R_1)\cos\theta + R_1\theta\sin\theta] \\ V'(\theta) &= mg[-h\sin\theta + R_1\theta\cos\theta] \\ V''(\theta) &= mg[(R_1-h)\cos\theta - R_1\theta\sin\theta] \end{aligned} \tag{11}$$

$V''(0) = mg(R_1 - h) > 0$ for $h < R_1$ $\Rightarrow$ The position $\theta = 0$ is stable if $h < R_1$ and unstable if $h > R_1$. For $h = R_1$, eq. (11) $\Rightarrow$

$$\begin{aligned} V''(\theta) &= -mgh\theta\sin\theta & &\Rightarrow \quad V''(0) = 0 \\ V'''(\theta) &= -mgh(\sin\theta + \theta\cos\theta) & &\Rightarrow \quad V'''(0) = 0 \\ V^{iv}(\theta) &= -mgh(2\cos\theta - \theta\sin\theta) & &\Rightarrow \quad V^{iv}(0) = -2mgh < 0. \end{aligned}$$

$\Rightarrow$ position $\theta = 0$ is unstable for $h = R_1$.

$$\begin{aligned} T &= \tfrac{1}{2}I_{zz}^D\dot{\theta}^2 = \tfrac{1}{2}m(k_C^2 + h^2 + R_1^2\theta^2)\dot{\theta}^2 \simeq \tfrac{1}{2}m(k_C^2 + h^2)\dot{\theta}^2 = \tfrac{1}{2}m_e\dot{\theta}^2 \;\Rightarrow \\ m_e &= m(k_C^2 + h^2), \qquad \omega^2 = V''(0)/m_e = g(R_1 - h)/(k_C^2 + h^2) \end{aligned}$$

(d) For $R_1 = \infty$, (Fig. E5.13d):

$$\begin{aligned} V(\theta) &= mg[R_2 + (h - R_2)\cos\theta], \qquad V'(\theta) = mg(R_2 - h)\sin\theta \\ V''(\theta) &= mg(R_2 - h)\cos\theta, \qquad V''(0) = mg(R_2 - h) \end{aligned}$$

$V''(0) > 0$ for $h < R_2$ $\Rightarrow$ The position $\theta = 0$ is stable if $h < R_2$ and unstable if $h > R_2$. It is also unstable for $h = R_2$ since $V(\theta) = mgR_2$ (a constant) for this case.

$$\begin{aligned} T &= \tfrac{1}{2}I_{zz}^D\dot{\theta}^2 \simeq \tfrac{1}{2}I_{zz}^A\dot{\theta}^2 = \tfrac{1}{2}m(k_C^2 + h^2)\dot{\theta}^2 = \tfrac{1}{2}m_e\dot{\theta}^2 \quad \Rightarrow \quad m_e = m(k_C^2 + h^2) \\ \omega^2 &= V''(0)/m_e = g(R_2 - h)/(k_C^2 + h^2) \end{aligned}$$

Thus, for cases in Figs E5.11a–e, the equilibrium is stable only if $h < R_1R_2/(R_1 + R_2)$. For the case in Fig. E5.11e: $R_2 = -R$, this condition of stable equilibrium reduces to $h < R_1/(1 - R_1/R)$. We can show for the case in Fig. E5.11f ($R_1 = -R$): $\theta = 0$ is stable only if $h < R_2/(1 - R_2/R)$ for $R \leq 2R_2$ and is stable if $h \leq R_2/(1 - R_2/R)$ for $R > 2R_2$.

These results apply to bodies with variable R_1, R_2 with R_1, R_2 as the values of the radii of curvature at the contact points in the equilibrium position.

Example 5.12: The bar AB is pinned to the support at A (Fig. E5.12a) and to the bar BC at B. The bar BC is pinned to a slider at D which is guided in a vertical slot. The bars are light and constrained by 5 springs which are undeformed for the given position.

Find the load P at which the vertical equilibrium configuration becomes unstable.

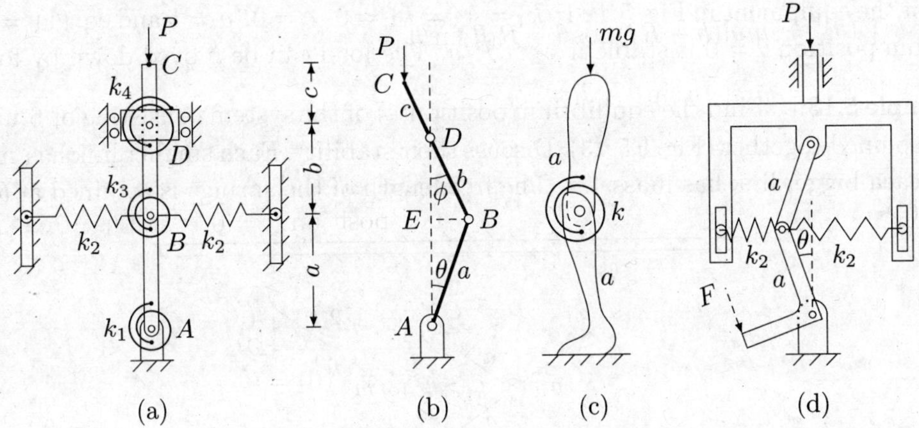

Fig. E5.12

Solution: In the displaced position (Fig. E5.12b), angles θ and ϕ are related by

$$BE = b\sin\phi = a\sin\theta \qquad \Rightarrow \qquad \text{up to 2nd order} \qquad \phi = a\theta/b$$

The relative rotation between the ends of the torsional spring k_3 equals $(\pi+\theta+\phi)-\pi = \theta+\phi$. The potential energy V of this conservative system, accurate up to 2nd order, is

$$\begin{aligned} V(\theta) &= \tfrac{1}{2}k_1\theta^2 + 2[\tfrac{1}{2}k_2(a\sin\theta)^2] + \tfrac{1}{2}k_3(\theta+\phi)^2 + \tfrac{1}{2}k_4\phi^2 + P[a\cos\theta + (b+c)\cos\phi] \\ &= \tfrac{1}{2}[k_1 + k_3(1+a/b)^2 + k_4a^2/b^2]\theta^2 + k_2a^2\sin^2\theta + P[a\cos\theta + (b+c)\cos(a\theta/b)] \\ V'(\theta) &= [k_1 + k_3(1+a/b)^2 + k_4a^2/b^2]\theta + k_2a^2\sin 2\theta \\ &\qquad -P[a\sin\theta + (b+c)(a/b)\sin(a\theta/b)] \\ V''(\theta) &= [k_1 + k_3(1+a/b)^2 + k_4a^2/b^2] + 2k_2a^2\cos 2\theta \\ &\qquad -P[a\cos\theta + (b+c)(a/b)^2\cos(a\theta/b)] \end{aligned}$$

$V'(0) = 0 \Rightarrow \theta = 0$ is an equilibrium configuration, which is stable if

$$V''(0) = [k_1 + k_3(1+a/b)^2 + k_4a^2/b^2] + 2k_2a^2 - P[a + (b+c)(a/b)^2] > 0$$

i.e., if $$P < \frac{k_1 + k_3(1+a/b)^2 + k_4a^2/b^2 + 2k_2a^2}{a + (b+c)(a/b)^2} = P_1 \text{ (say)} \qquad (1)$$

and unstable if $V''(0) < 0$, i.e., if $P > P_1$. The minimum value of P at which the equilibrium position becomes unstable is called the critical load P_{cr}, $P_{\text{cr}} = P_1$.

For a rigid link model of an artificial leg (Fig. 5.12c): $k_1 = k_4 = 0,\ k_2 = 0,\ k_3 = k,$

$c = 0$, $a = b$, $P = mg$. Equation (1) $\Rightarrow \theta = 0$ is stable if $mg < 2k/a$, i.e., if $k > \frac{1}{2}mga$.

For the equipment in Fig. 5.12d: $k_4 = k_3 = k_1 = 0$, $c = 0$, $a = b$ and eq. (1) $\Rightarrow$ equilibrium position $\theta = 0$ is stable if $k_2 > P/a$. The load can be moved down by force F.

Example 5.13: Find the equilibrium position(s) of the system consisting of 6 uniform links pinned together (Fig. E5.13). Discuss their stability. Each smaller link has mass m and each bigger link has mass $4m$. The free length of the springs is attained at $\theta = \theta_0$.

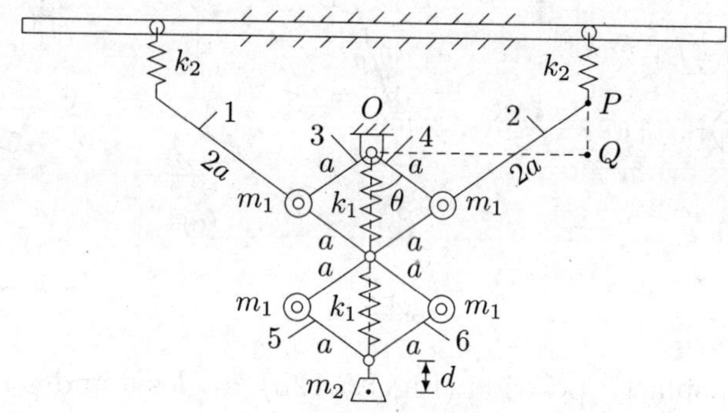

Fig. E5.13

Solution: In position θ, the length of the spring k_1 is $2a\cos\theta$ and $PQ = a\cos\theta$. The extensions of springs k_1 and k_2 are $(2a\cos\theta - 2a\cos\theta_0)$ and $(a\cos\theta_0 - a\cos\theta)$ respectively. The potential energy V with the datum for gravitational energy at O is

$$V(\theta) = 2[\tfrac{1}{2}k_1(2a\cos\theta - 2a\cos\theta_0)^2 + \tfrac{1}{2}k_2(a\cos\theta_0 - a\cos\theta)^2]$$
$$-2m_1g(a\cos\theta) - 2m_1g(3a\cos\theta) - m_2g(4a\cos\theta + d)$$
$$-2mg(0.5a\cos\theta) - 2mg(3.5a\cos\theta) - 2(4m)g(a\cos\theta)$$

$$V(\theta) = c_1(\cos\theta - \cos\theta_0)^2 - c_2\cos\theta - m_2gd \tag{1}$$

where $c_1 = (4k_1 + k_2)a^2$, $c_2 = 4(2m_1 + m_2 + 4m)ga$.

$$V'(\theta) = \sin\theta[c_2 - 2c_1(\cos\theta - \cos\theta_0)] \tag{2}$$

$$V''(\theta) = \cos\theta[c_2 - 2c_1(\cos\theta - \cos\theta_0)] + 2c_1\sin^2\theta \tag{3}$$

$$V' = 0 \quad \Rightarrow \quad \text{(i)}\ \sin\theta = 0 \qquad \text{(ii)}\ [c_2 - 2c_1(\cos\theta - \cos\theta_0)] = 0 \tag{4}$$

Thus, we consider the stability of equilibrium positions at: (1) $\theta = 0$, (2) $\theta = \pi$, (3) $\theta = \theta_3 = \cos^{-1}[c_2/2c_1 + \cos\theta_0]$ if $c_2/2c_1 + \cos\theta_0 < 1$ so that $\cos\theta_3 < 1$,

i.e., if $$c_2 - 2c_1(1 - \cos\theta_0) < 0 \tag{5}$$

(1) $\theta = 0$: $V''(0) = c_2 - 2c_1(1 - \cos\theta_0)$ and using eq. (5), we infer:

(a) if $V''(0) < 0$, i.e., if equilibrium position θ_3 exists, then position $\theta = 0$ is unstable.

(b) if $V''(0) > 0$, then $\theta = 0$ is a stable equilibrium position.

(c) if $V''(0) = c_2 - 2c_1(1 - \cos\theta_0) = 0$, then substituting $\cos\theta_0$ from this in eq. (3):

$$V''(\theta) = 2c_1(\sin^2\theta - \cos^2\theta + \cos\theta)$$

$$V'''(\theta) = 2c_1(2\sin 2\theta - \sin\theta), \qquad V'''(0) = 0$$

$$V^{iv}(\theta) = 2c_1(4\cos 2\theta - \cos\theta), \qquad V^{iv}(0) = 6c_1 > 0$$

$\Rightarrow$ $\theta = 0$ is a stable position.

(b), (c) $\Rightarrow$ if equilibrium position θ_3 does not exist, then $\theta = 0$ is a stable position.

(2) $\theta = \pi$: $V''(\pi) = -[c_2 + 2c_1(1 + \cos\theta_0)] < 0 \Rightarrow$ position $\theta = \pi$ is unstable.

(3) $\theta = \theta_3$: Using eqs (3), (4): $V''(\theta_3) = 2c_1 \sin^2\theta_3 > 0 \Rightarrow$ position $\theta = \theta_3$ is stable.

Example 5.14: *Stable Symmetric Point of Bifurcation.* A light rigid bar is hinged at one end and restrained by a torsional spring (Fig. E5.14a). Find its equilibrium positions θ under a vertical dead load P and discuss their stability, if the spring is undeformed for $\theta = \varepsilon$. Angles ε and θ may not be small.

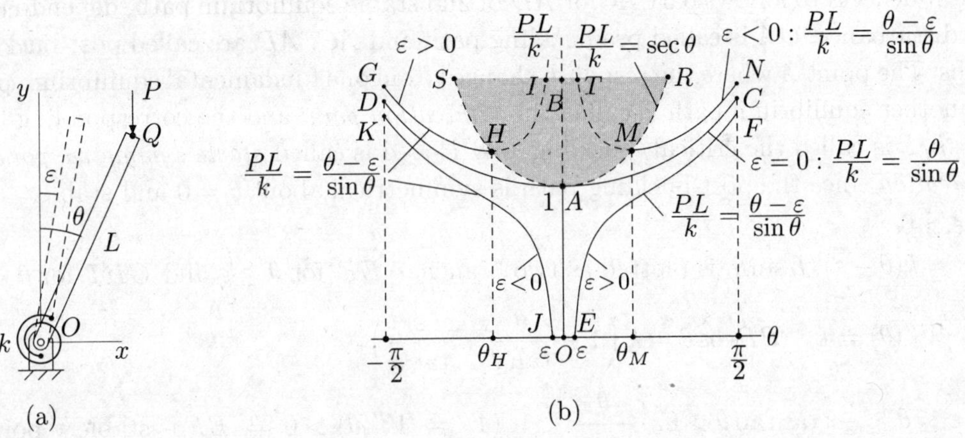

Fig. E5.14

Solution: $V(\theta) = \frac{1}{2}k(\theta - \varepsilon)^2 - (-P)y_Q = \frac{1}{2}k(\theta - \varepsilon)^2 + PL\cos\theta \quad \Rightarrow$

$$V'(\theta) = k(\theta - \varepsilon) - PL\sin\theta, \qquad V''(\theta) = k - PL\cos\theta$$

$$V'''(\theta) = PL\sin\theta, \qquad V^{iv}(\theta) = PL\cos\theta \tag{1}$$

Equilibrium eq. $V'(\theta) = 0 \Rightarrow$ $\quad k(\theta - \varepsilon) - PL\sin\theta = 0 \quad (2)$

$$P = k(\theta - \varepsilon)/L\sin\theta \tag{3}$$

The equilibrium position is stable if $V''(\theta) = k - PL\cos\theta > 0$, i.e., if $PL/k < \sec\theta$, and an unstable one if $V''(\theta) < 0$, i.e., if $PL/k > \sec\theta$. For the case $PL/k = \sec\theta$:

for $\theta \neq 0$: $\quad V'' = 0, \quad V''' \neq 0 \quad \Rightarrow$ unstable equilibrium;

for $\theta = 0,\ \varepsilon = 0$: $\quad V''(0) = 0,\ V'''(0) = 0,\ V^{iv}(0) = PL > 0 \Rightarrow$ stable equilibrium.

Hence, the unstable region is $PL/k \geq \sec\theta$ except for point $\theta = 0, PL/k = 1$, which is shown as shaded area in Fig. E5.14b. We consider 3 cases: $\varepsilon = 0,\ \varepsilon > 0,\ \varepsilon < 0$.

1. $\varepsilon = 0$:

$$\text{eq. (2)} \Rightarrow \text{equilibrium positions at:} \quad \theta = 0\ \forall\ P, \quad \text{or} \quad P = \frac{k\theta}{L\sin\theta} > \frac{k}{L} \quad (\theta \neq 0)$$

(a) As shown above, *fundamental path* $\theta = 0$ is stable for part OA ($P \leq k/L$) only.

(b) The second equilibrium path $P = k\theta/L\sin\theta$ is plotted as a curve DAC (excluding point A), which is symmetrical about $\theta = 0$. These positions are stable because

$$V'' = k - PL\cos\theta = k\left(1 - \frac{\theta}{\tan\theta}\right) > 0 \quad \text{since} \quad \frac{\tan\theta}{\theta} > 1 \quad \text{for} \quad 0 < |\theta| < \tfrac{1}{2}\pi$$

As P is increased from zero, the rod remains undeflected for the stable part OA and then it deflects to follows part AC or AD of 2nd stable equilibrium path, depending on the disturbance. OA is called pre-buckling path and AC, AD are called post-buckling paths. The point A where the response changes from one fundamental equilibrium path to another equilibrium path is called the *bifurcation point* and the corresponding load $P = k/L$ is called the critical/buckling load P_{cr}. A is called *stable symmetric point of bifurcation* since the post-buckling path is symmetrical about $\theta = 0$ and stable.

2. $\varepsilon > 0$:

$P = k(\theta - \varepsilon)/L\sin\theta$, is plotted as two branches EF for $\theta \geq \varepsilon$ and GHI for $\theta < 0$.

$$V''(\theta) = k - PL\cos\theta = k\left(1 - \frac{\theta}{\tan\theta} + \frac{\varepsilon}{\tan\theta}\right) \tag{4}$$

For $\varepsilon \leq \theta < \dfrac{\pi}{2}$: as $\tan\theta > 0$, $\dfrac{\theta}{\tan\theta} < 1$, (4) $\Rightarrow V''(\theta) > 0 \Rightarrow EF$ is stable $\forall$ points.

For $-\frac{1}{2}\pi < \theta < 0$, intersection H of curves GI and RAS: $P = k\sec\theta/L$, is given by

$$P = \frac{k(\theta - \varepsilon)}{L\sin\theta} = \frac{k\sec\theta}{L}, \quad \text{i.e.,} \quad \tan\theta - \theta + \varepsilon = 0 \quad \text{for } \theta_H \tag{5}$$

$$\text{At } P_{\min} \text{ for curve } GI: \quad \frac{dP}{d\theta} = \frac{k[\sin\theta - (\theta - \varepsilon)\cos\theta]}{L\sin^2\theta} = 0, \quad \text{i.e., at } \theta_H.$$

$$(4,)\,(5) \Rightarrow V''(\theta_H) = 0, \quad V''\left(-\frac{\pi}{2}\right) = k > 0 \quad \Rightarrow \quad V''(\theta) > 0 \quad \text{for } -\frac{\pi}{2} < \theta < \theta_H$$

i.e., the system is stable for this range of θ.

The configurations of GH and EF are stable (shown as solid lines) and those of HI are unstable. EF is a *natural equilibrium path* and GHI is a *complementary equilibrium path.* Only path EF is followed in a natural loading sequence from $P = 0$.

3. $\varepsilon < 0$:

Similar to case 2, the configurations of JK and MN are stable and those of TM are unstable. JK and TMN are the natural and the complementary equilibrium paths.

For an imperfect system ($\varepsilon \neq 0$): The natural equilibrium path is stable, runs close to that for the perfect system ($\varepsilon = 0$) and the rotation of the bar grows more rapidly as P_{cr} is approached. The effect of the imperfection is not significant.

Stable symmetric bifurcation occurs in columns and radially loaded circular plates.

Example 5.15: *Unstable symmetric point of bifurcation.* A rigid light rod is hinged at one end and restrained by a spring which always remains horizontal (Fig. E5.15a). Find the equilibrium configurations θ for a constant vertical load P and discuss their stability. The spring is undeformed for $\theta = \alpha$.

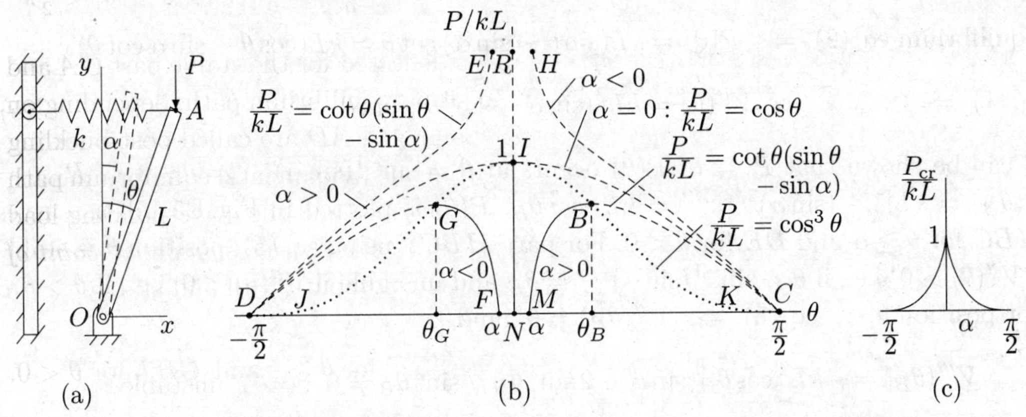

Fig. E5.15

Solution: $V(\theta) = \frac{1}{2}k(L\sin\theta - L\sin\alpha)^2 - (-P)L\cos\theta$

$$V'(\theta) = kL^2(\sin\theta - \sin\alpha)\cos\theta - PL\sin\theta$$

$$V''(\theta) = kL^2[\cos^2\theta - (\sin\theta - \sin\alpha)\sin\theta] - PL\cos\theta \tag{1}$$

Equilibrium eq. : $\quad V'(\theta) = kL^2(\sin\theta - \sin\alpha)\cos\theta - PL\sin\theta = 0 \quad (2)$

We consider 3 cases: $\alpha = 0,\ \alpha > 0,\ \alpha < 0$.

1. $\alpha = 0$:

eq. (2) $\Rightarrow$ $(kL^2\cos\theta - PL)\sin\theta = 0$ $\Rightarrow$ Two possible equilibrium configurations:

$$\theta = 0 \ \forall\, P, \qquad \text{and} \qquad P = kL\cos\theta \tag{3}$$

$$\text{eq. (1)} \Rightarrow \quad V''(\theta) = kL^2(\cos^2\theta - \sin^2\theta) - PL\cos\theta \tag{4}$$

$V''(0) = L(kL - P) > 0$ for $P < kL \Rightarrow$ The position $\theta = 0$ is stable if $P < kL$ and is unstable if $P > kL$. For $P = kL$, $V''(0) = 0$ and using eq. (4),

$$V'''(\theta) = -2kL^2\sin 2\theta + kL^2\sin\theta \quad \Rightarrow \quad V'''(0) = 0$$

$$V^{iv}(\theta) = -4kL^2\cos 2\theta + kL^2\cos\theta \quad \Rightarrow \quad V^{iv}(0) = -3kL^2 < 0 \Rightarrow \quad \theta = 0 \text{ unstable.}$$

For natural equilibrium path NIR in Fig. E5.17b, only part NI, excluding I, is stable.

For the 2nd set of equilibrium positions, $P = kL\cos\theta$ (curve DIC in Fig. E5.15b):

$$V''(\theta) \overset{(4)}{=} kL^2(\cos^2\theta - \sin^2\theta) - (kL\cos\theta)L\cos\theta = -kL^2\sin^2\theta < 0 \quad \Rightarrow \quad \text{unstable}$$

NIR and DIC intersect at I, which is an unstable symmetric point of bifurcation.

2. $\alpha > 0$:

Equilibrium eq. (2) $\Rightarrow$ $P(\theta) = kL(\sin\theta - \sin\alpha)\cot\theta = kL(\cos\theta - \sin\alpha\cot\theta)$,

$$\text{eq. (1)} \Rightarrow \qquad V''(\theta) = kL^2(\sin\alpha - \sin^3\theta)/\sin\theta \tag{5}$$

It can be shown that $P_{\max}$ of $P(\theta)$ occurs at $\theta = \sin^{-1}(\sin\alpha)^{1/3} = \theta_B$ with $P_{\max} = P(\theta_B) = kL[1 - (\sin\alpha)^{2/3}]^{3/2} = kL\cos^3\theta_B$. $P(\theta)$ is plotted in Fig. E5.15b as curves MBC for $\theta \geq \alpha$ and DE for $\theta < 0$. For part MBC: using eq. (5), position θ is stable if $V''(\theta) > 0$, i.e., if $\theta < \sin^{-1}(\sin\alpha)^{1/3} = \theta_B$ and unstable if $V''(\theta) < 0$, i.e., if $\theta > \theta_B$. For position θ_B: eq. (5) $\Rightarrow$ $V''(\theta_B) = 0$ and

$$V'''(\theta_B) = -kL^2\cos\theta_B(\sin\alpha + 2\sin^3\theta_B)/\sin^2\theta_B \neq 0 \quad \Rightarrow \quad \text{unstable}$$

Thus, $\theta = \theta_B$ is a limit point with $P_{cr} = P_{\max}$, since the positions $\theta \geq \theta_B$ are unstable.

The equilibrium positions of part MB, excluding B, are stable and those of BC are unstable. For positions of part DE: eq. (5) $\Rightarrow V''(\theta) < 0$ and hence these are unstable.

3. $\alpha < 0$:

The analysis is similar to case 2. The equilibrium path FG, excluding G, is stable and paths GD and CH are unstable. $\theta_G = -\sin^{-1}(\sin|\alpha|)^{1/3}$ and $P_{\max} = kL[1 - |\sin\alpha|^{2/3}]^{3/2} = kL\cos^3\theta_G$.

For cases 2 and 3, $P_{\max} = kL\cos^3\theta^*$, where θ^* is the equilibrium position for $P_{\max}$. The curve JIK in Fig. E5.15b is a plot of $P_{\max}$ versus θ^*. The critical load P_{cr} for perfect structure ($\alpha = 0$) is $P_{cr} = kL$, and for an imperfect one ($\alpha \neq 0$) it is

$$P_{cr} = P_{\max} = kL[1 - |\sin\alpha|^{2/3}]^{3/2} \tag{6}$$

$$\text{for } \alpha > 0: \quad \frac{dP_{cr}}{d\alpha} = -kL[1-(\sin\alpha)^{2/3}]^{1/2}(\sin\alpha)^{-1/3}\cos\alpha \;\to -\infty \quad \text{as } \alpha \to 0$$

The curve of $P_{cr}(\alpha)$, shown in Fig. E5.17c, has an infinite slope at $\alpha = 0$. Thus, the structure is extremely *sensitive to small imperfections* since these considerably reduce the maximum load which it can sustain.

Such behaviour occurs in deep arches and cylindrical panels. The exact solution for such problems is rare. Approximate analysis, based on a power series expansion of V in terms of θ and α, usually yields important information (see next exercise).

§ **Exercise 5.1:** For *approximate analysis* of Ex.5.15, with terms up to θ^4 in V, i.e., prove: $V(\theta) \simeq kL^2[\frac{1}{2}\alpha^2 - \alpha\theta + \frac{1}{2}\theta^2 - \frac{1}{6}(\alpha^4 - \alpha^3\theta - \alpha\theta^3 + \theta^4)] + PL(1 - \frac{1}{2}\theta^2 + \frac{1}{24}\theta^4)$ for $\alpha \ll 1$, where the first non-zero term of each type is included. Prove that $V' = 0 \Rightarrow$

$$P(\theta) = kL\Big[1 - \frac{\alpha - \alpha^3/6}{\theta} + \tfrac{1}{2}\alpha\theta - \tfrac{2}{3}\theta^2\Big]/(1 - \tfrac{1}{6}\theta^2) \simeq kL\Big(1 - \frac{\alpha - \alpha^3/6}{\theta} + \tfrac{1}{3}\alpha\theta - \tfrac{1}{2}\theta^2\Big]$$

and for this path $P(\theta)$: $\quad V''(\theta) \simeq kL^2[(\alpha - \theta^3) + \frac{1}{6}\alpha(\theta^2 - \alpha^2)]/\theta$

Show that this path is stable for $0 < \theta < \alpha^{1/3}$ with $P_{cr} \simeq P(\alpha^{1/3}) \simeq kL(1 - 1.5\alpha^{2/3})$ which are the two leading terms in the expansion of the exact expression [eq. (6)]. The approximate solution yields correct P_{cr} , correct leading terms of the post-buckling path for the perfect system, and correct leading terms of P_{cr} for the imperfect system⇒ *often critical load and imperfection sensitivity can be found by an approximate analysis.*

Example 5.16: *Asymmetric point of bifurcation.* The spring is undeformed for $\theta = 0$ (Fig. E5.16a). Find the equilibrium position θ for constant load P and discuss its stability.

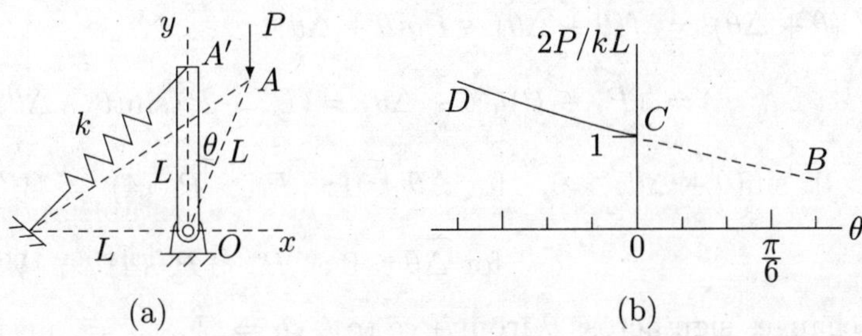

Fig. E5.16

Solution: $V(\theta) = \frac{1}{2}k[\{L^2(1+\sin\theta)^2 + L^2\cos^2\theta\}^{1/2} - \sqrt{2}L]^2 + Py_A$

$$\Rightarrow \quad V(\theta) = PL\cos\theta + kL^2[2 + \sin\theta - 2(1+\sin\theta)^{1/2}]$$

$$V'(\theta) = -PL\sin\theta + kL^2\cos\theta[1 - (1+\sin\theta)^{-1/2}] \tag{1}$$

Equilibrium equation: $V'(\theta) = 0 \;\Rightarrow$ two possible equilibrium positions are:

$$\theta = 0 \quad \forall P \quad \text{and} \quad P(\theta) = kL\cos\theta[1-(1+\sin\theta)^{-1/2}]/\sin\theta \tag{2}$$

$\theta = 0$ is a fundamental equilibrium path and eq. (2) describes the post-buckling path BCD (Fig. E5.16b) which is asymmetric about $\theta = 0$ as $P(\theta) \neq P(-\theta)$.

$$V''(\theta) = -PL\cos\theta - kL^2\sin\theta[1-(1+\sin\theta)^{-1/2}] + \tfrac{1}{2}kL^2\cos^2\theta(1+\sin\theta)^{-3/2} \tag{3}$$

$V''(0) = L(\frac{1}{2}kL - P) > 0$ for $P < 0.5kL$ $\Rightarrow$ position $\theta = 0$ is stable if $P < 0.5kL$ and is unstable if $P > 0.5kL$. For $P = 0.5kL$, $V''(0) = 0$ and

$$\begin{aligned} V'''(\theta) &= PL\sin\theta - kL^2\cos\theta[1-(1+\sin\theta)^{-1/2}] \\ &\quad -\tfrac{3}{2}kL^2\sin\theta(1+\sin\theta)^{-3/2}\cos\theta - \tfrac{3}{4}kL^2\cos^3\theta(1+\sin\theta)^{-5/2} \\ V'''(0) &= -\tfrac{3}{4}kL^2 \neq 0 \quad \Rightarrow \quad \theta = 0 \text{ is unstable.} \end{aligned}$$

$P(\theta)$ given by eq. (2) yields the post-buckling path BCD (Fig. E5.16b). It intersects the fundamental equilibrium path $\theta = 0$ at the bifurcation point C where critical load $P_{cr} = \frac{1}{2}kL$. C is called the asymmetric point of bifurcation since the post-buckling path BCD is asymmetric about the fundamental equilibrium path $\theta = 0$.

We decide the stability of equilibrium position θ under load $P = P(\theta)$, by studying the variation of V' across θ. Let $g(\theta) = L\sin\theta$, $f(\theta) = kL^2\cos\theta[1-(1+\sin\theta)^{-1/2}]$.

$$\begin{aligned} V'(\theta) &= f(\theta) - Pg(\theta), \qquad P = P(\theta) = f(\theta)/g(\theta) \\ P^* &= P(\theta+\Delta\theta) = f(\theta+\Delta\theta)/g(\theta+\Delta\theta) \\ \Rightarrow \quad V'(\theta+\Delta\theta) &= f(\theta+\Delta\theta) - Pg(\theta+\Delta\theta) \\ &= (P^*-P)g(\theta+\Delta\theta) = (P^*-P)\sin(\theta+\Delta\theta) \end{aligned} \tag{4}$$

For CB: $\theta > 0$, $\sin(\theta+\Delta\theta) > 0$, for $\Delta\theta < 0$: $P^* > P$, (4) $\Rightarrow V'(\theta+\Delta\theta) > 0$,

for $\Delta\theta > 0$: $P^* < P$, (4) $\Rightarrow V'(\theta+\Delta\theta) < 0$.

$\Rightarrow V'(\theta)$ changes sign across θ from +ve to −ve $\Rightarrow V_{max} \Rightarrow$ unstable position θ.

For CD: $\theta < 0$, $\sin(\theta+\Delta\theta) < 0$ $\Rightarrow$ $V'(\theta)$ changes across θ from −ve to +ve

$\Rightarrow V_{min} \Rightarrow$ stable position θ.

Hence, part BC of the post-buckling path BCD is unstable and part CD is stable.

Example 5.17: *Limit point and snap-through.* In model of a shallow arch (Fig. E5.17a), the springs are undeformed for $\theta = \alpha$. The pin A is loaded by a constant force P. For symmetric deformations, find the load P for equilibrium at θ and discuss its stability.

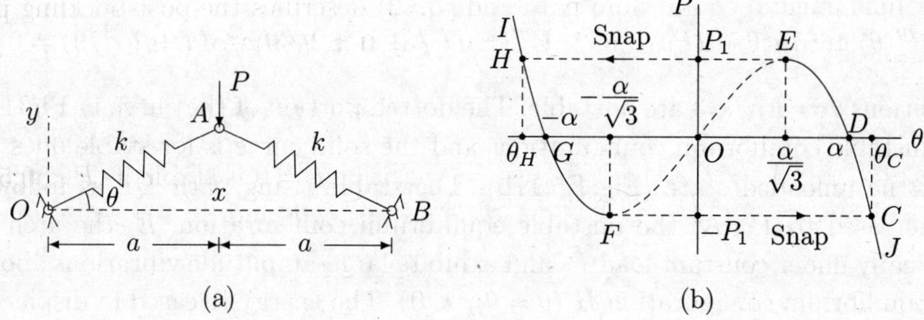

Fig. E5.17

Solution: The undeformed and deformed lengths (OA) of springs are: $a \sec\alpha$, $a\sec\theta$.

$$V(\theta) = 2[\tfrac{1}{2}k(a\sec\theta - a\sec\alpha)^2] - (-P)y_A = ka^2(\sec\theta - \sec\alpha)^2 + Pa\tan\theta$$

Since the arch is shallow, we use the approximation: $|\theta| \ll 1$, $|\alpha| \ll 1 \Rightarrow \sec\alpha \simeq 1 + \frac{1}{2}\alpha^2$, $\sec\theta \simeq 1 + \frac{1}{2}\theta^2$, $\tan\theta \simeq \theta + \frac{1}{3}\theta^3$, and retain terms up to θ^4 in V to get

$$V(\theta) \simeq \tfrac{1}{4}ka^2(\theta^4 - 2\theta^2\alpha^2 + \alpha^4) + Pa(\theta + \tfrac{1}{3}\theta^3)$$

$$\Rightarrow \quad V'(\theta) \simeq ka^2(\theta^3 - \theta\alpha^2) + Pa(1 + \theta^2)$$

$$V''(\theta) \simeq ka^2(3\theta^2 - \alpha^2) + 2Pa\theta \tag{1}$$

Equilibrium equation: $V'(\theta) = 0 \Rightarrow$

$$P = ka\theta(\alpha^2 - \theta^2)/(1 + \theta^2) \simeq ka\theta(\alpha^2 - \theta^2) \tag{2}$$

For $P = 0$, the 3 equilibrium configurations are: $\theta = 0, \alpha, -\alpha$. $P(\theta)$ is stationary for

$$\frac{dP}{d\theta} = ka(\alpha^2 - 3\theta^2) = 0, \qquad \text{for} \quad \theta = \pm\alpha/\sqrt{3}$$

$$P(\pm\alpha/\sqrt{3}) = \pm P_1 \qquad \text{with} \quad P_1 = 2ka\alpha^3/3\sqrt{3}$$

The P-θ curve is shown in Fig. E5.17b. For $-P_1 < P < P_1$, the cubic equation (2) has 3 real solutions and for $P < -P_1$ or $P > P_1$, eq. (2) has one real solution. Let us study the stability of the equilibrium configuration θ. Equations (1) and (2) $\Rightarrow$

$$V''(\theta) \simeq ka^2(3\theta^2 - \alpha^2) + 2Pa\theta = ka^2[3\theta^2 - \alpha^2 + 2\theta^2(\alpha^2 - \theta^2)]$$

$$\simeq ka^2(3\theta^2 - \alpha^2) \tag{3}$$

$\Rightarrow V''(\theta) > 0$ for $\theta^2 > \alpha^2/3 \Rightarrow$ equilibrium configuration θ is stable for $|\theta| > \alpha/\sqrt{3}$ and unstable for $-\alpha/\sqrt{3} < \theta < \alpha/\sqrt{3}$. Equation (1) $\Rightarrow$

$$V'''(\theta) \simeq 6ka^2\theta + 2Pa \quad \Rightarrow \quad V'''(\pm\alpha/\sqrt{3}) \simeq \pm 2\sqrt{3}ka^2\alpha(1 + 2\alpha^2/9) \neq 0$$

$\Rightarrow$ positions $\theta = \pm\alpha/\sqrt{3}$ are unstable. The dotted portion of the curve in Fig. E5.17b is for unstable equilibrium configurations and the solid curve is for stable ones.

D is the unloaded state (Fig. E5.17b). The stable rising path DE is followed as P is increased to P_1. At the unstable equilibrium configuration, E, the arch snaps dynamically under constant load P_1 and exhibits large-amplitude vibrations about the stable equilibrium configuration H ($\theta = \theta_H < 0$). The energy released in displacement corresponding to EH, gets dissipated in the presence of some positive damping and eventually, the arch takes the configuration $\theta = \theta_H$. The arch is now inverted and on further increase of load from P_1, it follows the stable rising path HI. On unloading, as P is decreased to $-P_1$, the arch remains inverted and follows the stable equilibrium configurations $IHGF$. At the unstable position of F, the arch snaps back dynamically from F to C. The arch follows the stable path CJ for further decrease of load. If the load is increased from J, then the arch follows the stable path $JCDE$. Thus, the system exhibits hysteresis. The points E and F, where snap-through into an inverted shape occurs, are called *limit points*. The corresponding load $P = P_1$ is the stationary value of $P(\theta)$ and is called the *limit load* or *critical load*.

Example 5.18: A rod is supported by two cables attached to a crane (Fig. E5.18a). It is slowly lifted up. Find the minimum value of length a of cables for stable equilibrium.

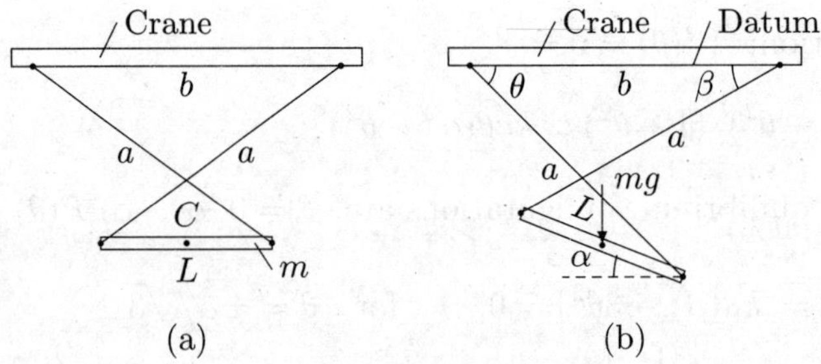

Fig. E5.18

Solution: For general position shown in Fig. E5.18b, α and β depend on θ by:

$$a \sin\theta - L \sin\alpha - a \sin\beta = 0 \tag{1}$$

$$a \cos\theta - L \cos\alpha + a \cos\beta = b \tag{2}$$

The potential energy V in the general position is:

$$V(\theta) = mg(-a\sin\theta + 0.5L\sin\alpha) \tag{3}$$

$$V'(\theta) = mg(-a\cos\theta + 0.5L\cos\alpha\,\alpha')$$

$$V''(\theta) = mg(a\sin\theta - 0.5L\sin\alpha\,{\alpha'}^2 + 0.5L\cos\alpha\,\alpha'') \tag{4}$$

Hence, we need to relate θ and α by eliminating β from eqs (1) and (2):

$$a^2 = (a\sin\beta)^2 + (a\cos\beta)^2 = (a\sin\theta - L\sin\alpha)^2 + (a\cos\theta - L\cos\alpha - b)^2 \tag{5}$$

Differentiating eq. (5) with respect to θ yields:

$$(a\sin\theta - L\sin\alpha)(a\cos\theta - L\cos\alpha\,\alpha') + (a\cos\theta - L\cos\alpha - b)(-a\sin\theta + L\sin\alpha\,\alpha') = 0$$

$$\Rightarrow \qquad a[b\sin\theta + L\sin(\theta-\alpha)] = L[b\sin\alpha + a\sin(\theta-\alpha)]\alpha' \tag{6}$$

Differentiating eq. (6) with respect to θ $\Rightarrow$

$$a[b\cos\theta + L\cos(\theta-\alpha)(1-\alpha')]$$
$$= L[b\cos\alpha\alpha' + a\cos(\theta-\alpha)(1-\alpha')]\alpha' + L[b\sin\alpha + a\sin(\theta-\alpha)]\alpha'' \tag{7}$$

For equilibrium position $\alpha = 0,\ \theta = \beta$: eqs (2), (6), (7) successively yield

$$2a\cos\theta = L+b, \qquad \alpha' = (b+L)/L, \qquad \alpha'' = -b(L+b)^2/2L^2 a\sin\theta \tag{8}$$

$$V''(\theta) \overset{(4)}{=} mg(a\sin\theta + 0.5L\alpha'') \overset{(8)}{=} [4La^2 - (L+b)^3]mg/4La\sin\theta$$

For stability: $\quad V''(\theta) > 0 \quad \Rightarrow \quad 4La^2 > (L+b)^3 \quad \Rightarrow \quad a_{\min} = [(L+b)^3/4L]^{1/2}$

5.5.3 Conditions of stability for 2-dof systems

Let q_1, q_2 be the generalised coordinates with $q_1 = q_2 = 0$ at an equilibrium position. Choose $(0,0)$ as datum for $V(q_1, q_2)$. This implies that

$$\left.\frac{\partial V}{\partial q_1}\right|_{(0,0)} = 0, \qquad \left.\frac{\partial V}{\partial q_2}\right|_{(0,0)} = 0, \qquad V(0,0) = 0$$

Let k_{ij}, elements of the matrix K, represent the second derivatives of V evaluated at the equilibrium configuration, $(0,0)$:

$$k_{11} = \left.\frac{\partial^2 V}{\partial q_1^2}\right|_{(0,0)}, \qquad k_{12} = k_{21} = \left.\frac{\partial^2 V}{\partial q_1 \partial q_2}\right|_{(0,0)}, \qquad k_{22} = \left.\frac{\partial^2 V}{\partial q_2^2}\right|_{(0,0)}$$

$$V(q_1, q_2) = V(0,0) + \left.\frac{\partial V}{\partial q_i}\right|_{(0,0)} q_i + \tfrac{1}{2}\left.\frac{\partial^2 V}{\partial q_i \partial q_j}\right|_{(0,0)} q_i q_j + \ldots \simeq \tfrac{1}{2} k_{ij} q_i q_j > 0$$

if the eigen-values λ_1, λ_2 of K are positive, since it ensures $V = \frac{1}{2}\lambda_1 q_1^{*2} + \lambda_2 q_2^{*2} > 0$.

The eigen-values λ of K satisfy $Kx = \lambda x \ \Rightarrow$

$$\begin{bmatrix} k_{11} & k_{12} \\ k_{21} & k_{22} \end{bmatrix} \begin{bmatrix} x_1 \\ x_2 \end{bmatrix} = \lambda \begin{bmatrix} x_1 \\ x_2 \end{bmatrix} \quad \Rightarrow \quad \begin{bmatrix} k_{11}-\lambda & k_{12} \\ k_{21} & k_{22}-\lambda \end{bmatrix} \begin{bmatrix} x_1 \\ x_2 \end{bmatrix} = \begin{bmatrix} 0 \\ 0 \end{bmatrix}$$

$$\Rightarrow \quad \begin{vmatrix} k_{11}-\lambda & k_{12} \\ k_{21} & k_{22}-\lambda \end{vmatrix} = \lambda^2 - (k_{11}+k_{22})\lambda + k_{11}k_{22} - k_{12}^2 = 0$$

$$\Rightarrow \quad \lambda_{1,2} = \tfrac{1}{2}(k_{11}+k_{22}) \pm [\tfrac{1}{4}(k_{11}-k_{22})^2 + k_{12}^2]^{1/2}$$

The two eigen-values λ_1 and λ_2 are positive if

$$\tfrac{1}{2}(k_{11}+k_{22}) > 0 \quad \text{and} \quad [\tfrac{1}{4}(k_{11}-k_{22})^2 + k_{12}^2]^{1/2} < \tfrac{1}{2}(k_{11}+k_{22})$$

$$\text{i.e.,} \quad k_{11}+k_{22} > 0 \quad \text{and} \quad k_{12}^2 - k_{11}k_{22} < 0$$

Note: (1) If $k_{12} = 0$, then the conditions reduce to $k_{11} > 0$, $k_{22} > 0$.

(2) If k_{11} or k_{22} is zero and $k_{12} \neq 0$, then the equilibrium position is unstable.

Hence, an isolated equilibrium configuration of a 2-dof conservative system in which not all second order derivatives of V at the equilibrium configuration are zero is stable

$$\text{iff} \quad \frac{\partial^2 V}{\partial q_1^2} + \frac{\partial^2 V}{\partial q_2^2} > 0 \quad \text{and} \quad \left(\frac{\partial^2 V}{\partial q_1 \partial q_2}\right)^2 - \frac{\partial^2 V}{\partial q_1^2}\frac{\partial^2 V}{\partial q_2^2} < 0 \tag{5.22}$$

If all the second order derivatives of V at the equilibrium configuration are zero, then it is necessary to examine the higher order derivatives to investigate stability.

It follows from the conditions of stability that if the potential energy of a 2-dof conservative linear system depends on a parameter P and if K is positive-definite for $P = 0$, then the critical value P_{cr} of P which makes the equilibrium configuration $q_1 = q_2 = 0$ unstable is the smallest value of P for which

$$k_{11}k_{22} - k_{12}^2 = 0 \tag{5.23}$$

i.e., for which one eigenvalue becomes zero. The displacement mode shape which the system follows on losing its stability is the solution of $Kq = 0$ for $P = P_{cr}$.

Example 5.19: Two uniform bars are hinged together and restrained by two torsion springs (Fig. E5.19a) of torsional stiffness k which are untwisted when the bars are vertical. Find the value of the load for which the vertical equilibrium position $\theta_1 = \theta_2 = 0$ becomes unstable for the two types of loading shown in Fig. E5.19b and find the buckling mode shapes. Neglect gravity.

Solution: (i) Let θ_1, θ_2 be generalised coordinates. F contributes $-(-F)y_B$ to V.

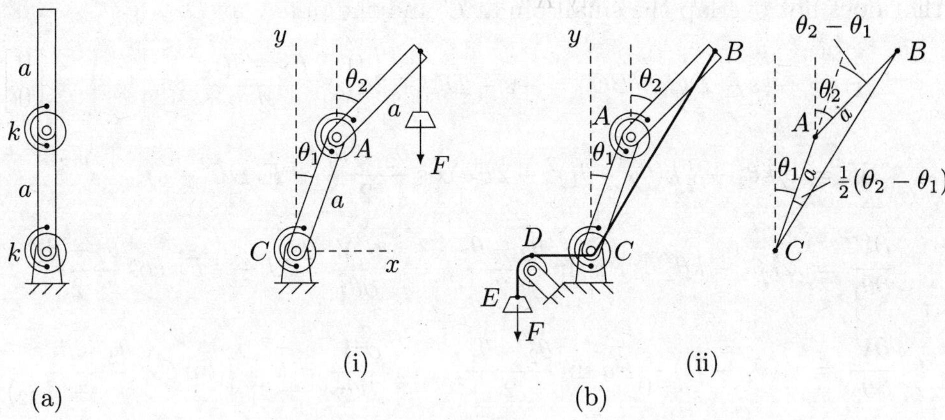

Fig. E5.19

$$V = \tfrac{1}{2}k\theta_1^2 + \tfrac{1}{2}k(\theta_2 - \theta_1)^2 - (-F)a(\cos\theta_1 + \cos\theta_2)$$

$$\Rightarrow \quad \frac{\partial V}{\partial \theta_1} = 2k\theta_1 - k\theta_2 - Fa\sin\theta_1, \qquad \frac{\partial V}{\partial \theta_2} = k(\theta_2 - \theta_1) - Fa\sin\theta_2$$

$$\frac{\partial^2 V}{\partial \theta_1^2} = 2k - Fa\cos\theta_1, \qquad \frac{\partial^2 V}{\partial \theta_2^2} = k - Fa\cos\theta_2, \qquad \frac{\partial^2 V}{\partial \theta_1 \partial \theta_2} = -k$$

The equilibrium equations $\partial V/\partial\theta_1 = 0$, $\partial V/\partial\theta_2 = 0$ are satisfied for the vertical position: $\theta_1 = \theta_2 = 0$. At this equilibrium position, second order derivatives of V are

$$\frac{\partial^2 V}{\partial \theta_1^2} = k_{11} = 2k - Fa = k(2 - c)$$

$$\frac{\partial^2 V}{\partial \theta_2^2} = k_{22} = k - Fa = k(1 - c), \qquad \frac{\partial^2 V}{\partial \theta_1 \partial \theta_2} = k_{12} = -k$$

where $Fa/k = c$. The critical values of c are obtained from eq. (5.23):

$$|K| = k_{11}k_{22} - k_{12}^2 = k^2(c^2 - 3c + 1) = 0$$

with roots $c_1 = 0.382$, $c_2 = 2.618$. The critical load F_{cr} corresponds to the smaller root c_1 with $F_{cr} = c_1 k/a = 0.382k/a$. The buckling mode shape is the solution of $Kx = 0$ with $k_{11} = k(2 - c_1) = 1.618k$, $k_{22} = k(1 - c_1) = 0.618k$, $k_{12} = -k$, i.e.,

$$\begin{bmatrix} 1.618k & -k \\ -k & 0.618k \end{bmatrix} \begin{bmatrix} \theta_1 \\ \theta_2 \end{bmatrix} = \begin{bmatrix} 0 \\ 0 \end{bmatrix} \quad \Rightarrow \quad 1.618k\theta_1 - k\theta_2 = 0$$

$\Rightarrow$ The buckling mode shape into which the system buckles at F_{cr} is $\theta_2 = 1.618\theta_1$.

(ii) The potential energy of F is $-(-F)y_E$. Let s be the length of the inextensible rope that does not overlap the small pin at C and the pulley at D.

$$y_E = -(s - DC - BC) = -s + DC + 2a\cos\frac{\theta_2 - \theta_1}{2}$$

$$V = \tfrac{1}{2}k\theta_1^2 + \tfrac{1}{2}k(\theta_2 - \theta_1)^2 + 2Fa\cos\frac{\theta_2 - \theta_1}{2} + F(DC - s)$$

$$\frac{\partial V}{\partial \theta_1} = 2k\theta_1 - k\theta_2 + Fa\sin\frac{\theta_2 - \theta_1}{2}, \qquad \frac{\partial^2 V}{\partial \theta_1^2} = 2k - \tfrac{1}{2}Fa\cos\frac{\theta_2 - \theta_1}{2}$$

$$\frac{\partial V}{\partial \theta_2} = k(\theta_2 - \theta_1) - Fa\sin\frac{\theta_2 - \theta_1}{2}, \qquad \frac{\partial^2 V}{\partial \theta_2^2} = k - \tfrac{1}{2}Fa\cos\frac{\theta_2 - \theta_1}{2}$$

$$\frac{\partial^2 V}{\partial \theta_1 \partial \theta_2} = -k + \tfrac{1}{2}Fa\cos\frac{\theta_2 - \theta_1}{2}$$

The equilibrium equations $\partial V/\partial\theta_1 = 0$, $\partial V/\partial\theta_2 = 0$ are satisfied for the vertical position: $\theta_1 = \theta_2 = 0$. At this equilibrium position, second order derivatives of V are

$$\frac{\partial^2 V}{\partial \theta_1^2} = k_{11} = 2k - \tfrac{1}{2}Fa = k\left(2 - \tfrac{1}{2}c\right)$$

$$\frac{\partial^2 V}{\partial \theta_2^2} = k_{22} = k - \tfrac{1}{2}Fa = k\left(1 - \tfrac{1}{2}c\right)$$

$$\frac{\partial^2 V}{\partial \theta_1 \partial \theta_2} = k_{12} = -k + \tfrac{1}{2}Fa = -k\left(1 - \tfrac{1}{2}c\right)$$

The vertical equilibrium position becomes unstable for

$$|K| = k_{11}k_{22} - k_{12}^2 = k^2(1 - \tfrac{1}{2}c) = 0, \qquad \text{i.e.,} \quad \text{for } c = 2$$

The critical load $F_{\text{cr}} = ck/a = 2k/a$. The buckling mode shape is the solution of $Kx = 0$ with k_{ij} evaluated for $c = 2$: $k_{11} = k$, $k_{22} = 0$, $k_{12} = 0$, i.e.,

$$\begin{bmatrix} k & 0 \\ 0 & 0 \end{bmatrix}\begin{bmatrix} \theta_1 \\ \theta_2 \end{bmatrix} = \begin{bmatrix} 0 \\ 0 \end{bmatrix} \quad \Rightarrow \quad k\theta_1 = 0$$

The system buckles at $F = F_{\text{cr}}$ with the buckling mode shape of $\theta_1 = 0$, $\theta_2 \neq 0$.

The drastic change in F_{cr} from $0.382k/a$ to $2k/a$ for two different ways of application of an end load to a system must be noted. It illuminates an *important aspect of stability studies that the precise definition of applied forces is very essential.*

We conclude that the investigation of stability should be performed with *extreme caution* taking into account the precise nature of the load, the effect of imperfections and mass distribution. Investigators should employ careful mathematics, sound intuition and common sense based on *experience in dealing with problems of stability.*

Review 5

1. A virtual displacement at time t is an infinitesimal displacement in zero time from a given admissible path C to a neighbouring admissible path C^*.
2. The virtual work, δW, is the work done by internal and external forces in a virtual displacement keeping the forces constant.
3. For a rigid body, subjected to a virtual displacement $\delta\vec{r}_A$ at point A and a virtual rotation $\overrightarrow{\delta\theta}$, we have, $\delta W = \vec{F}\cdot\delta\vec{r}_A + \vec{M}_A\cdot\overrightarrow{\delta\theta}$.
4. The work done by internal forces on a rigid body is zero.
5. The 'variation' δe of an entity 'e' for a given admissible path C is the difference in the value of e for the varied neighbouring admissible path C^* and for C.
6. The first variation $\delta^{(1)}e$ is the first order term in the variation δe.
7. $\frac{d}{dt}[\delta q(t)] = \delta\dot{q}(t)$, $\frac{d}{dt}[\delta f(q,\dot{q},t)] = \delta\dot{f}(q,\dot{q},t)$, $\int_{t_1}^{t_2}\delta h(q,\dot{q},t)\,dt = \delta\int_{t_1}^{t_2} h(q,\dot{q},t)\,dt$
8. Hamilton's principle: 'A geometrically admissible motion $q(t)$ of a system between prescribed $\hat{q}_1$ at t_1 and $\hat{q}_2$ at t_2 is the actual motion if $\int_{t_1}^{t_2}[\,\delta^{(1)}T+\delta W\,]dt = 0$ for any varied motion of the system $q^*(t) = q(t)+\delta q(t)$ with $\delta q(t_1) = \delta q(t_2) = 0$'.
9. Lagrange's equations for a system of rigid bodies with n independent generalised coordinates q_i, generalised forces Q_i, $\delta W = Q_i\delta q_i$, $L(\dot{q},q,t)\equiv T(\dot{q},q,t) - V(q)$:

$$\frac{d}{dt}\left(\frac{\partial T}{\partial\dot{q}_i}\right) - \frac{\partial T}{\partial q_i} + \frac{\partial V}{\partial q_i} = Q_i^{\rm nc}, \quad i = 1,\ldots,n \qquad \text{or} \qquad \frac{d}{dt}\left(\frac{\partial L}{\partial\dot{q}_i}\right) - \frac{\partial L}{\partial q_i} = Q_i^{\rm nc}$$

$$\text{For impulsive forces :}\quad \delta\hat{W} = \hat{Q}_i\delta q_i, \qquad \left.\frac{\partial T}{\partial\dot{q}_i}\right|_{t_1^+} - \left.\frac{\partial T}{\partial\dot{q}_i}\right|_{t_1^-} = \hat{Q}_i, \quad i = 1,\ldots,n$$

10. Principle of virtual work: If a system is in equilibrium, then the virtual work done by the internal and external forces in an arbitrary virtual displacement, δW, is zero. Conversely, a geometrically admissible configuration of a system is an equilibrium configuration if $\delta W = 0$ for an arbitrary virtual displacement, provided the system is initially in equilibrium.
11. Principle of potential energy: If a system, subjected only to conservative internal and external forces, is in equilibrium, then $\delta^{(1)}V = 0\ \forall\ \delta q$. Conversely, if $\delta^{(1)}V = 0\ \forall\ \delta q$ and if the system is initially in equilibrium, then it remains in equilibrium.
12. Principle of virtual rate of work: A system initially in equilibrium, will remain in equilibrium, if the virtual rate of work of all internal and external forces is zero for arbitrary virtual velocities. This is convenient for problems of mechanisms.

13. For discrete conservative system, the equivalent equilibrium equations are: (1) $\delta W = 0 \;\forall\, \delta q_i$, (2) $Q_i = 0 \;\forall\, i$, (3) $\partial V/\partial q_i = 0 \;\forall\, i$, (4) $\delta \dot{W} = Q_i \delta \dot{q}_i = 0 \;\forall\, \delta \dot{q}_i$, (5) $\vec{F} = \vec{0}$, $\vec{M}_A = \vec{0}$ for each rigid body.
14. For a discrete conservative system, if V has a local minimum at an equilibrium configuration, then the equilibrium configuration is stable.
15. For a one-dof conservative system, if the first non-zero derivative of V at an equilibrium position is of even order and positive, then this position is stable.
16. For a 2-dof conservative system, at an equilibrium position let $k_{ij} = \partial^2 V/\partial q_i \partial q_j$ with not all $k_{ij} = 0$. This position is stable iff $k_{11} + k_{22} > 0$, $k_{11}k_{22} - k_{12}^2 > 0$. If V depends on a parameter P and if K is positive-definite for $P = 0$, then P_{cr} is the smallest value of P for which $k_{11}k_{22} - k_{12}^2 = 0$ to initiate instability. The displacement mode shape post instability is the solution of $Kq = 0$ for $P = P_{\text{cr}}$.

5.6 Concept Review Questions 5

1. Find the virtual work done by force F acting at A at distance h (Fig. Q5.1) from the point of contact of a cylinder rolling without slip, in a virtual displacement for the variation $\delta\theta$ in the generalised coordinate θ. $[\delta\theta(R-r)hF/r]$

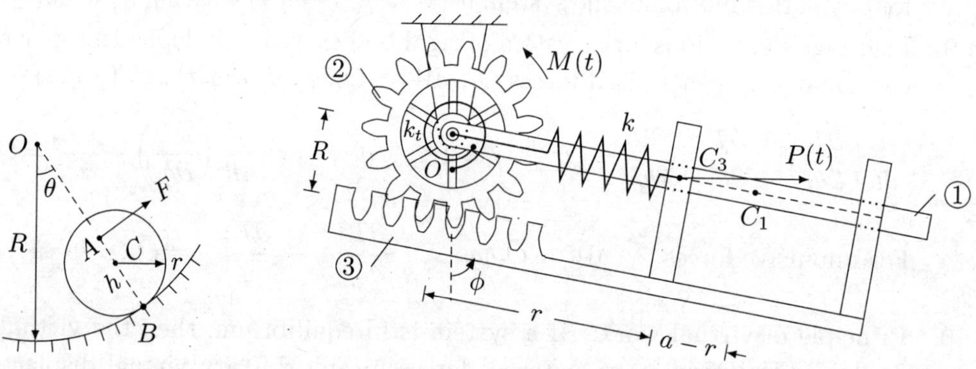

Fig. Q5.1 **Fig. Q5.2**

2*. A pendulum ① with mass m_1, centre of mass C_1 and $I_{zz}^O = I_1$ is suspended from a smooth hinge at O (Fig. Q5.2). A pinion ② having radius R and $I_{zz}^O = I_2$ is mounted on ① and is in mesh with a rack ③ which can slide on ①. The mass of the rack is m_3, with centre of mass at C_3 and $I_{zz}^{C_3} = I_3$. A torsional spring of stiffness k_t is connected between ① and ②. The springs are undeformed for $r = r_0$. The moment $M(t)$ and the force $P(t)$ are applied to the pinion and the rack. The frictional torque on the axle of the pinion is M_f. The generalised coordinates are r, ϕ. For $\dot{r} > 0$, prove that

$$\begin{aligned} T &= \tfrac{1}{2}(I_1 + I_3 + m_3 r^2)\dot{\phi}^2 + \tfrac{1}{2}I_2(\dot{\phi} + \dot{r}/R)^2 + \tfrac{1}{2}m_3\dot{r}^2, \\ V &= \tfrac{1}{2}k(r - r_0)^2 + \tfrac{1}{2}k_t(r - r_0)^2/R^2 - g\cos\phi(m_1 a + m_3 r), \\ \delta W^{\mathrm{nc}} &= (M + Pr\cos\phi)\delta\phi + (M/R - M_f/R + P\sin\phi)\delta r \end{aligned}$$

and set up the Lagrange's equations of motion.

$$[(m_3 + \frac{I_2}{R^2}\ddot{r} + \frac{I_2}{R}\ddot{\phi} - m_3 r\dot{\phi}^2 + k(r - r_0) + \frac{k_t(r - r_0)}{R^2} - m_3 g\cos\phi = P\sin\phi + \frac{M - M_f}{R}$$

$$(I_1 + I_2 + I_3 + m_3 r^2)\ddot{\phi} + 2m_3 r\dot{r}\dot{\phi} + \frac{I_2}{R}\ddot{r} + (m_1 a + m_3 r)g\sin\phi = M + Pr\cos\phi]$$

3. A bead of mass m moves under gravity on a smooth cycloidal guide $s = 4a\sin\phi$ in a vertical plane. The arc-length from the lowest point is s and ϕ is the angle between the tangent and the horizontal line. Derive the equation of motion with s as the generalised coordinate. Show that the exact solution is simple harmonic. $[\ddot{s} + (g/4a)s = 0]$

4*. Four rods, each of length $\sqrt{2}b$, form a rigid square frame of mass M (Fig. Q5.4). Four additional rods, each of length $2b$ and mass m, are hinged to this frame. Initially, the system is at rest on a smooth horizontal plane. A horizontal impulsive force $\hat{P}$ is applied. Find the angular velocity of the square frame just after the impulse has been applied. $[-1.5\hat{P}/b(2M + 3m)]$

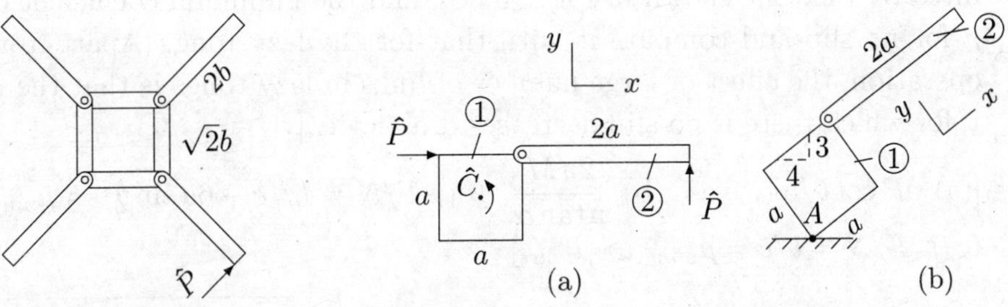

Fig. Q5.4 **Fig. Q5.5**

5. (a) A square plate ① of side a and mass m is hinged to a thin rod ② of length $2a$ and mass m (Fig. Q5.5a). The system is translating at $-v\vec{\jmath}$ when it is subjected to impulsive forces $\hat{P}\vec{\imath}$, $\hat{P}\vec{\jmath}$ and impulsive moment $\hat{C} = \hat{P}a$. Find the angular velocities just after the impact. (b) The system translating at $-v\vec{\jmath}$ makes impact at A (Fig. Q5.5b). The coefficient of restitution is e and the impact is smooth. Find the angular velocities just after the impact. (c) Rework part (b) for rough impact such that $\vec{v}'_A = \vec{0}$.

$[(a)\ (66,\ 54)\hat{P}/41ma\ \ (b)\ -(0.1011, 0.5225)(1 + e)\frac{v}{a}\ \ (c)\ -(0.3913, 0.4565)\frac{v}{a}]$

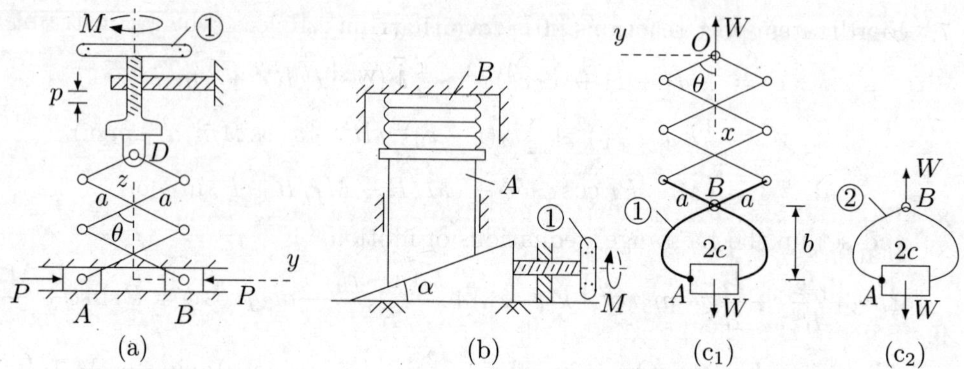

Fig. Q5.6

6. (a) Find the couple M applied to wheel ① to maintain equilibrium of sliders A and B against forces P each (Fig. Q5.6a). The pitch of the screw is p. Neglect friction. If the applied forces on the sliders are unequal, can equilibrium be maintained by a suitable value of M? (b) Find the force F with which block A presses on the bale B when a couple M is applied to the handle ① of a wedge press (Fig. Q5.6b). Neglect friction. The pitch of the screw is p. (c) A friction grip, in the form of lazy tongs, is used for lifting objects from a quenching tank (Fig. Q5.6c$_1$). Neglect weight of the tong. Find the relation between the normal force N and the frictional force F applied to the object at A. If the same block is lifted by the tong shown in Fig. Q5.6c$_2$, find the minimum coefficient of friction μ for no slip and compare it with that for the lazy tong. [Apart from ease of operation, the effect of large number of links in lazy tongs is that the minimum μ for which there is no slip decreases drastically.]

[(a) $pP\cot\theta/5\pi$, yes (b) $\dfrac{2\pi M}{p\tan\alpha}$ (c$_1$) $F/N = b/(c+6a\sin\theta) = \mu_{\min_1}$
(c$_2$) $F/N = b/c = \mu_{\min_2} \gg \mu_{\min_1}$]

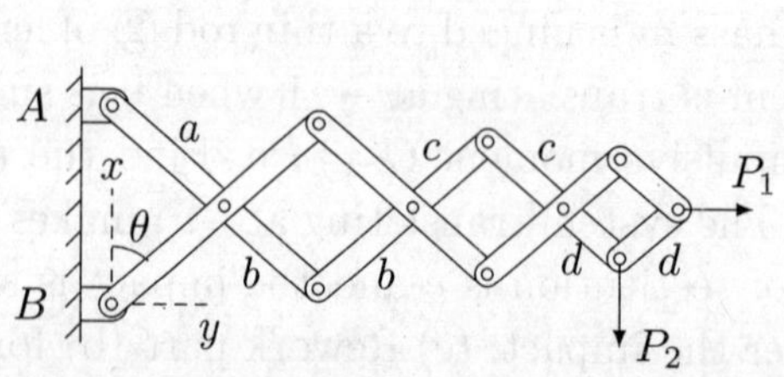

Fig. Q5.7

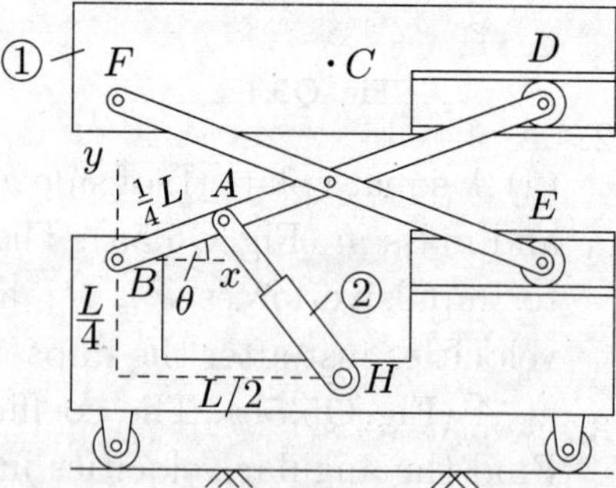

Fig. Q5.8

7. Find the support reactions in Fig. Q5.7. (Hint: Release the x-constraint at A.)

$$[\vec{R}_A = R_1\,\vec{\text{i}} + R_2\,\vec{\text{j}},\ \vec{R}_B = R_3\,\vec{\text{i}} + R_4\,\vec{\text{j}}\ \text{ with } R_3 = P_2 - R_1,\ R_4 = -P_1 - R_2,$$
$$R_1 = [(a+2b+2c+2d)P_1 \cot\theta + (a-d)P_2]/2a,$$
$$R_2 = -\{(a+2b+2c+d)P_2 \tan\theta + P_1 a\}/2a]$$

8. Find the force exerted by the hydraulic cylinder at A to hold in equilibrium the platform ①. The platform has weight W with centre of mass at C (Fig. Q5.8). The links are light. $\tan\theta = 0.75,\ BD = EF = L,\ BA = \frac{1}{4}L.$ [3.2W]

9. Force P is applied to the handle ③ of a hand shearing machine (Fig. Q5.9). Find the normal force applied on bar ① by member ② to shear it. The handle ③ is welded to the pinion. $[P(\cos\theta + a/R_2)(R_1 + R_2)/b]$

10. Find the relation between the loads on the mechanism (Fig. Q5.10) in equilibrium in the given position.
$[Pb\sin\theta + M_1 + aQ(.8\sin\theta + .6\cos\theta) + [R(d+e)(0.8\cos\theta - 0.6\sin\theta) + M_3]a/d = 0]$

11[d]. Find the conditions on the dimensions of the balance (Fig. Q5.11) so that the reading is independent of x. $[gc = bf]$

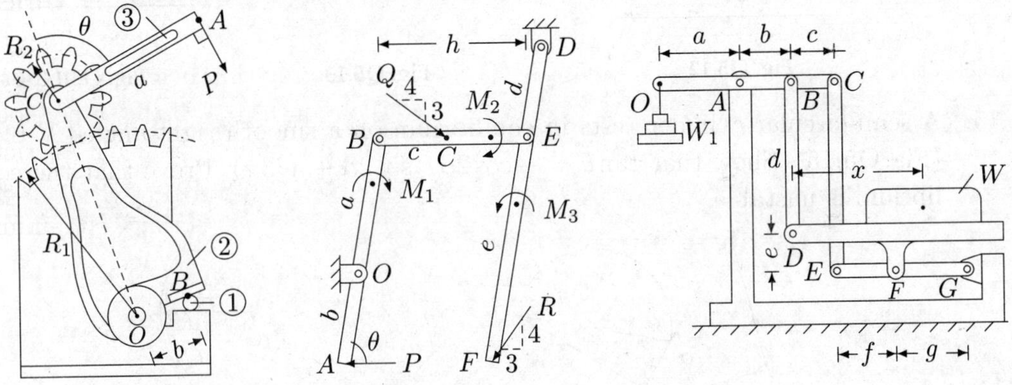

Fig. Q5.9 Fig. Q5.10 Fig. Q5.11

12. A plate of weight W is supported by three light rods (Fig. Q5.12). Find the value of θ $(0 < \theta < \pi/2)$ for the equilibrium configuration of the system under force $P > 0$ and couples $C > 0,\ M > 0$. What restrictions, if any, must the values of P and C satisfy for the equilibrium position to exist? Is this equilibrium position stable? $[\sin^{-1}[C/L(P^2 + W^2)^{1/2}] + \tan^{-1}(P/W),\ C^2 - L^2P^2 < L^2W^2$, stable]

13. A heavy bar AB is supported horizontally by two light vertical links. The bar is free to move in the vertical plane. Discuss the stability of the the equilibrium configuration in Fig. Q5.13 for (i) $b = 1.1a$, (ii) $b = 0.9a$ and (iii) $b = a$.

[(i) Stable (ii) Unstable (iii) Stable]

14. For the frictionless system shown in Fig. Q5.14, the free length of the spring is $2a$ and the weight of each roller and each rod is $W/2$. (a) Prove that $\theta = 0$ and $\theta = \pi$ are equilibrium positions for all values of W and $\theta = \cos^{-1}(2 - W/ka)$ is another equilibrium position if $3ka > W > ka$. (b) Determine the stability of these equilibrium positions. (c) Plot the W versus θ curve for equilibrium in the range $0 \leq \theta \leq \pi$ and mark stable portions of equilibrium on it.

[(b) $\theta = 0$ is stable for $W \leq ka$, else unstable;
$\theta = \pi$ is stable for $W \geq 3ka$, else unstable;
3rd position is stable.]

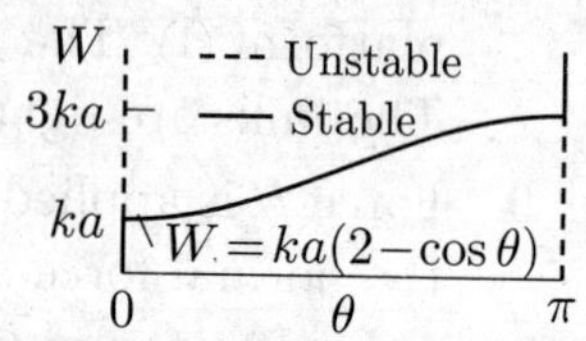

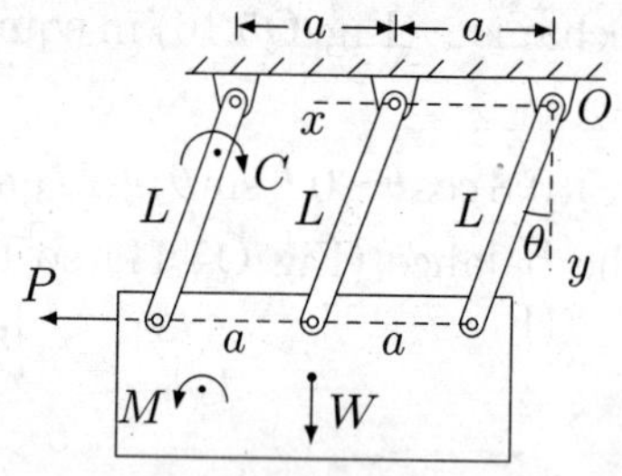

Fig. Q5.12

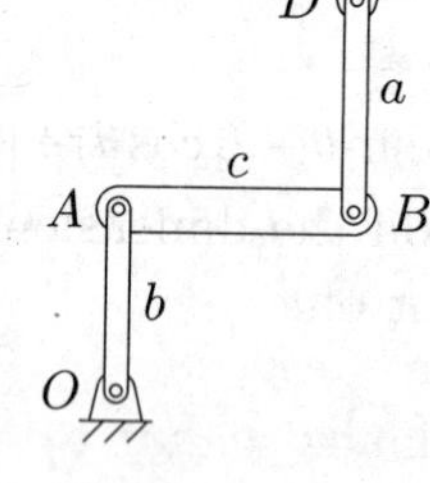

Fig. Q5.13

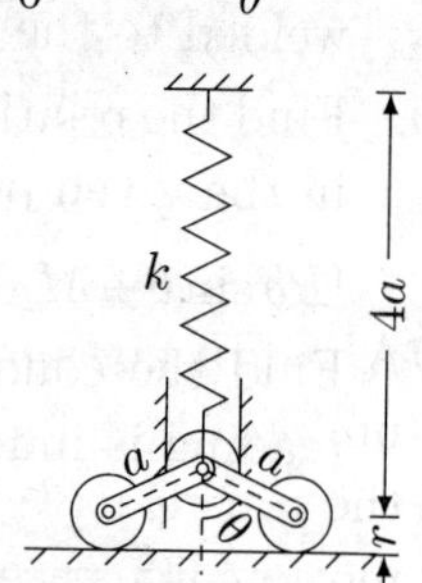

Fig. Q5.14

15. A semi-circular cylinder rests in equilibrium on a smooth right-angled V-block (Fig. Q5.15). Show that $\tan\theta = -\cos 2\alpha/(\sin 2\alpha + 4/3\pi)$. Prove that the equilibrium is unstable.

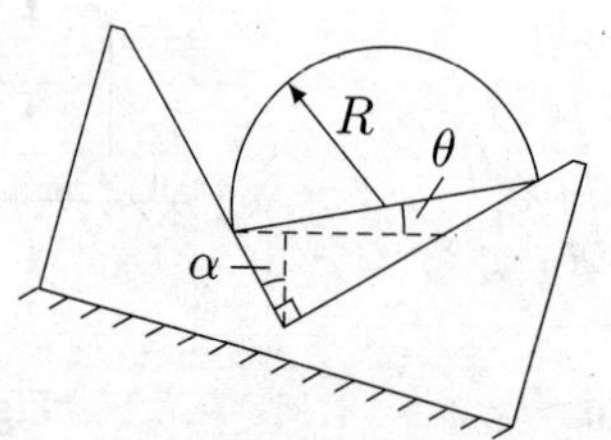

Fig. Q5.15

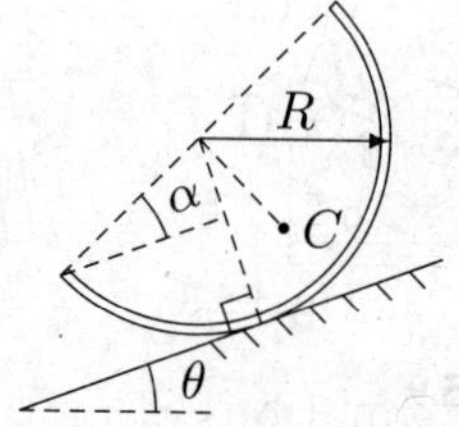

Fig. Q5.16

16. A thin semi-circular cylindrical shell of radius R rests on a rough inclined plane at angle θ to the horizontal (Fig. Q5.16). Prove that equilibrium is stable for $\theta < \sin^{-1}(2/\pi)$.

5.7 Practice Problems 5

1. Two sliders are pinned at A and C to a uniform rod AB of length $2L$ and mass m (Fig. P5.1). The resisting force on slider ① is cv_C and on slider ② is F. The

spring is unstretched for $\theta = 0$. Using θ as the generalised coordinate, set up the Lagrange's equation of motion.

$$[\{(m+m_1)\sin^2\theta + m_2\cos^2\theta + \tfrac{1}{3}m\}L^2\ddot{\theta} + (m+m_1-m_2)L^2\sin\theta\cos\theta\dot{\theta}^2$$
$$-(m+m_1)gL\sin\theta + kL^2\sin\theta\cos\theta = -M - cL^2\sin^2\theta\dot{\theta} - \left[P + F\frac{\cos\theta\dot{\theta}}{|\cos\theta\dot{\theta}|}\right]L\cos\theta]$$

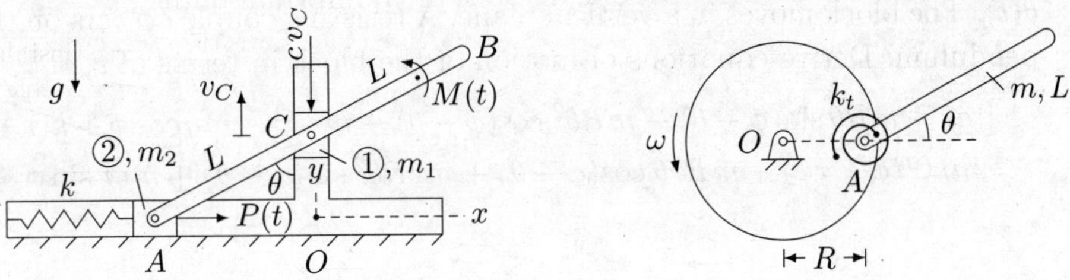

Fig. P5.1 Fig. P5.2

2. A rod of mass m and length L is pinned to a horizontal platform rotating with uniform angular velocity ω about a vertical axis (Fig. P5.2). The orientation of the rod relative to the rotating line OA is θ. Set up the Lagrange's equation of motion, linearise it for small θ and find the natural frequency of small oscillations. $[\frac{1}{3}mL^2\ddot{\theta} + k_t\theta + \frac{1}{2}m\omega^2 RL\sin\theta = 0, \;\; \sin\theta \simeq \theta, \;\; [3(k_t + \frac{1}{2}m\omega^2 RL)/mL^2]^{1/2}/2\pi]$

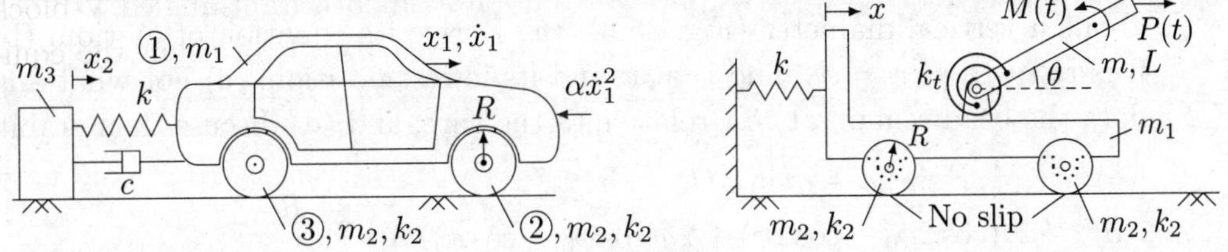

Fig. P5.3 Fig. P5.4

3. Derive the equations of motion of the system shown in Fig. P5.3 using x_1, x_2 as generalised coordinates. The driving torque on the rear wheels of the vehicle is M. Assume no slip of the wheels. The spring is undeformed for $x_1 = x_2 = 0$. The static and kinetic coefficients of friction for the ground are μ_s, μ_k. The axial radius of gyration of the wheels is k_2. The resistance offered by the damper is c times the relative velocity between the piston and cylinder.

$$[\{m_1 + 2m_2(1 + k_2^2/R^2)\}\ddot{x}_1 + k(x_1 - x_2) = -\alpha\dot{x}_1^2 - c(\dot{x}_1 - \dot{x}_2) + M/R,$$
$$m_3\ddot{x}_2 - k(x_1 - x_2) = c(\dot{x}_1 - \dot{x}_2) - \mu_k m_3 g]$$

4. Using the generalised coordinates x and θ (Fig. P5.4), set up the Lagrange's equations. The springs are unstretched for $\theta = 0, x = 0$. The axial radius of

gyration of the front and rear wheel assemblies is k_2.

$$[\{m_1 + 2m_2(1 + k_2^2/R^2) + m\}\ddot{x} - \tfrac{1}{2}mL(\ddot{\theta}\sin\theta + \dot{\theta}^2\cos\theta) + kx = P,$$
$$\tfrac{1}{3}mL^2\ddot{\theta} - \tfrac{1}{2}mL\ddot{x}\sin\theta + k_t\theta + \tfrac{1}{2}mgL\cos\theta = M - PL\sin\theta]$$

5. A small block of mass m is suspended from a spring of stiffness k and free length L (Fig. P5.5). The other end of the spring moves in a vertical circle at prescribed $\theta(t)$. The block moves in a vertical plane. A resisting couple $c\dot{\phi}$ acts on the spring pendulum. Derive equations of motion of the block in terms of r, ϕ.

$$[m\ddot{r} + mR\ddot{\theta}\sin(\phi - \theta) - mR\dot{\theta}^2\cos(\phi - \theta) - mr\dot{\phi}^2 - mg\cos\phi + k(r - L) = 0,$$
$$mr(2\dot{r}\dot{\phi} + r\ddot{\phi}) + mRr\ddot{\theta}\cos(\phi - \theta) + mrR\dot{\theta}^2\sin(\phi - \theta) + mgr\sin\phi + c\,\dot{\phi} = 0]$$

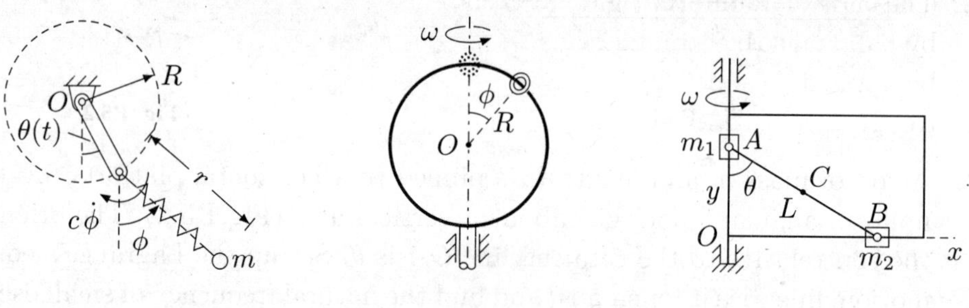

Fig. P5.5 Fig. P5.6 Fig. P5.7

6. A bead slides on a smooth wire which is rotated at constant angular velocity about a vertical diameter (Fig. P5.6). (a) Derive its equation of motion. (b) If it is released at $\phi = \phi_0$, find $\dot{\phi}$ at ϕ and its lowest position. (c) For what angles does the bead remain at rest relative to the wire; and (d) discuss their stability.

[(a) $\ddot{\phi} = \frac{1}{2}\omega^2\sin 2\phi + g\sin\phi/R$

(b) $[\frac{1}{2}\omega^2(\cos 2\phi_0 - \cos 2\phi) + 2g(\cos\phi_0 - \cos\phi)/R]^{1/2}$

$\phi_{\max} = \cos^{-1}[-(\cos\phi_0 + 2g/\omega^2 R)]$ if $\omega^2 > 2g/R(1 - \cos\phi_0)$, else $\phi_{\max} = \pi$

(c) $\phi_0 = 0, \pi$ and $\phi_0 = \cos^{-1}(-g/\omega^2 R)$ if $\omega^2 > g/R$

(d) $\phi_0 = 0$ is unstable, if $\phi_0 = \cos^{-1}(-g/\omega^2 R)$ exists then it is stable, $\phi_0 = \pi$ is stable if $\omega^2 \le g/R$ and unstable if $\omega^2 > g/R$]

7*. (a) The ends of a rod of mass m are connected to sliders which move on smooth arms of a rectangular frame (Fig. P5.7). The frame is rotated at constant angular velocity ω about the vertical axis. Derive the equation of motion of the rod. Find its rest position $\theta = \theta_0$ with respect to the rectangular frame and discuss its stability. (b) A rod AB of mass m and length L slides in a room such that A moves along a smooth straight vertical corner and B moves on the smooth floor. Derive equations of motion in terms of polar coordinate ϕ of B and inclination θ of AB to vertical. Obtain their first integrals.

[(a) $(m_1\sin^2\theta+m_2\cos^2\theta+\frac{1}{3}m)L^2\ddot{\theta}+\frac{1}{2}[(m_1-m_2)L^2\dot{\theta}^2-(m_2+\frac{1}{3}m)\omega^2L^2]\sin 2\theta$

$-(m_1+\frac{1}{2}m)gL\sin\theta=0$. If B moves on lower arm, then $\theta_0=0$ is unstable. If B moves on upper arm: (i) and if $\cos\theta_2=-(m_1+\frac{1}{2}m_2)g/\omega^2L(m_2+\frac{1}{3}m)>-1$, then $\theta_0=\theta_2$ exists and is stable; (ii) $\theta_0=\pi$ is unstable if θ_2 exists, else stable.

(b) $\ddot{\theta}-\dot{\phi}^2\sin\theta\cos\theta-3g\sin\theta/2L=0,\quad \frac{d}{dt}(\dot{\phi}\sin^2\theta)=0;$

$\dot{\theta}^2+\dot{\phi}^2\sin^2\theta+3g\cos\theta/L=c_2,\quad \dot{\phi}\sin^2\theta=c_1$]

8. Two hinged rods are initially at rest on a smooth horizontal table (Fig. P5.8). Find their angular velocities just after the application of an impulsive force $\hat{P}$ and an impulsive couple $\hat{P}L$, as shown. [$-10\hat{P}/3mL,\ \ 3\hat{P}/4mL$]

9. The ends of a uniform rigid bar ① of mass m_1 are connected to sliders ② and ③ by ball and socket joints (Fig. P5.9). OB is vertical. Find the force P that should be applied to slider ②, to maintain equilibrium in the given configuration. The free length of the spring is $\frac{2}{3}d$. [$\{(\frac{1}{2}m_1+m_3)g+\frac{1}{3}kd\}(L^2-a^2-d^2)^{1/2}/d$]

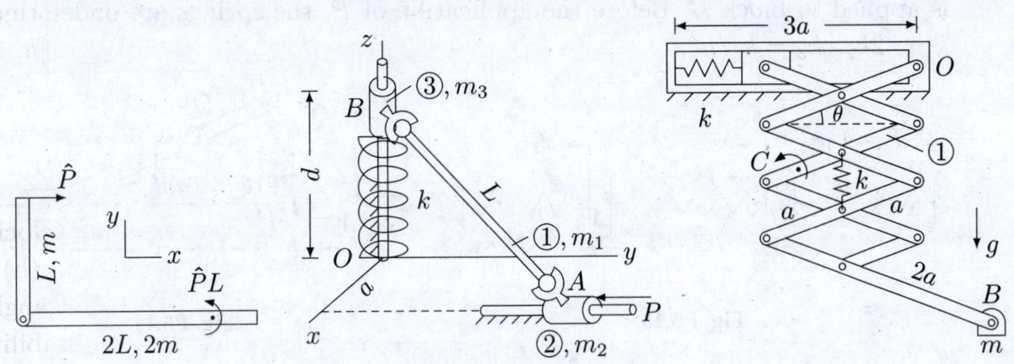

Fig. P5.8 **Fig. P5.9** **Fig. P5.10**

10. In the remote-control device shown in Fig. P5.10, the sensing device at B has mass m. The mass density of the links is λ kg/m. The free length of the springs is a. Find the couple C that should be applied to arm ① to maintain equilibrium in the given configuration. [$-(65\lambda a^2+9ma)g\cos\theta+ka^2(4\sin\theta-2\cos\theta)$]

11^d. The cross-section of a ventilating door in a roof is shown in Fig. P5.11. Its mass is m and the centre of mass is at C. One end of the door is pinned to a light rod which passes through a swivel block A and is restrained by a spring of stiffness k which is undeformed for $\theta=0$. Find the value of k so that the door is in equilibrium for any angle θ. Discuss the stability of the equilibrium configurations. [mgb/a^2, unstable]

12. In the planetary gear system (Fig. P5.12), a couple M_1 is applied to the sun gear ①. Find the couples M_2 and M_4 which must be applied to the spider ② and the

annular gear ④ for equilibrium. $[-2M_1(r_1+r_2)/r_1,\ \ M_1(r_1+2r_2)/r_1]$

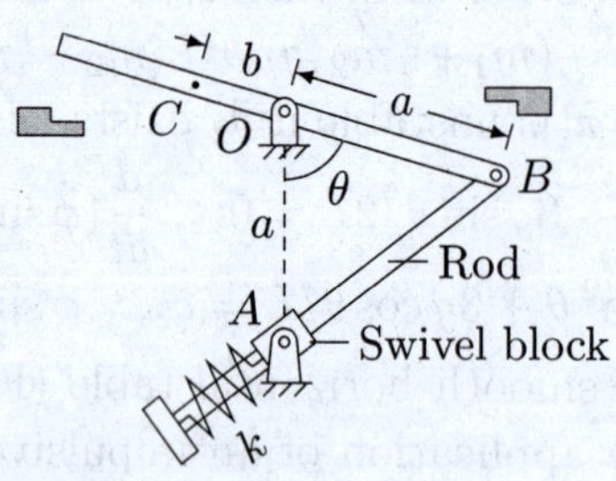

Fig. P5.11

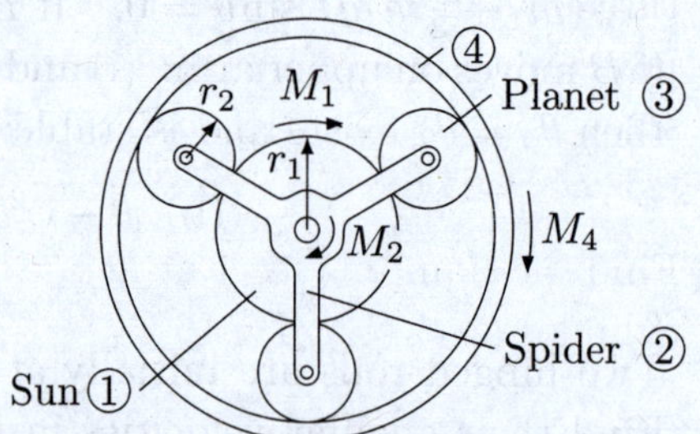

Fig. P5.12

13. Springs ① and ② (Fig. P5.13) have stiffness k_1 and k_2 respectively. Spring ③ is non-linear with force $F = k_3x_2^3$, where x_2 is its extension. Use the principle of stationary potential energy to find displacements x_1, x_2 when a force $P = 29k_2h$ is applied to block B. Before the application of P, the springs are undeformed. $k_1 = 2k_2,\ k_3 = k_2/h^2$. $[h,\ 3h]$

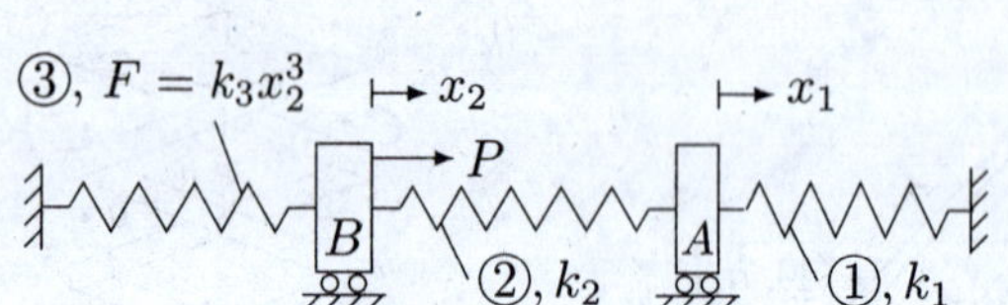

Fig. P5.13

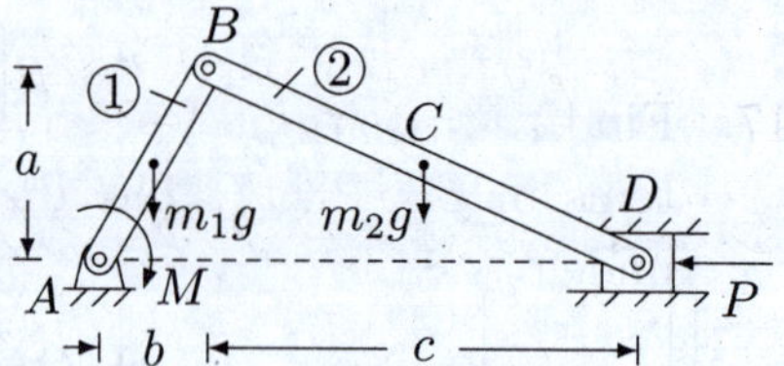

Fig. P5.14

14. The slider crank mechanism shown (Fig. P5.14) is in equilibrium. Neglect friction at A and D. The links have mass m_1 and m_2. The limiting frictional couple at hinge B is M_B. Find the force P for equilibrium.

$[P$ lies between $c[M + \frac{1}{2}(m_1+m_2)gb \mp M_B(1+b/c)]/a(b+c)]$

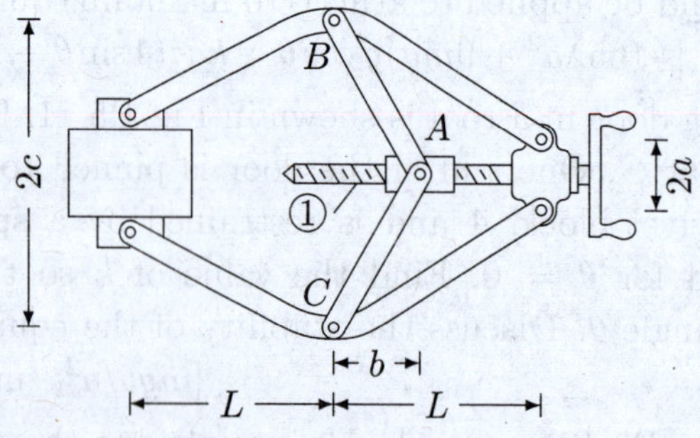

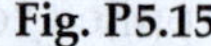

Fig. P5.15

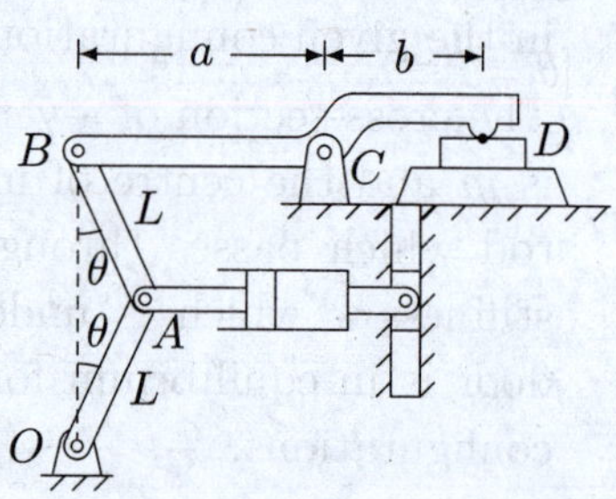

Fig. P5.16

15*. A clamp (Fig. P5.15) is operated by applying moment M to the handle of a threaded shaft ① which forces the threaded nut A to the right. Thereby, the jaws are tightened by the action of the toggle links AB and AC. The lead of the screw is l. Neglect friction. Find a relation between the clamping force P and M.

$[P = \pi M[cL - (c-a)b]/2lbL]$

16. Find the clamping force at D due to force P applied by the piston of the hydraulic cylinder at A (Fig. P5.16). $[aP\cot\theta/2b]$

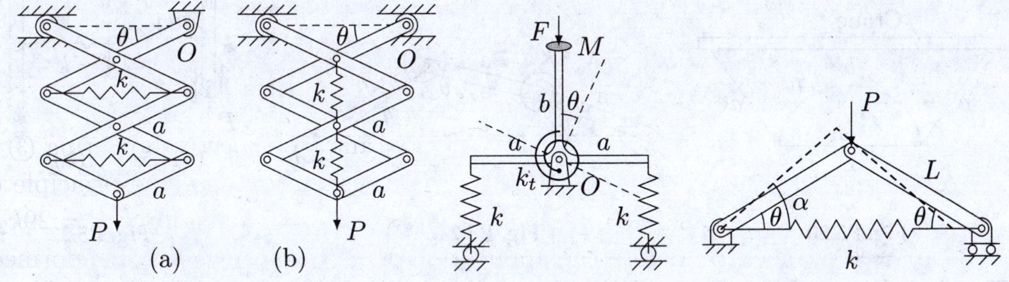

Fig. P5.17 Fig. P5.18 Fig. P5.19

17. Find the position(s) of equilibrium and discuss their stability for the two systems in Fig. P5.17. The linear density of the links is λ kg/m. The springs are undeformed for $\theta = \theta_0$. P acts vertically downwards and $0 \le \theta \le \pi/2$.

[(a) $\cos\theta(25a^2\lambda g + 5aP + 8ka^2\sin\theta) = 8ka^2\cos\theta_0\sin\theta$, stable

(b) $\pi/2$, $\theta = \theta_1 = \sin^{-1}\alpha$ if $\alpha = \sin\theta_0 + (25a\lambda g + 5P)/8ka < 1$;

θ_1 stable, $\pi/2$ stable if θ_1 does not exist, else unstable]

18. A uniform 'Tee'-shaped rod has a linear density of λ kg/m (Fig. P5.18). It is pivoted at O and carries a discrete mass M at the top. For $\theta = 0$, the initial compressive force in each spring is P and the torsional spring is undeformed. Find the equilibrium position(s) θ of the rod under a vertically downward load F and discuss their stability. Plot F versus θ curve(s) for small values of θ and mark the stable portions. Find the critical load. Discuss the effect of P.

[$\theta = 0\,\forall F$, stable for $F < F_{cr} = (k_t + 2ka^2)/b - (\frac{1}{2}\lambda b + M)g$, and for $F = F_{cr}$ if $k_t \ge 6ka^2$, else unstable. $F(\theta) = (k_t\theta + ka^2\sin 2\theta)/b\sin\theta - (M + \frac{1}{2}\lambda b)g$,

this θ is unstable for $F < F_{cr}$

and for $F = F_{cr}$ if $k_t \ge 6ka^2$, else stable.

P has no effect. $F(\theta) \simeq F_{cr} + (k_t - 6ka^2)\theta^2/6b$]

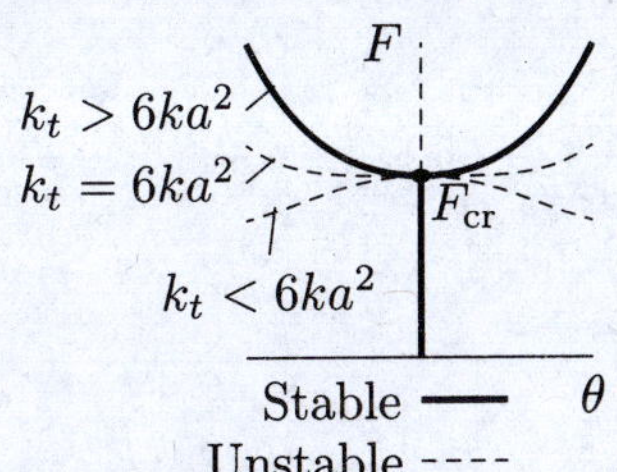

19*. Find the limit point and snap-through limit load for the shallow arch (Fig. P5.19). Also find the P-θ relation and discuss the stability of these equilibrium configurations. The spring is unstretched at $\theta = \alpha$. $\alpha \ll 1$.

$[P \simeq 2kL\theta(\alpha^2 - \theta^2)$ stable if $|\theta| > \dfrac{\alpha}{\sqrt{3}}$, limit point $\theta = \pm\dfrac{\alpha}{\sqrt{3}}$ with $P = \pm\dfrac{4kL\alpha^3}{3\sqrt{3}}]$

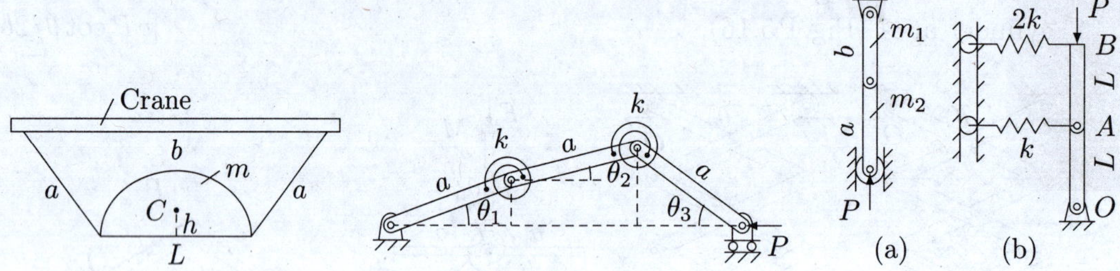

Fig. P5.20 Fig. P5.21 Fig. P5.22

20. An object of mass m, with centre of mass C, is supported by two cables of length a each, attached to a crane (Fig. P5.20). It is to be slowly lifted by the crane. Prove that the equilibrium position is stable if $h < L[4La^2 \sin^2\theta + b(b - L)^2]/4a(b-L)^2 \sin\theta$.

21. Three pinned light rods (Fig. P5.21) are subjected to a constant load P. The springs are undeformed when the bars are collinear. Discuss the stability of the equilibrium configuration $\theta_1 = \theta_2 = 0$. Find the buckling load and the buckling mode shape. [(a) Stable for $P < k/a$, $P_{\text{cr}} = k/a$, $\theta_2 = 0$, $\theta_3 = \theta_1$]

22. Two bars are hinged together and subjected to load P (Fig. P5.22). The springs are undeformed when the bars are vertical. (i) Find the critical value of the constant load P in case (a). (ii) Find the critical value of stiffness k in case (b) for which the vertical equilibrium configuration for light bars becomes unstable. Also find the buckling mode shape, θ_{AB}/θ_{OA}, where θ is the angle made with the vertical.

[(i) $g[\frac{1}{2}m_1 + m_2(1 + b/2a)]/(1 + b/a)$ (ii) $P/0.4384L$, $\theta_{AB}/\theta_{OA} = -1.281$]

CHAPTER

SPECIAL TOPICS

In this chapter central force motion, effect of earth's rotation, charged particle motion, systems with mass flow, vibration of 1 dof system and hydrostatics are covered.

A. CENTRAL FORCE MOTION

6.1 Motion Under a Central Force

Consider a body (Fig. 6.1a) of mass m moving under external force $\vec{F}$ which is always directed along a line from a fixed point O in inertial frame I to the centre of mass C of the body. Point O is called the *centre of force*, $\vec{F}$ is called a *central force* and the motion of C under $\vec{F}$ is called *central force motion*. The study of such systems has led to the explanation of planetary motion and helped physicists understand Rutherford's studies on the scattering of alpha particles.

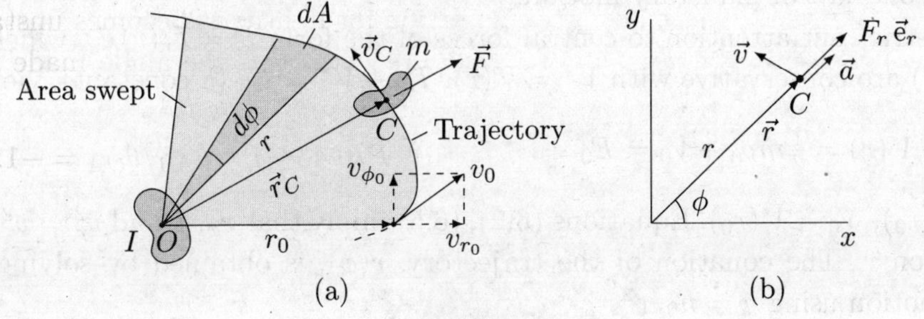

Fig. 6.1

We first consider general results which hold for all central forces. In Sections 6.1 and 6.2 all quantities are defined with respect to an inertial reference frame I and hence, for brevity, the subscript I is omitted. Thus, $\dot{(\ )} \equiv \dot{(\ )}_{|I}$ and $\vec{v} \equiv \vec{v}_{|I}$, $\vec{a} \equiv \vec{a}_{|I}$, etc.

By definition of central force motion,

$$\vec{F} \parallel \vec{r}_C \quad \Rightarrow \quad \vec{r}_C \times \vec{F} = \vec{0}$$

$$\Rightarrow \quad \vec{r}_C \times m\vec{a}_C = \vec{r}_C \times m\dot{\vec{v}}_C = \overline{(\vec{r}_C \times m\vec{v}_C)}^{\,\cdot} = \vec{0}$$

$$\Rightarrow \quad \vec{r}_C \times \vec{v}_C = \vec{H}^*_O/m = \vec{h}_0 = \text{constant} \tag{6.1}$$

where $\vec{H}^*_O$ would be moment of momentum, of the moving body, about O if all the mass were concentrated at C and $\vec{h}_0$ is its value per unit mass. For $\vec{h}_0 = \vec{0}$, eq. (6.1) $\Rightarrow$ $\vec{r}_C \parallel \vec{v}_C$ $\Rightarrow$ the path of C is a straight line through O. For $\vec{h}_0 \neq \vec{0}$, eq. (6.1) $\Rightarrow$ $\vec{r}_C \perp \vec{h}_0$, i.e., $\vec{r}_C$ is in a plane through O which is normal to $\vec{h}_0$. Hence, *the trajectory of C lies in a plane passing through the centre of force O*. Let us choose x and y axes in this plane with the origin at O (Fig. 6.1b). From equation (6.1) we have:

$$\vec{r}_C \times \vec{v}_C = r\,\vec{e}_r \times (\dot{r}\,\vec{e}_r + r\dot{\phi}\,\vec{e}_\phi) = r^2\dot{\phi}\,\vec{e}_z = \vec{h}_0 = h_0\,\vec{e}_z$$

$$\Rightarrow \quad r^2\dot{\phi} = rv_\phi = r_0 v_{\phi 0} = h_0 = \text{constant} \tag{6.2}$$

$$\Rightarrow \quad \dot{\phi} = h_0/r^2, \qquad v_\phi = h_0/r \tag{6.3}$$

Hence, $\dot{\phi}$ retains the same sign throughout the motion. The rate at which the position vector of C from O, sweeps area A (Fig. 6.1a), is given by

$$dA = \tfrac{1}{2}r^2 d\phi \quad \Rightarrow \quad \dot{A} = \tfrac{1}{2}r^2\dot{\phi} = \tfrac{1}{2}h_0 \tag{6.4}$$

hence, the rate at which the area is swept, i.e., *the areal velocity $\dot{A}$ is constant.* This is Kepler's second law of planetary motion.

We now restrict our attention to central forces of the form $\vec{F} = F_r(r)\,\vec{e}_r$, which (as if acting at C) are conservative with $V^* = V(r)$. $T_C + V^* = E_0$ (a constant), yields

$$\tfrac{1}{2}mv^2 + V(r) = \tfrac{1}{2}mv_0^2 + V_0 = E_0 \tag{6.5}$$

where $v_0 = v(r_0)$, $V_0 = V(r_0)$. Equations (6.2), (6.5) imply that v_ϕ, v and $v_r^2 = v^2 - v_\phi^2$ depend only on r. The equation of the trajectory, $r(\phi)$, is obtained by solving the equation of motion using $\dot{\phi} = h_0/r^2$:

$$ma_r = m(\ddot{r} - r\dot{\phi}^2) = m(\ddot{r} - h_0^2/r^3) = F_r$$

This non-linear equation is simplified by using $u = 1/r$: $\Rightarrow \quad \dot{\phi} = h_0/r^2 = h_0 u^2$,

$$\dot{r} = \frac{d}{dt}\left(\frac{1}{u}\right) = -\frac{\dot{u}}{u^2} = -r^2\,\frac{du}{d\phi}\,\frac{d\phi}{dt} = -h_0\,\frac{du}{d\phi}$$

$$\ddot{r} = -h_0 \frac{d^2u}{d\phi^2}\frac{d\phi}{dt} = -h_0^2 u^2 \frac{d^2u}{d\phi^2}$$

$$\Rightarrow \qquad \frac{d^2u}{d\phi^2} + u = -\frac{F_r(u)}{m h_0^2 u^2} \tag{6.6}$$

Example 6.1: The trajectory of a mass-point under central force is $r = r_0 \sec(n\phi)$, where $n \geq 0$. It is a circle $r = r_0$ for $n = 0$ and a straight line $x = r_0$ for $n = 1$. Find the nature of the central force and v_r, v_ϕ at any ϕ if the velocity is v_0 at $\phi = 0$.

Solution: $r^2\dot{\phi} = h_0 \quad \Rightarrow \quad \dot{\phi} = h_0 \cos^2(n\phi)/r_0^2, \qquad r = r_0 \sec(n\phi)$,

$$\Rightarrow \quad v_r \;=\; \dot{r} = r_0 \sec(n\phi)\tan(n\phi)\, n\dot{\phi} = n h_0 \sin(n\phi)/r_0 \tag{a}$$

$$\ddot{r} \;=\; n h_0 \cos(n\phi)\, n\dot{\phi}/r_0 = n^2 h_0^2 \cos^3(n\phi)/r_0^3 = n^2 h_0^2/r^3$$

$$F_r \;=\; m(\ddot{r} - r\dot{\phi}^2) = m\left(\frac{n^2h_0^2}{r^3} - \frac{rh_0^2}{r^4}\right) = m\left(n^2 - 1\right)\frac{h_0^2}{r^3} \quad \Rightarrow \quad F_r \propto \frac{1}{r^3}$$

Equation (a) $\Rightarrow$ at $\phi = 0$: $v_{r0} = 0 \Rightarrow v_{\phi 0} = v_0 \Rightarrow h_0 = r_0 v_{\phi 0} = r_0 v_0 \Rightarrow v_\phi = h_0/r = v_0 \cos n\phi \Rightarrow v_r = n v_0 \sin n\phi$.

§ **Exercise 6.1:** Consider the stability of a circular orbit $r = r_0$ with $r^2\dot{\phi} = h_0$ under central force motion. Let $r = r_0 + \xi$ be the perturbed motion with $r^2\dot{\phi} = h$. Show that $m(\ddot{r} - r\dot{\phi}^2) = F_r(r) \Rightarrow$

$$\ddot{\xi} = f(r) = F_r(r)/m + h^2/r^3 \simeq f(r_0) + f'(r_0)\xi \tag{1}$$

For the unperturbed motion: $\xi \equiv 0 \Rightarrow f(r_0) = 0$ and eq. (1) reduces to $\ddot{\xi} - f'(r_0)\xi = 0$. Its solution is bounded and the orbit is stable if

$$f'(r_0) \simeq F_r'(r_0)/m - 3h^2/r_0^4 \simeq [3F_r(r_0)/r_0 + F_r'(r_0)]/m < 0 \tag{6.7}$$

where the approximation $h^2 \simeq h_0^2$ is used, i.e., $f(r_0) \simeq F_r(r_0)/m + h_0^2/r_0^3 = 0$. In particular, for $F_r = -kr^n$ with $k > 0$, the circular orbit is stable if

$$3F_r(r_0)/r_0 + F_r'(r_0) = -(n+3)kr_0^{n-1} < 0 \qquad \text{i.e., if} \quad n > -3 \quad \Rightarrow$$

Stable circular orbits for linear ($n = 1$) and inverse square ($n = -2$) attractive forces.

6.1.1 Attractive linear central force

Consider the motion of a body of mass m under the central force $\vec{F} = -k\vec{r}$ (Fig. 6.2a), where k is a constant, $\omega^2 = k/m$ and the initial conditions are: $\vec{r}(0$ and $\vec{v}(0)$.

$$m\vec{a}_C = \vec{F} \quad \Rightarrow \quad m\ddot{\vec{r}} + k\vec{r} = \vec{0} \quad \Rightarrow \quad \vec{r}(t) = \vec{A}\cos\omega t + \vec{B}\sin\omega t$$

$$\vec{r}(0) = \vec{A}, \qquad \vec{v}(0) = \dot{\vec{r}}(0) = \omega\vec{B} \quad \Rightarrow$$

$$\vec{r}(t) = \vec{r}(0)\cos\omega t + \frac{\vec{v}(0)}{\omega}\sin\omega t \tag{6.8}$$

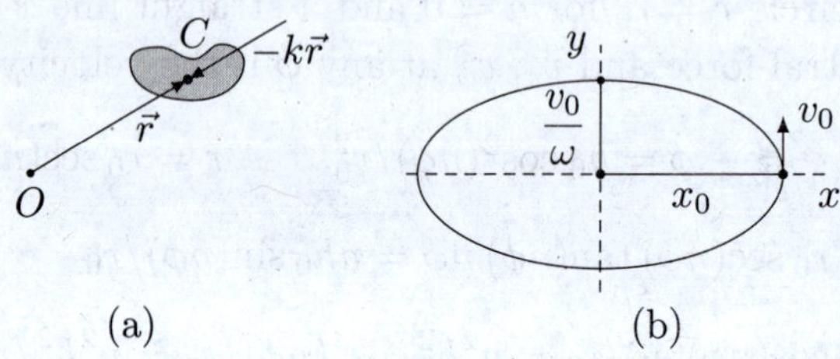

Fig. 6.2

The motion is simple harmonic and the orbit is bounded. Choose x-axis along the minimum radius $r_{\min} = x_0$ and y-axis along the velocity v_0 at this position (Fig. 6.2b). Then, at this position at $t = 0$, $\vec{r}(0) = x_0\,\vec{\imath}$, $\vec{v}(0) = v_0\,\vec{\jmath}$ and eq. (6.8) becomes

$$\vec{r}(t) = x_0\cos\omega t\,\vec{\imath} + \frac{v_0}{\omega}\sin\omega t\,\vec{\jmath}$$

$$\Rightarrow \quad x = x_0\cos\omega t, \qquad y = \frac{v_0}{\omega}\sin\omega t \qquad \Rightarrow \qquad \frac{x^2}{x_0^2} + \frac{y^2}{v_0^2/\omega^2} = 1$$

Hence, the orbit is an ellipse. $\vec{F}$ is a conservative force and energy conservation $\Rightarrow$ $E = \frac{1}{2}mv^2 + \frac{1}{2}kr^2 = E_0$ (a constant). This may easily be verified by substituting for the position and velocity from eq. (6.8) yielding $E \equiv \frac{1}{2}m(r_0^2\omega^2 + v_0^2) = E_0$.

6.2 Motion Under an Inverse Square Gravitational Force

Nearly all calculations for the trajectories of planets, satellites and missiles, based on the inverse square gravitational law, are in excellent agreement with observations. For the inverse square gravitational force $\vec{F} = -(GMm/r^2)\,\vec{e}_r$, the potential energy V^* of $\vec{F}$, as if it is acting at C, is $-GMm/r$. Hence, the work–energy relation for C moving at speed v is:

$$T_C + V^* = \text{constant} = E_0 \quad \text{(say)}$$

$$\Rightarrow \quad \tfrac{1}{2}mv^2 - \frac{GMm}{r} = \tfrac{1}{2}mv_0^2 - \frac{GMm}{r_0} = E_0$$

$$\Rightarrow \qquad v^2 = v_0^2 + 2GM\left(\frac{1}{r} - \frac{1}{r_0}\right) \qquad (a)$$

It follows that $v_{\max}$ occurs at $r_{\min}$ and $v_{\min}$ occurs at $r_{\max}$. From the Greek words, *peri* (near), *ap* (away from), *helios* (sun), *geos* (earth), the nearest position is called *pericentron (more specifically perigee, perihelion, periastron* for the earth, sun and another star respectively, as the attracting body) and the farthest position is called *apacentron (apogee, aphelion, apastron* as the case may be).

The body escapes to infinity if at $r = \infty$, the speed $v_\infty \geq 0$, i.e., if

$$v_\infty^2 = v_0^2 + 2GM\left(\frac{1}{\infty} - \frac{1}{r_0}\right) = v_0^2 - \frac{2GM}{r_0} \geq 0, \quad \text{i.e.,} \quad \text{if } v_0 \geq \left(\frac{2GM}{r_0}\right)^{1/2}$$

Hence, the *minimum escape speed* v_e at r_0, which is independent of its direction provided the trajectory does not hit body M, is:

$$v_e = \left(\frac{2GM}{r_0}\right)^{1/2} \qquad (6.9)$$

The negligible presence of hydrogen in the earth's atmosphere can be explained on the basis of v_e. Thermal and other excitations must have raised its velocity v above v_e since for a given kinetic energy $T = \frac{1}{2}mv^2 \Rightarrow v = (2T/m)^{1/2}$ is large for a small m. The Moon has no atmosphere since the gases have long since escaped. The more massive planets have atmosphere. The escape velocity from the sun is too great to allow even hydrogen, at a temperature of 6000^oC, to escape.

For a *circular orbit* of radius r_0, $\dot{r} = 0, \ddot{r} = 0$, $\vec{v} = r_0\dot{\phi}\,\vec{e}_\phi = v_c\,\vec{e}_\phi$ and the equation of motion yields the speed v_c:

$$m(\ddot{r} - r_0\dot{\phi}^2) = F_r = -\frac{GMm}{r_0^2} \quad \Rightarrow \quad v_c = r_0\dot{\phi} = \left(\frac{GM}{r_0}\right)^{1/2} = \frac{1}{\sqrt{2}}v_e \qquad (6.10)$$

6.2.1 Equation of the trajectory

For $F_r = -GMm/r^2 = -GMmu^2$, eq. (6.6) becomes

$$\frac{d^2u}{d\phi^2} + u = \frac{GM}{h_0^2}$$

Its complementary, particular and general solutions, u_c, u_p, and u respectively, are:

$$u_c = A\cos\phi + B\sin\phi = D\cos(\phi - \beta), \qquad u_p = \frac{GM}{h_0^2}$$

$$\frac{1}{r} = u = u_c + u_p = \frac{GM}{h_0^2} + D\cos(\phi - \beta)$$

where A, B, D, and β are constants.

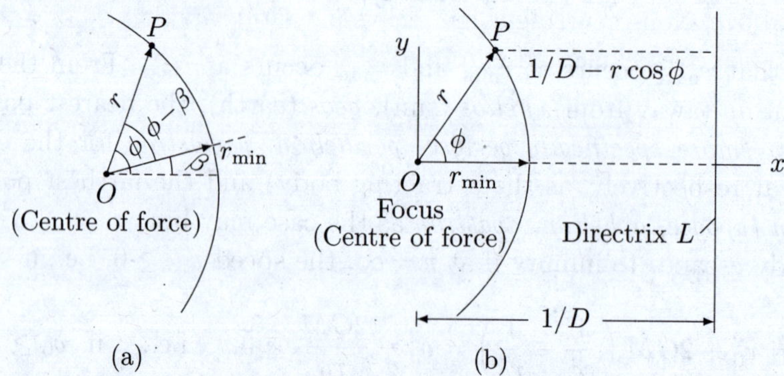

Fig. 6.3

Without loss of generality, let $D > 0$. Then $r_{\min}$ occurs at $\phi = \beta$ (Fig. 6.3a). Choose *the x-axis to pass through* $r_{\min}$, *(since by definition* ϕ *is measured from the x-axis), the equation of the trajectory* (Fig. 6.3b) is given by

$$\frac{1}{r} = \frac{GM}{h_0^2} + D\cos\phi \tag{6.11}$$

We get *radial velocity* v_r by differentiating (6.11) with respect to t and using $r^2\dot{\phi} = h_0$:

$$-\frac{\dot{r}}{r^2} = -D\sin\phi\,\dot{\phi} \quad \Rightarrow \quad v_r = \dot{r} = h_0 D\sin\phi \tag{6.12}$$

$$\text{Equation (6.11)} \Rightarrow \quad \frac{r}{1/D - r\cos\phi} = \frac{(\text{distance of } P \text{ from } O)}{(\text{distance of } P \text{ from line } L)} = \frac{h_0^2 D}{GM}$$

which is a constant. Hence, the orbit is a *conic section* (Fig. 6.3b) with *focus at the centre of force*, directrix L at a distance $1/D$ from it and its *eccentricity* e is given by

$$e = \frac{h_0^2 D}{GM} \tag{6.13}$$

This is Kepler's first law: planets describe ellipses with focus at the sun's centre.

Orbit	$e = \dfrac{h_0^2 D}{GM}$	v_0	$E_0 = \frac{1}{2}mv_0^2 - \dfrac{GMm}{r_0}$
Hyperbola	$e > 1$	$v_0 > (2GM/r_0)^{1/2}$	$E_0 > 0$
Parabola	$e = 1$	$v_0 = (2GM/r_0)^{1/2}$	$E_0 = 0$
Ellipse	$e < 1$	$v_0 < (2GM/r_0)^{1/2}$	$E_0 < 0, \quad E_0 = -GMm/2a$
Circle	$e = 0$	$v_0 = (GM/r_0)^{1/2}$	$E_0 < 0, \quad E_0 = -GMm/2r_0$

The dependence of the type of orbit on the value of each of e, v_0 and E_0 is given in the table above. Non-recurring comets have parabolic and hyperbolic paths. Recurring comets have narrow elongated elliptic paths with $e \simeq 1$.

Proof of $E_0 = -GMm/2a$ for an elliptic orbit of semi-major axis a:

At $r_1 = r_{\min}$ and $r_2 = r_{\max}$: $\dot{r} = 0 \Rightarrow v_{r_1} = v_{r_2} = 0 \Rightarrow v_{\phi_1} = v_1$ and $v_{\phi_2} = v_2$,

$$E_0 = \tfrac{1}{2}mv_1^2 - \frac{GMm}{r_1} = \tfrac{1}{2}mv_2^2 - \frac{GMm}{r_2} \qquad (a)$$

Put v_2 from $h_0 = r_2v_2 = r_1v_1$ in (a):

$$\tfrac{1}{2}mv_1^2 - \frac{GMm}{r_1} = \tfrac{1}{2}m\left(\frac{r_1v_1}{r_2}\right)^2 - \frac{GMm}{r_2} \quad \Rightarrow \quad v_1^2 = \frac{2GMr_2}{r_1(r_1+r_2)} \qquad (b)$$

Put v_1^2 from eq. (b) in eq. (a) to get:

$$E_0 = -\frac{GMm}{r_1+r_2} = -\frac{GMm}{2a} \tag{6.14}$$

The time of travel from ϕ_1 to ϕ_2 is obtained using the areal velocity and

$$\frac{d}{d\phi}\left[\frac{e\sin\phi}{1+e\cos\phi}\right] = \frac{1}{1+e\cos\phi} - \frac{1-e^2}{(1+e\cos\phi)^2}$$

$$t = \frac{A}{\dot{A}} = \frac{1}{h_0}\int_{\phi_1}^{\phi_2} r^2\,d\phi = \frac{h_0^3}{(GM)^2}\int_{\phi_1}^{\phi_2}\frac{1}{(1+e\cos\phi)^2}\,d\phi$$

$$\Rightarrow \quad t = \frac{h_0^3[I(\phi_2) - I(\phi_1)]}{(GM)^2} \tag{6.15}$$

with

$$I(\phi) = \begin{cases} \dfrac{2}{(1-e^2)^{3/2}}\tan^{-1}\left[\sqrt{\dfrac{1-e}{e+1}}\tan\dfrac{\phi}{2}\right] - \dfrac{e\sin\phi}{(1-e^2)(1+e\cos\phi)}, & \text{if } e<1 \\[2ex] \dfrac{1}{3}\tan\dfrac{\phi}{2}\left[\dfrac{2+\cos\phi}{1+\cos\phi}\right], & \text{if } e=1 \\[2ex] \dfrac{-2}{(e^2-1)^{3/2}}\tanh^{-1}\left[\sqrt{\dfrac{e-1}{e+1}}\tan\dfrac{\phi}{2}\right] - \dfrac{e\sin\phi}{(1-e^2)(1+e\cos\phi)}, & \text{if } e>1 \end{cases} \tag{6.16}$$

6.2.2 Time period for a closed orbit

The time period T for an elliptic orbit (Fig. 6.4), of semi-major and semi-minor axes a and b, is given by

$$T = \frac{\text{(Area of ellipse)}}{\text{(Areal velocity)}} = \frac{\pi ab}{\frac{1}{2}h_0} \tag{c}$$

$$\frac{1}{r_{\min}} = \frac{GM}{h_0^2} + D = \frac{GM}{h_0^2}(1+e), \qquad \frac{1}{r_{\max}} = \frac{GM}{h_0^2} - D = \frac{GM}{h_0^2}(1-e)$$

$$\Rightarrow \quad r_{\min} = \frac{h_0^2}{GM(1+e)}, \qquad r_{\max} = \frac{h_0^2}{GM(1-e)}$$

$$\Rightarrow \quad a = \tfrac{1}{2}(r_{\max} + r_{\min}) = \frac{h_0^2}{GM(1-e^2)}, \qquad b = a(1-e^2)^{1/2} \tag{d}$$

Equations (c), (d) $\Rightarrow$

$$T = \frac{2\pi ab}{h_0} = \frac{2\pi a^2(1-e^2)^{1/2}}{h_0} = \frac{2\pi a^2}{h_0}\left(\frac{h_0^2}{GMa}\right)^{1/2}$$

$$\Rightarrow \quad T = \frac{2\pi a^{3/2}}{(GM)^{1/2}} \tag{6.17}$$

i.e., $T^2 \propto a^3$, which is Kepler's third law of planetary motion. Equations (6.17) and (6.14) imply that all orbits having the same geometric parameter a or equivalently, the same dynamical constant E_0/m have the same periodic time. For example, electrons of an atom with the same energy level E_0/m move in elliptic orbits with the same semi-major axis a and the same periodic time.

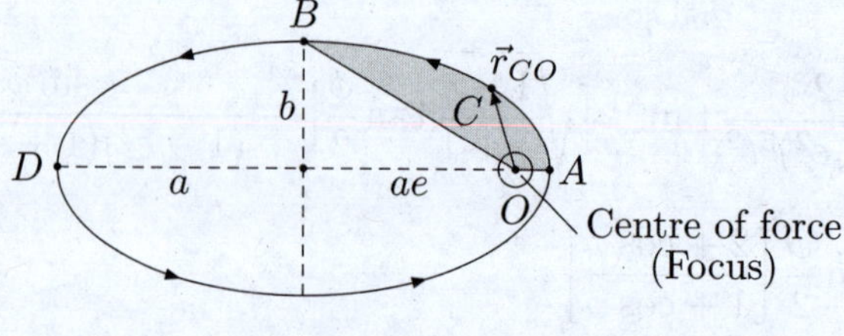

Fig. 6.4

For an elliptic orbit (Fig. 6.4), the time T_{AB} to traverse from A to B is not equal to the time T_{BD} to traverse from B to D, since the areal velocity corresponds to the area swept by the position vector with reference to the centre of force O:

$$T_{AB} = \frac{\text{Area } OAB}{\dot{A}} = \frac{\frac{1}{4}\pi ab - \frac{1}{2}b(ae)}{\frac{1}{2}h_0} = (1 - 2e/\pi)\frac{\frac{1}{4}\pi ab}{\frac{1}{2}h_0} = \tfrac{1}{4}T(1 - 2e/\pi)$$

Similarly, $T_{BD} = \frac{1}{4}T(1 + 2e/\pi)$.

§ **Exercise 6.2:** This problem was originally solved by Newton. Using Kepler's laws: $1/r = B(1 + e\cos\phi)$, $\dot{A} = \frac{1}{2}r^2\dot{\phi} = \text{constant} = \frac{1}{2}h_0$ (say), $T^2/a^3 = \text{constant} = 4\pi^2/D$ (say), prove that $\vec{F} = m\ddot{\vec{r}} = -(mD/r^2)\,\vec{e}_r$. Similarly, the force of a planet on its moon of mass m_1 is $-(m_1D_1/r^2)\,\vec{e}_r$. From the observed periods of the moon of a planet about it and of the planets about the sun, it was found that $D/M_{\text{sun}} = D_1/M_{\text{planet}} = \text{constant}$ (say G), i.e., $D = GM$. Hence, $\vec{F} = -(GMm/r^2)\,\vec{e}_r$. More precise measurements have shown that Kepler's laws are only an approximation. It was good for the development of mechanics that Kepler's laws were formulated from less accurate data, because otherwise the elegance of the law of gravitation would have been obscured.

6.2.3 Parameters of a satellite's orbit in terms of initial r_0, v_0, α_0

A body projected from the earth's surface can never become a satellite orbiting it. It will move like an unpowered missile and eventually crash or escape (see Fig. Ex6.4 below). To launch a satellite, it is necessary to accelerate it to a high-speed and take it outside the earth's atmosphere and then nudge it into the required orbit. Hence, a multi-stage rocket is required (see Fig. 6.5a). After the last stage of the rocket is fired (called *burnout*), the motion of the satellite (mass m) can be studied with reference to an (approximate) inertial frame I attached to earth's centre and with orientation fixed with reference to distant stars. Since the size of the satellite is very small compared to the distance from the centre of the earth (mass M) and other massive astronomical bodies are relatively far off, its motion can be obtained by modelling the net force on it by an inverse square gravitational force directed to the earth's centre fixed in I.

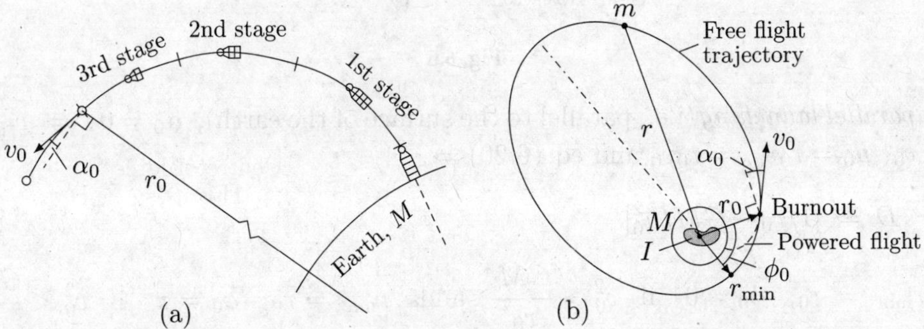

Fig. 6.5

The parameters GM, h_0, D of the satellite's orbit are computed from the velocity v_0 (*launching velocity*) at radius r_0 (at the end of the powered flight) at an angle α_0 (*launching angle*) with the circumferential direction $\vec{e}_\phi$ (Fig. 6.5b) with reference to frame I. The inclination ϕ_0 of the initial radial position from the minimum radial position is an unknown. GM is found from the acceleration g due to gravity at the surface of the earth (radius R): $GMm/R^2 = mg \Rightarrow GM = gR^2$. Equation (6.2) $\Rightarrow$

$$h_0 = rv_\phi = r_0 v_{\phi_0} = r_0 v_0 \cos\alpha_0 \tag{6.18}$$

Equations (6.11) and (6.12) for r_0 and v_{r_0} are solved for D and ϕ_0:

$$\frac{1}{r_0} = \frac{GM}{h_0^2} + D\cos\phi_0, \qquad v_{r_0} = v_0 \sin\alpha_0 = h_0 D \sin\phi_0 \Rightarrow$$

$$\tan\phi_0 = \frac{v_{r_0}/h_0}{1/r_0 - GM/h_0^2} = \frac{r_0 v_{r_0} v_{\phi_0}}{r_0 v_{\phi_0}^2 - GM} \tag{6.19}$$

$$D = \left[\left(\frac{1}{r_0} - \frac{GM}{h_0^2}\right)^2 + \left(\frac{v_{r_0}}{h_0}\right)^2\right]^{1/2} \tag{6.20}$$

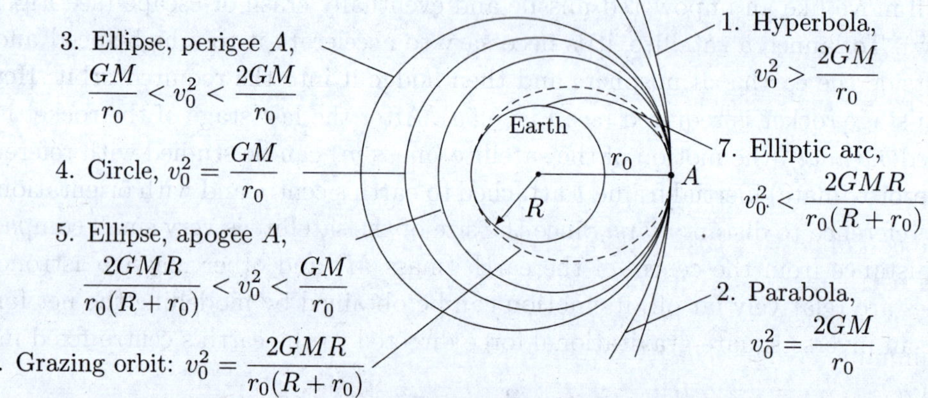

Fig. 6.6

For *parallel launching* (i.e., parallel to the surface of the earth), $\alpha_0 = 0$: $\Rightarrow v_{r_0} = 0$, $v_{\phi_0} = v_0$, $h_0 = r_0 v_{\phi_0} = r_0 v_0$ and eq. (6.20) $\Rightarrow$

$$D = |1/r_0 - GM/h_0^2| \tag{6.21}$$

$$r_{\min} = r_0, \ \ \phi_0 = 0 \ \text{ if } v_0^2 > \frac{GM}{r_0} \quad \text{and} \quad r_{\max} = r_0, \ \ \phi_0 = \pi \ \text{ if } v_0^2 < \frac{GM}{r_0}.$$

Once GM, h_0, D are obtained, then $e, r_{\min}, v_{\max}$ are computed from:

$$e = \frac{h_0^2 D}{GM}, \qquad \frac{1}{r_{\min}} = \frac{GM}{h_0^2} + D, \qquad v_{\max} = v_{|r_{\min}} = v_{\phi|r_{\min}} = \frac{h_0}{r_{\min}}$$

At a given radius r:

$$\cos\phi = [1/r - GM/h_0^2]\,/\,D, \qquad v_\phi = h_0/r, \qquad v_r = h_0 D \sin\phi$$

Alternatively, at first v^2 is obtained from the energy relation:

$$\tfrac{1}{2}mv^2 - \frac{GMm}{r} = \tfrac{1}{2}mv_0^2 - \frac{GMm}{r_0}, \qquad v_r^2 = v^2 - v_\phi^2$$

If $e < 1$: $$\frac{1}{r_{\max}} = \frac{GM}{h_0^2} - D, \qquad v_{\min} = v_{|r_{\max}} = v_{\phi|r_{\max}} = \frac{h_0}{r_{\max}}$$

$$\frac{1}{r_{\min}} = \frac{GM}{h_0^2} + D, \qquad v_{\max} = v_{|r_{\min}} = v_{\phi|r_{\min}} = \frac{h_0}{r_{\min}}$$

$$a = \tfrac{1}{2}(r_{\max} + r_{\min}), \qquad T = \frac{\pi ab}{\dot{A}} = \frac{\pi a^2(1-e^2)^{1/2}}{h_0/2}$$

If $e > 1$: $r_{\max} \to \infty$, $\qquad v_{\min} = (v_0^2 - 2GM/r_0)^{1/2}$

If $e = 1$: $r_{\max} \to \infty$, $\qquad v_{\min} = 0$

For parallel launching, the variation of orbits with the value of v_0 is shown in Fig. 6.6. The case of *radial launching* has been discussed in Example 1.6.

§ **Exercise 6.3:** Using eqs (6.13) and (6.20) show that $e = [1 + 2E_0 h_0^2/mG^2M^2]^{1/2}$.

The *best launching position* is at the equator with velocity v_{rel} relative to the surface of the earth in the direction west to east so that with reference to I, $v_0 = v_{\text{rel}} + \omega r_0 > v_{\text{rel}}$, where ω is the angular velocity of the earth. Neglecting air resistance, the minimum escape velocity at the surface of the earth, with radius $R = 6368$ km is $v_e^2 = (2GM/R)^{1/2} = (2gR)^{1/2} = 11.2$ km/s with reference to I.

6.2.4 Geostationary satellite

A geostationary communication satellite has a fixed position relative to the earth. Hence, its orbit is a *circle about* the N-S *axis traversed at the rate of rotation* ω of the earth. But the orbit is *coplanar with the centre of force O* at the earth's centre. Hence, it has a *circular orbit of radius* r_c *in the equatorial plane* and it moves from west to east at $\omega = 2\pi/[60 \times 60 \times 359/15]$ rad/s. (1 sidereal day= 23 h 56 min.= $\frac{359}{15}$ h) :

$$v_c = \omega r_c = \left(\frac{GM}{r_c}\right)^{1/2} \quad \Rightarrow \quad r_c = \left(\frac{GM}{\omega^2}\right)^{1/3} = \left(\frac{gR^2}{\omega^2}\right)^{1/3} \simeq 42140\,\text{km}$$

with $R = 6368\,\text{km}$ and $g = 9.81\,\text{m/s}^2$. Each satellite covers an angle 2α of the earth's surface (Fig. 6.7) with $\alpha = \cos^{-1}(R/r_c) = 81°$. Hence, the number of satellites to cover the earth (excluding small un-populated regions around the poles) is 3 since $360/2\alpha = 2.22$. Such high-altitude satellites cause an irritating echo due to the time gap of about 0.25 s for signal travel. To avoid this presently communication is achieved through multiple low-altitude satellites.

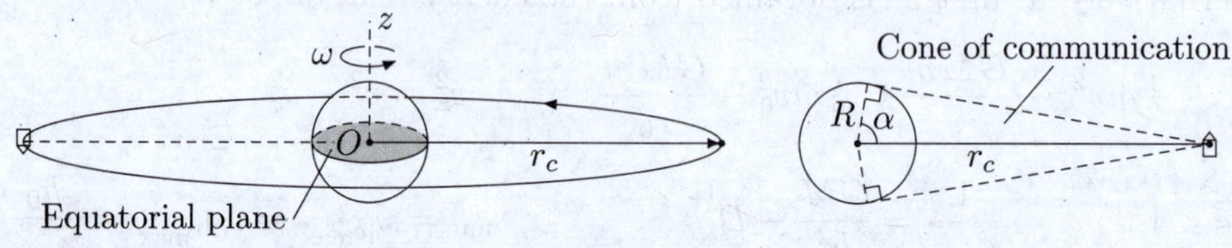

Fig. 6.7

6.2.5 Transfer orbit

Transfer from one type of orbit to another type of orbit requires a change in E_0. The transfer to a bigger orbit with greater E_0 should be executed at the optimum location in the orbit where the greatest increment in energy, ΔE_0, is achieved for the same change $|\Delta\vec{v}|$ in the magnitude of velocity (i.e., for the same expenditure of fuel):

$$\begin{aligned}
\Delta E_0 &= (\text{value of } E_0 \text{ in final orbit}) - (\text{value of } E_0 \text{ in initial orbit}) \\
&= \left(\tfrac{1}{2}mv_f^2 - \frac{GMm}{r}\right) - \left(\tfrac{1}{2}mv_i^2 - \frac{GMm}{r}\right) = \tfrac{1}{2}m(v_f^2 - v_i^2) \\
&= \tfrac{1}{2}m\left[(\vec{v}_i + \Delta\vec{v})^2 - \vec{v}_i^{\,2}\right] = \tfrac{1}{2}m\left[(\Delta\vec{v})^2 + 2\Delta\vec{v}\cdot\vec{v}_i\right]
\end{aligned}$$

(Note: In the analysis above, since the firing of the rockets takes place over a very small distance compared to the size of the orbit, r_i and r_f have both been taken as r.) Hence, $\Delta\vec{v}\cdot\vec{v}_i$ should be large, i.e., *$\Delta\vec{v}$ should be along $\vec{v}_i$ and at a location where $\vec{v}_i$ is maximum, i.e., at the location of the minimum radius r_{min}.*

6.2.6 Effect of atmospheric drag on the orbit of a satellite

Depending on the altitude of the satellite, we consider two cases.

1. The satellite is in an *elliptic orbit* and is in the atmosphere for a very short time when $r \simeq r_{\text{min}} = r_0$ (Fig. 6.8a). Hence, the atmospheric drag is modelled as an *instantaneous impulse* applied at $r = r_0$ which causes instantaneous change in speed from v_0 in the initial orbit to $v_0 + \Delta v_0$ in the next orbit ($\Delta v_0 < 0$).

$$E_0 = \tfrac{1}{2}mv_0^2 - \frac{GMm}{r_0} = -\frac{GMm}{2a} \quad\Rightarrow\quad \Delta E_0 = mv_0\Delta v_0 = \frac{GMm\Delta a}{2a^2}$$

$$\Rightarrow \quad \Delta a = \frac{2a^2 v_0 \Delta v_0}{GM} < 0$$

$$b^2 = a^2(1-e^2) = a(1-e)a(1+e) = r_{\max} r_{\min} = r_0(2a - r_0)$$

$\Rightarrow b$ also decreases. The semi-major axis of the successive orbits decreases and the trajectory tends to a circular orbit (Fig. 6.8a).

2. The satellite moves in a nearly elliptic orbit (Fig. 6.8b) of a semi-major axis a in atmosphere under drag force F. An approximate analysis is performed as follows.

$$E_0 \simeq -\frac{GMm}{2a}, \qquad \dot{E}_0 = \dot{T}_C + \dot{V}^* = \dot{W}^*_{nc}$$

$$\Rightarrow \frac{GMm\dot{a}}{2a^2} = -Fv \quad \Rightarrow \quad \dot{a} = -\frac{2Fva^2}{GMm} < 0 \tag{1}$$

Hence, the orbit continues to shrink and the satellite spirals slowly inwards (Fig. 6.8b). If the satellite moves in a nearly circular orbit of radius r in the upper highly rarefied layer of atmosphere, then the drag force is modelled as linear: $F = cv$ and with $v \simeq (GM/r)^{1/2}$, eq. (1) yields $\dot{r}/r = -2c/m$. Hence, r decreases and v increases, although there is a drag force. This occurs since the decrease in potential energy as r decreases is more than the negative work done by F.

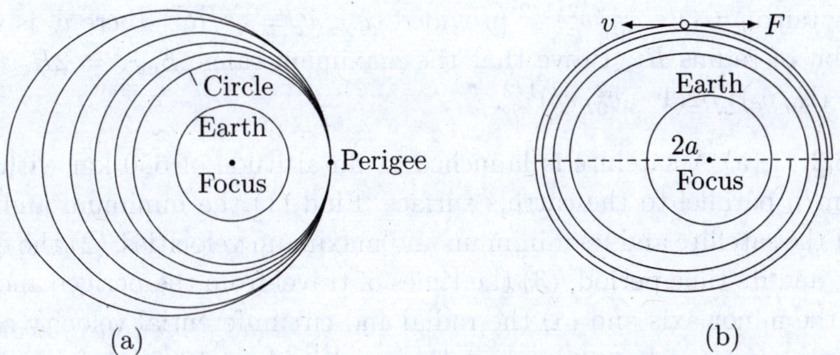

Fig. 6.8

In the altitude range 150 to 600 km, the drag force is modelled as $F = \alpha v^2$ with $\alpha = C_D A\rho$, where A is the cross-sectional area of the satellite, ρ is the atmospheric density and C_D is the drag coefficient. For nearly elliptic and circular orbits, eq. (1) $\Rightarrow$

$$\dot{a} = -\frac{2\alpha v^3 a^2}{GMm}, \qquad \dot{r} = -\frac{2\alpha(GMr)^{1/2}}{m} = -\frac{2C_D A\rho(GMr)^{1/2}}{m} \tag{6.22}$$

respectively. We have used $v \simeq (GM/r)^{1/2}$ for the circular orbit. Equation $(6.22)_2$ $\Rightarrow \dot{r} \propto \rho$, $\dot{r} \propto \sqrt{r}$. At low altitudes, $|\dot{r}|$ increases rapidly as ρ increases exponentially with decrease in height above the earth's surface. The orbit is completely unstable for altitudes less than 160 km. The useful life of a satellite can be found by integrating eq. (6.22) with the known variation of ρ with r.

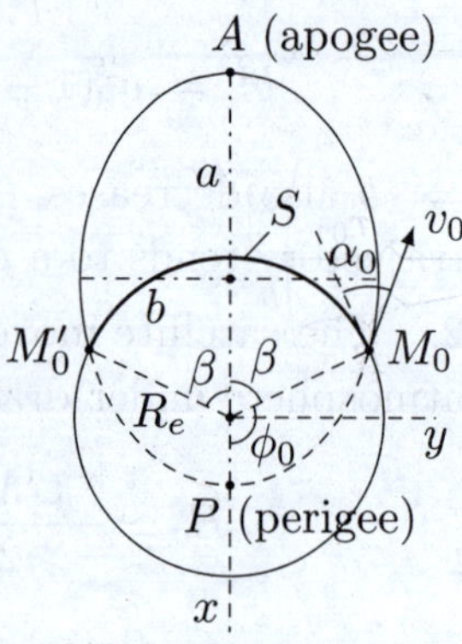

Fig. Ex6.4

§ **Exercise 6.4:** In the study of the path of a long-range intercontinental ballistic missile projected from the surface of the earth, the effect of earth's rotation should be included. If this effect and the drag force are both neglected, the path of the missile is an elliptic arc with its focus at centre of the earth and maximum altitude at $\phi = \pi$. A missile fired at M_0 at $r_0 \simeq R_e$ with v_0 at α_0 makes an impact at M_0' with range $\widehat{M_0M_0'} = S$ (Fig. Ex. 6.4). (a) Using eq. (6.19) with $\phi_0 = \pi - \beta$, show that the initial velocity v_0 for range S is given by

$$\tan\beta = \frac{v_0^2 \sin 2\alpha_0}{2(gR_e - v_0^2\cos^2\alpha_0)} \quad \Rightarrow \quad v_0 = \left[\frac{2gR_e\tan\beta}{\sin 2\alpha_0 + 2\cos^2\alpha_0\tan\beta}\right]^{1/2} \tag{6.23}$$

with $\beta = S/2R_e$. The time of flight is obtained from eq. (6.16) with $\phi_1 = \pi - \beta, \phi_2 = \pi + \beta$. (b) Prove that for a given range S, the minimum value of v_0 occurs for $\alpha_0 = \pi/4 - \beta/2$ with $v_{0_{\min}} = [2R_e g\sin\beta\,/\,(1+\sin\beta)]^{1/2}$. (c) For a given value of v_0, using $d(\tan\beta)/d\alpha_0 = 0$ from eq. (6.23), show that the range is maximum for optimum angle α_0 given by $\tan\alpha_0 = (1 - v_0^2/v_c^2)^{1/2}$ provided $v_0 < v_c = \sqrt{gR_e}$, where v_c is velocity for circular orbit of radius R_e. Prove that the maximum range $S_{\max} = 2R_e\beta_{\max}$, where $\tan\beta_{\max} = (v_0/v_c)^2\,/\,2(1 - v_0^2/v_c^2)^{1/2}$.

Example 6.2: (a) A satellite is launched at an altitude of 650 km with a velocity of 30000 km/h parallel to the earth's surface. Find (1) the minimum and maximum altitudes of the satellite and its minimum and maximum velocities, (2) the eccentricity of the orbit and its time period, (3) the times of travel from the perigee and apogee to the end of the minor axis and (4) the radial and circumferential velocity components and the speed when (i) the satellite reaches an altitude of 1600 km for the first time, and (ii) after traversing a polar angle of 60° at earth's centre from the launch position. (b) Work out part (a) if due to a malfunction, the velocity at launch is at an outward angle of 3° to the circumferential direction. (c) From the satellite in part (b), a space probe is to be launched to have a trajectory of eccentricity 1.2. Find the optimum location in the orbit where it should be launched and the change in its velocity at launch to achieve the objective. (d) At the maximum altitude of the satellite in part (b), an observation module is to be launched from it so that in its subsequent orbit, its minimum distance from the earth's surface is 300 km. Find the change in velocity required at launch (imparted by appropriate rockets) for: (1) circumferential launching, and (2) radial launching from the satellite. Take the radius R of the earth as 6400 km.

Solution: (a) We find GM, h_0, D from given data for *parallel launching* (Fig. E6.2a).

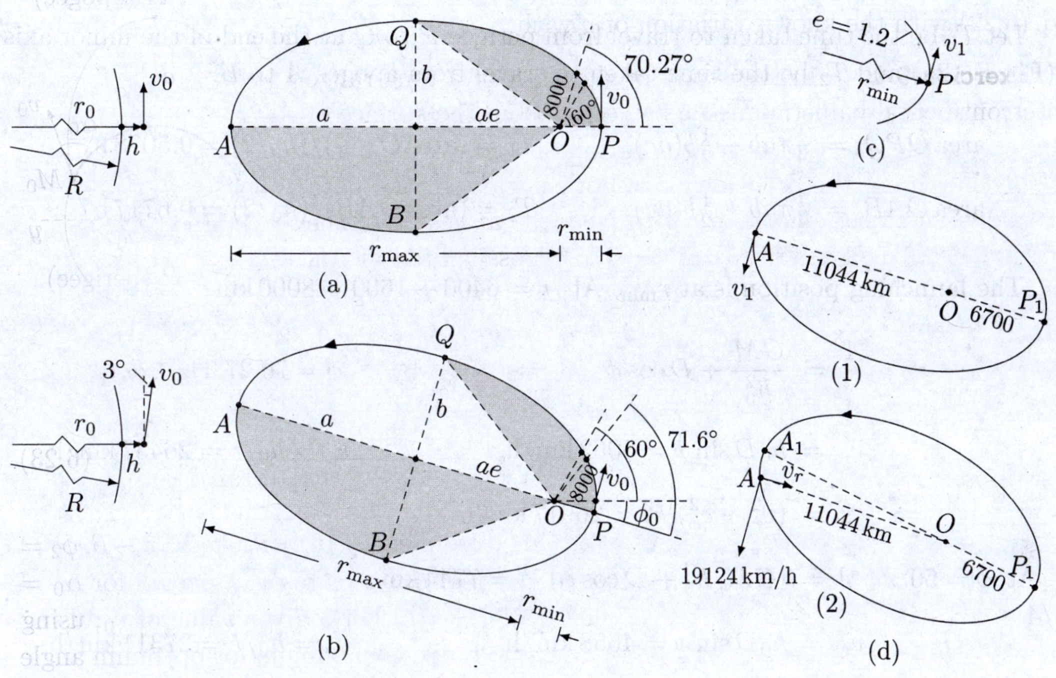

Fig. E6.2

$$GM = gR^2 = 9.81\,\frac{\text{m}}{\text{s}^2}\,(6400\,\text{km})^2 = 9.81\,\frac{10^{-3}\,\text{km}}{(1/3600)^2\,\text{h}^2}\,(6400\,\text{km})^2$$

$$= 5.20756 \times 10^{12}\,\text{km}^3/\,\text{h}^2$$

$$r_0 = 6400 + 650 = 7050\,\text{km}, \qquad v_0 = v_{\phi_0} = 30000\,\text{km/h}$$

$$h_0 = r_0 v_{\phi_0} = 211.5 \times 10^6\,\text{km}^2/\,\text{h}, \qquad GM/h_0^2 = 1.16416 \times 10^{-4}\,\text{km}^{-1}$$

$$D = \left|1/r_0 - GM/h_0^2\right| = 0.25428 \times 10^{-4}\,\text{km}^{-1}$$

$$r_{\min} = \left(\frac{GM}{h_0^2} + D\right)^{-1} = 7050\,\text{km}, \qquad r_{\max} = \left(\frac{GM}{h_0^2} - D\right)^{-1} = 10990\,\text{km}$$

$$h_{\min} = r_{\min} - R = 650\,\text{km}, \qquad h_{\max} = r_{\max} - R = 4590\,\text{km}$$

$$v_{\min} = h_0/r_{\max} = 19245\,\text{km/h}, \qquad v_{\max} = h_0/r_{\min} = 30000\,\text{km/h}$$

$$e = h_0^2 D/GM = 0.21842, \qquad a = \tfrac{1}{2}(r_{\min} + r_{\max}) = 9020\,\text{km}$$

$$b = a(1 - e^2)^{1/2} = 8802\,\text{km}, \qquad T = \pi ab/(h_0/2) = 2.359\,\text{h}$$

Let T_1 be the time taken to travel from perigee P to Q at the end of the minor axis (Fig. E6.2a) and T_2 be the time taken to travel from apogee A to B.

$$\text{area}\, OPQ = \tfrac{1}{4}\pi ab - \tfrac{1}{2}b(ae), \qquad T_1 = (\text{area}\, OPQ)/(h_0/2) = 0.5078\,\text{h}$$

$$\text{area}\, OAB = \tfrac{1}{4}\pi ab + \tfrac{1}{2}b(ae), \qquad T_2 = (\text{area}\, OAB)/(h_0/2) = 0.6717\,\text{h}$$

The launching position is at r_{min}. At $r = 6400 + 1600 = 8000\,\text{km}$:

$$\frac{1}{r} = \frac{GM}{h_0^2} + D\cos\phi \quad \Rightarrow \quad \phi = 70.27^\circ$$

$$v_r = h_0 D \sin\phi = 5062\,\text{km/h}, \qquad v_\phi = h_0/r = 26437\,\text{km/h}$$

$$\Rightarrow \quad v = (v_r^2 + v_\phi^2)^{1/2} = 26917\,\text{km/h}$$

$$\text{At } \phi = 60^\circ\text{:} \quad r = (GM/h_0^2 + D\cos\phi)^{-1} = 7744\,\text{km}$$

$$v_r = h_0 D \sin\phi = 4658\,\text{km/h}, \qquad v_\phi = h_0/r = 27311\,\text{km/h}$$

$$\Rightarrow \quad v = (v_r^2 + v_\phi^2)^{1/2} = 27705\,\text{km/h}$$

The speeds v could have been obtained using energy conservation: $\frac{1}{2}mv^2 - GMm/r = \frac{1}{2}mv_0^2 - GMm/r_0$.

(b) We first determine h_0, D for this *non-parallel launching* (Fig. E6.2b).

$$r_0 = 6400 + 650 = 7050\,\text{km}, \qquad v_0 = 30000\,\text{km/h}$$

$$v_{r0} = v_0 \sin 3^\circ = 1570.08\,\text{km/h}, \qquad v_{\phi 0} = v_0 \cos 3^\circ = 299589\,\text{km/h}$$

$$h_0 = r_0 v_{\phi 0} = 211.21 \times 10^6\,\text{km}^2/\,\text{h}, \qquad \frac{GM}{h_0^2} = 1.16736 \times 10^{-4}/\,\text{km}$$

$$D = \left[\left(\frac{1}{r_0} - \frac{GM}{h_0^2}\right)^2 + \left(\frac{v_{r0}}{h_0}\right)^2\right]^{1/2} = 0.261853 \times 10^{-4}\,\text{km}^{-1}$$

$$r_{\text{min}} = \left(\frac{GM}{h_0^2} + D\right)^{-1} = 6997\,\text{km}, \qquad r_{\text{max}} = \left(\frac{GM}{h_0^2} - D\right)^{-1} = 11044\,\text{km}$$

$$h_{\text{min}} = r_{\text{min}} - R = 597\,\text{km}, \qquad h_{\text{max}} = r_{\text{max}} - R = 4644\,\text{km}$$

$$v_{\text{min}} = \frac{h_0}{r_{\text{max}}} = 19124\,\text{km/h}, \qquad v_{\text{max}} = \frac{h_0}{r_{\text{min}}} = 30186\,\text{km/h}$$

$$e = \frac{h_0^2 D}{GM} = 0.22431, \qquad a = \tfrac{1}{2}(r_{\min} + r_{\max}) = 9020.5\,\text{km}$$

$$b = a(1-e^2)^{1/2} = 8790.6\,\text{km}, \qquad T = \frac{\pi ab}{h_0/2} = 2.359\,\text{h}$$

$$\text{area}\,OPQ = \tfrac{1}{4}\pi ab - \tfrac{1}{2}b(ae), \qquad T_1 = \frac{(\text{area}\,OPQ)}{h_0/2} = 0.5055\,\text{h}$$

$$\text{area}\,OAB = \tfrac{1}{4}\pi ab + \tfrac{1}{2}b(ae), \qquad T_2 = \frac{(\text{area}\,OAB)}{h_0/2} = 0.6740\,\text{h}$$

$$\tan\phi_0 = \frac{v_{r0}/h_0}{1/r_0 - GM/h_0^2} = 0.29607, \quad \phi_0 = 16.49^\circ$$

Except for round-off error, the values of a and T obtained for both parallel and non-parallel launchings are the same, since a depends on the initial energy level which is the same in both cases and $T^2 \propto a^3$. The launching position turns out to be at an angle $\phi_0 = 16.49^\circ$ to the position of $r_{\min}$ (Fig. E6.2b). At $r = 6400 + 1600 = 8000\,\text{km}$:

$$\frac{1}{r} = \frac{GM}{h_0^2} + D\cos\phi \quad \Rightarrow \quad \phi = 71.60^\circ$$

$$v_r = h_0 D\sin\phi = 5248\,\text{km/h}, \qquad v_\phi = \frac{h_0}{r} = 26401\,\text{km/h}$$

$$v = (v_r^2 + v_\phi^2)^{1/2} = 26917\,\text{km/h}$$

At $\theta = 60^\circ$, $\phi = 60^\circ + 16.49^\circ = 76.49^\circ$:

$$r = \left(\frac{GM}{h_0^2} + D\cos\phi\right)^{-1} = 8140\,\text{km}, \qquad v_r = h_0 D\sin\phi = 5378\,\text{km/h}$$

$$v_\phi = \frac{h_0}{r} = 25948\,\text{km/h}, \qquad v = (v_r^2 + v_\phi^2)^{1/2} = 26499\,\text{km/h}$$

The fact that the values of v at $r = 8000\,\text{km}$ are the same in parts (a) and (b) follows directly from the equation of conservation of energy.

(c) In order to achieve the objective with the minimum Δv, the probe should be launched in the circumferential direction at the position of $v_{\max}$ of the satellite, i.e., at $r_{\min} = 6997\,\text{km}$. This is also the minimum radius of the trajectory of the probe. Let h_1, D_1, e_1 refer to the trajectory of the probe (Fig. E6.2c).

$$e_1 = \frac{h_1^2 D_1}{GM} = 1.2 \quad \Rightarrow \quad \frac{GM}{h_1^2} = \frac{D_1}{1.2}$$

$$\frac{1}{r_{\text{min}}} = \frac{GM}{h_1^2} + D_1 \quad \Rightarrow \quad \frac{1}{6997} = \left(1 + \frac{1}{1.2}\right) D_1$$

$$\Rightarrow \quad D_1 = 0.779555 \times 10^{-4}\,\text{km}^{-1}$$

$$h_1 = \left(\frac{1.2GM}{D_1}\right)^{1/2} = 283.129 \times 10^6\,\text{km}^2/\text{h}$$

$$\Rightarrow \quad v_1 = v_{\phi 1} = \frac{h_1}{r_{\text{min}}} = 40464\,\text{km/h}$$

$$\Delta v = v_1 - v_{\text{max}} = 40464 - 30186 = 10278\,\text{km/h}$$

(d) (1) In this case, the maximum and minimum radii of the orbit of the module are 11044 km and 6700 km. Let h_0, D refer to the trajectory of the probe (Fig. E6.2d).

$$\frac{1}{6700} = \frac{GM}{h_0^2} + D, \quad \frac{1}{11044} = \frac{GM}{h_0^2} - D$$

$$\Rightarrow \quad \frac{GM}{h_0^2} = 1.199 \times 10^{-4}\,\text{km}^{-1} \quad \Rightarrow \quad h_0 = 208.404 \times 10^6\,\text{km}^2/\text{h}$$

$$v_1 = v_{\phi 1} = \frac{h_0}{11044} = 18870\,\text{km/h} \quad \Rightarrow \quad \Delta v = v_1 - 19124 = -254\,\text{km/h}$$

(2) Let v_r be the radial velocity imparted to the module at a launch radius of 11044 km, while its circumferential velocity v_ϕ remains unchanged at 19124 km/h. Hence, the velocity v_2 at the minimum radius $r_2 = 6700\,\text{km}$ of the module is given by

$$v_2 = v_{\phi 2} = \frac{h_0}{r_2} = \frac{11044 \times 19124}{6700} = 31523\,\text{km/h}$$

Conservation of energy:

$$\tfrac{1}{2}m(v_r^2 + 19124^2) - \frac{GMm}{11044} = \tfrac{1}{2}mv_2^2 - \frac{GMm}{r_2} \quad \Rightarrow \quad v_r = 4066\,\text{km/h}$$

which is very large compared to $|\Delta v|$ obtained in part (1).

Example 6.3: A space station S_1 is being assembled in its circular orbit of radius r_1 around the earth. The next payload S_2 of mass m is launched into a circular orbit of radius r_2 ($r_2 < r_1$). (a) Show that the transfer of payload to the space station would be with the least initial change in velocity in the circumferential direction if the transfer orbit is semi-elliptical. (b) Find the change in speed which must be imparted to the payload before the beginning and at the end of the semi-elliptical path. The carrier rocket of S_2 can develop a thrust P. Find the small times t_2 and t_1 for which it has to be fired to achieve these speed changes. (c) Find the relative angular positions of S_1 and S_2 at the time of launching.

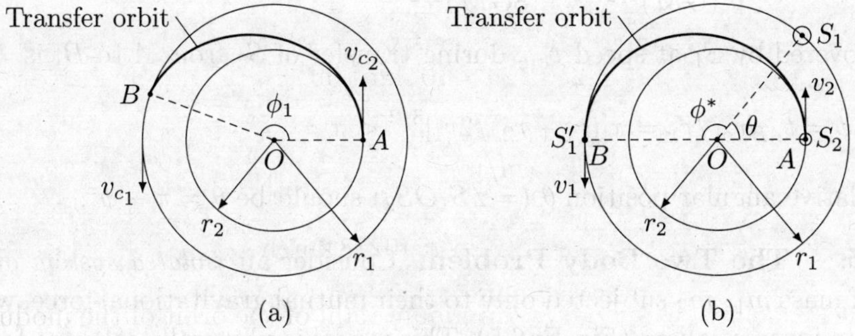

Fig. E6.3

Solution: (a) For the elliptic transfer orbit $\widehat{AB}$ (Fig. E6.3a), $r = r_2 = r_{\min}$ at $\phi = 0$ and $r = r_1$ at $\phi = \phi_1 \Rightarrow$

$$\frac{1}{r} = \frac{GM}{h_0^2} + D\cos\phi \quad \Rightarrow \quad \frac{1}{r_2} = \frac{GM}{h_0^2} + D, \qquad \frac{1}{r_1} = \frac{GM}{h_0^2} + D\cos\phi_1$$

$$\Rightarrow \quad h_0^2 = \frac{r_1 r_2 GM(1-\cos\phi_1)}{r_2 - r_1\cos\phi_1} \tag{1}$$

The least change in velocity corresponds to the least value of the function $f(\phi_1)$:

$$f(\phi_1) = \frac{1-\cos\phi_1}{r_2 - r_1\cos\phi_1} \quad \Rightarrow \quad \frac{df}{d\phi_1} = \frac{(r_2 - r_1)\sin\phi_1}{(r_2 - r_1\cos\phi_1)^2} = 0$$

$\Rightarrow \sin\phi_1 = 0 \Rightarrow \phi_1 = \pi$. Hence, the optimum transfer orbit is semi-elliptical.

(b) The speeds v_1 and v_2 in the transfer orbit (Fig. E6.3b) at r_1 and r_2 are obtained from h_0 given by eq. (1) for $\phi_1 = \pi$:

$$h_0 = r_1 v_1 = r_2 v_2 = \left[\frac{2r_1 r_2 GM}{r_1 + r_2}\right]^{1/2}$$

$$\Rightarrow \quad v_1 = \left[\frac{2r_2 GM}{r_1(r_1 + r_2)}\right]^{1/2}, \qquad v_2 = \left[\frac{2r_1 GM}{r_2(r_1 + r_2)}\right]^{1/2}$$

The speeds v_{c_1}, v_{c_2} for circular orbits of radii r_1, r_2 are $v_{c_1} = (GM/r_1)^{1/2}$, $v_{c_2} = (GM/r_2)^{1/2}$. The increase in speeds required at positions A and B are $\Delta v_A = v_2 - v_{c_2}$, $\Delta v_B = v_{c_1} - v_1$. The times t_1, t_2 of firing of engines at A and B are obtained from the impulse–momentum relations: $Pt_1 = m\Delta v_A$, $Pt_2 = m\Delta v_B$.

(c) The time of transfer:

$$t_{AB} = \tfrac{1}{2}(\text{period}) = \frac{\pi a^{3/2}}{(GM)^{1/2}} = \frac{\pi(r_1 + r_2)^{3/2}}{(8GM)^{1/2}}$$

The angle ϕ^* covered by S_1 at speed v_{c_1}, during transfer of S_2 from A to B, is

$$\phi^* = t_{AB}\,\dot{\phi} = t_{AB} v_{c_1}/r_1 = \pi[(r_1 + r_2)/2r_1]^{3/2} < \pi$$

Hence, their relative angular position θ $(= \angle S_1OS_2)$ should be $\theta = \pi - \phi^*$.

§ **Exercise 6.5: The Two Body Problem** Consider an *isolated system of two bodies* B_1, B_2 of mass m_1, m_2 subjected only to their mutual gravitational force, which is modelled as an inverse square (Fig. Ex6.5). This model is physically relevant for the motion of a double star system, the motion of the moon relative to the earth and the motion of satellites relative to earth. Prove that $\vec{R}_C(t) = \vec{R}_C(0) + \vec{v}_C(0)t$. Using the equations of motion of the centres of mass C_1, C_2 of individual bodies, prove that $m_1\ddot{\vec{r}}_{|I} = -\frac{Gm_1(m_1+m_2)}{r^2}\,\vec{e}$. Hence, the motion of C_1 relative to C_2 takes place as if the latter were fixed and its mass effectively increased from m_2 to $m_1 + m_2$. With this change, the results of the inverse square central force motion are applicable. Since C moves with uniform velocity with reference to I, a frame I_1 attached to C translating relative to I is itself an inertial frame. In this frame I_1, show that $\vec{r}_2 = -(m_1/m_2)\vec{r}_1$ and express the equation of motion of m_1 relative to I_1 in terms of $\vec{r}_1$ as:

$$m_1\ddot{\vec{r}}_1 = -\frac{Gm_1m_2^*}{r_1^2}\cdot\frac{\vec{r}_1}{r_1}, \qquad \text{with} \quad m_2^* = \frac{m_2}{(1 + m_1/m_2)^2}$$

Body B_1 moves relative to I_1 as if attracted to it by the gravitational force due to a mass, situated at fixed point C of I_1, equal to m_2^*.

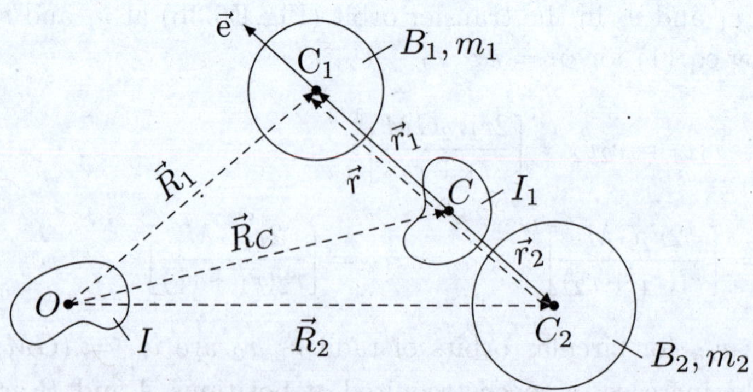

Fig. Ex6.5

B. MOTION OF A CHARGED 'PARTICLE'

6.3 Motion of a Charged 'Particle' in an Electromagnetic Field

The non-relativistic equation of motion of a point charge q of mass m (which is valid when its speed $v \ll c$, the speed of light) in an electromagnetic field $\vec{E}$ and $\vec{B}$ is

$$m\vec{a} = m\dot{\vec{v}} = \vec{F} = q\vec{E} + q\vec{v} \times \vec{B} \tag{6.24}$$

Let $V(x, y, z)$ be the potential of $\vec{E}$, then $\vec{E} = -\vec{\nabla}V$. $T = \frac{1}{2}mv^2 = \frac{1}{2}m\vec{v} \cdot \vec{v} \Rightarrow$

$$\begin{aligned} \dot{T} &= m\vec{a} \cdot \vec{v} = \vec{F} \cdot \vec{v} = (q\vec{E} + q\vec{v} \times \vec{B}) \cdot \vec{v} \\ &= -q\,\vec{\nabla}V \cdot \vec{v} = -q\left[\frac{\partial V}{\partial x}\frac{dx}{dt} + \frac{\partial V}{\partial y}\frac{dy}{dt} + \frac{\partial V}{\partial z}\frac{dz}{dt}\right] = -q\,\dot{V} \end{aligned}$$

$$\Rightarrow \quad T + qV = \tfrac{1}{2}mv^2 + qV = \text{constant} \tag{6.25}$$

For a purely magnetic field, $V = 0$ and eq. (6.25) implies that the speed v remains constant. For a purely electrical uniform field, eq. (6.24) $\Rightarrow$ $\vec{a} = q\vec{E}/m$ = constant. Hence, as in a uniform gravitational field, the particle describes a parabolic trajectory in the plane of $\vec{E}$ and the initial velocity $\vec{v}_0$, the trajectory being rectilinear if $\vec{v}_0 \parallel \vec{E}$. Unlike gravity, the acceleration and hence, the trajectory depends on the mass m.

6.3.1 Motion in a uniform magnetic field

Choose z-axis in the direction of $\vec{B} \Rightarrow \vec{B} = B\,\vec{k}$ and let $\omega = qB/m$. $m\vec{a} = \vec{F}$:

$$m(\dot{v}_x\,\vec{i} + \dot{v}_y\,\vec{j} + \dot{v}_z\,\vec{k}) = q(v_x\,\vec{i} + v_y\,\vec{j} + v_z\,\vec{k}) \times B\,\vec{k} = qB(-v_x\,\vec{j} + v_y\,\vec{i})$$

$$\vec{k}: \quad \dot{v}_z = 0 \qquad \Rightarrow \qquad v_z = v_z(0)$$

$$\vec{j}: \quad \dot{v}_y = -\omega v_x$$

$$\vec{i}: \quad \dot{v}_x = \omega v_y \qquad \Rightarrow \qquad \ddot{v}_x = \omega \dot{v}_y = -\omega^2 v_x$$

$$\Rightarrow \quad v_x = A\cos\omega t + D\sin\omega t, \qquad v_y = \dot{v}_x/\omega = -A\sin\omega t + D\cos\omega t$$

$$\Rightarrow \quad v_x^2 + v_y^2 = A^2 + D^2 = \text{constant}$$

A, D are found from $\vec{v}(0)$. Choose y-axis in the plane of $\vec{B}$ and $\vec{v}(0)$: $\vec{v}(0) = v_{y0}\,\vec{j} + v_{z0}\,\vec{k}$

$$\Rightarrow \quad v_x(0) = A = 0, \qquad v_y(0) = D = v_{y0}, \qquad v_z(0) = v_{z0}$$

$$\dot{x} = v_x = v_{y0} \sin\omega t, \qquad x(t) = x_0 + v_{y0}(1 - \cos\omega t)/\omega$$

$$\dot{y} = v_y = v_{y0} \cos\omega t, \qquad y(t) = y_0 + v_{y0} \sin\omega t/\omega \tag{6.26}$$

$$\dot{z} = v_z = v_{z0}, \qquad z(t) = z_0 + v_{z0} t$$

where initial position $\vec{r}(0) = x_0\,\vec{\mathrm{i}} + y_0\,\vec{\mathrm{j}} + z_0\,\vec{\mathrm{k}}$. Eliminating t from $x(t)$ and $y(t)$, yields

$$\left(x - x_0 - \frac{v_{y0}}{\omega}\right)^2 + (y - y_0)^2 = \left(\frac{v_{y0}}{\omega}\right)^2 \tag{6.27}$$

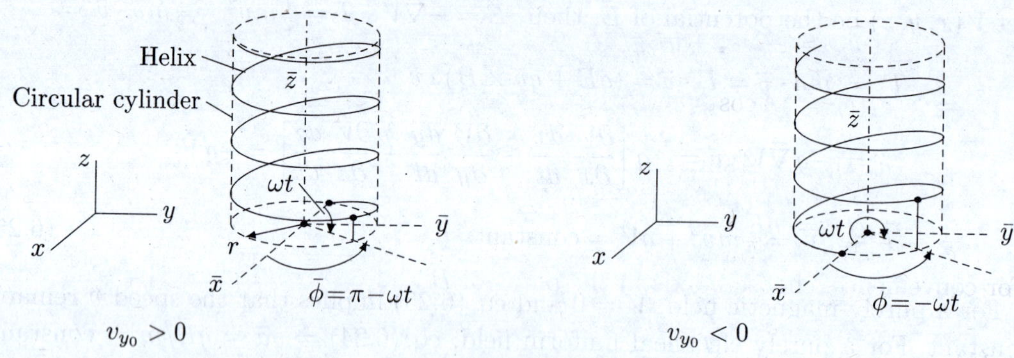

Fig. 6.9

The path is a helix on a right circular cylinder of radius $r = v_{y0}/\omega = mv_{y0}/qB$ with its axis parallel to $\vec{B}$ (Fig. 6.9) and passing through $(x_0 + v_{y0}/\omega,\ y_0,\ z_0)$. Consider axes $\bar{x}, \bar{y}, \bar{z}$ parallel to x, y, z, with origin at $(x_0 + v_{y0}/\omega,\ y_0,\ z_0)$ on the helix axis.

$$\text{eq. (6.26)} \ \Rightarrow\ \bar{x} = -\frac{v_{y0}}{\omega}\cos\omega t, \qquad \bar{y} = \frac{v_{y0}}{\omega}\sin\omega t,$$

$$\bar{z} = v_{z0} t, \qquad \bar{\phi} = \tan^{-1}\frac{\bar{y}}{\bar{x}} = \begin{cases} \pi - \omega t & \text{if} \quad v_{y0} > 0 \\ -\omega t & \text{if} \quad v_{y0} < 0 \end{cases} \tag{6.28}$$

i.e., constant $\dot{\bar{\phi}} = -\omega = -qB/m$. The path reduces to a circle described with constant speed if $\vec{v}(0) \perp \vec{B}$, i.e., if $v_{z0} = 0$. It reduces to the line $\bar{x} = \bar{y} = 0,\ \bar{z} = v_{z0} t$ if $v_{y0} = 0$.

The ratio q/m for an electron is obtained from its speed v in a circle of radius r:

$$v = |r\dot{\bar{\phi}}| = \left|-\frac{rqB}{m}\right| \quad \Rightarrow \quad \left(\frac{q}{m}\right)^2 = \left(\frac{v}{rB}\right)^2 \tag{a}$$

B is measured and r is measured from a photograph of the track of the electron, but the electrons move too fast for v to be measured directly. In Kaufman's method, the electron is accelerated by an electric field, starting from rest at potential V_0 to speed v at potential V_1 before entering the magnetic field. Using eq. (6.25) and eq. (a),

$$\tfrac{1}{2}mv^2 = q(V_0 - V_1) \quad \Rightarrow \quad \frac{|q|}{m} = \frac{2(V_0 - V_1)}{B^2 r^2}$$

6.3.2 Motion in an uniform electromagnetic field

Choose the z-axis in the direction of $\vec{B} \Rightarrow \vec{B} = B\,\vec{k}$ and let $\omega = qB/m$. $\quad m\vec{a} = \vec{F}$:

$$\begin{aligned}
m\dot{\vec{v}} &= q\vec{E} + q\vec{v}\times\vec{B} = q\vec{E} + q(v_x\,\vec{i} + v_y\,\vec{j} + v_z\,\vec{k})\times B\,\vec{k} \\
&= q\vec{E} + q(Bv_y\,\vec{i} - Bv_x\,\vec{j}) \\
\Rightarrow \quad \dot{v}_x &= \omega v_y + qE_x/m, \qquad\qquad \dot{v}_z = qE_z/m \\
\dot{v}_y &= -\omega v_x + qE_y/m \qquad \Rightarrow \qquad \ddot{v}_x + \omega^2 v_x = q\omega E_y/m \\
\Rightarrow \quad v_x(t) &= A\cos\omega t + B_1\sin\omega t + qE_y/\omega m \\
v_y(t) &= -A\sin\omega t + B_1\cos\omega t - qE_x/\omega m \qquad (1) \\
v_z(t) &= qE_z t/m + C_1
\end{aligned}$$

For convenience, choose y-axis in the plane of $\vec{B}$ and $\vec{v}(0)$: $\vec{v}(0) = v_{y0}\,\vec{j} + v_{z0}\,\vec{k}$.

$$\text{eq. (1)} \Rightarrow v_x(0) = A + \frac{qE_y}{\omega m} = 0, \qquad v_y(0) = B_1 - \frac{qE_x}{\omega m} = v_{y0}, \qquad v_z(0) = C_1 = v_{z0}$$

$$A = -\frac{qE_y}{\omega m} = -\frac{E_y}{B}, \qquad B_1 = \frac{qE_x}{\omega m} + v_{y0} = \frac{E_x}{B} + v_{y0}, \qquad C_1 = v_{z0}$$

$$\begin{aligned}
\Rightarrow \quad v_x &= -\frac{E_y}{B}\cos\omega t + (v_{y0} + E_x/B)\sin\omega t + \frac{E_y}{B} \\
v_y &= (v_{y0} + E_x/B)\cos\omega t + \frac{E_y}{B}\sin\omega t - \frac{E_x}{B} \qquad (6.29) \\
v_z &= \frac{q}{m}E_z t + v_{z0} \\
\Rightarrow \quad x &= -[E_y\sin\omega t + (Bv_{y0} + E_x)\cos\omega t]/\omega B + E_y t/B + C \\
&= -R\cos(\omega t - \psi) + E_y t/B + C \\
y &= [(Bv_{y0} + E_x)\sin\omega t - E_y\cos\omega t]/\omega B - E_x t/B + D \qquad (6.30) \\
&= R\sin(\omega t - \psi) - E_x t/B + D \\
z &= \frac{q}{2m}E_z t^2 + v_{z0}t + C_2
\end{aligned}$$

where $R^2 = [E_y^2 + (Bv_{y0} + E_x)^2]/\omega^2 B^2$, $\tan\psi = E_y/(Bv_{y0} + E_x)$ and $q/\omega m$ has been replaced by $1/B$. Eliminating $\sin(\omega t - \psi)$, $\cos(\omega t - \psi)$ from eqs $(6.30)_1$, $(6.30)_2$ $\Rightarrow$

$$\left(x - C - \frac{E_y t}{B}\right)^2 + \left(y - D + \frac{E_x t}{B}\right)^2 = R^2$$

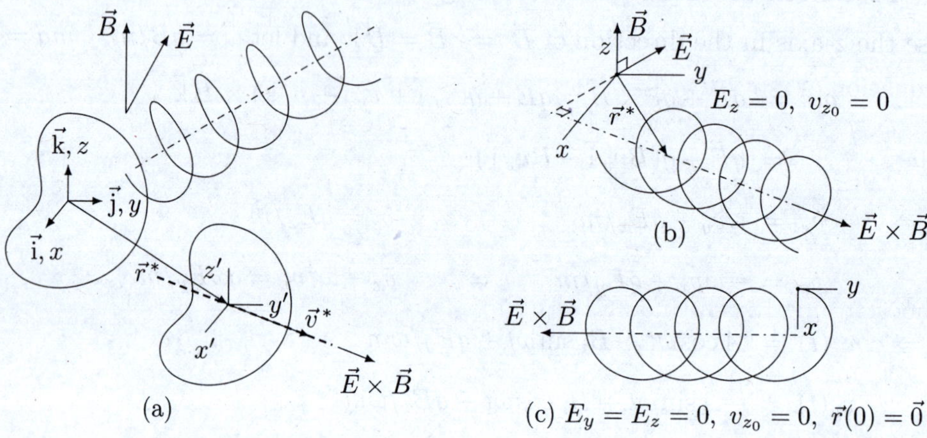

Fig. 6.10

The projection of motion on the xy-plane is a 'circle' of constant 'radius' R and centre O': $\vec{r}^* = (C + E_y t/B)\,\vec{\mathrm{i}} + (D - E_x t/B)\,\vec{\mathrm{j}}$. O' moves at $\vec{v}^* = \dot{\vec{r}}^* = (E_y\vec{\mathrm{i}} - E_x\vec{\mathrm{j}})/B = \vec{E} \times \vec{B}/B^2$, called the *Hall drift*, which is independent of the sign of the charge. The coordinates x', y', ϕ' (Fig. 6.10a) relative to a frame translating with O' are:

$$x' = -R\cos(\omega t - \psi), \qquad y' = R\sin(\omega t - \psi)$$

$$\phi' = \tan^{-1}(y'/x') = -\omega t + \psi \quad \Rightarrow \quad \dot{\phi}' = -\omega$$

The presence of $\vec{E}$ in addition to $\vec{B}$ causes the motion to shift across $\vec{B}$ in the direction of $\vec{E} \times \vec{B}$ and to have an additional acceleration in the direction of $\vec{B}$, apart from the spiralling around $\vec{B}$. We consider two special cases.

1. If $\vec{E} \perp \vec{B}$, i.e., $E_z = 0$ and $v_{z_0} = 0$, then the trajectory is a circle of constant radius R, relative to the frame of axes x', y' translating with uniform velocity relative to the frame of axes x, y (Fig. 6.10b), which is described with constant $\dot{\phi}' = -\omega = -qB/m$. The velocity of its centre $\vec{v}^* = \vec{E} \times \vec{B}/B^2$ is perpendicular to $\vec{E}$, i.e., on an average, surprisingly, the particle does not move along $\vec{E}$ but, perpendicular to it.

2. If $E_y = E_z = 0,\ v_{z_0} = 0$ and $x(0) = y(0) = z(0) = 0$, then eq. (6.30) yields the cycloidal trajectory (Fig. 6.10c):

$$x = A_1(1 - \cos\omega t), \qquad y = A_1 \sin\omega t - E_x t/B, \qquad z = 0 \quad \text{with } A_1 = \frac{Bv_{y_0} + E_x}{\omega B}$$

This motion of electrons is utilised in the *magnetron*, which is an electronic tube used to produce high-frequency radio waves, as in microwave ovens.

6.3.3 Effect of a weak magnetic field on motion in an electric field

For a frame 1 ($\vec{\mathrm{e}}_i$) rotating at $\vec{\omega}$ relative to an inertial frame I ($\vec{\mathrm{E}}_i$) [Fig. 6.11] we have,

$$\dot{\vec{r}}_{|I} = \vec{\omega}\times\vec{r} + \vec{v}_{|1}, \qquad \ddot{\vec{r}}_{|I} = \dot{\vec{\omega}}\times\vec{r} + \vec{\omega}\times(\vec{\omega}\times\vec{r}) + 2\vec{\omega}\times\vec{v}_{|1} + \vec{a}_{|1}$$

The equation of motion $m\vec{a} = q\vec{E} + q\vec{v}\times\vec{B}$ is expressed in terms of frames I and 1 as

$$\ddot{\vec{r}}_{|I} = \frac{q}{m}\,[\vec{E} + \dot{\vec{r}}_{|I}\times\vec{B}] \tag{6.31}$$

$$\Rightarrow\ \vec{a}_{|1} = \frac{q}{m}\,\vec{E} - \dot{\vec{\omega}}\times\vec{r} + \vec{v}_{|1}\times\left(2\vec{\omega} + \frac{q}{m}\,\vec{B}\right) + \frac{q}{m}(\vec{\omega}\times\vec{r})\times\vec{B} - \vec{\omega}\times(\vec{\omega}\times\vec{r}) \tag{6.32}$$

We choose frame 1 such that the term involving $\vec{v}_{|1}$ gets eliminated from eq. (6.32) $\Rightarrow$

$$2\vec{\omega} + \frac{q}{m}\,\vec{B} = 0 \quad\Rightarrow\quad \vec{\omega} = -\frac{q}{2m}\,\vec{B}, \qquad \dot{\vec{\omega}} = -\frac{q}{2m}\,\dot{\vec{B}} \tag{6.33}$$

$$\text{eq. (6.32)}\ \Rightarrow\qquad \ddot{\vec{r}}_{|1} = \frac{q}{m}\,\vec{E} + \frac{q^2}{4m^2}\,\vec{B}\times(\vec{B}\times\vec{r}) + \frac{q}{2m}\,\dot{\vec{B}}\times\vec{r} \tag{6.34}$$

$$\text{If}\quad \frac{q^2B^2r}{m^2} \ll \frac{qE}{m} \quad\text{and}\quad \frac{q\dot{B}r}{2m} \ll \frac{qE}{m}, \quad \text{i.e.,}$$

$$\text{if}\quad B \ll \left|\frac{mE}{qr}\right|^{1/2} \quad\text{and}\quad \dot{B} \ll \left|\frac{2E}{r}\right| \tag{6.35}$$

$$\text{then}\quad \text{eq. (6.34)} \quad\Rightarrow\quad \ddot{\vec{r}}_{|1} \simeq q\vec{E}/m \tag{6.36}$$

Fig. 6.11

The conditions for its validity are satisfied by a weak $\vec{B}$ for which both $\vec{B}$ and $\dot{\vec{B}}$ are small. The effect of a weak magnetic field is to make the trajectory under $\vec{E}$ alone to precess about $\vec{B}$ at a rate $\vec{\omega} = -q\vec{B}\,/2m$. This is known as the *Larmor precession.*

If an electron of charge $-e$ is in an orbit about a nucleus of charge e', then $\vec{E} = e'\vec{r}/r^3$ $\Rightarrow \ddot{\vec{r}} = -ee'\vec{r}/mr^3$. The trajectory is an ellipse under inverse square attractive central force. The addition of a weak $\vec{B}$ causes this ellipse to precess slowly with $\vec{\omega} = (e/2m)\vec{B}$ about an axis through the nucleus if the conditions of eq. (6.35) are satisfied, i.e.,

$$\text{if}\quad B = \frac{2m\omega}{e} \ll \sqrt{\frac{mE}{er}} = \sqrt{\frac{me'}{er^3}} \simeq \frac{m}{e}\sqrt{\frac{ee'}{ma^3}} = \frac{m}{e}\left(\frac{2\pi}{T}\right), \quad \text{i.e.,}\quad \omega \ll \frac{\pi}{T}$$

where $T = 2\pi\sqrt{ma^3/ee'}$ is the period of elliptic orbit of semi-major axis a, obtained from eq. (6.17) with GMm replaced by ee',

$$\text{and if}\ \ \dot{B} = \frac{2m\dot{\omega}}{e} \ll \frac{2E}{r} = \frac{2m}{e}\left(\frac{ee'}{mr^3}\right) \simeq \frac{2m}{e}\left(\frac{ee'}{ma^3}\right) = \frac{2m}{e}\left(\frac{4\pi^2}{T^2}\right), \quad \text{i.e.,}\quad \dot{\omega} \ll \frac{4\pi^2}{T^2}$$

C. EFFECT OF EARTH'S ROTATION

6.4 Motion Relative to a Non-Inertial Frame

Consider a non-inertial frame 1 having angular velocity $\vec{\omega}$ and angular acceleration $\dot{\vec{\omega}}$ with reference to inertial frame I and let a location A in it have acceleration $\vec{a}_{A|I}$ (Fig. 6.12). The equation of motion of the centre of mass C of a body of mass m is

$$m\vec{a}_{C|I} = m[\vec{a}_{A|I} + \dot{\vec{\omega}}\times\vec{r}_{CA} + \vec{\omega}\times(\vec{\omega}\times\vec{r}_{CA}) + 2\vec{\omega}\times\vec{v}_{C|1} + \vec{a}_{C|1}] = \vec{F}$$

$$m\vec{a}_{C|1} = \vec{F}\underbrace{-m\vec{a}_{A|I}-m\dot{\vec{\omega}}\times\vec{r}_{CA}-m\vec{\omega}\times(\vec{\omega}\times\vec{r}_{CA})-2m\vec{\omega}\times\vec{v}_{C|1}}_{\text{pseudo 'forces'}} \tag{6.37}$$

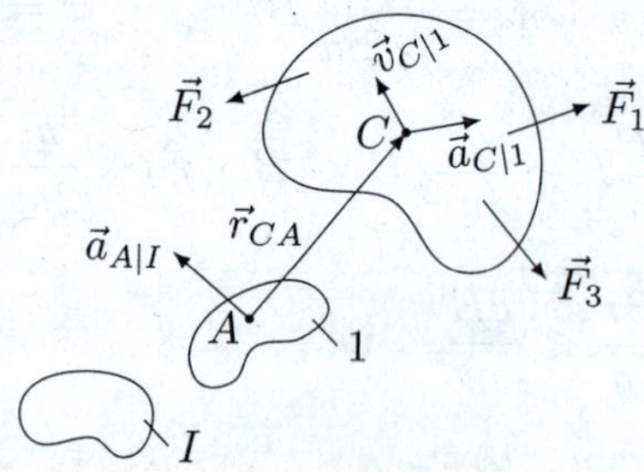

Fig. 6.12

Thus, $m\vec{a}_{C|1}$ equals the sum of the actual physical external force $\vec{F}$ and four pseudo 'forces' / inertia forces / fictitious forces which are in terms of the acceleration of the frame, angular acceleration of the frame, angular velocity of the frame and velocity relative to frame 1. The pseudo 'force', $-m\vec{\omega} \times (\vec{\omega} \times \vec{r}_{CA})$, is called the *centrifugal force* since *centium* and *fugere* in Latin mean 'centre' and 'to flee' and the pseudo force $-2m\vec{\omega} \times \vec{v}_{C|1}$ is called the *Coriolis force* after its discoverer. These are called pseudo forces since they are not due to physical interaction with other bodies and nor can we locate the other body on which opposite force acts. For a frame T translating relative to I, the only inertial force is $-m\vec{a}_{|T}$. For a frame rotating relative to I at constant $\vec{\omega}$ about a fixed axis, the only inertia forces are the centrifugal and the Coriolis forces.

6.5 Effect of Earth's Rotation

For most engineering systems, a frame fixed to the earth is a very good approximation to an inertial frame. When the distance covered is large, as in the flight of rockets, flow of rivers, movement of winds, formation of cyclones and ocean-currents, the frame attached to the earth can no longer be considered inertial and a frame I attached to the centre of the earth with its orientation fixed with reference to distant 'fixed' stars (Fig. 6.13a) is modelled as an inertial frame. In this section, we model the earth as a rigid sphere of radius $R = 6370\,\text{km}$, with a radially symmetric mass distribution, rotating at uniform angular velocity ω about its polar axis *fixed* in I. Its non-spherical shape, the local variation in its mass density and slight wobble (nutation) of its polar axis are neglected. Choose a non-inertial frame 1 attached to the earth at point O on the surface (Fig. 6.13a) with the axes x, y, z embedded in it. The frame 1 has $\vec{a}_{O|I}$, angular velocity $\vec{\omega} = \omega\,\text{E}$ and angular acceleration $\dot{\vec{\omega}} = \vec{0}$ relative to frame I.

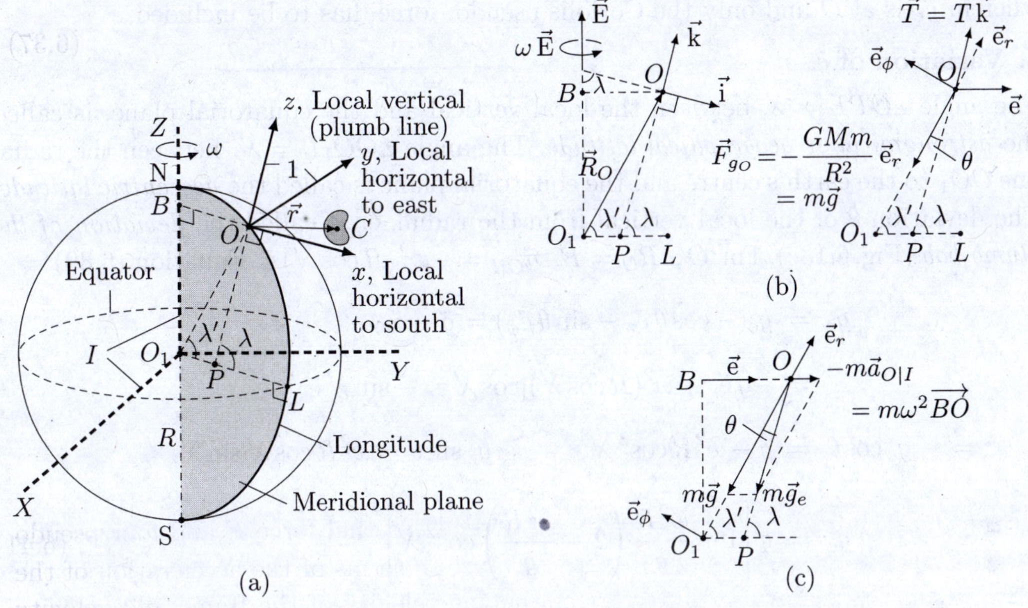

Fig. 6.13

Consider the motion of a body of mass m and size $\ll R$, under gravitational force $\vec{F}_g$ and other external forces $\vec{P}$. Let $\vec{r}_{CO} = \vec{r}$, $\vec{v}_{C|1} = \vec{v}_{|1}$, $\vec{a}_{C|1} = \vec{a}_{|1}$. Equation (6.37) $\Rightarrow$

$$m\vec{a}_{|1} = \vec{F}_g + \vec{P} - m\vec{a}_{O|I} - m\,\vec{\omega} \times (\vec{\omega} \times \vec{r}) - 2m\,\vec{\omega} \times \vec{v}_{|1} \tag{6.38}$$

We simplify this equation by including only the predominant effect as follows.

1. Effective Gravitational Acceleration g_e at O

For a stationary plumb-bob at O (Fig. 6.13b), $\vec{r} = \vec{0}$, $\vec{v}_{|1} = \vec{0}$, $\vec{a}_{|1} = \vec{0}$, the gravitational force $\vec{F}_g$ at O is $\vec{F}_{gO} = m\vec{g}$. The tension $\vec{T}$ is in the direction of the plumb line which is defined as the *local vertical* at O. Applying eq. (6.38) : $\vec{0} = \vec{F}_{gO} + \vec{T} - m\vec{a}_{O|I} \Rightarrow$

$$\vec{T} = -[\vec{F}_{gO} - m\vec{a}_{O|I}] = -m\vec{g}_e \text{ (say)} \qquad \Rightarrow \qquad \vec{g}_e = \vec{g} - \vec{a}_{O|I} \tag{6.39}$$

where $\vec{g}_e$ is the effective gravitational acceleration at O. In general, $\vec{g}_e$ differs from $\vec{g}$, except when O is on the polar axis.

2. Equation of Motion

For small values of $|\vec{r}| \ll R$, $\vec{F}_g - m\vec{a}_{O|I} \simeq \vec{F}_{gO} - m\vec{a}_{O|I} = m\vec{g}_e$. Since $\omega = 7.29 \times 10^{-5}$ rad/s $\Rightarrow \omega^2 r \ll \omega v$ holds in practice, we neglect the centrifugal term compared to the Coriolis term in eq. (6.38) and obtain

$$m\vec{a}_{|1} \simeq \vec{P} + m\vec{g}_e - 2m\vec{\omega} \times \vec{v}_{|1} \tag{6.40}$$

wherein $\vec{g}_e$ is at O and only the Coriolis pseudo 'force' has to be included.

3. Variation of $\vec{g}_e$

The angle $\angle OPL = \lambda$, between the local vertical and the equatorial plane, is called the *astronomical* or *geographical latitude.* The angle $\angle OO_1L = \lambda'$, between the radial line OO_1 to the earth's centre and the equatorial plane is called the *geocentric latitude.* The deviation θ of the local vertical from the radial $\vec{e}_r$ is called the *deviation of the plumb bob* (Fig. 6.13c). For O: $R_O = R$, $\vec{a}_{O|I} = -\omega^2(R\cos\lambda')\,\vec{e}$. Equation (6.39) $\Rightarrow$

$$\begin{aligned}
\vec{g}_e &= g_e(-\cos\theta\,\vec{e}_r - \sin\theta\,\vec{e}_\phi) = \vec{g} - \vec{a}_{O|I} \\
&= -g\,\vec{e}_r + \omega^2(R\cos\lambda')[\cos\lambda'\,\vec{e}_r - \sin\lambda'\,\vec{e}_\phi] \\
\Rightarrow \quad g_e\cos\theta &= g - \omega^2 R\cos^2\lambda', \qquad g_e\sin\theta = \omega^2 R\cos\lambda'\sin\lambda'
\end{aligned}$$

$$g_e = g\left[1 - \frac{\omega^2 R}{g}\left(2 - \frac{\omega^2 R}{g}\right)\cos^2\lambda'\right]^{1/2} \tag{6.41}$$

$$\tan\theta = \frac{(\omega^2 R/g)\cos\lambda'\sin\lambda'}{1 - (\omega^2 R/g)\cos^2\lambda'} \tag{6.42}$$

$$\begin{aligned}
g &= \frac{GM}{R^2} = \frac{6.670 \times 10^{-11}(5.976 \times 10^{24})}{(6370 \times 10^3)^2} = 9.823\,\text{m/s}^2 \\
\omega^2 R &= (7.292 \times 10^{-5})^2\, 6.370 \times 10^6 = 0.03387\,\text{m/s}^2 \\
\omega^2 R/g &= 0.003448 \ll 1
\end{aligned}$$

Equation (6.42) $\Rightarrow$ $\theta = 0$ for $\lambda' = 0$ (equator), and for $\lambda' = \pm\pi/2$ (poles); $\theta > 0$ for $0 < \lambda' < \pi/2$, $\theta < 0$ for $-\pi/2 < \lambda' < 0$, and $|\theta|_{\max} = 0.099°$ for $\lambda' = \pm 44.9°$.

$$g_0 = g_e(\lambda' = 0) = g_{e_{\min}} = 9.789\,\text{m/s}^2, \qquad g(\lambda' = \pm\pi/2) = g_{e\max} = g = 9.823\,\text{m/s}^2$$

The international gravity formula, based on spheroidal shape of the earth and its actual measurement, is

$$g_e = 9.780[1 + 0.005288 \sin^2 \lambda - 0.000\,006 \sin^2 2\lambda]\,\mathrm{m/s^2} \tag{6.43}$$

which shows that the earth's rotation [vide eq. (6.41), using $\omega^2/Rg \ll 1$ and binomial theorem] accounts for almost 2/3rd of the actual variation in g_e. Since $|\theta| < 0.099°$ and $(g_{\max} - g_0)/g_0 = 0.0034$ (0.34% variation), for most engineering applications we use $\theta \simeq 0$, $g_e \simeq \bar{g}$ (an average value of $9.806\,\mathrm{m/s^2}$).

4. Motion of a Fluid Parallel to the Earth's Surface

The effect of earth's rotation is important in meteorology and oceanography. Choose the z-axis along the *local vertical* at latitude λ at O and the axes x, y in the *local horizontal plane* with x-axis pointing south in the meridional plane O_1NOS (Fig. 6.14a). $\vec{\omega} = \omega\,\vec{\mathrm{E}} = \omega(-\cos\lambda\,\vec{\mathrm{i}} + \sin\lambda\,\vec{\mathrm{k}})$. Let $\vec{v}_{|1} = \vec{v}$, $\vec{a}_{|1} = \vec{a}$. The equation of motion (6.40)⇒

$$m\vec{a} = \vec{P} + m\vec{g}_e - 2m\,\vec{\omega} \times \vec{v} \tag{6.44}$$

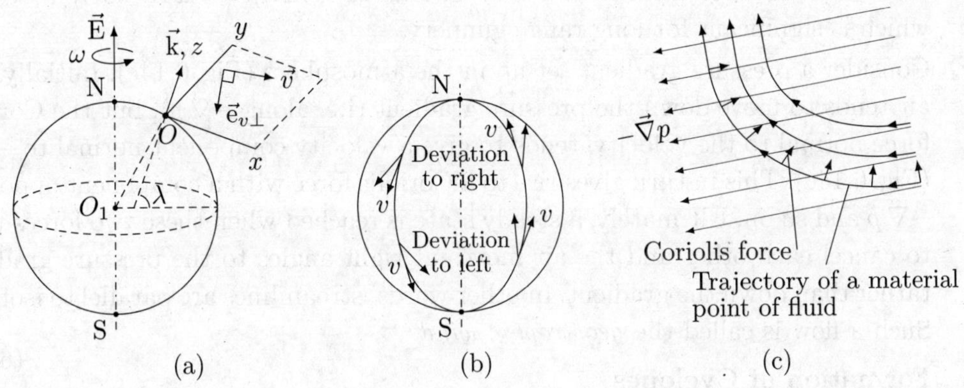

Fig. 6.14

For a fluid moving horizontally at latitude λ, $\vec{v} = v_x\,\vec{\mathrm{i}} + v_y\,\vec{\mathrm{j}}$ and

$$-2m\,\vec{\omega} \times \vec{v} = -2m\omega \sin\lambda\,\vec{\mathrm{k}} \times \vec{v} + 2m\omega\cos\lambda\,\vec{\mathrm{i}} \times \vec{v} \tag{a}$$

The second term on the right hand side of eq. (a) is in the vertical direction $\vec{\mathrm{k}}$. The contribution of the first term to the horizontal acceleration of the fluid is

$$-2\omega\sin\lambda\,\vec{\mathrm{k}} \times \vec{v} = 2\omega\sin\lambda\,\vec{v} \times \vec{\mathrm{k}} = 2\omega v \sin\lambda\,\vec{\mathrm{e}}_{v\perp} \tag{b}$$

where $\vec{\mathrm{e}}_{v\perp} = \vec{v} \times \vec{\mathrm{k}}\,/v$ is a unit vector normal to $\vec{v}$ and towards its right in the horizontal plane. The magnitude of this term in eq. (b), at a given location is independent of the direction of $\vec{v}$. Hence, the flow (Fig. 6.14b) deviates to its right in the northern hemisphere ($\lambda > 0 \Rightarrow \sin\lambda > 0$) and to its left in the southern hemisphere ($\lambda < 0 \Rightarrow \sin\lambda < 0$). This small deviation becomes noticeable over long distances. Based on these deviations, the following inferences can be drawn.

i. The air blowing towards the equator has a westward acceleration given by eq. (b). The wind blows as north-east trade wind in the northern hemisphere and as the south-east trade wind in the southern hemisphere.

ii. The right banks of the rivers flowing in the northern hemisphere and the left banks in the southern hemisphere are washed away, provided no strong local effects are present. This is known as *Ber's law*.

iii. The rightward/leftward (to the direction of motion) effect of the Coriolis force in the northern/southern hemispheres, explains the observed phenomenon of more wear on the right-hand rails than on the left-hand ones in the northern hemispheres on double-track railways wherein the train always goes in one direction.

iv. The deviation of the almost flat trajectory of a rifle bullet moving with speed v through distance H in time $t \simeq H/v$, is $\frac{1}{2}(2\omega v \sin\lambda)t^2 = \omega H^2 \sin\lambda/v \propto H^2$, which is significant for long range gunnery.

v. Consider a pressure gradient set up in the atmosphere (Fig. 6.14c). Initially, the air tends to move down the pressure gradient (i.e. along $-\vec{\nabla}\, p$), but the Coriolis force normal to the velocity, tends to give a velocity component normal to $-\vec{\nabla}\, p$ (Fig. 6.14c). This in turn gives rise to a Coriolis force with a component opposing $-\vec{\nabla}\, p$ and so on. Ultimately, a steady state is reached when these two forces tend to cancel each other and the air moves at right angles to the pressure gradient rather than down the gradient. In other words, streamlines are parallel to isobars! Such a flow is called the *geostrophic wind*.

vi. **Formation of Cyclones**
When a region of low pressure is formed in the atmosphere, may be due to differential heating, the wind blowing towards it along $-\vec{\nabla}\, p$ gets deflected sideways to the right in the northern hemisphere. This sets up a circulatory motion of the wind in addition to its motion along $-\vec{\nabla}\, p$ which generates a cyclone around the region of low pressure (Fig. 6.15a), with the wind circulating counter-clockwise along the isobars (as seen from above). Similarly, clockwise cyclones are generated around zones of low pressure in the southern hemisphere (Fig. 6.15b). The converse is true for regions of high pressure (Fig. 6.15c). No cyclonic wind systems form at the equator.

The navigation capability of migrating birds is amazing. In spite of the Coriolis drift and atmospheric disturbances, they cover large distances relative to the rotating earth to reach the desired spot. They seem to be more ingenious and knowledgeable than scientists in being able to 'monitor' their deviation from the desired path and in being able to incorporate a 'control mechanism' to correct it.

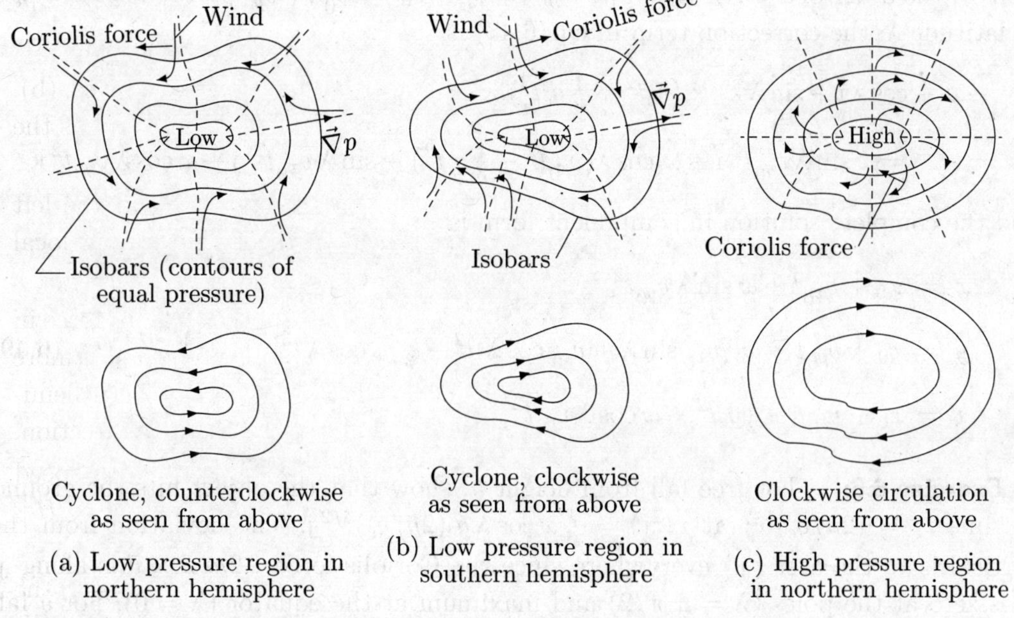

(a) Low pressure region in northern hemisphere
(b) Low pressure region in southern hemisphere
(c) High pressure region in northern hemisphere

Fig. 6.15

5. Approximate Solution for Free Fall with reference to Earth

For $\vec{P} = \vec{0}$, equation of motion (6.44) reduces to

$$\vec{a} = \vec{g}_e - 2\vec{\omega} \times \vec{v} \tag{6.45}$$

This equation is correct to the first order in ω. So, we solve it to the first order, for $\vec{r}(0) = \vec{r}_0$, $\vec{v}(0) = \vec{v}_0$, by successive approximations, using $\vec{v}$ from the previous approximation on the right hand side of eq. (6.45). For 1st approximation, using $\vec{v} = \vec{0}$ on right hand side of eq. (6.45), i.e., neglecting the effect of Coriolis force, we have:[1]

$$\vec{a} = \vec{g}_e \qquad \Rightarrow \qquad \vec{v} = \vec{g}_e t + \vec{v}_0 \tag{1}$$

For the 2nd approximation, $\vec{v}$ from eq. (1) is used on the right hand side of eq. (6.45):

$$\vec{a} = \vec{g}_e - 2\vec{\omega} \times (\vec{g}_e t + \vec{v}_0) \tag{6.46}$$

$$\Rightarrow \quad \vec{v} = \vec{v}_0 + \vec{g}_e t \underline{- \vec{\omega} \times (\vec{g}_e t^2 + 2\vec{v}_0 t)} \tag{6.47}$$

$$\vec{r} = \vec{r}_0 + \vec{v}_0 t + \tfrac{1}{2}\vec{g}_e t^2 \underline{- \vec{\omega} \times (\tfrac{1}{3}\vec{g}_e t^3 + \vec{v}_0 t^2)} \tag{6.48}$$

[1]It is easy to show that using $\vec{v} = \vec{0}$ as the initial guess does not affect the next approximation.

The underlined terms are the corrections, due to earth's rotation, to the usual results for uniform acceleration. The 3rd approximation is not needed since it involves corrections of the order of ω^2. For $\vec{r}_0 = x_0\,\vec{i} + y_0\,\vec{j} + z_0\,\vec{k}$, $\vec{v}_0 = v_{x0}\,\vec{i} + v_{y0}\,\vec{j} + v_{z0}\,\vec{k}$, $\vec{g}_e = -g_e\,\vec{k}$ at latitude λ, the correction term in eq. (6.48) is

$$-\omega(-\cos\lambda\,\vec{i} + \sin\lambda\,\vec{k}) \times (\vec{v}_0 t^2 + \tfrac{1}{3}\,\vec{g}_e t^3)$$

$$= \omega\sin\lambda\, v_{y0} t^2\,\vec{i} - \omega[\cos\lambda(v_{z0}t^2 - \tfrac{1}{3}\,g_e t^3) + \sin\lambda\, v_{x0} t^2]\,\vec{j} + \omega\cos\lambda\, v_{y0} t^2\,\vec{k}$$

and the complete solution in component form is

$$\begin{aligned} x &= x_0 + v_{x0}t + \omega\sin\lambda\, v_{y0}t^2 \\ y &= y_0 + v_{y0}t - \omega[(v_{x0}\sin\lambda + v_{z0}\cos\lambda)t^2 - \tfrac{1}{3}\,g_e\cos\lambda\, t^3] \\ z &= z_0 + v_{z0}t - \tfrac{1}{2}\,g_e t^2 + \omega\cos\lambda\, v_{y0}t^2 \end{aligned} \tag{6.49}$$

§ **Exercise 6.6:** For free fall from height h, show that the object hits the ground at time $t_1 = (2h/g_e)^{1/2}$ at $\vec{r}(t_1) = \frac{1}{3}\,\omega\cos\lambda\, g_e(2h/g_e)^{3/2}\,\vec{j}$. The deviation from the vertical is to the east ($\vec{j}$) everywhere since the Coriolis force $-2m\,\vec{\omega}\times\vec{v}$ is along $\vec{j}$. It is zero at the poles ($\lambda = \pm\pi/2$) and maximum at the equator ($\lambda = 0$). For a fall of 1000 m at $\lambda = 45°$, the deviation is only 49 cm which is difficult to separate from extraneous disturbances in the atmosphere.

§ **Exercise 6.7:** Prove that the maximum height h attained by a projectile thrown with $\vec{v}_0 = v_{x0}\,\vec{i} + v_{y0}\,\vec{j} + v_{z0}\,\vec{k}$ at $\vec{r}_0 = \vec{0}$, is $h = v_{z0}^2\,/\,(2g_e - 4\omega\cos\lambda\, v_{y0})$ which is greater (or less) than $v_{z0}^2/2g_e$ if v_{y0} is greater (or less) than 0.

§ **Exercise 6.8:** A projectile is fired at $\vec{r}_0 = \vec{0}$ with initial velocity v_0 in the x^*z-plane at an angle θ with the horizontal, where the x^*-axis is at an angle ϕ to the east of south (Fig. 6.16). Using $\vec{\omega} = \omega[-\cos\lambda(\cos\phi\,\vec{i}^* - \sin\phi\,\vec{j}^*) + \sin\lambda\,\vec{k}^*]$, $\vec{v}_0 = v_0\cos\theta\,\vec{i}^* + v_0\sin\theta\,\vec{k}^*$ and eq. (6.48), show that $\vec{r}(t)$ and the time of flight t_1 are given by

$$\begin{aligned} x^* &= v_0\cos\theta\, t - \omega\cos\lambda\sin\phi(v_0\sin\theta - \tfrac{1}{3}\,g_e t)t^2 \\ y^* &= -\omega\cos\lambda\cos\phi(v_0\sin\theta - \tfrac{1}{3}\,g_e t)t^2 - v_0\omega\cos\theta\,\sin\lambda\, t^2 \\ z^* &= v_0\sin\theta\, t - \tfrac{1}{2}\,g_e t^2 + v_0\omega\cos\lambda\sin\phi\cos\theta\, t^2 \\ t_1 &= (2v_0\sin\theta/g_e)(1 - 2v_0\omega\cos\lambda\cos\theta\sin\phi/g_e)^{-1} \\ &\simeq (2v_0\sin\theta/g_e)(1 + 2v_0\omega\cos\lambda\cos\theta\sin\phi/g_e) \\ &\simeq 2v_0\sin\theta/g_e \end{aligned}$$

Prove that at impact, the deviation from the vertical x^*z-plane through $\vec{v}_0$ is

$$y^*(t_1) \simeq -4v_0^3\omega \sin^2\theta(\cos\lambda\cos\phi\sin\theta + 3\cos\theta\sin\lambda)/3g_e^2$$

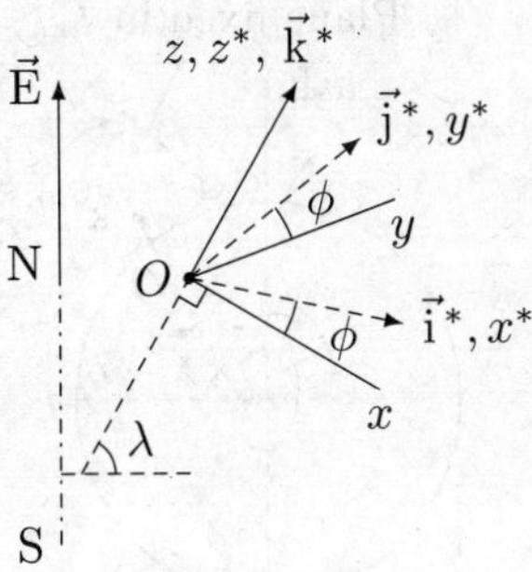

Fig. 6.16

[Hint: Use the value of t_1 to zero order in ω since $y^*(t_1)$ has a factor ωt_1^2 in its expression.] For the usual conditions of firing, the first term on the right hand side in the expression of $y^*(t_1)$ is smaller than the second. Hence, $y^* < 0$ for $\lambda > 0$ and $y^* > 0$ for $\lambda < 0$, i.e., *the deviation is to the right in the northern hemisphere and to the left in the southern hemisphere*. This is known as *Ferel's law*. Since there is a zero order term in ω in x^*, substitute the value of t_1 accurate up to first order in ω to prove that the range

$$x^*(t_1) \simeq (2v_0^2\cos\theta\sin\theta/g_e)\left[1 + \frac{2\omega v_0}{g_e}\cos\lambda\sin\phi(\cos\theta - \tfrac{1}{3}\sin^2\theta\sec\theta)\right]$$

The correction in the range due to earth's rotation vanishes for $\theta = 60°$. The effective g_e, time of flight, range and deviation from the vertical plane for $v_0 = 700\,\text{m/s}$, $\theta = 40°$, $\phi = 30°$, $\lambda = 45°$ are: $g_e = 9.8066\,\text{m/s}^2\,(9.81\,\text{m/s}^2)$,

$$t_1 = 92.03\,\text{s}\,(91.73\,\text{s}), \qquad x^*(t_1) = 49313\,\text{m}\,(49190\,\text{m}), \qquad y^*(t_1) = 289\,\text{m}\,(0)$$

The values listed in the parenthesis are the usual ones neglecting earth's rotation.

6. Foucault's Pendulum

After Copernicus presented his theory of planetary motion, scientists tried to devise an experiment on earth to demonstrate its daily rotation. Foucault (1819–1868) reasoned that if a plumb bob were swinging back and forth like a pendulum in a fixed plane in inertial frame I, then the earth would rotate under it. He publicly demonstrated his pendulum in 1851 in the Pantheon in Paris, with a steel wire 67 m long and a copper bob of mass 28 kg with a large period $[= 2\pi(L/g)^{1/2}]$ and small swing.

For a small swing, the position vector of the bob (Fig. 6.17a) suspended from B is

$$\vec{r} = r\,\vec{e}_r + z\,\vec{e}_z, \qquad \text{with } z, r \ll L, \qquad z = L - (L^2 - r^2)^{1/2} \simeq r^2/2L$$

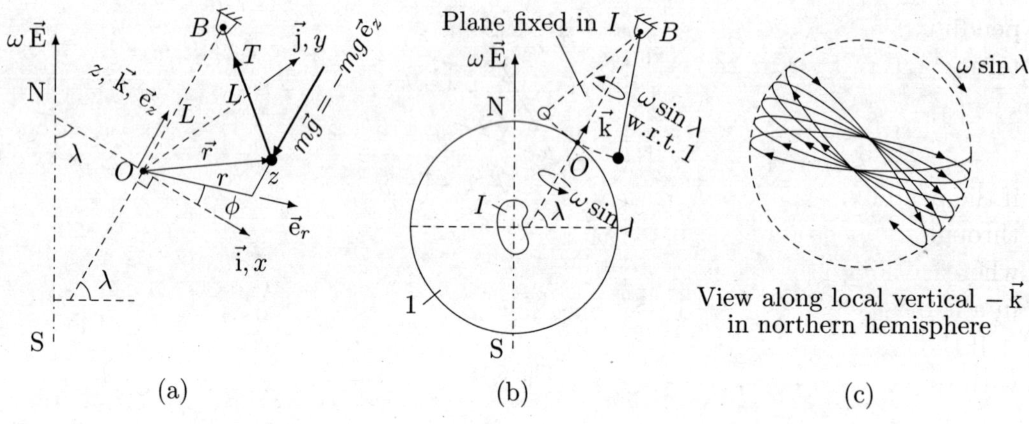

(a) (b) (c)

Fig. 6.17

Hence, if r is of first order, then z is of second order. The equation of motion (6.40) for the centre of mass of the bob, relative to frame 1 of the earth, is

$$m\vec{a}_{|1} = -mg\,\vec{e}_z + T[(L-z)\,\vec{e}_z - r\,\vec{e}_r]/L - 2m\vec{\omega}\times\vec{v}_{|1}$$

$$\Rightarrow \quad \ddot{\vec{r}} = (\ddot{r} - r\dot{\phi}^2)\,\vec{e}_r + \frac{1}{r}\frac{d}{dt}(r^2\dot{\phi})\,\vec{e}_\phi + \ddot{z}\,\vec{e}_z$$

$$= -g\,\vec{e}_z + [(L-z)\vec{e}_z - r\,\vec{e}_r]T/mL$$

$$-2\omega[-\cos\lambda(\cos\phi\,\vec{e}_r - \sin\phi\,\vec{e}_\phi) + \sin\lambda\,\vec{e}_z]\times(\dot{r}\,\vec{e}_r + r\dot{\phi}\,\vec{e}_\phi + \dot{z}\,\vec{e}_z)$$

$$\ddot{r} - r\dot{\phi}^2 = -2\omega(\dot{z}\cos\lambda\sin\phi - r\dot{\phi}\sin\lambda) - rT/mL \tag{1}$$

$$\frac{1}{r}\frac{d}{dt}(r^2\dot{\phi}) = -2\omega(\dot{z}\cos\lambda\cos\phi + \dot{r}\sin\lambda) \tag{2}$$

$$\ddot{z} = -g + 2\omega(r\dot{\phi}\cos\lambda\cos\phi + \dot{r}\cos\lambda\sin\phi) + T(L-z)/mL$$

$$\Rightarrow \quad (g - T/m) + [\ddot{z} - 2\omega\cos\lambda(r\dot{\phi}\cos\phi + \dot{r}\sin\phi) + Tz/mL] = 0 \tag{3}$$

For small swing, the expression within square brackets in eq. (3) is of second order and its value is much smaller compared to the other terms. Hence, $T \simeq mg$. Neglecting the second order $\dot{z}$ term compared to the $\dot{r}$ term on the right hand side of eq. (2):

$$\frac{d}{dt}(r^2\dot{\phi}) \simeq -2r\dot{r}\omega\sin\lambda = -\omega\sin\lambda\,\frac{dr^2}{dt}$$

$$\Rightarrow \quad r^2(\dot{\phi} + \omega \sin \lambda) = \text{constant} = C \quad \text{(say)} \tag{4}$$

If the initial conditions are such that the pendulum passes through the origin ($r = 0$), then eq. (4) yields $C = 0$ and $\dot{\phi} = -\omega \sin \lambda$, i.e., the vertical plane through the pendulum rotates at the rate $-\omega \sin \lambda$ with reference to the earth. Neglecting the $\dot{z}$ term compared to the $r\dot{\phi}$ term ($z \ll r$) in eq. (1) yields

$$\ddot{r} + (g/L)r = -r\omega^2 \sin^2 \lambda \simeq 0 \qquad \Rightarrow \qquad r = A \cos[(g/L)^{1/2} t]$$

if the pendulum is released from rest [$\dot{r}(0) = 0$] at $r(0) = A$. Hence, the vertical plane, through the pendulum, keeps rotating at the rate $-\omega \sin \lambda$ with reference to the earth, whereas the motion in this plane is simple harmonic. Note that the pendulum moves in a fixed plane (Fig. 6.17b) relative to inertial frame I attached to earth's centre.

If the initial conditions are such that the pendulum does not pass through the origin with the minimum distance from it $\ll L$, then eq. (4) implies

$$r^2 \dot{\theta} = C, \qquad \text{where} \qquad \dot{\theta} = \dot{\phi} + \omega \sin \lambda \tag{5}$$

Neglecting the $\dot{z}$ term compared to the $r\dot{\phi}$ term ($z \ll r$) in eq. (1), we get

$$\ddot{r} - r\dot{\phi}^2 \simeq 2r\dot{\phi}\omega \sin \lambda - rg/L$$

$$\text{and using eq. (5)} \Rightarrow \quad \ddot{r} - r\dot{\theta}^2 \simeq -r\omega^2 \sin^2 \lambda - rg/L$$

$$\Rightarrow \quad \ddot{r} - r\dot{\theta}^2 \simeq -rg/L \tag{6}$$

to first order. The equations of motion (5) and (6) in the r-θ plane are identical to those under an attractive central force varying linearly with the distance from the centre of force. Hence, the path in the r-θ plane is an ellipse. Equation (5) yields

$$\theta = \phi + \omega \sin \lambda \, t, \qquad \dot{\phi} = \dot{\theta} - \omega \sin \lambda$$

The effect of the earth's rotation on the instantaneous elliptic path of the spherical pendulum is to cause it to rotate at $-\omega \sin \lambda \, \vec{k}$ relative to the earth (Fig. 6.17c).

There is another much larger precession of the spherical pendulum when it moves as a conical pendulum. Such a motion can be made negligibly small by initially drawing the pendulum aside and letting it start from rest by burning the thread.

Review 6 A, B, C

1. For central force motion: the trajectory is coplanar with the centre of force with $rv_\phi = r_0 v_{\phi_0} = h_0$, areal velocity $= \frac{1}{2} h_0$.
2. For $\vec{F} = -\frac{GMm}{r^2} \vec{e}_r$: $\frac{1}{2}mv^2 - \frac{GMm}{r} = \frac{1}{2}mv_0^2 - \frac{GMm}{r_0} = E_0$, $v_e = (2GM/r_0)^{1/2}$,

$\dfrac{1}{r} = \dfrac{GM}{h_0^2} + D\cos\phi, \quad e = h_0^2 D/GM, \quad v_r = h_0 D \sin\phi, \quad v_c = (GM/r_0)^{1/2},$

$D = [(1/r_0 - GM/h_0^2)^2 + v_{r_0}^2/h_0^2]^{1/2}. \quad v_{\max}$ at $r_{\min}$, $v_{\min}$ at $r_{\max}$.

Hyperbola ($e > 1,\ E_0 > 0$), parabola ($e = 1,\ E_0 = 0$) or ellipse ($e < 1, E_0 < 0$).

Transfer to orbit of greater E_0, should be done at $r_{\min}$ with $\Delta\vec{v}$ along $\vec{e}_t$.

3. For a charged particle: $m\dot{\vec{v}} = q\vec{E} + q\vec{v}\times\vec{B}, \quad \frac{1}{2}mv^2 + qV =$ constant.

   (a) The trajectory for uniform $B\vec{k}$ is a circular helix with axis $\vec{k}$.

   (b) Motion under $B\vec{k}$ and $\vec{E}$ is circular helix of part (a) drifting across $\vec{B}$ (Hall drift) along $\vec{E}\times\vec{B}$ at $\vec{E}\times\vec{B}/B^2$ and additional acceleration of qE_z/m along $\vec{B}$.

4. $\vec{g}_e = \vec{g} - \vec{a}_{O|I}, \quad m\vec{a}_{|1} = \vec{P} + m\vec{g}_e - 2m\vec{\omega}\times\vec{v}_{|1}$.

   Fluid moving on earth deflects to right/left in northern/southern hemisphere.

   Free fall: $\vec{v} = \vec{v}_0 + \vec{g}_e t - \vec{\omega}\times(\vec{g}_e t^2 + 2\vec{v}_0 t),\ \vec{r} = \vec{r}_0 + \vec{v}_0 t + \frac{1}{2}\vec{g}_e t^2 - \vec{\omega}\times(\frac{1}{3}\vec{g}_e t^3 + \vec{v}_0 t^2)$.

6.6 Concept Review Questions 6 A, C

1. (a) Two satellites are in the same elliptic orbit at positions A and B at a small distance s ($s \ll R_1, s \ll R_2$) [Fig. Q6.1]. Find the distances s_1, s_2 between them when in positions A_1, B_1 and A_2, B_2. (b) Find the trajectory of a packet which falls off from a satellite with negligible velocity relative to it. The satellite is in an elliptical orbit under inverse square gravitational force.

   [(a) Rs/R_1, Rs/b (b) Same as satellite's orbit]

2. A satellite of mass m is in a circular orbit of radius R at an angle θ to the equatorial plane. Find the magnitude of the instantaneous impulse which should be applied to change it into an equatorial circular orbit.

   $[2m(GM/R)^{1/2}\sin\frac{1}{2}\theta]$

3. A point object (Fig. Q6.3) moves in the gravitational field of fixed spheres 1 and 2. Find the minimum speed v_e to escape from the given position on sphere 1 to reach sphere 2. $[[2G\{M/R + m/(d-R) - (\sqrt{m}+\sqrt{M})^2/d\}]^{1/2}]$

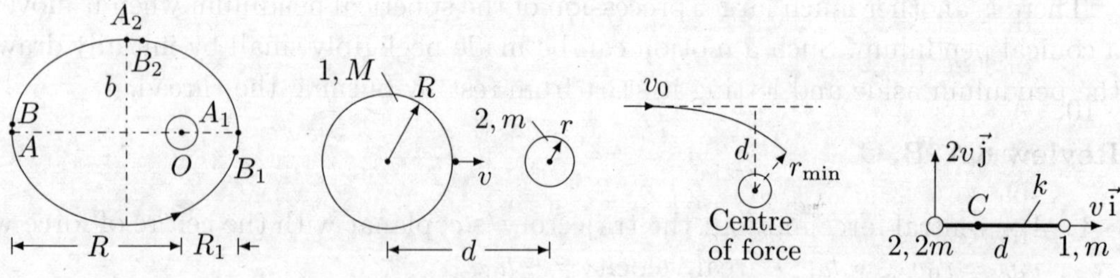

Fig. Q6.1 Fig. Q6.3 Fig. Q6.5 Fig. Q6.8

4. A ball of mass m is dropped down a fixed smooth vertical tube of height h at the equator. Find the contact force at time t after the ball is released from rest relative to the tube. Take $g_e = g - \omega^2 R$. Specify the side of the tube against which contact takes place. [$2m\omega g_e t$ to west, eastern side]
5. A mass-point moves in a repulsive central force field cmr^{-3}, where c is a constant. At a large distance from the centre of force, its speed is v_0 and the impact parameter is d (Fig. Q6.5). Find the closest distance of the mass-point from the centre of force. [$\sqrt{d^2 + c/v_0^2}$]
6. A satellite is in elliptic orbit around the earth with its maximum distance from the earth's centre being r_0. The eccentricity of the orbit is $\frac{1}{2}$. The universal gravitational constant is G and the mass of the earth is M. (a) Find the speed of the satellite when it is at the maximum distance r_0. (b) Find the minimum increment in speed that should be imparted to the satellite so that it escapes the earth's gravitational field. (c) Find the increment in velocity vector that should be given at the minimum radius so that the subsequent orbit is an ellipse with maximum radius r_0 and $e = 2/3$. Find the location of the minimum radial position in the new orbit from the starting position of the new orbit.
[$v = \sqrt{GM/r_0}$ (a) $0.707v$ (b) $0.3282v$ (c) $(-0.3893\vec{e}_\phi \pm 1.155\vec{e}_r)v$, $270°$ or $90°$]
7. Critics of Galileo argued that if earth rotates, then an object dropped from a high tower should fall to its west, but Galileo argued that it would fall to its east. Who was right? [Galileo]
8. Two skaters of mass m and $2m$ are connected by a rubber chord which is modelled to be a spring of zero length and stiffness k. The ground is smooth. (a) For separation d between them, each moves in a circle with reference to ground. Find the rate of rotation of the chord. (b) For the given initial conditions, find the motion of their centre of mass C and their motion with reference to the centre of mass frame (Fig. Q6.8). [(a) $\omega = (3k/2m)^{1/2}$
(b) $\vec{r}_C = \frac{1}{3}(vt\,\vec{\imath} + 4vt\,\vec{\jmath})$, $\vec{r}_1 = -2\vec{r}_2 = \frac{2}{3}d\cos\omega t\,\vec{\imath} + 2v(\vec{\imath} - 2\vec{\jmath})\sin\omega t/3\omega$]
9. Using the result for time period of an elliptical orbit under inverse square gravitational force GMm/r^2, obtain the result for the time of fall of an object from distance d from the centre of force. [Hint: The path is a degenerate elliptic orbit of semi-major axis $a = d$] [$\pi(0.5d)^{3/2}/\sqrt{GM}$]
10. A satellite in a circular orbit of radius r_0, under gravitational force and atmospheric drag force, is intermittently actuated by a thruster. The actuation is based on the motion of a small mass with reference to the satellite as the satellite changes its velocity. This is sensed and the thruster is actuated. The thruster provides a thrust of F for a total time of t_1 in n orbits. Find the drag force. [$Ft_1(GM/r_o^3)^{1/2}/2n\pi$]

11. Let r be the distance of a satellite from the centre of force. For a satellite in an elliptic orbit, find r and v at the end of the minor axis if the semi-major axis is a. (Hint : Use $E_0 = -GMm/2a$) $[a,\ (GM/a)^{1/2}]$

6.7 Practice Problems 6 A, C

1. A mass-point moves in a plane under a central force with velocity v_0 at radius r_0. Find the central force and the radial and circumferential components of the velocity in the general position if the trajectory is (a) $r = r_0 e^{\alpha\phi}$, (b) $r = r_0 \cos 2\phi$.
$[(a)\ -mv_0^2 \frac{r_0^2}{r^3},\ [\alpha, 1] v_0 r_0 / r(1+\alpha^2)^{1/2},\ (b)\ mv_0^2 r_0^2 (\frac{3}{r^3} - \frac{8r_0^2}{r^5}),\ -2\frac{r_0^2 v_0}{r^2} \sin 2\phi,\ \frac{v_0 r_0}{r}]$
2. A satellite to study Mars (mass M) has speed v at its nearest position distant r_0 from the centre of Mars. A probe is launched from it in this position by imparting it a radial velocity v_r relative to the satellite. Find v_r if the minimum distance of the probe is r_1 from the centre of Mars.
$[\{v^2(r_0^2/r_1^2 - 1) + 2GM(1/r_0 - 1/r_1)\}^{1/2}]$
3. A space shuttle is to dock with a space station which is in a circular orbit of radius $r_1 = 7000$ km around the earth. At burnout at A, the shuttle has a radial distance $r_0 = 6500$ km and its velocity v_0 forms an angle of 60° with the radial line. (a) Find v_0 so that the shuttle docks with the space station tangentially. (b) What impulse should the rockets impart to the shuttle of mass m just before docking? (c) If the shuttle is to be launched tangentially from another space station in a circular orbit of radius 8000 km, find the changes in its speed required at launching and docking and the relative positions of the orbiting space stations at the instant the shuttle is launched.
[(a) 17986 km/h (b) 12792m N.s (c) −864.1 km/h, −893.6 km/h, 19.62° behind]
4. A space vehicle approaches planet Jupiter with a trajectory of eccentricity 3 and comes to within 1600 km of its surface. Find the change in its circumferential velocity needed at its closest position so that it goes into an orbit around Jupiter with the minimum distance of 100 km from its surface. The radius of Jupiter is 69500 km and the acceleration of gravity at its surface is 27.67 m/s^2.
[−156918 km/h]
5. A satellite is launched parallel to the earth's surface at speed v_0 at radius r_0. Find (a) the eccentricity and the nature of the trajectory, (b) r_{max}, r_{min} and (c) v_{max}, v_{min}. (d) If the satellite goes into orbit, find the time taken (i) from the minimum altitude to the maximum altitude, (ii) for one orbit and (iii) from the maximum altitude to the end of the minor axis. (e) Find speed v at given radius r. What are the magnitudes of its radial and circumferential components?

Work out for three cases: (1) $v_0 = (1.5GM/r_0)^{1/2}$, (2) $v_0 = (2.5GM/r_0)^{1/2}$, (3) $v_0 = 25600\,\text{km/h}$, $r_0 = 7500\,\text{km}$, $R = 6400\,\text{km}$.

[(a) (1) 0.5 (2) 1.5 (3) 0.05477 (b) (1) $3r_0$, r_0 (2) ∞, r_0 (3) 7500, 6721 km

(c) (1), (2) : $(GM/r_0)^{1/2}[\sqrt{1.5},\ \sqrt{1/6};\ \ \sqrt{2.5},\ \sqrt{0.5}]$ (3) 28567, 25600 km/h

(d) (1) $(8.886,\ 17.77,\ 5.857)\sqrt{r_0^3/GM}$ (3) 0.826 h, 1.652 h, 0.4274 h

(e) $[v_0^2 + 2GM(1/r - 1/r_0)]^{1/2}$, $[v_0^2 + 2GM(1/r - 1/r_0) - h_0^2/r^2]^{1/2}$, h_0/r]

6. An earth satellite has to have an apogee and perigee of r_1 and r_2, respectively. Find its speed v_0 and the launching angle α_0 with the local horizontal if it is launched at radius r_0 from earth's centre.

$$[\{2GM(r_1 + r_2 - r_0)/r_0(r_1 + r_2)\}^{1/2},\ \ \cos^{-1}\{r_1 r_2/r_0(r_1 + r_2 - r_0)\}^{1/2}]$$

7. A satellite of mass m is in a circular orbit of radius r_0. Find the change in circumferential velocity that should be imparted to it if the time period of its subsequent orbit is T. Find the time of firing of engines of thrust P.

$$[\Delta v_\phi = (GM/r_0)^{1/2}[\sqrt{2}\{1 - (\pi^2 r_0^3/2T^2 GM)^{1/3}\}^{1/2} - 1],\ t = m\Delta v_\phi/P]$$

8. A satellite is in a circular orbit of radius r_1. Find the change in circumferential velocity required in the given position so that it lands at position θ with reference to this position. The radius of the earth is R.

$$[\{GM(1 - \cos\theta)R/r_1(r_1 - R\cos\theta)\}^{1/2} - (GM/r_1)^{1/2}]$$

9. (a) A train of mass m is moving towards the south at latitude $\lambda\,(> 0)$ with constant speed v_0. Find the force exerted on the track by the flanges on the inside of the wheels due to rotation ω of the earth. (b) Find the difference in the total vertical forces on the rails when the train travels to the east at latitude $\lambda\,(> 0)$ with constant speed v_0, as compared to when it travels to the west. (c) Rework part (b) for individual tracks. The height of centre of mass is h above the track. The distance between the wheels is b.

[(a) $2m\omega v_0 \sin\lambda$ to west, on western track (b) $\Delta(N_n + N_s) = -4m\omega v_0 \cos\lambda$

(c) $\Delta N_n,\ \Delta N_s = -2m\omega v_0 \cos\lambda \mp 4m\omega v_0 \sin\lambda\, h/b$]

10. At what angle to the vertical should a projectile be fired if its range in the vertical plane through the initial velocity is to be zero.

$$[\sin^{-1}(2\omega v_0 \cos\lambda \sin\phi/3g_e)]$$

11. An object is projected horizontally with velocity v_0 on the surface of a smooth horizontal plane at latitude λ. Show that owing to earth's angular velocity ω, it moves in a circle of radius $v_0/2\omega|\sin\lambda|$.

D. SYSTEMS WITH MASS FLOW

Euler's axioms $\vec{F} = \dot{\vec{p}}_{|I}$, $\vec{M}_O = \dot{\vec{H}}_{O|I}$ apply to any material system with a fixed well-identified quantity of matter—a solid, liquid, or gas. In fluid flow devices, rockets, blowers, jet aeroplanes, etc., flow of matter (mass) occurs across their boundaries. For such systems, it is convenient to consider mass flow across a well-defined boundary (a *control surface*) of a well-defined closed region of space (a *control volume*) which itself may be at rest or motion and may be changing in shape and size with time (Fig. 6.18a). We derive the relations for the rate of change of m, $\vec{p}$ and $\vec{H}_A$ for a control volume and apply them to several systems. All kinematics in Section D is with reference to an inertial frame I. This is not explicitly mentioned for brevity.

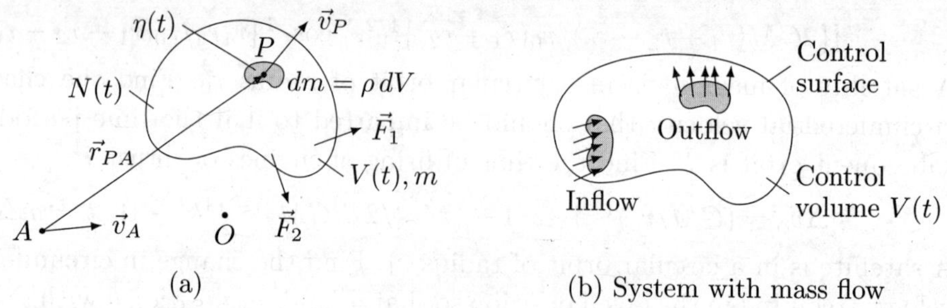

Fig. 6.18

6.8 Equations of Motion for a Control Volume

The conservation of mass and Euler's axioms involve the time derivative of an extensive property N (total mass, momentum and moment of momentum) of a well-identified material system of mass m which can be expressed in terms of the respective local intensive property η (which is the extensive property per unit mass):

$$N(t) = \int_m \eta(t) dm = \int_{V(t)} \eta(t) \rho \, dV$$

with $\eta = 1$, $\vec{v}_P$ and $\vec{r}_{PA} \times \vec{v}_{PA}$ for $N = m, \vec{p}$ and $\vec{H}_A$, respectively (Fig. 6.18b). Consider a control volume which at times t and $t + \Delta t$ occupies regions with volumes $V(t)$ and $V(t + \Delta t)$ of the inertial space with mass $m_V(t)$ and $m_V(t + \Delta t)$ within it (Fig. 6.19a,b). Let Δm_i be the mass that flows into the control volume and Δm_o be the mass that flows out of it in time Δt. Consider the material body B of mass m_B formed by the mass within the control volume $V(t)$ at time t, i.e., $m_B = m_V(t)$. By mass conservation, m_B is constant, hence, $\dot{m}_B = 0$.

Consider an entity $N_B(t)$ associated with body B and the same entity $N_V(t)$ associated with control volume $V(t)$. i.e. $N_B(t) = N_V(t)$. Using $m_V(t + \Delta t) = m_V(t) + \Delta m_i - \Delta m_0$, i.e., $m_B = m_V(t + \Delta t) - \Delta m_i + \Delta m_o$, we obtain for body B:

$$N_B(t+\Delta t) = N_V(t+\Delta t) - (N \text{ of } \Delta m_i \text{ at } t+\Delta t) + (N \text{ of } \Delta m_o \text{ at } t+\Delta t)$$

$$\Rightarrow \frac{N_B(t+\Delta t) - N_B(t)}{\Delta t} = \frac{N_V(t+\Delta t) - N_V(t)}{\Delta t} + \frac{(N \text{ of } \Delta m_o \text{ at } t+\Delta t) - (N \text{ of } \Delta m_i \text{ at } t+\Delta t)}{\Delta t}$$

$$\Rightarrow \quad \dot{N}_B(t) = \dot{N}_V(t) + (\text{net rate of flow of } N \text{ out of } V \text{ at } t) \tag{6.50}$$

in the limit as $\Delta t \to 0$. Applying eq. (6.50) to $m, \vec{p}$ and $\vec{H}_A$ yields

$$\begin{aligned} \dot{m}_B(t) &= \dot{m}_V(t) + (\text{net rate of flow of mass out of } V \text{ at } t) = 0 \\ \dot{\vec{p}}_B(t) &= \dot{\vec{p}}_V(t) + (\text{net rate of flow of } \vec{p} \text{ out of } V \text{ at } t) \\ \dot{\vec{H}}_{A_B}(t) &= \dot{\vec{H}}_{A_V}(t) + (\text{net rate of flow of } \vec{H}_A \text{ out of } V \text{ at } t) \end{aligned} \tag{6.51}$$

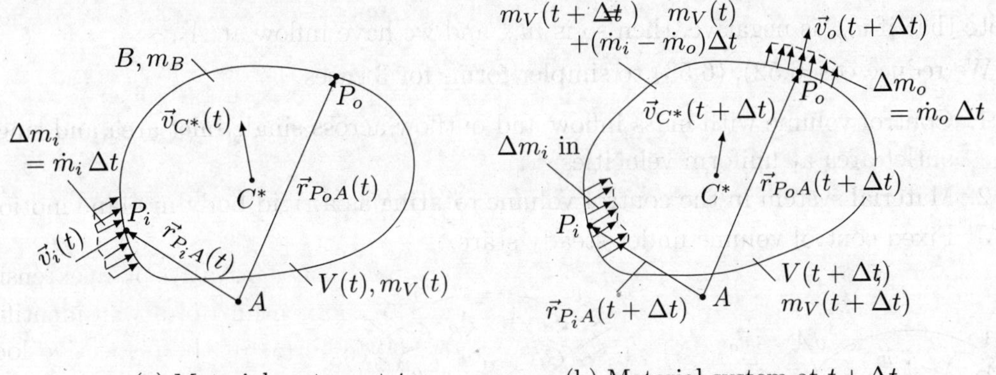

(a) Material system at t (b) Material system at $t+\Delta t$

Fig. 6.19

$\dot{m}_V(t) = \dot{m}_i - \dot{m}_o \Rightarrow$ eq. $(6.51)_1$, called the *continuity equation*, is satisfied. Let $\vec{F}$ be the total external force and $\vec{M}_A$ be the moment about a point A of the external loads on the body B at time t (i.e., on the control volume at t). For control volume $V(t)$, denote $m_V(t)$ as $m(t)$ with centre of mass C^* [*not the centroid of* $V(t)$] which coincides with the centre of mass C of body B at time t only, i.e., $\vec{v}_C = \vec{v}_{C^*}$, but in general $\vec{a}_C(t) \neq \vec{a}_{C^*}(t)$. For example, for steady flow in control volume of a fixed convergent nozzle, $\vec{v}_{C^*}(t) = \text{const.} \Rightarrow \vec{a}_{C^*} = \vec{0}$, but C accelerates since the outlet velocity is higher than that at the inlet. $p_V(t) = \int_{V(t)} \vec{v}_P \rho dV = m(t)\vec{v}_{C^*}(t)$. At time t, choose point $A \ni A \equiv C \equiv C^*$ or $\vec{a}_A(t) = \vec{0}$ or $\vec{a}_A(t)$ is along AC (AC^*). Using Euler's axioms for body B at time t ($\vec{F} = \dot{\vec{p}}_B,\ \vec{M}_A = \dot{\vec{H}}_{A_B}$) in eqs (6.51)$_{2,3}$ yields:

$$\dot{\vec{p}}_V = \frac{d}{dt}(m\vec{v}_{C^*}) = \vec{F} - (\text{net rate of flow of } \vec{p} \text{ out of } V) \tag{6.52}$$

$$\begin{aligned}\dot{\vec{H}}_{A_V} &= \frac{d}{dt}(\vec{H}_{C^*_V} + \vec{r}_{C^*A} \times m\vec{v}_{C^*A}) \\ &= \vec{M}_A - (\text{net rate of flow of } \vec{H}_A \text{ out of } V)\end{aligned} \tag{6.53}$$

Consider at time t, a fluid of density ρ_o moving with velocity $\vec{v}_0$ crosses an elementary area dA (with outward normal $\vec{n}_o$) of the area A_o on the control surface, at P_o whose velocity is $\vec{v}_{P_o}$ (Fig. 6.20). The mass outflow across dA depends on the component u_{on} of the velocity of the fluid relative to P_o in the direction of $\vec{n}_o$, i.e., $u_{on} = (\vec{v}_o - \vec{v}_{P_o}) \cdot \vec{n}_o$, since its tangential component does not cause any outflow. The mass outflow across dA in time dt is $\rho_o(u_{on}\, dt)\, dA$ and the entire rate of mass outflow $\dot{m}_o$ across A_o is:

$$\dot{m}_o = \int_{A_o} \rho_o u_{on}\, dA = \int_{A_o} \rho_o(\vec{v}_o - \vec{v}_{P_o}) \cdot \vec{n}_o\, dA \tag{6.54}$$

Note that if u_{on} is negative, then so is $\dot{m}_o$, and we have inflow at A_o.

We reduce eqs (6.52), (6.53) to simpler forms for 3 cases.

1. Control volume with mass inflow and outflow across single inlet area and single outlet area at uniform velocities.
2. Material system in the control volume rotating as a rigid body in plane motion.
3. Fixed control volume under steady state.

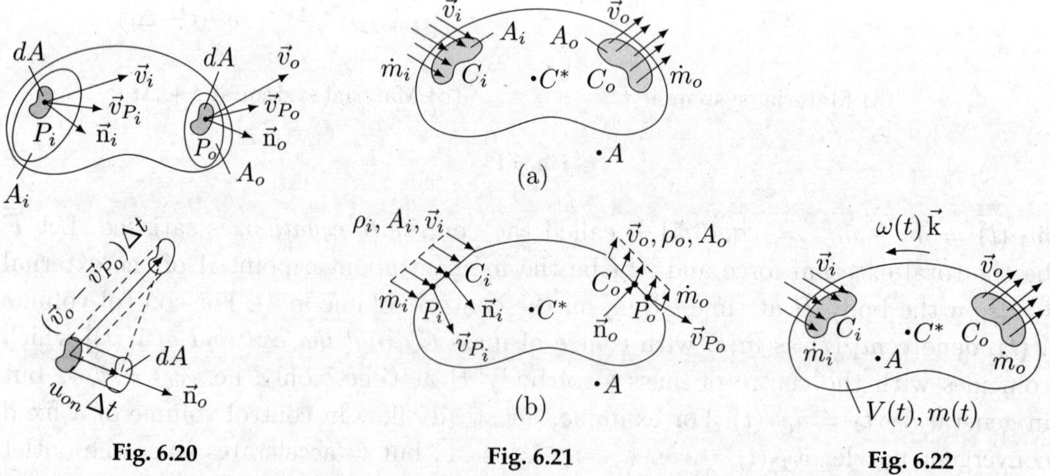

Fig. 6.20 **Fig. 6.21** **Fig. 6.22**

1. Consider a system with a single inlet of area A_i and a single outlet of area A_o. Let all mass inflow at rate $\dot{m}_i(t)$ at time t be with the same velocity $\vec{v}_i(t)$ and uniformly distributed across area A_i of the control surface; and all mass outflow

at rate $\dot{m}_o(t)$ be with the same velocity $\vec{v}_o(t)$ and uniformly distributed across area A_o of the control surface (Fig. 6.21a). The mass outflow rate $\dot{m}_o$ is obtained from eq. (6.54), while the mass inflow rate $\dot{m}_i$ is obtained from eq. (6.54) with subscript o replaced by i and $\vec{n}_i$ being the inward normal to the control surface at P_i. The net rate of *in*flow of mass $\dot{m}$, *out*flow of momentum $\dot{\vec{p}}^o$ and *out*flow of moment of momentum $\dot{\vec{H}}^o_A$ about point A are, respectively:

$$\begin{aligned} &\dot{m} = \dot{m}_i - \dot{m}_o, \qquad \dot{\vec{p}}^o = \dot{m}_o \vec{v}_o - \dot{m}_i \vec{v}_i \\ &\dot{\vec{H}}^o_A = \vec{r}_{C_oA} \times \dot{m}_o \vec{v}_{oA} - \vec{r}_{C_iA} \times \dot{m}_i \vec{v}_{iA} \end{aligned} \tag{6.55}$$

where C_i, C_o are the centroids of the projections of areas A_i, A_o normal to $\vec{v}_i$, $\vec{v}_o$, respectively and $\vec{v}_{iA} = \vec{v}_i - \vec{v}_A$, $\vec{v}_{oA} = \vec{v}_o - \vec{v}_A$. Using eqs (6.54), (6.55) and

$$\frac{d(m\vec{v}_{C^*})}{dt} = \dot{m}\vec{v}_{C^*} + m\vec{a}_{C^*} = (\dot{m}_i - \dot{m}_o)\vec{v}_{C^*} + m\vec{a}_{C^*}$$

eqs (6.52) and (6.53) yield

$$\vec{F} = \frac{d}{dt}(m\vec{v}_{C^*}) - \dot{m}_i \vec{v}_i + \dot{m}_o \vec{v}_o \tag{6.56}$$

$$\begin{aligned} \vec{M}_A = &\frac{d}{dt}(\vec{H}_{C^*} + \vec{r}_{C^*A} \times m\vec{v}_{C^*A}) \\ &- \vec{r}_{C_iA} \times \dot{m}_i(\vec{v}_i - \vec{v}_A) + \vec{r}_{C_oA} \times \dot{m}_o(\vec{v}_o - \vec{v}_A) \end{aligned} \tag{6.57}$$

Using $\dot{m} = \dot{m}_i - \dot{m}_o$, eq. (6.56) is expressed as

$$\vec{F} = m\vec{a}_{C^*} - \dot{m}_i(\vec{v}_i - \vec{v}_{C^*}) + \dot{m}_o(\vec{v}_o - \vec{v}_{C^*}) \tag{6.58}$$

For fluid inflow (outflow) (Fig. 6.21b) with uniform density ρ_i (ρ_o) and velocity $\vec{v}_i$ ($\vec{v}_o$) normal to a plane inlet area A_i (A_o) (of the control surface) with inward (outward) normal $\vec{n}_i$ ($\vec{n}_o$) and moving at $\vec{v}_{P_i}$ ($\vec{v}_{Po}$), the mass inflow (outflow) rates $\dot{m}_i$ ($\dot{m}_o$), obtained from eq. (6.54), are

$$\dot{m}_i = \rho_i A_i u_{in}, \qquad \dot{m}_o = \rho_o A_o u_{on} \tag{6.59}$$

where $\vec{u}_i = \vec{v}_i - \vec{v}_{P_i}$, $\vec{u}_o = \vec{v}_o - \vec{v}_{Po}$, $u_{in} = \vec{u}_i \cdot \vec{n}_i$, $u_{on} = \vec{u}_o \cdot \vec{n}_o$ are the velocities of fluid at inlet and exit relative to the inlet and exit areas of the control surface, and their components along the respective inward and outward normals.

2. If the control volume in case 1 is in *plane motion* with the material system of mass $m(t)$ in control volume $V(t)$ rotating as a rigid body, about a principal axis of inertia at A, with angular velocity $\vec{\omega}(t) = \omega(t)\vec{k}$ (Fig. 6.22) and A is a material point of $m(t)$, in addition to satisfying $A \equiv C \equiv C^*$, or $\vec{a}_A(t) = \vec{0}$ or $\vec{a}_A(t)$ along AC (AC^*), then $\vec{H}_{A_V}(t) = I^A_{zz}(t)\,\vec{\omega}(t)$ and eq. (6.57) $\Rightarrow$

$$\vec{M}_A = I_{zz}^A\,\dot{\vec{\omega}} + \dot{I}_{zz}^A\,\vec{\omega} - \vec{r}_{C_iA} \times \dot{m}_i(\vec{v}_i - \vec{v}_A) + \vec{r}_{C_oA} \times \dot{m}_o(\vec{v}_o - \vec{v}_A) \qquad (6.60)$$

3. **Fixed Control Volume Under Steady State**

For such a control volume fixed in I (Fig. 6.23a): (1) $m(t) = \text{constant}$, (2) $\vec{v}_{C^*}(t) = \text{constant} \Rightarrow \vec{p}_V(t) = m(t)\vec{v}_{C^*}(t) = \text{constant} \Rightarrow \dot{\vec{p}}_V = \vec{0}$, (3) $\vec{v}_P(t) = \text{constant} \Rightarrow \vec{H}_{C^*V}(t) = \text{constant} \Rightarrow \dot{\vec{H}}_{C^*V} = \vec{0}$,

$$\dot{\vec{H}}_{A_V} = \frac{d}{dt}(\vec{r}_{C^*A} \times m\vec{v}_{C^*A}) = \vec{v}_{C^*A} \times m\vec{v}_{C^*A} + \vec{r}_{C^*A} \times \dot{m}\vec{v}_{C^*A}$$

$$+\vec{r}_{C^*A} \times m\vec{a}_{C^*A} = -\vec{r}_{C^*A} \times m_V\vec{a}_A = \vec{0}$$

for $A \equiv C \equiv C^*$, or $\vec{a}_A = \vec{0}$, or $\vec{a}_A$ is along AC (AC^*). Hence, eqs (6.52) and (6.53) for case 1, with fixed control volume (Fig. 6.23b) under steady state, become

$$\vec{F} = \dot{\vec{p}}^o = \dot{m}_i(\vec{v}_o - \vec{v}_i) \qquad (6.61)$$

$$\vec{M}_A = \dot{\vec{H}}_A^o = [\vec{r}_{C_oA} \times (\vec{v}_o - \vec{v}_A) - \vec{r}_{C_iA} \times (\vec{v}_i - \vec{v}_A)]\dot{m}_i \qquad (6.62)$$

If point A is fixed in an inertial frame, then $\vec{v}_A = \vec{0}$ and eq. (6.62) reduces to

$$\vec{M}_A = [\vec{r}_{C_oA} \times \vec{v}_o - \vec{r}_{C_iA} \times \vec{v}_i]\dot{m}_i \qquad (6.63)$$

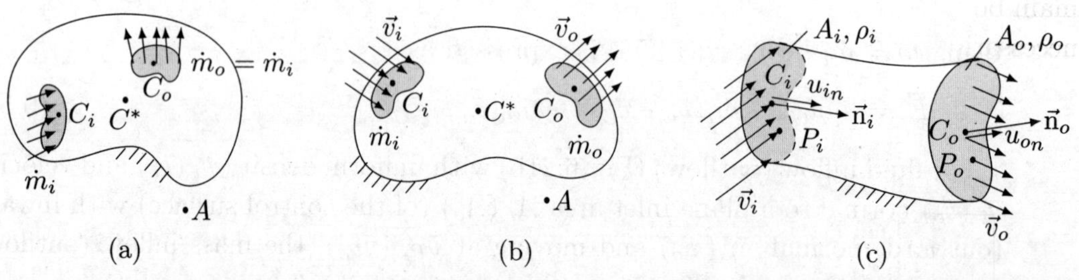

Fig. 6.23

For the fluid flow of Fig. 6.23c, the continuity equation for steady state mass flow across a fixed control volume, $\dot{m}_i = \dot{m}_o$ can be expressed as

$$\rho_i A_i |u_{in}| = \rho_o A_o |u_{on}| \qquad (6.64)$$

Flows of liquids and isothermal gas flows at low speeds (compared to the speed of sound) can be treated as incompressible. This is true for a *large* class of engineering problems. For an incompressible fluid flow, eq. (6.64) becomes

$$A_i |u_{in}| = A_o |u_{on}| \qquad (6.65)$$

Note: All round uniform atmospheric pressure p_a on the surface area A of control volume V does not contribute to $\vec{F}$ and $\vec{M}_A$, since it has null resultant at the origin O.

$$\vec{F}_R = \int_A -p_a\vec{n}\,dA = -p\int_A d\vec{A} = -p\vec{A} = \vec{0}$$

$$\vec{C}_{R_O} = \vec{M}_O = \int_A \vec{r}\times(-p_a\vec{n})dA = -p_a\int_A (x_i\,\vec{e}_i \times n_j\,\vec{e}_j)dA. \text{ Using divergence thm.,}$$

$$\vec{C}_{R_O} = -p_a\int_A \frac{\partial x_i}{\partial x_j}\,\vec{e}_i\times\vec{e}_j\,dA = -p_a\int_V \delta_{ij}\,\vec{e}_i\times\vec{e}_j\,dV = -p_a\int_V \vec{e}_j\times\vec{e}_j\,dV = \vec{0}$$

$\Rightarrow$ For pressure p at exit / entry, only gage pressure $p_g = p_e - p_a$ contributes to $\vec{F}$, $\vec{M}_A$.

6.9 Rocket

The fuel is burnt in the combustion chamber (Fig. 6.24a). The propellant is ejected at rate $\dot{m}_p$ with velocity $\vec{v}_o$. Let $\vec{v}$, $\vec{a}$ be the velocity and acceleration of main body of the rocket. Consider control volume to be the volume enclosed by the outer boundary of the rocket, having mass m at time t. Its centre of mass C^* shifts relative to main body at a very slow rate $\Rightarrow$ $\vec{v}_{C^*} \simeq \vec{v}$, $\vec{a}_{C^*} \simeq \vec{a}$. For $\dot{m}_i = 0$, $\dot{m}_o = \dot{m}_p$, eq. (6.58) becomes

$$\vec{F} = m\vec{a} + \dot{m}_o(\vec{v}_o - \vec{v}) = m\vec{a} + \dot{m}_p\vec{v}_{\text{rel}} \tag{6.66}$$

where $\vec{v}_{\text{rel}} = u_e\,\vec{e}$, with $\vec{e} = -\vec{v}/v$, is the velocity of the propellant relative to the main body of the rocket. Let p_e be the *gauge pressure* over the exit area A_e. p_e is not necessarily zero since the gases in the exhaust stream may not be fully expanded.

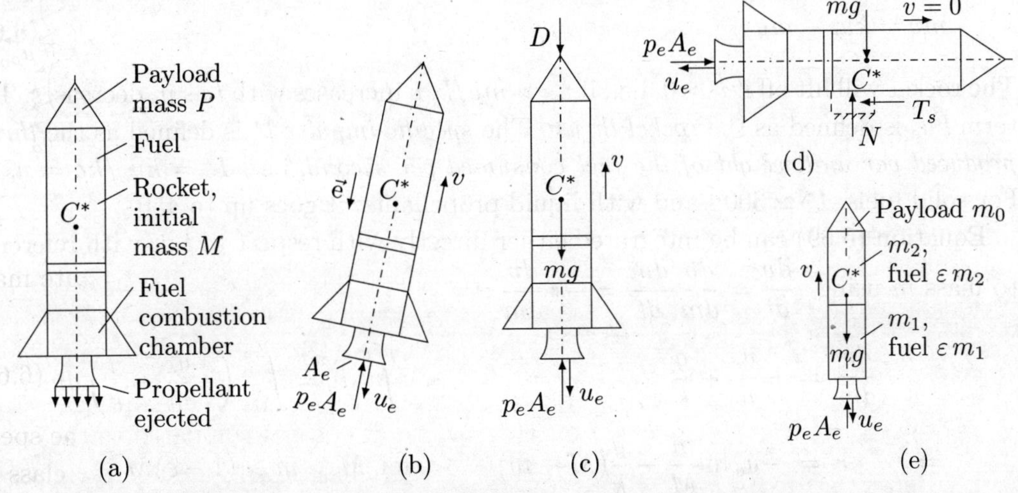

Fig. 6.24

We apply equation of motion eq. (6.66) to the 3 cases shown in Fig. 6.24b,c,d.

1. For a rocket in outer space (Fig. 6.24b), subjected to $F = p_e A_e$:
$$m\dot{v} = p_e A_e + \dot{m}_p u_e \tag{6.67}$$

2. For a rocket in the earth's atmosphere (Fig. 6.24c) under drag force D:
$$m\dot{v} = p_e A_e + \dot{m}_p u_e - mg - D \tag{6.68}$$

3. For a rocket in test stand (Fig.6.24d): $0 = p_e A_e + \dot{m}_p u_e - T_s \Rightarrow T_s = p_e A_e + \dot{m}_p u_e$. where T_s is the rocket's static thrust and $\dot{m}_p u_e$ is the jet reaction thrust.

In many cases, the pressure term $p_e A_e \ll \dot{m}_p u_e$ and may be omitted. The useful power P_1 of the jet reaction thrust, the power P_2 wasted in the kinetic energy of the exhaust gases moving at $v_2 = v - u_e$, and the efficiency η are given by

$$P_1 = (\dot{m}_p u_e)v, \qquad P_2 = \tfrac{1}{2}\dot{m}_p (v - u_e)^2, \qquad \eta = \frac{P_1}{P_1 + P_2} = \frac{2u_e v}{v^2 + u_e^2}$$

$\eta_{\max} = 1$ for $u_e = v$ since $P_2 = 0$ as $v_2 = v - u_e = 0$.

6.9.1 Single-stage rocket

Let the propellant be burnt at a constant rate k and be exhausted at the back at a constant backward velocity u_e relative to the rocket. Let M_C be the mass of the casing, P the mass of the payload, M the total initial mass of the rocket and εM that of the fuel, $0 < \varepsilon < 1$, $M = M_C + P + \varepsilon M$. Its mass m at time t is $m(t) = M - kt$, $\dot{m} = -k$, $\dot{m}_p = k$. The fuel gets burnt up in time $t = \tau = \varepsilon M/k$. Neglecting air resistance D and $p_e A_e$ as small, the equation of motion (6.68) for vertical lift-off becomes

$$m\dot{v} = k u_e - mg \tag{6.69}$$

The rocket will lift off if $\dot{v} > 0$, i.e., if $u_e > mg/k$. $\dot{v}$ increases with t as m decreases. The term $k u_e$ is defined as the *rocket thrust.* The *specific impulse* I^* is defined as the *thrust produced per unit weight of the fuel consumed per second,* i.e., $I^* = k u_e / kg = u_e/g$. For solid fuels, $I^* \simeq 300\,\text{s}$ and with liquid propellants, it goes up to 450 s.

Equation (6.69) can be integrated either directly with respect to t or with reference to mass m using $\dfrac{dv}{dt} = \dfrac{dv}{dm}\dfrac{dm}{dt} = -k\dfrac{dv}{dm}$:

$$\frac{dv}{dm} = -\frac{u_e}{m} + \frac{g}{k} \qquad \Rightarrow \qquad \int_0^v dv = \int_M^m \left(-\frac{u_e}{m} + \frac{g}{k}\right) dm$$

$$\Rightarrow \qquad v = -u_e \ln\frac{m}{M} - \frac{g}{k}(M - m) \qquad \text{for } M > m > (1-\varepsilon)M$$

$$\text{i.e.,} \qquad v = -u_e \ln\left(1 - \frac{kt}{M}\right) - gt \qquad \text{for } 0 < t < \tau \tag{6.70}$$

The maximum speed $v_{\max}$ occurs at $t = \tau = \varepsilon M/k$ when all the fuel is burnt off:

$$v_{\max} = u_e[-\ln(1-\varepsilon) - \varepsilon Mg/ku_e] \tag{6.71}$$

Interestingly, $v_{\max}$ is independent of the initial mass M, and depends only on u_e, ε and $\tau = \varepsilon M/k$. Since $ku_e > Mg$, the 2nd term in the expression of $v_{\max}$ in eq. (6.71) is negligible. For example, for $\varepsilon = 0.8$, $M = 10^5$ kg, $I^* = 300$ s, $k = 5 \times 10^3$ kg/s : $\tau = \varepsilon M/k = 16$ s, $u_e = I^* g \simeq 3.0$ km/s and the contributions of the two terms in eq. (6.71) are 4.83 km/s and 0.16 km/s yielding $v_{\max} = 4.67$ km/s. Thus, it follows that

$$v_{max} \simeq -u_e \ln(1-\varepsilon) \tag{6.72}$$

A single-stage solid-fuel rocket, even with 80% fuel, cannot place a satellite in a circular orbit at a height of $h = 100$ km, as required $v_c = [GM/(R+h)]^{1/2} = 7.8$ km/s $> v_{\max}$.

§ **Exercise 6.9:** If g is neglected, then the equation of motion of a single-stage rocket reduces to $dv/dt = ku_e/m$. Show that $v_{\max}$ is independent of whether the fuel is burnt at a constant or a variable rate.

In order to obtain a larger value of $v_{\max}$, u_e should be large and $\varepsilon \simeq 1$. Typically, $0.7 < \varepsilon < 0.9$. To escape from the earth using a single-stage rocket with $u_e = 3$ km/s, eq. (6.72) implies that $v_{\max} = (2GM/R)^{1/2} = 11.18$ km/s is obtained for $\varepsilon = 0.976$. This is structurally difficult to ensure as it leaves only 2.4% mass for the casing.

The height h at burn-off, when $m = M(1-\varepsilon)$, is obtained by integrating eq. (6.70):

$$v = \frac{dx}{dt} = \frac{dx}{dm}\frac{dm}{dt} = -k\frac{dx}{dm} = -u_e \ln\frac{m}{M} - \frac{g}{k}(M-m)$$

$$\Rightarrow \quad \int_0^x dx = \int_M^m \left[\frac{u_e}{k}\ln\frac{m}{M} + \frac{g}{k^2}(M-m)\right] dm$$

$$= \left[\frac{u_e}{k}\left(m\ln\frac{m}{M} - m\right) + \frac{g}{k^2}\left(Mm - \tfrac{1}{2}m^2\right)\right]\Bigg|_M^m$$

$$x = -\frac{g}{2k^2}(M-m)^2 + \frac{u_e}{k}\left(M - m + m\ln\frac{m}{M}\right)$$

$$\text{i.e.,} \quad x = -\tfrac{1}{2}gt^2 + u_e t + \frac{u_e}{k}(M-kt)\ln\left(1 - \frac{kt}{M}\right) \tag{6.73}$$

$$\Rightarrow \quad h = -\frac{g\varepsilon^2 M^2}{2k^2} + \frac{u_e}{k}[\varepsilon M + M(1-\varepsilon)\ln(1-\varepsilon)]$$

$$= u_e\tau\left[1 + \frac{1-\varepsilon}{\varepsilon}\ln(1-\varepsilon) - \frac{g\tau}{2u_e}\right] \tag{6.74}$$

Since $g\tau/2u_e = \varepsilon(Mg/2ku_e)$ is small compared to the other two terms,

$$h \simeq u_e\tau\left[1 + \frac{1-\varepsilon}{\varepsilon}\ln(1-\varepsilon)\right] \tag{6.75}$$

The height h depends on the time of burning $\tau = \varepsilon M/k$. An additional height $h_1 \simeq v_{\max}^2/2g$ is attained after burn-off. For the preceding data, $h \simeq 27.43\,\text{km}$ and $h_1 \simeq 1100\,\text{km}$, which is not small compared to the radius R of the earth and so is not valid. The maximum height is obtained from $\frac{1}{2}m^*v_{\max}^2 - GMm^*/(R+h) = -GMm^*/r_{\max}$.

6.9.2 Two-stage rocket

$v_{\max}$ can be increased considerably if the fuel casing is shed off at intervals so that m is reduced and the acceleration ku_e/m is increased. This concept is used in multi-stage rockets. A stage refers to an interval in which a certain amount of fuel is burnt after which the casing housing it, is shed off and the fuel of the next stage is ignited.

§ **Exercise 6.10:** Consider a 2-stage rocket with its payload, 1st stage and 2nd stage having mass m_0, m_1 and m_2, respectively. Let ε be the ratio of the mass of the fuel to that of the fuel+casing in each stage (Fig. 6.24e). (a) Using eq. (6.72) and neglecting the gravity terms, prove that the velocities v_1 and v_2 at the end of the 1st and 2nd stages are:

$$v_1 = -u_e \ln\left(1 - \frac{\varepsilon m_1}{m_0 + m_1 + m_2}\right)$$

$$v_2 = -u_e\left[\ln\left(1 - \frac{\varepsilon m_1}{m_0 + m_1 + m_2}\right) + \ln\left(1 - \frac{\varepsilon m_2}{m_0 + m_2}\right)\right]$$

(b) Prove that the optimum value of m_2 for which the final velocity v_2 is a maximum for given values of ε, m_0 and $M = m_1 + m_2$, is $m_2 = M[\alpha^{1/2}(1+\alpha)^{1/2} - \alpha] \simeq M\alpha^{1/2} = (m_0 M)^{1/2}$, where $\alpha = m_0/M \ll 1$. Show that the final velocity is, $v_2 \simeq -2u_e\ln[1 - \varepsilon(1-\alpha^{1/2})]$ with the velocity increment $\Delta v \simeq \frac{1}{2}v_2$ in each stage.

For $m_0/M = 0.01, \varepsilon = 0.8$, $v_2 \simeq 2.55u_e$ for an optimal 2-stage rocket with optimal $m_2 = (m_0 M)^{1/2} = M/10$. It is a huge improvement over $v_{\max} = -u_e\ln[1 - \varepsilon M/(M + m_0)] = 1.57u_e$ for a 1-stage rocket. Most of the mass of an optimal 2-stage rocket is in the first stage. For $I^* = 350\,\text{s}$ ($u_e \simeq 3.4\,\text{km/s}$), v_2 is about 8.67 km/s. Further improvement can be obtained by increasing the number of stages but with increasing cost of the rocket. Note the large ratio of fuel to pay load required, $\varepsilon M/m_0 = 80$.

Example 6.4: A pipe bend has fluid flowing at $Q\,\text{m}^3/\text{s}$. The density of the fluid is ρ. The static gauge pressure at B is p. Find the normal force N, shear force S and bending moment M across the pipe section at B for the two cases shown in Fig. E6.4a.

Solution: For fixed control volume of pipe bend under steady state (Fig. E6.4b),

$$\vec{v}_{C^*} = \text{const.},\ \ \vec{a}_{C^*} = \vec{0},\ \ \dot{\vec{H}}_O = \vec{0},\ \ v_i = \frac{Q}{A},\ \ v_o = \frac{Q}{0.5A} = \frac{2Q}{A},\ \dot{m}_i = \dot{m}_o = \rho Q,\ \vec{v}_O = \vec{0}$$

$$\vec{M}_O = \dot{\vec{H}}_O - \vec{r}_{P_iO} \times \dot{m}_i(\vec{v}_i - \vec{v}_O) + \vec{r}_{P_oO} \times \dot{m}_o(\vec{v}_o - \vec{v}_O) \tag{1}$$

$$\vec{F} = m\vec{a}_{C^*} - \dot{m}_i(\vec{v}_i - \vec{v}_{C^*}) + \dot{m}_o(\vec{v}_o - \vec{v}_{C^*}) \tag{2}$$

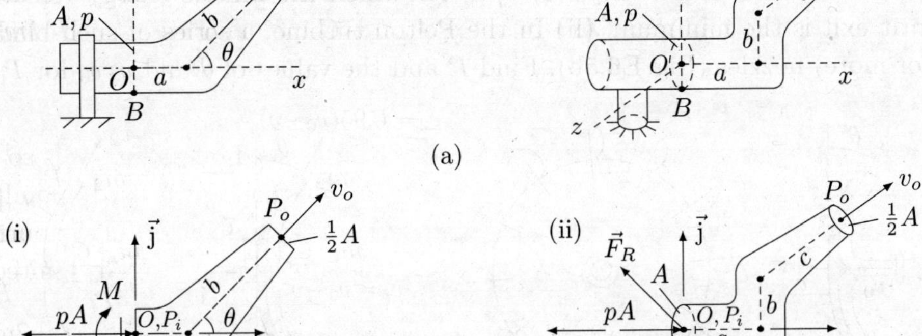

Fig. E6.4

(i) $\vec{r}_{P_iO} = \vec{0}, \quad \vec{r}_{P_oO} = (a + b\cos\theta)\,\vec{\mathrm{i}} + b\sin\theta\,\vec{\mathrm{j}}, \quad \vec{v}_i = \dfrac{Q}{A}\vec{\mathrm{i}}, \quad \vec{v}_o = \dfrac{2Q}{A}(\cos\theta\,\vec{\mathrm{i}} + \sin\theta\,\vec{\mathrm{j}})$

eq. (1) $\Rightarrow \vec{M}_O = -M\,\vec{\mathrm{k}} = [(a + b\cos\theta)\,\vec{\mathrm{i}} + b\sin\theta\,\vec{\mathrm{j}}] \times \dfrac{2\rho Q^2}{A}(\cos\theta\,\vec{\mathrm{i}} + \sin\theta\,\vec{\mathrm{j}})$

$\Rightarrow \ M = 2\rho Q^2 a \sin\theta / A$

eq. (2) $\Rightarrow \ \vec{F} = pA\,\vec{\mathrm{i}} - N\,\vec{\mathrm{i}} - S\,\vec{\mathrm{j}} = -\dfrac{\rho Q^2}{A}\,\vec{\mathrm{i}} + \dfrac{2\rho Q^2}{A}(\cos\theta\,\vec{\mathrm{i}} + \sin\theta\,\vec{\mathrm{j}})$

$\Rightarrow \ N = pA + \rho Q^2(1 - 2\cos\theta)/A, \qquad S = -2\rho Q^2 \sin\theta / A$

(ii) $\vec{r}_{P_iO} = \vec{0}, \quad \vec{r}_{P_oO} = a\,\vec{\mathrm{i}} + b\,\vec{\mathrm{j}} + c\,\vec{\mathrm{k}}, \quad \vec{v}_i = \dfrac{Q}{A}\,\vec{\mathrm{i}}, \quad \vec{v}_o = -\dfrac{2Q}{A}\,\vec{\mathrm{k}}, \quad \vec{n} = -\vec{\mathrm{i}}$

eq. (1) $\Rightarrow \vec{M}_O = \vec{C}_R = (a\,\vec{\mathrm{i}} + b\,\vec{\mathrm{j}} + c\,\vec{\mathrm{k}}) \times \dfrac{2\rho Q^2}{A}(-\vec{\mathrm{k}}) = \dfrac{2\rho Q^2}{A}(-b\,\vec{\mathrm{i}} + a\,\vec{\mathrm{j}})$

$\Rightarrow \ \vec{M}_t = -(2\rho Q^2 b/A)\,\vec{\mathrm{i}}, \qquad \vec{M}_b = (2\rho Q^2 a/A)\,\vec{\mathrm{j}}$

eq. (2) $\Rightarrow \ \vec{F} = pA\,\vec{\mathrm{i}} + \vec{F}_R = -\dfrac{\rho Q^2}{A}\,\vec{\mathrm{i}} + \dfrac{2\rho Q^2}{A}(-\vec{\mathrm{k}}) = -N\,\vec{\mathrm{i}} + \vec{S} + pA\,\vec{\mathrm{i}}$

$\Rightarrow \ N = pA + \rho Q^2/A, \qquad \vec{S} = -(2\rho Q^2/A)\,\vec{\mathrm{k}}$

Example 6.5: (i) A nozzle issues a jet of water with density ρ and cross-sectional area A (Fig. E6.5a). The jet strikes a moving curved blade, and is divided into two equal streams. The friction reduces the magnitude of the velocity of the water with reference to the blade by 5%. Find (a) the rate at which the kinetic energy emerges from the nozzle, (b) the force exerted on the blade due to flow of water, (c) the power P delivered to the blade by water and the ratio v/v_0 for which it is maximum, (d) the maximum mechanical efficiency, (e) v/v_0 for which the kinetic energy remaining in water at exit is the minimum. (ii) In the Pelton turbine, a series of such blades face one (or more) nozzles (Fig. E6.5b). Find P and the values of θ and v/v_0 for $P_{\max}$.

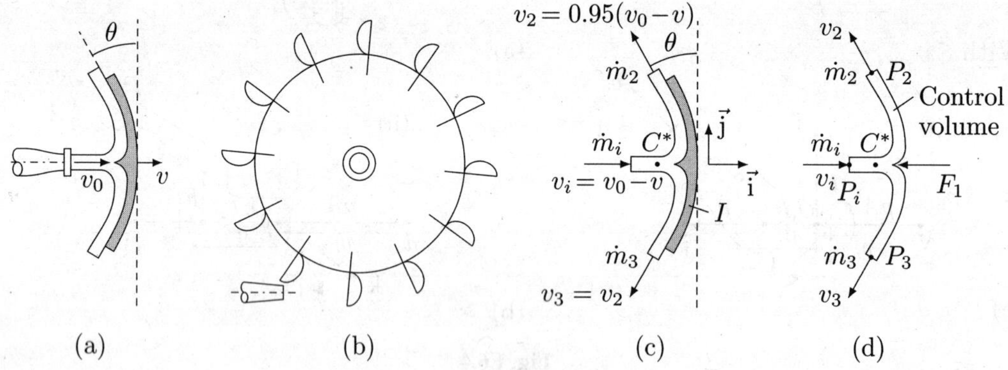

Fig. E6.5

Solution: (i) Let $\vec{F}_b$ be the force from blade on the control volume of water under steady state (Fig. E6.5d). Relative to the inertial frame I attached to the blade:

$$\vec{v}_{C*} = \text{constant}, \quad \vec{a}_{C*} = \vec{0}, \quad \dot{m}_i = 2\dot{m}_2 = 2\dot{m}_3 = \rho A(v_0 - v), \quad \vec{v}_i = (v_0 - v)\,\vec{\imath}$$

$$\vec{v}_2 = 0.95(v_0 - v)(-\sin\theta\,\vec{\imath} + \cos\theta\,\vec{\jmath}), \qquad \vec{v}_3 = 0.95(v_0 - v)(-\sin\theta\,\vec{\imath} - \cos\theta\,\vec{\jmath})$$

(a) From the nozzle: mass flow rate $\dot{m} = \rho A v_0$, $\dot{T} = \frac{1}{2}\dot{m}v_0^2 = \frac{1}{2}(\rho A v_0)v_0^2 = \frac{1}{2}\rho A v_0^3$.

(b) $$\vec{F} = m\vec{a}_{C*} - \dot{m}_i(\vec{v}_i - \vec{v}_{C*}) + \dot{m}_2(\vec{v}_2 - \vec{v}_{C*}) + \dot{m}_3(\vec{v}_3 - \vec{v}_{C*}) = \vec{F}_b + m\vec{g}$$

$$= \dot{m}_i(v_0 - v)[-\vec{\imath} + \frac{0.95}{2}(-\sin\theta\,\vec{\imath} + \cos\theta\,\vec{\jmath}) + \frac{0.95}{2}(-\sin\theta\,\vec{\imath} - \cos\theta\,\vec{\jmath})]$$

$$= -\rho A(v_0 - v)^2(1 + 0.95\sin\theta)\,\vec{\imath}$$

The force exerted by water on the blade is $-\vec{F}_b$ with force $\vec{F}_1$ due to flow of water as:

$$\vec{F}_1 = -\vec{F}_b - m\vec{g} = \rho A(v_0 - v)^2(1 + 0.95\sin\theta)\,\vec{\imath}$$

(c) $$P = \vec{F}_1 \cdot v\,\vec{\imath} = \rho A(v_0 - v)^2 v(1 + 0.95\sin\theta)$$

$$\frac{dP}{dv} = \rho A(1+0.95\sin\theta)(v_0^2 - 4v_0 v + 3v^2) = 0 \qquad \text{for} \quad v = v_0,\ \tfrac{1}{3}v_0$$

$$\Rightarrow \quad P_{\min} = 0 \quad \text{for } v = v_0 \quad \text{and } P_{\max} = \tfrac{4}{27}\rho A v_0^3(1+0.95\sin\theta) \quad \text{for } v = v_0/3.$$

(d) The mechanical efficiency η of the system is given by

$$\eta = \frac{\text{Power delivered to blade}}{\text{Rate of emission of kinetic energy from nozzle}}$$

$$= \frac{P}{\dot{T}} = 2\left(1 - \frac{v}{v_0}\right)^2 \frac{v}{v_0}(1+0.95\sin\theta)$$

With $\xi = v/v_0$,

$$\frac{d\eta}{d\xi} = 2(1 - 4\xi + 3\xi^2)(1+0.95\sin\theta) = 0 \qquad \text{for} \quad \xi = v/v_0 = 1,\ \tfrac{1}{3}$$

$$\Rightarrow \quad \eta_{\min} = 0 \quad \text{for } \xi = 1 \quad \text{and} \quad \eta_{max} = (1+0.95\sin\theta)8/27 \quad \text{for } \xi = v/v_0 = \tfrac{1}{3},$$

$$\Rightarrow \quad \eta_{\max} = 0.578 \quad \text{for } \theta = \pi/2.$$

(e)
$$\vec{v}_2|_{\text{ground}} = v\vec{\mathrm{i}} + 0.95(v_0 - v)(-\sin\theta\,\vec{\mathrm{i}} + \cos\,\vec{\mathrm{j}})$$

$$= [v - 0.95(v_0 - v)\sin\theta]\,\vec{\mathrm{i}} + 0.95(v_0 - v)\cos\theta\,\vec{\mathrm{j}}$$

The kinetic energy loss T^* per kg in water at exit is:

$$T^* = \tfrac{1}{2}(1)(\vec{v}_2|_{\text{ground}})^2 = \tfrac{1}{2}[v^2 + 0.95^2(v_0 - v)^2 - 1.9v(v_0 - v)\sin\theta]$$

$$= \tfrac{1}{2}v_0^2\,[\xi^2 + 0.95^2(1-\xi)^2 - 1.9\xi(1-\xi)\sin\theta]$$

$$= \tfrac{1}{2}v_0^2\,[(1.9025 + 1.9\sin\theta)\xi^2 - (1.805 + 1.9\sin\theta)\xi + 0.9025]$$

$$\frac{dT^*}{d\xi} = \tfrac{1}{2}v_0^2\,[2(1.9025 + 1.9\sin\theta)\xi - (1.805 + 1.9\sin\theta)] = 0$$

$$\text{for} \quad \xi = \xi_1 = (0.9025 + 0.95\sin\theta)/(1.9025 + 1.9\sin\theta)$$

$T^*_{\min}$ for $\xi = \xi_1$. For $\theta = \pi/2:$ $T^*_{\min} = 0$ for $\xi = 0.487 \simeq 0.5$, i.e., for $v \simeq 0.5v_0$.

(ii) For a single moving blade, the mass flow rate at the nozzle exit, $\dot{m} = \rho A v_0$, differs from $\dot{m}_i$ entering the control volume. However, for the blades arranged in Fig. E6.5b, the mass flow rate missed by one blade is intercepted by the next. Hence, for well-designed blade arrangement, the entire $\dot{m}$ contributes to useful output. The flow around each blade remains the same, the only change required is in $\dot{m}_i$: $\dot{m}_i = \rho A v_0 \Rightarrow$

$$\vec{F} = -\dot{m}_i(v_0 - v)(1 + 0.95\sin\theta)\,\vec{\mathrm{i}} = -\rho A v_0(v_0 - v)(1 + 0.95\sin\theta)\,\vec{\mathrm{i}}$$

The force F_1 on the blades is $\vec{F_1} = -\vec{F} = \rho A v_0(v_0 - v)(1 + 0.95\sin\theta)\,\vec{\imath}$.

$$P = \vec{F_1}\cdot\vec{v} = \rho A v_0(v_0 - v)v(1 + 0.95\sin\theta)$$

$$\eta = P/\dot{T} = 2(1-\xi)\xi(1 + 0.95\sin\theta)$$

P and η are maximum for $v = \frac{1}{2}v_0$, $\theta = 90°$ with $P_{\max} = 0.4875\rho A v_0^3$, $\eta_{\max} = 0.975$.

If blade friction is neglected, then $\eta_{\max} = 1$ for, $\xi = 0.5$, $\theta = 90°$, and the analysis of part (e) repeated for no friction yields $T^*_{\min} = 0$ for $\xi = 0.5$, consistent with $\eta_{\max} = 1$.

Example 6.6: A jet engine (Fig. E6.6a) scoops in air at A at the rate $\dot{m}_a$ and discharges the exhaust at B with a backward velocity u with reference to the airplane. (a) Find the propulsive thrust $\vec{F_T}$, when the airplane has constant speed v. $\dot{m}_a = 75\,\text{kg/s}$, $u = 300\,\text{m/s}$, $v = 360\,\text{km/h}$, $d = 3\,\text{m}$. (b) Solve part (a) taking into account the fuel consumed at the rate $\dot{m}_p = 1.2\,\text{kg/s}$. (c) Find the power output used to propel the airplane. (d) Find the power developed by the engine and its efficiency. (e) Assume that the drag is proportional to the square of the speed. Find the cruising speed of the airplane if the flow of air through the engine is decreased by 20%.

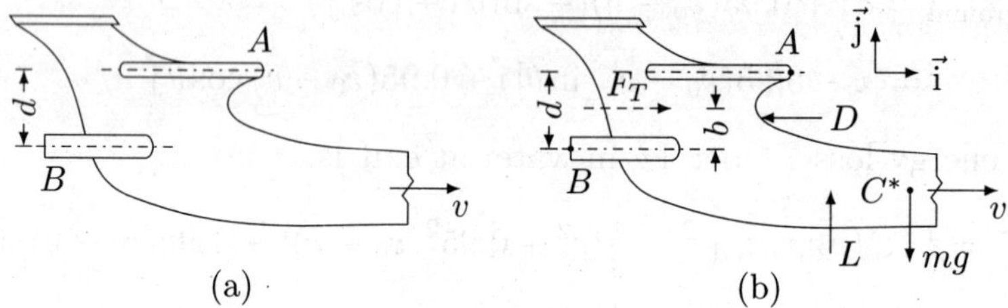

Fig. E6.6

Solution: Let m be the mass of the airplane and L, D be the lift and drag forces on it. $m_p \ll m$, $\dot{m}_p \ll \dot{m}_a$. For the control volume in Fig. E6.6b: $m \simeq$ constant and the small slow shift of its C^* due to $\dot{m}_p$ is neglected $\Rightarrow$ $\vec{v}_{C^*} \simeq v\,\vec{\imath}$ and

$$\dot{m}_i = \dot{m}_a, \qquad \vec{v}_{C^*} \simeq v\,\vec{\imath}, \qquad \vec{a}_{C^*} \simeq \dot{v}\,\vec{\imath}, \qquad \vec{v}_i = \vec{0}, \quad \vec{v}_o = -u\,\vec{\imath} + v\,\vec{\imath}$$

$$\vec{F} = m\vec{a}_{C^*} - \dot{m}_i(\vec{v}_i - \vec{v}_{C^*}) + \dot{m}_o(\vec{v}_o - \vec{v}_{C^*})$$

$$\Rightarrow \quad m\dot{v}\,\vec{\imath} = -mg\,\vec{\jmath} + L\,\vec{\jmath} - D\,\vec{\imath} + \dot{m}_i(\vec{v}_i - \vec{v}_{C^*}) - \dot{m}_o(\vec{v}_o - \vec{v}_{C^*})$$

$$\Rightarrow \quad m\dot{v} = \dot{m}_o u - \dot{m}_i v - D = F_T - D \tag{1}$$

where $F_T = F_i + F_o$, $\quad \vec{F_i} = F_i\,\vec{\imath} = -\dot{m}_i v\,\vec{\imath}$ at A, $\quad \vec{F_o} = F_o\,\vec{\imath} = \dot{m}_o u\,\vec{\imath}$ at B (2)

Let the line of action of the propulsive thrust $\vec{F_T}$ be at a distance b above B. It is resultant of parallel forces $\vec{F_i} = F_i\,\vec{\imath}$, $\vec{F_o} = F_o\,\vec{\imath}$. Hence,

$$b = \frac{F_i d + F_0 \times 0}{F_i + F_o} \tag{3}$$

(a) For $\dot{m}_i = \dot{m}_o = \dot{m}_a = 75\,\text{kg/s}, \quad u = 300\,\text{m/s}, \quad v = 100\,\text{m/s}, \quad d = 3\,\text{m},$

$$F_i \overset{(2)}{=} -\dot{m}_i v = -75(100) = -7500\,\text{N}, \qquad F_o \overset{(2)}{=} \dot{m}_O u = 75(300) = 22500\,\text{N}$$

$$b \overset{(3)}{=} \frac{-7500(3)}{22500 - 7500} = -1.5\,\text{m}$$

$\Rightarrow$ Resultant thrust $\vec{F}_T = (F_i + F_o)\,\vec{\imath} = 15000\,\vec{\imath}\ \text{N} = 15\,\vec{\imath}$ kN acts at 1.5 m below B.

(b) For $\dot{m}_i = \dot{m}_a = 75\,\text{kg/s}, \quad \dot{m}_o = \dot{m}_a + \dot{m}_p = 75 + 1.2 = 76.2\,\text{kg/s},$

$$F_i \overset{(2)}{=} -\dot{m}_i v = -75(100) = -7500\,\text{N}, \qquad F_o \overset{(2)}{=} \dot{m}_O u = 76.2(300) = 22860\,\text{N}$$

$$b \overset{(3)}{=} \frac{-7500(3)}{22860 - 7500} = -1.465\,\text{m}$$

$\Rightarrow$ Resultant thrust $\vec{F}_T = (F_i + F_o)\,\vec{\imath} = 15360\,\vec{\imath}\ \text{N} = 15.36\,\vec{\imath}$ kN acts at 1.465 m below B.

(c) The power, P_0, used to propel the plane is: $P_0 = F_T v = (F_i + F_o)v \overset{(2)}{=} \dot{m}_a(u - v)v$.

(d) The total power, P_1, developed by the engine is sum of P_0 and the rate of loss of kinetic energy in the exhaust gases:

$$\begin{aligned}
P_1 &= P_0 + \tfrac{1}{2}(\dot{m}_a + \dot{m}_p)v_2^2 \simeq P_0 + \tfrac{1}{2}\dot{m}_a(v - u)^2 \\
&= \dot{m}_a(u - v)v + \tfrac{1}{2}\dot{m}_a(v - u)^2 = \tfrac{1}{2}\dot{m}_a(u^2 - v^2) \\
\eta &= \frac{P_0}{P_1} = \frac{\dot{m}_a(u - v)v}{\tfrac{1}{2}\dot{m}_a(u^2 - v^2)} = \frac{2v}{u + v} \\
P_0 &= 15 \times 100 = 1500\,\text{kW} \\
P_1 &= \tfrac{1}{2} \times 75(300^2 - 100^2)\ \text{W} = 3000\,\text{kW} \\
\eta &= 2 \times 100/(300 + 100) = 0.5
\end{aligned}$$

(e) $v = 0$. Let $D = cv^2$. Equation (1) $\Rightarrow$

$$D = cv^2 = \dot{m}_a(u - v): \qquad c(100)^2 = 75(300 - 100) \quad \Rightarrow \quad c = 1.5$$

For $\dot{m}_a = 0.8 \times 75\,\text{kg/s}$, $u = 300\,\text{m/s}$, airplane's cruising speed v^* is obtained from

$$cv^{*2} = \dot{m}_a(u - v^*) \quad \Rightarrow \quad 1.5v^{*2} = 0.8 \times 75(300 - v^*)$$

$$\Rightarrow \quad v^* = 91.36\,\text{m/s}, \ -131.36\,\text{m/s} \quad \Rightarrow \quad v^* = 91.36\,\text{m/s} = 328.9\,\text{km/h}$$

Example 6.7: A cloud moves with velocity $\vec{v}$. A spherical raindrop has initial radius r_0 and velocity $\vec{v}$. It acquires mass by condensation at a rate $c_1 A$, where A is its cross-sectional area. The resistance force on it is $-c_2 A(\vec{v}_C - \vec{v})$, where $\vec{v}_C$ is the velocity of its centre of mass. The gravitational acceleration is $\vec{g}$. Find $\vec{v}_C$, $\vec{a}_C$ at t.

Solution: Let r be the radius of the rain drop at time t. Its mass m and area A are

$$m(t) = \frac{4}{3}\pi r^3 \rho, \qquad A(t) = \pi r^2 \quad \Rightarrow \quad \dot{m}_i = \dot{m} = 4(\pi r^2)\,\dot{r}\rho = c_1 A = c_1 \pi r^2$$

$$\Rightarrow \quad \dot{r} = c_1/4\rho \qquad\qquad \Rightarrow \quad r(t) = r_0 + c_1 t/4\rho$$

$\dot{m}_o = 0,\quad \vec{v}_i = \vec{v},\quad \vec{v}_{C*} = \vec{v}_C,\quad \vec{a}_{C*} \simeq \dot{\vec{v}}_C$. The equation of motion of the raindrop is:

$$m\vec{a}_{C*} - \dot{m}_i(\vec{v}_i - \vec{v}_{C*}) + \dot{m}_o(\vec{v}_o - \vec{v}_{C*}) = \vec{F} = m\vec{g} - c_2 A(\vec{v}_C - \vec{v})$$

$$\Rightarrow \quad \frac{4}{3}\pi r^3 \rho \frac{d\vec{v}_C}{dr}\frac{dr}{dt} - c_1 \pi r^2 (\vec{v} - \vec{v}_C) = \frac{4}{3}\pi r^3 \rho \vec{g} - c_2 \pi r^2 (\vec{v}_C - \vec{v})$$

$$\Rightarrow \quad r\frac{d\vec{v}_C}{dr} + 3\left(1 + \frac{c_2}{c_1}\right)\vec{v}_C = \frac{4}{c_1}\rho r \vec{g} + 3\left(1 + \frac{c_2}{c_1}\right)\vec{v} \qquad (1)$$

Equation (1) is of the Euler–Cauchy form: $\sum_{i=0}^{p} r^i [d^i(\,)/dr^i] = f(r)$. Substituting its complementary solution, $\vec{v}_C = \vec{B}r^n$, in the homogeneous part of eq. (1) yields

$$r(nr^{n-1})\vec{B} + 3(1 + c_2/c_1)\vec{B}r^n = 0 \qquad \Rightarrow \qquad n = -3(1 + c_2/c_1)$$

The particular solution of eq. (1) is of the form $\vec{v}_C = \vec{D}r + \vec{E}$. Substituting it in eq. (1) and comparing terms of the same degree in r, we get

$$r\vec{D}(4 + 3c_2/c_1) + 3(1 + c_2/c_1)\vec{E} = 4\rho r\vec{g}/c_1 + 3(1 + c_2/c_1)\vec{v}$$

$$\Rightarrow \quad \vec{D} = 4\rho\vec{g}/(4c_1 + 3c_2), \qquad \vec{E} = \vec{v}$$

$\Rightarrow$ General solution: $\quad \vec{v}_C = \vec{B}r^{-3(1+c_2/c_1)} + 4\rho r\vec{g}/(4c_1 + 3c_2) + \vec{v}$

The initial condition $\vec{v}_C = \vec{v}$ for $r = r_0$ yields: $\quad \vec{B} = -4\rho r_0^{(4+3c_2/c_1)}\vec{g}/(4c_1 + 3c_2)$.

$$\Rightarrow \quad \vec{v}_C = \frac{4\rho r_0}{4c_1 + 3c_2}\left[\left(\frac{r}{r_0}\right) - \left(\frac{r}{r_0}\right)^{-3(1+c_2/c_1)}\right]\vec{g} + \vec{v}$$

$$\vec{a}_C = \frac{d\vec{v}_C}{dt} = \frac{d\vec{v}_C}{dr}\dot{r} = \left[1 + 3\,(1 + c_2/c_1)\left(\frac{r}{r_0}\right)^{-(4+3c_2/c_1)}\right]\frac{c_1\vec{g}}{4c_1 + 3c_2}$$

which implies that as $r \to \infty$, $\vec{a}_C \to c_1\vec{g}/(4c_1 + 3c_2)$.

Example 6.8: Water flows into a rotary lawn sprinkler (Fig. E6.8a) at the rate of $Q\,\mathrm{m}^3/\mathrm{s}$. The cross-sectional area of flow of each arm is A. The density of water is ρ. The moment of inertia of the sprinkler, with the arms filled with water, about its vertical axis of rotation is I. (a) Find ω, if the frictional couple is M_f. (b) Find the couple C needed to hold the sprinkler stationary. (c) If the sprinkler starts from rest, find $\omega(t)$. (d) Find the torque to be applied to the sprinkler to maintain constant ω.

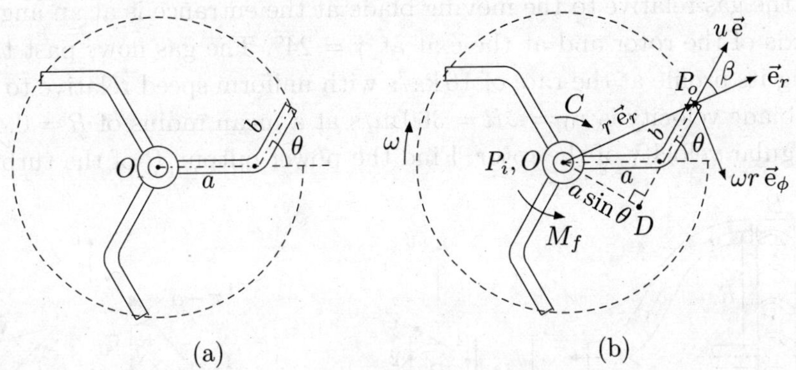

Fig. E6.8

Solution: For the control volume of the sprinkler (Fig. E6.8b), for each arm:

$$\dot m_i = \dot m_o = \rho Q/3, \qquad \vec r_{P_iO} = \vec 0, \qquad \vec r_{P_oO} = r\,\vec e_r, \qquad \vec v_O = \vec 0,$$

$$\vec v_o = \omega r\,\vec e_\phi + u\,\vec e, \qquad uA = Q/3, \qquad r = (a^2+b^2+2ab\cos\theta)^{1/2}$$

$$\frac{d}{dt}(I\omega)\,\vec e_z - 3\dot m_i \vec r_{P_iO}\times(\vec v_i - \vec v_O) + 3\dot m_o \vec r_{P_oO}\times(\vec v_o - \vec v_O) = M_{O_z}\,\vec e_z :$$

$$I\dot\omega\,\vec e_z + \rho Q r\,\vec e_r \times(\omega r\,\vec e_\phi + u\,\vec e) = (C - M_f)\,\vec e_z$$

$$I\dot\omega + \rho Q[(a^2+b^2+2ab\cos\theta)\omega - aQ\sin\theta/3A] = C - M_f \tag{1}$$

using $r\,\vec e_r \times \vec e = -r\sin\beta\,\vec e_z = -(OD)\,\vec e_z = -a\sin\theta\,\vec e_z$.

(a) For $\dot\omega = 0,\ \ C = 0$, eq. (1) $\Rightarrow\ \omega = \omega_\infty = \dfrac{aQ\sin\theta/3A - M_f/\rho Q}{a^2+b^2+2ab\cos\theta}$

(b) For $\omega = 0,\ \ \dot\omega = 0,\ \ M_f = 0$, eq. (1) $\Rightarrow\ C = -a\rho Q^2\sin\theta/3A$

(c) For $C = 0$, eq. (1) $\Rightarrow\ \dot\omega = c_1 - c_2\omega$ where,

$$c_1 = (\rho a Q^2\sin\theta/3A - M_f)/I, \qquad c_2 = \rho Q(a^2+b^2+2ab\cos\theta)/I$$

$$\int_0^\omega \frac{d\omega}{c_1 - c_2\omega} = \int_0^t dt, \qquad \frac{-1}{c_2}\ln\frac{c_1 - c_2\omega}{c_1} = t, \qquad \omega = \frac{c_1}{c_2}\left(1 - \mathrm{e}^{-c_2 t}\right)$$

$\Rightarrow$ as $t \to \infty$, $\omega \to c_1/c_2 = \omega_\infty$.

(d) For $\dot{\omega} = 0$, eq. (1) $\Rightarrow$ $C = M_f + \rho Q[(a^2 + b^2 + 2ab\cos\theta)\omega - aQ\sin\theta/3A]$

Example 6.9: A stationary nozzle diaphragm D and the rotor blades B of a gas turbine are shown in Fig. E6.9a. The combustion products leave the blades of D at an angle $\alpha = 28°$ and enter the moving rotor blades. For minimum turbulence, the velocity of the gas relative to the moving blade at the entrance is at an angle $\beta = 20°$ with the axis of the rotor and at the exit at $\gamma = 24°$. The gas flows past the moving blades along its profile at the rate of 16 kg/s with uniform speed relative to the blade. The mean blade velocity is $v_0 = \omega R = 360$ m/s at a mean radius of $R = 0.4$ m, where ω is the angular velocity of the rotor. Find the power output P of the turbine.

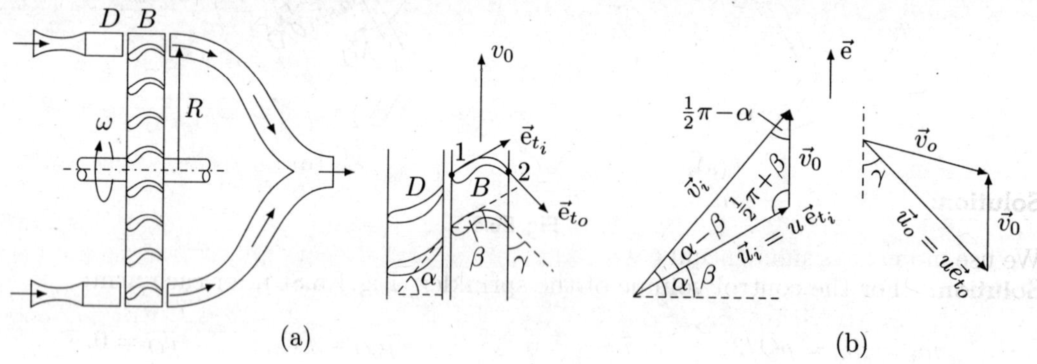

Fig. E6.9

Solution: Let u be the speed of the gas relative to the blade and $\vec{v}_i, \vec{v}_o$ be its velocities at entrance and exit: $\vec{v}_i = \vec{u}_i + \vec{v}_0$, $\vec{v}_o = \vec{u}_o + \vec{v}_0$, where $\vec{u}_i = u\,\vec{e}_{ti}$, $\vec{u}_o = u\,\vec{e}_{to}$, $\vec{v}_0 = v_0\,\vec{e}_\phi = v_0\,\vec{e}$, and $\vec{e}_{ti}$, $\vec{e}_{to}$ are the unit tangent vectors to the blade at entrance and exit. These vector relations are shown in Fig. E6.9b. From the velocity triangle,

$$\frac{v_i}{\sin(\pi/2 + \beta)} = \frac{u}{\sin(\pi/2 - \alpha)} = \frac{v_0}{\sin(\alpha - \beta)} \quad \Rightarrow \quad u = \frac{v_0 \cos\alpha}{\sin(\alpha - \beta)}$$

For control volume of gas stream on a blade: $\dot{m}_i = \dot{m}_o$, $\vec{a}_{C*} = -\omega^2 R\,\vec{e}_r$, $\vec{v}_{C*} = \vec{v}_0$,

$$\vec{F} = m\vec{a}_{C*} - \dot{m}_i(\vec{v}_i - \vec{v}_{C*}) + \dot{m}_o(\vec{v}_o - \vec{v}_{C*}) = \dot{m}_i(\vec{u}_o - \vec{u}_i) - m\omega^2 R\,\vec{e}_r$$

The total gas flow rate: $\dot{m}_i^t = 16$ kg/s. The force $\vec{F}_1$ on a blade and power P_1 are

$$\vec{F}_1 = -\vec{F} = \dot{m}_i(\vec{u}_i - \vec{u}_o) + m\omega^2 R\,\vec{e}_r$$

$$P_1 = \vec{F}_1 \cdot \vec{v}_0 = \dot{m}_i v_0(\vec{u}_i \cdot \vec{e} - \vec{u}_o \cdot \vec{e}) = \dot{m}_i v_0^2\,\frac{\cos\alpha(\sin\beta + \cos\gamma)}{\sin(\alpha - \beta)}$$

$$\Rightarrow \quad P = \dot{m}_i^t v_0^2 \cos\alpha(\sin\beta + \cos\gamma)/\sin(\alpha - \beta) = 16.52\,\text{MW}$$

Example 6.10: A wheel is wrapped on one face with a powder fuel in a long spirally wound tube of mass m forming a disc of radius a (Fig. E6.10a) with an axial moment of inertia $I_1 = \frac{1}{2}ma^2$. Its axial moment of inertia, without the charge, is I_0. The fuel burns at α kg/s and picks up gases from the still atmosphere at β kg/s. These are expelled tangentially at speed v relative to the wheel. Neglect gravity. Set up differential equation for ω. Assume that the configuration of the un-burnt fuel remains a circular disc. Find $\omega(r)$ if $I_0 \simeq 0$.

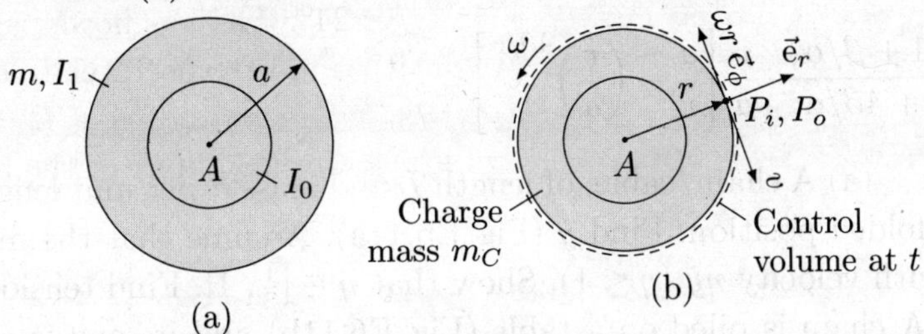

Fig. E6.10

Solution: $m_C = \dfrac{\pi r^2}{\pi a^2} m = \dfrac{r^2}{a^2} m \quad \Rightarrow \quad \dot{m}_C = \dfrac{2r\dot{r}}{a^2} m = -\alpha \quad \Rightarrow \quad \dot{r} = \dfrac{-a^2\alpha}{2rm}.$

We use moment of momentum equation about A for the control volume (Fig. E6.10b):

$$\dot{m}_i = \beta, \qquad \dot{m}_o = \alpha + \beta, \qquad \vec{r}_{P_iA} = \vec{r}_{P_oA} = r\,\vec{e}_r, \qquad \vec{v}_i = \vec{0},$$

$$\vec{v}_A = \vec{0}, \qquad \vec{v}_o = (\omega r - v)\,\vec{e}_\phi, \qquad I = I_0 + \tfrac{1}{2}m_C r^2 = I_0 + I_1 r^4/a^4$$

$$\frac{d}{dt}[I(t)\omega]\,\vec{e}_z - \dot{m}_i\vec{r}_{P_iA} \times (\vec{v}_i - \vec{v}_A) + \dot{m}_o\vec{r}_{P_oA} \times (\vec{v}_o - \vec{v}_A) = M_{A_z}\,\vec{e}_z = \vec{0}$$

$$\Rightarrow \quad \frac{d}{dt}(I\omega) = -(\alpha + \beta)(\omega r - v)r$$

$$\Rightarrow \quad \frac{d}{dr}(Iw)\frac{dr}{dt} = \frac{d}{dr}(I\omega)\left(-\frac{a^2\alpha}{2rm}\right) = -(\alpha+\beta)(\omega r - v)r$$

$$\Rightarrow \quad \frac{d}{dr}\left[\left\{I_0 + I_1\left(\frac{r}{a}\right)^4\right\}\omega\right] = \frac{2(\alpha+\beta)m}{\alpha a^2}(wr^3 - vr^2) \qquad (1)$$

For $I_0 \simeq 0$, using $I_1 = \frac{1}{2}ma^2$, the equation of motion (1) simplifies to

$$r\frac{d\omega}{dr} - \frac{4\beta}{\alpha}\omega = -4\left(1 + \frac{\beta}{\alpha}\right)\frac{v}{r} \qquad (2)$$

It is of the Euler–Cauchy form $\sum r^i[d^i(\;)/dr^i] = f(r)$, with complementary solution

$$\omega_C = Ar^n \quad \Rightarrow \quad (n - 4\beta/\alpha)Ar^n = 0 \quad \Rightarrow \quad n = 4\beta/\alpha$$

The particular solution of eq. (2) is $\omega_p = B/r$ since $f(r) = 1/r$. Put it in eq. (2):

$$(-1-4\beta/\alpha)B/r = -4(1+\beta/\alpha)v/r \quad \Rightarrow \quad B = 4(1+\beta/\alpha)v/(1+4\beta/\alpha)$$

$$\omega = \omega_C + \omega_p = Ar^{4\beta/\alpha} + \frac{4(1+\beta/\alpha)v}{(1+4\beta/\alpha)r}.$$

Constant A is determined from the initial condition: $\omega = 0$ for $r = a$:

$$0 = Aa^{4\beta/\alpha} + \frac{4(1+\beta/\alpha)v}{(1+4\beta/\alpha)a} \quad \Rightarrow \quad A = -\frac{4(1+\beta/\alpha)}{1+4\beta/\alpha}\cdot\frac{v}{a}a^{-4\beta/\alpha}$$

$$\omega = \frac{4(1+\beta/\alpha)}{1+4\beta/\alpha}\cdot\frac{v}{a}\left[\frac{a}{r} - \left(\frac{r}{a}\right)^{4\beta/\alpha}\right]$$

Example 6.11: (a) A chain/cable, of length L and mass σ per unit length, is released from its fully folded position. Find $\dot{y}$ (Fig. E6.11a). Assume that the mass leaves the moving part with velocity $\eta\dot{y}$ ($\eta \le 1$). Show that $\eta \in [\frac{1}{2}, 1]$. Find tension T_B at B for $\eta = 1, \frac{1}{2}$. (b) A chain is piled on a table (Fig. E6.11b) and its end is released in the hole. Find $\dot{y}$. Assume that the mass enters the moving part with velocity $\eta\dot{y}$ ($\eta \ge 0$).

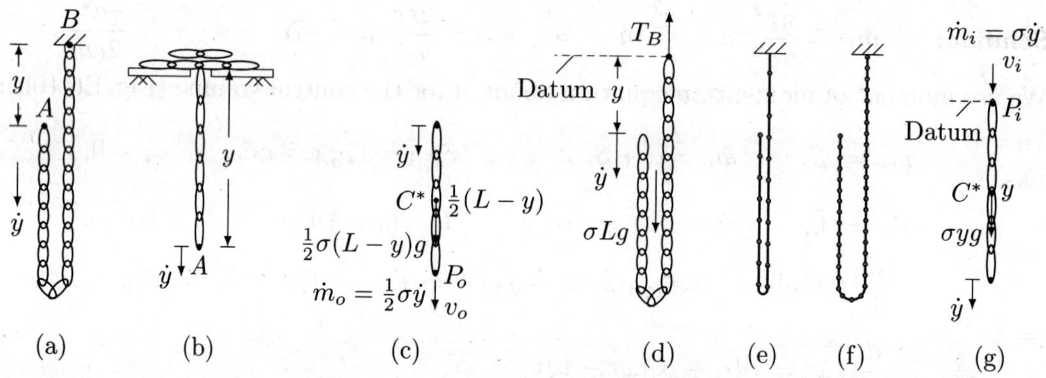

Fig. E6.11

Solution: (a) For control volume of part (Fig. E6.11c) moving at uniform speed $\dot{y}$:

$$m = \tfrac{1}{2}\sigma(L-y), \quad \dot{m}_i = 0, \quad \dot{m}_o = \tfrac{1}{2}\sigma\dot{y}, \quad v_o = \eta\dot{y}, \quad v_{C^*} = \dot{y}, \quad a_{C^*} = \ddot{y}, \quad T = \tfrac{1}{4}(L-y)\sigma\dot{y}^2$$

$$\vec{F} = m\vec{a}_{C^*} - \dot{m}_i(\vec{v}_i - \vec{v}_{C^*}) + \dot{m}_o(\vec{v}_o - \vec{v}_{C^*})$$

$$\Rightarrow \quad \tfrac{1}{2}\sigma(L-y)\ddot{y} + \tfrac{1}{2}\sigma\dot{y}(\eta-1)\dot{y} = \tfrac{1}{2}(L-y)\sigma g \tag{1}$$

Fig. E6.11d: $F = \sigma Lg - T_B = \dot{\overline{\tfrac{1}{2}(L-y)\sigma\dot{y}}} \Rightarrow T_B = \sigma Lg - \tfrac{1}{2}(L-y)\sigma\ddot{y} + \tfrac{1}{2}\sigma\dot{y}^2$ (2)

$$V = -\tfrac{1}{2}(L+y)\sigma g\tfrac{1}{4}(L+y) - \tfrac{1}{2}(L-y)\sigma g[y + \tfrac{1}{4}(L-y)] = -\tfrac{1}{4}(L^2+2Ly-y^2)\sigma g$$

$$\dot{T} + \dot{V} = \tfrac{1}{2}\sigma(L-y)\dot{y}\ddot{y} - \tfrac{1}{4}\sigma\dot{y}^3 - y\dot{y}\sigma g \overset{(1)}{=} \tfrac{1}{2}(\tfrac{1}{2}-\eta)\sigma\dot{y}^3 \le 0 \quad \Rightarrow \quad \eta \ge \tfrac{1}{2}$$

$\eta = 1:$ as in open-linked chain or case shown in Fig. E6.11e for which there is no tension at bottom string and mass-points exit at $\dot{y}$ and are brought to rest by tension in right part. Equation (1) $\Rightarrow$ $\ddot{y} = g \Rightarrow \dot{y}^2 = 2gy, \quad T_B \overset{(2)}{=} \frac{1}{2}\sigma Lg(1 + 3y/L)$.

$\eta = \frac{1}{2}:$ energy conservation as in cable (Fig. E6.11f). Each mass-point gradually comes to rest across the bend $\Rightarrow$ tension in string above it to retard it $\Rightarrow$ $\ddot{y} > g$.

$$T + V = \tfrac{1}{4}(L - y)\sigma\dot{y}^2 - \tfrac{1}{4}(L^2 + 2Ly - y^2)\sigma g = -\tfrac{1}{4}L^2\sigma g \quad \Rightarrow \quad \dot{y}^2 = \frac{gy(2L - y)}{L - y}$$

$$\ddot{\eta} \overset{(1)}{=} g + \frac{\dot{y}^2}{2(L - y)} \qquad \Rightarrow \qquad T_B \overset{(2)}{=} \frac{\frac{1}{2}\sigma Lg(1 + y/L - 3y^2/2L^2)}{1 - y/L}$$

(b) For control volume of part (Fig. E6.11g) moving at uniform speed $\dot{y}$: $m = \sigma y$,

$$\dot{m}_i = \sigma\dot{y}, \quad \dot{m}_o = 0, \quad v_i = \eta\dot{y}, \quad v_{C^*} = \dot{y}, \quad a_{C^*} = \ddot{y}, \quad T = \tfrac{1}{2}y\sigma\dot{y}^2, \quad V = -\tfrac{1}{2}y^2\sigma g$$

$$\vec{F} = m\vec{a}_{C^*} - \dot{m}_i(\vec{v}_i - \vec{v}_{C^*}) + \dot{m}_o(\vec{v}_o - \vec{v}_{C^*}) \quad \Rightarrow \quad \sigma y\ddot{y} - \sigma\dot{y}(\eta - 1)\dot{y} = y\sigma g \qquad (3)$$

$$\dot{T} + \dot{V} = \sigma y\dot{y}\ddot{y} + \tfrac{1}{2}\sigma\dot{y}^3 - \sigma gy\dot{y} \overset{(3)}{=} \tfrac{1}{2}(\eta - \tfrac{1}{2})\sigma\dot{y}^3 \le 0 \quad \Rightarrow \quad \eta \le \tfrac{1}{2} \Rightarrow \eta \in [0, \tfrac{1}{2}]$$

$\eta = 0:$ corresponds to open-linked chain for which possible relative motion between links ensures that the link entering the hole has no prior velocity. Put $y\dot{y} = z$ in eq. (3):

$$\dot{z} = \frac{dz}{dy}\dot{y} = gy \quad \Rightarrow \quad \int_0^z z\,dz = \int_0^y gy^2 dy \quad \Rightarrow \quad \tfrac{1}{2}z^2 = \tfrac{1}{3}gy^3 \quad \Rightarrow \quad \dot{y}^2 = \tfrac{2}{3}gy \Rightarrow \ddot{y} = \tfrac{1}{3}g$$

$\eta = \frac{1}{2}:$ is case of energy conservation: $\frac{1}{2}\sigma y\dot{y}^2 - \frac{1}{2}y^2\sigma g = 0 \Rightarrow \dot{y}^2 = gy \Rightarrow \ddot{y} = \frac{1}{2}g$.

Example 6.12: (i) An open-linked chain of mass σ per unit length and length L is suspended by a force P with the lower end just touching the floor. Find $P(t)$ and the normal force exerted by the ground on the chain if its top end is lowered at constant speed v (Fig. E6.12a). Assume that the links are loose so that the links piled on the ground exert negligible force on the suspended links. (ii) Rework for the case of a chain initially piled up on the ground with its end pulled up at constant speed v. (iii) Find $s(t)$ in part (ii) for the case of constant applied force P, for $t < 3\sqrt{P/\sigma}/g$.

Solution: (i) Consider equation of motion of length s of the chain (Fig. E6.12b) as the mass within the control volume at time t. It instantaneously has uniform speed v:

$$s = L - vt, \quad \dot{m}_i = 0, \quad \dot{m}_o = \sigma v, \quad m = \sigma s, \quad \vec{v}_o = -v\,\vec{\imath}, \quad \vec{v}_{C^*} = -v\,\vec{\imath} \Rightarrow \vec{a}_{C^*} = \vec{0}$$

$$\vec{F} = m\vec{a}_{C^*} - \dot{m}_i(\vec{v}_i - \vec{v}_{C^*}) + \dot{m}_o(\vec{v}_o - \vec{v}_{C^*})$$

$$\Rightarrow \qquad (P - \sigma sg)\,\vec{\imath} = \vec{0}, \qquad P = \sigma(L - vt)g$$

Equation of motion of the control volume around the chain on the floor (Fig. E6.12c):

$$m = \sigma(L - s), \quad \dot{m}_i = \sigma v, \quad \dot{m}_o = 0, \quad \vec{v}_i = -v\,\vec{i}, \quad \vec{v}_{C_1^*} = \vec{0}, \qquad \vec{a}_{C_1^*} = \vec{0}$$

$$\vec{F} = m\vec{a}_{C_1^*} - \dot{m}_i(\vec{v}_i - \vec{v}_{C_1^*}) + \dot{m}_o(\vec{v}_o - \vec{v}_{C_1^*})$$

$$\Rightarrow \quad [N - \sigma(L - s)g]\,\vec{i} = -\sigma v(-v\,\vec{i}), \qquad N = \sigma(v^2 + gvt)$$

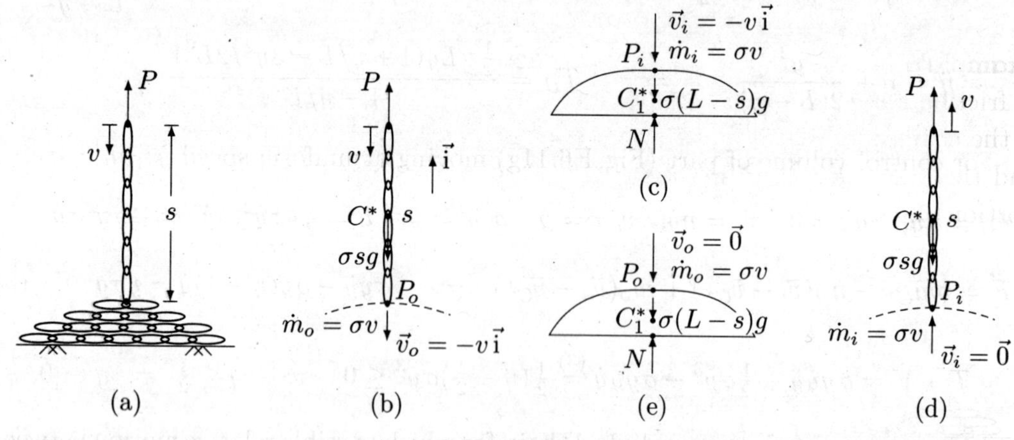

Fig. E6.12

(ii) For the control volume enclosing suspended part of chain going up (Fig. E6.12d):

$$s = vt, \quad m = s\sigma, \quad \vec{v}_{C^*} = v\,\vec{i}, \quad \vec{a}_{C^*} = \vec{0}, \quad \dot{m}_i = \sigma v, \quad \dot{m}_o = 0, \quad \vec{v}_i = \vec{0}$$

$$\vec{F} = m\vec{a}_{C^*} - \dot{m}_i(\vec{v}_i - \vec{v}_{C^*}) + \dot{m}_o(\vec{v}_o - \vec{v}_{C^*})$$

$$\Rightarrow \quad (P - \sigma sg)\,\vec{i} = -\sigma v(-v\,\vec{i}), \qquad P = \sigma(v^2 + gvt)$$

Consider the system enclosed in the control volume around the part of the chain piled upon the ground as shown in Fig. E6.12e. For this system, $m = (L - s)\sigma$, $\vec{v}_{C_1^*} = \vec{0}$, $\vec{a}_{C_1^*} = \vec{0}$, $\dot{m}_i = 0$, $\dot{m}_o = \sigma v$, $\vec{v}_o = \vec{0}$, and

$$\vec{F} = m\vec{a}_{C_1^*} - \dot{m}_i(\vec{v}_i - \vec{v}_{C_1^*}) + \dot{m}_o(\vec{v}_o - \vec{v}_{C_1^*})$$

$$\Rightarrow \quad [N - \sigma(L - s)g]\,\vec{i} = \vec{0}, \qquad N = \sigma(L - vt)g$$

(iii) For this case: $m = s\sigma$, $\vec{v}_{C^*} = \dot{s}\,\vec{i}$, $\vec{a}_{C^*} = \ddot{s}\,\vec{i}$, $\dot{m}_i = \sigma\dot{s}$, $\dot{m}_o = 0$, $\vec{v}_i = \vec{0}$.

$$\vec{F} = m\vec{a}_{C^*} - \dot{m}_i(\vec{v}_i - \vec{v}_{C^*}) + \dot{m}_o(\vec{v}_o - \vec{v}_{C^*})$$

$$\Rightarrow \quad (P - \sigma sg)\,\vec{i} = \sigma s\ddot{s}\,\vec{i} - \sigma\dot{s}(-\dot{s}\,\vec{i})$$

$$\Rightarrow \quad P - \sigma s g = \sigma s \ddot{s} + \sigma \dot{s}^2, \qquad \text{let } p = \sigma s \dot{s},$$

$$\Rightarrow \quad P - \sigma s g = \frac{d}{dt}(\sigma s \dot{s}) = \frac{dp}{dt} = \frac{dp}{ds}\dot{s} = \frac{p}{\sigma s}\frac{dp}{ds}$$

$$\Rightarrow \quad \int_0^p p\, dp = \int_0^s \sigma s (P - \sigma s g)\, ds \quad \Rightarrow \quad p^2 = \sigma^2 s^2 \dot{s}^2 = P\sigma s^2 - \frac{2}{3}\sigma^2 g s^3$$

$$\Rightarrow \quad \int_0^s \left(\frac{P}{\sigma} - \frac{2}{3} g s\right)^{-1/2} ds = \int_0^t dt \quad \Rightarrow \quad s = \frac{3}{2g}\left[P/\sigma - \left(\sqrt{P/\sigma} - \tfrac{1}{3} g t\right)^2\right]$$

Example 6.13: An open-linked chain in Fig. E6.13a is held at rest. The coefficient of friction between it and the surface BD is μ. BE is a smooth arc of a small radius. If the chain is released, find the speed of the chain when its end A has dropped by h and the pile at D is not exhausted. Neglect the motion of the chain inside the piled portion at D.

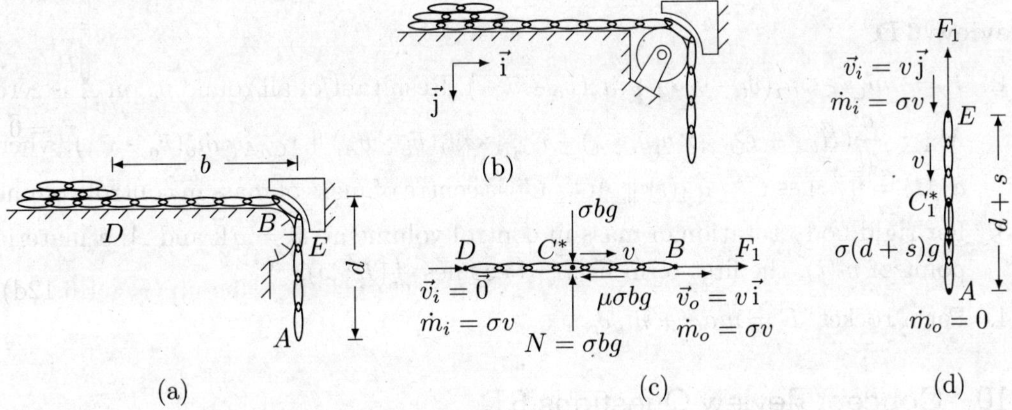

Fig. E6.13

Solution: An equivalent system is shown in Fig. E6.13b. Let σ be the mass per unit length of the chain. For the control volume of uncoiled chain of length b in Fig. 6.13c:

$$m = \sigma b, \quad v = \dot{s}, \quad \dot{m}_i = \dot{m}_o = \sigma v, \quad \vec{v}_i = \vec{0}, \quad \vec{v}_o = v\,\vec{i}, \quad \vec{v}_{C*} = v\,\vec{i}, \quad \vec{a}_{C*} = \dot{v}\,\vec{i}$$

$$\vec{F} = m\vec{a}_{C*} - \dot{m}_i(\vec{v}_i - \vec{v}_{C*}) + \dot{m}_o(\vec{v}_o - \vec{v}_{C*})$$

$$\Rightarrow \quad (F_1 - \mu\sigma b g)\,\vec{i} = (\sigma b \dot{v} + \sigma v^2)\,\vec{i} \quad \Rightarrow \quad F_1 = \mu\sigma b g + \sigma b \dot{v} + \sigma v^2 \tag{1}$$

For the control volume of the vertical portion of the chain shown in Fig. E6.13d:

$$m = (d+s)\sigma, \quad \dot{m}_i = \sigma v, \quad \dot{m}_o = 0, \quad \vec{v}_{C_1^*} = v\,\vec{j}, \quad \vec{a}_{C_1^*} = \dot{v}\,\vec{j}, \quad \vec{v}_i = v\,\vec{j}$$

$$\vec{F} = m\vec{a}_{C_1^*} - \dot{m}_i(\vec{v}_i - \vec{v}_{C_1^*}) + \dot{m}_o(\vec{v}_o - \vec{v}_{C_1^*})$$

$$\Rightarrow \quad [\sigma(d+s)g - F_1]\,\vec{j} = (d+s)\sigma\dot{v}\,\vec{j}$$

$$\Rightarrow \qquad F_1 = \sigma(d+s)g - (d+s)\sigma\dot{v} \qquad (2)$$

Let $(b+d+s)v = z$. Equations (1), (2) $\Rightarrow$

$$g(d+s-\mu b) = (b+d+s)\dot{v} + v^2 = \frac{d}{dt}[(b+d+s)v] = \frac{d}{ds}[(b+d+s)v]\frac{ds}{dt}$$

$$\Rightarrow \qquad z\frac{dz}{ds} = g(d+s-\mu b)(b+d+s) \Rightarrow$$

$$\int_0^{(b+d+h)v} z\,dz = \int_0^h g[s^2 + (b+2d-\mu b)s + (d-\mu b)(b+d)]\,ds$$

$$\Rightarrow \quad v^2 = \frac{2g[\frac{1}{3}h^3 + \frac{1}{2}(b+2d-\mu b)h^2 + (d-\mu b)(b+d)h]}{(b+d+h)^2}$$

$\dot{T} + \dot{W} \neq \dot{W}^{\text{nc}}$ as a small element in the pile has an instantaneous velocity change $\dot{s}$.

Review 6 D

1. $\vec{F} = m\vec{a}_{C^*} - \dot{m}_i(\vec{v}_i - \vec{v}_{C^*}) + \dot{m}_o(\vec{v}_o - \vec{v}_{C^*})$. Resultant of all round p_a on A is zero. $\vec{M}_A = \frac{d}{dt}(\vec{H}_{C^*} + \vec{r}_{C^*A} \times m\vec{v}_{C^*A}) - \vec{r}_{C_iA} \times \dot{m}_i(\vec{v}_i - \vec{v}_A) + \vec{r}_{C_oA} \times \dot{m}_o(\vec{v}_o - \vec{v}_A)$, where $\vec{a}_A(t) = \vec{0}\,/\,A \equiv C^*\,/\,\vec{a}_A(t) \parallel AC^*$, C^* is centre of mass of mass in control volume.
3. For rigid body rotation of mass in control volume at $\vec{\omega} = \omega\vec{\mathrm{k}}$ and A, a material point of $m(t)$, the first term in $\vec{M}_A$ becomes $\frac{d}{dt}(I_{zz}^A\,\vec{\omega})$.
4. For a rocket, $\vec{F} = m\vec{a}_C + \dot{m}_p\vec{v}_{\text{rel}}$.

6.10 Concept Review Questions 6 D

12. A table fan (Fig. Q6.12), having weight W and centre of mass C, has a slipstream of diameter d at front with velocity v_0 of blown air (density ρ). Find the minimum coefficient of friction μ and the minimum width b to ensure that the fan does not move. Note that our simplified model for the fan is that it moves air in the region inside the dotted lines shown in the figure, while the air outside is at rest. The streamlines that divide the two regions form the slipstream and are indicated with dotted lines in the figure. $[\pi\rho d^2 v_0^2/4mg,\ 2e + \pi\rho d^2 v_0^2 h/2mg]$
13. A boom consists of triangular units folded together in a canister (Fig. Q6.13). These are released during deployment such that the boom length at time t is $x(t)$. The mass of the boom per unit length is λ. Find the force–couple system exerted on the boom at B during the deployment from a vehicle which is (a) fixed, (b) translating at uniform velocity $v\,\vec{\mathrm{k}}$, (c) translating at uniform acceleration $a\,\vec{\mathrm{j}}$, (d) rotating at constant angular velocity $\omega\,\vec{\mathrm{k}}$ about the z-axis, and (e) rotating at constant angular acceleration $\dot{\omega}\,\vec{\mathrm{k}}$ about the fixed z-axis.

$[\vec{F}_R = \lambda x[g\vec{\mathrm{k}} + \vec{a}_A + \{\dot{\omega}(b+\frac{1}{2}x) + 2\omega\dot{x}\}\vec{\mathrm{j}} + \{\ddot{x} - \omega^2(b+\frac{1}{2}x)\}\vec{\mathrm{i}}] + \lambda\dot{x}^2\vec{\mathrm{i}} + \omega(\frac{1}{2}x)\lambda\dot{x}\vec{\mathrm{j}}$
$\vec{C}_R = \lambda(\dot{\omega}x^3 + 3x^2\dot{x}\omega)\vec{\mathrm{k}}/12 + \frac{1}{2}x\vec{\mathrm{i}} \times \vec{F}_R - \frac{1}{2}\lambda x\dot{x}\omega(\frac{1}{2}x)\vec{\mathrm{k}}$ with
$(a_A, \omega, \dot{\omega}) =$ (a) $(\vec{0}, 0, 0)$ (b) $(\vec{0}, 0, 0)$ (c) $(a\vec{\mathrm{j}}, 0, 0)$ (d) $(\vec{0}, \omega, 0)$ (e) $(\vec{0}, \omega, \dot{\omega})]$

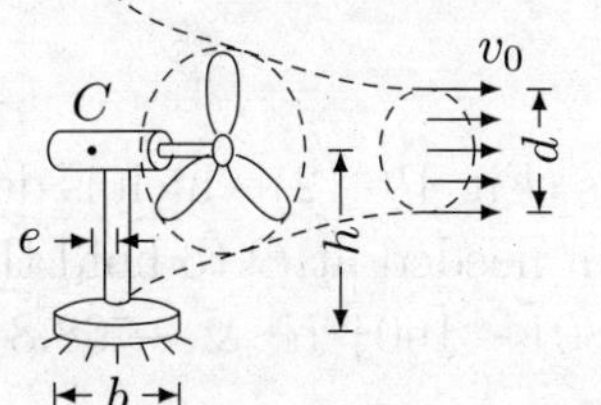

Fig. Q6.12

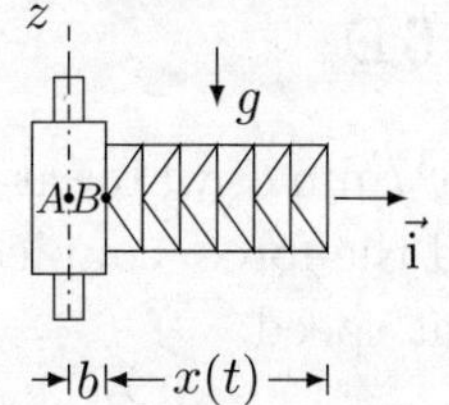

Fig. Q6.13

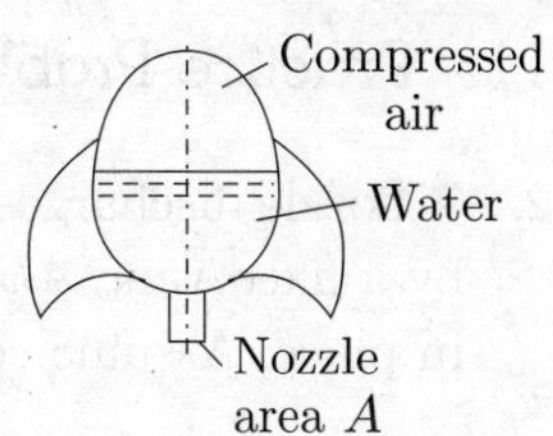

Fig. Q6.14

14. Water is injected into an air-filled body of a toy rocket, so that it is half filled with water and the rest with compressed air (Fig. Q6.14). The density of water is ρ, the nozzle area is A and the mass of the body and water is m. Find the speed v with which water must be ejected to lift the rocket. $[\sqrt{mg/\rho A}]$
15. A disc of density $\rho\,\mathrm{kg/m^2}$ rotates at $\omega\,\mathrm{rad/s}$ about its axis under axial torque M_0. It loses mass by disintegration of its rim such that its radius r decreases at a constant rate $\dot{r} = -c$. Is $(\rho\pi r^2)\frac{1}{2}r^2\dot{\omega} = M_0$ valid? [Yes]
16. A rocket car has mass m_0 when empty and carries fuel of mass m_f. The fuel is consumed at a constant rate of $\dot{m}$ and ejected at the back with velocity u relative to the car. The drag resistance at speed v is cv^2. Find the maximum speed v_m attained by the car, starting from rest. $[\sqrt{\dot{m}u/c}]$

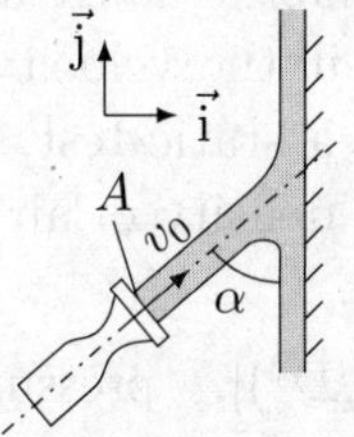

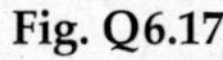
Fig. Q6.17

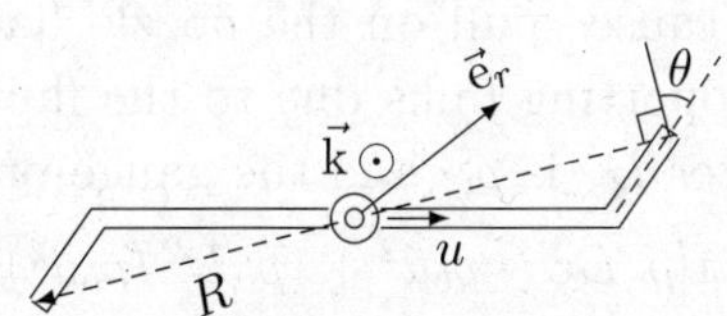

Fig. Q6.18

17. The nozzle of a fire hose of area of cross-section A directs water to a vertical screen at angle α and speed v_0 relative to the fire engine (Fig. Q6.17). The density of water is ρ. Find the normal force on the screen if (1) water flows along the screen which moves at velocity $v\,\vec{\mathrm{i}}$ and the fire engine moves forward at velocity $v_1\,\vec{\mathrm{i}}$, (2) the fire engine and the screen are at rest and water gets reflected from the screen at angle α_1 to it.

[(1) $\rho A(v_0 \sin\alpha + v_1 - v)[v_0 + (v_1 - v)\sin\alpha]$ (2) $\rho A v_0^2(\sin\alpha_1 + \sin\alpha)$]

18. Find the constant angular velocity of a lawn sprinkler (Fig. Q6.18) which discharges fluid in a horizontal plane with velocity u relative to nozzle. Neglect friction. Find the exit velocity of water with reference to ground.

$[-(u\cos\theta/R)\,\vec{k},\ u\sin\theta\,\vec{e}_r]$

6.11 Practice Problems 6 D

12. A nozzle discharges $1.2\,\text{m}^3/\text{min}$ of water at 40 m/s (Fig. P6.12) which is deflected by a fixed vane AB. Find the force–couple system needed at O to hold the vane in place. Assume constant speed. $[480\vec{i} - 160\vec{j}\ \text{N}\ \&\ -68.8\vec{k}\ \text{N.m}]$

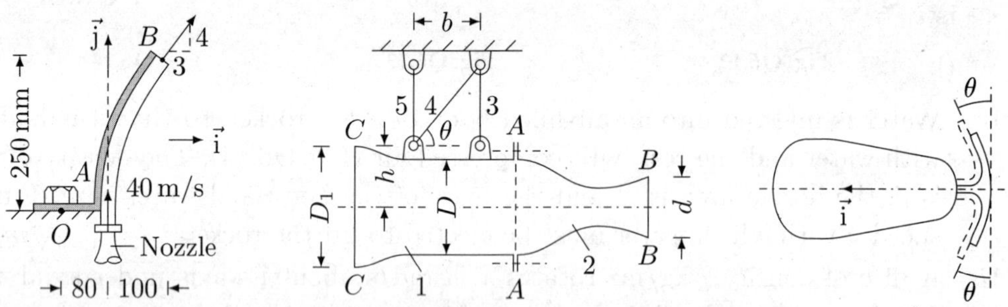

Fig. P6.12 Fig. P6.13 Fig. P6.14

13. An airstream is pumped through stationary duct 1 at speed v_0 across AA and is exhausted through the nozzle 2 (Fig. P6.13). The average static gauge pressures across the sections AA and BB are p_a and p_b. The densities of air at these sections are ρ_a and ρ_b. (a) Find the force exerted on the nozzle flange at AA by the bolts and the gasket to hold the nozzle in place. Is it the velocity change/pressure which causes pull on the nozzle flange? (b) In a static test, find the forces in the supporting links due to the fluid flow. The density of air at section CC of diameter D_1 is ρ_C and the gauge pressure is p_C.

[(a) $\frac{1}{4}\pi[p_aD^2 - p_bd^2 + (\rho_aD^2/\rho_bd^2)v_0^2(\rho_bd^2 - \rho_aD^2)]$, pressure

(b) $F_3 = 0.25\pi[D^4v_0^2\rho_a^2(1/d^2\rho_b - 1/D_1^2\rho_C) - (p_cD_1^2 - p_bd^2)]h/b$,

$F_4 = 0.25\pi[(p_bd^2 - p_cD_1^2) + D^4v_0^2\rho_a^2\left(\frac{1}{d^2\rho_b} - \frac{1}{D_1^2\rho_C}\right)\sec\theta,\ F_5 = -F_3 - F_4\sin\theta]$

14. A jet airplane is decelerated by deflecting the exhaust of the engines by a set of vanes (Fig. P6.14). Each engine scoops in air with intake area $A = 0.8\,\text{m}^2$ and discharges the combustion gases with velocity $u = 500$ m/s with reference to the engine. The density of air is $\rho = 1.2\,\text{kg/m}^3$. The speed of the airplane at the given instant is $v = 50$ m/s. (a) Find the reversed thrust provided by each engine ($\theta = 15°$). (b) Rework part (a) if the effect of the amount of fuel consumed at the rate of $\dot{m}_f = 0.5\,\text{kg}/s$ is also included. [(a) 8612 N (b) 8676 N]

15. Sand is dropped from a stationary bin into a hopper car of mass $m_0 = 8000\,\text{kg}$ when empty at the rate of $\dot{m}_1 = 200\,\text{kg/s}$ attaining a downward velocity of $v_1 = 8\,\text{m/s}$ before landing in the car (Fig. P6.15) and it leaks out at the rate of $\dot{m}_2 = 5\,\text{kg/s}$. (a) A horizontal force F is applied to give the car constant velocity $v = 3\,\text{m/s}$. Find F and the increase in the total reaction on the wheels due to sand falling in the car. (b) $F = 0$ and the velocity of the car is $v_0 = 3\,\text{m/s}$ to the right at $t = 0$ when loading begins. The car can move freely on its track. Find the position, velocity and acceleration of the car at time t. Find their values when $m_1 = 4000\,\text{kg}$ of sand has been loaded. (c) Find the force $F(t)$ needed to keep the car moving at constant acceleration $a_0 = 0.1\,\text{m/s}^2$ to the right. Its velocity is $v_0 = 3\,\text{m/s}$ at $t = 0$ when the loading begins. [(a) $600\,\text{N}$, $195gt + 1600\,\text{N}$ (b) $4800[1-(1+.024375t)^{-1/39}]$, $3(1+.024375t)^{-40/39}$, $-.075(1+.024375t)^{-79/39}$, $48.63\,\text{m}$, $1.996\,\text{m/s}$, $-0.03355\,\text{m/s}^2$ (c) $1400 + 39.5t\,\text{N}$]

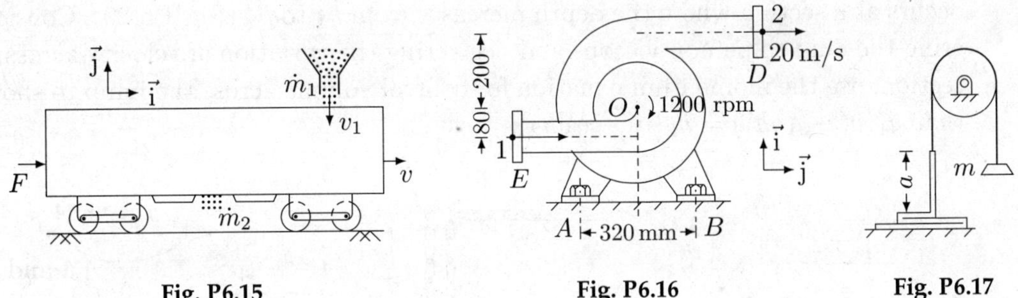

Fig. P6.15 Fig. P6.16 Fig. P6.17

16. A centrifugal pump, rotating at 1200 rpm, handles $25\,\text{m}^3$ of water per minute with inlet and outlet velocities of 20 m/s (Fig. P6.16). The motor delivers 50 kW to the impeller. The vertical reactions at the bearings A and B, when the pump is filled but not rotating, are 300 N each. Find the forces exerted by the foundation on the pump at A and B when it is running. To simplify the analysis you may assume that the tensions in connecting pipes at E and D are balanced by the respective forces due to the static pressure of water. [6348 N up, 5748 N down]

17. (a) An open-linked chain of mass σ per unit length is loosely coiled on the floor. Attached to its one end is a light cord which passes over a smooth pulley and is tied to a block of mass m (Fig. P6.17). A length a ($m > \sigma a$) of the chain is initially raised from the floor and the system is released from rest. Find the maximum height reached by the upper end of the chain. (b) Instead of an open-linked chain rework for a flexible cable for which there is no loss of energy.

[(a) $-\frac{1}{2}a + (3m^2/\sigma^2 - \frac{3}{4}a^2)^{1/2}$ (b) a or $2m/\sigma - a$]

18. A 1-stage rocket has mass $m_0 = 15\,\text{Mg}$ at take-off including mass $m_p = 12\,\text{Mg}$ of fuel. Each stage of a 2-stage rocket has mass $\frac{1}{2}m_0 = 7.5\,\text{Mg}$ at take-off including

mass $\frac{1}{2}m_p = 6\,\text{Mg}$ of fuel. The engines of both rockets consume fuel at the rate of $\dot{m}_p = 150\,\text{kg/s}$ exhausting them at relative velocity $u = 3\,\text{km/s}$. Find the maximum speed and height at burnout for the two rockets in vertical flight.

[4.044 km/s, 112.04 km; 5.576 km/s, 129.7 km]

19. A wheel of radius a can turn freely about a horizontal axis. A ribbon of charge is spread along its rim and ignited at $t = 0$. The rim loses mass at constant rate $\dot{m}_c$. Simultaneously, gas is taken from the atmosphere at rest at the constant rate $\dot{m}_a$. This total mass is discharged tangentially from the rim at velocity v with reference to the wheel. Find the angular velocity of the wheel and the angle turned by it at time t. The initial axial moment of inertia of the wheel is I. Neglect friction and gravity. $[\{1 - (1 - \dot{m}_c a^2 t/I)^{\dot{m}_a/\dot{m}_c}\}v(\dot{m}_a + \dot{m}_c)/a\dot{m}_a;$
$[(\dot{m}_c + \dot{m}_a)t + \{(1 - \dot{m}_c a^2 t/I)^{1+\dot{m}_a/\dot{m}_c} - 1\}I/a^2]v/a\dot{m}_a]$

20. For flow in an open channel of rectangular section of width w, a hydraulic jump occurs at a section where the depth increases from h_1 to h_2 (Fig. P6.20). Considering the fluid to be non-viscous and neglecting the variation of velocity across a section, use the momentum equation for control volume across the jump to show that $h_1(v_1^2 + \frac{1}{2}gh_1) = h_2(v_2^2 + \frac{1}{2}gh_2)$.

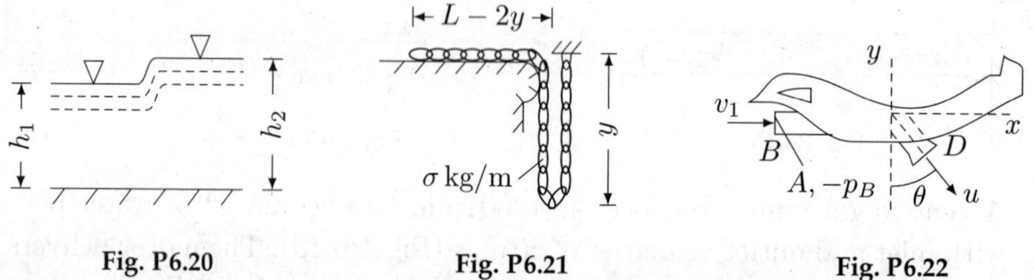

Fig. P6.20 Fig. P6.21 Fig. P6.22

21*. A chain is released from rest when $y = b$ (Fig. P6.21). Find the speed v_1 of its moving part when it just leaves the table and when its free end drops by $L/3$ from the table. The coefficient of friction of the table is μ. Work out for: (a) An open-linked chain. (b) A flexible cable/cycle chain with small sprockets at the bends, at the bottom and at the table, assuming conservation of energy.

$[(a)\ v_1^2 = 2(1+\mu)gL\ln\{2(1 - b/L)\} - (1 + 2\mu)g(L - 2b),\ \sqrt{v_1^2 + 2gL/3}$
$(b)\ [g\{(L^2 - 4b^2) - \mu(L - 2b)^2\}/L]^{1/2},\ [\{7L^2 - 18b^2 - 18\mu(0.5L - b)^2\}g/3L]^{1/2}]$

22. The vertical take-off and landing airplane (Fig. P6.22) sucks in air (ρ) at rate $\dot{m}_a$ at B across area A where the gauge pressure is $-p_B$. The fuel is consumed at the rate $\dot{m}_f$ with $\dot{m}_a/\dot{m}_f = n$. The exhaust at D is at speed of u with reference to the airplane. For forward flying, $\theta = 90°$. Find θ for vertical take-off with acceleration a. $[\tan^{-1}\{(\dot{m}_a^2/A\rho - p_B A)/m(a + g)\}]$

E. VIBRATION OF SINGLE D.O.F. LINEAR SYSTEMS

Vibratory motion is oscillation about an equilibrium position. Its harmful effects are: noise, passenger discomfort, wear, etc. Vibration is useful in musical instruments and audio equipment, in mechanical shakers for sieving and in fatigue testing machines. Ultrasonic methods are used in health care and for non-destructive testing. The control of vibrations requires careful analysis since intuition may sometimes be erroneous. We restrict to single degree of freedom linear system under a deterministic forcing function.

6.12 Equation of Motion

Consider a single dof system with a generalised coordinate x, shown schematically in Fig. 6.25a, where m is the *effective mass* (*inertia* of the system), k is the *effective spring constant (stiffness* due to the elastic *generalised restoring force* of the flexible elements) and c is its *effective coefficient of viscous damping* (the *energy loss mechanism*). F is the *effective generalised excitation force* representing the external time-dependent force.

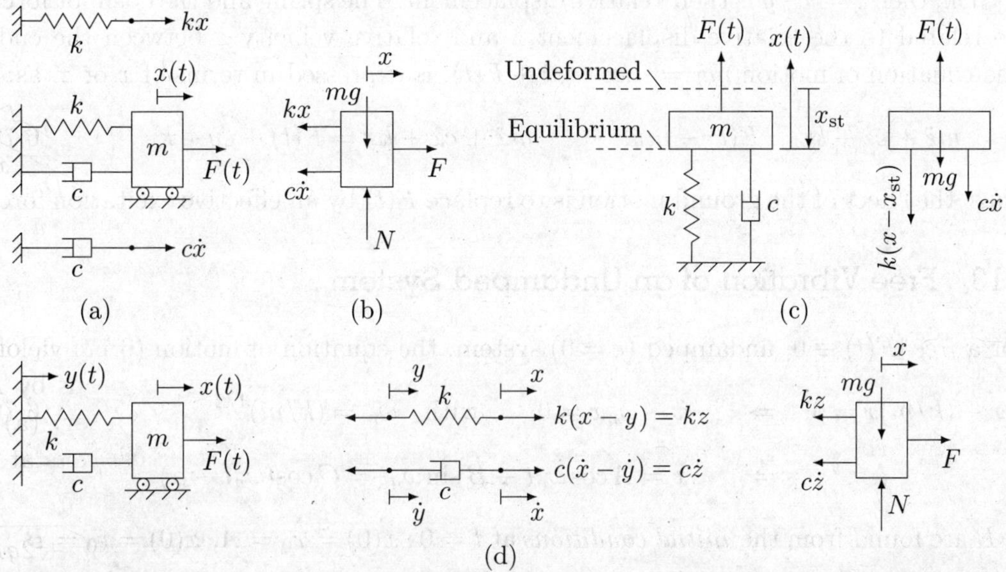

Fig. 6.25

The spring (massless, linear), damper/dashpot (massless, linear) and mass (rigid) represent effective k, c and m. The generalised restoring force due to k is kx, and due to damping is $c\dot{x}$, proportional to the generalised velocity $\dot{x}$ (v). A dashpot represents the source of viscous damping force which is proportional to the velocity of the piston relative to the cylinder. If x has unit of length, then c has unit of force per unit speed, i.e., N.s.m^{-1}. m, k, c and F are obtained from the kinetic energy T, potential energy V, the virtual work δW_d of the damping forces and the virtual work δW_e of the excitation forces:

$$T = \tfrac{1}{2}m\dot{x}^2, \quad V = \tfrac{1}{2}kx^2, \quad \delta W_d = (-c\dot{x})\delta x, \quad \delta W_e = F\delta x \tag{6.76}$$

Lagrange's equation or $m\ddot{x} = F_x$ for system (Fig. 6.25b), yields: $m\ddot{x} = F - kx - c\dot{x} \Rightarrow$

$$m\ddot{x} + c\dot{x} + kx = F(t) \tag{6.77}$$

It is *more convenient to take $x = 0$ at the equilibrium configuration.* For the spring–mass–damper system in Fig. 6.25c, $x_{\text{st}} = mg/k$ is the deflection in the equilibrium position under gravity load mg. Its equation of motion is

$$m\ddot{x} = F - k(x - x_{\text{st}}) - c\dot{x} - mg \qquad \Rightarrow \qquad m\ddot{x} + c\dot{x} + kx = F$$

Consider the effect of motion of the support on the system shown in Fig. 6.25d. Let x and y be the displacements of the system and the support from the equilibrium configuration, then $z = x - y$ is their relative displacement. The spring and the dashpot forces are related to the relative displacement z and relative velocity $\dot{z}$ between the ends. The equation of motion $m\ddot{x} = -kz - c\dot{z} + F(t)$, is expressed in terms of x or z, as:

$$m\ddot{z} + c\dot{z} + kz = F(t) - m\ddot{y}, \qquad m\ddot{x} + c\dot{x} + kx = F(t) + c\dot{y} + ky \tag{6.78}$$

Thus, the effect of the ground motion is to replace $F(t)$ by an effective excitation force.

6.13 Free Vibration of an Undamped System

For a *free* $[F(t) \equiv 0]$ undamped $(c = 0)$ system, the equation of motion (6.77) yields

$$\ddot{x} + (k/m)x = 0 \quad \Rightarrow \quad \ddot{x} + \omega_n^2 x = 0 \qquad \text{with} \quad \omega_n = (k/m)^{1/2} \tag{6.79}$$

$$\Rightarrow \quad x = A\cos\omega_n t + B\sin\omega_n t = C\cos(\omega_n t - \alpha)$$

A, B are found from the *initial conditions* at $t = 0$: $x(0) = x_0 = A,\ \dot{x}(0) = v_0 = B\omega_n$:

$$x = x_0\cos\omega_n t + \frac{v_0}{\omega_n}\sin\omega_n t = \left(x_0^2 + v_0^2/\omega_n^2\right)^{1/2}\cos(\omega_n t - \alpha) \tag{6.80}$$

where $\cos\alpha = \dfrac{x_0}{C}, \quad \sin\alpha = \dfrac{v_0}{\omega_n C}, \quad \tan\alpha = \dfrac{v_0}{\omega_n x_0}, \quad C = \left(x_0^2 + v_0^2/\omega_n^2\right)^{1/2}$

This *simple harmonic motion* repeats itself, i.e., it is periodic with a time period or *natural period T_n of free vibration*: $T_n = 2\pi/\omega_n$. C and α are called the *amplitude* and the *phase angle*, respectively. The *natural frequency*, f_n, is the number of periods (cycles) per unit time: $f_n = 1/T_n = (\omega_n/2\pi)$ Hz where Hz, is the abbreviation of the unit Hertz (cycles per second). The natural *circular frequency* is ω_n. Often, ω_n itself is called the natural frequency. It is remarkable that the *time period is independent of the initial conditions.* The x-t, v-t and the x-(v/ω_n) curves are plotted in Fig. 6.26.

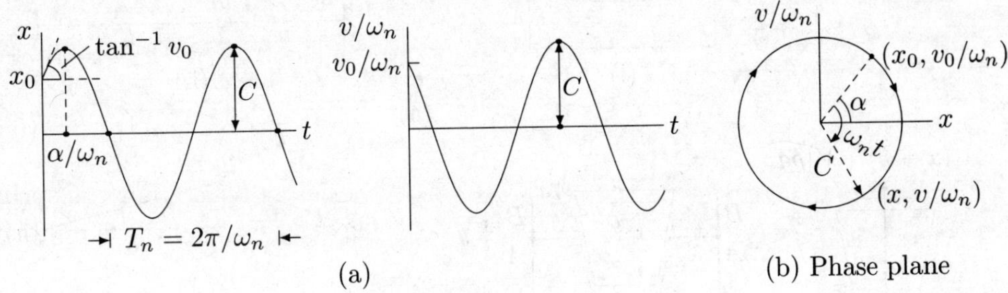

Fig. 6.26

The energy conservation relation for the conservative spring–mass system is

$$T + V = \tfrac{1}{2}m\dot{x}^2 + \tfrac{1}{2}kx^2 = \tfrac{1}{2}mv_0^2 + \tfrac{1}{2}kx_0^2 = E_0$$

$$\Rightarrow \quad x^2 + (v/\omega_n)^2 = 2E_0/k = C^2 \tag{6.81}$$

where mechanical energy $E = E_0$ is *constant.* When $x = x_{\text{max}} = C:\ \dot{x} = 0$ and $E_0 = V_{\text{max}} = \frac{1}{2}kC^2$. When $x = 0:\ \dot{x} = \pm\omega_n C = \pm v_{\text{max}},\ V = 0$ and $E_0 = T_{\text{max}} = \frac{1}{2}m(\omega_n C)^2$. Thus, with the equilibrium position as datum for V, $E_0 = V_{\text{max}} = T_{\text{max}}$. This is an alternative method (*Rayleigh's method*) *of finding* ω_n *by using* $V_{\text{max}} = T_{\text{max}}$:

$$V_{\text{max}} = \tfrac{1}{2}kx_{\text{max}}^2 = \tfrac{1}{2}kC^2 = T_{\text{max}} = \tfrac{1}{2}mv_{\text{max}}^2 = \tfrac{1}{2}m(\omega_n C)^2 \quad \Rightarrow \quad \omega_n = (k/m)^{1/2}$$

Example 6.14: Find the period of free vibration (Fig. E6.14a). (i) Vertical vibration of a ship floating in a canal lock of area A_1 with submerged volume V. (ii) A small sphere attached to a stretched cable under large tension T_0.

Solution: (i) Let x be the displacement of the ship from its equilibrium position. The additional volume of the ship, submerged below the water-line AA of equilibrium is A_2x. The water so displaced rises by y to the new water-line BB such that the total volume of water is the same: $(A_1 - A_2)y = A_2x$. The volume of the ship under the new water line is $V + (x+y)A_2$. The FBD is shown in Fig. E6.14b. $ma = F \Rightarrow$

$$m\ddot{x} = mg - \rho[V + (x+y)A_2]g \tag{1}$$

At equilibrium: $x = y = 0,\ \ddot{x} = 0 \Rightarrow mg = \rho Vg$ and $y = A_2x/(A_1 - A_2)$, eq. (1)$\Rightarrow$

$$\ddot{x} + \frac{A_1A_2g}{V(A_1 - A_2)}x = 0 \quad \Rightarrow \quad \omega_n^2 = \frac{A_1A_2g}{V(A_1 - A_2)} \quad \Rightarrow \quad T = \frac{2\pi}{\omega_n}$$

If $A_1 \gg A_2$, then $T \simeq 2\pi/[A_2 g/V]^{1/2}$, e.g., for a lactometer floating in a wide vessel.

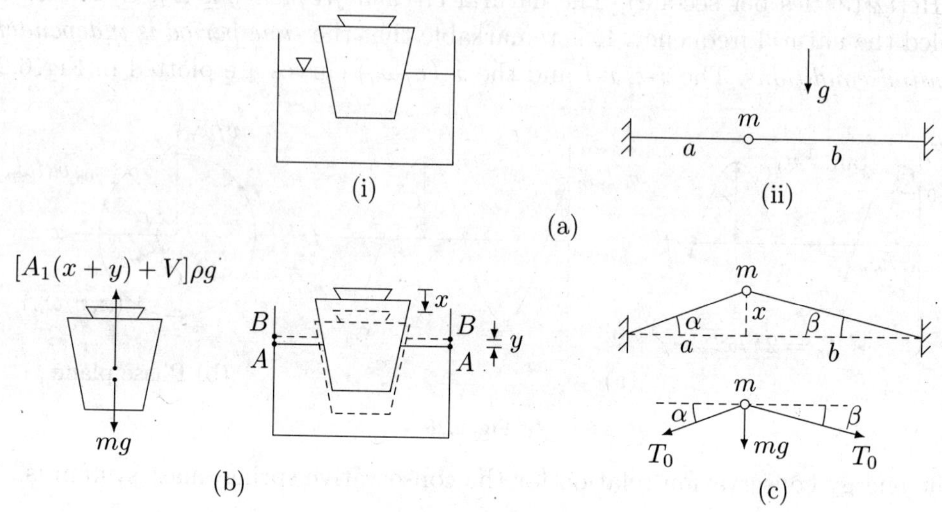

Fig. E6.14

(ii) Let x be the small displacement of the sphere (Fig. E6.14c), i.e., $x \ll a,\ x \ll b \Rightarrow$

$$\tan\alpha = \frac{x}{a} \ll 1, \qquad \tan\beta = \frac{x}{b} \ll 1 \quad \Rightarrow \quad \alpha \simeq \frac{x}{a} \ll 1, \quad \beta \simeq \frac{x}{b} \ll 1$$

We neglect the small increase in T_0 due to small additional stretching. $m\ddot{x} = F_x$:

$$m\ddot{x} = -mg - T_0\sin\alpha - T_0\sin\beta \simeq -mg - T_0(\alpha+\beta)$$

$$\Rightarrow \qquad \ddot{x} + \frac{T_0}{m}\left(\frac{1}{a}+\frac{1}{b}\right)x = -mg \qquad \Rightarrow \qquad T = 2\pi \Big/ \left[\frac{T_0}{m}\left(\frac{1}{a}+\frac{1}{b}\right)\right]^{1/2}$$

Example 6.15: (a) Find the natural frequency of small oscillations of the systems in Fig. E6.15 using Rayleigh's energy method. (i) The equilibrium position of the rod is vertical. (ii) A cylinder rolling without slip. (iii) A circular plate held by 3 wires and vibrating about its vertical axis. (iv) A block moving vertically and restrained by spring force $k_1x + k_2x^3$ for an extension x. (b) Find the minimum coefficient of friction to ensure no slip for case (ii) if the amplitude of motion is a large angle θ_0 $(<\frac{1}{2}\pi)$.

Solution: (a) (i) In the equilibrium position $\theta = 0$, let the extension of k_1 be x and the twist of k_2 be α. With the equilibrium position, $\theta = 0$, as the datum

$$V = -mga\theta - (Mg)\tfrac{1}{2}L(1-\cos\theta) + \tfrac{1}{2}k_1\left[(x+b\theta)^2 - x^2\right] + \tfrac{1}{2}k_2\left[(\theta+\alpha)^2 - \alpha^2\right]$$

$$\frac{dV}{d\theta} = -mga - \tfrac{1}{2}MgL\sin\theta + k_1(x+b\theta)b + k_2(\theta+\alpha)$$

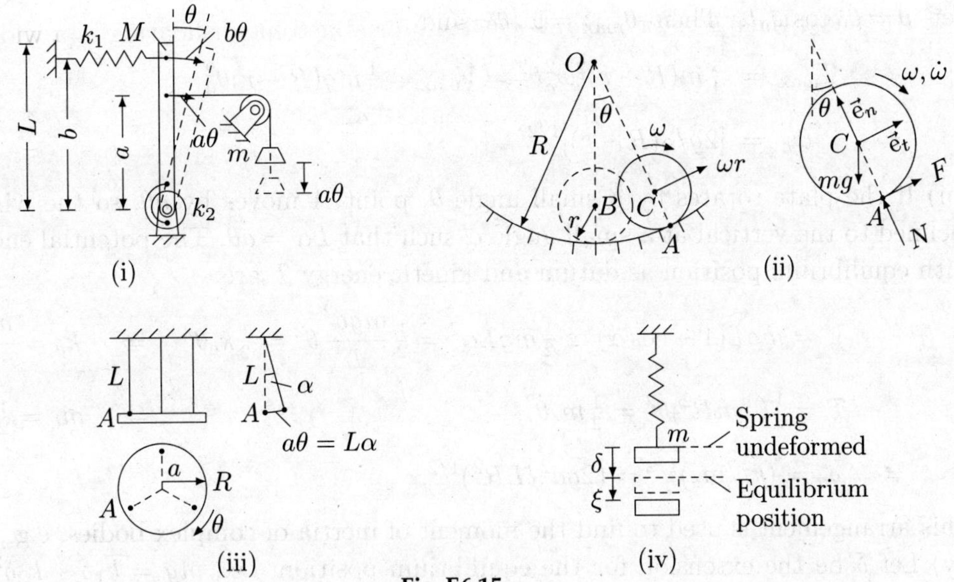

Fig. E6.15

Equilibrium at $\theta = 0 \;\Rightarrow\; (dV/d\theta)|_{\theta=0} = -mga + k_1xb + k_2\alpha = 0$. Hence,

$$\begin{aligned} V &= (-mga + k_1xb + k_2\alpha)\theta - MgL\sin^2\tfrac{1}{2}\theta + \tfrac{1}{2}(k_1b^2 + k_2)\theta^2 \\ &\simeq \tfrac{1}{2}\left(k_1b^2 + k_2 - \tfrac{1}{2}MgL\right)\theta^2 \\ T &= \tfrac{1}{2}(\tfrac{1}{3}ML^2)\dot{\theta}^2 + \tfrac{1}{2}m(a\dot{\theta})^2 = \tfrac{1}{2}\left(\tfrac{1}{3}ML^2 + ma^2\right)\dot{\theta}^2 \end{aligned}$$

Let $\theta = \theta_0\cos\omega_n t$. Then $\dot{\theta}_{\max} = \omega_n\theta_0$ and

$$\begin{aligned} T_{\max} &= \tfrac{1}{2}\left(\tfrac{1}{3}ML^2 + ma^2\right)\omega_n^2\theta_0^2 = V_{\max} = \tfrac{1}{2}\left(k_1b^2 + k_2 - \tfrac{1}{2}MgL\right)\theta_0^2 \\ \Rightarrow\quad \omega_n &= \left[\left(k_1b^2 + k_2 - \tfrac{1}{2}MgL\right)\Big/\left(\tfrac{1}{3}ML^2 + ma^2\right)\right]^{1/2} \end{aligned}$$

There will be vibration only if ω_n is real, i.e., if $k_1b^2 + k_2 > \frac{1}{2}MgL$.

(ii) C moves in a circle of radius $R - r$. Since the cylinder rolls without slip ($\vec{v}_A = \vec{0}$), the instantaneous centre is at A. If ω is the angular velocity of the cylinder, then

$$v_C = (R-r)\dot{\theta} = (AC)\omega = \omega r \quad\Rightarrow\quad \omega = (R-r)\dot{\theta}/r$$

Using $\cos\theta \simeq 1 - \frac{1}{2}\theta^2$ for small θ, V with datum at $\theta = 0$ (B) and kinetic energy T are:

$$\begin{aligned} V &= mg(R-r)(1-\cos\theta) \simeq \tfrac{1}{2}mg(R-r)\theta^2 \\ T &= \tfrac{1}{2}m(\omega r)^2 + \tfrac{1}{2}(\tfrac{1}{2}mr^2)\omega^2 = \tfrac{3}{4}mv_C^2 = \tfrac{3}{4}m(R-r)^2\dot{\theta}^2 \end{aligned}$$

Let $\theta = \theta_0 \cos \omega_n t$. Then $\dot{\theta}_{\max} = \omega_n \theta_0$ and

$$T_{\max} = \tfrac{3}{4} m(R-r)^2 \omega_n^2 \theta_0^2 = V_{\max} = \tfrac{1}{2} mg(R-r)\theta_0^2$$

$$\Rightarrow \quad \omega_n = [2g/3(R-r)]^{1/2}$$

(iii) If the plate rotates by a small angle θ, point A moves by $a\theta$, so the wire gets inclined to the vertical at a small angle α such that $L\alpha = a\theta$. The potential energy V with equilibrium position as datum and kinetic energy T are:

$$V = mgL(1-\cos\alpha) \simeq \tfrac{1}{2} mgL\alpha^2 = \tfrac{1}{2}\frac{mga^2}{L}\theta^2 = \tfrac{1}{2} k_e \theta^2 \quad \Rightarrow \quad k_e = \frac{mga^2}{L}$$

$$T = \tfrac{1}{2}(\tfrac{1}{2} mR^2)\dot{\theta}^2 = \tfrac{1}{2} m_e \dot{\theta}^2 \quad \Rightarrow \quad m_e = \tfrac{1}{2} mR^2$$

$$\Rightarrow \quad \omega_n = (k_e/m_e)^{1/2} = (2ga^2/LR^2)^{1/2}$$

This arrangement is used to find the moment of inertia of complex bodies, e.g., a car.
(iv) Let δ be the extension for the equilibrium position, i.e., $mg = k_1\delta + k_2\delta^3$. The potential energy V_s for the spring is $V_s = \frac{1}{2} k_1 x^2 + \frac{1}{4} k_2 x^4$. The potential energy V for extension $\delta + \xi$ with equilibrium position as datum and kinetic energy T are:

$$V(\xi) = \tfrac{1}{2} k_1(\delta+\xi)^2 + \tfrac{1}{4} k_2(\delta+\xi)^4 - mg\xi - \tfrac{1}{2} k_1\delta^2 - \tfrac{1}{4} k_2\delta^4$$

$$V(\xi) = V(0) + V'(0)\xi + \tfrac{1}{2} V''(0)\xi^2 + \ldots \simeq \tfrac{1}{2} V''(0)\xi^2$$

$$= \tfrac{1}{2}(k_1 + 3k_2\delta^2)\xi^2 = \tfrac{1}{2} k_e \xi^2 \quad \Rightarrow \quad k_e = k_1 + 3k_2\delta^2$$

$$T = \tfrac{1}{2} m\dot{\xi}^2 = \tfrac{1}{2} m_e \dot{\xi}^2 \quad \Rightarrow \quad m_e = m$$

$$\omega_n = (k_e/m_e)^{1/2} = [(k_1 + 3k_2\delta^2)/m]^{1/2}$$

(b) $T + V = T_0 + V_0$ yields v_C. F and N are found from the equations of motion.

$$T + V = T_0 + V_0: \qquad \tfrac{3}{4} mv_C^2 + mg(R-r)(1-\cos\theta) = mg(R-r)(1-\cos\theta_0)$$

$$\Rightarrow \quad v_C^2 = \tfrac{4}{3} g(R-r)(\cos\theta - \cos\theta_0)$$

$$F_n = ma_{C_n}: \qquad N - mg\cos\theta = mv_C^2/(R-r)$$

$$\Rightarrow \quad N = mg(7\cos\theta - 4\cos\theta_0)/3$$

$$F_t = ma_{C_t}: \qquad F - mg\sin\theta = m\dot{\omega}r \tag{1}$$

$$M_{A_z} = I_{zz}^A \dot{\omega}: \qquad \tfrac{3}{2} mr^2\dot{\omega} = -mgr\sin\theta \tag{2}$$

$$\text{Equations (1), (2)} \Rightarrow \quad F = \tfrac{1}{3} mg\sin\theta \quad \Rightarrow \quad \frac{|F|}{N} = \frac{\sin\theta}{7\cos\theta - 4\cos\theta_0}$$

Its maximum value occurs for $\theta = \theta_{\max} = \theta_0 \quad \Rightarrow \quad \mu_{\min} = (|F|/N)_{\max} = \frac{1}{3}\tan\theta_0$.

6.14 Free Vibration of a Damped System

The analysis is restricted to linear damping due to a viscous frictional force proportional to velocity. The equation of motion (6.77) for free vibration is

$$\ddot{x} + \frac{c}{m}\dot{x} + \frac{k}{m}x = 0 \tag{6.82}$$

where $(k/m) = \omega_n^2$. Its solution is $x = Ae^{\lambda t}$ with λ satisfying the characteristic equation

$$\lambda^2 + \frac{c}{m}\lambda + \frac{k}{m} = 0, \qquad \text{i.e.,} \qquad \lambda_{1,2} = -\frac{c}{2m} \pm \left(\frac{c^2}{4m^2} - \frac{k}{m}\right)^{1/2}$$

The solution depends on the nature of the roots λ_1, λ_2, i.e., on the sign of the discriminant which is zero for a value of $c = c_{\text{cr}}$, called the *critical damping coefficient*

$$c_{\text{cr}} = 2\sqrt{km} = 2m\omega_n = 2k/\omega_n \tag{6.83}$$

We define the *damping ratio* ξ as the ratio of the actual damping c to c_{cr}.

$$\xi = \frac{c}{c_{\text{cr}}} = \frac{c}{2\sqrt{km}} = \frac{c}{2m\omega_n} = \frac{c\,\omega_n}{2k} \tag{6.84}$$

Thus, $\lambda_{1,2} = -\xi\omega_n \pm \omega_n(\xi^2 - 1)^{1/2}$ and three cases aris: $\xi > 1,\ \xi = 1,\ \xi < 1$.

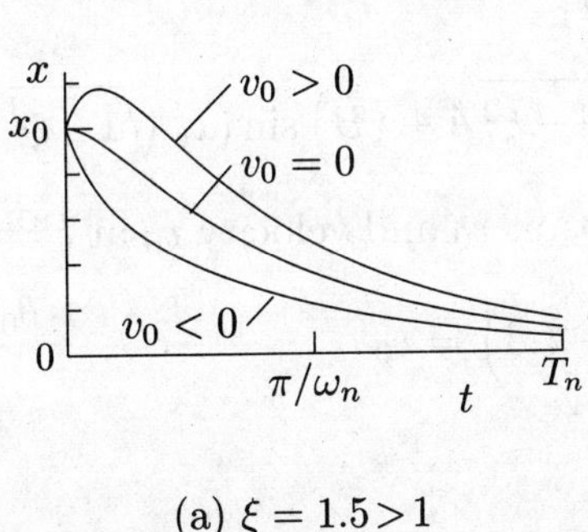

(a) $\xi = 1.5 > 1$

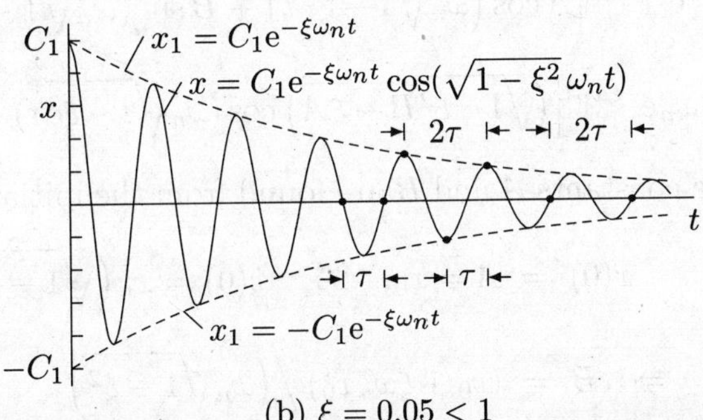

(b) $\xi = 0.05 < 1$

Fig. 6.27

1. Over-damped, $\xi > 1$.

The roots are real and negative: $\lambda_{1,2} = -\xi\omega_n \pm \omega_n\sqrt{\xi^2 - 1}$. The complete solution is

$$x = Ae^{\lambda_1 t} + Be^{\lambda_2 t} = e^{-\xi\omega_n t}\left[Ae^{\omega_n\sqrt{\xi^2-1}\,t} + Be^{-\omega_n\sqrt{\xi^2-1}\,t}\right]$$

$$\dot{x} = \omega_n e^{-\xi\omega_n t}\left[A\left(\sqrt{\xi^2-1}-\xi\right)e^{\omega_n\sqrt{\xi^2-1}\,t} - B\left(\sqrt{\xi^2-1}+\xi\right)e^{-\omega_n\sqrt{\xi^2-1}\,t}\right]$$

The velocity $\dot{x} = 0$ if there exists a time t such that

$$\exp\left(2\omega_n\sqrt{\xi^2-1}\,t\right) = B\left(\sqrt{\xi^2-1}+\xi\right)\Big/A\left(\sqrt{\xi^2-1}-\xi\right)$$

$\Rightarrow$ $\dot{x} = 0$ at most at one instant (when A and B are of opposite sign) (Fig. 6.27a). The motion is non-oscillatory or *dead-beat* and $x \to 0$ exponentially as $t \to \infty$. Such high damping is useful in the recoil mechanism of field guns to absorb shock.

2. Critically damped, $\xi = 1$.

The roots are $\lambda_1 = \lambda_2 = -\omega_n$. The complete solution is

$$x = (A + Bt)e^{-\omega_n t}, \qquad \dot{x} = [-\omega_n(A+Bt) + B]e^{-\omega_n t} \tag{6.85}$$

where $A = x_0$, $B = v_0 + \omega_n x_0$. $\dot{x} = 0$ if $t_1 = (B - \omega_n A)/B\omega_n = v_0/\omega_n(v_0 + \omega_n x_0) > 0$ exists, i.e., if $v_0 > 0$ or $v_0 < -\omega_n x_0$. The x-t curve may cross the x-axis at most once (if $v_0 < -\omega_n x_0$). The motion is non-oscillatory with $x \to 0$ as $t \to \infty$. We shall see next that for $\xi < 1$ motion is oscillatory. The motion changes from oscillatory to non-oscillatory for $c = c_{cr}$ ($\xi = 1$), which is aptly called the critical damping coefficient.

3. Under-damped, $\xi < 1$.

$\lambda_{1,2} = -\xi\omega_n \pm i\omega_n\sqrt{1-\xi^2}$. Using $e^{\pm i\theta} = \cos\theta \pm i\sin\theta$, the complete solution is

$$x = e^{-\xi\omega_n t}\left[A\cos\left(\omega_n\sqrt{1-\xi^2}\,t\right) + B\sin\left(\omega_n\sqrt{1-\xi^2}\,t\right)\right]$$

$$\dot{x} = \omega_n e^{-\xi\omega_n t}\left[\left(\sqrt{1-\xi^2}B - \xi A\right)\cos\left(\omega_n\sqrt{1-\xi^2}\,t\right) - \left(\sqrt{1-\xi^2}A + \xi B\right)\sin\left(\omega_n\sqrt{1-\xi^2}\,t\right)\right]$$

The constants A and B are found from the initial position x_0 and velocity v_0 at $t = 0$:

$$x(0) = A = x_0, \qquad \dot{x}(0) = \omega_n\left(\sqrt{1-\xi^2}B - \xi A\right) = v_0$$

$$\Rightarrow \quad B = (v_0 + \xi\omega_n x_0)\Big/\left(\omega_n\sqrt{1-\xi^2}\right)$$

$$x = e^{-\xi\omega_n t}\left[x_0\cos\left(\omega_n\sqrt{1-\xi^2}\,t\right) + \frac{v_0 + \xi\omega_n x_0}{\omega_n\sqrt{1-\xi^2}}\sin\left(\omega_n\sqrt{1-\xi^2}\,t\right)\right] \tag{6.86}$$

$$= Ce^{-\xi\omega_n t}\cos\left(\omega_n\sqrt{1-\xi^2}\,t - \phi\right)$$

The motion is not periodic. The amplitude decays exponentially with $x \to 0$ as $t \to \infty$. A suitable change of the instant from which t is measured, simplifies eq. (6.86) to

$$x = C_1 e^{-\xi\omega_n t}\cos\left(\omega_n\sqrt{1-\xi^2}\,t\right) \tag{6.87}$$

where C_1 is a constant. We compare x with $x_1,\, x_2 = \pm C_1 e^{-\xi\omega_n t}$ in Fig. 6.27b. Since $-1 < \cos(\omega_n\sqrt{1-\xi^2}\,t) < 1$, we have $x_2 < x < x_1$. The x-t curve touches the x_i-t curve, $i = 1$, 2, at $t = (2n-1+i)\pi/(\omega_n\sqrt{1-\xi^2})$, where n is an integer.

The maxima and the minima of x occurs at the instant when

$$\dot{x} = \omega_n C_1 e^{-\xi\omega_n t}\Big[-\xi\cos\left(\omega_n\sqrt{1-\xi^2}\,t\right) - \sqrt{1-\xi^2}\sin\left(\omega_n\sqrt{1-\xi^2}\,t\right)\Big] = 0$$

$$\Rightarrow \quad \tan\left(\omega_n\sqrt{1-\xi^2}\,t\right) = -\xi\Big/\sqrt{1-\xi^2}$$

The interval τ between successive maxima and minima is given by

$$\omega_n\sqrt{1-\xi^2}\,\tau = \pi, \quad \text{i.e.,} \quad \tau = \pi\Big/\omega_n\sqrt{1-\xi^2}$$

Thus, the time interval between two successive extreme positions on the same side of the equilibrium position is 2τ. Similarly, the mean position $x = 0$ occurs at the times t given by $\cos(\omega_n\sqrt{1-\xi^2}\,t) = 0$. The interval between two successive mean positions is also τ. The time interval between the successive passages through the equilibrium position in the same direction is 2τ. The motion is not periodic, but since the interval between successive mean and extreme positions is the same, we *may* refer to $2\tau = 2\pi/(\omega_n\sqrt{1-\xi^2})$ as the period T_d and $\omega_n\sqrt{1-\xi^2}$ as the *circular frequency* ω_d *of free damped* vibrations. $T_d > T_n = 2\pi/\omega_n$, but the increase is slight for small values of ξ. The damping ratio ξ for real structures is usually between 0.02 to 0.2. Even for $\xi = 0.2$, $\omega_d = \omega_n\sqrt{1-\xi^2} \simeq 0.98\omega_n$. Thus, in practice, for most cases $\omega_d \simeq \omega_n$.

Using eq. (6.87), successive extreme values, x_n at t_n and x_{n+1} at $t_{n+1} = t_n + 2\tau$ on the same side, are related by

$$\frac{x_{n+1}}{x_n} = \frac{C_1 e^{-\xi\omega_n t_{n+1}}}{C_1 e^{-\xi\omega_n t_n}} = e^{-\xi\omega_n(t_{n+1}-t_n)} = e^{-2\xi\omega_n\tau} = e^{-2\xi\pi/\sqrt{1-\xi^2}}$$

The logarithmic decrement, δ, is defined as the natural logarithm of the ratio of the successive maxima on the same side,

$$\delta = \ln\frac{x_n}{x_{n+1}} = \frac{2\pi\xi}{\sqrt{1-\xi^2}} \quad \Rightarrow \quad \frac{x_n}{x_{n+1}} = e^{\delta}, \quad \xi = \frac{\delta}{(4\pi^2+\delta^2)^{1/2}} \tag{6.88}$$

We find ξ and ω_n from measured T_d and δ, and c and k are obtained from given m.

For small values of ξ, $\xi^2 \ll 1$ and eq. (6.88) yields $\delta \simeq 2\pi\xi$, $\xi \simeq \delta/2\pi$. Experimentally, δ is evaluated by measuring maxima after p cycles:

$$\frac{x_n}{x_{n+p}} = (e^{\delta})^p = e^{p\delta} \quad \Rightarrow \quad p\,\delta = \ln\frac{x_n}{x_{n+p}}$$

The loss in V from one extreme x_n to the next x_{n+1} at the end of one cycle is $V_i - V_f = \frac{1}{2}k(x_n^2 - x_{n+1}^2)$, where V_i, V_f are potential energies at the beginning and at the end of a cycle. The fractional loss of potential energy per cycle is the same for all cycles:

$$\frac{V_i - V_f}{V_i} = \frac{x_n^2 - x_{n+1}^2}{x_n^2} = 1 - \mathrm{e}^{-2\delta} = 1 - (1 - 2\delta + \ldots) \simeq 2\delta \simeq 4\pi\xi, \qquad \text{if } \delta \ll 1$$

Thus, δ provides a measure of the rate of dissipation of energy by damping forces.

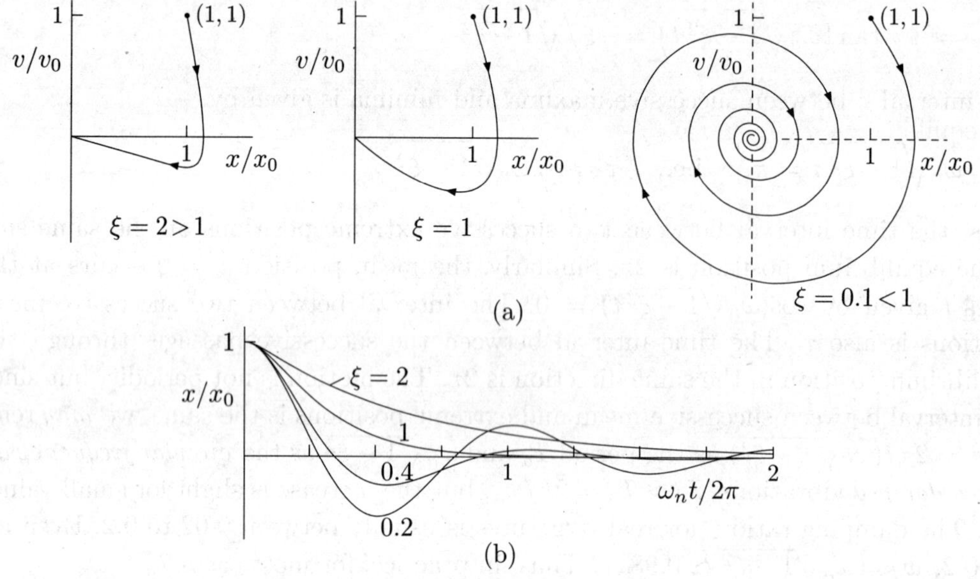

Fig. 6.28

The three cases of damped oscillations are illustrated in phase plane in Fig. 6.28a. The response for $x(0) = x_0$, $\dot{x}(0) = 0$, is shown in Fig. 6.28b for several values of ξ.

For given initial conditions, the critically damped system approaches the equilibrium position the fastest. Hence, such systems are used in door closers and dead-weight scales. In instruments which measure steady-state values, it is better to use $\xi = 0.7$, since with a little overshoot, the steady state value is attained faster.

Example 6.16: Find the critical damping coefficient for the system in Fig. E6.16a. If $c < c_{\mathrm{cr}}$ and the light rod AB is displaced by θ_0, find the time taken for the first return to the equilibrium position. The system is in equilibrium when AB is horizontal.

Solution: In the equilibrium position, let the extension of spring be δ and the twist of the torsional spring be α. The FBD is shown in Fig. E6.16b. $I_{zz}^O\ddot{\theta} = M_{Oz}$ yields

$$m(a^2 + b^2)\ddot{\theta} = mg(a\cos\theta + b\sin\theta) - k_1(\theta + \alpha) - (c\,d\,\dot{\theta})d - k_2(e\theta + \delta)e$$

For small oscillations, $\cos\theta \simeq 1,\ \sin\theta \simeq \theta$ and the above equation simplifies to

$$m(a^2+b^2)\ddot{\theta} + c\,d^2\dot{\theta} + (k_1 + k_2e^2 - mgb)\theta = mga - k_1\alpha - k_2\delta e \qquad (1)$$

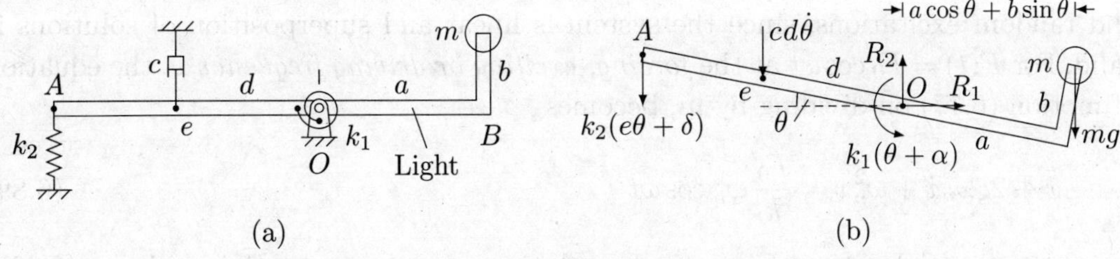

Fig. E6.16

At equilibrium, $\theta = \dot{\theta} = \ddot{\theta} = 0 \ \Rightarrow\ mga - k_1\alpha - k_2\delta e = 0$ and eq. (1) $\Rightarrow$

$$m(a^2+b^2)\ddot{\theta} + cd^2\dot{\theta} + (k_1 + k_2e^2 - mgb)\theta = 0$$

$$\Rightarrow \qquad m^*\ddot{\theta} + c^*\dot{\theta} + k^*\theta = 0$$

with $m^* = m(a^2+b^2),\ c^* = c\,d^2,\ k^* = k_1 + k_2e^2 - mgb$. The characteristic equation is $m^*\lambda^2 + c^*\lambda + k^* = 0$. Its discriminant $c^{*2} - 4m^*k^* = 0$ for $c^* = c^*_{\text{cr}} = 2\sqrt{m^*k^*} \Rightarrow$

$$\begin{aligned} c_{\text{cr}} &= 2[m(a^2+b^2)(k_1+k_2e^2-mgb)]^{1/2}/d^2 \\ \omega_n^2 &= \frac{k^*}{m^*} = \frac{k_1+k_2e^2-mgb}{m(a^2+b^2)} \\ \omega_d &= \omega_n\sqrt{1-\xi^2} = \omega_n\sqrt{1-(c/c_{cr})^2} \end{aligned}$$

For $\dot{\theta}(0) = 0,\ \theta(0) = \theta_0$, the solution is given by eq. (6.86) with x, x_0 and v_0 replaced by θ, θ_0 and $\dot{\theta}_0$, respectively. The system returns to $\theta = 0$ at $t = \tan^{-1}[-\sqrt{1-\xi^2}/\xi]/\omega_d$.

Example 6.17: A structure is modelled as a 1-dof system with $m = 1000\,\text{kg},\ \xi = 0.1$. A force of 4 kN causes a deflection of 0.2 cm. Find (a) ω_{n_d}, (b) δ and (c) the number of cycles for the amplitude to decrease to 10% and 1% of its initial value.

Solution: $k = P/\Delta = 4000/(0.2\times10^{-2}) = 2\times10^6\,\text{N/m},\ m = 1000\,\text{kg},\ \xi = 0.1$.

(a) $\omega_d = \sqrt{k/m}(1-\xi^2)^{1/2} = 44.50\,\text{rad/s}$

(b) $\delta = 2\pi\xi/(1-\xi^2)^{1/2} = 0.6315$

(c) $\dfrac{C_n}{C_0} = \dfrac{C_n}{C_{n-1}}\dfrac{C_{n-1}}{C_{n-2}}\cdots\dfrac{C_1}{C_0} = \left(\dfrac{C_1}{C_0}\right)^n \quad\Rightarrow\quad n = \dfrac{\ln(C_n/C_0)}{\ln(C_1/C_0)} = -\dfrac{\ln(C_n/C_0)}{\delta}$

$\Rightarrow C_n/C_0 = 0.1 \ \Rightarrow\ n = 3.65 \simeq 4$ cycles, $C_n/C_0 = 0.01 \ \Rightarrow\ n = 7.29 \simeq 8$ cycles.

6.15 Response to Harmonic Excitation

The source of periodic forcing function is some unbalance in the rotating parts. The response to harmonic excitation forms the basis of response to periodic, non-periodic and random excitations, since the system is linear and superposition of solutions is valid. For $F(t) = F_0 \cos \omega t$ at the *forcing, exciting or driving frequency* ω, the equation of motion (6.77) on dividing by m, becomes

$$\ddot{x} + 2\xi\omega_n\dot{x} + \omega_n^2 x = \frac{F_0}{k}\,\omega_n^2 \cos\omega t \tag{6.89}$$

Let $\beta = \omega/\omega_n$. Substitute the particular solution $x_p = A\cos\omega t + B\sin\omega t$ in eq. (6.89):

$$[(\omega_n^2 - \omega^2)A + 2\xi\omega_n\omega B]\cos\omega t + [-2\xi\omega_n\omega A + (\omega_n^2 - \omega^2)B]\sin\omega t = (F_0/k)\,\omega_n^2 \cos\omega t$$

$$\Rightarrow \quad (\omega_n^2 - \omega^2)A + 2\xi\omega_n\omega B = (F_0/k)\,\omega_n^2$$

$$-2\xi\omega_n\omega A + (\omega_n^2 - \omega^2)B = 0$$

$$\Rightarrow \quad A = (1-\beta^2)D^2\,\frac{F_0}{k}, \qquad B = 2\xi\beta D^2\,\frac{F_0}{k}$$

$$D = \left[(1-\beta^2)^2 + (2\xi\beta)^2\right]^{-1/2} \tag{6.90}$$

$$\Rightarrow \quad x_p = D^2\,\frac{F_0}{k}\,[(1-\beta^2)\cos\omega t + (2\xi\beta)\sin\omega t]$$

$$x_p = \frac{F_0/k}{[\,(1-\beta^2)^2 + (2\xi\beta)^2\,]^{1/2}}\cos(\omega t - \phi)$$

$$= X_s\cos(\omega t - \phi) \tag{6.91}$$

with

$$X_s = D\,\frac{F_0}{k}, \qquad \tan\phi = \frac{2\xi\beta}{1-\beta^2} \tag{6.92}$$

$$\cos\phi = \frac{1-\beta^2}{[\,(1-\beta^2)^2 + (2\xi\beta)^2\,]^{1/2}}, \qquad \sin\phi = \frac{2\xi\beta}{[\,(1-\beta^2)^2 + (2\xi\beta)^2\,]^{1/2}}$$

$D = X_s/(F_0/k)$ is the *dynamic magnification factor* defined as the ratio of the steady state amplitude X_s to the static deflection F_0/k. D depends on ξ and β.

The complete solution of eq. (6.89) is $x(t) = x_c(t) + x_p(t)$. For $0 < \xi < 1$,

$$x(t) = \mathrm{e}^{-\xi\omega_n t}(A_1\cos\omega_d t + B_1\sin\omega_d t) + D(F_0/k)\cos(\omega t - \phi) \tag{6.93}$$

A_1 and B_1 are obtained from $x(0) = x_0$, $\dot{x}(0) = v_0$ *using the complete equation* (6.93).

For large t, $x_c(t) \to 0$, $x(t) \to x_p(t)$ and so the complementary solution $x_c(t)$ is called a *transient solution.* The particular solution $x_p(t)$ does not vanish for large t and is called the *steady-state solution.*

Alternative solution: We present an elegant approach for response to harmonic excitation in terms of complex numbers, which is very convenient to find the response to arbitrary loads using frequency domain analysis. We write the force and x_p in terms of the real parts of the complex functions $F_0 e^{i\omega t}$ and $X e^{i\omega t}$ as

$$
\begin{aligned}
F_0 \cos\omega t &= \mathrm{Re}(F_0 e^{i\omega t}), \qquad x_p = \mathrm{Re}(X e^{i\omega t}) \\
\Rightarrow \quad \dot{x}_p &= \mathrm{Re}(i\omega X e^{i\omega t}), \qquad \ddot{x}_p = \mathrm{Re}(-\omega^2 X e^{i\omega t})
\end{aligned}
$$

Substituting these in eq. (6.89) yields

$$
\begin{aligned}
&\{(\omega_n^2 - \omega^2) + i2\xi\omega\omega_n\}X = F_0\omega_n^2/k \\
\Rightarrow \quad X &= F_0/k(1-\beta^2+i2\xi\beta) \\
\Rightarrow \quad x_p &= \mathrm{Re}\left[\frac{F_0}{k}\left(1-\beta^2+i2\xi\beta\right)^{-1} e^{i\omega t}\right] \\
&= \mathrm{Re}\left[\frac{F_0}{k}\left[\{(1-\beta)^2+(2\xi\beta)^2\}^{1/2}e^{i\phi}\right]^{-1} e^{i\omega t}\right] \\
&= \frac{F_0}{k}\left[(1-\beta)^2+(2\xi\beta)^2\right]^{-1/2} \cos(\omega t-\phi)
\end{aligned}
$$

The ratio of X to F_0/k is called the *complex frequency response* function $H(\omega)$:

$$
\begin{aligned}
H(\omega) &= \frac{X}{F_0/k} = [(1-\beta^2)+i2\xi\beta]^{-1}, \qquad |H(\omega)| = D \qquad (6.94) \\
x_p(t) &= \mathrm{Re}[H(\omega)(F_0/k)e^{i\omega t}]
\end{aligned}
$$

Some salient features of the parametric plots of D and ϕ (Figs 6.29 and 6.30), are:

1. For $\beta = 0$: $D = 1$, $\phi = 0$, the response is a static one in phase with the force. For $\beta \to 0$, $D \to 1$, $\phi \to 0$ irrespective of ξ. For $\xi = 1/\sqrt{2}$, $D = (1+\beta^4)^{-1/2}$ does not contain the β^2 term. Hence, for $\xi = 1/\sqrt{2}$, $D \simeq 1$ to within 1% error for β up to 0.4 and to within 5% error for β up to 0.6. Moreover, for $\xi = 1/\sqrt{2}$, $\tan\phi = \sqrt{2}\beta/(1-\beta^2) \Rightarrow \phi$ is approximately a linear function of β for $\beta < 1$. Thus, the time lag for different harmonic components of the output would be constant implying no distortion.

   Hence, in vibration measuring instruments, $\xi \in (0.65,\ 0.7)$ and the operating range is restricted to $\omega/\omega_n \leq 0.4$ to ensure no distortion in the measured entity.

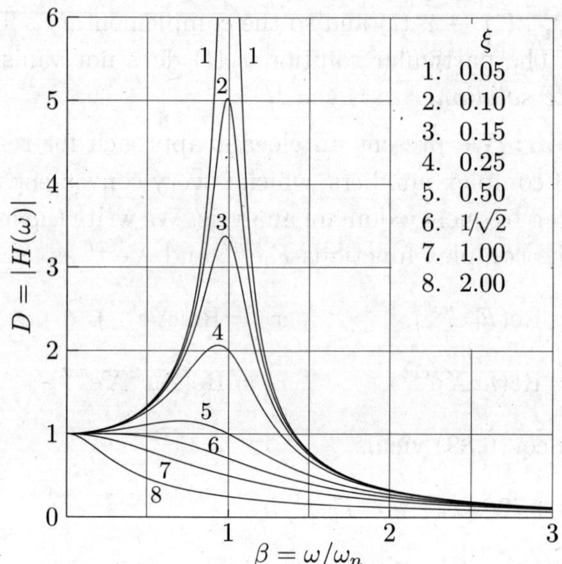

Fig. 6.29

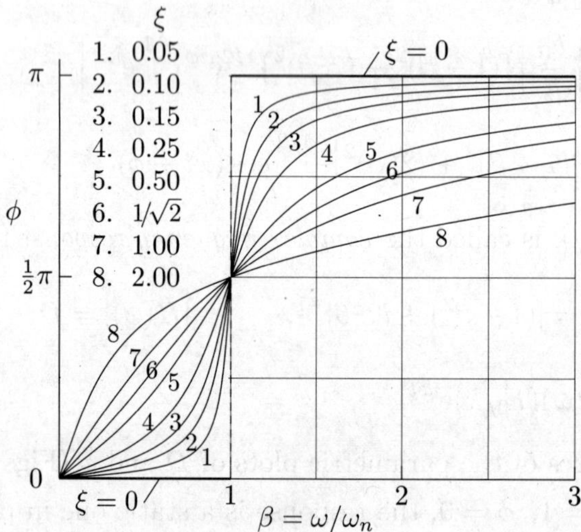

Fig. 6.30

2. For $\beta \gg 1$, i.e., $\beta \to \infty$: $D \to 0$, $\phi \to \pi$, $\forall\ \xi$. The response is very small and out of phase with the forcing function, i.e., when the force is maximum on the positive side, the response is maximum on the negative side and vice versa.
3. At *resonance*, $\omega = \omega_n$: $\beta = \omega/\omega_n = 1$, $D = 1/2\xi$, $\phi = \pi/2$ (if $\xi \neq 0$). The response is large for small values of ξ. When the force has maximum magnitude, the displacement has a zero magnitude and vice versa.

4. For a given β, D decreases as damping increases.
5. For a given ξ, D is maximum when $dD/d\beta = 0 \Rightarrow$ for either

$$\beta = \frac{\omega}{\omega_n} = \left(1 - 2\xi^2\right)^{1/2} \quad \text{with} \quad D = \frac{1}{2\xi\,(1-\xi^2)^{1/2}} \quad \left(= 1 \text{ if } \xi = \frac{1}{\sqrt{2}}\right)$$

$$\text{or} \quad \beta = \frac{\omega}{\omega_n} = 0 \qquad\qquad \text{with} \quad D = 1$$

$$\Rightarrow D_{\text{max}} = \begin{cases} 1/\,2\xi(1-\xi^2)^{1/2} & \text{if } \xi < 1/\sqrt{2} \\ 1 & \text{if } \xi \geq 1/\sqrt{2} \end{cases} \tag{6.95}$$

D_{max} occurs at $\omega < \omega_n$, but for light damping, the value of $D = 1/2\xi$ at $\omega = \omega_n$ does not differ much from $D_{\text{max}} = 1\,/\,2\xi(1-\xi^2)^{1/2}$. $D_{\text{max}} > 1$ if $\xi < 1/\sqrt{2}$.

6. For light damping, large values of D occur near $\beta = 1$.
7. For $\xi = 0,\ \beta \neq 1$: $D = 1/|1-\beta^2|$ has discontinuity at $\beta = 1$. $D \to \infty$ as $\beta \to 1$.
8. For $\beta \ll 1$, eq. (6.92) $\Rightarrow$ $\sin\phi \simeq 2\xi\beta \ \Rightarrow \phi \simeq 2\xi\beta$, i.e., $\phi \propto \beta$.
9. For $\xi = 0,\ \beta \neq 1$: eq. (6.89) $\Rightarrow$ $\ddot{x} + \omega_n^2 x = \dfrac{F_0}{k}\omega_n^2 \cos\omega t \ \Rightarrow \ x_p = C\cos\omega t \ \Rightarrow$

$$x_p(t) = \frac{F_0/k}{1-\beta^2}\cos\omega t \tag{6.96}$$

$\Rightarrow$ $x_p(t)$ is in phase with $F(t)$ if $\beta < 1$ and 180° out of phase for $\beta > 1$.

10. For steady state: $x(t) = D(F_0/k)\cos(\omega t - \phi) \quad \Rightarrow$

$$\dot{x}(t) = -\omega D\,\frac{F_0}{k}\,\sin(\omega t - \phi)$$

$$\Rightarrow \qquad v_{\text{amp}} = \omega D\,\frac{F_0}{k} = \beta\omega_n D\,\frac{F_0}{k} = \omega_n\,\frac{F_0}{k}\left[(\beta^{-1} - \beta)^2 + 4\xi^2\right]^{-1/2}$$

v_{amp} is maximum for $\beta^{-1} - \beta = 0$, i.e., for $\beta = 1$ $(\omega = \omega_n)$ $\Rightarrow$ ω_n is the *velocity resonant frequency* for harmonic load. The *amplitude resonant frequency* is $\omega = (1-2\xi^2)^{1/2}\omega_n$, if $\xi < 1/\sqrt{2}$ and differs from ω_n significantly for large ξ.

11. The maximum response of an undamped system under step load F_0 is $2F_0/k \Rightarrow$ $D = 2$. The maximum steady state response under harmonic load is greater if

$$D = \left[(1-\beta^2)^2 + (2\xi\beta)^2\right]^{-1/2} > 2 \quad \Rightarrow \quad \text{for a narrow range } \ \beta_2 < \beta < \beta_1$$

$$\text{where,} \quad \beta_1, \beta_2 = \left[1 - 2\xi^2 \pm (4\xi^4 - 4\xi^2 + 0.25)^{1/2}\right]^{1/2} \qquad \text{if } \ \xi < 0.259$$

$$(\xi, \beta_1, \beta_2) = (0, .7071, 1.225),\ (.1, .7220, 1.199),\ (.2, .7807, 1.109),\ (.25, .8660, 1)$$

The response is large at resonance ($\omega/\omega_n = 1$). Organ pipes and industrial sieving machines use it effectively. A radio/TV receiver tunes into a specific frequency by adjusting the natural frequency of the circuit to be exactly equal to the frequency of the transmitter. A major concern in many problems of dynamics is to avoid resonance, e.g., in bridges, aircraft structures, propeller shafts for ships, rotating machinery, etc.

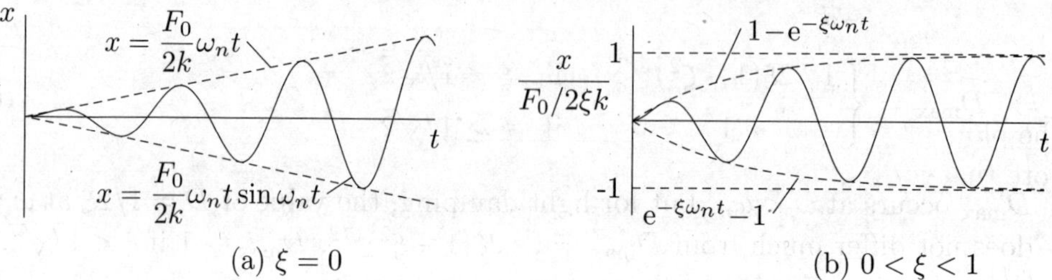

Fig. 6.31

For an undamped system at resonance ($\omega = \omega_n$), the equation of motion (6.89) becomes

$$\ddot{x} + \omega_n^2 x = (F_0/k)\omega_n^2 \cos \omega_n t \qquad \Rightarrow \qquad x_p(t) = (F_0/2k)\omega_n t \sin \omega_n t$$

This particular solution, can be verified by substituting it into the governing equation. The amplitude of this oscillatory response increases linearly with time (Fig. 6.31a). It deviates from the actual response as x increases since the assumption of linear restoring force is not valid for large deformations. The response $x(t)$ for $\xi \neq 0$ under $F = F_0 \cos \omega_n t$, with $x(0) = 0,\ \dot{x}(0) = 0$, is:

$$x(t) = \frac{F_0}{2\xi k}\left[\sin \omega_n t - \frac{1}{\sqrt{1-\xi^2}}\, e^{-\xi\omega_n t} \sin\left(\omega_n \sqrt{1-\xi^2}\, t\right)\right]$$

For comparison, this response is also plotted in Fig. 6.31b. It takes *time to build up*, but soon enough in 2 to 3 cycles, it approaches $x_p(t)$ which is simple harmonic.

6.15.1 Response of a rotating unbalanced rotor

Consider a rotor with an unbalanced mass m at an eccentricity e (Fig. 6.32a), rotating at ω. The total mass of the system is M. Let $\delta = Mg/k$ be the static deflection and $x(t)$ be the displacement of the main body, having mass $(M-m)$, from the equilibrium position. The vertical displacement of the unbalanced mass is $x + e \sin \omega t$. The FBD of the system is shown in Fig. 6.32b. The equation of motion for vertical displacement is

$$\begin{aligned}(M-m)\ddot{x} + m\overline{(x + e\sin\omega t)}^{\,\cdot\cdot} &= -c\dot{x} - k(x-\delta) - (M-m)g - mg \\ M(\ddot{x} + 2\xi\omega_n\dot{x} + \omega_n^2 x) &= me\omega^2 \mathrm{Im}(e^{i\omega t}) \qquad (6.97)\end{aligned}$$

with $\omega_n^2 = k/M,\ \xi = 0.5c/\sqrt{Mk}$. Using eq. (6.94), the steady state response x(t) is

$$x(t) = \text{Im}\left[\frac{me\omega^2}{k}\,|H(\omega)|\,\mathrm{e}^{i(\omega t-\phi)}\right] = \frac{m}{M}\,e\beta^2\,D\,\sin(\omega t-\phi)$$

where ϕ and $H(\omega)$ are given by eqs (6.92) and (6.94). The amplitude X of $x(t)$ is

$$X = \frac{m}{M}\,e\beta^2 D \quad \Rightarrow \quad \frac{MX}{me} = \beta^2 D = \beta^2\,|H(\omega)| \tag{6.98}$$

The plots of MX/me versus $\beta = \omega/\omega_n$ with ξ as a parameter are shown in Fig. 6.32c. Note that for $\beta = 0,\ \beta^2 D = 0$; for $\beta = 1,\ \beta^2 D = 0.5/\xi$; and as $\beta \to \infty,\ \beta^2 D \to 1$.

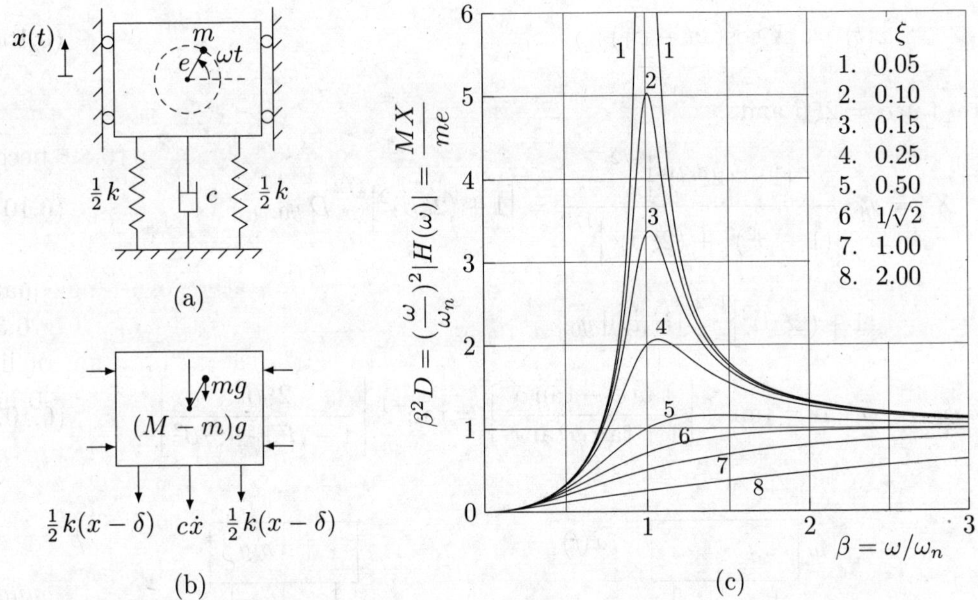

Fig. 6.32

6.15.2 Response to the harmonic motion of the support

Consider the problem of isolating the equipment from the vibration of the support. Let the support of a 1-dof system (Fig. 6.33a) be subjected to $y(t) = y_0\cos\omega t = \text{Re}(y_0\mathrm{e}^{i\omega t})$. Let $x(t)$ be the displacement from equilibrium position $x = y = 0$, where the springs have compression δ. The FBD is shown in Fig. 6.33b. The equation of motion is

$$m\ddot{x} = -c(\dot{x}-\dot{y}) - k(x-y-\delta) - mg$$

$$\Rightarrow \qquad m\ddot{x} + c\dot{x} + kx = c\dot{y} + ky + (k\delta - mg) \qquad (k\delta = mg)$$

$$\Rightarrow \qquad \ddot{x} + 2\xi\omega_n\dot{x} + \omega_n^2 x = 2\xi\omega_n\dot{y} + \omega_n^2 y \tag{6.99}$$

Let the steady state solution of eq. (6.99) be $x(t) = \text{Re}(Be^{i\omega t})$. Equation (6.99) $\Rightarrow$

$$\Big[(\omega_n^2 - \omega^2) + i2\xi\omega_n\omega\Big]B = \Big(\omega_n^2 + i2\xi\omega_n\omega\Big)y_0$$

$$\Rightarrow \quad B = \frac{(1 + i2\xi\beta)y_0}{1 - \beta^2 + i2\xi\beta}$$

$$\Rightarrow \quad x(t) = \text{Re}\left[\frac{1 + i2\xi\beta}{1 - \beta^2 + i2\xi\beta}\, y_0\, e^{i\omega t}\right] = \text{Re}\Big[(1 + i2\xi\beta)\, H(\omega)\, y_0\, e^{i\omega t}\Big] \qquad (a)$$

$$= \text{Re}\left[\frac{\{1 + (2\xi\beta)^2\}^{1/2}\, e^{i\alpha}\, y_0\, e^{i\omega t}}{\{(1 - \beta^2)^2 + (2\xi\beta)^2\}^{1/2}\, e^{i\phi}}\right]$$

$$x(t) = X\cos(\omega t - \phi_1) \tag{6.100}$$

where $\tan\alpha = 2\xi\beta$ and

$$X = y_0 \frac{\Big[1 + (2\xi\beta)^2\Big]^{1/2}}{\Big[(1 - \beta^2)^2 + (2\xi\beta)^2\Big]^{1/2}} = \Big[1 + (2\xi\beta)^2\Big]^{1/2} D\, y_0 \tag{6.101}$$

$$= \Big[1 + (2\xi\beta)^2\Big]^{1/2} |H(\omega)|\, y_0$$

$$\phi_1 = \phi - \alpha = \tan^{-1}\left[\frac{\tan\phi - \tan\alpha}{1 + \tan\phi\tan\alpha}\right] = \tan^{-1}\left[\frac{2\xi\beta^3}{1 - \beta^2 + 4\xi^2\beta^2}\right] \tag{6.102}$$

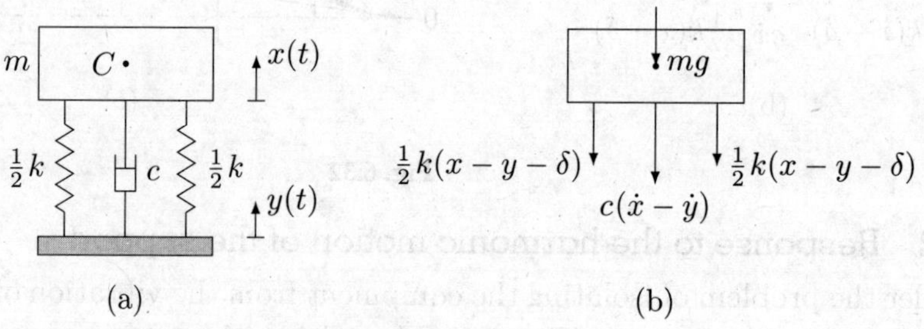

Fig. 6.33

The non-dimensional ratio (TR) of the amplitudes of the system and the support is

$$\text{TR} = \frac{X}{y_0} = \Big[1 + (2\xi\beta)^2\Big]^{1/2} |H(\omega)| = \frac{\Big[1 + (2\xi\beta)^2\Big]^{1/2}}{\Big[(1 - \beta^2)^2 + (2\xi\beta)^2\Big]^{1/2}} \tag{6.103}$$

The transmissibility ratio TR represents the degree of isolation of the system from the vibration of the support. Plots of X/y_0 and ϕ_1 versus β, with ξ as a parameter, are given in Fig. 6.34.

$$\text{For } \beta = 0, \sqrt{2}: \quad \frac{X}{y_0} = 1 \ \forall\ \xi; \qquad \text{for } \beta = 1: \quad \frac{X}{y_0} = \frac{(1+4\xi^2)^{1/2}}{2\xi}$$

For $\beta < \sqrt{2}$, X/y_0 decreases with increase in ξ, whereas for $\beta > \sqrt{2}$, X/y_0 increases with increase in ξ. As $\beta \to \infty$, $X/y_0 \to 0$.

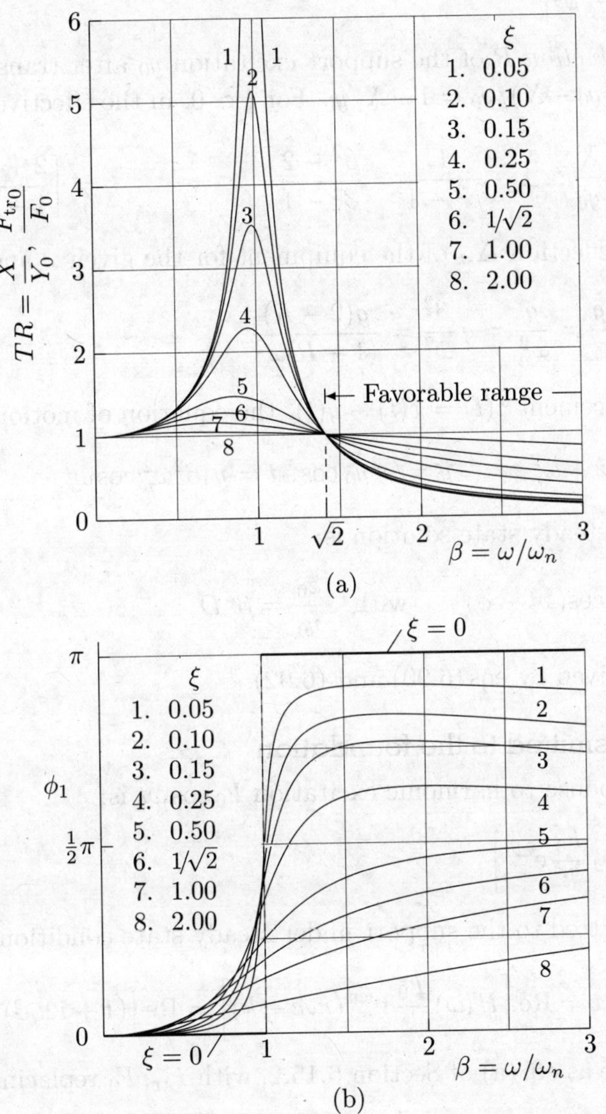

Fig. 6.34

Isolation of the equipment from the support is possible only if $\beta > \sqrt{2}$, i.e., with $\omega_n = \sqrt{k/m} < \omega/\sqrt{2}$ which can be ensured by a large m or a small k. In the range $\beta > \sqrt{2}$, less damping is more beneficial than large damping. Some damping is needed to limit the large amplitude at resonance if ω varies through $\omega = \omega_n$. If very small damping is used, then the larger amplitude at resonance can be limited by stops. In the effective range $\beta > \sqrt{2}$, we want $\xi \ll 1$. For $\xi = 0$, eq. (6.103) simplifies to

$$\frac{X}{y_0} = \frac{1}{|1-\beta^2|} \tag{6.104}$$

The *fractional reduction* R of the support excitation y_0 after transmission as X to the system is: $R = (y_0 - X)/y_0 = 1 - X/y_0$. For $\xi \simeq 0$, in the effective range $\beta > \sqrt{2}$,

$$R = 1 - \frac{X}{y_0} \simeq 1 - \frac{1}{\beta^2 - 1} = \frac{\beta^2 - 2}{\beta^2 - 1} \qquad \Rightarrow \qquad \beta = \left[\frac{2-R}{1-R}\right]^{1/2}$$

The desired static deflection Δ_{st} of the equipment for the given ω and R is

$$\Delta_{st} = \frac{mg}{k} = \frac{g}{\omega_n^2} = g\,\frac{\beta^2}{\omega^2} = \frac{g(2-R)}{(1-R)\omega^2}$$

For relative displacement $z(t) = x(t) - y(t)$, the equation of motion (6.99) yields

$$\ddot{z} + 2\xi\omega_n\dot{z} + \omega_n^2 z = -\ddot{y} = \omega^2 y_0 \cos\omega t = y_0\beta^2\omega_n^2 \cos\omega t$$

As for eq. (6.89), its steady state solution is:

$$z(t) = z_0\cos(\omega t - \phi) \qquad \text{with} \quad \frac{z_0}{y_0} = \beta^2 D$$

where D and ϕ are given by eqs (6.90) and (6.92).

6.15.3 Force transmitted to the foundation

The steady state response to harmonic excitation $F_0\cos\omega t$ is:

$$x(t) = \mathrm{Re}\left[H(\omega)\,\frac{F_0}{k}\,\mathrm{e}^{i\omega t}\right]$$

The force F_{tr} transmitted to the support under steady state conditions is

$$F_{\mathrm{tr}}(t) = c\dot{x} + kx = \mathrm{Re}\left[H(\omega)\,\frac{F_0}{k}\,\mathrm{e}^{i\omega t}(i\omega c + k)\right] = \mathrm{Re}\left[(1 + i2\xi\beta)\,H(\omega)\,F_0\mathrm{e}^{i\omega t}\right]$$

This equation is same as eq. (a) of Section 6.15.2, with F_{tr}, F_0 replacing x, y_0. Hence,

$$F_{\mathrm{tr}}(t) = F_{\mathrm{tr}_0}\cos(\omega t - \phi_1) \qquad \text{with} \quad \mathrm{TR} = \frac{F_{\mathrm{tr}_0}}{F_0} \tag{6.105}$$

given by eq. (6.103) and ϕ_1 given by eq. (6.102). TR $= F_{\text{tr}0}/F_0$ is called the *transmissibility* of force from the structure to the foundation and equals the TR of motion from the foundation to the structure. Fig. 6.34 also represents force transmissibility ratio.

Example 6.18: The frame consists of two light cantilevers (Fig. E6.18a) supporting a rigid platform of mass 5000 kg pinned to it. For a cantilever under end load P, $\delta = PL^3/3EI$, where E is Young's modulus and I is the area moment of inertia of the cross-section about the bending axis. For bending moment M, the maximum stress $\sigma_{\max} = Mc/I$, where c is the maximum distance of a point from the bending axis. Assume that $\xi = 0.1$. (a) The equipment on the platform subjects it to a force $F(t) = 1.8\cos 9.6t$ kN. Find the steady state amplitude and phase angle of: (i) the displacement of the platform, (ii) the force transmitted to the ground and the transmissibility of force to the ground, and (iii) $\sigma_{\max}$. (b) $F(t)$ does not act, but the ground motion is $y(t) = 5\cos 13t$ mm. Find the steady state amplitude and phase angle of: (i) the displacement of the platform and the transmissibility of motion to the platform, (ii) the displacement of the platform relative to the ground; (iii) the force transmitted to the ground, and (iv) $\sigma_{\max}$. $E = 2\times 10^5$ MPa, $I = 2800\,\text{cm}^4$, $c = 10$ cm, $L = 4$ m.

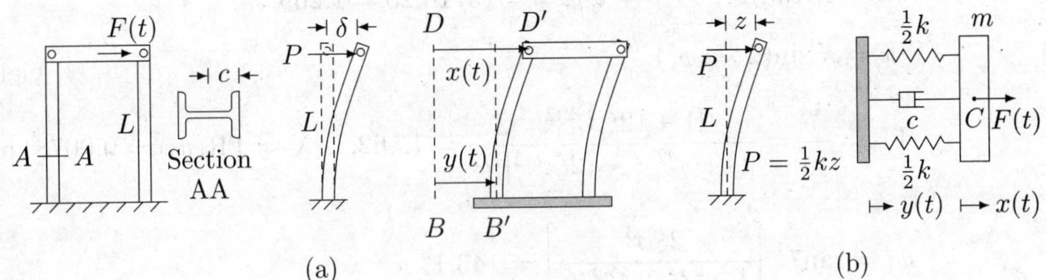

Fig. E6.18

Solution: In the displacement of a cantilever from the equilibrium position BD to position $B'D'$, the deflection $z(t)$ of its tip D' relative to its base B' is $z(t) = x(t) - y(t)$. Its stiffness for 1–dof model (Fig. E6.18b) is $k/2 = P/\delta = 3EI/L^3 \Rightarrow$

$$m = 5000\,\text{kg},\qquad \xi = 0.1,\qquad E = 2\times 10^{11}\,\text{N/m}^2,$$

$$I = 28\times 10^{-6}\,\text{m}^4,\qquad L = 4\,\text{m},\qquad c = 0.1\,\text{m},$$

$$k = 6EI/L^3 = 525\times 10^3\,\text{N/m},\qquad \omega_n = (k/m)^{1/2} = 10.25\,\text{rad/s}.$$

(a) $F(t) = F_0\cos\omega t = 1800\cos 9.6t$ N $\Rightarrow$ $\omega = 9.6$ rad/s, $F_0 = 1800$ N,

$$\beta = \omega/\omega_n = 0.9369,\qquad F_0/k = 3.429\times 10^{-3}\,\text{m} = 3.429\,\text{mm}.$$

(i) The response is $x = X_s\cos(\omega t - \phi)$. $D = [(1-\beta^2)^2 + (2\xi\beta)^2]^{-1/2} = 4.470,$

$$\phi = \tan^{-1}\left[\frac{2\xi\beta}{1-\beta^2}\right] = 56.89°, \qquad X_s = D\frac{F_0}{k} = 0.01533\,\text{m}$$

(ii) $\quad F_{tr} = F_{tr_0}\cos(\omega t - \phi_1)$. We get F_{tr_0} and ϕ_1 from eqs (6.103) and (6.102):

$$\text{TR} = \frac{F_{tr_0}}{F_0} = \frac{[1+(2\xi\beta)^2]^{1/2}}{[(1-\beta^2+(2\xi\beta)^2]^{1/2}} = 4.548, \qquad F_{tr_0} = F_0(\text{TR}) = 8186\,\text{N}$$

$$\phi_1 = \tan^{-1}\left(\frac{2\xi\beta^3}{1-\beta^2+4\xi^2\beta^2}\right) = 46.27°$$

(iii) $P_{max} = \frac{1}{2}kz_{max}$ and σ_{max} is at the base B: $M_{max} = LP_{max} = \frac{1}{2}Lkz_{max} \quad \Rightarrow$

$$\sigma_{max} = \frac{M_{max}c}{I} = \frac{3Ecz_{max}}{L^2} = \frac{3EcX_s}{L^2} = 57.48\,\text{MPa} \tag{1}$$

The amplitude of the stress is very large because ω is very near to ω_n.

(b) $\quad y(t) = y_0\cos\omega t = 0.005\sin 13t\,\text{m}, \qquad y_0 = 0.005\,\text{m},$

$$\omega = 13\,\text{rad/s}, \qquad \beta = \omega/\omega_n = 13/10.25 = 1.269.$$

(i) $\quad x = X\sin(\omega t - \phi_1),$

$$\text{TR} = \frac{X}{y_0} = \frac{[1+(2\xi\beta)^2]^{1/2}}{[(1-\beta^2)^2+(2\xi\beta)^2]^{1/2}} = 1.562, \quad X = \text{TR}(y_0) = 0.00781\,\text{m},$$

$$\phi_1 = \tan^{-1}\left[\frac{2\xi\beta^3}{1-\beta^2+4\xi^2\beta^2}\right] = 143.1°$$

(ii) $\quad z(t) = z_0\cos(\omega t - \phi),$

$$\frac{z_0}{y_0} = \beta^2 D = \frac{\beta^2}{[(1-\beta^2)^2+(2\xi\beta)^2]^{1/2}} = 2.436 \Rightarrow z_0 = 12.18\,\text{mm} = 0.01218\,\text{m}$$

$$\phi = \tan^{-1}\left[\frac{2\xi\beta}{1-\beta^2}\right] = 157.4°$$

(iii) $\quad F_{tr} = kz + c\dot{z}. \qquad m\ddot{x} + c\dot{z} + kz = 0$

$$\Rightarrow \quad F_{tr} = -m\ddot{x} = \omega^2 mX\sin(\omega t - \phi_1)$$

$$\Rightarrow \quad F_{tr_0} = \omega^2 mX = (13)^2 \times 5000 \times 0.00781 = 6599\,\text{N}.$$

(iv) Equation (1) $\Rightarrow \quad \sigma_{max} = \frac{3Ecz_{max}}{L^2} = \frac{3Ecz_0}{L^2} = 45.68 \times 10^6\,\text{N/m}^2 = 45.68\,\text{MPa}$

Review 6 E

1. $T = \frac{1}{2} m_e \dot{q}^2, \quad V = \frac{1}{2} k_e q^2, \quad \omega_n = (k_e/m_e)^{1/2}, \quad \dot{q}_{\max} = \omega_n q_{\max},$
$T_{\max} = V_{\max}, \quad q(t) = q_0 \cos \omega_n t + (\dot{q}_0/\omega_n) \sin \omega_n t.$

2. $c_c = \sqrt{4km}, \quad \xi = c/c_c, \quad \omega_{n_d} = \omega_n \sqrt{1-\xi^2}, \quad \delta = \ln(x_n/x_{n+2}) = 2\pi\xi/\sqrt{1-\xi^2},$

$$x = e^{-\xi\omega_n t}\Big[x_0 \cos \omega_{n_d} t + \frac{v_0 + \xi\omega_n x_0}{\omega_{n_d}} \sin \omega_{n_d} t\Big]$$

3. For $F = F_0 \cos \omega t: \ \beta = \omega/\omega_n, \ x = X_s \cos(\omega t - \phi), \ H(\omega) = (1 - \beta^2 + i2\xi\beta)^{-1},$

$$D = \frac{X_s}{F_0/k} = [(1-\beta^2)^2 + (2\xi\beta)^2]^{-1/2}, \quad \tan\phi = \frac{2\xi\beta}{1-\beta^2}, \quad |H(\omega)| = D,$$

$$\mathrm{TR} = X/y_0 = F_{\mathrm{tr}_0}/F_0 = [1 + (2\xi\beta)^2]^{1/2} \,/\, [(1-\beta^2)^2 + (2\xi\beta)^2]^{1/2}.$$

4. Unbalanced rotor M with m at eccentricity e rotating at ω.

$$x = X \sin(\omega t - \phi), \quad \frac{MX}{me} = \beta^2 D$$

5. For support motion $y = y_0 \cos \omega t: \ x(t) = X \cos(\omega t - \phi_1), \quad \dfrac{X}{y_0} = \mathrm{TR},$

$$\phi_1 = \tan^{-1} \frac{2\xi\beta^3}{1 - \beta^2 + 4\xi^2\beta^2}.$$

6.16 Concept Review Questions 6 E

19. State the order relation for natural frequencies ω_1 and ω_2 for each case. (a) For the system in Fig. Q6.19a, for oscillations in (i) horizontal plane and (ii) in the vertical plane. (b) For the system in Fig. Q6.19b for oscillations in the (vertical) xy-plane for orientations (i) and (ii). (c) For a spring–mass system on the earth and the moon. (d) A ship has a rolling frequency ω_1 (Fig. Q6.19c). Cargo is shifted horizontally and symmetrically from near the plane of symmetry of the ship to the sides as shown in Fig. Q6.19c and the rolling frequency becomes ω_2.

[(a) $\omega_1 = \omega_2$ (b) $\omega_1 > \omega_2$ (c) $\omega_1 = \omega_2$ (d) $\omega_2 < \omega_1$]

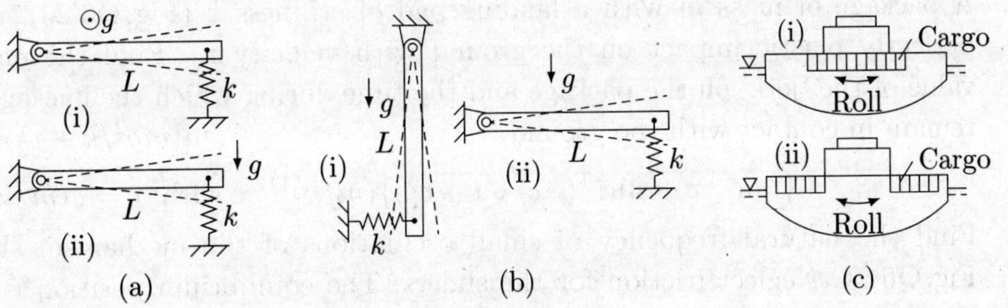

Fig. Q6.19

20. (a) For each of the following cases, find the steady stable position and the period of small oscillation about it, for a simple pendulum of length L suspended in: (i) a train accelerating at a to the right, (ii) a train moving at constant velocity on an inclined plane at an angle θ to the horizontal, (iii) an elevator going up with an acceleration a. (b) Rework part (a) for axial oscillation of a spring–mass (k, m) system suspended from the ceiling. [(a) (ii) vertical, $2\pi\sqrt{L/g}$
(i) $\tan^{-1}(a/g)$ to left of vertical, $2\pi\sqrt{L}/(g^2+a^2)^{1/4}$ (iii) vertical, $2\pi\sqrt{L/(g+a)}$
(b) $x_0 = (\sqrt{g^2+a^2},\ g,\ g+a)\frac{m}{k}$, orientation as in (a), $T = 2\pi\sqrt{m/k}\ \forall$ cases]

21. Find the natural circular frequency ω_0 of small oscillations for the following systems. (a) A pendulum with a thin disc hinged to it (Fig. Q6.21a). (b) A plate pendulum shown in Fig. Q6.21b. The springs are undeformed for $\theta = 0$. (c) A body of mass m and axial moment of inertia I about OB (Fig. Q6.21c), which is oscillating with small values of θ. [(a) $[(m_1+\frac{1}{2}m)gL+k]^{1/2}/(m_1+\frac{1}{3}m)^{1/2}L$
(b) $[\{2k_t + kL^2 + (m+m_1)gL\}/(m_1+2m/3)L^2]^{1/2}$ (c) $(mge\sin\alpha/I)^{1/2}$]

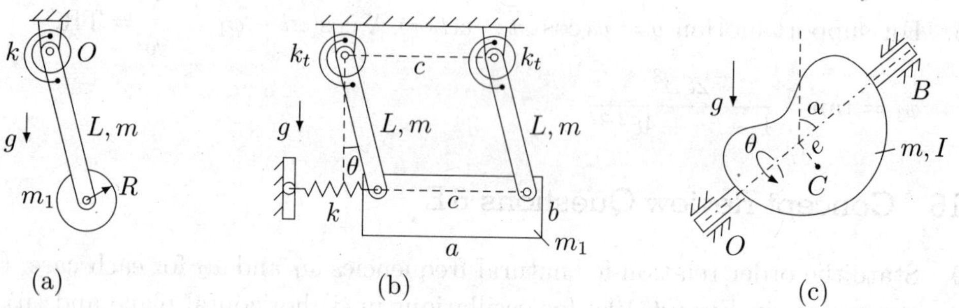

Fig. Q6.21

22. Find the natural circular frequency ω of small oscillations of the spring and pulley system of Fig. Q6.22. Each pulley has axial radius of gyration k_1. Work out for (a) smooth pulleys ($\mu = 0$) and (b) for perfectly rough pulleys (no slip).
[(a) $[16k/(16m+5m_1+m_2)]^{1/2}$ (b) $\{16k/(16m+5m_1(1+k_1^2/r^2)+m_2)\}^{1/2}$]

23. A package of mass m with a landing pad of stiffness k (Fig. Q6.23), makes a perfectly plastic impact on the ground with velocity v_0. Find the maximum value of the 'jerk' on the package and the time during which the package would remain in contact with the ground. [$(mg^2/k+v_0^2)^{1/2}k/m$,
$\{\tan^{-1}c + \sin^{-1}(-c/\sqrt{1+c^2})\}(m/k)^{1/2}$ with $c = g(m/k)^{1/2}/v_0$]

24*. Find the natural frequency of small oscillations of the mechanism shown in Fig. Q6.24. Neglect friction for the sliders. The equilibrium position is shown. Work out for (a) smooth pulley and (b) for perfectly rough pulley.

[(a) $[12k/(3m + 3m_1 + m_2 \operatorname{cosec}^2\theta + 3m_3 \cot^2\theta)]^{1/2}/2\pi$
(b) $[12k/(4.5m + 3m_1 + m_2 \operatorname{cosec}^2\theta + 3m_3 \cot^2\theta)]^{1/2}/2\pi$]

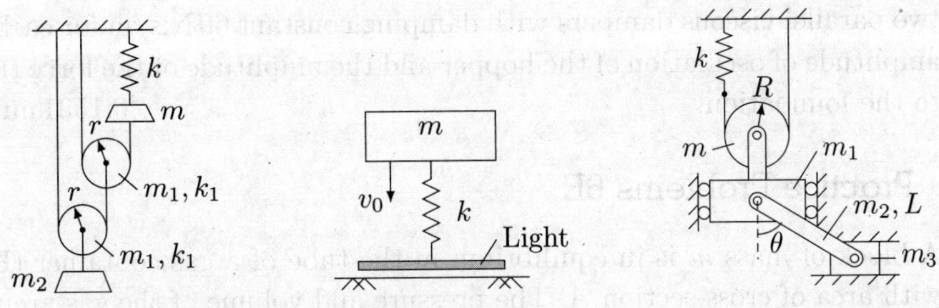

Fig. Q6.22 **Fig. Q6.23** **Fig. Q6.24**

25. Find the natural frequencies of the systems shown in Fig. Q6.25. Neglect the inertia of the beam in case (i) and of the rigid rod in case (ii). The central deflection δ of the beam (without the springs) shown in Fig. Q6.25(i) under a central load P is $\delta = PL^3/48EI$.

[(i) $[k_1(k_2 + 48EI/L^3)/(k_1 + k_2 + 48EI/L^3)m]^{1/2}/2\pi$
(ii) $[k_1(k_2 + 9k_3)/m(4k_1 + k_2 + 9k_3)]^{1/2}/2\pi$ (iii) $\sqrt{2k/m}\sin\phi/2\pi$]

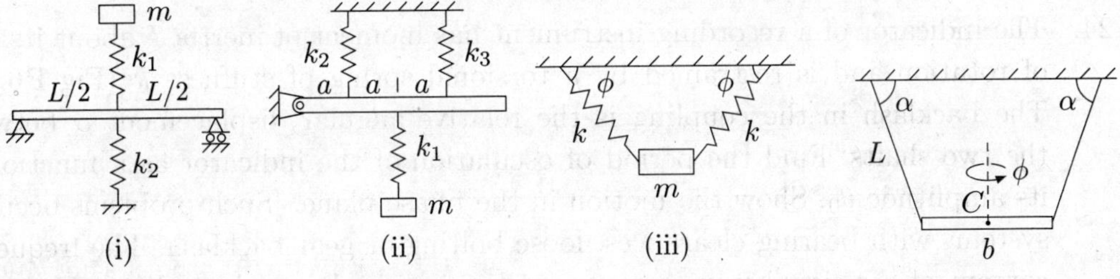

Fig. Q6.25 **Fig. P6.26**

26. A body of mass m, centre of mass C and moment of inertia I about the vertical axis through the centre of mass, is suspended by two ropes (Fig. Q6.26). Find the natural frequency of small oscillations about the vertical axis through C.

[$\{mg(\frac{1}{2}b + L\cos\alpha)b/2IL\sin\alpha\}^{1/2}/2\pi$]

27. A single degree of freedom spring–mass–damper system has a stiffness of 4 kN/cm. The damping element experiences a relative velocity of 40 cm/s if a force of 0.6 kN is applied to it. The mass of the system is 3000 kg. Find the damped natural frequency, the damping ratio, the logarithmic decrement, the ratio of successive amplitudes and the number of cycles after which the amplitude reduces to 5% of its initial value. [1.837 Hz, 0.02165, 0.1361, 0.8728, 22 cycles]

28. To facilitate flow of material in a hopper, a shaker imparts it a harmonic force of amplitude 100 N at a frequency of 50 Hz. The mass of the hopper and its contents is 500 kg. It is restrained by two parallel springs each of stiffness 25 MN/m and two parallel viscous dampers with damping constant 60 N.s/m for each. Find the amplitude of oscillation of the hopper and the amplitude of the force transmitted to the foundation. [0.1531 mm, 7656 N]

6.17 Practice Problems 6E

23. A block of mass m is in equilibrium in the tube of a gas container (Fig. P6.23) with area of cross-section A. The pressure and volume of the gas are p_0 and V_0. Find the natural frequency of small vertical oscillations of the block. Assume isothermal conditions for the gas inside the container. $[(p_0 A^2/mV_0)^{1/2}/2\pi]$

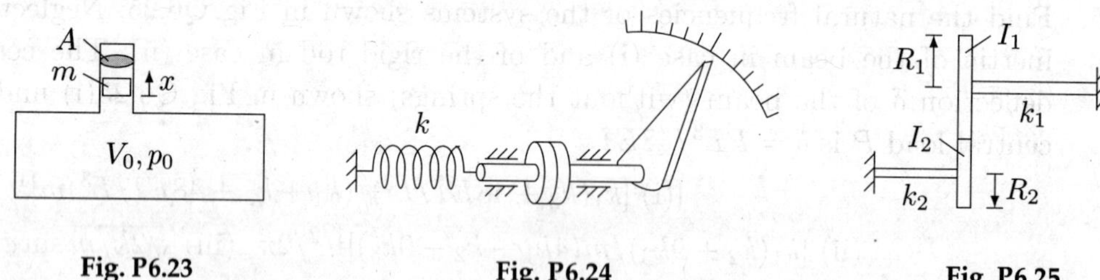

Fig. P6.23 Fig. P6.24 Fig. P6.25

24. The indicator of a recording instrument has moment of inertia I about its axis of rotation and is restrained by a torsional spring of stiffness k (Fig. P6.24). The backlash in the coupling is the relative angular displacement α between the two shafts. Find the period of oscillation of the indicator as a function of its amplitude θ_0. Show the motion in the phase plane. [Such problems occur in systems with bearing clearances, loose bolting or gear backlash. The frequency content of the signal from a *worn out bearing* can be used by the maintenance engineer to monitor wear and effect timely replacement of worn out parts.]

$[2\{\pi + 2\alpha/(2\theta_0 - \alpha)\}/\sqrt{k/I}]$

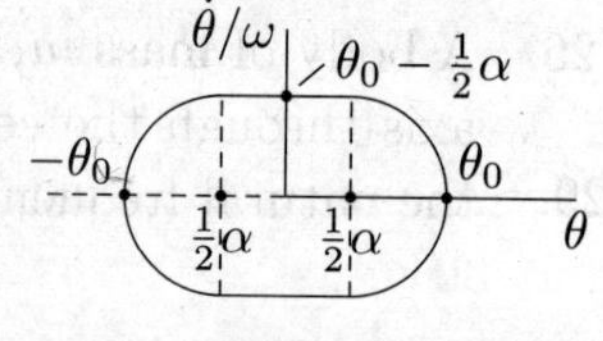

25. Find the natural frequency of oscillation of the geared system in Fig. P6.25. The torsional stiffness of the shafts are k_1 and k_2. Neglect their inertia. The axial moments of inertia of the gears are I_1 and I_2.

$[\{(k_1 + k_2 R_1^2/R_2^2)/(I_1 + I_2 R_1^2/R_2^2)\}^{1/2}/2\pi]$

26. Find the natural frequency of the system shown in Fig. P6.26. The T shaped member is hinged at O. The system is in equilibrium and the springs are undeformed when OA is vertical. The length and mass of each of the three segments of the uniform T shaped member are given.

$[\{k_2 + (k_1 + k_3)b^2 - (P + \frac{1}{2}m_1 g)a\}^{1/2}/(\frac{1}{3}m_1 a^2 + \frac{2}{3}m_2 b^2)^{1/2} 2\pi]$

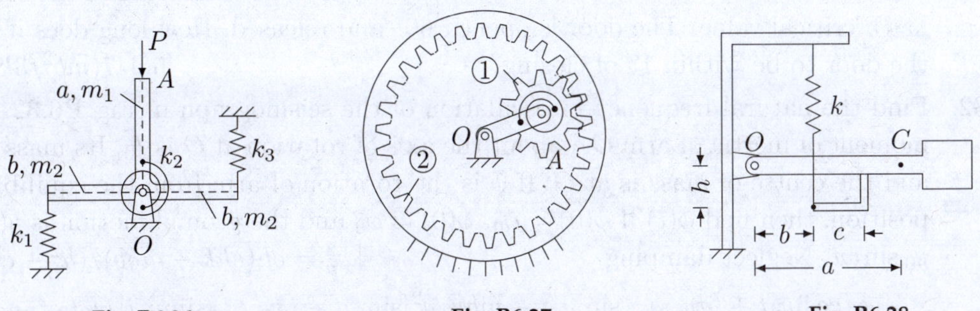

Fig. P6.26 **Fig. P6.27** **Fig. P6.28**

27. Gear ① of radius r shown in Fig. P6.27 is mounted on a crank OA of length R. It is in mesh with a fixed gear ②. A torsional spring connecting the gear to the crank has stiffness k. The axial moment of inertia of gear ① is I_1 and its mass is m_1. The moment of inertia of the crank about its axis of rotation is I. Find the natural frequency of vibration of the crank. The axis of rotation of the crank is vertical. $[\{k(1 + r/R)^2/(I_1 + m_1 r^2 + Ir^2/R^2)\}^{1/2}/2\pi]$

28*. A lever has mass m, centre of mass at C and moment of inertia I about the axis of rotation at O. It is restrained by a spring (Fig. P6.28). OC is horizontal in the equilibrium position. (a) Find the natural frequency of oscillations of the lever in the vertical plane. (b) Find the minimum value of k for the equilibrium position to be stable. [(a) $\{k(b^2 - mgah/kb)/I\}^{1/2}/2\pi$ (b) $mgah/b^3$]

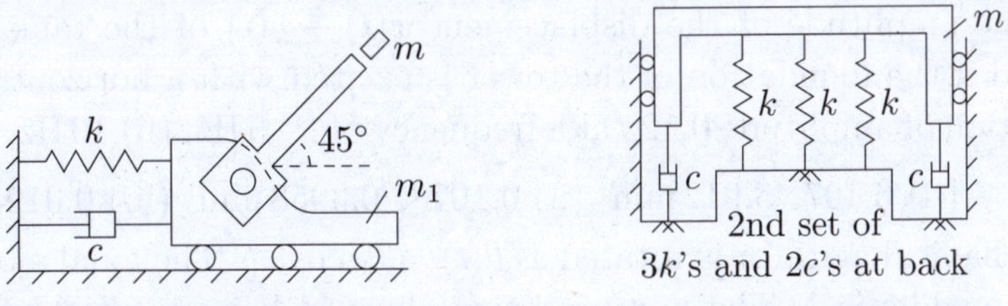

Fig. P6.29 **Fig. P6.30**

29. A gun carriage is restrained by a spring–dashpot system (Fig. P6.29) with $k = 30\,\text{kN/m}$, $c = 10^4\,\text{N.s/m}$. The mass m_1 of the gun is 800 kg and a mortar of mass $m = 4\,\text{kg}$ is fired with velocity 240 m/s relative to the barrel. Find the maximum recoil movement of the gun and the time taken for that. [5 cm, 0.1622 s]

30^d. A platform balance (Fig. P6.30) should have a damped natural frequency of f Hz and its amplitude should decrease to 0.4% of its initial value in 4 cycles, for the unloaded condition. Design appropriate values of k and c. Find the damped natural frequency and the ratio of successive amplitudes after a body of mass $M = 10m$ is placed on it. [$6.897f^2m$, $0.6902fm$, $0.3081f$, 0.6654]

31. A uniform thin door of mass m, width b and height h is fitted with an automatic door closer. The torsional stiffness of the spring is k. The damping in the system has a critical value. The door is opened 60° and released. How long does it take the door to be within 1° of closing? $[6.047(mb^2/3k)^{1/2}]$

32. Find the natural frequency of oscillation of the seismograph in Fig. P6.32. The moment of inertia of arm OA about the axis of rotation at O is I_0. Its mass is m and the centre of mass is at C. If ϕ is the rotation of arm from the equilibrium position, then find $\phi(t)$ if $\phi(0) = \phi_0$, $\dot{\phi}(0) = \dot{\phi}_0$ and the ground motion is $y(t) = y_0 \sin \omega t$. Neglect damping. $[\omega_n^2 = eh(ehk - mgb)/(h^2 + e^2)I_0,$ $\phi_0 \cos \omega_n t + (\dot{\phi}_0/\omega_n) \sin \omega_n t - mby_0\omega^2[\sin \omega t - (\omega/\omega_n) \sin \omega_n t]/I_0(\omega_n^2 - \omega^2)]$

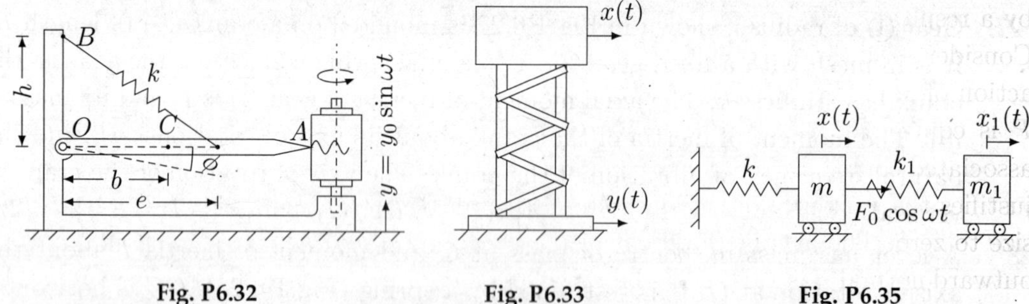

Fig. P6.32

Fig. P6.33

Fig. P6.35

33. A water tank tower shown in Fig. P6.33 is modelled as a single degree of freedom system with a mass of 40 Mg, a stiffness of 40 MN/m for transverse displacement and a damping ratio of 0.1. Find the amplitude of displacement $x(t)$ of the tank and the amplitude of the displacement $x(t) - y(t)$ of the tank relative to the ground if the foundation of the tower is excited with a horizontal harmonic acceleration of amplitude $0.12g$ at a frequency of (i) 5 Hz, (ii) 10 Hz, and (iii) 15 Hz.

[(i) 6.107, 5.912 mm (ii) 0.1078, 0.3958 mm (iii) 0.01952, 0.1489 mm]

34. A mechanical agitator is rotated as $\theta(t) = \theta_0 \cos \omega t$. The axial moment of inertia of the paddle is I. The mixture being churned has the effect of increasing the moment of inertia to $1.8I$. For the shaft: $k_t = 350$ N.m/rad. Find the amplitude of the angular displacement of the paddle and the amplitude of the twist in the shaft. $\theta_0 = 20°$, $\omega = 20$ rad/s, $I = 0.12$ kg.m^2. [26.56°, 6.555°]

35. A spring–mass (k, m) system is subjected to a harmonic force $F_0 \cos \omega t$. A vibration absorber [Fig. P6.35] to completely absorb the vibration of m $(x \equiv 0)$ consists of a mass m_1, attached to m through a spring of stiffness k_1. To achieve this, prove that $k_1/m_1 = \omega^2$ and show that the amplitude of m_1 is $x_{1_0} = F_0/k_1$.

F. HYDROSTATICS

The concept of stress vector at a point is introduced. The concepts of fluid and solid are mechanically differentiated. Pascal's law is proved. The equation of motion at a point of a non-viscous fluid is derived. The pressure variation in a fluid in equilibrium is discussed. The resultant force on a surface in a fluid in equilibrium and location of centre of pressure are covered. Buoyant force, equilibrium conditions of floating bodies, metacentre and stability of partially submerged bodies are discussed.

6.18 Concept of Stress Vector

While dealing with bars, we replaced the internal forces between parts of a body by a resultant force system. We now pursue further the discussion of internal forces. Consider an interior point P of a body and a surface π through it (Fig. 6.35a). The action of part ② of the body on part ① across the area element ΔA of π containing P is equivalent to a force-couple resultant at P consisting of a force $\overrightarrow{\Delta F}$ and an associated couple $\overrightarrow{\Delta C}$ (Fig. 6.35b). The continuum hypothesis described in Section 1.4 justifies the existence of the internal elemental surface ΔA and permits shrinking its size to zero. The average force or *average traction* at P on an area element ΔA with outward normal $\vec{n}$ is $\overrightarrow{\Delta F}/\Delta A$, the *average moment* being $\overrightarrow{\Delta C}/\Delta A$. *Cauchy and Euler postulated an axiom* regarding their limits as ΔA shrinks to P, keeping $\vec{n}$ constant:

'The limit of average traction as $\Delta A \to 0$, keeping $\vec{n}$ constant, exists and is independent of the shape of ΔA and the way in which $\Delta A \to 0$. This limit is called the *traction vector* or *stress vector* $\vec{T}^n$ at P on an area element at P with outward normal $\vec{n}$:

$$\vec{T}^n(\vec{r}) = \lim_{\substack{\Delta A \to 0 \\ \vec{n} = \text{const.}}} \frac{\overrightarrow{\Delta F}}{\Delta A} \tag{6.106}$$

It is also usually postulated that $|\overrightarrow{\Delta C}|$ is of the same order as $|\overrightarrow{\Delta F}|a$ where length a lies within the shrinking region $\Delta A \Rightarrow \lim_{\Delta A \to 0} \frac{\overrightarrow{\Delta C}}{\Delta A} = \vec{0}$.'

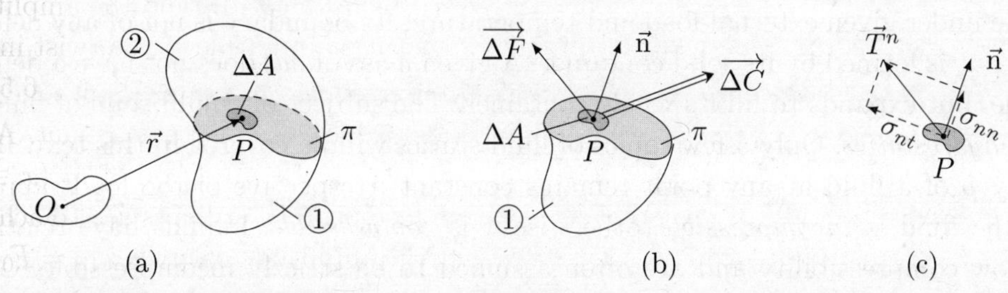

Fig. 6.35

The stress vector has dimensions of $[FL^{-2}] = [ML^{-1}T^{-2}]$. The S.I. unit is N/m^2

which is also denoted as Pascal (abbreviated as Pa). The stress vector $T^n(\vec{r})$ is a function of $\vec{r}$ and $\vec{n}$. Unlike a field of force, *state of stress* within a body is not defined by a vector function of position.

The forces between atoms and electrons at the microscopic level of a physical body are idealised into the concept of stress-vector at the macroscopic level in the mathematical model of a physical body as a continuous body. The axiom of Euler and Cauchy is justified on the experimental validity of the results of mechanics of deformable bodies.

The component of $\vec{T}^n$ (Fig. 6.35c) in the direction of $\vec{n}$ is called the *normal stress* σ_{nn}

$$\sigma_{nn} = \vec{T}^n \cdot \vec{n} = T_i^n n_i \qquad \text{where} \qquad \vec{T}^n = T_i^n \vec{e}_i, \quad \vec{n} = n_i \vec{e}_i$$

It is a *tensile/compressive stress* according as σ_{nn} is positive or negative. The component of stress vector in the tangent plane to the area element at P is called the *shear stress*, σ_{nt} (Fig. 6.35c). The shear stress as a vector ($\vec{T}_t^n$) is obtained by subtracting the normal component $\vec{T}_n^n = \sigma_{nn}\vec{n}$ from the stress vector $\vec{T}^n$.

6.19 Concept of a Fluid

The concepts of solids and fluids have been developed by us for the sake of convenience. Substances like tooth paste, paints, paraffin, polymers, etc., act somewhat like solid and fluid. But most substances in engineering can be classified as solid or fluid.

A *fluid* is mechanically differentiated from a solid by the definition that a shear stress, *however small*, applied to the surface of a fluid at rest will cause it to move indefinitely as long as the stress is applied. A *solid*, however, is defined as a material which has at most a definite additional deformation for a given additional shear stress applied on its surface. It follows that for all fluids in equilibrium, shear stress at all points of the body is zero, i.e., $\vec{T}^n = \sigma_{nn}\vec{n}$ (no sum on n). The normal stress is compressive to avoid separation, i.e., $\vec{T}^n = \sigma_{nn}\vec{n} = -p\,\vec{n}$, where pressure $p = p(\vec{r}, \vec{n})$ is the compressive normal stress at position $\vec{r}$ on an area element whose outward normal is $\vec{n}$. A liquid free of dissolved gases can withstand small tension.

Fluids are classified into *liquids* and *gases*. A given mass of *liquid* occupies a definite volume under given external load and temperature. Its boundary is not of any definite shape but is formed by its solid container. A given mass of *gas* does not have a definite volume, but expands to fill its whole container. The subject of equilibrium of fluids is called *hydrostatics*. Only a few topics of fluid statics will be covered in this text. If the density ρ of a fluid at any point remains constant irrespective of the loads applied, then the fluid is *incompressible*, otherwise it is *compressible*. Liquids have relatively very low compressibility and are often assumed to be strictly incompressible. Gases are compressible. In dynamics, the terms *compressible* and *incompressible* are used to qualify the flow. Gases in certain flow regimes can be modelled as incompressible while propagation of sound waves in liquids, at finite speed, indicates their compressibility.

6.20 Pascal's Law

Pascal's Law states that for (i) any fluid at rest, (ii) a non-viscous fluid in motion, and (iii) a viscous fluid having rigid body motion, the pressure p for all area elements through a point is the same. In all these cases, the shear stresses are zero.

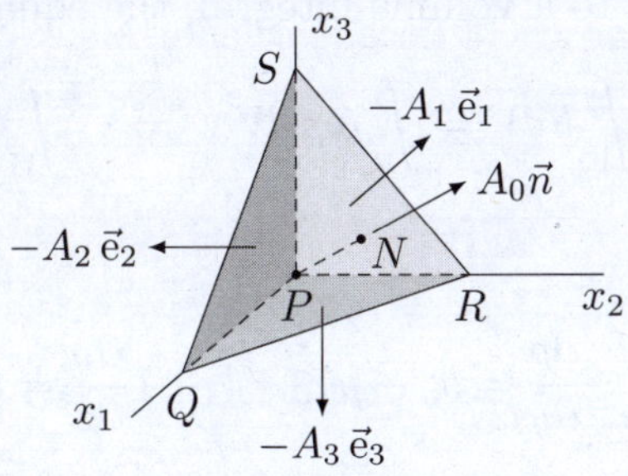

Fig. 6.36

Proof: Consider a tetrahedron $PQRS$ at a point P of the fluid (Fig. 6.36). Three of its faces are parallel to the coordinate planes with outward normals in $-\vec{e}_i$ directions and the fourth face SQR has outward normal $\vec{n} = n_i\,\vec{e}_i$. Let the body force density be $\vec{b}(\vec{r})$ per unit volume. Let A and V be the surface area and volume of the tetrahedron. Let area $\triangle QRS = A_0$. $V = \frac{1}{3}\,A_0 h$ where $h = PN$ is the distance of the vertex P from the base QRS. The density is ρ and $\vec{a}$ is the acceleration of a typical point in V. Let p_i and p be pressure on area elements with normals in $\vec{e}_i$ and $\vec{n}$ directions, respectively. From Ex.1.2d, the area A_i of the face with normal $-\vec{e}_i$ is:

$$A_i = n_i A_0$$

Applying Euler's 1st axiom to the tetrahedron and using mean value theorem, we get

$$\vec{F} = \sum_{i=1}^{3}\int_{A_i} -p_i(-\vec{e}_i)\,dA + \int_{A_0}(-p\,\vec{n}\,)dA + \int_V \vec{b}\,dV = \int_V \vec{a}\,\rho\,dV$$

$$\sum_{i=1}^{3}(p_{i\mathrm{av}}n_i\,\vec{e}_i A_0 - p_{\mathrm{av}}n_i\,\vec{e}_i A_0) + (\vec{b} - \vec{a}\,\rho)_{\mathrm{av}}(\tfrac{1}{3}A_0\,h) = \vec{0}$$

As $h \to 0$, keeping $\vec{n}$ constant, areas shrink to P, average values become values at P:

$$\sum_{i=1}^{3}(p_i - p)n_i\,\vec{e}_i = \vec{0} \qquad \Rightarrow \qquad p_i = p \qquad \Rightarrow \qquad p_1 = p_2 = p_3 = p$$

Hence, pressure at a point is independent of $\vec{n}$ and we can speak of *the* pressure at point P. Thus, pressure is a scalar function of position and $\vec{T}^n(\vec{r}) = -p\,(\vec{r}\,)\,\vec{n}$.

6.21 Equations of Motion

The shear stresses are zero in an in-viscid fluid in arbitrary motion and in a viscous fluid in rigid body motion. Hence, $\vec{T}^n = -p\vec{n}$. Consider an element of such a fluid having volume V and surface area A subjected to body force $\vec{b}$. Using Gauss's theorem to convert the area integral to a volume integral, the equation of motion is:

$$\vec{F} = \int_A -p\,\vec{n}\,dA + \int_V \vec{b}\,dV = \int_V \vec{a}\,\rho\,dV \quad \Rightarrow \quad \int_V (-\vec{\nabla}p + \vec{b} - \rho\,\vec{a})\,dV = \vec{0}$$

$$\Rightarrow \quad \vec{\nabla}p = \vec{b} - \rho\,\vec{a}$$

$$\Rightarrow \quad \frac{\partial p}{\partial x} = b_x - \rho a_x, \qquad \frac{\partial p}{\partial y} = b_y - \rho a_y, \qquad \frac{\partial p}{\partial z} = b_z - \rho a_z \tag{6.107}$$

Equations (6.107) are the *differential equations of motion at a point of* (i) *an in-viscid fluid in arbitrary motion and* (ii) *a viscous fluid having rigid body motion.*

6.21.1 Equations of equilibrium of fluids

From eq. (6.107), we obtain the *differential equations of equilibrium* ($\vec{a} = \vec{0}$) *at a point of any fluid (viscous or non-viscous)*:

$$\vec{\nabla}p = \vec{b}, \qquad \text{i.e.,} \qquad \frac{\partial p}{\partial x} = b_x, \qquad \frac{\partial p}{\partial y} = b_y, \qquad \frac{\partial p}{\partial z} = b_z \tag{6.108}$$

$$\Rightarrow \quad \vec{\nabla}\times\vec{b} = \begin{vmatrix} \vec{\mathrm{i}} & \vec{\mathrm{j}} & \vec{\mathrm{k}} \\ \dfrac{\partial}{\partial x} & \dfrac{\partial}{\partial y} & \dfrac{\partial}{\partial z} \\ b_x & b_y & b_z \end{vmatrix} = \left(\frac{\partial b_z}{\partial y} - \frac{\partial b_y}{\partial z}\right)\vec{\mathrm{i}} + \ldots = \left(\frac{\partial^2 p}{\partial y \partial z} - \frac{\partial^2 p}{\partial z \partial y}\right)\vec{\mathrm{i}} + \ldots = \vec{0}$$

Hence, a fluid can remain in equilibrium only if $\vec{b}$ is ir-rotational, i.e., $\vec{\nabla}\times\vec{b} = \vec{0}$. $\Rightarrow$ A potential V exists for the body force $\vec{b}$ such that $\vec{b} = -\vec{\nabla}V$. Equation (6.108) becomes

$$\vec{\nabla}(p + V) = \vec{0} \qquad \Rightarrow \qquad p + V = \text{constant} = C \quad \text{(say)} \tag{6.109}$$

It is the integrated form of the equilibrium equation. It implies that the surfaces of constant p are the level surfaces of V. A free surface is a surface of constant p and thus corresponds to a level surface of V. If $\vec{b}$ can be neglected (as in gases), then eq. (6.109) implies that the pressure in a fluid in equilibrium is the same everywhere.

Let $p = p_1(\vec{r})$ be a solution of eq. (6.108) for a given $\vec{b}$. Then $p = p_1(\vec{r}) + p_0$ also satisfies eq. (6.108), where p_0 is an arbitrary constant which represents a constant increase of pressure at every point. Thus, in equilibrium, if the pressure at one point is changed by p_0, then the pressure at every point changes by p_0 when equilibrium is restored. This finds use in hydraulic jack, hydraulic press and hydraulic brake.

6.22 Pressure Variation in a Fluid in Equilibrium

The gravitational force on a volume element dV is $-\frac{GM}{r^2}\rho\, dV\, \vec{\mathrm{e}} = \vec{b}\, dV$, i.e., $\vec{b} = -\rho g\, \vec{\mathrm{e}}$ where ρ is the density of the fluid and r is its distance from the earth's centre. $\vec{\mathrm{e}}$ is the local upward vertical unit vector and $g = GM/r^2$ is the local gravitational acceleration. g is modelled as uniform for small vertical and horizontal distances, and as variable for pressure variation through large vertical distances in the earth's atmosphere.

Equation (6.108) is to be solved subject to boundary conditions. We need constitutive relation for the fluid material which relates pressure and density. This is a *thermodynamic equation of state* which relates p to the temperature $T(t)$ of the fluid:

$$F(p, \rho, T) = 0 \tag{6.110}$$

For isothermal processes of constant T, eq. (6.110)) is an equation relating p and ρ. We consider the equation of state for incompressible fluids and compressible ideal gases.

6.22.1 Incompressible fluids

The equation of state for incompressible fluids[2] is ρ = constant. This idealization is used for liquids for a wide range of temperature and pressure. The gravitational body force
is in the direction of the local vertical (z-axis vertically down), $\vec{b} = \rho g\, \vec{\mathrm{k}} \;\Rightarrow\; b_x = b_y = 0,\; b_z = \rho g$. $\vec{b}$ is almost constant if the variations in z are small. Equation (6.108) $\Rightarrow$

$$\frac{\partial p}{\partial x} = 0, \qquad \frac{\partial p}{\partial y} = 0, \qquad \frac{\partial p}{\partial z} = \rho g \tag{a}$$

Equations $(\mathrm{a})_{1,2} \Rightarrow p$ is independent of x, y, i.e., *p is the same at any given horizontal level.* Hence, the free surface is normal to the local vertical direction since p at the free surface is constant. Moreover, eq. $(\mathrm{a})_3$ becomes an *ordinary differential equation*:

$$\frac{dp}{dz} = \rho g > 0 \tag{6.111}$$

Hence, *p increases with depth* and $\rho = g^{-1}\frac{dp}{dz}$ implies that *for a fluid in equilibrium under gravitational force, the density is independent of x, y and can depend only on the depth z.* It follows that the surface separating two fluids of different densities, which do not mix and are in equilibrium in a container, is a horizontal plane.

Equation (6.111) $\Rightarrow$ $\qquad p(z) = p(0) + \int_0^z \rho g dz$

if g and ρ are constant, then $\qquad p = p_0 + \rho g z \qquad (6.112)$

[2]Strictly, incompressibilty is a property of the flow and not the fluid as discussed on p. 794 (bottom).

where p_0 is the pressure at the level $z = 0$. The pressure in an incompressible fluid subject to uniform gravity increases linearly with depth. This is obvious from the FBD of a cylinder of fluid from the free surface to a depth z (Fig. 6.37). Choose $z = 0$ at the free surface between the air and the fluid. p_0 = (atmospheric pressure p_a at $z = 0$),

$$p_g = p - p_a = \rho g z \tag{6.113}$$

where p_g is the *gage pressure* $= (p - p_a)$ at depth z. At sea level, $p_a \simeq 100\,\text{kPa}$. For water, $\rho g \simeq 10^4\,\text{N/m}^3$ and eq. (6.113) implies that the pressure is about 100 kPa at the free surface and increases by the same amount for each additional 10 m depth.

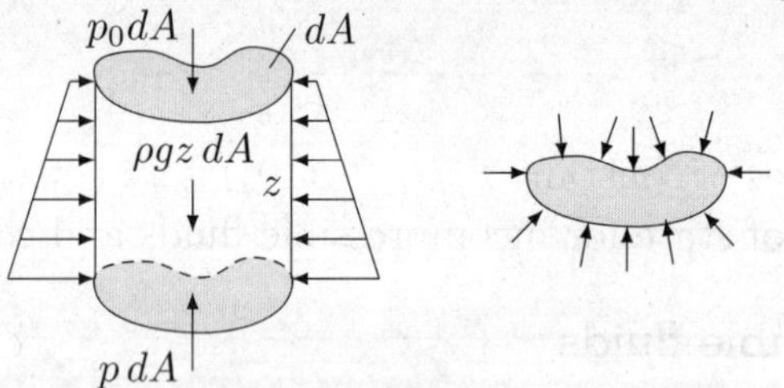

Fig. 6.37

By a pressure of x cm of water or y mm of mercury we mean that the pressure is $\rho_w g(10^{-2}x)$ Pa and $\rho_m g(10^{-3}y)$ Pa respectively, where $\rho_w = 1000\,\text{kg/m}^3$, $\rho_m = 13600\,\text{kg/m}^3$, $g = 9.81\,\text{m/s}^2$. The variation of pressure with depth is significant in liquid containers, but it is insignificant in gas containers since $\rho_{\text{gas}} \ll \rho_{\text{liquid}}$.

Example 6.19: (a) Discuss the variation in the pressure of blood while standing and while lying down. (b) Why do pilots have 'black out' during sharp manoeuvres of the aircraft?

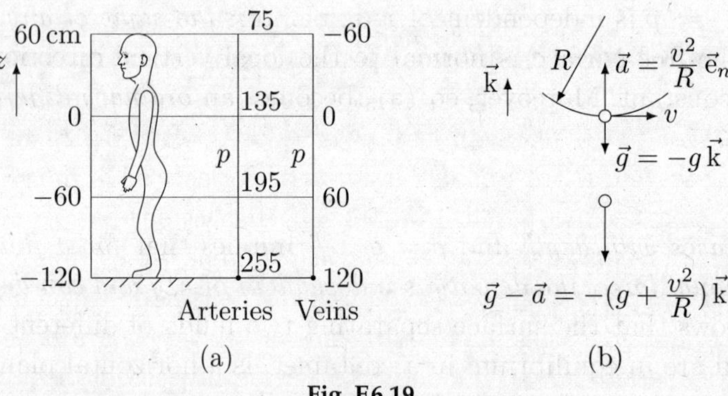

Fig. E6.19

Solution: (a) The blood pressure in the arteries varies between 120 mm and 80 mm of mercury with the mean at the level of the heart being 100 mm of mercury $\simeq$ 135 cm of water. This is also the average pressure difference between the blood in the arteries and the veins at any level. The specific density of blood is almost 1. Hence, a 1.8 m tall

person while standing has a pressure difference of 180 cm of water between the brain and the feet. The mean arterial and venous pressures at different levels are shown in Fig. E6.19a in centimeters of water. While lying down, the mean pressure everywhere is about 135 cm of water in the arteries and almost zero in the veins. On rising suddenly, the increase of blood pressure in the veins expands them and the blood 'pools' there. The venous return of the blood to the heart drops drastically. Consequently, the blood circulation to the brain falls resulting in dizziness or a feeling of faintness. The arterial pressure should be measured with the arm held at the level of the heart.
(b) For a pilot in an aircraft moving at $v = 720\,\text{km/h} = 200\,\text{m/s}$ at the bottom of a vertical loop of radius $R = 2\,\text{km} = 2000\,\text{m}$ (Fig. E6.19b), $\vec{a} = (v^2/R)\,\vec{\text{k}}$. Hence,

$$\vec{\nabla}\, p = \vec{b} - \rho\vec{a} = \rho(\vec{g} - \vec{a}) = -\rho\left(g + \frac{v^2}{R}\right)\vec{\text{k}} \quad \Rightarrow \quad \frac{dp}{dz} = -\rho\left(g + \frac{v^2}{R}\right)$$

instead of $dp/dz = -\rho g$ for statics. $v^2/R = 20\,\text{m/s}^2 \simeq 2g$ and $dp/dz \simeq -3\rho g$. Hence, referring to Fig. E6.19a, the arterial pressure in the brain is reduced to $135 - 3 \times 60 = -45$ cm of water. The additional pressure in the veins, near the feet, distends them causing 'pooling' of blood. This reduces return of blood from the veins and supply to the brain falls markedly. This may lead to 'black out', i.e., loss of vision/consciousness.

Example 6.20: In the system shown (Fig. E6.20) find the minimum pressure in the water tank. The densities of air, oil, mercury and water are ρ_a, ρ_0, ρ_m and ρ_w.

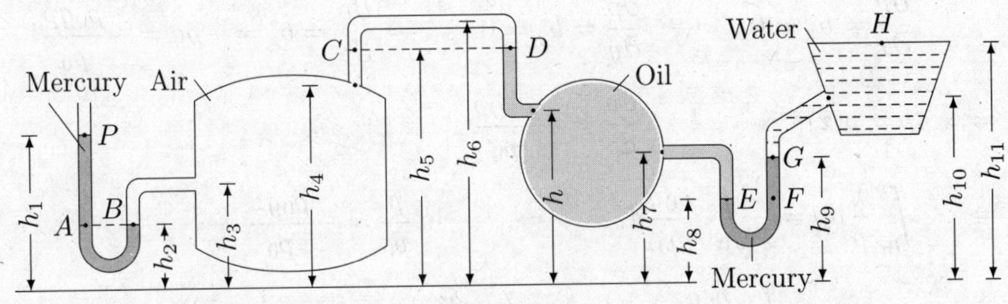

Fig. E6.20

Solution: For a fluid in equilibrium, Δp for increase of h in depth is $\rho g h$, and pressure p at the same level is the same. Hence, $p_P = p_a$,

$$p_A = p_P + \rho_m g(h_1 - h_2) \quad \Rightarrow \quad p_A = p_a + \rho_m g(h_1 - h_2) \tag{1}$$

$$p_A = p_B = p_C + \rho_a g(h_5 - h_2) \quad \Rightarrow \quad p_C = p_A - \rho_a g(h_5 - h_2) \tag{2}$$

$$p_C = p_D = p_E - \rho_0 g(h_5 - h_8) \quad \Rightarrow \quad p_E = p_C + \rho_0 g(h_5 - h_8) \tag{3}$$

$$p_E = p_F = p_G + \rho_m g(h_9 - h_8) \quad \Rightarrow \quad p_G = p_E - \rho_m g(h_9 - h_8) \tag{4}$$

$$p_H = p_G - \rho_w g(h_{11} - h_9) \tag{5}$$

The minimum pressure in the water tank is at H. Equations (1) to (5) $\Rightarrow$

$$p_H = p_a + \rho_m g(h_1 - h_2) - \rho_a g(h_5 - h_2) + \rho_0 g(h_5 - h_8) - \rho_m g(h_9 - h_8) - \rho_w g(h_{11} - h_9)$$

6.22.2 Compressible ideal gas

For *an ideal gas* we have the following *axiom* relating its pressure p, volume V, absolute temperature T (Kelvin), density ρ and mass m $(= \rho V)$:

$$pV = mRT \qquad \Rightarrow \qquad p = \rho RT \tag{6.114}$$

where R is a material constant of a particular gas. R is related to the universal gas constant $R^* = 8.31\,\text{J.mol}^{-1}.\text{K}^{-1}$ by $R = R^*/[6.02 \times 10^{23} m_1] = 1.38 \times 10^{-23}/m_1$ where m_1 is the mass of one molecule of the gas. This axiom is valid for air or its constituents for large range of p and T. It is not valid for some gases, e.g., steam. We consider 2 cases.

1. Uniform temperature distribution T

$$\text{Equation (6.114)} \quad \Rightarrow \quad \frac{p}{\rho} = \frac{p_0}{\rho_0} = RT \quad \Rightarrow \quad \rho = \frac{p\rho_0}{p_0}$$

where p_0 and ρ_0 are at $z = 0$. If the gas is subjected only to gravitational body force, then with $\vec{k}$ vertically up $b_x = 0$, $b_y = 0$, $b_z = -\rho g$. Equilibrium equations $\Rightarrow$

$$\frac{\partial p}{\partial x} = b_x = 0, \qquad \frac{\partial p}{\partial y} = b_y = 0, \qquad \frac{\partial p}{\partial z} = b_z = -\rho g = -\frac{\rho_0 g p}{p_0}$$

$$\Rightarrow \quad p = p(z), \qquad \frac{dp}{dz} = -\frac{\rho_0 g p}{p_0}$$

$$\Rightarrow \quad \int_{p_0}^{p} \frac{1}{p}\, dp = -\int_0^z \frac{\rho_0 g}{p_0} dz \quad \Rightarrow \quad \ln\frac{p}{p_0} = -\frac{\rho_0 g z}{p_0}$$

$$\Rightarrow \quad p = p_0 \exp\left(-\frac{\rho_0 g z}{p_0}\right) = p_0 \exp\left(-\frac{gz}{RT}\right) = p_0 \exp\left(-\frac{z}{h_0}\right)$$

where $h_0 = RT/g$. In an isothermal atmosphere (near earth's surface) in equilibrium, p and hence, ρ decrease exponentially with z. At heights h_0 and $2h_0$, p/p_0 falls to $1/\text{e}$ and $1/\text{e}^2$. For air at $T = 273\,\text{K}$, $m_1 = 4.75 \times 10^{-26}\,\text{kg}$, $R = 290.6\,\text{J.kg}^{-1}.\text{K}^{-1}$, $h_0 = RT/g \simeq 8.1\,\text{km}$ for which $p \simeq 0.37 p_0$ with associated physical effects. Atop mount Everest: $z = 8.84\,\text{km}$, $p/p_0 \simeq 0.34$; and at the height of jet travel: $z \simeq 12\,\text{km}$, $p/p_0 \simeq 0.23$.

2. Linear variation of temperature with elevation

Let $T = T_0 + cz$ where T_0 is the temperature at $z = 0$ and $c \simeq -0.0065\,\text{K/m}$. Hence,

$$\frac{dp}{dz} = -\rho g = -\frac{pg}{RT} = \frac{-pg}{R(T_0 + cz)}$$

$$\Rightarrow \quad \int_{p_0}^{p} \frac{1}{p}\, dp = -\frac{g}{R}\int_0^z \frac{1}{T_0 + cz} dz \quad \Rightarrow \quad \frac{p}{p_0} = \Big(\frac{T_0}{T_0 + cz}\Big)^{g/cR}$$

The temperature and pressure outside a jet airplane are much smaller than the normal pressure and temperature maintained inside for comfort. The fuselage is designed to withstand this large pressure difference which alters as the plane ascends/descends.

6.23 Resultant Force on a Surface in a Fluid in Equilibrium

We consider a plane surface and a curved surface.

1. Plane surface

Consider a plane area A with centroid C (Fig. 6.38a) having a fluid of density ρ in equilibrium on one side and atmosphere on the other side. At C choose z-axis normal to the plane A and x-axis parallel to the line of intersection of the plane of the area A with the free surface. Let θ be the inclination of the plane surface with the horizontal. An area element dA at (x, y) has a depth $h = PQ = OP \sin\theta = (y'_C + y)\sin\theta$.

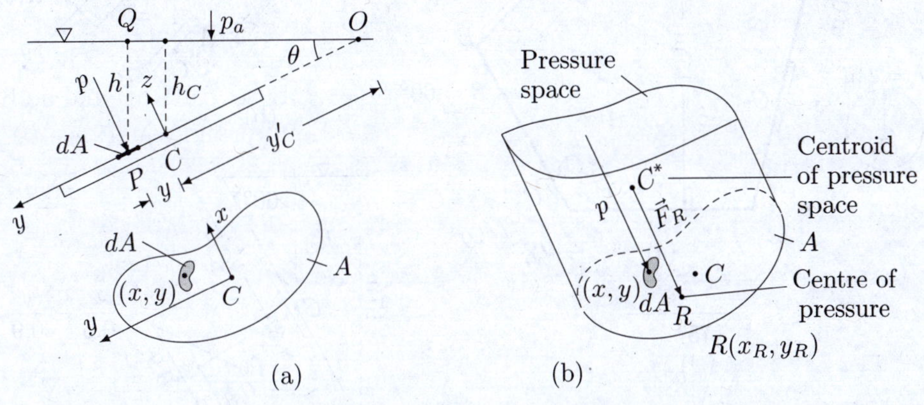

Fig. 6.38

The net pressure p on dA is $p = (\rho g h + p_a) - p_a = \rho g h$. It is a parallel force distribution on A in the same direction (Fig. 6.38b). Hence, its simplest resultant is a single force acting at the centre of pressure R at (x_R, y_R):

$$\vec{F}_R = -\int_A p\,\vec{\mathrm{k}}\, dA = -\int_A \rho g h\, dA\,\vec{\mathrm{k}} = -(\rho g h_C)A\,\vec{\mathrm{k}} = -p_C A\,\vec{\mathrm{k}} \tag{6.115}$$

where $p_C = \rho g h_C$ is the pressure at the centroid C of the area at depth $h_C = y'_C \sin\theta$ below the free surface. R is the centre of parallel force distribution with

$$\vec{r}_R = \frac{\sum \vec{r}_i F_i}{\sum F_i} = \frac{\int_A (x\,\vec{\mathrm{i}} + y\,\vec{\mathrm{j}})\rho g(y'_C + y)\sin\theta\, dA}{\rho g y'_C \sin\theta\, A} = \frac{\int \left[(xy'_C + xy)\,\vec{\mathrm{i}} + (yy'_C + y^2)\,\vec{\mathrm{j}}\right]\, dA}{A y'_C}$$

$$\vec{r}_R = \frac{-I^C_{xy}\,\vec{\mathrm{i}} + I^C_{xx}\,\vec{\mathrm{j}}}{Ay'_C}, \qquad x_R = -\frac{I^C_{xy}}{Ay'_C}, \qquad y_R = \frac{I^C_{xx}}{Ay'_C} > 0 \tag{6.116}$$

since $\int_A x\,dA = \int_A y\,dA = 0$, $I^C_{xx} = \int_A y^2\,dA$, $I^C_{xy} = -\int_A xy\,dA$. I^C_{xx} and I^C_{xy} are the area moment of inertia about x-axis and the area product of inertia of the area A about x, y-axes at C. R is always below C. If y is an axis of symmetry of area A, then $I^C_{xy} = 0 \Rightarrow x_R = 0 \Rightarrow R$ lies on the symmetry axis. For an area A at greater depth, y'_C increases and eq. (6.116) $\Rightarrow x_R \to 0$, $y_R \to 0$, i.e., R gradually shifts towards C. For $y'_C \gg$ (dimensions of area A): $p \simeq \rho g h_C \Rightarrow R$ is almost coincident with C.

Example 6.21: Find the resultant force on rectangular, circular and triangular windows (Fig. E6.21a) in a submersible body used for exploration. Assume that the pressure inside is atmospheric. The density of liquid outside is $\rho = 1000\,\mathrm{kg/m^3}$.

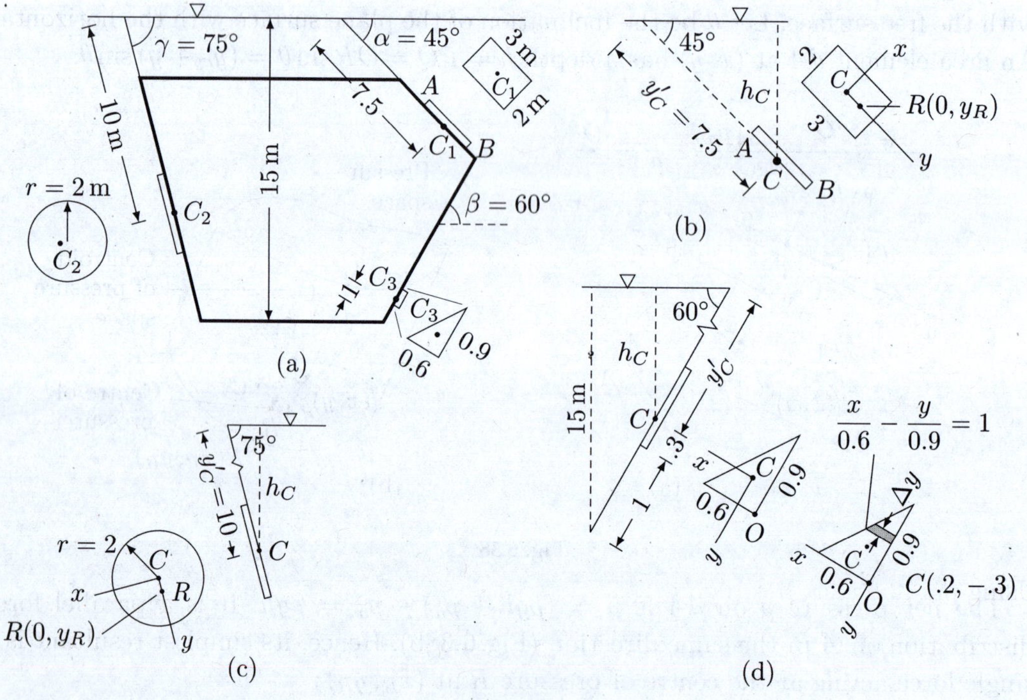

Fig. E6.21

Solution: (i) Rectangular window (Fig. E6.21b): $y'_C = 7.5\,\mathrm{m}$. The depth of the centroid C is $h_C = 7.5 \sin 45° = 5.303\,\mathrm{m}$. Area $A = 6\,\mathrm{m^2}$, $\rho = 1000\,\mathrm{kg/m^3}$. The area is symmetrical about the y-axis. The resultant force F_R acts at $R_1(0, y_R)$ where

$$F_R = (\rho g h_C)A = 312.1\,\mathrm{kN}, \qquad y_R = \frac{I^C_{xx}}{Ay'_C} = \frac{2 \times 3^3/12}{6 \times 7.5} = 0.1\,\mathrm{m}$$

(ii) Circular window (Fig. E6.21c) : y-axis is its axis of symmetry. $\vec{F}_R$ acts at $R(0, y_R)$.

$$r = 2\,\text{m}, \quad A = \pi(2)^2 = 12.57\,\text{m}^2, \quad y'_C = 10\,\text{m}$$

$$h_C = 10\sin 75^\circ = 9.659\,\text{m}, \quad I_{xx}^C = \tfrac{1}{4}\pi r^4$$

$$F_R = (\rho g h_C)A = 1191\,\text{kN}, \quad y_R = \frac{I_{xx}^C}{A y'_C} = \frac{\pi(2)^4/4}{\pi(2)^2(10)} = 0.1\,\text{m}$$

(iii) Triangular window (Fig. E6.21d) : y-axis is not its symmetry axis.

$$A = \tfrac{1}{2}(0.6 \times 0.9) = 0.27\,\text{m}^2$$

$$y'_C = 15\,\text{cosec}\,60^\circ - (1 + 0.3) = 16.02\,\text{m}$$

$$h_C = y'_C \sin 60^\circ = 13.87\,\text{m}$$

$$I_{xy}^O = -\int xy\,dA = -\int_{-0.9}^{0}\left[y\int_0^{0.6(1+y/0.9)} x\,dx\right]dy$$

$$= -\int_{-0.9}^{0}(y)\tfrac{1}{2}\left[0.6\left(1 + \frac{y}{0.9}\right)\right]^2 dy$$

$$= -\int_{-0.9}^{0} 0.18\left[y + \frac{2}{0.9}y^2 + \frac{1}{0.81}y^3\right]dy = 1.215 \times 10^{-2}\,\text{m}^4$$

$$I_{xx}^O = \int y^2\,dA = \int_{-.0.9}^{0}\left[y^2\int_0^{0.6(1+y/0.9)} dx\right]dy$$

$$= \int_{-0.9}^{0} 0.6\left(y^2 + \frac{y^3}{0.9}\right)dy = 0.03645\,\text{m}^4$$

Using parallel axes theorems with $x_C = 0.2,\ y_C = -0.3$,

$$I_{xy}^O = I_{xy}^C - A x_C y_C \quad \Rightarrow \quad I_{xy}^C = -0.00405\,\text{m}^4$$

$$I_{xx}^O = I_{xx}^C + A y_C^2 \quad \Rightarrow \quad I_{xx}^C = 0.01215\,\text{m}^4$$

The resultant force is $F_R = (\rho g h_C)A = 36.74\,\text{kN}$ at R with $\overrightarrow{CR} = x_R\vec{\imath} + y_R\vec{\jmath}$ where,

$$x_R = -\frac{I_{xy}^C}{A y'_C} = 9.36 \times 10^{-4}\,\text{m}, \quad y_R = \frac{I_{xx}^C}{A y'_C} = 0.00281\,\text{m}$$

R almost coincides with C since for any point of the window, the depth $h \simeq h_C$.

2. Curved surface

The simplest resultant of spatial pressure force distribution on a curved surface area A is, in general, a wrench. The force on an area element dA is $-p\vec{n}\,dA$ at $\vec{r}$ (Fig. 6.39a). The resultant at origin O consists of a force $\vec{F}_R$ and a moment $\vec{C}_{R_O}$ given by

$$\vec{F}_R = -\int_A p\vec{n}\,dA = -\int_A p(n_x\vec{\mathrm{i}} + n_y\vec{\mathrm{j}} + n_z\vec{\mathrm{k}})\,dA$$

$$= -\int_{A_x} p\,dA_x\vec{\mathrm{i}} - \int_{A_y} p\,dA_y\vec{\mathrm{j}} - \int_{A_z} p\,dA_z\vec{\mathrm{k}}$$

$$\vec{C}_{R_O} = -\int_A \vec{r}\times p\vec{n}\,dA$$

where $dA_x = n_x\,dA$, $dA_y = n_y\,dA$, $dA_z = n_x\,dA$ and A_x, A_y, A_z are projected components of the area element dA and area A on planes perpendicular to x, y, z axes. The simplest resultants F_{Rx}, F_{Ry}, F_{Rz}, of force distribution $p\,dA_x$ on A_x, $p\,dA_y$ on A_y and $p\,dA_z$ on A_z, are

$$F_{Rx} = -\int_{A_x} p\,dA_x, \qquad F_{Ry} = -\int_{A_z} p\,dA_y, \qquad F_{Rz} = -\int_{A_z} p\,dA_z$$

The pressure on dA, dA_x, dA_y is the same as these are at the same depth. Hence, the forces F_{Rx}, F_{Ry} on A are the same as those on projected vertical surfaces A_x, A_y.

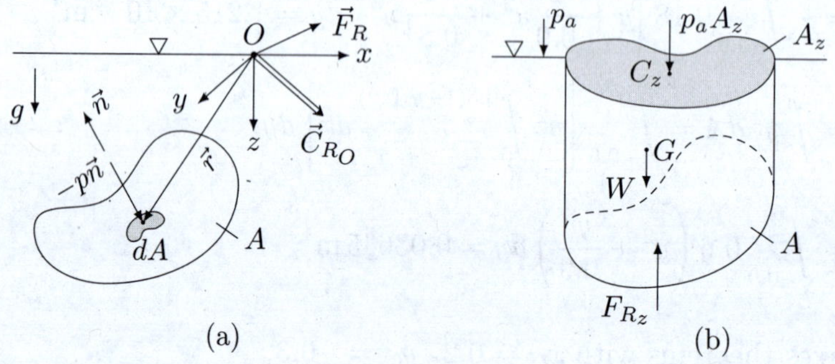

Fig. 6.39

Consider a cylindrical column of fluid with base as the given curved surface, lateral surface parallel to z-axis and top-surface at the free surface of the fluid (Fig. 6.39b). Let its volume and weight be V and W. The vertical forces on this column are shown in Fig. 6.39b. Hence, $F_{Rz} = W + p_aA_z$ and its line of action is along the resultant of force W at the centre of gravity G of the fluid column and the force p_aA_z at the centroid C_z of area A_z. The forces F_{Rx}, F_{Ry}, F_{Rz} along with their lines of action are equivalent to the distributed pressure force on the curved area. The simplest resultant of these forces is, in general, not a single force since they may not be concurrent.

However, for some curved surfaces the simplest resultant is a single force.

(i) The curved area has a vertical plane of symmetry. The resultant force on a horizontal strip of depth dz is a force in the plane of symmetry since the pressure on it is uniform. The pressure force distribution is equivalent to a coplanar force system in the vertical plane of symmetry, whose simplest resultant is a single force.

(ii) For spherical surface, the resultant of concurrent forces at the centre is a force.

(iii) Cylindrical surface with generator parallel to the free surface

Consider a cylindrical surface $ABDE$ with horizontal generators. AB is parallel to the free surface (Fig. 6.40a) at a depth of h_1. The parallel pressure distribution on strip $L \times ds$ has a resultant $pLds$ on GF. Thus, there is coplanar line force on the curve PQ (Fig. 6.40b). Its simplest resultant is a force which can be found by integration.

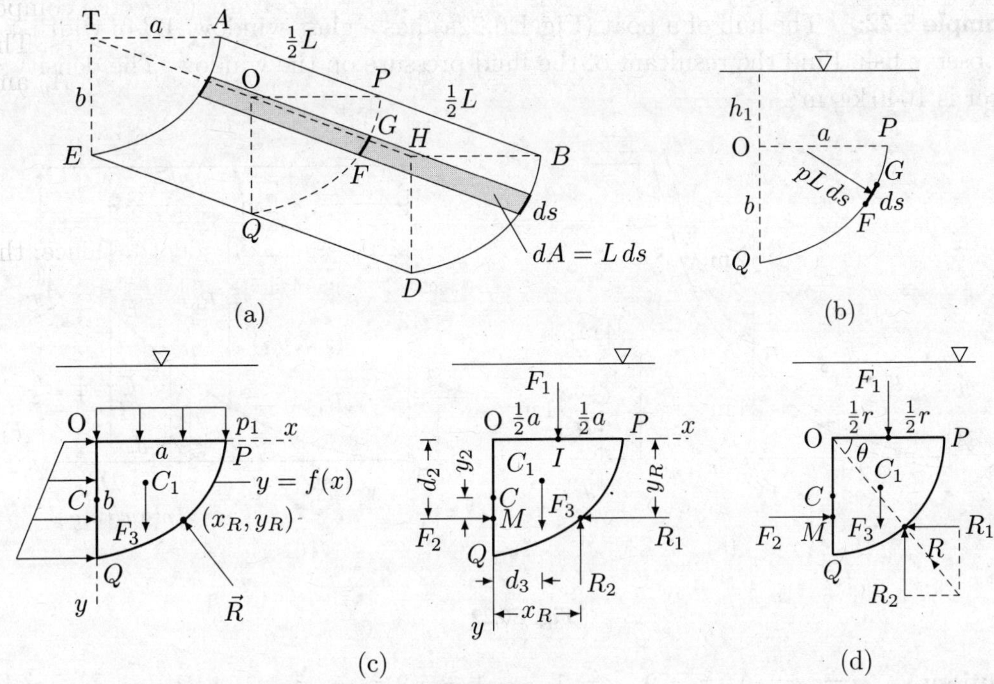

Fig. 6.40

An *alternative method* is to consider the equilibrium of fluid contained in a sector $HDBATE$ of a cylinder (Fig. 6.40a). The forces acting on it are shown in Fig. 6.40c.

(1) Resultant force $F_1 = p_1 La$ at I, due to pressure p_1 on $L \times a$ face $HBAT$.

(2) Resultant force F_2 acting on $L \times b$ face $DHTE$, at centre of pressure M at y_2.

(3) Weight F_3 of the cylinder of fluid of length L and area A_1 = (area OPQ) acting through its centre of gravity C_1. For a uniform fluid of density ρ, $F_3 = \rho(A_1 L)g$.

(4) The pressure on the curved surface $DBAE$. Its resultant is a force $\vec{R}$ acting through

the centre of pressure at $[x_R,\ y_R = f(x_R)]$ where the curve PQ is defined by $y = f(x)$.

The forces R_1, R_2 and x_R are found from the equilibrium equations of $HDBATE$.

$$F_x = -R_1 + F_2 = 0 \quad \Rightarrow \quad R_1 = F_2$$

$$F_y = -R_2 + F_1 + F_3 = 0 \quad \Rightarrow \quad R_2 = F_1 + F_3$$

$$M_{Oz} = \tfrac{1}{2}aF_1 + d_3F_3 - d_2F_2 - x_RR_2 + y_RR_1 = 0 \tag{1}$$

Using $y_R = f(x_R)$ we solve eq. (1) for x_R.

If it is a circular cylindrical surface, then the distributed load on PQ is concurrent at O (Fig. 6.40d). Its simplest resultant $\vec{R}$ passes through O. Instead of using eq. (1), the centre of pressure is obtained directly at $(r\cos\theta, r\sin\theta)$ with $\theta = \tan^{-1}(R_2/R_1)$.

Example 6.22: The hull of a boat (Fig. E6.22a) has a glass window AB of width 4 m to observe fish. Find the resultant of the fluid pressure on the window. The density of water is 1040 kg/m^3.

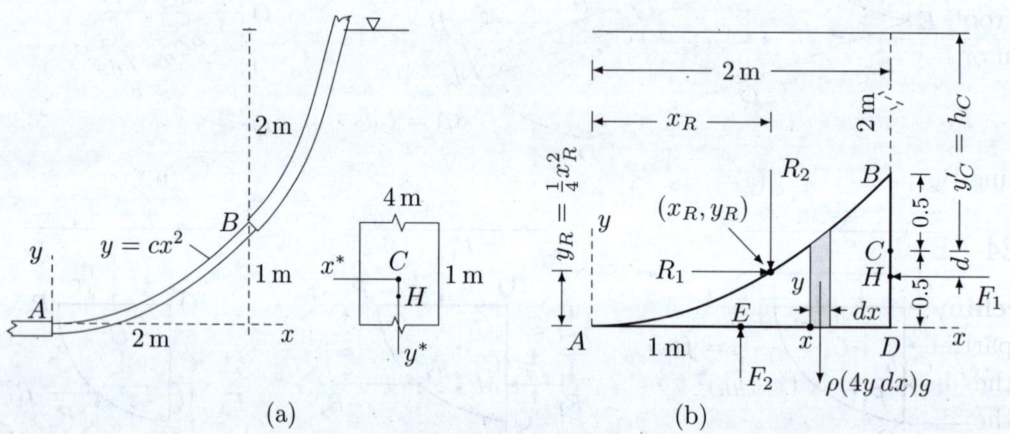

Fig. E6.22

Solution: $y = cx^2$. At $x = 2,\ y = 1 \Rightarrow 1 = c(2)^2 \Rightarrow c = 1/4$. Hence, $y = x^2/4$. Consider the equilibrium of fluid contained within the window and its horizontal and vertical projected areas. The cross-section of this fluid is ABD and its width is 4 m. The FBD of this fluid is shown in Fig. E6.22b. The area of face BD is $A = 1 \times 4\ \text{m}^2$ with centroid C. The resultant force F_1 on face BD acts at H with

$$F_1 = \rho g h_C A = 1040 \times 9.81(2 + 0.5)(1 \times 4) = 102.0 \times 10^3\ \text{N} = 102.0\ \text{kN}$$

$$d = CH = \frac{I_{xx}^C}{Ay'_C} = \frac{4(1)^3/12}{(1 \times 4)2.5} = 0.0333\ \text{m}$$

The resultant force F_2 on face AD (area A_1) acts at its centroid E ($AE = 1\,\text{m}$) and

$$F_2 = \rho g h_E A_1 = 1040 \times 9.81 \times 3 \times (2 \times 4)\,\text{N} = 244.9\,\text{kN}$$

Let the resultant reaction $\vec{R}$ of the window on water have horizontal and vertical components R_1 and R_2 and act at (x_R, y_R) with $y_R = x_R^2/4$. The gravity force on the elementary volume of the fluid shown as shaded in Fig. E6.22b is $\rho(4y\,dx)g = 10.2x^2\,dx\,\text{kN}$ where x is in meters. The equations of equilibrium of the fluid are:

$$F_x = R_1 - F_1 = 0 \qquad \Rightarrow \qquad R_1 = F_1 = 102.0\,\text{kN}$$

$$F_y = F_2 - R_2 - \int_0^2 10.2x^2\,dx = 0 \qquad \Rightarrow \qquad R_2 = 217.7\,\text{kN}$$

$$M_{A_z} = (0.5 - d)F_1 + 1(F_2) - \int_0^2 x(10.2x^2)\,dx - (x_R^2/4)R_1 - x_R R_2 = 0$$

$$\Rightarrow \qquad 25.5x_R^2 + 217.7x_R - 251.7 = 0$$

Its roots are $x_R = 1.032,\ -9.57\,\text{m}$. The first root corresponds to the point of intersection of $\vec{R}$ with the window. Hence, the resultant force $\vec{F}_R$ on window is $-\vec{R}$.

$$\vec{F}_R = -\vec{R} = -R_1\,\vec{\imath} + R_2\,\vec{\jmath} = -102\,\vec{\imath} + 217.7\,\vec{\jmath}\ \text{kN}$$

acting through $(x_R, y_R) = (1.032\,\text{m}, 0.258\,\text{m})$.

6.24 Buoyancy

Archimedes' Principle: 'The resultant force due to fluid pressure, on a body wholly or partially immersed in a *fluid in equilibrium*, is the negative of the weight force vector of the 'displaced fluid''. This force is called the *buoyant force* and the centre of gravity of the 'displaced fluid' is the *centre of buoyancy.*

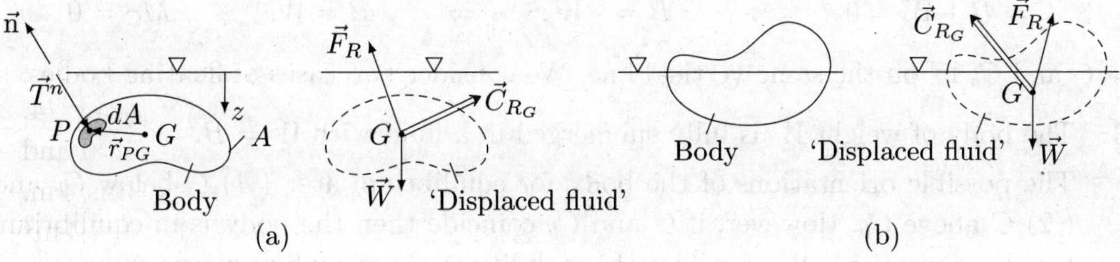

Fig. 6.41

Proof: Consider the body shown in Fig. 6.41 which may be fully immersed in one fluid (Fig. 6.41a) or partially submerged in two fluids (Fig. 6.41b). The resultant force of the fluid on the submerged body, being an integral of the pressure distribution over the surface of the body depends only on the shape of that surface and its location in the fluid and not on the material of the body. Remove the body and replace it by a fluid which occupies the same volume as the body and whose $\rho(z)$ is identical to that of the surrounding fluid. This mass of fluid- the 'fluid displaced' by the body, is in equilibrium under its own weight and the pressure of the fluid surrounding it on its boundary surface A. The 'displaced fluid' is in equilibrium since the satisfaction of the equation of equilibrium, $\vec{\nabla} p = \rho\vec{g}$, in the surrounding fluid, automatically ensures satisfaction in the interior of the 'displaced fluid' as $\rho(z)$ for both is the same.

The weight $\vec{W}$ of the 'displaced fluid' acts at its centre of gravity G. In general, the resultant of the pressure distribution on surface area A of the 'displaced fluid'/ submerged body, at G is a force $\vec{F}_R$ and a couple $\vec{C}_{R_G}$. The equilibrium equations of the 'displaced fluid' are

$$\vec{F} = \vec{F}_R + \vec{W} = \vec{0} \qquad \Rightarrow \qquad \vec{F}_R = -\vec{W} \quad \text{and} \quad \vec{M}_G = \vec{C}_{R_G} = \vec{0}$$

Hence, the resultant of the fluid pressure on the surface A of the submerged body is a single force $\vec{F}_R = -\vec{W}$ acting through the centre of gravity of the 'displaced fluid'.

A fluid translates, as a rigid body, with acceleration $\vec{a}$. For a body submerged in it, the equation of motion of its 'displaced fluid' of mass m is:

$$\vec{F} = \vec{F}_R + \vec{W} = \vec{F}_R + m\vec{g} = m\vec{a}, \qquad \vec{M}_G = \vec{C}_{R_G} = \vec{0}$$

$\Rightarrow \vec{F}_R = -m(\vec{g} - \vec{a}) \Rightarrow$ buoyant force is $m|\vec{g} - \vec{a}|$ acting through G opposite to $(\vec{g} - \vec{a})$.

6.24.1 Equilibrium conditions of floating bodies

A body floating in a fluid (compressible or incompressible) is in equilibrium under buoyancy force $\vec{B}$ and its weight $\vec{W}$ (Fig. 6.42a). $\vec{B}$ acts vertically upwards at centre of gravity C_b of the 'displaced fluid'. $\vec{W}$ acts vertically downwards at the centre of gravity C of the body. The conditions of equilibrium of the floating body are

$$\vec{F} = \vec{B} + \vec{W} = \vec{0} \qquad \Rightarrow \qquad \vec{B} = -\vec{W}, \qquad \Rightarrow \qquad B = W, \qquad \vec{M}_C = \vec{0}$$

$\Rightarrow C$ and C_b lie on the same vertical line. We consider two cases of floating bodies.

1. The body of weight W is fully submerged in a fluid with $W = B$.

   The possible orientations of the body for equilibrium are: (1) C below C_b and (2) C above C_b. However, if C and C_b coincide then the body is in equilibrium for all orientations. We consider the stability of these configurations:

   (i) C below C_b: $\vec{B}$, $\vec{W}$ form restoring couple in disturbed position (Fig. 6.42b).

   (ii) C above C_b: $\vec{B}$, $\vec{W}$ form overturning couple in disturbed position (Fig. 6.42c).

   (iii) C at C_b: $\vec{B}$, $\vec{W}$ do not form couple in disturbed position (Fig. 6.42d).

Hence, an equilibrium configuration is stable only if C is below C_b.

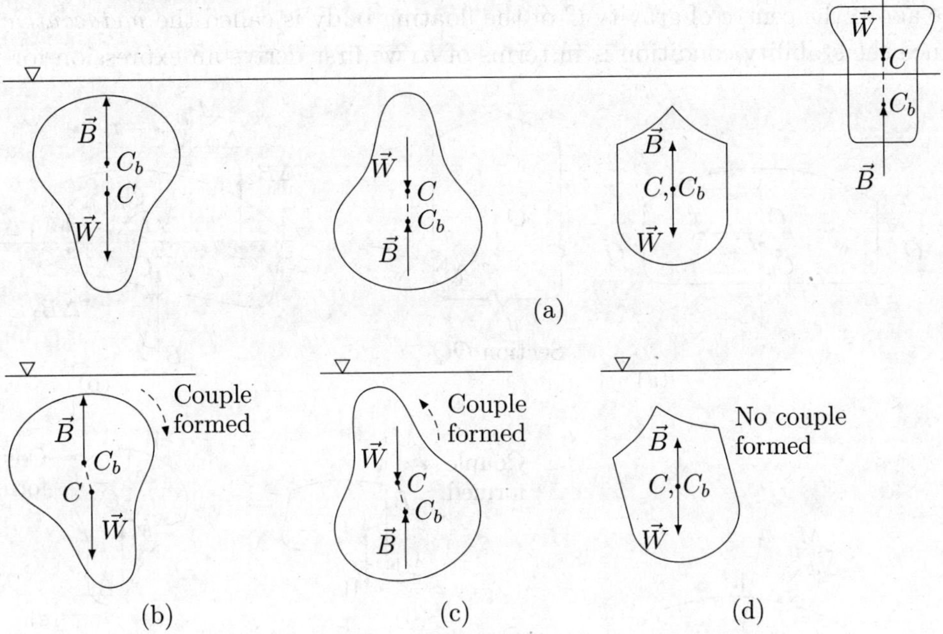

(a)

(b) (c) (d)

Fig. 6.42

2. If the weight W of a body is less than the buoyancy force B had it been fully submerged, then the body will float partially submerged. The buoyancy force B^* in the partially submerged condition in a liquid is

$$B^* = B_{\text{liq}} + B_{\text{air}} = \text{Weight of liquid displaced} + \text{Weight of air displaced} \simeq B_{\text{liq}}$$

The depth of immersion is found from $W = B^*$. C and C_b must lie on the same vertical. The investigation of stability becomes more complex since the boundary of the displaced fluid depends on the orientation of the submerged body.

6.24.2 Stability of partially submerged floating bodies

The stability of partially submerged floating bodies in liquids are important for boats and ships. We restrict this study to bodies that have large dimension normal to the plane of the diagram of Fig. 6.43a and a vertical plane of symmetry when floating in the rest position. If the body experiences an infinitesimal rotation $\Delta\phi\,\vec{\jmath}$ about the y-axis through O (Fig. 6.43b), then the fluid displacement increases on the left side and decreases on the right side of the plane of symmetry. The position of the centre of gravity of the displaced fluid thus shifts to the left from C_b to C_b'. The point of intersection of the vertical buoyancy force through C_b' in the displaced configuration with the plane of symmetry is called the *metacentre* M. The distance h (positive upward) of the metacentre above the centre of gravity C of the floating body is called the *metacentric height.* Since the stability condition is in terms of h, we first derive an expression for it.

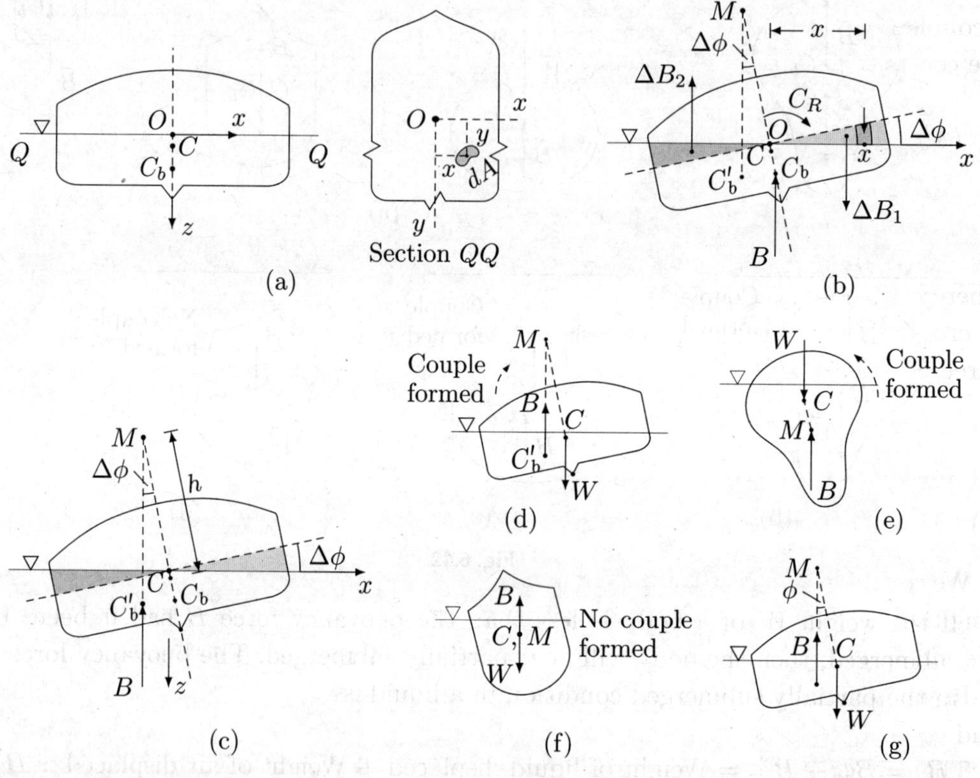

Fig. 6.43

Let A be the area of cross-section of the body at the water level QQ, with axis of symmetry y and centroid at O. The density of water is ρ and the volume of displaced water is V. The buoyancy force consists of B up at C_b and the resultant of the incremental buoyancy force. The area dA at (x, y) moves up by $x\Delta\phi \Rightarrow$ volume $x\Delta\phi\, dA$ moves out $\Rightarrow$ incremental elementary buoyancy force at $x\vec{\mathrm{i}} + y\vec{\mathrm{j}}$ is $\rho g(x\,\Delta\phi)dA\,\vec{\mathrm{k}}$ (a decrease for $x > 0 \Rightarrow$ in downward $\vec{\mathrm{k}}$ direction). Total

$$\Delta\vec{B} = \rho g \Delta\phi \int_A x\, dA\,\vec{\mathrm{k}} = \vec{0} \qquad \text{since symmetry} \Rightarrow \int x\, dA = 0$$

The decrease in buoyancy ΔB_1 on the right is balanced by the increase ΔB_2 $(= \Delta B_1)$ on the left side. Thus, the resultant of the incremental buoyancy force is a couple

$$\vec{C}_R = \vec{M}_O = \int_A (x\,\vec{\mathrm{i}} + y\,\vec{\mathrm{j}}) \times \rho g(x\,\Delta\phi)dA\,\vec{\mathrm{k}} = \rho g\,\Delta\phi\left[-\int_A x^2 dA\,\vec{\mathrm{j}} + \int_A xy\, dA\,\vec{\mathrm{i}}\right]$$

$$= \rho g\,\Delta\phi\left[-I^O_{yy}\vec{\jmath} - I^O_{xy}\vec{\imath}\right] = -\rho g\,\Delta\phi I^O_{yy}\vec{\jmath}$$

as $I^O_{xy} = 0$ since x-axis is principal axis for the symmetric area A at O.

The resultant of buoyancy force in displaced position is a force $B = \rho g V$ at C_b and a couple $C_R = \rho g\,\Delta\phi\, I^O_{yy}$ (Fig. 6.43b). Their simplest resultant is the force B through the centre of gravity C_b' of the displaced submerged volume (Fig. 6.43c) such that

$$(M_{C_b})_y = -[(C_bM)\,\Delta\phi]B = C_{Ry} = -\rho g\,\Delta\phi\, I^O_{yy} \quad\Rightarrow\quad C_bM = \frac{\rho g\Delta\phi\, I^O_{yy}}{\rho g V\,\Delta\phi} = \frac{I^O_{yy}}{V}$$

$$\Rightarrow \qquad h = CM = C_bM - C_bC = \frac{I^O_{yy}}{V} - a \tag{6.117}$$

where a is the height of C *above* C_b and I^O_{yy} is the area moment of inertia of the area of cross-section of the ship at the water line about the longitudinal axis. There are three possibilities:

(i) M above C $(h > 0)$: when disturbed, $\vec{B}$, $\vec{W}$ form restoring couple (Fig. 6.43d).

(ii) M below C $(h < 0)$: when disturbed, $\vec{B}$, $\vec{W}$ form overturning couple (Fig. 6.43e).

(iii) M at C $(h = 0)$: when disturbed, $\vec{B}$, $\vec{W}$ do not form couple (Fig. 6.43f).

We conclude that an equilibrium configuration of a partially submerged body having longitudinal plane of symmetry *is stable if the metacentric height h is positive, i.e., if the metacentre is above the centre of mass.*

In particular, if the centre of buoyancy C_b is above the centre of mass C then $a < 0$ and eq. (6.117) yields $h > 0$. Hence, the body is definitely stable if C_b is above C.

The equation of motion of the body in small rolling motion (Fig. 6.43g) is:

$$I^C_{yy}\ddot{\phi} = M_{Cy} = -W(CM)\sin\phi = -mgh\sin\phi \simeq -mgh\phi \qquad (B = W)$$

$$\Rightarrow \qquad \ddot{\phi} + (mgh/I^C_{yy})\phi = 0$$

where I^C_{yy} is the mass moment of inertia of the body about the longitudinal y-axis through C. If $h > 0$ then the rolling motion of the body is simple harmonic, bounded and stable with frequency $(mgh/I^C_{yy})^{1/2}/(2\pi)$. We conclude that though we need $h > 0$ for stability, yet h should not be made too large since that would increase the frequency of its rolling in the water waves, leading to discomfort.

Example 6.23: A long prismatic block (Fig. E6.23a) has triangular cross-section. The densities of the block and water are ρ_1 and ρ_2. Find $\alpha_{\min}$ for stability.

Solution: Let L and h_1 be the length and height of the prismatic block. Let it float with depth of immersion h_2 as shown in Fig. E6.23b. The volume of fluid displaced is $V = \frac{1}{2}(2h_2\tan\alpha)h_2L = h_2^2L\tan\alpha$. The buoyant force on the block is $h_2^2L\rho_2 g\tan\alpha$.

The weight of block is $[\frac{1}{2}(2h_1 \tan\alpha)h_1]L\rho_1 g = h_1^2 L\rho_1 g \tan\alpha$. For equilibrium:

$$h_1^2 L\rho_1 g \tan\alpha = h_2^2 L\rho_2 g \tan\alpha \quad \Rightarrow \quad h_1 = h_2(\rho_2/\rho_1)^{1/2}$$

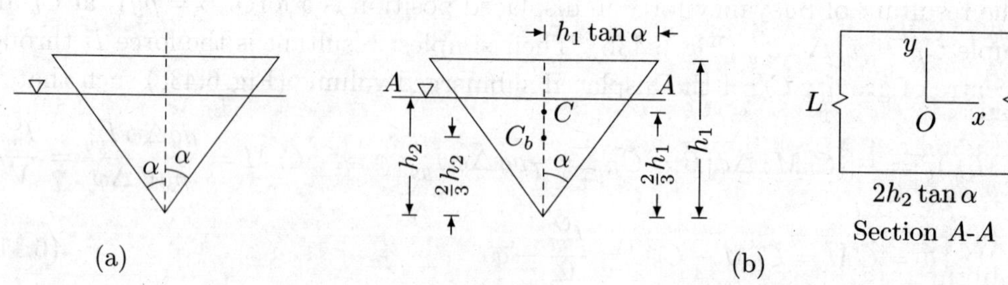

Fig. E6.23

The cross-section of the block at water level AA is a rectangle with sides $2h_2 \tan\alpha$ and L. Its $I_{yy}^O = (2h_2 \tan\alpha)^3 L/12 = \frac{2}{3}h_2^3 L \tan^3\alpha$. The height a of C above C_b is $a = C_bC = \frac{2}{3}(h_1 - h_2)$ [Fig. E6.23b]. The metacentric height h is given by eq. (6.117):

$$h = \frac{I_{yy}^O}{V} - a = \frac{2}{3}[h_2 \tan^2\alpha - (h_1 - h_2)]$$

Hence, the equilibrium configuration of the floating block is stable if $h > 0$, i.e., if

$$\tan^2\alpha > (h_1/h_2 - 1) = [(\rho_2/\rho_1)^{1/2} - 1]$$

$$\Rightarrow \quad \text{if} \quad \alpha > \tan^{-1}[(\rho_2/\rho_1)^{1/2} - 1]^{1/2}$$

$\Rightarrow$ The minimum angle for which the block is stable is $\alpha_{\min} = \tan^{-1}[(\rho_2/\rho_1)^{1/2} - 1]^{1/2}$.

Review 6 F

1. Equation of motion of a non-viscous fluid: $\vec{\nabla}p = \vec{b} - \rho\vec{a}$
2. Equation of equilibrium of a fluid: $\vec{\nabla}p = \vec{b}$.

   Incompressible fluid: $\frac{dp}{dz} = \rho g, \quad p = p_0 + \rho g z$, where z points down.

   Compressible ideal gas: $\frac{p}{\rho} = \frac{p_0}{\rho_0} = RT \quad \Rightarrow \quad \frac{dp}{dz} = -\frac{\rho_0 g p}{p_0}$, where z points up.
3. On a plane area A with x, y, z axes at centroid C and z-axis normal to plane:

   $\vec{F}_R = -\rho g h_C A \vec{k}$ at $\vec{r}_R = (-I_{xy}^C \vec{i} + I_{xx}^C \vec{j})/Ay_C'$.
4. Floating body is stable if metacentric height $h = I_{yy}^O/V - a > 0$ where a is the height of C *above* C_b and I_{yy}^O is the area moment of inertia of the area of cross-section of the body at the water line about the longitudinal axis.

6.25 Concept Review Questions 6 F

29. Find the magnitude of pressure gradient within oil of density ρ in a tanker translating (i) horizontally with acceleration a, (ii) vertically up with acceleration a, and (iii) in free vertical fall. [(i) $\rho\sqrt{g^2+a^2}$ (ii) $\rho(g+a)$ (iii) 0]

30. A tube completely filled with a fluid is rotated at high angular speed ω. Neglect gravity. The pressure and density at $r=0$ are p_0 and ρ_0. Find pressure p at radius r if (i) the fluid is incompressible and (ii) the fluid is compressible with $p/p_0 = \rho/\rho_0$. [(i) $p_0 + \frac{1}{2}\rho\omega^2 r^2$ (ii) $p_0 \exp(\rho_0\omega^2 r^2/2p_0)$]

31. (a) For which of the following fluids is $\vec{T}^n = -p_n\vec{n}$? (1) In-viscid fluid in equilibrium, (2) in-viscid fluid moving as a rigid body, (3) in-viscid fluid in general motion, and (4) viscous fluid in general motion. (b) List the cases given in part (a) for which $\vec{\nabla} p = \vec{b} - \rho\vec{a}$ where $\vec{b}$ is the body force density, $\vec{a}$ is the acceleration and ρ is the density of the fluid. [(a), (b) For all cases except 4]

32. An arch dam (Fig. Q6.32) has a cylindrical surface. The density and depth of water are ρ and h. Find the simplest resultant force of the water pressure on the curved arch of the dam. [$\rho g h^2 R \sin\theta$ at $h/3$ from base]

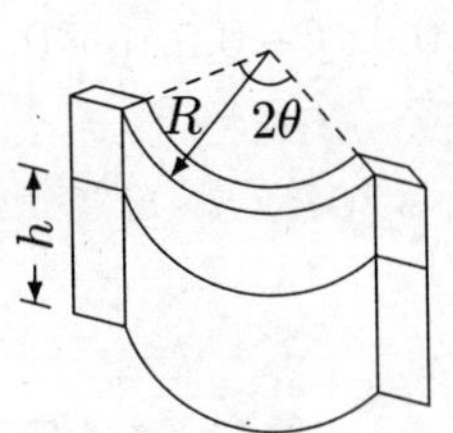

Fig. Q6.32

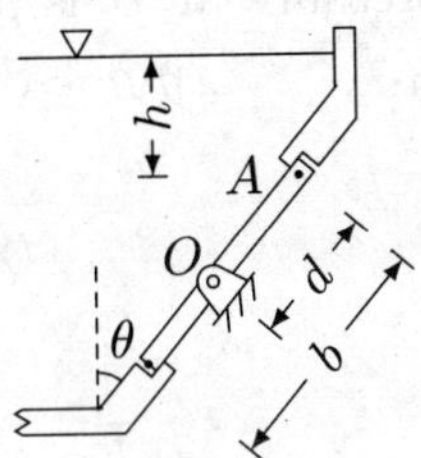

Fig. Q6.33

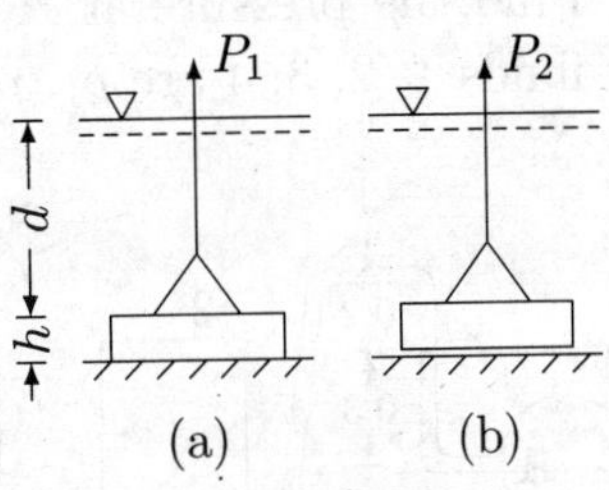

Fig. Q6.34

33. Find d so that the rectangular gate (Fig. Q6.33) will open when the water level reaches a height h above A. [$\frac{1}{2}b + b^2/6(b+2h/\cos\theta)$]

34. A cuboidal slab $a \times b \times h$ of weight W rests at the bottom of a pool (Fig. Q6.34a). The density of water is ρ. There is no water under the slab. Find the pull P_1 exerted on the rope to just lift it off and pull P_2 to support it just after lift off (Fig. Q6.34b). The atmospheric pressure is p_a. [$W + ab(p_a + \rho g d)$, $W - abh\rho g$]

35. A block with vertex h_1 below the base has density ρ_1. It floats in water of density ρ_2 with submerged depth h_2 of the vertex. Find h_2/h_1 if the block is (i) a wedge, and (ii) a cone. [(i) $\sqrt{\rho_1/\rho_2}$ (ii) $(\rho_1/\rho_2)^{1/3}$]

36. The buoyancy force on a fully submerged body of volume V in a fluid is F when the atmospheric pressure is p_0. Find the increase in the buoyancy force if atmospheric pressure increases to $2p_0$. [No increase]

37. How does Archimedes principle get modified, if at all, for a body of volume V submerged in a fluid of density ρ contained in a vessel under the following motions? The vessel is (i) on a lift moving up with acceleration a, (ii) on a lift moving down with acceleration a, (iii) on a vehicle moving horizontally with acceleration a, (iv) on a satellite moving in a circular orbit of radius R, and (v) falling freely under gravity. [(i) $\rho V(a+g)$ (ii) $\rho V(g-a)$ (iii) $\rho V\sqrt{g^2+a^2}$ (iv) 0 (v) 0]
38. The pressure in a satellite orbiting the earth in a circle of radius R is p_a. What is the height of the mercury in a mercury barometer in the satellite? [full tube]
39. Why does a sinking ship often turn over as it becomes immersed in water?

[As metacentric height h decreases and becomes less than 0 with sinking of ship]

6.26 Practice Problems 6 F

36. The equation of state of an atmosphere is $p/p_0 = (\rho/\rho_0)^\gamma$ where γ is adiabatic index and $p_0 = R\rho_0 T_0$ at $z = 0$. Prove that the pressure and temperature (under constant gravity) at height z are: $p = p_0[1-(\gamma-1)\rho_0 gz/(\gamma p_0)]^{\gamma/(\gamma-1)}$, $T/T_0 = 1-(\gamma-1)\rho_0 gz/(\gamma p_0)$.
37. Find the pressure at A if the pressure at B is p_B (Fig. P6.37). The densities of fluids 1, 2, 3, 4 are ρ_1, ρ_2, ρ_3, ρ_4. [$p_B + (0.3\rho_1 - 0.6\rho_2 + 0.5\rho_3 + 0.46\rho_4)g$]

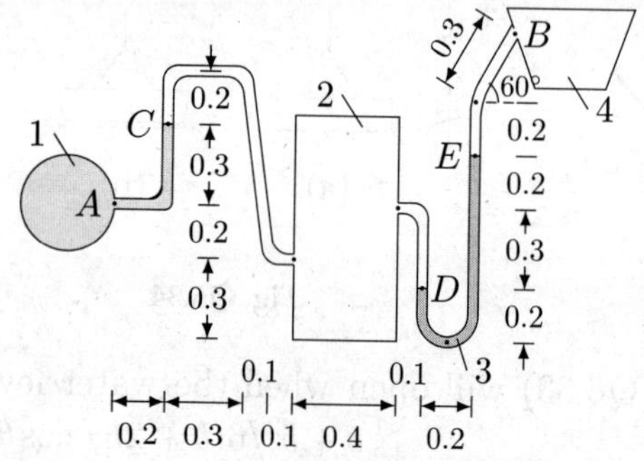

Fig. P6.37

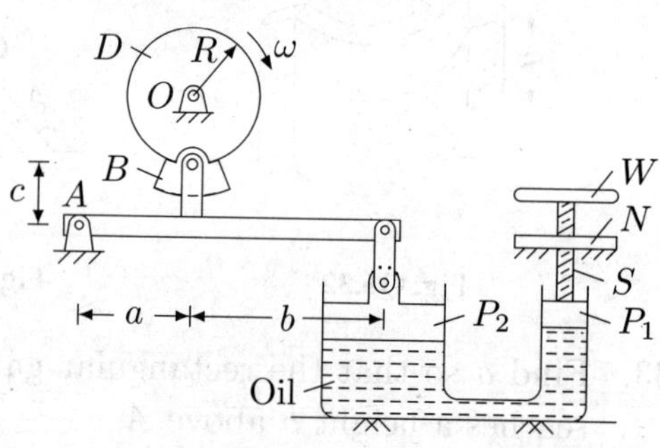

Fig. P6.38

38. A torque M_0 is applied to a wheel W of the brake system (Fig. P6.38). The screw S has single square thread with a helix angle α and mean radius r. The coefficient of friction between the nut N and the screw S is $\tan^{-1}\phi$. The radii of pistons P_1 and P_2 are R_1 and R_2. The coefficient of friction between the brake shoe B and the brake drum D is μ. The contact between the screw S and piston P_1 is smooth. Assume that the contact force on brake shoe acts in the middle. Find the braking torque on the brake drum. [$\mu RM_0(a+b)R_2^2/R_1^2 r(a-\mu c)\tan(\alpha+\phi)$]
39. A small body of mass m and density ρ is submerged in a fluid chamber of an ultra

centrifuge at distance r from its vertical axis of rotation $\vec{e}_z$. The angular velocity ω is such that $g_e = \omega^2 r \gg g$. The density of the fluid is ρ_1. (a) Show that the buoyant force is $m\rho_1(g^2+\omega^4 r^2)^{1/2}/\rho$. (b) Show that the body will be accelerated towards/away from the axis of rotation according as $\rho < \rho_1$ or $\rho > \rho_1$. Hence, show that the density of a rotating fluid can only increase with r. (c) The density ρ_1 of the fluid in the centrifuge varies as $\rho_1 = \frac{1}{2}\rho_0(1 + r/r_0)$, e.g., a salt solution of increasing concentration. The macromolecules (proteins) or virus particles of density $\rho > \frac{1}{2}\rho_0$ are initially uniformly distributed in the fluid. Find their location after extended centrifugation. (d) The viscous drag on a macromolecule of mass m and density ρ is c times its velocity relative to the fluid. Show that its equation of motion is $\ddot{r} = \omega^2 r(1-\rho_1/\rho) - c\dot{r}/m$ if ρ_1 is constant. Since $\omega^2 r \gg g$, the separation rate of the molecules from the main solution in a centrifuge is very much enhanced compared to gravitational separation even if $\rho \simeq \rho_1$. Show that if the terminal speed of the molecules is v, then $m \simeq vc/[g_e(1-\rho_1/\rho)]$. [(c) $r = r_0(2\rho/\rho_0 - 1)$]

40. Find the resultant force on the following windows in a vertical side of a vessel filled with a fluid of density ρ. (i) An isosceles triangular window with base b at depth h from the free surface and vertex d below the base. (ii) A semicircular windows of radius R with diameter at the top at a depth h below the free surface.

[(i) $\frac{1}{2}\rho g(h+\frac{1}{3}d)bd$ at $y_R = d^2/(18h+6d)$ below C
(ii) $\frac{1}{2}\pi\rho g R^2(h+4R/3\pi)$ at $(9\pi^2-64)R^2/12\pi(3\pi h+4R)$ below C]

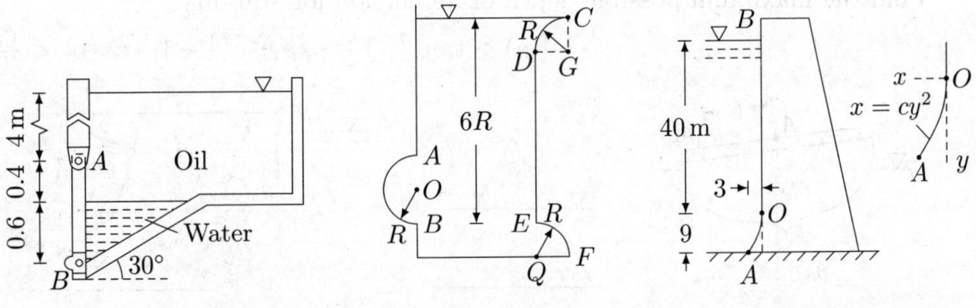

Fig. P6.41 Fig. P6.42 Fig. P6.43

41. Water has accumulated in an oil-storage tank (Fig. P6.41). The specific gravity of oil is 0.85. For water, $\rho = 1000\,\text{kg/m}^3$. A door AB of 0.5 m width is hinged at A and has a latch at B. Find the force exerted by the latch on the door. [9834 N]

42. Find the simplest resultant of the fluid pressure on curved cylindrical gates (a) AB, (b) CD, (c) EF (Fig. P6.42). The width of each gate is b. The density of water is ρ.

[(a) $10.12\rho g R^2 b$ at $8.927°$ below horizontal through O
(b) $0.544\rho g R^2 b$ at $23.23°$ below horizontal through G

[(c) $8.993\rho g R^2 b$ at $43.71°$ above horizontal through Q]

43. Find the resultant force system at A due to the fluid pressure acting on face AB of a dam of width 20 m shown in Fig. P6.43. The density of water is 1000 kg/m^3.

[$F_R = 237.09$ MN at $6.557°$ to horizontal and $C_R = 3887$ MN.m]

44. A simple experiment is conducted on a loaded ship to find the position of its centre of mass C. A known additional load of mass m_0 is placed at a distance b from the plane of symmetry and the angle of tilt θ_0 is measured by a plumb bob (Fig. P6.44). The tilt θ_1 for the magnitude m_1 of another additional load is also measured. The mass of the ship is M. Find the distance d of C from A, and the metacentric height.

[$b\{(1+\frac{M}{m_1})/\theta_0-(1+\frac{M}{m_0})/\theta_1\}/M(1/m_0-1/m_1)$, $b(1/\theta_0-1/\theta_1)/M(1/m_0-1/m_1)$]

45. The material of a spherical balloon is such that its un-stretched radius r increases by $\alpha r p$ when an internal gage pressure p is applied by the helium gas. The density of air is ρ_a and that of the helium gas at the given temperature and gage pressure p is $\rho_g = \beta(p + p_a)$ where p_a is the atmospheric pressure. Find the pressure p for which the lift on the balloon is the maximum. [$\frac{1}{4}(3\rho_a/\beta - 3p_a - 1/\alpha)$]

46. A beam of density ρ has an isosceles triangular section (Fig. P6.46). Its length is large compared to its cross-sectional dimensions. It floats in water of density ρ_1. (a) Find the range of α for which the equilibrium configuration is stable. (b) If $\alpha = 60°$, find the range of ρ/ρ_1 for which the equilibrium configuration is stable. Find the maximum possible depth of immersion for stability.

[(a) $> \tan^{-1}[(1-\rho/\rho_1)^{-1/2} - 1]^{1/2}$ (b) $< \frac{15}{16}$, $\frac{3}{4}h$]

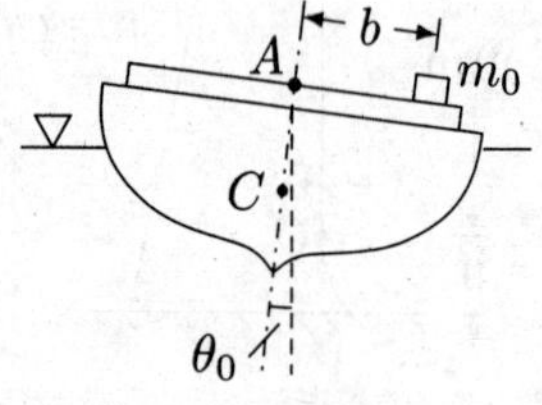

Fig. P6.44

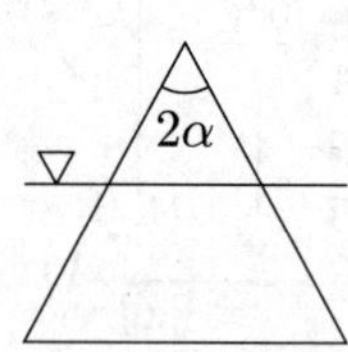

Fig. P6.46

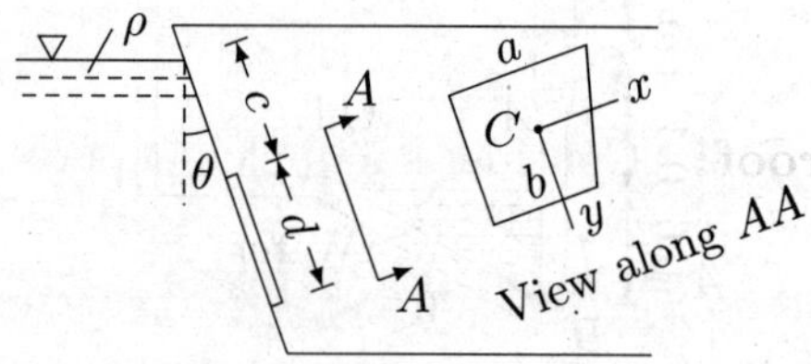

Fig. P6.48

47*. A bar has a square section and its density is ρ. It can float in equilibrium configuration in a liquid of density ρ_1 with a diagonal of the square horizontal. Find the range of ρ/ρ_1 for which it is stable. [$9/32 < \rho/\rho_1 < 23/32$]

48*. Find the resultant force and the centre of pressure due to hydrostatic pressure on the window shown in Fig. P6.48.

[$\frac{1}{6}\rho g d[3(b+a)c + (2b+a)d]\cos\theta$ at (x_R, y_R) with $y'_C = c + (2b+a)d/3(a+b)$
$x_R = -(a^3+3a^2b-3ab^2-b^3)d/36(a+b)^2 y'_C$, $y_R = (a^2+b^2+4ab)d^2/18(a+b)^2 y'_C$]

APPENDIXES

APPENDIX A: CENTRE OF MASS OF BODIES

A.1 Theorems of Pappus and Guldinus

Consider a plane curve AB of length L (Fig. A.1a) and a plane surface of area A_1 (Fig. A.1b) and an axis of revolution in their plane which does not intersect them. The area A of the surface of revolution and the volume V of the body of revolution developed by revolving them about the axis of revolution through angle θ radians are given as follows in terms of the distances y_C and y_{C_1}, of the centroid C of the curve and the centroid C_1 of the area A_1, from the axis of revolution x:

$$A = \theta\, y_C L, \qquad V = \theta\, y_{C_1} A_1 \tag{A.1}$$

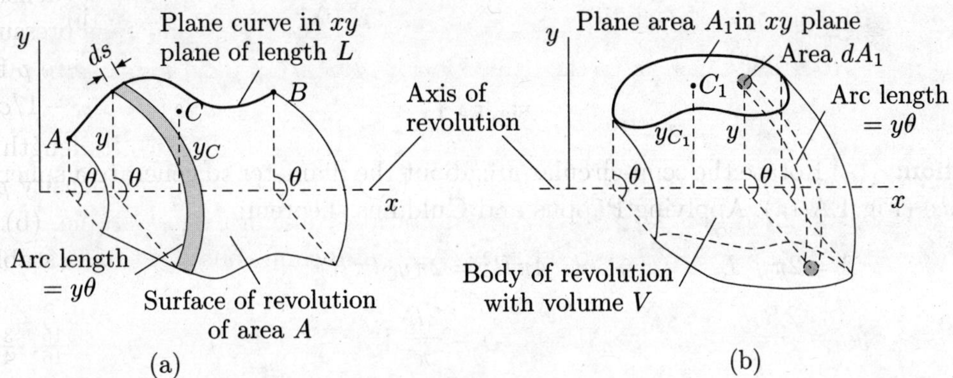

Fig. A.1

Proof: Consider small elements ds and dA_1 of the curve AB and of the area A_1.

$$A = \int_s (y\theta)\, ds = \theta \int_s y\, ds = \theta\, y_C L$$

$$V = \int_{A_1} (y\theta)\, dA_1 = \theta \int_{A_1} y\, dA_1 = \theta\, y_{C_1} A_1$$

For $\theta = 2\pi$, we get the area of a complete surface of revolution and volume of a complete body of revolution:

$$A = 2\pi y_C L, \qquad V = 2\pi y_{C_1} A_1 \tag{A.2}$$

If a generating curve or a generating area intersects the axis of revolution, then these are decomposed into appropriate parts by the intersection with this axis and the theorem applied to each part separately to find their contribution. The theorems are useful for finding the distance of the centroid of the generating curve or the generating area from the axis of rotation from the known value of the surface area or the volume and vice-versa.

Example A.1: (a) Find the centre of mass of a uniform (i) semi-circular wire, (ii) quarter-circular wire. (b) Find the centre of mass of a uniform (i) semi-circular plate, (ii) quarter-circular plate.

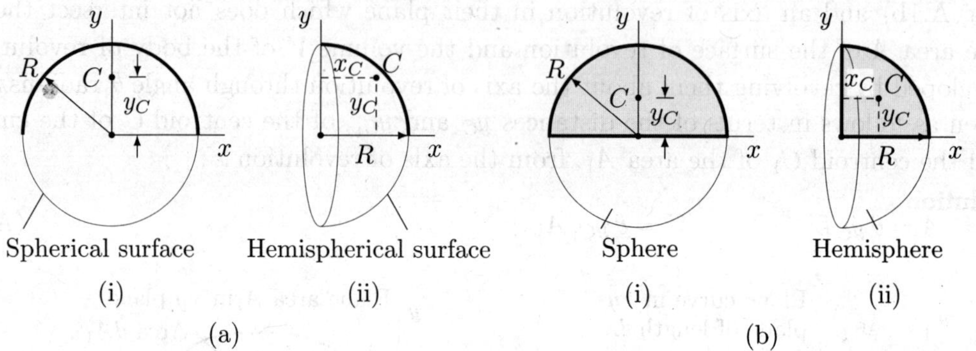

Fig. EA.1

Solution: (a) Rotate the semi-circular arc about the diameter to generate a spherical surface (Fig. EA.1a). Applying Pappus and Guldinus theorem:

$$A = 2\pi y_C L \quad \Rightarrow \quad 4\pi R^2 = 2\pi y_C(\pi R)$$

$$\Rightarrow \quad y_C = \frac{2R}{\pi} \quad \Rightarrow \quad \vec{r}_C = \frac{2R}{\pi}\vec{j}$$

For a quarter circle, rotations about x and y-axes yield similarly $y_C = x_C = 2R/\pi \Rightarrow \vec{r}_C = (2R/\pi)(\vec{i} + \vec{j})$.

(b) Rotate the semi-circular area about the diameter to generate a sphere (Fig. EA.1b). Applying Pappus and Guldinus theorem:

$$V = 2\pi y_C A \quad \Rightarrow \quad \tfrac{4}{3}\pi R^3 = 2\pi y_C(\tfrac{1}{2}\pi R^2)$$

$$\Rightarrow \quad y_C = \frac{4R}{3\pi} \quad \Rightarrow \quad \vec{r}_C = \frac{4R}{3\pi}\vec{j}$$

It implies that $y_C = 4R/3\pi$ for a quarter-circular area also, and $x_C = y_C$ yields $\vec{r}_C = (4R/3\pi)(\vec{i} + \vec{j})$.

These results are applicable to thin uniform semi-circular and quarter-circular cylindrical shells; and uniform semi-circular and quarter-circular cylinders with the mid-plane as xy-plane.

Example A.2: Find the surface area A and volume V of a bell shaped body of revolution shown in Fig. EA.2a using Pappus and Guldinus theorem.

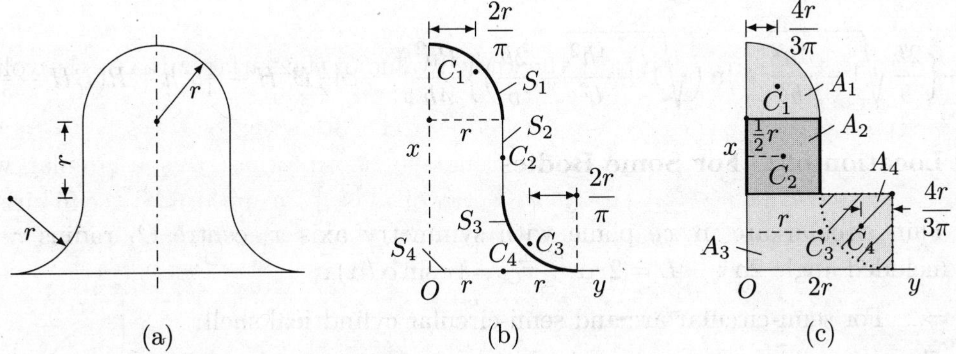

Fig. EA.2

Solution: The generating curve for the surface is decomposed as in Fig. EA.2b:

$$y_{C_1} = \frac{2r}{\pi}, \quad s_1 = \tfrac{1}{2}\pi r, \quad y_{C_2} = r, \quad s_2 = r, \quad y_{C_3} = 2r - \frac{2r}{\pi}, \quad s_3 = \tfrac{1}{2}\pi r$$

$$y_{C_4} = r, \quad s_4 = 2r, \qquad A = 2\pi y_C L = 2\pi \sum_{i=1}^{4} y_{C_i} s_i = 2\pi(\pi + 3)r^2$$

The generating area for the volume is decomposed as in Fig. EA.2c where A_3 is a rectangle $(2r \times r)$ and A_4 (a quarter circle of radius r) is a deletion.

$$y_{C_1} = \frac{4r}{3\pi}, \quad A_1 = \frac{\pi r^2}{4}, \quad y_{C_2} = \tfrac{1}{2}r, \quad A_2 = r^2$$

$$y_{C_3} = r, \quad A_3 = 2r^2, \quad y_{C_4} = 2r - \frac{4r}{3\pi}, \quad A_4 = -\frac{\pi r^2}{4}$$

$$V = 2\pi y_C A = 2\pi \sum_{i=1}^{4} y_{C_i} A_i = (19/3 - \pi)\pi r^3$$

§ Exercise A.1: Find the surface area and volume of the rim of a V-belt steel pulley, an arch dam and a steel collar (Fig. ExA.1).

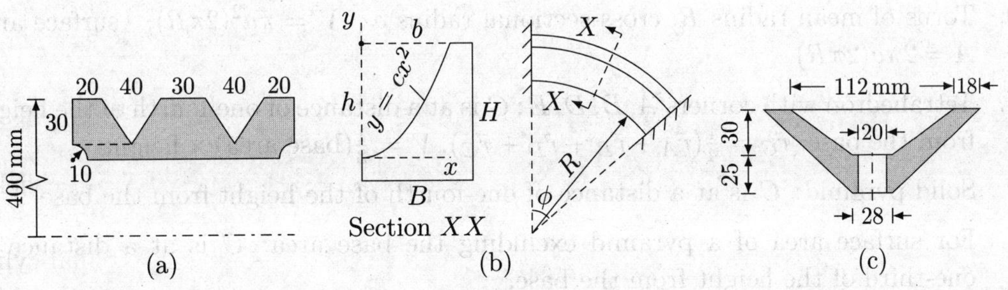

Fig. ExA.1

[(a) $1.036\,\text{m}^2$, $0.01103\,\text{m}^3$ (c) $433.8\,\text{cm}^2$, $249.2\,\text{cm}^3$

(b) $\phi[R(2H - h + 2B - b) + B^2 + HB - \frac{1}{2}b^2 + \frac{1}{12}\{(1 + 4h^2/b^2)^{3/2} - 1\}b^4/h^2$

$$+\left\{\frac{2h}{b}\sqrt{1+\frac{4h^2}{b^2}}+\ln\left(\sqrt{1+\frac{4h^2}{b^2}}+\frac{2h}{b}\right)\right\}\frac{Rb^2}{4h}\Big], \quad \phi[\tfrac{1}{2}B^2H-\tfrac{1}{4}b^2h+R(BH-\tfrac{2}{3}bh)]$$

A.2 Location of C For Some Bodies

1. Thin circular arc in xy plane with symmetry axis x, centre O, radius r and included angle 2α: $L = 2r\alpha$, $\vec{r}_C = (r\sin\alpha/\alpha)\,\vec{\imath}$

   $\Rightarrow$ For semi-circular arc and semi-circular cylindrical shell:

   $\vec{r}_C = (2r/\pi)\,\vec{\imath}$, where x-axis is the symmetry axis in the mid-plane of shell.

2. Triangular plate with corners at $(0,0)$, $(b,0)$, (a,h) in xy plane: $A = \frac{1}{2}bh$, $\vec{r}_C = \frac{1}{3}(a+b)\,\vec{\imath} + \frac{1}{3}h\,\vec{\jmath}$.

3. Triangular prism of length L from $z = -\frac{1}{2}L$ to $z = \frac{1}{2}L$, with cross-section as in case 2: $V = \frac{1}{2}bhL$, $\vec{r}_C = \frac{1}{3}(a+b)\,\vec{\imath} + \frac{1}{3}h\,\vec{\jmath}$.

4. Circular sector in xy plane with symmetry axis x, centre O, radius R and included angle 2α: $A = \alpha R^2$, $\vec{r}_C = (2R\sin\alpha/3\alpha)\,\vec{\imath}$

   $\Rightarrow$ For semi-circular disc and semi-circular cylinder: $\vec{r}_C = (4R/3\pi)\,\vec{\imath}$ where x-axis is the symmetry axis in the mid-plane of the cylinder.

5. Elliptic disc with axes $2a$ and $2b$ along x and y axes: $A = \pi ab$.

6. Right circular cone of height h, base radius R and axis z has C at a distance $h/4$ from the base and $V = \frac{1}{3}\pi R^2 h$.

7. Right circular conical shell of height h, base radius R, has C at a distance $h/3$ from the base and: $A = \pi R(R^2 + h^2)^{1/2}$.

8. Solid hemisphere of radius R: C is at distance $3R/8$ from the base.

9. Hemispherical shell of radius R: C is at distance $\frac{1}{2}R$ from the base.

10. Torus of mean radius R, cross-sectional radius a: $V = \pi a^2(2\pi R)$, surface area $A = 2\pi a(2\pi R)$.

11. Tetrahedron with corners A, B, D, E: C is at a distance of one-fourth of the height from the base, $\vec{r}_C = \frac{1}{4}(\vec{r}_A + \vec{r}_B + \vec{r}_D + \vec{r}_E)$, $V = \frac{1}{3}$(base area) $\times$ height.

12. Solid pyramid: C is at a distance of one-fourth of the height from the base.

13. For surface area of a pyramid excluding the base area: C is at a distance of one-third of the height from the base.

APPENDIX B: INERTIA ELEMENTS OF BODIES

1. Thin rod of length L and axis x : $I^C_{xx} = 0$, $I^C_{yy} = I^C_{zz} = mL^2/12$.
2. Thin circular arc in xy plane with axis x, centre O, radius r included angle 2α:
   $I^O_{zz} = mr^2$, $I^O_{xx}, I^O_{yy} = mr^2(1 \mp \sin 2\alpha/2\alpha)/2$. $\Rightarrow$ Thin ring: $I^O_{xx} = \frac{1}{2}I^O_{zz} = \frac{1}{2}mr^2$.
3. Triangular plate with corners at $(0,0)$, $(b,0)$, (a,h) in xy plane:
   $I^C_{xx} = mh^2/18$, $I^C_{yy} = m(b^2 + a^2 - ab)/18$, $I^C_{xy} = -mh(2a - b)/36$.
4. Thin triangular plate with sides a, b along x, y axes:
   $I^C_{xx} = mb^2/18$, $I^C_{yy} = ma^2/18$, $I^C_{xy} = mab/36$.
5. Circular sector in xy plane with axis x, centre O, radius R, included angle 2α:
   $I^O_{xx}, I^O_{yy} = \frac{1}{4}(1 \mp \sin 2\alpha/2\alpha)mR^2$, $I^O_{zz} = \frac{1}{2}mR^2$. Disc: $I^O_{xx} = \frac{1}{2}I^O_{zz} = \frac{1}{4}mR^2$.
6. Elliptic disc with axes $2a$ and $2b$ along x and y axes:
   $I^O_{xx} = mb^2/4$, $I^O_{yy} = ma^2/4$, $I^O_{zz} = m(a^2 + b^2)/4$.
7. Cuboid with edges a, b, c along axes x, y, z: $I^C_{xx} = m(b^2 + c^2)/12, \ldots$
8. Right circular cone of height h, base radius R and axis z has C at $h/4$ from base,
   $I^C_{xx} = I^C_{yy} = 3m(4R^2 + h^2)/80$, $I^C_{zz} = 3mR^2/10$.
9. Right circular conical shell: height h, base radius R, axis z, C at $h/3$ from base,
   $I^C_{xx} = I^C_{yy} = m(R^2/4 + h^2/18)$, $I^C_{zz} = mR^2/2$.
10. Right circular solid cylinder of radius R, length L and axis z:
    $I^C_{xx} = I^C_{yy} = m(R^2/4 + L^2/12)$, $I^C_{zz} = mR^2/2$.
11. Circular cylindrical shell of radius R, length L and axis z:
    $I^C_{xx} = I^C_{yy} = m(R^2/2 + L^2/12)$, $I^C_{zz} = mR^2$.
12. Solid sphere of radius R: $I^C_{xx} = I^C_{yy} = I^C_{zz} = 2mR^2/5$.
13. Spherical shell of radius R: $I^C_{xx} = I^C_{yy} = I^C_{zz} = 2mR^2/3$.
14. Elliptical cylinder of length L and axes $2a$ and $2b$ along x and y:
    $I^C_{xx} = m(b^2/4 + L^2/12)$, $I^C_{yy} = m(a^2/4 + L^2/12)$, $I^C_{zz} = m(a^2 + b^2)/4$.
15. Elliptical paraboloid: $x^2/a^2 + y^2/b^2 - z/c = 1$ with base $z = 0$ has C at $-\frac{1}{3}c\vec{k}$,
    $I^C_{xx} = m(b^2 + c^2/3)/6$, $I^C_{yy} = (a^2 + c^2/3)/6$, $I^C_{zz} = m(a^2 + b^2)/6$.
16. Ellipsoid: $\dfrac{x^2}{a^2} + \dfrac{y^2}{b^2} + \dfrac{z^2}{c^2} = 1$: $[I^O_{xx}, I^O_{yy}, I^O_{zz}] = \dfrac{m}{5}[(b^2 + c^2),\ (c^2 + a^2),\ (a^2 + b^2)]$.
17. Torus with axis z, mean radius R in xy plane and cross-sectional radius a:
    solid: $I^C_{xx} = m(\frac{1}{2}R^2 + \frac{5}{8}a^2)$, $I^C_{zz} = m(R^2 + \frac{3}{4}a^2)$
    shell: $I^C_{xx} = m(\frac{1}{2}R^2 + \frac{5}{4}a^2)$, $I^C_{zz} = m(R^2 + \frac{3}{2}a^2)$.
18. Tetrahedron with edges a, b, c along axes x, y, z at O:
    $I^O_{xx} = m(b^2 + c^2)/10$, $I^O_{yy} = m(c^2 + a^2)/10$, $I^O_{zz} = m(a^2 + b^2)/10$

APPENDIX C: INERTIA MATRIX FOR AREA

C.1 Inertia Matrix For Area

From the material concept of inertia matrix, we abstract the geometrical concept of the *inertia matrix of an area* at point B with respect to axes $\vec{e}_i$ at B (Fig. C.1a) by replacing mass element dm by the area element dA:

$$I_{ij}^B = \int_A (r^2\delta_{ij} - x_i x_j)\, dA \tag{C.1}$$

where the position vector $\vec{r}_{PB}$ of a point P in dA relative to B is $\vec{r}_{PB} = x_i \vec{e}_i$ with $r^2 = x_k x_k$. Explicitly

$$\begin{aligned} I_{11}^B &= \int_A (x_2^2 + x_3^2)\, dA, & I_{12}^B &= -\int_A x_1 x_2\, dA \\ I_{22}^B &= \int_A (x_3^2 + x_1^2)\, dA, & I_{23}^B &= -\int_A x_2 x_3\, dA \\ I_{33}^B &= \int_A (x_1^2 + x_2^2)\, dA, & I_{31}^B &= -\int_A x_3 x_1\, dA \end{aligned} \tag{C.2}$$

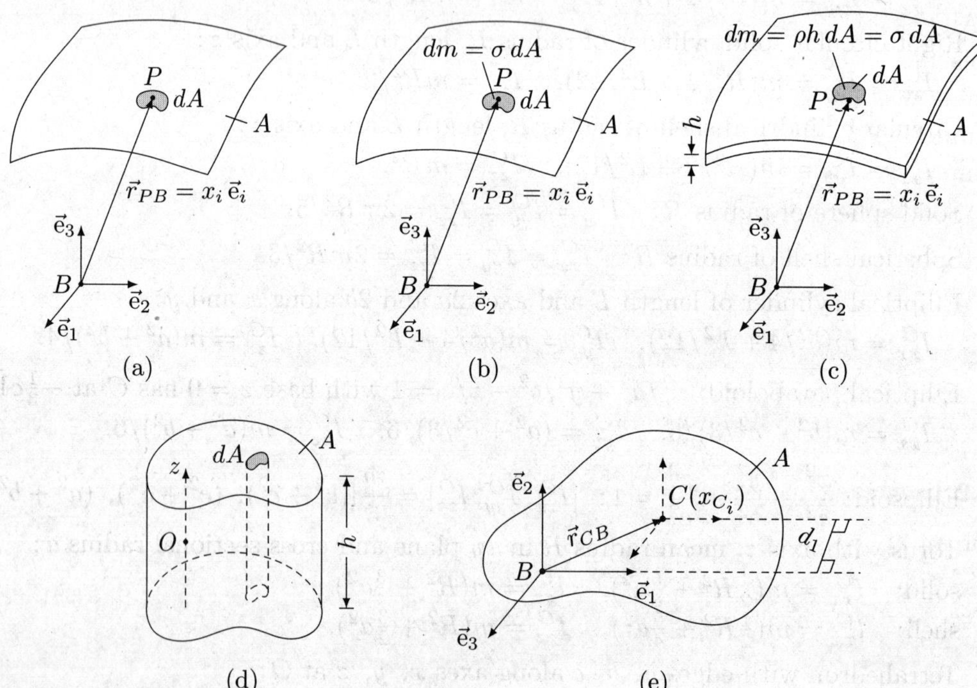

Fig. C.1

The word inertia in the name is meaningless for an area, but traditionally the diagonal elements $I_{11}^B, I_{22}^B, I_{33}^B$ are called *moments of inertia of area* and the off-diagonal elements $I_{12}^B = I_{21}^B, I_{23}^B = I_{32}^B, I_{31}^B = I_{13}^B$ are called the *products of inertia of the area.* The elements I_{ij}^B are also called *second moments of area.* The moments of inertia are non-negative, but the products of inertia can be of any sign.

The inertia matrix $[I_{\text{mass}}^B]$ of a uniform distribution of mass over an area with mass density σ (Fig. C.1b), $dm = \sigma\, dA$, is related to the inertia matrix $[I_{\text{area}}^B]$ by

$$[I_{\text{mass}}^B] = \sigma[I_{\text{area}}^B] \tag{C.3}$$

For a thin shell of density ρ and thickness h (Fig. C.1c), eq. (C.3) is approximately valid provided $\rho h = \sigma$ ($dm = \sigma\, dA = \rho h\, dA$) is constant. The mass moment of inertia of a homogeneous right cylinder (Fig. C.1d) about an axis parallel to the generators is exactly related to the area moment of inertia of the cross-section

$$(I_{zz}^O)_{\text{mass}} = \int_m (x^2 + y^2)\, dm = \rho h \int_A (x^2 + y^2)\, dA = \rho h (I_{zz}^O)_{\text{area}} \tag{C.4}$$

The inertia matrix of an area is also useful in the study of bending of elastic beams, torsion of elastic circular rods, the hydrostatic force on a plane surface in a fluid, stability of floating bodies, etc.

All results of inertia matrix of mass are applicable to inertia matrix of area with dA replacing dm, viz., transformation relations for rotation of axes, results for principal axes and principal moments of inertia, directions of principal axes of inertia for certain symmetries, parallel axes theorems, radius of gyration $k_n^B = (I_{nn}^B/A)^{1/2}$ and the decomposition theorem. For example, with $\vec{r}_{CB} = x_{C_i}\, \vec{e}_i = x_C\, \vec{i} + y_C\, \vec{j} + z_C\, \vec{k}$ (Fig. C.1e), the parallel axes theorem for centroid C of the area is

$$\begin{aligned} I_{ij}^B &= I_{ij}^C + (r_{CB}^2 \delta_{ij} - x_{C_i} x_{C_j}) A \\ I_{11}^B &= I_{11}^C + A(x_{C_2}^2 + x_{C_3}^2) = I_{11}^C + A d_1^2, \qquad I_{12}^B = I_{12}^C - A x_{C_1} x_{C_2} \\ I_{xx}^B &= I_{xx}^C + A(y_C^2 + z_C^2) = I_{xx}^C + A d_x^2, \qquad I_{xy}^B = I_{xy}^C - A x_C y_C \end{aligned} \tag{C.5}$$

C.2 Inertia Matrix for a Plane Area

For an area in xy-plane ($z = 0$) (Fig. C.2a), for point B *in the plane of the area*, $I_{xz}^B = I_{yz}^B = 0$ and

$$I_{xx}^B = \int_A y^2\, dA, \qquad I_{yy}^B = \int_A x^2\, dA, \qquad I_{xy}^B = -\int_A xy\, dA \tag{C.6}$$

$$I_{zz}^B = \int_A (x^2 + y^2)\, dA = I_{xx}^B + I_{yy}^B \tag{C.7}$$

I_{zz}^B is called *polar moment of inertia* about z-axis at B. If an axis x or y through B is an axis of symmetry of the area, then x, y, z-axes are the principal axes of inertia at B (Fig. C.2b).

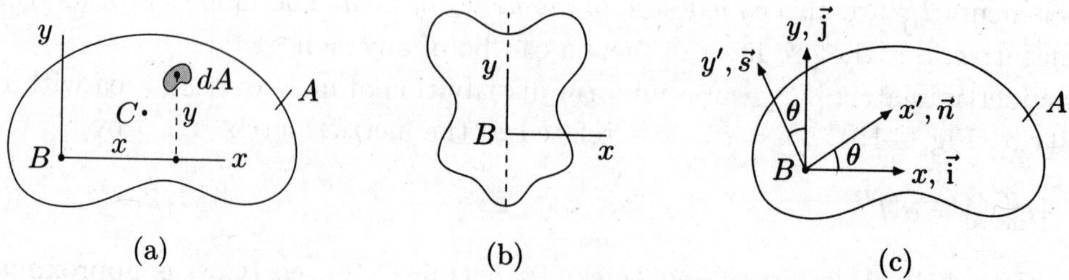

Fig. C.2

The transformation relations of elements of the inertia matrix at B in the plane of the area with respect to axes x, y, z and x', y', z' (Fig. C.2c) are obtained from eqs (3.10) and (3.11) with $\vec{n} = (\cos\theta, \sin\theta, 0)$, $\vec{s} = (-\sin\theta, \cos\theta, 0)$:

$$\begin{aligned} I_{xx}^{B\prime} &= I_{xx}^B \cos^2\theta + I_{yy}^B \sin^2\theta + 2I_{xy}^B \cos\theta \sin\theta \\ &= \tfrac{1}{2}(I_{xx}^B + I_{yy}^B) + \tfrac{1}{2}(I_{xx}^B - I_{yy}^B)\cos 2\theta + I_{xy}^B \sin 2\theta \end{aligned} \tag{C.8}$$

$$\begin{aligned} I_{xy}^{B\prime} &= -(I_{xx}^B - I_{yy}^B)\sin\theta\cos\theta + I_{xy}^B(\cos^2\theta - \sin^2\theta) \\ &= -\tfrac{1}{2}(I_{xx}^B - I_{yy}^B)\sin 2\theta + I_{xy}^B \cos 2\theta \end{aligned} \tag{C.9}$$

$$I_{yy}^{B\prime} = I_{xx}^B + I_{yy}^B - I_{xx}^{B\prime}$$

$$I_{zz}^{B\prime} = I_{zz}^B, \qquad I_{xz}^{B\prime} = I_{yz}^{B\prime} = 0$$

(using invariant $I_{xx}^B + I_{yy}^B + I_{zz}^B = I_{xx}^{B\prime} + I_{yy}^{B\prime} + I_{zz}^{B\prime}$). z-axis is a principal axis at B. Equation (C.9) implies that x'-axis is a principal axis at B if $I_{xy}^{B\prime} = 0$, i.e., if

$$\tan 2\theta = \frac{2I_{xy}^B}{I_{xx}^B - I_{yy}^B} \tag{C.10}$$

Substituting from eq. (C.10) in eq. (C.8) yields the principal moments of inertia:

$$I_{11}^{B*}, I_{22}^{B*} = \tfrac{1}{2}(I_{xx}^B + I_{yy}^B) \pm [\{\tfrac{1}{2}(I_{xx}^B - I_{yy}^B)\}^2 + {I_{xy}^B}^2]^{1/2} \tag{C.11}$$

Using eqs (C.10) and (C.11), we can show that the principal axis $\vec{e}_1^*$ for I_{11}^{B*} is inclined to $\vec{i}$ in the xy-plane at an angle θ_1 given by

$$\tan\theta_1 = \frac{I_{11}^{B*} - I_{xx}^B}{I_{xy}^B} \tag{C.12}$$

The principal axis $\vec{e}_2^*$ for I_{22}^{B*} is inclined to $\vec{i}$ in the xy-plane at an angle $\pi/2+\theta_1$.

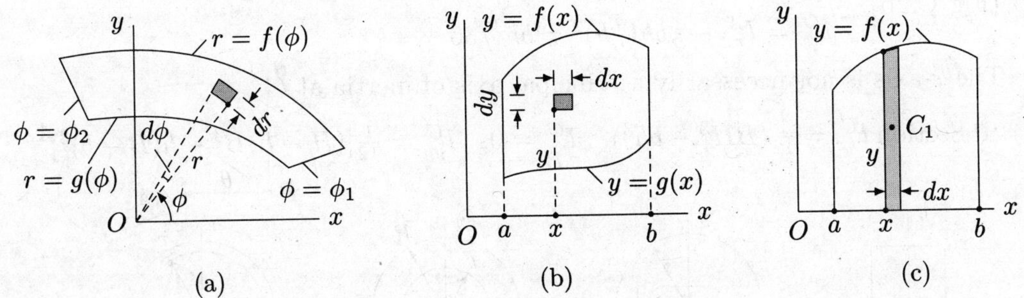

Fig. C.3

For computing $[I^B]$, if the boundary of area is described easily in terms of r, ϕ co-ordinates, then take $dA = r\,d\phi\,dr$, $x = r\cos\phi$, $y = r\sin\phi$ (Fig. C.3a). If the boundary is described as $y = f(x)$, then take $dA = dx\,dy$ (Fig. C.3b). For the area in Fig. C.3c, choose rectangular strip as area element: $dA = f(x)\,dx$,

$$dI_{xx}^O = \frac{1}{3}y^3\,dx = \frac{1}{3}f^3(x)\,dx, \qquad dI_{yy}^O = dI_{yy}^{C_1} + x^2\,dA = x^2 f(x)\,dx$$

$$dI_{xy}^O = dI_{xy}^{C_1} - x(\tfrac{1}{2}y)\,dA = -\tfrac{1}{2}xf^2(x)\,dx$$

$$I_{xx}^O = \tfrac{1}{3}\int_a^b f^3(x)\,dx, \qquad I_{yy}^O = \int_a^b x^2 f(x)\,dx, \qquad I_{xy}^O = -\tfrac{1}{2}\int_a^b xf^2(x)\,dx$$

Some results of inertia matrix of plane areas of Figs C.4a to C.4i, based on the mass inertia matrix of thin bodies of the same shape are as follows.

(a) Rectangle : $I_{xy}^C = 0, \quad I_{xx}^C = \tfrac{1}{12}m(h^2+0^2) = \tfrac{1}{12}Ah^2 = \tfrac{1}{12}bh^3 \;\Rightarrow\; I_{yy}^C = \tfrac{1}{12}hb^3,$
$I_{zz}^C = I_{xx}^C + I_{yy}^C = \tfrac{1}{12}(bh^3+hb^3), \qquad I_{xx}^A = \tfrac{1}{12}bh^3 + bh(\tfrac{1}{2}h)^2 = \tfrac{1}{3}bh^3$

(b) Circle : $I_{xx}^C = I_{yy}^C = \tfrac{1}{4}mr^2 = \tfrac{1}{4}Ar^2 = \tfrac{1}{4}\pi r^4$
$I_{zz}^C = \tfrac{1}{2}mr^2 = \tfrac{1}{2}Ar^2 = \tfrac{1}{2}\pi r^4, \qquad I_{xy}^C = 0$

(c) Annulus : $I_{xx}^C = I_{yy}^C = \tfrac{1}{4}\pi(R^4 - r^4), \qquad I_{zz}^C = \tfrac{1}{2}\pi(R^4-r^4), \qquad I_{xy}^C = 0$

(d) Thin ring : $A = 2\pi rt, \qquad I_{zz}^C = Ar^2, \qquad I_{xx}^C = I_{yy}^C = \tfrac{1}{2}I_{zz}^C = \tfrac{1}{2}Ar^2$

(e) Parallelogram : $I_{xx}^C =$ (same as for rectangle in Fig. C.4e) $= \tfrac{1}{12}bh^3$

The x-axis is not necessarily a principal axis of inertia at C.

(f) Triangle : $I_{xx}^{A} = \frac{1}{2}[I_{xx}^{A}$ of the parallelogram in Fig. C.4f$] = bh^3/24$,

$$I_{xx}^{C} = I_{xx}^{A} - \tfrac{1}{2}bh(\tfrac{1}{6}h)^2 = bh^3/36$$

The x-axis is not necessarily a principal axis of inertia at C.

(g) 'I' Section $I_{xx}^{C} = \frac{1}{12}(BH^3 - bh^3)$, $I_{xy}^{C} = 0$, $I_{yy}^{C} = \frac{1}{12}[(H-h)B^3 + h(B-b)^3]$

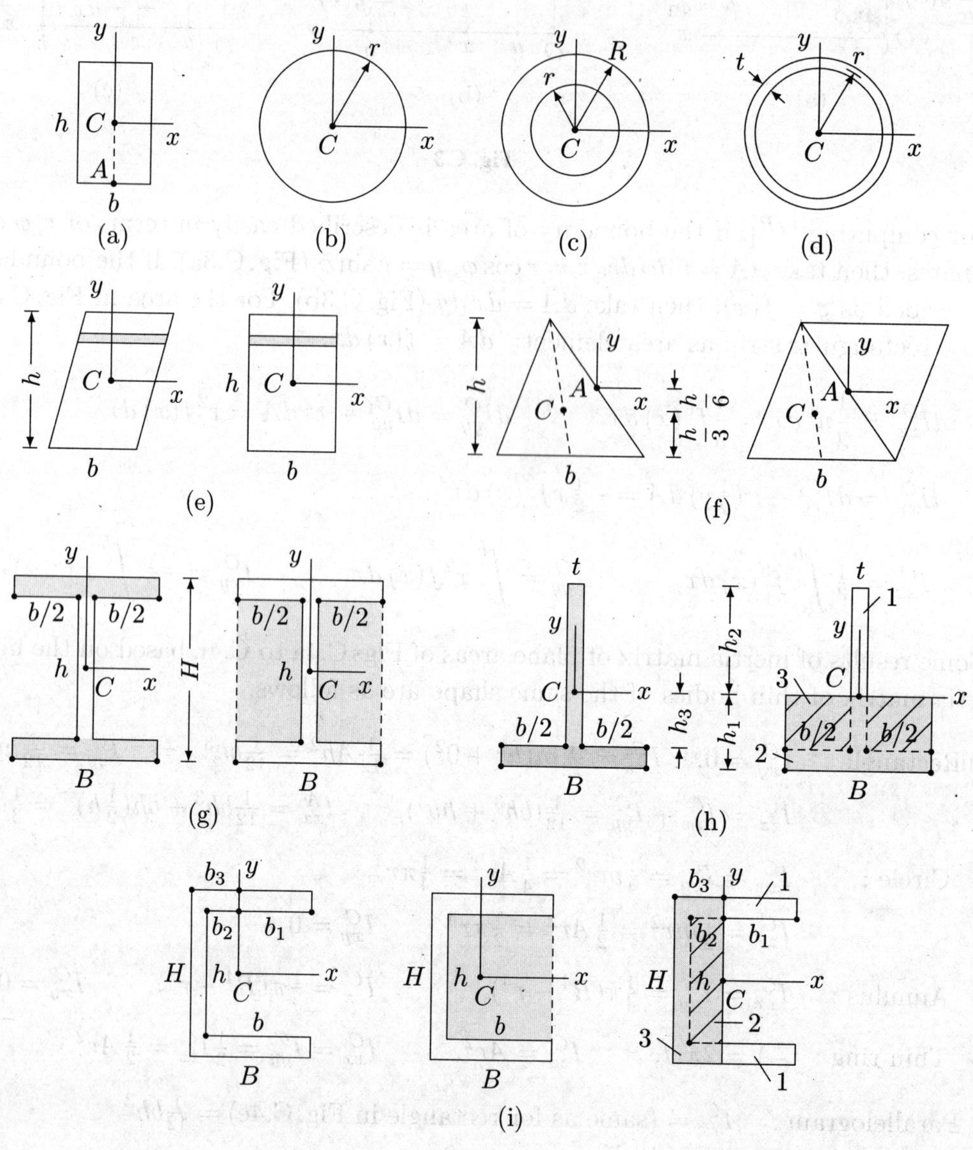

Fig. C.4

(h) 'T' Section : $I_{xx}^C = \frac{1}{3}[th_2^3 + Bh_1^3 - bh_3^3], \quad I_{xy}^C = 0$

$$I_{yy}^C = \tfrac{1}{12}[(h_2 + h_3)t^3 + (h_1 - h_3)B^3]$$

(i) Channel Section : $I_{xx}^C = \frac{1}{12}(BH^3 - bh^3), \quad I_{xy}^C = 0$

$$I_{yy}^C = \tfrac{1}{3}[Hb_3^3 - hb_2^3 + (H - h)b_1^3]$$

Example C.1: The cross-section of an angle section beam is shown in Fig. EC.1a. Find the inertia matrix of the area at its centroid C and find the principal axes and principal moments of inertia at C.

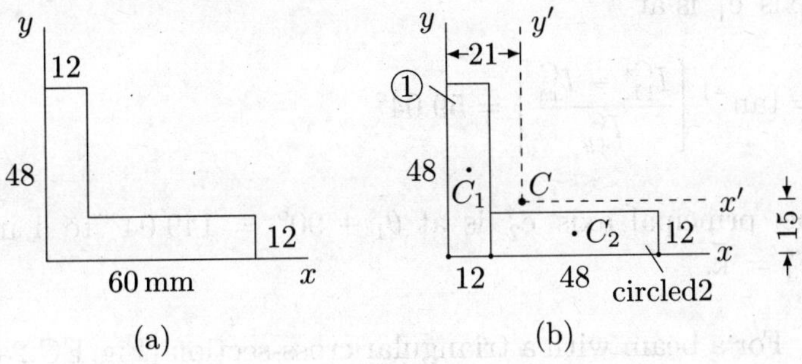

Fig. EC.1

Solution: The areas and centroids of the rectangles in Fig. EC.1b are:

$$A_1 = A_2 = 48 \times 12 = 576\,\text{mm}^2, \qquad \vec{r}_{C_1} = 6\vec{\imath} + 24\vec{\jmath}, \qquad \vec{r}_{C_2} = 36\vec{\imath} + 6\vec{\jmath}\ \text{mm},$$

$$\vec{r}_C = (A_1\vec{r}_{C_1} + A_2\vec{r}_{C_2})/(A_1 + A_2) = 21\vec{\imath} + 15\vec{\jmath}\ \text{mm},$$

$$\vec{r}_{C_1C} = -15\vec{\imath} + 9\vec{\jmath}, \qquad \vec{r}_{C_2C} = 15\vec{\imath} - 9\vec{\jmath}$$

Using parallel axes theorem:

$$I_{xx}^C = [(I_{xx}^C)_1 + A_1(9)^2] + [(I_{xx}^C)_2 + A_2(9)^2]$$

$$= \frac{12 \times 48^3}{12} + 576 \times 9^2 \times 2 + \frac{48 \times 12^3}{12} = 21.08 \times 10^4\,\text{mm}^4$$

$$I_{yy}^C = [(I_{yy}^C)_1 + A_1(15)^2] + [(I_{yy}^C)_2 + A_2(15)^2]$$

$$= \frac{48 \times 12^3}{12} + 576 \times 15^2 \times 2 + \frac{12 \times 48^3}{12} = 37.67 \times 10^4\,\text{mm}^4$$

$$I_{xy}^C = [(I_{xy}^C)_1 - A_1(-15)(9)] + [(I_{xy}^C)_2 - A_2(15)(-9)]$$

$$= -576(-15)(9) - 576(15)(-9) = 15.55 \times 10^4\,\text{mm}^4$$

$$I_{zz}^C = I_{xx}^C + I_{yy}^C = 58.75\,\text{mm}^4, \qquad I_{xz}^C = I_{yz}^C = 0$$

The principal moments of inertia at C are given by

$$I_{11}^{C*}, I_{22}^{C*} = \tfrac{1}{2}(I_{xx}^C + I_{yy}^C) \pm [\{\tfrac{1}{2}(I_{xx}^C - I_{yy}^C)\}^2 + I_{xy}^{C\,2}]^{1/2} = 29.375 \pm 17.624$$

$$I_{11}^{C*} = 47.00 \times 10^4, \qquad I_{22}^{C*} = 11.75 \times 10^4$$

$$I_{33}^{C*} = I_{zz}^C = 58.75 \times 10^4\ \text{mm}^4$$

The principal axis $\vec{e}_1^*$ is at

$$\theta_1 = \tan^{-1}\left[\frac{I_{11}^{C*} - I_{xx}^C}{I_{xy}^C}\right] = 59.04°$$

to $\vec{i}$ in xy-plane, principal axis $\vec{e}_2^*$ is at $\theta_1 + 90° = 149.04°$ to $\vec{i}$ in xy-plane and principal axis $\vec{e}_3^* = \vec{k}$.

Example C.2: For a beam with a triangular cross-section (Fig. EC.2a), find the area moment of inertia about axis AD. Find the inertia matrix of the area at the centroid C and find the principal axes and principal moments of inertia at C.

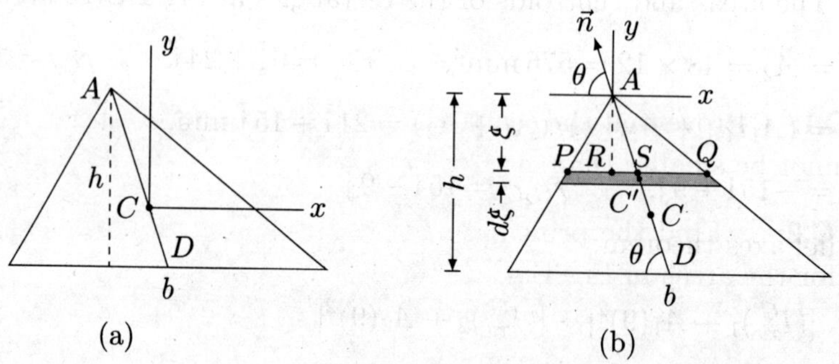

Fig. EC.2

Solution: Divide $\triangle$ into strips parallel to the base (Fig. EC.2b). It is easier to first find $[I^A]$ and then use parallel axes theorem to find $[I^C]$. $x_C = \frac{2}{3}h\cot\theta,\ y_C = -\frac{2}{3}h,\ z_C = 0.$

For the strip $PQ = b\xi/h,\ RS = \xi\cot\theta,\ dA = (b\xi/h)\,d\xi$ and using parallel axes theorem for A and C' $(\xi\cot\theta, -\xi, 0)$,

$$I_{xx}^A = \int_0^h \xi^2\left(\frac{b}{h}\,\xi\,d\xi\right) = \frac{bh^3}{4}$$

$$I^A_{xy} = \int_0^h -x_{C'}y_{C'}\left(\frac{b}{h}\,\xi\,d\xi\right) = bh\cot\theta\int_0^h \xi^3\,d\xi = \frac{bh^3}{4}\cot\theta$$

$$I^A_{yy} = \int_0^h \left[\frac{1}{12}\left(\frac{b}{h}\,\xi\right)^2 + (\xi\cot\theta)^2\right]\left(\frac{b}{h}\,\xi\,d\xi\right) = \frac{b^3h}{48} + \frac{bh^3}{4}\cot^2\theta$$

$I^C_{xz} = I^C_{yz} = 0$. Using parallel axes theorem for points A and C with area $A = bh/2$:

$$I^C_{xx} = I^A_{xx} - A(y^2_C + z^2_C) = bh^3/36$$

$$I^C_{xy} = I^A_{xy} + Ax_Cy_C = (bh^3/36)\cot\theta$$

$$I^C_{yy} = I^A_{yy} - A(z^2_C + x^2_C) = b^3h/48 + (bh^3/36)\cot^2\theta$$

$$I^C_{zz} = I^C_{xx} + I^C_{yy} = b^3h/48 + (bh^3/36)\operatorname{cosec}^2\theta$$

The principal moments of inertia at C are given by

$$I^{C*}_{11}, I^{C*}_{22} = \tfrac{1}{2}(I^C_{xx} + I^C_{yy}) \pm [\{\tfrac{1}{2}(I^C_{xx} - I^C_{yy})\}^2 + {I^C_{xy}}^2]^{1/2}, \qquad I^{C*}_{33} = I^C_{zz}$$

The principal axes are: $\vec{e}^*_1$ at $\theta_1 = \tan^{-1}[(I^{C*}_{11} - I^C_{xx})/I^C_{xy}]$ to $\vec{i}$ in xy-plane, $\vec{e}^*_2$ at $\theta_1 + 90°$ to $\vec{i}$ in xy-plane and $\vec{e}^*_3 = \vec{k}$.

Unit vector $\vec{n}$ along AD is $\vec{n} = (-\cos\theta, \sin\theta, 0)$,

$$I_{AD} = I^A_{ij}n_in_j = I^A_{xx}\cos^2\theta + I^A_{yy}\sin^2\theta - 2I^A_{xy}\sin\theta\cos\theta = (hb^3/48)\sin^2\theta$$

§ **Exercise C.1:** Prove that the centroidal polar moment of inertia of a plane surface of area A cannot be smaller than $A^2/2\pi$.

§ **Exercise C.2:** Find the principal axes of inertia and the principal moments of inertia at A for the areas in Fig. ExC.2.

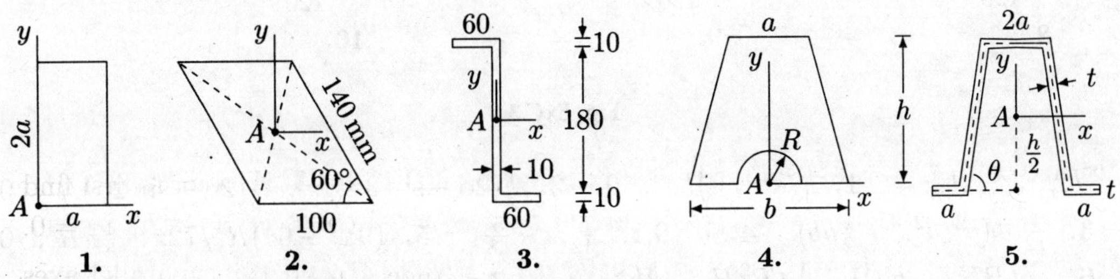

Fig. ExC.2

[**1.** $3.081a^4$, $0.2525a^4$, $-22.5°$, $67.5°$ **2.** 2353, $637.7\,\text{cm}^4$, 45.35°, 135.4°

3. $1624,\quad 58.75\,\text{cm}^4,\quad 10.68°,\quad 100.68°$

4. $(b+3a)h^3/12-\pi R^4/8,\quad h(a^2+b^2)(a+b)/48-\pi R^4/8,\quad 0°,\quad 90°$

5. $ath^2+th^3/6\sin\theta,$

$\frac{5}{6}a^3t+2at(1.5a+h\cot\theta)^2+[2ht(a+.5h\cot\theta)^2+\frac{1}{6}h^3t\cot^2\theta]/\sin\theta,\quad 0°,\quad 90°]$

§ **Exercise C.3:** Find the inertia matrix of area at point A with respect to the given axes for the areas in Fig. ExC.3.

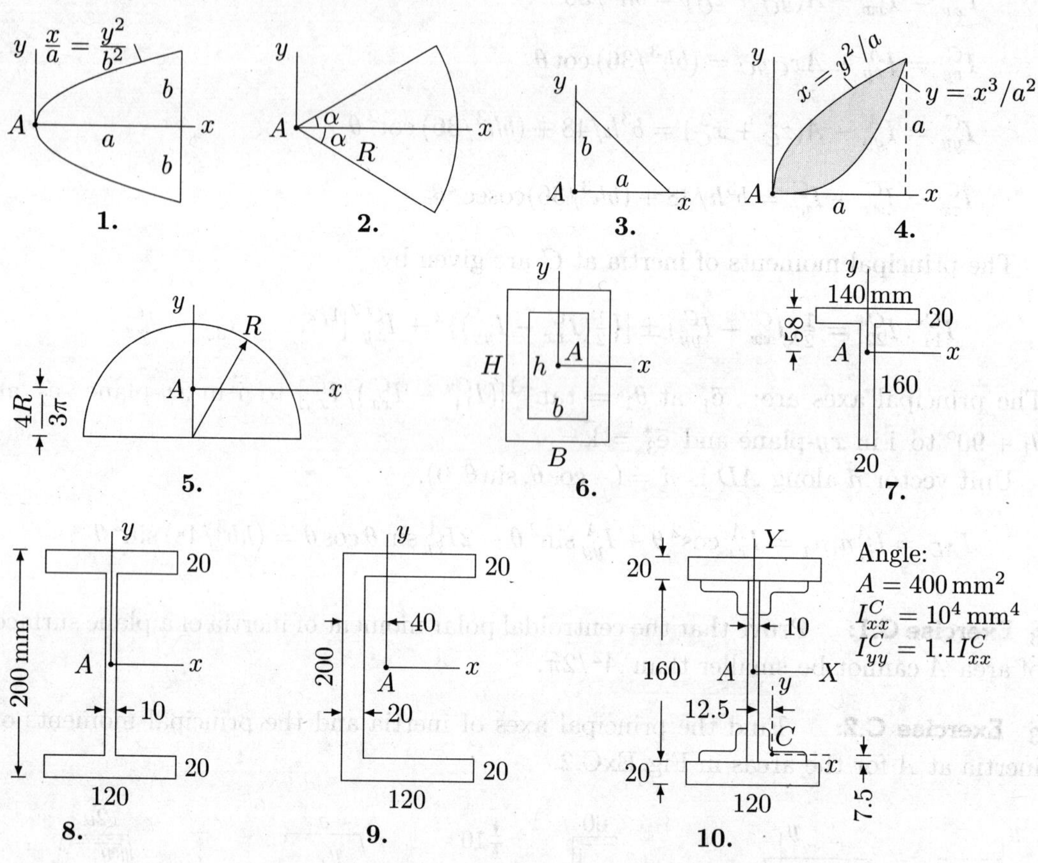

Fig. ExC.3

$[I^A_{xx},\ I^A_{yy},\ I^A_{xy}=$ **1.** $\frac{4}{15}ab^3,\ \frac{4}{7}ba^3,\ 0$ **2.** $\frac{1}{8}(2\alpha-\sin 2\alpha)R^4,\ \frac{1}{8}(2\alpha+\sin 2\alpha)R^4,\ 0$
3. $\frac{1}{12}ab(b^2, a^2, -\frac{1}{2}ab)$ **4.** $a^4(0.1,\ \frac{5}{42},\ -\frac{5}{48})$ **5.** $(9\pi^2-64)R^4/72\pi,\ \frac{1}{8}\pi R^4,\ 0$
6. $\frac{1}{12}(BH^3-bh^3),\ \frac{1}{12}(B^3H-b^3h),\ 0$ **7.** $(1902,\ 468,\ 0)\,\text{cm}^4$
8. $(4245,\ 577.3,\ 0)\,\text{cm}^4$ **9.** $(4587,\ 1067,\ 0)\,\text{cm}^4$ **10.** $(5090,\ 630.7,\ 0)\,\text{cm}^4]$